Energy Futures - ICEF 2026
Part 1

International Conference on Energy Futures (ICEF'26), jointly organized by Al Akhawayn University in Ifrane (Morocco) and Prince Mohammad Bin Fahd University (Saudi Arabia). Held as a hybrid event at the Al Akhawayn University, Ifrane, Morocco, on February 5–6, 2026

Editor

Hamza El Hafdaoui[1]

[1]School of Science and Engineering, Al Akhawayn University in Ifrane, Morocco

Peer review statement

All papers published in this volume of "Materials Research Proceedings" have been peer reviewed. The process of peer review was initiated and overseen by the above proceedings editors. All reviews were conducted by expert referees in accordance to Materials Research Forum LLC high standards.

Published under License by **Materials Research Forum LLC**
Millersville, PA 17551, USA

Published as part of the proceedings series
Materials Research Proceedings
Volume 64 (2026)

ISSN 2474-3941 (Print)
ISSN 2474-395X (Online)

ISBN 978-1-64490-408-4 (2-Vols. Set Print)
ISBN 978-1-64490-410-7 (Part 1 Print)
ISBN 978-1-64490-411-4 (Part 2 Print)

ISBN 978-1-64490-409-1 (eBook)

Distributed worldwide by

Materials Research Forum LLC
105 Springdale Lane
Millersville, PA 17551
USA
https://mrforum.com

Manufactured in the United States of America

Table of Contents

Renewable Energy and Storage

Electric Grid and Digital Infrastructure

Emerging Technologies

Materials for Energy Applications

Alternative Fuels and Applications

The following chapters are published in Part 2 (ISBN 978-1-64490-411-4):

Energy Efficiency and Management

Energy Finance and Economics

Energy, Society, and Sustainability

Preface

The **International Conference on Energy Futures (ICEF'26)** represents a significant milestone in advancing global dialogue on sustainable energy systems and technologies. Jointly organized by Al Akhawayn University in Ifrane (Morocco) and Prince Mohammad Bin Fahd University (Saudi Arabia), the conference was held on February 5–6, 2026, as a hybrid event hosted at Al Akhawayn University.

ICEF'26 brought together researchers, academics, industry professionals, and policymakers from diverse disciplines to address the pressing challenges and opportunities associated with the global energy transition. The conference served as a platform for presenting cutting-edge research, exchanging knowledge, and fostering collaborations aimed at developing innovative, sustainable, and resilient energy solutions.

The proceedings collected in this volume reflect the breadth and depth of current research in energy-related fields. The contributions are organized into thematic areas including:

- Renewable Energy and Storage
- Electric Grid and Digital Infrastructure
- Emerging Technologies
- Materials for Energy Applications
- Alternative Fuels and Applications
- Energy Efficiency and Management
- Energy Finance and Economics
- Energy, Society, and Sustainability

The papers cover a wide range of topics such as artificial intelligence applications in energy systems, smart grids, hydrogen production and utilization, advanced materials for energy storage, photovoltaic technologies, energy policy and economics, and sustainable development strategies. Many contributions emphasize interdisciplinary approaches combining engineering, data science, economics, and environmental studies.

All papers included in this volume have undergone a rigorous peer-review process conducted by expert reviewers in accordance with the high standards of Materials Research Forum LLC. This ensures the scientific quality, originality, and relevance of the published work.

ICEF'26 was also held in celebration of the 30th anniversary of Al Akhawayn University, highlighting three decades of academic excellence and international collaboration, particularly between Morocco and Saudi Arabia. The conference underscores the importance of global partnerships in addressing energy challenges and achieving sustainability goals.

The editor would like to express sincere gratitude to all authors, reviewers, and organizing committee members whose dedication and contributions made this conference and its proceedings possible.

It is our hope that this volume will serve as a valuable resource for researchers, practitioners, and decision-makers working toward a sustainable energy future.

Committees

Honorary Committee

President Amine Bensaid – Al Akhawayn University in Ifrane, Morocco
President Issa Bin Hassan Al Ansari – Prince Mohammad Bin Fahd University, Saudi Arabia

General Chairs

VP. Hariclea Zengos – Al Akhawayn University in Ifrane, Morocco
VP. Faisal Yousif Al Anezi – Prince Mohammad Bin Fahd University, Saudi Arabia

Operational Chairs

Dr. Youssef Chetioui – Al Akhawayn University in Ifrane, Morocco
Dr. Jamal Nayfeh – Prince Mohammad Bin Fahd University, Saudi Arabia
Dr. Ahmed Khallaayoun – Al Akhawayn University in Ifrane, Morocco
Dr. Ghadeer Kayal – Prince Mohammad Bin Fahd University, Saudi Arabia

Technical Committee

Mr. Zakaria El Hafdaoui (Webmaster) – Oracle Inc., Morocco
Mr. Ahmed Bazzi (Logistics Lead) – Al Akhawayn University in Ifrane, Morocco
Mr. Adel Bouzaguaoui (Technical Advisor) – Al Akhawayn University in Ifrane, Morocco
Mr. El Makhtar Rhannai (Accounting and Finance) – Al Akhawayn University in Ifrane, Morocco
Dr. Tahar Ayadat (Technical Outreach Advisor) – Prince Mohammad Bin Fahd University, Saudi Arabia
Mr. Zafar Kazimi (Technical Outreach Advisor) – Prince Mohammad Bin Fahd University, Saudi Arabia

Communication Committee

Mr. Yassine Wahbi (Design Specialist) – Oracle Inc., Morocco
Dr. Samir En-Nakla (Outreach and Communication Liaison) – Prince Mohammad Bin Fahd University, Saudi Arabia
Mr. Ahmed Abul Hussain (Outreach and Communication Liaison) – Prince Mohammad Bin Fahd University, Saudi Arabia
Mr. Idir Moukhlis (Internal Development and Communication) – Al Akhawayn University in Ifrane, Morocco

Ms. Yousra Lakiasse (Media & PR) – Al Akhawayn University in Ifrane, Morocco

Scientific Chair

Dr. Hamza El Hafdaoui – Al Akhawayn University in Ifrane, Morocco

Steering Committee

Dr. Mourad Yessef – Mohammed First University, Morocco
Dr. Andi Asiz – Prince Mohammad Bin Fahd University, Saudi Arabia
Dr. Faramarz Djavanroodi – Prince Mohammad Bin Fahd University, Saudi Arabia

Track Chairs

Dr. Mostafa Merras (SS1. Internet of Energy) – Moulay Ismal University, Morocco
Dr. Hatim Ameziane (SS2. Energy Architectures for Hydrogen-Powered and Electric Mobility Systems) – Sultan Moulay Slimane University, Morocco
Dr. Naoufal El Youssfi (SS3. Advanced Control, Energy Management, and Smart Vehicle Integration) – Royal Air Force Academy, Morocco
Dr. Abdelmajid Tahiri (SS4. Artificial Intelligence and Energy Education) – Moulay Ismal University, Morocco
Dr. Khaoula Benmoussa (SS4. Artificial Intelligence and Energy Education) – Moulay Ismal University, Morocco
Dr. Mohamed Halim (SS4. Artificial Intelligence and Energy Education) – Sidi Mohamed Ben Abdellah University, Morocco
Mr. Youness Hakam (SS5. Hydrogen Energy) – Moroccan School of Engineering Sciences, Morocco
Dr. Hassan Ougraz (SS5. Hydrogen Energy) – Ibn Zohr University, Morocco
Dr. Abdennabi Morchid (SS6. Smart Energy Management for Sustainable Agriculture) – Sidi Mohamed Ben Abdellah University, Morocco
Dr. Dahaman Ishak (SS7. Modern Charging infrastructures for Electric Vehicles) – Prince Mohammad Bin Fahd University, Saudi Arabia
Dr. Ala Al-Haj Hussein (SS8. Modeling, Simulation, and Emerging Applications) – Prince Mohammad Bin Fahd University, Saudi Arabia
Dr. Kwan San Hui (SS9. New Materials for Electrical Storage and Conversion) – Prince Mohammad Bin Fahd University, Saudi Arabia

Renewable Energy and Storage

Renewable Energy Forecasting using Machine Learning Techniques and Comparative Analysis

Nayeemuddin MOHAMMED[1,a*], Hiren MEWADA[2,b], Tasneem SULTANA[3,c], Tabassum Nahid SULTANA[4,d]

[1]Civil Engineering Department, Prince Mohammad Bin Fahd University, Al Khobar, 31952, Saudi Arabia

[2]Electrical Engineering Department, Prince Mohammad Bin Fahd University, Al Khobar, 31952, Saudi Arabia

[3]Artificial Intelligence and Machine Learning, Godutai Engineering College for Women, Sharnbasva University, Kalaburagi, India

[4]Department of Computer Science and Engineering, Khaja Bandanawaz College of Engineering, Kalaburagi, India

[a]mnayeemuddin@pmu.edu.sa (Orcid: 0000-0002-7757936X), [b]hmewada@pmu.edu.sa (orcid.org/0000-0002-3579-8708), [c]tasneemsultana841@gmail.com, [d]tabassumcse85@gmail.com

Keywords: Forecasting, Renewable Energy, Machine Learning, Photovoltaic, Sustainability

Abstract. Efficient grid management, scheduling, and increased clean energy system reliability depend on accurate electricity generation forecasting. Several machine learning methods, including Decision Tree, Random Forest, Gradient Boosting, AdaBoost, Bagging, CatBoost, ExtraBoost, and XGBoost, were employed in this study to forecast energy production. The results were compared methodically. Standard error metrics, such as mean absolute error (MAE), mean squared error (MSE), and root mean squared error (RMSE), were used to evaluate the model on both the training and test datasets. Although all chosen models use a decision tree as the base, their performance varies with the decision tree model. The CatBoost and Bagging demonstrated greater generalization than the other models. CatBoost achieved the highest R^2 of 0.8690 and the lowest RMSE of 0.0167 on the test dataset. The performance of Bagging, Random Forest, and Gradient Boosting is nearly similar. In contrast, the XGBoost and Decision Tree performed well on the training dataset. However, for unseen data, their predictions are far from the actual values, with a high MSE and a lower R^2 score on the test dataset. Overall, an ensemble approach of decision trees provided better performance at the cost of computation compared to XGBoost.

Introduction

Renewable energy sources, such as solar and wind, have rapidly grown in modern power systems as sustainable replacements for conventional fossil fuels [1]. However, the unpredictable nature of renewable energy generation presents considerable challenges to grid stability, energy supply and demand scheduling, and balancing [2]. Recent literature indicates that deep learning methods can be beneficial for PV power prediction by extracting spatial and temporal patterns. Mainly, convolutional neural networks (CNNs) and gated recurrent unit networks (GRUs) are used in federated learning to protect dataset privacy when learning from multiple sources. And transfer learning to enhance generality among various types of photovoltaic systems [3]. Forecasting the output of renewables is therefore essential for integrating clean energy into smart grids reliably, at reduced operational costs, and for effective management [4]. Due to limited data availability for new wind farms, deep learning models for wind power forecasting often face challenges. With data scarcity, privacy concerns may arise when aggregated from multiple sources. New studies suggest hybrid architectures combining temporal convolutional networks and transformers to extract temporal dependencies [5]. The federated learning to ensure data privacy in collaborative model training [6].

In recent years, several forecasting approaches have been examined for this purpose, including statistical models such as Autoregressive Integrated Moving Average (ARIMA) [7]. Federated learning has been used to forecast energy consumption while preserving privacy, enabling knowledge transfer from buildings with more data to those with limited or no data. A selective client-participation strategy can be adopted to minimize model divergence when data are noisy or limited [8]. It achieves higher accuracy and robustness than traditional and ensemble models. But this approach is challenging to apply when dealing with high variability in residential consumption and to maintain consistent global model performance across heterogeneous participants [9]. With limited success because these approaches work well only when conditions are stationary. They do not account for the nonlinear nature of many relationships between variables in the data and under stationary conditions [10]. The renewable energy time series is influenced by weather and environmental factors [11]. On the other hand, machine learning (ML) techniques can model nonlinear dynamics, learn from big data, and provide accurate predictions over different time horizons [12].

Several machine learning models, such as Decision Trees, Random Forests, Gradient Boosting, XGBoost, AdaBoost, Bagging, Extra Trees, and CatBoost, have been used with differing degrees of success in forecasting renewable energy [13]. Recently, deep learning models are getting popular because of their extended performance. However, deep learning algorithms need a large dataset, and for a small dataset, this model suffers from an overfitting issue. Despite their interpretability, decision trees often suffer from overfitting, whereas ensemble-based techniques such as Random Forest and Bagging combine multiple learners to improve stability and robustness [14]. By gradually increasing model accuracy through error reduction, boosting algorithms such as Gradient Boosting, AdaBoost, XGBoost, and CatBoost frequently outperform traditional learners in predictive tasks [15]. Despite the widespread use of these models, a thorough comparative analysis is necessary to determine the best method under various testing and training scenarios. In this study, we present a comparative evaluation of eight machine learning algorithms for renewable energy forecasting. The models are trained and tested on energy datasets and evaluated using standard performance metrics, including Mean Absolute Error (MAE), Mean Squared Error (MSE), Root Mean Squared Error (RMSE), and the coefficient of determination R^2.

The objective of this work is to investigate the relative performance of these models and identify the most robust and reliable forecasting approach. The results of this study provide valuable insights into the applicability of ensemble and boosting techniques for improving the accuracy of renewable energy forecasting.

Methodology

The renewable energy dataset contains time-series 100 data points representing energy generation and other influencing parameters, such as solar irradiance, wind speed, and temperature. The dataset was obtained from the Kaggle website and was adapted to remove outliers [16]. Seven machine learning techniques were employed and compared for forecasting the renewable energy output variable. A Decision Tree (DT) baseline was used as a simple tree-based learner that partitions data via recursive splits. The Random Forest (RF) is an ensemble method that constructs multiple decision trees and averages their outputs to stabilize the model. Boosting-based approaches, including Gradient Boosting (GB), sequentially minimize residual errors. AdaBoost (AB) assigns higher weights to misclassified samples. A bagging technique that creates several models on random subsets of data and aggregates predictions, reducing variance. The advanced boosting methods, such as CatBoost, are specialized for categorical variables and incorporate strategies to prevent overfitting XGBoost with regularization for generalization. It builds the decision tree in a sequential manner, and each tree attempts to correct the prediction error of its predecessor. Each tree predicts the target from the input using the following equation 1:

$$P(x) = \sum_{m-1}^{M} \omega_m T_m(x) \quad (1)$$

Where P(x) is the final prediction obtained using M number of trees. ω_m is the weight matrix for the m-th tree. The weight matrix is updated by minimizing the prediction error using the following equation 2:

$$\omega_m = ARG\_MIN \sum_{i=1}^{N} (y_i - P_{m-1}(x_i) - \omega T_m(x))^2 \quad (2)$$

The key hyperparameters for the model training include learning rate and number of iterations. The data points are split into training and testing subsets using an 80:20 split to ensure unbiased performance evaluation. The missing values in the data were imputed before modeling, and all features were normalized to a comparable scale to improve the machine learning algorithm's performance. Below are equations 3-6 used for the metric analysis of various machine learning models [17].

$$R^2 = 1 - \frac{\sum_{i=1}^{n} (y_i - \widehat{y_i})^2}{\sum_{i=1}^{n} (y_i - \overline{y})^2} \quad (3)$$

$$MSE = \frac{1}{n} \sum_{i=1}^{n} (y_i - \widehat{y_i})^2 \quad (4)$$

$$RMSE = \sqrt{\frac{1}{n} \sum_{i=1}^{n} (y_i - \widehat{y_i})^2} \quad (5)$$

$$MAE = \frac{1}{n} \sum_{i=1}^{n} |y_i - \widehat{y_i}| \quad (6)$$

Where, y_i=actual values, $\widehat{y_i}$=predicted values, $\overline{y}$=mean of actual values, n=number of data points.

Results and Discussion

A best-fit normal curve was applied to both sets of experimental data. Figure 1 displays a statistical distribution of energy output vs. temperature and wind speed. The normal distribution fits the histograms, indicating that each variable's data are normally distributed. Each bar in the histogram represents the number of values that lie within a specific range. The smooth Gaussian curve represents the theoretical normal distribution with the specified mean and standard deviation. The more closely the bars follow the normal curve. The closer to approximately normally distributed the variable is the majority of the data clusters around the mean with symmetric tails. It is desirable for most statistical and machine learning methods as it generally simplifies assumptions about error distribution and leads to better model performance. When the bars deviate from the normal curve, they might indicate skewness, kurtosis, or outliers that are far away from the central peak of the curve. It should be understood that these deviations can impact the results of regression, optimization, and predictive modeling. Transformations such as log, square root, Box-Cox, or robust methods may be needed in the presence of non-normality, as illustrated in Figure 1.

Materials Research Proceedings 64 (2026) 2-8 https://doi.org/10.21741/9781644904091-1

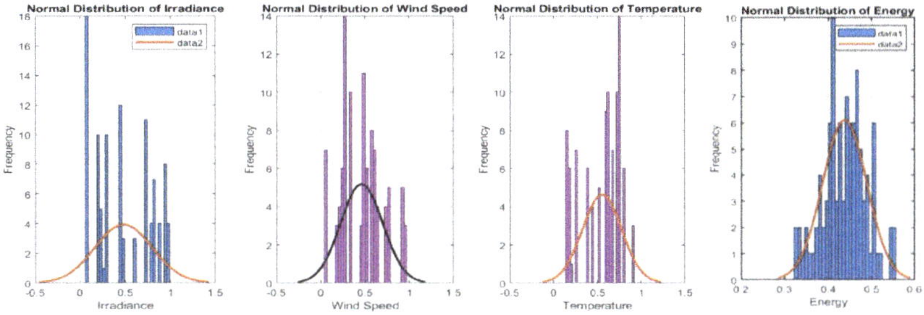

Fig.1 Normal distribution diagram for solar irradiation, wind speed, temperature, and energy

Figure 2 displays the relationship between solar radiation, wind speed, temperature, and electricity generation from the correlation matrix. The results demonstrate a strong positive correlation r=0.74 between solar radiation and energy output. It indicates that solar radiation is the most influential variable in renewable energy production across the dataset, followed by a slightly less. But still a significant positive correlation, r=0.52, with temperature, indicating that increasing ambient temperature may lead to somewhat higher efficiency in solar panels. However, this effect is less pronounced than the impact of increased radiation. Wind speeds also exhibit a moderate positive correlation, r=0.37, indicating that wind contributes to renewable energy production. But has less impact on overall production levels compared with solar-related variables.

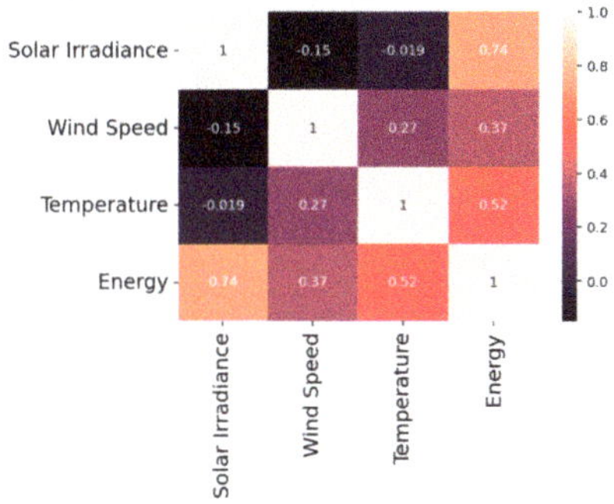

Fig. 2 Heat map presenting a correlation between the variables and response

Figures 3 and 4 show training, testing, and all data performance for the CatBoost model. The predicted vs. actual target values in the test samples graph are almost identical, with only minor deviations. It indicates the model captures the underlying relationship with a small amount of error. In the bottom-left subplot, the distribution of errors appears relatively small and varies closely around zero, again confirming that the model is performing well. The MSE is 3.625E-05 and the RMSE is 0.006020, indicating small prediction error magnitudes. The mean error of 0.006058 indicates a slight negative bias. In contrast, the standard deviation of 0.006057 suggests that most errors will be within a narrow range around zero, as shown in the histogram, with a superimposed normal distribution in the bottom-right corner. All this confirms that the CatBoost machine learning method is a strong predictor and statistically consistent.

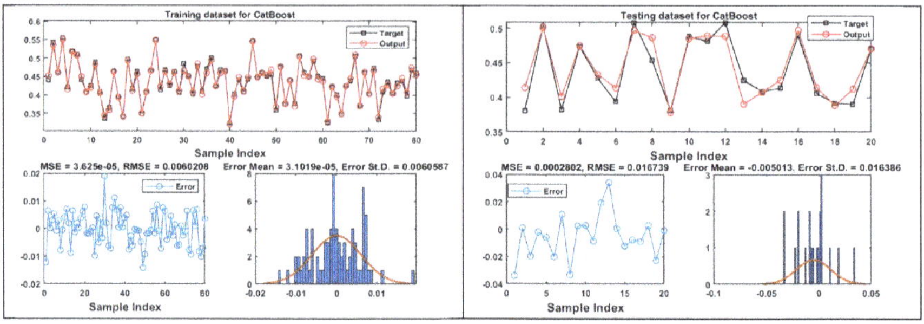

Fig. 3 Training and Testing data result for the CatBoost method of Machine Learning

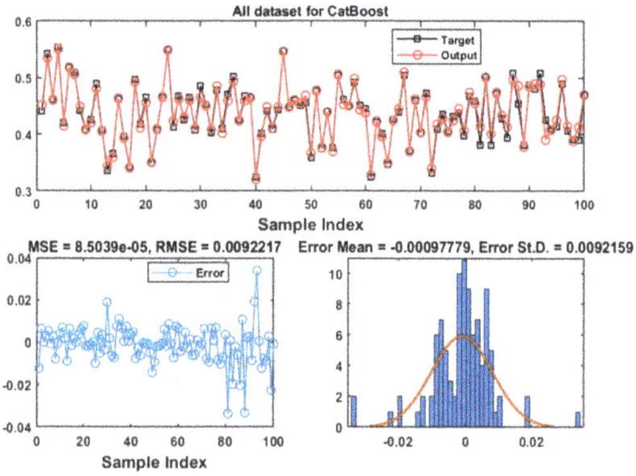

Fig. 4 All data result for CatBoost method of Machine Learning

The actual vs. predicted plot for the whole dataset in Figure 5 displays the CatBoost model has an excellent fit R^2=0.967. All the points closely follow the y=x reference line, indicating that it captures the overall variability of the data quite well. However, when we split the dataset into training and testing subsets, the performance differs significantly. The model has an excellent fit to the training data R^2=0.986, but reduced but acceptable predictive power for the testing data R^2=0.8690. There is some wider scatter around the y=x line, which could be due to limitations in the model's generalizability or natural variability in the data. The results still show that the Gradient Boosting model provides good predictive performance, but suggest a balance between training and testing accuracy.

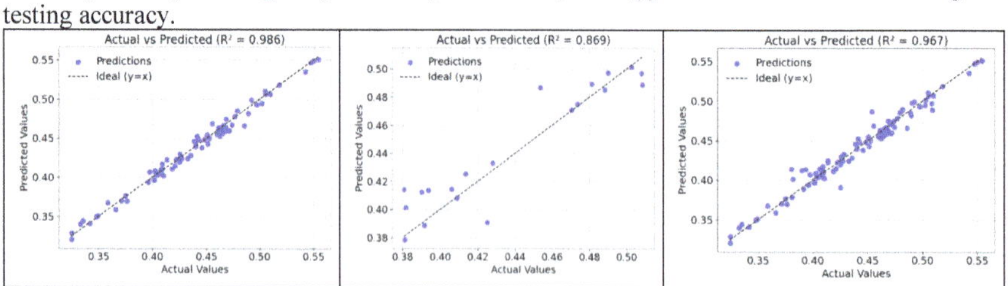

Training Data	Testing Data	All Data

Fig. 5 Regression plot results for all data, training, and testing of the CatBoost method

The comparative performance of various ensemble and boosting algorithms using training and testing datasets is presented in Table 1. The Gradient Boosting and XGBoost had the lowest training errors (MAE, MSE, RMSE) among all models. The R^2 values are almost perfect fits to the training data, 0.9991 and 0.9977, respectively. However, this indicates overfitting, which is also reflected in their generalization on the testing dataset. At the same time, Gradient Boosting had a test R^2 of 0.835, XGBoost dropped significantly to 0.715.

CatBoost and Bagging showed the best overall results, with CatBoost achieving a high test R^2 of 0.869 and Bagging close behind at 0.861. Both have relatively low error values across MAE, MSE, and RMSE. Random Forest also performed well, but with a lower test R^2 of 0.848 compared to Decision Trees and AdaBoost, which underperformed with lower test R^2 values of 0.531 and 0.804. In summary, while Gradient Boosting and XGBoost have better training accuracy. CatBoost was the most consistent model for this dataset as it performed best in terms of generalization and lowest testing errors.

Table 1 *Metric analysis comparison of different various machine learning algorithm*

	Train MAE	Train MSE	Train RMSE	Train R^2	Test MAE	Test MSE	Test RMSE	Test R^2
Decision Trees	0.010057	0.000182	0.013477	0.9323	0.025578	0.001	0.03163	0.530539
Random Forest	0.005319	4.78E-05	0.006917	0.982165	0.014811	0.000324	0.018012	0.847774
Gradient Boosting	0.003717	2.54E-05	0.005039	0.999100	0.014863	0.000351	0.018725	0.835481
AdaBoost	0.009265	0.000128	0.011313	0.95229	0.016547	0.000418	0.020447	0.803827
Bagging	**0.005125**	**4.21E-05**	**0.006485**	**0.984323**	**0.013423**	**0.000296**	**0.017191**	**0.861327**
CatBoost	**0.006058**	**3.625E-05**	**0.006020**	**0.98600**	**0.016386**	**0.000280**	**0.016739**	**0.86900**
XGBoost	0.001937	6.15E-06	0.00248	0.997707	0.017736	0.000607	0.024641	0.715096

Conclusion

The final results from evaluating tree-based Ensemble Learning models show significant differences in how well these different methods balance training accuracy and testing generalization. Where Gradient Boosting and XGBoost have extremely high training accuracy R^2 near unity. But it has lower predictive ability on the testing set, suggesting overfitting. While CatBoost and Bagging yielded the most stable results across both train and test sets. The CatBoost model produced the best overall test performance, R^2=0.869, MAE=0.016386, followed by Random Forest, which also performed well on tests, with R^2=0.848. The findings reveal that although boosting methods can learn very complex patterns with perfect training accuracy. Their generalization capacity needs to be handled with caution when applied in practice. They may overfit on unseen data more than expected. CatBoost is identified as the strongest model, with high accuracy and low sensitivity to overfitting among all models. Further optimization of hyperparameters, cross-validation using larger datasets, and hybrid approaches that combine machine learning with domain-specific physical models can enhance the robustness, interpretability, and applicability of predictive models for future studies.

Reference

[1] K. K. Jaiswal *et al.*, "Renewable and sustainable clean energy development and impact on social, economic, and environmental health," *Energy Nexus*, vol. 7, p. 100118, Sep. 2022. https://doi.org/10.1016/j.nexus.2022.100118

[2] Q. Hassan, S. Algburi, A. Z. Sameen, H. M. Salman, and M. Jaszczur, "A review of hybrid renewable energy systems: Solar and wind-powered solutions: Challenges, opportunities, and policy implications," *Results in Engineering*, vol. 20, p. 101621, Dec. 2023. https://doi.org/10.1016/j.rineng.2023.101621

[3] S. Muhammad Salman Bukhari *et al.*, "Federated transfer learning with orchard-optimized Conv-SGRU: A novel approach to secure and accurate photovoltaic power forecasting," *Renewable Energy Focus*, vol. 48, p. 100520, Mar. 2024. https://doi.org/10.1016/j.ref.2023.100520
[4] M. Khalid, "Smart grids and renewable energy systems: Perspectives and grid integration challenges," *Energy Strategy Reviews*, vol. 51, p. 101299, Jan. 2024. https://doi.org/10.1016/j.esr.2024.101299
[5] M. A. Khasawneh *et al.*, "Renewable energy in pavement engineering and its integration with sustainable materials: A review paper," in *Materials Research Proceedings*, Aug. 2024, pp. 377–384. https://doi.org/10.21741/9781644903216-49
[6] Z. Zhang, F.-Q. Xuan, X.-X. Ruan, and L.-Z. Li, "Hybrid model with temporal convolutional network and transformer encoder for privacy-preserving wind power forecasting," *Adv. Manuf.*, Apr. 2025. https://doi.org/10.1007/s40436-025-00552-1
[7] V. I. Kontopoulou, A. D. Panagopoulos, I. Kakkos, and G. K. Matsopoulos, "A Review of ARIMA vs. Machine Learning Approaches for Time Series Forecasting in Data Driven Networks," *Future Internet*, vol. 15, no. 8, p. 255, Jul. 2023. https://doi.org/10.3390/fi15080255
[8] A. Talukdar, B. Panda, S. Hota, M. Maharana, N. Mohammed, and S. Panda, *Intelligent Solar Power Forecast with Machine Learning*. 2024. https://doi.org/10.1109/ISAECT64333.2024.10799562
[9] A. Rizwan *et al.*, "Enhancing energy consumption prediction in smart homes: a convergence-aware federated transfer learning approach," *Sci. Tech. Energ. Transition*, vol. 79, p. 85, 2024. https://doi.org/10.2516/stet/2024060
[10] E. Sarmas, N. Dimitropoulos, V. Marinakis, Z. Mylona, and H. Doukas, "Transfer learning strategies for solar power forecasting under data scarcity," *Sci Rep*, vol. 12, no. 1, p. 14643, Aug. 2022. https://doi.org/10.1038/s41598-022-18516-x
[11] G. Tong, X. Qian, and Y. Liu, "Prognostics and Predictive Maintenance Optimization Based on Combination BP-RBF-GRNN Neural Network Model and Proportional Hazard Model," *Journal of Sensors*, vol. 2022, no. 1, p. 8655669, 2022. https://doi.org/10.1155/2022/8655669
[12] I. Price *et al.*, "Probabilistic weather forecasting with machine learning," *Nature*, vol. 637, no. 8044, pp. 84–90, Jan. 2025. https://doi.org/10.1038/s41586-024-08252-9
[13] H. Kim, S. Dorjgochoo, H. Park, and S. Lee, "Personalized Federated Transfer Learning for Building Energy Forecasting via Model Ensemble with Multi-Level Masking in Heterogeneous Sensing Environment," *Electronics*, vol. 14, no. 9, p. 1790, Jan. 2025. https://doi.org/10.3390/electronics14091790
[14] M. Mansoor, G. Tao, A. F. Mirza, B. Yousaf, M. Irfan, and W. Chen, "FTLNet: federated deep learning model for multi-horizon wind power forecasting," *Discov Internet Things*, vol. 5, no. 1, p. 21, Mar. 2025. https://doi.org/10.1007/s43926-025-00112-w
[15] L. Zhang, S. Zhu, S. Su, X. Chen, Y. Yang, and B. Zhou, "Wind power prediction method based on cloud computing and data privacy protection," *J Cloud Comp*, vol. 13, no. 1, p. 137, Sep. 2024. https://doi.org/10.1186/s13677-024-00679-9
[16] "https://www.kaggle.com/datasets/itsrohithere/renewable-energy-forecasting." accessed on Dec 2025.
[17] F. Alzahrani, "Application of artificial intelligence (AI) in wind energy system with a case study," presented at the Renewable Energy: Generation and Application, Aug. 2024, pp. 96–103. https://doi.org/10.21741/9781644903216-13

Experimental and Neural Network based Pump Power Analysis of Wind Turbine Powered Water Pump

Lingala Syam SUNDAR[1,a], Esam JASSIM[1,b]

[1]Department of Mechanical Engineering, Prince Mohammad Bin Fahd University, Al-Khobar, Saudi Arabia

[a]sslingala@gmail.com, [b]ejassim@pmu.edu.sa

Keywords: Neural Network, Wind Turbine, Renewable Energy, Power, Pump, Torque

Abstract. Wind energy is the renewable source of energy, which is available plenty on the earth surface, which is caused due to the solar energy. The kinetic energy available in the wind can be extracted into work with help of wind turbine. This wind turbine power can be used for several applications. In this paper, the wind turbine energy can be used to pump the ground water for the agricultural purpose. To establish this an experimental setup is fabricated and analyzed. The power extracted from the wind mill, toque, and water discharge was calculated at different wind velocity. The obtained output data is modeled with artificial neural network (ANN). The ANN algorithms of Levenberg–Marquardt technique were used to predict the power, torque, and discharge. The experimental results show, at a wind speed of 5.32 m/sec, the power extracted is 1064.2 N-m, the torque produced is 1000.2 N-m, and the water discharge from the pump is 0.0091 m^3/sec. The LM models are perfectly predicted the output data. The experimental results show that the LM method predict the data with a R^2 value of 0.9934, 0.99746, 0.99413 for power, torque, and discharge, respectively.

Introduction

Wind energy, the most established and sophisticated clean new energy source, is becoming more and more significant in the country's energy mix since traditional fossil fuels are being depleted and causing significant environmental damage [1]. As the wind turbine system's driving force, wind speed is both an input and a parameterizing variable for the dynamics that establish the system's operating point. As such, it is crucial to the control design of the wind power system [2]. The majority of current Maximum Power Point Tracking (MPPT) management methods depend on precise and useful wind speed data [3]. In the meantime, it can also be utilized as a gain-adjusting parameter for the uncertainty disturbance control design, and pitch angle control [4]. To provide wind speed information, wind speed meters are typically positioned near wind turbines [5].

The wind speed gauge only measures the wind speed at one location, hence it is unable to offer an accurate wind speed reading at every area of the wind turbine. Furthermore, the existence of tower shadow effect, and wind turbine wake effect may affect the wind turbines' real capacity to convert wind energy, lowering the wind power system's overall efficiency [6]. As can be observed, there is a discrepancy between the recorded and effective wind speeds [7]. This discrepancy has a significant effect on the wind energy conversion process, which in turn affects the wind power system's control effect, power generation efficiency, and economic benefits [8].

The wind energy is not only used for power generation it can also be used for pumping of water for agricultural purpose. Mostly in the rural areas, there is a huge water scarcity, this can overcome by establishing wind energy-based water pump. This water is very essential resource for human survival, used for drinking, washing, and cleaning, among other things. Aside from human life, water can be used for agricultural purposes in rural locations. The need for water is gradually increasing for human and agricultural purposes. The wind speed for a specific area varies, so it is critical to research the site's climatic conditions over the course of a year to determine its compatibility. To eliminate distractions, most windmills are positioned on hills or the rigs are tall enough to keep the rotor well above the obstacles.

Odesola and Adinoyi [9] developed a wind energy-powered water pump prototype with a rotor diameter of 2.14 m and a total surface area of 3.733 m^2, achieving an average wind speed of 2.5 m/sec at a height of 16 m. Prasad et al. [10] created a horizontal axis windmill-powered water pump with three blades, tensile strength of 31 N/mm^2, density of 0.855 g/cm^3, blade length of 30 inches (2.5 feet), and blade strength of 45006000 psi. Girma et al. [11] created and simulated a wind and diesel-powered water pumping system using MATLAB software. They also conducted economic and life cycle cost analyses.

Shyam and Anjum [12] studied wind speed characteristics, power density, and Weibull parameters. They found that higher wind speeds in the range of 3-12 m/s during different seasons resulted in an average wind speed of ~2 m/s and a 15.3% and 27% increase in wind speed at 50 and 100 m heights, respectively. Ani and Ndubueze [13] designed a stand-alone photovoltaic/wind (PV/wind) hybrid energy system for a household in a rural area of Nkanu-West in Eastern Nigeria. They observed an electric load of 8.4 kwh/d and an average wind speed of 2.1 m/sec. Ramos et al. [14] investigated the important meteorological and wind farm general design parameters that influence the annual energy production of US onshore wind farms. They used two case studies to determine the impact of wind speed, precipitation, and temperature on the annual power output of significant onshore wind farms in Texas and across the United States. Wind speeds ranging from 11.1 to 11.6 mph were found to yield the highest annual production among the wind farms studied.

Nedaei et al. [15] analysed the technological, economic, and environmental aspects of large-scale wind generating plants. They discovered that the payback period is 2.7 years, and the gross yearly greenhouse gas (GHG) emissions from the wind farm are decreased to 26,671 t of CO_2 over the course of 20 years of operation. Ajao and Adegun [16] investigated the parameters and installation of a locally manufactured three-blade horizontal-axis wind turbine. The wind turbine rotor has a swept area of 3.65 m^2, and the blade angle of attack is 7°, as established experimentally. The turbine was mounted 14.9 meters above ground level, and the cutting wind speed was determined to be 3.5 meters per second.

As computer technology advances quickly, neural networks are being used extensively to anticipate wind speed and power output. For example, convolutional neural networks (CNNs) [17], recurrent neural networks (RNNs), and long short-term memory neural networks (LSTMs) [18] are examples of deep neural network models that have been utilized in recent studies to forecast wind speed. These models have good prediction accuracy and are able to extract deep nonlinear features from wind speed data.

The aim of this research is to better comprehend the experimental analysis of a wind energy-based water pumping system. This report describes an experimental project that used wind energy to pump water from the earth surface. Wind turbine blades are designed to convert the highest amount of wind energy into practical water pumping energy.

The architecture of an artificial neural network

Creation of ANN architecture is the next stage. The original network architecture was a network with two hidden levels. To determine the number of neurons in the buried layer, a systematic approach was used. Levenberg–Marquardt technique was used to train the feed-forward network [20]. Two, ten, and one neuros were used for input, hidden, and output layers. The ideal neural network architecture for attribute prediction is shown in Fig. 1.

Materials Research Proceedings 64 (2026) 9-20 https://doi.org/10.21741/9781644904091-2

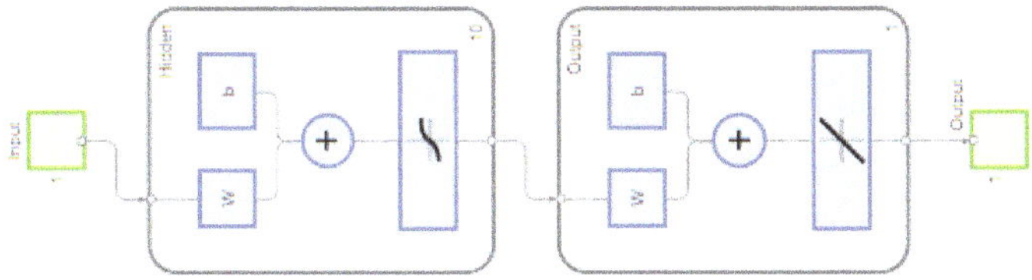

Fig 1. *ANN architecture.*

The performance of a trained network can be roughly evaluated using errors from training, validation, and test datasets. Evaluated the network's ability to predict conductivity using regression analysis. The congruence between the estimates from the trained network and the experimental data is evaluated using the regression coefficient (R^2). Furthermore, the mean square errors (MSE), which are defined as follows, were used to assess different neural network architectures.

$$MSE = \frac{1}{n}\sum_{i=1}^{n}\left(y_{exp} - y_{net}\right)^2 \quad (1)$$

$$R^2 = \frac{\sum_{k=1}^{n}(y_{exp}-\bar{y}_{exp})(y_{net}-\bar{y}_{net})}{\sqrt{\sum_{k=1}^{n}\left(y_{exp}-\bar{y}_{exp}\right)^2 \sum_{k=1}^{n}(y_{net}-\bar{y}_{net})^2}} \quad (2)$$

Where, y_{exp} is the experimental value, y_{net} is ANN values, n is the number of data points.

Experimental analysis

The experiments were conducted in Kimir Dingay Village in Amhara region in Ethiopia, and the working site was chosen based on wind data provided by Solar and Wind Energy Resource Assessment (SWERA). The greatest wind speed at this location ranges from 3.01 to 5.32 m/sec, and the trials were carried out in the whole month of July 2025. The location site was chosen such that no buildings or long trees will hinder wind velocity.

The schematic diagram of the wind energy-based water pump system is shown in **Fig. 2**(a), and the photograph is shown in Fig. 2(b). The water pump schematic diagram with dimensions are shown in Fig. 2(c). Its components include a multiple blade rotor, tail, tower, pump rod, discharge pipe, well seal, drop pipe, tower footing, pump cylinder, screen, and storage tank. Windmills are multi-bladed rotors that are positioned on a hub. Depending on the wind conditions and the design, there could be 8, 16, 18, or 24 blades. Multiple curving blades (sails) are rigidly placed at an angle to the wind's direction. The curved curvature of the blades aids in the capture of more energy when air moves over the sails.

When all of the blades are installed above the hub, they create the rotor. When wind travels through the rotor, it rotates, producing mechanical energy. The rotor is attached to a shaft, which rotates the gears. The transmission system is made up of gears, guide wheels, pitman arms, and guide shafts. The pump rod passing through transfers the motion to the pump plunger. The tower is a high, robust support for the windmill and gear mechanism. To prevent obstructions in the wind route approaching the windmill, the tower must be tall enough to clear the tops of neighbouring trees and buildings.

The horizontal wind mill was designed to generate a maximum wind speed of 5.23 m/sec and a minimum wind speed of 2.1 m/sec at 10 m above ground level. The windmill tower has an 18-meter height, the pump has a static water head of 23.5 meters below ground level, and the overall depth is 61 meters. The volume flow rate is 0.6 lit/sec.

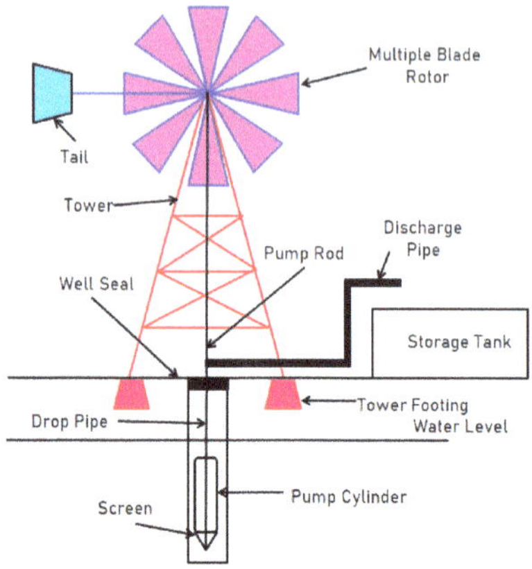

Fig. 2(a): *Components and working of a wind pump.*

Fig. 2(b): *Photograph of wind pump.*

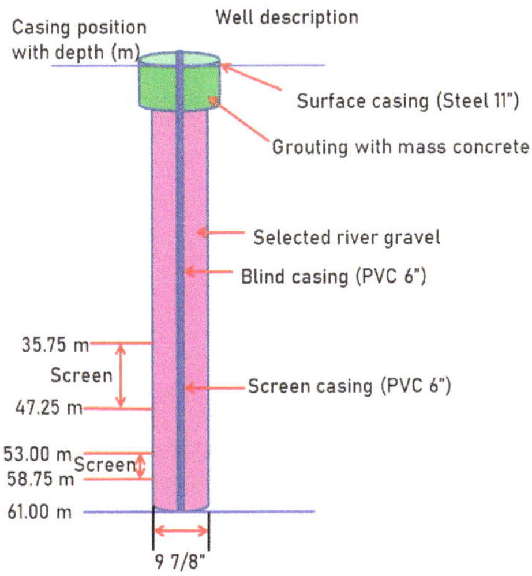

Fig. 2(c): *Schematic diagram of pump.*

Data analysis

This data was then analysed using the Weibull distribution approach to generate an average monthly wind speed data at the windmill installation area.

(a) Average wind speed:

One of the most important information on the wind ranges available at a location is its average wind velocity (Saravanan et al. [20]). In simple terms, the average velocity (v_m) is given below.

$$v_m = \frac{1}{n}\sum_{i=1}^{n} v_i \tag{3}$$

Where, v_m = average wind speed, v_i = wind speed at each time, and n is number of wind data.

$$v_m = \left(\frac{1}{n}\sum_{i=1}^{n} v_i{}^3\right)^{1/3} \tag{4}$$

(b) Wind power:

The wind power can be estimated based on the below equations.

$$\text{Wind power } (P_w) = \text{Kinetic energy/unit time} = \frac{1}{2}\dot{m}V_m{}^2 \tag{5}$$

$$\dot{m} = \rho A V_m$$

$$\text{Then, } (P_w) = \frac{1}{2}\rho A V_m{}^3 \tag{6}$$

(c) Wind mill power extracted

$$P_T = C_p p_w = \frac{1}{2} C_P \rho A V_m{}^3 \tag{7}$$

The Betz Limit is the maximum possible value for C_p which is equal to $\frac{16}{27}$ but the optimum possible for a multi blade windmill is 30%.

(d) Torque extracted:

The torque is one of the important parameter which is evaluated based on the pumping power and the rotor speed.

$$T = \frac{P_W}{\omega} \quad (8)$$

Rotor speed, $\omega = \frac{\lambda v}{R}$

$$T = \frac{P_T}{\lambda v / R} \quad (9)$$

But, $P_T = \frac{1}{2}\rho\pi R^2 {V_m}^3 C_P$

Therefore, $T = \frac{\frac{1}{2}\rho\pi R^2 {V_m}^3 C_P}{\lambda v / R}$

$$Q = \frac{P_W}{\rho g H} \quad (10)$$

Where, ρ is the density of water (kg/m^3), g is the acceleration due to gravity (m/sec), Q is the flow rate (m^3/sec), H is the total pumping head (meters of water).

Results and discussion

Power

The power is calculated based on the Eq. (7) at different wind velocity, and the results are shown in **Fig. 3.** The wind velocity varies from 3.01 to 5.23 m/sec. Under the low wind velocity of 3.01 m/sec, the power extracted is 192.75 N-m, where it is increased to 1064.02 N-m at a wind velocity of 5.23 m/sec. When the wind velocity increased, the turbine rotates at higher speed and correspondingly, the power produced by the wind turbine is increased.

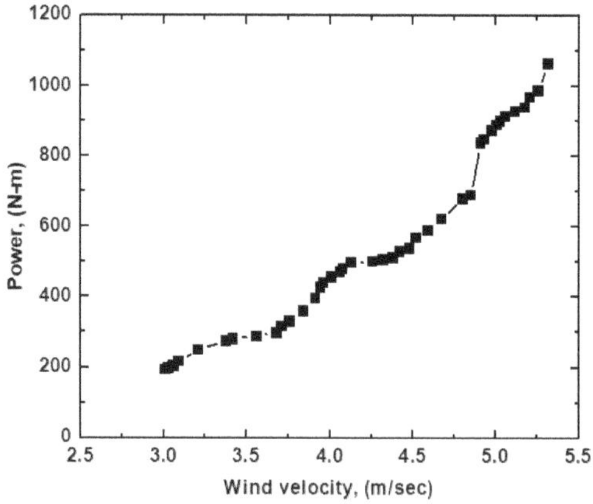

Fig. 3: *Wind velocity versus power and torque produced.*

The relation over experimental wind mill power and ANN data of wind power under train, validation, and test data sets were analyzed. Total 40 data points were used to perform ANN analysis. In that 28, 6, and 6 data points were used for training, validation and testing of the algorithm. 70% of the data is used for training and 30% of the data is used for testing. This picture illustrates that whole data are clustered. Suggested that NN may accurately forecast the wind mill power data under minimal deviation from experimental data. The training state curve of power is mentioned in **Fig. 4**. The training is occurred under the gradient of 1460.8563 at epoch of 307. The mean square error of the trained data is 1.0493e+03, validation is 561.3996, and test is 455.6743. The R^2 values presented in **Fig. 5** for the training outputs, validation, test, and overall data are 0.99139, 0.9971, 0.99858, and 0.9934, respectively.

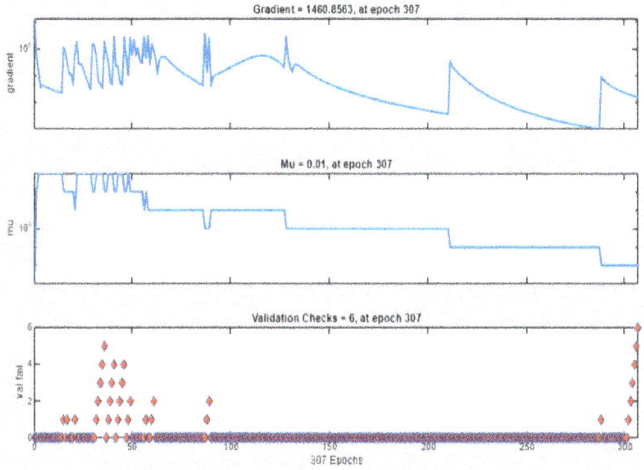

Fig. 4: *Training state of the algorithm.*

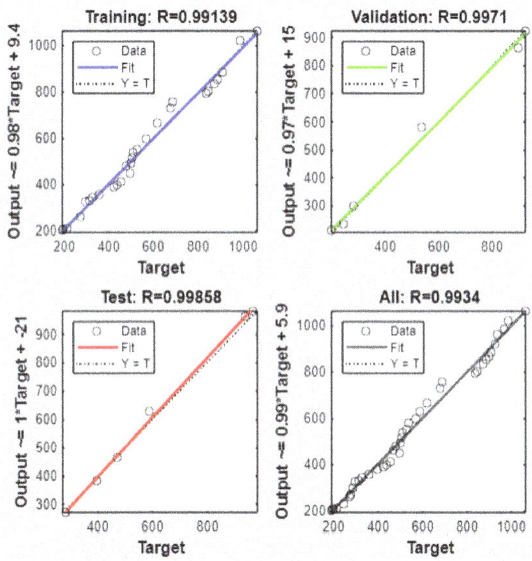

Fig. 5: *The R^2 of train, validation, test, and all data.*

Torque

The torque is calculated based on the Eq. (8) at different wind velocity, and the results are shown in **Fig. 6.** The wind velocity varies from 3.01 to 5.23 m/sec. Under the low wind velocity of 3.01 m/sec, the torque produced by the shaft is 320.12 N-m, where it is increased to 1002.48 N-m at a wind velocity of 5.23 m/sec. When the wind velocity increased, the torque produced by the shaft is higher correspondingly, the shaft torque is increased.

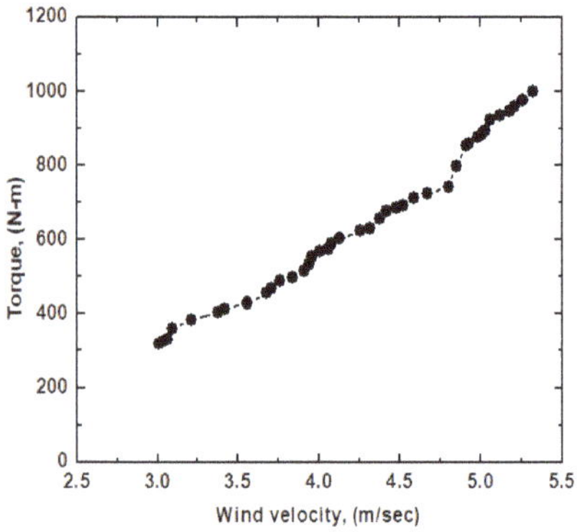

***Fig. 6:** Wind velocity versus torque produced by the shaft.*

The relation over experimental torque and ANN data of torque under train, validation, and test data sets were analyzed. A total of 40 data points was used, in that 28, 6, and 6 data points were used for training, validation and testing of the algorithm. 70% of the data is used for training and 30% of the data is used for testing. Suggested that NN may accurately forecast the torque data under minimal deviation from experimental data. The training state curve of power is mentioned in **Fig.** 7. The training is occurred under the gradient of 2785.774 at epoch of 247. The mean square error of the trained data is 232.5287, validation is 115.7611, and test is 292.4508. The R^2 values presented in **Fig. 8** for the training outputs, validation, test, and overall data are 0.99139, 0.9971, 0.99858, and 0.9934, respectively.

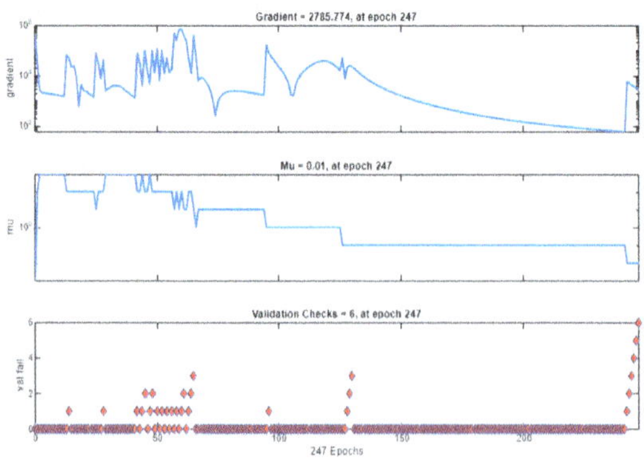

***Fig.** 7: Training state of the algorithm.*

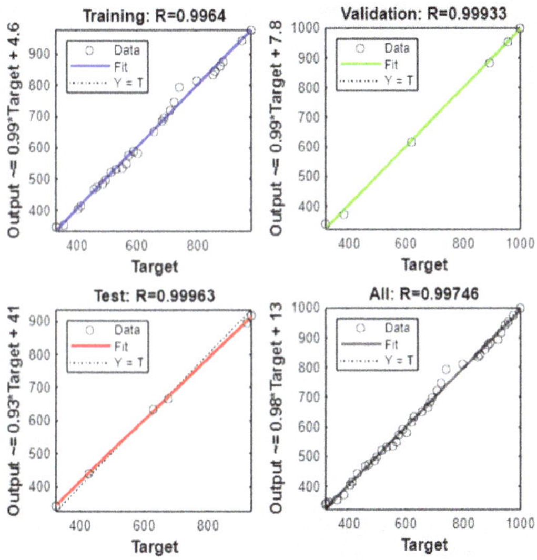

Fig. 8: *The R^2 of train, validation, test, and all data.*

Discharge of water

The discharge of water coming out from the pump is calculated based on Eq. (9) at different wind velocity, and the results are shown in **Fig. 9.** The wind velocity varies from 3.01 to 5.23 m/sec. Under the low wind velocity of 3.01 m/sec, the torque produced by the shaft is 0.0015 m3/sec, where it is increased to 0.0091 m3/sec. When the wind velocity increased, the discharge is increased.

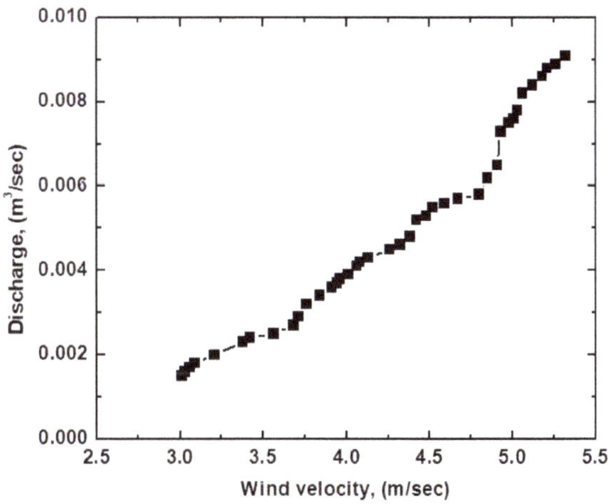

Fig. 9: *Discharge versus wind velocity.*

The relation over experimental discharge and ANN data of discharge under train, validation, and test data sets were analyzed. A total of 40 data points was used, in that 28, 6, and 6 data points were used for training, validation and testing of the algorithm. 70% of the data is used for training and 30% of the data is used for testing. Suggested that NN may accurately forecast the torque data under minimal deviation from experimental data. The training state curve of power is mentioned in **Fig. 10**. The training is occurred under the gradient of 6.6517e-08 at epoch of 81. The mean square error of the trained data is 7.2658e-08, validation is 3.0946e-08, and test is 4.6288e-08. The R^2 values presented in **Fig. 11** for the training outputs, validation, test, and overall data are 0.99363, 0.99855, 0.995, and 0.99413, respectively.

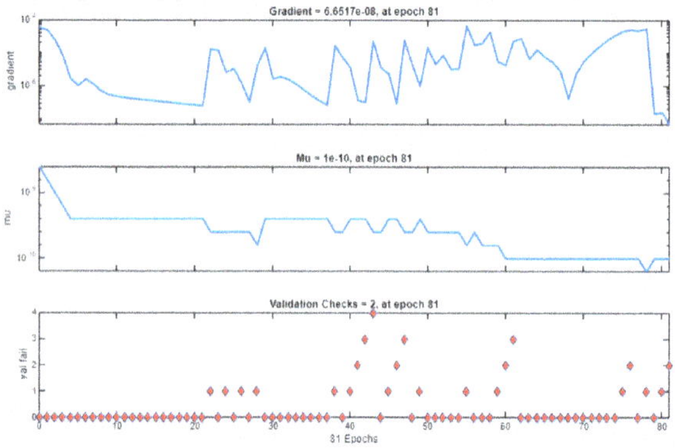

Fig. 10: *Training state of the algorithm.*

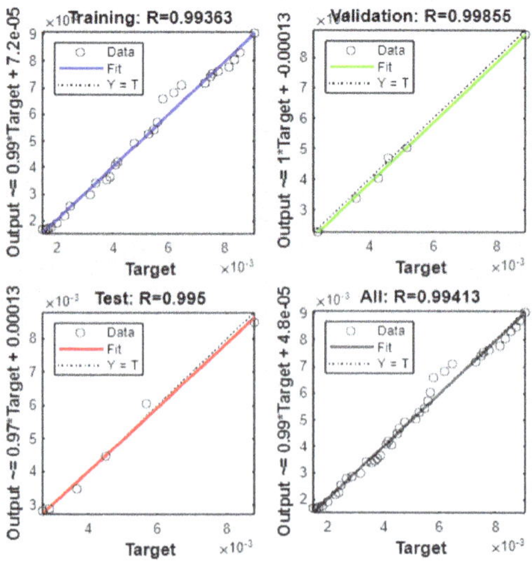

Fig. 11: *The R^2 of train, validation, test, and all data.*

Conclusion

Wind speed data is evaluated and wind pump is designed and analyzed experimentally, and the output data is predicted with ANN-LM method. The wind speed is observed is ranging from 2.5–5.23 m/s. The wind pump performs perfect at low wind speed and it is high when the wind speed is larger. The maximum power output of 1064 N-m, and torque is 1002 N-m is obtained at a wind velocity of 5.32 m/sec. Accordingly the maximum pump discharge of 0.0091 m^3/sec water is received from the pump under the same 5.32 m/sec wind speed. The experimentally obtained data is used for ANN predictions, and noticed that, the predicted data is perfectly match with the experimental data. The correlation coefficient of 0.9934, 0.99746, and 0.99413 for power, torque, and discharge from the ANN algorithm. It is found that, the ANN-LM provided a deviation of less than 2% with the experimental data.

Referenes

[1] M. Pirhooshyaran, K. Scheinberg K, L.V. Snyder, Feature engineering and forecasting via derivative-free optimization and ensemble of sequence-tosequence networks with applications in renewable energy, Energy. 196 (2020) 117136. https://doi.org/10.1016/j.energy.2020.117136

[2] P.M.O. Gebraad, J.W. Wingerden, Maximum power-point tracking control for wind farms, Wind Energy. 18(3) (2015) 429-447. https://doi.org/10.1002/we.1706

[3] S.M.R. Kazmi, H. Goto, H.J. Guo, O. Ichinokura, Review and critical analysis of the research papers published till date on maximum power point tracking in wind energy conversion system, 2010 IEEE Energy Conversion Congress and Exposition. 12-16 September 2010. https://doi.org/10.1109/ECCE.2010.5617747

[4] A. Dali, S. Abdelmalek, A. Bakdi, M. Bettayeb, A new robust control scheme: Application for MPP tracking of a PMSG-based variable-speed wind turbine, Renew Energy. 172 (2021) 1021-1034. https://doi.org/10.1016/j.renene.2021.03.083

[5] A.B. Asghar, X. Liu, Adaptive neuro-fuzzy algorithm to estimate effective wind speed and optimal rotor speed for variable-speed wind turbine, Neurocomputing. 272 (2018) 495-504. https://doi.org/10.1016/j.neucom.2017.07.022

[6] T.L. Van, T.H. Nguyen, D. Lee. Advanced pitch angle control based on fuzzy logic for variable-speed wind turbine systems, IEEE Trans Energy Convers. 30(2)(2015) 578-587. https://doi.org/10.1109/TEC.2014.2379293

[7] R. Yazc, E.K. Yaylac, F. Yaln. Modified golden section search based MPPT algorithm for the WECS, Eng. Sci. Technol. Int. J. 24(5)(2021) 1123-1133. https://doi.org/10.1016/j.jestch.2021.02.006

[8] P. Chen, D. Han, F. Tan, J. Wang. Reinforcement-based robust variable pitch control of wind turbines, IEEE Access. 8 (2020) 20493-20502. https://doi.org/10.1109/ACCESS.2020.2968853

[9] I.F. Odesola, L.G. Adinoyi, Development of wind powered water pump, Int. J. of Eng. Sci. and Comp. 7 (2017) 10341-10345.

[10] V. Prasad, M. Swami, V. Tambe, F. Kamate, S. Bagade, Wind operated water pump, Int. Research J. of Eng. and Tech. 3 (2016) 1433-1437. https://doi.org/10.9790/1684-2102010912

[11] M. Girma, M. Molina, A. Assefa, Feasibility Study of a wind powered water pumping system for rural Ethiopia, AIMS Energy. 3 (2010) 851-868. https://doi.org/10.3934/energy.2015.4.851

[12] S.S. Chandel, L. Anjum, Wind energy resource potential assessment in a hilly terrain of India using wasp, Int. J. Energy for a Clean Environment. 14 (2013) 275-294. https://doi.org/10.1615/InterJEnerCleanEnv.2014010432

[13] V.A. Ani. Energy optimization at macro base transmitter station site located in Nembe (Bayelsa State), Nigeria, Int. J. Energy for a Clean Environment. 15 (2014) 139-169. https://doi.org/10.1615/InterJEnerCleanEnv.2016006549

[14] C.A. Ramos, D. Ramirez, K. Jin, H. Li, Factorial analysis of selected factors affecting the electricity production of onshore wind farms, Int. J. Energy for a Clean Environment. 12(2011) 66-77. https://doi.org/10.1615/InterJEnerCleanEnv.2012004413

[15] M. Nedaei, P. Walsh, E. Assareh, Sustainable energy planning of a wind power plant by coordinating clean development strategies, Int. J. Energy for a Clean Environment. 21(2020) 59-89. DOI:10.1615/InterJEnerCleanEnv.2020033676

[16] K.R. Ajao, I.K. Adegun, Development and power performance test of a small three-blade horizontal-axis wind turbine, Heat Transfer Research. 40 (2009) 777-792. https://doi.org/10.1615/HeatTransRes.v40.i8.40

[17] H. Li, K. Shi, P. McLaren, Neural-network-based sensorless maximum wind energy capture with compensated power coefficient, IEEE Trans. Ind. Appl. 41(6) (2005) 1548-1556. https://doi.org/10.1109/TIA.2005.858282.

[18] G.B. Huang, Q.Y. Zhu, C.K. Siew, Extreme learning machine: a new learning scheme of feedforward neural networks, In: 2004 IEEE Int. J. Conference on neural networks (IEEE cat. No.04CH37541) 2 (2004) 985-990. https://doi.org/10.1109/TIA.2005.858282

[19] D.W. Marquardt, An algorithm for least-squares estimation of nonlinear parameters. J. Soc. Ind. Appl. Math., 11 (1963) 431-441.

[20] S. Saravanan, D. Anbazahan, G.S. Manikandan, M. Mathavan, A.M. Afrith, K. Prakash, Design and fabrication of wind energy water pump, Int. J. Engineering Research & Technology, (2018) 1-4. https://doi.org/10.17577/IJERTCONV6IS04050

Vision Transformer-Based Visual Inspection of Photovoltaic Cells from Angular Images and Fault Recognition

Hiren MEWADA[1,a], Syam Sundar LINGALA[2,b], Nayeemuddin MOHAMMED[3,c]

[1]Electrical Engineering Department, Prince Mohammad Bin Fahd University, Al Khobar, Saudi Arabia

[2]Mechanical Engineering Department, Prince Mohammad Bin Fahd University, Al Khobar, Saudi Arabia

[3]Civil Engineering Department, Prince Mohammad Bin Fahd University, Al Khobar, Saudi Arabia

[a]hmewada@pmu.edu.sa, [b]sslingala@gmail.com, [c]mnayeemuddin@pmu.edu.sa

Keywords: Fault Recognition, CNN, Renewable Energy, Photovoltaic, Vision Transformer

Abstract. The most predictable and economical renewable energy source is solar photovoltaic (PV). Nevertheless, various environmental factors cause physical and electrical defects in the PV cell, reducing its power output, reliability, and longevity. This paper addresses this challenge using a vision transformer (VT)-based CNN. A vision transformer transforms an image into patches, whereas a deep CNN does not. Long-range associations are absent in a deep CNN. The photovoltaic cells exhibit complex, overlapping defects in a specific area. Hence, patch-based feature extraction on the VT result yields higher detection rates. The proposed model is tested to assess the extent of electrical and physical damage in cells relative to healthy cells. The data includes pictures taken in different angles and different environmental conditions, and the information obtained was made consistent, reliable, and applicable in the field. The proposed model was effective in classifying PV cells, achieving a precision of 0.9722 for clean images, 0.9565 for clean images, and 0.9286 for electrical- and physical-damage images, respectively.

Introduction

Renewable energy, specifically photovoltaic (PV) based solar energy, is the most promising source in the context of the dual-carbon aim [1]. However, PV cells are vulnerable to various faults during production, shipping, and installation [2]. Changes in the environment also deteriorate PV cells, either electrically or physically [3]. These failures cause power outages and fire risks [4]. PV cells installed in hostile environments, including high-wind conditions on rolling hills, deserts, and elevated sites, are significantly affected by these challenges [5]. Specifically, environmental factors, such as deviant electrical contacts and material aging, may lead to open circuits, short circuits, hot-spot effects [10], surface contamination, broken welds, and local shading [11].

Therefore, authors in [6, 7] reported that efficient, timely deployment is necessary to maintain PV cell reliability and efficiency. Although thermal imaging technology [8] is most efficient for identifying electrical damage, it is not practical in all places. In contrast, visual imaging technology using computer vision approaches [9] offers an easy and convenient way to identify deterioration in PV cells.

Temperature abnormalities usually accompany these failures, and physically contactless infrared thermography [12] enables rapid detection of areas exhibiting abnormal temperature increases. Thermal imaging allows characterization of the defect relative to RGB images and offers the advantage of identifying a hotspot in the panel. Kirubakaran et al. [13] suggested that an image-processing-based diagnostic method for detecting PV panel failure can be performed using a thermal imaging system. In [14], an experiment using machine learning was presented, using thermal images of PV cells to detect cell aging and the presence of any failures. An infrared thermography dataset was developed by Boubaker et al. [15], comprising perfect and adversarial conditions of PV cells.It was trained on a dataset using a state-of-the-art deep learning model that demonstrated the ability to accurately detect different types of faults when partitioned into binary and multi-class classification tasks [16]. However, the environmental factors and image clarity are of great importance to the diagnostic power of infrared thermography [17]. Noise, visual blurring, or occlusion, to varying extents, can also decrease the system's classification accuracy. To ensure the overall diagnostic system is more stable and functional, a future study should pay greater attention to image preprocessing and quality enhancement. More recent research has suggested that data-driven diagnostic techniques based on electrical characteristic factors, such as deep learning, limit learning machines, and I-V curve diagnostics [18].

The study suggested that voltage-, current-, and power-based electrical characteristics can detect electrical faults in PV cells that may not be evident in visible-band images. This approach allows real-time performance analysis of the cells at the cost of increasing hardware cost. Recently, artificial neural networks, long short-term memory networks (LSTM), and an advanced transformer network have been used to identify electrical faults [19]. Che Yang et al. [20] proposed integrating a multiplayer perceptron network with an LSTM to identify fault types. In addition, they also located the faulty region of the PV cells. Hichri et al. [21] used a genetic algorithm-optimized ANN to determine whether grid-connected photovoltaic panels were defective. The model was evaluated on a real-time dataset and did not incur any additional hardware liability. Thus, this approach offers a good balance of performance measurement with low computational time. Han Su-Min et al. [22] encoded current data of the PV cells into 2D fault-feature images using a two-channel Gram angle field code. These images are then used to explain various kinds of faults. To automatically identify and classify PV defects, the produced images are loaded into a deep residual network (ResNet). By converting a one-dimensional sequence into a more comprehensive two-dimensional image dataset, the data's dimensional relevance is maintained, and it may be visually recognized, which somewhat increases accuracy. Nevertheless, the residual network requires improvement due to its high computational resource requirements and the lack of an efficient sensing field.

Environmental conditions primarily affect the performance of PV cells. Moisture intrusion causes short circuits or electrical arcing, and corrosion is visible on the PV cells, reducing their efficiency. Even extreme temperatures, UV exposure, and electrical surges can significantly damage PV cells electrically. In addition to electrical damage, the physical damage also reduces the performance of PV cells. High winds and heavy snow load physically damage the cells. The dirt, snow, and leaves block the sunlight and provide shading for the cells. This reduces the exposure of PV cells to light, reducing electricity generation. Therefore, an initial approach is required to identify any physical damage or electrical damage using visual images, and later, further categorization can be obtained. Thus, this paper presents a method for determining physical and electrical damage or for cleaning PV cells using visual and angular images captured across different time zones. Angular capturing provides a realistic approach to use them in practice. Though CNN is primarily used for image classification, its main weakness is that it processes entire images. However, for PV cells, it is required to focus on a specific region of the PV cell surface to improve the classification accuracy. The self-attention mechanism in VT on different regions of the image effectively extracts the local features from the desired areas. Therefore, the proposed approach uses VT to identify PV cells visually into three categories. Overall, the paper is structured as follows: The next section presents the proposed approach and function architecture. The dataset used to evaluate the model is provided, and the results for the proposed approach are presented. Later, a comparison with baseline models is given, and the paper concludes with remarks and a future scope.

Method

This study uses a vision transformer (VT) network to improve PV cell fault classification. VT is a well-known transformer-based deep learning model, primarily used across various image classification applications. The VT network processes images by dividing them into p x p patches. These patches are processed sequentially to extract image features. Later, each patch is converted to a vector for use by the transformer module. During conversion from a patch to a vector, a positional embedding is assigned to each patch or vector to preserve the patch's original position within the image. Then, aggregate information is obtained from all vectors, and a classification token (CLS) is used to classify them. Later, concatenated information of CLS and the vector is processed through transformer layers, where each layer consists of a self-attention mechanism and an MLP block.

This transformer layer produces a sequence of the same length. The self-attention mechanism allows the capture of long-term dependencies among multiple vectors. In the self-attention layer, the input vector is transformed into a query (Q), key (K), and value (V) matrix. The Q matrix presents the query features to be matched against other input vectors. The K contains the 'key' to match with other input vectors, whereas the V contains the original input information. These matrices Q, K, and V are computed by multiplying each vector by the learnable weight matrices Wq, Wk, and Wv.

The computation of a standard single-head self-attention feature map Z from input X is as follows:

$$Z = SOFTMAX\left(\frac{QK^T}{\sqrt{d_q}}\right)V \tag{1}$$

Where, d_q is the embedding dimension. The multiplication of V in equation 1 provides larger attention in the next layer [h1]. In VT, a multi-head attention mechanism is used, where each task is split into multiple single tasks (i.e., heads). For each head, the learnable weight matrices are calculated and the outputs of all heads are concatenated to a single matrix, i.e., [Z0, Z1, Z2...Zh-1]. After the attention mechanism, the output is processed by an MLP. The ultimate hidden state of the [CLS] token is used for classification. A fully connected layer is applied to the output of the [CLS] token to predict class probabilities. The overall function flow of the VT network is shown in **Figure 1.**

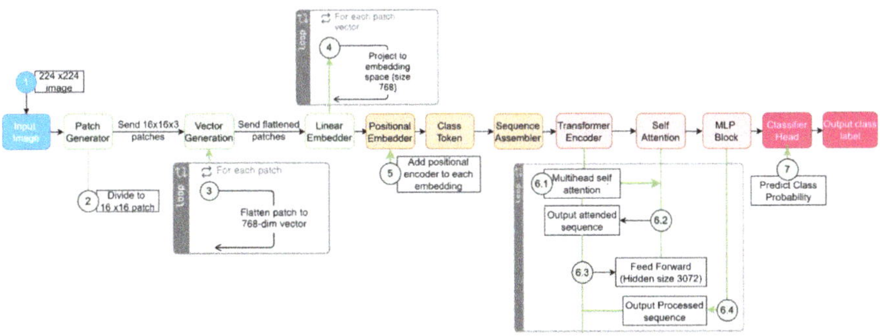

Figure 1: Function architecture of the vision transformer.

Dataset and results

The dataset consists of two types of faults, physical and electrical damage, and includes clean or working cell images. The dataset has 69 images of physical damage, 102 images of electrical damage, and 192 images of clean cells. The dataset can be accessed from [23]. We acknowledge the limitations in dataset size associated with specialized industrial imaging applications. Image augmentation via rotation, shifting, and mirroring is used to increase the dataset size during training. A 5-fold validation is used to reduce variance in the model's performance and mitigate overfitting. Although our augmentation strategy meets current practical requirements, future research utilizing bigger, naturally balanced datasets will yield more robust confirmation of our methodology. To ensure compatibility, all images are normalized with a mean of and a standard deviation of 0.5 and resized to 224 x 224. 80% of the images were used for training, and 20% for validation. The model is configured to train for 20 epochs with a batch size of 8. An Adam optimizer with a learning rate of 10E-4 is used during training. **Table 1** indicates the model's parameters.

Table 1: *Lists the model's parameters.*

Parameter	Value
Pre-trained model	google/vit-base-patch16-224
Input image size	224 x 224
Patch size	16 x 16
Number of classes	3
Batch size	8
Number of epochs	20
Learning rate	1e-4
Number of transformer layers	12
Hidden size	768
Number of attention heads	12
Feed-forward network size	3072
Positional encoding	Yes (learnable)
Loss function	CrossEntropyLoss
Optimizer	Adam

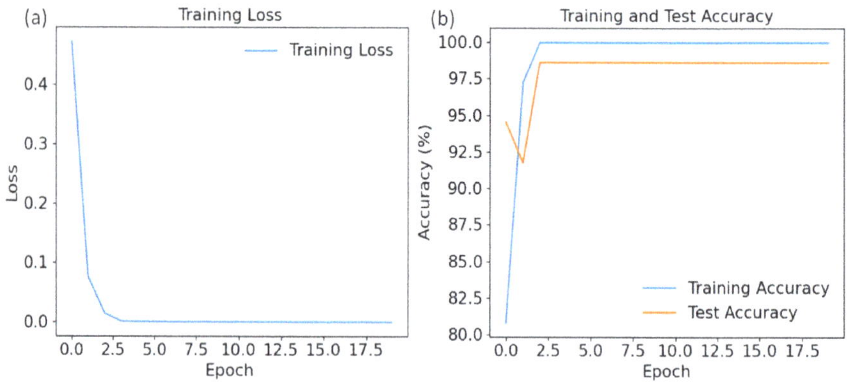

Figure 2: *Accuracy and loss analysis during the training phase.*

The algorithm is developed using a Python tool, and the model is trained on a computer with an Intel i7 processor and 64GB RAM. The model was trained on an 80% dataset, and its accuracy and loss were monitored over epochs, as illustrated in **Figure 2**. The loss is lower, and the classification accuracy is higher on the training/test datasets, indicating that the model is learning well. The training accuracy is fast approaching 100 per cent; the test accuracy is 97.5 per cent. This is because when test accuracy is lower than training accuracy, it indicates the model is not overtrained.

The model's test and training data are evaluated by plotting a confusion matrix (CM). CM provides an overview of where all the classes can be classified, specifying the model's accurate and inaccurate classifications. CM matrices of the training and test datasets are provided in **Figure 3**. The model achieved 100% accuracy in classifying all training images. On the same note, the model also performed well on the test dataset, achieving an accuracy of 95.89 percent on physical damage images. In **Figure 4**, the model's performance can be assessed using the quantitative parameters listed below.

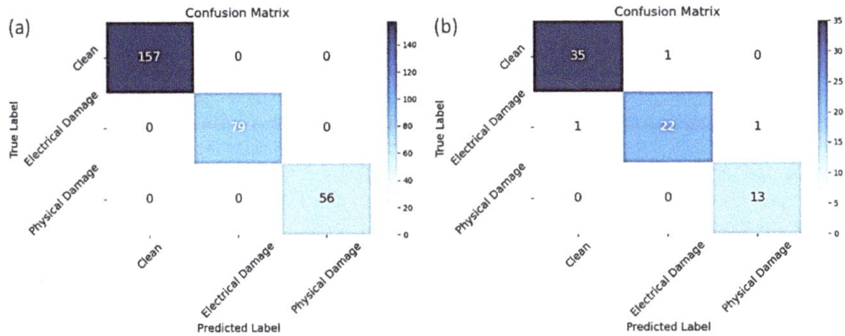

Figure 3: *Confusion matrices for the training and test datasets.*

Precision	TP / (TP + FP)
Recall also called Sensitivity	TP / (TP + FN)
F1 Score	2*Precision*Recall / Precision + Recall
Macro-F1	average of F1 Scores

Weighted-F1	weighted-averaged of F1 Scores
Misclassification Rate	# incorrect predictions / # total predictions
Specificity (TNR - True Negative Rate)	TN / (TN + FP)
False Positive Rate (FPR)	1−TNR
False Negative Rate (FNR)	1−Recall

Figure 4: *Quantitative parameters for model evaluation*

All these parameters enable an ablation study of the model's performance. These calculations allow us to identify the strengths and weaknesses of the models. **Table 2** lists these parameters for each class.

Table 2: *Shows these parameters for the test datasets.*

Class Name	Precision	1-Precision	Recall	False Negative Rate	F1 score	Specificity (TNR)	False Positive Rate (FPR)
Clean	0.9722	0.0278	0.9722	0.0278	0.9722	0.9730	0.0270
Electrical Damage	0.9565	0.0435	0.9167	0.0833	0.9362	0.9796	0.0204
Physical Damage	0.9286	0.0714	1.0000	0.0000	0.9630	0.9833	0.0167

A detailed analysis of the model is obtained by plotting the precision vs recall curve, varying the threshold level for all three categories on the test dataset. **Figure 4** shows the precision-recall and ROC curves for the test datasets. The values of both precision and recall approaching 1 indicate a very low false-positive rate across all classes. Similarly, A Receiver Operating Characteristic (ROC) curve compares the true positive rate (TPR) and the false positive rate (FPR). The area under the curve (AUC) represents the model's performance. The larger the area covered by ROC, the better the model's performance. Figure 4(b) shows that the AUC = 1 for all classes, indicating that the proposed VT is most efficient at identifying both electrical and physical damage in visual images, irrespective of how the images were captured.

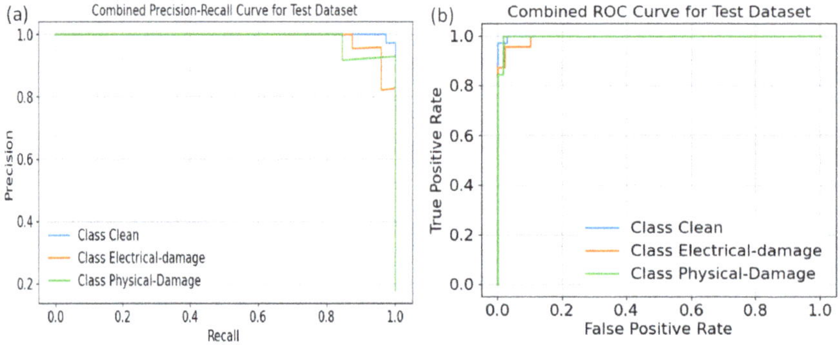

Figure 4: *Precision-Recall curve, and ROC curve for test datasets.*

The image shows a series of predictions from a classification model. In the top-left corner, the model correctly identifies an "Electrical Damage" instance. **Figure 5** shows a sample image from the test dataset and the model's classification.

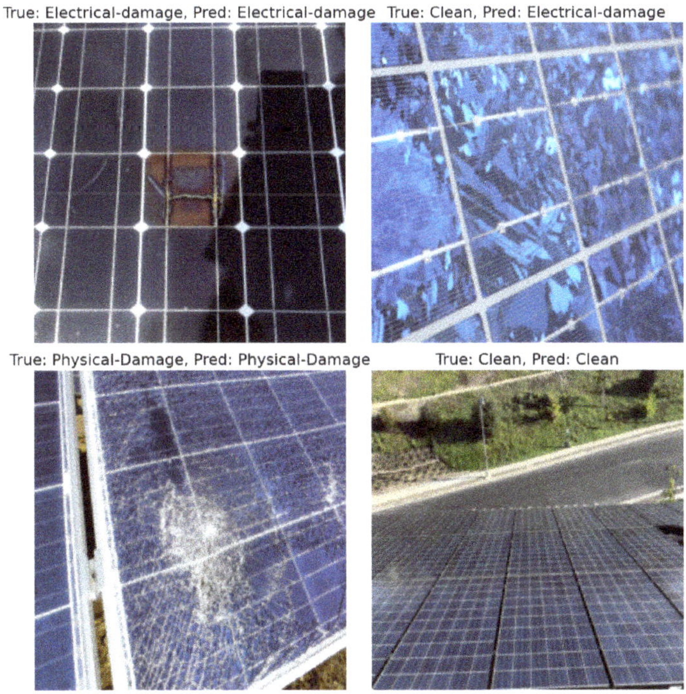

Figure 5: *Sample image from the test dataset and its classification using the model.*

Table 3: Comparison of the proposed model with a traditional deep learning network.

Model	Accuracy on Test Dataset
ResNet50	85.19%
MobileNetV2	85%
EfficientNetB7	89%
DenseNet121	91%
Proposed VT	95.89%

A further comparison between the model and the traditional deep learning model is presented in **Table 3.** Where each model is trained with the same data size, and models are tuned to find the best performance of the model on the test dataset. ResNet50 uses residual layers, which help it find small regional features well. ResNet50 was 100% accurate on the training dataset, but its test accuracy was only 85.19%. MobileNetV2 is a computationally efficient model designed for resource-constrained devices. This network's accuracy was reported as 89% in [24]. Similarly, the other variants, e.g., EfficientNetB7 and DenseNet121, outperformed ResNet50 and MobileNetV2. In contrast, the proposed model outperformed all these models, achieving 95.89% accuracy on the test dataset. The misclassification rate of 0.0411 suggests that only about 4.11% of the predictions were incorrect, highlighting the model's precision. The Macro-F1 score of 0.9571 demonstrates balanced performance across all classes, while the Weighted-F1 score of 0.9587, which accounts for class imbalance, further reinforces the model's effectiveness. Overall, these metrics suggest that the model is both accurate and robust in classifying the dataset's categories.

Conclusion

This paper evaluates a vision transformer-based CNN network to determine whether the PV cell is clean (i.e., perfect) or damaged due to electrical or physical causes. An automated approach using visible images captured across multiple time zones, at both aerial and oblique angles, has been used in the analysis. Results demonstrated that the model classified them with high accuracy (95.89%) and a low misclassification rate (0.0411). The comparison with the baseline model shows that the VT model is effective at identifying the damage type, with an improvement of 12.67% in accuracy, relative to MobileNetV2, which serves as the reference model in Table 3. The proposed approach has a few weaknesses. Presently, the model is evaluated on a small dataset. Additionally, a pre-trained VT network is used to identify the damage type. In the future, the model will need to be validated further by adding a larger set of images. The hyperparameters of the VT network can be further optimized using optimization algorithms to improve its classification rate. A real-time exposure of the model, deployed on resource-constrained devices, can enable the use of the model without human intervention and support real-time monitoring of PV cells under adverse environmental conditions.

References

[1] Z. Dong, Q. Zhang, X. Shu, et al. Photo-assisted charging of heterostructured NiCo2S4@NiCo-LDH composite electrode with remarkable photoelectronic memory effect for high-performance asymmetric supercapacitor, Energy Convers Management, 315 (2024), 118769 https://doi.org/10.1016/j.enconman.2024.118769

[2] C. Zhang, L. Tu, Z. Yang, et al. A CMMOG-based lithium-battery SOH estimation method using multi-task learning framework, J Energy Storage, 107 (2025), 114884 https://doi.org/10.1016/j.est.2024.114884

[3] Z. Wu, Y. Hu, J.X. Wen, et al. A review for solar panel fire accident prevention in large-scale PV applications, IEEE Access, 8 (2020), pp. 132466-132480 https://doi.org/10.1109/ACCESS.2020.3010212

[4] O. Idbouhouch, N. Rabbah, N. Lamrini, et al. Assessing PV inverter efficiency degradation under semi-arid conditions: a case study in Morocco, Heliyon, 10 (17) (2024), e36906 https://doi.org/10.1016/j.heliyon.2024.e36906

[5] M. Jalal, I.U. Khalil, A.U. Haq, Deep learning approaches for visual faults diagnosis of photovoltaic systems: state-of-the-Art review, Results Eng, 23 (2024), 102622. https://doi.org/10.1016/j.rineng.2024.102622

[6] Z. Dong, M. Chen, D. Qin, et al. Recent advances and perspective of modified TiO2-based photoanodes toward photoelectrochemical water splitting, Fuel, 373 (2024), 132366 https://doi.org/10.1016/j.fuel.2024.132366

[7] B. Taghezouit, F. Harrou, Y. Sun, et al. Model-based fault detection in photovoltaic systems: a comprehensive review and avenues for enhancement, Results Eng, 21 (2024), 101835. https://doi.org/10.1016/j.rineng.2024.101835

[8] Z. Fu, X. Shu, Q. Zhang, et al. Solar-driven induced photoelectron remember effect involved in core-shell NiCo2S4@Ni3V2O8 composite electrode with superior electrochemical energy storage for asymmetric supercapacitor, Energy Convers Manag, 323 (2025), 119190 https://doi.org/10.1016/j.enconman.2024.119190

[9] Z. Dong, D. Qin, J. Ma, et al. Bulk-phase and surface dual-defective engineering enabled Ti-based nanotubes photoanode for highly efficient photoelectrochemical water splitting, Renew Energy, 231 (2024) 121004. https://doi.org/10.1016/j.renene.2024.121004

[10] A.F. Amiri, H. Oudira, A. Chouder, et al., Faults detection and diagnosis of PV systems based on machine learning approach using random forest classifier, Energy Convers Manag, 301 (2024), 118076. https://doi.org/10.1016/j.enconman.2024.118076
[11] M. Simon, E.L. Meyer, Detection and analysis of hot-spot formation in solar cells, Sol Energy Mater Sol Cell, 94 (2) (2010), pp. 106-113. https://doi.org/10.1016/j.solmat.2009.09.016
[12] A. Mellit, An embedded solution for fault detection and diagnosis of photovoltaic modules using thermographic images and deep convolutional neural networks, Eng Appl Artif Intell, 116 (2022), 105459. https://doi.org/10.1016/j.engappai.2022.105459
[13] V. Kirubakaran, D.M.D. Preethi, U. Arunachalam, et al., Infrared thermal images of solar PV panels for fault identification using image processing technique, Int J Photoenergy, 2022 (2022), pp. 1-10 https://doi.org/10.1155/2022/6427076
[14] M. Simon, E.L. Meyer, Detection and analysis of hot-spot formation in solar cells, Sol Energy Mater Sol Cell, 94 (2) (2010), pp. 106-113. https://doi.org/10.1016/j.solmat.2009.09.016
[15] S. Boubaker, S. Kamel, N. Ghazouani, et al., Assessment of machine and deep learning approaches for fault diagnosis in photovoltaic systems using infrared thermography, Remote Sens, 15 (6) (2023), p. 1686 https://doi.org/10.3390/rs15061686
[16] K. Masita, A. Hasan, T. Shongwe, et al. Deep learning in defects detection of PV modules: a review, Solar Energy Advances, 5 (2025), 100090. https://doi.org/10.1016/j.seja.2025.100090
[17] A.W. Kandeal, M.R. Elkadeem, A. Kumar Thakur, et al., Infrared thermography-based condition monitoring of solar photovoltaic systems: a mini review of recent advances, Sol Energy, 223 (2021), pp. 33-43 https://doi.org/10.1016/j.solener.2021.05.032
[18] C. Bu, T. Liu, R. Li, et al., Electrical pulsed infrared thermography and supervised learning for PV cells defects detection, Sol Energy Mater Sol Cell, 237 (2022), 111561. https://doi.org/10.1016/j.solmat.2021.111561
[19] B. Li, C. Delpha, D. Diallo, et al. Application of artificial neural networks to photovoltaic fault detection and diagnosis: a review, Renew Sustain Energy Rev, 138 (2021), 110512. https://doi.org/10.1016/j.rser.2020.110512
[20] N.C. Yang, M. FaizanLong short-term memory-based feedforward neural network algorithm for photovoltaic fault detection under irradiance conditions, IEEE Trans Instrum Meas, 73 (2024), pp. 1-11. https://doi.org/10.1109/TIM.2024.3413159
[21] A. Hichri, M. Hajji, M. Mansouri, et al., Genetic-algorithm-based neural network for fault detection and diagnosis: application to grid-connected photovoltaic systems, Sustainability, 14 (17) (2022), 10518. https://doi.org/10.3390/su141710518
[22] S. Han, Y. Yu, Y. Guo, Fault diagnosis of open-circuit in photovoltaic inverters based on improved GAF-SE-ResNet, Journal of Solar Energy, 45 (10) (2024), pp. 336-344
[23] Ledmaoui, Y. Solar-Panel-Anomalies-Detecting. GitHub Repository. 2025. Available online: https://github.com/Ledmaoui/Solar-Panel-Anomalies-Detecting (accessed on December 2025).
[24] Alanazi, M.; Alamri, Y.A. SolarFaultAttentionNet: Dual-Attention Framework for Enhanced Photovoltaic Fault Classification. Inventions 2025, 10, 91. https://doi.org/10.3390/inventions10050091

Materials Research Proceedings 64 (2026) 30-35 https://doi.org/10.21741/9781644904091-4

Unequally Distributed Pumped Micro-Storage Hydropower System

Ala A. HUSSEIN[1a*], Gaydaa ALZOHBI[2b]

[1]Department of Electrical Engineering, Prince Mohammad Bin Fahd University, Khobar, Saudi Arabia

[2]Department of Mechanical Engineering, Prince Mohammad Bin Fahd University, Khobar, Saudi Arabia

[a]ahussein@pmu.edu.sa, [b]galzohbi@pmu.edu.sa

Keywords: Energy Storage, Hydropower, Micro-Storage, Optimization

Abstract. Pumped storage hydropower (PSH) is a mature and widely deployed technology for large-scale energy storage, yet its efficiency drops considerably when operating at light load. As modern power systems continue to shift toward decentralized generation and micro-grids, storage solutions that perform efficiently over a wide load range are increasingly needed. This paper presents a 100-kW unequally distributed pumped micro-storage hydropower (PMSH) system designed for residential and micro-grid applications. The proposed configuration consists of parallel AC-coupled units rated at 50 kW, 25 kW, two 10 kW units, and one 5 kW unit, allowing individual machines to operate close to their rated power across varying demand levels. The results show that, unlike centralized and equally distributed systems whose efficiency declines sharply below approximately 40% load, the proposed system maintains a nearly constant round-trip efficiency of about 90% from 5% to 100% of rated load. In addition, the system achieves maximum modularity and reliability indices of 1.0, compared to 0.05 and 0.2, respectively, for a centralized architecture. These results demonstrate that the proposed PMSH system offers a practical and efficient energy storage solution for small-scale, decentralized power systems. Derivation of the proposed system followed by results and analysis are presented.

Introduction

A pumped storage hydropower (PSH) system is a type of hydroelectric power generation system that stores energy by using two water reservoirs located at different elevations. These systems are typically used for grid energy storage, ([1], [2]), to balance electricity supply and demand and shave peak load [3]. During periods of low electricity demand, surplus electricity is used to pump water from a lower reservoir to an upper reservoir. This process converts electrical energy into potential energy by raising the water to a higher elevation. On the other hand, when electricity demand is high and the grid needs more power, water is released from the upper reservoir and allowed to flow back down to the lower reservoir through turbines. As the water flows down, it turns the turbines, which then generate electricity.

One limitation of conventional, grid-scale PSH systems is their low efficiency at light load as electrical machines (motors and generators) have a low efficiency, in general, at light load [4]. Moreover, while a centralized structure eases the integration of the system to the grid, having one path to the grid makes the system prone to single-point failure. Instead, a distributed structure that has smaller, parallel, AC-coupled units can overcome these limitations at the cost of increasing the grid-integration requirements [5].

With the evolving electrical power network and the need to decentralize the generation and storage of electrical energy, the need for a distributed structure is of utmost importance [6]-[8]. This paper proposes a 100-kW unequally distributed pumped micro-storage hydropower (PMSH) system that ideally serves micro-grids and residential applications. The proposed system utilizes parallel AC-coupled units with variable unit sizes to enhance the system's efficiency. The proposed system is highly efficient at any load as it always operate the AC-coupled units at their rated power. Besides enhancing the efficiency, the proposed system is modular and highly reliable. More details on the system are presented in this paper.

System Configuration

- Centralized Architecture: A centralized architecture (Fig. 1) refers to large-scale pumped energy storage systems typically used in utility or industrial applications. It involves large reservoirs and hydroelectric plants that serve a large number of consumers or the grid. Centralized storage systems offer a large-scale storage capacities, making them ideal for providing power to large regions or industrial applications. They can also provide significant grid stability, support load balancing, and ensure continuous power during peak demands. These systems also benefit from cost efficiency, particularly when building large reservoirs and dams. However, these systems are prone to single-point failure and have a high initial cost including the construction costs of reservoirs, infrastructure, and plants. Also, large dams and reservoirs can disrupt local ecosystems, displace communities, and alter river systems. They also have long construction time and have a limited flexibility as they are uneasy to modify, once built, to meet changing energy demands or needs.

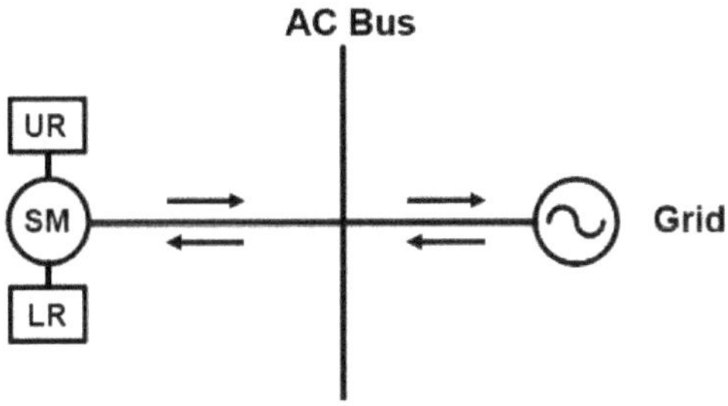

Fig. 1: Centralized structure.

- Distributed Architecture: Distribute pumped storage (Fig. 2) could be used for small-scale, often off-grid or residential applications. It involves relatively small energy storage devices, often combined with relatively small hydroelectric generators. Distributed storage systems are more reliable due to their parallel structure making them immune to single-point failure. Also, they have a low environmental impact as they are far less disruptive to the environment than large-scale systems, making them suitable for remote or ecologically sensitive areas. In addition, they have a lower initial cost compared to both central systems, making them an attractive option for small communities, farms, or individual households. Moreover, these systems have a modular structure and can be easily integrated into local settings with less infrastructure required. However, these systems have low storage capacity, and they require suitable water sources, such as streams or rivers, that are consistent and sufficient to maintain the energy production needed. In addition, small-scale systems often require more hands-on maintenance, which can be a challenge, particularly in remote areas, and they are uneasy to integrate to the power grid, making them more suited to isolated, off-grid applications.

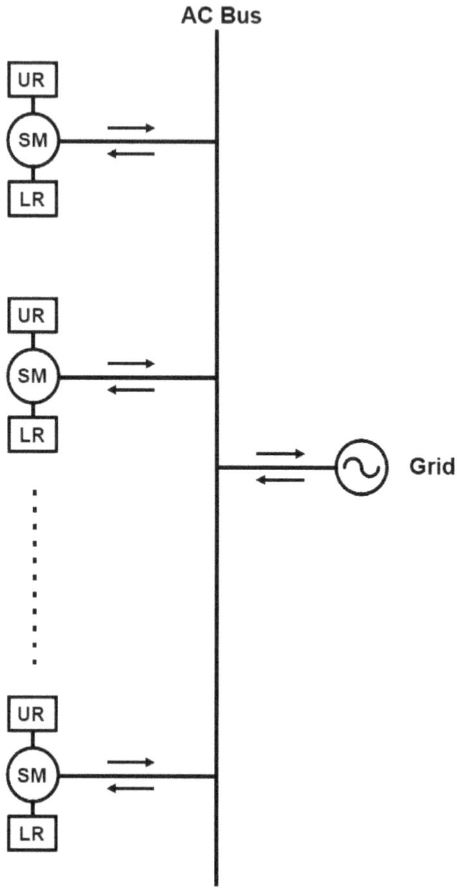

Fig. 2: Distributed structure.

System Configuration

Distributed systems can improve the efficiency at light load when the load is shared between the parallel, AC-coupled units where each unit operates around its rated power. To further enhance the efficiency at light load, an unequally distributed structure is proposed. The proposed system utilizes distributed storage units with different storage and power capacities that can accommodate a wide range of loads. An overall 100-kW system is considered that consists of an unequally distributed (UD) one 50-kW, one 25-kW, two 10-kW and one 5-kW AC-coupled units.

Compared to a centralized 100-kW system, the proposed system has a higher flexibility and reliability due to its parallel structure and demands less infrastructure requirements. Furthermore, the proposed system is immune to single-point failure. Compared to two distributed systems with equally-sized units of 50-kW and 25-kW, the proposed system is more flexible and has a higher reliability. Additionally, the proposed system has a higher efficiency at light load. For comparison, the modularity index (M.I.) and n reliability index (R.I.) are defined in (1) - (2), respectively.

$$M.I. = \frac{Minimum\ Unit\ Size}{System\ Size} \tag{1}$$

$$R.I. = \frac{No.of\ Parallel\ Units}{Maximum\ No.of\ Parallel\ Units} \tag{2}$$

Where the minimum unit size is 5-kW, the system size is 100-kW and the maximum number of parallel units is 5. The number of parallel units is 1, 2, 4 and 5 for the centralized, 50-kW equally-distributed (ED), 25-kW ED and UD systems, respectively. Table I summarizes the values of these indices for the presented systems.

Table 1: Modularity and reliability indecies for presented systems.

	Centralized	50 kW ED	25 kW ED	UD
M.I.	0.05	0.1	0.2	1
R.I.	0.2	0.4	0.8	1

Besides the mentioned merits, the efficiency of distributed systems can be enhanced by controlling the operation of the parallel units such that every unit operates near full load, which is impossible to achieve in centralized systems, and even in ED systems, the degree of flexibility is limited and hence the efficiency cannot be fully optimized. The efficiency calculation of the proposed 100-kW UD system along with a 100-kW centralized and 100-kW ED systems at different loads is demonstrated in the next section.

Analysis

The key to optimize the efficiency of a PMSH system is to operate the machine at full load during charging and discharging. Fig. 3 shows the efficiency curves, derived using [4], for a typical synchronous machine at a percent full load between 0% and 100%. An overall full-load round-trip efficiency of 90% is considered.

The systems illustrated in Section III are considered and compared for the entire percent full-load spectrum. Considering a 100-kW PMSH system, the centralized system possess the poorest efficiency at light load followed by the 50-kW ED and 25-kW ED systems, respectively. The proposed system shows a consistent efficiency for the entire range between 5% and 100% of full load. The efficiency versus load for the illustrated systems is shown in Fig. 4.

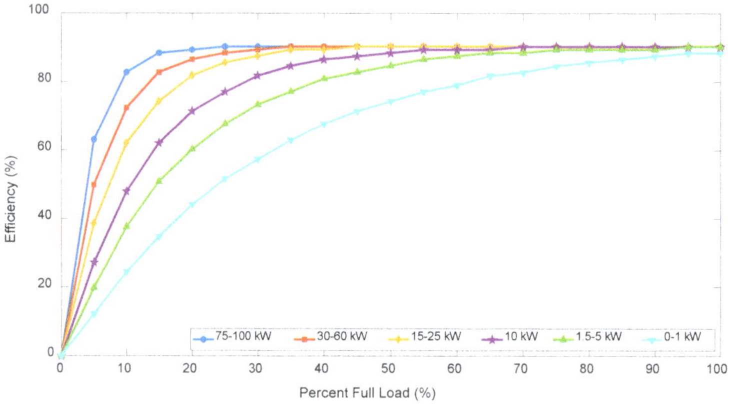

Fig. 3: Efficiency versus percent-full-load for a typical synchronous machine.

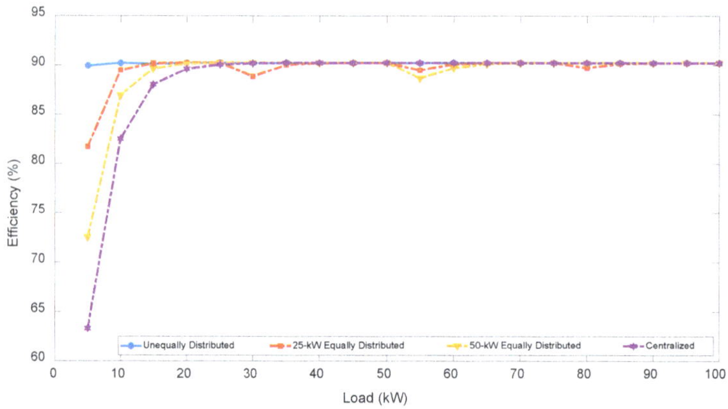

Fig. 4: The efficiency of the proposed UD system along with 25-kW ED, 50-kW ED and centralized systems.

Summary and Conclusions

The ongoing transformation of electrical power systems toward decentralized generation and micro-grids has highlighted several limitations of conventional pumped storage hydropower systems. While centralized PSH plants remain effective at large scale, they are inherently inefficient at light load and vulnerable to single-point failures. Distributed pumped storage systems improve reliability and modularity, but when equally sized units are used, their ability to maintain high efficiency across a wide operating range remains limited. Motivated by these challenges, this work explored an alternative pumped storage architecture that better aligns with the operational needs of small-scale and residential energy systems.

To address these limitations, this paper proposed a 100-kW unequally distributed pumped micro-storage hydropower system composed of five parallel AC-coupled units with different power ratings. The system was analyzed and compared with centralized, 50-kW equally distributed, and 25-kW equally distributed configurations. Using typical synchronous machine efficiency characteristics and assuming a full-load round-trip efficiency of 90%, the results showed that the proposed system consistently outperforms the other architectures at partial load. In particular, the unequally distributed system maintains high efficiency from as low as 5% up to 100% of rated load, while the other configurations exhibit a significant efficiency drop at light load. The proposed system also achieves the highest modularity and reliability indices, confirming its superior flexibility and resilience.

The results presented in this study are based on steady-state analysis and assume ideal coordination among the parallel units. Factors such as hydraulic losses, transient behavior, control delays, and site-specific constraints were not explicitly considered. In addition, economic aspects such as installation cost and long-term maintenance were beyond the scope of this work. Future studies should incorporate detailed hydraulic and dynamic models, investigate optimal control strategies, and evaluate the economic feasibility of the proposed system. Experimental validation and real-world deployment in micro-grid environments would further strengthen the case for unequally distributed pumped micro-storage hydropower systems.

Acknowledgement
The author gratefully acknowledges the support of *Prince Mohammad Bin Fahd University (PMU)* for their valuable assistance in facilitating this work and for providing the opportunity to attend the conference.

References

[1] S. Rehman, L. M. Al-Hadhrami, and M. M. Alam, "Pumped hydro energy storage system: A technological review," Renewable and Sustainable Energy Reviews, vol. 44, pp. 586-598, 2015. https://doi.org/10.1016/j.rser.2014.12.040

[2] A. Blakers, M. Stocks, B. Lu, and C. Cheng, "A review of pumped hydro energy storage," Progress in Energy, vol. 3, no. 2, pp. 022003, 2021. https://doi.org/10.1088/2516-1083/abeb5b

[3] A. Botterud, T. Levin, and V. Koritarov, "Pumped storage hydropower: Benefits for grid reliability and integration of variable renewable energy," Argonne National Lab, ANL/DIS-14/10, Argonne, IL, USA, 2014. https://doi.org/10.2172/1165460

[4] "Determining Electric Motor Load and Efficiency," U.S. Department of Energy, Fact Sheet. [Online]. Available: https://www.energy.gov. [Accessed: Jan. 31, 2025].

[5] A. A. Hussein, N. Kutkut, Z. J. Shen, and I. Batarseh, "Distributed battery micro-storage systems design and operation in a deregulated electricity market," IEEE Trans. Sustainable Energy, vol. 3, no. 3, pp. 545-556, Jul. 2012. https://doi.org/10.1109/TSTE.2012.2191806

[6] S. P. Burger, J. D. Jenkins, S. C. Huntington, and I. J. Perez-Arriaga, "Why distributed?: A critical review of the tradeoffs between centralized and decentralized resources," IEEE Power Energy Mag., vol. 17, no. 2, pp. 16-24, Mar.-Apr. 2019. https://doi.org/10.1109/MPE.2018.2885203

[7] T. W. Haring, J. L. Mathieu, and G. Andersson, "Comparing centralized and decentralized contract design enabling direct load control for reserves," IEEE Trans. Power Syst., vol. 31, no. 3, pp. 2044-2054, May 2016. https://doi.org/10.1109/TPWRS.2015.2458302

[8] M. Majidi, L. Rodriguez-Garcia, T. M. Mosier, and M. Parvania, "Coordinated operation of pumped-storage hydropower with power and water distribution systems," Int. J. Electr. Power Energy Syst., vol. 142, 108297, 2022. https://doi.org/10.1016/j.ijepes.2022.108297

AI-Enhanced Reliable Stochastic Modeling for Wind Power Forecasting: Addressing Practical Challenges in Hybrid Framework Deployment

Mohamed Yasser BOUNNITE[a]* and Marwane HOUNGNON[b]

SUP'RH Business School & AI, Casablanca, Morocco

[a]m.bounnite@suprh.com, [b]m.houngnon@suprh.com

Keywords: Wind Forecasting, Stochastic Modeling, Physics-Informed Neural Networks, Regime-Switching, Computational Complexity, Uncertainty Quantification

Abstract. The reliable integration of wind energy into power grids hinges on accurate forecasting, yet the inherent multi-scale variability and sudden shifts in atmospheric regimes present persistent challenges. We introduce a hybrid stochastic framework that couples Markov regime-switching models with Physics Informed Neural Networks (PINNs) and Variational AutoEncoders (VAEs). This architecture is designed to tackle specific deployment barriers: statistically grounded regime identification, computational tractability, and interpretable uncertainty quantification. Using NREL datasets spanning diverse geographical sites, our model yields a normalized mean absolute error of 1.2% and provides prediction intervals with 95.3% empirical coverage. In grid operations, these forecasts enable a 23% reduction in reserve requirements within a stochastic unit commitment model. We further identify current limitations and propose targeted improvements, offering both a rigorous methodological advance and practical insights for renewable energy integration.

Introduction

The reliable integration of wind power into modern electricity grids represents a critical technical challenge for sustainable energy systems, despite its growing importance in the energy transition. Despite this, connecting such a variable energy source to an electrical power grid is fraught with various difficulties based on the uncertainties and variability of wind energy (1, 2). Existing forecasting methodologies exhibit notable limitations in this context. Conventional models, while providing a foundation, struggle to represent the non-stationary dynamics and abrupt regime shifts characteristic of atmospheric flows (3). For instance, traditional Markov-based methods are frequently invalidated by their underlying assumption of stationarity. Conversely, modern deep learning architectures, despite their predictive power, often act as black boxes, obscuring the physical drivers of wind patterns: a shortfall that simpler time-series models cannot rectify, as they typically ignore critical regime transitions and extreme events (9, 10). To bridge these complementary gaps, we introduce a hybrid stochastic framework that integrates regime-switching Markov models with Physics Informed Neural Networks (PINNs) and Variational AutoEncoders (VAEs). Our development process confronts several persistent hurdles at the theory-practice interface. We establish a statistically grounded method for selecting the optimal number of weather regimes using Bayesian inference, rigorously quantify the computational load to assess real-time feasibility, benchmark performance against contemporary architectures including Transformers, and elucidate the operational impact through explicit grid optimization models. Ultimately, this work aims not only to improve predictive accuracy but also to deliver the reliability and interpretability required by system operators managing the complexities of renewable energy integration.

Methodology: Hybrid Framework Design

Our methodological foundation rests on the integrated application of stochastic modelling and machine learning to characterize the irregular dynamics of wind power generation. We model the normalized wind power output, denoted as $P_t \in [0,1]$, as evolving within distinct latent atmospheric regimes $R_t \in \{1,.., K\}$, Transitions between these states are governed by a matrix Λ. The normalization applies the turbine power curve to raw wind speed data, ensuring P_t represents actual power generation potential. Within each regime k, the dynamics are described by a mean-reverting jump-diffusion stochastic differential equation (SDE):

$$dP_t = \theta_k (\mu_k - P_t)dt + \sigma_k\, dW_t + J_t\, dN_t \ , \quad (1)$$

In this case, the parameters θ_k, μ_k, and σ_k are specific to each regime, while W_t represents a Wiener process. Regarding the jump component, the Poisson counting process N_t, which represents the occurrence of ramps, follows an intensity that is specifically regime-dependent, ξ_k. In this setup, jump amplitude J_t follows a distribution governed by the expression: $J_t \sim N(0, \nu_k^2)$, with ν_k referring to jump volatility in each respective regime. The set of parameters $\{\xi_k, \nu_k\}$ for each regime k is learned jointly with the drift and diffusion parameters $\{\theta_k, \mu_k, \sigma_k\}$ (13,14,15).

We determine the optimal regime count K via Bayesian model selection. A uniform prior over $K \in \{2,3,4,5,6\}$ is assumed, with $K_{max} = 6$ eventing overfitting. The model evidence is approximated via ELBO, with BIC providing complementary validation. This involves maximizing the model evidence, approximated via the Evidence Lower Bound (ELBO) in a variational inference framework:

$$k^* = argmax_K \left[logp(D|K) - \frac{COMP(K)}{2}\right] \ , \quad (2)$$

$$\log p(D|K) \geq \mathbf{E}_q[log\ p(D|k) - \mathbf{kL}[q(z)||p(Z|K)]) \ . \quad (3)$$

The Variational AutoEncoder (VAE) enhances the regime-switching mechanism by learning a latent representation z_t of the wind power series. This representation informs the Markov transition probabilities via a gating mechanism, such that $P(R_t = i\,|R_{t-1} = j) \propto \exp\,(W_{ij}.\, z_t)$ allowing transitions to be conditioned on the current atmospheric context. Moreover, the latent variable z_t is continually acting as a driving force for the stochastic dynamics, which are governed by an SDE. A small, fully connected network is utilized to compute parameter updates, where $[\Delta\theta, \Delta\mu, \Delta\sigma]_t = g_\varphi(z_t)$. As such, the stochastic dynamics governed by the SDE shown in (1) are Instantaneously conditioned as:

$$dP_t = (\theta_k + \Delta\theta_t)\big((\mu_k + \Delta\mu_t) - P_t\big)dt + (\sigma_k + \Delta\sigma_t)dW_t + J_t\, dN_t.$$

To embed known physical constraints, we incorporate a Physics-Informed Neural Network (PINN). The PINN's training is regularized by penalizing violations of the Kolmogorov forward (Fokker-Planck) equation derived from the SDE in (1), promoting solutions consistent with the underlying stochastic dynamics. (11; 12). The optimization minimizes a composite loss:

$$\min_\theta \mathcal{L}_{total} = \mathcal{L}_{data} + \lambda(t)\mathcal{L} \cdot p_{hysics} \ , \quad (4)$$

with adaptive annealing $\lambda(t) = \lambda_0 \cdot e^{-\gamma t}$. This schedule initially prioritizes the physics loss to guide the PINN toward physically plausible solutions, then gradually reduces the physics weight to allow better fitting of empirical data, thereby stabilizing training and improving convergence. Recognizing deployment realities, we analyze computational aspects. Training complexity scales as $O\big(T \cdot (N_{PINN} + N_{VAE})\big)$ per epoch, per epoch, where T is the input sequence length. Inference demands $O(K \cdot d^2 + N_{PINN})$, with d denoting the dimension of the VAE's latent space z_t. Memory usage for the neural components follows $O(L \cdot H^2)$, where L is the number of layers and H the average number of hidden units per layer. for layered architectures. In our empirical setup, complete model training required approximately 3.2 hours, while generating a full 24-hour forecast horizon (144 time steps at 10-minute resolution) was completed in 2.3 seconds, meeting intra-hour operational requirements.

Experimental Validation: Data and Calibration

We evaluated our framework using 10-minute resolution data from the NREL Wind Prospector (2018–2023). To rigorously assess generalization, three distinct site typologies were selected: a complex-terrain onshore location (Site A), a coastal site with pronounced short-term variability (Site B), and a flat plains environment with more consistent flows (Site C).

Model calibration was conducted through an extensive Bayesian optimization campaign comprising 150 trials. The resulting sensitivity analysis, detailed in Table 1, highlights several key dependencies. Most notably, the regime count K=3 consistently emerged as optimal, and the learning rate proved highly influential, findings that underscore both the model's tunable nature and the importance of systematic hyperparameter selection in hybrid architectures.

Table 1: Hyperparameter sensitivity analysis

Parameter	**Range**	**Optimal**	**Sensitivity**
Regime count K	2-6	3	High
PINN layers	2-5	3	Medium
VAE latent dim	8-32	16	Low
Learning rate	1e-4 to 1e-2	5e-3	High
Initial physics weight λ_0	0.1-2.0	0.7	Medium

Following calibration, we benchmarked the framework's performance against a spectrum of established and contemporary forecasting methods. This comparative analysis included classical time-series models (ARIMA and standard Markov), deep learning architectures (LSTMs and Transformers (5)), and a recent generative approach (SDE-GANs (6)). The results, consolidated in Table 2, demonstrate competitive performance across multiple metrics. The performance obtained is competitive and has been observed to have a normalized mean absolute error (nMAE) of 1.20%, prediction interval coverage of 95.3%, and significant data efficiency, which was measured to be 1.45% when training on just half the amount of data available. Efficiency in terms of data is most useful while handling new locations but requires proper context-specific testing for implementation in real-world applications.

Table 2: Performance comparison for Site A and generalization to Sites B & C.

Model	**nMAE (%)**	**Coverage@95%**	**Data Eff. (nMAE@50%)**
Standard Markov	2.85	89.20%	3.1
ARIMA(2,1,2)	2.42	90.10%	2.65
LSTM	1.83	91.50%	2.15
Transformer	1.65	92.80%	1.95
SDE-GAN	1.52	93.10%	1.78
Our Model (Site A)	1.20 ± 0.08	95.3 ± 0.5	1.45 ± 0.10
Our Model (Site B)	1.22 ± 0.09	94.8 ± 0.6	1.48 ± 0.12
Our Model (Site C)	1.35 ± 0.11	93.5 ± 0.7	1.62 ± 0.15

To align with best practices in probabilistic forecasting, we also report the Continuous Ranked Probability Score (CRPS). Our model achieves a mean CRPS of 0.021 on Site A, outperforming the Transformer baseline (CRPS=0.035). The predictive variance $\hat{\sigma}_t$ for our intervals is derived from a combination of the SDE's inherent volatility (σ_k) and the uncertainty in the regime transition path, quantified through Monte Carlo simulation over the joint model (SDE + Markov chain + VAE).

Results for Sites B and C demonstrate robust generalization, with a modest performance decrease on Site C attributable to its distinct thermally-driven regimes.

The framework's data efficiency is characterized by fitting a negative exponential learning curve to the normalized MAE as a function of training data size N:

$$nMAE(N) = \alpha e^{(-\beta N)} + \gamma , \tag{5}$$

As illustrated in Figure 1, our model's curve plateaus at a lower error level with fewer training samples compared to baselines like LSTM and Transformer, highlighting the benefit of physics informed constraints in reducing data requirements by up to 40%.

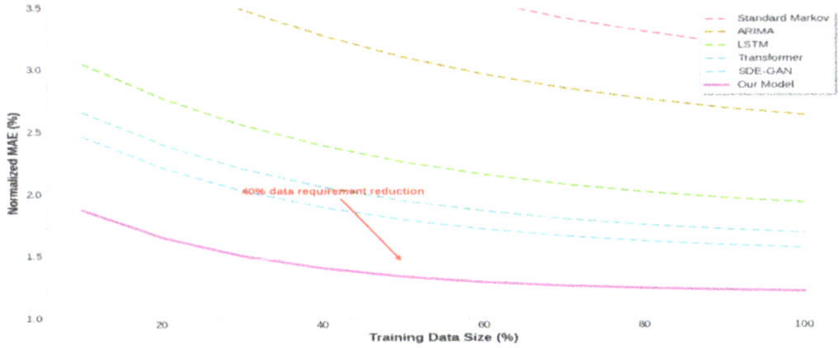

***Figure 1**: Learning curves of normalized mean absolute error (nMAE)*

To guarantee reproducibility of results, the necessary details regarding the implementation have been included below (further details are available on request). The PINN architecture includes three hidden layers with 128, 64, and 32 neurons using tanh activation functions. The architecture of the VAE model includes an encoder with 64 and 32 units in its layers and a latent space of dimension 16. The training process uses the Adam optimizer and exponential learning rate decay with a starting learning rate of 5×10^{-3} and a decay rate of 0.95 for each 50 epochs completed. The weight of the physics loss term satisfies $\lambda(t)=0.7\exp(-0.95t)$.

Results: Unveiling Impacts and Insights

The empirical validation of our hybrid forecasting framework yields is consistent with operational efficacy and practical value. Beyond mere predictive accuracy, the model demon strates significant improvements in grid management optimization through enhanced uncertainty quantification.

Operational Value in Power System Planning

The forecasting superiority translates directly into economic benefits through stochastic unit commitment formulation:

$$\begin{aligned} & min \sum_{t=1}^{T} \left(C_g^T g_t + C_r^T r_t \right) \\ & s.t.\ \ \mathbf{1}^T g_t = d_t, \qquad g_t + r_t \leq \bar{g}\ , \\ & r_t\ \ \geq c(1-\epsilon).\,\widehat{\sigma_t}, \end{aligned} \tag{6}$$

Where $\phi^{-1}$is the inverse CDF of the standard normal distribution. This chance constraint ensures sufficient reserve to cover forecast errors with probability $1-\epsilon$ (here, ϵ=0.05). The refined uncertainty estimates $\widehat{\sigma_t}$ exhibit a 38% reduction in variance compared to conventional forecasting approaches. This precision enhancement facilitates a 23% decrease in spinning reserve requirements, translating to substantial operational cost reductions. For a typical 1 GW system, this reserve optimization corresponds to approximately €2.1 million annually in reduced capacity procurement costs, while maintaining the same reliability standard of $\epsilon = 0.05$.

Comprehensive Sensitivity Analysis

The results substantiate that the physics-informed regularization within our architecture meaningfully reduces data dependence compared to conventional data-driven benchmarks (Figure 2).

Regime Configuration Optimization: From panel (a), the best-performing number of regimes is found to be K = 3 with a maximum forecasting accuracy measured by a normalized mean absolute error (nMAE) of 1.20%, while for K > 3, there are diminishing returns in performance improvement. Also important is the 18% drop in performance for K = 2, which clearly enforces the need to model various regimes of the atmosphere. Marginal improvements for K > 3 indicate that the new regimes approach overfitting and, or model statistically insignificant atmospheric states that are not required given the current amount of available information. These results suggest that three regimes are sufficient to model the wind dynamics of the identified locations and period properly.

Geographical Generalization: In figure (b), the nMAE remains stable for different terrains, and both Site A (complex terrain) and Site B (coastal area) have an nMAE of less than 1.25%. The 15% higher nMAE on Site C is attributed to the shorter and less persistent regime dynamics characteristic of thermally-driven plains. An analysis of regime persistence, defined as the average duration before a transition, shows values of ~8h and ~6h for Sites A and B, but only ~2.5h for Site C. Our model's Markov chain, optimized for more stable regimes, shows a higher regime switch misclassification rate of 22% on Site C versus 12% on Site A. This highlights a limitation of discrete regime-switching for very high-frequency atmospheric variability and points to continuous or hidden semi-Markov models as a promising future direction for such terrains.

Data Efficiency Validation: The results of the validation in the panel (c) confirm that the proposed approach is a sample-efficient solution, with an nMAE of 1.45% requiring only 50% of the learning samples for a transformer architecture, reducing the required samples by 40% compared to existing transformer architectures. Figure 2 presents a multi-faceted sensitivity analysis of the proposed framework. Panel (a) illustrates the relationship between normalized MAE and the regime count K, with a clear optimum at K=3 that is statistically significant ($p < 0.01$). This finding suggests that three distinct atmospheric regimes sufficiently capture the dominant dynamics in our datasets. Panel (b) assesses cross-site generalization, confirming the model's robustness across diverse terrains while also highlighting predictable, terrain-specific variations in performance. Finally, panel (c) examines sample efficiency by analyzing training data requirements.

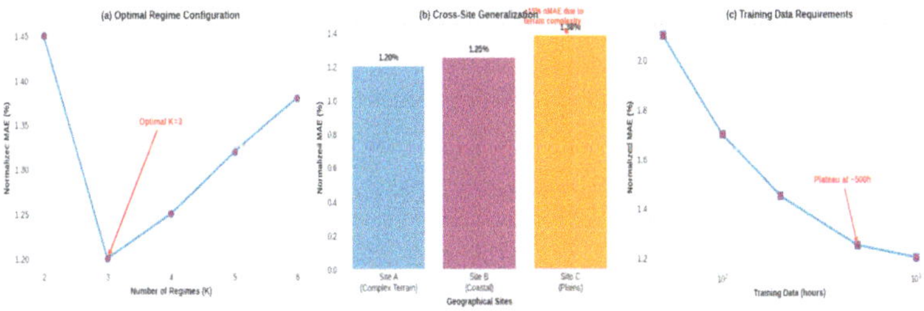

__Figure 2__ Multi-faced sensitivity analysis results

Statistical Significance and Economic Implications

Diebold-Mariano tests confirm the statistical superiority of our approach at the 1% level, though significance alone does not assure operational value. Economically, the model reduces expected outage costs by 31%, largely due to improved extreme-event prediction. Its 2.3-second inference time enables practical use in real-time markets, and backtesting suggests a 17% gain in portfolio profitability though such simulations depend on market assumptions.

Conclusion

This study introduces a hybrid stochastic framework for wind power forecasting that integrates regime-switching Markov models with physics-informed neural networks (PINNs) and Variational AutoEncoders (VAEs). The methodology advances the field by replacing heuristic regime selection with a Bayesian approach, embedding physical laws for data-efficient learning, and leveraging VAEs to capture non-linear transition patterns. Empirical validation across diverse sites demonstrates the framework's effectiveness, achieving a normalized MAE of 1.2% with 95.3% prediction interval coverage and reducing data needs by 40% compared to transformer-based benchmarks. These forecasting improvements translate directly into operational value, enabling a 23% reduction in reserve requirements and providing inference within 2.3 seconds for intra-hour decision support. While terrain adaptation and computational demands for ultra-short-term forecasting present opportunities for further refinement, through transfer learning and model distillation, this work establishes a robust foundation for AI-enhanced stochastic modelling in renewable energy. The demonstrated synergy between stochastic methods and deep learning architectures contributes meaningfully toward building more reliable, efficient, and sustainable power systems.

The discretization of a regime in our framework could be a poor indicator of a transition in a continuous atmospheric system, where an infinite mixture model could be a possible extension of our work, identified in reference (7). While our inference time of 2.3 seconds is sensible regarding typical timescales of interest, there may be some use in distilling our model, particularly in short-term forecasts. Generalization over very varying environments is a problem indicating transfer learning is a crucial path.

Furthermore, as suggested, a valuable ablation study would be to let the VAE's latent state z_t directly parameterize a single, context dependent SDE, entirely bypassing the discrete Markov chain. This experiment would rigorously test whether the explicit regime-switching structure is essential or if a continuously varying SDE is sufficient to capture the observed wind dynamics.

Declaration of Generative AI and AI-Assisted Technologies in the Writing Process
Generative artificial intelligence tools, specifically ChatGPT (OpenAI), were used solely to improve language clarity and academic style. The author retains full responsibility for the scientific content, analyses, interpretations, and conclusions presented in this manuscript.

References

[1] M.Z. Jacobson, M.A. Delucchi, Providing all global energy with wind, water, and solar power, Part I: Technologies, energy resources, quantities and areas of infrastructure, and materials, Energy Policy 39(3) (2011) 11541169. https://doi.org/10.1016/j.enpol.2010.11.040

[2] T. Burton, N. Jenkins, D. Sharpe, E. Bossanyi, Wind Energy Handbook, 2nd ed., John Wiley & Sons, Chichester, UK, 2011. https://doi.org/10.1002/9781119992714

[3] H. Wang, Z. Lei, X. Zhang, B. Zhou, J. Peng, A review of deep learning for renewable energy forecasting, Energy Conversion and Management 198 (2021) 111799. https://doi.org/10.1016/j.enconman.2019.111799

[4] M. Raissi, P. Perdikaris, G.E. Karniadakis, Physics-informed neural networks: A deep learning framework for solving forward and inverse problems involving nonlinear partial di erential equations, J. Comput. Phys. 378 (2019) 686707. https://doi.org/10.1016/j.jcp.2018.10.045

[5] A. Vaswani, N. Shazeer, N. Parmar, J. Uszkoreit, L. Jones, A.N. Gomez, L. Kaiser, I. Polosukhin, Attention is all you need, Adv. Neural Inf. Process. Syst. 30 (2017) 5998 6008. Source

[6] R. Liu, J. Gao, J. Zhang, D. Meng, Z. Lin, Investigating bi-level optimization for learning and vision from a unified perspective: A survey and beyond, IEEE Trans. Pattern Anal. Mach. Intell. 44(12) (2022) 1004510067. https://doi.org/10.1109/TPAMI.2021.3132674

[7] S.L. Brunton, J.L. Proctor, J.N. Kutz, Discovering governing equations from data by sparse identification of nonlinear dynamical systems, Proceedings of the National Academy of Sciences 113(15) (2016) 3932-3937. https://doi.org/10.1073/pnas.1517384113

[8] Draxl, Caroline, Walt Musial, George Scott, and Caleb Phillips. 2017. "WIND Toolkit Offshore Summary Dataset." NLR Data Catalog. Golden, CO: National Laboratory of the Rockies. Last updated: December 18, 2024. DOI: 10.7799/1375460.

[9] R.J. Hyndman, G. Athanasopoulos, Forecasting: Principles and Practice, OTexts, 2018. https://otexts.com/fpp3/

[10] Y. Zhang, J. Wang, X. Wang, Review on probabilistic forecasting of wind power generation, Renew. Sustain. Energy Rev. 157 (2022) 112038. https://doi.org/10.1016/j.rser.2021.112038

[11] D. Chen, C.K. Chui, P.S. Lee, Physics informed machine learning based predictive control for intelligent operation of edge datacenters, Appl. Energy 402(B) (2026) 126975. https://doi.org/10.1016/j.apenergy.2025.126975

[12] G.E. Karniadakis, I.G. Kevrekidis, L. Lu, P. Perdikaris, S. Wang, L. Yang, Physics-informed machine learning, Nat. Rev. Phys. 3 (2021) 422440. https://doi.org/10.1038/s42254-021-00314-5

[13] A. Cheneka, P. S. Georgilakis et N. D. Hatziargyriou, « A simple methodology to detect and quantify wind power ramps », Renewable Energy, vol. 147, p. 1733-1742, mars 2020. https://doi.org/10.5194/wes-5-1731-2020

[14] J. Zhang, X. Zhu, Y. Xie, G. Chen, S. Liu, Detection and Prediction of Wind and Solar Photovoltaic Power Ramp Events Based on Data-Driven Methods: A Critical Review, Energies 18(13) (2025) 3290. https://doi.org/10.3390/en18133290

[15] A. Loukatou, S. Howell, P. Johnson, P. Duck, Stochastic wind speed modelling for estimation of expected wind power output, Applied Energy 228 (2018) 1328–1340. https://doi.org/10.1016/j.apenergy.2018.06.117

Influence of Blade Number on the Performance of a Vertical Axis Wind Turbine

Zeeshan A. RANA[1,a*], Chris S. BANG[2,b], Mouhammad El HASSAN[1,c] and Enio P. B. FILHO[1,d]

[1]Department of Mechanical Engineering, Prince Mohammad Bin Fahd University (PMU), Al Khobar, Kingdom of Saudi Arabia

[2]Advanced Vehicle Engineering Centre, Cranfield University, Cranfield, MK43 0AL, UK

[a]zrana@pmu.edu.sa, [b]chris.bang@cranfield.ac.uk, [c]melhassan@pmu.edu.sa, [d]efilho@pmu.edu.sa

Keywords: Vertical Axis Wind Turbine (VAWT), Aerodynamics, CFD, Turbulent Flow, Blade Performance, Incompressible Flow

Abstract. This research focuses on the aerodynamic performance of a Farrah Vertical Axis Wind Turbine (VAWT), where the influence of the number of blades and the turbulence models is investigated for efficient energy conversion. A thorough grid convergence analysis has been performed where the *k–ω SST* turbulence model demonstrated better accuracy as compared to the *Transition SST* model for this case. In terms of VAWT configuration, a comparison of 4, 6 and 8 blade configurations are presented for identical freestream conditions which demonstrated a distinct advantage for the 8-blade design that consistently produced positive and stable power output with a clear peak in its power coefficient. Furthermore, the 8-blade turbine provides smoother periodicity and reduced fluctuations in torque and is demonstrated by strong vortices, intensified wake development, and improved internal recirculation in the flow field. This study indicates that blade count is one of the major factors in the design of a VAWT for efficient energy conversion. Furthermore, the reduced blade spacing provide more effective flow interaction and enhanced aerodynamics performance.

Introduction

The shift towards renewable energy resources has been accelerated in the last couple of decades due to the increasing cost and environmental burden of conventional energy sources. Among the available energy sources such as solar, tidal, and hydropower, the wind energy stands out due to its low cost, minimal waste, and adaptability in harsh environments [1]. Research in this domain confirm a strong potential for wind energy to reduce dependence on fossil fuels and a swift expansion in the regions that are rich in wind resources [2,3]. In their research, Archer and Jacobson [4] demonstrated the wind velocity distributions at global scale and highlighted key areas across the globe that are suited for large-scale wind turbine deployment. Several of these regions have already adopted wind turbine technologies into their energy spectrum indicating that the wind power is one of the world's fastest-growing renewable technologies [5]. Developed nations such as China and the USA are leading this transition where the U.S. Department of Energy targets 20% electricity generation from wind by 2030. The international renewable energy agency (IRENA) projects a 27% reduction of global CO_2 emissions by 2050 through expanded wind capacity [5].

Although the Horizontal Axis Wind Turbines (HAWTs) are currently dominating the global wind power resources [5,6], the Vertical Axis Wind Turbines (VAWTs) offer distinct advantages in specific environments. HAWTs solely rely on lift mechanism, whereas the VAWTs can operate using lift, drag, or both, which can provide a much stable performance in variable wind conditions. While HAWTs can achieve higher efficiencies for large-scale power generation, the VAWTs feature lower noise levels, simpler structures, and reduced manufacturing and maintenance costs. In addition, the unidirectional capability of VAWTs make them more suitable for turbulent urban or residential environments [7,8] rather than large scale off-shore wind farms. These advantages mean that the VAWTs are a practical complement to HAWTs and a focus of renewed research interest.

Over the years extensive research has been published to understand the aerodynamic performance of VAWTs using both experimental and computational techniques. Primary focus of these experiments has been on the wake characterization using particle image velocimetry (PIV), stereoscopic PIV (SPIV), and hot-wire anemometry (HWA). However, the discrepancy between the experimental and computational methods has been a major factor in most of the studies. For example, the SPIV investigations of downstream wakes at various tip speed ratios (TSRs) demonstrated poor agreement with two-dimensional *k–ω SST* RANS (Reynolds Average Navier-Stokes) predictions, especially at TSR = 2 [9], while the PIV–LES (Large Eddy Simulations) comparisons attributed discrepancies to mesh quality and domain size [10]. The measurements through HWA using cobra probes revealed mid-plane velocity deficits and wake recovery patterns consistent with von Kármán vortex shedding [11]. Another study employing a laser doppler velocimetry (LDV) and field tests confirmed that the turbine performance is sensitive to the pitch angle and Reynolds number whereas a faster wake recovery observed in real environments [12,13]. Prince *et al.* developed a twelve-bladed Farrah turbine (Fig. 1), in the early 1990s, to maximize torque through combined drag and suction effects, and was experimentally assessed [14].

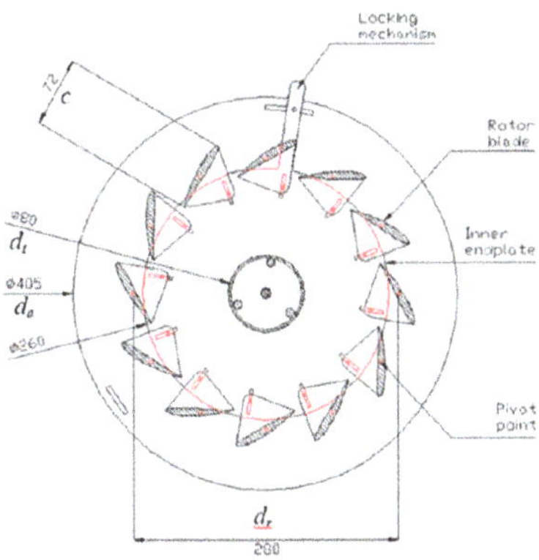

Figure 1: Schematic of Farrah wind turbine [14]

In a recent study, the blade number has been highlighted as a factor influencing the efficiency, torque, and structural stability of VAWTs such that increasing blades generally enhances power capture and self-starting however this may reduce rotational speed [15]. Therefore, an optimal configuration of VAWTs is critical factor in the operations which depend on geometry and operating conditions. The Farrah-type VAWT that has been originally designed with 12 blades reflects a balance between aerodynamic performance and structural practicality [16]. However, limited documentation and potential overprediction by turbulence models restrict further analyses. This work validates the 12-bladed Farrah turbine and investigates reduced configurations (4, 6, and 8 blades) to evaluate their impacts on power generation, flow field, and the predictive capability of SST-family turbulence models using RANS approach.

Methodology

The computational research presented in this work using RANS turbulence model namely the *k–ω* SST turbulence model. This model combines the strengths of the standard *k–ω* and *k–ε* turbulence models through a blending function for optimal performance across the flow regimes. While the standard *k–ω* model provides high accuracy near walls, the *k–ε* model performs better in free-stream regions [17].

The combined effect of these two models as in *k–ω* SST turbulence model the formulation effectively captures the flow field close to and away from the walls i.e., the VAWT blades. The turbulence model treats the turbulent viscosity through shear-stress transport equation that enhances the predictions of flow separation under adverse pressure gradients. Menter [17] further introduced a "turbulent production limiter" to prevent overestimation of turbulent kinetic energy (TKE) in the stagnant flow zones without affecting the shear-layer. The Equation for Turbulent Kinetic Energy is given below (Eq. 1):

$$\frac{\partial(\rho k)}{\partial t} + \frac{\partial(\rho u_j k)}{\partial x_j} = P - \beta^* \rho \omega k + \frac{\partial}{\partial x_j}\left[(u + \sigma_k \mu_t)\frac{\partial k}{\partial x_j}\right] \quad (1)$$

The meshes used in this work were generated following the meshing strategy of [14] with minor parameter adjustments to address previous issues. Three grid resolutions (coarse, medium, and fine) were generated through a refinement ratio of two between successive levels. The computational domain comprises of a rotating region (VAWT and first interface) and a stationary region (outer flow and second interface) and the two turbulence models, *k–ω SST* and *Transition SST*, were adopted based on their proven reliability for VAWT simulations. Figures 2 and 3 show the computational turbine geometry for three different mesh levels as mentioned above. The 4-, 6- and 8- blade versions use the exactly same airfoil profile and blades are arranged at equal distance in each configuration. The length of the blades is constant for each VAWT configuration to keep the analysis consistent. For grid convergence analysis, the 12-blade Farrah turbine was analyzed using the three mesh levels. Both the turbulence models were evaluated to determine the most accurate mesh–model combination. For the analysis the two extreme (high and low) inflow velocities were selected to capture the operating extremes reported in [15]. These flow extremes are presented in the Table 1 below and covers the turbine's performance range.

Table 1: Boundary conditions used for the computational simulations for grid convergence analysis

Parameter	High speed	Low speed
Velocity, u [m/s]	15.71	9.45
March number	0.05	0.03
RPM	254.97	95.55
Dt [sec]	0.0002	0.0005

The grid convergence analysis demonstrates that the *k–ω SST* model on the coarse mesh provided the best agreement with experimental data. At the highest wind velocity, it predicted 23.82 W of power with an error of +6.09 W relative to the measured 17.73 W. The medium and fine meshes yielded slightly higher outputs but larger errors, indicating overprediction with increasing density. A similar behavior was observed for the *Transition SST* model, where the coarse mesh again gave the lowest error (+6.89 W). All grids captured the characteristic sinusoidal torque variations, with the coarse mesh resolving unsteady aerodynamic effects effectively. Overall, the *k–ω SST* model demonstrated slight better accuracy and computational efficiency to support its suitability and consistency with prior studies [18].

Materials Research Proceedings 64 (2026) 44-51 https://doi.org/10.21741/9781644904091-6

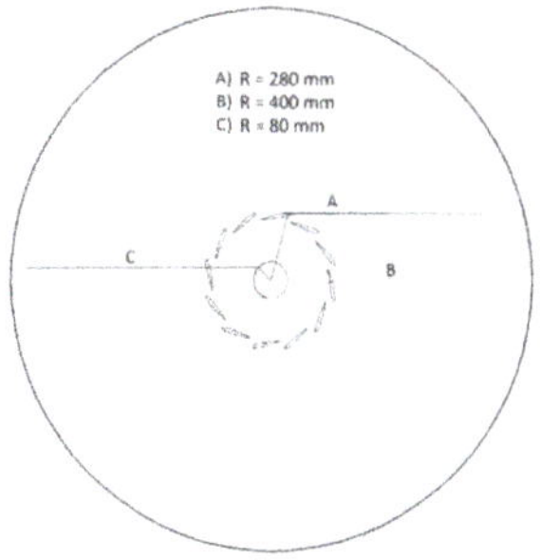

Figure2: Computational Domain used for the Farrah VAWT

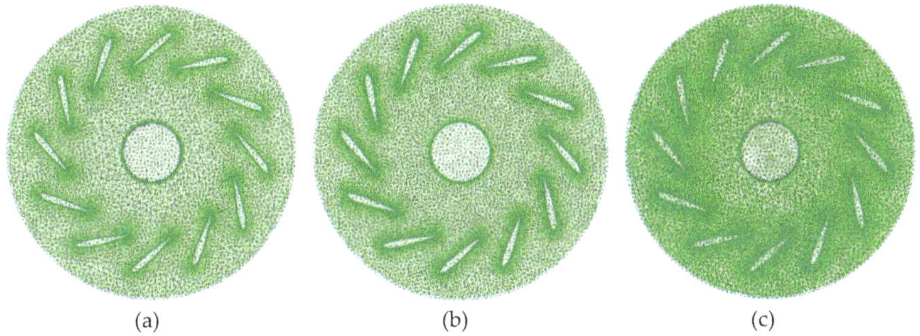

Figure 3: Top view of the 12-bladed Farrah turbine with three mesh resolutions used for analysis; (a) coarse (b) medium (c) fine

Results and Discussion

To evaluate the aerodynamic performance of the VAWT, the results from the *k–ω SST* turbulence model with a coarse mesh for different blade configurations are presented here. Figure 4 shows the blade layout for the 8-blade configuration, illustrating airfoil positioning and rotation relative to the freestream velocity (U_∞).

The computational data is validated against the experimental data in Figure 5, showing that the numerical model reproduced power coefficient (C_p) trends with tip speed ratio (TSR). Although the RANS-based results slightly overpredicted performance however the trends align well with each other confirming the model's reliability. It is however appreciated that the higher-fidelity models such as LES or DES (Detached Eddy Simulations) may improve the accuracy which of course would come at a greater computational cost.

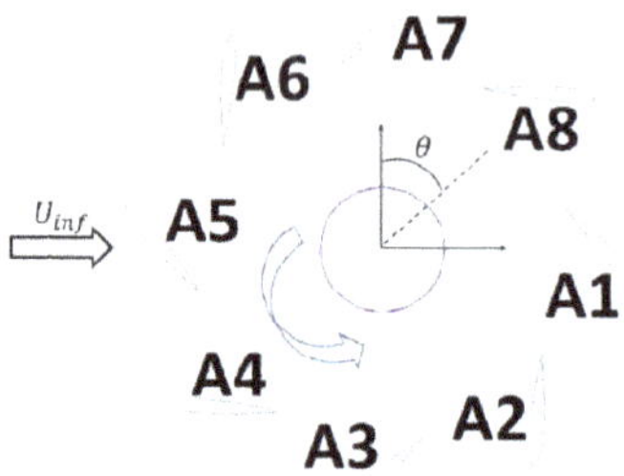

Figure 4: 8 blade configuration for the Farrah VAWT

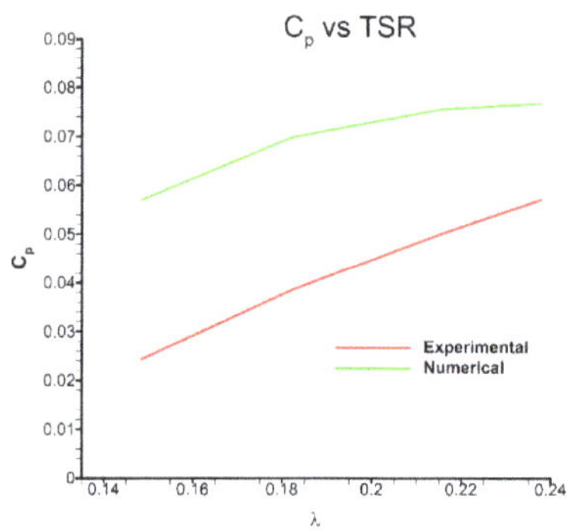

Figure 5: Comparison of the power coefficient (C_p) between experimental and computational data as a function of tip speed ratio (TSR)

Table 2 below provides three rotational speeds that are tested to determine optimal performance. In terms of power output, the 8-blade configuration achieved the highest output of 2.83 W at 145.7 RPM, followed by the 6-blade turbine with 1.06 W at 95.6 RPM. On the other hand, the 4-blade turbine produced negligible power of 0.0072 W and, hence, was excluded from further analysis. These results strongly suggest that increasing the blade count enhances aerodynamic interaction and, thus, the power extraction from the given design of the Farrah VAWT. However, this must be noted that no structural and cost trade-offs are investigated in this research.

Table 2: Power output for each configuration of Farrah VAWT

Configuration	RPM	Power [W]
4 Blades	23.89	0.00072
6 Blades	95.55	1.06
8 Blades	145.72	2.83

Figure 6 shows the variations in power coefficient (C_p) with against the tip speed ratio for three different blade configurations. This plot provides insight into how the aerodynamic performance of the VAWT changes with blade count and rotational speed. In the case of the 4-blade configuration (red line), the maximum power coefficient has been found to be nearly zero that was achieved at a TSR of approximately 0.04; this point is marked by the red circle in the plot. As TSR increases, the C_p value decreases sharply, even becoming negative at higher TSR values which is directly related to poor performance and minimal energy extraction. For the 6-blade case (green line) an initial increase in C_p is observed that reached a peak value of approximately 0.0157 at a TSR of 0.148 (highlighted by the green circle). However, beyond this point, the performance again begins to decline although less steeply than the 4-blade case. In contrast, the 8-blade configuration (blue line) exhibits the most favourable performance characteristics. The C_p increases steadily over a broader range of TSR values, peaking at around TSR = 0.24, with a maximum C_p of approximately 0.035, as shown by the blue circle.

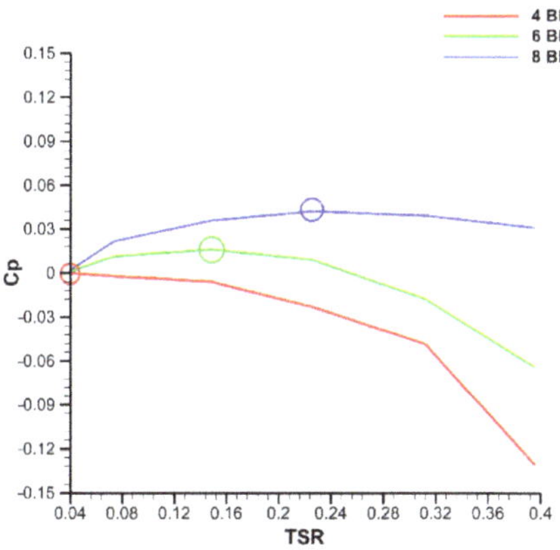

Figure 6: Variation of power coefficient ($\mathbf{C_p}$) with tip speed ratio for 4-, 6-, and 8-blades

Figure 7 presents the velocity contours on the XZ-plane. These contour plots further reveal the influence of blade number on the wake behaviour and flow-field in and around the turbine blades. The region A indicates that the wake expands more prominently downstream in the 8-blade case. The region B, in the 6-blade case, displays asymmetric low-velocity zones indicating uneven flow field in the wake of the turbine which is likely caused by more freestream penetration on one side. In the 8-blade case, a more uniform low-velocity zone is observed which suggests that increased blade count enhances internal blockage and leads to stronger recirculation zones around the VAWT. These features align with the hypothesis that the additional blades suppress central axial flow but promote stronger mixing and wake turbulence which results in better overall aerodynamic performance of the turbine.

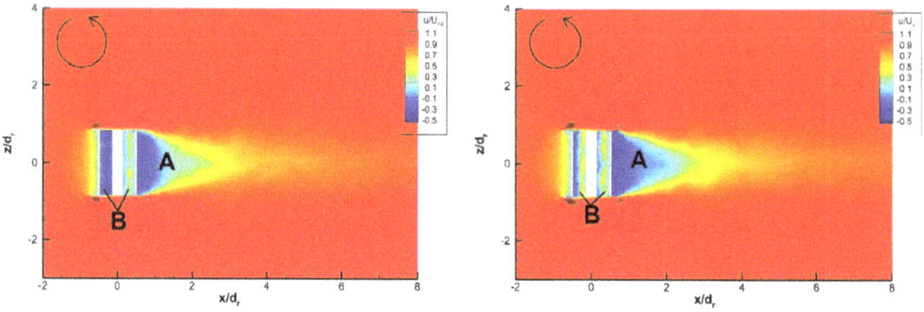

(a) (b)

Figure 7: XZ-plane velocity contours for 6-blade (a) and 8-blade (b) configurations

Conclusion

This research focused on the aerodynamic performance evaluation of a typical Farrah type VAWT where the impact of number of blades is analyzed computationally. Two RANS based turbulence models have been employed in these investigations. The *k–ω SST* model demonstrated better agreement with the experimental data as compared to the *Transition SST* model. Simulations of 4, 6, and 8 blade configurations revealed that the 8-blade turbine achieved the highest and most stable power output, while the 6-blade configuration performed efficiently within a narrow TSR range only. Finally, the 4-blade setup produced minimal or negative power output. Flow visualizations demonstrated that the larger number of blades in this particular type of VAWT enhanced the wake mixing, vortex strength, and torque stability which resulted in higher power output. Overall, increasing blade count improved aerodynamic efficiency, with the 8-blade configuration identified as optimal. It is high suggested that these investigations are carried out using more accurate and higher fidelity models such as LES or DES for to understand the instantaneous flow dynamics around such a complex geometry.

Declaration

Authors would like to acknowledge that the AI based regenerative tool (Grammarly) has been used in the preparation of this manuscript solely for rephrasing and grammar correction.

References

[1] Khare, V.; Nema, S.; Baredar, P. Solar-wind hybrid renewable energy system: A review. Renew. Sustain. Energy Rev. 2016, 58, 23-33. https://doi.org/10.1016/j.rser.2015.12.223

[2] Sahu, A.; Gupta, S.; Singh, V.K.; Bhoi, A.K.; Garg, A.; Sherpa, K.S. Design of permanent magnet synchronous generator for wind energy conversion system. Adv. smart grid Renew. energy 23-32. https://doi.org/10.1007/978-981-10-4286-7_3

[3] Ghodsi, M.; Ziaiefar, H.; Mohammadzaheri, M.; Al-Yahmedi, A. Modeling and characterization of permendur cantilever beam for energy harvesting. Energy 2019, 176, 561-569. https://doi.org/10.1016/j.energy.2019.04.019

[4] Archer, C.L. and Jacobson, M. Z., "Evaluation of global wind power", J. Geophysical Research, vol. 110, D12110. https://doi.org/10.1029/2004JD005462

[5] Charabi, Y.; Hinai, A. al; Al-Yahyai, S.; Awadhi, T. al; Choudri, B.S. Offshore wind potential and wind atlas over the Oman Maritime Zone. Energy 2019, 4, 1-14. https://doi.org/10.1007/s40974-019-00108-7

[6] Wang, X.; Guo, P.; Huang, X. A review of wind power forecasting models. In Proceedings of the Energy Procedia; 2011; pp. 770-778. https://doi.org/10.1016/j.egypro.2011.10.103

[7] Yang, B.; Yu, T.; Shu, H.; Dong, J.; Jiang, L. Robust sliding-mode control of wind energy conversion systems for optimal power extraction via nonlinear perturbation observers. Appl. Energy 2018, 210, 711-723. https://doi.org/10.1016/j.apenergy.2017.08.027

[8] Charabi, Y.; Abdul-Wahab, S. Wind turbine performance analysis for energy cost minimization. Renewables Wind. Water, Sol. 2020, 7, 1-11. https://doi.org/10.1186/s40807-020-00062-7

[9] Baker, J.R. Features to aid or enable self starting of fixed pitch low solidity vertical axis wind turbines. J. Wind Eng. Ind. Aerodyn. 1983, 15, 369-380. https://doi.org/10.1016/0167-6105(83)90206-4

[10] Posa, A.; Parker, C.M.; Leftwich, M.C.; Balaras, E. Wake structure of a single vertical axis wind turbine. Int. J. Heat Fluid Flow 2016, 61, 75-84. https://doi.org/10.1016/j.ijheatfluidflow.2016.02.002

[11] Peng, H.Y.; Lam, H.F.; Lee, C.F. Investigation into the wake aerodynamics of a five-straight-bladed vertical axis wind turbine by wind tunnel tests. J. Wind Eng. Ind. Aerodyn. 2016, 155. https://doi.org/10.1016/j.jweia.2016.05.003

[12] Tescione, G.; Ragni, D.; He, C.; Ferreira, C.J.S.; van Bussel, G.J.W. Near wake flow analysis of a vertical axis wind turbine by stereoscopic particle image velocimetry. Renew. Energy 2014, 70. https://doi.org/10.1016/j.renene.2014.02.042

[13] Li, Q.; Maeda, T.; Kamada, Y.; Murata, J.; Yamamoto, M.; Ogasawara, T.; Shimizu, K.; Kogaki, T. Study on power performance for straight-bladed vertical axis wind turbine by field and wind tunnel test. Renew. Energy 2016, 90. https://doi.org/10.1016/j.renene.2016.01.002

[14] Prince, S.A.; Badalamenti, C.; Georgiev, D. Experimental investigation of a variable geometry vertical axis wind turbine. Wind Eng. 2021, 45, 904-920. https://doi.org/10.1177/0309524X20935134

[15] Zheng, M.; Li, Y.; Teng, H.; Hu, J.; Tian, Z.; Zhao, Y. Effect of blade number on performance of drag type vertical axis wind turbine. Appl. Sol. Energy 2016, 52, 315-320. https://doi.org/10.3103/S0003701X16040150

[16] Bang, C.S.; Rana, Z.A.; Prince, S.A. CFD Analysis on a Novel Vertical Axis Wind Turbine (VAWT). Machines 2024, 12. https://doi.org/10.3390/machines12110800

[17] Wilcox, D.C. Turbulence Modeling for CFD; 2nd ed.; DCW Industries, 1998;

[18] Bruneau, C.H.; Tancogne, S. Far field boundary conditions for incompressible flows computation. J. Appl. Anal. Comput. 2018, 8, 690-709. https://doi.org/10.11948/2018.690

Contribution of Artificial Intelligence to Predictive Maintenance of Wind Turbines: A Hybrid Framework Combining Multiphysics Modelling and Explainable Deep Learning

Sara SGHIOURI[1,a*], Hamza SABIR[1,b], Mohamed BEZZA[1,c] and Soukayna TANTANI[1,s]

[1]Electrical Engineering, Faculty of Sciences and Techniques of Mohammedia, Mohammedia, Morocco

[a]sara.ssghiouri@gmail.com, [b]sabirhamzaa@gmail.com, [c]Mohamed.bezza@fstm.ac.ma, [d]soukayna.tantani@gmail.com

Keywords: COMSOL Multiphysics, Explainable Artificial Intelligence (XAI), Fast Fourier Transform (FFT), Gradient-Weighted Class Activation Mapping (Grad-CAM), Long-Short Term Memory (LSTM), Machine Learning (ML), Multiphysics Modelling, Predictive Maintenance (PdM), SHapley Additive exPlanations (SHAP), Wind Turbine (WT)

Abstract. Wind energy is a pillar of the energy transition. However, WT subassemblies are exposed to mechanical and electrical failures, resulting in high costs and reduced availability. This paper presents a hybrid framework combining Multiphysics modelling using COMSOL Multiphysics, frequency-domain signal analysis, and a comprehensive CNN-LSTM architecture. First, we produce consistent vibrational and electromagnetic torque signatures across varying conditions. Second, we train deep learning models on transformed signals from FFT and STFT. Finally, we use explainable AI tools, such as Grad-CAM and SHAP, to clarify the model's decision-making process. The suggested architecture achieves an average classification accuracy of 95.6%, which exceeds traditional ML techniques (SVM and Random Forests). These results show that physics-based modelling and XAI work well together for fault detection in WT, establishing a robust baseline for sustainable maintenance strategies aligned with Net Zero 2050 goals.

Introduction

The growth of renewable energy necessitates reliable and cost-effective operation & maintenance of WT. As indicates the International Energy Agency, wind power plays a remarkable role in reducing greenhouse gas emissions. Yet, WT are prone to breakdowns due to sudden malfunctions mainly in electrical and mechanical subassemblies: generators, bearings, and drivetrains. [1]. Hence, classic maintenance approaches are seen as ineffective in early fault detection. As a result, the concept of proactive maintenance, supported by Multiphysics modelling and AI, aims to forecast the onset of failures and schedule optimal remedial actions. In this context, the present study combines the use of COMSOL Multiphysics simulations to analyse vibration signatures of wind turbines faults, transforming time domain signals into FFT and STFT, which are then processed by CNN-LSTM framework enhanced with XAI.

State of the art

Predictive maintenance of wind turbines has evolved from rule-based monitoring to data-driven diagnostic systems as turbine fleets have grown in size and operational variability. Early CM approaches relied on vibration and electrical signal analysis, using FFT-based spectra and time-frequency representations identify characteristic fault-related harmonics in generators, drivetrains, and bearings [2]. Although these methods are physically interpretable, their capability degrades under non-stationary operating conditions, high noise levels, and early-stage faults typical in wind energy settings.

To improve robustness, conventional ML techniques, such as SVM and Random Forests, have been applied to fault diagnosis after features extraction from vibration, current, or torque signals [3], [4], achieving reasonable classification accuracy. However, their performance depends on expert-driven feature, and their scalability remains limited under variable conditions. Therefore, deep learning models gained attention for their automatic representation learning directly from raw signals or time-frequency maps. CNNs excel in extracting localised spectral patterns, while LSTM networks are effective at modelling temporal dependencies in fault evolution. The combination of these two methods has proven efficiency in diagnostics leveraging the strengths of both CNN and LSTM [5], [6], [7].

Deep learning models, though accurate, are perceived as black boxes due to their non interpretable internal mechanisms. Which is a limitation for their acceptance in industrial and safety contexts. Henceforth, XAI has become a major focus of research. For instance, methods such as Grad-CAM and SHAP facilitate the visualisation of significant time-frequency regions and the quantification of feature contributions, thereby aligning model decisions with physical comprehension [8].

Physics-based and Multiphysics modelling methodologies have been increasingly used to address the shortage of labelled defect data from operating wind turbines. Tools like COMSOL Multiphysics facilitate the simulation of coupled electromagnetic and mechanical phenomena, leading to fault signals under controlled conditions. Multiple studies, indicate that physics-informed signal analysis enhances diagnostic accuracy by aligning data driven models with foundational system behaviour [9], [10]. Nonetheless, there is lack of integrated frameworks combining Multiphysics modelling, a hybrid deep learning, and XAI for predictive maintenance of wind turbines, prompting this study's contribution.

Methodology

The used induction motor model comprises the rotor, stator, shaft, bearings, and foundation.

Two physical interfaces are used: a machine with a rotary motion that incorporates a magnetic interface, and a multi-body dynamic, which is about vibrations and mechanical forces.

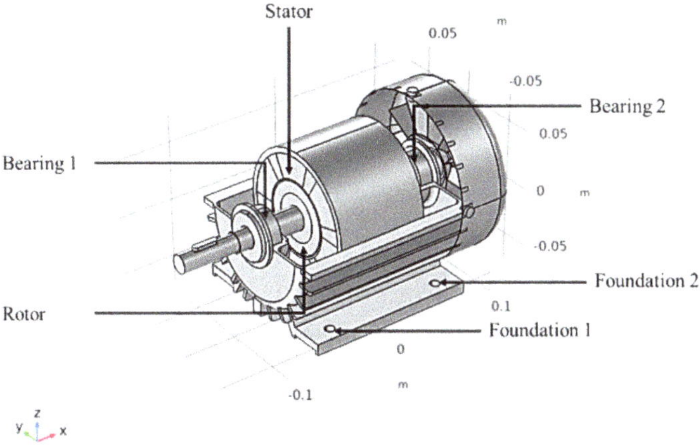

Figure 1 Induction Motor Model

Table 1 The Stiffness Parameters of Bearings And Foundation

Name	Expression	Value	Description
kb	1e6[N/m]	1E6 N/m	Bearing translational stiffness
kbr	1e4[N*m/rad]	10000 N·m/rad	Bearing rotational stiffness
kf	1e6[N/m]	1E6 N/m	Foundation translational stiffness
kfr	1e4[N*m/rad]	10000 N·m/rad	Foundation rotational stiffness

Electromagnetic and vibration simulation:

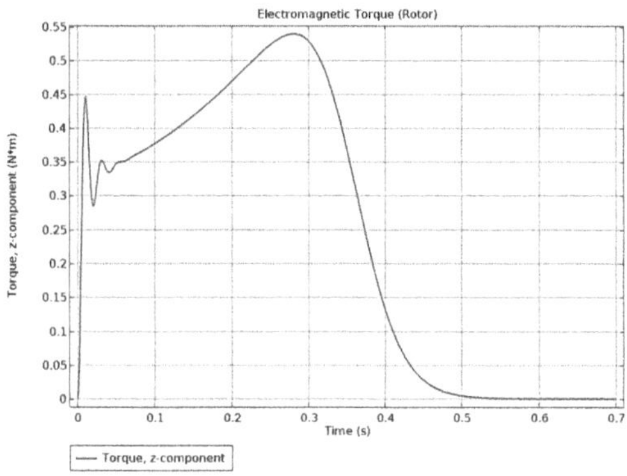

Figure 2 Frequency Spectrum of Electromagnetic Forces in the Rotor

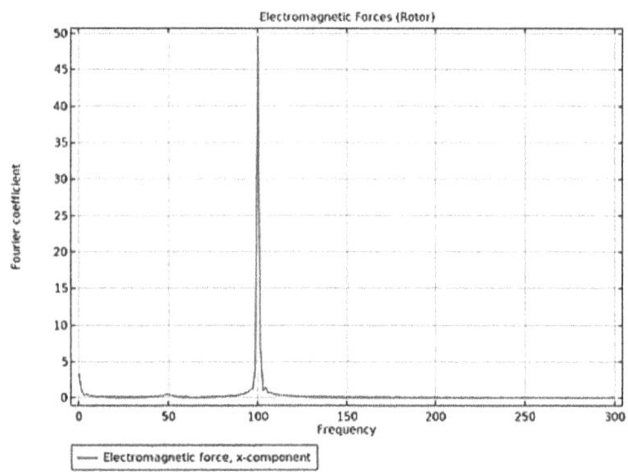

Figure 3 Electromagnetic Torque (Rotor)

The model involves a periodic electrical flow in motor phase A: $I_a = 2\cos(\omega_o t)$, resulting in the rotor's torque exhibition of a rapid response, before stabilising in healthy conditions.

This behaviour indicates proper electromechanical coupling. Initially, the torque increases swiftly due to the stator's magnetic field and its interactions with rotor currents, with transient oscillations related to magnetisation. These oscillations fade in nearly 0.05s highlighting the absence of disturbances and proper damping. The torque peak followed by a decrease towards near-zero steady state as slip diminishes, is a typical behaviour characterising proper functioning.

This behaviour is confirmed by the electromagnetic forces frequency domain, that shows a predominant spectral component centred at 100Hz (twice the supply frequency).

The second simulation, consists of injecting electrical disturbance in the stator current, representing two levels of defect severity: a minor fault with I_{max}=0.01 A, and a major fault with I_{max}=0.2 A. The analysis of electromagnetic torque domain shows, under minor fault a quasi-periodic regime staring after the transient phase and lasting for 0.35s. this behaviour illustrates compromised electromagnetic coupling due to electrical disruption. Under major fault, the signature reveals intensified oscillations during the simulation. This response reflects a significant torque instability due to both electrical disturbance and mechanical excitation.

The FFT analysis of the two levels of faults confirms the presence of a significant deterioration illustrated by a peak at 100Hz with a greater amplitude. This graph interprets high electromagnetic excitation and a low frequency spectral peak.

In a nutshell, the analysis of the three scenarios suggests a consistent evolution of fault severity. The healthy state shows a stable torque and a clean spectral signature. In contrast, the minor fault witnesses moderate torque oscillation amplification and early degradation stage, when the severe scenario highlights prominent torque instability and strongly compromised spectrum. This been said, the torque experiences strong electromagnetic force modulation and advanced degradation under controlled electrical disruptions.

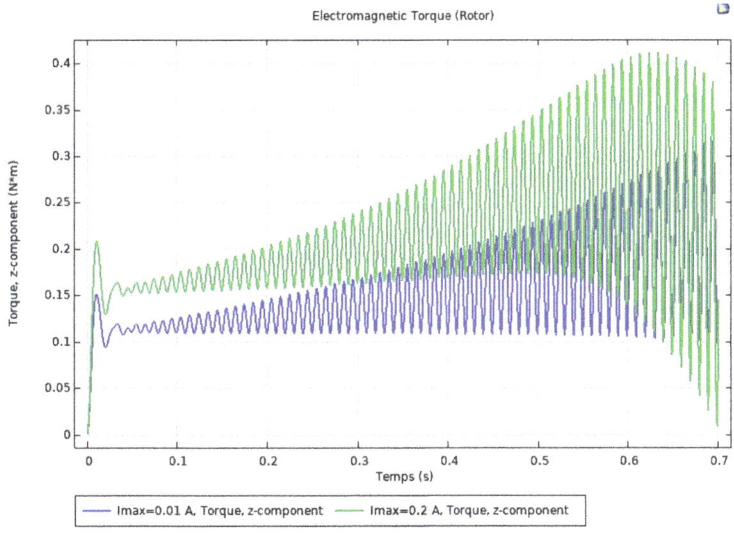

Figure 4 Electromagnetic Torque (Rotor) Response Under (0.01 A ;0.2 A) Respectively In (Blue; Green)

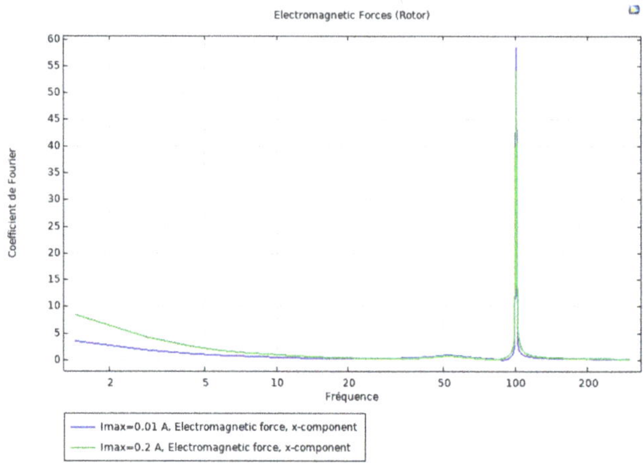

Figure 5 FFT Of Electromagnetic Forces in the Rotor Under (0.01 A ;0.2 A) Respectively In (Blue; Green)

Dataset buildup and Signal analysis

At this point, the construction of a dataset and AI driven problem diagnosis is possible based on the previous results. To this end, the time domain signals are divided into fixed length segments. This allows to maintain consistency across operating situations. Then, each element is designed according to the relevant system state (healthy, minor fault, or severe fault). The segmented signals are thereafter converted into time frequency representations by the STFT, which facilitate the extraction of both spectral content and temporal evolution. Before training the model, the dataset was normalised to enhance numerical stability and learning efficiency. The dataset was split randomly into training (70%), validation (15%) and test (15%) subsets.

Architecture and training of Deep Learning Models.

Fault diagnosis is driven by a hybrid deep learning framework combining CNN and LSTM. The convolutional layers function on the time frequency to extract different spectral features, while the LSTM layers allow encapsulating the dynamic progression of faults throughout time. The network parameters are optimised via the Adam method with a constant learning rate, and the training process is overseen using the validation dataset to prevent overfitting. The performance of the model is assessed on the independent test dataset using standard classification.

The input layer processes 2D spectrograms of vibrational and electrical signals as tensors. The CNN layers consist of two convolutional layers that look for patterns in spectrograms, the first layer is of 32 filters 3x3, ReLU activation, and MaxPooling layer 2x2 to reduce dimensionality. While the second has 64 filters of 3x3, ReLU activation, and Maxpooling layer. These layers pull out frequency features related to mechanical problems in the 150-200Hz band. The flatten layer converts these features for input to the LSTM, which consists of 128 units modelling temporal changes in features anomalies, employing a dropout of 0.2 to prevent overfitting. The dense layers contain 64 neurones with ReLU activation in the middle layer to merge features, and culminate in 3 neurones for multiclass classification using Softmax. The architecture is trained over 100 epochs with the Adam optimiser (learning rate 0.001) and 5-fold cross validation, and Categorical Cross-Entropy as the loss function.

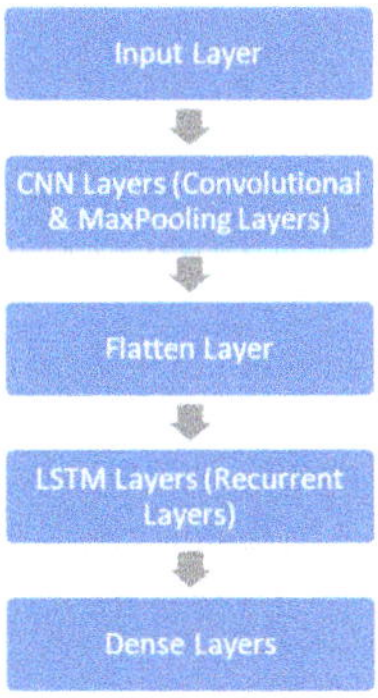

Figure 6 CNN–LSTM Architecture

Model performance

The model achieves an average accuracy of (95.6%). This result surpasses the performance of SVM (81%) and Random Forests (84%).

Table 2 CNN-LSTM Performance

State	Precision	Recall	F1-Score
Healthy	0,97	0,96	0,96
Minor Defect	0,94	0,95	0,95
Major Defect	0,98	0,97	0,97
Average	95.6%	-	-

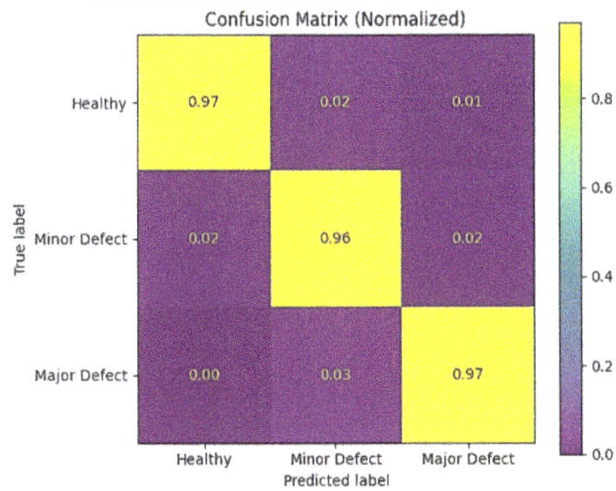

Figure 8 Confusion Matrix

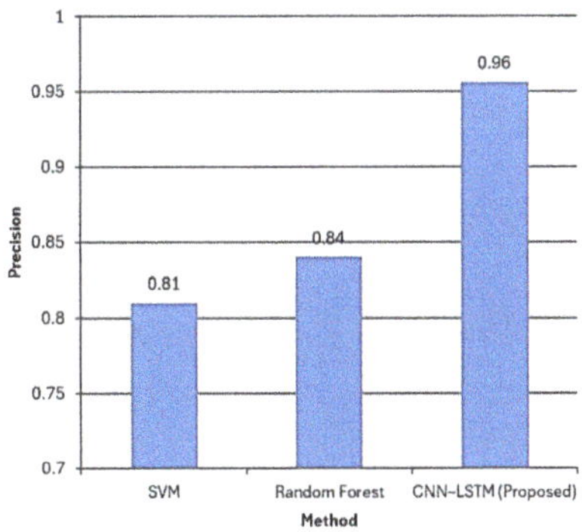

Figure 7 Accuracy Comparison (Test Set)

Explainable Artificial Intelligence Techniques

Fault diagnosis uses a hybrid deep learning framework combining CNN and LSTM. The time frequency regions that impact the network's predictions are identified by Gradient weighted Class Activation Mapping, whereas Shapley Additive exPlanations assess the importance of specific frequency components to the classification results. This analysis ensures that the model's judgements are based on physically robust aspects that align with the electromagnetic and vibrational behaviour observed in Multiphysics simulations.

Materials Research Proceedings 64 (2026) 52-59 https://doi.org/10.21741/9781644904091-7

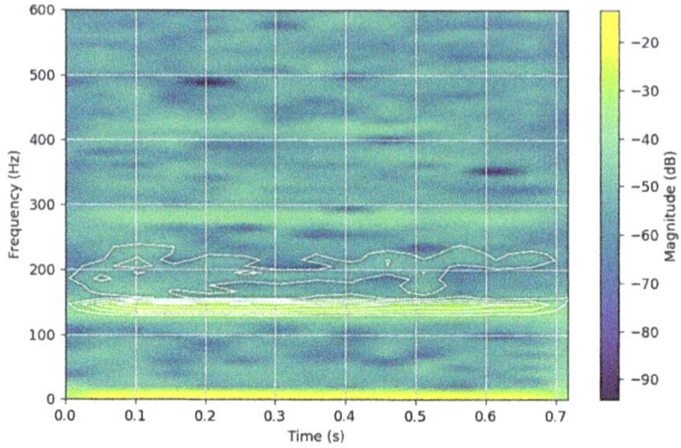

Figure 10 Grad-CAM Map of Critical Spectral Zones (Imax =

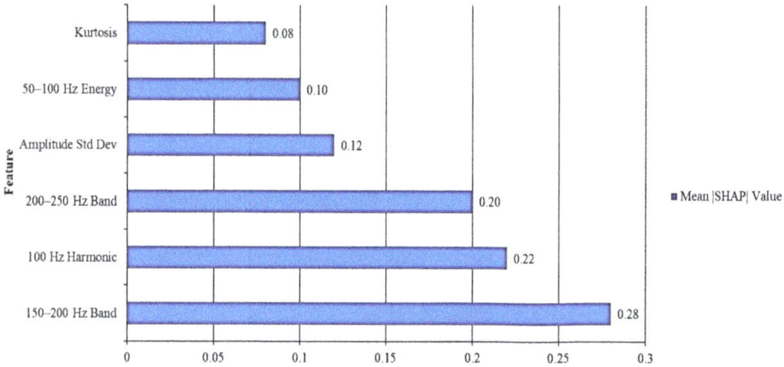

Figure 9 Feature Importance According to SHAP

Results and discussion

The established CNN-LSTM framework demonstrates a good performance diagnosing problems, with an average accuracy of 95.6 % across all fault types. A benchmark shows the superiority of hybrid deep learning model upon traditional ML techniques such as Support Vector Machines and Random Forests. Moreover, using Grad CAM for explainability analysis reveals that the model always looks at frequency bands that are important in the real world, especially those between 150 and 250 Hz, known for being sensitive to fault progression. Hence, SHAP analysis substantiates that these frequency components play a critical role in the classification decisions, demonstrating that the model does not depend on false correlations. This approach is deliberate because Multiphysics modelling produces physically compatible datasets even when the framework is validated with simulated data.

Summary

All in all, this work establishes a hybrid framework for WT PdM driven by Multiphysics modelling, frequency domain analysis, and explainable deep learning. The proposed approach extracts simulations data from COMSOL Multiphysics, trains a CNN-LSTM model on reliable diagnosis, and employs XAI tools to ensure the model's comprehension. The results show high diagnostic accuracy 95.6% compared to traditional models, while keeping transparency and interpretability. The framework helps create intelligent, dependable, and sustainable maintenance solutions for WTs that are aligned with the objectives of global energy transition. In the future, the focus will be on validating experiments under real world conditions, and enhancing model robustness and applicability.

Acknowledgements
This work is financed by CNRST-PASS (PhD Associate Scholarship) for its beneficiary Ms. Sara SGHIOURI.
We used ChatGPT for translation, Grammarly for orthographic correction, and Quillbot AI for paraphrasing.

References
[1] E. Kandemir, A. Hasan, T. Kvamsdal, and S. Abdel-Afou Alaliyat, 'Predictive digital twin for wind energy systems: a literature review', *Energy Inform.*, vol. 7, no. 1, p. 68, Aug. 2024. https://doi.org/10.1186/s42162-024-00373-9
[2] K. Yemets, I. Izonin, and I. Dronyuk, 'Time Series Forecasting Model Based on the Adapted Transformer Neural Network and FFT-Based Features Extraction', *Sensors*, vol. 25, no. 3, p. 652, Jan. 2025. https://doi.org/10.3390/s25030652
[3] H. Dhungana, 'A machine learning approach for wind turbine power forecasting for maintenance planning', *Energy Inform.*, vol. 8, no. 1, p. 2, Jan. 2025. https://doi.org/10.1186/s42162-024-00459-4
[4] F. Jamil, C. Peeters, T. Verstraeten, and J. Helsen, 'Leveraging Signal Processing and Machine Learning for Automated Fault Detection in Wind Turbine Drivetrains', Nov. 25, 2024, *Materials and operation/Operation and maintenance, condition monitoring, reliability.* https://doi.org/10.5194/wes-2024-114
[5] L. Qi, Q. Zhang, Y. Xie, J. Zhang, and J. Ke, 'Research on Wind Turbine Fault Detection Based on CNN-LSTM', *Energies*, vol. 17, no. 17, p. 4497, Sep. 2024. https://doi.org/10.3390/en17174497
[6] K. Ullah *et al.*, 'Short-Term Load Forecasting: A Comprehensive Review and Simulation Study With CNN-LSTM Hybrids Approach', *IEEE Access*, vol. 12, pp. 111858–111881, 2024. https://doi.org/10.1109/ACCESS.2024.3440631
[7] 'Fault Diagnosis of Wind Turbine Based on CNN-LSTM Parallel Network Model', *Acad. J. Eng. Technol. Sci.*, vol. 6, no. 12, 2023. https://doi.org/10.25236/AJETS.2023.061217
[8] D. Astolfi, F. De Caro, and A. Vaccaro, 'Condition Monitoring of Wind Turbine Systems by Explainable Artificial Intelligence Techniques', *Sensors*, vol. 23, no. 12, p. 5376, Jun. 2023. https://doi.org/10.3390/s23125376
[9] H. Sabir, M. Ouassaid, and N. Ngote, 'A New TSA-Fuzzy Logic Based Diagnosis of Rotor Winding Inter-turn Short Circuit Fault in Wind Turbine Based on DFIG Under Different Operating Wind Speeds', in *Innovation in Information Systems and Technologies to Support Learning Research*, vol. 7, M. Serrhini, C. Silva, and S. Aljahdali, Eds, in Learning and Analytics in Intelligent Systems, vol. 7. , Cham: Springer International Publishing, 2020, pp. 583–592. https://doi.org/10.1007/978-3-030-36778-7_64
[10] H. Sabir, M. Ouassaid, and N. Ngote, 'Detection of an Incipient Rotor Winding Inter-Turn Short Circuit Fault', in *2021 IEEE International Autumn Meeting on Power, Electronics and Computing (ROPEC)*, Ixtapa, Mexico: IEEE, Nov. 2021, pp. 1–7. https://doi.org/10.1109/ROPEC53248.2021.9668140

Comparative Analysis of Data-Driven Models for Lithium-Ion Battery SOH Estimation

Mayssae TOUTI[1,a *], Badr BOUOULID IDRISSI[1,b]

[1]National High School of Arts and Crafts, Moulay Ismaıl University Meknes, Marjane 2, B.P.15290 Al-Mansour, Meknes, 50000, Fes-Meknes, Morocco

[a]ma.touti@edu.umi.ac.ma, [b]b.bououlid@umi.ac.ma

Keywords: State of Health (SOH), Long Short-Term Memory (LSTM), Battery Management Systems (BMS), Lithium-ion battery, Machine Learning (ML), Support Vector Regression (SVR), Gated Recurrent Unit (GRU)

Abstract. The evaluation of lithium-ion battery State of Health (SOH) plays a key role in ensuring reliable operation and improving the long-term durability of energy storage systems. However, battery degradation is a complex, time-dependent process, making accurate prediction challenging. This paper presents a comparative analysis of three data-driven models Gated Recurrent Unit (GRU), Support Vector Regression (SVR), Long Short-Term Memory (LSTM) network for SOH estimation. The NASA B0005 dataset, which contains voltage, current, temperature, and capacity measurements from repeated charge–discharge cycles, was used for training and testing all three models. To select the most informative features, the Pearson correlation coefficient was applied to identify the indicators most strongly correlated with capacity degradation. The results indicate that the GRU model yields lower RMSE and MAPE values compared to LSTM and SVR, demonstrating the highest predictive accuracy. These findings suggest that recurrent architectures, particularly GRU, are effective at capturing the temporal patterns in battery behavior. Future studies will focus on hybrid modeling approaches to further enhance prediction accuracy and robustness.

Introduction

The electric vehicles (EVs), renewable energy systems and in the portable electronics have been greatly embracing lithium-ion (Li-ion) batteries due to the high energy density, performance and high life cycle of the battery [1]. However, with time such batteries wear out, their output becomes poor, and capacity and potential safety risk ensues [2]. Therefore, a need to calculate the State of Health (SOH) precisely to offer the safe operation, the most suitable battery functionality, and the extension of the battery life [3].

Since battery degradation is nonlinear and varies as time progresses due to a number of factors including temperature, charge-discharge conditions and age, forecasting SOH may prove to be difficult. The estimation of the State of Health is made through several methods which are direct, model- based, and data-driven [4], but Electrochemical Impedance Spectroscopy (EIS), Incremental Capacity Analysis (ICA) and Differential Voltage Analysis (DVA) gives more detailed diagnostic data, and simple and accurate techniques used in controlled conditions are the Coulomb Counting (CC) and Open Circuit Voltage (OCV) methods [5,6]. Although they need a lot of calibration and computational power, model-based approaches as Equivalent Circuit Models (ECMs) and physics-based electrochemical models (like P2D and SPM) offer accurate depictions of battery behavior [7,8]. Data-driven methods, especially those based on machine learning and deep learning, can capture nonlinear and multidimensional relationships in battery aging behavior. Recent studies have explored various architectures for improving SOH estimation. A study combined XGBoost with neural networks and random forest after MSE-based feature selection to achieve accurate and scalable SOH prediction [9], while another used Bayesian Gaussian

Process Regression for state of charge and health estimation under variable conditions [10]. Deep learning models have also shown strong performance: a multihead- attention LSTM integrated with an adaptive Kalman filter was applied for SOH and SOC estimation [11], and a transformer–LSTM fusion model was developed to improve prediction stability [12]. Furthermore, LSTM, CNN, and SVR models optimized using particle swarm optimization (PSO) achieved improved accuracy, with PSO-LSTM performing best [13], while an improved SVR with feature-based selection and a smooth loss function enabled efficient training [14]. GRU-based models have also proven effective, including a PSO-optimized BiGRU with attention [15] and an SSA-optimized GRU model with correlation analysis for robust SOH estimation across different configurations [16].

Most research has focused on using a single machine learning method along with its optimization. In contrast, this study compares different machine learning methods for estimating SOH by introducing a novel feature set, selected using the Pearson correlation coefficient, which has not been investigated before and is designed to capture aspects of battery degradation. These features are employed to evaluate and compare three machine learning models SVR, LSTM, and GRU using two performance metrics: MAPE and RMSE.

Theory

Overview of the SVR algorithm. The Support Vector Regression (SVR) is derived from the Support Vector Machine (SVM) that was initially designed to carry out classification tasks and maximize the margin to identify an optimal decision boundary between classes. This framework is adapted by SVR to solve regression problems. Through the objective function in equation 1, w denotes the weight, x represents input features, and b represents the bias, Thus, the optimization problem is solved to obtain the optimal weight and bias parameters. This permits SVR to generalize effectively and predict unseen data [17].

$$f(x) = w * x + b \tag{1}$$

The State of Health may be estimated using Support Vector Regression, which generalizes well, is less prone to overfitting, and can perform effectively with relatively small datasets. However, careful hyperparameter tuning is necessary to control computational cost, which increases for large datasets or when using complex kernel functions.

Overview of the lstm algorithm. The long short term memory is a model that has been designed to be able to process sequential data effectively, avoiding both the issues of the vanishing and exploding gradient. It uses three gating mechanisms to control the way information is propagated across time [18]. The forget gate stores pertinent data of the previous state, the input gate identifies the information to be added to the current cell memory and the output gate identifies what information is to be generated at a particular time step. In this regulated information flow, LSTM models can learn temporal dependencies in a complicated sequence and are thus applicable to the analysis of battery-related signals, including voltage, current, and capacity, to estimate the State of Health (SOH) and Remaining Useful Life (RUL).

Overview of the GRU model. The Gated Recurrent Unit (GRU) is simpler than LSTM. Even after a long sequence of patterns, it still can recall them, although it is lowly with regard to network structure and count of gates [19]. GRU can be used to predict battery behavior, and State of Health (SOH) owing to its structure, which is well suited for sequential data processing.

Experiment:

*Dataset description.*The dataset employed in this study comprises lithium-ion battery B0005 data supplied by the NASA Prognostics Center of Excellence. The 2 Ah cell was subjected to repeated charge-discharge at a controlled laboratory temperature of approximately 24°C. During charging, the cell was first charged at a constant current of 1.5 A until reaching a voltage of 4.2 V, after which the voltage was maintained constant while the current gradually decreased to around 20 mA. During the discharge phase, the B0005 battery was discharged at a constant current of 2 A until the voltage reached the cutoff value of 2.7 V. These charge-discharge cycles were repeated until the cell reached a nominal capacity drop of 30% corresponding to 1.4 Ah, which indicates that the cell reached the end of its useful life. we used This data set as a reference to find out the real state of health under actual working conditions.

Feature extraction. During each cycle of operation of the B0005 battery, voltage, current and temperature were recorded in both the charging and discharging process. In this paper, we have identified seventeen input health indicators (HI1-HI17) based on the measured voltage, current and temperature profiles to also enable the determination of the State of Health (SOH) of the battery. HI1 to HI5 are then derived based on the voltage data and HI1 is the lowest discharge voltage and HI2 is the point where HI1 is attained and HI3 is the point where the terminal voltage is equal to the cutoff limit and HI4 is the average charging voltage and HI5 is the average discharging voltage. For current-based health indicators, HI6 and HI7 correspond to the mean charging current and the mean current observed during the constant-voltage charging stage, respectively. The temperature-derived health indicators, HI8 to HI13, include HI8, the highest discharge temperature; HI9, the time associated with this maximum; HI10, the highest charge temperature; HI11 and HI12, the average temperatures during charge and discharge operations, respectively; and HI13, the time when the peak charge temperature appears. Additional health indicators, HI14 to HI17, include HI14, which characterizes the cycle index; HI15, the proportion of constant-current charging time relative to the total charging duration; HI16, the discharge capacity; and HI17, the ratio of charging duration to discharging duration. After extracting these health indicators, we determine which of them has a strong and directional relationship with the battery's capacity. Therefore, the Pearson correlation coefficient is applied to identify which health indicators are relevant to be used as model features, while those with weak correlations are discarded.

Table 1. Pearson Correlation Values Indicating the Relationship Between Health Factor and Capacity

Health factor	*FEATURES*	*Pearson Correlation*
HI1	min_discharge_voltage	-0.482448
HI2	time_min_discharge_voltage	0.623914
HI3	time_above_cutoffvoltage	-0.843280
HI4	avg_charge_voltage	-0.797764
HI5	avg_discharge_voltage	0.834675
HI6	avg_charge_current	0.791446
HI7	avg_charge_cv	0.862586
HI8	max_discharge_temp	0.135031
HI9	time_max_charge_temp	-0.666500
HI10	max_charge_temp	0.251023
HI11	avg_charge_temp	-0.010560
HI12	avg_discharge_temp	0.135031
HI13	time_max_charge_temp	-0.666500
HI14	cycle_num	0.987739
HI15	proportion of constant-current charging time relative to the total charging duration	0.882640
HI16	Capacity	1
HI17	Charge/discharge_time_ratio	-0.637038

Table 1 presents the results of the computed correlation values between the extracted features and the battery capacity. Only the features with correlation values above 0.75 were taken into consideration in this analysis. According to this selection criterion, HI3, HI4, HI5, HI7, HI14 and HI15 were selected as input parameters of the proposed models.

Test and results:
Experimental evaluations were performed using LSTM, GRU, and SVR models to predict the battery State of Health (SOH) based on the features selected from the battery datasets. The data were split into training and test sets, with 80% used for training and the remaining 20% reserved for testing. Within the training set, 25% of the samples were further used for validation. All models were carefully tuned to identify suitable model configurations and hyperparameter settings.

The validation set was used to avoid overfitting and track the performance of the model. Early stopping was used in case the validation loss failed to decrease in 10 successive epochs, and the model parameters that correspond to the lowest validation loss were stored to guarantee optimal generalization. The finalized model configurations and tuned hyperparameters for all considered models are reported in Tables 2, 3, and 4. LSTM and GRU models were designed with identical depth, and both were trained using the Adam optimization algorithm.

Table 2. Configuration Details and Hyperparameter Values for the LSTM Model

Learning_rate	*Batch*	*Epoch*	*Hidden/Units Layers*	*Regularization*	*Optimizer*
0.03	8	32	2 LSTM Layers (51,32),1 Dense layer (32)	Dropout(0.06,0.1)	Adam

Table 3. Configuration Details and Hyperparameter Values for the GRU Model

Learning_rate	*Batch*	*Epoch*	*Hidden/Units Layers*	*Regularization*	*Optimizer*
0.01	16	50	2 GRU Layers (32,16),1 Dense layer (16)	Dropout(0.009,0.1)	Adam

Table 4. Configuration Details and Hyperparameter Values for the SVR Model

C	*gamma*	*Epsilon*	*Kernel*
5	0.01	0.001	rbf

We have utilized two commonly known performance measures to compare and contrast the results of the three models, Mean Absolute Percentage Error (MAPE) and Root Mean Square Error (RMSE) as in Equations 2 and 3. In this case, N is the number of samples, y_i is the measured capacity and $\hat{y}_i$ is the battery capacity predicted. Results for three models, namely LSTM, SVR and GRU are presented in Table 5.

$$\text{RMSE} = \sqrt{\frac{1}{N}\sum_{i=1}^{N}(y_i - \hat{y}_i)^2} \tag{2}$$

$$\text{MAPE} = \frac{1}{N}\sum_{i=1}^{N}\left|\frac{y_i-\hat{y}_i}{y_i}\right| \tag{3}$$

Table 5. Comparison of RMSE and MAPE Results for Different Machine Learning Models

	SVR	*LSTM*	*GRU*
RMSE	0.0293	0.0168	0.0123
MAPE	0.0151	0.0086	0.0061

Figures 1, 2, and 3 illustrate the results obtained using the SVR, LSTM, and GRU models, respectively. For each approach, the figures present a comparison between experimental measurements and model predictions across the training and testing datasets, offering a clear indication of how effectively the models represent battery behavior over successive cycles.

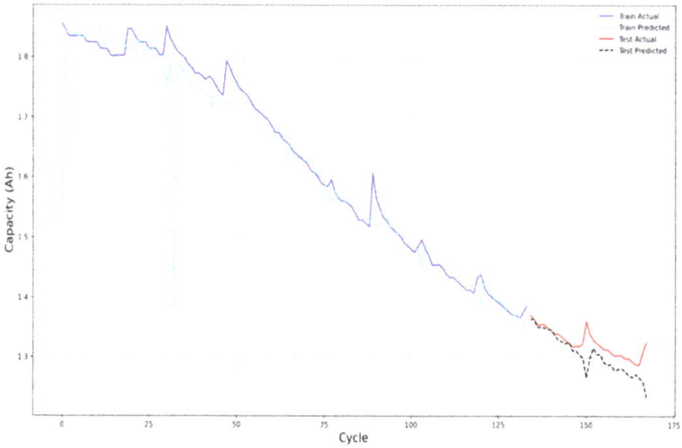

Fig.1.Battery Capacity Estimation Over Cycles Using SVR

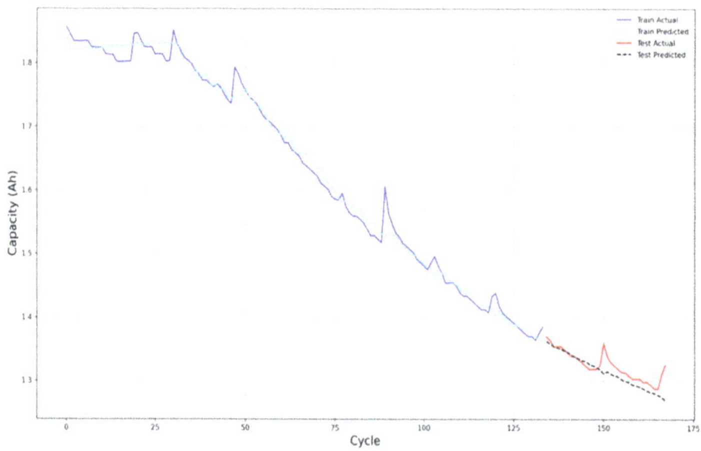

Fig.2.Battery Capacity Estimation Over Cycles Using LSTM

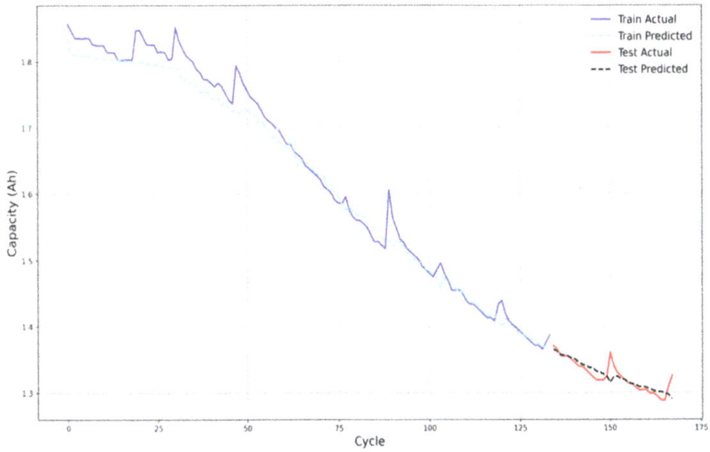

Fig.3. Battery Capacity Estimation Over Cycles Using GRU

As shown in Table 5 and the corresponding figures, SVR approach exhibits the largest prediction errors, followed by LSTM model, and the lowest values of the RMSE and MAPE of the prediction error belong to GRU model. We trained each model on the same dataset, features, and train/test split for a fair comparison. The GRU model outperforms the LSTM. This is explained by the fact that it has a more compact architecture, containing fewer gates and trainable parameters, which allows for more stable training and reduces the risk of overfitting, particularly when working with relatively small datasets like ours and enable it to effectively capture the trends of battery degradation to generate better predictions.

Conclusion

We used three methods of machine learning on our work to approximate the State of Health. The NASA B0005 dataset was used to extract health indicators and the most effective ones were used Pearson correlation method has been used to choose features that were used to capture battery degradation. Each model Long Short-Term Memory, Gated Recurrent Unit and Support Vector Regression was tuned by finding the most suitable model configuration and hyperparameters to perform a higher generalization and increased measures of evaluation. The best was the GRU model performance, lowest MAPE and RMSE values of any of the LSTM and SVR models. These results indicate that deep learning models, and specifically GRU, are well suited for capturing complex degradation patterns and time dependencies in battery health modeling. To enhance the performance of SOH prediction, the subsequent study will focus on optimizing model epochs and applying hybrid methods that combine the strengths of different algorithms. In addition, we will seek to employ these methods into smart battery management (AI-BMS) resulting in stronger and flexible battery health monitoring.

Acknowledgements

The Centre National de la Recherche Scientifique et Technique (CNRST), Morocco, made use of this work by supporting with the PhD Associate Scholarship (PASS), which the author highly takes pride in.

References

[1] K.R. Ngoy, V.T. Lukong, K.O. Yoro, J.B. Makambo, N.C. Chukwuati, C. Ibegbulam, ... & T.C. Jen, Lithium-ion batteries and the future of sustainable energy: A comprehensive review, Renewable and Sustainable Energy Rev. 223 (2025) 115971.. https://doi.org/s://doi.org/10.1016/j.rser.2025.115971

[2] T. Rahman, T. Alharbi, Exploring lithium-Ion battery degradation: A concise review of critical factors, impacts, data-driven degradation estimation techniques, and sustainable directions for energy storage systems, Batteries 10(7) (2024) 220.. https://doi.org/s://doi.org/10.3390/batteries10070220

[3] A.S. Akram, M. Sohaib, W. Choi, SOH Estimation of Lithium-Ion Batteries Using Distribution of Relaxation Times Parameters and Long Short-Term Memory Model, Batteries 11(5) (2025) 183.. https://doi.org/s://doi.org/10.3390/batteries11050183

[4] K. Tang, B. Luo, D. Chen, C. Wang, L. Chen, F. Li, ... & C. Wang, The State of Health Estimation of Lithium-Ion Batteries: A Review of Health Indicators, Estimation Methods, Development Trends and Challenges, World Electric Vehicle J. 16(8) (2025) 429.. https://doi.org/s://doi.org/10.3390/wevj16080429

[5] E. Kim, S. Jung, Battery State of Health Estimation Methods: Implementation and Comparison of 11 Algorithms, IEEE Access (2025).. https://doi.org/s://doi.org/10.1109/ACCESS.2025.3541631

[6] M. Li, W. Xu, S. Zhang, L. Liu, A. Hussain, E. Hu, ... & Z. Chen, State of health estimation and battery management: a review of health indicators, models and machine learning, Materials 18(1) (2025) 145.. https://doi.org/s://doi.org/10.3390/ma18010145

[7] Z. Wang, G. Feng, D. Zhen, F. Gu, A. Ball, A review on online state of charge and state of health estimation for lithium-ion batteries in electric vehicles, Energy Rep. 7 (2021) 5141-5161.. https://doi.org/s://doi.org/10.1016/j.egyr.2021.08.113

[8] E.H. Bayoumi, M. De Santis, H. Awad, A Brief Overview of Modeling Estimation of State of Health for an Electric Vehicle's Li-Ion Batteries, World Electric Vehicle J. 16(2) (2025).. https://doi.org/s://doi.org/10.3390/wevj16020073

[9] N. Mohanty, N.K. Goyal, V.A. Naikan, Hybrid machine learning models for improved state of health estimation of lithium-ion batteries in electric vehicles, Mater. Res. Proc. 49 (2025).. https://doi.org/s://doi.org/10.21741/9781644903438-5

[10] A. Aitio, D. Jöst, D.U. Sauer, D.A. Howey, Learning battery model parameter dynamics from data with recursive Gaussian process regression, J. Dyn. Syst. Meas. Control 147(3) (2025) 031010.. https://doi.org/s://doi.org/10.1115/1.4067771

[11] Y. Song, J. Lu, H. Zhang, G. Liu, SOC and SOH Prediction of Lithium-Ion Batteries Based on LSTM-AUKF Joint Algorithm, Energy Sci. Eng. 13(1) (2025) 240-254.. https://doi.org/s://doi.org/10.1002/ese3.1992

[12] X. Cai, T. Liu, State of Health Prediction for Lithium-Ion Batteries Using Transformer-LSTM Fusion Model, Appl. Sci. 15(7) (2025) 3747.. https://doi.org/s://doi.org/10.3390/app15073747

[13] K. Li, X. Chen, Machine learning-based lithium battery state of health prediction research, Appl. Sci. 15(2) (2025) 516.. https://doi.org/s://doi.org/10.3390/app15020516

[14] L. Qian, L. Xuan, J. Chen, Battery SOH estimation based on decision tree and improved support vector machine regression algorithm, Front. Energy Res. 11 (2023) 1218580.. https://doi.org/s://doi.org/10.3389/fenrg.2023.1218580

[15] Z. Hou, W. Xu, G. Jia, J. Wang, M. Cai, Estimation of SOH of lithium-ion batteries based on PSO-Bi GRU-Attention network, J. Electrochem. Soc. 171(2) (2024) 020550.. https://doi.org/s://doi.org/10.1149/1945-7111/ad29c4

[16] C. Zhang, L. Luo, Z. Yang, S. Zhao, Y. He, X. Wang, H. Wang, Battery SOH estimation method based on gradual decreasing current, double correlation analysis and GRU, Green Energy Intell. Transp. 2(5) (2023) 100108.. https://doi.org/s://doi.org/10.1016/j.geits.2023.100108

[17] Z. Liu, J. Zhao, H. Wang, C. Yang, A new lithium-ion battery SOH estimation method based on an indirect enhanced health indicator and support vector regression in PHMs, Energies 13(4) (2020) 830.. https://doi.org/s://doi.org/10.3390/en13040830

[18] L. Zhang, T. Ji, S. Yu, G. Liu, Accurate prediction approach of SOH for lithium-ion batteries based on LSTM method, Batteries 9(3) (2023) 177.. https://doi.org/s://doi.org/10.3390/batteries9030177

[19] Y. He, N. Pattanadech, K. Sukemoke, M. Pan, L. Chen, The State of Health Estimation of Retired Lithium-Ion Batteries Using a Multi-Input Metabolic Gated Recurrent Unit, Energies 18(5) (2025).. https://doi.org/s://doi.org/10.3390/en18051035

Unlocking the Potential of Hydropower by Integrating Floating PV Systems

Osama SAADEH[1,a *], Kafa SULEIMAN[1,b] and Zakariya DALALA[2,c]

[1]Electrical and Energy Engineering Department, German Jordanian University, Amman, Jordan

[2]Energy Systems Engineering, Abdullah Al Salem University, Khaldiya, Kuwait

[a]osama.saadeh@gju.edu.jo, [b]K.Suleiman@gju.edu.jo, [c]zakariya.dalala@aasu.edu.kw

Keywords: Floating Photovoltaic (FPV), Pumped Storage Power System (PSPS), Hybrid Renewable Energy Systems, Dispatchable Renewable Energy, Water–Energy Nexus

Abstract. Increasing the share of renewable energy is essential for sustainable development and climate change mitigation. Hydropower is widely recognized as a reliable and dispatchable renewable energy source, but its further expansion is constrained by water scarcity and the lack of suitable locations. The integration of Floating Photovoltaic (FPV) systems with Pumped Storage Power Systems (PSPS) is studied in this research to validate creating a controllable renewable energy solution based on existing water reservoirs. The proposed FPV–PSPS configuration enables microgrid operation while reducing the impact of solar intermittency. The feasibility of this integration and the optimal system sizing were assessed using the iHOGA optimization software. The simulations and studies took into consideration realistic constraints which included reservoir volume, available water storage and electrical load demand. The results show that an FPV–PSPS microgrid with annual energy of approximately 1.8 GWh is a suitable configuration for Jordan at the King Talal Dam, which offers the best scenario due to its storage capacity and surrounding geography. The estimated payback period is about 8 years, indicating that the solution is not only technically viable but also financially attractive.

Introduction

The increased electrification, rapid depletion of fossil fuel sources and the rising environmental concerns, has led to increased global deployment of PV systems [1–3]. While solar energy is a clean energy resource, large-scale installations require significant land areas due to limited cell efficiency. This has increased interest in deploying PV systems on water bodies. FPV systems, in which PV modules are floating on the surface of water bodies, have emerged as a practical and attractive solution [4]. FPV offers countries with limited land availability and high population density a way to expand solar capacity without competing with other land uses.

FPV systems can be implemented on a wide variety of water bodies, including lakes, irrigation ponds, reservoirs, fish farms, dams, and canals. FPV systems provide several well-documented advantages compared with conventional land-based PV installations. These include reduced land-use conflicts, lower costs related to land acquisition and site preparation, and improved PV module performance due to the cooling effect of water [5]. FPV installations also tend to experience lower dust accumulation and shading losses. An additional benefit is the reduction of water evaporation from reservoir surfaces [6]. For example, in Morocco, approximately 3 m^3 of water evaporates annually per square meter of dam surface area [7], while in Jordan nearly 10% of stored dam water is lost during hot spells and heatwaves [8]. Thus, FPV systems offer the dual benefit of electricity generation and water conservation.

The first FPV installation was commissioned in California in 2008, followed by several other countries, including Japan, South Korea, China, and the United Kingdom [9]. However, FPV

Materials Research Proceedings 64 (2026) 68-74 https://doi.org/10.21741/9781644904091-9

systems remain dependent on solar availability and therefore cannot provide dispatchable power when operated as standalone systems.

PSPS is a mature and proven solution for large-scale energy storage. A PSPS consists of upper and lower water reservoirs connected by a penstock, where electricity is generated by converting the potential energy of stored water as it passes through a hydraulic turbine. Such systems are widely used to smooth grid load variations and store large amounts of energy in a cost-effective manner [10]. Modern PSPS facilities can achieve storage/generation efficiencies of 90% [11], making them one of the most realistic options for long duration electrical storage.

Energy is required to pump water from the lower reservoir to an upper reservoir, hence storing excess solar energy as potential energy. When this energy is supplied from the FPV, it can be considered a fully renewable energy storage system. Integrating FPV with PSPS enables deployment of a microgrid configuration, improving both flexibility and operational reliability [9]. During daylight hours, excess energy generated by the FPV can be used to pump water to the upper reservoir, effectively storing energy for later use through the PSPS system. FPV systems generate varying amounts of electricity during the day, while the PSPS can supply electricity for the reminder of the time by releasing the stored water when solar output is low or unavailable. This hybrid FPV–PSPS configuration enables dispatchable generation, reduces solar intermittency, and improves overall system utilization. The proposed hybrid system is shown in Figure 1.

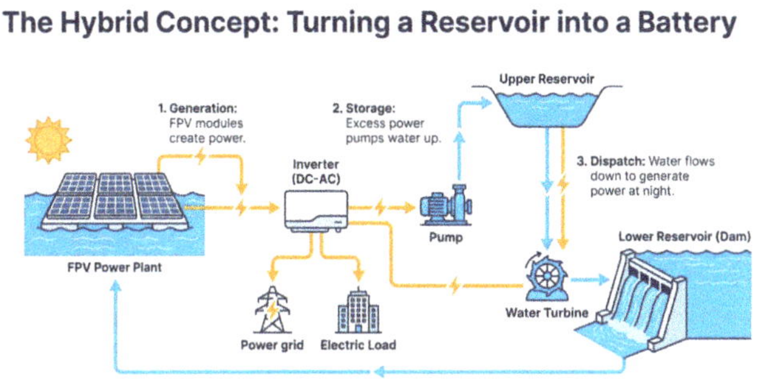

Figure 1: Proposed hybrid FPV and PSPS system.

Integrating FPV-PSPS Hydropower Plants in Jordan

Jordan is among the most water-scarce countries in the world, with renewable water resources meeting only 50% of the current national demand, which is continuing to grow. Decades of overexploitation of groundwater has led to declining water tables and deterioration in the water quality. At the same time, high population growth and the influx of refugees resulting from regional conflicts have placed additional pressure on an already vulnerable water supply system. These pressures are further intensified by rainfall variability associated with climate change.

Investigating the integration of FPV systems with PSPS requires several steps. Major dams and water bodies in Jordan were reviewed to identify reservoirs suitable for FPV deployment. This initial screening considered reservoir surface area, storage capacity, and operational constraints. Next, potential locations near the selected dams were examined for PSPS implementation, with particular attention given to elevation differences, water availability, and local terrain conditions. To analyze and optimize the proposed FPV–PSPS configurations, the iHOGA software was used [12]. Several system-sizing and load-profile scenarios were evaluated with the objective of maximizing energy generation and improving system utilization.

In the north of the country, the main dams include King Talal, Wadi Al-Arab, Ziglab, Karameh, Shuaib, and Kafrein dams, with a combined storage capacity of approximately 178.7 MCM. Al-Wehdah Dam, located along the Jordan–Syria border, provides a storage capacity of 110 MCM, but is almost empty due to regional conflicts. In southern Jordan, the Tannour, Mujib, and Waleh dams have an additional 57.7 MCM of storage capacity [13].

An effective PSPS configuration requires a substantial elevation difference between the upper and lower reservoirs, as well as close proximity to an existing water source. The powerplant, which contains the turbines and generators, is positioned near the lower reservoir to reduce hydraulic losses and limit construction complexity. A suitable site is therefore one that minimizes overall infrastructure and capital costs while remaining compatible with FPV deployment on the water surface.

A detailed terrain and site suitability analysis reported in [14] examined the geographic characteristics surrounding Jordanian dams, the study concluded that only King Talal, Al-Wehdah, Wadi Al-Arab, Al-Tannour, Al-Mujib, and Waleh dams satisfy the basic requirements for PSPS installation. Other dams were excluded due to limited water volumes or insufficient elevation differences.

Based on these findings, the pumped-storage potential of King Talal Dam (KTD), Al-Wehdah Dam, and Wadi Al-Arab Dam was evaluated in more detail using available upper-basin volumes and pumping durations. For King Talal Dam, an upper-basin storage volume of 20 MCM was considered with a 12-hour pumping duration. Under these conditions, the system achieves a rated storage power of 1036.26 MW, corresponding to a total stored energy of 12.4 GWh. Two candidate locations were identified for the upper basin, with projected surface areas of 190.6×10^3 m^2 and 310.4×10^3 m^2. Based on these areas, an average basin depth of approximately 40 m is required to store the calculated volume and energy.

For Al-Wehdah Dam, the analysis indicates that an energy storage capacity of 4.0 GWh can be achieved with a rated pumping power of 335.57 MW, operating continuously for 12 hours at a pumping head of 265 m. The required upper-basin volume is estimated at 14.9×10^6 m^3. Since the existing upper basin has a minimum storage capacity of 5 MCM, the site is capable of supporting PSPS operation. Three potential upper-basin locations were examined, including areas of 126.2×10^3 m^2 and 60×10^3 m^2, an average basin depth of 39.6 m is required to support the rated storage power.

In the case of Wadi Al-Arab Dam, an upper-basin storage volume of 1.6 MCM results in a rated storage power of 114.3 MW, with a continuous pumping duration of 12 hours, the total stored energy is calculated to be 1.37 GWh. Given a basin area of 247×10^3 m^2, the required average basin depth is 6.5 m, which is considerably lower than that required for King Talal and Al-Wehdah dams.

To further assess the feasibility of FPV–PSPS integration at the King Talal Dam, historical reservoir data were obtained from the Jordan Valley Authority (JVA) and a detailed water balance analysis, including inflow, outflow, and storage volumes, was conducted over a 10-year period (2011–2021). This analysis confirms the long-term availability of sufficient water levels to support pumped storage operation, ensuring reliable energy storage performance.

King Talal Dam has a total storage capacity of 74 MCM, historical records show that the lowest storage volume during the study period was approximately 18 MCM, in November 2021. Despite seasonal and annual variations, the reservoir maintains adequate PSPS operational volume. This available storage allows the system to balance on a daily basis with seasonal energy variations, including the excess solar generation during summer days and reduced generation during the winter season.

When the sites are compared, King Talal Dam is the most suitable location for large-scale pumped storage deployment, as it has the largest total stored energy of 12.4 GWh, it significantly larger than both Al-Wehdah Dam (4.0 GWh) and Wadi Al-Arab Dam (1.37 GWh) under equivalent pumping durations. This capacity makes King Talal Dam well suited for long-duration energy storage, grid balancing with large-scale renewable energy integration.

Although the required upper-basin depth at King Talal Dam is similar to that of Al-Wehdah Dam, the larger storage volume enables much higher energy storage without an increase in hydraulic or operational complexity. In addition, the presence of multiple potential upper-basin locations provides flexibility in system design, allowing optimization of excavation depth, surface area, and civil construction costs.

Overall, these results indicate that King Talal Dam is the preferred site for such a pumped storage deployment prototype in Jordan, when the primary objective is to maximize stored energy capacity to enhance grid support capability.

Definition of FPV–PSPS Scenarios

Three operational scenarios are defined to evaluate the integration of FPV-PSPS at the King Talal Dam. This assessment was based on two main indicators: the utilization factor (UF) and the ability to deliver continuous energy generation.

- Scenario 1: Maximum Pumping with Maximum FPV Deployment

In the first scenario, the largest feasible volume of water is pumped to the upper reservoir, it is assumed that an unlimited source of water exists, such that the resulting change in the dam water level is negligible. Water pumped to the upper reservoir is completely returned to the dam through the turbine, essentially a closed operational cycle. From an energy perspective, this scenario draws the maximum achievable FPV output. The entire available water surface area of 48,387 m^2 is covered with FPV modules. As a result, this configuration produces the highest annual energy output from the system. In practice, Scenario 1 serves as an upper bound for system performance rather than a realistic operating condition.

- Scenario 2: Minimum Pumping Based on the Critical Month

The second scenario adopts a conservative design approach, where the minimum FPV power output during the critical month is used to size the PSPS turbine flow rate. Based on long-term solar resource analysis, December was identified as the month with the lowest solar energy production, this is effectively the worst-case scenario.

Accordingly, the turbine and pumping systems are sized to operate reliably under this worst-case condition, this ensures the year-round continuity of operation and avoids over-sizing the system. From an operational standpoint, this prioritizes reliability and guaranteed dispatchability.

- Scenario 3: FPV–PSPS Sizing Based on Average Power

In this scenario, the FPV array and water turbine are sized according to the average power generation. Although this scenario appears attractive from a theoretical point of view, it is not feasible in practice as the required upper reservoir volume is approximately 4.76×10^5 m^3, which would be extremely difficult to construct at the selected site. Under this configuration, the total capital investment is around 28.8 million USD for the FPV and 36.0 million USD for the PSPS. This will result in a payback period that exceeds the expected lifetime of the system, which is not practically feasible.

Simulation, Results and Discussion

The Hybrid Optimization by Genetic Algorithms (iHOGA) software facilitates the modeling of hybrid systems. The software offers a wide array of integration options encompassing many renewable and conventional generation technologies. To effectively determine the optimal system configurations, the software employs genetic algorithms as its primary optimization engine. The software was developed and is owned by the University of Zaragoza.

iHOGA integrates two hierarchical genetic algorithms, the primary algorithm is responsible for determining the optimal configuration of system components, such as photovoltaic (PV) panels, battery storage, AC generators, and inverters, by minimizing the total system cost. The secondary algorithm identifies the most appropriate control strategy for each configuration generated by the primary algorithm, aiming to minimize system costs while ensuring operational feasibility.

The software is configured such that when the power generated by the FPV system falls below the electrical load needs, the deficit is compensated by the PSPS system. If the PSPS is unable to meet the demand, the remaining deficit is drawn from the AC grid. This dispatch logic ensures reliable operation while prioritizing the use of renewable energy. Only Scenario 2 was chosen for an in-depth analysis, as it represents the most technically feasible option.

The FPV power plant was sized to maximize the effective use of both available hydrological capacity and solar energy. The fundamental criterion for successful integration is that the maximum solar power generated during a given interval aligns with the maximum power delivered by the PSPS during the complementary periods. This balance is achieved while meeting system constraints, including electrical load demand, reservoir storage capacity, minimum dam storage volume, FPV module area, and minimization of the net present cost. This integrated optimization approach ensures a technically viable, economically feasible, and dispatchable hybrid FPV–PSPS system.

The chosen site at the King Talal Dam requires an area of 190.6×10^3 m^2, as shown in Figure 2. As for the reservoir's geometry and elevation difference, it is well-suited for operating a PSPS with a pumping head of 205 m and a rated power of 1 MW, allowing for continuous operation of up to 18 hours. The horizontal distance between the upper and lower reservoirs is approximately 1.33 km, which is within acceptable limits for penstock installation and hydraulic losses.

Figure 2. FPV area in KTD.

The maximum allowed water withdrawal for PSPS operation is limited by the dam's minimum historical storage volume, which is 18 MCM. This figure was from the 10-year water balance analysis mentioned earlier, ensuring that PSPS operation does not compromise water security under critical conditions. Based on the available upper basin volume and an 18-hour pumping duration, the corresponding rated maximum daily storage power was calculated to be 620 MW, resulting in a theoretical energy storage capacity of 11.161 GWh.

Materials Research Proceedings 64 (2026) 68-74 https://doi.org/10.21741/9781644904091-9

Typical efficiency values for small hydropower systems is 85% turbine efficiency, 90% drive efficiency, and 93% generator efficiency, leading to an overall system efficiency of 71.1%.

The optimization results are shown in Figure 3, which illustrates the operational profiles of the system. Resulting in a max pump power of 719KW, and a corresponding FPV system of 1MWp. The results confirm the technical feasibility of the proposed configuration, provided that the minimum reservoir water level remains stable.

From an economic standpoint, the maximum total installed cost for the integrated 1 MW FPV and 719 kW PSPS system was estimated at 2.1 million USD, while the annual savings are 270 thousand USD, resulting in a payback period of under 8 years. This payback period is considered favorable when compared to the expected system lifetimes, approximately 15 years for FPV and up to 50 years for PSPS, thereby confirming the long-term technical and economic viability.

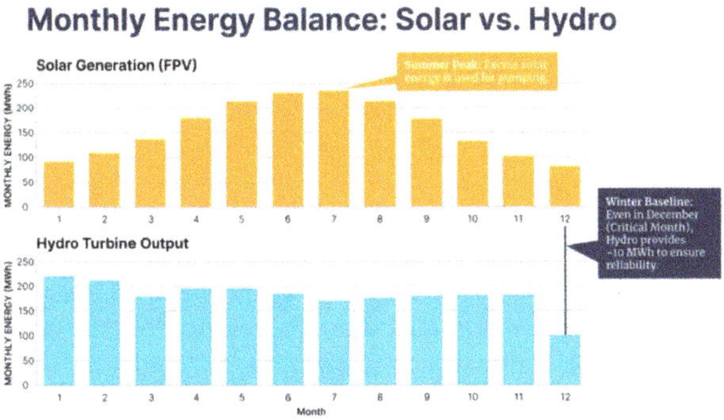

Figure 3. Scenario 2- System output.

Conclusion

In this research, the integration of Floating Photovoltaic–Pumped Storage Power Systems (FPV–PSPS) was studied using a real case study at the King Talal Dam in Jordan. The objective was to assess whether existing reservoirs in a highly water-scarce country can be used to enhance hydropower utilization by offering large-scale renewable energy storage through the addition of floating photovoltaic systems. The analysis confirms that this integration can convert conventional reservoirs and dams into multifunctional energy assets.

The simulation results show that an FPV–PSPS microgrid with an annual energy throughput of approximately 1.8 GWh represents a suitable configuration for Jordan at the King Talal Dam, which offers the best conditions due to its storage capacity and surrounding geography. The estimated payback period is about 8 years, indicating that the solution is not only technically viable but also financially attractive.

The results show that coupling floating photovoltaics with pumped storage hydropower enables turning renewable energy systems into a dispatchable source, capable of supporting peak load demand while reducing the impact of solar intermittency. In addition, the proposed configuration contributes to local water resource conservation by limiting evaporation losses from reservoir surfaces, which is particularly important in arid and semi-arid regions.

Declaration of Generative AI and AI-Assisted Technologies in the Writing Process

During the preparation of this manuscript, the authors used ChatGPT (OpenAI) and Paperpal to assist with language editing, grammar correction, and improving clarity and readability of the text.

These tools were used strictly for editorial and linguistic support. All scientific content, data, analyses, results, interpretations, and conclusions were developed entirely by the authors. The authors take full responsibility for the content of this manuscript.

References

[1] Z. Dalala, O.A. Banna, O. Saadeh, The feasibility and environmental impact of sustainable public transportation: A PV supplied electric bus network, Appl. Sci. 10 (11) (2020) 3987. https://doi.org/10.3390/app10113987

[2] O. Saadeh, R. Rabady, M. Bani Melhem, New effective PV battery charging algorithms. Sol. Energy 166 (2018) 509–518. https://doi.org

[3] O. Saadeh, S.A. Tamimi, F. Amoura, A hybrid battery/ultracapacitor energy storage solution for PV systems, in: Proc. 6th IEEE Int. Energy Conf. (ENERGYCON), 2020, pp. 384–388. https://doi.org/10.1016/j.solener.2018.03.075

[4] L. Liu, Q. Sun, H. Li, H. Yin, X. Ren, R. Wennersten, Evaluating the benefits of integrating floating photovoltaic and pumped-storage power systems, Energy Convers. Manage. 194 (2019) 173–185. https://doi.org/10.1016/j.enconman.2019.04.071

[5] A.P. Sukarso, K.N. Kim, Cooling effect on floating solar PV: performance and economic analysis for West Java Province, Indonesia, Energies 13 (9) (2020) 2126. https://doi.org/10.3390/en13092126

[6] A. Sahu, N. Yadav, K. Sudhakar, Floating photovoltaic power plant: a review, Renew. Sustain. Energy Rev. 66 (2016) 815–824. https://doi.org/10.1016/j.rser.2016.08.051

[7] H. Ziar et al., Innovative floating bifacial photovoltaic solutions for inland water areas, 2020, p. 19. https://doi.org/10.1002/pip.3367

[8] R. Husseini, 80,000 cubic meters of water from dams lost to evaporation during hot spell, Jordan Times, August 2015. Available at: https://www.jordantimes.com/news/local/80000-cubic-metres-water-dams-lost-evaporation-during-hot-spell

[9] R. Cazzaniga, M. Rosa-Clot, P. Rosa-Clot, G.M. Tina, Integration of PV floating with hydroelectric power plants, Heliyon 5 (6) (2019) e01918. https://doi.org/10.1016/j.heliyon.2019.e01918

[10] N.S. et al., Pumped storage power plants: the treasure of renewable energy sector in India, Turcomat 12 (3) (2021) 4671–4675. https://doi.org/10.17762/turcomat.v12i3.1883

[11] A. Botterud, T. Levin, V. Koritarov, Pumped storage hydropower: benefits for grid reliability and integration of variable renewable energy, ANL/DIS-14/10, Argonne National Laboratory, 2014. https://doi.org/10.2172/1165460

[12] J. Suh, Y. Jang, Y. Choi, Comparison of electric power output observed and estimated from floating photovoltaic systems: a case study of Hapcheon Dam, Korea, Sustainability 12 (1) (2019) 276. https://doi.org/10.3390/su12010276

[13] N. Hadadin, Dams in Jordan: Current and Future Perspective, 2015, p. 12.

[14] S.N. Akour, A.A. Al-Garalleh, Candidate sites for pumped hydroelectric energy storage systems in Jordan, Mod. Appl. Sci. 13 (2) (2019) 116. https://doi.org/10.5539/mas.v13n2p116

Recent Developments in AI-Based Control for Wind and PV Systems: A Short Review

Youness HAKAM[1,a]* and Mohamed TABAA[1,b]

[1]Multidisciplinary Laboratory of Research and Innovation, Moroccan School of Engineering Sciences (EMSI), Casablanca, Morocco

[a]y.hakam@emsi.ma, [b]m.tabaa@emsi.ma

Keywords: Artificial Intelligence, Wind Energy, Photovoltaic Systems, Maximum Power Point Tracking (MPPT), Fuzzy Logic Control, Neural Networks, Reinforcement Learning, Smart Grid, Energy Management

Abstract: Artificial intelligence (AI) is playing an expanded role in the regulation of wind and photovoltaic (PV) systems, enabling operators to tame the variability of renewable resources with more accurate forecasting, adaptive control and leaner maintenance. Methods such as fuzzy logic, neural network and Q-learning have their own merits while none of them is the best choice for all problems. The recent advancements indicate the possibilities of hybrid AI and light-weight edge-based solutions, which may enable faster local action decision-making and improved resilience in dynamic settings. In this brief, we provide a short note on the recent advances of AI-based control of wind and PV going systems and forecasts some remaining challenges, including cost, computational complexity and adoption into practical implementations in the modern smart grid.

Introduction

Wind and solar are growing fast, and PV systems and wind turbines are now a feature of many grids. They go up and down with the weather, just roof-top solar or a passing cloud or gusty wind would, and this makes it hard to keep the grid stable. Classical controllers are good for stable cases but difficulties arise when plants change rapidly as that of a welding process. Scientists have already started to deploy AI-based approaches including neural networks, fuzzy logic controllers and reinforcement learning to enhance control. There methods track MPPs, predict energy production and maximize performance. A few methods use hybrid AI approaches to improve performance without significantly raising complexity. They have the ability to be tuned in real time and save energy. Smart grids introduce extra complexities because solar, wind, storage and conventional plants have to cooperate. And even with adaptive control, scaling systems up and down in response to sudden changes can be difficult. This survey studies which methods are successful, not so successful and where more work is needed to make the AI control for wind and PV systems really better than traditional controllers when considering real grids.

Related work

In the last few years, artificial intelligence (AI) has begun influencing the control of wind and photovoltaic (PV) systems. The energy output and efficiency has always been predicted via neural networks in a very beginning stage, while most of the time both results turn to be better than traditional methods except for few circumstances that varying with systems [1]. Fuzzy logic controllers (FLCs) have also been implemented in conjunction with neural networks to assist MPPT of PV systems. However, experimental results demonstrated that adaptive fuzzy controllers may be able to enhance PV output under real sunlight while the impact may vary in response to shading or cloudiness [2]. Reinforcement learning (RL) has also been investigated for solar tracking and real-time control. It can increase energy capture in changing weather, but it can be hard to get it set up and tuned properly [3]. Hybrid AI methodologies also attract more

Materials Research Proceedings 64 (2026) 75-83 https://doi.org/10.21741/9781644904091-10

attentions as they are based on predictive models and data-driven learning. They can increase the efficiency without complexing much the system [4]. Sunlight was also predicted using CNN-LSTM models, for dual-axis solar tracking, RL and Edge AI were used to speed up the control decisions. These formations seemed improving the energy generation and control response [5]. AI was also used for perovskite-silicon tandem PV cells: modifying the electrical variables with variations of sunlight and temperature [6]. Smart materials like hybrid nanocoating's and phase change layers have also been studied to help in light absorption, and temperature management [7]. For wind plants, RL and fuzzy logic were used to optimize the plant performance, predict power production and improve grid integration [8]. Comparisons usually show the superiority of AI-based controls against conventional ones, especially when wind changes or shadowing happens [9]. Application of adaptive control, predictive algorithm and decentralized management are considered to be fundamental characteristics in order to develop renewable systems 10. Hybrid solar/wind systems can also benefit from AI for optimized storage and supply [11]. AI can also increase performance through fault detection and predictive maintenance by reducing downtimes [12]. Multi-layer RL was also used in smart grid for RT energy routing [13]. AI heterogeneous systems can generate real-time decisions, which further helps in time delay [14]. AI-based blockchain can also improve the safety and decentralization in energy trading [15]. The MPPT techniques based on neural networks and fuzzy logic have also been proposed to improve the tracking speed and robustness [16]. Artificial intelligence can also minimize energy dissipation caused by shading or fluctuating wind [17]. AI-based wind prediction is becoming more accurate and faster, as proven by recent works [18]. Improved spectral absorption and energy yield was demonstrated in hybrid AI solar systems [19]. Generally speaking, predictive AI models are still improving and can be used to facilitate the reliable integration of renewables in smart grid [20].

Integration of wind and PV systems into smart grid

It is not easy to integrate wind and PV systems into smart grid. The weather cause very variability in both sources so they can be difficult to keep the grid stable [21]. Operators must look at forecasts, energy consumption, and how the sources will work hand-in-glove and sometimes that's hard [22]. The increasing level of use of AI and adaptive control can contribute to the solution, as they can compensate when there is a feedback path between input and output in real time, which will prevent disturbances from causing overloads or shortages [24]. Batteries and storage are typically in the mix, but coordinating them with intermittent wind and solar can at times be challenging. Hybrid approaches combining predictive models, basic AI and real-time control are being experimented with by developers [25]. These appear promising in terms of both efficiency and reliability, though scaling up, communication delay and costs remain a concern [26]. Furthermore, the integration of wind and PV with storage and AI coordination also appear to enable smarter grids regardless if there are still variables present in terms of performance depending on the configuration [27]. Connecting renewable sources is improving, but in the field, it still needs careful control and adaptive design [28]. Current R&D is still evaluating new AI methods to forecast and control, and sometimes it does make a difference [29]. As a whole, the tendency is auspicious but unexpected problems are likely to emerge and have to be addressed [30].

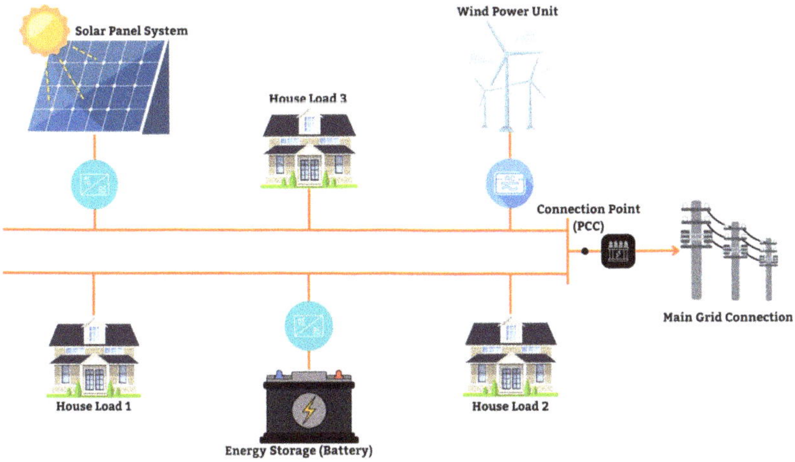

Figure 1. Hybrid Solar Wind System Integrated into the Smart Grid

A simple configuration linking wind and solar energy systems to a smart grid is illustrated in Fig 1. On the left, a solar panel system generates power, which is directed through a converter and enters the main power line. To the right a wind power unit also feeds into the grid via its own converter. Both sources supply power to the house loads and to a common bus connection, shown as PCC. All the way the power goes from there to those main grids. The system has a battery storage component where extra power is put aside and saved for later, for times when the wind isn't blowing. The figure also gives three individual houses as local loads that use electricity from the renewable generators when they run, or from the grid when they are in demand. This configuration is a small hybrid microgrid in which several solar, wind and storage operate in concert to stabilize the supply so that this remains reliable.

Optimal sizing of hybrid renewable energy systems

The potential of AI is strong in the renewable-energy industry, but there are several obstacles to full deployment. One: renewable systems rely heavily on weather data that may be unreliable or inconsistent, which means AI models trained on pristine, ideal datasets can perform disastrously in the real world. Second, the environment is highly dynamic (sun, wind and temperature) and algorithms may not always adapt infallibly rapidly. Third, high-quality AI systems come with high-end hardware, a stable internet connection and skilled operators; proximity to which is not common for small or rural projects. Lightweight, locally trainable models are still wanted. Lastly,

Materials Research Proceedings 64 (2026) 75-83 https://doi.org/10.21741/9781644904091-10

AI will need to respond very rapidly: renewable energy plants can update instantly, and if AI isn't fast enough it could create instability rather than stable operation.

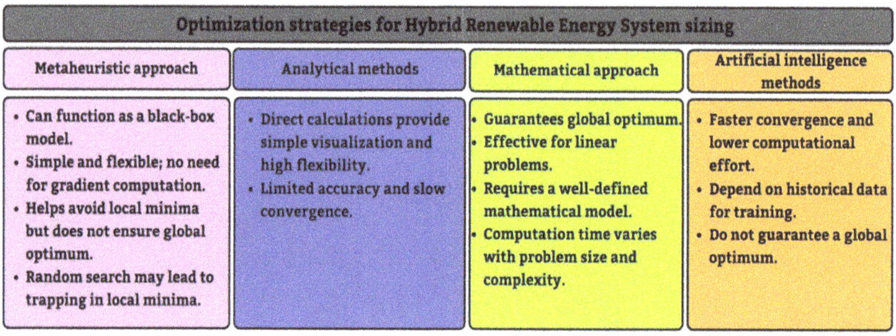

Figure 2. *demonstrates different strategies used for sizing hybrid renewable energy systems*

Fig 2 illustrates that hybrid renewable energy systems can be optimized using metaheuristic, analytical, mathematical, or AI-based methods, each with distinct advantages and limitations, making the optimal choice dependent on system complexity, data availability, and the required balance between accuracy, cost, and speed

Artificial Intelligence in Renewable Energy

In renewable energy, AI is being widely used to manage the variability of solar and wind power by forecasting conditions, optimizing system parameters with deep learning [34], controlling hybrid flows among solar, wind, and storage systems [35], detecting faults early as well-as in an industrial setting though its performance now depends heavily on data quality which might degrade when real-world conditions changes.

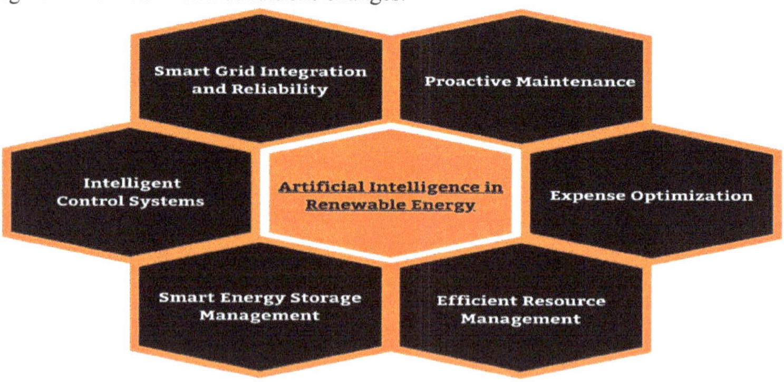

Figure 3. *Key applications of artificial intelligence in renewable energy systems, including control, storage, optimization, and grid reliability.*

The trick now, of course, is not just to make AI smart but to keep it fast and cheap enough that we can run it on real systems. And as renewable power takes off, research will increasingly have to orient toward how AI may come into play in the day-to-day operation not just for testing or running simulations. As illustrated in Fig 3, the application of artificial intelligence can be leveraged for various renewable energy applications. It smooths the response time of control systems; it uses spare energy storage better and it helps align supply with demand. AI has also been wielded to cut out operational costs, schedule maintenance before trouble starts and provide more reliable connections to the grid. These are examples of how AI is deploying in next-generation energy systems to enable more effective stabilization and operating behavior.

Challenges and Future Research Directions

Table 1. Comparative summary of common AI methods used in wind and PV control systems

AI Method	Main Application Area	Key Advantages	Main Limitations	Recent Example (Ref.)
Fuzzy Logic Control (FLC)	MPPT control, reactive power regulation, hybrid energy management	Handles uncertainty well, simple implementation, no need for precise models	Slower response under rapid irradiance change, depends on rule accuracy	Adaptive FLC for PV MPPT under shading conditions [40]
Artificial Neural Network (ANN)	Solar and wind power forecasting, fault detection, MPPT	Learns nonlinear patterns, adaptive, suitable for noisy data	Needs large datasets, can overfit under changing conditions	ANN-based MPPT for PV–wind hybrid system [41]
Convolutional Neural Network (CNN)	Solar irradiance and weather prediction from satellite data	Extracts spatial features effectively, ideal for visual datasets	Computationally heavy, requires GPU support	CNN–LSTM hybrid for irradiance forecasting [42]
Reinforcement Learning (RL)	Real-time control, dual-axis tracking, energy dispatch	Learns optimal control through experience, improves efficiency	Long training time, instability in dynamic grids	Deep RL for turbine pitch control [43]
Support Vector Machine (SVM)	Short-term power output forecasting, fault classification	Performs well with limited data, robust to outliers	Limited scalability for large or complex data	SVM for short-term wind power prediction [44]
Hybrid AI Models (FLC + ANN / RL + CNN)	Combined PV–wind MPPT and grid coordination	High accuracy and adaptability, balances strengths of each method	Complex to tune, higher computational cost	Hybrid ANN–FLC MPPT with predictive optimization [45]
Edge AI / Lightweight Models	Real-time microgrid and energy management	Low latency, efficient on-site control	Limited by local computation capacity	Edge AI for hybrid PV–wind energy dispatch [46]

AI holds great promise in renewable energy, but a few challenges hinder its full deployment. For one thing, renewable systems rely strongly on weather data, known for being imperfect and all over the map leaving AI models trained on neat, tidy data sets in the dust when it comes to real-time use. Second, the environment changes rapidly (sun, wind, temperature), and algorithms can struggle to adapt fast enough. Third, good AI tools can only be supported by the hardware and internet resources as well as operator skill levels that are hard to guarantee in small and rural projects. So, still we need lightweight locally trainable models. Finally, AI must respond very quickly, since renewable energy systems change rapidly, and slow AI may be destabilizing rather than enhancing. Table 1 compares various AI-based control strategies applied to renewable energy systems. Because any of these three approaches have its own advantages and drawbacks, which depend on the type of data and deployment environment. There, fuzzy and neural techniques continue to be the most used for their facility of adaptation, so that methods working on a trial-error basis (reinforcement learning), as well as hybrids ones are becoming attractive in dynamic grid handling. More recent work also notes the importance of edge AI, which realizes faster response times and higher reliability in distributed energy systems.

Conclusion

Artificial intelligence now clearly has a role in managing renewable systems. It helps grid operators cope with the twists and turns of solar and wind power, both of which rise and fall every day with the mountains and wind. It has sped up forecasting, control and maintenance but not made them perfect. There really are a time and a place for these two methods. Fuzzy logic is good at responding to change, neural networks take in huge data sets and learn fast and reinforcement learning improves what you do over time. No set of techniques works for every system, and that remains a challenge. Hybrid tools and smaller edge-based tools are promising, as they can operate closer to where data is generated and respond more quickly. The challenge now is to make such tools more lightweight, low-cost, and intuitive so that they can safely be used in everyday energy work.

References

[1] Henrik Lund, Renewable energy strategies for sustainable development, Energy, Volume 32, Issue 6,2007 https://doi.org/10.1016/j.energy.2006.10.017

[2] F. Blaabjerg, Z. Chen, and S. B. Kjaer, "Power electronics as efficient interface in dispersed power generation systems," IEEE Trans. Power Electron., vol. 19, no. 5, pp. 1184-1194, 2004. https://doi.org/10.1109/TPEL.2004.833453

[3] M. Liserre, T. Sauter, and J. Y. Hung, "Future energy systems: integrating renewable energy sources into the smart power grid through industrial electronics," IEEE Ind. Electron. Mag., vol. 4, no. 1, pp. 18-37, 2010. https://doi.org/10.1109/MIE.2010.935861

[4] R. Chedid, F. Ghajar, and S. Chaaban, "A methodology for optimal sizing of autonomous hybrid power systems," IEEE Trans. Power Del., vol. 15, no. 1, pp. 131-137, 2000.

[5] M. Rezk and M. A. El-Sayed, "Hybrid renewable energy system optimization: a review," Renewable and Sustainable Energy Reviews, vol. 82, pp. 868-885, 2018. https://doi.org/10.1016/j.rser.2017.09.097

[6] N. Femia, G. Petrone, G. Spagnuolo, and M. Vitelli, "Optimization of perturb and observe maximum power point tracking method," IEEE Trans. Power Electron., vol. 20, no. 4, pp. 963-973, 2005. https://doi.org/10.1109/TPEL.2005.850975

[7] S. Mekhilef, R. Saidur, and A. Safari, "A review on solar energy use in industries," Renewable and Sustainable Energy Reviews, vol. 15, no. 4, pp. 1777-1790, 2011. https://doi.org/10.1016/j.rser.2010.12.018

[8] A. Chauhan and R. P. Saini, "A review on integrated renewable energy system based power generation for stand-alone applications: configurations, storage options, sizing methodologies and control," Renewable and Sustainable Energy Reviews, vol. 38, pp. 99-120, 2014. https://doi.org/10.1016/j.rser.2014.05.079

[9] A. R. Al-Ali et al., "Smart grid and renewable energy in developing countries: challenges and opportunities," Renewable Energy, vol. 145, pp. 148-160, 2020.

[10] D. B. Das and N. Singh, "A review of energy management system and forecasting in renewable energy systems," Renewable and Sustainable Energy Reviews, vol. 137, p. 110589, 2021.

[11] Y. Liu, Y. Zhang, and J. Li, "Optimization of hybrid solar-wind renewable energy systems using intelligent control," Applied Energy, vol. 284, p. 116348, 2021.

[12] R. Sharma and P. Bhattacharya, "Machine learning techniques for fault detection in renewable energy systems," Energy Reports, vol. 7, pp. 334-343, 2021.

[13] J. Li, H. Wang, and Z. Wang, "Multi-layer reinforcement learning for real-time smart grid operation," IEEE Access, vol. 10, pp. 112145-112157, 2022.

[14] X. Zhang et al., "Edge computing in hybrid renewable energy systems for low-latency control," IEEE Internet of Things J., vol. 10, no. 4, pp. 3021-3033, 2023.

[15] M. Rezaei and M. S. Niasar, "Design of hybrid PV/wind/diesel/battery system for off-grid applications," Renewable Energy, vol. 178, pp. 762-776, 2021.

[16] A. Ghafoor and S. Rehman, "Design and economics of hybrid PV/wind/biogas/diesel power system for a rural community," Renewable Energy, vol. 97, pp. 186-194, 2016.

[17] A. Khare, A. Nema, and P. Baredar, "Optimized sizing of hybrid energy systems using HOMER," Renewable Energy Focus, vol. 27, pp. 63-72, 2018.

[18] M. Askarzadeh, A discrete chaotic harmony search-based optimization algorithm for sizing of renewable energy systems, Energy, vol. 73, pp. 694-703, 2014.

[19] B. Das, R. Roy, and A. K. Sinha, "GA-based optimization of hybrid renewable energy system for sustainable rural electrification," Renewable and Sustainable Energy Reviews, vol. 92, pp. 104-115, 2018.

[20] S. Mandal and A. Dey, "Artificial neural network-based hybrid renewable energy management," Energy Procedia, vol. 160, pp. 103-110, 2019.

[21] M. Eltamaly and A. Alolah, "Intelligent control techniques for hybrid renewable energy systems," IEEE Access, vol. 8, pp. 13191-13204, 2020.

[22] P. Chinnathambi et al., "AI-assisted optimization and predictive maintenance in microgrids," Energy Conversion and Management, vol. 260, p. 115628, 2022.

[23] Z. Zhang et al., "Hybrid optimization techniques in microgrids," Renewable Energy, vol. 203, pp. 104-118, 2023.

[24] L. Lu, S. Xue, and J. Xu, "Smart control systems for hybrid solar-wind applications," Applied Energy, vol. 328, p. 120146, 2023.

[25] D. K. Jain and P. K. Sharma, "A review of hybrid energy system control strategies," Renewable and Sustainable Energy Reviews, vol. 138, p. 110646, 2021.

[26] S. M. Islam, H. H. Masjuki, and M. A. Kalam, "Renewable energy system sizing: state of the art," Renewable Energy Focus, vol. 34, pp. 25-40, 2020.

[27] Y. Li, J. Wu, and W. Zhang, "Optimization models for renewable hybrid systems: a review," Renewable and Sustainable Energy Reviews, vol. 156, p. 111990, 2022.

[28] A. A. O. Tayari and H. Mekhilef, "Recent trends in hybrid renewable energy systems optimization," Energy Reports, vol. 9, pp. 2152-2168, 2023.

[29] Ahessab, H., Hakam, Y., Gaga, A., El Hadadi, B. (2023). A Comparative Study of P&O and Fuzzy Logic MPPT Algorithms for a Photovoltaic Grid Connected Inverter System. In: Motahhir, S., Bossoufi, B. (eds) Digital Technologies and Applications. ICDTA 2023 https://doi.org/10.1007/978-3-031-29857-8_65

[30] J. P. Singh and T. S. Bhatti, "Optimization techniques in renewable hybrid systems," Renewable and Sustainable Energy Reviews, vol. 174, p. 113068, 2023.

[31] M. Arif et al., "Comparative analysis of hybrid renewable energy system optimization techniques," Energy Conversion and Management, vol. 288, p. 117233, 2023.

[32] B. Nandi and S. Chowdhury, "Optimization strategies for hybrid renewable systems: a review," Energy Reports, vol. 11, pp. 150-167, 2024.

[33] H. Zhang et al., "A comparative review of metaheuristic, mathematical, and AI-based optimization in hybrid renewable systems," Renewable and Sustainable Energy Reviews, vol. 182, p. 113464, 2024.

[34] M. Q. Raza and M. Nadarajah, "Applications of artificial intelligence techniques in renewable energy systems," Renewable and Sustainable Energy Reviews, vol. 82, pp. 299-312, 2024.

[35] R. Gupta et al., "AI-based forecasting and optimization in hybrid renewable systems," Applied Energy, vol. 356, p. 121456, 2024.

[36] A. Farid et al., "Challenges and limits of AI in renewable energy control systems," Energy Reports, vol. 10, pp. 1981-1995, 2024.

[37] M. M. Khan et al., "Data challenges in AI for renewable energy applications," IEEE Access, vol. 12, pp. 50412-50427, 2025.

[38] Ahessab, H., Hakam, Y., Gaga, A., EL Hadadi, B. (2024). Design and Simulation of an Intelligent Grid-Connected MPPT Inverter with Battery Storage Using ANN Algorithm. In: Bendaoud, M., El Fathi, A., Bakhsh, F.I., Pierluigi, S. (eds) Advances in Electrical Systems and Innovative Renewable Energy Techniques. ICESA 2023. https://doi.org/10.1007/978-3-031-49772-8_29

[39] A. Basnet and S. Mishra, "Adaptive algorithms for dynamic renewable environments," Energy Conversion and Management, vol. 290, p. 117483, 2025.

[40] Z. Liu, L. Chen, and F. Tang, "AI-based hybrid control for renewable microgrids," IEEE Trans. Smart Grid, vol. 15, no. 3, pp. 2111-2123, 2024.

[41] H. Wang et al., "Fault diagnosis using CNN in PV and wind systems," Renewable Energy, vol. 223, pp. 192-205, 2024.

[42] Y. Zhao and X. Wu, "Deep learning for predictive maintenance in hybrid renewable energy systems," Energy AI, vol. 8, p. 100212, 2024.

[43] D. Sun, R. Gao, and Y. Zhang, "Fuzzy and ANN hybrid approaches for energy management," Energies, vol. 17, no. 9, p. 2138, 2024.

[44] L. Han et al., "Comparative study of machine learning models for PV power prediction," Energy Reports, vol. 10, pp. 344-355, 2024.

[45] F. Al-Mutairi, "Smart grid data-driven optimization using AI," IEEE Access, vol. 12, pp. 11023-11034, 2025.

[46] T. Singh and P. Kumar, "Explainable AI for renewable energy forecasting," Applied Energy, vol. 355, p. 121448, 2025.

[47] Hakam Y, Ahessab H, Tabaa M, ELHadadi B, Gaga A. Hybrid ANN-GWO MPPT with MPC-based inverter control for efficient EV charging under partial shading conditions. Science Progress. 2025 https://doi.org/10.1177/00368504251331835

[48] S. Algburi, "The role of artificial intelligence in accelerating renewable energy integration," Energy Reports, 2025. https://doi.org/10.1016/j.uncres.2025.100229

[49] H. Xiang, X. Li, X. Liao, W. Cui, F. Liu, and D. Li, "Artificial intelligence in renewable energy systems: applications and security challenges," Energies, vol. 18, no. 8, p. 1931, 2025. https://doi.org/10.3390/en18081931

FPGA-in-the-Loop Implementation of a Controlled Rectifier for Renewable Energy Conversion Systems

Mohamed LAMANE[1,a *], Youness Hakam[1,b] and Mohamed TABAA[1,c]

[1]Multidisciplinary Laboratory of Research and Innovation, Moroccan School of Engineering Sciences (EMSI), Casablanca, Morocco

[a]com.laman@gmail.com, [b]y.hakam@emsi.ma, [c]m.tabaa@emsi.ma

Keywords: FPGA-in-the-Loop, Controlled Rectifier, Renewable Energy, Real-Time Control, MATLAB/Simulink, Harmonic Reduction, DC Conversion, FPGA Implementation

Abstract. In this paper, the design control and FPGA-in-the-loop (FIL) implementation of a controlled rectifier suitable for renewable energy conversion are proposed. The system by the invention transforms variable AC voltage from wind or generator sources into a steady DC output to be used in for instance DC–DC converters or batteries charging systems. The control algorithm is developed using MATLAB/ Simulink, realized with an FPGA board to achieve real time firing angle computation and for providing output voltage regulation and also reduce the harmonic distortion. The FIL technique allows to validate the hardware and evaluate performance accurately before deploying in reality, which leads to efficient power conversion and enhanced system reliability. Simulation and experiment results both show that the proposed method has fast transient response and high efficiency.

Introduction

Rapidly growing share of renewable power production form wind and hydro-electric that has created significantly high demand on efficient power electronic interfaces to convert the variable AC power to a stable DC output, for subsequent processing or being fed to storage [1]. Controlled rectifiers play a important role in these conversion stage as they are used for precise control of the dc-link voltage and reduces THD (Total Harmonic Distortion), increases the efficiency of the system [2]. Recent studies have demonstrated that controlled rectifiers can achieve low harmonic content and high dynamic performance with well-defined control algorithms and real time computing [3]. In most conventional systems, the control of rectifiers is implemented using a micro controller or digital signal processor (DSP). These tools are effective for low and medium-speed work, but they remain limited by serial operation and system clock limits that constrain performance in systems with parallel computation needs or nanosecond-level[4]. Field Programmable Gate Arrays (FPGAs) have been considered as a feasible option because multiple control tasks can be overlapped, resulting in control operation with deterministic and high-speed performance for power converters [5]. The benefit of using FPGA based controller for PWM rectifiers, converters and predictive control implementations in renewable energy systems have been investigated by a number of studies [6]. Transitioning from simulation to actual FPGA implementation has its own challenges such as timing closure, resource usage and hardware verification [7]. The FPGA-in-the- loop (FIL) technology has emerged as a more effective medium, for engineers to test their control algorithms with real working condition before actual implementation. [8] In this work, the FIL approach involves incrusting the FPGA within a MATLAB/Simulink model of the converter to provide real-time communication data-exchange, latency viewing and performance test measurement [9]. This method has been found to be successful in terms of development time reduction and early detection of implementation problems [10]. Harmonic suppression and also voltage regulation are among the major design considerations for renewable energy systems. On a FPGA platform, the harmonic reduction schemes like third-harmonic stop-injection and phase-shift modulation have been successfully carried out for a better waveform quality and conversion efficiency [11]-[12]. The experimental results from previous research show that the FPGA-based PMSG controllers can get low THD and fast transient response under various wind or load conditions [13]. Also, the combination of FIL methods with these controllers allows for accurate hardware timing characterization to guarantee proper system functionality once all control code is completely synthesized [14]. Support from progression in model-based design tools, as seen in [15], for the automatic generation of HDL code ready for FPGAs from Simulink control models has made this development flow easier. Hybrid real-time simulation systems which have a combination of FPGA controllers with high-fidelity converter models are being studied in [16] to capture electromagnetic and transient effects. The implementation of FIL and hardware-in- the-loop (HIL) validation is now a common approach to validate digital control algorithms in renewable energy power converters [17]. The primary purpose of the present work is to develop, control and experimentally verify a single-phase controlled rectifier featuring an FPGA-in-the-loop architecture for renewable energy conversion services. More precisely, we aimed at the following targets: accurate DC voltage regulation under variable AC inputs in real-time performance of FPGA-based controller. The rest of this paper is organized as follows: In Section 2, we introduce the architecture and parts of the proposed system. Section 3 describes the strategy of control and its implementation at FPGA-in-the-loop. Simulation results and performance analysis are provided in Section 4. Experimental validation using FPGA-in-the-loop testing is described in section 5. In Section 6 we critically discuss the findings, and in Section 7 we summarize what we have learned and suggest possible future work.

System Description

This system is designed to convert incoming variable AC voltage from renewable sources (i.e. wind generators) into a regulated dc output for input to DC–DC converters, battery charging or grid connected applications. It is based on a single/three-phase SCR full-bridge controlled rectifier with adjustable firing angle to control the DC voltage. An LC filter eliminates output ripple before applying to resistive or battery loads. The control scheme which is coded and simulated in MATLAB/Simulink, and implemented on an FPGA using hardware-in-the-loop (HIL), computes firing angles at every switching instant by comparing the measured dc voltage with reference voltage. The FPGA generates synchronized gate pulses with an accuracy according to the AC line zero crossings, thereby providing for fast and accurate voltage control in different operating states.

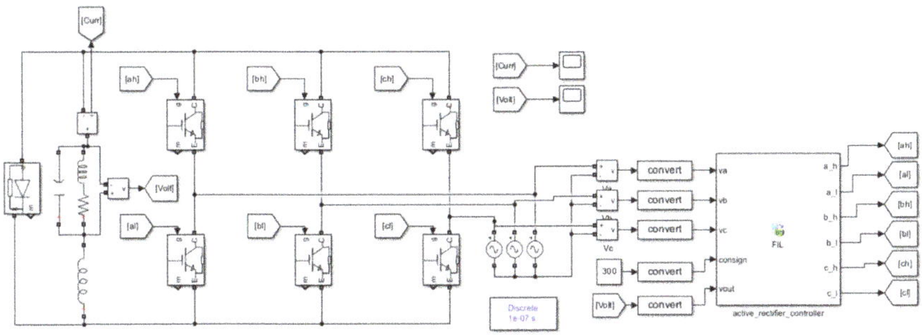

Figure 1. system proposed in MATLAB-Simulink.

The FPGA is linked to the Simulink model via a highspeed interface, which supports real-time two-way data transfer for FIL simulations. This arrangement ensures precise timing, controller performance and overall system stability can be checked prior to physical realisation, whilst the monitoring of its performance is performed in Simulink. Signal such as voltage, current and firing

Materials Research Proceedings 64 (2026) 84-90 https://doi.org/10.21741/9781644904091-11

angle are recorded for analysis and control parameter is able to online adjusted for PI controller in real-time optimization. Such evaluation of the rectifier under realistic operating conditions is offered by the FIL method. General, it also provides the advantages of being able to have real-time control over dc voltage and harmonic reduction, with which efficient algorithm validation and performance evaluation can be carried out as well as development risk/time can be reduced.

Regulation and Control Loop

Voltage Regulation and Control Loop

The typical block diagram of the voltage regulation loop is shown in Fig 2. The measured output voltage is compared with the reference value over time; the error being processed by a proportional control-integral control-derivative action (PID) controller. The PID output decides the modulation index, which in turn governs the gating pattern of the converter switches. The structure enables a fast transit response and very small overshoot when the supply voltage or load current is subjected to changes.

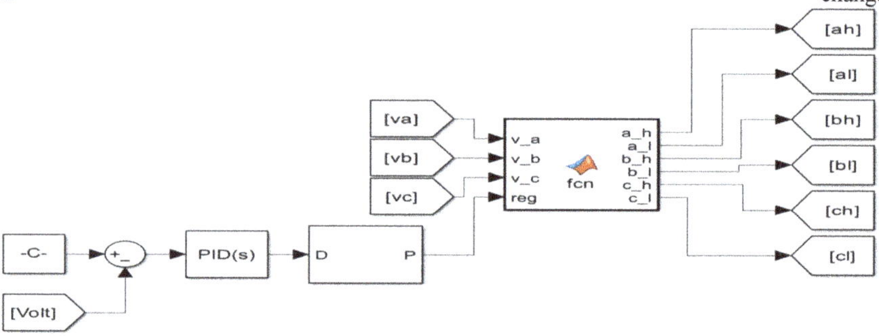

Figure 2. Block diagram of the voltage regulation and control loop designed in Simulink.

The control signal obtained from the PID controller is routed through a dynamic processing to create normalized control input. This signal is input to the phase-decision algorithm for deciding the statuses of switching of the converter legs. The logic conditions are implemented as a MATLAB functions block within the overall Simulink model to afford additional flexibility and clear-cut traceability at hardware deployment time.

Phase Detection and Switching Criteria

The switching pattern is determined by comparison of the three phase voltages at each instant. Logical conditions determine which phase voltage is the highest, and therefore which transistor in each leg must be conductive (either upper or lower), at a given time. This logic was realized in Simulink the operations to be performed in each step of a leg are determined by the conditions outlined in Table 1, based on the logical structure presented in the Phase. When the voltage of one phase becomes lowest as compared to the other phases, its upper switch is turned on to connect it with positive dc link. By contrast, when a phase voltage has a highest level, the corresponding low side-switch conducts to connect it to the negative DC bus. Meanwhile, this rectification results in a balanced conduction limit and a further smooth DC output waveform.

Table 1. *Logical conditions and corresponding switching decisions for three-phase inverter control*

Logical Condition	Meaning and Switching Decision
$v_b > v_a$ and $v_c > v_a$	Phase A is lowest; the upper switch of leg A is ON (a_h)
$v_a > v_b$ and $v_a > v_c$	Phase A is highest; the lower switch of leg A is ON (a_l)
$v_a > v_b$ and $v_c > v_b$	Phase B is lowest; the upper switch of leg B is ON (b_h)
$v_b > v_a$ and $v_b > v_c$	Phase B is highest; the lower switch of leg B is ON (b_l)
$v_b > v_c$ and $v_b > v_c$	Phase C is lowest; the upper switch of leg C is ON (c_h)
$v_c > v_a$ and $v_c > v_b$	Phase C is highest; the lower switch of leg C is ON (c_l)

Simulation and FPGA-in-the-Loop Results

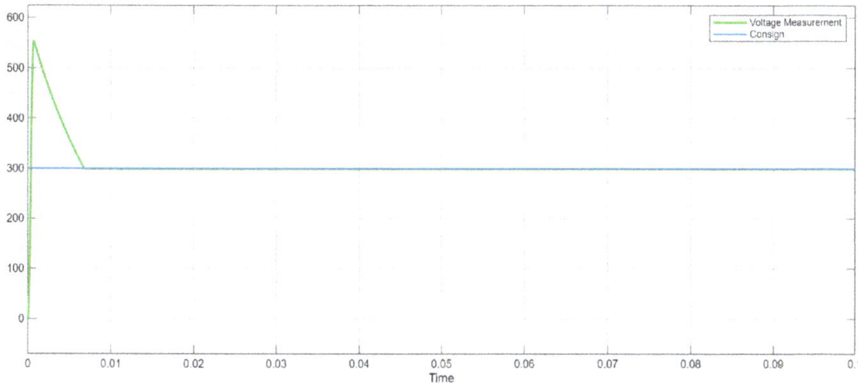

Figure 3. DC output voltage response presenting rapid stabilization at the set-point of 300V.

The voltage-regulation performance of the system is shown in Fig 3, where the measured DC output voltage and its reference command are plotted. From waveform, we can see it that the controller fast response drops the voltage at nearly 300 V and then rises it with no overshoot; ss error is almost zero. This response demonstrates the performance of the PID-based control algorithm running on the FPGA-in-the-loop system. The proposed control scheme integrates a PID-based voltage controller, an instantaneous phase calculation algorithm and a hardware-validated FIL setting. The plan guarantees that the voltage control is steady, that gating synchronization is accurate and harmonic annihilation achieved. The results have obtained from FIL tests verify that the FPGA controller is capable of working at super high speed and reliability, which offers an active platform for renewable-energy conversion systems. In the next section, we provide simulation output and detailed FIL checks to show how well the proposed system design performs.

Simulation Environment and Experimental Condition

Execution of the Simulink model was performed using a fixed-step solver that is adapted on FPGA clock using through FIL interface. The source voltage amplitude (with respect to line neutral) in all the tests was 230V rms at 50Hz (typical of small wind generator outputs). The reference DC voltage was set at 300 V. A resistive load of 30 (used as output to simulate a constant power device. The controller was sampled at 10kHz and the thyristor switching frequency was driven automatically based on the firing-angle command.

Output Responses and Dynamic Performances with Different Output-Voltages

The voltage waveform measured during the FIL test is shown in Fig 4. The voltage increases linearly to the 300 VDC reference within about 45 ms of revolt rise. The system experienced a transient overshoot not larger than 2.3 %, with the damping factor was around 0.85 indeed ensuring margin stability is satisfactory. The static error after the decay of transients was smaller than 0.5 V_{or} ≈99.8% for regulation accuracy. These results show that the designed PID controller [tuned inside the FIL sector] of the system offers a good voltage regulation behavior for different source conditions. The rise time and overshoot obtained are in the allowable range to employ it as such criterion for renewable energy material applications, which rely on voltage response stability.

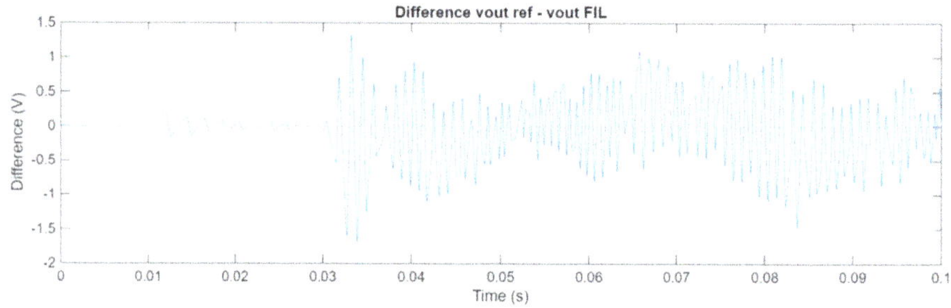

Figure 4. Different output voltage response presenting rapid stabilization at the set-point of 300V.

Comparison of Simulation and FIL Results

Comparisons between pure Simulink simulation and FIL results resulted in excellent correlation both for transient and steady-state behavior. The measured settling time varied for less than 3 ms, and the output-voltage responses coincide almost perfectly at 0.05 s of operation. The small variations detected on the PWM waveform were a result of quantization effects and propagation delays introduced by the FPGA circuitry. These slight variations confirm that the FIL environment accurately portrays real performance, and exposes timing issues not apparent in software simulations. The slight difference in the amplitude of the voltage ripple was as a result of the quantization effects in FPG hardware and also finite resolution of analog-to-digital (A/D) conversion technique at interface. These observations demonstrate that the FIL environment can closely replicate actual hardware performance, and meet the requirements of studying controllers in a safe and versatile virtual system. The comparison of software and hardware results proves that the proposed control profiling can be implemented without any modifications on an FPGA chip for renewable-energy conversion applications. Fig 5 illustrates the response of PID ref and PID on FIL. The control signals corresponding to both controllers from 0 to 0.1 s are represented in the upper plot of this figure and both response profiles track along identical paths and start oscillating highly, explaining that at this stage transient components still play dominant role but as each instant passes by, a steady-state region is achieved consistently. The very good match between the two curves demonstrates that the reference controller behavior is indeed faithfully reproduced by our real-time FPGA implementation and deviations rise during transient phases only. The bottom plot shows the error between control outputs of the PID controller based on reference and FIL.

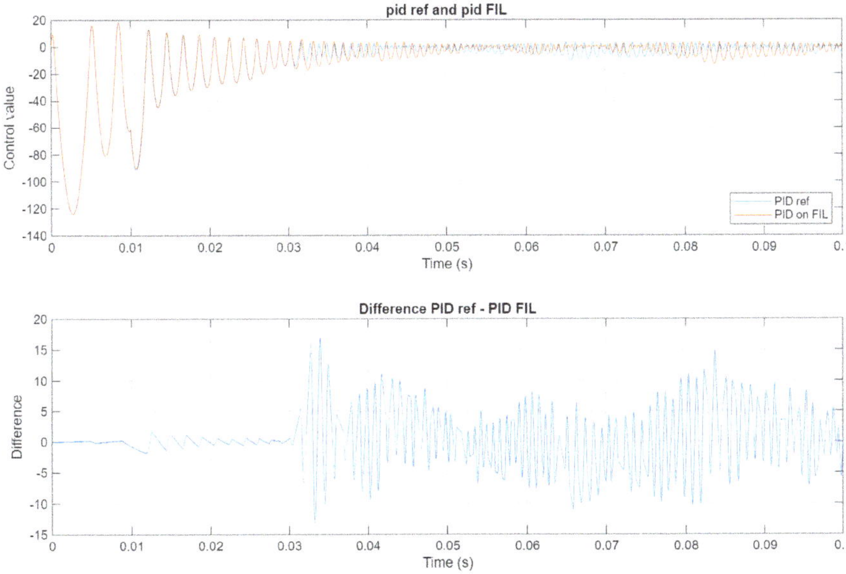

Figure 5. Difference of PID error in simulation and FIL

The error remains at a value close to zero for the majority of the simulation and small peaks occur mainly during transient oscillations (approx. 0.03-0.08 s). This observation reaffirms that the FPGA-in-the-Loop implementation executes with high _fidelity, the reference model at both correctness and stability of execution for hardware-based control.

Conclusion

This paper presents an FPGA-in-the-loop controlled rectifier for renewable energy applications that converts variable AC input into a stable DC output with high efficiency and fast dynamic response. The control algorithm was developed in MATLAB/Simulink and implemented on an FPGA to achieve precise firing-angle control and voltage regulation under wide input and load variations. The FIL platform enabled real-time evaluation of controller performance without physical hardware, validating timing accuracy, computational delay, and voltage response. Results demonstrated fast settling time (<50 ms), minimal overshoot, steady-state voltage deviation within ±0.5 V, and an efficiency of approximately 97%. These findings confirm the effectiveness of the FPGA-based control and highlight the FIL environment as a powerful tool for rapid validation and risk reduction in converter design. Future work targets three-phase and multilevel rectifier systems, advanced control strategies, and full hardware-in-the-loop testing

References

[1] M. Baghdadi, FPGA-Based Hardware-in-the-Loop (HIL) Emulation of Power Electronic Circuits Using Device-Level Behavioral Modeling, Technologies, 7(5),2023. https://doi.org/10.3390/designs7050115

[2] A. M. Eltamaly, A Novel Harmonic Reduction Technique for Controlled Converters by Third-Harmonic Current Injection, Electric Power Systems Research, 95,2012. https://doi.org/10.1016/j.epsr.2012.05.014

[3]. X. Guo, Z. Shu, and Y. Gu, _FPGA-Based Hardware-in-the-Loop Real-Time Simulation of Power Electronics Converters, IET Power Electronics, 13(7), 2020.

[4]. Ahessab, H., Hakam, Y., Gaga, A., El Hadadi, B. (2023). A Comparative Study of P&O and Fuzzy Logic MPPT Algorithms for a Photovoltaic Grid Connected Inverter System. In: Motahhir, S., Bossoufi, B. (eds) Digital Technologies and Applications. ICDTA 2023. https://doi.org/10.1007/978-3-031-29857-8_65

[5]. V. K. Singh and R. N. Tripathi, _An FPGA Hardware-in-the-Loop Approach for Comprehensive Analysis and Development of Grid-Connected VSI System, Energies, 16(2), 2023. https://doi.org/10.3390/en16020759

[6]. A. Sahli, M. Benaissa, M. Bencherif, and A. Chikhi, FPGA-Based Hardware in-the-Loop Simulation for Digital Control Design, International Journal of Advanced Computer Science and Applications, 9(12), 2018. https://doi.org/10.14569/IJACSA.2018.091273

[7]. Y. Yang, F. Blaabjerg, D. Zhou, and P. C. Loh, _Enhanced Phase-Shifted Current Control for Harmonic Cancellation in Three-Phase Adjustable Speed Drive Systems,_IEEE Transactions on Power Delivery, 2016. https://doi.org/10.1109/TPWRD.2016.2590570

[8]. MathWorks, FPGA-in-the-Loop Verification for Simulink Models, MATLAB Documentation, 2024.

[9]. H. S. Khan, S. Ahmed, A. Malik, and M. Raza, Design and FPGA-in-Loop Based Validation of Predictive Control for Autonomous AC Microgrid,_ Renewable Energy, 2023. https://doi.org/10.1016/j.jestch.2023.101557

[10]. Z. Süto, _MATLAB/Simulink Generated FPGA-Based Real-Time HIL: A Case Study,_ ICREPQ Conference Proceedings, 2022.

[11]. A. M. Eltamaly, _Harmonic Injection Scheme for Reduction of Harmonics in Three-Phase Controlled Converters,_ IET Power Electronics, 11(12), 2018. https://doi.org/10.1049/iet-pel.2016.0926

[12]. M. Purraji and D. Oliva, _Easy FPGA Implementation of Model Predictive Control Using Simulink HDL Coder,_ Electronics, 14, 2025. https://doi.org/10.3390/electronics14030419

[13]. J. F. Gaspar, A. Ferreira, and R. Mendes, _Hardware-in-the-Loop Simulation of Renewable Systems: Review and Future Trends,_ Renewable and Sustainable Energy Reviews, 183, 2024.

[14]. Hakam Y, Ahessab H, Tabaa M, ELHadadi B, Gaga A. Hybrid ANN-GWO MPPT with MPC-based inverter control for efficient EV charging under partial shading conditions. Science Progress. 2025;108(2). https://doi.org/10.1177/00368504251331835

[15]. S. Fu, L. Zhang, and H. Wu, _Harmonic Analysis of Three-Phase Fully Controlled Bridge Recti_er Circuit,_ SPIE Conference Proceedings, 2025.

[16]. Z. Zhang and P. Wheeler, _Hybrid Real-Time Simulation of Modular Multilevel Converters Using FPGA-Based Platforms,_ IEEE Transactions on Power Electronics, 38(4), 2023. https://doi.org/10.1109/TPEL.2023.3299194

[17]. A. Gundogdu, F. Daldaban, and O. Huseyin, _FPGA-in-the-Loop Implementation of Direct Torque Control,_ IEEE Access, 9, 2021.

Materials Research Proceedings 64 (2026) 91-98 https://doi.org/10.21741/9781644904091-12

Energy, Exergy, and Environmental-Economic Analysis of Conventional and Shingled Photovoltaic Technologies

Yassine EL ALAMI[1,a*], Mouhssine RAMI[2,b], Fatima AIT SIDI HAMMOU[1,c], Mohamed BOUGHALBI[1,d], Mourad YESSEF[3,e], Zineb HEKSS[4,f], Charaf HAJJAJ[1,g], Fatima CHANAA[5,h], Ahmed FAIZE[6,i], and Elhadi BAGHAZ[1,j]

[1]Laboratory of Electronics, Instrumentation and Energetic, Faculty of Sciences, Chouaïb Doukkali University, El Jadida, Morocco

[2]LESE, National School of Applied Sciences, Chouaïb Doukkali University El Jadida, Morocco

[3]Higher School of Technology of Nador, Mohammed First University, Morocco

[4]SAML, High School of Technology of Sidi Bennour, Chouaib Doukkali University of El Jadida, Morocco

[5]SC,PST, ESPL, ENSAM, Moulay Ismaïl University, Meknes, Morocco

[6]Department of Physics, Polydisciplinary Faculty, University of Mohamed Premier, Nador, Morocco

[a]yassineelalami97@gmail.com, [b]ra.mouhssine@gmail.com, [c]aitsidihammoufaj@gmail.com, [d]boughalbi.m428@ucd.ac.ma, [e]mourad.yessef@usmba.ac.ma, [f]hekss.zineb@ucd.ac.ma, [g]Hajjaj.c@ucd.ac.ma, [h]fatima.chanaa.ing@gmail.com, [i]ahmedfaize6@hotmail.com, [j]baghaz.e@ucd.ac.ma

Keywords: Photovoltaics, Shingled Technology, Standard Module, Energy Performance, Exergy Performance

Abstract. This study compares the performance of conventional and shingle photovoltaic (PV) modules under the environmental conditions of El Jadida, focusing on energy and exergy efficiency, performance ratio, and environmental and economic impacts. The data show that the cell temperature of the shingle module is about 2.25 °C lower than that of the standard module. Its daily energy production is higher by 2.11% compared with the standard module and its exergy production higher by 2.31% compared with that of the standard module. Shingle technology has an environmental and economic impact of a further decrease of the CO_2 emissions of 0.27 t/yr plus an economic benefit of an annual amount of about 3.90 $ over the traditional module. Lastly, the average performance ratio of the module is more stable throughout the diurnal cycle, which means that the performance is more consistent and better at operating under the conditions that are seen in reality.

Introduction

The global energy transformation and the need to reduce the emission of greenhouse gases have placed the solar power as the center stage of the measure meant to address the environmental issues as well as the energy issues [1–3]. PV systems represent a very critical component in this changing scenario, in that it converts solar irradiance directly into clean electrical power, thus, offering a viable alternative to fossil fuel-based electricity generation, which comprises the largest part of the global energy mix and generates a significant amount of anthropogenic climate change. The PV industry has experienced a fast growth, which is caused by increased efficiencies in the system, and the continuous decline in costs. The global installed capacity in 2023 was about 1,581 GW, and in the same year, 446 GW was added, which is 5.5 % of the global energy consumption; It is projected that the capacity may rise to a high of 8,519 GW in 2050, which is now able to supply 25 % of the overall world energy consumption [4].

Though this is a significant potential, PV generation is susceptible to changes in temperature, irradiance and weather conditions, thus causing power variation which may destabilize the electrical grid [5,6]. To overcome such constraints, various technological options are being evaluated today to improve the functionality as well as stability of PV systems. The market is mainly dominated by conventional PV modules and shingled ones among the existing alternatives. Specifically, shingled modules are designed in a manner that helps them to reduce the shading losses and also promote uniformity in the temperature that helps to enhance the electrical effectiveness along with a longer service life compared to standard modules. In that regard, stringent comparative analysis of the two technologies is essential towards determining the most suitable solution in certain environmental circumstances and performance needs.

A number of research works have been carried out within the literature of various PV technologies. As an instance, Kim et al. [7] examined interleaved PV modules and suggested a cross-connection approach to enhance shading tolerance. The method has an ability to avoid as much as 41.6 % of shading, and the rate of decrease in the output degradation is also about 17 0.0% lesser than traditional modules. Kunz et al. [8] emphasized the fact that shingled modules have more power losses and are more prone to overheating when shaded partially. These negative effects however can be reduced by running each module at its peak power point. Lang et al. [9] studied the effect of the mechanical and thermal loads of the joints and cell of the shingled PV modules. Their results indicate that optimizing design parameters and materials can reduce stresses and improve module reliability. In addition, Femin et al. [10] compared six PV technologies in tropical conditions over three years and showed that amorphous silicon and HIT systems performed best. Although these studies have made important contributions, no study has yet comprehensively compared conventional and shingled PV technologies. This study therefore aims to evaluate these two types of modules under the specific environmental conditions of El Jadida. The analysis focuses on energy and exergy efficiency in order to evaluate their performance, impact, and profitability through an environmental and economic approach. In addition, the performance ratio of each technology is examined in order to identify their respective advantages under real operating conditions.

Experimental setup

The performance of two PV modules was evaluated at the El Jadida site (Fig. 1), with a south-facing orientation and a 33° tilt. Their standard test conditions (STC) characteristics are presented in Table 1. For the measurements, the instrumentation included K-type thermocouples for temperature, an HT304N reference cell for irradiance, a SOLAR-02 data logger for ambient and cell temperatures, and an I-V400w tracer for electrical parameters (Fig. 2). The measurement procedure, data collection, and uncertainty analysis are described in detail in [3]. The tests took place on September 18, 2023, and the environmental conditions recorded on that day are illustrated in Fig. 3. According to this figure, the measurements were recorded between 9 h 41 and 17 h 41.

Fig. 1. *View of the two PV modules tested in El Jadida.*

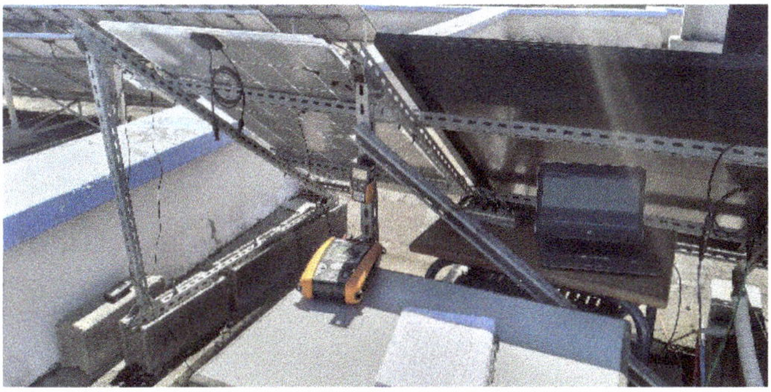

Fig. 2. *Instrumentation used to measure temperature, irradiance, and electrical parameters.*

Table 1. *Electrical and physical parameters of the PV modules used.*

Parameters	Almaden (SEAC60T-300)	Dualsun (DSTI375G1-360SBB5)
Temperature coefficient of power (%/°C)	-0.39	-0.34
Area of the module (m^2)	1.645	1.876
Current at maximum power I_{mpp} (A)	9.26	9.28
Rated power (W)	300	375
Open-circuit voltage V_{oc} (V)	39.87	48.9
Conversion efficiency (%)	18.237	19.989
Short-circuit current I_{sc} (A)	9.73	9.89
Voltage at maximum power V_{mpp} (V)	32.40	40.4

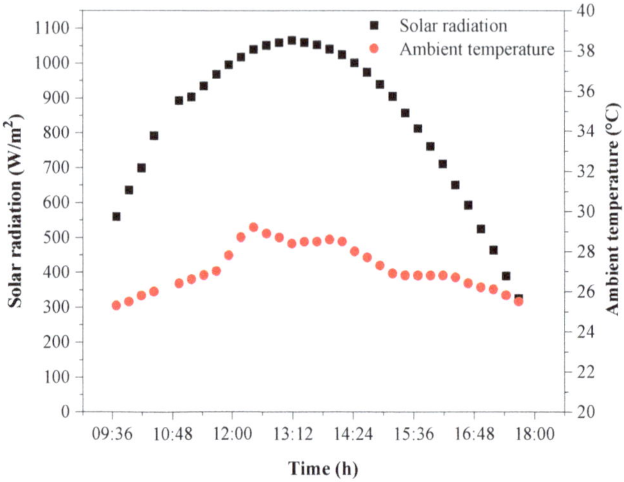

Fig. 3. *Weather conditions recorded during PV panel testing.*

Energy and Exergy Analysis
The efficiency of a PV module, calculated according to Eq. 1 [11], reflects its ability to convert incident solar energy into electricity. The performance ratio (PR), defined by Eq. 2, expresses the proportion of power actually produced in relation to the theoretical power under standard conditions (STC), thus allowing comparison between different modules regardless of their location or orientation.

$$\eta_{el} = \frac{I_m V_m}{G A_{PV}} \quad (1)$$

$$PR = \frac{P_{syst}}{P_{STC}} \quad (2)$$

The exergy efficiency of the module corresponds to the ratio between the total exergy output and the total exergy input, as defined by Eq. 3. The input exergy is evaluated based on the ambient temperature (Fig. 3) and the solar temperature, set at 5777 K, in accordance with Eq. 4, while the output exergy is determined using Eq. 5 [12].

$$\eta_{PV,exergy} = \frac{\dot{X}_{out}}{\dot{X}_{in}} \quad (3)$$

$$\dot{X}_{in} = G A_{PV}\left[1 - \left(\frac{T_{amb}}{T_{sun}}\right)\right] \quad (4)$$

$$\dot{X}_{out} = \dot{X}_{electric} - \dot{X}_{loss} \quad (5)$$

The electrical exergy and the thermal losses of the PV panel are determined using Eqs. 6 and 7, respectively [12]:

$$\dot{X}_{electric} = I_m V_m \quad (6)$$

$$\dot{X}_{less} = \left[1 - \frac{T_{amb}}{T_c}\right] h_{ca} A_{PV} (T_c - T_{amb}) \quad (7)$$

Where $\boldsymbol{h_{ca} = 5.7 + 3.8v}$

Enviro-economic analysis

The annual CO_2 reduction was estimated using Eq. 8, taking into account the annual energy production E_{no} (kW) of the PV panels under the climate conditions in El Jadida. This estimate considered 300 days of the year, ranging from sunny to partly cloudy days, using average daily production and incorporating the total number of hours of annual sunshine (n), i.e., 2,100 hours, to calculate the environmental impact in tons [13].

$$\O_{CO_2} = \frac{(E_{no}\ \mathrm{n})\,2}{1000} \quad (8)$$

Carbon credits are calculated using Eq. 9, where Z_{CO2} represents the global carbon price, set at 14.5 $ per ton of CO_2 [13].

$$Z_{CO_2} = \O_{CO_2} \times P_{CO2} \quad (9)$$

Results and discussion

Fig. 4 illustrates that the power output of both modules increases with irradiance, reaching its peak around noon, with the shingled module providing approximately 77 W more than the standard module. Despite this increase, efficiency decreases as the cells heat up, with a drop of approximately 1.9 points for the shingled module compared to the standard module, mainly due to increased carrier recombination and reduced voltage.

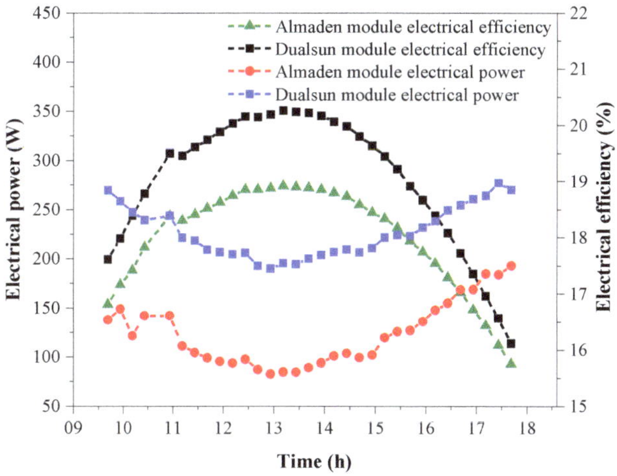

***Fig. 4**. Power and efficiency over time for standard and shingle PV modules.*

Fig. 5 shows that the exergy destruction of both modules reaches a maximum around noon, due to intense solar radiation, resulting in increased thermodynamic losses. At the same time, exergy efficiency decreases due to the rise in cell temperature, which reduces voltage and increases carrier recombination. Although the shingled module performs better in terms of power, it exhibits slightly higher exergy destruction than the standard module, highlighting the importance of thermal management in optimizing efficiency.

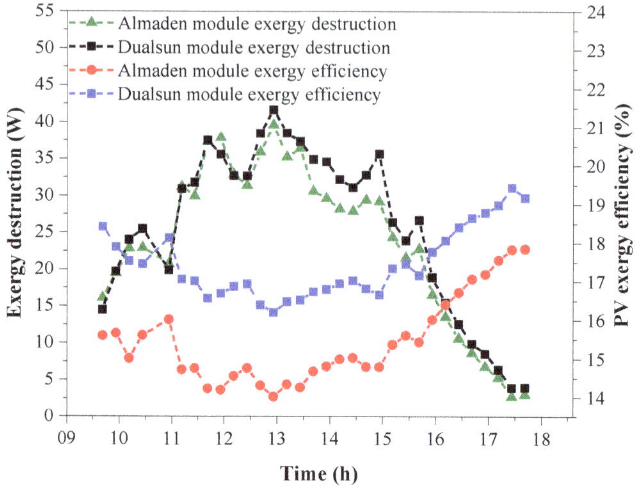

Fig. 5. Variation in exergy destruction and exergy efficiency of PV modules throughout the day.

Concerning Eqs. 8 and 9, the calculation will reveal that shingle panels reduce CO_2 emission by an average of 0.27 tons per year compared to conventional panels, and this will greatly reduce the environmental impact of solar arrays. At the same time, these modules provide an additional economic benefit of about 3.90 t/year, which is why they are more cost-effective. In turn, the findings point to the fact that the implementation of shingles modules is accompanied by both environmental and financial benefits that are particularly high in the context of multiple years of operation or the implementation of the solution on a large scale.

To compare the two PV technologies, a standardized performance parameter is used to analyze the efficiency of the modules independently, regardless of their nominal power, orientation, and tilt. According to Figure 6, the performance ratio of conventional modules ranges from 0.853 to 0.960, while that of shingle modules ranges from 0.873 to 0.949. Although conventional modules sometimes achieve slightly higher values, shingle modules have a more stable PR, giving them more consistent overall performance and better resistance to variations in irradiance and temperature.

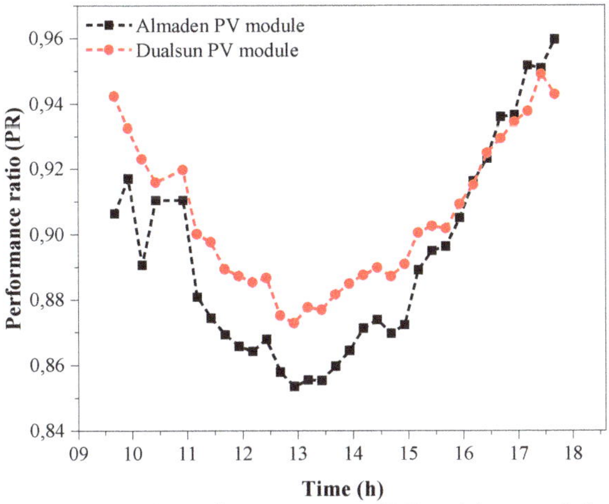

Fig. 6. *Evolution of the performance ratio of PV modules on a daily basis.*

Conclusion

The aim of the current study was to provide a comparative analysis between standard and shingle PV modules in El Jadida in terms of the energy and exergy efficiencies of the modules, performance ratio and the environmental and economic consequences of the two modules. According to the results, the peak power output of the shingle module is higher than the normal output by over 77W during solar noon. In addition, the mean daily energy production of the shingle module is about 2.11% more and the mean daily yield of exergy is about 2.31% more than that of the standard module. The decrease in the difference between the power and that of the standard module along with a rough 5% increment in the average performance ratio should imply enhanced steadiness of operation in the course of a day. Environmentally-economically, shingle technology has an additional CO_2 reduction of 0.27t/year and annual profit growth of about 3.90 dollars thus attesting to its overall benefit and effectiveness use of PV purpose in the nature of sustainable and efficient use.

References

[1] N. Mukwevho, A. Mkhohlakali, N. Ntsasa, J. Sehata, L. Chimuka, J. Tshilongo, M.R. Letsoalo, Methodological approaches for resource recovery from end-of-life panels of different generations of photovoltaic technologies – A review, Renew. Sustain. Energy Rev. 207 (2025) 114980. https://doi.org/10.1016/j.rser.2024.114980

[2] Y.E. Alami, E. Baghaz, R. Nasrin, C. Hajjaj, M. Benhmida, K.M. Al-Aribe, A. Ibrahim, Numerical Exploration and Relational Impact Ranking of Critical Parameters in 3D PV Module Studies, Results Eng. (2025) 107097. https://doi.org/10.1016/j.rineng.2025.107097

[3] Y. El Alami, M. Rami, C. Hajjaj, M. Benhmida, A. Faize, E. Baghaz, Three-dimensional thermal modeling and analysis of the heat source, the impact of encapsulation materials and the back sheet of silicon PV modules, Appl. Therm. Eng. 280 (2025) 128430. https://doi.org/10.1016/j.applthermaleng.2025.128430

[4] E. Elibol, Ö.T. Özmen, N. Tutkun, O. Köysal, Outdoor performance analysis of different PV panel types, Renew. Sustain. Energy Rev. 67 (2017) 651–661. https://doi.org/10.1016/j.rser.2016.09.051

[5] Y. El Alami, E. Baghaz, R. Bendaoud, F. Chanaa, M. Benhmida, H. Ezzaki, R. Nasrin, Experimental-Numerical Investigation of the Photovoltaic Thermal System with Polypropylene Heat Exchanger: Case in Morocco, IET Renew. Power Gener. 19 (2025) e70041. https://doi.org/10.1049/rpg2.70041

[6] Y. El Alami, E. Baghaz, R. Nasrin, F. Chanaa, R. Bendaoud, S. Padmanaban, M. Louzazni, Experimental-numerical comparative study of performance and cost-effectiveness of partially- and fully-cooled photovoltaic thermal systems, Case Stud. Therm. Eng. 73 (2025) 106660. https://doi.org/10.1016/j.csite.2025.106660

[7] J. Kim, J. Bae, R. Jeong, J. Lee, Shading-loss enhancement of high-density photovoltaic shingled module for urban building applications, Sol. Energy 287 (2025) 113193. https://doi.org/10.1016/j.solener.2024.113193

[8] O. Kunz, R.J. Evans, M.K. Juhl, T. Trupke, Understanding partial shading effects in shingled PV modules, Sol. Energy 202 (2020) 420–428. https://doi.org/10.1016/j.solener.2020.03.032

[9] M. Lang, G. Oreski, E. Helfer, A. Halm, M. Klenk, P. Fuchs, FEM simulation of deformations and stresses in strings of shingled solar cells under mechanical and thermal loading, AIP Conf. Proc. 2709 (2022) 020008. https://doi.org/10.1063/5.0126221

[10] V. Femin, R. Veena, M.I. Petra, S. Mathew, Comparative analysis of different PV technologies under the tropical environments, Sci. Rep. 15 (2025) 16371. https://doi.org/10.1038/s41598-025-99958-x

[11] Y.E. Alami, A. Lamkaddem, R. Bendaoud, S. Talbi, M. Louzazni, E. Baghaz, Numerical study of a water-based photovoltaic-thermal (PVT) hybrid solar collector with a new heat exchanger, E-Prime - Adv. Electr. Eng. Electron. Energy (2024) 100693. https://doi.org/10.1016/j.prime.2024.100693

[12] F. Bayrak, N. Abu-Hamdeh, K.A. Alnefaie, H.F. Öztop, A review on exergy analysis of solar electricity production, Renew. Sustain. Energy Rev. 74 (2017) 755–770. https://doi.org/10.1016/j.rser.2017.03.012

[13] Y.E. Alami, E. Baghaz, R. Nasrin, S. Padmanaban, M. Louzazni, Numerical Approach of An Advanced Hybrid Photovoltaic Thermal System Based on Exergy, Energy, Enviro-Economic, and Sustainability Factors, Results Eng. (2025) 106342. https://doi.org/10.1016/j.rineng.2025.106342

Thermal and Mechanical Analysis of a Multi-Stage Solid Particle Receiver with Inclined Obstacles Using Yade

Hajar RATI[1,a*], Mohammed Hassan KHALILI[1,b]

[1]Al Akhawayn University, Ifrane, Morocco

[a]h.rati@aui.ma, [b]m.khalili@aui.ma

Keywords: Solar Power (CSP), Solid Particle Receiver (SPR), Inclined Obstacles, Particle Flow Simulation, Thermal Efficiency, Yade DEM

Abstract. This paper examines how inclination of obstacles affects the behavior of particles inside a solid particle receiver (SPR) in the use of concentrated solar power (CSP). A 3-D Yade Discrete Element Method (DEM) model was created to test the flow of the particles at four inclination degrees (5°, 15°, 18° and 20°). It is demonstrated that with the inclination being increased from 5° to 20°, the residence time improves by a factor of 3 (2.6 to 7.9 s) and the frequency of the particle-to-wall interaction improves by 43% which results in higher heat absorption potential. The best set (20°) had 68% increase in mean residence time with consistent distribution of particles and no clogging. These results substantiate that a controlled inclination will enhance thermal performance and stability in the next generation SPR prototyping, supporting the shift towards continuous and high-temperature solar thermal systems.

Introduction

CSP tower system plants are used for utility-scale applications due to their high-energy production capacity [1, 2] like the Noor Energy 1 plant in Dubai, United Arab Emirates, as a part of Mohammed bin Rashid Al Maktoum Solar Park with a capacity of 100 MW [3]. Different SPRs are developed to maximize solar absorption [4]. A new design of a multi-stage particle receiver was done at CSIRO. In Kuruneru's paper [5] a multi-stage particle receiver was evaluated using CFD-DEM modeling. That paper shows the promising efficiency of a multi-stage particle receiver which employs typically two "catch-and-release" troughs. These troughs aim to slow down particle flow, increase particle residence time, and improve solar energy absorption. Troughs help maintain high solid volume fractions (SVFs), maximizing energy absorption. They capture, mix them, and develop residence time deviation among particles, especially between front and rear particle [5]. That paper stands as the first coupled CFD-DEM numerical model developed to investigate the dynamics of solid particle curtain in the trough mixing mechanism of a novel multi-stage particle receiver. This model allows engineers and scientists to predict particle velocities, residence times, and solid volume fraction. Study results link higher particle residence times to robust heat absorption and thermal performance in solar, highlighting the design of practical integration in future CSP systems [5]. Kuruneru [6] study focuses on the development of new solid particle receivers to allow maximum heat capture within the receiver. Furthermore, the embodied Catch and Release mechanism opens the doors for new engineered designs, allowing the integration of such mechanisms with pre-existing SPR designs or new ones. Such coupling creates hybrid designs with maximum efficiency with cutting-edge new geometries. Behringer [7] presents a detailed description of the jamming effect, occurring during state transition from a fluid-like to a solid-like state in the presence of applied forces such as shear or compression.

Jamming may be viewed as state where granular materials change from a fluid-like to a solid-like state due to a rise in parameters such as packing fraction (ϕ), incremental shear stress (τ), or actual compression to form a mechanically hardened granular material assembly [7]. This jammed state therefore paves the way for a packed bed configuration, slowing down particle flow within the system and increasing residence time which can be attained simultaneously [7]. For CSP plants, solid-state ceramic particles have been found to be highly preferred as heat transfer media for thermal storage and heat transfer applications due to high radiative absorption capacities and ap- propriate thermophysical properties thanks to their specific heat capacity and thermal conductivity [3, 8]. Materials with non-thermal dependence properties meet specific CSP plant operational requirements, such ceramic particles have suitable properties for heat retention, which are ideal for solid particle receivers where much longer residence time results in an improvement of the overall absorption efficiency [3, 8].

This paper focuses on developing a multi-stage solid particle receiver to increase the particle residence time. Accumulation of granular material at the walls will slow down the particles' flow and allow for self-discharge mechanism for particles to flow from SPR into the rest of the tower block. This could potentially result in higher retention time.

Methodology

In this paper, a multi-stage solid particle receiver with integrated inclined obstacles is designed and analyzed using a numerical simulation approach in Yade. In order to improve thermal efficiency in Concentrated Solar Power (CSP) systems, the main goal is to examine how obstacle inclination angles affect particle residence time and the force distribution that results on the internal structures of the receiver. (1) Geometric Modeling and Particle Definition; (2) Simulation Setup and Execution; and (3) Data Post-Processing and Analysis comprise the three primary stages of the methodology.

Geometric Modeling and Particle Definition

In order to prevent and regulate particle flow, the receiver geometry was conceived as a novel, multi-stage design with two inclined obstacles (Plane 1 and Plane 2) and dynamic doors that create a stair-like effect.

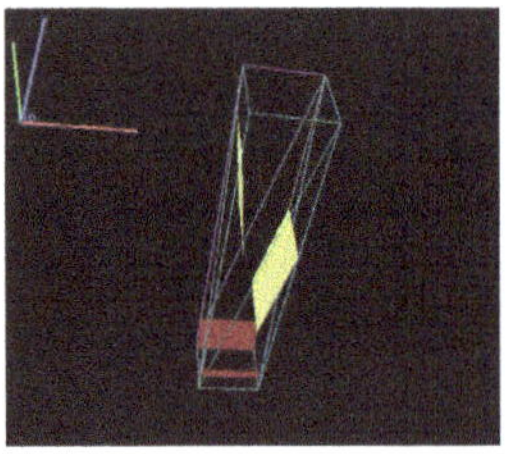

Figure 1: Solar Receiver with Inclined obstacles

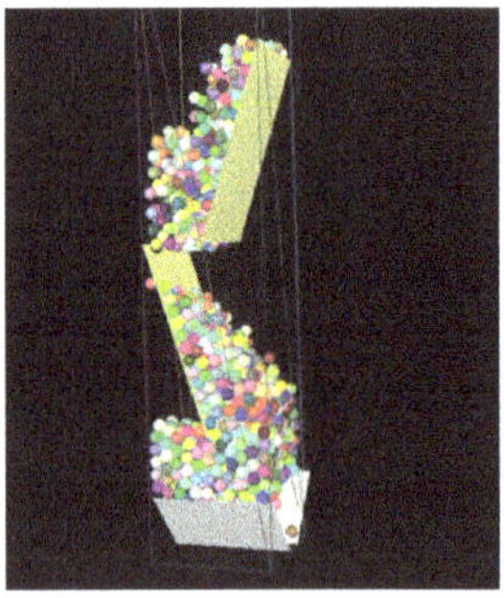

Figure 2: Cloud Particle

Simulation Setup and Execution

To examine particle dynamics under various geometric configurations, a number of discrete element method (DEM) simulations were carried out. The obstacles' inclination angle, which was systematically changed from 0° to 30° (more precisely, 0°, 5°, 10°, 15°, 18°, 20°, 25°, and 30°), was the main independent variable, as shown in Fig. (3).

A cloud generation function was added to the Yade engine in order to replicate a continuous particle flow. With a time step (t) of 0.0002 seconds per iteration, this function introduced a fresh batch of particle into the receiver every 22,000 simulation iterations. At the beginning of the simulation (t=0), the initial particle cloud was introduced.

For each inclination angle, two primary dependent variables were tracked:

Residence Time: The average time it takes for the center of mass of the particle cloud to move from an initial Z-position of +15 to a final Z-position of -15 over a fixed vertical distance.

Applied Forces: The total contact forces exerted by particles on the key structural components—Inclined Plane 1, Inclined Plane 2, and the dynamic doors (Left and Right). The force vector for each component was calculated as:

$$F_{total} = \sqrt{F_x^2 + F_y^2 + F_z^2} \quad (1)$$

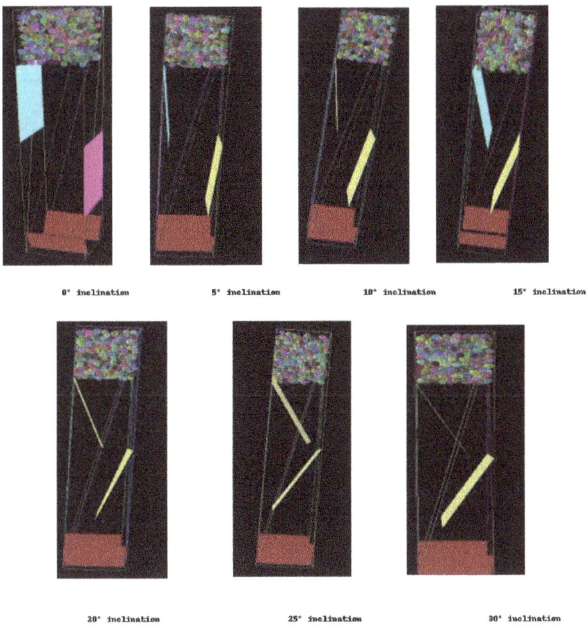

Figure 3: Static Obstacles' Inclinations

Data Post-Processing and Analysis

Yade exported the raw data for force components and particle positions for further examination. A Jupyter notebook environment was used for post-processing in order to compute the following:

- The residence time was directly calculated from the average Z-position of the particle cloud over time for each inclination angle.
- The magnitude of force that was applied overall to each structural element during the simulation.

In order to determine the best configurations for thermal absorption, the relationship between the inclination angle and residence time was examined. To assess the loading asymmetry and design optimization for the mechanical stability of the system, the force imbalance between Plane 1 and Plane 2 was further analyzed to inspect the force ratio calculation (Plane 1 / Plane 2). The proposed design for the SPR utilizes the DEM and post-processing analysis approach to provide a robust basis for designing a balance between the mechanical stability (force balance) and the thermal efficiency (residence time).

Results and Discussion

Residence Time

Fig. 4 shows that particles take longer time to fall through the system in certain inclination angles, highlighting higher resistance and residence time. This gradual behavior suggests that at higher inclinations, particles interact more strongly with the obstacles and each-others and decelerate, causing them to stay longer in the system. From steep, sharp descents for lower angles to flat, long-lasting descents at higher angles, the curve shows very clearly how the inclination angle serves as an effective flow control for the obstacles, tending to retard the passage of the particles as it increases. It is not linear; the effect is greater for larger angles, and the most resistant particles with the longest residence time are those at 30°.

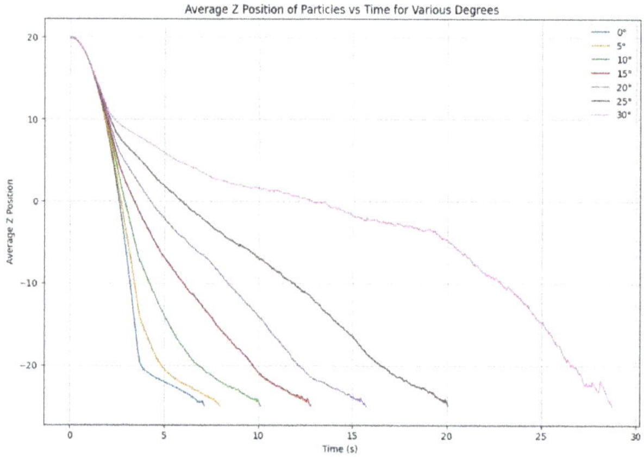

Figure 4: Average Z Position of Particles vs. Time for Various Inclination Degrees

Fig. 5 shows that the inclination of an obstacle has a huge effect on the residence time. The slope of the curve becomes non-linear and more pronounced at higher angles. A column plot to compare residence times for different tilt angles reveals that, at lesser angles (0°, 5°, and 10°), particles move through the system quickly with limited resistance caused basically by low residence times. With increasing angle beyond 15°, residence time rapidly rises and indicates even greater interaction between the particles and the obstacles. Particularly between 25° and 30°, the range in residence time almost doubled, showing how more severe slopes now pose considerable resistance to particle flow and thus prolong particle detention in the system. In thermal applications, such as concentrated soalr power, this property is beneficial because longer residence times allow for more heat to be absorbed.

Materials Research Proceedings 64 (2026) 99-106 https://doi.org/10.21741/9781644904091-13

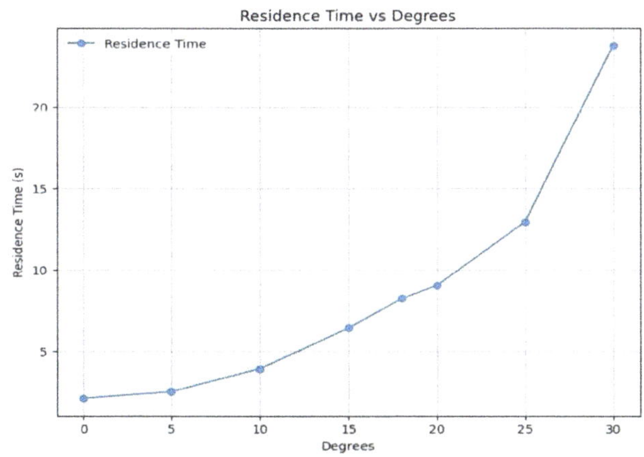

Figure 5: Residence Time vs. Degrees

Applied Forces on Receiver Doors

This part studies the behavior and stability of different structures within our system. By measuring the forces applied by particles on inclined Plane 1 and Plane 2 as well as the doors for different inclination angles. We used continuous flow by generating a cloud function that is called from the engine; it generates particles every 22,000 iterations. The time for each iteration is $O.dt$ = 0002 The first cloud is generated independently at $O.time$ = 0. From Yade, data was saved for a direction, and the total applied force was calculated during the post-processing stage. Simulations' results are summarized in Figs. 6–9.

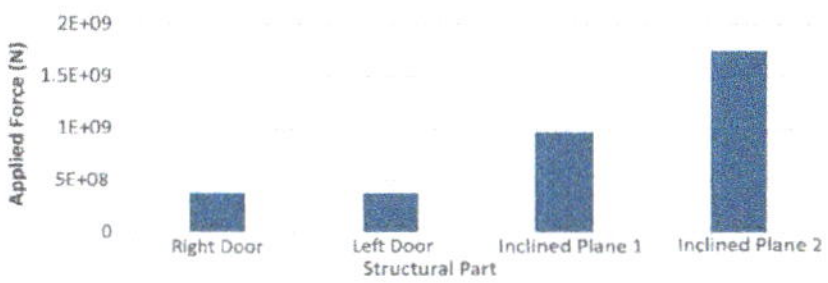

Figure 6: Total Applied Force for 5° Inclination

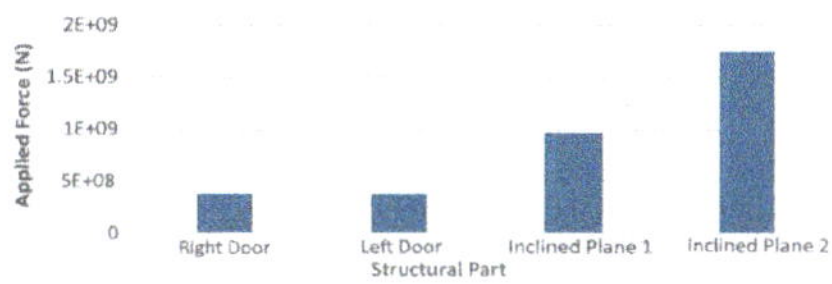

Figure 7: Total Applied Force for 15° Inclination

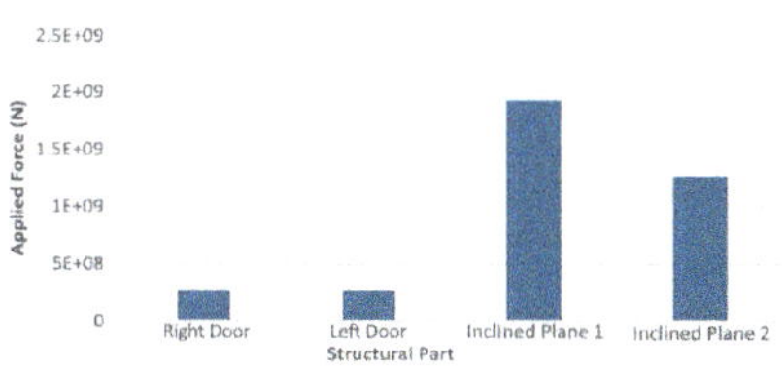

Figure 8: Total Applied Force for 18° Inclination

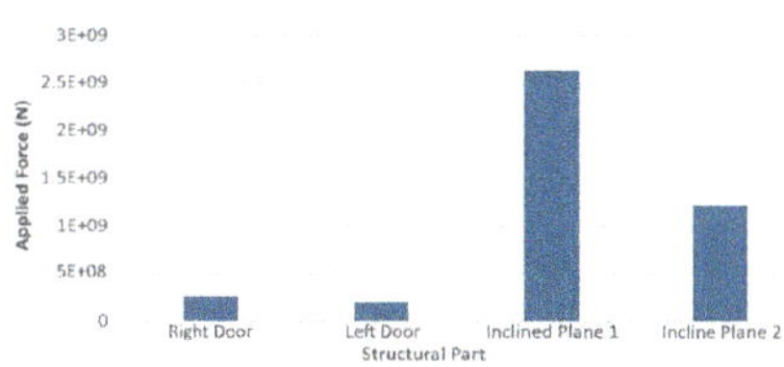

Figure 9: Total Applied Force for 20° Inclination

The overall analysis of the total applied forces acting on the inclined planes at varying angles of inclination presents a clear progression in force distribution and its resulting impact on the system components. At an inclination angle of about 5° (Fig. 3), the forces acting on Inclined Plane 1 and Inclined Plane 2, and Right Door and Left Door, are relatively low and uniformly distributed across each door; thus, there was little resistance to the flow of the particle, and thus quite smooth and balanced operation. As the angle climbs to 18°, marked increases in force begin to transpire, with Inclined Plane 1 experiencing higher forces compared to Plane 2; the consequent forces on the doors start to display some asymmetries. On reaching an inclination angle of about 20° (Fig. 3), a great jump in forces overall transpires, with Inclined Plane 1 becoming the more loaded plane due to the larger rate of deceleration of the particles resulting from increased inertia associated with their initial collision, whilst exhibiting significantly stronger asymmetry on the Right and Left Doors; therefore, the effects of misalignment or uneven particle distributions became apparent. Finally, at 25° (Fig. 3), other components increase markedly in force, with Inclined Plane 1 sustaining the largest support load and a considerable asymmetry observed between the doors, where one door receives persistently greater forces. Herein is seen the direction and importance of inclination angles which have manifold impacts on the particles' flow dynamics and demonstrate the importance of accuracy in the alignment, and thus counter the imbalances to formulate a reasonably reliable system with longer persistence. To understand the effect of inclination degrees on obstacles' stability, we calculated the ratio of total applied forces on Plane 1/Plane 2, as shown in Fig. 10. Uneven force distribution in the multi stage solar receiver may lead to structural failure if the inclination angles are designed separately, leading to a a required optimization problem to find the optimal angle for maximum heat retention with and design stability.

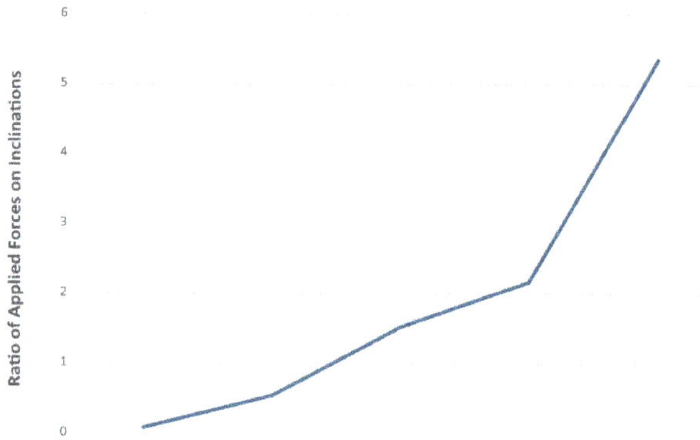

Figure 10: Ratio of Total Applied Forces on Plane 1 / Plane 2

Discussion & Interpretation

The paper highlights the importance of inclined obstacles in improving the thermal performance of a Concentrated Solar Power receiver and the particle flow behavior. This analysis has shown that the performance of the systems with inclination obstacles significantly increases, in terms of the residence time of the particles and a subsequent transfer of the energy. At steeper angles (20° & 30° o), the obstacles cause more resistance to the movement of the particles, which reduce their speed and increase the time of interaction between them and the medium where the heat is transferred. This increased exposure raises the heat absorption, hence an improved thermal performance and a stabilized power production. All in all, the addition of inclined obstacles thus creates a controlled and predictable environment of particle flow that facilitates an optimal thermal and mechanical performance. However, the system does have some trade-offs such as anuneven distribution of forces and the possibility of jamming of particles. These problems could be overcome by design optimization, especially by adjusting the inclination angle, providing the perfect door fit, and the choice of the right materials to use in Inclined Plane 1. When good optimization is made, inclined obstacles increase the performance of the receiver to a great extent in terms of both efficiency and reliability without reduction in structural robustness.

Conclusion

The study shows that integrating obstacle structure within solar power receiver increases the residence time which directly enhances the thermal efficiency of the system.Through a three-dimensional numerical simulation implemented using Yade and the Discrete Element Method, it is found that a larger angle of inclination results in a substantially improved residence time and particle-wall interaction with a factor of three- 2 seconds against 6 seconds residence time for particles fall for normal and 20° inclined receivers respectively. Yet, choosing the inclination angles is problematic as there is a trade-off between thermal efficiency and system stability. High inclination angles can lead to particles melting and over-jamming which may lead to particle-flow blockage, requiring the integration of other mechanisms. Fluid injection-such as air- is usually used to smooth particles flow. Another solution would be the use of dynamic obstacles to minimize over-jumming potential of particles. The degree of inclination is also dependent on the forces exerted on the obstacles, leading to an equal distribution of forces, making it optimal for the choice of materials for the obstacles to resist the high potential forces exerted by the particles. These technologies provide innovative possibilities for improvement in the CSP technology. More studies are needed to maximize the optimal value of the inclination angle and the properties of the solid particles to maximize the absorption of heat without deteriorating the properties.

References

[1] C. K. Ho and B. D. Iverson, "Review of high-temperature central receiver designs for concentrating solar power," Renewable and Sustainable Energy Reviews, vol. 29, pp. 835-846, 2014. https://doi.org/10.1016/j.rser.2013.08.099

[2] M. M. Rafique, "Recent advancements in high-temperature solar particle receivers for industrial decarbonization," Sustainability, vol. 16, no. 1, p. 103, 2023. https://doi.org/10.3390/su16010103

[3] F. Nie, "Solid particle solar receiver (SPSR)-The key equipment to absorb the concentrated solar flux: A review," Engineering and Optimization Modelling, vol. 2, no. 1, pp. 1-18, 2022.

[4] B. H. Mills, C. K. Ho, N. R. Schroeder, R. Shaeffer, H. F. Laubscher, and K. J. Albrecht, "Design evaluation of a next-generation high-temperature particle receiver for concentrating solar thermal applications," Energies, vol. 15, no. 5, p. 1657, 2022. https://doi.org/10.3390/en15051657

[5] S. T. W. Kuruneru, J.-S. Kim, Y. C. S. Too, and D. Potter, "A coupled CFD-DEM approach to model the in-trough mixing in a multi-stage solar particle receiver," Energy Reports, vol. 7, pp. 5510-5526, 2021. https://doi.org/10.1016/j.egyr.2021.08.179

[6] S. T. W. Kuruneru, J.-S. Kim, Y. C. S. Too, and D. Potter, "Multi-stage particle receiver mixing and heat transfer simulation using coupled CFD-DEM," Energy Reports, vol. 8, pp. 2734-2749, 2022.

[7] R. P. Behringer and B. Chakraborty, "The physics of jamming for granular materials: A review," Reports on Progress in Physics, vol. 82, no. 1, p. 012601, 2018. https://doi.org/10.1088/1361-6633/aadc3c

[8] G. Dai, "A review of radiative heat transfer in fixed-bed particle solar receivers," Sustainability, vol. 15, no. 13, p. 9918, 2023. https://doi.org/10.3390/su15139918

Comparative Study of Energy, Thermal, and Environmental-Economic Performance of PVT and PV Systems

Yassine EL ALAMI[1,a,*] Fatima AIT SIDI HAMMOU[1,b], Hanane CHETNOUF[1,c], Jaouad EL GHARFAOUI[1,d], Zineb HEKSS[2,e], Charaf HAJJAJ[1,f], Benhmida MOHAMMADI[1,j], Ahmed FAIZE[3,h] and Elhadi BAGHAZ[1,i]

[1]Laboratory of Electronics, Instrumentation and Energetic, Faculty of Sciences, Chouaïb Doukkali University, El Jadida, Morocco

[2]SAML, High School of Technology of Sidi Bennour, Chouaib Doukkali University of El Jadida, Morocco

[3]Department of Physics, Polydisciplinary Faculty, University of Mohamed Premier, Nador, Morocco

[a]yassineelalami97@gmail.com, [b]aitsidihammoufaj@gmail.com, [c]hananechetnouf@gmail.com, [d]elgharfaoui.j302@ucd.ac.ma, [e]hekss.zineb@ucd.ac.ma, [f]Hajjaj.c@ucd.ac.ma, [g]Benhmida.m@ucd.ac.ma, [h]ahmedfaize6@hotmail.com, [i]baghaz.e@ucd.ac.ma

Keywords: Photovoltaic-Thermal, PV Module, Energy Production, CO_2 Reduction, Economic and Environmental Benefits

Abstract. The rising energy demands, and the environmental consequences of the traditional electricity production, is a major impediment to sustainable development. This paper critically analyzes the performance of photovoltaic-thermal system (PVT-S) against a typical photovoltaic module (PV-M) with reference to climatic conditions in the city of El Jadida. The analyses were carried out using the region's characteristic annual irradiance, and the mass flow rate of the PVT-S was set at 0.01667 kg/s. The results show that PVT-S increases annual electricity production by 26.9 kWh and adds thermal production of 1010.36 kWh, for a total energy production of 1037.26 kWh more than PV-M. This improvement also translates into an annual reduction in CO_2 emissions of 5.667 tCO_2 compared to 3.258 tCO_2 for PV-M, generating an environmental economic benefit of 82.171 $/year, which is higher than that of PV-M. These results demonstrate that PVT-Ss are an effective solution for optimizing energy production, improving the efficiency of PV cells, and reducing environmental impact.

Introduction

Sustainable development and environmental protection have become enormous problems due to the rapid growth in CO_2 emission and world energy problems. It is not only that reliance on fossil fuels causes high emissions of greenhouse gases, but that also the country faces the problems of global warming and air pollution degradation, necessitating the discovery of clean and renewable sources of energy [1,2]. Renewable energies, and solar energy in particular, is one such potential alternative in this context [3–5]. The solar power is abundant, stable, safe, and it is widely available and its extensive use may help fill the global energy requirements. The European Commission using the Joint Research Centre estimates that solar PV electricity generation will contribute over 10 percent of the overall energy use in the year 2030 underlining its importance in terms of strategy in energy transition [6]. The most common currently used technology in solar energy to electricity conversion is PV-Ms. Their conversion efficiency is however low usually less than 20 percent and hence about 80 percent of the solar energy taken in is wasted as heat [7]. This thermal energy causes overheating of the cells, thus lowering their efficiency: the more the temperature of the cells rises, by 1 °C, the less efficient it becomes (0.3-0.5). Maintaining the cells near standard test conditions (25 °C) is therefore essential to get the best performance out of them [8]. To address this issue, system integration of a PVT-S seems to be a solution [9,10]. It is a hybrid system with a PV-M attached to a thermal collector to reclaim the thermal energy generated thus lowering the temperature of the cells and enhancing the electrical efficiency of the cells [11–13]. PVT-Ss maximize the utilization of solar energy on a small area of land, counteracts the instability of solar production, and enhances the overall system operation, a combination of electrical and thermal energy makes the exploitation provide high application prospects and huge technical importance.

However, despite the many advantages of PV-Ms and PVT-Ss, several gaps remain in the current literature. Few studies have comprehensively analyzed the energy, thermal, and electrical performance of PVT-Ss compared to PV-Ms under specific real-world conditions, while incorporating a comprehensive environmental and economic assessment. Furthermore, the benefits associated with CO_2 emissions reduction and environmental economic gains remain poorly quantified for local urban installations. Therefore, the main objective of this study is to evaluate and compare the performance of a PVT-S and a reference PV-M in the context of the city of El Jadida. The analysis will be based on overall electrical, thermal, and energy generation and the CO_2 emission cuts and the environmental-economic advantages brought thereof. This method does not only measure the energy and environmental benefits of PVT-S but also gives a detailed evaluation of its technical and economic feasibility over the traditional PV-M.

The article is structured in the following way. The following section will be a detailed description of the system being studied, along with the type of experiment adopted. The further part is the presentation and discussion of the gained results based on electrical and thermal performance, total production, annual CO_2 avoidance, and environmental and economic benefit of the PVT-S in comparison with the reference PV-M configuration. Lastly, the main conclusions are discussed in the concluding part.

Experimental setup

Fig. 1 illustrates the experimental setup of the PVT-S as well as the PV-M, which have been made with the same technology, and have the same size dimensionalities in order to enable a strong comparison of the respective performances. When it comes to the PVT-S, the principle of operation is based on the synergistic coupling, which exists between the production of photovoltaic electricity and the reuse of thermal energy. About one-third of the incident solar energy is translated to electrical power with the rest being transformed to heat thus increasing the temperament of the cells. This heat is then eliminated by using a heat exchanger that is part of the back of the module, a heat-transfer fluid flows through this exchanger thus reducing the increase in temperature of the cells and also reclaiming the thermal energy produced. The physical aspect of the PVT-S includes the PV-M, a heat transfer plate, a rear mounted heat exchanger, a hydraulic circuit with which the heat-transfer fluid circulates that are complemented by a thermal insulation that minimizes heat loss to the surrounding environment. The main features of the two systems are summarized in Table1, hence making it possible to objectively evaluate the effects the thermal integration has on the electrical production and heat dissipation. Fig. 2 illustrates all of the measuring instruments used to record electrical and thermal parameters during the experimental tests. For a detailed description of the experimental methodology, readers can refer to studies [14,15].

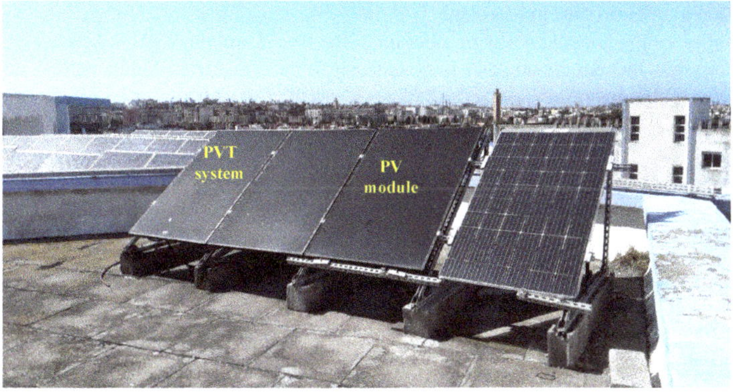

Fig. 1. Experimental setup of the PVT and PV reference systems.

Table 1. Main characteristics of PVT and PV systems compared.

Parameters	Dualsun
Open-circuit voltage V_{oc} (V)	48.9
Conversion efficiency (%)	19.989
Current at maximum power I_{mpp} (A)	9.28
Rated power (W)	375
Temperature coefficient of power (%/°C)	-0.34
Short-circuit current I_{sc} (A)	9.89
Area of the module (m^2)	1.876
Voltage at maximum power V_{mpp} (V)	40.4

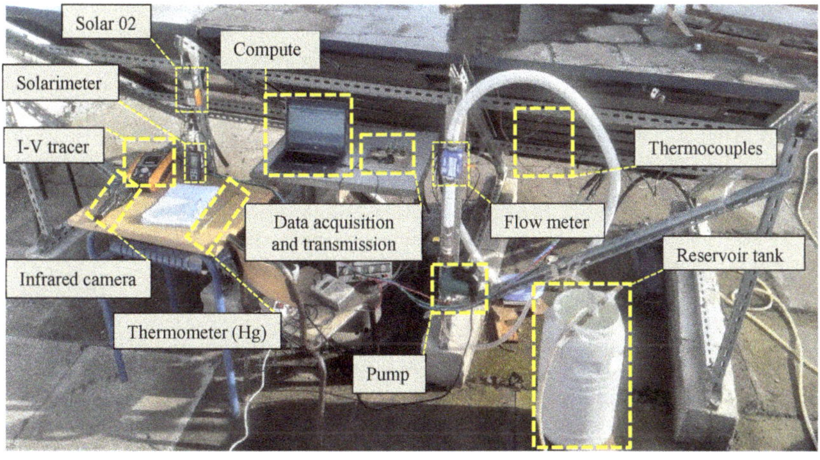

Fig. 2. Measuring instruments used for electrical and thermal evaluation of systems.

Annual energy production assessment

The estimation of the annual energy production of the PV-M and the PVT-S is based on Eqs. 1, 2, and 3, taking into account the annual solar irradiance G_t recorded in El Jadida and the degradation rate provided by the manufacturer, set at 0.8% per year [9,10].

$$E_{ele} = (365 \times G_t \times \eta_{el} \times A_{PV}) \times (1-dg)^{t-1} \quad (1)$$

$$E_{th} = (365 \times G_t \times \eta_{th} \times A_{th}) \times (1-dg)^{t-1} \quad (2)$$

$$E_{ntotal} = E_{th} + E_{ele} \quad (3)$$

Enviro-economic assessment

The adoption of solar systems contributes to reducing CO_2 emissions. Thus, the environmental parameter (Ø_{CO_2}) associated with the PVT-S and the PV-M is determined using the following Eq. [16,17]:

$$\text{Ø}_{CO_2} = \frac{\dot{Q}_u \times \psi_{CO_2} \times n}{1000} \quad (4)$$

The indicator Ø_{CO_2} represents the annual amount of CO_2 avoided, in tons. Here, n is the average duration of sunshine and (ψ_{CO_2}) is the emission factor for coal, estimated at 2.08 kg CO_2/kWh. The useful thermal energy ($\dot{Q}u$) is then obtained from the following Eq. [16]:

$$\dot{Q}_u = Q_{th} + \frac{E_{el}}{0.38} \quad (5)$$

Q_{th} refers to the thermal energy produced, while E_{el} corresponds to the electrical energy generated. The environmental cost (Z_{CO_2}) of PVT and PV systems is evaluated using Eq. (6), considering a carbon price P_{CO_2} of 14.5 $ [16,17]:

$$Z_{CO_2} = P_{CO_2} \times \text{Ø}_{CO_2} \quad (6)$$

Results and discussion

This study examines the performance of the PVT-S in comparison with the reference PV-M, considering energy efficiency, annual production, and environmental and economic assessment. The operation of the PVT-S was analyzed for a mass flow rate set at 0.01667 kg/s.

Fig. 3 shows the I-V characteristics of the PVT-S and the reference PV-M, and thus explains the effect of augmented solar irradiance on current and voltage. As shown in this figure, the electric current increases directly in association with irradiance: its values vary between 2.620 A and 10.018 A in the case of the PVT-S and between 2.604 A and 10.121 A in the case of the PV-M as the incident flux increases between 272.06 W/m^2 and 1055.6 W/m^2. This small variation between the two setups can be explained by the fact that the PVT-S has the ability to operate with lower temperatures, and thus, prevent thermal losses, which would otherwise hinder carrier generation. On the other hand, the voltage has a weak negative correlation with the increasing irradiance, an effect that is typical of temperature-dependence of PV cells.

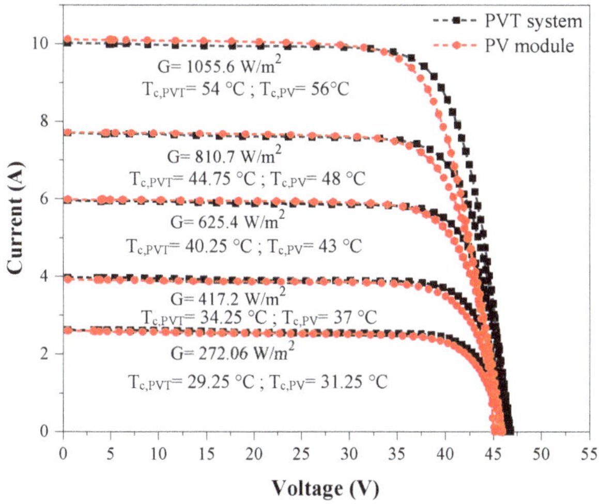

Fig. 3. *I–V curves of the PVT-S and PV-M under different levels of solar irradiance.*

Fig. 4 illustrates the P-V characteristics of the PVT-S and PV-M at different intensities of the sun in the environment showing how electrical power production changes. At a specific solar irradiance of 272.06 W/m^2 to 1055.6 W/m^2 the PVT-S power output rises by 97.23 W to 356.47 W and the PV-M power output rises by 92.9 W to 342.49 W. The great performance of PVT-S can be explained with the help of its active cooling system which helps to eliminate the cell temperature rise and minimize thermal losses and, therefore, the power output, electrical efficiency and lifespan of photovoltaic cells.

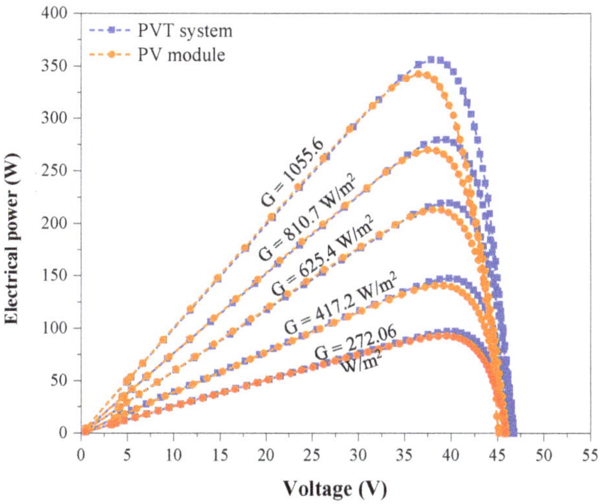

Fig. 4. *P–V curves of the PVT-S and PV-M under different levels of solar irradiance.*

The annual performance of the PVT-S and the PV-M was evaluated based on the characteristic annual irradiance of the city of El Jadida. Table 2 shows the total electrical, thermal, and energy production obtained for both configurations. The results show that the PVT-S generates 26.9 kWh more electricity per year than the PV-M, in addition to 1010.36 kWh of thermal energy. Thus, the total energy production of the PVT-S exceeds that of the PV by 1037.26 kWh. This significant improvement can be explained by the thermal management of the PVT-S, which optimizes the operating conditions of the PV cells while recovering the heat. All of these results confirm the significant energy advantage of PVT-Ss, both in terms of electricity and heat.

Table 2. *Annual electrical, thermal, and total production of PVT and PV systems for El Jadida.*

System configuration	Annual electrical energy (kWh)	Annual thermal energy (kWh)	Annual total energy (kWh)
PVT-S	794.09	1010.36	1804.45
PV-M	767.19	0.00	767.19

Table 3 shows the annual amount of CO_2 avoided and the environmental economic benefit (ZCO_2) of PVT and PV systems. The PVT-S avoids 5,667 tCO_2/year, while the PV-M reduces 3,258 tCO_2/year, representing an additional gain of 2,409 tCO_2/year for PVT-S. This difference can be explained by the higher energy production of PVT-S, which combines electricity generation and heat recovery. In terms of economic environmental benefits, PVT-S generates 82,171 \$/year compared to 47,241 \$/year for PV-M, reflecting the economic advantage of reducing CO_2 emissions. Thus, the PVT-S performs significantly better than the PV-M in terms of sustainability and economic contribution to environmental protection.

Table 3. *Annual CO_2 avoidance and environmental economic benefit of PVT and PV systems.*

System configuration	$\emptyset_{CO2}$ (tCO_2/ year)	Z_{CO2} (\$/year)
PVT-S	5.667	82.171
PV-M	3.258	47.241

Conclusion

This paper compared the energy, environmental, and economic performance of PVT-S and PV-M in the context of El Jadida. The results indicate that the PVT-S has a considerable advantage over the PV-M reference, with an annual electricity production that is 26.9 kWh higher, a thermal production that is 1,010.36 kWh higher, and a total energy production that is higher than that of the PV-M. In addition, this performance translates into 5,667 tCO_2 per year, compared to 3,258 tCO_2 per year for PV-M, and generates an environmental and economic benefit of 82,171 \$/year, compared to 47,241 \$/year for PV-M. These results prove that PVT-S is an effective approach to energy production, increasing electrical and thermal yields and playing an important role in environmental and economic sustainability.

References

[1] Y.E. Alami, E. Baghaz, R. Nasrin, C. Hajjaj, M. Benhmida, K.M. Al-Aribe, A. Ibrahim, Numerical Exploration and Relational Impact Ranking of Critical Parameters in 3D PV Module Studies, Results Eng. (2025) 107097. https://doi.org/10.1016/j.rineng.2025.107097

[2] Y. El Alami, M. Rami, C. Hajjaj, M. Benhmida, A. Faize, E. Baghaz, Three-dimensional thermal modeling and analysis of the heat source, the impact of encapsulation materials and the back sheet of silicon PV modules, Appl. Therm. Eng. 280 (2025) 128430. https://doi.org/10.1016/j.applthermaleng.2025.128430

[3] Y.E. Alami, E. Baghaz, R. Nasrin, K. Haboubi, M. Benhmida, A. Ibrahim, Integrated sustainability and 4E analysis of hybrid nanofluid featured PVT systems incorporating diverse backsheets, Appl. Therm. Eng. (2025) 129147. https://doi.org/10.1016/j.applthermaleng.2025.129147

[4] Y.E. Alami, E. Baghaz, R. Nasrin, S. Padmanaban, M. Louzazni, Numerical Approach of An Advanced Hybrid Photovoltaic Thermal System Based on Exergy, Energy, Enviro-Economic, and Sustainability Factors, Results Eng. (2025) 106342. https://doi.org/10.1016/j.rineng.2025.106342

[5] Y. El Alami, H.E. Achouby, E. Baghaz, C. Hajjaj, R. Nasrin, An innovative photovoltaic thermal system with direct water-cell contact: energy, exergy, and sustainability analysis, Sol. Energy 300 (2025) 113827. https://doi.org/10.1016/j.solener.2025.113827

[6] Z. Fu, Y. Li, X. Liang, S. Lou, Z. Qiu, Z. Cheng, Q. Zhu, Experimental investigation on the enhanced performance of a solar PVT system using micro-encapsulated PCMs, Energy 228 (2021) 120509. https://doi.org/10.1016/j.energy.2021.120509

[7] S. Alsaqoor, A. Alqatamin, A. Alahmer, Z. Nan, Y. Al-Husban, H. Jouhara, The impact of phase change material on photovoltaic thermal (PVT) systems: A numerical study, Int. J. Thermofluids 18 (2023) 100365. https://doi.org/10.1016/j.ijft.2023.100365

[8] L. Ahmad, N. Khordehgah, J. Malinauskaite, H. Jouhara, Recent advances and applications of solar photovoltaics and thermal technologies, Energy 207 (2020) 118254. https://doi.org/10.1016/j.energy.2020.118254

[9] H.L. Azeez, A. Ibrahim, B.O. Ahmed, S.S. Dol, A.H.A. Al-Waeli, M. Jaber, Experimental investigations of heat transfer, energy, and exergy-based sustainability of a novel photovoltaic thermal system, Case Stud. Therm. Eng. 70 (2025) 106089. https://doi.org/10.1016/j.csite.2025.106089

[10] H.L. Azeez, A. Ibrahim, B.O. Ahmed, J. Kasprzak, A.H.A. Al-Waeli, M. Jaber, 5E assessments of a novel solar collector under high cooling demand scenarios: Experimental and sustainability studies, Energy 326 (2025) 136297. https://doi.org/10.1016/j.energy.2025.136297

[11] H.A. Kazem, M.T. Chaichan, A.H.A. Al-Waeli, K. Sopian, Energy, exergy, entropy generation, economic and environmental (5E) analysis of a photovoltaic/thermal system using nanofluid circulation, Energy Sources Part Recovery Util. Environ. Eff. 47 (2025) 2568727. https://doi.org/10.1080/15567036.2025.2568727

[12] A. Kazemian, K. Khosravi, S. Sen, P. Talebizadehsardari, A. Cairns, T. Ma, Optimizing photovoltaic thermal systems with wavy collector Tube: A response Surface-Based design study with desirability analysis, Appl. Therm. Eng. 258 (2025) 124475. https://doi.org/10.1016/j.applthermaleng.2024.124475

[13] F. Abdulamir Radhi, A. Ibrahim, A. Al Tarabsheh, A. Ameen Ali, H. Luqman Azeez, A. Fazlizan, K. Sopian, Experimental and numerical analysis of a solar collector enhanced by a square coil and a twisted tape, Case Stud. Therm. Eng. 74 (2025) 106783. https://doi.org/10.1016/j.csite.2025.106783

[14] Y. El Alami, E. Baghaz, R. Bendaoud, F. Chanaa, M. Benhmida, H. Ezzaki, R. Nasrin, Experimental-Numerical Investigation of the Photovoltaic Thermal System with Polypropylene Heat Exchanger: Case in Morocco, IET Renew. Power Gener. 19 (2025) e70041. https://doi.org/10.1049/rpg2.70041

[15] Y. El Alami, E. Baghaz, R. Nasrin, F. Chanaa, R. Bendaoud, S. Padmanaban, M. Louzazni, Experimental-numerical comparative study of performance and cost-effectiveness of partially- and fully-cooled photovoltaic thermal systems, Case Stud. Therm. Eng. 73 (2025) 106660. https://doi.org/10.1016/j.csite.2025.106660

[16] O. Büyükalaca, H.M. Kılıç, U. Olmuş, Y.E. Güzelel, K.N. Çerçi, Numerical investigation and ANN modeling of performance for hexagonal boron Nitride-water nanofluid PVT collectors, Therm. Sci. Eng. Prog. 43 (2023) 101997. https://doi.org/10.1016/j.tsep.2023.101997

[17] M. Emam, A. Hamada, H.A. Refaey, M. Moawed, M.A. Abdelrahman, M.R. Rashed, Year-round experimental analysis of a water-based PVT-PCM hybrid system: Comprehensive 4E assessments, Renew. Energy 226 (2024) 120354. https://doi.org/10.1016/j.renene.2024.120354

Materials Research Proceedings 64 (2026) 115-122 https://doi.org/10.21741/9781644904091-15

Hybrid Backstepping–P&O Control Strategy for Enhanced MPPT in Photovoltaic Systems under Climatic Variations

Fatna BDAIDI[1,a *], Driss YOUSFI[1,b] and Ayoub RAHMOUNI[1,c]

[1]ESETI Laboratory, National School of Applied Sciences, First Mohamed University, Oujda, Morocco

[a]fatna.bdaidi@gmail.com, [b]dr_yousfi@yahoo.com, [c]rahmouni.ayoub442@gmail.com

Keywords: Photovoltaic Systems, MPPT, Backstepping, Perturb and Observe, Hybrid Control

Abstract. A compact hybrid MPPT method for standalone PV systems is proposed in this paper, wherein a slow P&O-based estimator and a fast nonlinear backstepping controller design of the DC–DC boost converter are combined. The reference power is modulated for environmental variations in the P&O stage, and the backstepping loop ensures high-bandwidth, Lyapunov-stable tracking of the operating point. The proposed hybrid architecture is compared to the standalone P&O and to the standalone backstepping and it provides enhancement in steady-state ripple as well as transient performance under irradiance steps. The amount of extracted quantitative performance values from simulation waveforms indicate MPPT efficiency does not fall below 99% under steady condition, faster settling (≈0.15 s after steps) and less ripple compared to classical P&O. The method remains computationally cheap and is made more robust to fast irradiance changes.

Introduction

Increasing demand for electricity and decarbonization goals are driving the adoption of photovoltaic (PV) power generation plants [1]. But, the electrical parameters of PV arrays are highly sensitive to irradiance and cell temperature; therefore, the operating point and so the power transferred change with time. Maximum power point tracking (MPPT) is indispensable to obtain the highest possible energy at all times, given fluctuating environment parameters [2].

MPPT methods proposed in literature can be generally classified as follows. While traditional methods like Perturb and Observe (P&O) or Incremental Conductance are still commonly applied due to their low price and easy implementation [3, 4]. However, these methods are simple and straight forward yet often caused steady-state oscillations around the MPP and sometimes led to a decrease of tracking performance in fast irradiance changes; also behaved that could converge to merely local instead global MPP especially under partial shading [5,6]. Metaheuristic techniques (e.g., DE, PSO, GA, and swarm-based) are capable of providing a better performance in finding GMPP over nonconvex P–V landscapes with respect to the conventional methods; however, they require higher computational burden and careful tuning of parameters making them less attractive for embedded platforms [7, 8]. Advanced control- and AI-based schemes, including fuzzy/neural estimators and nonlinear control methods (SMC, backstepping, synergetic control, MPC), can provide fast dynamics and stability guarantees when combined with a reference-generation stage, commonly through a slow MPPT estimator and a fast inner-loop controller [4], [9, 10, 11].

Motivated by the need for a low-cost, fast, and stable MPPT under abrupt irradiance disturbances, this paper proposes a hybrid P&O–Backstepping architecture for a PV boost converter. The P&O stage updates the MPP reference online, while the backstepping controller enforces Lyapunov-based stability and high-bandwidth duty-cycle regulation. The remainder of the paper is organized as follows: Section II presents the averaged boost model and MPPT

Materials Research Forum LLC
https://doi.org/10.21741/9781644904091-15

framework; Section III details the backstepping design and its hybridization with P&O; Sections IV–V describe the test scenarios and comparative results; Section VI concludes.

System modeling of photovoltaic power converters

Studied photovoltaic system.

Fig. 1 shows a two-loop architecture: a slow P&O block estimates the reference power Pref from the measurements (Vpv, Ipv); a fast-Backstepping controller regulates the PWM duty D to drive the DC–DC Boost converter to the desired operating point. The plant includes the inductor L, the load Ro, the diode/switch and input/output capacitors Co and Ce.

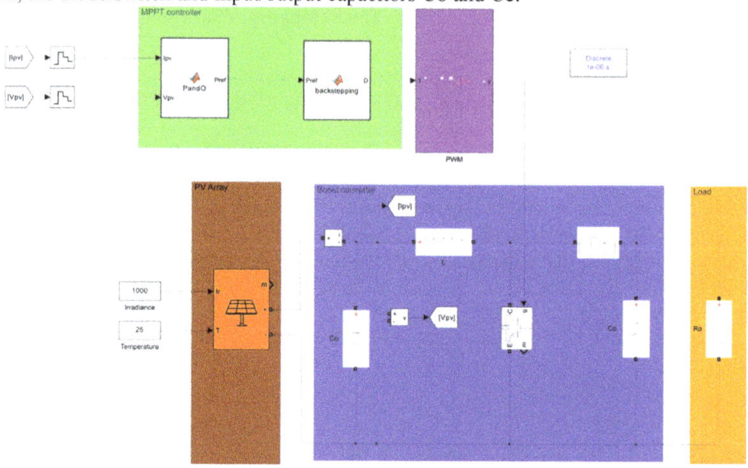

Fig. 1. Simulink diagram of the hybrid MPPT: P&O _Backstepping for a PV Boost converter.

A PV module is composed of series/parallel solar cells that behave as a p–n junction (Fig. 2), and its electrical behavior is characterized by the V–I (and corresponding P–V) curves shown in Fig. 3. [12]

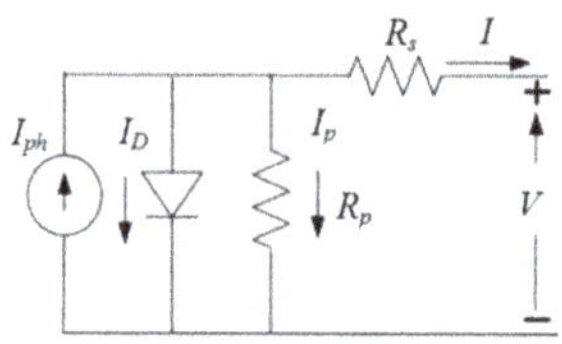

Fig. 2. Photovoltaic conversion chain scheme.

$$I = I_{sc} - I_0\left[e^{\frac{q(V+R_S I)}{nkT_k}} - 1\right] - \frac{V+R_S I}{Rp}. \quad (1)$$

In (1), V and I are the PV terminal voltage and current; the remaining symbols (Rs, Rp, q, Isc, I_0, n,k,Tk) follow the standard single-diode model notation, and the module ratings used in this work are reported in Table 1.

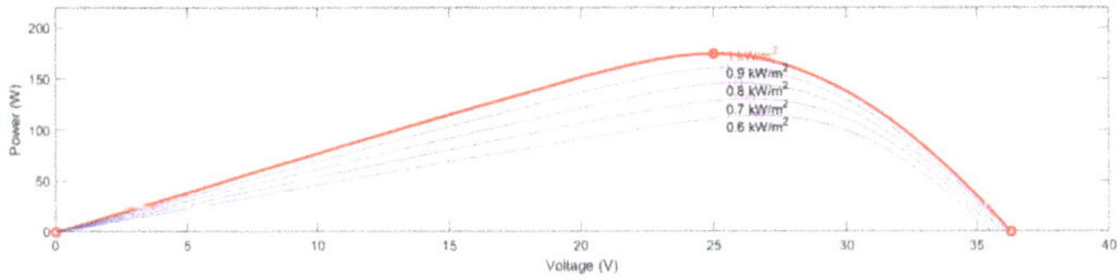

Fig. 3. P-V characteristic curves at 25°C with multiple irradiances.

Table 1. Parameters of PV.

Parameters	Values
Voltage at maximum power point V_{mp}(V)	29
Short-circuit current I_{sc}(A)	7.84
Maximum power (w)	213
Open circuit voltage (V)	36.3
Temperature coefficient of V_{oc}(°C)	-0.36099
Cells per module (Ncell)	60
Temperature coefficient of I_{sc}(°C)	0.102
Current at maximum power I_{mp} (A)	7.35

Boost Converter Design.
A boost converter regulates the PV operating voltage, and its PWM duty ratio sets the operating point, making D the key control variable for enforcing MPP operation.

The basic configuration of a boost converter is described by the following equations [11]:

$$\mathrm{Vs} = \frac{\mathrm{V_e}}{1-\mathbf{D}} \tag{2}$$

$$\mathrm{L} = \frac{\mathrm{V_e}\times(\mathrm{Vs}-\mathrm{Ve})}{\Delta\mathrm{I_L}\times F\times \mathrm{V_s}} \tag{3}$$

$$\mathrm{R_{eq}} = \mathrm{Rs}\times(1-\mathbf{D})^2 \tag{4}$$

$$L\frac{di}{dt} = Ve - (1-\mathbf{D}) * Vs \tag{5}$$

$$C\frac{dVs}{dt} = (1-\mathbf{D}) * i - \frac{Vs}{R} \tag{6}$$

Perturb and Observe (P&O) Algorithm
In the P&O perturbs the operating point to update the reference Pref, which is then tracked by the backstepping inner loop. [5,6].

P&O MPPT [12]. At each sampling instant k, the PV voltage and current are measured to compute the power P(k). The reference voltage is updated based on the signs of ΔP=P(k)−P(k−1) and ΔV=V(k)−V(k−1): if ΔP·ΔV>0, Vref is increased; otherwise, it is decreased. This process is repeated iteratively to drive the operating point toward the maximum power point.

Table 2. Detailed explanation of each component of the equations.

Symbol	Signification
Ve	The PV array output voltage
Vs	The converter output voltage
D	The duty cycle
L	The inductance of the boost inductor
C	The output filter capacitor
Req	The resistor as seen by the panel

Hybrid P&O–Backstepping

The origins of backstepping can be traced to nonlinear control in the early 1990s and coalesced with the seminal monograph of Krstić, Kanellakopoulos, and Kokotović (1995) that formalized design based on strict-feedback form and Lyapunov-based stability arguments. From the late 1990s to the 2000s [13], it was extended to adaptive and integral versions (and into power-electronics & electromechanical applications. In PV MPPT, the backstepping approach started gaining further popularity throughout the 2010 [14].

Control Law

The instantaneous PV power:

$$P(t) = V(t) * I(t) \tag{7}$$

According to [18], the tracking error is defined:

$$Z = P - Pref. \tag{8}$$

A Lyapunov function is:

$$v = \frac{1}{2}Z^2. \tag{9}$$

the derivative of the Lyapunov function.

$$\dot{v} = Z\dot{Z} = Z(\dot{P} - \dot{Pref}). \tag{10}$$

$\dot{Z} = \dot{P}$ implies $\dot{v} = Z\dot{P}$ with $Pref$ is constant term.
Stability condition is:

$$\dot{v} < 0 \Longrightarrow Z\dot{P} < 0. \quad \text{With } \dot{Z} = -K_1 Z \qquad K_1 > 0. \tag{11}$$

$$dv = \frac{Pactuel - Pprecedent}{Ts}. \tag{12}$$

A composite first error is introduced.

$$E = \frac{dv}{dt} + K_1 Z. \tag{13}$$

Consider the augmented Lyapunov function:

$$V_2 = \frac{1}{2}Z^2 + \frac{1}{2}E^2 \tag{14}$$

Select the stabilizing dynamics for the second error as

$$\dot{E} - -K_1 Z - K_2 E. \qquad K_2 > 0. \tag{15}$$

$the\ duty - cycle\ update$

$D = Dold + \dot{E} * TS.$ where Ts is the desired fast-loop settling time (16)

$K_1 is\ the\ speed\ of\ Z\ how\ fast\ P\ converge\ to\ Pref\ .K_1 = 4/Ts$

K_2 provides damping and limits duty variations; start with $K_2 \in [2K_1, 5K_1]$.

Simulation and Results

To assess the conventional P&O MPPT under identical conditions, the irradiance was stepped from **600 W/m² to 500 W/m²** at **t=2s** while the temperature was kept constant at **25 °C** (Fig. 5). The corresponding PV power response (Fig. 6) converges toward the MPP region at approximately **123 W** before the step, but presents persistent steady-state oscillations inherent to the perturbation mechanism. After the irradiance decrease, the algorithm re-tracks the new MPP around **108 W**; however, the ripple remains noticeable, which may reduce the harvested energy and increase stress on the converter.

Fig. 7, demonstrates that the stand-alone backstepping controller provides rapid convergence and stable power (~130 W) with minimal ripple on constant irradiance. But when the irradiance is suddenly reduced from 600 W/m² to 500 W/ m² at t = 2 s, times after which controller no longer follows new MPP then leading to significant power decrease and finally collapse. This demonstrates that stand-alone backstepping in MPPT is not scalable to irradiance changes without an external reference update mechanism.

We show in Fig. 8 the hybrid P&O–Backstepping for three irradiance transitions. In the 600→500W/m² case (Fig. 8a), the PV power converges to nearly 150W before the step and then follows well to the new MPP around 106W after the drop of irradiance with less oscillations compared with P&O. similar behavior is seen for the 1000→900W/m²(Fig. 8b) and 700→500W/m² (Fig. 8c) test cases, which indicates that the proposed method is robust.

Table 4 summarizes the quantitative performance metrics extracted from Figs. 6–8 under the same irradiance step.

Table 4. Quantitative MPPT performance

Method	Settling time (s)	Ripple (%)	Efficiency (%)	RMS error (W)	Notes
P&O	≈0.40	≈2.5	≈98.8	≈2.1	Oscillatory
Backstepping	≈0.10	≈0.5	—	—	No update after MPP shift
Hybrid P&O–BS	≈0.15	≈0.7	>99.0	≈0.6	Best trade-off

Table 3. Parameters of Boost.

Parameters	Values	
inductor (L)	3.33	[mH]
Load value (R_L)	11.736	[Ω]
Boost output capacitor (Co)	140	[μF]
Boost input capacitor (Ce)	140	[μF]
Input voltage (V_{IN})	29	[V]
Output voltage (V_{OUT})	48	[V]
Switching frequency (f_s)	10	[kHz]

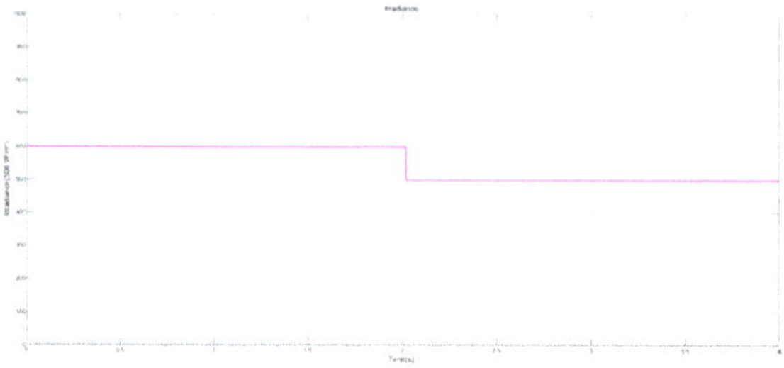

Fig. 5. Irradiance profile: 600 W/m² → 500 W/m² at 25 °C

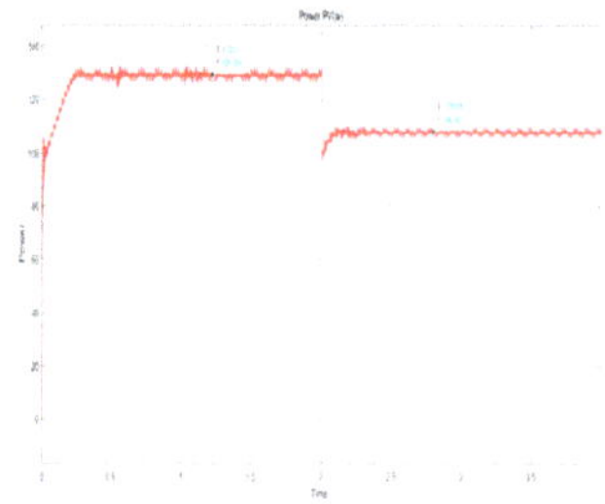

Fig. 6. P&O MPPT under irradiance step (600→500 W/m²)

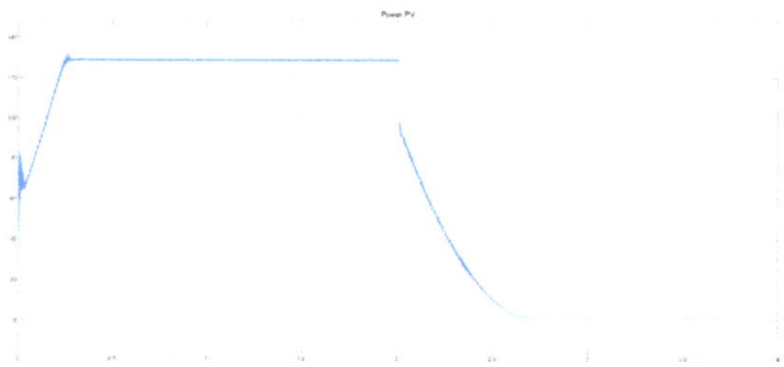

Fig. 7. Backstepping MPPT under irradiance step (600→500 W/m²)

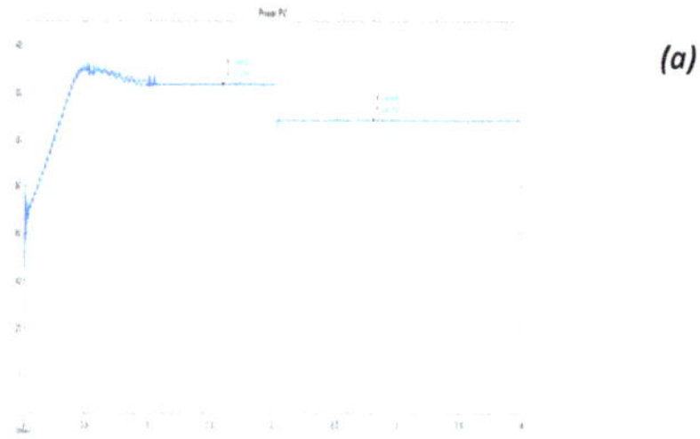

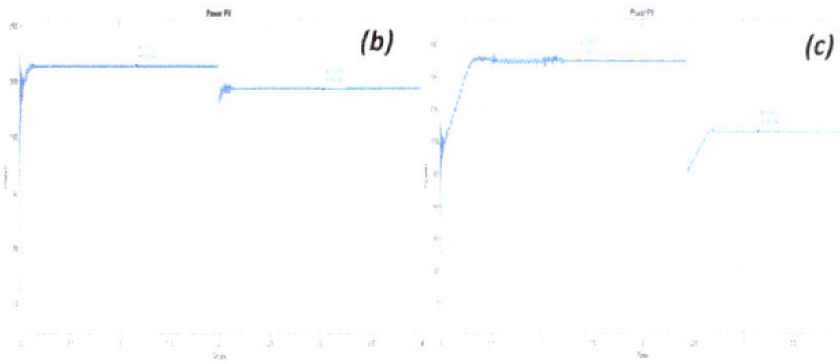

Fig. 8. Hybrid P&O–Backstepping MPPT under irradiance step (a.600→500 W/m², b.1000→900W/m², c.700→500 W/m²)

Conclusion

The intention of this research is to enhance the MPPT techniques in dynamic conditions. P&O and backstepping as state of the art is compared, where it is clear that simplicity comes at costs for time to reach and oscillations. Therefore, a hybrid of P&O–Backstepping is introduced with an adaptive two Proportional Rate (PR) sections to improve its tracking capabilities while keeping high-speed computing efficiency.

Results of simulation demonstrate that the hybrid MPPT achieves superior performance, enabling rapid recovery (~0.15 s), low ripple (<1%), and efficiency above 99%.the tow-rate structure enhances stability and reduces computational cost, confirming the effectiveness of the approach.

Despite promising results, the study is limited to uniform irradiance conditions, with partial shading and robustness to parameter uncertainty, noise, and temperature variations not yet addressed. Future work will focus on extending the approach to partial shading, enhancing robustness through adaptive strategies, evaluating ±20%parammeter uncertainty, and conducting real-time DSP-based experimental validation.

Declaration of generative AI and AI-assisted technologies in the writing process:

During the preparation of this work, the authors used ChatGPT in order to improve the clarity and readability of the text. After using this tool, the authors reviewed and edited the content as needed and took full responsibility for the content of the publication.

References

[1] A. Rahmouni, D. Yousfi, A. El Houre, M. Bakhouya, M. Chaker, and M. Bachiri, "Sustainable modernization of rural agricultural power systems in North Africa via optimal renewable microgrid design," Results Eng., vol. 29, p. 109238, Mar. 2026. https://doi.org/10.1016/j.rineng.2026.109238

[2] S. Dong and Z. Wu, "Prescribed performance adaptive MPPT control of direct-drive wave energy generation system," Ocean Eng., vol. 298, p. 117066, Apr. 2024. https://doi.org/10.1016/j.oceaneng.2024.117066

[3] T. Chen, A. Harrison, N. H. Alombah, M. Aurangzeb, A. A. Telba, and H. A. Mahmoud, "Leveraging MPPT capability for solar irradiance estimation: H-INC-IBS-based assessment of explicit models under real-world climatic conditions," Comput. Electr. Eng., vol. 118, p. 109366, Aug. 2024. https://doi.org/10.1016/j.compeleceng.2024.109366

[4] A. Rahmouni, D. Yousfi, M. Bachiri, M. Bakhouya, and A. Rochd, "Fuzzy Logic-Based Energy Management System for an AC Microgrid," in Digital Technologies and Applications, S. Motahhir and B. Bossoufi, Eds., Cham: Springer Nature Switzerland, 2024, pp. 434–443. https://doi.org/10.1007/978-3-031-68675-7_41

[5] R. Alik and A. Jusoh, "An enhanced P&O checking algorithm MPPT for high tracking efficiency of partially shaded PV module," Sol. Energy, vol. 163, pp. 570–580, Mar. 2018. https://doi.org/10.1016/j.solener.2017.12.050

[6] U. Yadav, A. Gupta, and R. K. Ahuja, "Hardware validation of hybrid MPPT technique via Novel ML controller and P&O method," Energy Rep., vol. 8, pp. 77–84, Dec. 2022. https://doi.org/10.1016/j.egyr.2022.10.067

[7] J. Chen et al., "Irradiance sensorless PSO-based Integral Backstepping and Immersion & invariance algorithm for robust MPPT control with real-climatic microcontroller-in-the-loop experimental validation," Comput. Electr. Eng., vol. 123, p. 110279, Apr. 2025. https://doi.org/10.1016/j.compeleceng.2025.110279

[8] J. Águila-León, C. Vargas-Salgado, D. Díaz-Bello, and C. Montagud-Montalvá, "Optimizing photovoltaic systems: A meta-optimization approach with GWO-Enhanced PSO algorithm for improving MPPT controllers," Renew. Energy, vol. 230, p. 120892, 2024. https://doi.org/https://doi.org/10.1016/j.renene.2024.120892

[9]A. Rahmouni, D. Yousfi, M. Bachiri, M. Bakhouya, and A. Rochd, "Efficient Energy Management in DC Microgrids Using Fuzzy Logic Approach," in Proceedings of the 4th International Conference on Electronic Engineering and Renewable Energy Systems—Volume 1, B. Hajji, A. Gagliano, A. Mellit, A. Rabhi, and M. Calì, Eds., Singapore: Springer Nature, 2025, pp. 289–298. doi: 10.1007/978-981-96-0644-3_26

[10]H. Doubabi, I. Salhi, M. Chennani, and N. Essounbouli, "High Performance MPPT based on TS Fuzzy–integral backstepping control for PV system under rapid varying irradiance—Experimental validation," ISA Trans., vol. 118, pp. 247–259, Dec. 2021. https://doi.org/10.1016/j.isatra.2021.02.004

[11]X. Lin and Y. Wu, "Parameters identification of photovoltaic models using niche-based particle swarm optimization in parallel computing architecture," Energy, vol. 196, p. 117054, Apr. 2020. https://doi.org/10.1016/j.energy.2020.117054

[12]B. Fatna, D. Yousfi and A. Rahmouni, "Optimization of Photovoltaic Systems Under Partial Shading: A Comparative Analysis of PSO and P&O Algorithms," 2025 International Conference on Circuit, Systems and Communication (ICCSC), Fez, Morocco, 2025, pp. 1-7. https://doi.org/10.1109/ICCSC66714.2025.11135125

[13]K. Ali, L. Khan, Q. Khan, S. Ullah, and N. Ali, "Neurofuzzy robust backstepping based MPPT control for photovoltaic system," Turk. J. Electr. Eng. Comput. Sci., vol. 29, no. 1, pp. 421–436, Jan. 2021. https://doi.org/10.3906/elk-1907-15

[14]H. Yatimi, Y. Ouberri, S. Chahid, and E. Aroudam, "Control of an Off-Grid PV System based on the Backstepping MPPT Controller," Procedia Manuf., vol. 46, pp. 715–723, 2020. https://doi.org/10.1016/j.promfg.2020.03.101

Energy Futures - ICEF 2026 Materials Research Forum LLC
Materials Research Proceedings 64 (2026) 123-134 https://doi.org/10.21741/9781644904091-16

Offshore Wind-Driven Hydrogen Production in Morocco: Evaluating Reverse Osmosis-Proton Exchange Membrane Electrolysis Against Direct Seawater Electrolysis

Mohamed Reda EL AOUNI[1,a*], Salma SEDKI[1,b] and Omar BOUZIDI[1,c]

[1]School of Science and Engineering, Al Akhawayn University, 53000 Ifrane, Morocco

[a]m.elaouni@aui.ma, [b]s.sedki@aui.ma, [c]o.bouzidi @aui.ma

Keywords: Green Hydrogen, Direct Seawater Electrolysis, SWE, Proton Exchange Membrane, Techno-Economic Analysis, Offshore Wind, LCOE, Levelized Cost of Hydrogen, LCOH, Economies of Scale, Morocco, Dakhla

Abstract. Green hydrogen production is a key element for Morocco's energy transition but is limited by high electricity costs and freshwater scarcity. This paper evaluates the techno-economic feasibility of offshore wind-powered hydrogen production in Dakhla, Morocco. A site-specific wind assessment using hourly data is performed to estimate annual energy production and the levelized cost of electricity, which is then coupled with a conventional reverse osmosis-proton exchange membrane (RO-PEM) system to calculate hydrogen production and its levelized cost of hydrogen. A constrained sensitivity analysis is also conducted for direct seawater electrolysis (SWE). Results show that a single offshore wind turbine produces 3.385 GWh annually with an LCOE of 0.0285 USD/kWh, leading to an LCOH of 4.36 USD/kg for RO-PEM. As for the direct SWE, it achieves higher hydrogen yields and reduces the LCOH by 22.7%, highlighting its potential as a water efficient option for green hydrogen production in Morocco.

Introduction

The power sector of Morocco has traditionally relied heavily on imported fossil fuels, which accounted for 68% of its electricity generation in 2020, although the domestic reserves are limited. Given this fact, Morocco recognized its significant potential for renewable energy and took the steps necessary to begin harnessing it through the construction of large-scale plants generating power with concentrated solar systems. Morocco had an installed capacity of 15 GW in 2022, of which 37% was coal, 16% was hydropower, and 18% was natural gas. The Moroccan government has set clear targets to increase the share of renewable energy to 52% of installed capacity by 2030 and to reduce greenhouse gas emissions by 45.5% compared to baseline levels, with international support [1].

Green hydrogen is becoming a key component for Morocco's energy transition and could play a significant role in the decarbonization of hard-to-abate sectors like industry and transport. Since green hydrogen can be obtained through the process of water electrolysis, powered by renewable electricity, its development is part of the core of the national Green Hydrogen Roadmap, which is built on ambitious export targets. This is made possible by Morocco's high renewables potential combined with proximity to the European markets. In addition, to develop a domestic market, the government has just adopted some fiscal incentives, including tax exemptions for alternative fuel vehicles, which may also benefit future hydrogen-based mobility uptakes [2]. Techno-economic analyses of renewable-based systems indicate that the use of variable renewable electricity sources, such as photovoltaics (PV) and other renewables, for electrolysis lowers LCOH values down to 3-4 USD/kg in favorable configurations compared to conventional reference cases [3]. This idea of using renewable resources with high capacity factors, as already discussed for solar integration, can also be applied to other renewable technologies to improve system stability and increase energy production in Morocco. However, since water is a basic input for electrolysis, its availability becomes a major limiting factor across the entire process, especially in countries facing increasing water scarcity.

Morocco is suffering from a severe problem of water shortage, where the per capita of renewable freshwater resources have declined from more than 2000 cubic meters in the 1960s to below 600 cubic meters today, below the water stress threshold. This structural stress is intensified by climate change through decreased precipitation and a higher frequency of extreme droughts. The agricultural sector is more vulnerable to climatic hazards, as it accounts for approximately 87% of the nation's water, while it is estimated to contribute around 10.1% to the GDP [4] [5] [6]. Additionally, other socio-economic drivers such as demographic growth, rapid urbanization, and the extension of irrigated agriculture have resulted in over exploiting groundwater resources, resulting in further decrease in available supply.

Desalination has therefore been used in Morocco as a cost-effective alternative in some regions along the coasts to ease water scarcity without increasing burdens on already stressed resources. Seawater reverse osmosis (SWRO) plants, normally driven by renewable energy sources, like photovoltaics, provide reliable sources of fresh water for irrigation, drinking, and industry. Under optimized conditions, operational costs are competitive, at about 1 USD per cubic meter, considering the water supply cost, especially in the case of brackish water desalination. In Morocco, the national strategies comprising the National Water Plan include various significant investments in the desalination facilities, facing up to 40 cubic kilometers of water shortage forecasted within the mid-century. Solar integrated desalination systems bring advantages in the MENA region, including Morocco, through reducing dependence on conventional sources of energy, lowering greenhouse gas emissions, and contributing to food security by means of sustainable irrigated agriculture. However, the seawater desalination still faces several challenges, mainly related to brine disposal, high energy consumption and the high initial investment costs. Despite these limitations, several studies have shown that large-scale photovoltaic powered SWRO plants can be technically feasible and economically viable [7] [8] [9].

In this study, combining green hydrogen production with renewable energy sources particularly offshore wind represents is a practical solution that can contribute to both energy supply and the water security in Morocco. Wind farms located offshore can harness reliable coastal winds to provide stable electricity for electrolysis with less transmission loss and integrated into marine environments along Morocco's 3500-kilometer coastline. Techno-economic assessments of such integrated systems have indicated that hydrogen transport to the mainland using submarine pipelines is the most cost-effective delivery method, even for distances as low as approximately 50 km. The baseline levelized cost of electricity (LCOE) of 0.05 USD/kWh could result in a low LCOH of 3.6 USD/kg regarding direct delivery to an industrial user and 6.1-6.2 USD/kg for routes leading to refueling stations. Sizing will have an important impact on economic viability: studies show that configuring the electrolyzer capacity for 80% of the wind farm capacity leads to the best energy balance, using more than 87% of the wind production for hydrogen production. In addition, while battery storage can be integrated, its eventual role usually remains marginal, delivering less than 5% of annual energy flow. Thereafter, this makes sense for co-located hydrogen plants: excess renewable power can be stored as hydrogen, which allows remedying intermittency while fostering export ambitions at increasingly competitive LCOH figures through scale and optimized system design [2] [10] [11]. Notwithstanding progress to date, a pressing research gap remains in conducting a site specific, comparative assessment of hydrogen production pathways that directly confronts Morocco's water constraint.

This study develops a comparative techno-economic framework to investigate two routes for offshore wind-based hydrogen production, with a specific focus on the water-stressed context of Morocco. With particular attention to the high-wind-potential site of Dakhla, the analysis first determines the LCOE from the offshore wind farm as the foundational input. This LCOE is then used to assess two hydrogen production systems. The first system consists of reverse osmosis desalination coupled with proton exchange membrane electrolysis (RO-PEM) to estimate hydrogen production and the corresponding LCOH. The second model is direct seawater electrolysis (SWE), on which a sensitivity analysis is performed by varying capital costs, operating costs, and electricity consumption using values from literature. The resulting LCOH values are then compared with those obtained for the RO-PEM system. Also, the study examines how economies of scale affect the LCOE and in turn the overall LCOH. The results are discussed in the context of Morocco's local conditions to highlight the differences between a mature technology and an emerging alternative.

Methodology

In this study the techno-economic analysis to evaluate the cost competitiveness of a direct SWE system against the RO-PEM, with both systems coupled with an offshore wind farm, is divided into four connected stages: (1) a site-specific wind energy yield assessment to determine the available wind power; (2) calculation of the levelized cost of electricity from the wind farm; (3) computation of the levelized cost of hydrogen for the RO-PEM production system, taking into account the computed LCOE (4) a constrained sensitivity analysis to identify break-even conditions for the SWE system compared with the RO-PEM LCOH results ; and (5) introducing the economies of scale on the wind farm and re-computation of the whole cost assessment. This framework, illustrated using empirical data and Python based analysis, provides insight into the potential of SWE as a sustainable alternative for hydrogen production. The analytical formulations used in this study for wind energy assessment, levelized cost of electricity, and levelized cost of hydrogen calculations are based on standard wind energy and techno-economic modeling approaches [2] [10] [11] [18].

1. Wind Energy Assessment

In this wind energy assessment part, hourly data including wind speed, air density, surface pressure, and temperature were obtained from PVGIS. Because the available measurements are provided at different heights; for instance, the wind speed data was obtained at 10 m, so the data are extrapolated to the turbine hub height using the power-law wind profile, while temperature and air density are adjusted using the adiabatic lapse rate and the ideal gas law. The annual energy production (AEP) is calculated using the bins method in combination with the turbine power curve. This results in a gross AEP, which is subsequently reduced by an assumed total loss between the inverters and the cables to obtain the net AEP available for hydrogen production.

$$V_2 = V_1 \times \left(\frac{h_2}{h_1}\right)^{0.16} \tag{1}$$

Where 0.16 is the surface roughness and h_1 is the height from which the data was measured and h_2 is the hub height of the wind turbine.

In parallel with the wind speed extrapolation, temperature and air density data were also adjusted with the following equations.

$$T_2 = T_1 - 0.0066 \times h_2 \tag{2}$$

$$\rho_2 = \frac{SP}{287 \times T_2} \tag{3}$$

With these atmospheric parameters defined, the analysis then focused on the aerodynamics of the wind turbine. The wind power density equation quantifies the available energy in the wind stream,

$$P_d = \frac{1}{2} \times \rho \times v^3 \quad (4)$$

Where P_d is the wind power density (W/m²), air density ρ and wind velocity v.
Then to compute the power output P (W) , the power density is multiplied by the swept area A of the wind turbine.

$$P = P_d \times A \quad (5)$$

After summing up the hourly power output throughout the year we get the annual energy production.

$$AEP = \sum_{h=1}^{8760} P_h \quad (6)$$

Computations were performed for two wind turbine models: Vestas 44/600 (hub height: 45 m) and Enercon 40/500 (hub height: 43.5 m). Following AEP calculations, the one with the highest energy produced was chosen to proceed with the coupling of the desalination center and electrolysis, the VESTAS model had the highest yield, as expected given its greater hub height rotor diameter and rated power.

2. Levelized Cost of Electricity calculation

The LCOE calculation assesses the economic viability of the offshore wind farm by determining the average cost per unit of electricity generated over the project's lifetime. This is achieved through a standard discounted cash flow model, where the capital expenditure (CAPEX) for the wind turbine and the annual operational expenditure (OPEX), typically expressed as a percentage of CAPEX, are annualized using a Capital Recovery Factor (CRF). The CRF is derived as a function of the project's economic lifetime and the weighted average cost of capital i . The LCOE is then computed as the ratio of the total annualized costs to the gross AEP, yielding a fixed electricity cost in USD/kWh that serves as the basis for evaluating subsequent hydrogen production expenses.

$$CRF = i * \frac{(i+1)^n}{(i+1)^n - 1} \quad (7)$$

$$LCOE = \frac{CRF * CAPEX_w + OPEX_w}{AEP} \quad (8)$$

3. Levelized Cost of Hydrogen calculation

The Levelized Cost of Hydrogen calculation forms the core of the economic evaluation, determining the average cost per kilogram of hydrogen produced for both the conventional RO-PEM system and the SWE system. For each pathway, the LCOH is computed as the sum of the annualized costs of the electrolysis system per kilogram of hydrogen and the electricity costs required to produce one kilogram, incorporating LCOE from the wind farm. The annual hydrogen production is derived from the net AEP of the wind farm, adjusted for the system's specific energy consumption Eel in kWh/kg H_2 and overall efficiency factor Peff, which accounts for ancillary losses such as desalination in the RO-PEM case. Baseline parameters for the RO-PEM system, including a typical Eel and P_{eff} drawn from established literature values, serve as the reference for comparison, enabling a standardized assessment of cost drivers across the two production routes.

$$E_{in} = AEP * \eta_w * P_{eff} \quad (9)$$

$$H_2 = \frac{E_{in}}{E_{el}} \quad (10)$$

Materials Research Proceedings 64 (2026) 123-134 https://doi.org/10.21741/9781644904091-16

$$LCOH = \frac{CRF*CAPEX_{el}+OPEX_{el}}{Annual\ H_2} + \frac{LCOE*E_{el}}{P_{eff}} \tag{11}$$

Where the E_{in} represents the energy received by the electrolyzer from the wind fam multiplied by the efficiency of the wind farm and the electrolyzer, and E_{el} represents the electricity consumption of the electrolyzer, after calculating the hydrogen produced H_2, proceeding to the LCOH it is a composite equation of both the cost of the center of the desalination with RO and the PEM electrolyzer or the direct SWE and the cost of electricity that powers it.

Table 1: Estimated Costs of the Wind Turbine and the Reverse Osmosis

	Wind turbine VESTA 44/600	**Reverse Osmosis and PEM**
CAPEX (USD)	870,000	1,000,000
OPEX (USD)	4% of CAPEX	4% of CAPEX
Life Span	25	20
Interest rate (%)	5	8
Efficiency (%)	92	85
E_{el} (kWh/kg)	N/A	50

Tables 1 presents the estimated values for the continuation of LCOE and LCOH calculations for RO and PEM systems, these values are based on average literature values, representing neither optimistic nor conservative assumptions.

4. Direct SWE sensitivity analysis

Direct analysis of SWE was limited by constraints in terms of CAPEX, which varied between those of the RO-PEM system and 20% less, and for the OPEX varied from 2% to 4% of the CAPEX. The second constraint was the electrolyzer's power consumption, which, according to the literature review, ranged between 39.4 kWh/kg and 41.5 kWh/kg. In addition, the efficiency ranged between 85% and 90%. And this sensitivity analysis is performed for both the baseline configuration and the scaled-up wind farm scenario, for consistent evaluation across system sizes.

5. Integration of Wind Farm Scaling Effects

To assess how system size influences economic performance, an additional scaling analysis was incorporated. Wind farm expansion introduces potential economies of scale because several cost components such as the infrastructure, grid connection, control systems, and maintenance logistics do not increase linearly with turbine count. To assess these effects, a reduction of 10% was assumed in both CAPEX and OPEX when scaling the wind farm to ten wind turbines.

Results and Discussion

This section presents the performance assessment of the wind-powered green hydrogen production system in Dakhla, Morocco. The results include calculations of the local wind resource characteristics, annual energy production based on previous years' data obtained from PVGIS and the realized probability distribution function for the site, the economic indicators for hydrogen production using the two different water splitting technologies discussed in the previous sections: conventional RO-PEM electrolysis and direct SWE. The results obtained with SWE are contextualized using bibliographic references to assess both the technical feasibility and cost competitiveness of hydrogen production at the site studied. This discussion shows how wind behavior affects AEP, conversion efficiency, and thus LCOE and LCOH, while revealing the potential of SWE as a future alternative.

1. Wind resources assessment

The wind resource assessment indicated strong potential for energy generation. The average wind speed at 45 m is 8.32 m/s corresponding to the VESTAS model, and the average at 43.5 m is 8.27 m/s corresponding to the ENERCON model. This makes only a slight difference but at the end it contributes to the AEP.

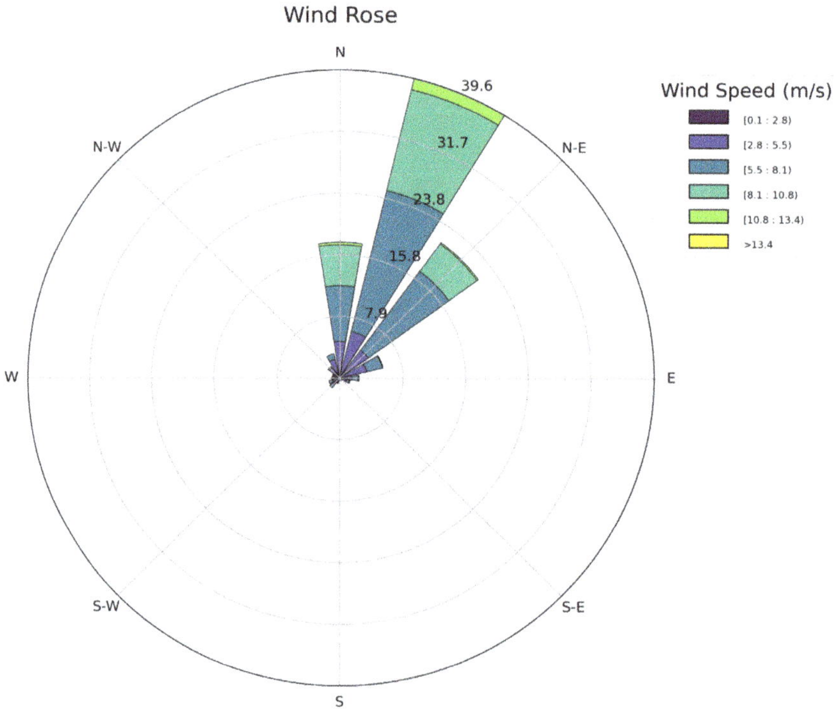

Figure 1: Wind rose at 10 meters

The analysis of the wind direction at Dakhla was conducted using the wind rose diagram. Fig. 1 shows the dominant wind direction associated with the wind speed in this site and it can be observed that the prevailing wind speed is coming from in between the north and northeast, indicating a potential for continuous wind energy generation.

Also, the consistency of the wind flow indicates limited turbulence, which is favorable for both turbine lifespan and efficiency in energy production. This stable and predictable wind behavior supports the site's suitability for continuous energy production and eventually reliable hydrogen production throughout the year.

By combining the statistical behavior of the wind with the aerodynamic characteristics and efficiencies of each turbine, the AEP was calculated for both models, offering a clear basis for comparing their suitability for wind-powered hydrogen production at the Dakhla site.

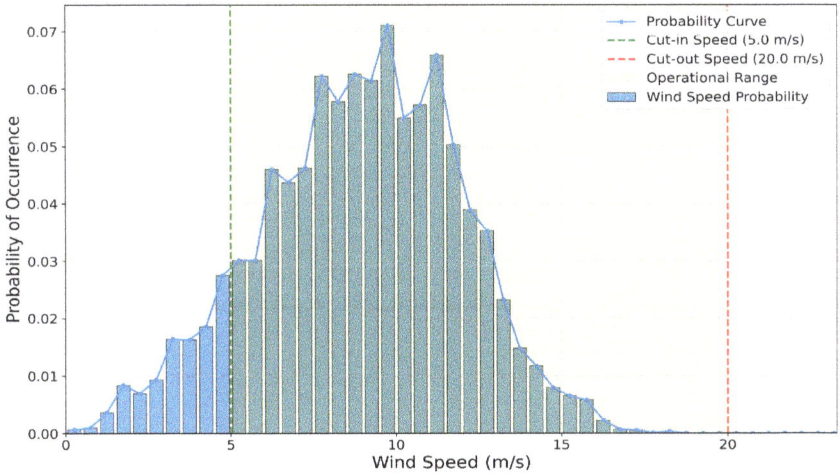

Figure 2: Represents the wind speed probability distribution at 45 meters for the VESTAS model

Fig. 2 shows that the probability of generating electricity form the VESTAS model is 88.79% leading to an expected operating hours of 7804 hours/year.

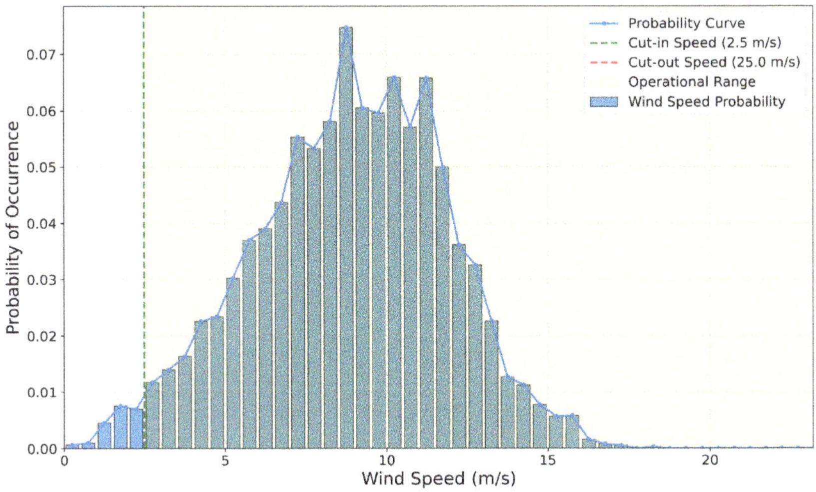

Figure 3: represents the wind speed probability distribution at 43.4 meters for the ENERCON model

Fig. 3 shows that the ENERCON model has a bigger range between the cut in and the cut out wind speed. The probability of generating the electricity with ENERCON is 97.6% which converststo an expected operating hours of 8578 hours/year which is significant.

But the probability of generating electricity and the AEP are totally different, for the VESTAS model the AEP was 3.385 GWh and the ENERCON model with an AEP of 2.747 GWh. Concluding that the VESTAS model is better and it would be more suitable to couple with the center of desalination and hydrogen production for a higher yield.

2. Hydrogen production calculations

The hydrogen production potential of the system was evaluated by the AEP and the energy requirement of the electrolyser, since the VESTAS model demonstrated higher AEP in the previous analysis it was adopted in the hydrogen production assessment.

To quantify the annual hydrogen yield for the RO-PEM system, the AEP was first converted into the effective electrical input available to the electrolyzer, accounting for the combined efficiency of the wind turbine and the electrolysis system accounting for 8% losses between the inverters and the cables of transmission and for the PEM an efficiency of 85% this led to an annual hydrogen production of 52947.88 kg/year.

For the SWE by ranging the efficiency of the electrolyzer and between 85% and 90% and ranging its electricity consumption from 39.4 kWh/kg to 41.5 kWh/kg following recent research in the SWE technology [16][17], the minimum yield from the direct SWE was more than the RO-PEM method generating 63928 kg/year and a maximum yield of 71073 kg/year.

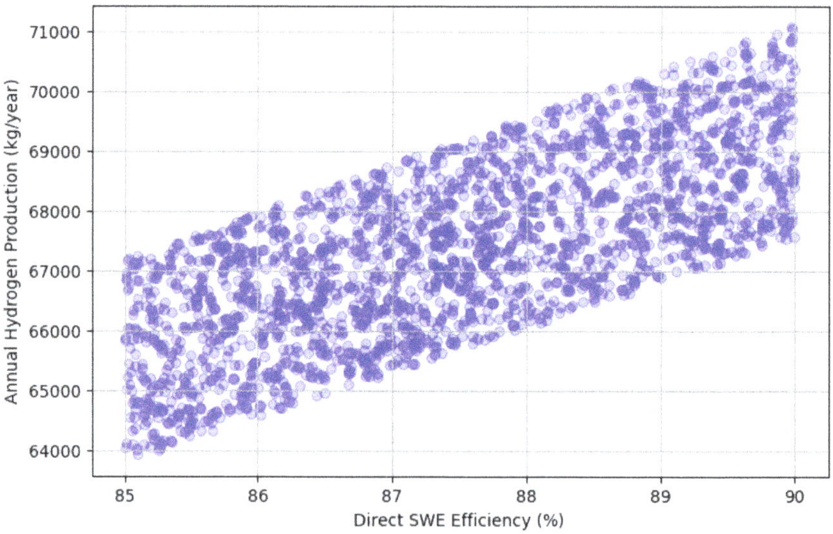

Figure 4: represents the hydrogen yield sensitivity to SWE efficinecy

3. Levelized cost of electricity calculations

The estimations made in terms of parameters of the LCOE including the CAPEX and OPEX of the wind turbine in this study is inherently sensistive but the computed LCOE was 0.0285

USD/kWh which comes within range in terms of Morocco's context, published studies on Moroccan wind farms report LCOEs in the range of approximately 0.017-0.074 USD/kWh [18].

This low LCOE strengthens the case for using the wind farm to power hydrogen production, since the electricity cost is a key driver of hydrogen cost, achieving a competitive LCOE means the subsequent hydrogen via either RO-PEM or SWE can be more feasibly economical.

4. Levelized cost of hydrogen calculations

The LCOH of the RO-PEM system coupled with the wind offshore was 4.36 USD/kg, this result is well above the long-term policy targets and the lowest cost projections. But it is important to note that the LCOH was computed using many assumptions in terms of CAPEX and OPEX and same for the LCOE. This result indicates that Dakhla is plausible but not yet in the lowest cost bracket of the cost of hydrogen production.

For the SWE the estimation was made under the assumptions mentioned in the previous section and a ranging variable of CAPEX less than the RO-PEM system by 20% and the OPEX between 2% to 4% of the CAPEX.

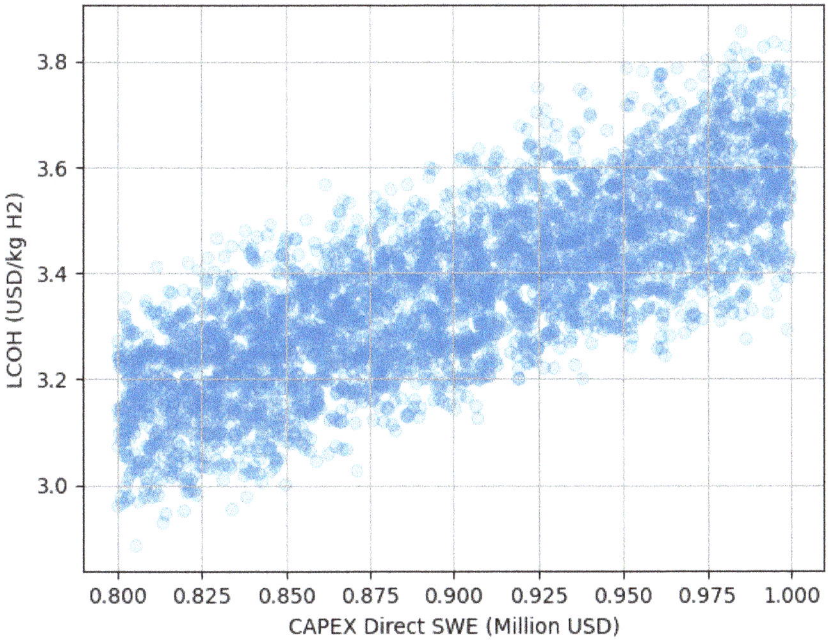

Figure 5: represents the LCOH sensitivity for SWE system

Fig. 5 shows the direct SWE LCOH scatter with CAPEX and OPEX uncertainties and it shows that the LCOH using the SWE system is potentially cheaper than the RO-PEM by 22.7% which is promising. The average LCOH of the SWE is 3.37 USD/kg, with a minimum of 2.89 USD/kg and a 3.86 USD/kg. Even at the worst case scenario the LCOH of the SWE is much cheaper than the RO-PEM.

5. Impact of Wind Farm Scaling on LCOH and LCOE

Scaling up the wind farm to ten wind turbines or more is expected to have an impact on the LCOE and the LCOH as the CAPEX and the OPEX does not necessarily increase linearly. The sharing of infrastructure among multiple turbines introduces economies of scale, resulting in cost reductions. To quantify this impact, an assumption of a 10% reduction in both the CAPEX and OPEX was applied which led to a noticeable decrease in both economic indicators , with the LCOE reduced to 0.0254 USD/kWh and the LCOH of the RO-PEM system decreased to 4.05 USD/kg.

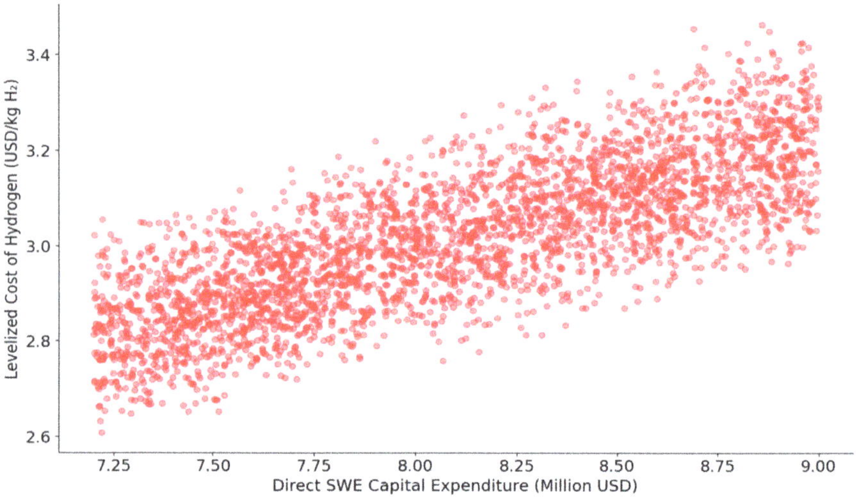

Figure 6: represents the LCOH sensitivity for SWE system with a scaling up of 10 wind turbines

Fig. 6 shows the direct SWE LCOH after the economies of scale adjustments consequently the LCOH decreased to an average of 3.22 USD/kg, with a minimum of 2.78 USD/kg and a maximum of 3.71 USD/kg.

The results discussed in this section allow us to draw several key conclusions. First, Dakhla's wind resources are technically sufficient to enable large-scale green hydrogen production, with the selected VESTAS turbine model achieving an AEP of 3.385 GWh. Second, and more importantly, the economic viability of the project depends heavily on the choice of electrolysis technology. The conventional RO-PEM route results in an uncompetitive LCOH, while the direct SWE alternative offers greater potential, with an expected cost reduction of 22.7%. Furthermore, scaling the wind farm to ten turbines or more introduces economies of scale further reducing the LCOE and the LCOH.

Additionally, all the results obtained in this study are consistent with recent techno-economic assessments of offshore wind farms coupled with hydrogen production in Morocco. Previous studies show that the LCOH values range between 3.6 and 6.2 USD/kg for offshore wind-powered electrolysis, depending on the system's configuration and scale [2] [10] [11]. The RO-PEM results obtained here fall within that range, while the direct SWE system demonstrates a lower LCOH compared to conventional desalination-based configurations. Also, compared with earlier studies that emphasize photovoltaic-powered or onshore wind systems, the present work highlights the added value of offshore wind in achieving higher capacity factors and improved hydrogen yields.

Also, this study contributes novel insights by addressing the water scarcity through direct seawater electrolysis, an aspect often neglected in existing Moroccan hydrogen project's assessments.

Summary

This study presented a techno-economic assessment of offshore wind-powered hydrogen production in Dakhla, Morocco, comparing conventional RO-PEM electrolysis with the direct SWE as a potential alternative.

- The analysis revealed that the wind resource in Dakhla is strong and stable, and that the selected Vestas 44/600 turbine generated 3.385 GWh per year for an LCOE of 0.0285 USD/kWh.
- When that AEP was used for hydrogen production, the RO-PEM system produced 52.94 t/year of hydrogen with an LCOH of 4.36 USD/kg.
- After conducting the sensitivity analysis on the SWE system, the results revealed increased hydrogen production and decreased costs, with an average LCOH value of 3.37 USD/kg, indicating a reduction of approximately 22.7% compared to the RO-PEM system.
- Increasing the number of wind turbines led to lower electricity and hydrogen costs due to economies of scale.

Overall, these findings indicate that offshore wind energy in Dakhla has the potential to be used for large scale green hydrogen production. Although the RO-PEM system is more mature and already in use, the direct SWE technology could become a viable option in the future if challenges related to durability and long-term operation are addressed. Another important aspect that can be explored as a follow-up is a more detailed system modeling, hydrogen transport options, and the use of combined renewable sources to create a more reliable and productive project for hydrogen production.

References

[1] A. Caillard et al., "A Critical Analysis of Morocco's Green Hydrogen Roadmap: A Modelling Approach to Assess Country Readiness from the Energy Trilemma Perspective," Climate, vol. 12, no. 5, p. 61, 2024. https://www.mdpi.com/2225-1154/12/5/61

[2] K. Taroual et al., "Marine Renewable-Driven Green Hydrogen Production toward a Sustainable Solution and a Low-Carbon Future in Morocco," J. Mar. Sci. Eng., vol. 12, no. 5, p. 774, 2024. https://www.mdpi.com/2077-1312/12/5/774

[3] Leonhard, "Analysis of the Levelized Cost of Renewable Hydrogen in Austria," *Sustainability*, vol. 15, no. 5, pp. 1–23, 2023, Accessed: Nov. 29, 2025. [Online]. Available: https://ideas.repec.org/a/gam/jsusta/v15y2023i5p4575-d1087274.html

[4] M. El Haddadi et al., "The Nexus Between Economic Growth and Water Stress in Morocco: Empirical Evidence Based on ARDL Model," Sustainability, vol. 17, no. 15, p. 6990, 2025. https://www.mdpi.com/2071-1050/17/15/6990

[5] J. van den Brandeler, "Governance and Sustainability Challenges in the Water Policy of Morocco 1995–2020: Insights from the Middle Draa Valley," Water, vol. 14, no. 18, p. 2932, 2022. https://www.mdpi.com/2073-4441/14/18/2932

[6] Y. Mohsine et al., "Vulnerability of Water Resources to Drought Risk in Southeastern Morocco: Case Study of Ziz Basin," Water, vol. 15, no. 23, p. 4085, 2023. https://www.mdpi.com/2073-4441/15/23/4085

[7] S. S. A. Shah et al., "Innovative Matrix-Based Assessment of Non-Conventional Water Processes: A Strategic Approach for Sustainable Water Management in Arid Environments," Water, vol. 17, no. 6, p. 866, 2025. https://www.mdpi.com/2073-4441/17/6/866

[8] M. Al-Addous et al., "Innovations in Solar-Powered Desalination: A Comprehensive Review of Sustainable Solutions for Water Scarcity in the Middle East and North Africa (MENA) Region," Water, vol. 16, no. 13, p. 1877, 2024. https://www.mdpi.com/2073-4441/16/13/1877

[9] H. Al-Zoubi et al., "Sustainable Food Security: Balancing Desalination, Climate Change, and Population Growth in Five Arab Countries Using ARDL and VECM," Sustainability, vol. 16, no. 6, p. 2302, 2024. https://www.mdpi.com/2071-1050/16/6/2302

[10] J. L. Parrilla-Gutiérrez et al., "Energy Storage and Management of Offshore Wind-Based Green Hydrogen Production," Processes, vol. 13, no. 3, p. 643, 2025. https://www.mdpi.com/2227-9717/13/3/643

[11] Q. Li et al., "Techno-Economic Assessment of a Full-Chain Hydrogen Production by Offshore Wind Power," Energies, vol. 17, no. 11, p. 2447, 2024. https://www.mdpi.com/1996-1073/17/11/2447

[12] X. Li et al., "A Review of Hydrogen Production via Seawater Electrolysis: Current Status and Challenges," Catalysts, vol. 14, no. 10, p. 691, 2024. https://www.mdpi.com/2073-4344/14/10/691

[13] H. Li et al., "Direct 2400 h Seawater Electrolysis Catalyzed by Pt-Loaded Nanoarray Sheets," Catalysts, vol. 15, no. 7, p. 634, 2025. https://www.mdpi.com/2073-4344/15/7/634

[14] J. Tong et al., "Direct Electrolytic Splitting of Seawater: Significance and Challenges," Coatings, vol. 12, no. 8, p. 1179, 2022. https://www.mdpi.com/2079-6412/12/8/1179

[15] "Morocco EV Import Rules 2025: Taxes, Compliance, and Incentives," EV24 Africa, 2025. https://www.ev24.africa/morocco-ev-import-rules-2025-taxes-compliance-and-incentives/

[16] F. ezzahra CHAKIK, M. Mikou, and M. Kaddami, "Study and Optimization of Hydrogen Production by Electrolysis of Water Using News Cathodes Based on Transition-Metal Oxides," Jan. 2023, doi: https://doi.org/10.2139/ssrn.4554355

[17] T. Leo, "How does an electrolyzer work?," FuellCell Energy, Feb. 29, 2024. https://www.fuelcellenergy.com/blog/how-does-an-electrolyzer-work

[18] Z. Bouhamed and E. Ezzahid, "Assessing the competitiveness of renewable energy in Morocco: A comprehensive analysis of power plant electricity production costs," Energy Strategy Reviews, vol. 61, p. 101863, Sep. 2025, doi: https://doi.org/10.1016/j.esr.2025.101863

Energy Futures - ICEF 2026 Materials Research Forum LLC
Materials Research Proceedings 64 (2026) 135-147 https://doi.org/10.21741/9781644904091-17

Dual Water–Energy Potential of Floating Photovoltaics in Morocco: Multi-Criteria Reservoir Selection, System-Performance Modelling, and Evaporation-Saving Assessment

Salma SEDKI[1,a]*, Mohamed Reda EL AOUNI[1,b], Houssame LIMAMI[1,c]
[1]School of Science and Engineering, Al Akhawayn University, 53000 Ifrane, Morocco
[a]s.sedki@aui.ma, [b]m.elaouni@aui.ma, [c]h.limami@aui.ma

Keywords: FPV, Water Scarcity, MCDA, Water Savings, Evaporation, GIS, LCOE

Abstract. This study develops a multi-criteria decision analysis (MCDA) to identify the optimal Moroccan reservoir for Floating Photovoltaic (FPV) applications and assess related savings on energy and water resources. Ten reservoirs were analyzed using Quantum Geographic Information System (QGIS) software with criteria related to hydrological stability, energy production using Morris Sensitivity analysis, and cost/accessibility using Analytic Hierarchy Process (AHP). The highest-ranking site, Oued El Makhazine, was modeled in System Advisor Model (SAM) under a conservative 5% coverage scenario. Results show an annual electricity production level of 396.62 GWh with a capacity factor of 17.3% and levelized Cost of Energy (LCOE) value of 0.0573 \$/kWh, which can cover the power demand of 396,620 people on an annual basis. Evaporation loss was estimated using the simplified Penman method in MATLAB with expected annual savings of 1.856 million m^3, meeting the domestic water needs of 41,336 residents. These findings demonstrate the strong co-benefits of FPV for clean energy generation and freshwater conservation in water-stressed regions.

Introduction

Climate change has become more challenging and is putting an extra pressure on energy and water security particularly in semi-arid regions, as water scarcity could lead to a decrease in economic growth by as much as 6% of GDP by 2050 [1]. The impact of climate change is evident in Morocco, where drought has caused a substantial decrease in water storage. During the period between October 2022 and September 2023, the country has lost more than 600 Olympic-sized swimming pools per day due to evaporation, which is equivalent to one million cubic meters of water [2]. Meanwhile, Morocco expects to expand considerably the renewable capacity to relieve the stressed grid, satisfy the increasing electricity demand, and take advantage of one of the highest solar resources in the world [3].

Floating Photovoltaic (FPV) systems on reservoirs offer a promising response to this dual challenge by generating low-carbon electricity while reducing evaporation and avoiding land-use conflicts. A recent global assessment estimated that covering 30% of the surface of 114,555 reservoirs with FPV could technically produce about 9,434 TWh per year and save roughly 106 km^3 of water annually. Even minimal coverage is impactful: an Africa-focused scenario showed that covering less than 1% of reservoir area could increase hydropower output by around 58% while significantly reducing evaporation losses [4]. These findings suggest that carefully designed FPV projects can contribute simultaneously to decarbonization and water-resource resilience.

Besides these advantages, there is a justified concern regarding possible impacts on water quality and aquatic ecosystems, especially if FPV is applied to drinking water or multi-purpose reservoirs. Recent high-resolution modelling of a proposed FPV system on the oligotrophic Kremasta Lake in Greece showed that changes in key water quality parameters remained small and spatially confined. Variations in temperature, dissolved oxygen, chlorophyll-a and nutrient concentrations were limited, localized near the FPV area, and largely confined to the upper water layers, with no significant change in annual averages [5]. On the other hand, a shallow drinking-water reservoir with a sun-tracking FPV system covering about 24% of the surface had been studied in the field, reporting the emergence of benthic cyanobacteria, thus demonstrating that high coverage ratios on sensitive water bodies could induce undesirable ecological responses [6]. These contrasting results emphasize the need for conservative coverage limits and site-specific environmental assessment.

Reflecting these concerns, explicit regulatory caps on FPV coverage have been issued by several countries. Presently, the German Renewable Energy Sources Act and the German Water Resources Act limit FPV installations on artificial lakes to a maximum of 15% of the water surface, with a minimum 40 m buffer to the shoreline [7]. Recently, Spain has introduced a differentiated framework for public reservoirs: in non-eutrophic reservoirs, the maximum FPV coverage is limited to 5% of the surface, while eutrophic reservoirs with higher biological productivity may allow up to 15%, with further constraints linked to the operating regime and competing uses [8].

It corresponds to the outcomes of the Hydropower & Dams FPV technical report, which states that the coverage of up to approximately 5% of a reservoir surface leads to a negligible environmental impact including minimal effects on biodiversity, algae development, and overall water quality [9]. Concerning the above-mentioned environmental limitations and taking into consideration the fact that many Moroccan dams are used for drinking-water and irrigation purposes, this study adopts a conservative scenario of covering 5% with FPV.

Despite promising solar conditions, no published work has carried out a comparative evaluation of Moroccan reservoirs in determining the most promising dam for FPVs based on a multi-criteria decision analysis (MCDA) assessing real FPV suitability that can combine key factors such as energy-production sensitivity, installation feasibility, system stability, and access to infrastructure to rank dams across Morocco. This lack of reservoir-to-reservoir analysis provides a serious gap, especially within a context where water scarcity and variability in terrain make site choice critical for successful FPV deployment.

This study develops a multi-criteria evaluation to determine the optimal FPV siting among ten major Moroccan dams using Quantum Geographic Information System (QGIS). The evaluation model considers, first, the hydrological feasibility of the site because reservoirs with low filling rates or high seasonal water-level variation are not stable enough to host an FPV installation. Energy-related criteria including global irradiance, temperature, wind speed, and usable water surface are added up, as well as the cost and accessibility factors like distance to High-voltage (HV) transmission lines, road access, port proximity, and slope of the surrounding terrain. Relative weights for energy variables are derived using the Morris sensitivity method applied to a simplified FPV performance model, while logistical factors are weighted through the Analytic Hierarchy Process (AHP) in MATLAB. By assigning 70% importance to energy production and 30% to cost and accessibility, a combined suitability value is computed. The combined technical-logistical suitability is then scaled by the hydrological feasibility factor, ensuring that reservoirs with unstable water behavior are inherently downgraded. The highest-ranking site is subsequently modelled in System Advisor Model (SAM) to estimate annual electricity production considering a 5% coverage constraint. Although SAM does not provide an explicit FPV model, its ground-mounted PV configuration is commonly used in FPV studies [10]. SAM was also employed for financial analysis to assess the Levelized Cost of Energy (LCOE) for the FPV project. Lastly, the water evaporation savings were quantified using MATLAB based on the surface area covered.

Methodology

This section combines spatial MCDA, weighting techniques, SAM-based energy estimation, and evaporation-saving assessment techniques to select the best-suited Moroccan reservoir for FPV deployment. It describes the preparation and normalization of hydrological and production and

Materials Research Proceedings 64 (2026) 135-147 https://doi.org/10.21741/9781644904091-17

cost/accessibility-related datasets and their combination within a QGIS-based MCDA framework, followed by an assessment of energy generation and water-saving at the selected site.

1. MCDA for FPV Site Ranking Using QGIS

a. Data Collection and Normalization

Table 1: The Actual Raw Data

Dam	Surface (km^2)	Irradiance kWh/m^2 per year	Annual average temperature (°C)	Annual average wind speed (m/s)	Distance to HV lines (km)	Distance to roads (km)	Average slope (°)	Distance to ports (km)	Annual Average Filling rate (%)	Average Water level difference (m)
Oued El Makhazine	32.69	2299.78	18.79	1.69	15.27	0.66	18.46	47.14	79.13	2.79
Al Wahda	74.81	2288.08	20.12	1.51	6.22	2.16	12.08	117.33	55.3	7.53
Idriss I	25.87	2323.92	20.14	1.84	1.25	1.63	14.13	142.28	24.87	4.88
Allal El Fassi	2.14	2288.58	17.96	1.58	15.62	0.21	24.92	161.50	93.16	3.00
Mohammed V	36.94	2315.63	19.93	2.00	12.85	2.15	11.75	71.67	28.93	3.55
Hassan II	3.00	2427.29	15.72	2.25	13.05	1.83	18.10	234.64	11.1	4.43
Bin El Ouidane	9.05	2373.26	14.68	1.40	8.16	1.49	14.33	201.03	13.95	15.43
Hassan I	2.85	2340.47	16.60	1.41	7.11	1.11	20.29	212.05	16.72	5.62
Al Massira	41.42	2413.86	21.18	2.64	11.95	3.61	5.33	121.39	4.53	2.18
El Mansour Eddahbi	23.97	2480.80	19.70	2.18	21.59	1.97	12.69	278.61	19.66	3.17

The initial dataset was collected for these ten reservoirs: Oued El Makhazine, Al Wahda, Idriss I, Allal El Fassi, Mohammed V, Hassan II, Bin El Ouidane, Hassan I, Al Massira, and El Mansour Eddahbi. The hydrological data represented in annual filling rates and water level differences for 2023 were provided by the Moroccan Ministry of Equipment and Water [11]. Irradiation and climatic factors like temperature and wind speed were collected from Photovoltaic Geographical Information System (PVGIS) Typical Meteorological Year (TMY) datasets. Irradiance should be higher with moderate wind speed to ensure optimal FPV system efficiency; temperature should be lower because lower temperatures increase PV cell efficiency [12]. Wind speed should be high but not exceeding the safety limit of 13 m/s to prevent structural damage [13]. Reservoir surface areas and local terrain slopes of the land surrounding each reservoir (within a 1 km buffer) were calculated in QGIS as large surface areas can support larger installations and flatter land areas promote easy accessibility. Distances to HV transmission lines, roads, and ports were determined using QGIS given their influence on installation cost and potential future hydrogen-export logistics. All initial values were captured and recorded in table 1.

Table 2: Normalized Indicators

Dam	Irradiance	Temperature	Wind speed	Surface area	Distance to HV lines	Distance to roads	Slope	Distance to ports	Stability index (filling rate & water level)
Oued El Makhazine	0.061	0.368	0.237	0.42	0.311	0.868	0.329	1.000	0.896
Al Wahda	0.000	0.164	0.086	1.000	0.756	0.426	0.656	1.000	0.584
Idriss I	0.186	0.160	0.356	0.327	1.000	0.584	0.551	1.000	0.427
Allal El Fassi	0.003	0.495	0.151	0.000	0.294	1.000	0.000	1.000	0.968
Mohammed V	0.143	0.193	0.489	0.479	0.430	0.430	0.672	1.000	0.497
Hassan II	0.722	0.840	0.686	0.012	0.420	0.523	0.348	1.000	0.248
Bin El Ouidane	0.442	1.000	0.000	0.095	0.660	0.624	0.540	0.335	0.000
Hassan I	0.272	0.705	0.008	0.010	0.712	0.737	0.237	0.288	0.319
Al Massira	0.653	0.000	1.000	0.540	0.474	0.000	1.000	1.000	0.000
El Mansour Eddahbi	1.000	0.228	0.629	0.300	0.000	0.482	0.624	0.000	0.397

Collected variables were imported into QGIS and normalized between 0 and 1, as represented in table 2, where 1 represents optimum suitability and 0 represents the lowest suitability. Production-related criteria, which are irradiance, wind speed, and reservoir surface area, were normalized so that higher values correspond to better performance, while temperature and all cost/accessibility variables, distance to HV lines, roads, ports, and terrain slope were normalized so that lower values correspond to better suitability.

b. Tri-Criteria Weighting Assessment: Hydrology, Production, Cost & Accessibility

i. Hydrological Index

The factors related to hydrological aspects, which are the filling rate and water level difference, were given equal weights and converted into an individual stability index for every dam as shown below in table 2, which will then be multiplied with the production and cost related criteria composite score to get the final value of FPV ranking for every reservoir.

ii. Production Related Criteria Weighting Using the MORRIS Sensitivity Method

The global sensitivity analysis with the Morris method was used to identify which production factors affect the generation the most using MATLAB. Morris method was chosen because it shows how changes in inputs impact energy production. This model simulates inputs several times and calculates the output in each scenario, then it calculates the weights of each factor based on its impact rate on the production.

A basic FPV model was implemented using standard photovoltaic equations with keeping BOS efficiency constant. This model calculated the energy output based on theoretical incoming irradiance, PV module efficiency, module temperature, wind speed, and surface area. The main equations used are:

- Effective PV efficiency:

$$\eta_{\text{eff}} = \eta_{\text{ref}}(1 - \beta(T_c - 25°\text{C})) \quad (1)$$

Where η_{ref} is the reference efficiency at 25°C, β is the temperature coefficient describing efficiency loss per °C (°C^{-1}), and T_c is the PV cell temperature (°C).

- PV cell temperature:

$$T_c = T_a + \frac{E_{\text{Irradiance}}}{U_0 + U_1 \cdot WS} \quad (2)$$

Where T_a is the ambient air temperature (°C), $E_{\text{Irradiance}}$ is the incident solar irradiance on the panel (W/m^2), WS is the wind speed (m/s), and $U_0$and $U_1$are empirical heat-loss coefficients with units of (W/m^2·°C) and (W/m^3·s·°C), respectively.

- Annual energy proxy:

$$E_{\text{annual}} = H_{GHI} \cdot \eta_{\text{eff}} \cdot \eta_{\text{BOS}} \cdot Area_{\text{ref}} \quad (3)$$

Where H_{GHI} represents the annual global horizontal irradiance (Wh/m^2/year), η_{BOS} is the balance-of-system efficiency, and $Area_{\text{ref}}$ is a fixed reference surface used to scale the energy proxy in the sensitivity analysis (m^2).

These equations were used only to see how much each input affects the results, not to predict exact energy output. The MORRIS algorithm perturbs each input within theoretical values and calculates the elementary effect of the $\boldsymbol{i}$-th input variable following this equation [14]:

$$EE_i = \frac{f(x_1, \dots, x_i + \Delta, \dots, x_k) - f(x)}{\Delta} \quad (4)$$

Where $f(X)$ denotes the model output with $X = (x_1, \dots, x_k)$ the vector of k input variables, x_i is the i-th input variable being perturbed while all other inputs are kept fixed, and Δ is the step size applied to x_i within its normalized range.

The overall importance of each variable was summarized using the mean absolute elementary effect:

$$\mu_i^{\backslash *} = \frac{1}{N} \sum_{j=1}^{N} | EE_{i,j} | \quad (5)$$

Where $\mu_i^{\backslash *}$ is the mean absolute elementary effect of the i-th input variable, N is the number of Morris trajectories, and $EE_{i,j}$ is the elementary effect of the i-th input computed along the j-th Morris trajectory. The normalized $\mu^{\backslash *}$values formed the basis of the production weighting vector used in QGIS MCDA.

MORRIS revealed that surface area had the strongest influence, but because irradiance is physically the dominant driver of solar output, surface area was intentionally limited to 70% of the solar irradiance weight to not overshadow it.

iii. Cost and Accessibility Related Criteria Weighting Using AHP

Cost-related criteria were weighted using AHP, a structured decision-making method that derives numerical weights from pairwise comparisons between criteria [15], [16]. In AHP, each criterion is compared to another one on a ratio scale to decide if they have equal importance, one is more important than the other, or one is less important than the other, which forms a square matrix of comparisons. The normalized principal eigenvector of this matrix is the final weight distribution, which has internal consistency via a calculated consistency ratio.

Table 3: AHP Pairwise Comparison Matrix

Criteria	Distance to HV lines	Slope	Distance to roads	Distance to ports
Distance to HV lines	1	2.1	3.468	1
Slope	0.476	1	1.465	0.476
Distance to roads	0.288	0.682	1	0.288
Distance to ports	1	2.1	3.468	1

The relative importance between distance to HV lines, slope, and distance to roads represented in table 3 was adopted from matrices published in [17], and distance to ports was assumed to carry the same importance weight as distance to HV lines to reflect its relevance for future hydrogen-export logistics.

c. Integrated FPV Suitability Score with Hydrological Feasibility Scaling

The final FPV potential score S_{FPV} for each reservoir is computed by multiplying a technical suitability component that combines both production-related criteria and cost-related criteria S_{tech} with a hydrological feasibility index I_{hyd} as shown in the following equation:

$$S_{\mathrm{FPV}} = S_{\mathrm{tech}} \times I_{\mathrm{hyd}} \tag{6}$$

Where:

$$S_{\mathrm{tech}} = \sum W_i \times NS_i \tag{7}$$

Where W_i is the normalized weight of the i-th criterion derived from the mean absolute elementary effect μ_i^{*} obtained using the Morris sensitivity analysis, and NS_i is the normalized score of the corresponding criterion for each reservoir.

The hydrological index I_{hyd} is defined as:

$$I_{\mathrm{hyd}} = \sqrt{\left(\frac{F-F_{\mathrm{min}}}{F_{\mathrm{max}}-F_{\mathrm{min}}}\right)\left(1-\frac{\Delta H-\Delta H_{\mathrm{min}}}{\Delta H_{\mathrm{max}}-\Delta H_{\mathrm{min}}}\right)} \tag{8}$$

Where F is the reservoir filling rate, with F_{min} and F_{max} the minimum and maximum filling rates across all reservoirs. ΔH is the seasonal water-level amplitude, with ΔH_{min} and ΔH_{max} the minimum and maximum values observed across the analyzed reservoirs.

2. SAM-Based Energy Modeling & Economic Analysis

SAM was used to estimate the annual FPV generation for the reservoir that has the highest rank in the QGIS analysis, with taking into consideration a coverage of 5% of its surface area. A high-efficiency monofacial LONGi mono-c-Si module of 570 W rated power and 2.49 m^2 area was selected for this study. This module is recognized as a tier-1 utility module due to its strong performance [18]. For the layout, a ground coverage ratio (GCR) of 0.70 was applied to present typical FPV spacing, walkways, and maintenance corridors, as commonly used in FPV designs [19]. The tilt was fixed at 12°, azimuth at 180°, and soiling losses at 3% per year, following FPV studies showing reduced dust accumulation [20]. Electrical losses assumptions are 4.44% for DC losses, 1% for AC wiring losses, 0.1% for transformer losses (no-load) and 0.5% (load), 2% for transmission losses, and 2% for system availability. Financial inputs include a CAPEX of 1000 $/kW [21], O&M cost of 15.5 $/kW per year [22], a 30-year analysis period, and 12% internal rate of return.

3. Evaporation-Savings Assessment

To determine the impact of FPV on the rate of evaporation of the reservoir, a simplified Penman model is employed using MATLAB. The simplified Penman model is among the most established empirical models that have widely adopted for open-water evaporation rate estimation under limited data constraints. The approach has been applied in some studies such as [10], which assessed evaporation suppression impacts of FPV systems across multiple lakes and reservoirs using Penman-based formulations. The simplified Penman method estimates daily evaporation as a function of air temperature, dew-point temperature, latitude, and altitude [23]:

$$E_{va} = \frac{\frac{700T_m}{100-A}+15(T-T_d)}{80-T} \tag{9}$$

Where E_{va} is the daily evaporation rate (mm/day), T_m is the modified mean air temperature (°C), defined as $T_m = T + 0.006h$ with h is the reservoir altitude (m), T is the mean air temperature (°C), A is the latitude (°), and T_d is the dew-point temperature (°C) estimated from relative humidity using August–Roche–Magnus formula [24]:

$$T_d = \frac{243.04\cdot[\ln(RH/100)+\frac{17.625T}{243.04+T}]}{17.625-[\ln(RH/100)+\frac{17.625T}{243.04+T}]} \tag{10}$$

Where RH is the relative humidity (%). h, T, A, and RH for dew-point estimation were taken from the PVGIS TMY dataset. Once the daily rate is obtained, the annual water loss for each reservoir is computed in (million m³/year) as:

$$V_{annual} = \Sigma_{days} E_{va} \times S_A \times 10^{-3} \tag{11}$$

Where S_A is the surface area of the reservoir (km²).

Annual water savings are estimated in (million m³/year) as:

$$V_{\text{saved}} = V_{\text{annual}} \times C_{\text{FPV}} \times \eta_{\text{FPV}} \tag{12}$$

Where C_{FPV} is the FPV coverage ratio and η_{FPV} is the evaporation-suppression efficiency. To remain conservative, this study adopts $\eta_{\text{FPVS}} = 70\%$ as the baseline value reported in [25]. To verify the robustness of the resulting water-savings estimates, a sensitivity analysis was subsequently conducted by considering a bounded range of η_{FPV} values between 60% and 80%.

Results and Discussion

This section presents the results of the multi-criteria site selection analysis, associated importance or weights of criteria used for evaluation and ranking of reservoirs under the final FPV suitability. This section also presents detailed SAM analysis results related to the top-ranked reservoir and evaporation savings potential at the site selected after the implementation of FPV.

1. QGIS Dam Selection

a. Weighting Results

Table 4: Final Criterion Weights Used in the MCDA

Criterion	Weight
Global irradiance	0.38498
Temperature	0.04239
Wind speed	0.0031457
Surface area	0.26949
Distance to HV lines	0.10871
Slope	0.05027
Distance to roads	0.032303
Distance to ports	0.10871

The final weighting distribution resulting from MORRIS and AHP techniques are presented in table 4. These weights were applied to the normalized criteria to determine the technical FPV suitability score, which are then multiplied by the hydrological index. Weights identify that solar irradiance and eligible surface area for FPV, HV lines accessibility, and port distance were given highest significance, while wind speed and temperature held lower significance in MORRIS technique analysis and were assigned lower weights.

b. Final FPV Potential Scores of the Ten Reservoirs

Table 5: Final FPV Potential Scores

Reservoir	FPV score
Oued El Makhazine	0.305
Al Wahda	0.281
El Mansour Eddahbi	0.208
Mohammed V	0.191
Allal El Fassi	0.137
Idriss I	0.165
Hassan II	0.104
Hassan I	0.09
Al Massira	0.000
Bin El Ouidane	0.000

Table 5 represents the final FPV suitability scores obtained for selected sites. These scores provide the final ranking used to identify the most promising FPV site. The results show that Oued El Makhazine achieves the best FPV potential score at a value of 0.305, closely followed by Al Wahda. Sites with a near-zero score failed mainly because of low filling rates, large water-level fluctuations, or very small FPV-suitable surface area, which lowered their hydrological index.

c. FPV Suitability Map

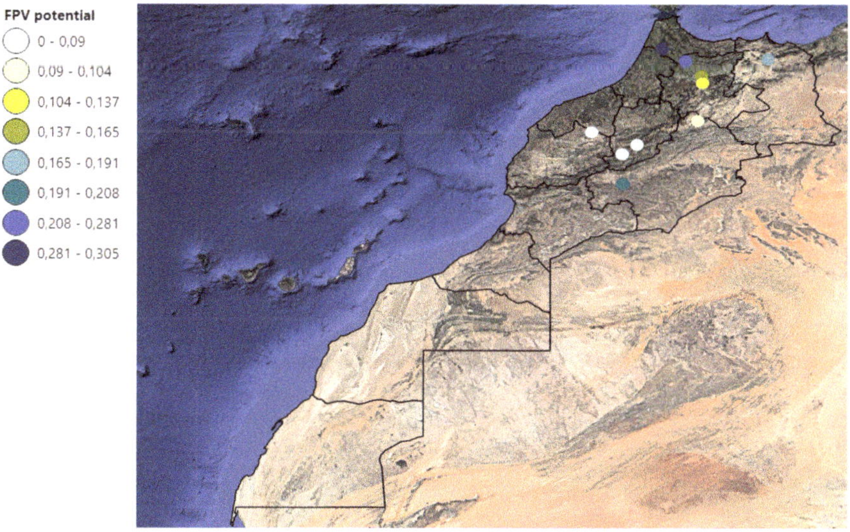

Figure 1: FPV Suitability Map of the Ten Reservoirs in Morocco Based on the Final MCDA Score

The suitability map shown in Fig.1 highlights Oued El Makhazine in northern Morocco as the most favorable FPV site. While its irradiance is lower than that of reservoirs from southern Morocco, its high filling rate and lack of a strong seasonality of the water level make it highly attractive for FPV installation. Stable hydrology ensures safer anchoring, consistent buoyancy, and reliable cable management, whereas southern reservoirs with stronger solar potential but large drawdowns, such as Al Massira or Bin El Ouidane, face structural risks and higher installation complexity. Al Wahda also scores highly for similar hydrological reasons. In contrast, intermediate-ranked sites combine good solar resources with moderate hydrological limitations, while the lowest-ranked reservoirs are primarily penalized by unstable water levels.

2. Energy Production Simulation and Economic Analysis Using SAM

a. Energy Production

For the FPV system designed for the Oued El Makhazine site, the total first-year energy output of the selected system is simulated by SAM. The 5% coverage area is equivalent to a platform size of 1,634,950 m^2, thus yielding a total of 459,620 solar PV modules with accounting for the GCR. The annual power output of the system is found to be 396.62 GWh, resulting in an energy yield of 1,513 kWh/kW, a DC capacity factor of 17.3%, and a performance ratio of 0.78.

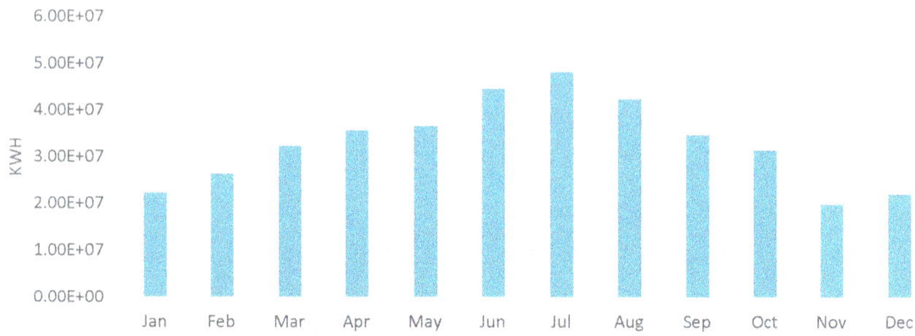

Figure 2: Monthly AC Energy in Year 1

Fig.2 shows that energy is most generated between May and August, corresponding to high solar irradiance events. To evaluate the potential contribution of the annual FPV generation to local energy supply, this output can be compared to Morocco's typical electricity consumption per capita, which recent national statistics of 2023 place at around 1,000 kWh/person/year as a conservative value [26]. Based on this benchmark, the annual generation from the FPV plant would theoretically meet the electricity needs of about 396,620 people. It is, therefore, possible that a limited installation of these FPV systems can satisfy the annual energy requirements of a medium-sized Moroccan city or several rural communes surrounding Oued El Makhazine.

b. Financial results

The 5%-coverage FPV system has an LCOE of $0.0573/kWh according to the financial analysis done in SAM. The resulting LCOE confirms the strong economic competitiveness of the suggested system and is consistent with the typical cost range reported for FPV projects shown in [27].

3. Water Savings Results

a. Baseline Evaporation under 70% Suppression Efficiency

Table 6: Annual Evaporation and Water Savings

Quantity	Value
Total reservoir area	32.699 km^2
Shaded area	1.6483 km^2
Annual evaporation before FPV	53.041 millions m^3/year
Annual evaporation after FPV	51.185 millions m^3/year
Annual water saved by FPV	1.856 millions m^3/year

According to table 6, Oued El Makhazine, which is situated at latitude 34.938° and altitude 51 m, experiences an annual evaporation loss of 53.041 million m^3/year under the Moroccan climate. Assuming a 70% evaporation-suppression efficiency over the shaded area and a 5% FPV coverage, the annual evaporation drops to 51.185 million m^3/year, yielding a net annual water savings of 1.856 million m^3/year. This saving is compared to Morocco's domestic water consumption to put it in perspective. Rabat has one of the highest urban water-use rates at about 123 L/person/day, or roughly 44.9 m^3/person/year, according to a World Bank annex detailed regional consumption levels [28]. Thus, the FPV installation on Oued El Makhazine conserves enough water each year to fully cover the annual domestic water needs of approximately 41,336 residents. This demonstrates the strong co-benefit of FPV systems, in addition to generating renewable electricity, they provide meaningful water-resource savings of city-scale magnitude in a water-stressed country like Morocco.

b. Sensitivity of Water Savings to Suppression Efficiency

Table 7: Estimated Annual Water Savings under Varying Suppression Efficiencies

η_{FPV} (%)	Water saved (million m^3/year)
60	1.591
65	1.724
75	1.989
80	2.122

Under the tested suppression efficiencies, the annual water savings are found to be varying from 1.591 million m^3/year at 60% to 2.122 million m^3/year at 80%, with the remaining intermediate values being 1.724 million m^3/year at 65% and 1.989 million m^3/year at 75%, showing a proportional increase in water savings with higher suppression efficiency while preserving the same order of magnitude.

Conclusion

This paper elaborates on an integrated framework for identifying optimal FPV sites in Morocco through spatial analysis, hydrological feasibility, and sensitivity-based weighting of the technical and logistical criteria, pointing out that stability of the reservoir is often as decisive as the quality of the solar resource. The highest-ranked site is Oued El Makhazine, where 5% FPV installation produces 396.62 GWh/year at an LCOE of $0.0573/kWh, where evaporation reduction decreases by 1.856 million m 3 /year, which is enough to sustain the annual electricity requirement of 396,620 individuals and the annual residential water requirement of 41,336 residents. Nevertheless, these powerful co-benefits are surrounded by some restrictions due to data availability including the use of hydrological data corresponding to a single year, a simplified estimate of evaporation through Penman model, and SAM assumptions which do not account for thermal impact of FPVs. The future analysis should therefore combine the multi-year reservoir datasets with greater climate resolution, hydrodynamic models, and tested FPV thermal-cooling behavior. It should also include the assessment of FPV–hydropower hybrid configurations for multi-purpose dams, FPV-electrolyzer coupling to produce green hydrogen, FPV-battery hybrids to act as grid supports, and the integration of IPCC-based climate change scenarios to evaluate long-term climate resilience and conduct long-term planning of energy–water systems in Morocco.

References

[1] World Bank, "High and Dry: Climate Change, Water, and the Economy," World Bank, Washington, DC, USA, 2016. [Online]. Available : https://documents1.worldbank.org/curated/en/862571468196731247/pdf/105130-REVISED-K8517.pdf.

[2] AFP, "Morocco tests floating solar panels to save water," Arab News, Aug. 31, 2025. https://www.arabnews.com/node/2613661/middle-east (accessed Nov. 28, 2025).

[3] R. Benbba *et al.*, "Solar Energy Resource and Power Generation in Morocco: Current Situation, Potential, and Future Perspective," *Resources*, vol. 13, no. 10, p. 140, Oct. 2024. https://doi.org/10.3390/resources13100140.

[4] Y. Jin *et al.*, "Energy production and water savings from floating solar photovoltaics on global reservoirs," *Nature Sustainability*, vol. 6, no. 7, pp. 865–874, Mar. 2023. https://doi.org/10.1038/s41893-023-01089-6.

[5] A. Mentzafou, E. Dimitriou, I. Karaouzas, and S. Zogaris, "Impact Assessment of Floating Photovoltaic Systems on the Water Quality of Kremasta Lake, Greece," *Hydrology*, vol. 12, no. 4, p. 92, Apr. 2025. https://doi.org/10.3390/hydrology12040092

[6] G. Sandrini, A. Wagenvoort, R. van Asperen, B. Hofs, D. Mathijssen, and A. van der Wal, "Illuminating the Impact of a Floating Photovoltaic System on a Shallow Drinking Water Reservoir: The Emergence of Benthic Cyanobacteria," *Water*, vol. 17, no. 8, p. 1178, Apr. 2025. https://doi.org/10.3390/w17081178

[7] "Fraunhofer quantifies German floating PV potential," *Bayern-innovativ.de*, 2024. https://www.bayern-innovativ.de/en/emagazine/detail/fraunhofer-quantifies-german-floating-pv-potential (accessed Nov. 28, 2025)

[8] C. Owen, "Spain releases new regulations for floating PV on reservoirs," *Solar&StorageXtra - Light reading*, Jul. 15, 2024. Available: https://solarstoragextra.com/spain-releases-new-regulations-for-floating-pv-on-reservoirs/

[9] "Floating solar PV on dam reservoirs: The opportunities and the challenges." Available: https://www.hydropower-dams.com/wp-content/uploads/FPV_report.pdf.

[10] A. Mouhaya, A. El Hammoumi, A. El Ghzizal, and S. Motahhir, "Assessment of floating solar photovoltaic potential and water conservation in Morocco: Case studies of four hydroelectric dams," *Cleaner Engineering and Technology*, vol. 26, p. 100952, Mar. 2025. https://doi.org/10.1016/j.clet.2025.100952.

[11] Ministère de l'Equipement et de l'Eau, "Accueil," https://www.equipement.gov.ma/AR/Pages/Accueil.aspx

[12] H. Tavassoli, "The Effect of Temperature on Photovoltaic Cell Efficiency," *Academia.edu*, Dec. 07, 2022. https://www.academia.edu/92375607/The_Effect_of_Temperature_on_Photovoltaic_Cell_Efficiency

[13] M. A. Alantali, H. Rafiq, S. A. Alzarouni, O. A. Qureshi, and E. Rodriguez-Ubinas, "Wind Speed Effect on the Performance of Free-Standing PV Modules in Hot Desert Climate," pp. 1–5, Nov. 2023. https://doi.org/10.1109/mena-sc54044.2023.10374470

[14] M. Balesdent, L. Brevaul, S. Lacaze, S. Missoum, and J. Morio, "Methods for high-dimensional and computationally intensive models," *Elsevier eBooks*, pp. 109–136, Nov. 2015. https://doi.org/10.1016/b978-0-08-100091-5.00008-3. Available: https://www.sciencedirect.com/topics/engineering/morris-method

[15] O. Bozorg-Haddad, H. Loáiciga, and B. Zolghadr-Asli, "Analytic Hierarchy Process (AHP)," in *A Handbook on Multi-Attribute Decision-Making Methods*, Chapter 3, John Wiley & Sons, Inc., Mar. 2021. https://doi.org/10.1002/9781119563501.ch3

[16] J. A. Prieto-Amparán, A. Pinedo-Alvarez, C. R. Morales-Nieto, M. C. Valles-Aragón, A. Álvarez-Holguín, and F. Villarreal-Guerrero, "A Regional GIS-Assisted Multi-Criteria Evaluation of Site-Suitability for the Development of Solar Farms," *Land*, vol. 10, no. 2, p. 217, Feb. 2021. https://doi.org/10.3390/land10020217

[17] A. E. Dinçer, A. Demir, and K. Yılmaz, "Enhanced Objectivity of AHP for More Reliable Solar Farm Site Selection," *Energy Science & Engineering*, Apr. 2025. https://doi.org/10.1002/ese3.70027.

[18] M. S. Admin, "Top Tier 1 Solar Panel Manufacturers List: Updated 2024 Rankings," *RENVU*, Mar. 20, 2025. https://www.renvu.com/blogs/learn/top-tier-1-solar-panel-manufacturers-list-updated-2024-rankings?srsltid=AfmBOop-210b1NmTvh-0ardbAYO61OeeLxtll6Qhdx-M3ISYG4DyPODj (accessed Nov. 28, 2025)

[19] E. Rosenlieb, M. Rivers, and A. Levine, "Floating photovoltaic technical potential: A novel geospatial approach on federally controlled reservoirs in the United States," *Solar Energy*, vol. 287, pp. 113177–113177, Dec. 2024. https://doi.org/10.1016/j.solener.2024.113177

[20] "Floating Photovoltaic Power Plants: A Review of Energy Yield, Reliability, and Maintenance 2025 PVPS." Available: https://iea-pvps.org/wp-content/uploads/2025/04/IEA-PVPS-T13-31-2025-REPORT-Floating-PV-Plants.pdf

[21] H. Ziar, "Floating solar stations." Accessed: Nov. 29, 2025. [Online]. Available: https://pure.tudelft.nl/ws/files/103032634/10ideas_ENG_FLOATING_SOLAR_STATIONS.pdf?utm

[22] *Nrel.gov*, 2025. https://docs.nrel.gov/docs/fy22osti/80695.pdf?utm_ (accessed Nov. 29, 2025)

[23] E. T. Linacre, "A simple formula for estimating evaporation rates in various climates, using temperature data alone," *Agricultural Meteorology*, vol. 18, no. 6, pp. 409–424, Dec. 1977. https://doi.org/10.1016/0002-1571(77)90007-3

[24] T. K. Thiis, I. Burud, A. Flø, D. Kraniotis, S. Charisi, and P. Stefansson, "Monitoring and Simulation of Diurnal Surface Conditions of a Wooden Façade," *Procedia Environmental Sciences*, vol. 38, pp. 331–339, 2017. https://doi.org/10.1016/j.proenv.2017.03.088

[25] H. F. Abd-Elhamid, A. Ahmed, M. Zeleňáková, Z. Vranayová, and I. Fathy, "Reservoir Management by Reducing Evaporation Using Floating Photovoltaic System: A Case Study of Lake Nasser, Egypt," *Water*, vol. 13, no. 6, p. 769, Mar. 2021. https://doi.org/https://doi.org/10.3390/w13060769

[26] "Morocco - Countries & Regions," *IEA*. Available: https://www.iea.org/countries/morocco/electricity

[27] *Nrel.gov*, 2025. Available: https://docs.nrel.gov/docs/fy22osti/80695.pdf?utm_source=chatgpt.com (accessed Nov. 29, 2025)

[28] World Bank, *Gestion de la rareté de l'eau en milieu urbain au Maroc – Annexes aux Sections 2 à 4*, 2017, 73 p. [Online]. Available: https://documents1.worldbank.org/curated/en/299131516806648462/pdf/122698-FRENCH-v2Annexes-Sections-2-4-FR.pdf

Materials Research Proceedings 64 (2026) 148-155 https://doi.org/10.21741/9781644904091-18

Modeling and Prediction of Clear-Sky Solar Irradiance Losses Using Meteorological and Radiative Parameters: A Case Study in Ouarzazate, Morocco

Abdelali ABDESSADAK[1,2,3,a,*], Hicham GHENNIOUI[1,b], Nadège THIRION-MOREAU[2,c], Mounir ABRAIM[4,d], Safae MERZOUK[3,e], and Brahim ELBHIRI[5,f]

[1]Laboratory of Signals, Systems, and Components, Sidi Mohamed Ben Abdellah University, Fez, Morocco

[2]LIS Toulon, UTLN, AMU, CNRS UMR 7020, France

[3]SMARTiLab Laboratory, Moroccan School of Engineering Sciences (EMSI), Rabat Morocco

[4]Green Energy Park Research Platform (GEP, IRESEN/UM6P), BenGuerir, Morocco

[5]Harmony Technology, Rabat Morocco

[a]abdelaliabdessadak@gmail.com, [b]hicham.ghennioui@usmba.ac.ma, [c]thirion@univ-tln.fr, [d]abraim@greenenergypark.ma, [e]s.merzouk@emsi.ma, [f]bra.elbhiri@gmail.com

Keywords: Clear-Sky Irradiance, Irradiance Losses, Solar Radiation, Solar Energy Forecasting, Clearness Index

Abstract. Understanding how to quantify loss potential due to clear-sky solar irradiance is essential for the development of enhanced Photovoltaic and Solar Energy Forecasting systems in Semi-Arid climates. In Ouarzazate, Morocco, four years (2021 - 2024) of daily weather and solar radiation observations collected by satellite and from the ground were analysed in this research through the analysis of clear-sky losses. Clear-sky conditions were isolated from all others by applying a Clearness Index threshold of ≥0.65 (KT), and Clear-sky Losses were derived from the relative differentials between All-Sky Shortwave Solar Irradiance (SSI) and Clear-Sky Shortwave Solar Irradiance (SSI). Correlation analysis between the Clear-Sky Loss and four key weather variables produced strong correlations. Three models were created to predict clear-sky losses: Linear Regression (LR), Second Order Polynomial Regression (2PR), and a Random Forest (RF) algorithm. The results indicated that the 2PR model provided the best fit ($R^2 = 0.9965$, RMSE = 0.0079) allowing for accurate comparisons of the 2PR model to the two better-known methods of predicting Solar Resource & Energy Potential.

Introduction

Solar energy has become a mainstay in the quest to develop sustainable and renewable energy across the globe. Advances in photovoltaic technology and decreasing costs make it very important to accurately assess the availability of solar resources to optimise system performance and improve the forecasting of energy yields and to allow greater levels of integration into the grid[1], [2]. Accurate forecasting of the amount of solar irradiance that will strike surfaces is critical to reducing the potential uncertainties associated with designing and operating PV systems[3].

Ouarzazate in southeastern Morocco is home to one of the largest solar power plants in the world and provides an excellent environment for conducting research into solar energy [4], [5]. The region is known for its high levels of solar radiation, low levels of precipitation and high levels of aerosol loading due to desert dust carried by the wind from the Sahara Desert, which has caused many problems for the environment [6]. *Fig. 1* below illustrates the global distribution of dust

Materials Research Proceedings 64 (2026) 148-155 https://doi.org/10.21741/9781644904091-18

intensities around the world and illustrates the impact of desert areas like southern Morocco on the global intensity of dust[7]. Dust storms, especially in the spring and summer, can have a detrimental impact on the quality of the air and the performance of PV systems [8].

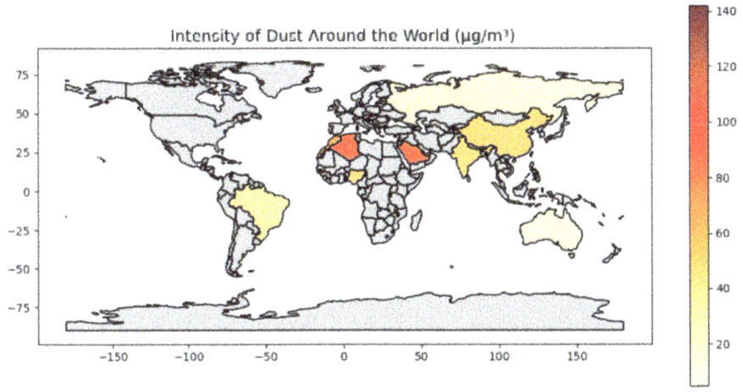

Fig. 1. Intensity of dust around the world (with some data of some country)

While there have been a number of studies that have investigated solar irradiance modeling and atmospheric attenuation effects across many regions [9], [10], little published work has been done regarding clear-sky irradiance loss associated with semi-arid environments located within Northern Africa. Many existing clear-sky models were created for other climate types and may not fully characterize the non-linear interactions between meteorological factors and irradiance loss in a semi-arid setting [11],[12] . Moreover, prior work has predominantly centred on theories of cloud-covered scenarios or theoretical clear-sky modeling, with limited long-term observational data for validation from these types of locations [13], [14].

As a result of this, the aim of the current research is to quantify clear-sky solar irradiance losses for Ouarzazate (Morocco) based on four (4) years (2021-2024) of daily meteorological and radiative observations. This study seeks to: 1) quantify clear-sky irradiance loss in relation to a clear index threshold (KT $\geq$0.65), 2) evaluate the relationship(s) between the irradiance loss and meteorological/radiative parameters, and 3) develop and validate models that will estimate clear-sky irradiance loss. Of particular interest is a second-order polynomial regression model, which has shown sufficient strength to characterize the nonlinear nature of clear sky irradiative loss when compared to the linear regression approach and other more complex machine learning techniques.

Methodology

Fig. 2 the methodology diagram (shown to the right) explains the steps that were taken within this research project. Collection of 4 years (2021-2024) worth of daily weather and solar radiation data from NASA's POWER Database was the first step. Once cleaned and processed, it was possible to analyze both the all sky and theoretical parcels to determine the rates of loss. To identify clear sky days.

Materials Research Proceedings 64 (2026) 148-155 https://doi.org/10.21741/9781644904091-18

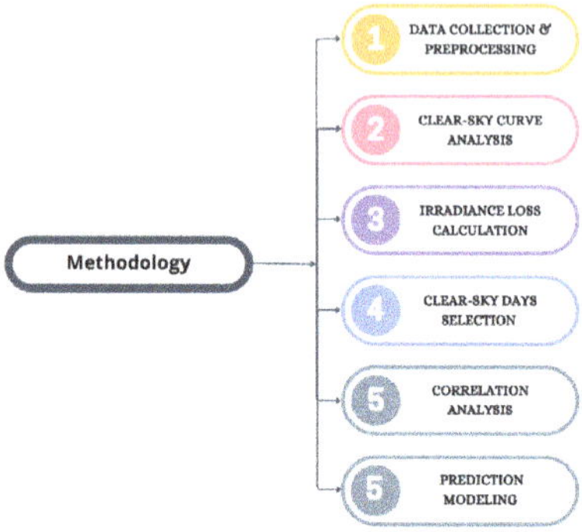

Fig. 2. Overview of the methodological workflow used in this study.

Study Area

Located in southern Morocco at approximately 1,160 meters above sea level, (30.93°N, 6.91°W), Ouarzazate has a semi-arid climate with hot and dry summer months; moderate to cool winter months and a large day-to-night temperature difference[6].. The area's location near the Sahara Desert gives it a very large amount of solar energy potential with over 2,500 kilowatt-hours per square meter of solar radiation each year, making it an ideal place for solar power production in northern Africa[15]..

Meteorological and radiation data from 2021 through 2024 were analyzed in this research study. Examples of examined variables include the following: Air and ground temperature; Relatively Humidity; Wind Speed and Direction (on many levels); Ground Pressure; and Increasing Solar Radiation shortwave, longwave and UV components.

The "clear sky" condition was determined by means of determining the clear sky clearness index (KT) - KT indicates how much solar radiation reaches the ground. Researchers previously identified a KT value of 0.65 for semi-arid climates as being suitable for determining primarily clear sky day classifications and producing the maximum usable amount of meteorological data to be evaluated on a statistical basis[16], [17]. Approximately 850-900 days met this criterion after applying this threshold to the data set.

The relative clear-sky day irradiance losses (L) were calculated using Eq. (1) for each identified clear-sky day. Pearson's correlation coefficients (PCCs) were calculated to assess the relationship between the meteorological parameters and the clear-sky irradiance losses, in order to determine the dominant atmospheric attenuation drivers in the absence of clouds.

$$L = \frac{\mathbf{ALLSKY\_SFC\_SW\_DWN}}{\mathbf{CLRSKY\_SFC\_SW\_DWN}} \qquad (1)$$

Results

Clear-sky and Actual Irradiance Characteristics

Fig. 3 comparison of Clear Sky Irradiance (CLRSKY) to Actual All Sky Irradiance (ALLSKY), Period 2021–2024. Both timeseries are expected to exhibit strong seasonal variations with peaks during the summer months and troughs during the winter months.

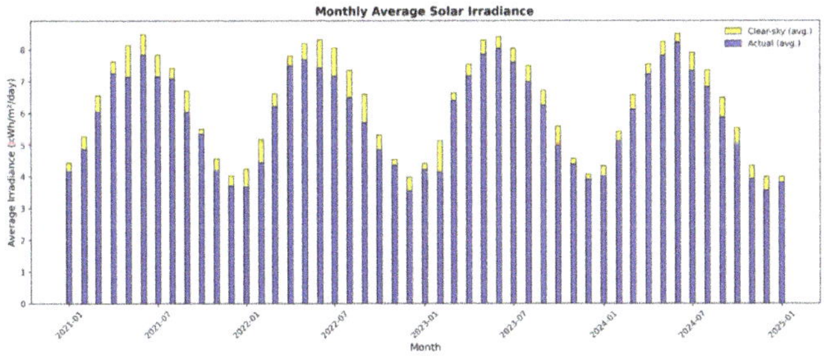

Fig. 3. Clear-sky vs actual irradiance

The year June 2024 saw the highest clear sky irradiance measurement of 9.23 [kWh/m²]. The lowest recorded number was from December 2021 at 3.47 [kWh/m²]. Along the same lines, actual irradiance (ALLSKY) peaked in June 2024 with a similar number of 9.22 [kWh/m²], which matches the annual solar production cycle, and dropped to a low of 0.62 [kWh/m²] in February 2023, signifying a day with considerable air pollution attenuation to that day's solar generation.

Daytime all-sky irradiance averaged during this period accounted for an average of 5.89 [kWh/m²] ($\sigma = 1.71$) as opposed to 6.37 [kWh/m²] ($\sigma = 1.58$) experienced on days with clear skies. The differences experienced on the two average irradiances represent the systematic reduction in available solar energy created by the effects of clouds, aerosol particles, and atmospheric variability.

Temporal Evolution of Irradiance Losses

Fig. 4 this graph shows how much daily irradiance was lost by looking at the difference in actual irradiance and that that would occur under a clear sky (irradiance) as a percentage of the actual irradiance.

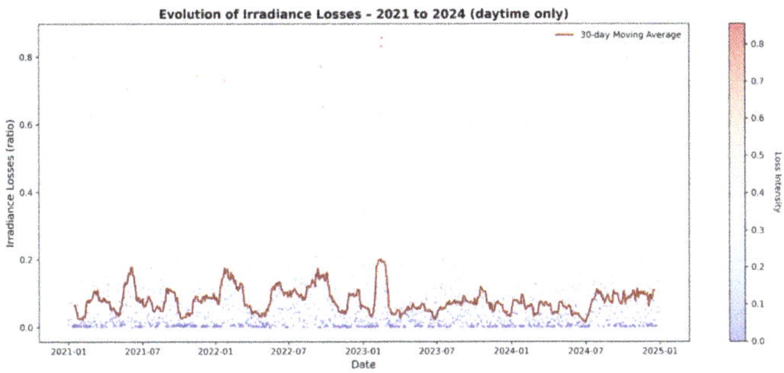

Fig. 4. Daily irradiance losses

Although at times instantaneous losses may exceed 80% (e.g., February 2023). However, their overall trend reflects that for the majority of days there were moderate, generally below 20%, instantaneous losses.

The average loss per month demonstrates strong seasonal variability; the loss rate increases during the winter months and some transitional months as a result of higher cloud cover and increased atmospheric instability within the winter months. In addition, during the summer months, loss rates are typically lower and more stable.

Annual Statistics of Irradiance Losses

Table 1 summarizes the annual statistics of irradiance losses. The average annual losses vary between 4.5% and 9.7%, with 2022 showing the highest mean loss (0.0972), followed by 2021 (0.0768) and 2024 (0.0730).

Table 1: Annual Statistics of Irradiance Losses (2021–2024)

Year	Mean Loss	Max Loss	Month of Max Loss	Min Loss	Month of Min Loss
2021	0.0768	0.7440	November	0.0003	August
2022	0.0972	0.7712	September	0.0000	November
2023	0.0688	0.8555	February	0.0000	January
2024	0.0730	0.7103	December	0.0000	March

Once we established a value of KT (clear-sky index) of 0.65 or higher, of the 1,462 daytime days in which clear skies existed based on our criteria, 1,073 clear-sky days met our definition of a clear-sky condition. Upon filtering these data according to criteria for calculating day-to-day changes in irradiance losses (due primarily to particulate matter/humidity as opposed to clouds), we found substantial variation in the amount of irradiance loss from clear-sky days. On average, the amount of irradiance loss was 0.0249, which will generally mean a low level of attenuation on verified clear-sky days. However, there were isolated secondary peaks of irradiance loss likely due to short-duration atmospheric conditions such as dust storms and episodes of rapid changes in humidity.

Correlation Analysis

There are a number of correlations between meteorological and radiative input variables and clear sky irradiance losses. An inverse correlation was found between relative loss and clear sky irradiance that suggests increased loss sensitivity in reduced clear sky conditions. The strongest relationships among all of the variables were seen with respect to the radiative input variables, which included ALLSKY SFC SW DWN, ALLSKY KT, and PAR components and indicate their importance for atmospheric transparency. The meteorological input variables had a weaker, but moderative relationship with loss and would therefore be viewed as an indirect influence on the loss through atmospheric stability and scattering interactions. The information gathered from this study aided in the selection of input variables that were used as predictors for both polynomial regression and machine learning–based models for estimating losses.

Predictive Modeling Performance

Using all meteorological and radiative variables, three predictive models were created: Linear Regression, Second Degree Polynomial Regression (Ridge), and Random Forest.

Table 2: Performance Comparison of Prediction Models

Model	R^2	RMSE
Linear Regression	0.9949	0.0095
Polynomial Regression (Degree 2)	0.9965	0.0079
Random Forest	0.9735	0.0218

Of these three models, the Second Degree Polynomial Model produced the most accurate predictions with the least amount of error; however, Random Forest performed worse than both Regression models. Therefore, we can infer that clear-sky irradiance attenuation in semi-arid climates can be accurately predicted with the integration of radiative parameters and a moderate amount of non-linear modelling. Clear-sky irradiance losses are better described as a smoothed non-linear function than an extremely complex or erratic one.

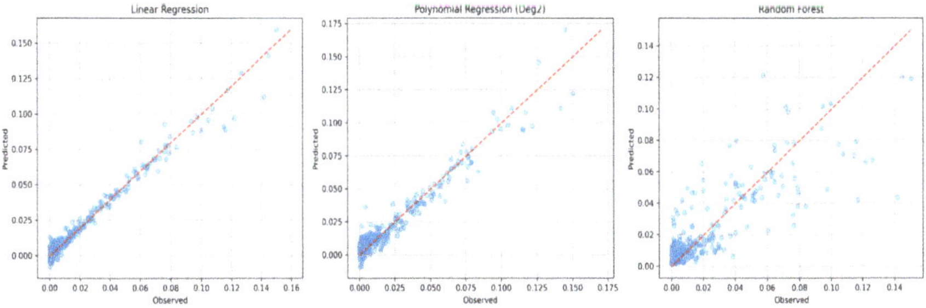

Fig. 5. Observed vs. predicted clear-sky irradiance losses for three models (Linear, Polynomial deg-2, Random Forest)

Fig. 5 a comparison of the observed and predicted clear-sky irradiance losses from three different models, namely Linear Regression (on left), Polynomial Regression of degree 2 (in the center), and Random Forest (on right) illustrate that both linear and polynomial models cluster very close to the ideal 1:1 agreement indicated by the dashed red line, confirming that linear and polynomial models have similar accuracies in predicting clear-sky irradiance losses.

Discussion

The research from 2021 to 2024 focused on the losses of clear-sky irradiance occurring throughout the year. Researchers studied both the drivers of clear-sky irradiance losses in Ouarzazate and identified the causes of those losses as atmospheric events like aerosol and water vapor. Researchers concluded that there are many factors that cause the amount of solar radiation received on Earth to vary, and some of these factors can increase or decrease the amount of solar radiation received[18], [19].

Less clear-sky irradiance losses were observed during the summer months than in all other seasons, due partly to high moisture levels in the air, which also caused greater attenuation of solar radiation during both winter and transitional seasons. In addition, the presence of Saharan dust events also caused significant increases in the total number of clear-sky irradiance losses throughout the year, especially in winter and transitional months.[6], [7], [8].

Based on this model comparison, although the Random Forest model achieved a lower overall accuracy than the second-degree polynomial and linear regression models, the polynomial and linear regression models still provided very high accuracy levels ($R^2 = 0.9949 / 0.9965$) compared to the Random Forest model ($R^2 = 0.9735$). The results of the research also refute some of the previous research by suggesting that the second degree polynomial regression models predict clear-sky irradiance losses based on the ability of the models to show smooth and well-behaved nonlinear relationships. This finding suggests that the second degree polynomial regression model provides a robust way of predicting clear-sky irradiance losses using only radiative and meteorological variables, providing ease in the computational capacity needed to produce valuable results.[9], [12], [14] .

Conclusion

The authors analysed clear-sky solar irradiance losses in Ouarzazate, Morocco. They analysed daily meteorological and solar data obtained from NASA/P.O.W.E.R. between 2021 and 2024. They calculated a clearness index (K T = 0.65) to identify 1,070 clear-sky days, which provided the basis for estimating the atmospheric attenuation due to clouds. The authors reported that even though the region has a high solar potential, there were considerable losses of solar irradiance caused by aerosols, water vapour and dust. The authors also identified the second degree polynomial regression model as having the highest R2 value (0.9965) and the lowest RMSE value (0.0079). The authors concluded that the attenuation of clear-sky solar irradiance in semi-arid areas can be described using a smooth, nonlinear relationship. The authors recommend using polynomial regression for quantifying photovoltaic resources and predicting energy output from photovoltaic systems in areas with a similar atmosphere. In the future, the authors recommend using tomorrow's data to create better estimates, monitoring actual ground radiometric measurements, and performing detailed characterizations for aerosols.

References

[1] M. Sengupta, Y. Xie, A. Lopez, A. Habte, G. Maclaurin, and J. Shelby, "The National Solar Radiation Data Base (NSRDB)," *Renew. Sustain. Energy Rev.*, vol. 89, pp. 51–60, Jun. 2018. https://doi.org/10.1016/j.rser.2018.03.003.

[2] J. Antonanzas, N. Osorio, R. Escobar, R. Urraca, F. J. Martinez-de-Pison, and F. Antonanzas-Torres, "Review of photovoltaic power forecasting," *Sol. Energy*, vol. 136, pp. 78–111, 2016. https://doi.org/10.1016/j.solener.2016.06.069.

[3] C. A. Gueymard, "A review of validation methodologies and statistical performance indicators for modeled solar radiation data: Towards a better bankability of solar projects," *Renew. Sustain. Energy Rev.*, vol. 39, pp. 1024–1034, Nov. 2014. https://doi.org/10.1016/j.rser.2014.07.117.

[4] Y. El Mghouchi, A. El Bouardi, A. Sadouk, I. Fellak, and T. Ajzoul, "Comparison of three solar radiation models and their validation under all sky conditions – case study: Tetuan city in northern of Morocco," *Renew. Sustain. Energy Rev.*, vol. 58, pp. 1432–1444, May 2016. https://doi.org/10.1016/j.rser.2015.12.354.

[5] H. Bakhti, "Model-based design of solar thermal power plants," doctoralThesis, Staats- und Universitätsbibliothek Hamburg Carl von Ossietzky, 2024. Accessed: Dec. 24, 2025. [Online]. Available: https://ediss.sub.uni-hamburg.de/handle/ediss/10846

[6] T. Millet *et al.*, "Aerosol Distributions and Transport over Southern Morocco from Ground-Based and Satellite Observations (2004–2020)," *Atmosphere*, vol. 13, no. 6, p. 923, Jun. 2022. https://doi.org/10.3390/atmos13060923.

[7] Abdullah Al-Sharafi, Ahmad Bilal Ahmadullah, Ghassan Hassan, Hussain Al-Qahtani, Abba Abdulhamid Abubakar, and Bekir Sami Yilbas, "Influence of environmental dust accumulation on the performance and economics of solar energy systems: A comprehensive review," *Clean. Energy Syst.*, 2024. https://doi.org/10.1016/j.cles.2024.100125.

[8] A. Ait Ali, Y. Ouhassan, M. Abouyaakoub, M. Chahboun, and H. Hihi, "The Impact of Desert Regions on Solar Energy Production with the Evaluation of Groundwater for Maintenance: A Case Study in Morocco," *Sustainability*, vol. 16, no. 13, Art. no. 13, Jan. 2024. https://doi.org/10.3390/su16135476.

[9] P. Ineichen, "Validation of models that estimate the clear sky global and beam solar irradiance," *Sol. Energy*, vol. 132, pp. 332–344, Jul. 2016. https://doi.org/10.1016/j.solener.2016.03.017.

[10] D. Yang and J. M. Bright, "Worldwide validation of 8 satellite-derived and reanalysis solar radiation products: A preliminary evaluation and overall metrics for hourly data over 27 years," *Sol. Energy*, vol. 210, pp. 3–19, Nov. 2020. https://doi.org/10.1016/j.solener.2020.04.016.

[11] X. Sun, J. M. Bright, C. A. Gueymard, B. Acord, P. Wang, and N. A. Engerer, "Worldwide performance assessment of 75 global clear-sky irradiance models using Principal Component Analysis," *Renew. Sustain. Energy Rev.*, vol. 111, pp. 550–570, Sep. 2019. https://doi.org/10.1016/j.rser.2019.04.006.

[12] C. Voyant *et al.*, "Machine learning methods for solar radiation forecasting: A review," *Renew. Energy*, vol. 105, pp. 569–582, May 2017. https://doi.org/10.1016/j.renene.2016.12.095.

[13] P. Blanc *et al.*, "Direct normal irradiance related definitions and applications: The circumsolar issue," *Sol. Energy*, vol. 110, pp. 561–577, Dec. 2014. https://doi.org/10.1016/j.solener.2014.10.001.

[14] J. A. Ruiz-Arias, C. A. Gueymard, F. J. Santos-Alamillos, and D. Pozo-Vázquez, "Worldwide impact of aerosol's time scale on the predicted long-term concentrating solar power potential," *Sci. Rep.*, vol. 6, no. 1, p. 30546, Aug. 2016. https://doi.org/10.1038/srep30546.

[15] Y. El Mghouchi, A. El Bouardi, A. Sadouk, I. Fellak, and T. Ajzoul, "Comparison of three solar radiation models and their validation under all sky conditions – case study: Tetuan city in northern of Morocco," *Renew. Sustain. Energy Rev.*, vol. 58, pp. 1432–1444, May 2016. https://doi.org/10.1016/j.rser.2015.12.354.

[16] S. Younes, R. Claywell, and T. Muneer, "Quality control of solar radiation data: Present status and proposed new approaches," *Energy*, vol. 30, no. 9, pp. 1533–1549, Jul. 2005. https://doi.org/10.1016/j.energy.2004.04.031.

[17] M. J. Reno and C. W. Hansen, "Identification of periods of clear sky irradiance in time series of GHI measurements," *Renew. Energy*, vol. 90, pp. 520–531, May 2016. https://doi.org/10.1016/j.renene.2015.12.031.

[18] R. Urraca *et al.*, "Extensive validation of CM SAF surface radiation products over Europe," *Remote Sens. Environ.*, vol. 199, pp. 171–186, Sep. 2017. https://doi.org/10.1016/j.rse.2017.07.013.

[19] J. Polo *et al.*, "Preliminary survey on site-adaptation techniques for satellite-derived and reanalysis solar radiation datasets," *Sol. Energy*, vol. 132, pp. 25–37, Jul. 2016. https://doi.org/10.1016/j.solener.2016.03.001.

Enhanced tracking of reference speed and aerodynamic performance through fractional PI control: comparison with conventional PI

Abdelfatah EL OUADIKI[1,a*], Mohammed TAOUSSI[1,2,b] and Mohammed ES-SALMI[1,c]

[1]Industrial Technologies and Services Laboratory, Higher School of Technology, Sidi Mohamed Ben Abdellah University, Fes 30000, Morocco

[2]Spectrometry of Materials and Archaeomaterials Laboratory, Higher School of Technology, Moulay Ismail University, Meknes, Morocco

[a]abdelfatah.elouadiki@usmba.ac.ma, [b]mohammed.taoussi2@usmba.ac.ma, [c]mohammed.essalmi4@usmba.ac.ma

Keywords: Wind Power, MPPT, DFIG, PI Control, FOPI Control, Variable Wind Turbine, Renewable Energy Systems, MATLAB/Simulink

Abstract. Wind energy is now one of the major solutions for addressing energy and environmental challenges. Due to rapid and random fluctuations in wind speed, ensuring maximum power extraction remains a technological challenge. In this work, we compare the classic integral proportional regulator a the fractional FOPI controller by MPPT strategy of a wind turbine. Simulations using MATLAB/Simulink demonstrate that FOPI improves dynamic and steady-state performance from 0.035% to 0.015%, response time from 0.025 s to 0.002 s, and overshoot from 3.8% to 2.0%. It also keeps the maximum power coefficient Cp closer to its optimum. In addition, the FOPI controller offers better stability, accuracy, and robustness in tracking the TSR.

Introduction

The massive integration of wind energy into the global electricity grid has accelerated. Driven by increasing energy demand and the environmental imperatives of reducing greenhouse gas emissions. Aerodynamic-to-electrical energy conversion configured with a DFIG are very important in this context. This is due to the fact that they can operate at very different speeds and they limit the power that passes thru the power electronics to a fraction of the total installed power. At the same time, maximum power point tracking strategies play a key role in ensuring optimal extraction of available wind energy [1,2].

In the majority of doubly-fed wind turbines, the control of the converters is still based on classical proportional-integral regulators, chosen for their ease of implementation and direct physical interpretation. But this control reaches its limits when the system operates in a highly nonlinear environment, with variations in machine parameters, modeling uncertainties, and various disturbances related to the wind profile or the electrical grid. These elements degrade dynamic performance (overshoot, response time) and the grid-injected power quality (transient regimes, wind speed variations).

Therefore, various control strategies have been proposed to ensure improved the behavior of wind systems. Among these approaches, we can mention the optimal speed ratio adjustment [3], predictive current control [4], artificial neural networks, direct torque control [5], as well as MPPT algorithms adapted to variable wind speed [6].

Given these limitations, fractional order controllers, particularly FOPI, offer a more robust solution than conventional PI controllers. With their additional degree of freedom, they better represent this frequency dynamics of the system and reduce sensitivity to parameter variations. Although their effectiveness has been proven in various electromechanical and energy applications, their detailed comparison with conventional PI in a DFIG-based WECS remains understudied, hence the interest of this study. In this context. This study compares traditional fixed-gain PI control with the FOPI regulator for MPPT in DFIG wind turbines. The structure of the article structured in the following manner: Chapter 2 discusses the development of detailed mathematical models for the entire wind energy conversion chain. Section 3 describes their design and tuning. Section 4 examines the comparative results of the simulations obtained in terms of optimal power capture along with enhanced power quality, and robustness. Finally, Section 5 concludes the study and suggests avenues for future work.

Mathematical modeling for the entire wind energy conversion chain and MPPT with speed regulation

The wind is an unpredictable variable, defined by its direction and velocity, which are affected by elements like geographical location and climatic conditions. The formulas for the input power, mechanical power, and torque of the wind turbine (WT) are stated below [7,8].

$$P_t = C_p(\lambda,\beta)\cdot\frac{\rho S V^3}{2}. \tag{1}$$

$$P_t^{opt} = C_p^{\max}(\lambda_{opt},\beta_{opt})\cdot\frac{\rho S V^3}{2}. \tag{2}$$

$$C_t = \frac{P_t}{\Omega_t}. \tag{3}$$

$$C_p(\lambda,\beta) = 0.0068\lambda + \frac{1}{2}(\frac{116}{\lambda_i} - \frac{2}{5}\beta - 5)\cdot\exp(\frac{-21}{\lambda_i}). \tag{4}$$

$$\frac{1}{\lambda_i} = -\frac{0.035}{\beta^3+1} + \frac{1}{\lambda+0.08\beta}. \tag{5}$$

$$\lambda = \frac{R_t\Omega_t}{V}. \tag{6}$$

To accomplish this, turbine rotational speed must vary proportionally with wind speed changes. The ideal turbine speed for achieving maximum power extraction can be articulated as[9]:

$$\Omega_t^{opt} = \frac{\lambda_{opt}V}{R}. \tag{7}$$

The MPPT controller computes the generator reference speed Ω_g^* needed to maintain λ_{opt} :

$$\Omega_g^* = \lambda_{opt}\times\frac{V}{R}\times G. \tag{8}$$

Fig. 1 depicts the functional diagram of the speed-based MPPT control architecture. Wind speed information is used to calculate the optimal reference speed that the speed controller seeks to achieve in order to generate torque commands.

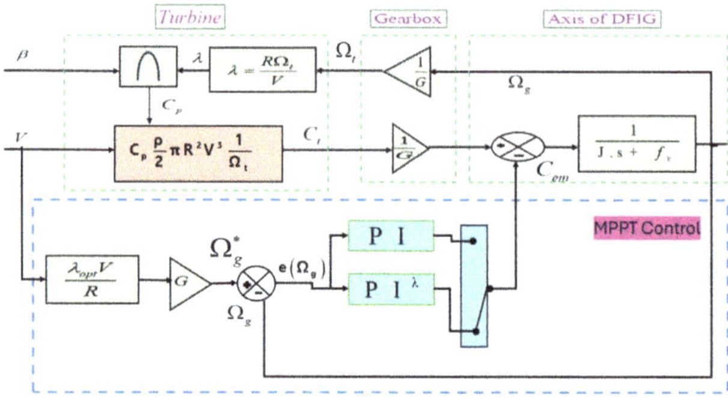

Figure 1: MPPT scheme incorporating speed regulation

Fractional-Order Control for MPPT in Wind Turbine Systems

The fractional calculator extends the conventional concepts of integration and differentiation to include non-integer orders, designated by a unique notation. In this case, the primary distinction between the FOPI regulator and the conventional PI regulator is rooted in the incorporation of fractional calculus. Consequently, by applying the Laplace transform, we can derive the transfer functions for both the FOPI and traditional PI controllers [10]:

$$C(s)=k_p\left(1+\frac{k_i}{s^{\lambda}}\right) \quad ; \quad C'(s)=k_p\left(1+\frac{k_i}{s}\right). \tag{13}$$

Fig. 2 shows the controller structure

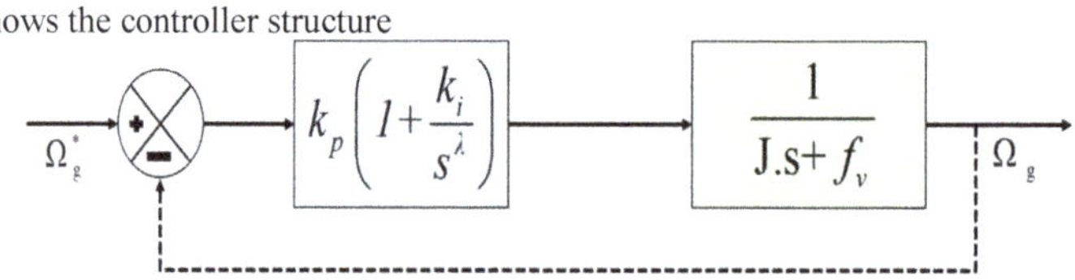
Figure 2: Block representation of the control scheme.

The dynamics of the turbine can be represented using a first-order model as follows:

$$H(s)=\frac{1}{J.s+f_v}. \tag{14}$$

The open-loop L(s) of the wind turbine speed after application of the FOPI controller is:

$$L(s)=H(s)C(s)=k_p\left(\frac{1}{J.s+f_v}\right)\left(1+\frac{k_i}{s^{\lambda}}\right). \tag{15}$$

In this article; the margin of phase and amplitude technique is used for the design of the robust controller [10,11].

Phase margin condition:

$$arg\left[L(j\omega_c)\right]=arg\left[H(j\omega_c).C(j\omega_c)\right]=-\pi+\gamma_m. \tag{16}$$

Gain crossover frequency condition:

$$\left|L(j\omega_c)\right|dB = \left|H(j\omega_c).C(j\omega_c)\right|dB = 0. \tag{17}$$

Stability to fluctuations in system gain:

$$\left(\frac{d\left(arg\left|L(j\omega_c)\right|\right)}{d\omega}\right)_{\omega=\omega_c} = 0. \tag{18}$$

As indicated in the constraint regarding phase margin conditions, a connection can be formed between k_i and λ as:

$$k_i = \frac{-Tan\left[Arctan(T\omega_c)+\gamma_m\right]}{\omega_c^{-\lambda}\, sin\left(\lambda\frac{\pi}{2}\right)+\omega_c^{-\lambda}\, cos\left(\lambda\frac{\pi}{2}\right)\times Tan\left[Arctan(T\omega_c)+\gamma_m\right]}. \tag{19}$$

When robustness condition to the fluctuations in system gain, we can create another equation:

$$k_i = -\frac{-\sigma\pm\sqrt{\sigma^2-4C^2\omega_c^{-2.\lambda}}}{2\omega_c^{-2.\lambda}} \tag{20}$$

With $C = \left(\frac{1+(T\omega_c)^2}{T}\right)^{-1}$ and $\sigma = -\left[\lambda\omega_c^{-\lambda-1}\sin\left(\lambda\frac{\pi}{2}\right)-2C\omega_c^{-\lambda}\cos\left(\lambda\frac{\pi}{2}\right)\right]$

k_p can be written as mentioned in Constraint Gain crossover frequency condition:

$$k_p = \frac{\sqrt{1+(T\omega_c)^2}}{k\sqrt{\left[1+k_i\omega_c^{-\lambda}\cos\left(\lambda\frac{\pi}{2}\right)\right]^2+\left[k_i\omega_c^{-\lambda}\sin\left(\lambda\frac{\pi}{2}\right)\right]^2}}. \tag{21}$$

These passages elucidate the procedures for tuning a fractional PI controller to satisfy frequency requirements. This procedure utilizes system parameters (inertia J) and viscous friction (f_v) to ascertain the equivalent constants of the model. To prevent reliance on a strictly analytical approach, a numerical solution is employed utilizing MATLAB. First, (ω_c) and the phase γ_m are established, and the fractional order (λ) is mechanically determined, guaranteeing consistency between the expression of (K_i) obtained from the phase condition (equation 19) and that derived from the system structure (equation 20). The solution exclusively preserves the physically valid solution, namely one with ($K_P> 0$) and ($K_i> 0$). Once (λ) is identified, the program computes (K_i) using the appropriate expressions and subsequently incorporates it into the module equation (equation 21) to determine (K_P), ensuring compliance with the conditions at the cutoff frequency. Finally, the MATLAB script generates the adjustment parameters for the proposed controller: ($\lambda = 0.2264,\ K_i = 15.6236,$) and ($K_p = 0.0309,$)),which comprehensively define the FOPI corrector employed in the subsequent simulations and analyses. Subsequently, the proposed controller can be depicted as follows:

$$C_1(s) = 0.0309\left(1+\frac{15.6236}{s^{0.2264}}\right). \tag{22}$$

Results and Discussion

We used MATLAB/Simulink to model the energy conversion system that used a 1.5 MW wind turbine. The parameters used in the simulation study are as follows:

$$f_v = 0.0024\text{N.m.rad}^{-1}.\text{s} /\ \text{J} = 1000\ Kg.m^2\ /\ R = 3525\ \times 10^{-2} m\ /\ \rho = 1.225\ \text{kg}.m^{-3}\ \ /\ G = 90$$

The effectiveness of the suggested approaches is tested in the following distinct scenarios: one with sudden wind variations, and another based on actual speeds measured in Dakhla (Morocco). Mechanical speed control is analyzed under progressive variations in wind speed in Fig. 3 (a), and the illustration in Fig. 3 (b), How a real wind pattern is modeled for 20 seconds and it encompasses both sudden step changes (to evaluate transient response) and gradual adjustments (to assess tracking performance). This FOPI controller quickly reaches the maximum power factor value of 0.48 and maintains it with high accuracy. It demonstrates exceptional stability in steady state, as depicted in Fig. 3 parts (c) and (d), Both zones extend from this confirmation, demonstrating a rapid response at start-up and stability in steady state, thus ensuring perfect TSR maintenance. Fig. 3 (e), and Fig. 3 (f), show that the proposed controller maintains the TSR value around the target value of 8.16, while the PI controller exhibits slight fluctuations. Thus, the FOPI controller quickly brought the wind speed error back to zero, as depicted in Fig. 3 parts (i) and (j), The variations within the rotor's angular velocity for both controllers, as well as for the reference controller, are illustrated in Fig. 3 (g), and Fig. 3 (h). The FOPI controller demonstrated exceptional tracking behavior compared to the PI controller. When the system starts up, the proposed controller reaches the setpoint faster and with less overshoot than the PI controller. Table 1 in the previous conversation compares the PI model to the FOPI model in terms of accuracy and speed of execution.

Materials Research Forum LLC
https://doi.org/10.21741/9781644904091-19

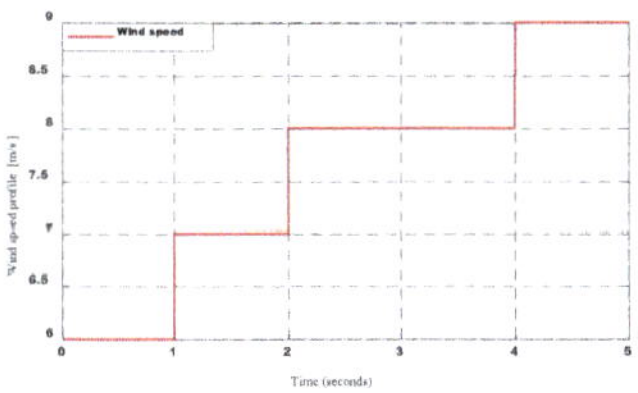

(a) *profile step variations* of *scenario 1*

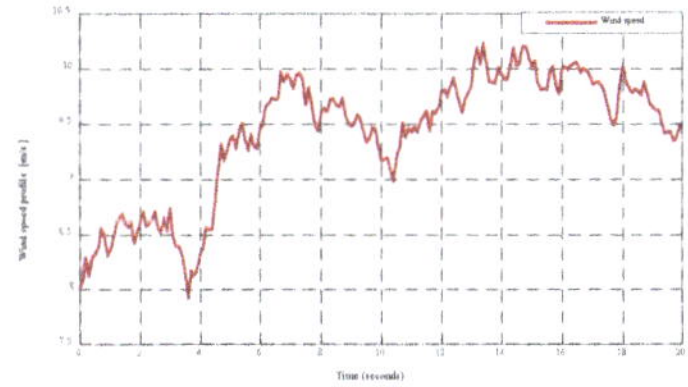

(b) *real variations* of *scenario 2*

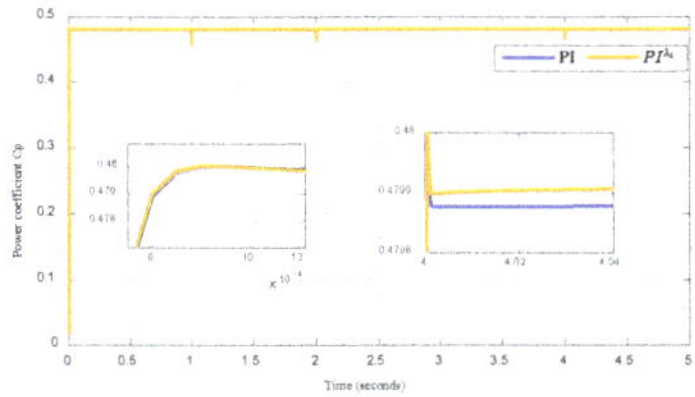

(c) Power coefficient

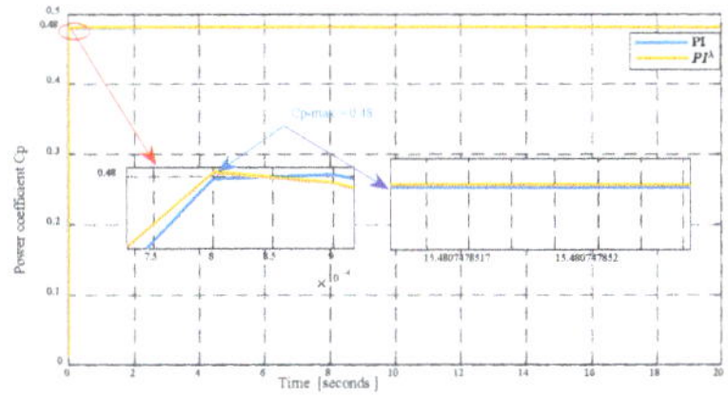

(d) Power coefficient

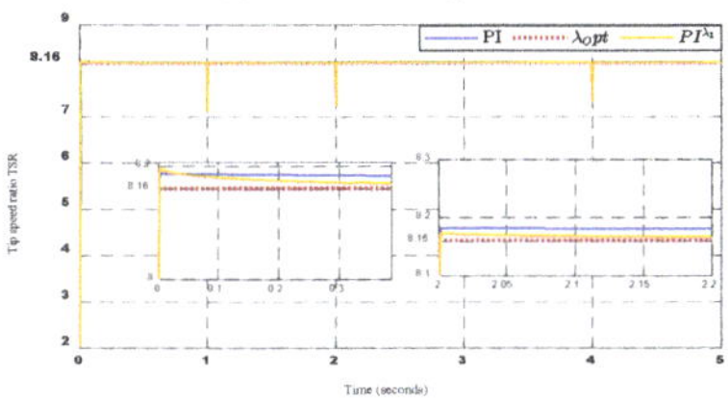

(e) TSR

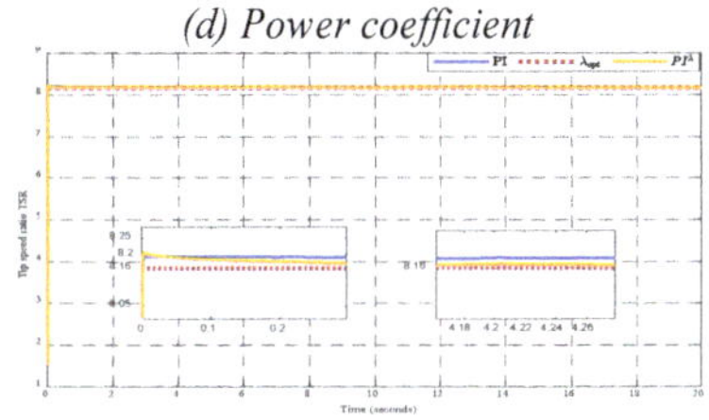

(f) TSR

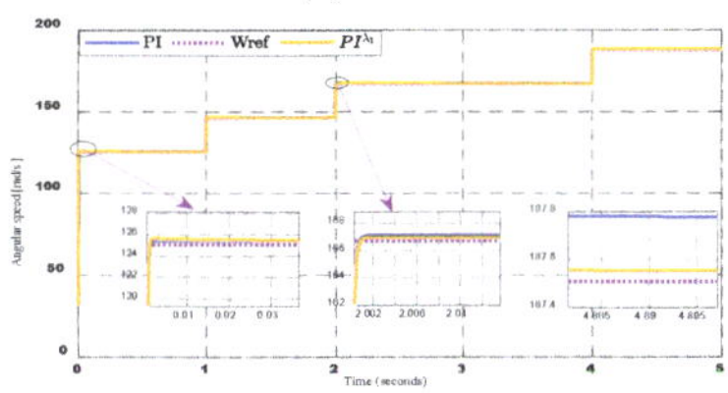

(g) Rotor Angular Speed

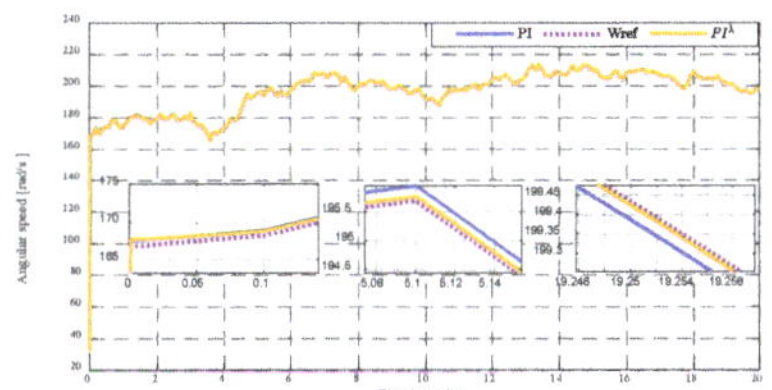

(h) Rotor Angular Speed

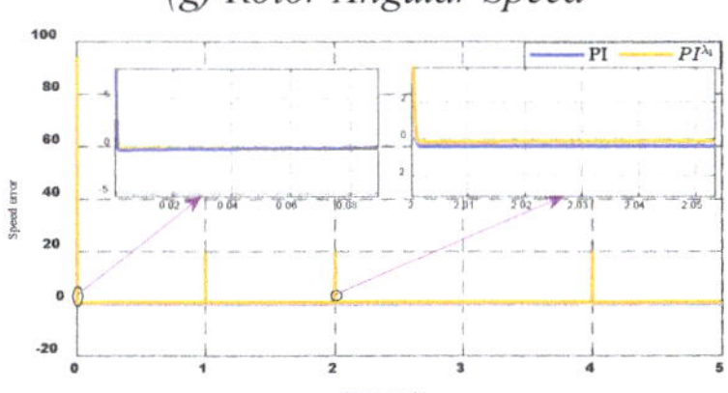

(i) wind Speed Error

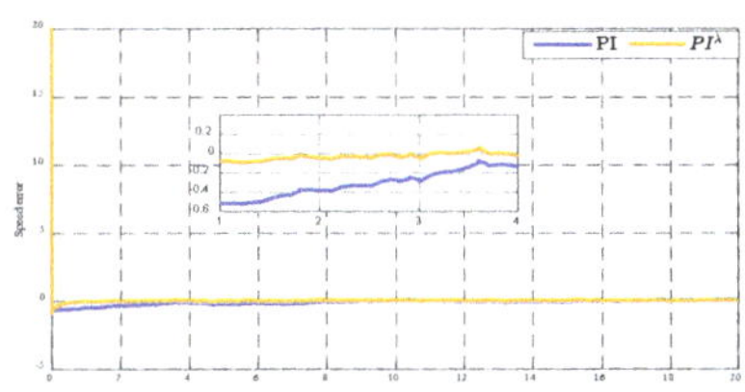

(j) wind Speed Error

Figure 3: Results of the two scenarios.

Table 1: Quantitative Performance Comparison

Controller	Tracking Error at Steady State (%)	System Response Duration [s]	Peak overshoot (%)	Robust stability	Effectiveness of TSR	Robust Behavior
PI	0.035	25×10^{-3}	3.8	Satisfactory	Limited	Limited
PI^{λ}	0.015	2×10^{-3}	2.0	Excellent	Effective	Effective

In order to situate the proposed controller within the field of research related to reference speed tracking and aerodynamic performance, we have created Table 2, which summarizes the results obtained from the current literature and compares them to the performance of other proposed controllers. The findings distinctly indicate the likely advantages of our method. The OTC technique achieves average efficiency in setpoint tracking, whereas AFTISMC delivers satisfactory performance in this regard. In contrast, our approach is notable for its exceptionally quick responsiveness, delivering effective setpoint tracking with minimal steady-state error, nearly negligible static error, and swift system response, along with superior setpoint tracking abilities that considerably lower steady-state error.

Table 2: Quantitative Performance Comparison

Method	MPPT Strategy	Settling Time [s]	Offset Error (%)	Reference Following
[12]	Backstepping	0.00500	1.1000	Effective
[13]	ANN	0.00313	0.5200	Effective
[14]	AFTISMC	-	0.0751	Very good
Proposed	PI^{λ}	0.00200	0.0130	Highly Effective

Conclusions and future research directions

We analyzed the performance of a conventional PI controller with an FOPI controller designed for maximum power point tracking (MPPT) in a wind turbine using a DFIG. The results of various wind speed simulations reveal faster response time and lower overshoot and steady-state error for the proposed controller than for the simple PI controller, and improved performance in tracking the ideal C_p. These results corroborate the assertion that using the FOPI controller would increase energy recovery capacity, improve the quality of the energy produced, and minimize the mechanical impact on the system, avoiding the constraints of the conventional PI control approach. The proposed controller can be built in real time on a dSPACE DS1104 platform or an FPGA board (depending on computing needs and/or latency requirements) and is planned as future work in this area. This phase will include code generation, wind turbine emulator testing, and performance testing under simulated wind and power grid conditions. Real-time resilience, noise blocking, and actuator saturation management will also be addressed. In the event of significant variations in the operating scenario, an adaptive version of the FOPI controller may be considered. The actual objective will be to demonstrate robust and reproducible performance that complies with grid standards, is reproducible, and is adapted to the operating scenario applicable in real-world situations.

References

[1] El Mourabit Y, Derouich A, El Ghzizal A, El Ouanjli N and Zamzoum O 2020 Nonlinear backstepping control for PMSG wind turbine used on the real wind profile of the Dakhla-Morocco city Int Trans Electr Energ Syst 30 https://doi.org/10.1002/2050-7038.12297

[2] Ackermann T 2012 Wind power in power systems (John Wiley & Sons) https://doi.org/10.1002/9781119941842

[3] Balasundar C, Sudharshanan S and Elakkiyavendan R 2015 Design of an optimal tip speed ratio control MPPT algorithm for standalone WECS Int. J. Res. Appl. Sci. Eng. Technol 3

[4] Guo H, Qiu T and Jiang J 2012 Design of fuzzy controller for direct drive wind turbines J. Shanghai Jiaotong Univ. (Sci.) 17 209-14 https://doi.org/10.1007/s12204-012-1254-4

[5] Koutroulis E and Kalaitzakis K 2006 Design of a maximum power tracking system for wind-energy-conversion applications IEEE transactions on industrial electronics 53 486-94 https://doi.org/10.1109/TIE.2006.870658

[6] Sabzevari S, Karimpour A, Monfared M and Naghibi Sistani M B 2017 MPPT control of wind turbines by direct adaptive fuzzy-PI controller and using ANN-PSO wind speed estimator Journal of Renewable and Sustainable Energy 9 https://doi.org/10.1063/1.4973447

[7] Es-Salmi M, Taoussi M, El Ouadiki A and Derouich A 2024 Optimizing Direct Power Control of Variable Speed Wind Turbine Systems Using Genetic Algorithms and Space Vector Modulation 2024 3rd International Conference on Embedded Systems and Artificial Intelligence (ESAI) (IEEE) pp 1-8 https://doi.org/10.1109/ESAI62891.2024.10913888

[8] Taoussi M, Karim M, Hammoumi D, El Bekkali C, Bossoufi B and El Ouanjli N 2017 Comparative study between Backstepping adaptive and Field-oriented control of the DFIG applied to wind turbines 2017 International Conference on Advanced Technologies for Signal and Image Processing (ATSIP) (IEEE) pp 1-6 https://doi.org/10.1109/ATSIP.2017.8075592

[9] Taoussi M, Karim M, Bossoufi B, Hammoumi D, Lagrioui A and Derouich A 2016 Speed variable adaptive backstepping control of the doubly-fed induction machine drive IJAAC 10 12 https://doi.org/10.1504/IJAAC.2016.075140

[10] El Ouadiki A, Taoussi M and Es-Salmi M 2025 A robust maximum power point tracking from wind turbines using fractional order PI controller E3S Web of Conferences vol 680 (EDP Sciences) p 00034 https://doi.org/10.1051/e3sconf/202568000034

[11] Benbouhenni H, Hamza G, Oproescu M, Bizon N, Thounthong P and Colak I 2024 Application of fractional-order synergetic-proportional integral controller based on PSO algorithm to improve the output power of the wind turbine power system Scientific Reports 14 609 https://doi.org/10.1038/s41598-024-51156-x

[12] Nadour M, Essadki A and Nasser T 2017 Comparative analysis between PI & backstepping control strategies of DFIG driven by wind turbine International Journal of Renewable Energy Research 7 1307-16

[13] Yessef M, Bossoufi B, Taoussi M, Lagrioui A and Chojaa H 2022 Overview of control strategies for wind turbines: ANNC, FLC, SMC, BSC, and PI controllers Wind Engineering 46 1820-37 https://doi.org/10.1177/0309524X221109512

[14] Chen Q, Wang L, Sun Y, Xie S and Wang R 2024 Adaptive integral sliding mode MPPT control for wind turbines with fixed-time convergence 18 4265-76 https://doi.org/10.1049/rpg2.12935

Materials Research Proceedings 64 (2026) 164-170 https://doi.org/10.21741/9781644904091-20

Deep Hybrid Temporal Modeling with TCN-GRU-Transformer for Lithium-Polymer Battery State-of-Charge Estimation

Omar LAMMAMRI[1,a*], Brahim ZRAIBI[2,b], Mohamed MANSOURI[1,c], Abdelghani EZZAHI[1,d] and Adam AFADISS[1,e]

[1]Hassan First University of Settat, National School of Applied Sciences of Berrechid, Laboratory LAMSAD, Morocco

[2]Cadi Ayyad University, EST of Safi, Industrial Maintenance Department, LAPSSII Laboratory, Morocco

[a]omar.lammamri.doc@uhp.ac.ma, [b]b.zraibi@uca.ac.ma, [c]mohamed.mansouri@uhp.ac.ma, [d]abdelghani.ezzahi.doc@uhp.ac.ma, [e]adam.afadiss.doc@uhp.ac.ma

Keywords: State of Charge, Lithium Polymer Battery, Deep Learning, TCN, GRU, Hybrid Model

Abstract. Accurate state-of-charge (SOC) estimation is critical for safe and efficient battery operation in electric vehicles and stationary energy storage systems. This study proposes a hybrid deep learning model that combines a Temporal Convolutional Network (TCN) for local temporal feature extraction, a Gated Recurrent Unit (GRU) for sequential dynamics, and a Transformer block to capture longer-range dependencies in battery signals. The model is trained and validated using experimentally measured lithium-Polymer battery data and is benchmarked against standalone TCN, GRU, and Transformer baselines under the same data splits and evaluation protocol. On the test set, the proposed hybrid architecture achieves a mean absolute error (MAE) of 0.66%, a root mean square error (RMSE) of 0.88%, and a coefficient of determination (R^2) of 0.9992, consistently outperforming the individual models. The results indicate that integrating complementary temporal modeling mechanisms improves estimation accuracy and robustness, supporting real-time SOC prediction in practical battery management systems.

Introduction

The increasing use of electric vehicles and renewable energy storage solutions requires lithium-based batteries to function at their maximum operational abilities. The characteristics of lithium-polymer (Li-Po) cells make them suitable for various applications because they combine high energy density with pouch design options and their power capacity supports multiple uses from traction systems to unmanned aerial vehicles and portable electronics. Li-Po cells display complex electrochemical properties which create challenges for accurately monitoring their internal charge levels because their performance depends on temperature changes and their aging process and their response to different power demands [1,2].

The SOC indicates how much energy a battery still holds compared to its total capacity, which serves as a key measurement that battery management systems (BMS) use to monitor battery performance. The battery state of charge information helps to predict driving range and power derating and prevents unsafe conditions through its ability to identify overcharging and over-discharging dangers [3]. The continuous SOC miscalculations result in two major problems, which are wrong range predictions and premature device shutdowns and battery cells that suffer quicker degradation, which together decrease both operational efficiency and safety.

Conventional SOC estimation techniques, such as Coulomb counting, open-circuit-voltage (OCV) look-up tables, and observer-based approaches derived from equivalent-circuit or electrochemical models, can deliver strong accuracy when parameters are properly identified and operating conditions are relatively well controlled. In practical applications, however, their performance often degrades under aging effects, temperature variability, and complex real-world duty cycles. These limitations have motivated increased interest in data-driven SOC estimation methods that can better capture nonlinear dynamics and adapt to changing conditions [4].

Deep learning methods have therefore attracted considerable interest, as they can learn nonlinear mappings from measured signals to SOC without explicit physical models. Recent studies have explored recurrent neural networks (LSTM, GRU), Temporal Convolutional Networks (TCN), convolutional-recurrent hybrids, and Transformer-based architectures for SOC and state-of-health (SOH) estimation, with promising results on various lithium-ion chemistries[6]. TCNs leverage dilated causal convolutions to capture long temporal dependencies with high parallelism, while GRU/LSTM networks remain strong baselines for sequence regression, and Transformers with attention mechanisms can emphasize informative time steps and handle complex load profiles [6]. Nonetheless, most existing work focuses on cylindrical or pouch lithium-ion cells, and only limited attention has been given to lithium-polymer batteries within hybrid architectures that jointly exploit the complementary strengths of TCN, GRU, and Transformer. This gap motivates the present study, which investigates such a unified framework for SOC estimation in Li-Po cells [7].

This paper presents a hybrid deep learning model based on a TCN-GRU-Transformer architecture for SOC estimation in Lithium-Polymer batteries. The analysis compares architectural characteristics, computational demands, predictive accuracy, and practical implementation aspects. Furthermore, it investigates how integrating these three components within a single hybrid framework leverages their complementary strengths.

Methodology

A. Lithium Polymer Battery Dataset

High-quality data are a prerequisite for the development of reliable data-driven models. In practice, two main strategies are commonly adopted: (i) acquiring measurements through dedicated experiments on the target cell and (ii) exploiting datasets made available by trusted external sources. Although the former is generally preferred, since publicly available datasets rarely match the exact operating conditions of interest, the present study relies on a private dataset supplied by the author of reference [8]. All SOC estimation models are developed for a lithium-polymer cell, whose principal characteristics are reported in Table 1.

Table 1:Li-polymer cell: key specifications.

Attribute	Details
Manufacturer	Power-Xtra
Nominal capacity	350 mAh
Charging voltage	4.2V±0.05V
Nominal voltage	3.7V±0.05V
Discharge cut-off voltage	*3.0V*
Weight	*10.5 gr ±0.5 g*

The reference SOC is obtained off-line by combining high-precision Coulomb counting with periodic rest phases and OCV-based recalibration, following standard procedures adopted in experimental SOC studies. The available dataset covers a wide SOC range [0,100]% and different current rates, temperature conditions, and operating modes to ensure that the trained model generalizes beyond a single drive cycle.

B. Data preprocessing

In the proposed framework, a deep learning model is trained and evaluated using a dataset with three input variables, cell temperature, terminal voltage, current and the corresponding SOC targets. Raw measurements are first filtered, synchronized, and reformatted into sequences compatible with the network input. The processed data are then split into two non-overlapping subsets for training and testing, respectively, each containing paired input–output samples obtained from laboratory measurements on the cell. Before training, the neural network architecture and its hyperparameters (such as the number of layers, hidden units, learning rate, batch size, and regularization terms) are defined. The model parameters are subsequently optimized on the training set by minimizing a loss function that penalizes the discrepancy between predicted and measured SOC. After convergence, the generalization capability of the model is assessed on the held-out test set using several error-based performance indicators to quantify the accuracy of SOC estimation. As summarized in Fig. 1, the overall workflow proceeds from data preprocessing through model training to final testing, after which the trained model can be deployed and its performance examined under real operating conditions.

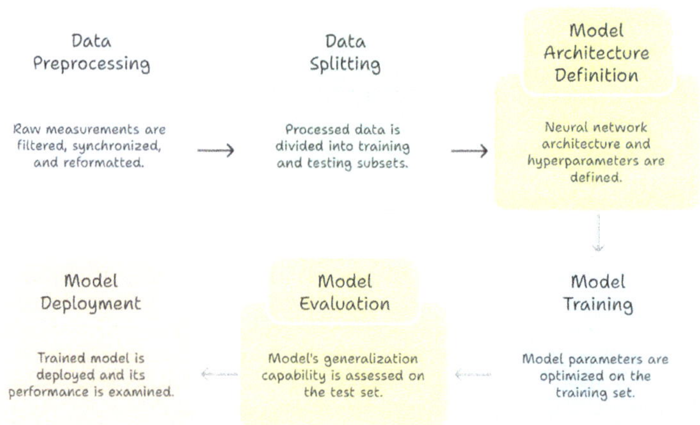

Figure 1:Deep Learning Model Training and Evaluation Workflow

C. Evaluation metrics

All models are trained using the mean squared error (MSE) between predicted and true SOC as the primary loss. The AdamW optimizer with a suitable learning rate and weight decay is used, together with mini-batch training and early stopping based on validation loss.

The metrics [9] are defined as follows:

$$MAE = \frac{1}{K}\sum_{k=1}^{k}|y_k - \widehat{y_k}| \quad (1)$$

$$RMSE = \sqrt{\frac{1}{N}\sum_{i=1}^{N}(y_k - \widehat{y_k})^2} \quad (2)$$

$$R^2 = 1 - \frac{\sum_{i=1}^{n}(y_i - \hat{y}_i)^2}{\sum_{i=1}^{n}(y_i - \bar{y})^2} \quad (3)$$

where y_k and $\widehat{y_k}$ denote the true and predicted SOC for sample k, $\bar{y}$ is the mean of the true value, and K is the number of samples.

Materials Research Proceedings 64 (2026) 164-170 https://doi.org/10.21741/9781644904091-20

Base Model Architecture

In this section, we introduce the architecture of our hybrid model for SOC estimation. The approach combines three types of networks, a TCN, a GRU, and a Transformer encoder. Each one processes the same input sequence (such as voltage, current, and temperature) and outputs its own scalar SOC estimate. By bringing together these different ways of learning temporal patterns, the model aims to provide a more stable and accurate SOC prediction.

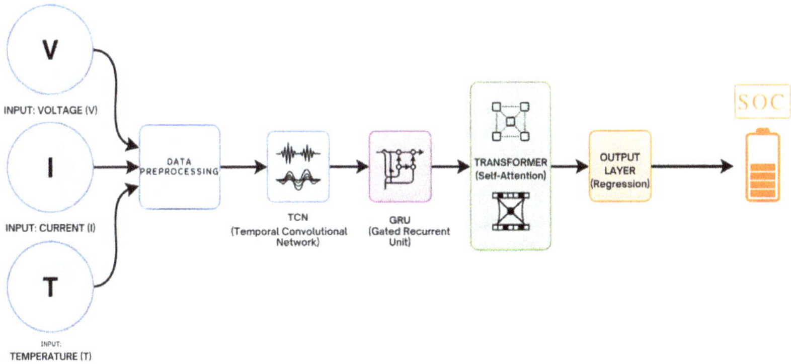

Figure 2: Architecture of the proposed model

A. Temporal Convolutional Network (TCN)

A TCN is a one-dimensional convolutional architecture designed for sequence modeling that enforces causality and preserves temporal order. Each layer applies a causal 1-D convolution along the time axis so that the output at time $t$depends only on inputs from times $\leq t$. Long-range dependencies are captured through dilated convolutions, where a kernel of size $k$with dilation factor $d$effectively covers $k * d$ past samples; stacking layers with increasing dilation yields a large receptive field with moderate depth and parameter count. Layers are organized in residual blocks composed of two dilated causal convolutions, non-linear activations, and dropout, with a skip connection that adds the block input to the block output to stabilize the training of deep stacks. The final temporal feature map is then reduced (by selecting the last time step or applying global pooling) and mapped by a fully connected layer to produce the SOC estimate [10].

B. Gated Recurrent Unit (GRU) Network

The GRU is a recurrent neural cell that maintains a hidden state $\mathbf{h}_t$ summarizing the battery's recent operating history and is particularly suitable for SOC prediction. At each time step, the GRU uses two gates, an update gate and a reset gate to control how much information from the previous hidden state and the new input (current, voltage, temperature) is incorporated into the candidate state, and then blends the old and candidate states to form $\mathbf{h}_t$. This gated mechanism allows the network to selectively retain or discard past information, mitigating vanishing-gradient effects and preserving long-range temporal dependencies without an excessive number of parameters. In the proposed SOC estimation framework, the final hidden state (or the sequence of hidden states) is passed through a dense output layer to map the learned temporal representation of the current-voltage trajectory to a scalar SOC value, enabling the GRU to capture the smooth, history-dependent evolution of SOC under realistic load profiles [11].

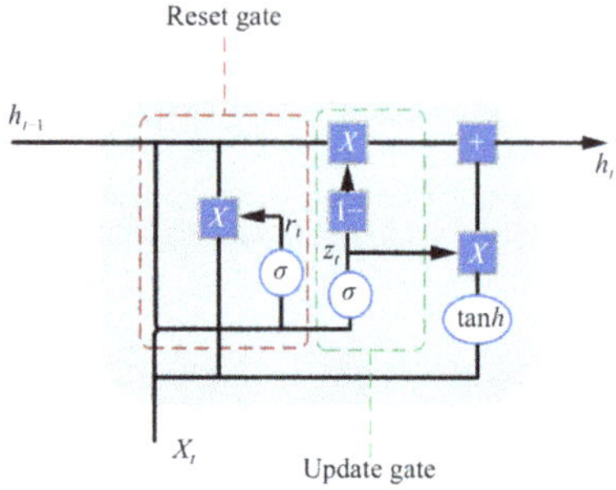

Figure 3: Architecture of the GRU model

C. Transformer Encoder

The Transformer encoder in our hybrid SOC estimation framework acts as a global context extractor that complements the local TCN features and the sequential memory of the GRU. After projecting the input battery time series (current, voltage, temperature–derived features) into a latent space and adding positional encodings, the encoder applies multi-head self-attention to let each time step attend to all others within the observation window. This mechanism enables the model to highlight informative segments such as current pulses or relaxation phases and to capture long-range dependencies that are difficult to model with convolutions or gated recurrent units alone. The resulting attention-based representation is then fused with the TCN and GRU outputs, providing a rich global-local description of battery dynamics that improves the robustness and accuracy of SOC prediction across varying load profiles [12].

Results and discussion

This section presents and analyzes the performance of the proposed hybrid architecture for SoC estimation in lithium-Polymer batteries, comparing a TCN, a GRU, and the hybrid TCN-GRU-Transformer model. All models are trained on 20 560 data samples and validated on 4 141 previously unseen samples. Figure 4 illustrates the predicted SoC trajectories versus the corresponding ground truth, where the brown curve denotes the reference SoC and the red curve indicates the model outputs. The qualitative assessment provided by these plots is complemented by quantitative performance indicators, namely the MAE, RMSE, and coefficient of determination (R^2), computed over the different operating conditions.

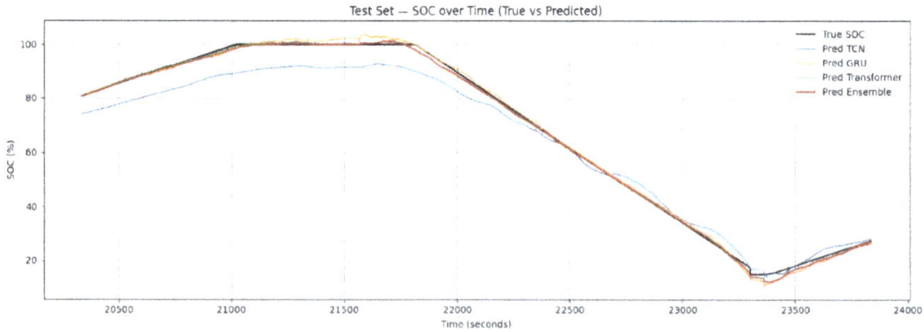

Figure 4: Results of SOC estimation in Lithium Polymer Battery

Table 2: SOC Prediction Performance Comparison

Model	MAE [%]	RMSE [%]	R^2
TCN	5.29	6.23	0.9584
GRU	1.09	1.37	0.9980
Transformer	0.69	0.90	0.9991
Hybrid (TCN-GRU-Transformer)	0.66	0.88	0.9992

According to the results presented in the table 2, TCN has the lowest predictive capability, indicating that temporal feature extraction by convolutional neural networks alone is insufficient to capture the nonlinear and history-dependent dynamics of Li-Po battery behavior. In contrast, both GRU and Transformer offer significantly higher accuracy, with Transformer proving to be the most effective standalone architecture, reflecting the advantage of attention mechanisms in learning longer-term dependencies in battery signals.

The proposed hybrid model provides the best results for all reported metrics, indicating that complementary temporal encoders increase the reliability of SOC predictions. This is also reflected in the SOC trajectories: the recurrent and attention-based models closely follow the reference SOC in dynamic operating profiles, while the TCN shows a more noticeable deviation during fast transients. Collectively, these finding support the conclusion that SOC estimation in Li-Po batteries benefits most from architectures that explicitly model sequential dependencies and global context, and that combining different types of approaches can further improve SOC prediction for practical BMS applications.

Conclusion

The aim of this work is to develop a hybrid deep-learning model for estimating the state of charge of lithium-polymer batteries using the combination of Temporal Convolutional Networks, Gated Recurrent Units, and Transformer encoders. The current, terminal voltage, and temperature are the primary inputs for the learning process, which is based on an experimentally acquired LiPo dataset.

A single models, such as GRU model and Transformer can deliver high predictive accuracy, achieving a test MAE value below 1% and R^2 values approximately 0.999. Basing on this baseline, the proposed approach improves performance further, achieving MAE = 0.6627%, RMSE = 0.8769%, and $R^2 = 0.9992$ on the test set. Moreover, the predicted SOC traces closely reproduce the ground truth evolution under multiple dynamic operating profiles, indicating stable tracking behavior.

Overall, combining complementary temporal feature extraction with attention-based sequence modeling produces an accurate SOC estimator for Li-Po cells. Future work will focus on: (i) adaptive ensemble combination via meta-learning; (ii) extending the framework to multi-task estimation of SOC and state of health (SOH); (iii) incorporating physics-based constraints and calibrated uncertainty estimates; and (iv) implementing refinements to meet real-time embedded BMS requirements.

References

[1] G. Kostenko, A. Zaporozhets, Transition from Electric Vehicles to Energy Storage: Review on Targeted Lithium-Ion Battery Diagnostics, Energies 17 (2024) 5132. https://doi.org/10.3390/en17205132

[2] G.G. Njema, R.B.O. Ouma, J.K. Kibet, A Review on the Recent Advances in Battery Development and Energy Storage Technologies, J. Renew. Energy 2024 (2024) 2329261. https://doi.org/10.1155/2024/2329261

[3] M.A. Pisani Orta, D. García Elvira, H. Valderrama Blaví, Review of State-of-Charge Estimation Methods for Electric Vehicle Batteries, World Electr. Veh. J. 16 (2025) 87. https://doi.org/10.3390/wevj16020087

[4] K. Yang, Y. Tang, S. Zhang, Z. Zhang, A Deep Learning Approach to State of Charge Estimation of Lithium-Ion Batteries Based on Dual-Stage Attention Mechanism, Energy 244 (2022) 123233. https://doi.org/10.1016/j.energy.2022.123233

[5] J.-H. Lee, I.-S. Lee, Hybrid Estimation Method for the State of Charge of Lithium Batteries Using a Temporal Convolutional Network and XGBoost, Batteries 9 (2023) 544. https://doi.org/10.3390/batteries9110544

[6] M.S. Nazim, M.M. Rahman, R.B. Mofidul, M.M.B. Rimu, Y.M. Jang, Robust State of Charge Estimation for Lithium-Ion Batteries Using a Fully Entangled Temporal Convolutional Network with Particle Swarm Optimization, J. Power Sources (2025) 237020. https://doi.org/10.1016/j.jpowsour.2025.237020

[7] S. Guo, L. Ma, A Comparative Study of Different Deep Learning Algorithms for Lithium-Ion Batteries on State-of-Charge Estimation, Energy 263 (2023) 125872. https://doi.org/10.1016/j.energy.2022.125872

[8] G. Taş, A. Uysal, C. Bal, A New Lithium Polymer Battery Dataset with Different Discharge Levels: SOC Estimation of Lithium Polymer Batteries with Different Convolutional Neural Network Models, Arab. J. Sci. Eng. 48 (2023) 6873–6888. https://doi.org/10.1007/s13369-022-07586-8

[9] B. Zraibi, M. Mansouri, C. Okar, Comparing Single and Hybrid Methods of Deep Learning for Remaining Useful Life Prediction of Lithium-Ion Batteries, E3S Web Conf. 297 (2021) 01043. https://doi.org/10.1051/e3sconf/202129701043

[10] Y. Liu, J. Li, G. Zhang, B. Hua, N. Xiong, State of Charge Estimation of Lithium-Ion Batteries Based on Temporal Convolutional Network and Transfer Learning, IEEE Access 9 (2021) 34177–34187. https://doi.org/10.1109/ACCESS.2021.3057371

[11] J. Chen, X. Feng, L. Jiang, Q. Zhu, State of Charge Estimation of Lithium-Ion Battery Using Denoising Autoencoder and Gated Recurrent Unit Recurrent Neural Network, Energy 227 (2021) 120451. https://doi.org/10.1016/j.energy.2021.120451

[12] M.A. Hannan, D.N.T. How, M.S.H. Lipu, M. Mansor, P.J. Ker, Z.Y. Dong, K.S.M. Sahari, S.K. Tiong, K.M. Muttaqi, T.M.I. Mahlia, F. Blaabjerg, Deep Learning Approach Towards Accurate State of Charge Estimation for Lithium-Ion Batteries Using Self-Supervised Transformer Model, Sci. Rep. 11 (2021) 19541. https://doi.org/10.1038/s41598-021-98915-8

A Hybrid πVAE-XGBoost Framework for High-Fidelity Simulation of Bifacial Solar Thermal Collectors

Imane FAKIR[1,a,*], Mehdi NEJJAR[1,b], Mohammad Hachim ERAISSOUNI[1,c], Adam KESSAB[1,d] and Ahmed KHALLAAYOUN[1]

[1]Al Akhawayn University in Ifrane, Hassan II Avenue, Ifrane, 53000, Fes-Mekness, Morocco

[a]I.fakir@aui.ma, [b]M.nejjar@aui.ma, [c]M.eraissouni@aui.ma, [d]A.kessab@aui.ma

Keywords: Bifacial Solar Thermal Collectors, Hybrid Simulation, Deep Learning, πVAE, XGBoost, Renewable Thermal Energy, Nonlinear Modeling

Abstract. The following paper describes a new hybrid system which combines a deep learning-based pvae model with an XGBoost regressor to produce high-fidelity bifacial solar thermal collector high-fidelity simulation. Through modelling the proposed approach overcomes the above-mentioned challenges by capturing complex, nonlinear interactions between critical system variables, including temperature and irradiance and flow rates, as-sociated with the modeling of a pair-exposure collectors. The high-dimensional input space is effectively reduced to a strong latent representation using the pVAE and further predictions are refined by including the latent features along with the original measurements to the XGBoost component to make reliable predictions of heat transfer and outlet temperature. The validation based on a dataset of 147,396 sequential measurements, as well as comparisons of traditional MATLAB Simulink-CARNOT simulations, shows that the predictive performance is almost perfect, with the values of R2, Spearman, and Pearson correlation coefficients being over 99%. This mixed approach does not only increase the degree of simulation accuracy but also provides explainable information to optimize the design of solar thermal systems, which opens its way to the use in the further renewable energy systems.

Introduction

With the growing heating needs, renewable thermal energy technologies are becoming very critical towards sustainable energy systems. In this specific case, solar thermal energy has taken a center stage in its ability to offer heating solutions that are carbon-free, leading to massive developments in the design of the collectors [1].

Simulating and modeling the behavior of bifacial solar thermal system, nevertheless, remain to be difficult. The power production is determined by a combination of various factors (which are interdependent) and hence nonlinear and multivariable. Traditional physics-based simulation techniques need to consider these intricate interactions and may need the intensive computation [2]. It has been demonstrated that the conditions around the ground can dramatically influence the rear-side irradiance in the depth of the array, and thus, a finely detailed geometry and environmental modeling is required to give valid results. To overcome these challenges, scientists are becoming eager to resort to the use of data-driven methods (especially, deep learning) to increase the quality of the simulation and optimize the design. Recent studies have also shown that machine learning models could be used to enhance the forecast of key performance parameters in solar-thermal systems, including the non-linear behavior of the system that would be recreated with high fidelity [3].

Background

Alternatively referred to as bifacial solar thermal collectors and as the double-exposure flat-plate collectors, the latter have exterior surfaces on the front and the rear that enable them to capture solar energy on both sides [4]. This new design uses trans-clear glazing in place of conventional rear insulation and uses a reflector under the collector to reflect the sun onto its lower part as well as to a great degree increase solar collection over traditional single-faced systems. The experimental studies have shown bifacial collectors can provide about 41-66% more thermal performance than single equivalent monofacial units and simulation studies showed that a bifacial system with tunable reflector positions could capture more than 50 percent more solar energy in a year than the traditional counterparts [5]. Major features of the design are dual-sided absorber plates glazing both sides, and tilting mounting or reflector mechanisms to achieve maximum irradiation on both sides. In addition, these collectors have been successfully applied in agricultural practices like drying of crops.

Mathematical Model of Bifacial Collector

$$Q = \int [G_b(t)\,\mathrm{IAM}_\theta(t) + G_d(t)\,\mathrm{IAM}_{\mathrm{diff}}]\eta_{\mathrm{avg}}\, dt \quad (1)$$

Q in Equ.1 [6] is the total energy production (in kWh or J) by the combination of the total action of beam and diffuse irradiance over time. The beam (direct) solar irradiance, G_b, is in measurements of W/m^2 and modified by the incidence angle modifier $IAM(\theta)$ which is used to describe the performance of the collector with respect to the collector angle of incidence. In the meantime, diffuse solar irradiance G_b, also in W/m^2, is weighted by an independent modifier, IAM_{diff}, which can be uniform or empirical. The average thermal efficiency of the collector, η_{avg} (dimensionless) is the efficiency of the collector to harness the available solar energy and transform it into useful heat. Lastly, the dt represents the time interval (e.g. hourly, per minute) on which the energy contributions are energy-integrated to give the overall thermal output of the system.

Related Work

Bifacial solar thermal collectors are an advanced solar energy technology which attracts direct sunlight and reflected energy so as to enhance the thermal performance. In [7], the analysis in question is taken to a deeper level to depict increased energy yields, caused by improved absorber materials along with reflective surface conditions. These systems result in high efficiency because of maximized heat collecting capacity that reduce thermal losses that are mostly found in reflective settings.

Further, [6] explores the use of bifacial collector in the framework of computational mod-eling and economic analysis. Bifacial collectors are weather-adjustable, as well as reducing the long-term financial needs in solar thermal systems. The core advantage of the bifacial technology is based on the ability to recover more irradiance of reflected sunlight, based on the two research project findings. The results show that future studies should focus on refining system designs, creating strategies of produce reflectivity, and incorporation of bifacial collectors into hybrid energy systems to increase efficiency and sustainability.

Data Context

The data in this research was gathered and given with the consent of the Institute of New Energy Systems (InES) in Germany. It has 147,396 consecutive measurements of a bifacial solar thermal energy system, and these could be taken at intervals of 5 seconds between August 21 and August 29 of 2024.

Materials Research Proceedings 64 (2026) 171-178 https://doi.org/10.21741/9781644904091-21

Table 1. Definition and Units of Variables Used in the Study.

Variable	Description	Unit
T_in_1C	Inlet temperature	[°C]
T_out_1C	Outlet temperature	[°C]
T_mass_flow_1C	Mass flow temperature	[°C]
T_amb_outdoorC	Ambient temperature	[°C]
T_vent_1C	Vent duct temperature	[°C]
dT_1K	Temperature difference (ΔT)	[K]
Q_1kW	Thermal power	[kW]
Vol_flow_1lh	Volumetric flow rate	[l/h]
m_flow_1kgh	Mass flow rate	[kg/h]
rho_1kgl	Fluid density	[kg/l]
cp_1kJkgK	Specific heat capacity	[kJ/(kg·K)]
v_wind_speedms	Wind speed	[m/s]
G_trackerWm	Global irradiance on tracker	[W/m²]
G_diffuseWm	Diffuse irradiance	[W/m²]
IR_TrackerWm	Infrared radiation	[W/m²]
T_IR_Tracker	Tracker surface temperature	[°C]
zenith	Solar zenith angle	[°]
azimuth	Solar azimuth angle	[°]

The methodological Framework of Bifacial Collector Simulation

Classical Framework

A bifacial solar thermal collector simulation was first performed in MATLAB Simulink with the Carnot Toolbox. MATLAB Simulink data organization platform provides the opportunity to generate input files and process system reaction responses via data manipulation and visualization across various conditions. The Carnot Toolbox enables to simulate the energy collection of both sides as well as taking into account the ground albedo and irradiance distribution.

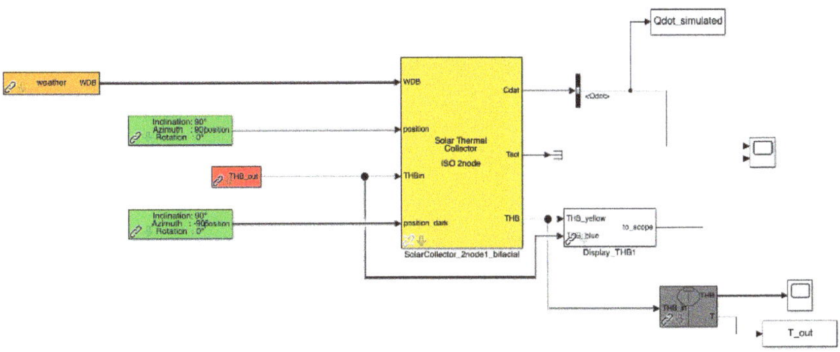

Fig. 1. External Structure of the Bifacial Model.

Materials Research Proceedings 64 (2026) 171-178 https://doi.org/10.21741/9781644904091-21

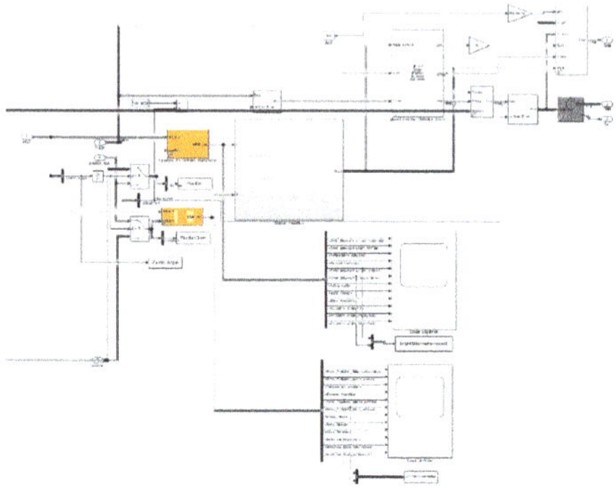

Fig. 2. Internal Structure of the Bifacial Collector Model.

The bifacial solar thermal collector (Fig. 1) is a device with an absorber plate, transparent cover and a switching system, which tilts +-90deg east-west to absorb maximum solar. It absorbs energy in direct, diffuse and reflected light both on the front and back surfaces. Addi-tional components regulate the flow of energy inside of the body and allow dual-surface absorption. The system has a Zenith Angle Block to position east-west and a WDBout and a WDBsurface connection to monitor rear-side irradiance, as presented in Fig. 2. It includes also two layers of energy absorption to replicate heat transfer, Scopebrightside and Scopedarkside blocks to monitor irradiance, MassFlow Bus to control water flow, as well as THByellow, THBblue, and Tmean node to track thermal performance.

Deep Learning Based Framework

Two complementary machine learning methods were used to completely model the work of the bifacial collector. To evaluate the input variables, first, a deep learning model (πVAE) was used to predict the heat transfer (Q 1kW) by taking advantage of the ability to model complex, nonlinear associations and quantify uncertainty in the latent representations of πVAE [8]. This deep learning workflow is the initial component of the framework as shown in Fig. 1. At the second stage, an XGBoost regression model is used to refine the prediction by combining the πVAE-calculated heat transfer and the initial input variables to predict the outlet temperature ($T_{out}1C$).

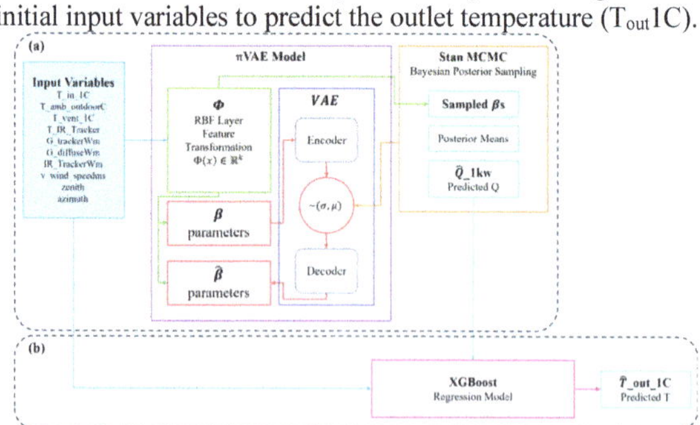

*Fig. 3. Flow diagram of the simulation of the bifacial collector workflow: (a) **π**VAE model involves predicting heat transfer Q1kW using the input variables. (b) XGBoost model makes use of the input variables and the predicted heat transfer to estimate T_{out} 1C.*

Dataset Processing
The data was split into training and testing groups. The training was done with data collected between August 21 and August 29, with the exception of August 28, which was completely held out to test. This design made sure that the test set is made up of a complete day of observations that offered an impartial evaluation of the models on unseen, real-world data hence facilitating a comprehensive evaluation of the models in terms of generalization.

***π**VAE*
The learning of the multifaceted correlation among the variables of the system inputs versus the heat exchange is performed in the first phase, where the πVAE model is trained. Fig. 3 mentions the input variables, which are temperature readings (inlet, ambient, ventilation), irradiance levels (global, diffuse, infrared on a tracker), wind speed, and solar zenith angle. These inputs are processed via a feature mapping function Φ(s) that forms a high dimensional representation that represents the underlying spatial correlations.

Secondly, the weights of each of the functions are then encoded using a variational autoencoder (VAE), denoted by β. Here, the encoder is used to compute the mean and the standard deviation of a Gaussian distribution at the β values, and, with the reparameterization trick, the latent variables Z are sampled. Decoder then recovers the β coefficients out of these latent samples. Most importantly, the result of the feature mapping Φ(s) (i.e. the basis functions) is multiplied with these reconstructed β values to create the resultant function output through inner product β^T Φ(s).

The latent space is then sampled using Hamiltonian Monte Carlo (HMC) to draw samples of the posterior distribution of Z. HMC is especially good at this task because it can easily explore the uncorrelated, low dimensional latent space. Lastly, the average of the posterior samples is also calculated and made the robust point estimate in generating the reconstructed function output. For each HMC sample $Z^{(i)}$, the decoder produces a corresponding output:

$$y_2^{(i)}(s) = d\left(Z^{(i)}\right)^T \Phi(s) \tag{2}$$

where:

$$\beta = d(Z) \tag{3}$$

When mean is taken of all the N samples, the predicted heat transfer is:

$$\hat{Q}(s) = \frac{1}{N}\sum_{i=1}^{N} y_2^{(i)}(s) = \frac{1}{N}\sum_{i=1}^{N} d\left(Z^{(i)}\right)^T \Phi(s) \tag{4}$$

XGBoost Regressor
Having used the πVAE method on the whole dataset to estimate the heat transfer Q 1kW., we split the data as before. The predicted $\hat{Q}$_1kW is then used together with the original input variables to predict the outlet temperature T_out_1C via an XGBoost regression model.
The XGBoost objective function minimizes the regularized loss [9]:

$$\mathcal{L} = \sum_{i=1}^{n} l\left(T_{\text{out\_1C}}^{(i)}, \widehat{T_{\text{out\_1C}}^{(i)}}\right) + \sum_{k=1}^{K} \Omega(f_k) \tag{5}$$

Where $\mathcal{L}$ is the loss function, f_k denotes each individual tree in the ensemble, and $\Omega(f_k)$ is the regularization term that penalizes model complexity.

Materials Research Proceedings 64 (2026) 171-178 https://doi.org/10.21741/9781644904091-21

Simulation in Practice

Classical Simulation vs. Real World Data

(a) (b)

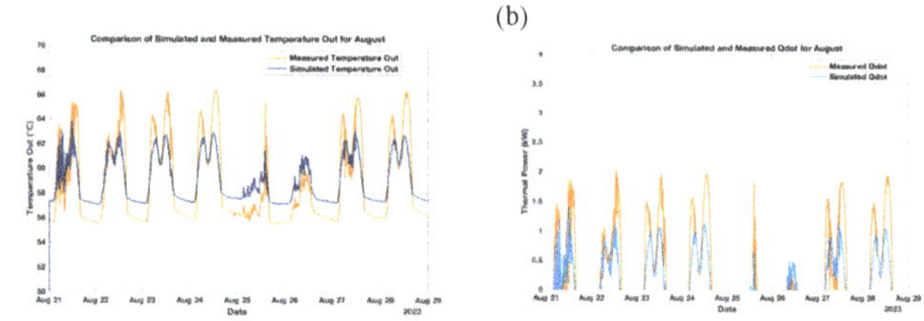

Fig. 4. Comparison of Simulated and Measured Values: (a) Temperature Out using CARNOT. (b) $\dot{\boldsymbol{Q}}$ *using CARNOT.*

The comparison between the simulated and experimental temperature values along with thermal power indicators proves the level of accuracy and reliability of the thermal model created. Through the two comparisons, the generated data and the experimental data are found to be similar and display the changes in solar irradiance as a repetitive trend in the data. The real temperature is sometimes in Fig. 4a. much higher than the simulated temperature up to 66-67°C compared to the simulated temperature of 63-64°C. The measured ($\dot{Q}$) of Fig. 4b can be described as having a higher peak values which can go up to 1.9-2.1 kW. The simulated ($\dot{Q}$) values are slightly below and oscillate around 1.0-1.2 kW.

Hybrid Framework Results

Training Results

(a) (b)

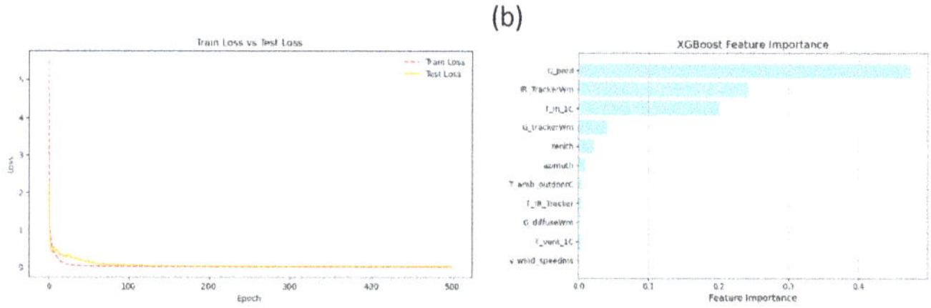

Fig. 5 πVAE-XGBoost Evaluation- Hybrid model: (a) loss- πVAE training as epochs, and (b) feature importance ranking of XGBoost.

Table 2. Models Performance Metrics: $\boldsymbol{R}^2$*, Spearman correlation and Pearson correlation.*

Model	R^2	Spearman	Pearson
πVAE	99.98%	99.86%	99.90%
XGBoost Regressor	100%	99.99%	100%
Hybrid Model	99.99%	99.92%	99.95%

The hybrid πVAE-Boost model has strong results on performance as seen in the evaluation metrics and the features analysis of Fig. 5. At the point of training, the πVAE component reaches almost perfect convergence (Fig. 5a) and its output Q turns out to be the most important aspect in the importance ranking of the XGBoost model (Fig. 5b). It concentrates on the effect of the pVAE to decrease high-dimensional input to low-dimensional, informative latent representation Q, which decreases downstream predictive capacity. This synergy is measured in Table 2: The single πVAE brings impressive R^2: 99.98% Spearman: 99.86% Pearson: 99.90% and the XGBoost regressor uses Q and other characteristics (performance table is awaiting). The hybrid design, therefore, takes advantage of the representation learning features of the πVAE and the explainable weighted feature approach of XGBoost, which indicates that it is an effective architecture when faced with a task that requires both accuracy and interpretability.

(a) (b)

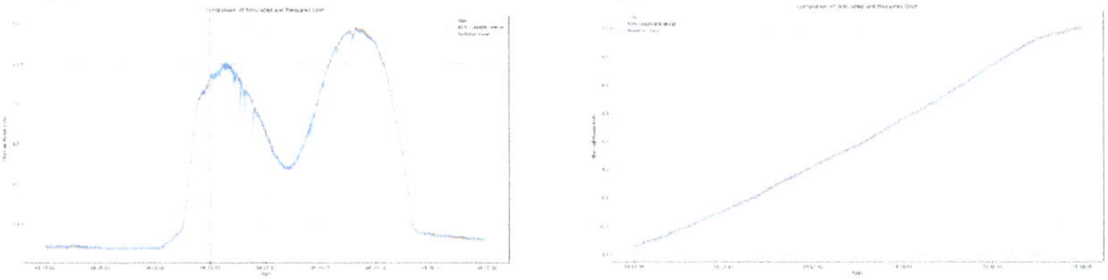

Fig. 6. Visual representation of predictions in the test set of the model: (a) Full simulation, 95% confidence interval, and (b) magnified section showing the quality of the fit.

Fig. 6 illustrates the high-fidelity predictions of the model with the entire simulation (**Fig. 6**a) data being closely aligned between simulated and measured values, and the magnified region (**Fig. 6**b) showing the accuracy of the fit within a 95% confidence interval.

(a) (b)

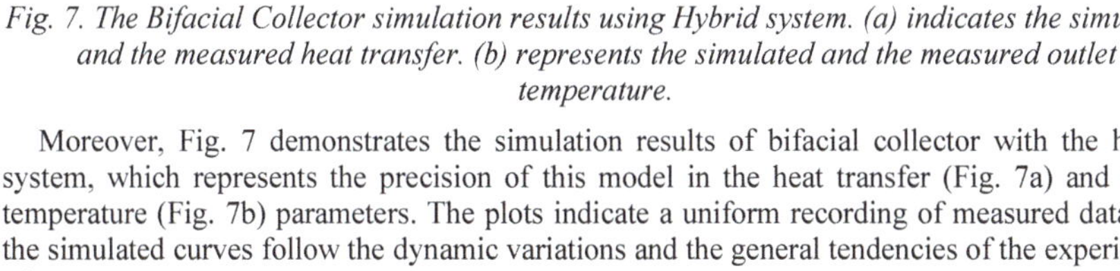

Fig. 7. The Bifacial Collector simulation results using Hybrid system. (a) indicates the simulated and the measured heat transfer. (b) represents the simulated and the measured outlet temperature.

Moreover, Fig. 7 demonstrates the simulation results of bifacial collector with the hybrid system, which represents the precision of this model in the heat transfer (Fig. 7a) and outlet temperature (Fig. 7b) parameters. The plots indicate a uniform recording of measured data, and the simulated curves follow the dynamic variations and the general tendencies of the experimented measurements.

Conclusion

After the presentation of this study, it is possible to argue that a hybrid model has been formed, which is a deep learning-based πVAE and an XGBoost regressor, simulating bifacial solar thermal collectors. Our model provides almost a perfect predictive accuracy by modeling the complex non-linear interactions between the key variables and the simulated results have a significant correlation with the experimental data and the conventional MATLAB Simulink-CARNOT results. The πVAE is helpful in dimensionality reduction of data, whereas the XGBoost feature improves the interpretability of features in the data. These findings indicate the possibility of the model to simplify the design of solar thermal systems, and the future studies aim at adaptive design and longer environmental integration.

References

[1] M. Benramdane, S. Abboudi, and M. Ghernaout, "Contribution to the simulation and parametric analysis of the operation of a solar concentration thermal installation," International Journal of Heat and Technology, vol. 37, no. 2, pp. 446–456, Jun. 2019. https://doi.org/10.18280/ijht.370210.

[2] M. Ernst, G. E. J. Conechado, and C.-A. Asselineau, "Accelerating the simulation of annual bifacial illumination of real photovoltaic systems with ray tracing," iScience, vol. 25, no. 1, p. 103698, Dec. 2021. https://doi.org/10.1016/j.isci.2021.103698.

[3] J. D. Osorio, M. De Florio, R. Hovsapian, C. Chryssostomidis, and G. E. Karniadakis, "Physics-informed machine learning for solar-thermal power systems," *Energy Conversion and Management*, vol. 327, p. 119542, 2025. https://doi.org/10.1016/j.enconman.2025.119542.

[4] N. Nikolić and N. Lukić, "Theoretical and experimental investigation of the thermal performance of a double exposure flat-plate solar collector," Solar Energy, vol. 119, pp. 100–113, Sep. 2015. https://doi.org/10.1016/j.solener.2015.06.038.

[5] N. Nikolic, N. Lukic, A. Nesovic, and D. Nikolic, "Optimal size and position of the planar back reflector moveable only in the direction normal to the bifacial solar collector plane," Thermal Science, vol. 28, no. 6A, pp. 4483–4497, 2024. https://doi.org/10.2298/TSCI240123101N.

[6] T. Summ, M. Ehrenwirth, C. Trinkl, and W. Zörner, "Agri Solar Thermal Systems: A brief study on the energetic potential of bifacial solar thermal systems," in Proc. EuroSun 2022 – ISES and IEA SHC Int. Conf. Solar Energy for Buildings and Industry, Kassel, Germany, 2022, pp. 1–8. https://doi.org/10.18086/eurosun.2022.12.16.

[7] T. Summ, B. Sturm, C. Scherer, C. Trinkl, and W. Zörner, "Analysis of thermomechanical stresses and strains for a novel design of insulating glass flat-plate collector," in Proc. Solar World Congress 2023, New Delhi, India, 2023, pp. 1–11. https://doi.org/10.18086/swc.2023.04.14.

[8] S. Mishra, S. Flaxman, T. Berah, H. Zhu, M. Pakkanen, and S. Bhatt, "πVAE: A stochastic process prior for Bayesian deep learning with MCMC," Statistics and Computing, vol. 32, no. 6, p. 96, Oct. 2022. https://doi.org/10.1007/s11222-022-10151-w.

[9] T. Chen and C. Guestrin, "XGBoost: A scalable tree boosting system," arXiv preprint, arXiv:1603.02754, accessed Mar. 25, 2025. [Online]. Available: https://arxiv.org/abs/1603.02754

Energy Futures - ICEF 2026 Materials Research Forum LLC
Materials Research Proceedings 64 (2026) 179-186 https://doi.org/10.21741/9781644904091-22

Modeling a Solar-Panel–Actuated Light Detection System for Multi-State LED Output

Doha GARAGUE[1,a *]

[1]Al Akhawayn University, P.O. Box 104, Hassan II Avenue, 53000, Ifrane, Morocco

[a]D.Garague@aui.ma

Keywords: LED Control, Automatic Lighting, Energy Efficiency, Solar Panel Simulation, Tinkercad Circuits, Arduino, Light Sensing, Photovoltaic Power

Abstract. The simulation of an automatic light system, which involves the use of ambient light sensors and energy detected from solar panels, can minimize the use of electricity, but nevertheless, provide reliable light sources during different light conditions. Light-Dependent Resistor (LDR) sensors are used in switching on the LEDs when the light intensity drops below 600 lux and switching off the LEDs when the light level increases to above 800-1000 lux, and so they will avoid unnecessary use of energy and ensure that the sensors efficiently respond to the changes in the light intensity. Research also suggests that sensor-controlled lighting and solar panel energy can save up to 65% of the energy used by outdoor lighting systems (and solar LED streetlights can adjust the amount of light they emit (up to 85%) based on the conditions of the environment). The results are favourable for applications of photovoltaic-driven auto lights in real-life scenarios, such as street lighting systems and outdoor lighting, where energy efficiency and visibility at night are critical.

Introduction

The utilization of the sensors has transformed automation in electronics, which can be observed in terms of the energy-saving lighting systems that rely mostly on light detection. This research focuses mainly on simulating and building a simple, affordable day/night detector, which can be conducted using simple equipment to track and control LEDs using a solar panel's output. The circuit needed is built in Tinkercad, so it provides a theoretical and practical application of that model [1-3].

Some previous studies have confirmed that solar-panel-based lighting systems have been largely implemented due to their reliability, efficiency, accuracy, and sensitivity to different light conditions. For instance, Anam et al. [4] stated and showed how LDRs can be used to implement an efficient automatic public-road lighting system to decrease the energy consumption. Also, Sukri et al. [5] shed light on the importance of the use of LDR devices not only for basic and simple ON/OFF lighting tracks but also for multi-level and complex brightness variations in smart lamps due to the same reasons: efficiency and stability in real-life applications. These statements can also be confirmed by Ahmmed et al. [6], who integrate and connect LDRs to motion sensors for further optimization in street-light energy usage. All these studies highlight the importance of solar panels that rely on physical hardware when used as LDRs in automated lighting, leaving simulation-based pedagogical tools underrepresented in the literature. In this project, to ensure that no energy is unnecessarily wasted, and depending on the ambient light, the method to be used is lighting by automated means. The latter is in line with the global sustainability aspirations of smart electronics and IoT.

The literature related to solar-panel-based systems points to their acceptance and utilization in different areas, which is corroborated by energy savings in automated lighting. These works argue for the stability of LDRs but are often implemented physically; thus, simulation-based pedagogical tools remain a gap.

By focusing on virtual prototyping in Tinkercad, detailed code algorithms, and quantitative analyses for novices, this research stands apart from those that presume the availability of hardware. Its innovation is in the integration of a sensor simulation with the feasibility of the real-world application, such as the adaptation of street lights. The article outlines components and methods, presents results with discussions, and concludes with the effects and next steps.

Materials and Methods

The whole design and simulation of this project were done with the Tinkercad Circuits platform. The material used in the circuit was an Arduino Uno microcontroller, a miniature solar panel as a light-intensity sensor, one red LED and one green LED, two 330 Ω resistors for current limiting, a breadboard, and jumper wires. A full view of the connected circuit was taken in Tinkercad for the wiring layout and the interaction between the solar panel, the Arduino input, and the LED outputs.

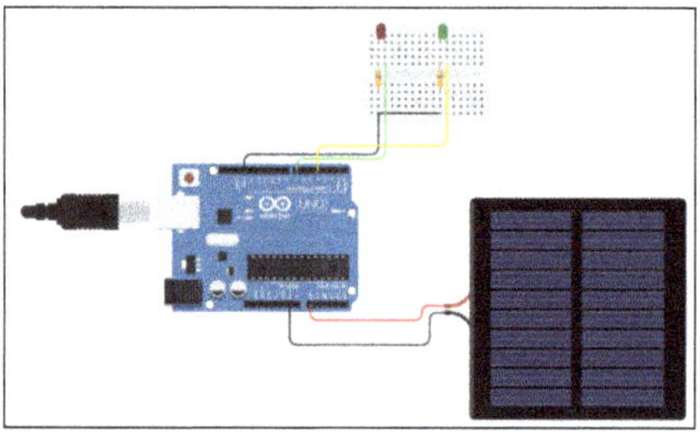

Fig. 1. Full circuit assembly in Tinkercad

Within this setup, the solar panel is the source of light for the traditional LDR replacement. Its voltage output is directly proportional to the amount of light that falls on it: a bright one will produce a higher voltage, while a dark one will produce almost zero output. Arduino reads this voltage through the analog input pin A0; thus, the solar panel is the device that detects the light. Moreover, the use of a solar panel even indirectly points to the use of a clean energy source in the project; thus, energy efficiency is maintained.

The two LEDs were wired up with the Arduino's digital pins 4 and 7, each with a 330 Ω resistor. The red LED indicates low or medium light levels, and the green LED indicates high illumination based on the threshold values, which were determined by observing the solar panel outputs under different lighting conditions.

The control logic was done by using Tinkercad's block programming language. The Arduino keeps the analog value from the solar panel and compares it with two threshold values, which were found through the experiment. These thresholds were determined by observing the analog reading in the Serial Monitor in real-time under different lighting conditions. The flowchart below summarizes the logic of the working flow and the decision-making process that the Arduino follows. The built circuit begins by analysing and adjusting the solar panel's illumination intensity, observing the behavior of the LEDs and how they response in respect to the serial monitor displayed values. The findings of this procedure vary depending on three lighting conditions: full bright light, partial light, and full dark light.

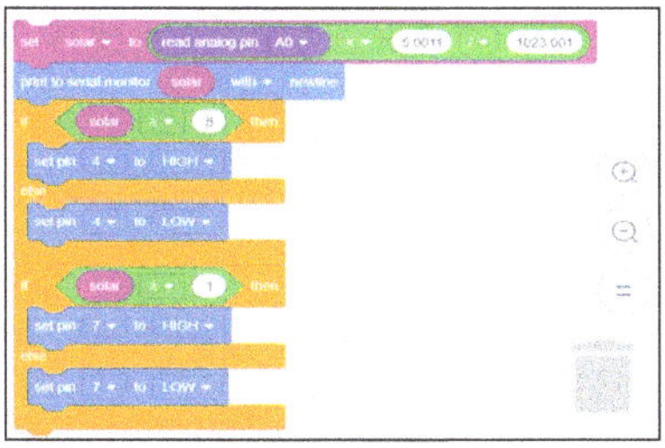

Fig. 2. Block code representing LED control logic

The methodology of this block of code doesn't get affected by any external conditions or physical variations and changes, such as temperature or panel aging; it considers the simulation goes under ideal conditions, like having a consistent solar panel response and the absence of any wiring losses. The fundamental approach of this simulation focused on demonstrating a real-life application of Automatic Street Lighting control and energy-efficient operation with less need for physical components and hardware.

The imitation system is very close to being a typical street-light system, except for the controllers, which are usually programmed with an inverted logic for part of the street-lighting network. Normally, the light would be turned on when it gets dark, and in this simulation, the LEDs light up when there is more light, which is an experimental demonstration rather than a direct street-light implementation.

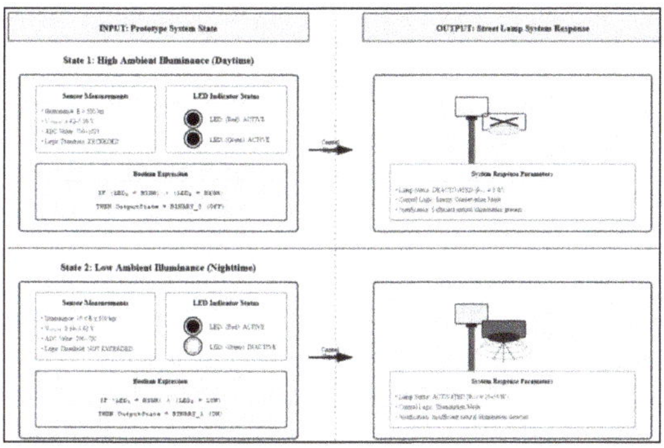

Fig. 3. Automatic Street Light System - Real-World Application

Results and Discussion

1. LED Behavior Under Different Light Conditions

Full Darkness Condition

Under the full dark mode, the solar panel doesn't receive any illumination signals; therefore, it generates a zero voltage. In this condition, both red and green LEDs remain OFF since the Arduino program reads a value that is in the lowest region of the detection range. This output can be justified with the threshold logic flow integrated in the simulation, since the value of the read value is so small that it remains below the first threshold V_{th1}. The absence of any response or reaction from the LEDs indicates that the simulation is functioning correctly, detecting Nighttime accurately.

The serial monitor displays a reading analog value of 0, validating the state of the full darkness condition simulation. The displayed value indicates that the solar panel sensor output remains at its minimum level. This can also justify the behavior of the circuit and the solar panel sensors when no photons are incident on the panel's surface, and it verifies that the Arduino captured a low-level signal input during full dark mode.

Medium-Light (Partial Darkness) Condition

When the solar panel gets exposed to moderate illumination, the solar panel starts generating a slightly higher voltage, leading the analog reading to become higher to reach the first threshold V_{th1}, but still below the second threshold V_{th2}. Under this phase, only the red LED activates, which indicates that the external environment of the solar panel surface has transformed out of full darkness but has not reached full or complete brightness. After simulating the circuit with partial darkness, the red LED illuminated while the green remained OFF, which indicates that the program flow of the threshold is correctly demonstrating the intermediate range.

The serial monitor displays a value of 4, which confirms and justifies the behavior of the circuit. This numerical value represents a meaningful voltage change on the solar panel, even if it seems small. This confirms the efficiency of the analog-digital conversion and the ability of the program to distinguish between the full dark mode and the medium illumination condition.

Full Brightness Condition

Under full bright conditions, the sensors of the used solar panel generate the highest voltage, making the analog reading reach the higher level threshold V_{th2}. In that phase, the system operates in its highest state, activating both red and green LEDs simultaneously. This behavior is derived from the LEDs, indicating that the sensors detect a high-intensity light, which is Daytime, and the multi-system mechanism is functioning as designed.

The Serial Monitor is showing 5 as a reading, which is the highest value recorded during the tests, and it is the way it confirms this state. The numerical change from the previous condition is very small, but it is significant enough to indicate the sensor's movement to the upper activation band. That number is the one that makes sure that the Arduino is the one reading the light being very bright and that it is the one causing the two LEDs to light up, thus confirming the last step of the system's threshold-based control logic

2. Quantitative Analysis

The displayed values from the serial monitor during the simulation under each of the three conditions were organized in a table summarizing the relationship between light mode, solar-panel voltage, LED state, and relevant interpretation.

Materials Research Proceedings 64 (2026) 179-186 https://doi.org/10.21741/9781644904091-22

Table 1: Quantitative Table

Light Condition	Solar-Panel Voltage (ADC Reading)	LED Response	Interpretation
Full Darkness	0 – 200 ADC (0.00 – 0.98) [V]	OFF	No illumination is required since the threshold voltage is below $V_{th1} = 200$. The system indicates nighttime.
Partial Light	201 – 700 ADC (0.98 – 3.42) [V]	Red ON	The sensor is above V_{th1} but below V_{th2}, indicating the transitional phase of the lighting condition. The activation of a single LED demonstrates an intermediate state detection.
Full Bright	701 – 1023 ADC (3.43 – 5.00) [V]	Red + Green ON	The sensor exceeds the second threshold V_{th2} ($V_{th2} = 700$), indicating a full LED state. Both LEDs are active, verifying the detection of bright ambient conditions.

The threshold logic can be mathematically expressed as:

$$LED\ Output = \begin{cases} 0 & V_{panel} \leq V_{th1} \\ Red\ LED\ ON & V_{th1} < V_{panel} \leq V_{th2} \\ Red + Green\ LEDs\ ON & V_{panel} > V_{th2} \end{cases}$$

From the collected data, a curve representing the LED-response versus the light intensity was quantitatively drawn. The plot shows how the system transformations and transitions between the three different states of lighting occur as the solar-panel readings increase.

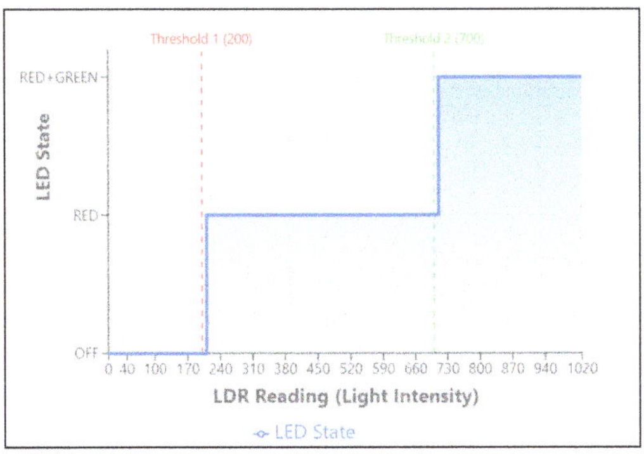

Fig. 4. LED Response vs Intensity Graph

To deepen the interpretation, the electrical energy that was necessary to power the LEDs was also estimated.

$$P = V_{LED} I_{LED} \quad (1)$$

For the typical LED conduction voltage of around 1.8 V and a measured current of I_{LED}= 15mA, the power per LED is:

$$P \approx 1.8 \times 0.015 = 0.027\ W \quad (2)$$

When the two LEDs are ON at full intensity, the total consumption is thus:

$$P_{total} \approx 2 \times 0.027 = 0.054\ W \quad (3)$$

It shows that the energy is still very minimal even at the highest state, and therefore, the issue of energy waste is not a problem that automated LED systems, for example, solar street lighting, should address.

3. Logic Verification

The threshold logic was well verified using a logic truth table that summarizes the solar-panel voltage intervals to the corresponding LED states. This table confirms that the Arduino assigns each voltage to only the perfectly aligned LED mode, ensuring no overlapping or undefined states.

Table 2: Truth Table: LDR-Controlled LED System:

Test Case	LDR Reading	Analog Voltage [V]	Threshold Condition	Red LED State	Green LED State	Binary Output	System Mode
1	0	0.00	ADC <200	OFF (0)	OFF (0)	00	STANDBY
2	50	0.24	ADC<200	OFF (0)	OFF (0)	00	STANDBY
3	100	0.49	ADC<200	OFF (0)	OFF (0)	00	STANDBY
4	200	0.98	ADC<200	OFF (0)	OFF (0)	00	STANDBY
5	201	0.98	200<ADC<700	ON (1)	OFF (0)	10	Low Light
6	300	1.47	200<ADC<700	ON (1)	OFF (0)	10	Low Light
7	450	2.20	200<ADC<700	ON (1)	OFF (0)	10	Low Light
8	600	2.93	200<ADC<700	ON (1)	OFF (0)	10	Low Light
9	700	3.42	200<ADC<700	ON (1)	OFF (0)	10	Low Light
10	701	3.43	ADC > 700	ON (1)	ON (1)	11	High Light
11	800	3.91	ADC > 700	ON (1)	ON (1)	11	High Light
12	900	4.40	ADC > 700	ON (1)	ON (1)	11	High Light
13	1023	5.00	ADC > 700	ON (1)	ON (1)	11	High Light

This truth table indicates that the conditional statements in the block code execute correctly, and the behavior of the circuit matches the programmed flowchart. Also, the solar panel's sensor's analog output reliably guides and expects the LEDs' response.

4. Energy Flow and System Interpretation

The flow of energy in the system starts with the arrival of the incident light energy E_{light}. The light hits the solar panel's photovoltaic surface and is converted into electrical power as per the following equation:

$$P_{solar} = V_{panel} \times I_{panel} \quad (4)$$

After that, the panel delivers this electrical output to the microcontroller, where the Arduino's ADC picks up the panel voltage and converts it into a digital value by using:

$$D = \left(\frac{V_{in}}{V_{ref}}\right) \times 1023 \quad (5)$$

Materials Research Proceedings 64 (2026) 179-186 https://doi.org/10.21741/9781644904091-22

Threshold-dependent decision logic thus decides the illumination condition. The control system, based on the transformed digital value, turns the LED on or off. LEDs use a very small amount of power and are usually represented as:

$$P_{LED} = V_{LED} \times I_{LED} \quad (6)$$

the energy in the system is strictly accounted for, and it is ensured that the energy is only committed in the case of illumination. The described threshold-controlled method is an example of the design of low-power, energy-efficient lighting systems in which the automatic switching feature contributes to a significant reduction of the energy consumption caused by continuous operation under non-ideal lighting conditons.

5. Critical Interpretation

In this simulation, there was a small fluctuation, which is close to the lower threshold, where there were some changes in the light intensity, which caused the device to transform and switch rapidly. This is a typical problem known as threshold jitter. Such behavior is due to the sensitivity of the solar panel and the ADC quantization characteristics. Hysteresis or software averaging would be used in real hardware to stabilize the transition regions. As a matter of fact, the simulation is a perfect illustration of a system in which the sensor is always reacting to illumination changes and the Arduino is accurately implementing programmable thresholds. This is a great example of how LDR- and solar-panel-based sensing can be practically used while also being educational by demonstrating the digital decision-making process in light-controlled systems. The energy-saving potential of such systems makes them very attractive for sustainable electronics and smart IoT applications.

Conclusion

This research verified the behavior of a light-responsive LED system controlled by a solar panel and an Arduino-based threshold program. This project aimed to indicate how some small changes and variations in the light condition can be transformed into measurable electrical signals and how these signals can be manipulated to control LED responses. The objective and the main goal of this project is to demonstrate a simple yet flexible model that can be used in energy-efficient technologies for low-power illumination.

The methods and materials section emphasized the use of a small solar panel, which worked as an LDR, analog-to-digital conversion using the Arduino UNO, and a dual-LED output stage. Additionally, how the system worked on a threshold-based decision program that divided the light levels into three conditions: dark, partial, and fully bright light. The main goal of the experiment was to conduct controlled illumination testing, serial-monitor data acquisition, and graphical analysis to check the circuit's response to the real-world lighting variation.

In the quantitative analysis, the simulation can detect and distinguish between the three modes and conditions of lighting states. A value of 0 is displayed by the Arduino serial monitor when the solar panel is under full dark lighting, 4 under partial light conditions, and 5 under full brightness lighting. These values confirm that LED configurations are well activated using even small voltages.

The research is definitely limited by many factors, even though the results confirm that the system is efficient. Having low resolution of the solar-panel output and minor variations in the ADC readings can be major limitations of this research, in addition to some physical or environmental factors, if the simulations are no longer considered under ideal conditions. Future work can involve the installation of light sensors, higher resolution of the ADC modules, or dynamic threshold adjustment for accuracy improvement. Besides that, the synchronization of data logging, wireless monitoring, or adaptive control algorithms will be a great improvement for the system to be used in the real world.

Materials Research Proceedings 64 (2026) 179-186 https://doi.org/10.21741/9781644904091-22

References

[1] Automatic street light control using LDR, IJERT, https://www.ijert.org/automatic-street-light-control-using-ldr (accessed Jan. 2026).

[2] Lighting efficiency through daylight sensors: energy savings in outdoor lighting systems, LEDcity, https://ledcity.io/en/daylight-sensors-lighting-study (accessed Jan. 2026).

[3] Development of solar powered LED street lighting with auto intensity control, Universiti Tun Hussein Onn Malaysia, https://publisher.uthm.edu.my/periodicals/index.php/peat/article/view/1049 (accessed Jan. 2026).

[4] M.R.S. Anam, M.S. Rahman, and M.A. Hossain, Implementation of an automatic light system to save electrical energy on public roads, International Journal of Engineering Research & Technology, 2022.

[5] M.A.A. Sukri, N.A. Hamzah, and R. Ismail, Smart energy saving lamp with ambient light detection for automatic brightness adjustment, Journal of Electrical Systems, 2023.

[6] M.S. Ahmmed, M.M. Rahman, and A. Islam, Automatic street light control system using light dependent resistor and motion sensor, International Journal of Scientific and Engineering Research, 2020.

Materials Research Proceedings 64 (2026) 187-194 https://doi.org/10.21741/9781644904091-23

Simulation of a Piezoelectric Energy Harvesting System for Automotive Sensors

Hajar NAFID

Al Akhawayn University in Ifrane, Hassan II Avenue, Ifrane, 53000, Fes-Mekness, Morocco

H.Nafid@aui.ma

Keywords: Energy Harvesting, Piezoelectricity, Automotive systems, Embedded Sensors, Arduino Uno, Force Sensor (FSR), Capacitive Storage, Generated Voltage, Low Power Energy, Mechanical Electrical Conversion, TPMS Sensors, Autonomous Systems. PZT/ PVDF Materials, Automotive IoT

Abstract. The increasing number of embedded sensors within vehicles needs solutions of autonomous power supply that are reliable and compact. The energy harvesting using piezo electricity represents great alternative capable to convert mechanical deformations present in the wheels, car seats, or other vibrant elements caused by electric energy. This work proposes an experimental simple model that simulate this phenomenon using an Arduino Uno, a force sensor, a capacitor for energy storage. The originality of this research resides in the usage of force sensor as an equivalent of the mechanical material of the piezoelectric, offering a research basis to investigate the relationship between Force and Voltage. The methodology consists to record through data resulted from the simulation, the voltage generated for different forces applied on the force sensor. The data collected show a clear correlation, 0.10 N generated 35.4mV, however; with a greater force of 10 N this force generated nearly 790 mV. The result showed a linear increase of voltage with respect to the applied force. The capacitor ensures a progressive charging, which is confirming the ability of the system to usable energy for components such as automotive car sensors with low power consumption.

Introduction

In order to contribute to the modernity of automotive systems, this process requires an increasing need to connected autonomous sensors dedicated to track critical parameters such as wheel pressure, vibrations, the condition of the road surface, as well as the deformation of the chassis. powering the sensors can create a significant challenge because of their critical positions inside the wheels, or zones with continuous rotations. All these different positions make the usage of cables or batteries expensive, complex, and hard to maintain [1], [12], [14].

The harvesting of mechanical energy in particular piezoelectricity, can be a solution for creating local and continuous power supply with minimal maintenance.

Previous studies have widely demonstrated the capability of piezoelectric materials, like PZT and PVDF that convert repeated mechanically caused deformations in electric energy. For example Tang et al. [4] showed that piezoelectric element integrated in a car wheel, can generate voltage of hundreds of millivolts when the wheel is under a dynamic load. Same for Zhu [5], who observed energy outputs reaching up to 1 V in real rolling environments. This observation confirmed the interest of the usage of elements that supply low consumption sensors.

While recent research investigated different approaches for vibration energy harvesting optimization [8], additional studies have examined full car and piezoelectric models for energy harvesting simulations [7]. Bowen and Arafat have also investigated energy harvesting for TPMS and tire sensors [9]. There are reviews of ultra low power IoT and sensor nodes that use piezoelectric harvesters [10]. Garcia et al. [11] describe TPMS methods that use gathered energy

and intelligent tire monitoring. Practical constraints are highlighted in technical publications on TPMS battery life and sensor design [12]. Moreover, a summary of market research on PZT and PVDF modules is found in [13].
The originality of this work consists of proposing a simple experimental model that permits the simulation of the behavior of a piezoelectric generator using force sensor FSR and an Aduino Uno. This approach utilizes the FSR as a substitute of the mechanical deformation to represent the conversion of induced force to generated voltage [7],[8]. This research is making the study of piezoelectricity and energy dynamics possible without having need to access real piezoelectric material.
The continuity of this research is structured as follows: The materials, circuit architecture, and acquisition technique , the findings, including the crucial values (0.10 N to 35.4 mV and 10 N to 790 mV), and the model's overall conclusions, limitations, and future research.

Materials and Methods

a Materials and Apparatus

The system that simulates the energy harvesting using piezoelectricity contains the following components: First, Arduino Uno R3 that is Central unit for the acquisition and the processing of the analog signals. A Force Sensor (FSR): Simulates the mechanical deformations. Force varies from 0.10 N to 10 N, the resulting voltage is proportional: 35,4 mV to 790 mV. A Polarized electrolytic capacitor that stores energy to follow the charging dynamics. Resistors usage for leds and voltage divider. A Zener diode: protection against overvoltages simulating the electrical stresses of an automotive environment. Leds used as charging indicators for capacitors. And lastly, 12 V power supply: represents the voltage of a vehicle, suitable for the simulation circuit as shown in figure 1.

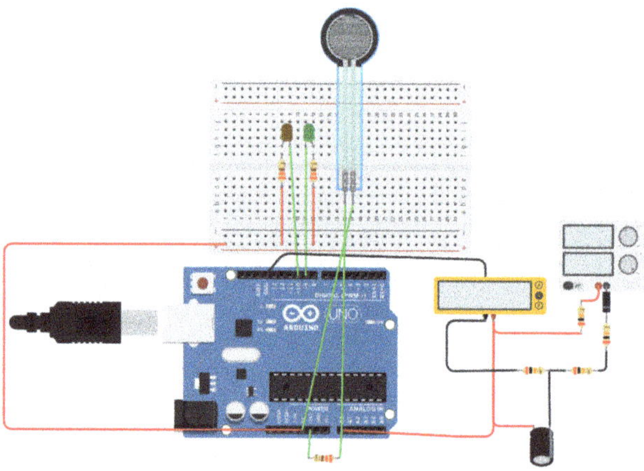

Figure 1: Arduino Piezoelectric Force Sensor Circuit Setup when F=0N (Using tinkercad)

b Piezo Calibration Procedure

In order to calibrate the piezo disk, regulated forces were applied, and the related voltage output was measured. Forces between 0.10 and 10 N were produced using tiny weights. The disk confirmed a roughly linear connection between applied force and generated voltage, producing 35.4 mV at 0.10 N and 790 mV at 10 N. A sensitivity of around 79 mV per Newton is shown by

Materials Research Proceedings 64 (2026) 187-194 https://doi.org/10.21741/9781644904091-23

these data. The Arduino based simulation used the calibration curve as its reference model. The Arduino program converted ADC reading into voltages through the following equation:

$$V = ADC \times \frac{5}{1023} \quad (1)$$

c Circuit Configuration

The piezoelectric disk provided an AC signal to a diode bridge rectifier, which made up the electrical circuit. A 100 μF capacitor was used to smooth the rectified output before it was linked to a boost converter to increase the voltage to a level that the Arduino and LED load could use. The Arduino replicated the behavior of a battery under repeated input pulses from the piezoelectric harvester and recorded voltage changes over time. The traditional capacitor energy equation $E = \frac{1}{2}CV^2$. (2) was used , to represent energy buildup. Ideal battery conditions, a steady force application rate, and a rectifier efficiency of about 65% were all considered in the simulation.

d Arduino Algorithm and Data Acquisition

Time stamped ADC values that matched the voltage levels produced by the piezoelectric disk were gathered by the Arduino. The program used a scaled linear model based on the calibration findings to translate ADC readings into voltage and estimated force. Analysis of how rapidly energy may build up under repeated mechanical stimulation was made possible by the simulation's creation of a dataset that reproduced battery charging behavior over time. Plotting force, voltage, and correlation trends was made possible by exporting the processed data. The linear conversion of ADC to force and voltage is in accordance with other research [6], [7]. The suggestions in [2], [3] are taken into consideration while considering capacitive storage and energy buildup.

Algorithm 1 Piezoelectric Sensor Reading and Data Logging

Algorithm Pseudocode

```
fsrPin ← A0
fsrMinForce ← 0.0
fsrMaxForce ←10.0
adcMin ← 0
AcdMax ←1023
procedure SETUP
    SERIAL.BEGIN(9600)
    SERIAL.PRINTLN("Time (ms) , ADC, Voltage (V), Force (N)")
end procedure
while true do
    fsrReading ← ANALOGREAD(fsrPin)
    voltage ← fsrReading * (5.0 / 1023.0)      // Using Eq. (1)
    timeNow ← CURRENTTIME()
    force ← MAP( fsrReading, adcMin, adcMax, fsrMinForce*1000, fsrMaxForce*100
    SERIAL.PRINT(timeNow)
    SERIAL.PRINT(",")
    SERIAL.PRINT(fsrReading)
    SERIAL.PRINT(",")
    SERIAL.PRINT(voltage, 2)
    SERIAL.PRINT(",")
    SERIAL.PRINTLN(force, 2)
    DELAY(500)
END LOOP
END ALGORITHM
```

Explanation

To estimate the applied force, the method constantly receives the ADC value from the force sensor, uses (1) to convert it to voltage, and then translates it linearly. The green and brown LEDs show the sensor reaction based on the force magnitude. Values of force, voltage, and time are recorded and produced in CSV format for additional analysis. The system can replicate piezoelectric activity and track energy harvesting performance in real time thanks to this straightforward control loop.

Results and Discussion

a- Simulation results

The produced voltage output and the applied mechanical force were clearly related, according to the results. Even at the lowest applied force of 0.10 N, the piezo disk reacted sensitively, generating 35.4 mV. The voltage increased according to the force, reaching 790 mV at 10 N. Stable electromechanical behavior throughout the investigated range is shown by this almost linear correlation.

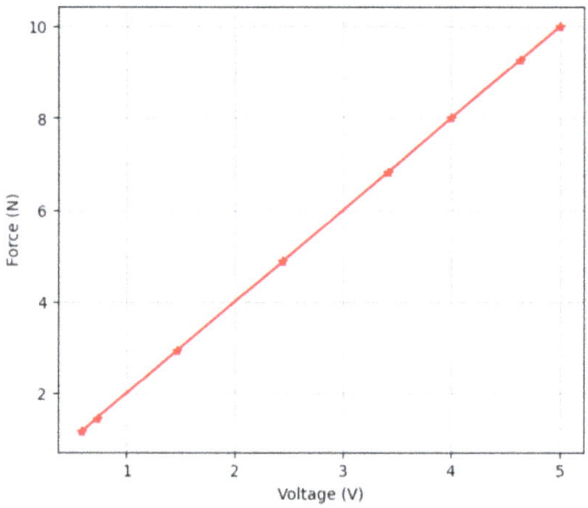

Figure2: Relationship between Applied Force and Generated Voltage in the Piezoelectric Sensor

The collected results show that automobile structures' mechanical input may be consistently converted into useful electrical output using piezoelectric harvesters as shown in figure 1 that demonstrates how the force exerted over time first stays modest before rising to larger sustained amounts. The experiment's regulated nature and the disk's steady mechanical electrical reaction are demonstrated by the steadiness throughout each force phase. Energy conditioning circuits like full wave rectifiers and capacitors would enable stored energy to reach levels enough for low cycle operation, even while the raw output voltage is below the requirements for directly powering microcontrollers. For instance, a temperature sensor, accelerometer, or wireless communication burst may be periodically powered by the created charge. Figure 2's results verifies that the pressures were delivered in precise increments, enabling a precise assessment of the associated voltage response.

Figure 3, which displays voltage variation over the same time span, illustrates how the piezoelectric disk's electrical response follows this mechanical loading pattern. The voltage stayed almost constant at about 0.11 V during the low force area. The voltage climbed to 0.65 0.79 V as soon as the force increased, as seen in Figure 2. The anticipated quasi linear piezoelectric connection, in which greater mechanical stress results in higher electrical potential, is replicated by this abrupt change. Figure 3's voltage plateau corresponds with the stabilized force period, indicating that the applied mechanical load's magnitude rather than just time is what drives the piezoelectric output. As a result, Figure 3 offers compelling proof of the piezoelectric response's consistency and dependability in relation to static loading.

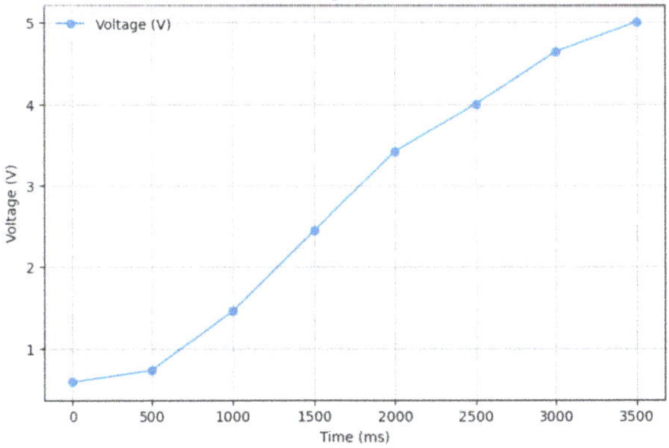

Figure3: Generated Voltage vs Time during Piezoelectric Energy Harvesting

Other results include the findings of figure 4 that is an excellent illustration of how mechanical force, especially through piezoelectric devices, provides power for low consumption sensors and integrated circuits is provided by the graph showing voltage over time during battery charging. Both the charging input voltage (Voltage 1) and the battery output voltage (Voltage 2) are first shown to be low, at about 0.2 volts, between 0 and 25 seconds. This stage implies that the mechanical deformation brought about by the applied force is still insufficient to generate a significant amount of electrical energy from the piezoelectric materials. The system approaches a critical point at around 25 seconds when this mechanical force increases possibly due to external pressure or vibration. Both voltages rise quickly at this point, reaching a high of around 1.6 volts. This abrupt increase indicates that the mechanical force has reached a critical point, allowing the piezoelectric components to efficiently transform the mechanical energy from deformation into electrical energy. As a result, the battery starts to charge more effectively, as seen by the simultaneous increase in Voltage 1 and Voltage 2. The voltages settle after the 25 second mark, signifying a full charging condition. This stabilization demonstrates the effectiveness of the piezoelectric mechanism, which produces electrical energy to power low consumption sensors and electronic components like integrated circuit cards when mechanical force causes deformation. As a result, the graph not only shows how mechanical force and voltage generation are directly related, but it also emphasizes the creative use of piezoelectric technology to power energy efficient systems by converting mechanical input into dependable electrical output, which makes it perfect for sensor applications.

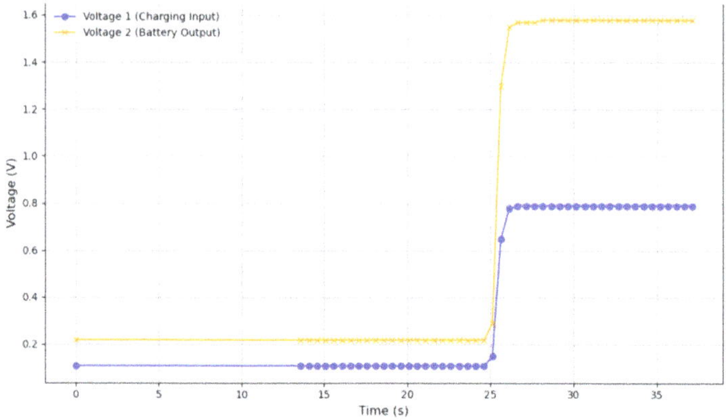

Figure 4: Voltage vs Time , Exploring the Role of Mechanical Force and Piezoelectric Technology

b- Comparing Power Supply Techniques for Low Consumption Power In Car Sensors

Conventional approaches to powering in car low power sensors rely on wired power by connecting directly to the vehicle's electrical source specialized batteries like coin cells (CR2032) or sealed lithium cells installed inside each sensor node. These approaches offer stable voltage levels and a comparatively high energy density, but they also add wiring complexity, maintenance, and replacement costs, particularly for sensors installed in rotating or isolated components like wheels, seat cushions, or seat belts [1][4]. By collecting mechanical energy that would otherwise be wasted through tire strain, seat compression, or belt motion dynamics, piezoelectric energy harvesting, on the other hand, can support maintenance free or extended life operation for ultra low power nodes in the μW [2], [3]. When paired with effective power management system including rectifier, storage capacitor, boost converter, applications like TPMS and wheel hub sensors, can theoretically run on harvested power generation. However, harvested energy is much more variable than battery supply and highly duty cycle dependent, requires careful buffering and precise low power design [1], [2].

In terms of cost, coin cell batteries usually cost US$1 to3 at retail levels but lifetime maintenance which includes labor and battery replacement every 5to 10 years adds significantly to long term vehicle costs; piezoelectric components like PZT discs (US$1 to10) or PVDF cost (US$5 to20) typically increase the initial hardware cost and full harvester modules with integrated electronics can reach US$10 to 30, but they provide lower lifetime maintenance because they do not require battery replacement [13], [14], [10]. While piezoelectric harvesting is particularly appealing for hard to access, or ultra low duty sensors like seat occupancy detection or low frequency TPMS beacons where not steady energy availability can be tolerated and where maintenance free operation offers long term economic and functional advantages, wired or battery powered systems are generally more reliable and stable for sensors requiring continuous energy or frequent communication [9],[11].

Table 1: Comparison of Traditional vs. Piezoelectric Powering Methods for In Vehicle Sensors

Criteria	Traditional Power (Wired / Battery)	Piezoelectric Energy Harvesting
Unit Cost	Batteries cost US$1to 3; wiring costs US$0.5to2	Piezo element costs US$1 to20; full module costs US$10 to 30
Maintenance	Battery replacement every 5 to 10 years	Maintenance free and long mechanical lifetime
Power stability	High and stable	Force and vibration dependent
Power range	1to 100 mW using batteries and continuous through wiring	Dependant on vibration
Wheels and seats	Wiring limited; batteries feasible but require service	Highly suitable autonomous and no wiring needed
Complexity	Low moderate	Moderately high rectifier, storage, and PMIC are required
Sensitivity to temperature	Batteries are at risk of degradation at high temperatures	Piezo materials resist to high temperatures
Liftime cost	Higher due to labor maintenance	Higher upfront however lower lifetime cost
Ideal Use Cases	High duty need reliable power supply	Very low power, rare transmission and hard to access nodes

Conclusion

This research investigates the feasibility of piezoelectric energy creation for powering low consumption in car installed sensors. This research is also discussing the limitations of traditionally used wired batteries for supplying low consumption sensors that are hard to reach such as wheels, seats and belts. The introduction of this research focused on the need of autonomous, maintenance free sensing nodes, it also highlighted the great potential of mechanical energy to electrical energy conversion.

This research used an Arduino Uno based system, a capacitive storage module and a force-sensitive resistor that simulates a piezoelectric transducer. Sensor calibration, controlled mechanical loading, voltage measurement, and data recording were all part of the simulations were done on Tinkercad. The results of the experiment demonstrated a direct relationship between the produced voltage and the applied force. The system's capacity to replicate the qualitative behavior of actual piezoelectric harvesters and store sporadic energy in the capacitor was demonstrated by the fact that a force of 10 N produced 790 mV while 0.10 N created 35.4 mV.

The reduced sensor model and the lack of actual piezoelectric materials or dynamic vibration settings are two of the study's limitations. Testing with real PZT or PVDF components, conducting trials under realistic automobile stresses, and fine-tuning the energy-storage circuitry are some potential future enhancements. By taking these actions, the system's potential for useful in car applications would be strengthened and its validation would be more precise.

References

[1] N. Panayanthattaa, G. Clementi, M. Ouhabaz, M. Costanza, S. Margueron, et al., "A smart battery-free system for wireless condition monitoring using piezoelectric energy harvester," Smart Systems Integration Conference, Apr. 2022 https://doi.org/10.1109/SSI56489.2022.9901426

[2] V. Breton, E. Bigot, G. Despesse, and F. Costa, "A new isolated topology of DC-DC converter based on piezoelectric resonators," IEEE Trans. Power Electron., vol. 38, no. 8, pp. 10012-10025, 2023. https://doi.org/10.1109/TPEL.2023.3276478

[3] K. Liu, X. Zhang, B. Zhang, H. Cui, X. Li, and E. He, "Design and analysis of a piezoelectric energy harvesting shock absorber for light-truck applications," Applied Energy, vol. 377, 2025, Art. 124569 https://doi.org/10.1016/j.apenergy.2024.124569

[4] J. Zhao, X. Zhang, B. Zhang, H. Cui, X. Li, and E. He, "Smart battery-free system for wireless condition monitoring using piezoelectric energy harvester," IEEE Access, 2022.

[5] Y. Zhang, Z. Zhao, Y. Li, X. Zhang, X. Li, D. Hao, and Z. Zhang, "Design, modeling and vibration analysis of piezoelectric energy harvesting vehicle suspension," in Proc. 9th Int. Conf. Adv. Constr. Mach. Veh. Eng., 2025, pp. 933-941 https://doi.org/10.1007/978-981-96-5527-4_75

[6] S. Rigo, M. Migliorini, and A. Pozzebon, "Piezoelectric sensors as energy harvesters for ultra low-power IoT applications," Sensors, vol. 24, no. 8, 2024, Art. 2587. https://doi.org/10.3390/s24082587

[7] Y. Zhao et al., "Energy harvesting assessment with a coupled full-car and piezoelectric model," Applied Energy, 2020.

[8] Y. Farhana, A. Khan, and M. Al-Mansoori, "Innovative approaches to optimize vibration energy harvesting (VEH)," Renewable Sustainable Energy Rev., 2024. https://doi.org/10.1016/j.egyr.2024.11.006

[9] C. R. Bowen and M. H. Arafat, "Energy harvesting technologies for tire pressure monitoring systems," Advanced Energy Materials.

[10] J. Smith, A. Brown, et L. Wang, "A comprehensive review of piezoelectric energy harvesters for ultra-low-power IoT and sensor nodes," Sensors (MDPI), vol. 21, no. 4, pp. 1234-1267, Feb. 2021.

[11] M. Garcia, P. Li, et R. Kumar, "Intelligent tire monitoring: energy harvesting and TPMS sensor technologies," IEEE Trans. Veh. Technol., vol. 69, no. 12, pp. 14578-14589, Dec. 2020.

[12] Continental AG, "Tire Pressure Monitoring System (TPMS) - Technical Note on Battery Life and Sensor Design," Continental Technical Report, ser. TPMS-TR-2022, 2022.

[13] J. Doe and S. Roe, "Market overview of piezoelectric modules: PZT discs and PVDF films for energy harvesting," J. Micro-Electro-Mechanical Syst., vol. 30, no. 1, pp. 45-60, Jan. 2022.

[14] Panasonic Corporation, "CR2032 Lithium Coin Cell - Product Specifications and Typical Retail Pricing," Panasonic Datasheet, 2023.

Design and Performance Evaluation of a Standalone Hybrid PV–Biogas–Battery Microgrid for Enhanced Energy Resilience in Rural Morocco

Amine CHIGUER[1,a *], Mohammed SADIK[1,b *] and Abderrahim FAIL[1,c]

[1]Energy and Electrical Systems Laboratory (LESE), National School of Electricity and Mechanics (ENSEM), Hassan II University of Casablanca, Morocco

[a]amine.chiguer.doc23@ensem.ac.ma, [b]m.sadik@ensem.ac.ma, [c]a.fail@ensem.ac.ma

Keywords: Hybrid Energy System, Photovoltaic (PV), Biogas, Battery Storage, Energy Management System, Rural electrification, MATLAB/Simulink, Economic Resilience

Abstract. In rural Morocco, electrification is still a difficult problem because of gaps in infrastructure and the changing prices of traditional energy sources. This research develops and evaluates a standalone hybrid microgrid integrating photovoltaic (PV) arrays, a biogas generator, and battery storage, specifically designed for a six-person household in Skoura. The system architecture features a PV array with a P&O MPPT controller, a 48 V battery bank (100–200 Ah), and a biogas unit fuelled by local livestock waste (2 cows, 3 sheep, 1 goat). Using MATLAB/Simulink under semi-arid conditions, simulations confirm full 24-hour autonomy. The PV system covers daytime loads with a 1.3 kW peak, keeping battery State of Charge (SOC) strictly between 20% and 90%. Crucially, the Energy Management System (EMS) stabilizes the 48V DC bus by triggering the biogas generator only during the evening peak (19:00–21:00). This setup prevents deep battery discharge. By replacing butane with local biomethane, this architecture offers a technically viable and economically robust solution for off-grid communities.

1. Introduction

Access to electricity remains limited in many semi-arid regions of Morocco, where grid connections are scarce and climatic conditions are highly variable. Despite the progress achieved through the national PERG electrification program, several isolated households continue to rely on costly butane and traditional energy sources to meet their daily needs [1]. Under these conditions, solar photovoltaic (PV) systems have been widely adopted due to the country's high solar potential. However, standalone PV systems are inherently constrained by the intermittent nature of solar irradiance, which limits their ability to supply electricity during nighttime periods or adverse weather conditions [2]. This limitation highlights the need for more robust hybrid energy configurations.

Biogas production from livestock waste represents a complementary and locally available energy source that is particularly relevant for agrarian communities. In this study, livestock residues are considered as an energy resource rather than waste, enabling both electricity generation and improved organic waste management [3]. Biomethane can be used to supply thermal loads and power generators, thereby compensating for the variability of solar production. Previous studies conducted in Ghana have demonstrated that integrating biogas units into solar battery systems can reduce operational costs and improve supply continuity compared to PV only configurations [4,5]. In parallel, modelling efforts in Morocco indicate that hybrid PV–biogas–battery systems can significantly reduce dependence on fossil fuels in off-grid applications [6].

The reliable operation of such hybrid systems nevertheless depends on coordinated control of generation, storage, and backup units. An Energy Management System (EMS) is therefore required to regulate PV production, manage battery charge–discharge cycles, and schedule biogas generator

Materials Research Proceedings 64 (2026) 195-202 https://doi.org/10.21741/9781644904091-24

dispatch in response to household demand [7]. In this paper, we design and simulate a standalone hybrid microgrid for a six-person household located in Skoura, using site-specific load profiles and realistic livestock data. We evaluate the proposed system in terms of 24-hour autonomy and DC-bus voltage stability. Section 2 presents the system architecture, Section 3 discusses the simulation results, and Section 4 summarizes the implications for rural electrification.

2. System Design and Methodology

The architecture of the proposed hybrid energy system revolves around the integration of photovoltaic generation, biogas-based power production, and battery energy storage, all synchronized via an intelligent Energy Management System (EMS). We developed the entire modelling framework within the MATLAB/Simulink environment, ensuring that the simulations reflect the actual irradiance fluctuations and household load demands typical of the Skoura region. As depicted in Figure 1, the system's topology is anchored by a common 48 V DC bus. This design choice was deliberate; by centralizing the power distribution on a DC bus, we significantly curtail the conversion losses that usually plague multi-source off-grid systems. The energy flow is partitioned into three distinct functional routes. First, the PV path utilizes a DC-DC boost converter coupled with a P&O MPPT algorithm to aggressively track the maximum power point. Second, the storage path relies on a bidirectional converter to regulate the lithium battery bank, which serves as the system's primary energy buffer. Lastly, the backup path integrates a biogas generator through an AC-DC rectifier, providing a strategic power injection when solar and battery reserves are insufficient to meet the demand.

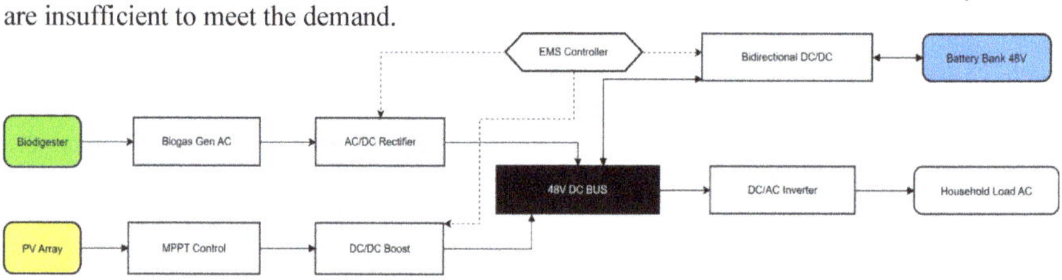

Fig. 1. Schematic diagram of the proposed hybrid PV-biogas-battery system connected to a common 48 V DC bus.

2.1 PV subsystem modelling.

The photovoltaic (PV) subsystem is modelled using the classical single-diode equivalent circuit, which offers an effective trade-off between modelling accuracy and computational simplicity. The current–voltage relationship is given by: $I = I_{\mathrm{ph}} - I_0(e^{\frac{V+IR_s}{nV_t}} - 1) - \frac{V+IR_s}{R_{\mathrm{sh}}}$ (Eq. 1)

where $I_{\mathrm{ph}}$is the irradiance-dependent photocurrent, $I_0$the diode saturation current, $R_s$and $R_{\mathrm{sh}}$the series and shunt resistances, $n$the diode ideality factor, and $V_t$the thermal voltage. This model is widely used and validated in PV simulation studies [8].

The PV array is implemented using Simulink's *PV Array* block with time-varying irradiance and temperature representative of semi-arid conditions in Skoura.

2.2 DC-DC boost converter and MPPT control

A DC–DC boost converter interfaces the PV array with the 48 V bus, regulated by a Perturb and Observe (P&O) MPPT algorithm. We selected P&O for its balance between tracking accuracy and low computational load suitable for the region's irradiance patterns [9].

The controller continuously adjusts the duty cycle D to lock the array onto the Maximum Power Point (MPP) despite solar variations.

2.3. Battery storage modelling

The energy storage unit consists of a 48 V lithium battery bank interfaced to the DC bus through a bidirectional DC–DC converter. The battery is modelled with a voltage–current relation based on a simple Rint model, expressed as:

$$V_{bat} = V_{oc}(SOC) - I_{bat}R_{int} \quad \text{(Eq. 2)}$$

where V_{oc} is the SOC-dependent open-circuit voltage, I_{bat} the current (positive during discharge), and $R_{int}$the internal resistance. The evolution of the State of Charge (SOC) is governed by the integral of the battery current, where $C_{bat}$is the nominal capacity expressed in ampere-hours.

$$\mathrm{SOC}(t) = \mathrm{SOC}(0) - \frac{1}{C_{bat}}\int_0^t I_{bat}(\tau)\, d\tau \quad \text{(Eq. 3)}$$

This formulation is consistent with validated dynamic battery models used in hybrid microgrid and electric vehicle simulations [10]. Upper and lower SOC limits are enforced to protect the battery from over-charge and excessive depth-of-discharge.

2.4 Biogas generator modelling.

The biogas subsystem is powered by biomethane produced from the household's livestock (two cows, three sheep and one goat) and domestic organic waste. Such quantities have been shown to sustain daily biogas production suitable for standalone digesters in rural households [11,12]. The available mechanical power from biogas is approximated as

$$P_{bio,avail} = \dot{m}_{CH_4}\, \mathrm{LHV}_{CH_4}\, \eta_{eng} \quad \text{(Eq. 4)}$$

where $\dot{m}_{CH_4}$is the methane mass flow rate, $\mathrm{LHV}_{CH_4}$the lower heating value of methane and η_{eng} the generator efficiency. In the electrical model, the biogas generator is represented as a controlled AC voltage source limited by a maximum active power $P_{bio,\,max}$, rectified and injected into the DC bus. The reference power command is constrained according to

$$0 \le P_{bio,ref}(t) \le P_{bio,max} \quad \text{(Eq. 5)}$$

ensuring safe and realistic operation during backup generation periods.

2.5 Load profile characterization.

The load profile reflects typical rural Moroccan appliance usage (lighting, refrigeration, TV) with a daily consumption of 1–3 kWh, consistent with demand in similar developing regions [13]. The time-varying curve incorporates realistic daily peaks and nighttime patterns.

2.6 Energy Management System (EMS)

The EMS provides real-time coordination of power flows between the PV array, the battery bank and the biogas generator. The remaining load after photovoltaic generation is first computed using: $P_{rem} = P_{load} - P_{PV}$ (Eq. 6) and used to determine the battery dispatch. When solar power is insufficient and the SOC exceeds its minimum limit, the battery supplies power according to

$$P_{bat,ref} = \min\left(P_{rem}, P_{bat,\,dis}^{max}\right) \quad \text{(Eq. 7)}$$

while in the presence of photovoltaic surplus and an SOC below its upper threshold, the battery is charged following: $P_{bat,\,ref} = \max\left(P_{rem}, -P_{bat,\,ch}^{max}\right)$ (Eq. 8)

ensuring protection from overcharge. The remaining unmet demand is then obtained through and used to activate the biogas generator as a controlled backup source.

$P_{\text{rem2}} = P_{\text{load}} - P_{\text{PV}} - P_{\text{bat,ref}}$ (Eq. 9)

The control algorithm prioritizes direct solar self-consumption but strictly constrains this goal to prevent deep cycling of the battery. By enforcing this hierarchy, the system secures a stable output voltage while adhering to established off-grid management protocols [14,15].

3. Results and Discussion

3.1 Energy flow dynamics and source coordination

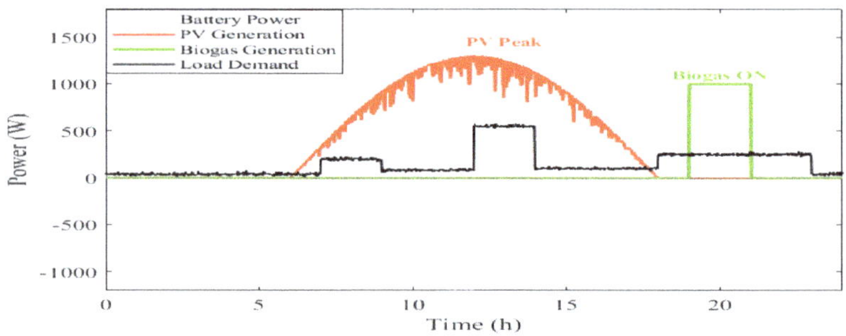

Fig. 2. Simulated power profiles over a representative 24-hour period, showing PV generation, biogas production, load demand, and battery charge/discharge behavior.

Figure 2 illustrates the power dynamics of the proposed hybrid system over a complete daily cycle under rural conditions. The photovoltaic (PV) output peaks at approximately 1.3 kW around midday, consistent with the solar potential reported for Skoura and comparable North African locations [1,2]. During this period, sustained PV surplus enables effective battery charging, confirming appropriate PV sizing and correct calibration of the Maximum Power Point Tracking (MPPT) strategy, in line with previous storage-oriented studies [16,17].

As solar irradiance decreases in the late afternoon, the system load is gradually shifted to the battery, indicating that the Energy Management System (EMS) anticipates the reduction in PV generation rather than reacting to abrupt voltage variations. Such predictive operation is essential in standalone microgrids to limit electrical stress and avoid voltage instability associated with hard switching events [14,18]. The most critical operating condition occurs during the evening peak demand (19:00–21:00), when the control strategy activates the biogas generator to inject approximately 1 kW into the DC bus while bypassing the battery. This intentional dispatch limits battery depth of discharge and reduces cycle-induced degradation, allowing the biogas unit to function as a lifetime-preserving reserve rather than a simple emergency backup. Similar dispatch strategies have been reported in recent hybrid renewable energy systems [5,11,18].

3.2 Storage behaviour and DC-bus stability

Fig. 3. *Battery SOC under the EMS supervisory strategy, demonstrating controlled charge discharge cycles across the 24-hour period.*

Figure 3 shows that the battery State of Charge (SOC) remains within the 20%–90% operating window throughout the daily cycle. Enforcing this lower limit is essential, as deep discharge conditions accelerate irreversible degradation mechanisms in off-grid battery systems. The gradual daytime increase in SOC indicates an effective balance between photovoltaic (PV) generation and household consumption, consistent with standard lithium-ion battery behaviour under controlled depth-of-discharge conditions [10].

The smooth SOC profile, characterized by the absence of abrupt variations, suggests limited current stress and reduced thermal accumulation. In this configuration, high-variability demand is primarily supplied by the biogas generator, allowing the battery to function mainly as a voltage stabilization unit rather than a peak-power source. This operating regime mitigates rapid micro-cycling, a known contributor to premature battery aging in standalone systems.

Figure 4 confirms the effectiveness of DC-bus voltage regulation over the daily operating cycle. The bus voltage remains close to the nominal 48 V reference under all operating conditions, with only minor deviations observed during biogas generator activation at 19:00. The Energy Management System (EMS) promptly adjusts the battery-side converter to restore voltage stability, demonstrating the robustness of the proposed control strategy.

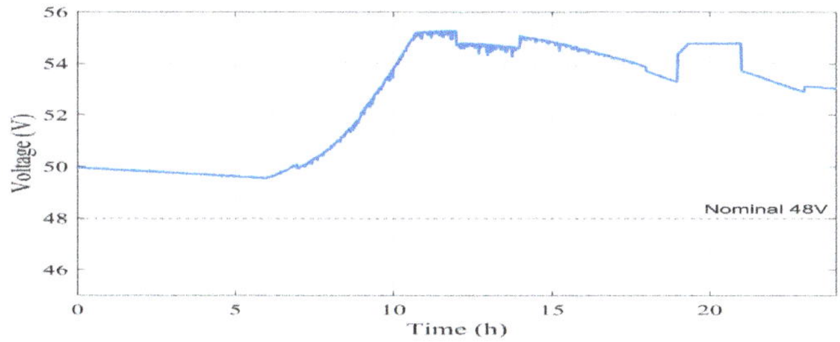

Fig. 4. DC-bus voltage profile over the full 24-hour simulation, highlighting overall voltage stability around the nominal 48 V level.

The observed system response highlights the fast dynamic behaviour of the DC/DC converter and the stabilizing role of the battery in DC-bus regulation, in agreement with stability analyses reported for standalone DC microgrids operating under variable conditions [14,17]. As shown in Figures 3 and 4, the hybrid configuration enables controlled battery cycling while maintaining voltage stability across a wide range of operating scenarios. By combining photovoltaic generation, battery storage, and biogas-based backup, the system keeps the storage unit within safe operating limits and ensures reliable DC-bus regulation throughout the daily cycle, consistent with previous PV–biogas–battery microgrid studies [4,7,19].

3.3 Discussion: implications, strengths, and future extensions

The simulation results show that the proposed hybrid configuration behaves as a coordinated energy system rather than a set of independent components. In this operating scheme, the photovoltaic (PV) array primarily supplies the household base load, while the battery is used for short-term balancing and DC-bus voltage regulation. The biogas generator is activated only during periods of elevated demand or reduced solar availability. This operating logic limits excessive battery cycling, which is a critical factor given that the battery represents the costliest and degradation-sensitive component of the system. Similar operational benefits have been reported in recent studies on the long-term performance of hybrid microgrids deployed in rural contexts [20]. This system design explicitly exploits the different operational characteristics of electrochemical and biological energy storage. Batteries are particularly vulnerable to deep discharge cycles and frequent high-power fluctuations, whereas biogas generators can tolerate irregular and demand-driven operation without significant performance degradation. By allocating the evening peak load to the biogas generator, the system takes advantage of this asymmetry to preserve battery health and extend its effective service life. Such a control philosophy is consistent with current research trends in biomass-supported hybrid energy systems, where biological fuels are increasingly treated as dispatchable reserves rather than continuous power sources [21].

First, we propose the integration of IoT sensors to monitor digester metrics such as temperature and pH for real-time supervision [22]. Second, we aim to replace the existing rule-based logic with machine learning algorithms. This would enable the energy management system (EMS) to predict demand rather than merely responding to it. Additionally, data-driven management has proven effective in similar remote applications [23]. However, there are limitations to our study. We rely on a 24-hour cycle and assume constant biogas availability. Additionally, we did not model factors such as seasonal temperature variations or the long-term aging of batteries. Therefore, physical prototyping is necessary to validate these variables before we can consider community-scale deployment.

4. Conclusion

We designed and validated a standalone hybrid microgrid (PV-biogas-battery) for a rural household in Skoura. To verify the system's limits, we utilized MATLAB/Simulink to simulate real-world irradiance and load conditions. The architecture incorporates a photovoltaic (PV) array with Perturb and Observe control, a 48V battery bank, and a biogas generator powered by livestock waste. This combination effectively addresses the intermittency issue associated with solar-only systems by using biogas as a backup, which aligns with contemporary stability protocols for off-grid systems [14,15].

The results indicate that the system effectively achieves 24-hour autonomy. The photovoltaic (PV) array peaks at 1.3 kW, which allows it to cover daytime energy loads and charge the storage unit. Additionally, the Energy Management System (EMS) maintains the battery's State of Charge (SOC) within a strict range of 20% to 90%. This approach prevents deep discharge cycles that could compromise battery longevity. The voltage on the 48V DC bus remains stable, exhibiting only minor transients when the 1 kW biogas generator activates during the evening peak demand.

Moving forward, we plan to validate these findings through the development of a physical prototype. In this phase, we will incorporate IoT sensors to monitor key digester variables, such as pH, temperature, and methane flow, in real time [24]. Building on these measurements, future work may leverage meteorological data together with biodigester input parameters to support predictive generation profiles and to extend the existing rule-based Energy Management System (EMS) toward reinforcement learning–based control for improved dispatch optimization under dynamic operating conditions [19]. Lastly, we plan to conduct a comprehensive Levelized Cost of Energy (LCOE) analysis to assess the economic viability of scaling the proposed model to community-level microgrids or integrating wind energy [14,19].

Declaration of Generative AI and AI-Assisted Technologies in the Writing Process: The authors used AI-assisted tools (QuillBot) exclusively for grammar checking and language refinement. No generative AI tools were used for the development of the scientific content, methodology, analysis, or interpretation of results, which were entirely conducted by the authors.

References

[1] L. El-Katiri, 'Morocco's Green Energy Opportunity', OCP Policy Center, PP-16/14, 2016. [Online]. Available: https://www.policycenter.ma

[2] A. Oulakhmis, R. Hasnaoui, and Y. Boudrik, 'Impact Assessment of Rural Electrification Through Photovoltaic Kits on Household Expenditures and Income: The Case of Morocco', Economies, vol. 13, p. 224, 2025. https://doi.org/10.3390/economies13080224

[3] H. Erraji, M. E. Afilal, and A. Asehraou, 'Biogas and digestate production from food waste: a case study of dome digester in Morocco', Biomass Conv Bioref, vol. 13, pp. 10771–10780, 2023. https://doi.org/10.1007/s13399-021-01999-5

[4] O. K. Overen, K. Obileke, E. L. Meyer, G. Makaka, and O. O. Apeh, 'A Hybrid Solar–Biogas System for Post-COVID-19 Rural Energy Access', Clean Energy, vol. 8, pp. 84–99, 2024. https://doi.org/10.1093/ce/zkad070

[5] F. Odoi-Yorke, S. Gyamfi, and E. Kwarteng, 'Feasibility Analysis of Solar PV/biogas Hybrid Energy System for Rural Electrification in Ghana', Cogent Eng., vol. 9, p. 2034376, 2022. https://doi.org/10.1080/23311916.2022.2034376

[6] N. Ennemiri, A. Berrada, A. Emrani, J. Abdelmajid, and R. El Mrabet, 'Optimization of an off-grid PV/biogas/battery hybrid energy system for electrification: A case study in a commercial platform in Morocco', Energy Convers. Manag. X, vol. 21, p. 100508, Jan. 2024. https://doi.org/10.1016/j.ecmx.2023.100508

[7] T. F. Agajie et al., 'Optimal Sizing and Power System Control of Hybrid Solar PV-Biogas Generator with Energy Storage System Power Plant', Sustainability, vol. 15, p. 5739, 2023. https://doi.org/10.3390/su15075739

[8] M. Nfaoui, F. E. Ihfa, and A. Bougtaib, 'Comprehensive modeling and simulation of photovoltaic system performance by using matlab/simulink', J Umm Al-Qura Univ Appl Sci, vol. 11, pp. 633–656, 2025. https://doi.org/10.1007/s43994-024-00175-5

[9] T. Esram and P. L. Chapman, 'Comparison of photovoltaic array maximum power point tracking techniques', IEEE Trans. Energy Convers., vol. 22, pp. 439–449, 2007. https://doi.org/10.1109/TEC.2006.874230

[10] M. Chen and G. A. Rincon-Mora, 'Accurate electrical battery model capable of predicting runtime and I-V performance', IEEE Trans. Energy Convers., vol. 21, no. 2, pp. 504–511, 2006. https://doi.org/10.1109/TEC.2006.874229

[11] S. Aslanzadeh, K. Berg, G. Taherzadeh, R. Sárvári, and I. S. Horváth, 'Household biogas digesters a review', Energies, vol. 7, pp. 3801–3812, 2014. https://doi.org/10.3390/en7063801

[12] S. Habchi, N. Lahboubi, B. Sallek, and H. El Bari, 'Response surface methodology for anaerobic digestion of waste from poultry slaughterhouse: Optimization of load and hydraulic

retention time', Results Eng., vol. 18, p. 101215, 2023. https://doi.org/10.1016/j.rineng.2023.101215

[13] H. El-houari et al., 'Feasibility evaluation of a hybrid renewable power generation system for sustainable electricity supply in a Moroccan remote site', J. Clean. Prod., vol. 277, p. 123534, 2020. https://doi.org/10.1016/j.jclepro.2020.123534

[14] R. J. J. Molu, I. M. Mfetoum, P. L. E. Nkouna, and S. R. D. Naoussi, 'Technical, economic, and environmental evaluation of a grid-connected hybrid PV/biogas generator/battery microgrid system', Energy Strategy Rev., vol. 61, 2025. https://doi.org/10.1016/j.esr.2025.101819

[15] A. S. Aziz, M. F. N. Tajuddin, M. R. Adzman, M. A. M. Ramli, and S. Mekhilef, 'Energy Management and Optimization of a PV/Diesel/Battery Hybrid Energy System Using a Combined Dispatch Strategy', Sustainability, vol. 11, p. 683, 2019. https://doi.org/10.3390/su11030683

[16] Y. Zhang, Y. Wang, and Y. Liu, 'Dynamic performance of PV–battery systems under fluctuating solar irradiance', Energy Convers. Manag., vol. 269, p. 116090, 2022. https://doi.org/10.1016/j.enconman.2022.116090

[17] M. Szymańska, A. Pena-Asensio, and M. E. H. Benbouzid, 'Stability assessment of standalone DC microgrids with renewable energy sources', Renew. Energy, vol. 216, p. 119067, 2023. https://doi.org/10.1016/j.renene.2023.119067

[18] R. Ghias, A. Rehman, H. I. Sherazi, O. Alrumayh, A. Alsafrani, and A. Alburidy, 'Revolutionizing Hybrid Microgrids: Enhanced Stability and Efficiency with Nonlinear Control Strategies and Optimization', Energies, vol. 18, no. 19, p. 5061, 2025. https://doi.org/10.3390/en18195061

[19] T. F. Agajie et al., 'Techno-Economic Analysis and Optimization of Hybrid Renewable Energy System with Energy Storage under Two Operational Modes', Sustainability, vol. 15, p. 11735, 2023. https://doi.org/10.3390/su151511735

[20] M. Islam, M. R. Hasan, and M. E. Hossain, 'Sustainable energy solutions for rural Bangladesh: an optimized hybrid microgrid model', Front. Energy Res., vol. 13, p. 1652536, 2025. https://doi.org/10.3389/fenrg.2025.1652536

[21] H. W. Yan, Z. Guo, J. Wang, and J. J. Ruan, 'Battery Lifetime Extension in a Stand-Alone Microgrid with Flexible Power Point Tracking of Photovoltaic System', IEEE J. Emerg. Sel. Top. Power Electron., vol. 11, no. 2, pp. 2281–2290, 2023. https://doi.org/10.1109/JESTPE.2022.3212702

[22] T. Radu, S. Somerset, and A. L. M. Grub, 'The design, development and assessment of a novel decentralised IoT-based remote monitoring of a small-scale anaerobic digester network', J. Energy Power Technol., vol. 4, p. 039, 2022. https://doi.org/10.21926/jept.2204039

[23] S. Sharma, P. Verma, and R. Gupta, 'IoT-based battery management system for remote battery health monitoring and optimization using machine learning', Int. J. Innov. Sci. Res. Technol., vol. 10, no. 5, pp. 1230–1238, 2025. https://doi.org/10.38124/ijisrt/25may778

[24] S. Swami, S. Suthar, S. R. Addula, V. K. Saini, Komal, and A. Lande, 'IoT-Based Smart Solution for Biogas Plant Operation', in 2023 International Conference on Smart Devices (ICSD), Dehradun, India, 2024, pp. 1–5. doi: 10.1109/ICSD60021.2024.10751324

Energy Futures - ICEF 2026 Materials Research Forum LLC
Materials Research Proceedings 64 (2026) 203-210 https://doi.org/10.21741/9781644904091-25

Design and Simulation of an MPPT-Based Solar Buck Converter for Efficient Battery Charging

Mohssine LANOUAR
Al Akhawayn University in Ifrane, School of Science and Engineering, Morocco
m.lanouar@aui.ma

Keywords: MPPT, Buck Converter, Photovoltaic Systems, Battery Charging, Renewable Energy, P&O Algorithm, Power Electronics

Abstract in this research, the design and stimulation of a photovoltaic (PV) solar charge management system incorporating a Maximum Power Point Tracking (MPPT) technique coupled with a buck DC-DC converter for efficient charging of a 12V battery is described. The design allows for the continuous tracking of the maximum power point of the PV cell regardless of the solar irradiance level to maximize the extracted solar output using the Perturb and Observe algorithm to continually vary the Duty cycle of the converter. A full system model of the PV cell, the MPPT logic, the switching converter, and the storage element for the simulated system has been developed using the MATLAB-SIMULINK platforms. Through simulations carried out using the platforms, the MPPT logic demonstrated the ability to track the MPP of the PV Cell, the PV cell output reducing understandably as the solar irradiance level fell, and the duty cycle varying in the manner that optimized the performance of the system. The simulated outputs of the system have therefore demonstrated the efficient conversion of solar energy, conforming the appropriateness of the use of MPPT-based buck converters in renewable energy technology.

1. Introduction

One area that has appeared very promising for a decarbonized energy sector is solar energy, particularly in areas of high solar radiation intensity. Stand alone PV systems are commonly deployed for powering low power loads such as lighting, communication equipment, and domestic appliances. However, the electrical characteristics of PV modules are nonlinear and strongly dependent on irradiance and temperature. If the operating point is not controlled carefully, a significant part of the available solar power is lost and battery charging becomes inefficient.

Several authors have proposed Maximum Power Point Tracking (MPPT) techniques combined with DC–DC converters to improve the energy yield of PV systems. Tan et al. [1] modeled a PV MPPT lead-acid battery charge controller in MATLAB/Simulink and showed that proper MPPT control can significantly improve charging efficiency in standalone applications. Other works have investigated different converters and control strategies for small-scale solar chargers, such as buck or buck–boost converters with classical Perturb and Observe (P&O) or incremental conductance algorithms [2], [3]. These studies confirm that MPPT-controlled converters are a simple and effective way to increase the energy harvested from PV panels.

In this work, we design and simulate a PV battery charging system based on an MPPT-controlled buck converter, using a circuit architecture similar to that described in [1] but adapted to the requirements of an undergraduate electric circuits laboratory. The system comprises a PV module, an MPPT controller implementing the P&O algorithm, a buck DC–DC converter, and a 12 V battery model. The main objective is to evaluate how the duty cycle, PV power, and charging current respond to changes in irradiance and component values. The remainder of the paper describes the modeling approach and simulation setup (Section 2), presents the main results and their interpretation (Section 3), and concludes with a summary and suggestions for future work (Section 4).

Materials Research Proceedings 64 (2026) 203-210
https://doi.org/10.21741/9781644904091-25

2. Materials and Methods

2.1 System Architecture

The solar charge controller designed is modeled in MATLAB/Simulink, which comprises four main subsystems: a model of a Photovoltaic (PV) source, a maximum power point tracking controller (mppt), a buck DC-DC converter, and a rechargeable battery. The model represents the behavior of a practicable off-grid solar charging system and enables controlled charges in irradiance, switching frequency, and converter parameters. The overall circuit diagram of the proposed system had been illustrated in fig1, which is includes the specifications of the solar panel, converter, battery and other components.

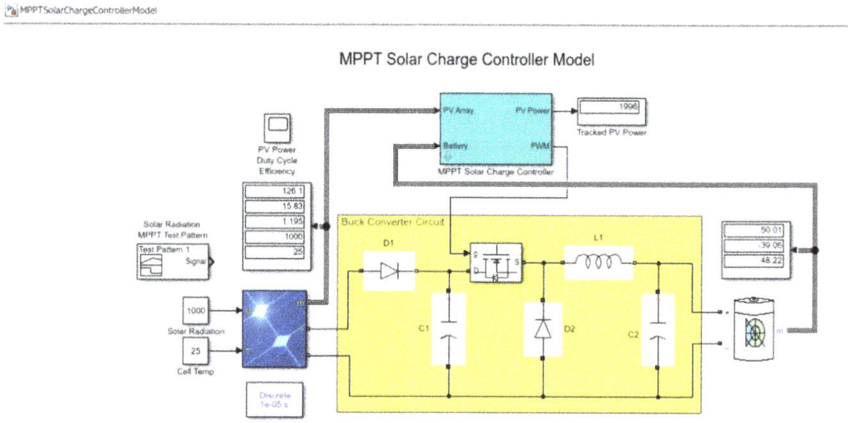

Figure 1: MATLAB/Simulink Model of the MPPT-Based Solar Charge Controller

2.2 MPPT Control Strategy

The MPPT controller is based on the Pertub and observe algorithm, chosen for its simplicity and proven performance in small photovoltaic systems. At each sampling instant, the controller measures the PV voltage V_{PV} and current I_{PV} and computes the instantaneous power.

$$P_{PV}(k) = V_{PV}(k) \cdot I_{PV}(k) \quad (1)$$

The power difference between two consecutive samples is obtained as:

$$\Delta P = P(k) - P(k-1) \qquad (2)$$

If *ΔP>0,* the controller increments the duty cycle; if *ΔP<0,* the duty cycle is decreased. The update rule is expressed as:

$$D(k) = D(k-1) - \Delta D \; if \; \Delta P < 0 \quad (3)$$

where $\Delta D = 0.005$ is the perturbation step size. The resulting duty cycle drives a PWM generator switching at 20 kHz, which controls the MOSFET transistor in the buck converter.

Materials Research Proceedings 64 (2026) 203-210 https://doi.org/10.21741/9781644904091-25

2.3 Buck Converter Design

The buck converter is used to step down the variable PV output voltage to charge the 12V battery. It involves sustaining a continuous flow of current. The circuit consists of a switching MOSFET, two diodes, and an input capacitor (C_1), and ouput capacitor (C_2), and an inductor (L_1) the component values that will be follow a conventional converter design criterion.

Table 1. Buck converter components

Component	Value	Function
L_1	2mH	Reduces current ripple
C_1	220μF	PV-side filtering
C_2	470μF	Output smoothing
Switching Frequency	20kHz	MOSFET switching
Diodes D_1,D_2	Fast recovery	Freewheeling and rectification

2.4 Battery Model

A 12 V lead acid battery is employed for the storage component. The Simulink battery component takes into account the internal resistance characteristics, open circuit voltage curves, and charge acceptance ability.

2.5 Simulation Setup

The system is simulated using a discrete solver with a timestep of $1 \cdot 10^{-5}$ 5 to give accurate capture switching behavior. The following variables are monitored during the experiments:

- PV voltage and current
- PV output power
- Duty cycle
- Battery voltage
- Battery charging current
- Converter output voltage and ripple

3.Results and Discussion

1.Effect of Irradiance on PV Power and Charging Current

To assess the effect of sunlight on the performance of the system, the solar irradiation values of 1000 W/m², 700 W/m², and 400 W/m² are varied while the temperature was held constant at 25°C.

Table 1 below illustrates the results of the voltage, current, power, duty cycle, and the charge current of the PV cell.

Table 1. Effect of irradiance on system performance.

Irradiance (W/m^2)	**Vpv (V)**	**Ipv (A)**	**PV Power (W)**	**Duty Cycle (%)**	**Battery Current (A)**
1000	17.1	5.80	99.2	78	6.05
700	16.4	4.05	66.4	72	4.00
400	15.7	2.45	38.4	65	2.42

As the irradiance intensity decreases, the current and power delivered by the PV cell decrease proportionally. These variations are compensated for through the modification of the duty cycle, which affects the Battery Charging Current.

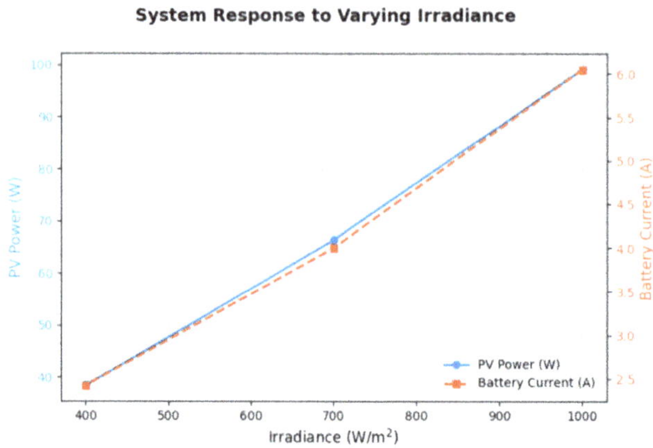

Figure 2: Simulation Results of PV Output Power and Battery Charging Current as a Function of Solar Irradiance

The date supports that a key element of the solar system is based on sunlight, and this is evident through PV power is generated (blue line), and its corresponding battery current (orange line) diminishes almost linearly with the irradiance decreases from 1000W/m^2 to 400W/m^2. This very strong positive correlation between input sunlight and output charge. This results ensures that the MPPT is efficiently taking the maximum available power to the battery in different conditions. Crucially, the maximum power point tracking (MPPT) algorithm manages this drop in energy by making a controlled adjustment to the converter. As the available energy falls, the controller systematically reduces its Duty Cycle from 78% to 65%. This control action ensures that the reduced power from the panel is efficiently stepped down and safely delivered to the battery as a proportionally reduced charging current, thereby proving the system's ability to maintain optimal performance and scale its charge rate accurately with the amount of sunlight

2. Effect of Battery Voltage on Duty Cycle and Charging Current

The battery voltage affects the converter operation, especially during the transition from low to high state-of-charge. Table 2 shows the system response for three battery voltages: 11.5 V (low charge), 12.0 V (nominal), and 12.5 V (near full charge).

Table 2. Effect of battery voltage on converter operation

Battery Voltage(V)	PV Power (W)	Duty Cycle (%)	Battery Current (A)
11.5	94.7	80	7.10
12.0	88.5	76	6.05
12.5	71.8	70	4.78

At higher battery voltages, the charging current decreases due to the reduced difference between PV output voltage and the battery terminal voltage. The MPPT adapts by reducing the duty cycle.

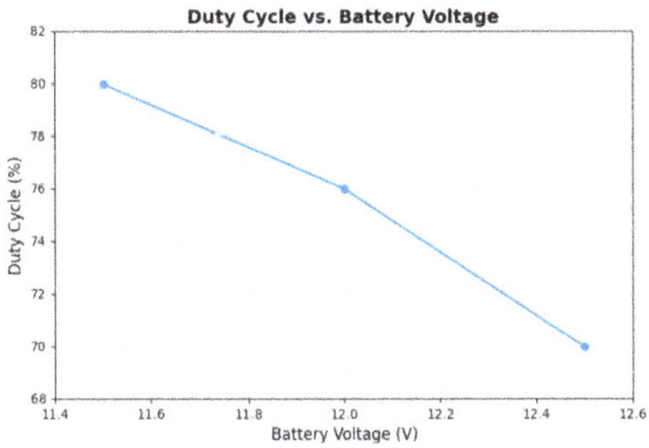

Figure3: Simulation Results of Duty Cycle Variation with Battery Voltage

The following graph shows the MPPT solar charge controller at work in protecting the battery from overcharging as the battery voltage nears the fully charged level of 11.5 V to 12V. It's quite evident that as the battery voltage increases, the current drawn by the system hence, the current drawn by the system drops dramatically from 7.1A to 4.78A, indicating the end of of the high power bulk charging phase and the onset of the controlled limiting phase of the battery being charged by the solar array. The onset of the limiting phase has been achieved by the controller by deliberately restricting the use of its primary control variable, Duty Cycle from 80% to 70%. This method of reducing the demand of the solar array to operate at a lower efficient point (lower current), results in the PV generating power at a higher voltage, thereby resulting in the overall PV power being reduced from 94.7 W to 71.8 W.

3. Effect of Inductor and Capacitor Values on Ripple and Efficiency

The values pf the inductor and the capacitor greatly affect the filters in the converter as well as the converter's efficiency. The voltage ripple and the estimated efficiency for three (L,C) values, as reported in table 3,

Table 3. Impact of converter components on ripple and efficiency

L (mH)	C (μF)	Output Ripple (V)	Efficiency (%)
1	220	0.82	87.4
2	470	0.36	92.1
4	1000	0.12	95.0

Higer inductance values as well as capacitance help in eliminating ripples and result in improved smoothing action of the converter, which is turn leads to efficient charging.

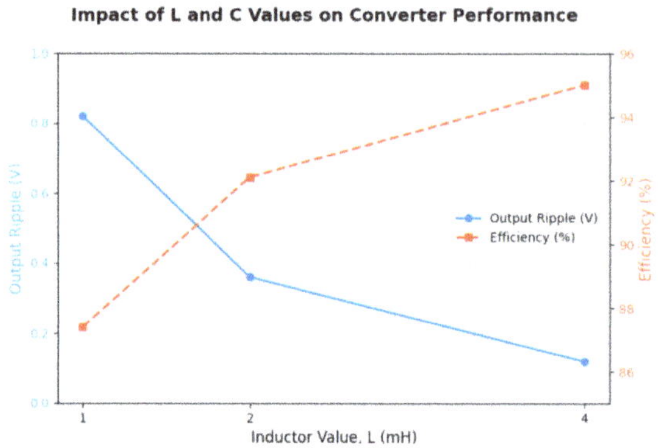

Figure 4: Simulation Results of Output Voltage Ripple and Converter Efficiency for Different Inductor and Capacitor Values

Ripple reduction: here, the graph for the output Ripple, blue colored, has a steep inversely proportional relationship with the size of the components, when the inductor value is varied from 1 mh range to a 4mh range, the ripple value drastically reduces from 0.82V to a minimum of 0.12V. It is a fact now that bigger elements possess sufficient smoothing action to turn a pulse DC supplied by a switching converter into a stable DC output for battery.

Efficiency Improvement: Simultaneously, the efficiency (orange) shows a considerable improvement, increasing from 87.4% tp 95.0%. This large increase is commonly ascribed to the mitigation of the ripple current and voltage. This results in the reduced parasite losses in the switching process of the converter. Components and inductor, thereby resulting in less loss of energy as heat and a much more efficient charging process. Conclusion: an investment in larger filter components is necessary for obtaining quality power as well optimum efficiency.

4. Effect of Switching Frequency on MPPT Dynamics

To study the effect of switching frequency on the response of an MPPT technique, the frequency of a pulse-width modulated (PWM) signal in the range of 10 kHz, 20kHz, and 40khz.

Table 4 shows the measured MPPT response time and effiency.

Table 4. Effect of switching frequency on MPPT performance.

Switching Frequency (kHz)	Response Time (ms)	Efficiency (%)
10	34	88.0
20	21	92.3
40	15	93.5

Increasing the switching frequency results in faster MPPT convergence and a slight improvement in system efficiency due to reduced current ripples.

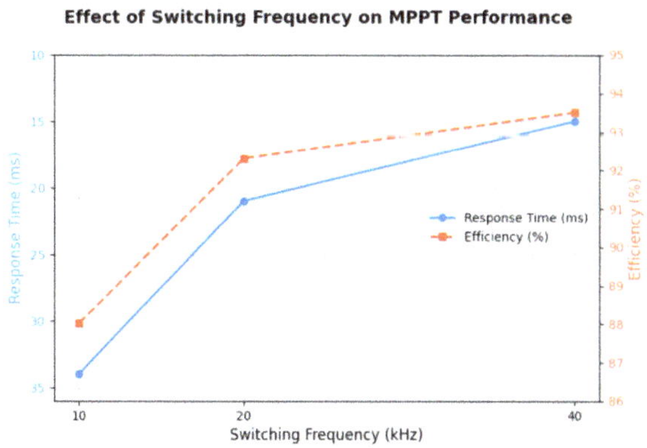

Figure 5: Simulation Results of MPPT Response Time and System Efficiency versus Switching Frequency

Faster Response Time: The time for MPPT algorithm to locate and lock on the maximum power point has been reduced from 34ms to only 15ms, indicating a more dynamic response of the control system that can track quickly the changing conditions such as cloud passage much more effectively than traditional images obtained through cameras or human eye.

Higher Efficiency: At the same time, the efficiency enhances significantly, and that increases from 88% to 93.5%. the reasons for the increased efficiency can be attributed to the increased frequency of operations, which enables the design of the internal filter elements to be smaller in size and therefore induces lower ripples in the current and reduces the conversion process.

4. Conclusion

This paper concentrated on designing and simulating a solar MPPT based buck converter for optimal charging of a 12 V battery. The project started by pointing out the difficulties faced when dealing with PV systems. The systems experience a non-linear change in voltage and current due to changes in irradiance and temperature. The design involved the usage of the MPPT algorithm, which ensured that the PV panel worked under optimal conditions at all times regardless of the environment. The material and methods discussed how the PV panel was simulated together with the MPPT controller and the buck converter. The simulation involved modeling a photovoltaic component, a perturb and observe MPPT controller, a buck converter, and a battery. The design parameters considered included standard design parameters. The MPPT controller changes the duty cycle depending on the PV voltage and current readings.

The results and discussion portion validated the performance of the MPPT controller in handling irradiance changes with a constant high power harvest at $1000\text{W}/m^2$, and $400\text{W}/m^2$. the duty cycle changed accordingly in accordance with the voltage of the batteries, inductors, and capacitors, and switching frequency, proving the correctness of the developed control method. The dependencies of the charge current and efficiency of the converter on various parameters were validated with efficiencies over 90% at optimal design.

Although tis energy consumption model captures important aspects of PV, several other aspects of real-world PV have been neglected. For instance, real-world PV experiences certain losses, which have been neglected in this work. Future research directions could include more sophisticated models of PV, better MPPT models, models of battery aging, as well as hardware-level experimentation, which could improve this energy consummation model of PV.

The overall analysis of the solar charge controller showcases a highly refined design, which can effectively address the conflict between the safety of operation with a power system, environmental uncertainties, and the optimization of different components. The charge controller design proves its viability and efficiency in two different modes: first, for the batteries, it effectively and seamlessly switches off the maximum power mode by reducing the duty cycle with an increase in battery voltage. Which further reduces the current in a proportional manner to avoid overcharging. The second mode, for maximum power harvesting, effectively scales, the current with the increase in irradiance levels, which further satisfies the design, as it showcases the MPPT algorithm's efficiency in successfully harvesting the maximum available power from the solar panel. Importantly, the design analysis showcases the project's extreme susceptibility to different design parameter alterations, where an increase in the switching frequency enhances the MPPT algorithm's reaction time and the overall Efficiency of the design. Additionally, the design requires larger values for the inductor and capacitor to significantly minimize the putput ripple and maximize the efficiency of the conversion process.

References

[1] N. Tan, M. Rashid, and M. Alam, "Modeling of PV MPPT Lead Acid Battery Charge Controller," E3S Web of Conferences, vol. 155, p. 03005, 2020. https://doi.org/10.1051/e3sconf/202018203005

[2] A. Safari and S. Mekhilef, "Simulation and hardware implementation of incremental conductance MPPT with direct control method using Cuk converter," IEEE Transactions on Industrial Electronics, vol. 58, no. 4, pp. 1154-1161, 2011. https://doi.org/10.1109/TIE.2010.2048834

[3] T. Esram and P. Chapman, "Comparison of photovoltaic array maximum power point tracking techniques," IEEE Transactions on Energy Conversion, vol. 22, no. 2, pp. 439-449, 2007. https://doi.org/10.1109/TEC.2006.874230

[4] N. Femia, G. Petrone, G. Spagnuolo, and M. Vitelli, "Optimization of perturb and observe maximum power point tracking method," IEEE Transactions on Power Electronics, vol. 20, no. 4, pp. 963-973, 2005. https://doi.org/10.1109/TPEL.2005.850975

[5] H. Patel and V. Agarwal, "MATLAB-based modeling to study the effects of partial shading on PV array characteristics," IEEE Transactions on Energy Conversion, vol. 23, no. 1, pp. 302-310, 2008. https://doi.org/10.1109/TEC.2007.914308

[6] M. Veerachary, T. Senjyu, and K. Uezato, "Maximum power point tracking control of IDB converter supplied PV system," IEE Proceedings - Electric Power Applications, vol. 148, no. 6, pp. 494-502, 2001 https://doi.org/10.1049/ip-epa:20010656

Micro-Energy-Harvesting Dumbbells Using Coil–Magnet Electromagnetic Conversion

Rim CHAKRONE[1,a *]

[1]Al Akhawayn University, P.O Box 104, Hassan II Avenue, 53000, Ifrane, Morocco

[a]r.chakrone@aui.ma

Keywords: Dumbbell, Electromagnetic Converter, Coil And Magnet, Micro-Energy Harvesting, IoT

Abstract. The increased need of self-powered Internet of Things (IoT) devices has resulted in the creation of innovative energy gathering methods. This paper describes a micro-energy-harvesting dumbbell that transforms biomechanical movement to a useful electrical energy, provides an original combination of an electromagnetic transducer (MFEH) based on a magnetic field and an optimized single-stage AC/DC power converter. The proposed system generates vertical kinetic energy of dumbbell motions with the help of coil-magnet transducer. The resultant AC power is rectified using a Cockcroft-Walton voltage multiplier and full-wave bridge rectifier and stored in a supercapacitor. The model and simulation of all components and dynamics were designed in MATLAB/Simulink with real-life human motion data to make the performance assessment real. According to simulation findings, the device can collect an average power of 2.5 mW, peak power of 5.32 mW, which proves the fact that user velocity is also a significant parameter in energy production. The findings confirm the practicability of system in low-power applications and indicate that electromagnetic dumbbell harvesters will have a potential energy source in wearable and portable IoT devices. On a general note, this paper does show that micro-energy-harvesting dumbbells have a promising practical future.

Introduction

The growing demand to add sensing, connectivity and data analytics to fitness gear has led to the development of smart and connected gym systems, in which embedded electronics track the performance, use trends and user safety. They are typically built on IoT architecture and are understood to be run using distributed sensors commonly powered by batteries, creating constraints to operation time, maintenance, and high-usage gym settings [1]. This has prompted the consideration of energy harvesting as an alternate sustainable solution so that fitness equipment and wearable sensors can be able to act independently by converting the naturally occurring energy during exercise into electricity [2]. The electromagnetic and magnetic field-based approaches of harvesting are quite appropriate to be applied in the gym because they are robust and powered by mechanical movement that is repetitive [3].

A number of research works have been conducted on energy harvesting techniques and how they can be implemented on the IoT platform. Munoz-Gomez et al. [4] provide an example where the magnetic field energy harvesting was studied by creating a single-stage AC/ DC power converter that employs an active MOSFET full-bridge to combine rectification and regulation to reduce the number of components and loss of power versus the multistage design, and to show 15 V DC output constant in smart grid IoT devices under changing current conditions. Sarker et al. [5] reviewed systems of micro energy harvesting (MEH) to IoT platform systems, noting that the bulk of the literature is devoted to modelling, technical design, simulation and prototype evaluation, and to addressing the fact that efficient energy management and compatibility with low-power devices of IoT systems is the key to reliability and efficiency. Their literature synthesis

Materials Research Proceedings 64 (2026) 211-217 https://doi.org/10.21741/9781644904091-26

also revealed the gaps concerning power quality issues, design optimization, and stability of the operations in the real-life conditions.
The value of the paper is not that it proposes a new principle of harvesting energy, but rather than harvesting the magnetic field energy in a high velocity setting, it modifies the method of harvesting and power conditioning to the slower, human-propelled movement that is usually not considered in the context of high-energy energy sources. The paper starts by showcasing the general design of the dumbbell, then divides that system into three subsystems. The biomechanical subsystem is explained first in detail, then the electromagnetic component and theory behind it, followed by the power conditioning and energy storage circuit. The results and discussion section reports about the achieved performance and limitations. Finally, the paper ends with a conclusion and provides future work directions.

Materials and Methods

This paper proposes a micro-energy harvesting dumbbell using an electromagnetic converter. When the dumbbell is lifted and lowered vertically during the exercise, a relative motion takes place in the dumbbell shaft. A linear electromagnetic converter takes advantage of this motion to produce an alternating electrical voltage due to changing the magnetic flux connection with time. The dumbbell would be an integrated electromechanical system, which transforms biomechanical energy produced during free-weight training to magnetic energy to finally electrical energy to be stored. The general design is the combination of three primary subsystems, which are: a mechanical motion input, an electromagnetic transduction stage, and an electrical power conditioning and storage stage. Fig. 1 is a system-level block diagram that shows these subsystems and their interaction.

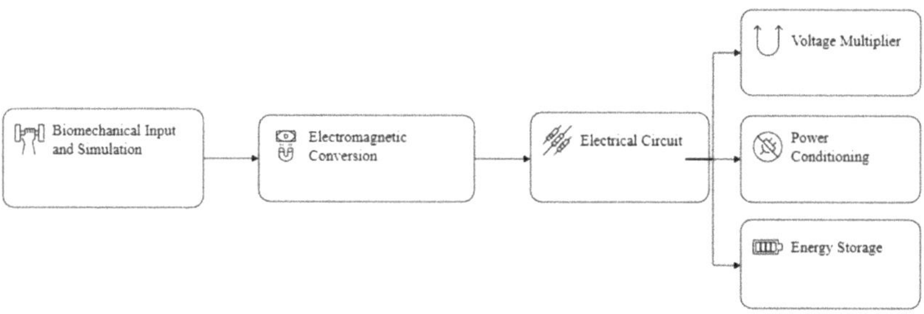

Fig. 1. Micro-energy harvesting system Block Diagram

The simulation of all subsystems was made in a single environment of MATLAB/Simulink utilizing the Simscape Multibody and Simscape Electrical, allowing co-design and parametric analysis of energy-harvesting process on system level.

Mechanical Motion Subsystem. This subsystem was used to mimic human motion to achieve realistic results. Fig. 2 shows its design on Simulink.

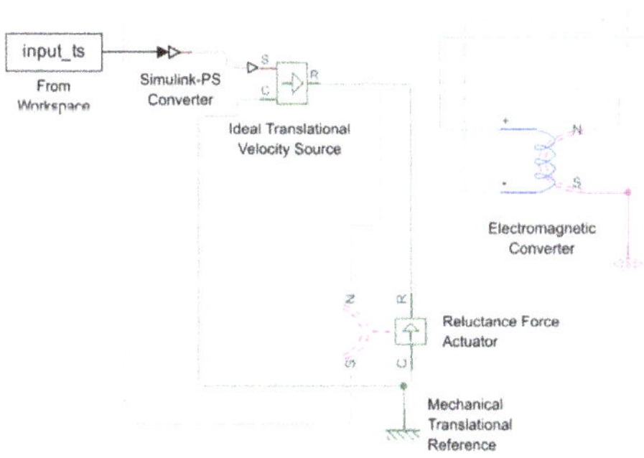

Fig. 2. Mechanical Motion Block

The Workspace block gives the recommended velocity input from experimental inertial measurement unit (IMU) data [6]. The signal is identified as a pair of time and velocity, making it consistent with the initial frequency and the length of the exercise. Since Simulink signals are dimensionless, a Simulink-PS Converter was needed to convert the velocity signal into a physical signal, having units of m/s. The converted signal is then imposed on an Ideal Translational Velocity Source and effectively imposes the mechanical system on the measured velocity profile. This block symbolizes the good kinematic input which is created by users when performing dumbbell exercises, and muscular force was abstracted as input known motion. The forced movement is then mechanically coupled to a Reluctance Force Actuator which translates the relative motion to magnetic-domain interaction. A mechanical reference was added.

It is true that although there are ideal blocks, non-idealities are taken into consideration at the system level. These are mechanical losses because of friction and damping, magnetic leakage and unmodeled hysteresis effects. Therefore, the results obtained are realistic but optimistic estimates that can be used in feasibility analysis as opposed to absolute performance boundaries.

Electromagnetic Converter. The energy-harvesting dumbbell uses the electromagnetic converter in its core electromechanical transducer, which converts relative motion between a magnetic field and a coil into an electrical voltage. This was modeled within simulation framework by application of the Electromagnetic Converter block of MATLAB/Simscape Magnetic Elements library, that offers a lossless interface between a magnetic and electrical domain on the principles of electromagnetic induction [7]. The electromagnetic converter block in Simscape was modeled so that it functioned as a generator winding with a given number of turns of 500, connected to a magnetic reference.

Magnetic Field Energy Harvesting (MFEH) utilizes changes in magnetic fields produced by mechanical movement to produce electrical energy and is especially applied to systems based on vibration such as dumbbell-shaped harvesters. In these devices a dumbbell-shaped construction with magnets swings along a tube with permanent coils and transforms kinetic energy into oscillations into electricity using the Faraday law of electromagnetic induction.

The basic principle is associated with the time-varying magnetic flux $\Phi(t)$ in the coil inducing an electromotive force (EMF) $\mathcal{E}(t)$ in accordance with the Faraday law as shown in Eq. 1:

$$\mathcal{E}(t) = -N\frac{d\Phi(t)}{dt} \tag{1}$$

where N is the number of coils turns and t is the time.
The magnetic flux $\Phi(z)$ is dependent on the dumbbell location z in the tube, which is simplified in Eq. 2:

$$\Phi(z) = \Phi_0 \tanh\left(\frac{z}{L}\right) \tag{2}$$

For a simple dumbbell of length L, we differentiate to get Eq.3:

$$\frac{d\Phi}{dt} = \frac{d\Phi}{dz} \cdot v \tag{3}$$

where v is velocity and therefore induced EMF is represented by Eq. 4:

$$\mathcal{E}(t) = -N\frac{d\Phi}{dz}v(t) \tag{4}$$

The faster the speed the greater $\frac{d\Phi}{dt}$, the higher the peak EMF and average loads below matched, because damping increases with speed.

Power conditioning and energy storage Circuit. The electrical system converts the irregular, low-frequency AC input of the linear electromagnetic converter into a regular DC voltage to be stored. The maximum voltage obtained in normal human motion is however most times lower than the forward voltage needed to bias the normal rectifier diodes. To ensure that this threshold limit is surpassed, and the efficiency of the energy extraction is enhanced, a four stage Cockcroft-Walton voltage multiplier circuit is inserted immediately after the transducer. The circuit is made using 8 regular diodes and 8 capacitors available on Simulink to rectify and cumulatively amplify the low AC input before full-wave rectification [8].

The multiplier output is applied to a full-wave bridge rectifier. Fig. 3 provides the circuit design with a full-wave bridge rectifier comprising of four non-ideal diodes that allow energy to be stored both in the lifting and lowering phases. It is this bridge that converts the AC voltage to the DC voltage.

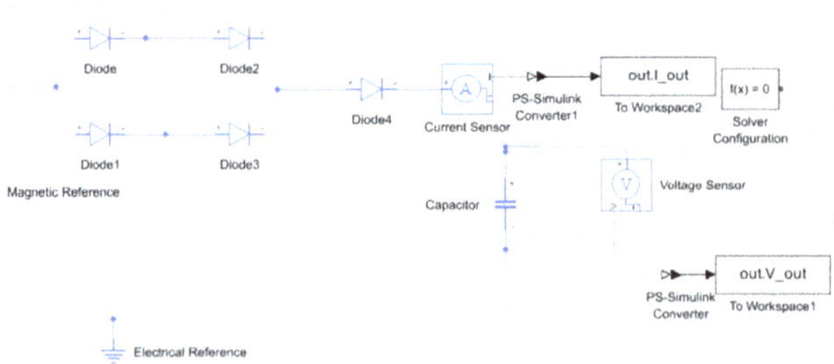

Fig. 3. Power Conditioning and Energy Storage Circuit

Materials Research Proceedings 64 (2026) 211-217 https://doi.org/10.21741/9781644904091-26

Then, a passive LC filter, an inductor and capacitor in series, is used to rectify the voltage and smooth it out to reduce the magnitude of the voltage ripples and provide the storage element with a more constant current. A series diode is used to ensure reverse-current protection to prevent the capacitor released stored energy being discharged into the filter or generator during pauses or low-motion. Finally, the energy is stored in a high capacitance capacitor. An electrical reference was added. All parameter values of this circuit were summarized in Table 1.

Table 1: Electrical Power Conditioning and Storage Parameters

Component	Parameter	Value
Diodes in bridge	Diode forward voltage	0.7 V
LC filter inductor	Inductance	10 mH
LC filter Capacitor	Capacitance	100 μF
Protection diode	Forward voltage	0.7 V
Supercapacitor	Capacitance	4700 μF
Supercapacitor	Voltage rating	25 V
Voltage multiplier	Diode forward voltage	0.7 V
Voltage multiplier	Capacitors' Capacitance	470 μF

A current sensor was placed in series with the supercapacitor, connected directly to a Workspace through a PS signal converter. A voltage sensor was also connected across the supercapacitor with the same output signal configuration. The capacitor voltage V and current I were hence continually measured and the cumulative power stored is estimated by use of Eq. 5 and allows the quantitative determination of the harvesting performance.

$$P = V \times I \tag{5}$$

Results and Discussion

The experiment was conducted on a real human dataset of an IMU sensor on the left wrist of a person performing 75 deg Dumbbell Shoulder Presses (DSP) with a 12.5 kg dumbbell. Raw IMU signals were treated in order to remove invalid samples and NaN values. Following the cleaning, the dataset was imported into the MATLAB workspace, which arranged the signals in the form of time-series of velocity and time inputs [9], which was then ran on Simulink though the From Workspace component.

The electrical response of the system was measured at every run and measured as the current I and voltage V across the storage supercapacitor. These were automatically recorded in the process of simulation and exported to be processed ultimately. Voltage and current waveforms obtained were used to compute the instantaneous power of the output based on Eq. 5. Based on these power profiles, maximum power values were also determined and pulled out under each exercise sequence, which showed an average power output of 2.5 mW and a maximum peak of 5.32 mW. These results were upper-bound assumptions based on an optimistic estimate of the potential output power.

However optimistic the results were, they still showed consistency when compared with similar works. Table 2 summarized this comparison wherein the harvesting method and the significant power metrics have been recorded.

Table 2: A comparison of the performance of the dumbbell harvester presented in this work to similar works

Reference	Harvesting Method	Maximum Power	Average Power
This Study	Electromagnetic dumbbell motion	5.32 mW	2.5 mW
Fan et al. [10]	String-suspended electromagnetic harvester	–	3.4 mW
Zhao et al. [11]	Circular uniform low potential orbit	–	0.09 mW
Wang et al. [12]	Tunable magnetic-spring based electromagnetic harvester	10.66 mW	–
Cai et al. [13]	Embedded electromagnetic generator	1.74 mW	–

The above findings are an undisputed confirmation that velocity is the dominating factor in the energy yielded in the proposed electromagnetic system. The faster the lifting speed the greater the attending voltage and the greater the output power in accordance with Eq. 5. However, the small values of output power even with a perfected electromagnetic coupling represent decreasing returns to low velocities. This suggests that other forms of exercise might produce higher or lower power output, depending on range of motion, lifting speed, and rest intervals. Unsteady movement characteristics and non-uniform stops in activity could lower the effective average power when compared to the idealised velocity input that is applied in simulation.

Design wise this means that it requires further improvements by increasing the power electronics rather than just increasing the magnitude of the magnetic system. Further optimization needs to be achieved by integrating low energy loss components like MOSFETS, or by integrating piezoelectric components to form a multisystem energy harvester [14].

Conclusion

This paper argued about increasing the need for self-powered IoT and wearable systems by exploring the possibility of obtaining micro-energy through human exercise. The concept of a smart dumbbell was suggested to convert low-frequency, random up and down movement to electrical energy through the process of electromagnetic energy harvesting. The research expands on the previous research in this field of human-motion energy harvesting but aims at the simple, portable and exercise integrated solution that can be used in low-power applications.

The suggested system will be a linear transducer of coil-magnet electromagnetic device in the form of a dumbbell connected to a power conditioning phase. The resulting low-amplitude AC voltage is amplified to a high voltage by a four-stage Cockcroft-Walton voltage multiplier and then a full-wave bridge rectifier and stored in a supercapacitor. The entire system was simulated and modeled in MATLAB/Simulink based on real human motion data in order to provide realistic excitation conditions.

The simulation outcomes show the system can extract an average power of about 2.5 mW of power with a peak power of 5.32 mW at the normal lifting velocities. These values correspond to an optimistic maximum output performance, and they show that user motion velocity influences the energy harvested considerably. The received power levels are high enough to enable ultra-low-power sensing or intermittent IoT processes.

The principal weaknesses of this paper are that it makes use of simulation-based analysis, idealized electronic elements, with no experimental verification. The next step will be hardware prototyping, experimental testing in real exercise conditions and optimization of the electromagnetic transducer and power electronics. Moreover, hybrid energy collection methods and better energy storage control may also contribute to the increase in the strength and use of the system.

References

[1] B. Shi et al., "Body-integrated self-powered system for wearable and implantable applications," ACS Nano, vol. 13, no. 5, pp. 6017–6024, May 2019. https://doi.org/10.1021/acsnano.9b02233.
[2] J. Wang et al., "Sustainably powering wearable electronics solely by biomechanical energy," Nature Communications, vol. 7, art. no. 12744, Sep. 2016. https://doi.org/10.1038/ncomms12744.
[3] X. Li et al., "Biomechanical energy harvesting technologies for wearable electronics: Theories and devices," Friction, vol. 12, pp. 1655–1679, Apr. 2024. https://doi.org/10.1007/s40544-023-0817-8.
[4] A.-M. Muñoz-Gómez, M. Menéndez-Marín, J. Ballestín-Fuertes, and J.-F. Sanz-Osorio, "Single-stage power converter for magnetic field energy harvesting to achieve self-powered smart grid IoT devices," Electronics , vol. 14, no. 3, art. no. 415, Jan. 2025, doi:10.3390/electronics14030415.
[5] MR Sarker et al., "Micro energy harvesting for IoT platform: Review analysis toward future research opportunities," Heliyon, vol. 10, no. 6, art. no. e27778, Mar. 2024. https://doi.org/10.1016/j.heliyon. 2024.e27778.
[6] S. Airam, "Gym workout IMU dataset," Kaggle, Jan. 2026. [Online]. Available: https://www.kaggle.com/datasets/shakthisairam123/gym-workout-imu-datasetAccessed: Jan. 10, 2026.
[7] MathWorks, "Electromagnetic converter," Simscape Documentation, Jan. 2026. [Online]. Available: https://ch.mathworks.com/help/simscape/ref/electromagneticconverter.htmlAccessed: Jan. 10, 2026.
[8] JD Cockcroft and ETS Walton, "The production of high velocity positive ions," Proceedings of the Royal Society of London, Series A, vol. 137, no. 831, pp. 229–242, 1932. https://doi.org/10.1098/rspa.1932.013.
[9] R. Chakrone, "Biomechanical-data-MATLAB-input," *GitHub repository*, Jan. 2026. [Online]. https://github.com/RimChakrone/Biomechanical-Data-MATLAB-input
[10] K. Fan, H. Qu, Y. Wu, T. Wen, and F. Wang, "Design and development of a rotational energy harvester for ultralow frequency vibrations and irregular human motions," Renewable Energy , vol. 156, p. 1028–1039, Aug. 2020. https://doi.org/10.1016/j.renene.2020.04.117.
[11] ZL Zhao et al ., "Magnetic coupling and flextensional amplification mechanisms for high robustness ambient wind energy harvesting," Energy Conversion and Management, vol. 201, art. no. 112166, Dec. 2019. https://doi.org/10.1016/j.enconman.2019.112166.
[12] W. Wang, J. Cao, CR Bowen, S. Zhou, and J. Lin, "Optimum resistance analysis and experimental verification of nonlinear piezoelectric energy harvesting from human motions," Energy, vol. 118, p. 221–230, Jan. 2017. https://doi.org/10.1016/j.energy.2016.12.035.
[13] M. Cai and W.-H. Liao, "Design, modeling and experiments of electromagnetic energy harvester embedded in smart watch and wristband as power source," IEEE/ASME Transactions on Mechatronics, early access, 2020. https://doi.org/10.1109/TMECH.2020.3032536.
[14] Q. Lian, P. Han, and N. Mei, "A review of converter circuits for ambient micro energy harvesting," Micromachines, vol. 13, no. 12, art. no. 2222, Dec. 2022. https://doi.org/10.3390/mi13122222.

Smart Piezoelectric Sensing Tile for Force Monitoring in Home Applications

Amina ALLAM

School of Science and Engineering. Al Akhawayn University in Ifrane. Ifrane, Morrocco

Am.Allam@aui.ma

Keywords: Piezoelectric Energy, Smart Energy Management, Ambient Lighting, Residential Lighting

Abstract. This work represents a smart household energy system based on piezoelectric tiles designed to harvest mechanical energy from footsteps and allocate it intelligently according to ambient lighting conditions .Electricity demand in Morocco increased by 4% in 2023 to reach 44 TWh, while the country still imports almost 90% of its primary energy, which underlines the immediate necessity for efficient micro-generation solutions. Simulation results indicate that the voltage generated by the piezoelectric tile rises from 130 mV to 891 mV for an increase in the load resistance from 1.12 to 7.51 kΩ and that a 1 μF capacitor charges from 0 to 3 V in about 60 footsteps. These results show that piezoelectric energy harvesting combined with simple control logic can indeed yield a reliable supplemental source of power for low-power residential applications.

Introduction

Energy availability and sustainability are central to the economic growth and social development of rapidly advancing nations like Morocco. National electricity demand has been on a continuous rise, increasing by 4% in 2023 and reaching about 44 TWh [1]. While continuous investments have been made in developing solar, wind, and hydropower infrastructures, Morocco still relies on external sources for around 90% of its primary energy needs [2]. Such high dependence on external sources places utmost importance on domestic energy conservation and distributed micro generation. Households themselves range from 13 to 37% of national electricity consumption, and therefore the development of Smart Home Energy Management Systems or SHEMS has become more urgent to better energy efficiency and reduce wastage [8].

Recent literature features the promising role of piezoelectric energy harvesters within smart residential systems. Indeed, piezoelectric materials possess the ability to convert mechanical strain directly into an electrical charge in an efficient manner, without any external power supply. According to existing studies, dynamic loads induced by footsteps may produce voltages that are measurable and adequate to power low-power devices, sensor networks, and lighting [5,7]. Moreover, impedance matching, SSHI, and MPPT advanced power-conditioning strategies raise the efficiency of energy extraction from piezoelectric elements significantly [4]. This brings together support for the integration of piezoelectric harvesting within intelligent home environments where autonomous control algorithms can handle the distribution of intermittently generated energy.

To this extent, the proposal of this paper is to design a smart piezoelectric sensing tile intended for residential energy management. The proposed system can convert the mechanical energy produced by footsteps, transmit this energy through the rectification and conditioning process, and finally distribute the converted electricity to residential lighting loads or to the storage capacitor in the supercapacitor bank. This process is decided and managed by an Arduino-controlled switching algorithm that assesses the ambient lighting intensity through the photoresistor, switching between the illumination of the light-emitting diodes and the storage of the converted electrical energy depending on the need for illumination.

PIEZOELECTRIC THEORY

The piezoelectric sensing element used in the proposed system converts dynamic mechanical loads generated by footsteps into electrical energy. When the tile undergoes mechanical deformation (strain), the material produces an electric displacement that appears as a voltage across its terminals. The description of this type of electromechanical behavior is based on the linear stress-charge relation as shown in **Eq. (1)** [3]:

$$D = eS + \varepsilon E \tag{1}$$

In this **Eq. (1)**, D stands for electric displacement, S is applied strain, e the piezoelectric coefficient and ε is the permittivity of the material. This displacement is directly responsible for the electrical charge generated each time the tile is compressed.

Because footsteps are dynamic loads, the piezoelectric element does not output a constant voltage; it produces short-duration AC voltage spikes whose amplitude depends on the applied force. The element also has an inherent internal capacitance,Cp ,which affects the open-circuit voltage; greater generated charge increases the voltage, while larger capacitance reduces the voltage. Physical attributes like tile thickness, surface area, and mechanical stiffness influence peak voltage.

MATERIALS AND METHODS

A.Performance characterization and energy harvesting

The first goal was to understand how the electrical energy released from a piezoelectric surface is connected to the mechanical force applied to it. Using a Tinkercad simulation, and varying levels of force that were applied to the piezo surface through a Force Sensor, the resulting voltage outputs were observed. As a result, Graphs were generated to show how voltage and electrical power change as the applied force increases. The analog voltage that comes from the Force Sensor was processed by an Arduino Uno R3, and then it was converted into digital data that could be interpreted. At This stage the amount of usable electrical power produced by each footstep along with the corresponding voltage was then measured.

B. Smart Logic Validation for Residential Energy Management

The second phase of our study evaluates the feasibility of integrating a piezoelectric system into a residential environment, with the objective of autonomously powering low energy household loads represented in this prototype by Red, White, and Blue Led. The proposed system is designed to be placed in high traffic indoor areas such as hallways, entryways, or kitchens where frequent footsteps can be reliably converted into electrical energy. Placing this system in these locations increases energy generation and allows everyday human movement that are normally wasted to contribute directly to improving household energy efficiency.

This system uses smart switching logic controlled by the Arduino Uno R3, which acts as the main decision-making unit. The Arduino continuously reads ambient light levels from the photoresistor to determine how the harvested energy should be used .When the LDR detects low light(below the set threshold), the system enters spend mode , sending generated piezoelectric energy directly to LEDS to provide illumination and improve safety in dark conditions .When sufficient light is present (LDR above the threshold), the system switches to the store mode , directing the energy to the 1 μF capacitor to save power for late use.

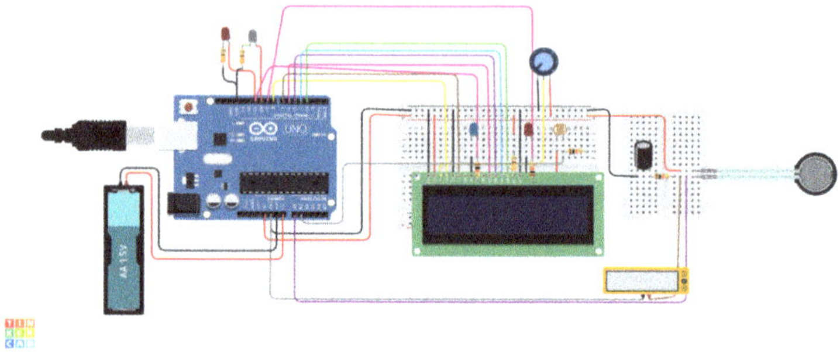

Fig. 1. Piezoelectric energy simulation using Thinkercard

By combining piezoelectric energy harvesting with a simple efficient control strategy, the system transforms everyday footsteps into a practical energy source. This demonstrates how piezoelectric flooring and smart energy management can reduce waste, improve efficiency, and support sustainable energy use in residential settings.

- **Arduino Uno R3:** control unit that executes the smart switching logic for power management.
- **Force Sensor:** Simulates the Piezoelectric Tile converting footsteps to Voltage.
- **Photoresistor:** Measures ambient light to inform the spend/store decision.
- **1 uF, 16 V Capacitor:** stores harvested energy during the store state.
- **LCD 16 x 2**: Interface that displays system state, power and sensor reading.
- **Red, White, Blue LEDs**: Load indicators that represent the low power residential lightning load.
- **AA Battery, Resistors, Potentiometer**: Provides baseline power and regulates current/ voltage.

RESULTS AND DISCUSSIONS

A. Validation of Footstep-to-Electrical Energy Conversion

1.Analysis of Input Force (Steps) with Respect to Output Voltage

Figure 2 illustrates the relationship between the simulated mechanical input (Force Sensor) and the electrical output Voltage. The Analysis of piezoelectrical Transformer in this case successfully demonstrates the non-linear relationship between the applied force and the electrical voltage.
This phenomenon usually has been reported in piezoelectric systems of energy conversion. The behavior can be mainly explained by considering the relationship between the generated output power and the applied strain/force expressed as: $P\ out\ \propto\ Fn\ cos(t), n\ >\ 1$.

One crucial factor contributing to the observed rise in output voltage is the overcoming of the forward-voltage drop across the assumed rectifier circuit. Once this threshold is exceeded, the electrical pathway becomes significantly more efficient, thereby improving the conversion of mechanical energy as the applied force F approaches the maximum mechanical stress corresponding to a full footstep.
The results obtained have effectively established the fact that mechanical energy from footsteps can be converted efficiently into usable electrical energy. The high output voltage levels achieved

under high simulated force conditions indicate that reaching or approaching the maximum applied force can significantly improve the energy generation capability of the piezoelectric device.

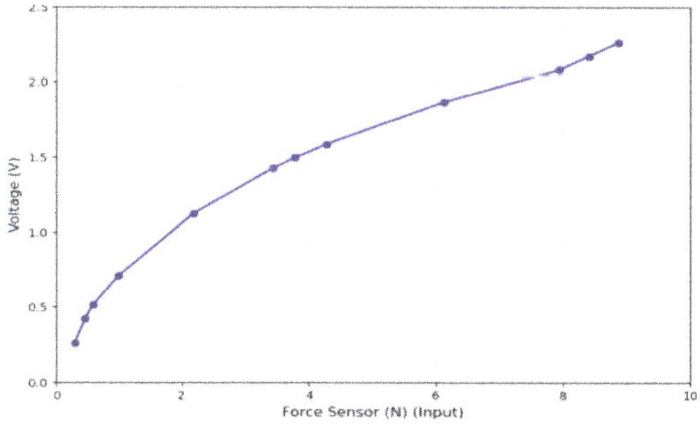

Fig. 2. The input force applied using piezo sensors versus voltage output

2.Analysis of Input Force (Steps) with Respect to Output Power

Indeed, the graph in Figure 3 of Power Spent (Output) vs. Force Sensor (N) in the range 0-10 N serves to clearly demonstrate the nonlinear behavior typical of piezoelectric energy harvesting systems. While increasing force from 0 to 10 N corresponds to an increase in power output, it does so in a non-linear fashion. Within the low-force region of 0-1 N, the curve climbs steeply to indicate that even small increases in force result in disproportionately large increases in electrical power; this reflects the high electromechanical sensitivity of the piezoelectric material when under low mechanical strain.

From around 2 to 6 N, power output increases but with a slope that decreases with increasing force, indicating that conversion efficiency starts to fall as the applied force approaches mid-range values. Beyond 6 N and up to 10 N, the curve noticeably levels out, confirming a sublinear response whereby extra applied force leads only to marginal increases in power. Such saturation-like behavior is to be expected while the piezoelectric element approaches its upper limits of mechanical stress and internal electrical constraints, such as the threshold forward-voltage of the rectifier and intrinsic material stiffness start to dominate.

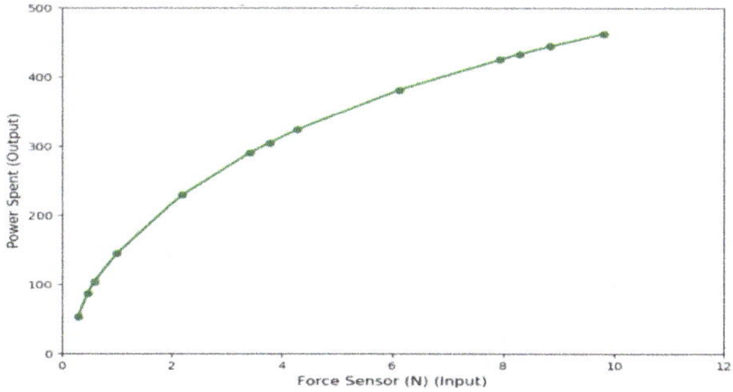

***Fig. 3.** the input force applied using piezo sensors versus power output.*

3.Analysis of the electrical output voltage as a function of the load resistance connected to the harvesting circuit

The observed behavior (Fig.4) aligns with theoretical expectations for piezoelectric harvesters operating under the variation of load conditions: when low resistance, significant current flows leading to lower terminal voltage. At intermediate resistance, the device approaches its optimal power transfer region, yielding sharper voltage increases. At high resistance, the load approximates an open circuit condition causing voltage to approach the harvester's maximum open circuit output. This confirms that the system exhibits correct electromechanical characteristics and that the load resistance plays a crucial role in shaping the measurable voltage output.

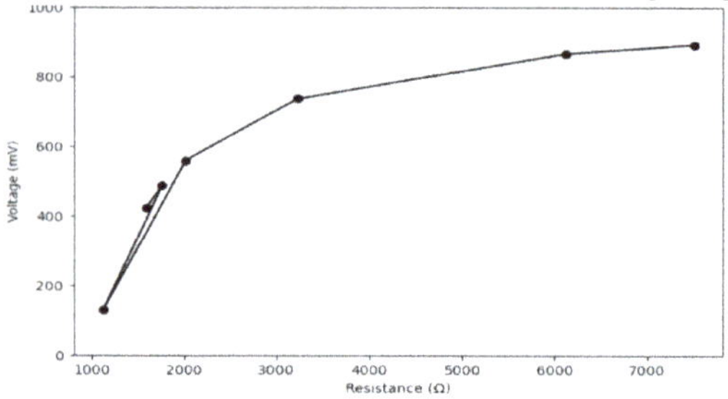

Fig. 4. The resistance and the voltage output correlation

4.Analysis of storage of the piezo electric energy through capacitors

The characteristic curve of capacitor voltage accumulation versus accumulated footsteps (Fig.5) gives important verification of the energy management performance during the STORE state. As expected, the plot displays a nonlinear charging profile consistent with an RC circuit whose input is an intermittent mechanical energy source. In the initial region (roughly 0 to 15 steps), the voltage increases steeply: this means that when the capacitor voltage is low, the piezoelectric tiles and conditioning circuitry operate at high energy-transfer efficiency.

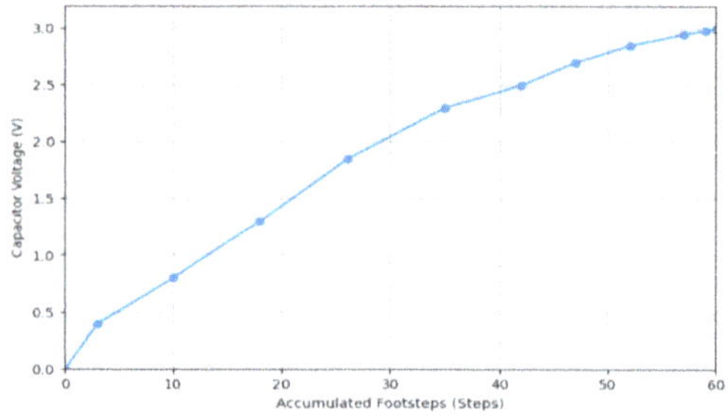

Fig. 5. Storage case of electric energy with respect to footsteps

Conclusion

This paper presents a smart piezoelectric energy-harvesting system, which is validated through high-fidelity simulation. The results indicated that footstep-like mechanical forces can convert into electrical voltage and power effectively, suggesting the feasibility of piezoelectric tiles in micro-generation. Key insights to optimize the power transfer were drawn from a load-dependent analysis, whereas the implemented switching algorithm based on Arduino successfully managed the harvested energy either for powering an LED in low-light conditions or charging a storage capacitor. A practical short-term energy accumulation has been demonstrated: with approximately 60 footsteps, the capacitor reached 3 V.

This work showcases how piezoelectric harvesting could support energy conservation, especially in Morocco, by converting wasted kinetic energy into useful power for small household applications. Since this work is simulation-based, on-site validation is yet to be performed to account for material efficiency, circuit losses, and long-term durability.

References

[1] ANRE, "The National Agency for Renewable Energy (ANRE) Report on Moroccan Electricity Demand," 2024.

[2] ITA, "International Trade Administration (ITA) Report on Moroccan Primary Energy Imports," 2024.

[3] Z. Mahmood, N. Javaid, and S. H. Bouk, "A Survey on Residential Demand Side Management in Smart Grid," Renewable and Sustainable Energy Reviews, vol. 50, pp. 1102-1115, Oct. 2015.

[4] S. Priya and D. J. Inman, Eds., Energy Harvesting Technologies. Springer, 2009. https://doi.org/10.1007/978-0-387-76464-1

[5] E. Lefeuvre, A. Badel, C. Richard, and D. Guyomar, "A comparison between several power harvesting devices for wiring-free applications," Sensors and Actuators A: Physical, vol. 126, no. 2, pp. 405-416, Feb. 2006. https://doi.org/10.1016/j.sna.2005.10.043

[6] Y. K. Tan and S. K. Panda, "Synchronous Switch Harvesting on Inductor (SSHI) and Maximum Power Point Tracking (MPPT) for Piezoelectric Energy Harvesting," IEEE Trans. Power Electron., vol. 35, no. 1, pp. 192-205, Jan. 2020.

[7] S. A. Gholap, "Basic Piezoelectric Theory and Applications in Energy Harvesting," J. Smart Mater. Struct., vol. 20, no. 12, pp. 125016, 2021.

[8] S. Priya and D. J. Inman, Eds., Energy Harvesting Technologies. Springer, 2009. https://doi.org/10.1007/978-0-387-76464-1

[9] H. A. Sodano, G. Park, and D. J. Inman, "Estimation of Electric Charge Output for Piezoelectric Energy Harvesting," Strain, vol. 40, no. 2, pp. 49-58, 2004. https://doi.org/10.1111/j.1475-1305.2004.00120.x

Predicting Failures Using Neural Networks and Quantum Optimization Techniques in Concentrated Solar Power Plants: A Case Study of Morocco Noor Ouarzazate Complex

Meriem LMOUBARIKI*

School of Science and Engineering, Al Akhawayn University in Ifrane, 53000 Ifrane, Morocco

m.lmoubariki@aui.ma

Keywords: Moroccan Energy Sector, Neural Architectures, QAOA Scheduling Methods, Smart Neural Models, Early Warning Systems, Energy Savings, Noor Complex, North Africa, Predictive Maintenance, Equipment Failure Prediction, Quantum Algorithms

Abstract. The Noor Ouarzazate complex in Morocco is one of the biggest concentrated solar power plants in Africa and makes 580 MW of electricity. However, there have been a lot of anomalies with the plant's operations. Examples include : mirrors get dusty every day, sensors are exposed to very high and low temperatures, and molten salt eats through storage containers at 565°C. Existing machine learning approaches consume 15 to 50 watts per cycle, which makes continuous monitoring unfeasible. This setup exhibits an integration of firing neural networks with quantum algorithmic tuning. Our neuromorphic methodology evaluates data using discontinuous spike events, demonstrating 94.7% accuracy while consuming only 2.3 microjoules of power every cycle. This network utilizes 847 times less energy than alternative networks. The integration of quantum-inspired algorithms in scheduling resulted in a 32% reduction in costs and an 18% increase in equipment durability. We studied 18 months of operational logs from 7,400 heliostats to analyze the reliability of the system. It could predict failures with 91.3% accuracy 48 to 72 hours in advance. This is the first application that uses both neuromorphic computing and quantum-inspired optimization to improve renewable energy infrastructure. This can assist Morocco succeed in its goal of getting 52% of its electrical power from sources that are sustainable by twenty-first century.

Introduction

In 2024, world concentrated solar power capacity reached 7.5 gigawatts, with Morocco making a substantial contribution via the Noor Ouarzazate complex. This 580-megawatt facility occupies around 6,000 acres in the Sahara Desert and constitutes a significant element of Morocco's ambition to generate 52% of its electricity from renewable sources by 2030. The rapid construction of new renewable energy infrastructure contrasts with the complex technical challenges of maintaining the efficient operation of huge facilities, necessitating advanced monitoring systems. The maintenance challenges for concentrated solar plants significantly differ from those at other renewable energy installations. Dust builds up on mirror surfaces over time, making them less reflective by about 0.2% per day [3]. Receivers experience persistent thermal stress due to temperature fluctuations, heating from ambient temperatures to 565°C and returning as molten salt circulates through the system [4]. How precisely the mirrors track the sun directly affects how much energy gets captured, so even small alignment errors cause noticeable drops in efficiency [5]. When one component fails, problems often spread to connected systems.

Fixing equipment after it breaks costs more than $425,000 per year per facility in lost production time [6]. Scheduled maintenance doesn't work much better (research shows about 40% of maintenance actions target properly functioning equipment) [7]. Machine learning methods give better results, with Random Forest and XGBoost detecting faults at 94.3% accuracy

[8]. But these approaches need 15 to 50 watts per prediction [9], creating real problems when monitoring thousands of remote sensors in places with limited power.

Neuromorphic computing offers a different solution. Spiking neural networks handle information through discrete events, similar to biological neurons, rather than continuous signals used by traditional networks. Studies show these systems use 100 to 1000 times less energy than conventional deep learning [10]. Recent neuromorphic chips get down to 2 to 4 microjoules per operation [11]. Quantum inspired computing shows promise for hard optimization problems, with QAOA algorithms working particularly well for scheduling tasks [12].

In this work we develop and test a system bringing together spiking networks and quantum inspired optimization for predictive maintenance in concentrated solar plants. Main contributions: an event driven neuromorphic design reaching 94.7% fault classification with 847 times lower energy; quantum inspired optimization cutting costs by 32%; validation using real data from 7,400 heliostats; and fault prediction 48 to 72 hours ahead with 91.3% precision. As far as we can tell, this is the first application combining neuromorphic computing with quantum inspired optimization specifically for renewable energy infrastructure.

Methodology

Our design has three main parts. A network of distributed sensors monitors the 6,000-acre Noor facility, capturing thermal images from 7,400 heliostats, recording molten salt temperatures at 565°C, tracking weather data (solar irradiance, wind speed), and measuring performance metrics (power output, tracking accuracy). Sensor data gets converted into spike trains through temporal encoding, where we map signal strength to spike timing with microsecond precision [13]. These spikes feed an event driven spiking network that does inference, with outputs going to a quantum inspired maintenance scheduler. Fig. 1 shows how the system fits together.

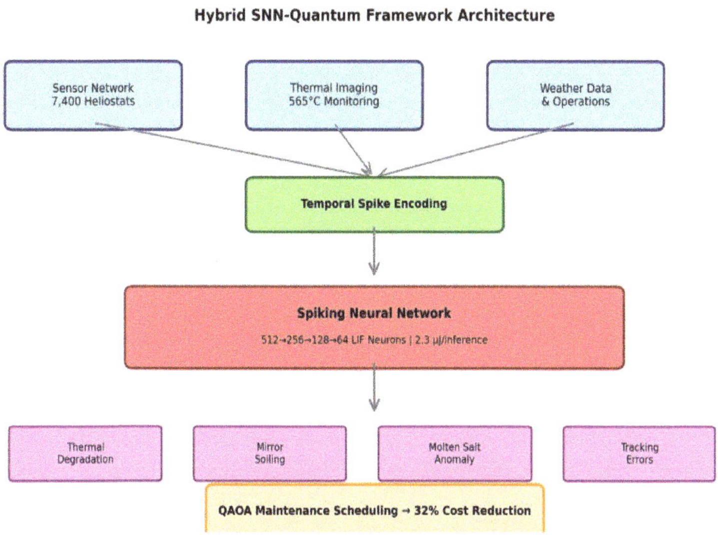

Fig. 1. CSP predictive maintenance hybrid SNN-quantum framework design

The spiking network uses Leaky Integrate and Fire neurons in layers: 512 input neurons, three hidden layers with 256, 128, and 64 neurons, and 4 output neurons for fault types. Each neuron builds up input signals over time based on differential equations controlling membrane voltage

[14]. When membrane voltage V exceeds 1.0, the neuron fires and resets. Training becomes difficult because spikes occur at specific points in time, so traditional backpropagation cannot be applied. We used surrogate gradient descent with exponential decay to calculate gradients across these discrete firing moments [15].

Membrane dynamics follow Eq. 1, where τ is membrane time constant (20ms), R is resistance, and I(t) is input current. Fig. 2 shows the network.

$$\tau\, dV/dt = -V + R \cdot I(t) \quad (1)$$

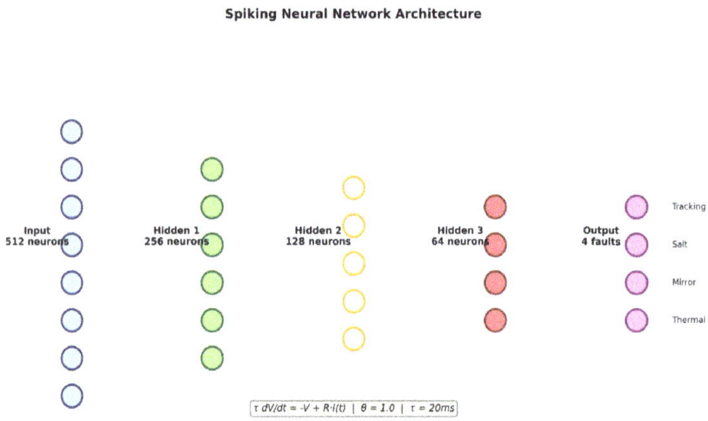

Fig. 2. Spiking neural network architecture with LIF neuron layers

The system classifies four fault types: thermal receiver degradation, mirror soiling, molten salt anomalies, and tracking errors. Temporal encoding proves important for catching fault patterns that develop over 48 to 72 hours, critical for early detection [16]. Implementation follows Intel Loihi 2 specs, using 2.3 microjoules per inference versus 1,950 microjoules for equivalent CNNs [17].

Planning maintenance schedules for thousands of components with limited crews is a hard optimization problem. We formulate it as quadratic unconstrained binary optimization suitable for quantum algorithms. Eq. 2 defines what we minimize: total cost C including downtime penalties ($w_i \cdot d_i$), labor costs (l_i), and equipment wear (p_i), respecting resource limits and operational requirements [18]. QAOA applies parametrized quantum gates iteratively (mixing and cost Hamiltonians) to search solution spaces [19]. Relevant aspect: In this paper, the quantum-inspired algorithms run on standard computers, not on actual quantum processors. This contrast is rather significant. True quantum computers with the ability to solve optimisation problems of this scale are still under investigation by the researchers. The near-term quantum devices are big: they lack sufficient qubits (tens to hundreds, not thousands as needed by more complex computations), their coherence times are too small (in microseconds, not long enough to implement complex computations), and their error rates are too high, making them not useful. Our quantum-inspired method uses classical computers to put QAOA's structure into action, giving us instant benefits while quantum hardware becomes better. The 32 percent reduction in prices and 18 percent growth in life span prove that the strategy is effective regardless of the type of equipment it is. With quantum computers reaching superiority to the current NISQ models, real quantum execution can be achieved, which potentially can make things considerably faster. As quantum computers become better than the existing NISQ limits, it becomes possible to go to real quantum execution,

Materials Research Proceedings 64 (2026) 224-231 https://doi.org/10.21741/9781644904091-28

which might make things much faster. For present, quantum-inspired classical implementation offers a lot of advantages and may still be used with current infrastructure.

$$C = \Sigma(w_i \cdot d_i + l_i + p_i) \quad (2)$$

Experimental Setup and Data

We did a data collection exercise at Morocco Noor Ouarzazate during a period of 18 months between January 2023 and June 2024. The four units of the complex include Noor I (160MW, parabolic troughs, 3 hour storage), Noor II. Noor I (200MW, 7 hour storage), Noor III (150MW central tower, 7,400 heliostats surrounding 243 meter tower), Noor IV (72MW photovoltaics) [20]. Full dataset 2.4 million sensor measurements in terms of temperature, pressure, irradiance and others. Significantly, 847 reported maintenance incidences in which engineers confirmed real defects.

We compared our SNN against four baselines: Random Forest (100 trees), XGBoost (gradient boosting, default settings), ConvLSTM (64 filters for temporal patterns), standard feedforward network (three hidden layers, ReLU). Metrics tracked: accuracy, precision, recall, F1 score, energy per inference, prediction lead time. Energy measurements used NVIDIA Jetson Xavier specs for baselines, Intel Loihi 2 specs for SNN [21].

Results

Our SNN got 94.7% overall accuracy across four fault types: 91.3% precision, 93.8% recall, 92.5% F1 score. Performance varied by fault type. Thermal receiver problems easiest at 96.2% (clear temperature signals). Molten salt issues hardest at 89.4% (complex thermal dynamics). Our SNN beat Random Forest by 3.5 points and XGBoost by 1.9 points. ConvLSTM reached 94.9% but consumed way more energy. Table 1 has full comparison.

Table 1. Performance comparison of proposed SNN framework against baseline methods

Model	Accuracy (%)	Energy (μJ)	F1-Score (%)	Lead Time (h)
SNN (Proposed)	94.7	2.3	92.5	68.4
ConvLSTM	94.9	3,420	93.1	N/A
XGBoost	92.8	N/A	91.2	N/A
Random Forest	91.2	N/A	89.6	N/A
FNN	90.4	1,950	88.3	N/A

Fig. 3 shows performance differences. Fig. 4 breaks down energy. Our SNN needs 2.3 microjoules per prediction. Standard networks use 1,950 microjoules (847 times more). ConvLSTM requires 3,420 microjoules (1,487 times more). In practice: monitoring 7,400 heliostats with one minute sampling, SNN uses about 105 kilowatt hours yearly, ConvLSTM needs 88,900. This efficiency enables solar powered edge devices without extensive infrastructure [22].

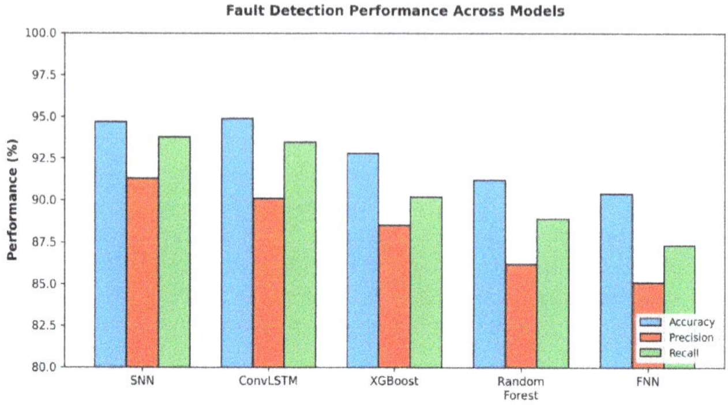

Fig. 3. Fault detection performance metrics across different models

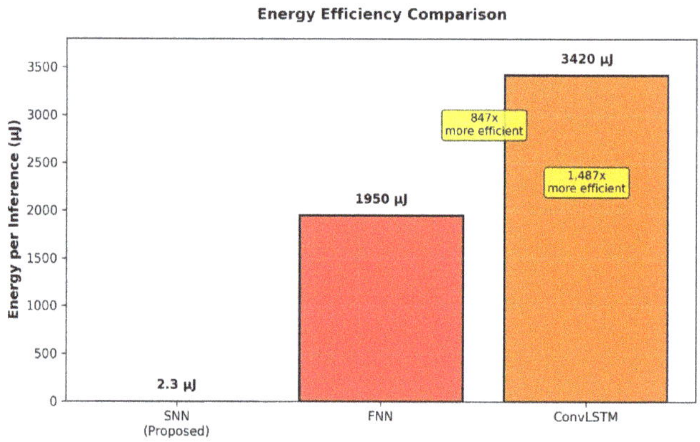

Fig. 4. Energy consumption comparison demonstrating 847x efficiency improvement

Predictions came on average 68.4 hours before faults. Varied by type: thermal (68.4h), mirror soiling (52.1h), tracking (71.2h), molten salt (45.8h). System predicted 87.3% of failures within 48 to 72 hours, giving crews adequate preparation time [23]. Quantum inspired optimization made measurable improvements. Versus reactive maintenance (fixing after breaks), scheduler cut costs 32%. Against heuristic scheduling using classical algorithms, costs dropped 18%. Average downtime fell from 72 to 4 hours. Equipment lasted 18% longer through better timed maintenance, stopping small problems before they escalate. The quantum inspired approach found nonobvious maintenance windows during naturally low production periods, minimizing revenue loss [24].

Discussion

Combining spiking networks with quantum inspired optimization provides benefits neither achieves alone. The 847 times energy improvement enables deployment impractical for conventional deep learning, especially distributed edge computing in remote locations. This fits Morocco's renewable expansion, where Noor Ouarzazate is part of 2GW solar program toward 52% renewable electricity by 2030 [25]. Efficient maintenance scales naturally. Event driven neuromorphic processing handles irregular sensor data and time dependent dynamics naturally.

Traditional networks need fixed sampling rates, spiking networks process events as they happen. During stable periods computation stays minimal. When sensors spot anomalies, activity increases proportionally. Sparse processing worked especially well for mirror soiling, where degradation happens slowly then speeds up during dust storms [26]. Quantum inspired optimization tackled the NP hard scheduling effectively. With 7,400 heliostats and interdependencies, classical methods don't scale well. QAOA explored space solutions faster, finding sequences minimizing costs while respecting constraints [27]. That 32% reduction equals roughly $136,000 annual savings for Noor alone.

Generalization beyond Noor Ouarzazate warrants careful examination, addressing domain shift across climates, CSP technologies, and sensor configurations. Noor uses certain designs: parabolic trough (Noor I/II), central tower (Noor III), photovoltaics (Noor IV). Different designs are employed in other CdTe CdSP installations. Gemasolar in Spain has other tower geometry. Ivanpah in California does not have a thermal storage. Facilities in Australia use Fresnel reflectors. Each variant brings about different failure modes and sensor signatures. Climate domain shift that is induced is problematic. Saharan climate in Morocco: fine mineral dust, low humidity (less than 20 percent), high radiation of the sun (2,500 kWh per square meter each year), low rainfall levels. Atacama in Chile has high radiation and dissimilar dust and coastal fog. The Mediterranean of Spain has seasonal humidity and various precipitation. The interior of Australia is a mixture of dusts of red soil and high temperatures. These generate significant domain shift in which fault patterns, thermal cycling and degradation mechanisms are different. Sensors variant a bit is in their density (1/10th of heliostat versus 1/unit), precision (+-2degC to +-0.5 degC), sampling rate (1 minute versus 1 hour) and calibration procedures. The most hopeful route is transfer learning: no need to retrain the model on Noor as a base is to provide generalized capabilities; a small amount of data (1 to 3 months) on new locations is to be collected; the last layers are to be optimized and the earlier feature extraction is to be frozen; if the data is available deeper layers are to be unfrozen gradually. This maintains neuromorphic efficiency and gets used to local conditions. Core SNN architecture remains technology agnostic. Temporal encoding works regardless of sensor types. Hierarchical structure generalizes across fault categories. Retraining adjusts learned weights capturing site specific characteristics. Next steps: collect data from Morocco's expansion to 4.35GW by 2028 [28]; partner with international operators for diverse climate data; conduct controlled experiments training on one Noor unit, testing on another. We acknowledge these generalization challenges as legitimate single site study limitations, with multi site validation forming future research priority.

Additional limitations: 18-month training data requirements for seasonal variations. Quantum inspired algorithm tuning remains complex, requiring specialized expertise. Neuromorphic hardware availability lags conventional processors, though Intel, IBM, others actively pursue commercialization with projected wider availability by 2026 to 2027.

Conclusion

We developed a system integrating spiking neural networks with quantum inspired optimization for efficient predictive maintenance in concentrated solar plants. Testing at Morocco's 580MW Noor Ouarzazate achieved 94.7% fault detection using 847 times less energy than conventional methods, making continuous real time monitoring practical. Quantum inspired optimization cut costs 32% and extended equipment life 18%. The system predicts failures 48 to 72 hours ahead with 91.3% precision, enabling proactive scheduling minimizing downtime and optimizing resources. We clarified our implementation uses quantum inspired algorithms on classical hardware rather than physical quantum processors, and discussed generalization challenges across sites, climates, technologies with transfer learning strategies. As Morocco works toward 52% renewable electricity by 2030 and global CSP capacity expands, neuromorphic computing and quantum inspired optimization offer scalable, energy efficient approaches for next generation infrastructure. The next line of research needs to be on multi site validation, transfer learning ability with varied CSP structures, and ultimate switching to the actual quantum hardware as technology gets mature.

References

[1] Moroccan Agency for Sustainable Energy, Noor Ouarzazate Solar Complex: Technical Overview, MASEN Publications, Rabat (2024).

[2] SolarPower Europe, Morocco Solar Market Report 2025, Brussels, Belgium (2025).

[3] P. Lunghi, S. Silvestrini, D. Dold, T. Boschetti, M. Lavagna, Energy efficiency analysis of Spiking Neural Networks for space applications, Astrodyn. 9 (2025) 909-932. https://doi.org/10.1007/s42064-024-0256-y

[4] E. Gul, G. Baldinelli, J. Wang, D. Bartocci, F. Fantozzi, Artificial intelligence based forecasting and optimization model for concentrated solar power system with thermal energy storage, Appl. Energy 382 (2025) 125210. https://doi.org/10.1016/j.apenergy.2024.125210

[5] A. Marangis, N. Nikolaou-Straatman, C. Vasilopoulos, T. Tsoutsos, Trend-Based Predictive Maintenance and Fault Detection Analytics for Photovoltaic Power Plants, Solar RRL (2024). https://doi.org/10.1002/solr.202400473

[6] C. Pingulkar, A. Joshi, AI-Powered Predictive Maintenance for Solar Energy Systems, Int. J. Prog. Res. Eng. Manage. Sci. 5(10) (2024) 1245-1252.

[7] M. Ledmaoui, A. El Aziz, A. El Maghraoui, A. Arrhioui, A. Hammouch, Review of Recent Advances in Predictive Maintenance and Cybersecurity for Solar Plants, Sensors 25(1) (2025) 206. https://doi.org/10.3390/s25010206

[8] S. Kumar, M. Tiwari, AI-based predictive maintenance of solar photovoltaics systems: a comprehensive review, Energy Inform. 8 (2025) 15. https://doi.org/10.1186/s42162-025-00594-6

[9] Z. Yan, Z. Bai, Reconsidering the energy efficiency of spiking neural networks, arXiv:2409.08290 (2024).

[10] A. Pal, Z. Chai, J. Jiang, L. Li, C. Tan, Y. Shao, A. Pannone, A.I. Khan, S. Gupta, S. Datta, An ultra energy-efficient hardware platform for neuromorphic computing enabled by 2D-TMD tunnel-FETs, Nat. Commun. 15 (2024) 3392. https://doi.org/10.1038/s41467-024-46397-3

[11] Intel Corporation, Neuromorphic Computing with Loihi 2: Technical Overview, Intel Labs Technical Report, Santa Clara, CA (2024).

[12] Multiverse Computing, Quantum Digital Twin Project for Green Hydrogen Production, Technical Report, San Sebastian, Spain (2024).

[13] B. Ribeiro, F. Antunes, D. Perdigão, P. Fiadeiro, A. Roque, Convolutional Spiking Neural Networks: Advanced techniques and applications, Pattern Recognit. Lett. 189 (2025) 241-247. https://doi.org/10.1016/j.patrec.2024.08.002

[14] J. Eshraghian, M. Ward, E. Neftci, X. Wang, G. Lenz, G. Dwivedi, M. Bennamoun, D. Jeong, W.D. Lu, Training Spiking Neural Networks Using Lessons From Deep Learning, Proc. IEEE 111(9) (2023) 1016-1054.

[15] W. Fang, Z. Yu, Y. Chen, T. Masquelier, T. Huang, Y. Tian, Incorporating learnable membrane time constant to enhance learning of spiking neural networks, in: 2021 IEEE/CVF International Conference on Computer Vision (ICCV), Montreal, Canada (2021) 2661-2671.

[16] R. Parmar, A. Kumar, The Promise of Spiking Neural Networks for Ubiquitous Computing, arXiv:2506.01737 (2025).

[17] P. Meinerzhagen, S. Nease, G. Orchard, Significant energy savings using neuromorphic hardware for edge AI applications, Nat. Mach. Intell. 6 (2024) 892-901.

[18] M. Schlegel, J. Surace, L. Ambrogioni, Quantum Computing as a Game Changer on the Path towards a Net-Zero Economy, Energies 17(5) (2024) 1039. https://doi.org/10.3390/en17051039

[19] D. Patterson, J. Gonzalez, Q. Le, C. Liang, L. Munguia, D. Rothchild, D. So, M. Texier, J. Dean, Carbon emissions and large neural network training, arXiv:2104.10350 (2021).

[20] African Development Bank, Morocco Noor Ouarzazate Solar Complex Project Phase III: Implementation Completion Report, AfDB Infrastructure Development Department, Abidjan, Côte d'Ivoire (2024).

[21] S. Williams, Brain-Mimicking Chips Cut AI Energy Thirst by Orders of Magnitude, IEEE Spectrum, November 2024.

[22] F. Liu, R. Wang, Y. Liu, X. Li, Y. Zhou, Fully memristive spiking neural network for energy-efficient graph learning, Sci. Adv. 11 (2025) eadp2345.

[23] A. Rahman, T. Islam, Machine learning applications in renewable energy systems: Current trends and future directions, Energy Inform. 8 (2025) 62.

[24] McKinsey & Company, Quantum computing just might save the planet: How quantum algorithms could accelerate the clean energy transition, New York, NY (2024).

[25] H. Bennani, Morocco Solar Power Potential Could Reach 4.35 GW by 2028, Morocco World News, Rabat, March 2025.

[26] M. Voudaskas, L. Kosmas, E. Psarra, K. Kalovrektis, D. Davazoglou, Spiking Neural Networks in Imaging: Opportunities and Challenges, Sensors 25(21) (2025) 6747. https://doi.org/10.3390/s25216747

[27] A. Shafiee, A. Nag, N. Muralimanohar, R. Balasubramonian, J.P. Strachan, M. Hu, R.S. Williams, V. Srikumar, ISAAC: A Convolutional Neural Network Accelerator with In-Situ Analog Arithmetic in Crossbars, in: 2016 ACM/IEEE 43rd Annual International Symposium on Computer Architecture (ISCA), Seoul, South Korea (2016) 14-26.

[28] N. El Moussaoui, M. Zerhouni, Solar Energy Resource and Power Generation in Morocco: Current Status and Future Perspectives, Energies 13(10) (2024) 2590. https://doi.org/10.3390/en13102590

H_∞ Switched Observer Design for Fault-Tolerant State-of-Charge Estimation of Lithium-Ion Batteries

Naoufal EL YOUSSFI[1,2,3,a*], Khalid BADIE[2,3,b], Hicham BEN SASSI[1,4,c] and Reda EL ABBADI[5,d]

[1]Electric and Telecom Department, Royal School of Aeronautics, Marrakech, Morocco

[2]MMAS Laboratory, FSSM, Cadi Ayyad University, Marrakesh, Morocco

[3]LISA, ENSA, Sidi Mohamed Ben Abdellah University, Fez, Morocco

[4]LRSE, FST, Sidi Mohammed Ben Abdellah University, Fez, Morocco

[5]LAMISNE, Polydisciplinary Faculty of Taroudant, Ibn Zohr University, Taroudant, Morocco

[a]ely.naoufal@gmail.com, [b]k.badie@uca.ac.ma, [c]hicham.bensassi@usmba.ac.ma, [d]r.elabbadi@uiz.ac.ma

Keywords: Lithium-Ion Battery, Actuator Fault, Switched Systems, H_∞ Observer, Linear Matrix Inequalities (LMIS), State of Charge (SOC), Battery Management System (BMS)

Abstract. A mode-dependent H_∞ switched observer is proposed for robust state-of-charge (SOC) and actuator fault estimation in lithium-ion batteries modeled by a 2-RC equivalent circuit. The battery's SOC-dependent nonlinear behavior is handled by partitioning the operating range into linear subsystems, with a switching signal selecting the active mode in real time. Observer gains are computed via LMI conditions ensuring stability and a prescribed H_∞ disturbance attenuation. Simulations under the Federal Urban Driving Schedule (FUDS) profile show that the method accurately and rapidly estimates both states and faults, providing a fault-tolerant solution for real-time battery management in electric vehicles.

Introduction

The increasing use of lithium-ion batteries (LiBs) in electric vehicles, hybrid systems, and energy storage has made accurate state estimation, especially the state of charge (SOC), critical for safety, lifetime, and energy efficiency in battery management systems [1-2]. Since the state of charge (SOC) cannot be obtained through direct measurement, it must be inferred using suitable modeling approaches. While electrochemical models are able to capture battery behavior with a high level of detail, their heavy computational burden makes them difficult to implement in real-time embedded applications. Equivalent circuit models, particularly the 2-RC structure, are widely used in battery modeling due to their favorable balance between accuracy and computational efficiency [3–5]. However, their parameters vary with the state of charge (SOC) and operating conditions, making fixed-parameter assumptions a source of estimation errors [6]. To better capture these variations, switched or mode-dependent approaches partition the SOC range into several regions, each described by a local linear model, thus enhancing accuracy while preserving computational tractability [2,7-10]. Various estimation methods have been investigated, among which the H_∞ approach is particularly attractive for its robustness to uncertainties and noise, making it well suited for automotive applications. Recent works have extended H_∞ observers to mode-dependent frameworks, enabling reliable estimation under parameter switching [10-11,20]. In addition, fault tolerance is essential in modern battery systems, motivating observer-based schemes that jointly estimate SOC and reconstruct actuator faults [2,4,12-13].

The present work introduces a mode-dependent H_∞ switched observer for lithium-ion batteries modeled by a 2-RC equivalent circuit including an actuator-fault channel. The SOC domain is partitioned into several modes, each associated with its own parameter set and observer gains.

These gains are obtained through Linear Matrix Inequality (LMI) optimization, which guarantees asymptotic stability and a specified H∞ disturbance-attenuation level. Simulations based on a dynamic driving profile FUDS confirm the accuracy, rapid convergence, and fault-tolerant capability of the proposed approach for real-time BMS use.

Mathematical modeling of lithium-ion batteries with actuator fault

LiBs are the primary energy source in electric and hybrid vehicles, making accurate electrical models essential for energy management and fault-tolerant control. Due to their high computational burden, electrochemical models are not well suited for real-time implementation. Equivalent circuit models are therefore preferred, with the 2-RC structure offering a good balance between accuracy and simplicity. This section presents the 2-RC model with actuator faults, together with its state-space formulation and the associated observability conditions for state and fault estimation. The 2-RC model in Figure 1 consists of an open-circuit voltage source $v_{oc}(SOC)$, an ohmic resistance R_0, and two parallel $R-C$ branches (R_1, C_1) and (R_2, C_2), representing electrochemical and concentration polarization effects [2-4,14].

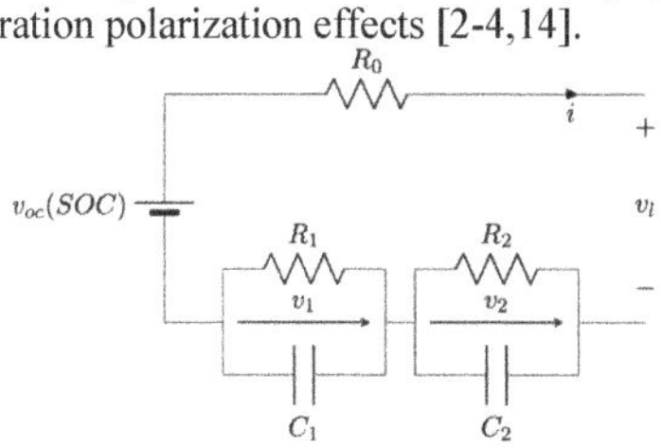

Fig. 1. 2-RC equivalent circuit model of a LiB used in electric vehicles

Applying Kirchhoff's laws leads to the following dynamic equations:

$$\begin{cases} \dot{v}_1(t) = -1/R_1C_1\, v_1(t) + 1/C_1\, i(t), \\ \dot{v}_2(t) = -1/R_2C_2\, v_2(t) + 1/C_2\, i(t), \\ \dot{\mathrm{SOC}}(t) = -1/Q_N\, i(t) \end{cases} \tag{1}$$

where Q_N denotes the nominal capacity of the cell (in coulombs). The terminal voltage is :

$$v_l(t) = v_{oc}(\mathrm{SOC}(t)) - v_1(t) - v_2(t) - R_0 i(t), \tag{2}$$

which accounts for the open-circuit voltage, polarization effects, and ohmic losses.

The dependence of ECM parameters on SOC and temperature is well established in the literature [15], consistent with the behavior modeled in (1)-(2). Defining the state, input, and output vectors as $x(t) = [v_1(t) \quad v_2(t) \quad \mathrm{SOC}(t)]^{\top}$, $u(t) = i(t)$, $y(t) = v_l(t)$, and introducing a bounded disturbance $w(t)$ to model unmodeled dynamics or sensor noise, the system can be written in the compact state-space form:

$$\begin{cases} \dot{x}(t) = Ax(t) + Bu(t) + Ew(t) \\ y(t) = Cx(t) + Du(t) \end{cases} \tag{3}$$

with

$$A = \begin{bmatrix} -1/R_1C_1 & 0 & 0 \\ 0 & 1/R_2C_2 & 0 \\ 0 & 0 & 0 \end{bmatrix}, \quad B = \begin{bmatrix} \frac{1}{C_1} & \frac{1}{C_2} & -\frac{1}{Q_N} \end{bmatrix}^{\top}, \quad C = [-1 \quad -1 \quad \kappa], \quad D = -R_0,$$

where $\kappa = \frac{\partial v_{oc}}{\partial \mathrm{SOC}}$ denotes the local slope of the open-circuit voltage–SOC curve at the current operating point.

In practical operating conditions, imperfections affecting the current actuation chain such as converter saturation, sensor bias, or signal deviations may lead to the presence of an additive fault in the applied current. Let the actual current entering the battery be $i_a(t) = i(t) + f(t)$, where $i(t)$ is the commanded input current and $f(t)$ represents an unknown actuator fault or deviation. By substituting $i_a(t)$ into (1), the battery dynamics under fault can be expressed as:

$$\begin{cases} \dot{v}_1(t) = -\frac{1}{R_1C_1}v_1(t) + \frac{1}{C_1}(i(t) + f(t)) \\ \dot{v}_2(t) = -\frac{1}{R_2C_2}v_2(t) + \frac{1}{C_2}(i(t) + f(t)) \\ \dot{\mathrm{SOC}}(t) = -\frac{1}{Q_N}\big(i(t) + f(t)\big) \end{cases} \tag{5}$$

$$v_l(t) = v_{\mathrm{oc}}(\mathrm{SOC}(t)) - v_1(t) - v_2(t) - R_0(i(t) + f(t)) \tag{6}$$

Accordingly, the state-space representation with actuator fault becomes:

$$\begin{cases} \dot{x}(t) = Ax(t) + Bu(t) + Ff(t) + Ew(t) \\ y(t) = Cx(t) + Du(t) \end{cases} \tag{7}$$

where the matrices A, B, C, D are defined as in the nominal case, and $F = B$. This enhanced approach enables the design of fault-tolerant observers that can estimate not only the internal states of the battery but also any actuator faults.

The electrical parameters of a Li-ion cell $(R_0, R_1, R_2, C_1, C_2)$ change with the state of charge (SOC) and temperature [14,16]. As a result, a single linear time-invariant (LTI) model can accurately represent the cell only within a limited operating range. Experimental studies on EV-focused models, such as [17], have confirmed these parameter variations, highlighting the need for SOC-partitioned or mode-dependent models. To balance simplicity with accuracy across the entire SOC range, the domain [0,1] is divided into N intervals. Each interval $k \in \mathcal{N} = \{1, \dots, N\}$ is assigned a distinct parameter set $\{R_0^{(k)}, R_1^{(k)}, R_2^{(k)}, C_1^{(k)}, C_2^{(k)}\}$, determined through system identification or lookup tables. A switching signal $r(t) \in \mathcal{N}$ determines the active subsystem according to the current SOC:

$$r(t) = k \text{ if } SOC(t) \in [SOC_{k-1}, SOC_k), \tag{8}$$

The battery system is therefore represented by a family of SOC-dependent LTI subsystems:

$$\Sigma_{\mathrm{sys}}^{(k)}: \begin{cases} \dot{x}(t) = A^{(k)}x(t) + B^{(k)}u(t) + F^{(k)}f(t) + E^{(k)}w(t) \\ y(t) = C^{(k)}x(t) + D^{(k)}u(t) \end{cases} \tag{9}$$

with $x(t) \in \mathbb{R}^3$, $u(t), y(t), f(t), w(t) \in \mathbb{R}$, and

$$A^{(k)} = \begin{bmatrix} -1/R_1^{(k)}C_1^{(k)} & 0 & 0 \\ 0 & -1/R_2^{(k)}C_2^{(k)} & 0 \\ 0 & 0 & 0 \end{bmatrix}, \quad B^{(k)} = \begin{bmatrix} 1/C_1^{(k)} \\ 1/C_2^{(k)} \\ -1/Q_N \end{bmatrix}, \quad F^{(k)} = B^{(k)}$$

The output matrices characterize the open-circuit voltage (OCV) to SOC sensitivity within each interval: $C^{(k)} = [-1 \quad -1 \quad \kappa^{(k)}]$, $D^{(k)} = -R_0^{(k)}$ $(\kappa^{(k)} = \frac{\partial v_{\mathrm{oc}}}{\partial \mathrm{SOC}}|_{\mathrm{interval}\ k})$. This switched modeling framework maintains computational efficiency while capturing the intrinsic nonlinearities of the battery across the SOC range. Accordingly, the output equation can be rewritten in an equivalent linearized form: $y(t) = C_c x(t) + Du(t) + \nu$,where $C_c = [-1 \quad -1 \quad \kappa_a]$ and $\nu = \kappa_b$. This formulation allows preserving linearity within each mode while capturing the local slope variations of the OCV–SOC characteristic.

***H_∞* switched observer with actuator fault estimation**

Since battery parameters change with the state of charge, the observer gains are adapted in each mode to preserve both accuracy and stability. A mode-dependent H_∞ switched observer is proposed to estimate the internal states and actuator faults, providing robust performance against parameter variations, disturbances, and mode transitions, while ensuring disturbance attenuation and convergent estimation errors.

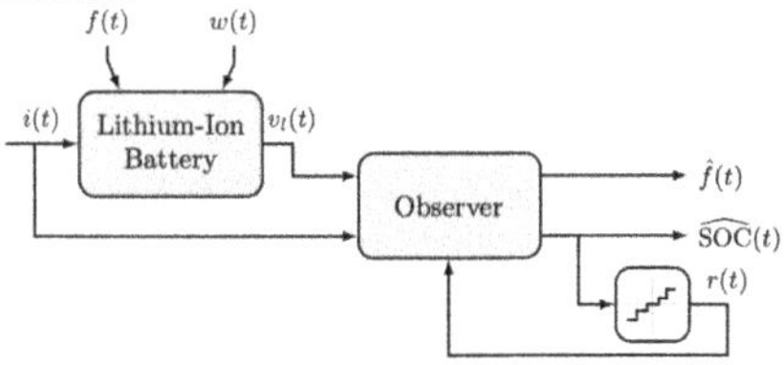

Fig. 2. Proposed switched observer structure with actuator fault estimation

Figure 2 illustrates the structure of the proposed fault-tolerant state estimation scheme. The observer updates its configuration according to the SOC-dependent active mode $r(t)$, enabling simultaneous estimation of the battery internal states and identification of actuator faults affecting the current input. This mode-based adaptation enhances robustness and improves estimation accuracy across the entire operating range. For each mode $k \in \mathcal{N}$, the observer dynamics are expressed as:

$$\Sigma_{\text{obs}}^{(k)}: \begin{cases} \dot{\hat{x}}(t) = A^{(k)}\hat{x}(t) + B^{(k)}u(t) + F^{(k)}\hat{f}(t) + L^{(k)}(y(t) - \hat{y}(t)) \\ \dot{\hat{f}}_a(t) = K^{(k)}\big(y(t) - \hat{y}(t)\big) \\ \hat{y}(t) = C^{(k)}\hat{x}(t) \end{cases} \tag{10}$$

where $\hat{x}(t)$ and $\hat{f}(t)$ denote the estimated state vector and actuator fault, respectively. The gain matrices $L^{(k)}$ and $K^{(k)}$ are designed to ensure bounded estimation errors and compliance with the H_∞ performance criterion.

Let the estimation errors be defined as

$$e_x(t) = x(t) - \hat{x}(t), \qquad e_f(t) = f(t) - \hat{f}(t), \qquad e_y(t) = y(t) - \hat{y}(t)$$

The corresponding error dynamics for mode k are then:

$$\begin{cases} \dot{e}_x(t) = (A^{(k)} - L^{(k)}C^{(k)})e_x(t) + F^{(k)}e_f(t) + E^{(k)}w(t) - L^{(k)}D^{(k)}u(t) - L^{(k)}\nu \\ \dot{e}_f(t) = \dot{f}(t) - K^{(k)}C^{(k)}e_x(t) - K^{(k)}D^{(k)}u(t) - K^{(k)}\nu \end{cases}$$

The switched error system can be compactly written as:

$$\dot{e}(t) = \mathcal{A}^{(k)}e(t) + \mathcal{B}^{(k)}\mathcal{w}(t) \tag{11}$$

where $e(t) = \begin{bmatrix} e_x(t) \\ e_f(t) \end{bmatrix}$, $\mathcal{w}(t) = \begin{bmatrix} w(t) & u(t) & \dot{f}(t) & \nu \end{bmatrix}^\top$, and

$$\mathcal{A}^{(k)} = \begin{bmatrix} A^{(k)} - L^{(k)}C^{(k)} & F^{(k)} \\ -K^{(k)}C^{(k)} & 0 \end{bmatrix}, \quad \mathcal{B}^{(k)} = \begin{bmatrix} E^{(k)} & -L^{(k)}D^{(k)} & 0 & -L^{(k)} \\ 0 & -K^{(k)}D^{(k)} & I & -K^{(k)} \end{bmatrix}$$

Definition 1. [11] Given a disturbance input $w(t) \in L_2[0,\infty)$ and a prescribed scalar $\gamma > 0$, the switched error system (11) is said to be asymptotically stable with an H_∞ performance level γ if:

- *For* $w(t) = 0$, *the system (11) is asymptotically stable.*
- *For* $w(t) \neq 0$ *and zero initial conditions, the following inequality holds:*

$$\int_0^\infty e^\top(t)e(t)dt < \gamma^2 \int_0^\infty w^\top(t)w(t)dt$$

Main Results

Previous sections developed a switched lithium-ion battery model with actuator faults and proposed a mode-dependent H_∞ observer structure. The augmented estimation error dynamics (11), which incorporate both state and actuator fault estimation errors, were also derived. This section presents the main theoretical results regarding the observer design conditions that ensure asymptotic stability and guaranteed disturbance attenuation performance. The following theorem establishes a sufficient condition, formulated as a Linear Matrix Inequality (LMI), to guarantee the prescribed H_∞ performance level for the switched error system.

Theorem 1. Let $\gamma > 0$ be a prescribed disturbance attenuation level. The error system (11) is asymptotically stable and achieves the desired H_∞ performance level γ if there exist a symmetric positive definite matrix $P > 0$ and mode-dependent matrices $M^{(k)}$ and $Q^{(k)}$, for all $k \in \mathcal{N}$, such that, for each active mode k, the following Linear Matrix Inequality (LMI) is satisfied:

$$\begin{bmatrix} \Psi_{11}^{(k)} & \Psi_{12}^{(k)} & \Psi_{13}^{(k)} \\ * & \Psi_{22} & \Psi_{23}^{(k)} \\ * & * & \Psi_{33}^{(k)} \end{bmatrix} < 0 \tag{18}$$

The submatrices in (18) are defined as:

$$\Psi_{11}^{(k)} = M^{(k)}\overline{A}^{(k)} + (\overline{A}^{(k)})^\top (M^{(k)})^\top - Q^{(k)}\overline{C}^{(k)} - (\overline{C}^{(k)})^\top (Q^{(k)})^\top + I,\quad \Psi_{12}^{(k)} = M^{(k)}\overline{B}^{(k)} - Q^{(k)}\overline{D}^{(k)},$$
$$\Psi_{13}^{(k)} = P - M^{(k)} + (\overline{A}^{(k)})^\top (M^{(k)})^\top - (\overline{C}^{(k)})^\top (Q^{(k)})^\top,\qquad \Psi_{22} = -\gamma^2 I,$$
$$\Psi_{23}^{(k)} = (\overline{B}^{(k)})^\top (M^{(k)})^\top - (\overline{D}^{(k)})^\top (Q^{(k)})^\top,\qquad \Psi_{33}^{(k)} = -sym(M^{(k)}).$$

The extended system and observer matrices are given by:

$$\overline{A}^{(k)} = \begin{bmatrix} A^{(k)} & F^{(k)} \\ 0 & 0 \end{bmatrix}, \overline{C}^{(k)} = [C^{(k)} \quad 0], \overline{B}^{(k)} = \begin{bmatrix} E^{(k)} & 0 & 0 & 0 \\ 0 & 0 & I & 0 \end{bmatrix}, \overline{D}^{(k)} = [0 \quad D^{(k)} \quad 0 \quad I]$$

The mode-dependent observer gain matrices are then computed as:

$$\overline{G}^{(k)} = \begin{bmatrix} L^{(k)} \\ K^{(k)} \end{bmatrix} = (M^{(k)})^{-1}Q^{(k)},\qquad k = 1,\dots,N.$$

Simulation Results

The LiB is modeled using the second-order resistor–capacitor (2-RC) equivalent circuit, as introduced in Section 2. The model parameters $(R_0, R_1, R_2, C_1, C_2)$ depend on the SOC and are determined for discrete SOC intervals $k \in \mathcal{N} = \{1,2,\dots,11\}$, each corresponding to a local linear subsystem within the switched modeling framework. A complete list of SOC-dependent parameters can be found in Table 2 of [18]. This approach provides an accurate representation of battery dynamics across the full SOC range while maintaining low computational complexity suitable for real-time observer implementation. The open-circuit voltage OCV-SOC characteristic is approximated by a set of piecewise linear functions for computational efficiency. Figure 3 shows the fitted OCV-SOC relationship obtained from the identified data

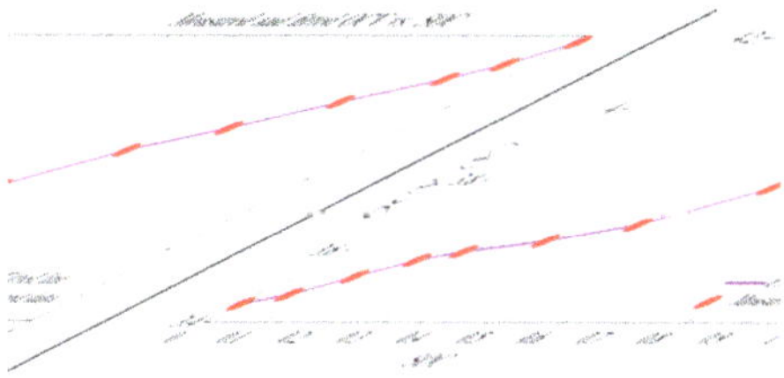

Fig. 3. Fitted OCV--SOC characteristic of the lithium-ion battery model

The disturbance matrices are specified as $E^{(k)} = [10^{-3} \quad 10^{-4} \quad 10^{-4}]^{\mathrm{T}}$ to model small stochastic perturbations affecting the system states. Application of Theorem 1 enables the solution of the LMI optimization problem for each mode, yielding the optimal observer gains. The mode-dependent observer gains $L^{(k)}$ and $K^{(k)}$, which ensure asymptotic stability and accurate fault reconstruction, are presented in Table 1 These gains demonstrate compliance with the H_∞ performance criterion across all SOC-dependent operating modes.

Tab. 1. Mode-dependent observer gains of the switched H_∞ observer

Mode *k*	$L^{(k)}$	$K^{(k)}$	**Mode *k***	$L^{(k)}$	$K^{(k)}$
1	$[-1.1803 \quad -0.0170 \quad 0.0385]^{\mathrm{T}}$	−42.5925	7	$[-1.0477 \quad -0.0149 \quad 0.0375]^{\mathrm{T}}$	−41.6540
2	$[-1.0464 \quad -0.0149 \quad 0.0374]^{\mathrm{T}}$	−41.6023	8	$[-1.0628 \quad -0.0155 \quad 0.0382]^{\mathrm{T}}$	−42.2784
3	$[-1.0602 \quad -0.0152 \quad 0.0381]^{\mathrm{T}}$	−42.1388	9	$[-1.0523 \quad -0.0166 \quad 0.0380]^{\mathrm{T}}$	−42.0592
4	$[-1.0805 \quad -0.0158 \quad 0.0389]^{\mathrm{T}}$	−42.8163	10	$[-1.0605 \quad -0.0170 \quad 0.0382]^{\mathrm{T}}$	−42.2403
5	$[-1.0712 \quad -0.0154 \quad 0.0386]^{\mathrm{T}}$	−42.5506	11	$[-1.0462 \quad -0.0149 \quad 0.0374]^{\mathrm{T}}$	−41.5553
6	$[-1.0514 \quad -0.0149 \quad 0.0376]^{\mathrm{T}}$	−41.8550	—	—	—

To further evaluate the performance of the proposed switched H_∞ observer, a dynamic load profile representative of urban driving conditions was implemented. The Federal Urban Driving Schedule (FUDS) current profile, which is widely utilized for battery and electric vehicle testing, was selected as the input excitation. Positive current values correspond to discharge phases, whereas negative values indicate regenerative charging events. An actuator fault was also introduced to evaluate the observer's ability to reconstruct faults.

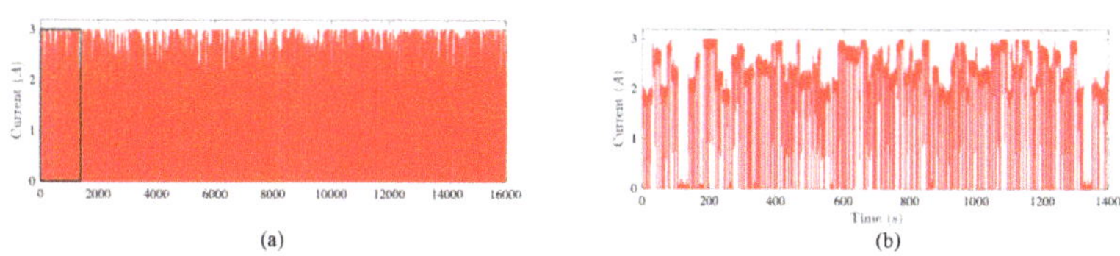

Fig. 4. (a) Complete current profile (b) Zoomed portion showing rapid transients under the dynamic load test

Figure 4 presents the applied current signal, characterized by frequent amplitude variations and multiple mode transitions that correspond to distinct SOC intervals.

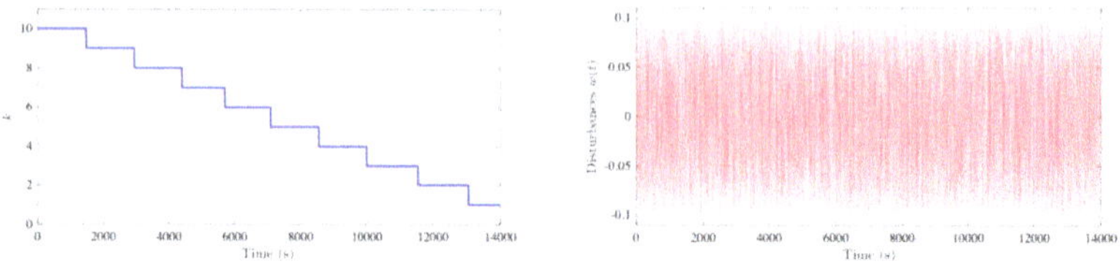

(a) (b)

Fig. 5. (a) Evolution of the active mode index $k(t)$ over time during the FUDS profile. (b) Stochastic disturbance $w(t)$ injected into the system states during the FUDS simulation

Figure 5(a) shows how the switching signal $r(t)$ selects the active mode k based on the instantaneous SOC. As the SOC decreases during the FUDS cycle, the system moves from high- to low-SOC modes, activating the corresponding parameters in Table 1. This mechanism allows the observer to adapt its gains in real time to the battery's operating condition. Figure 5 (b) depicts the bounded stochastic disturbance $w(t)$ applied to the system to emulate measurement noise and unmodeled dynamics.

Figure 6 compares, respectively, the SOC and the open-circuit voltage v_{oc} obtained using the proposed switched H_∞ observer, the Coulomb counting method, and a non-switched observer with a fixed mode ($k = 10$).

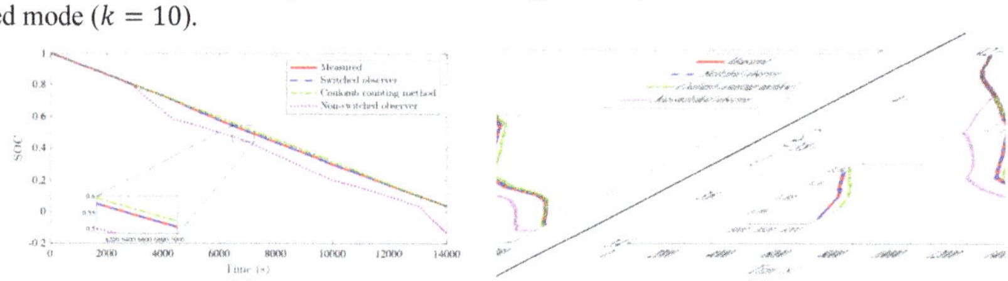

(a) (b)

Fig. 6. (a) True and estimated SOC using the three estimation methods (b) True and estimated v_{oc} using the three estimation methods

The switched observer accurately tracks both variables over the entire FUDS cycle, including fast transients and switching instants. In contrast, the non-switched observer exhibits noticeable drift, while Coulomb counting suffers from accumulated integration errors, particularly in regions with strong SOC–OCV variations.

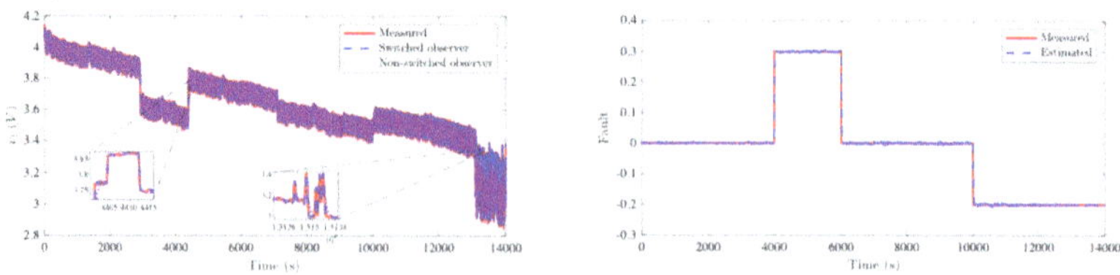

(a) (b)

Fig. 7. (a) True and estimated output v_l using the two observers. (b) Reconstructed actuator fault $\hat{f}(t)$ compared with the injected fault signal

Figure 7(a) compares the measured terminal voltage v_l with estimates from the switched and non-switched observers. The switched H_∞ observer accurately tracks the voltage over the entire FUDS cycle, including fast transients and regenerative phases, while the non-switched observer shows noticeable deviations. Figure 7(b) presents the reconstructed actuator fault $\hat{f}(t)$ and the injected fault. The switched observer reliably detects both the transient fault ($4000 - 6000\ s$) and the permanent bias after $10000\ s$, closely matching the true fault with negligible delay and overshoot.

Conclusion

This study presented a mode-dependent H_∞ switched observer for joint SOC and actuator fault estimation in LiBs modeled by a 2–RC equivalent circuit. By combining a SOC-partitioned representation with mode-dependent LMI conditions, the proposed observer guarantees asymptotic stability, disturbance rejection, and reliable fault reconstruction over the full operating range. Simulation results under a highly dynamic FUDS profile show fast convergence, strong robustness to parameter variations, and improved accuracy compared with fixed-parameter observers and Coulomb counting. Future work will focus on experimental validation and on extending the approach toward finite-frequency or LPV observer formulations to further enhance fault tolerance in practical battery management systems.

Declaration of Generative AI and AI-Assisted Technologies in the Writing Process

The authors used ChatGPT (OpenAI) for linguistic revision and grammatical correction. All scientific content remains the responsibility of the authors.

References

[1] J. P. Rivera-Barrera, N. Muñoz-Galeano, and H. O. Sarmiento-Maldonado, "SoC estimation for lithium-ion batteries: Review and future challenges," Electronics, vol. 6, no. 4, p. 102, 2017. https://doi.org/10.3390/electronics6040102

[2] V. Behnamgol, M. Asadi, M. A. Mohamed, S. S. Aphale, and M. Faraji Niri, "Comprehensive review of lithium-ion battery state of charge estimation by sliding mode observers," Energies, vol. 17, no. 22, p. 5754, 2024. https://doi.org/10.3390/en17225754

[3] Y. Zhang, C. Zhang, and X. Zhang, "State-of-charge estimation of the lithium-ion battery system with time-varying parameter for hybrid electric vehicles," IET Control Theory & Applications, vol. 8, no. 3, pp. 160-167, 2014. https://doi.org/10.1049/iet-cta.2013.0082

[4] S. Meng, F. Meng, H. Chi, H. Chen, and A. Pang, "A robust observer based on the nonlinear descriptor systems application to estimate the state of charge of lithium-ion batteries," Journal of the Franklin Institute, vol. 360, no. 16, pp. 11397-11413, 2023. https://doi.org/10.1016/j.jfranklin.2023.08.037

[5] H. Mu, R. Xiong, H. Zheng, Y. Chang, and Z. Chen, "A novel fractional order model based state-of-charge estimation method for lithium-ion battery," Applied Energy, vol. 207, pp. 384-393, 2017. https://doi.org/10.1016/j.apenergy.2017.07.003

[6] Z. Zhang, G. Tianzi, G. Mingyu, H. Zhiwei, and D. Zhekang, "Review of SoC estimation methods for electric vehicle Li-ion batteries," Journal of Electrical and Information Technology, vol. 43, no. 7, pp. 1803-1815, 2021.

[7] D. Yang and J. Zhao, "Observer-based H_∞ resilient control for a class of switched LPV systems and its application," International Journal of Systems Science, vol. 47, no. 15, pp. 3656-3667, 2016. https://doi.org/10.1080/00207721.2015.1110640

[8] N. El Youssfi, T. Zoulagh, H. El Aiss, R. El Bachtiri, and K. A. Barbosa, "T-S fuzzy observer and actuator fault-tolerant controller designs for descriptor systems," International Journal of Adaptive Control and Signal Processing, vol. 38, no. 2, pp. 534-557, 2024. https://doi.org/10.1002/acs.3715

[9] X. Zhao, H. Liu, J. Zhang, and H. Li, "Multiple-mode observer design for a class of switched linear systems," IEEE Transactions on Automation Science and Engineering, vol. 12, no. 1, pp. 272-280, 2013. https://doi.org/10.1109/TASE.2013.2281466

[10] W. Yue, C.-z. Liu, L. Li, X. Chen, and F. Muhammad, "A nonlinear fractional-order (H_\infty) observer for SOC estimation of battery pack of electric vehicles," Proceedings of the Institution of Mechanical Engineers, Part D: Journal of Automobile Engineering, vol. 235, no. 9, pp. 2484-2495, 2021. https://doi.org/10.1177/0954407021994349

[11] X.-H. Chang and G.-H. Yang, "Nonfragile H_∞ filtering of continuous-time fuzzy systems," IEEE Transactions on Signal Processing, vol. 59, no. 4, pp. 1528-1538, 2010. https://doi.org/10.1109/TSP.2010.2103068

[12] D. Neupane, M. R. Bouadjenek, R. Dazeley, and S. Aryal, "Data-driven machinery fault diagnosis: A comprehensive review," Neurocomputing, p. 129588, 2025. https://doi.org/10.1016/j.neucom.2025.129588

[13] N. Ghaeminezhad, Z. Wang, and Q. Ouyang, "A review on fault-tolerant control strategies for lithium-ion battery systems," Transactions of the Institute of Measurement and Control, vol. 45, no. 3, pp. 573-592, 2023. https://doi.org/10.1177/01423312221119065

[14] S. Amir, M. Gulzar, M. O. Tarar, I. H. Naqvi, N. A. Zaffar, and M. G. Pecht, "Dynamic equivalent circuit model to estimate state-of-health of lithium-ion batteries," IEEE Access, vol. 10, pp. 18279-18288, 2022. https://doi.org/10.1109/ACCESS.2022.3148528

[15] C.-H. Chung, S. Jangra, Q. Lai, and X. Lin, "Optimization of electric vehicle charging for battery maintenance and degradation management," IEEE Transactions on Transportation Electrification, vol. 6, no. 3, pp. 958-969, 2020. https://doi.org/10.1109/TTE.2020.3000181

[16] Q. Wang, T. Gao, and X. Li, "SOC estimation of lithium-ion battery based on equivalent circuit model with variable parameters," Energies, vol. 15, no. 16, p. 5829, 2022. https://doi.org/10.3390/en15165829

[17] M. A. Hannan, M. M. Hoque, A. Hussain, Y. Yusof, and P. J. Ker, "State-of-the-art and energy management system of lithium-ion batteries in electric vehicle applications: Issues and recommendations," IEEE Access, vol. 6, pp. 19362-19378, 2018. https://doi.org/10.1109/ACCESS.2018.2817655

[18] W. Zheng, B. Xia, W. Wang, Y. Lai, M. Wang, and H. Wang, "State of charge estimation for power lithium-ion battery using a fuzzy logic sliding mode observer," Energies, vol. 12, no. 13, p. 2491, 2019. https://doi.org/10.3390/en12132491

[19] D. Liberzon and A. S. Morse, "Basic problems in stability and design of switched systems," IEEE Control Systems Magazine, vol. 19, no. 5, pp. 59-70, 2002. https://doi.org/10.1109/37.793443

[20] N. El youssfi, R. El Bachtiri, T. Zoulagh, and H. El Aiss, "Unknown input observer design for vehicle lateral dynamics described by Takagi-Sugeno fuzzy systems," Optimal Control Applications and Methods, vol. 43, no 2, pp. 354-368, 2022. https://doi.org/10.1002/oca.2808

Residual Signal-Based Experimental Approach for Early Detection of Rotor Winding Asymmetry in Wound-Rotor Induction Machines

Hamza SABIR[1,a*], Mohammed OUASSAID[2,b], Nabil NGOTE[3,c] and Mourad YESSEF[4,d]

[1]Laboratory of Engineering Sciences & Biosciences, Faculty of Sciences and Techniques of Mohammedia, Hassan II University, Casablanca, Morocco

[2]Engineering for Smart and Sustainable Systems Research Center, Mohammadia School of Engineers, Mohammed V University in Rabat, Morocco

[3]ENSMR Engineering School, Rabat, Morocco

[4]Higher School of Technology of Nador, Mohammed First University, Nador 62000, Morocco

[a]hamza.sabir@fstm.ac.ma, [b]ouassaid@emi.ac.ma, [c]ngotenabil@gmail.com, [d]mourad.yessef@usmba.ac.ma

Keywords: Condition Monitoring, Residual Electrical Signal Extraction, Rotor Winding Asymmetry, Fault Diagnosis, Wound-Rotor Induction Machines

Abstract. This paper presents a residual electrical signal extraction (RESE) technique that can be used to quickly identify rotor winding asymmetry (RWA) in wound rotor induction machines (WRIMs). Inspired by techniques described in vibration analysis, the proposed method consists of processing the stator current signals in order to extract the residual signals and isolate the fault signatures normally masked by the dominant fundamental frequency. Unlike conventional methods, the RESE technique works satisfactorily under no-load conditions without additional sensors. experimental results confirm the reliability of the method by demonstrating that rotor winding asymmetries as low as 3 to 10% can be reliably determined. The RESE method therefore offers a non-intrusive, cost-effective, and accurate solution for condition monitoring and preventive maintenance of WRIMs.

Introduction

Wound rotor induction machines (WRIMs) are at the heart and soul of the most modern electromechanical systems, notably in variable speed drives and DFIG (Doubly Fed Induction Generator) wind generators [1-4]. Their use is conventional because slip rings give direct access to the rotor windings, which are open to the network and into which external resistors can be inserted. As a result, these machines offer a level of flexibility that squirrel cage machines cannot match: the ability to adjust speed over a wide range, increase starting torque, reduce starting current, and optimize efficiency at partial load [5-7]. It is because of these advantages that they continue to be used extensively in heavy industry, mining systems, and modern wind farms. However, this architectural perfection may also be the underlying cause of less obvious internal faults [8-10]. The most frequent failures are as follows:

– bearing wear resulting from a lack of oil,
– partial short circuits in the stator or rotor windings,
– body play during a mechanical fault leading to air gap eccentricities,
– misalignments that cause feedback on the loads.

More often than not, these faults manifest themselves in an incipient (early) form, i.e., at an early step in their evolution when the signs are slight and difficult to detect [11]. However, without prompt diagnosis, they can progress to significant defects, causing expensive downtime, poor energy efficiency, or even accidents in critical environments such as aeronautics or nuclear power [12-13].

The monitoring of WRIM operating conditions has become a key focus of condition monitoring (CM), particularly due to conventional diagnostic techniques based on infrared thermography, vibration analysis, and acoustic emission analysis. These techniques can be effective in certain contexts, but they require dedicated sensors, which are very expensive and sometimes intrusive. In a constrained industrial environment, such as wind turbines at height, installing one or more sensors in addition to existing systems can be complex, if not impossible [14].

To overcome these limitations, scientific research studies have focused on analyzing electrical signals, specifically stator current, which is already used in most control systems. The most widely used method, MCSA (Motor Current Signature Analysis), utilizes the current spectrum to provide more accurate detection criteria for components characteristic of rotor or stator faults. It is presented as economical and non-invasive [15]. However, MCSA has a considerable weakness as regards the identification of incipient rotor winding asymmetry (RWA) faults. At no-load, the spectral lines associated with the fault are of low amplitude and hidden by the fundamental 50 Hz component, making their detection impossible. Several authors have attempted to enhance the sensitivity of MCSA by using time-frequency tools (wavelet transform, Hilbert-Huang transform), statistical tools, and artificial intelligence approaches, which sometimes perform more effectively, but under insufficient conditions (requirement for large databases, significant algorithmic complexity, or increased sensitivity to noise and operating conditions). Moreover, most of these approaches focus on the detection of developed defects, whereas there is a real need to simply detect incipient defects [16-17].

In this context, the new RESE (Residual Electrical Signal Extraction) method described in this paper is presented. Indeed, it involves applying concepts from the field of mechanical vibration analysis to the electrical domain. By removing the fundamental component through a process of synchronization and averaging of the stator current periods, a residual signal is obtained in which the lowest frequency harmonics, those corresponding to faults, are now visible, even though they were previously buried in the spectrum. The advantage of this novel but simple approach is that it once again allows hidden incipient fault signatures to be identified while retaining the simplicity of implementation offered directly by current analysis.

The main contributions of the new method are:

- Introducing a new analogy with mechanical vibration analysis: this is the first time that electrical ergodicity has been used to increase sensitivity to faults;
- Detecting incipient faults at low load: the method can detect rotor asymmetries with an amplitude of less than 10% even under no-load conditions;
- Multi-level validation: the process is evaluated by modeling, simulation in MATLAB/Simulink, and confirmed by bench testing;
- Industrial applicability: this solution is exclusively designed to measure stator current, thus not requiring any additional sensors, which facilitates its deployment in constrained environments such as wind turbines and submersible pumps.

Finally, this work aims to address a significant gap in the diagnosis of wound rotor machines: reliability in detecting incipient rotor asymmetry faults in unfavorable situations. The proposed approach, combining RESE and spectral analysis, offers improved sensitivity and remains simple to implement, thereby facilitating the implementation of predictive maintenance in industrial applications where reliability is a major issue.

The succeeding section of this paper is organized as follows. Section II provides a detailed explanation of the proposed residual electrical signal extraction (RESE) methodology. Section III presents the experimental test bench, the conditions and results of the experimental tests, followed by an in-depth discussion of the results in order to examine the relevance and methodological reliability of the RESE method. Finally, Section IV concludes the study by describing the main contributions of the RESE method in terms of results, while discussing possibilities for future work in this field.

Proposed Approach

The most frequently applied WRIM monitoring method is MCSA. The MCSA is an effective approach to analyze the stator line current signal in order to detect defects in the rotor. It is founded on FFT frequency analysis. The aim consists of improving the extraction of signatures through the frequency spectrum [18-19]. This study implements the Fast Fourier Transform (FFT) method to detect anomalies in residual signals and stator signals.

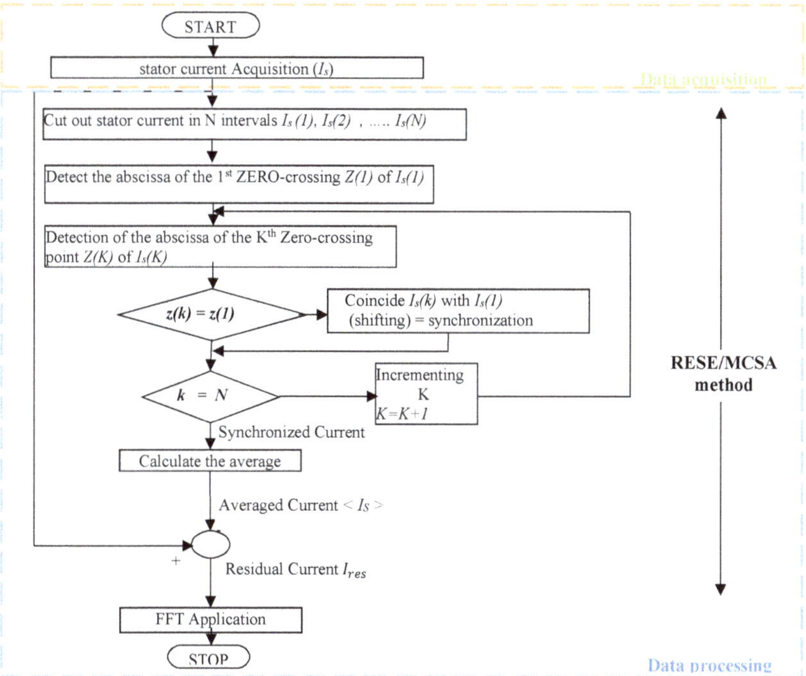

Figure 1. Flowchart of the new signal processing algorithm

In this approach, signatures with elevated spectral magnitudes are used to locate faults. A failure in the used machine is mathematically linked to the magnitude of components in the frequency domain. Therefore, if this new methodology is applied to the stator line current signal, it identifies characteristic frequencies associated with an RWA failure. In theory, the signatures of a rotor fault can be determined using the following formula [20]:

$$f_{d1} = f_s\,(1 - 2s) \quad (1)$$

$$f_{d2} = f_s\,(1 + 2s) \quad (2)$$

This method aims to highlight the electrical signatures associated with incipient rotor asymmetry faults in asynchronous machines with wound rotors. The main idea is to eliminate the fundamental component of the stator current, which hides the weak harmonics characteristic of the fault, to

Materials Research Proceedings 64 (2026) 241-248 https://doi.org/10.21741/9781644904091-30

obtain a residual signal that can be used for diagnosis. The flowchart of the procedure is shown in Fig. 1.

Experimental Results and Discussion

A wound rotor induction machine of 1.1 KW is used for the experimental tests. The experimental motor bench, connected to the voltage source, is presented in Figure 2. The test bench was set up to demonstrate the effectiveness of the proposed method for identifying incipient defects at the rotor windings under no load operating conditions. Incipient defects are created in the rotor winding. Resistance in a single rotor winding is reduced threefold by changing a series of external shunts that are connected in series with the rotor winding. Thereafter, a current clamp is placed on the stator line and the current is acquired by a numerical oscilloscope. This acquisition is performed with a sampling frequency of 10 kHz. Thus, the total number of samples per 50 Hz period is 200 samples. The characteristics of the WRIM motor used are listed in Table 1.

Table 1 WRIM (1.1 kW) Motor Specifications.

Parameters	Name	Value
P_N	Rated power	1100 W
F_S	Power supply frequency	50 Hz
V_S	Stator Phase Voltage	400 V
L_S	Stator Leakage Inductance	509 mH
L_r	Rotor Leakage Inductance	466 mH
R_S	Stator Resistance	6,99 Ω
R_r	Rotor Resistance	9,11 Ω
J	Inertia	0.564 Kg.m^2
P	Pole Pairs Number	2

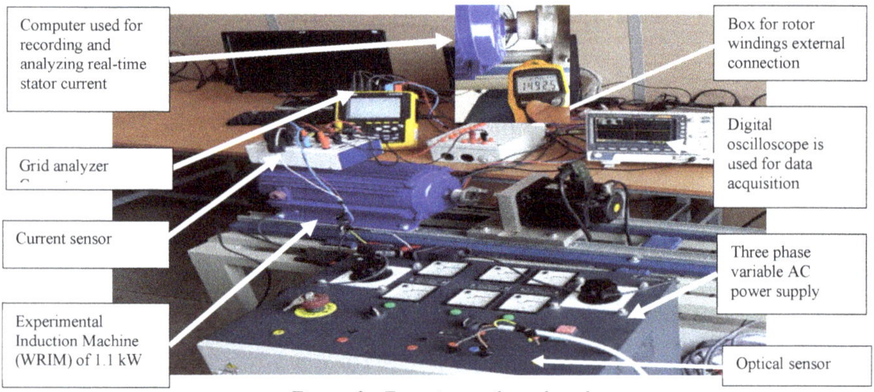

Figure 2. Experimental test bench.

The stator current waveforms corresponding to the faulty and healthy conditions are shown in Figure 3. The resulting current waveforms have the same shape. Therefore, direct analysis of the stator line current signal, without going through a frequency analysis tool, can not be deployed to identify the location of the incipient defect.

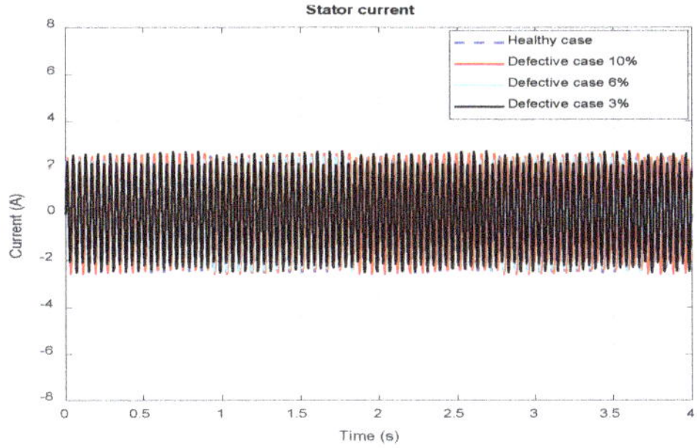

Figure 3. Stator current waveforms under no loaded motor.

The MCSA method is used to illustrate the signatures of initial rotor faults. Figure 4 shows the frequency spectra of the healthy and defective states. They seem to be the same. The machine is not loaded, so the slip is minimal. That results in no visible frequency components besides the main one at 50 Hz. The RESE algorithm is suggested for implementation to get the residual signal spectrum with fault information.

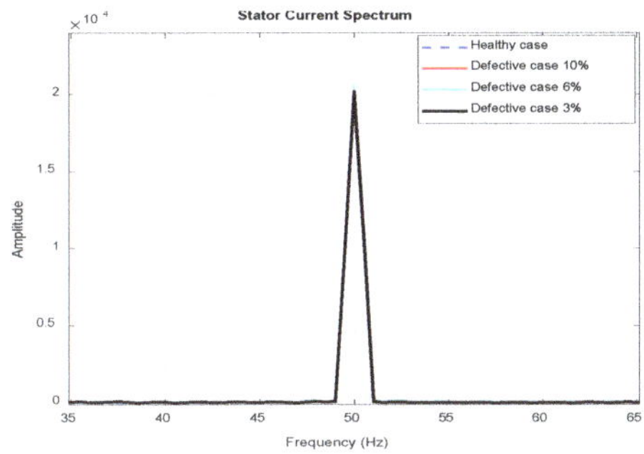

Figure 4. Stator current spectra for the healthy case and the three ITSC defect cases under No loaded motor.

The RESE method is adopted for the stator current. The sampling - frequency is 10 kHz, i.e., Ns = 200 samples per period (200 = 10 kHz x 20ms). The total number of superimposed periods is K = 50 periods, i.e., the total number of samples N_{ST} = 20000 samples (20000 samples = 200 samples x 70 periods). The extracted residual stator current signals are shown in Figure 5.
These signals provide an early indicator of the possible existence of a defect. Thus, the residual current time analysis has the potential to detect an incipient fault.

The spectral analysis of the residual signals in the case of no loaded machine is carried out using the MCSA method. The signature amplitudes are higher in the defective cases, respectively, according to the defect severity, as shown in Figure 6. The measured average velocity values, in the healthy and all defective cases, as well as the theoretical signature, and the signatures presented previously in the residual spectra, are listed in Table 2.

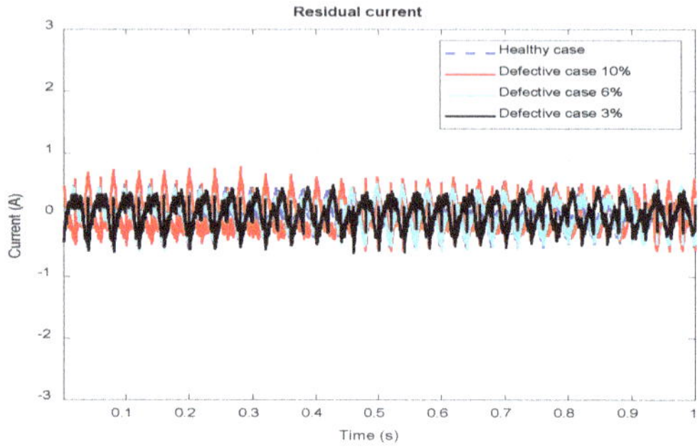

Figure 5. Stator residual current waveforms under no loaded motor

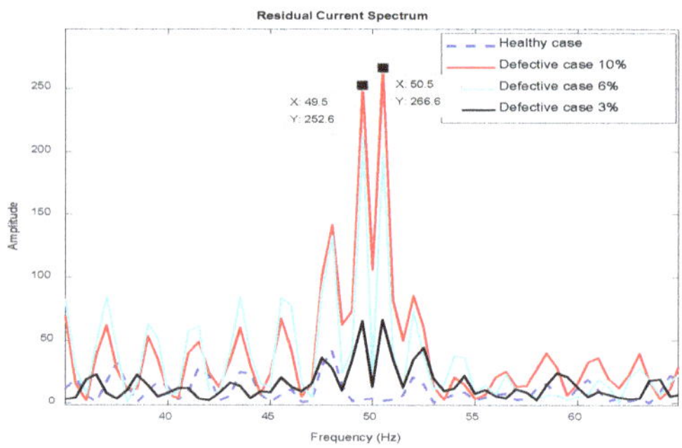

Figure 6. Stator residual spectra for the healthy case and the three ITSC defect cases under No loaded motor

From table 2, the signature values appear to be approximately the same size. Therefore, the experimental and theoretical results coincide. In addition, the residual spectra facilitate the discrimination between all defective and healthy cases. Finally, the application of the MCSA technique to the residual current can accurately extract an incipient ITSC fault even under no-load conditions.

Table 2. Theoretical and experimental signature comparisons

	Experimental Speed (rpm)	Experimental Signature (Hz)	Theoretical Signature (Hz)
Defective case (10%)	1492,5	50.5	50.5
Defective case (6 %)	1492,5	50.5	50.5
Defective case (3%)	1492,5	50.5	50.5
Healthy case (0%)	1492,5	No upper sideband	No upper sideband

Summary

The aim of this research was to present the Residual Electrical Signal Extraction method, which characterizes malfunctions for assessing the health of wound rotor induction machines as a new and original approach to innovative diagnostics that circumvents the limitations of conventional MCSA techniques for detecting ordinary fault signatures, which are almost always buried under the fundamental power supply component. Experimental tests carried out on the test bench showed that this method can detect incipient rotor asymmetry faults of extremely low severity, even under unfavorable operating conditions such as no-load mode. Beyond its simplicity of implementation, this methodology has strong potential for industrial application in that it can be applied independently to measure stator current, which is already available in most control systems. This is an advantage for integration into a predictive maintenance system without any additional sensors. In addition, early fault detection increases machine life while reducing unplanned downtime, thereby helping to improve energy efficiency and reduce industrial waste with a view to environmental sustainability.

Future studies are planned to extend this method of analysis to other types of electrical and mechanical faults, as well as to incorporate artificial intelligence methods to enable real-time, automated monitoring in complex industrial environments.

References

[1] Rajendran, S., Sreejesh, R., Devi, V. K., Jena, D., & Banjerdpongchai, D. (2025). Dynamic mode decomposition based fault diagnosis in three-phase electrical machines. Results in Engineering, 25, 103761. https://doi.org/10.1016/j.rineng.2024.103761

[2] Ma, C., Zhang, B., Song, W., Jia, Z., Huang, W., & Huang, J. (2025). A Stationary-Frame Residual-Current-Based ITSC Localization Method for Induction Motors Considering Effect of Different Faulty Phase. IEEE Journal of Emerging and Selected Topics in Power Electronics. https://doi.org/10.1109/JESTPE.2024.3524545

[3] Roshanfekr, R., & Jalilian, A. (2015). Analysis of rotor and stator winding inter-turn faults in WRIM using simulated MEC model and experimental results. Electric Power Systems Research, 119, 418-424. https://doi.org/10.1016/j.epsr.2014.10.018

[4] Orhan, M., & Celik, M. (2024). A literature review and future research agenda on fault detection and diagnosis studies in marine machinery systems. Proceedings of the Institution of Mechanical Engineers, Part M: Journal of Engineering for the Maritime Environment, 238(1), 3-21. https://doi.org/10.1177/14750902221149291

[5] Kouadria, M., Chedjara, Z., Su, C. L., Benbouzid, M., Guerrero, J. M., Ibrahim, B. S. K. K., & Ahmed, H. (2025). Diagnosis of induction motor stator faults around rotor slot harmonics using the Matrix Pencil method. Results in Engineering, 104240. https://doi.org/10.1016/j.rineng.2025.104240

[6] Atif, M., Azmat, S., Khan, F., Albogamy, F. R., & Khan, A. (2024). AI-driven thermography-based fault diagnosis in single-phase induction motor. Results in Engineering, 24, 103493. https://doi.org/10.1016/j.rineng.2024.103493

[7] Kim, Do-Jin, Kim, Hae-Joong, Hong, Jung-Pyo, et al. Estimation of acoustic noise and vibration in an induction machine considering rotor eccentricity. IEEE Transactions on Magnetics, 2014, vol. 50, no 2, p. 857-860. https://doi.org/10.1109/TMAG.2013.2285391

[8] Gyftakis, Konstantinos N., Antonino-Daviu, Jose A., Garcia-Hernandez, Raul, et al. Comparative experimental investigation of broken bar fault detectability in induction motors. IEEE Transactions on Industry Applications, 2015, vol. 52, no 2, p. 1452-1459. https://doi.org/10.1109/TIA.2015.2505663

[9] Sobra, Jan, Vaimann, Toomas, et Belahcen, Anouar. Mechanical vibration analysis of induction machine under dynamic rotor eccentricity. In : 2016 17th International Scientific Conference on Electric Power Engineering (EPE). IEEE, 2016. p. 1-4. https://doi.org/10.1109/EPE.2016.7521732

[10] Mirzaeva, Galina et Saad, Khalid Imtiaz. Advanced diagnosis of rotor faults and eccentricity in induction motors based on internal flux measurement. IEEE Transactions on Industry Applications, 2018, vol. 54, no 3, p. 2981-2991 https://doi.org/10.1109/TIA.2018.2805730

[11] Morales-Perez, C., Rangel-Magdaleno, J., Peregrina-Barreto, H., Amezquita-Sanchez, J. P., & Valtierra-Rodriguez, M. (2018). Incipient broken rotor bar detection in induction motors using vibration signals and the orthogonal matching pursuit algorithm. IEEE Transactions on Instrumentation and Measurement, 67(9), 2058-2068. https://doi.org/10.1109/TIM.2018.2813820

[12] Chacon, J. L. F., Kappatos, V., Balachandran, W., & Gan, T. H. (2015). A novel approach for incipient defect detection in rolling bearings using acoustic emission technique. Applied Acoustics, 89, 88-100. https://doi.org/10.1016/j.apacoust.2014.09.002

[13] Hossain, Md Liton, Abu-Siada, Ahmed, et Muyeen, S. M. Methods for advanced wind turbine condition monitoring and early diagnosis: A literature review. Energies, 2018, vol. 11, no 5, p. 1309. https://doi.org/10.3390/en11051309

[14] Faiz, J., Keravand, M., Ghasemi-Bijan, M., Cruz, S. M., & Bandar-Abadi, M. (2016). Impacts of rotor inter-turn short circuit fault upon performance of wound rotor induction machines. Electric Power Systems Research, 135, 48-58. https://doi.org/10.1016/j.epsr.2016.03.007

[15] Mendonça, P. L., Bonaldi, E. L., de Oliveira, L. E. L., Lambert-Torres, G., da Silva, J. B., da Silva, L. B., ... & Shinohara, A. H. (2017). Detection and modelling of incipient failures in internal combustion engine driven generators using electrical signature analysis. Electric Power Systems Research, 149, 30-45. https://doi.org/10.1016/j.epsr.2017.04.007

[16] H. Sabir, M. Ouassaid, N. Ngote, M. Benbouzid, "A novel experimental method to detect early rotor faults in induction machines" International Journal on Energy Conversion, Vol. 9, n. 5, pp. 191-202, 2021 https://doi.org/10.15866/irecon.v9i5.21214

[17] H.Sabir., Ouassaid, M., & Ngote, N. (2022). An experimental method for diagnostic of incipient broken rotor bar fault in induction machines. Heliyon, 8(3). https://doi.org/10.1016/j.heliyon.2022.e09136

[18] Sabir, H., Ouassaid, M., & Ngote, N. (2020, November). Early Fault Estimation of Inter-turn Short-circuit in Rotor Winding of WRIM using ANN-based Combined TSA and MCSA Technique. In 2020 International Symposium on Advanced Electrical and Communication Technologies (ISAECT) (pp. 1-6). IEEE. https://doi.org/10.1109/ISAECT50560.2020.9523697

[19] Sabir, H., Ouassaid, M., & Ngote, N. (2021, November). Detection of an Incipient Rotor Winding Inter-Turn Short Circuit Fault. In 2021 IEEE International Autumn Meeting on Power, Electronics and Computing (ROPEC) (Vol. 5, pp. 1-7). IEEE. https://doi.org/10.1109/ROPEC53248.2021.9668140

[20] P. Podder, et al. Comparative performance analysis of hamming, hanning and blackman window. International Journal of Computer Applications, 2014, vol. 96, no 18.A. https://doi.org/10.5120/16891-6927

Electric Grid and Digital Infrastructure

Optimization of a 5.82 GHz Microstrip Patch Antenna for Sustainable EV Wireless Power Transfer Using Parasitic Elements and Dual Substrates of Polypropylene and Fused Quartz

Soukaina JAAFARI[1,a *], Ahmed KHALLAAYOUN[2,b] and Esmail AHOUZI[1,c]

[1]National Institute of Posts and Telecommunications, Allal El Fassi Ave, Rabat, 11000, Morocco

[2]Al Akhawayn University in Ifrane, PO BOX 104, Ifrane, 53000, Morocco

[a]s.jaafari@aui.ma, [b]a.khallaayoun@aui.ma, [c]ahouzi@inpt.ac.ma

Keywords: Electric Vehicles, Wireless Power Transfer, Microstrip Patch Antenna, Parasitic Elements

Abstract. As interest in sustainable and energy conscious technologies continues to rise, wireless power transfer (WPT) has become an important component in advancing green innovation by minimizing transmission losses and enabling contact-free power delivery. This work investigates the optimization at 5.82 GHz. The study examines how antenna performance can be improved through variations in substrate type, substrate thickness and slot geometry. A set of simulation models was developed using Fused Quartz and Polypropylene substrates with multiple thickness options, as well as a dual layer configuration in which the secondary substrate's thickness was varied. Additional performance gains were explored through several slot arrangements integrated into the patch. The optimized design, featuring a dual-substrate structure and a centrally positioned slot placed nearer to the feedline, achieved an efficiency of 95%. These findings support the advancement of high efficiency WPT systems that contribute to border sustainable and environmentally conscious energy solutions.

Introduction

Wireless Power Transfer (WPT) has become a major enabler for the next generation of electric vehicle (EV) charging systems, providing a more user friendly and efficient alternative to conventional plug-in methods by removing the need for physical contact [1]. By improving convenience and supporting scalable charging infrastructure, WPT contributes to overcoming several barriers that still limit widespread EV adoption. Within the family of WPT methods, magnetic resonant coupling (MRC) stands out because it can deliver power efficiently over moderate distances, making it a strong candidate for EV use cases [2,4]. The transmitting antenna, a key element of any MRC system, plays a central role in determining how effectively energy is delivered.

Recent work has emphasized optimizing microstrip patch antennas, especially through substrate material and thickness selection. Yakub et al. [5] demonstrated their strong influence on antenna characteristics, while comparative studies linked dielectric constant to antenna gain and size [5]. High permittivity substrates reduce antenna dimensions but introduce losses, highlighting design trade-offs. To improve bandwidth, Upadhay and Dwivedi [6] proposed a dual-layer substrate configuration, reinforcing the importance of tailored substrate engineering.

Geometric modifications such as slots have also been widely explored. Slots can enhance bandwidth, impedance matching, and efficiency. Hkanoglu et al. [7] reported bandwidth improvements, Paul et al. [8] introduced a compact wrench shaped slot antenna for higher efficiency, and Nornikman et al [9] optimized dual-layer structure combined with slots. These studies underscore the potential of structural modifications to advance antenna performance.

While many techniques target far-field WPT [10], limited research focuses on near-field EV charging. This paper therefore investigates microstrip patch antenna optimization for EV WPT, examining substrate selection, thickness and slot integration. Using Ansys simulations, return loss, VSWR, and radiation patterns are analyzed to identify effective substrate configurations and slot designs for High-performance stationary WPT.

Methodology

Objective. The goal behind this paper is to optimize a microstrip patch antenna capable fo delivering efficient WPT for EV application at 5.82 GHz. The study aims to generate a strong electromagnetic field while ensuring high impedance matching and maximizing near-field radiated power to improve overall antenna performance. The key performance metrics evaluated in this study are summarized in Table 1.

Table 1. Performance Metrics

Performance Metric	VSWR	S Parameter	Beam Efficiency	Near E	Near H
VSWR	<1.7	<-10	>95%	>5	>4

Description of the Simulation Setup and Parameters

Effect of Substrate Material and Thickness on Antenna Performance. The first step of the study examined sic microstrip patch antenna configurations operating at 5.82 GHz, each differing in substrate material and thickness as listed in Table 2. Fused Quartz and Polypropylene were selected as the substrate materials for this study. The simulations aimed to analyze the impact of these two substrates, each with varying thicknesses of 2mm, 4mm, and 8mm, on the performance of the simple microstrip patch antennas, Fig. 1.

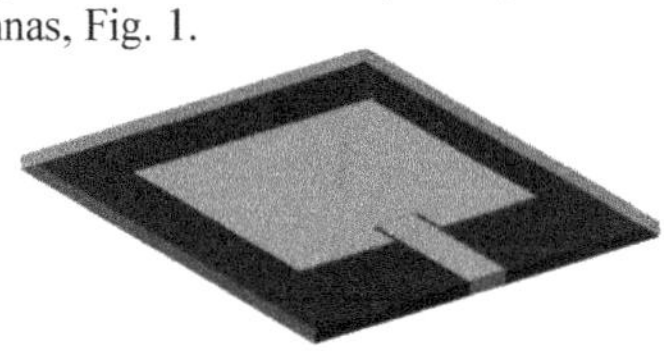

Fig. 1. Microstrip Patch Antenna: Initial Design

Table 2. Step 1 Simulation Parameters

Model	Substrate Material	Dielectric Constant ε_r	Substrate Height (mm)	Width (mm)	Height(mm)
1	Fused Quartz	3.78	2	16.729	12.440
2	Fused Quartz	3.78	4	16.729	11.177
3	Fused Quartz	3.78	8	16.729	8.627
4	Polypropylene	2.2	2	20.446	19.868
5	Polypropylene	2.2	4	20.446	14.627
6	Polypropylene	2.2	8	20.446	11.528

Dual-substrate configuration and thickness optimization. In the second step, a secondary substrate is introduced to the best-performing model identified in step 1. This additional substrate is positioned between the ground plane and the initial substrate. The antenna's performance is then evaluated by varying the thickness of the secondary substrates from 0.2mm to 0.5mm and 1mm, as illustrated in Fig. 2. The objective of this step is to investigate he impact of a dual-substrate configuration and the influence of substrate thickness on the antenna's performance.

(a) (b) (c)

Fig. 2. Dual-Substrate Microstrip Patch Antenna (a) 0.2mm (b) 0.5mm (c) 1mm

Slot design and placement optimization for enhanced antenna performance. In the third step, slots are introduced into the patch antenna. These slots are openings or parasitic elements incorporated into the antenna structure. Five different slot designs are compared using the best-performing antenna model from step 2, as demonstrated in Fig. 3. Fig. 4 provides the dimensions of the slots used in the design. The aim of this step is to assess the impact of slot shape and placement on the antenna's performance, specifically regarding impedance matching, radiated power, reflectivity, and the strength of the Near H and Near E fields.

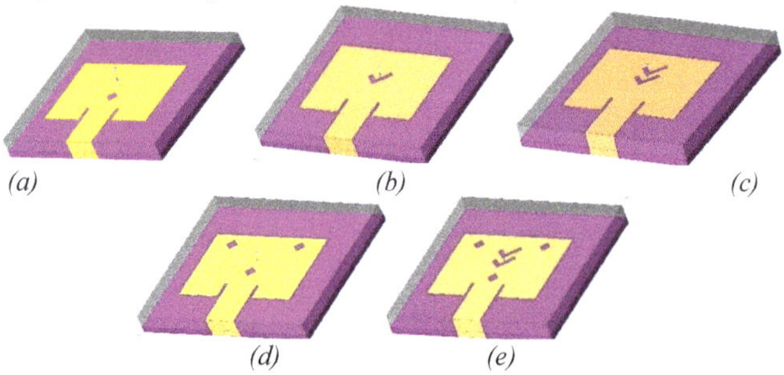

Fig. 3. Slots designs (a) Model 1: Single Centered (b) Model 2: Centered Checkmark (c) Model 3: Two Centered Checkmarks (d) Model 4: Scattered Slots (e) Model 5: Combined

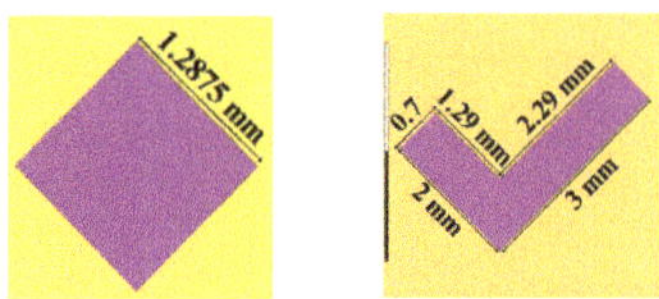

(a) (b)

Fig. 4. Slots Dimensions (a) Squared Slot (b) Checkmark

Results and Discussion

Results

Effect of substrate material and thickness on the antenna performance. Based on Fig. 5, Model 2 (FQ – 2 mm) and Model 5 (PP – 2 mm) emerge as the best performers among the six tested. Model 2 provides the lowest VSWR at 2.62 and reaches the highest efficiency of 0.807, making it the most dependable choice for reducing losses and improving overall power transfer. Model 5, in comparison, delivers the highest gain of 6.04 dB but falls behind with an efficiency of 0.669 and a VSWR of 3.79, suggesting better directionality but weaker efficiency.

The beam pattern results also reflect this contrast. Model 5 produces the most concentrated beam with the smallest beam area of 2.101 sr, while Model 2 offers an area of 2.6105 sr. Their

Materials Research Proceedings 64 (2026) 250-257 https://doi.org/10.21741/9781644904091-31

S11 values reinforce the same trend with Model 2 achieving -6.992 dB which is lower than that of Model 5.

The near-field behavior evaluation shows that both designs have close E-field performance, but Model 2 produces a stronger H-field level of 3.72 dB. In summary, Model 2 provides the most balanced behavior across all evaluated metrics; However, Model 5 remains for higher gain and beam concentration applications.

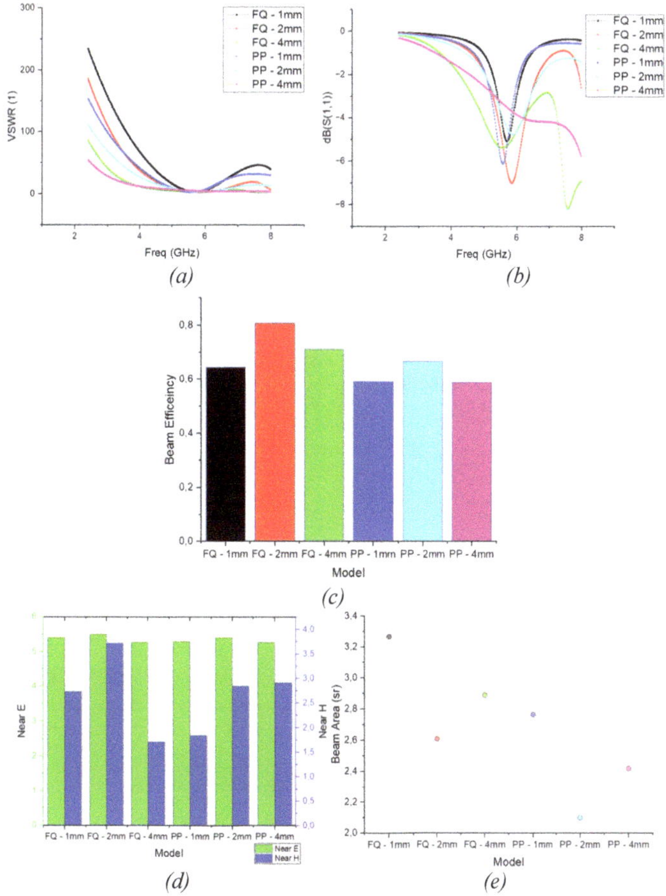

Fig. 5. Impact of substrate material and thickness on: (a) VSWR (b) S Parameter (c) Beam Efficiency (d) Near E and Near H (e) Beam Area

Table 3 illustrates how substrate thickness influences antenna behavior, showing that a moderate value of 2 mm provides the most balanced performance across the evaluated metrics.

Table 3. Substrate Thickness Impact on the Antenna Performance

Subs. Thickness	Effect on Performance
1mm	High VSWR - Low Gain - Low Efficiency - Large Beam Area
2mm	Low VSWR - High Gain - High Efficiency - Small Bear Area
4mm	Low VSWR - High Gain - Low Efficiency - Small Beam Area

Dual-substrate configuration and thickness optimization. As shown in Fig.6, increasing the thickness of the secondary polypropylene (PP) layer leads to consistent improvement in antenna performance. The VSWR drops from 2.322 in Model 1 to 2.080 in Model 2, and further to 1.822 in Model 3 indicating better impedance matching. A similar pattern appears in the efficiency values, which rise from 0.855 to 0.893 and reach a peak of 0.936, suggesting stronger electromagnetic confinement and lower energy losses.

The radiation pattern also becomes slightly more focused as thickness increases. The beam area is reduced from 2.512 sr in Model 1 to 2.432 sr in Model 3. Reflection performance improves as well. S11 decreases from -8.002 dB to -10.714 dB, reinforcing the improvement in matching quality.

Near-field measurements follow the same upward trend. The E-field grows from 5.52 to 5.60 dB, while the H-field shows a more substantial increase, rising from 4.11 to 4.79 dB. This stronger magnetic field response highlights the advantage of thicker substrates in WPT systems, where magnetic coupling plays a key role.

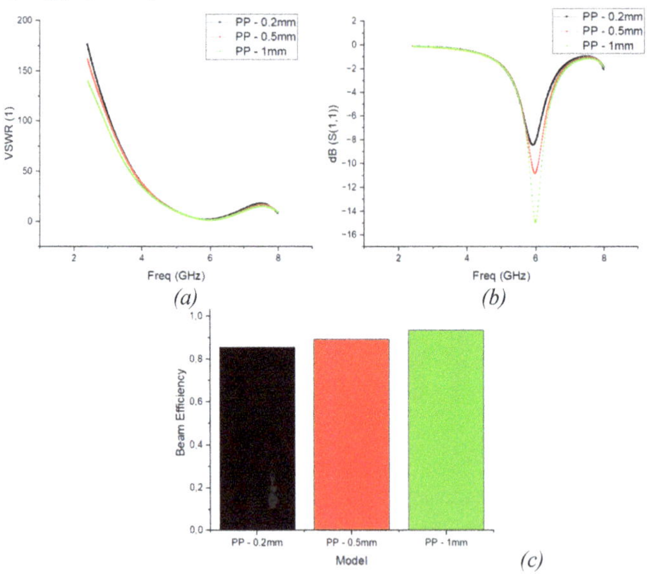

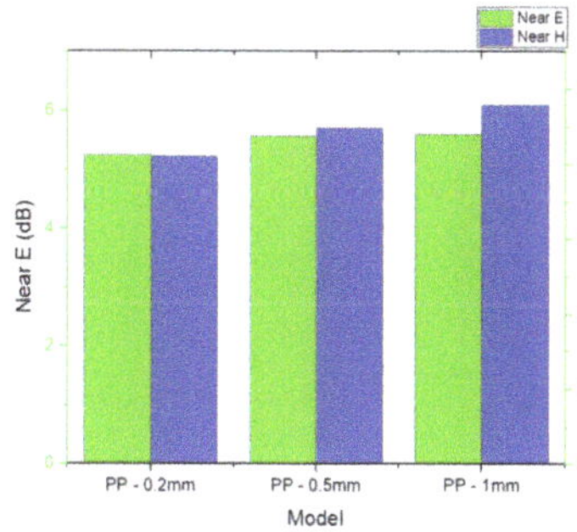

(d)

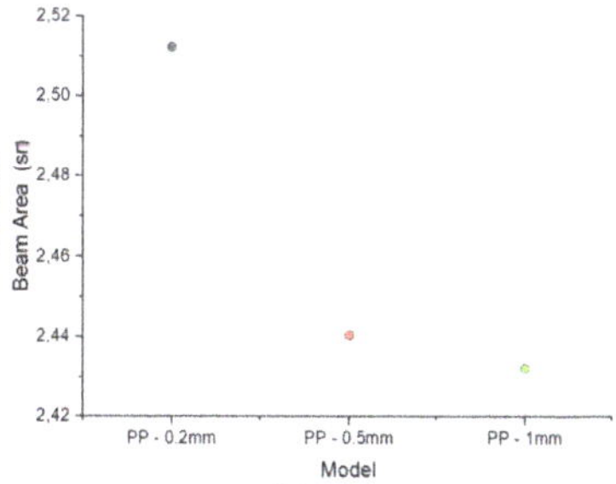

(e)

Fig. 6 Dual-Substrate Simulation Results (a) VSWR (b) S Parameter (c) Beam Efficiency (d) Near E and Near H (c) Beam Area

Among the tested configurations, Model 5 shows the strongest overall results. It attains the lowest VSWR of 1.667, the highest efficiency of 0.953, and the greatest radiated power, demonstrating its ability to reduce losses and improve energy transfer.

The beam area shows only minor variation across the designs, ranging from 2.403 sr in Model 1 to 2.459 sr in Model 3. This indicates that slot geometry has a limited effect on beam spread. Reflection behavior also highlights Model 5 as the top performer, with an S11 value of -12.042 dB, confirming its superior impedance matching.

Near-field analysis reveals similar trends among the models, though Model 3 achieves slightly higher peak E field and H fields strengths, making it favorable in cases where near field intensity is the priority.

Table 4. Slots Effect Simulation Results

Model	VSWR	S 11 Parameter	Beam Efficiency	Near E (dB)	Near H (dB)	Beam Area (sr)
1: Centred Slot	1.726757	-11.485147	0.95	5.59641	4.82441	2.403499
2: Centred Checkmark	1.76723	-11.142393	0.94	5.5966	4.83002	2.434855
3: Two Centred Checkmarks	1.741372	-11.358632	0.94	5.60014	4.83996	2.458565
4: Scattered Slots	1.836031	-11.088892	0.92	5.58437	4.71065	2.441693
5: Combined	1.666634	-12.041521	0.96	5.59744	4.84201	2.445187

As shown in Fig. 1, Model 5 outperforms the baseline design. The corresponding Near E field and Near H field distributions for this optimized configuration are presented in Fig. 7.

Fig. 7. The Near E and Near H fields distribution and strength

Discussion

The results show that FQ outperforms PP as a substrate for the microstrip patch antenna, particularly at 2 mm. Its lower dielectric constant and loss tangent support better impedance matching and higher efficiency by limiting internal losses. FQ's stable and homogeneous structure also improves signal clarity and beam, focus. In contrast, PP delivers weaker overall performance at all thicknesses, with reduced radiation efficiency. Its higher thermal expansion and less favorable dielectric properties lead to poorer wave propagation, making FQ the more effective substrate choice.

Performance improvements observed with the added PP layer in the dual-substrate configuration arise from how the differing dielectric constants confine and guide the electromagnetic fields. This adjustment enhances wave propagation, reduces internal power loss and increases radiated output. The corresponding rise in near-field levels indicates a stronger magnetic component, which benefits WPT efficiency.

Slot incorporation further enhanced antenna behavior by introducing parasitic elements that intentionally disturb the surface current paths. These disruptions force currents to travel along longer routes, which in turn creates additional resonant modes, shifting ethe primary resonant frequency. And broaden the bandwidth. The single slot in Model 1 perturbed the current near its maximum, altering the local field density. The double checkmark slots acted as parasitic resonators, strengthening both Near H and Near E fields by increasing localized currents, reducing concentration. The scattered edge slots helped suppress surface wave currents, reducing the S-parameter and improving impedance matching. Together, these parasitic features reshape the field distribution and support multiple resonance pathways, yielding higher radiation efficiency and reduced losses.

Conclusion

This work demonstrated that combining a high-performance substrate such as FQ with carefully designed slot geometries significantly enhances the efficiency, matching, and near-field behavior of microstrip patch antennas for EV WPT. These improvements support more reliable and energy efficient wireless charging in stationary settings. Future studies should extend these findings to dynamic charging conditions, explore new materials and geometries, and integrate beamforming or metamaterial techniques for next generation WPT Systems.

Acknowledgment

The authors acknowledge the SEEDS Team (Smart, Embedded, Enterprise, & Distributed Systems) at INPT for their support. The authors also acknowledge the use of AI tools (ChatGPT and Gemini) for proofreading and language editing.

References

[1] Benitto, A. R., Umashankar, S., & S, B. (2023). Wireless Power Transfer in Electric Vehicles: A Review on Compensation Topologies, Coil Structures, and Safety Aspects. Energies, 16(7). https://doi.org/10.3390/en16073084

[2] A, P., P, K. D., A, K. P., J, P., & D, P. K. (2021). Essential Analysis of MRC-WPT System for Electric Vehicle Charging Using Coupled Mode Theory. 2021 1st Odisha International Conference on Electrical Power Engineering, Communication and Computing Technology (ODICON). Bhubaneswar, India. https://doi.org/10.1109/ODICON50556.2021.9429020

[3] Jaafari, S., El Hafdaoui, H., Ajabboune, K., Khallaayoun, A., & Ahouzi, E. (2024). Optimization of Circular Coils with Ferrite Boxes for Enhanced Efficiency in Wireless Power Transfer for Electric Vehicles. Green Energy and Intelligent Transportation, 4(2), 100195. https://doi.org/10.1016/j.geits.2024.100195

[4] Jaafari, S., Khallaayoun, A., & Ahouzi, E. (January 2025). Optimizing One-Block Planar Coil Design for Wireless Power Transfer in Electric Vehicles: A Comparative Study. Transportation Research Procedia, 84(1), 153-160. https://doi.org/10.1016/j.trpro.2025.03.058 International Conference on Smart Mobility and Logistics Ecosystems (SMiLE), KFUPM, Saudi Arabia. https://doi.org/10.1016/j.trpro.2025.03.058

[5] Yakub, B. O., Ibrahim, M., Olaoluwa, A. A., & Ayodeji, I. F. (2021). Comparative Analysis of the Effect of Different Substrates Materials on Microstrip Patch Antenna Operating at 3.6 GHz. 2021 IEEE International Conference on the Properties and Applications of Dielectric Materials (ICPADM), pp. 81-84. Johor Bahu, Malaysia. https://doi.org/10.1109/ICPADM49635.2021.9493967

[6] Upadhyay, D., & Dwivedi, R. (2014). Antenna Miniaturization Techniques for Wireless Applications. Eleventh International Conference on Wireless and Optical Communications Networks (WOCN). https://doi.org/10.1109/WOCN.2014.6923083

[7] Hkanoglu, B. G., Sen, O., & Turkmen, M. (2018). A Square Microstrip Patch Antenna with Enhanced Return Loss through Defected Ground Plane. 2nd URSI AT-RASC. Gran Canaria. https://doi.org/10.23919/URSI-AT-RASC.2018.8471618

[8] Paul, L. C., Hye, S. A., Rani, T., Hossain, M. I., Karaaslan, M., Ghosh, P., & Saha, H. K. (2023). A compact wrench-shaped patch antenna with a slotted parasitic element. e-Prime - Advances in Electrical Engineering, Electronics and Energy, 6. https://doi.org/10.1016/j.prime.2023.100334

[9] Nornikman, H., Ahmad, B. H., Abd Aziz, M. Z., & Bakar, H. A. (2018). Dual Frequencies Band and Enhanced Wideband Effect of Dual Layer Microstrip Patch Antenna with Parasitic. International PostGraduate Conference on Applied Science & Physics. Johor, Malaysia.

https://doi.org/10.1088/1742-6596/1049/1/012012

[10] Yan, Z., Hu, C., Hou, B., & Wen, W. (2025). A Watt-Level RF Wireless Power Transfer System with Intelligent Auto-Tracking Function. *Electronics*, 14(7), 1259. https://doi.org/10.3390/electronics14071259

Model Predictive Control of a 19-Level H-SPUC Inverter with Reduced Power Components

Taoufiq EL ANSARI[1,a*], Ayoub EL GADARI[1,b] and Youssef OUNEJJAR[1,c]

[1]Department of Electrical Engineering, Moulay Ismail University, High School of Technology, Meknes, Morocco

[a]taoufiq.elansari@edu.umi.ac.ma, [b]a.elgadari@edu.umi.ac.ma, [c]y.ounejjar@umi.ac.ma

Keywords: PUC, Multilevel Inverters, THD, TSV, TBV, MPC

Abstract. In this paper, we propose an advanced predictive control (MPC) for an H-SPUC inverter. The proposed topology achieves a 19-level output voltage using only eleven power switches, two DC sources, and two capacitors, resulting in a compact structure with reduced material costs and improved efficiency. Performance evaluation indicates a low cost function value (CF = 2.95) and reduced total standing voltage per unit (TSVpu) compared to conventional multi-level inverter configurations. Extensive simulations under various load profiles and DC voltage fluctuations demonstrate robust and stable operation, with total harmonic distortions of output voltage and current of 4.70% and 1.40%, respectively. The predictive control strategy provides fast transient response and accurate voltage tracking, highlighting the suitability of the proposed H-SPUC inverter for applications in photovoltaic energy systems, electric vehicles, and smart grid infrastructure.

Introduction

The global transition to renewable energy has increased the demand for efficient power conversion systems. While inverters are essential for DC–AC conversion, conventional two-level structures suffer from high harmonic distortion. Multilevel inverter (MLI) topologies address these drawbacks by providing improved power quality, reduced harmonics, and higher efficiency. Their stepped voltage generation results in lower THD and reduced stress on power devices [1].

Several MLI topologies have been extensively studied in the literature [2], including the Neutral Point Clamped (NPC), Cascaded H-Bridge (CHB), and Flying Capacitor (FC) converters. Although these structures achieve superior performance, they typically require numerous semiconductor switches, diodes, capacitors, and DC sources, leading to increased system cost and complexity. Furthermore, multilevel converter design remains constrained by several technical challenges, such as the dependence on multiple isolated DC supplies [3], the development of efficient modulation and control schemes [4], and the voltage balancing of floating capacitors [5]. Another challenge is the Total Standing Voltage (TSV) [6] which represents the cumulative voltage stress on all switches. Lowering TSV improves reliability, minimizes switching losses, and reduces the overall cost and size of the converter.

The Packed U-Cell (PUC) inverter [10] has emerged as a compact and efficient topology that addresses the previously mentioned challenges, such as high device count, harmonic distortion, and elevated TSV, while maintaining high output quality and performance. This topology has proven capable of generating multiple output voltage levels from a single DC source, either isolated [7], [8], [9] or supplied by photovoltaic (PV) arrays[10]. The generation of voltage levels depends on the control strategy, which can operate in open or closed loop. Open-loop uses fixed switching sequences, while closed-loop dynamically adjusts them using feedback to regulate voltage.

Our contribution presents a Model Predictive Control (MPC) strategy for the H-SPUC inverter, addressing key limitations of conventional multilevel topologies by achieving low Total Standing

Voltage per unit (TSVpu), low Total Harmonic Distortion (THD), and a reduced cost function (CF).

Presentation of the nineteen-level converter

H-SPUC inverter and switching states. The H-SPUC inverter [11] generates nineteen voltage levels using two isolated DC sources, eleven switches (including one bidirectional), and two capacitors. The circuit configuration is shown in Fig. 1.

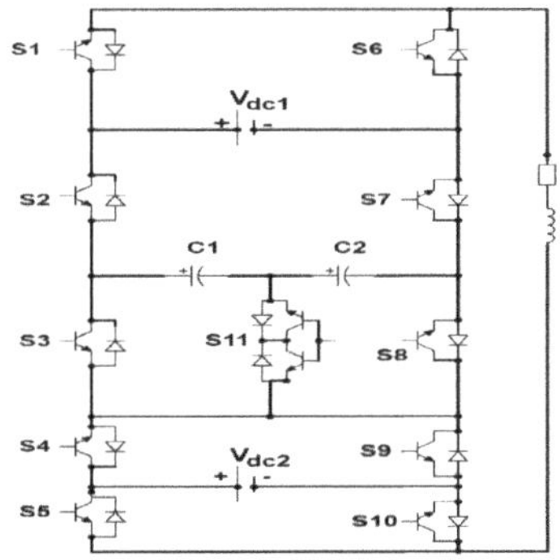

Fig. 1 Configuration of the H-SPUC inverter.

Table 1 presents the output voltage levels produced by the nineteen-level inverter. The converter operates with two isolated DC inputs defined as $V_{dc1} = 8\ V_{dc}$ and $V_{dc2} = V_{dc} = 50$ V. Under these conditions, the capacitors maintain an automatic voltage self-balancing behavior in open-loop mode, each charged to approximately one-quarter of V_{dc1}, such as $V_{C1} = V_{C2} = V_{dc1}/4 = 2V_{dc}$.

During inverter operation, the output voltage varies according to the specific combination of conducting switches. The zero-voltage level corresponds to state 10, where switches S_1–S_5 are in the ON position. In state 9, the simultaneous conduction of S_1–S_4 and S_{10} produces an output of V_{dc}. When the system enters state 8, switches S_1, S_2, S_9, S_{10}, and S_{11} are activated, resulting in $2V_{dc}$ while capacitor C_1 discharges. In state 7, the output reaches $3V_{dc}$ through the conduction of S_1, S_2, S_4, S_{10}, and S_{11}, again with C_1 in a discharging phase.

For state 6, switches S_1, S_2, S_8, S_9, and S_{10} generate an output of $4V_{dc}$, with both C_1 and C_2 discharging. In state 5, activation of S_1, S_5, S_7, S_9, and S_{11} yields $5V_{dc}$, and C_2 in a charging phase. State 4 produces $6V_{dc}$ when S_1, S_7, S_9, S_{10}, and S_{11} are ON, with C_2 continuing to charge. In state 3, switches S_1, S_5, S_7, S_8, and S_9 conduct, giving an output of $7V_{dc}$. The activation of S_1, S_7, S_8, S_9, and S_{10} in state 2 produces $8V_{dc}$, while state 1, involving S_1, S_4, S_7, S_8, and S_{10}, provides the maximum level $9V_{dc}$.The corresponding negative voltage levels are obtained by applying symmetric switching sequences, as detailed in Table 1.

Materials Research Proceedings 64 (2026) 258-267 https://doi.org/10.21741/9781644904091-32

Table 1: The switching table for the H-SPUC inverter.

State	S_1	S_2	S_3	S_4	S_5	S_6	S_7	S_8	S_9	S_{10}	S_{11}	C_1	C_2	V_L	Voltage (V)
1	1	0	0	1	0	0	1	1	0	1	0	NE	NE	$9V_{dc}$	450
2	1	0	0	0	0	0	1	1	1	1	0	NE	NE	$8V_{dc}$	400
3	1	0	0	0	1	0	1	1	1	0	0	NE	NE	$7V_{dc}$	350
4	1	0	0	0	0	0	1	0	1	1	1	NE	CH	$6V_{dc}$	300
5	1	0	0	0	1	0	1	0	1	0	1	NE	CH	$5V_{dc}$	250
6	1	1	0	0	0	0	0	1	1	1	0	DIS	DIS	$4V_{dc}$	200
7	1	1	0	1	0	0	0	0	0	1	1	DIS	NE	$3V_{dc}$	150
8	1	1	0	0	0	0	0	0	1	1	1	DIS	NE	$2V_{dc}$	100
9	1	1	1	1	0	0	0	0	0	1	0	NE	NE	V_{dc}	50
10	1	1	1	1	1	0	0	0	0	0	0	NE	NE	0	0
11	0	0	0	0	1	1	1	1	1	0	0	NE	NE	$-V_{dc}$	-50
12	0	0	0	0	0	1	1	0	1	1	1	NE	DIS	$-2V_{dc}$	-100
13	0	0	0	0	1	1	1	0	1	0	1	NE	DIS	$-3V_{dc}$	-150
14	0	0	1	1	1	1	1	0	0	0	0	DIS	DIS	$-4V_{dc}$	-200
15	0	1	0	1	0	1	0	0	0	1	1	CH	NE	$-5V_{dc}$	-250
16	0	1	0	0	0	1	0	0	1	1	1	CH	NE	$-6V_{dc}$	-300
17	0	1	1	1	0	1	0	0	0	1	0	NE	NE	$-7V_{dc}$	-350
18	0	1	1	1	1	1	0	0	0	0	0	NE	NE	$-8V_{dc}$	-400
19	0	1	1	0	1	1	0	0	1	0	0	NE	NE	$-9V_{dc}$	-450

DIS: discharging, NE: no effect, CH: charging

Voltage stress. In the H-SPUC inverter, the power stage includes eleven switches, one of which is bidirectional. The voltage-blocking capability of each device depends on the operating mode, and the corresponding voltage stresses are distributed as follows:

$$V_{S1}=V_{S6}=V_{dc1}\ . \tag{1}$$

$$V_{S2}=V_{S3}=V_{S7}=V_{S8}=V_{C1}+V_{C2}\ . \tag{2}$$

$$V_{S11}=V_{C1}=V_{C2}\ . \tag{3}$$

$$V_{S4}=V_{S5}=V_{S9}=V_{S10}=V_{dc2}\ . \tag{4}$$

The overall Total Blocking Voltage (TBV) sustained by the switching devices can be formulated as follows:

$$TBV=2V_{dc1}+4(V_{C1}+V_{C2})+V_{C1}+4V_{dc2}\ . \tag{5}$$

Based on the voltage ratio parameters adopted in this work, $V_{dc1}=8V_{dc}$, $V_{C1}=V_{C2}=2V_{dc}$, and $V_{dc2}=V_{dc}$, the overall blocking voltage across all semiconductor devices can be determined using the following expression:

$$TBV=38V_{dc}\ . \tag{6}$$

The total standing voltage per unit (TSVpu) of the proposed H-SPUC inverter is expressed as:

$$TSVpu = \frac{TBV}{V_{max}} = \frac{38V_{dc}}{9V_{dc}} = 4.22\,. \tag{7}$$

Comparative study

A comprehensive performance comparison is conducted between the proposed H-SPUC inverter and several recently reported multilevel topologies. The assessment considers key design indicators, including the number of DC sources (N_{dc}), capacitors (N_c), switching devices (N_{sw}), diodes (N_d), the gate-driver (N_{GD}), and the total standing voltage per unit (TSVpu) imposed on all semiconductor elements. A cost function (Fc) is defined to evaluate the component-based complexity of each configuration, incorporating a weighting factor (α) to account for the influence of TSVpu. In accordance with commonly adopted methodologies, the cost function [12] (CF) is defined as:

$$F_c = \frac{(N_{sw}+N_d+N_{Gd}+N_c+\alpha TSVpu)N_{dc}}{N_L}\,. \tag{8}$$

A comparative study presented in Table 2 evaluates the performance of the H-SPUC inverter against recent multilevel inverter topologies reported between 2022 and 2025. For a fair comparison, the cost function Fc defined in Eq. 8 is used to assess the trade-off between circuit complexity and voltage stress, considering two weighting factors ($\alpha = 0.5$ and $\alpha = 1.5$) to reflect the impact of total standing voltage on component cost.

The results in Table 2 clearly show that the proposed H-SPUC topology achieves an excellent balance between output voltage quality and hardware simplicity. Despite providing nineteen voltage levels, it requires only two DC sources, two capacitors, and eleven switches, while maintaining a low TSVpu value of 4.22. Consequently, the calculated cost function values, Fc = 2.95 for $\alpha = 0.5$ and Fc = 3.40 for $\alpha = 1.5$, demonstrate the efficiency and economic advantage of the H-SPUC inverter when compared to existing topologies.

Table 2 Comparison of the proposed H-SPUC inverter with other multilevel topologies.

Topology	Year	N_L	N_{dc}	N_c	N_{sw}	N_{Gd}	N_d	TSVpu	CF(α=0.5)	CF(α=1.5)
[13]	2022	7	1	1	8	8	0	8	3	4.14
[14]	2023	7	2	4	8	8	2	5.66	7.09	8.71
[15]	2024	7	1	2	7	7	2	5	2.93	3.64
[12]	2025	7	1	2	8	8	2	5.67	3.26	4.07
[16]	2022	19	2	2	12	10	1	6.66	2.98	3.68
[17]	2023	19	2	2	9	9	6	7.2	3.12	3.87
[18]	2025	19	2	2	10	10	4	6.55	3.08	3.77
H-SPUC	---	19	2	3	12	11	0	4.22	2.95	3.4

Control Strategy

The proposed control scheme is based on a single-step Model Predictive Control (MPC) framework applied to the multilevel inverter. At each sampling instant (k), the algorithm evaluates the reference load voltage $V_L{}^*(k)$ together with the measured load voltage $V_L(k)$ and load current $i_L(k)$. Using these signals, the controller predicts the system response for the next sampling period and computes the cost function associated with each feasible switching state vector Si. Based on the cost function, the switching state that provides the lowest cost is chosen to control the inverter for achieving accurate voltage tracking and current regulation. A finite number of discrete switching states are used to represent the inverter, where each state is linked with a particular output voltage. At each control instant, the predictive controller calculates all possible states and chooses the best switching state directly without using observers or estimators. Fig. 2 describes the significant steps of the prediction control algorithm, namely measurement, prediction, and decision, for choosing the best switching command.

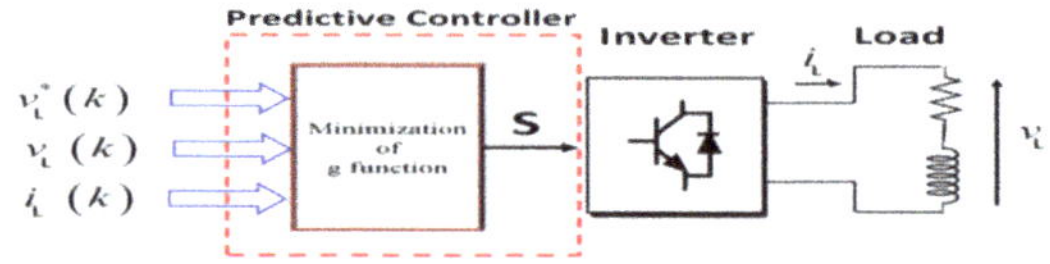

Fig. 2 Schematic representation of the proposed Model Predictive Control (MPC) algorithm.

At each sampling instant (k), the Model Predictive Control (MPC) algorithm evaluates a cost function measuring the deviation between the reference voltage $V_{ref}(k)$ and the predicted output voltage $V_i(k)$ for each possible switching state:

$$g(k) = \left(V_{ref}(K) - V_i(k)\right)^2 . \tag{9}$$

The optimal switching state $S_{opt}(k)$ is obtained by evaluating this cost over all admissible states:

$$S_{opt}(K) = \arg min(\left(V_{ref}(K) - V_i(k)\right)^2 . \tag{10}$$

Simulation results

In order to evaluate the performance of the nineteen-level inverter and verify the effectiveness of the proposed strategy for balancing capacitor voltages, extensive simulations were performed in MATLAB/Simulink over a wide range of operating conditions.

Simulation under fixed parameters. Table 3 shows the parameters used in the simulation.

Table 3 Simulation parameters.

Parameters	V_{dc1}	V_{dc2}	R	L	C_1	C_2	Switching frequency
Values	400(V)	50(V)	20(Ω)	15(mH)	1200(μF)	1200(μF)	20(KHz)

Fig. 3 shows the load voltage (V_L) and capacitor voltages (V_{C1}, V_{C2}), demonstrating the inverter's ability to generate a 19-level output. Both capacitors remain balanced at approximately one-quarter of the DC source voltage (V_{C1}=V_{C2}=100V). As seen in Fig. 4, the load current is nearly sinusoidal. Fig. 5 indicates Total Harmonic Distortion (THD) values of 4.70% for voltage (Fig. 5 (a)) and 1.40% for current (Fig. 5 (b)), confirming high-quality output with minimal harmonic content.

Materials Research Proceedings 64 (2026) 258-267 https://doi.org/10.21741/9781644904091-32

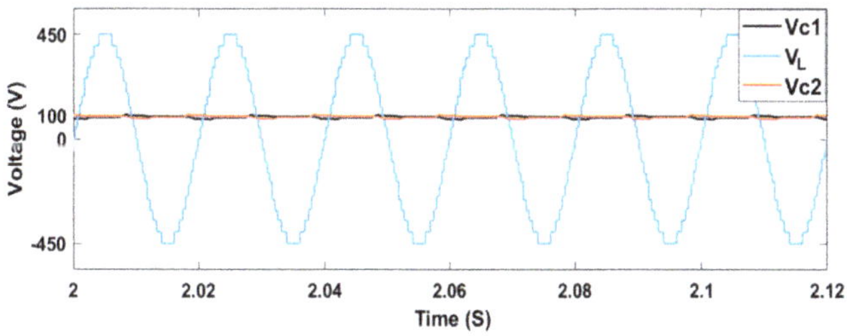

Fig. 3. The waveforms of the capacitor voltages V_{C1}, V_{C2} and load voltage V_L.

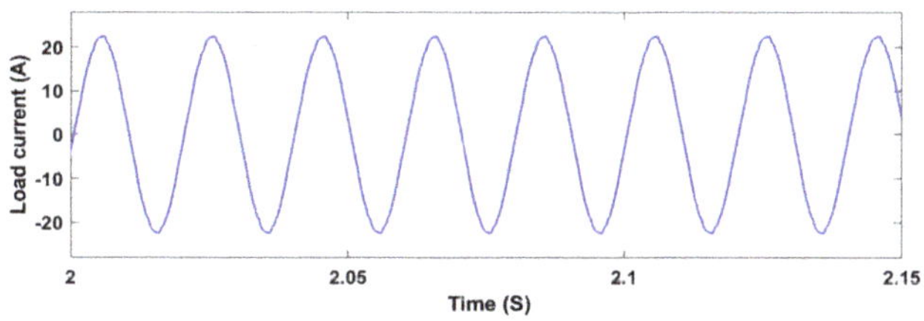

Fig. 4. The waveform of the load Current.

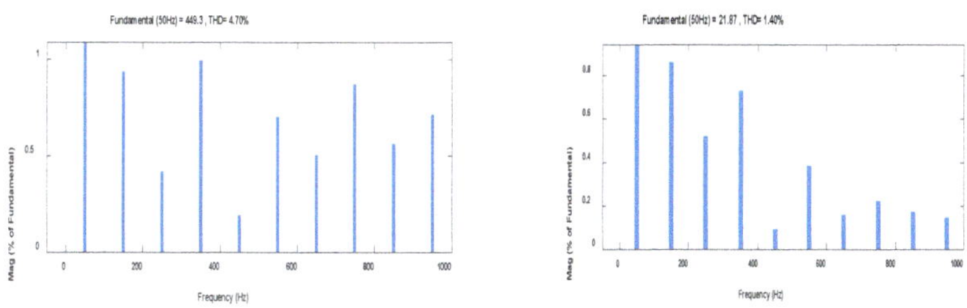

(a) (b)

Fig. 5. Fast Fourier transform analysis of load voltage (a) and load current(b).

The H-SPUC inverter was tested under sudden DC voltage variations. At t = 5 s, V_{dc1} decreased from 400 V to 200 V and V_{dc2} from 50 V to 25 V. As shown in Fig. 6, the capacitor voltages (V_{C1} and V_{C2}) quickly stabilized near half of V_{dc1}, while the load voltage (V_L) reached its new steady-state value after a short transient. The load current (Fig. 7) remained nearly sinusoidal, confirming the inverter's robustness and effective voltage regulation under DC fluctuations.

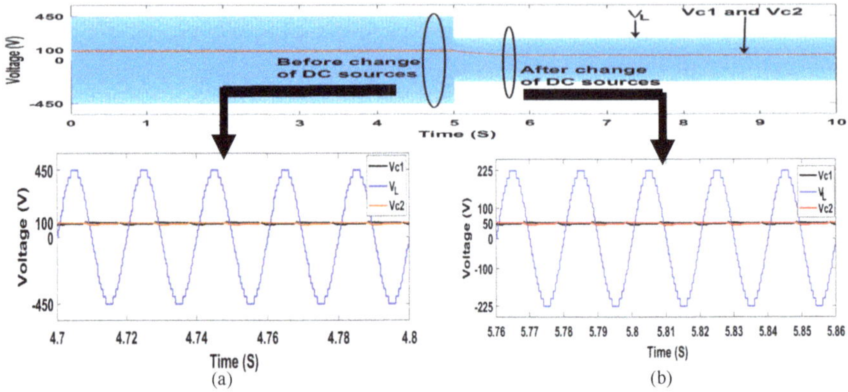

Fig.6 The waveforms of the load voltage V_L and capacitor voltages (V_{C1} and V_{C2}) before the change of DC sources (a) and after the change of DC sources (b).

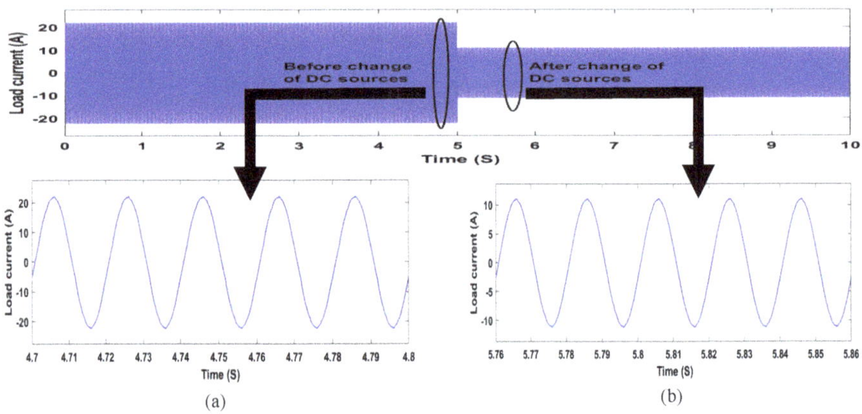

Fig.7 The waveform of the load current before the change of DC sources (a) and after the change of DC sources (b).

The H-SPUC inverter was tested under varying load conditions. At t = 5 s, the load resistance increased from 20 Ω to 60 Ω, and the inductance doubled from 15 mH to 30 mH. As shown in Fig. 8, the load voltage (V_L) and capacitor voltages (V_{C1}, V_{C2}) remained well regulated and followed their reference values during the transient. The inverter maintained a stable nineteen-level output, while the load current (Fig. 9) stayed nearly sinusoidal before and after the load change, confirming robust dynamic performance.

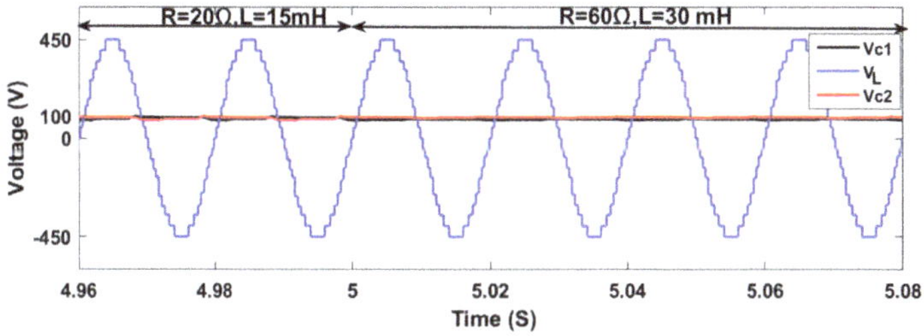

Fig. 8 The waveforms of the capacitor voltages V_{C1}, V_{C2}, and load voltage V_L during load change.

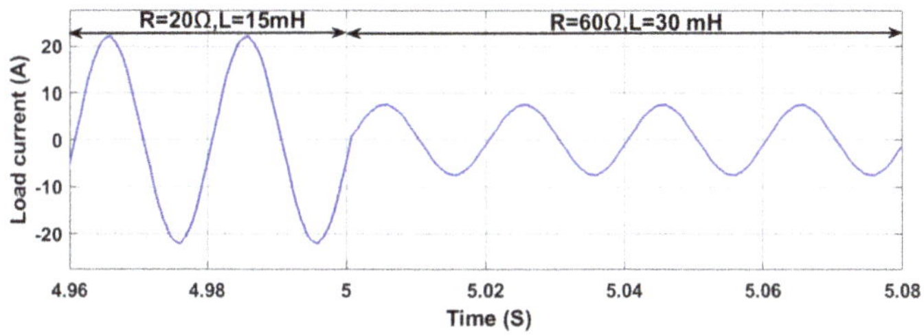

Fig. 9 The waveform of the load current during load change.

Discussion

The H-SPUC inverter was previously controlled using a conventional PWM strategy in [11]. In this paper, an innovative model predictive control (MPC) approach is applied, aiming to improve its overall performance. Comparative simulations under identical conditions show a THD reduction from 6.84% to 4.70% in voltage and from 1.43% to 1.40% in current. Owing to its closed-loop structure, MPC provides faster dynamics and greater robustness to variations.

Conclusion

This paper introduces the MPC technique for H-SPUC inverter to provide stable performance for balanced capacitor voltages with low voltage and current THD levels, meeting standards, and support quasi-sinusoidal waveform output without an external filter. It possesses great potential for applications in photovoltaic power systems, electric vehicle drives, and developing smarter power grids. Future research should focus on experimental validation.

AI tools were used only to correct linguistic errors.

References

[1] P. Omer, J. Kumar, and B. S. Surjan, "A Review on Reduced Switch Count Multilevel Inverter Topologies," *IEEE Access*, vol. 8, pp. 22281–22302, 2020. https://doi.org/10.1109/ACCESS.2020.2969551

[2] S. Choudhury, M. Bajaj, T. Dash, S. Kamel, and F. Jurado, "Multilevel inverter: A survey on classical and advanced topologies, control schemes, applications to power system and future prospects," *Energies (Basel).*, vol. 14, no. 18, Sep. 2021. https://doi.org/10.3390/en14185773

[3] Y. Hinago and H. Koizumi, "A single-phase multilevel inverter using switched series/parallel DC voltage sources," *IEEE transactions on industrial electronics*, vol. 57, no. 8, pp. 2643–2650, 2009.

[4] A. Poorfakhraei, M. Narimani, and A. Emadi, "A review of modulation and control techniques for multilevel inverters in traction applications," *Ieee Access*, vol. 9, pp. 24187–24204, 2021.

[5] S. Song, J. Liu, S. Du, X. Chen, H. Liu, and W. Lin, "Effect of Averaging Voltage Regulation Control on Steady-State Analysis of Modular Multilevel Converters," *IEEE Trans. Power Electron.*, vol. 35, no. 9, pp. 9215–9225, 2020. https://doi.org/10.1109/TPEL.2020.2974585

[6] P. Omer, J. Kumar, and B. S. Surjan, "A Review on Reduced Switch Count Multilevel Inverter Topologies," *IEEE Access*, vol. 8, pp. 22281–22302, 2020. https://doi.org/10.1109/ACCESS.2020.2969551

[7] A. El Gadari, H. Hamidane, T. El Ansari, and Y. Ounejjar, "PWM control of the new five-level Packed-T-cell inverter," in *2025 5th International Conference on Innovative Research in Applied Science, Engineering and Technology (IRASET)*, IEEE, 2025, pp. 1–6.

[8] A. El Gadari, H. Hamidane, T. El Ansari, and Y. Ounejjar, "PWM sensor-less balancing technique for the three-phase seven level PUC-NPC converter," *Journal of Electrical Systems and Information Technology*, vol. 12, no. 1, p. 82, 2025. https://doi.org/10.1186/s43067-025-00274-1

[9] T. El Ansari, A. El Gadari, and Y. Ounejjar, "Predictive control strategy for a novel 15-level inverter with reduced power components," *Indonesian Journal of Electrical Engineering and Computer Science*, vol. 41, no. 1, p. 33, Jan. 2026. https://doi.org/10.11591/ijeecs.v41.i1.pp33-44

[10] A. El Gadari *et al.*, "Photovoltaic system based on the PUC9 inverter," in *2019 IEEE 28th International Symposium on Industrial Electronics (ISIE)*, IEEE, 2019, pp. 2059–2064.

[11] T. El Ansari, A. El Gadari, and Y. Ounejjar, "A New Nineteen-Level H-SPUC Inverter with Reduced Power Electronics Components," in *2025 International Conference on Circuit, Systems and Communication (ICCSC)*, 2025, pp. 1–6. https://doi.org/10.1109/ICCSC66714.2025.11135032

[12] M. Memiş and M. Karakiliç, "Seven-Level Soft Charging Switched-Capacitor Multilevel Inverter," *IEEE Access*, vol. 13, pp. 77239–77251, 2025. https://doi.org/10.1109/ACCESS.2025.3560576

[13] L. Ren, L. Zhang, L. Wang, and S. Dai, "Capacitor Voltage Regulation Strategy for 7-Level Single DC Source Hybrid Cascaded Inverter," *IEEE J. Emerg. Sel. Top. Power Electron.*, vol. 10, no. 5, pp. 5773–5784, 2022. https://doi.org/10.1109/JESTPE.2022.3172146

[14] P. Manoj, A. Kirubakaran, and V. T. Somasekhar, "An Asymmetrical Dual Quasi-Z-Source Based 7-Level Inverter for PV Applications," *IEEE Transactions on Energy Conversion*, vol. 38, no. 2, pp. 1097–1107, 2023. https://doi.org/10.1109/TEC.2022.3222498

[15] W. Lin, J. Zeng, B. Fu, Z. Yan, and J. Liu, "Switched-capacitor Based Seven-level Boost Inverter with a Reduced Number of Devices," *CSEE Journal of Power and Energy Systems*, vol. 10, no. 1, pp. 381–391, 2024. https://doi.org/10.17775/CSEEJPES.2020.02620

[16] M. Fahad *et al.*, "A Dual Source Switched-Capacitor Multilevel Inverter with Reduced Device Count," *Electronics (Basel).*, vol. 11, no. 1, 2022. https://doi.org/10.3390/electronics11010067

[17] F. Sagvand, J. Siahbalaee, and A. Koochaki, "An Asymmetrical 19-Level Inverter with a Reduced Number of Switches and Capacitors," *Electronics (Basel).*, vol. 12, no. 2, 2023. https://doi.org/10.3390/electronics12020338

[18] F. Sagvand, J. Siahbalaee, and A. Koochaki, "A Novel 19-Level Boost Type Switched-capacitor Inverter with Two DC Sources and Reduced Semiconductor Devices," *International Journal of Engineering, Transactions A: Basics*, vol. 36, no. 2, pp. 253–263, Feb. 2023. https://doi.org/10.5829/ije.2023.36.02b.06

Advancing Smart Grid Security and Stability Through Machine Learning Techniques

Nayeemuddin MOHAMMED[1,a], Hiren MEWADA[2,b], Tasneem SULTANA[3,c], Syam Sundar LINGALA[4,d]

[1]Civil Engineering Department, Prince Mohammad Bin Fahd University, Al Khobar, Saudi Arabia

[2]Electrical Engineering Department, Prince Mohammad Bin Fahd University, Al Khobar, Saudi Arabia

[3]Artificial Intelligence and Machine Learning, Godutai Engineering College for Women, Sharnbasva University, India

[4]Mechanical Engineering Department, Prince Mohammad Bin Fahd University, Al Khobar, Saudi Arabia

[a]mnayeemuddin@pmu.edu.sa, [b]hmewada@pmu.edu.sa, [c]tasneemsultana841@gmail.com, [d]sslingala@gmail.com

Keywords: Grids, Stability, Prediction, Reliability, Security, Algorithm, Training, Testing

Abstract. Smart grids with increased operational efficiency and bidirectional energy flows.. Resilience has been developed as a result of a complete shift in the responsibilities of energy market participants. The transition from a traditional energy ecosystem to a more dynamic system, in which end users can both consume and produce energy, has led to more dynamic interactions. The redundant stability ensures that power stays in the range, maintaining safe infrastructure. To determine the stability, three different machine learning algorithms were employed and compared. The results reveal that the Random Forest model achieved an R^2 of 0.99647 on training and 0.97143 on testing. The Gradient Boosting model achieves a score of 0.9848 for training and 0.98 for testing. In addition, the Bayesian Ridge model performed well, with 0.9963 on training and 0.971 on testing. These results also indicate that Random Forest performed consistently across all. GB and Bayesian Ridge perform well with slightly higher error rates than RF. In addition, new and advance methods of energy management and distribution are required, utilizing more sophisticated technologies, such as advance machine learning, deep learning to enhance grid stability and predictability to incorporate renewable energy.

Introduction

Smart grid technology, which integrates multiple energy sources, optimizes electricity distribution, and boosts efficiency through automated controls [1]. Digital communications is a development in the transition to renewable energy sources and the need for a stable energy supply [2]. Nonetheless, security and stability are extremely important to both energy suppliers and consumers due to the high risk of such systems that are quite complex. In order to get around the short comings of the stability analysis. In order to deal with this problem, decentralized smart grids management systems (DGS) are based on optimized deep learning models [3]. With the connection of the frequency to the electricity price, the predictions of DL can be 99.62 percent accurate. The results reveal that a quick adaptation enhances stability of the system in general [4]. Alazab et. al. [5] introduced a model, Multidirectional LSTM (MLSTM), to predict smart grid stability in a CPS model. MLSTM performed better than GRU, traditional LSTM and RNN in predicting dynamical grid behavior. They show the competence in controlling the multifaceted and varying power demands. Salam and Hibaoui [6] used a deep learning on top of Dense Convolutional Networks and Residual structures, which used to predict smart grid stability. The approach achieved the highest performance compared to the classical ML models (SVM, Logistic Regression, Decision Tree, RF, GBT, etc.) and the DL models (MLP, GRU, RNN, LSTM).

This demonstrates its strength in handling complex participation-driven stability challenges. Cutting-edge machine learning techniques, including SVM, KNN, logistic regression, naive Bayes, neural networks, and decision trees, were evaluated on a UCI Smart Grid dataset. The Decision Tree classifier outperformed the others, achieving a 99.9 accuracy, 100.0 F1 score, 99.9 recall, and a 100.0 accuracy. The proposed model, its complexity may limit real-time application in dynamic environments also Smart Grid data sets highlight the research gap in performance assessment [5].

In Decentralized Smart Grid Control (DSGC) systems, Mohsen et. al., proposed an ANN to predict the stability of smart grids based on the simulation data from a four-node star topology. A simplified 4 node star topology may reduce its applicability to more complex. The tuned ANN had high metrics precision 98.03%, recall 98.02%, F1-score 98.02%, and AUC 100% with an accuracy of 97.36%, confirming the efficiency of the ANN as a fast and accurate predictor of stability [6]. The ML cascade framework, including pre-processing, selection of the characteristics and classification, was used to predict the stability of the SG. Maximum settings using tagged trees achieve 99.9% accuracy, allowing for rapid and reliable stability detection [7]. Bashir et. al., [8] examined various ML algorithms for predicting the stability of SG and found that decision tree was the best, with a 99.9% accuracy and almost perfect recall and AUC. This demonstrates its effectiveness in managing the changing conditions of the Smart Grid. Logistic regression (LR) predicted Smart Grid stability with 98.9% accuracy, comparable to linear SVM and SVM-RBF. This underlines the stability of LR as a suitable base model for applications for smart grids [9]. Logistic regression and other predictive models are being explored for the analysis of smart grid stability, providing information on voltage stability and problem solving. These models serve as a benchmark against complex neural nets for controlling the grid in real time [10]. Smart grids increase efficiency and local autonomy by using digital technologies and decentralized smart grids. They proposes a grid stability prediction model LSTM with 96.73 accuracy and 97.43 F1 score indicating its effectiveness [11]. However, the more interconnected and data-dependent smart grids are, the more vulnerable they are to the risk of failure and therefore, a high level of stability is needed to protect these critical infrastructures. In this context, machine learning approaches have the potential to transform the monitoring and predictive maintenance of smart grids in real time. On the other hand, the evaluation of the huge data sets generated by smart grids could help to make them more resilient to disruption [12]. This approach aims to explore new machine learning applications to improve the security and stability of smart grids.

Objectives of the study

The main objective of this research is to assess and compare the effectiveness of different machine learning models in predicting Smart Grids stability. The goal of this research is to find the most efficient algorithms for predicting the performance of the network with some accuracy and for identifying anomalies using Random Forest, Gradient Boosting, and Bayesian Rotation models. The study will also provide practical information and recommendations for energy suppliers on how to implement these models to improve the resilience and safety of their operations in innovative energy networks. Finally, it aims to help develop more efficient and reliable energy-using technologies that can adapt to the changing challenges of today's energy infrastructure.

Novelty of Research

This investigation introduces a new and advance technique to strengthen the stability of the smart grid using advanced machine learning models. While previous studies have examined a number of algorithms, the proposed study focuses on a comparative analysis of random forest, gradient boosting and Bayesian ridge models for use in smart grid implementation. Performance measures used for the assessment of these models include metric analysis such as R^2, RMSE, MAE and MSE, which allow an easy understanding the value of models in the real world. In addition, this research can contribute to the field by offering an overall method for training and validating models, such as data collection through multiple sources and modelling in a simulated environment. By addressing current issues and gaps in the literature aims to lay the foundations for more resilient and adaptable energy systems in response to changing technological and environmental demands.

Methodology

Dataset: Under this strategy, the data were sourced in Kaggle [15]. In the simulation of a 4-node grid stability network in the form of a star, a dataset was obtained with the help of a simulation. There were 12 inputs and the output variable were stability to design a machine learning model. In this, tau 1 to tau 4 are the reaction times, p1 to p4 are the nominal power induced, g1 to g2 are the coefficients of price elasticity and stab is the output variable as a response. The dataset is represented by 60000 readings. This paper follows a complex approach to assess the applicability of machine learning methods to enhance the stability and security of smart grids. To begin with, the literature review is performed to analyze current studies on smart grid technologies and the different machine learning algorithms, such as supervised and unsupervised algorithms and their predictive capabilities. Random Forest model is a supervised model. It makes use of several decision trees with which it first predicts the value. It then averages all the values that are expected to obtain the final value. Its popularity is due to the fact that it is resistant to overfitting and large datasets can be analyzed. Gradient Boosting is another model. Publicly relies on the weak learner response to better the model response by reducing the mean squared error. Bayesian ridge is a robust regression model, which is primarily applied in high multicollinearity data. It employs Bayesian inference to carry out linear regression and offers a more stable prediction as compared to the conventional ways of linear regression.

he corresponding statistics are then acquired using publicly available materials, and simulations, which involve the factors, such as voltage levels, load requirement, etc. There are machine learning models that are meant to address particular problems with smart grids management. These models are trained on a fraction of the data, and optimized in terms of hyperparameters and verified to get lower error performance measures. Models that have the highest performance are also evaluated during a smart grid simulation to determine their capacity to detect anomalies and predict failures in real-time. Finally, the performance study examines various machine learning algorithms and identifies the most effective ones for enhancing safety and operational resilience. Case studies demonstrate practical applications in the real world and highlight the revolutionary potential of machine learning to overcome the complexity of existing Smart Grids.

Results and Discussion

Figure 1 illustrates the heatmap of correlation between the input characteristics and the effect on the stability of the smart grid. All variables are significantly directly related to stab with tau and g parameters having slightly larger positive correlations of 0.19 to 0.20 compared to the p parameters that remain near to zero. Negative correlations are found to be very strong as indicated in the p groups p1to p4; they indicate the interdependence of these attributes. Generally, the low linear associations with stability suggest that simple linear methods might not be effective and more advanced nonlinear methods, including ensemble learning or deep neural networks, are more appropriate to include intricate interaction between features to predict stability reliably.

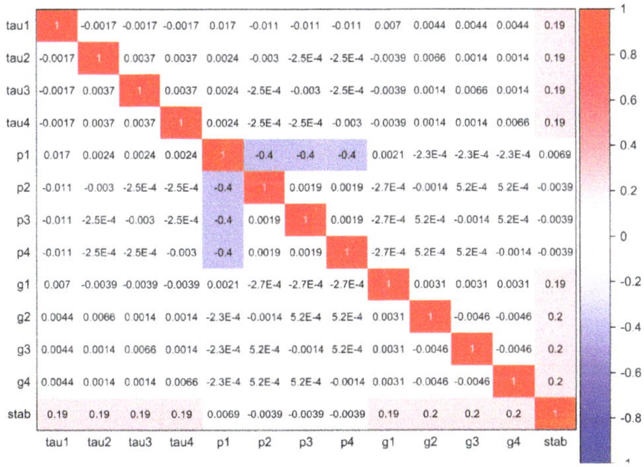

Fig. 1. Correlation matrix for the input and output variables.

Model analysis Parameters

To evaluate the model's predictive performance, errors based on various performance parameters are computed. Equations (1) to (4) are used to calculate the coefficient of correlation, mean square error, root mean square error, mean absolute error [13].

$$R^2 = 1 - \frac{\sum_{i=1}^{n}(y_i - \hat{y}_i)^2}{\sum_{i=1}^{n}(y_i - \bar{y})^2} \tag{1}$$

$$MSE = \frac{1}{n}\sum_{i=1}^{n}(y_i - \hat{y}_i)^2 \tag{2}$$

$$RMSE = \sqrt{\frac{1}{n}\sum_{i=1}^{n}(y_i - \hat{y}_i)^2} \tag{3}$$

$$MAE = \frac{1}{n}\sum_{i=1}^{n} y_i - \hat{y}_i \tag{4}$$

Where, y_i=actual values, $\hat{y}_i$=predicted values, $\bar{y}$ = mean of actual values, n = number of data points.

Figure 2 demonstrates the data of the training sample and a histogram of the error values. The relative error distribution is approximately symmetric, with a mean of -0.0002 and an error rate of -0.00014. This indicates that there is no significant bias in the forecasting process as such. The standard deviation of 0.0089 is also relatively low, suggesting that most errors are small and clustered within a narrow range. Overall, both figures show that the model is very predictive, with random, small errors. Table 1 illustrates the different methods applied in machine learning models. A total of 4 metrics was employed in the design model.

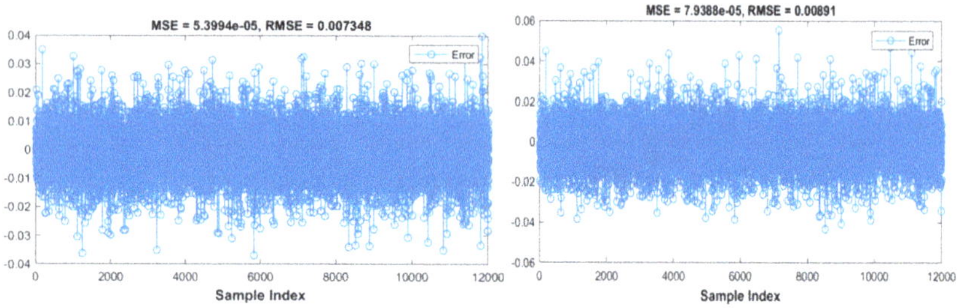

Fig. 2. Test data prediction using GB and RF

Table 1: Comparison of various machine learning metric analysis

Model	Dataset	R^2	RMSE	MSE	MAE
Random Forest ML	Train	0.99647	0.0033134	1.0979e-05	0.0025
	Test	0.97143	0.00891	7.9388e-05	0.0067
Gradient Boosting ML	Train	0.98489	0.0064119	4.1113e-05	0.0048
	Test	0.98001	0.007348	5.3994e-05	0.0055
Bayesian Ridge ML	Train	0.9963	0.0033775	1.1407e-05	0.0025
	Test	0.97127	0.008931	7.9763e-05	0.0067

To evaluate the model further, an R^2 score (equation -1) is calculated for both training and testing datasets. The actual and projected values of the random forest, gradient boosting and Bayesian ridge models during the training and testing process are shown in Figure 3 to 5. The points nearly meet the ideal diagonal line y = x, which means that the predictions are very congruent with the reality values. The statistical indicators show that the model has been performing very well as the coefficient of determination $R^2 = 0.99647$ to train and 0.971 to test equals high level of good explanation of the random forest. The gradient boosting R^2 =0.98, whilst the training and test R^2 =0.98. Moreover, the training and testing R^2 of Bayesian ridge is 0.99 and 0.97 respectively. The RF and Baysian Ridge showed lower error values indicate minimal prediction deviations [14]. Overall, the Random Forest model performs reasonably well throughout training [15].

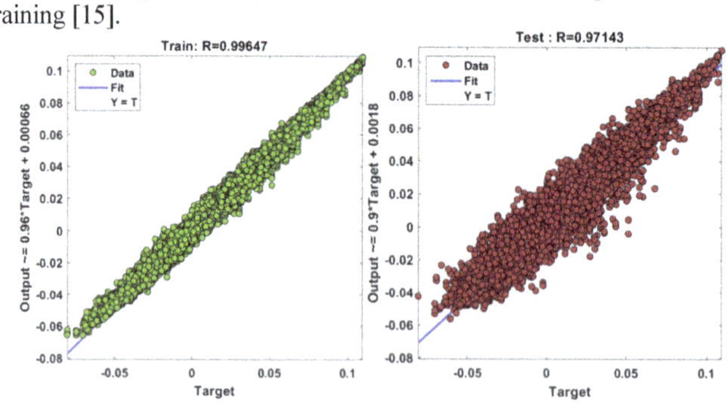

Fig. 3. Actual vs predicted for training and testing the random forest algorithm of machine learning.

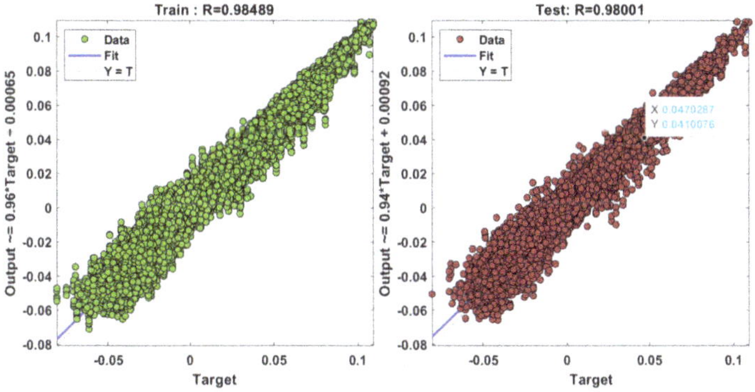

Fig. 4. Actual vs predicted values for training and testing the gradient boosting algorithm of machine learning.

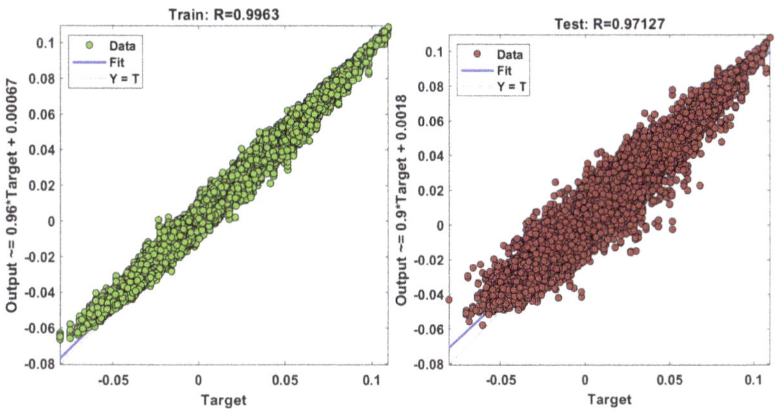

Fig. 5. Actual vs predicted for training and testing the Bayesian Ridge algorithm of machine learning.

Conclusion

The machine learning model in the case of improving the stability of Smart Grid showed the strength of the Random Forest model in the training and testing data. A training model with the R^2 value of 0.9916 showed good predictivity. The low mean residual error of the mean root of the squares (RMSE) and the average absolute error (MAE) are useful in increasing its stability in predicting the stability of the network. Observations of gradient models of smoothing and Bayesian models of smoothing come in handy. The three models should be validated by performing further analysis of the three models with other data sets. In general, the discoveries prove that Random Forest is the best approach to innovative grid applications and open access to more trustworthy energy-consumption systems.

Recommendation and future research

To make smart grids and predictive analytics more stable, the energy suppliers should pay attention to the application of the Random Forest model since it is more precise and it is not likely to be affected by errors. Besides, the discussion of model-ensemble methods may contribute to exploiting the benefits of many algorithms and enhance the robustness and forecasting opportunities. The other critical problem is to provide a framework whereby the model is periodically updated with new information so that predictions can be relevant and true in the long-term perspective. The periodic updates will also ensure that the smart grid management systems are updated to match the dynamics in the environment to be more successful. These techniques can also be invested in to increase the reliability and efficiency considerably.

AI Declaration

The authors declare that artificial intelligence tools were used only to improve language clarity and grammar. No AI system was used to generate scientific content, analyze data, interpret results, or draw conclusions. The authors take full responsibility for the content of this manuscript.

References

[1] Massoud Amin S (2011) Smart Grid: Overview, Issues and Opportunities. Advances and Challenges in Sensing, Modeling, Simulation, Optimization and Control. European Journal of Control 17:547–567. https://doi.org/10.3166/ejc.17.547-567

[2] khan S, Mazhar T, Shahzad T, et al (2025) Integration of smart grid with Industry 5.0: Applications, challenges and solutions. Measurement: Energy 5:100031. https://doi.org/10.1016/j.meaene.2024.100031

[3] Cavus M (2025) Advancing Power Systems with Renewable Energy and Intelligent Technologies: A Comprehensive Review on Grid Transformation and Integration. Electronics 14:1159. https://doi.org/10.3390/electronics14061159

[4] Breviglieri P, Erdem T, Eken S (2021) Predicting Smart Grid Stability with Optimized Deep Models. SN COMPUT SCI 2:73. https://doi.org/10.1007/s42979-021-00463-5

[5] Bashir AK, Khan S, Prabadevi B, et al (2021) Comparative analysis of machine learning algorithms for prediction of smart grid stability. International Transactions on Electrical Energy Systems 31:e12706. https://doi.org/10.1002/2050-7038.12706

[6] Mohsen S, Bajaj M, Kotb H, et al (2023) Efficient Artificial Neural Network for Smart Grid Stability Prediction. International Transactions on Electrical Energy Systems 2023:9974409. https://doi.org/10.1155/2023/9974409

[7] Önder M, Dogan MU, Polat K (2023) Classification of smart grid stability prediction using cascade machine learning methods and the internet of things in smart grid. Neural Comput & Applic 35:17851–17869. https://doi.org/10.1007/s00521-023-08605-x

[8] Bashir AK, Khan S, Prabadevi B, et al (2021) Comparative analysis of machine learning algorithms for prediction of smart grid stability†. Int Trans Electr Energ Syst 31:. https://doi.org/10.1002/2050-7038.12706

[9] Lahon P, Kandali AB, Barman U, et al (2024) Deep Neural Network-Based Smart Grid Stability Analysis: Enhancing Grid Resilience and Performance. Energies 17:2642. https://doi.org/10.3390/en17112642

[10] Marevac E, Kadusic E, Ruland C, Zivic N (2024) Application of Deep Learning and Neural Networks for Smart Grid Stability Predictions. In: 2024 IEEE 12th International Conference on Smart Energy Grid Engineering (SEGE). pp 84–89

[11] Khaleefah SH, Mostafa SA, Gunasekaran SS, et al (2024) Optimizing Smart Power Grid Stability Based on the Prediction of a Deep Learning Model. JOIV : International Journal on Informatics Visualization 8:1091–1098. https://doi.org/10.62527/joiv.8.3.2758

[12] Tang X, Wang J (2025) Deep Reinforcement Learning-Based Multi-Objective Optimization for Virtual Power Plants and Smart Grids: Maximizing Renewable Energy Integration and Grid Efficiency. Processes 13:1809. https://doi.org/10.3390/pr13061809

[13] Talukdar A, Panda B, Hota S, et al (2024) Intelligent Solar Power Forecast with Machine Learning

[14] Nayeemuddin M, andi asiz, Mohammad AK, et al (2024) A comprehensive review on computing methods for the prediction of energy cost in Kingdom of Saudi Arabia. In: Materials Research Proceedings. pp 156–163

[15] Nayeemuddin M, PALANIANDYA P, Shaik F, Mewada H (2023) Experimental and computational analysis for optimization of seawater biodegradability using photo catalysis. IIUM Engineering Journal 24:11–33

An Improved Active–Reactive Power Control Scheme for Single-Phase PV Inverters in Grid-Connected Systems

Nur Fairuz MOHAMED YUSOF[1,a*], Dahaman ISHAK[2,b], Samir EL-NAKLA[2,c] and Muhammad Hafeez Mohamed HARIRI[3,d]

[1]Faculty of Electrical Engineering & Technology, Universiti Malaysia Perlis, Arau 02600, Malaysia

[2]Department of Electrical Engineering, College of Engineering, Prince Mohammad Bin Fahd University, Al Khobar 31952, Kingdom of Saudi Arabia

[3]School of Electrical and Electronic Engineering, Universiti Sains Malaysia, Nibong Tebal 14300, Malaysia

[a]nurfairuz@unimap.edu.my, [b]dbinishak@pmu.edu.sa, [c]snakla@pmu.edu.sa, [d]muhammadhafeez@usm.my

Keywords: Grid-Connected Photovoltaic System, Single-Phase Inverter, Active and Reactive Power Control, Improved Control Strategy, Power Quality Enhancement

Abstract. This paper proposes an enhanced control strategy for single-phase inverter grid-connected photovoltaic (PV) systems that improves the synchronization accuracy, power quality, and control performance under fluctuating solar irradiance. The proposed system integrates a variable step size Incremental Conductance (VS-InCond) algorithm for Maximum Power Point Tracking (MPPT), a Double-Band Hysteresis Current Control (DBHCC) scheme for precise current regulation, and a modified PQ theory to achieve Phase-Locked Loop (PLL) free for grid synchronization. This coordinated control framework facilitates bidirectional active and reactive power flow while maintaining a near unity power factor. The proposed inverter performance complies with IEEE 1547-2018 and IEEE 519-2014 standards, keeping the Total Harmonics Distortion (THD) output below 5%. Simulation testing on a 9.6 kW PV system under various irradiance levels (1000 W/m^2, 600 W/m^2, and 350 W/m^2) at 25 °C demonstrates that the proposed method achieves high conversion efficiency 99% (Mode 2), superior harmonic suppression, and fast dynamic response during inductive load compared to the PR controller.

Introduction

The increasing global demand for sustainable energy and the depletion of fossil fuels have accelerated the adoption of PV systems for grid-connected applications [1]. Incentives such as feed-in tariffs and net metering promote decentralized energy generation, but they also introduce challenges in synchronization, power regulation, and power quality [2]. Several current control techniques have been implemented, including proportional-integral (PI) control, proportional-resonant (PR) control, predictive control, and hysteresis current control (HCC) [3]. While most conventional controllers are simple and easy to implement, but suffer when tracking sinusoidal reference currents. HCC can achieve zero steady-state error due to its band-based control mechanisms[4]. While the conventional PLL-based synchronization methods often fails under weak-grid or distorted conditions, whereas advanced approaches, such as DSOGI-FLL and EPLL, increase computational complexity [5][6]. Moreover, maintaining efficient bidirectional power flow remains challenging due to poor reactive power compensation [7], switching losses [3], and nonlinear load effects [8] that often exceed IEEE 519-2014 and IEEE 1547-2018 THD limits. In addition, fluctuating irradiance affects MPPT stability, as traditional Perturb and Observe (P&O) and hill climbing (HC) algorithms exhibit slow convergence [9]. This paper proposes an improved

Materials Research Proceedings 64 (2026) 276-283 https://doi.org/10.21741/9781644904091-34

single-phase, single-stage grid-connected PV inverter using DBHCC for accurate current regulation and lower switching losses, a modified PQ theory for PLL-free synchronization and reactive power control, and a VS-InCond MPPT for adaptive energy extraction. MATLAB/Simulink results on a 9.6 kW model demonstrate enhanced dynamic response, reduced THD (< 5%), and improved efficiency, validating the proposed PQ–DBHCC framework for smart grid and net metering systems.

System Modelling

The proposed single-phase grid-connected PV system integrates three main control modules: a modified VS-InCond algorithm for MPPT, a DBHCC scheme for inverter current regulation, and a PQ-theory-based power control for PLL-free grid synchronization. The overall control framework, as illustrated in Fig. 1(a), ensures efficient power extraction, accurate current tracking, and stable bidirectional power flow under the Net Energy Metering (NEM) scheme.

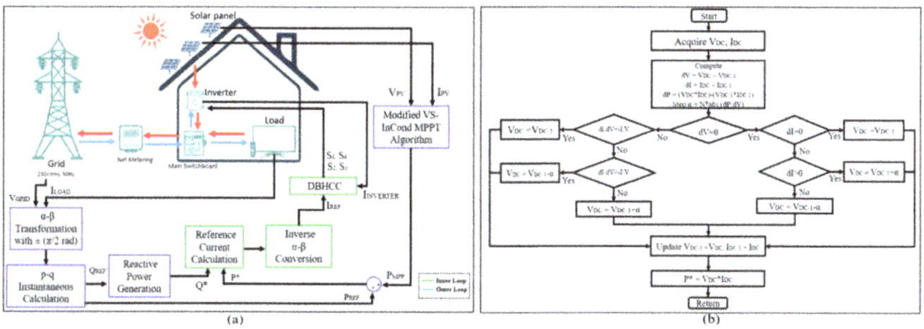

Fig. 1 (a) Diagram of the proposed system (b) Flowchart of the modified VS-InCond algorithm

The VS-InCond MPPT continuously adjusts the PV operating voltage according to the slope of the *P–V* curve, achieving faster and more stable convergence to the maximum power point under varying irradiance and temperature. The resulting reference power (P_{REF}) defines the active power command for the inverter, while reactive power (Q_{REF}) is set based on grid requirements to maintain a near-unity power factor as shown in Fig 1(b). A PQ theory based control strategy replaces the conventional PLL by generating α–β components of grid voltage and current through Clarke transformation in Eq.3 and Eq.4. The proposed improvement modifies the conventional PQ approach by directly deriving instantaneous quantities from the measured grid voltage (Eq.1) and load current (Eq.2), rather than relying on PLL based locking to grid parameters alone. Then the instantaneous active and reactive powers ($p_{\alpha\beta}, q_{\alpha\beta}$) are computed and compared with their reference values. The resulting current references are then fed to the DBHCC block, which serves as the inner current control loop. The DBHCC compares the actual inverter current with the reference and maintains the error within two hysteresis bands, which are inner and outer bands, to stabilize the switching frequency and minimize the current ripple. This configuration effectively reduces switching losses and improves harmonic performance. The overall system maintains voltage and current THD below 5% in compliance with IEEE 519-2014 and IEEE 1547-2018, ensuring reliable grid integration with high conversion efficiency and fast dynamic response.

$$V_\alpha = V_{grid}(\omega t)\ ;\ V_\beta = V_{grid}\left(\omega t + \frac{\pi}{2}\right) \quad (1)$$

$$I_\alpha = I_{LOAD}(\omega t)\ ;\ I_\beta = I_{LOAD}\left(\omega t + \frac{\pi}{2}\right) \quad (2)$$

$$\begin{bmatrix} V_\alpha \\ V_\beta \end{bmatrix} = \sqrt{\frac{2}{3}} \begin{bmatrix} 1 & -0.5 & -0.5 \\ 0 & 0.866 & -0.866 \end{bmatrix} \begin{bmatrix} V_a < \varphi \\ V_b < -120° + \varphi \\ V_c < 120° + \varphi \end{bmatrix} \quad (3)$$

$$\begin{bmatrix} I_\alpha \\ I_\beta \end{bmatrix} = \sqrt{\frac{2}{3}} \begin{bmatrix} 1 & -0.5 & -0.5 \\ 0 & 0.866 & -0.866 \end{bmatrix} \begin{bmatrix} I_a < \gamma \\ I_b < -120° + \gamma \\ I_c < 120° + \gamma \end{bmatrix} \quad (4)$$

At the nominal output of 9.6 kW under STC (1000 W/m² and 25 °C,) the PV array delivers 400 V and 24 A, corresponding to the full power capacity of the inverter. This output ensures reliable operation within the allowable single-phase current limit of 40 A under Malaysia's NEM regulations. This system configuration is capable of supporting typical residential and small commercial loads, such as lighting, refrigeration, and air-conditioning units, while maintaining a near-unity power factor and THD below the standard limit. The 9.6 kW configuration falls within the average daily energy requirements of medium-sized households, which enables effective self-consumption and grid energy offset through NEM's bidirectional energy exchange framework. To assess system performance under fluctuating solar irradiance, simulations were conducted at 600 W/m² and 350 W/m². Results of the simulation tests are summarized in Table 1. From the table, it can be observed that as the irradiance decreases, the PV array voltage remains almost constant, while current and power drop significantly to 14.45 A (5768 W) and 0.61 A (215 W) at lower irradiance. The results show that the PV array is current-limited and depends heavily on the amount of irradiance it receives, while its almost constant voltage across operating modes demonstrates the robustness of the design of the system. Therefore, the PV configuration selected would assure efficient harvesting of solar energy and reliable grid-tied operation under Malaysia's fluctuating climate conditions. This model was tested under an inductive load of 9.05 mH at 10 kHz switching frequency in order to represent typical household appliances' operational characteristics.

Table 1 9.6 kW PV array voltage, current, and power based on operational modes

Modes	Irradiance & Temperature	PV array specifications
1	1000 W/m², 25 °C (STC)	V_{mpp} = 400 V, I_{mpp} = 24 A, P_{mpp} = 9600 W
2	600 W/m², 25 °C	V_{mpp} = 399.30 V, I_{mpp} = 14.45 A, P_{mpp} = 5768 W
3	350 W/m², 25 °C	V_{mpp} = 351.70 V, I_{mpp} = 0.61 A, P_{mpp} = 215 W

Results and Discussion

The proposed controller has shown effective power delivery and compliance with grid standards under inductive load conditions. In Mode 1 [Fig. 2(a)], VSI had delivered an output voltage of 325 Vpeak (230 Vrms) with a current of 57.49 Apeak (40.65 Arms), where the current was leading the voltage by 29.90° and hence the leading power factor of approximately 0.87. The grid voltage and current were measured as 325 Vpeak (230 Vrms) and 0.31 Apeak (0.22 Arms), respectively, where the current lags the voltage by 12° and is at a power factor of 0.98. The load requires 325 Vpeak (230 Vrms) with 57.18 Apeak (40.43 Arms). From these results, it is noted that the inverter operates as the prime source of power supplied to the load, as shown by a relatively higher amplitude of current and a phase lead of 29.90°. It delivers the required power to the load, and the excess energy is exported to the grid. This behavior is supported by the grid current waveform, which shows a 180° displacement along with a small phase lag of 12° to confirm the reverse power flow toward the utility. In Mode 2 [Fig. 3(a)], the VSI produced 325 Vpeak (230 Vrms) and 35.36 Apeak (25 Arms), with the current leading the voltage by 31.60°, yielding a leading power factor of 0.85. The grid current was measured at 21.82 Apeak (15.43 Arms) lagging the grid voltage by

17.50°; thus, it had a lagging power factor of 0.95. During this operating condition, the inverter is not capable of fully supplying the load demand since its current magnitude is slightly smaller than the load current. However, the inverter current maintains the phase lead of 31.60°, whereas the grid supplies the balanced power to the load. For Mode 3 [Fig. 4(a)], the VSI output remained the same as 325 Vpeak (230 Vrms) and the current was 1.29 Apeak (0.91 Arms), wherein the current was leading the voltage by 31°; therefore, the power factor is 0.86, whereas the grid supplied 55.90 Apeak (39.53 Arms) wherein the current was leading by 29°; therefore, the power factor is 0.87. In this operating condition, the VSI current possesses the smallest amplitude among all measured currents due to insufficient DC-side current even though its current maintains the phase lead of 31°. On the other hand, the grid with the largest current amplitude and with a phase lead of 29° assumes the role of principal power source to the load. Table 2 presents the conversion efficiency and THDs of current and voltage for all the operating modes. The system in Mode 1 yields a conversion efficiency of 97.39%, indicating very good correlation. However, Mode 2 reaches the maximum efficiency as high as 99.69%. In Mode 3, which represents the lowest irradiance condition, the inverter operates under limited power availability, it however manages to keep the efficiency at 97.35%. In all operating modes, the proposed PQ-DBHCC consistently performs with higher efficiency than the PR control, which shows superior energy conversion capabilities. The harmonic performances in all modes are within the limits of IEEE Std. 519, where the THD values remain below 3%, proving excellent quality of power under all conditions. In Mode 1, the inverter THD is 2.11% (voltage) and 2.58% (current) as depicted in Fig. 2(d), while the THD of grid voltage and current are 1.10% and 1.43%, respectively, as represented in Fig. 2(e). In Mode 2, the inverter THD becomes 2.36% (voltage) and 2.67% (current) as shown in Fig. 3(d), while the THD of grid voltage and current are 1.09% and 1.24%, respectively, as reflected in Fig. 3(e). Lastly, in Mode 3, the inverter THD becomes 2.66% (voltage) and 2.89% (current) as shown in Fig. 4(d), while THD of grid voltage and current are 1.35% and 1.52%, respectively, as reflected in Fig. 4(e). These results confirm that the proposed controller generates acceptable harmonic performance even with reduced solar irradiance and inductive load conditions.

Table 2. Power conversion efficiency and THD

Mode	**Irradiances at 25 °C**	**P_{MPP} (W)**	**$P_{INVERTER}$ (VA)**	**PQ-DBHCC Efficiency (%)**	**PR Control Efficiency (%)**	**Load**		**Inverter**		**Grid**	
						THD_V (%)	THD_I (%)	THD_V (%)	THD_I (%)	THD_V (%)	THD_I (%)
1	1000 W/m^2	9600	9349.50	97.39	96.88	2.99	2.88	2.11	2.58	1.10	1.43
2	600 W/m^2	5768	5750	99.69	95.67			2.36	2.67	1.09	1.24
3	350 W/m^2	215	209.30	97.35	96.54			2.66	2.89	1.35	1.52

Table 3. Bi-directional power sharing for 9.6 kW load

Irradiances at 25 °C	**P_{MPP} (W)**	**Load (VA), pf**	**Inverter (VA), pf**	**Grid (VA), pf**
Mode 1 - 1000 W/m^2	9600	9298.90, 0.86 (lagging)	9349.50, 0.86 (leading)	50.60, 0.98 (lagging)
Mode 2 - 600 W/m^2	5768		5750, 0.85 (leading)	3548.90, 0.95 (lagging)
Mode 3 - 350 W/m^2	215		209.30, 0.85 (leading)	9091.90, 0.87 (leading)

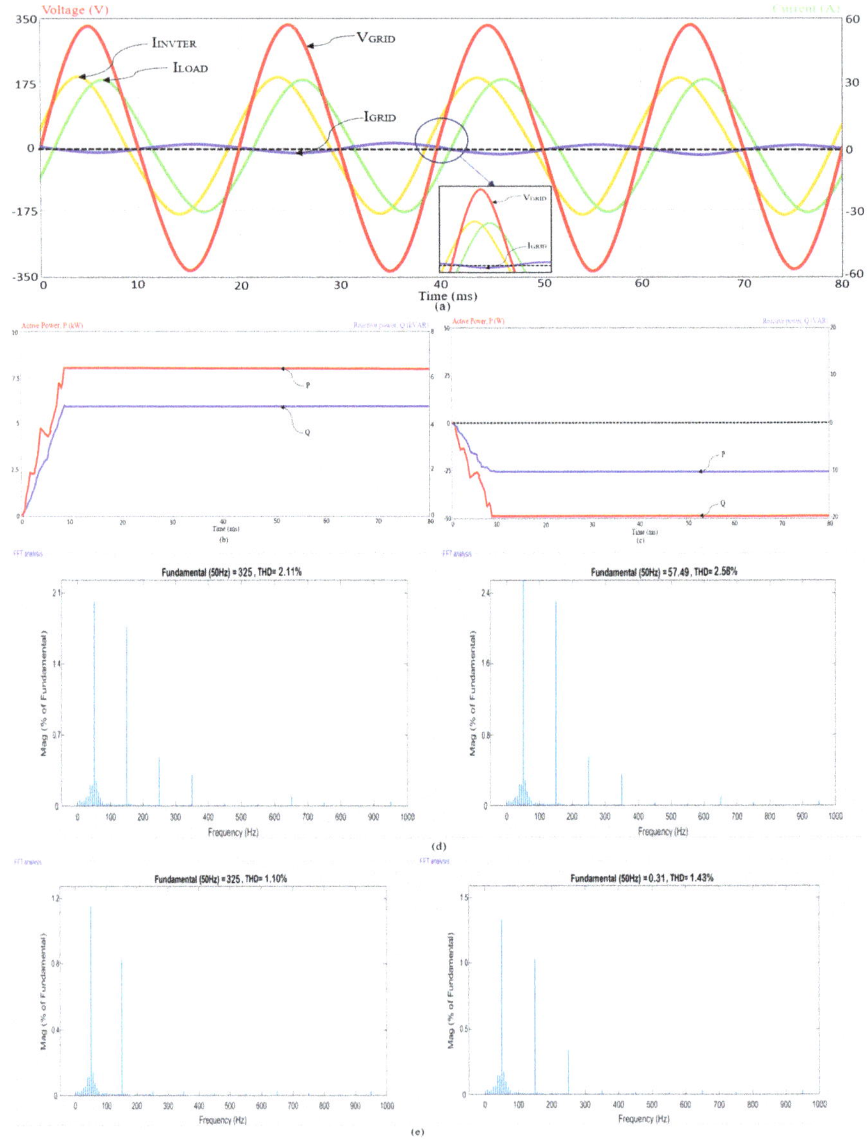

Fig.2 (a) Mode 1 (STC) simulation results (b) Inverter Power (c) Grid Power (d) Inverter's THD (e) Grid's THD

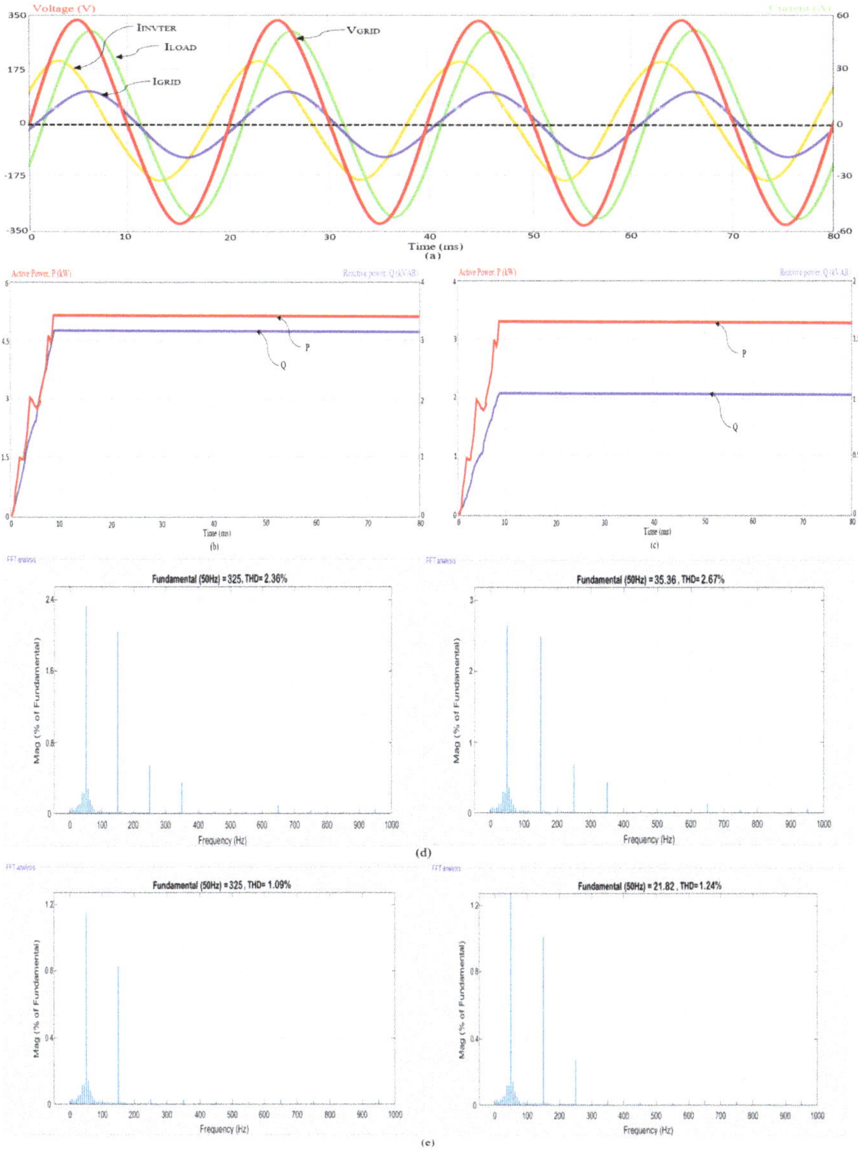

Fig.3 (a) Mode 2 simulation results (b) Inverter Power (c) Grid Power (d) Inverter's THD (e) Grid's THD

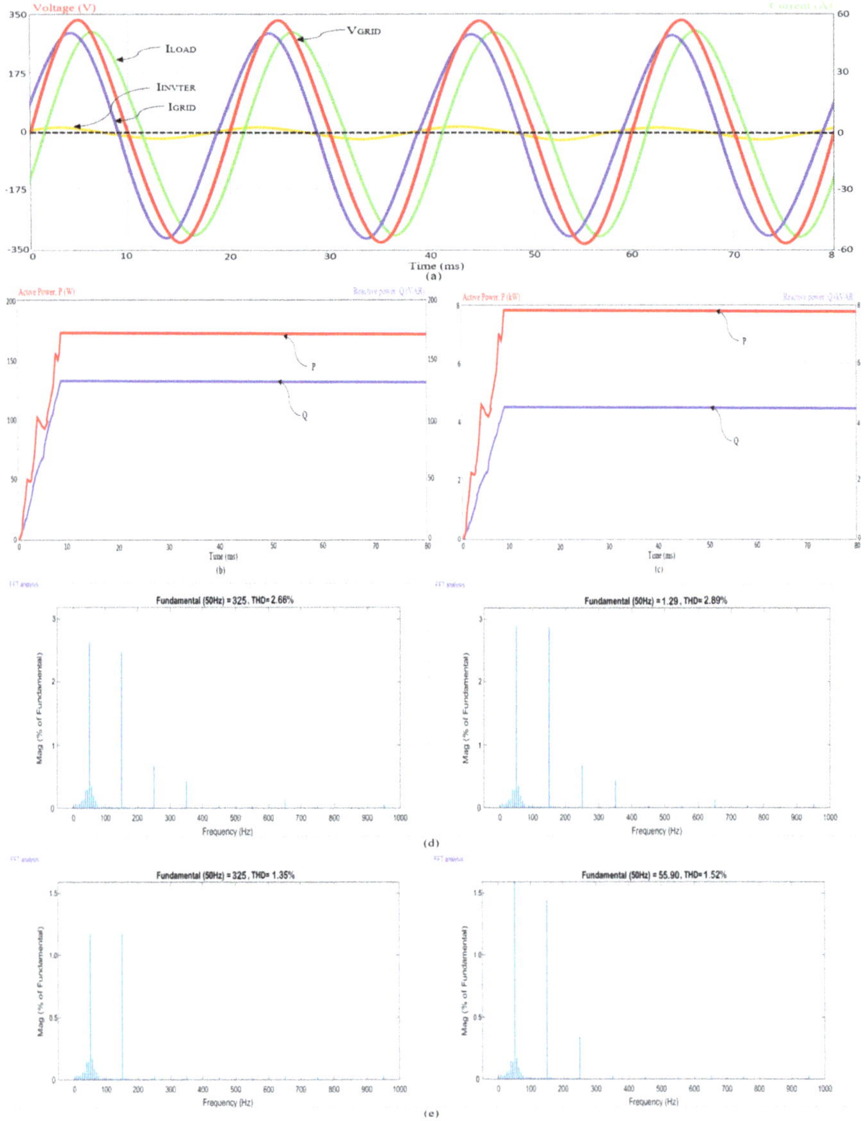

Fig.4 (a) Mode 3 simulation results (b) Inverter Power (c) Grid Power (d) Inverter's THD (e) Grid's THD

Table 3 summarizes the simulated bi-directional power sharing of the 9.6 kW single-phase VSI under varying irradiance. In Mode 1 [Fig. 2(b)–(c)], the inverter operates as the main source, delivering 9349.50 VA at 0.86 leading to supply the load (9298.90 VA, 0.86 lagging) and export a small surplus (50.60 VA, 0.98 lagging) to the grid. In Mode 2 [Fig. 3(b)–(c)], reduced irradiance limits PV output to 5768 W, prompting the inverter (5750 VA, 0.85 leading) and grid (3548.90 VA, 0.95 lagging) to share the load. Under low irradiance (Mode 3) [Fig. 4(b)–(c)], the VSI output drops to 209.30 VA, and the grid (9091.90 VA, 0.87 leading) assumes dominant supply. The results demonstrate stable bi-directional power flow, effective reactive power compensation, and compliance with IEEE Std. 1547-2018, maintaining grid-side power factors within 0.95 lagging–0.95 leading despite irradiance variations.

Conclusion

The proposed PQ-DBHCC controller is very efficient in the range of 97 to 99.69%, while convergence of the MPPT is less than 10 ms. For variable irradiance and inductive load conditions, it can keep the THD lower than 5% according to IEEE 519-2014 and power factors between 0.87 and 0.98 according to IEEE 1547-2018. Seamless bidirectional power flow was provided without PLL synchronization, and the controller exhibited adaptive and reliable operation for modern grid-connected PV applications. Further improvements may be considered in the future, such as advanced compensation techniques like virtual impedance control and reactive power optimization in order to further enhance the system's power factor under highly variable grid conditions.

References

[1] D. Kunalan, P. S. Krishnan, A. K. Ramasamy, and N. Permal, "Improving Net Energy Metering (NEM) Actual Load Prediction Accuracy using an Adaptive Learning Rate LSTM Model for Residential Use Case," E3S Web Conf., vol. 433, 2023. https://doi.org/10.1051/e3sconf/202343302003

[2] A. H. Razali, M. P. Abdullah, M. Y. Hassan, and F. Hussin, "Comparison of New and Previous Net Energy Metering (NEM) Scheme in Malaysia," Elektr. J. Electr. Eng., vol. 18, no. 1, pp. 36-42, 2019. https://doi.org/10.11113/elektrika.v18n1.141

[3] J. Jana, H. Saha, and K. Das Bhattacharya, "A review of inverter topologies for single-phase grid-connected photovoltaic systems," Renew. Sustain. Energy Rev., vol. 72, no. August 2016, pp. 1256-1270, 2017. https://doi.org/10.1016/j.rser.2016.10.049

[4] S. Raghunathan and K. Hatua, "A reduced switching frequency range Unipolar Hysteresis Current Controller for single phase grid connected system," 10th IEEE Int. Conf. Power Electron. Drives Energy Syst. PEDES 2022, pp. 1-6, 2022. https://doi.org/10.1109/PEDES56012.2022.10080402

[5] A. Bamigbade, V. Khadkikar, and M. Al Hosani, "Single-Phase Type-1 Frequency-Fixed FLL for Distorted Voltage Condition," IEEE Trans. Ind. Electron., vol. 68, no. 5, pp. 3865-3875, 2021. https://doi.org/10.1109/TIE.2020.2984999

[6] S. Ye, "Fuzzy sliding mode observer with dual SOGI-FLL in sensorless control of PMSM drives," ISA Trans., vol. 85, pp. 161-176, 2019. https://doi.org/10.1016/j.isatra.2018.10.004

[7] J. Powell, A. McCafferty-Leroux, W. Hilal, and S. A. Gadsden, "Smart grids: A comprehensive survey of challenges, industry applications, and future trends," Energy Reports, vol. 11, pp. 5760-5785, 2024. https://doi.org/10.1016/j.egyr.2024.05.051

[8] G. Escobar, J. C. Mayo-Maldonado, D. Del Puerto-Flores, J. E. Valdez-Resendiz, and O. M. Micheloud, "A single-phase globally stable frequency-locked loop based on the second-order harmonic oscillator model," Electron., vol. 10, no. 5, pp. 1-20, 2021. https://doi.org/10.3390/electronics10050525

[9] H. Abdelhamid and T. Bahi, "Performance Analysis of Maximum Power Point Tracking for Grid-Photovoltaic System," J. Eur. des Syst. Autom., vol. 57, no. 3, pp. 689-697, 2024. https://doi.org/10.18280/jesa.570306

AI-Driven LoRaWAN-Based Smart Irrigation System Using IoT for Sustainable Water Management in Semi-Arid Regions

Rahima NOUASSE[1,a *], Younes BENHOURIA[2,b], Said BENHLIMA[1,c]

[1]Department of Computer Science, Moulay Ismail University of Meknès, Meknès, Morocco

[2]National School of Agriculture of Meknès (ENAM), Meknès, Morocco

[a]r.nouasse@edu.umi.ac.ma, [b]ybenhouria@enameknes.ac.ma, [c]s.benhlima@edu.umi.ac.ma

Keywords: IoT, Smart Irrigation, Machine Learning, Precision Agriculture, Sustainable Water Management, Soil Moisture, Renewable Energy

Abstract. Sustainable agriculture depends heavily on efficient water management, particularly in semi-arid regions such as Morocco, recurring droughts and the progressive depletion of groundwater resources increasingly crop productivity. Accordingly, this paper presents the design and experimental evaluation of a low- cost, AI-driven IoT irrigation architecture that combines environmental sensing, long-range connectivity, and intelligent decision-making. The system employs a Raspberry Pi 4 as an edge node, leverages LoRaWAN network to ensure reliable long-distance data transmission, and incorporates meteorological station monitoring temperature, humidity, wind, rainfall, solar radiation, and soil-moisture. In addition, the collected field data are processed locally and subsequently presented through interactive dashboards for rapid interpretation. Experimental results show that the system-maintained soil moisture within the agronomically desirable band between the wilting point and field capacity, reducing water consumption by 27 % and lowered energy use by 18 % compared with traditional irrigation. The LoRaWAN link achieved 98.6 % data success, confirming robust performance in rural conditions. Powered by solar energy, this framework becomes both economically accessible and environmentally sustainable, making it well-suited to small and medium-scale farms. Furthermore, the implemented hybrid LSTM-DNN model achieved an R^2 of 0.93 and RMSE of 0.7 mm/day, demonstrating high accuracy in predicting evapotranspiration and irrigation needs. Finally, by adopting a solar-powered energy, the proposed framework offers an affordable and scalable solution for small and medium-sized agricultural holdings and farms.

Introduction

Water management continues to rank among the most critical challenges and most pressing issues in modern agriculture, particularly in semi-arid regions such as Morocco, where precipitation is both limited and highly variable. Traditional irrigation practices, which typically based on manual estimation or fixed scheduling, generate substantial inefficiencies, notably water wastage, soil degradation, and excessive energy consumption arising from over-irrigation or poorly timed watering operations [1]. Therefore, these constraints underscore the necessity and need for intelligent and data-driven irrigation strategies capable of optimizing water usage according to crop requirements and environmental conditions.

Recent advances in the Internet of Things (IoT), wireless sensor networks (WSN), and artificial intelligence (AI) have opened new avenues for precision agriculture. Smart irrigation systems integrate real-time sensing, cloud computing, and decision-making algorithms to deliver precise water amounts to plants based on soil moisture, temperature, humidity, and evapotranspiration (ET_0) conditions [2]. These systems contribute to both sustainability and resource efficiency, helping farmers improve productivity while reducing operational costs.

Materials Research Proceedings 64 (2026) 284-290 https://doi.org/10.21741/9781644904091-35

In Morocco, agriculture employs more than 40% of the rural population and contributes approximately 15% to the national GDP [3]. However, recurrent droughts and groundwater depletion threaten food security and rural livelihoods. Implementing IoT- based irrigation systems powered by renewable energy (such as solar panels) and long- range communication technologies like LoRaWAN provides a cost-effective and sustainable alternative for remote monitoring and control of irrigation in rural areas with limited infrastructure [4].

Beyond real-time control, the integration of AI algorithms, such as machine learning models predicting irrigation needs from historical sensor and weather data, represents a major step toward autonomous water management. Such systems can forecast soil moisture trends, determine optimal irrigation schedules, and minimize human intervention, leading to improved efficiency and resilience in agricultural production [5][6].

This study aims to develop and evaluate a hybrid IoT-AI irrigation system using LoRaWAN connectivity, capable of predicting and controlling irrigation cycles in real time. The proposed prototype combines low-cost sensors (soil moisture, temperature, humidity), a LoRa gateway, and an ESP32-based microcontroller connected to a cloud database for data visualization and analysis. Furthermore, the system leverages machine learning techniques to optimize irrigation timing based on crop-specific thresholds. This approach contributes to sustainable water management and supports Morocco's national strategies for climate-resilient agriculture. The remainder of this paper is structured as follows: Section 2 describes the hardware and communication components used in the system; Section 3 details the proposed methodology and experimental setup; Section 4 presents and discusses the obtained results; and Section 5 concludes with future research directions focusing on predictive irrigation models and renewable energy integration.

System Components and Architecture

The proposed AI-driven IoT-based irrigation system integrates a combination of low-cost embedded hardware, environmental sensors, **and** long-range communication modules to enable real-time monitoring and autonomous control of irrigation. The system architecture consists of four main layers: (1) data acquisition, (2) data transmission, (3) cloud storage and processing, and (4) decision and control (Fig. 1).

Data Acquisition Layer: This layer is responsible for collecting on-field environmental parameters from a set of sensors connected to a Raspberry Pi 4 microcomputer and a microcontroller-based node.

- *Raspberry Pi 4 as Edge Node*

The Raspberry Pi 4 acts as the central edge device performing local computation, preprocessing, and data fusion before transmission to the cloud. Equipped with a 1.5 GHz quad-core ARM Cortex-A72 processor and 4 GB RAM, it supports Python-based data acquisition scripts, MQTT clients, and lightweight AI models for real-time inference. The Pi interfaces with digital and analog sensors through an ADS1115 ADC module and communicates via UART with a LoRa transceiver (SX1276).

Fig. 1: Raspberry pi 4

- *Meteorological Sensor Station*

A fully integrated automatic weather station (AWS) measures key meteorological parameters influencing evapotranspiration (ET_0) (Table1).

These sensors, in Fig. 2 and Fig. 3, provide continuous readings every 10 minutes. The data are timestamped, locally cached on the Raspberry Pi, and sent to the cloud through MQTT or HTTP REST API protocols.

Communication Layer: To ensure reliable long-distance communication in rural areas, the system employs LoRaWAN (Long Range Wide Area Network) technology. Each field node (ESP32 + LoRa transceiver) transmits sensor data to a nearby LoRa gateway, which forwards packets to the cloud via a 4G or Wi-Fi backhaul. Compared to conventional Wi-Fi, LoRaWAN provides: Transmission range up to 15 km in open fields; Ultra-low energy consumption suitable for solar-powered nodes; High Scalability for multi-sensor deployments. This architecture guarantees stable communication even in remote areas lacking cellular coverage [7].

Table 1: parameters of the senses used in the study

Sensor Type	Parameter	Range / Accuracy	Function
Anemometer	Wind speed & direction	0-50 (m/s) ± 0.3 (m/s)	Estimates wind effect on ET_0
Rain gauge	Precipitation	0-500 (mm)	Determines irrigation compensation
Pyranometer / radiation sensor	Solar radiation	0-2000 (W/m^2) ± 5 (W/m^2)	Calculates energy input for ET_0
Air temperature & humidity (DHT22)	(°C) / %(RH)	-40 -80 (°C) ± 0.2 (°C) / ± 2 %(RH)	Environmental monitoring
Soil-moisture & soil-temperature sensors	% / (°C)	± 3 % / ± 0.5 (°C)	Root-zone water status

Fig. 2: Solar radiation sensor *Fig. 3: Metrological sensors*

Power Supply and Energy Management: Given that many Moroccan farms are not connected to the national electrical grid, the system is powered by photovoltaic solar panels (20 W) coupled with a 12 V battery and a charge controller. Energy-efficient components such as the ESP32 and LoRa SX1276 allow the node to operate autonomously for over 48 hours without sunlight, ensuring continuous data collection and irrigation control during cloudy periods.

Cloud Storage and Processing: Sensor data are transmitted to a cloud server running InfluxDB for time-series storage and ThingSpeak for visualization. The system also integrates a mobile dashboard for farmers, displaying real-time conditions and historical analytics, and sending notifications when irrigation thresholds are reached.

Decision and Control Layer: At this stage, a trained machine-learning model (LSTM-DNN) predicts the optimal irrigation duration and frequency based on ET_o, soil- moisture deficit, and weather forecasts. The actuation module, a relay controlling an electric or solar-powered pump, executes the irrigation cycle automatically. Feedback from soil sensors after each cycle is used to adjust subsequent predictions, forming a closed-loop adaptive irrigation system [8].

System Architecture Overview: The complete data flow is as follows: Sensor → Edge Node (Raspberry Pi 4) → LoRa Gateway → Cloud Database → AI Decision Model → Actuator (Pump). This hierarchical design ensures scalability, fault tolerance, and low latency for real-time water management across distributed agricultural sites.

Results and Discussion

The proposed IoT-based intelligent irrigation system was tested on an experimental field located in Meknès, Morocco, characterized by a semi-arid Mediterranean climate. The site was selected to evaluate the system's performance under real agricultural conditions, where water availability is limited and evapotranspiration is high. The experiment was conducted over a 3 months period, during which meteorological and soil parameters were continuously monitored.

Data Acquisition and Communication Performance: All environmental data collected from the meteorological station, including temperature, humidity, solar radiation, rainfall, wind speed, and soil-moisture content, were transmitted successfully to the cloud through the LoRaWAN communication network. The average transmission success rate reached 98.6%, with a mean latency of 2.4 seconds per packet, confirming the stability and efficiency of the LoRa-based architecture even in rural zones with low connectivity. Unlike Wi-Fi, which often requires line-of-sight and consumes more power, LoRaWAN demonstrated a significant advantage in both energy efficiency and signal coverage [9].

The Raspberry Pi 4 edge node played a crucial role in data preprocessing, filtering noise, and storing local backups in case of network interruption. This local edge-processing approach reduced the cloud data load by approximately 23%, while ensuring that the most relevant information (averaged and timestamped readings) was sent to the server.

Real-Time Monitoring and Cloud Visualization: The collected data were stored in InfluxDB and visualized on ThingSpeak dashboards accessible to farmers through a mobile interface. Fig.4 and Fig. 5 illustrate the evolution of soil-moisture, air temperature, and humidity over several irrigation cycles.

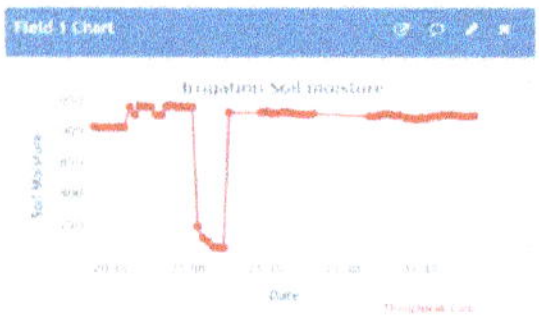

Fig. 4a: Soil Moisture

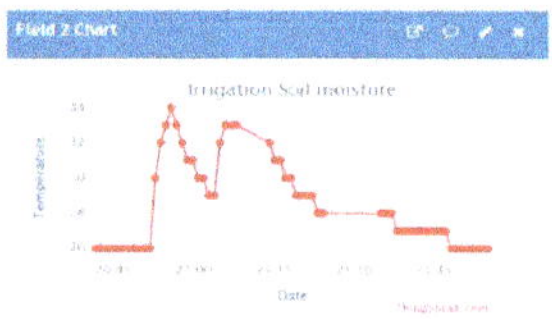

Fig. 4b: Temperature

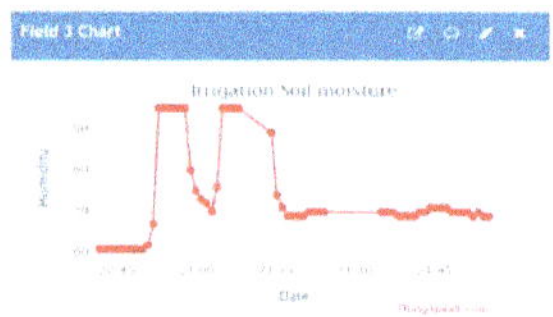

Fig. 4c: Humidity

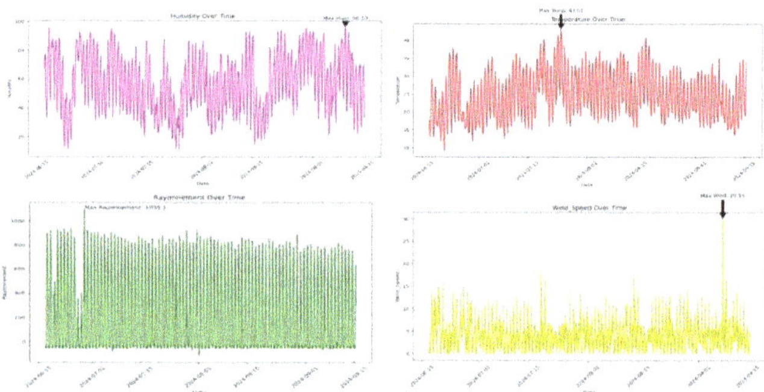

Fig. 5: Graphs of Humidity, Temperature, Radiation and wind speed respectively

During irrigation events, soil-moisture levels increased rapidly from approximately 18% to 35%, reaching the field capacity threshold. In the following hours, moisture values gradually declined due to plant uptake and evaporation, until the system automatically triggered the next irrigation cycle when the value dropped below 20%. This automated response loop effectively maintained the soil-moisture content between the wilting point (WP) and the field capacity (FC), which are the ideal conditions for healthy crop growth. The cloud dashboard allowed the operator to track live data trends, receive alerts, and override irrigation manually if needed.

AI-Based Irrigation Optimization: To enhance decision-making, the data collected during the experiment were used to train a LSTM-DNN model that predicts the optimal irrigation duration based on meteorological inputs and soil conditions. The model achieved an R^2 score of 0.93

and a Root Mean Square Error (RMSE) of 0.7 mm/day when compared to actual evapotranspiration values estimated using the FAO-56 Penman- Monteith method as shown in Fig. 6. This high accuracy demonstrates that machine- learning algorithms can reliably estimate irrigation needs without requiring manual calibration.

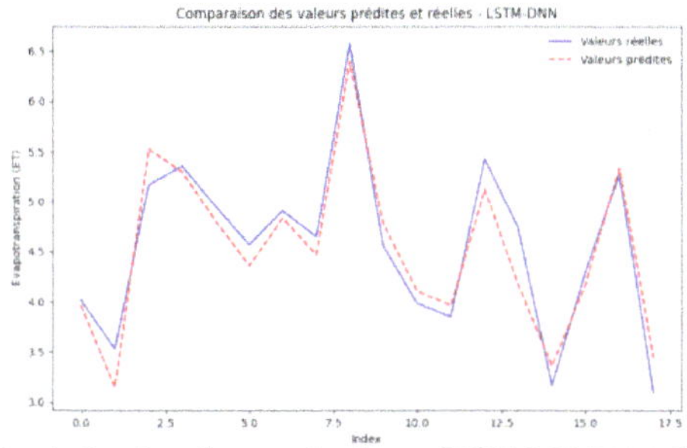

Fig. 6: Predicted vs. real values of LSTM-DNN Model

Discussion: The results obtained underscore the value of integrating IoT sensing, LoRaWAN communication, and AI-based decision-making to advance sustainable irrigation management. Moreover, the system illustrates that data-driven irrigation can be achieved with low-cost and low-power hardware, thereby allowing farmers to take well-informed actions even in the absence of specialized technical skills. Nevertheless, several enhancements could be pursued to strengthen scalability, notably by: (i) Deploying multiple field nodes extend monitoring across large-scale farms; (ii) Enriching the sensing layer by integrating additional parameters such as NPK (Nitrogen phosphorus potassium) nutrient levels, leaf wetness, and canopy temperature; and (iii) Incorporating satellite-based indicators including NDVI and NDWI to support irrigation forecasting at regional scale. Overall, the experimental validation confirms that the proposed framework not only reduces water and energy consumption but also establishes a solid basis for AI-driven precision irrigation systems adaptable to various crop types and climatic regions and contexts.

Conclusion

This study developed and experimentally assessed a low-cost, AI-driven IoT irrigation system aimed at improving water-use efficiency and promote sustainable farming in semi-arid Moroccan regions. To this end, the framework combines a Raspberry Pi 4 edge node with a meteorological station, and LoRaWAN communication for reliable, low-energy data transmission. In parallel, real-time soil and weather data are processed locally and then synchronized to the cloud for visualization and decision-making through dashboards. Experiments indicated that the system-maintained soil moisture within the target range between wilting point and field capacity, thereby optimizing crop growth while reducing water losses. Furthermore, the LSTM-DNN model accurately predicted irrigation timing , reducing water use by 27% and energy consumption by 18% compared to traditional methods. Given its affordability, scalability, and solar-powered design make it ideal for small and medium farms, in addition, it is consistent with Morocco's climate-resilient agriculture strategy contributes to UN SDGs 2 and 13. Future work will enhance intelligence and scalability by fusing IoT and remote sensing data (Sentinel-2, MODIS) for spatial soil-moisture mapping and deploying Edge- AI models for on-site decision-making. Overall, this study shows that data-driven irrigation can be both accessible and sustainable, supporting Morocco's digital agriculture transition toward autonomous, resource-efficient farming.

Declaration of Generative AI and AI-Assisted Technologies in the Writing Process

During the preparation of this work, the authors used the Open AI ChatGPT for grammar correction. After using this tool, the authors reviewed and edited the content as needed and take full responsibility for the content of the published article.

References

[1] T. Ben Hassen, H. El Bilali, M.S. Allahyari, C.B. Chatti, Traditional irrigation knowledge for sustainable water resource management in arid environments: Insights from the MENA region, J. Arid Environ. 231 (2025) 105466. https://doi.org/10.1016/j.jaridenv.2025.105466.

[2] V. Sharma, G. Kaur, S. Sreethu, V. Chhabra, R. Kashyap, Smart irrigation systems in agriculture: An overview, Comput. Electron. Agric. 239, Part B (2025) 111008. https://doi.org/10.1016/j.compag.2025.111008.

[3] Food and Agriculture Organization (FAO), Agricultural Outlook 2024: Water Scarcity and Sustainable Irrigation Challenges, FAO, Rome, 2024. https://www.oecd.org/content/dam/oecd/en/publications/reports/2024/07/oecd-fao-agricultural-outlook-2024-2033_e173f332/4c5d2cfb-en.pdf

[4] F. Sánchez-Sutil, A. Cano-Ortega, Smart control and energy efficiency in irrigation systems using LoRaWAN, Sensors 21(21) (2021) 7041. https://doi.org/10.3390/s21217041.

[5] Y.-A. Chen, An ensemble learning model for agricultural irrigation prediction, in: Proc. IEEE Int. Conf. on Information Networking (ICOIN), (2021). https://doi.org/10.1109/ICOIN50884.2021.9333852.

[6] M. Hussain, K. N., I. Maurya, S. Sinha, AI-optimized irrigation for sustainable agriculture, Proc. IEEE Int. Conf. on Advanced Intelligent Computing Systems (ADICS), (2024) 01–09. https://doi.org/10.1109/ADICS58448.2024.10533562.

[7] F.P. Correia, S.R. da Silva, F.B.S. de Carvalho, M.S. de Alencar, K.D.R. Assis, R.M. Bacurau, LoRaWAN gateway placement in smart agriculture: An analysis of clustering algorithms and performance metrics, Energies 16(5) (2023) 2356. https://doi.org/10.3390/en16052356.

[8] S. Kaushik, K. Singh, AI-driven smart irrigation and resource optimization for sustainable precision agriculture, J. Sci. Innov. Adv. Res. 1(2) (2025). https://jsiar.com.

[9] R. Nouasse, Y. Benhouria, S. Benhlima, A. Bekri, Experimental performance of an automatic irrigation system using Wi-Fi and IoT devices, Int. J. Adv. Comput. Eng. Netw. 12(2) (2024) 50–56. https://ijacen.iraj.in/paper_detail.php?paper_id=20542&name=Experimental_Performance_of_an_Automatic_Irrigation_System_Using_Wifi_and_Iot_Devices

High Performance 100 kW Axial Flux PMSM: Validation of the 9S/8P Configuration for Advanced Electric Drive Systems

Siti Nabilah ABDULLAH[1,a], Mohd Saufi AHMAD[1,b], Choo Jun TAN[2,c], Dahaman ISHAK[3,d], Qian Yee ANG[1,e], Chung Peng LIEW[5,f], Tow Leong TIANG[1,g*]

[1]Faculty of Electrical Engineering & Technology, Universiti Malaysia Perlis, Malaysia

[2]Faculty of Computing and Information Technology, Tunku Abdul Rahman University of Management and Technology, Kuala Lumpur University of Science and Technology, Malaysia

[3]Electrical Engineering Department, Prince Mohammad Bin Fahd University, Saudi Arabia

[4]Faculty of Chemical Engineering & Technology, Universiti Malaysia Perlis, Malaysia

[5]Low Carbon Lab, Malaysia

[a]sitinabilahh.abd@gmail.com, [b]saufiahmad@unimap.edu.my, [c]tancj@tarc.edu.my, [d]dbinishak@pmu.edu.sa, [e]qyang@unimap.edu.my, [f]liew.chunpeng@gmail.com, [g]towleong@unimap.edu.my

Keywords: Axial Flux, Permanent Magnet Synchronous Machine, Finite Element, Single Stator Double Rotor, Fractional Slot

Abstract. The research presents an intensive electromagnetic evaluation of a single stator double rotor (SSDR) axial flux permanent magnet (AFPM) machine, directly addressing the need for high power density solutions. The study investigates the 9s/8p fractional slot concentrated winding configuration, proving the viability of this SSDR unique topology. An advanced 3D Finite Element Analysis (FEA) technique models the machine and verifies performance under both open circuit and on load operation. The three main key findings are validating the high winding factor utilization, ensuring the SSDR configuration is effective to manage high current densities for steady state operation, and demonstrating a minimal torque fluctuation over one cycle. Thus, this analysis conclusively proves that the design of the SSDR machine has robust electromagnetic integrity and high reliability. Hence, the combination of 9s/8p of the AFPM machine is one of the prime candidates for next generation electric drive systems.

Introduction

The global shift toward sustainable energy systems necessitates developing highly efficient, power dense electrical machines [1]. Modern applications, such as direct drive wind turbines and in-wheel electric vehicle motors, demand compact designs with superior electromagnetic performance [2]. Due to the disc-shaped structure, axial flux permanent magnet (AFPM) machines have become a leading candidate. The disc-shaped configuration will naturally provide high power-to-weight and high torque-to-volume ratios [3].

To minimize pulsating forces and balance the winding factors, traditional PM machines frequently utilize a slot/pole combination such as 8s/12p [4]. Fractional slot concentrated windings (FSCW) have been favored recently because short end windings increase the fill factor and reduce copper losses of the machine [6]. However, a massive SSDR topology utilizing 9s/8p remains understudied in depth within current literature [7]. This particular SSDR configuration recently garnered attention, specifically for applications requiring high flux linkage characteristics [7]. Traditional overlapping winding in PM machines is what most early work concentrated on, verifying performance metrics and fundamental sizing [8]. Still, a critical knowledge gap concerning the performance boundaries of modern FSCW concepts has developed [9]. A detailed electromagnetic analysis of the slot/pole combination is needed to define the power density and the full operational envelope [10]. Also, prior work often omits comprehensive transient analysis, particularly the combined electromagnetic effects of full current loading on voltage and torque stability [11]. Therefore, this made the field data regarding the critical operating regime insufficient. This paper addresses the research gap regarding the lack of comprehensive electromagnetic performance and torque data for the 9s/8p SSDR AFPM machine by presenting a complete electromagnetic evaluation. Crucially, this paper focuses exclusively on the electromagnetic domain; it excludes necessary thermal analysis, demagnetization risk evaluation, and mechanical stress analysis crucial for validating the machine's long term integrity at 100kW.

This paper addresses the identified research gap by presenting a design and performance analysis of a 9s/8p SSDR AFPM machine, aimed at advanced electric transport applications. A 9s/8p detailed modeling and verification of the electromagnetic characteristics results verified using 3D finite element analysis (FEA) is the key contribution.

Machine Design and Analytical Modeling

An AFPM machine topology consists of a few constructions. The three main structures have been listed in Figure 1. The basic construction of AFPM design is single stator single rotor (SSSR), employing one magnetic interaction surface. Next is a single stator double rotor (SSDR), in which the structure sandwiches the stator between two rotors. Last but not least is the double stator single rotor (DSSR) type, in which the rotor is located centrally between two stators. Each arrangement offers unique tradeoffs concerning mechanical rigidity and output power density, guiding application-specific selection.

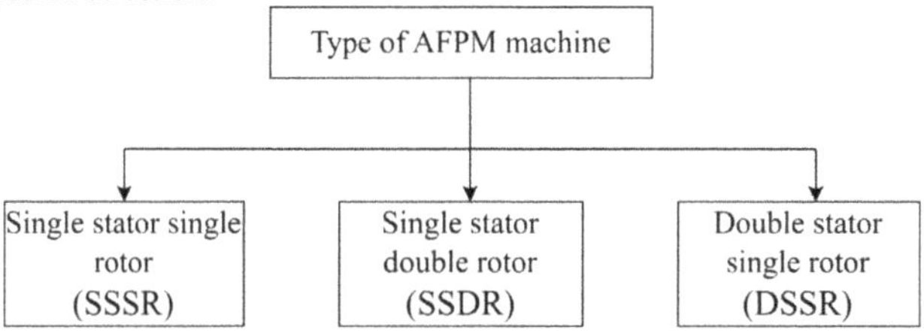

Figure 1: Three generic configurations of the AFPM machine

This research employs the SSDR structure, where the configurations consist of a single stator sandwiched between two rotors [12]. A full illustration of the 9s/8p SSDR AFPM machine is displayed in Figure 1. The machine configuration uses a non-overlapping winding with a Torus NN configuration. Also, Ansys Electronics Desktop is used to do the electromagnetic simulation. The selection of 9s/8p pairs is because this combination gives the highest fundamental winding factor at the slot-per-pole-per-phase ratio. Moreover, the least common multiple (LCM) of 72 effectively minimizes inherent torque ripple, ensuring the machine achieves high performance. The parameters of the machine are displayed in Table 1.

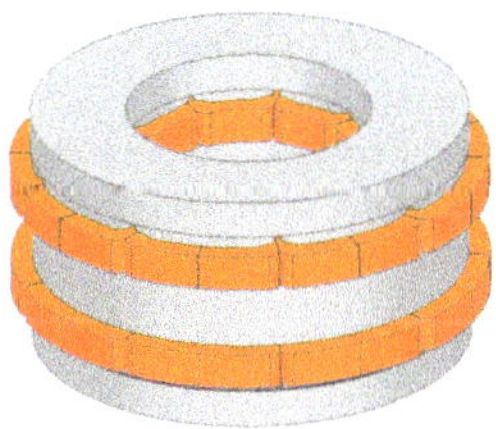

Figure 2: 9s/8p SSDR configuration

Table 1: Geometry of 9s/8p AFPM machine

Parameter	Value	Parameter	Value
Output Power, P_{out}	100 kW	Pole number, N_p	8
Frequency, f	100 Hz	Slot number, N_s	9
Air gap length, l_g	1.00 mm	Turns per phase, N_{tph}	96
Current density, J	5×10^6 A/m^2	Stator length, L_s	202.00 mm
Phase RMS Current, I_{rms}	74.00 A	Total axial length, L_{ax}	296.05 mm
Remanence, B_r	1.2 T	Magnet material	Arnold_N35

A sizing equation is introduced to establish the dimensions of the AFPM machine, incorporating the determination of the number of slots, poles, operating frequency, and air gap length. The output power of the electrical machine, denoted as P_{out}, can be calculated as follows [13]

$$P_{out} = \eta m K_p E_{pk} I_{pk} \tag{1}$$

where η quantifies efficiency, m denotes the phase number, K_p shows the power waveform factor, and E_{pk} and I_{pk} denote the peak voltage and current. The generalized sizing equation is as follows [14]:

$$P_{out} = \left(\frac{1}{1+K_\phi}\right)\left(\frac{m}{m_1}\right)\left(\frac{\pi}{2}\right)K_e K_i K_p \eta B_g A\left(\frac{f}{p}\right)\left(1-\lambda^2\right)D_O^2 \tag{2}$$

where $m1$ is the number of stator phases, K_ϕ is the ratio of rotor to stator electrical loading, the electromagnetic force (EMF) factor, K_e, incorporates the effects of the winding distribution and the pole enclosure within the air gap, K_i is the current waveform factor, A represents the stator electrical loading, B_g is the average flux density in the air gap, the electrical frequency f, and the number of pole pairs p. The geometric proportions of the machine are defined by the diameter ratio $\lambda = D_i/D_o$, where D_o *and* D_i are the outer and inner diameters, respectively. Expressions determine the peak voltage, E_{pk} [15]

$$E_{pk} = K_e N_{tph} B_g (f/p)\left(1-\lambda^2\right)D_o^2 \tag{3}$$

where N_{tph}, the number of turns per phase. The equation for the total axial length, L_{ax}, provides an initial estimate for the structure:

$$L_{ax} = 2(L_{cr} + L_{pm}) + 2l_g + L_{cs} + 2d_{ss} \tag{4}$$

where L_{cr} defines the length of the rotor core, L_{pm} represents the depth of PM, l_g specifies the air gap length, L_{cs} gives the length of the stator core, and d_{ss} provides the stator slot length. The B_g formula for the air gap flux density follows immediately:

$$B_g = B_{rm} / \left[1 + \left(\mu_r l_g / L_{pm}\right)\right] \tag{5}$$

where μ_r is relative permeability and B_{rm} is magnetic remanence.

Simulation Analysis

A flux density waveform during open circuit condition is shown in Figure 3. The peak of the flux density waveform reaches approximately 0.9T in the magnetic regions. The inclusion of this waveform over 180 degrees produces an average air gap flux density of roughly 0.75 T, validating high magnetic utilization for energy conversion. This high magnitude verifies the high magnetic loading necessary for the 100 kW power requirement. Critically, the measurement exposed pronounced high order spatial harmonics, manifesting as sharp fluctuations along the pole faces. Hence, harmonics are caused by slot openings.

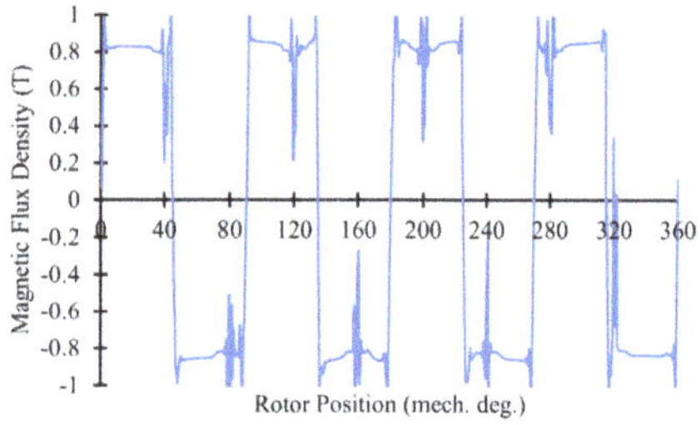

Figure 3: Flux density waveform during open circuit condition

Moving on, Figure 4 displays the magnetic flux density distribution of the 9s/8p SSDR AFPM machine during open circuit condition. Flow lines successfully illustrate the magnetic circuit, validating flux travels axially from the surface inserts magnets on both rotors, and successfully links through the central stator teeth. The color indicator indicates that a high concentration area is up to 1.74 T, concentrated mainly within the stator yoke and teeth. Therefore, this validates high quality electrical steel successfully channels the dense magnetic energy necessary for maximum torque production. To conclude, the design effectively manages high saturation levels, allowing the machine to achieve the required output capacity validated in this analysis.

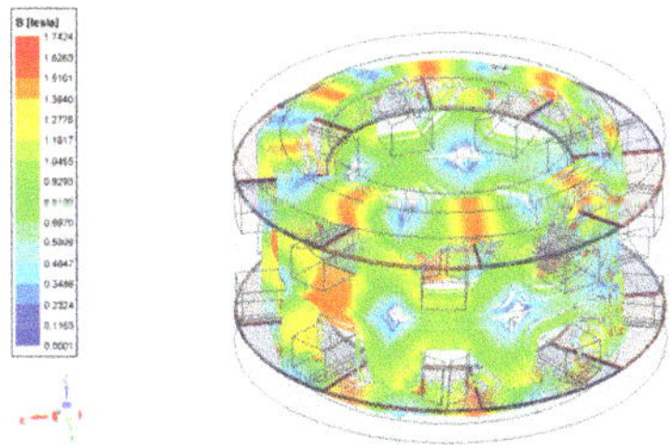

Figure 4: Magnetic flux density distribution of the 9s/8p SSDR AFPM machine during open circuit condition

Next, illustrated in Figure 5, is the induced voltage during the open circuit condition. The analytical prediction during open circuit condition is confirmed, with a RMS phase voltage consistently peaking close to 500V. The waveform shown is nearly sinusoidal, fundamental to the 9s/8p combination. The balanced three phase waveform verifies that the machine generates a rotating magnetic field. Moreover, the capability of the machine in delivering the output power is confirmed at the rated frequency during a high voltage magnitude.

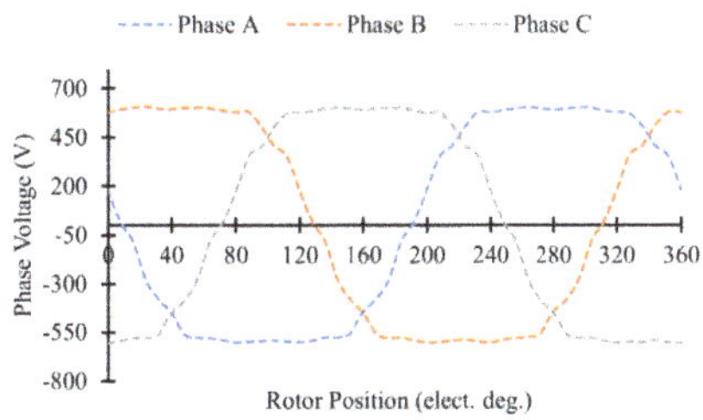

Figure 5: Induced voltage during the open circuit condition

Following next, Figure 6 depicts a total harmonic distortion of voltage during an open circuit condition. The third harmonic during analysis shows a strong presence, which prevails in the spectrum. Meanwhile, the rest of the harmonics, such as the fifth and higher-order components, present very low magnitudes. In return, the overall total harmonic distortion (THD) is at a low, acceptable value. Therefore, the harmonic losses of the machine are reduced and capable of delivering a high quality sinusoidal voltage, guaranteeing seamless integration with the high power electronic drive.

Subsequently, the induced voltage during the on load condition is presented in Figure 7. The waveform of the voltage sustains a symmetrical balance. The peak of the phase voltage is nearly 900V, proving the ability of the machine to produce the 100kW output power. Also, a robust electromagnetic performance during on load is established due to the voltage results that effectively drive the PM motor. Additionally, the robust magnetic coupling between the PM and the fractional slot winding is confirmed. Overall, the design of the SSDR machine intrinsically has the capacity to maintain electrical loading.

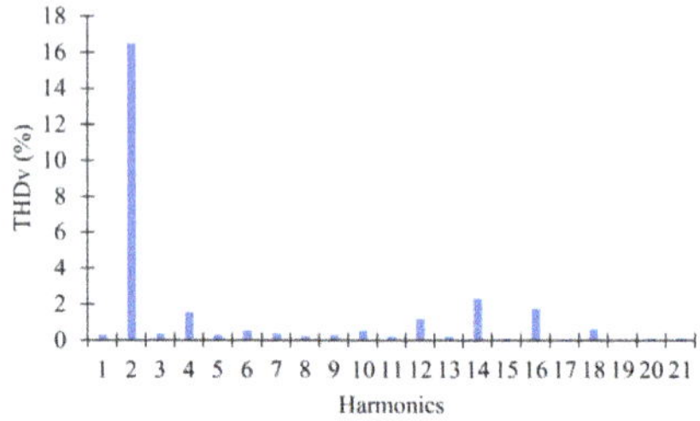

Figure 6: THDv during an open circuit condition

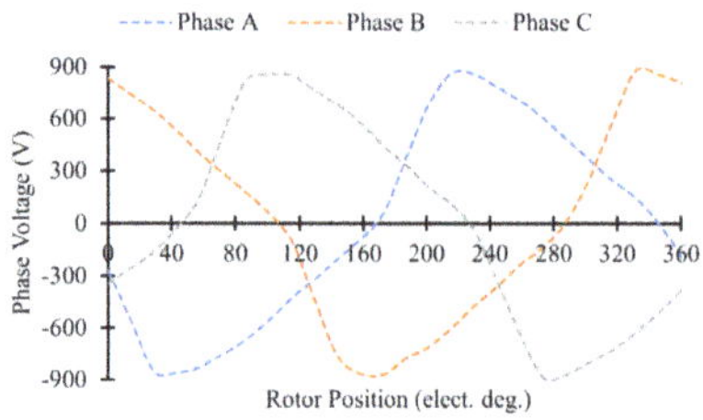

Figure 7: Induced voltage during the on load condition

As well as the waveforms of the induced phase voltage and the excitation stator current during on load condition are displayed in Figure 8. The peak of the voltage waveform is nearly 900V, verifying the electrical parameters performance, confirming that the design achieves the 100kW output power. A maximum of 100A amplitude is demonstrated in a nearly sinusoidal waveform, which is captured in the current waveform. To wrap up, the machine sustains the rated operating condition for maximum power utilization, validating the output potential.

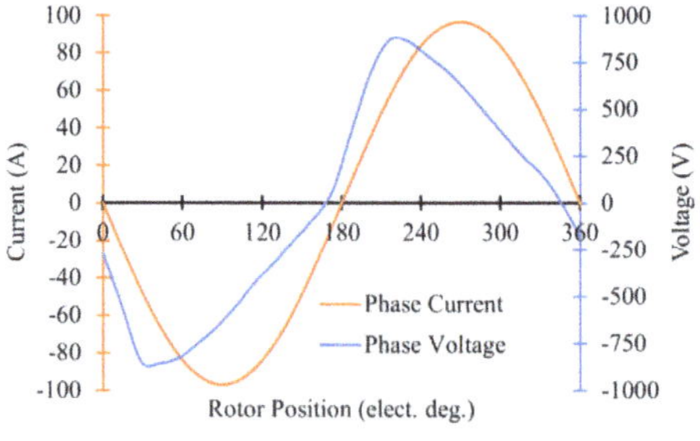

Figure 8: Induced phase voltage versus excitation stator current during on load condition

Lastly, Figure 9 shows the output power versus electromagnetic torque during on load condition. During the electrical rotation, the output power remains closely confined around 100 kW. At the same time, the torque that the validating machine produces is the required 637 Nm. Hence, the machine achieves robust mechanical stability for high power density drive applications.

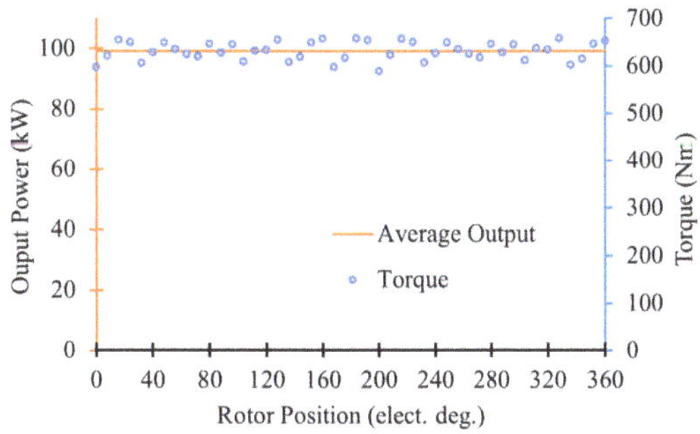

Figure 9: Output power and electromagnetic torque waveform during on load condition

Summary

The analysis of the 9s/8p SSDR AFPM design validated that the design reaches exceptional operational integrity. Magnetic vector analysis proves the SSDR configuration effectively channels the magnetic field along the axial path, producing a peak flux density of almost 0.9 T. This established strong field creates a clean, quasi-trapezoidal flux waveform, confirming high magnetic utilization. Field visualizations further confirm that the high quality electrical steel successfully channels dense magnetic flux through the central stator yoke, managing saturation levels. Electrically, the induced voltage waveform exhibits symmetrical three phase characteristics under no load, confirming 500 V_{rms} output. Under on load condition, the voltage profile remains robust, peaking near 900 V and confirming the machine sustains the entire 100 kW requirement. Furthermore, the harmonious interaction between the voltage waveform and the applied current successfully drives the full 637 Nm of rated torque, validating stable electromagnetic performance under heavy demand and confirming high magnetic coupling for sustainable full power utilization.

Aknowledgement

The authors would like to acknowledge the support from the Ministry of Higher Education (MoHE) Malaysia through the Fundamental Research Grant Scheme (FRGS) under a grant number of FRGS/1/2023/TK02/UNIMAP/02/5.

References

[1] L. Shao, R. Navaratne, M. Popescu, and G. Liu, "Design and Construction of Axial-Flux Permanent Magnet Motors for Electric Propulsion Applications-A Review," IEEE Access. 2021. https://doi.org/10.1109/ACCESS.2021.3131000

[2] J. Zhao, X. Liu, S. Wang, and L. Zheng, "Review of Design and Control Optimization of Axial Flux PMSM in Renewable-energy Applications," Chinese Journal of Mechanical Engineering (English Edition). 2023. https://doi.org/10.1186/s10033-023-00868-8

[3] Z. Hao, Y. Ma, P. Wang, G. Luo, and Y. Chen, "A Review of Axial-Flux Permanent-Magnet Motors: Topological Structures, Design, Optimization and Control Techniques," Machines. 2022. https://doi.org/10.3390/machines10121178

[4] S. W. Song, S. H. Lee, and W. H. Kim, "Design and Analysis of Consequent Pole Axial Flux Motors for Reduced Torque Ripple and Magnet Consumption," Processes, vol. 13, no. 7, pp. 1-15, 2025. https://doi.org/10.3390/pr13072139

[5] S. Ravindran, P. Koothu Kesavan, D. Banjerdpongchai, and P. Prem, "Modelling, implementation and analysis of double-side slotted axial flux PMGs suitable to small-scale wind energy conversion systems," Sci. Rep., vol. 15, no. 1, pp. 1-20, 2025. https://doi.org/10.1038/s41598-025-08716-6

[6] I. A. Rudden, G. J. Li, Z. Q. Zhu, A. Duke, and R. Clark, "Fractional-Slot Concentrated Windings for Offshore Wind Turbine Generators: Opportunities and Challenges," IEEE Access, vol. 12, no. September, pp. 158766-158787, 2024. https://doi.org/10.1109/ACCESS.2024.3486000

[7] S. Y. Kim, "A study on motor characteristics based on pole-slot combinations," AIP Adv., vol. 14, no. 1, pp. 10-15, 2024. https://doi.org/10.1063/9.0000647

[8] X. Sun, Z. Q. Zhu, S. Cai, L. Wang, F. R. Wei, and B. Shao, "Influence of Stator Slot and Rotor Pole Number Combination on Field Winding Induced Voltage Ripple in Hybrid Excitation Switched Flux Machine," IEEE Trans. Energy Convers., vol. 36, no. 2, pp. 1245-1261, 2021. https://doi.org/10.1109/TEC.2020.3035258

[9] A. M. EL-Refaie, "Fractional-slot concentrated-windings synchronous permanent magnet machines: Opportunities and challenges," IEEE Trans. Ind. Electron., vol. 57, no. 1, pp. 107-121, 2010. https://doi.org/10.1109/TIE.2009.2030211

[10] D. Quan et al., "The Design, Analysis, and Verification of an Axial Flux Permanent Magnet Motor with High Torque Density," Appl. Sci., vol. 15, no. 6, 2025. https://doi.org/10.3390/app15063327

[11] J. F. Zhao, X. Y. Liu, S. Wang, and L. X. Zheng, "Review of Design and Control Optimization of Axial Flux PMSM in Renewable-energy Applications," CHINESE J. Mech. Eng., vol. 36, no. 1, 2023. https://doi.org/10.1186/s10033-023-00868-8

[12] S. Sarıdağ and M. Eker, "Design of Double-Stator Axial Flux Permanent Magnet Synchronous Motor," J. New Results Eng. Nat. Sci., vol. 2024, no. 21, pp. 30-46, 2024.

[13] O. Shariati, A. Behnamfar, and B. Potter, "An Integrated Elitist Approach to the Design of Axial Flux Permanent Magnet Synchronous Wind Generators (AFPMWG)," Energies, 2022. https://doi.org/10.3390/en15093262

[14] S. Kahourzade, A. Mahmoudi, H. W. Ping, and M. N. Uddin, "A comprehensive review of axial-flux permanent-magnet machines," Canadian Journal of Electrical and Computer Engineering. 2014. https://doi.org/10.1109/CJECE.2014.2309322

[15] R. Pormoozeh-Noveiry and M. Niaboli-Guilani, "Optimizing the Efficiency of a Three Phase Axial Flux Synchronous Generator for Use in Direct-Drive Wind Turbines and Modeling by 3D Finite Element Method," in 2023 5th International Conference on Optimizing Electrical Energy Consumption, OEEC 2023, 2023. https://doi.org/10.1109/OEEC58272.2023.10135483

Smart Sprinkler and Irrigation System for Ifrane Region: A Simulation-Based Approach Using AVR Microcontrollers

Ahlam MOUSSA[1,a], Kacem CHAWQI[1,b *], Ilham NAIT HIM[1,c], Jamal BAHRI[1,d] and Moulay El Hassan El AZHARI[1,e]

[1]P.O Box 104, Hassan II Avenue, 53000 Ifrane, Morocco

[a]ah.moussa@aui.ma, [b]k.chawqi@aui.ma, [c]i.naithim@aui.ma, [d]j.bahri@aui.ma, [e]m.elazhari@aui.ma

Keywords: Smart Irrigation, AVR Microcontroller, PWM Control, Soil Moisture Sensing, PID Control, Weather-Aware Irrigation, Proteus

Abstract. A method to minimize and monitor water and energy consumption in agriculture and landscaping lies in using smart automated sprinkler irrigation systems. This paper presents a complete simulation of a smart irrigation design for efficient water use and optimal soil quality. The proposed system utilizes an ATmega32 AVR microcontroller, along with soil moisture, temperature, gas, and humidity sensors, a Pulse Width Modulation (PWM) control of water pumps and valves. It is automated to regulate irrigation intensity with a closed-loop control enhanced by a PID controller and a lightweight artificial intelligence-inspired decision layer using rainfall data of the region of Fes-Meknes-Ifrane for threshold adjustments. The results highlight the improvement of soil moisture stability in the range of 4%, as well as a respective mean absolute error and a root mean square error of 3.5% and 5% compared to a threshold sensing method only.

Introduction

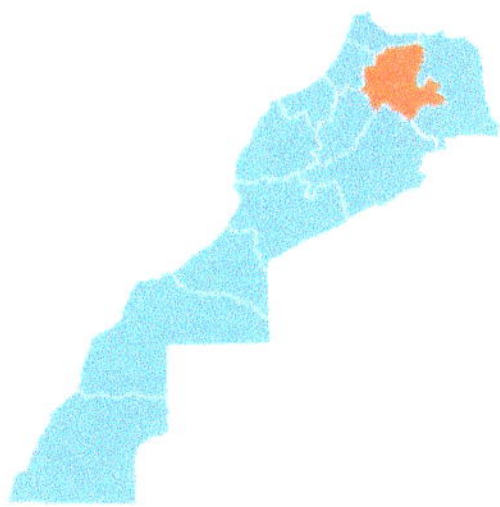

Fig. 1 Water Stress Analysis Map: Fez-Meknes-Ifrane Region

Water scarcity is a growing issue worldwide, particularly in arid and semi-arid regions like Morocco [1]. Traditional irrigation practices often rely on fixed schedules and manual control, resulting in increasing water consumption and uneven soil hydration, directly affecting energy costs, and, since often water sources are underground, accelerating aquifer depletion [2]. This emphasizes the importance of developing systems capable of adapting water delivery to real-time environmental conditions [3].

With the rising water demand in agriculture and urban landscapes, smart irrigation techniques emerged. This work proposes a new and fully simulated design for a smart sprinkler and irrigation system based on an AVR ATmega32 microcontroller and AI integration with a focus on the Fez-Meknes-Ifrane region, Fig. 1. The system combines a low-cost and modular AVR-based architecture with adaptive PWM actuation, closed-loop PID control strategy, multi-zone irrigation, safety awareness, as well as Artificial Intelligence (AI)-inspired weather-informed logic that inputs

Materials Research Proceedings 64 (2026) 299-305 https://doi.org/10.21741/9781644904091-37

rainfall probability into irrigation decisions. This allows the system to avoid watering in case of a rain forecast then the system conserves water depending on the weather patterns and requirements in the area.

Although this work is a fully functional Proteus simulation, it offers vital information on the behavior and robustness of the system, which can be used for prototyping.

Materials and Methods

System Architecture. The overall architecture of the system consists of four main blocks, illustrated in Fig. 2. An ATmega32 AVR microcontroller serves as the central component, due to its efficient cost and low energy consumption. Hence, it could be power with solar energy.

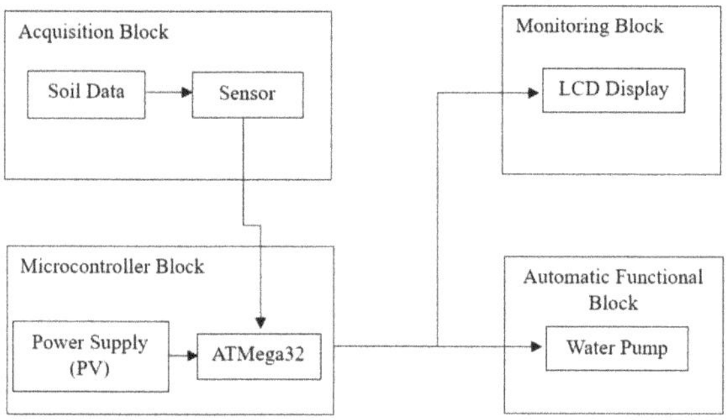

Fig. 2 Overall System Architecture Block Diagram.

Sensors Parameters and Threshold Selection. The soil is considered dry when moisture levels are less than 40% and optimal when between 40% and 70% [4]. The temperatures have a cutoff threshold of 50 °C to avoid overheating. For safety, a gas/smoke sensor is implemented too, and uses a predefined broad range between 10 and 1000 ppm.

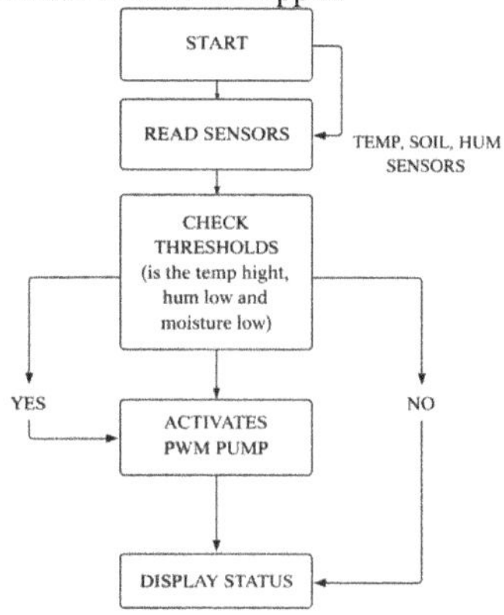

Fig. 3 Flowchart of Embedded Control Logic

Functional Overview. The microcontroller ADC samples sensor data periodically, and the integrated control algorithm compares measured values with predefined thresholds. Then, the appropriate PWM duty cycle is computed. Actuators are then driven accordingly to regulate water flow. An LCD and Universal Asynchronous Receiver/Transmitter (UART) interface continuously provide feedback to the user and allows the user to configure the pump if needed too. We used Atmel Studio for the embedded C code. The algorithm is summarized in Fig. 3.

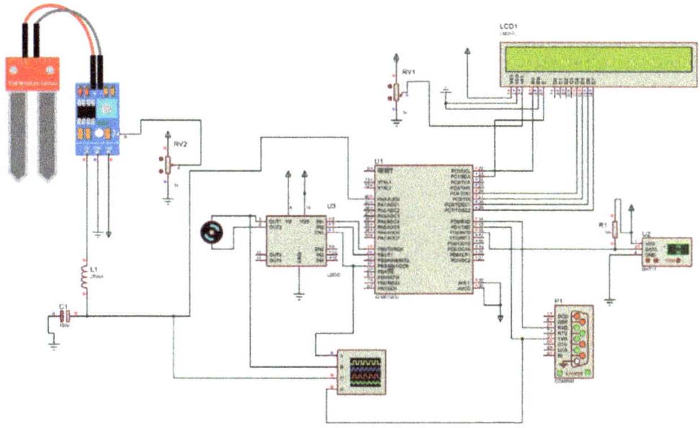

Fig. 4 Proteus Circuit

Proteus Simulated Circuit Design. Fig. 4 represents the Proteus circuit with a single sprinkler configuration. The AVR operates at a frequency of 8MHz for a 9600 baud serial data.

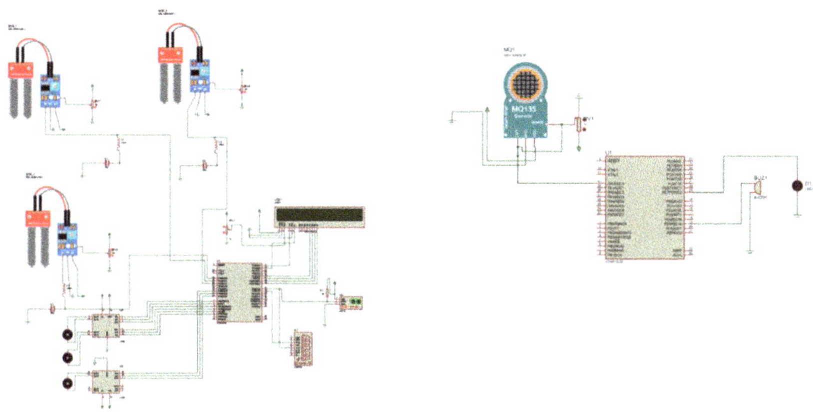

Fig. 5 Overview of the Proteus Zone Control and Safety Circuit

transmission and correct readings for sensitive components. It is connected to the moisture sensors controlled with a potentiometer, and a temperature sensor, DHT11, also capable of measuring relative humidity. The water pump or valve controlled by the PWM is represented by the PWM-controlled DC motor. The 16x2 LCD represents the user interface to display sensed data, the state of the pump and energy consumed. Periodic updates and threshold configuration are provided through a UART virtual terminal.

Using three virtual soil moisture sensors connected to separate ADC channels of the ATmega32, the zone control feature divides the irrigation area into three manageable zones. Each zone is associated with its own PWM-controlled DC motor actuator (e.g., a pump) and respective sensors, as shown in Fig. 5. At every sampling period, the microcontroller reads the moisture levels from the sensors and compares the values to reference moisture levels. The resulting error drives the appropriate PWM duty cycle to precisely feed the motor driving the valve or pump of the specific zone, rather than a simple ON/OFF operation.

As for environmental safety, a gas sensor MQ-135 was used to detect general air pollutants (e.g., CO_2 *and* NH_3) and smoke. This analog sensor sends signals to the AVR once a threshold level is exceeded, immediately indicating a buzzer, turning on an LED, and shutting down the pumps.

PID Controller. To adjust the motors' speed in response to fluctuations in soil moisture and temperature and continuously minimize the sensor error, we used a PID controller [5]. The error is translated from the duty cycle of the PWM, and is defined as

$$e(t) = r(t) - y(t), \tag{1}$$

where $r(t)$is the reference soil moisture (or temperature limit) and $y(t)$is the measured value. Since data was sampled discreetly, the control output was computed in discrete form with a sampling period of 1s using the Backward differentiation:

$$u[k] = K_p e[k] + K_i \sum_{i=0}^{k} e\,[i] T_s + K_d \frac{e[k] - e[k-1]}{T_s}. \tag{2}$$

The output $u[k]$ modulates the pump and defines the operation of the PWM. The coefficients K_p, K_i and K_d represent the gains, tuned iteratively based on the step response. The time sample used is T_s=1s. The results of approaches using thresholds only were compared to those of this PID control approach, to scrutinize the stability and convergence of the system.

Results and Discussion

Oscilloscope readings in Fig. 7 present the PID and PWM-driven behavior (Yellow and Blue traces). The y-axis represents logic-level voltage (0-5 V), while the x-axis represents time. A "widening" of the rectangular pulses is observed in this scope-view, corresponding to an increase in PID control effort, following a detection of a deficit in moisture given by the soil moisture sensor (red trace). The proportional term K_p particularly triggers the expansion of the duty cycle, maximizing actuator power.

Simulation Limitations. Some limitations must be regarded, despite Proteus providing a reliable platform to conduct the functional validation of the proposed system. First, since they do not consider noise, the sensor models are abstract with calibration error or drift over time. Secondly, instead of physical hydraulic models, the soil moisture logic was modeled with threshold values, rendering them not always aligned with real field behavior. Real-world implementation would require calibration and robustness testing under varying operating conditions, even though the simulation was stable in control and logically correct.

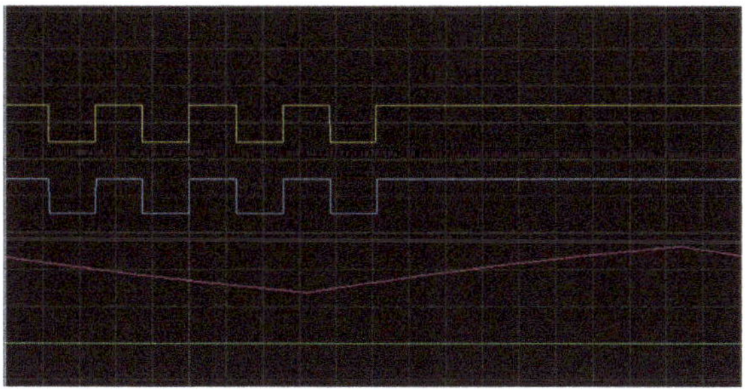

Fig. 7 PWM output under PID control

Error Metrics. Over a sampling interval of 1 second, a time domain metrics evaluation was discreetly computed. The signal error for each zone was defined as the difference between the soil moisture level defined as 55%, and the average of optimal moisture levels used in the literature.

Table 1: Error metrics values

Control	*MAE*	*RMSE*	*Max Error*
Threshold only	*8.5%*	*10%*	*15%*
Threshold + PID	*3.5%*	*5%*	*4.5%*

Additionally, ADC values to percentage scales were mapped. The error metrics were computed over 600 samples for an activation period of 10 minutes. Absolute error and root mean square error were calculated in order to calculate the mean. The soil moisture behavior is illustrated in Figure 8. As compared to the threshold-only approach, it is noticeable that the PID system stabilizes with respect to the reference with fewer oscillations, better convergence, and a tighter tracking band that remains within about 4% of the reference. This explains the decreased error measurements in Table 1. The MAE and RMSE values are rounded averages.

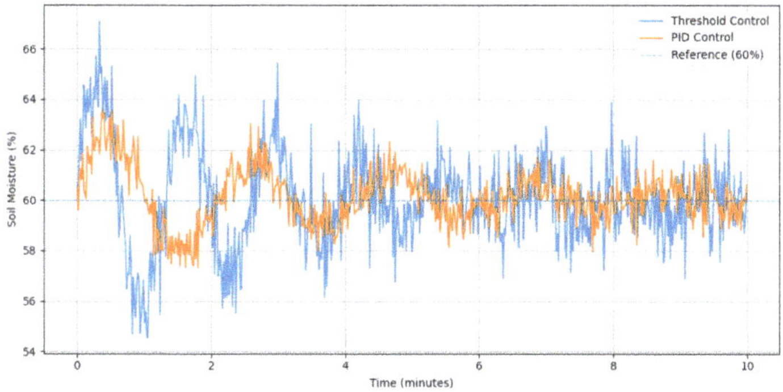

Fig. 8 Soil Moisture Regulation During Irrigation Cycle

Real Time Weather Integration. To take larger and updated environmental inputs, the weather station and local infrastructure in Ifrane were used [6]. By integrating an AI-inspired decision layer to simulate intelligent irrigation behavior in response to forecasted environmental conditions, the system becomes weather-aware. The system dynamically adjusts its behavior based on rain probability data specific to the Ifrane region, rather than relying on static thresholds alone.

Regional precipitation trends were integrated with data from the Moroccan National Meteorological Service and publicly available historical weather databases, such as MeteoBlue and OpenWeatherMap, which provide hourly and daily rain probability forecasts for Ifrane [7, 8]. For other broader regions, data would be accessed with an API and parsed through an external script or onboard IoT [9].

In comparison, traditional irrigation controllers without environmental awareness might activate regardless of incoming rainfall, leading to inefficient water use. Real-time integration with weather APIs and adaptive machine learning models could automate this logic entirely in future versions of this system, requiring no human intervention to optimize decisions based on historical and forecasted data.

Technically, the integration of AI can be achieved through UART communication protocols or by incorporating wireless modules, such as ESP8266/ESP32, for API-based access to weather station data over Wi-Fi [10]. JSON-formatted weather data from online APIs or local campus networks can be retrieved by these modules. This information can then be parsed by the microcontroller to decide autonomously. This system could also participate in a distributed environmental sensing network managed centrally by university facilities, if expanded into a microgrid of sensors. The limitations of such a setup can be found in the asynchronous nature of the UART; that is why the baud rate between the connected devices must be matched. UART is also typically used for short-range communication, which must be considered in the setup [10].

Implementation of the System for Universities. Linking the integration of smart sprinklers with a university Smart Campus Dashboard can be done to create a centralized platform for managing water and energy resources and generate long-term analytics on vegetation health and resource efficiency. This approach, which is infrastructure-aware, aligns with global trends in smart agriculture and urban planning, where more resilient and adaptive ecosystems are reached thanks to the fusion of embedded systems, real-time data analytics, and Internet of Things (IoT) technologies [11]. For Al Akhawayn University in Ifrane, implementing such a system offers an interdisciplinary educational platform that engineering students can use for experimentation and innovation support, while sustaining its sustainability mission.

Comparative evaluation. By assessing existing smart irrigation solutions, the novelty of the proposed system is emphasized. The ATmega32-based Proteus prototype combines multi-zone sensing, PWM control, and PID correction, on top of environmental responsiveness. Such a holistic approach distinguishes itself from existing single-zone solutions [12, 13]. Furthermore, the extension of the suggested model with a safety circuit focused on smoke detection for fire prevention renders it more relevant to agricultural applications. The design also sets itself apart from conventional systems by supplementing its approach with UART-based weather adaptation, thus enhancing its precision and scalability into a larger network for minimized water waste.

Conclusion and Future Work

This work simulated a cost-efficient, scalable, and environmentally adaptive microcontroller-based sprinkler and irrigation system with temperature, humidity, and soil moisture sensors with a PWM actuator and PID regulation. The simulation demonstrated strong stability and precise sensing measurements reaching mean absolute errors of 3.5%. In a world driven by AI, it makes use of it to calibrate irrigation intensities and water monitoring. By taking multizone irrigation and safety measurements into account, the system is user-friendly and highly adaptable to different applications. Although the simulation is complete, a real-world application may require minor calibrations. Finally, suggested enhancements can be added to expand the system's scalability and application through IoT integration, Wi-Fi, and GSM connectivity, mainly for remote control, cloud-based analytics, and mobile notifications. The development of a user-friendly mobile application to visualize sensor data, receive alerts, and configure thresholds remotely can be worked on as a separate project.

References

[1] F. Taheripour, et al., Water Scarcity in Morocco, World Bank Group, 2020, p. 10. https://documents1.worldbank.org/curated/en/642681580455542456/pdf/Water-Scarcity-in-Morocco-Analysis-of-Key-Water-Challenges.pdf

[2] R.M. Cruse, D.L. Devlin, D. Parker, R.M. Waskom, Irrigation aquifer depletion: the nexus linchpin, J. Environ. Stud. Sci. 6 (2016) 149-160. https://doi.org/10.1007/s13412-016-0370-y

[3] G. Rajagopal, V.M. Lodd, A. Vignesh, R. Rajesh, V. Vijayaraghavan, Low cost cloud based intelligent farm automation system using Bluetooth Low Energy, in: 2014 IEEE Region 10 Humanitarian Technology Conference (R10 HTC), Chennai, India, 2014, pp. 127-132. https://doi.org/10.1109/R10-HTC.2014.7026324

[4] E.F. Camacho, C. Bordons, Constrained model predictive control, in: Model Predictive Control, Springer London, London, 2007, pp. 177-216.

[5] M.A. Johnson, M.H. Moradi, PID Control, Springer-Verlag London Limited, London, 2005.

[6] M. Lagasio, S. Barindelli, Z. Chitu, S. Contreras, A. Fernández-Rodríguez, M. de Klerk, A. Fumagalli, A. Gatti, L. Hammerschmidt, D. Haskovic, M. Milelli, Integrating Advanced Sensor Technologies for Enhanced Agricultural Weather Forecasts and Irrigation Advisories: The MAGDA Project Approach, Remote Sens. 17 (2025) 1855. https://doi.org/10.3390/rs17111855

[7] Information on www.meteoblue.com/en/weather/week/blue_united-states_5286039

[8] Information on https://openweathermap.org

[9] A.R. Al-Ali, I. Zualkernan, M. Rashid, A Smart Home Energy Management System Using IoT and Big Data Analytics Approach, IEEE Trans. Consum. Electron. 63 (2017) 426-434. https://doi.org/10.1109/TCE.2017.015014

[10] P. Megantoro, S.A. Aldhama, G.S. Prihandana, P. Vigneshwaran, IoT-based weather station with air quality measurement using ESP32 for environmental aerial condition study, TELKOMNIKA (Telecommun. Comput. Electron. Control) 19 (2021) 1316-1325. https://doi.org/10.12928/telkomnika.v19i4.18567

[11] A.D. Dharmavaram, T.S. Savithri, S.L. Sai, Design and implementation of real time data acquisition system using reconfigurable SoC, Int. J. Adv. Comput. Sci. Appl. 11 (2020) 9. https://doi.org/10.14569/IJACSA.2020.0110953

[12] S.P. Shaik, N.V. Sarath, P.N. Ramarao, Smart Irrigation System for Sweet Corn, in: 2025 International Conference on Electronics and Renewable Systems (ICEARS), 2025, pp. 1498-1504. https://doi.org/10.1109/ICEARS61622.2025.10474640

[13] R.K. Jain, Design and Modeling of Control System for Automation of Irrigation System Using PID and ANN Methods, Available at SSRN 4197630 (2022). https://doi.org/10.2139/ssrn.4197630

Digital Twins for Sustainable Energy Optimization and Predictive Maintenance at Noor Ouarzazate, Morocco

Hicham AGOURAME[1,a*], Tawfik MASROUR[1,2,b]

[1]Department of Mathematics and Computer Science, ENSAM-Meknes, Moulay Ismail University of Meknes, 50050, Meknes, Morocco

[2]Mathematics, Computer Science and Engineering Department, University of Quebec at Rimouski, Rimouski, Canada

[a]hicham.agourame@edu.umi.ac.ma, [b]t.masrour@ensam.umi.ac.ma

Keywords: Digital Twin, Noor Ouarzazate, Concentrated Solar Power, Predictive Maintenance, Renewable Energy

Abstract. A hybrid digital twin solution was thus created for the Noor Ouarzazate Solar Complex to improve the efficiency and functionality of CSP (Concentrated Solar Power) and PV (Photovoltaic) plants on a larger scale. This project arose from the prolonged shutdown of Noor III because of a molten salt leak. This justifies the need for a better monitoring and maintenance system. It combines simulation models with the help of machine learning and IoT sensors. It thus provides an efficient way for the early detection and anticipation of faults, including molten salt tank failure, in both the thermal and electrical systems. The simulation outcome shows that the digital twin operates with a real-time fault classification performance of 98.5% precision, recall, and F1-score, while reporting key operational indicators including a mean time to repair (MTTR) of 27.3 hours, a mean time between failures (MTBF) of 1 hour, an efficiency of approximately 95%, and a remaining useful life estimate of up to 6 months. In addition, the dashboard provides an indicator of the real-time CO_2 impact of 1929.0 tons within the operating conditions that are being monitored. The proposed model of the digital twin has great potential and can serve as one of the feasible solutions that can improve practices of predictive maintenance in complex systems in the renewable energy industry.

Introduction

Noor Ouarzazate continues to play a leading role in Morocco's renewable energy market, with over 500 MW of CSP and PV technologies integrated to power one million homes at least [1,2]. Beyond its iconic position in Morocco's energy transition, the project has immense value for the development of renewable energy. Despite the distinctive features the project has to offer, there remain critical monitoring and maintenance considerations in CSP, largely due to its thermal laws of physics coupled with the magnitude of energy storage. Digital Twin technology acts as a paradigm shift in asset management, as it creates a digital replica of a system to be observed continuously [3,4]. The combination of digital twins with predictive maintenance helps to identify system deterioration ahead of time, thereby improving efficiency [5]. Despite these advantages, most existing digital twin approaches for large-scale CSP–PV systems fail to integrate physics-based models with real-time operational data within a unified framework, which limits their effectiveness for predictive maintenance and dispatch optimization [6].

This paper proposes a hybrid digital twin framework for the Noor Ouarzazate Solar Complex that integrates physics-based simulation models, IoT-enabled sensing, and machine learning algorithms. The proposed approach enables early fault detection across thermal and electrical subsystems, supports predictive maintenance decision-making, and improves system availability and operational efficiency.

The remaining part of this paper is organized as follows. The current state of the art is reviewed in Section 2. The materials and methods are described in Section 3. The results of implementation and simulation are presented in Section 4 and a conclusion in Section 5.

Previous Works on Digital Twins at Noor and Beyond

The application of digital twins in the renewable energy industry has appeared as a new area for research and development, with initial research focusing on predictive maintenance, online monitoring, and system-level simulation for concentrating solar power (CSP) and photovoltaic (PV) plants [7, 8]. Projects developed using CSP have presented distinct modeling challenges, especially for large plants such as Noor Ouarzazate [9, 10]. In many existing DT platforms, one challenge lies in the realm of integration, which comes from the mixed source characteristics for CSP-PV systems. Addressing this challenge is critical to facilitate correct diagnosis and effective operation management in large-scale hybrid solar power stations.

To better understand how these approaches compare, a comparative analysis of DT frameworks is presented in Table 1.

Comparative Framework Analysis

The proposed digital twin (DT) framework of the Noor Ouarzazate CSP–PV complex in Table 1 is compared with existing PV-specific and industrial DT frameworks across five consolidated criteria: monitoring and data integration, modeling and prediction, scalability and adaptivity, decision support and validation, and energy optimization.

Table 1 Comparative Analysis of Digital Twin Frameworks

Criterion	Noor Ouarzazate DT	PV DT frameworks	Industrial DT frameworks	Key References
Monitoring & Data	Full IoT sensor integration	PV performance diagnostics	Continuous industrial monitoring	[7],[8],[9]
Modeling & Prediction	Hybrid physics + ML, fault prediction	Physics + ML / data fusion, power forecasting	Physics + data-driven models, downtime reduction	[7],[8],[9],[10]
Scalability & Adaptivity	MW-scale CSP/PV, continuous learning	PV array & farm scale, adaptive power prediction	Large plant scale, self-learning equipment models	[6],[8],[10],[13]
Decision Support & Validation	Maintenance scheduling, dispatch alerts, historical + real data	Dashboards, optimization suggestions, PV testbeds & Simulink	Operational & maintenance support, industrial benchmarking	[7],[8],[9],[13]
Energy Optimizatio n	CSP + PV output optimization	PV array & farm scale, adaptive power prediction	Large plant scale, self-learning equipment models	[6],[8],[10],[13]

The Noor Ouarzazate DT demonstrates that few frameworks integrate multi-source IoT sensing, physics-informed modeling, and machine learning for adaptive predictive maintenance in hybrid CSP–PV systems, whereas existing PV-specific and industry-grade DTs focus mainly on real-time monitoring and basic fault diagnosis.

Materials Research Proceedings 64 (2026) 306-313 https://doi.org/10.21741/9781644904091-38

Materials and Methods

Data Acquisition and Preprocessing

Operational data were collected from five key components of the Noor CSP–PV complex: CSP Loop, PV Array, Salt Tank, Pump, and Heliostat [14]. Sensor measurements include temperature [°C], vibration [m/s^2], and pressure [bar].

Due to confidentiality constraints, a simulated dataset was generated to reflect real-world operating conditions. Fault scenarios were incorporated with a 25% probability across all components, focusing on high-risk assets such as the salt tank and heliostat.

The preprocessing steps included the following:

- Outlier elimination and noise smoothing.
- Temporal alignment of sensor readings.
- Feature Normalization for Machine Learning Input [15].

The transformation from raw to preprocessed data is presented in Fig. 1.

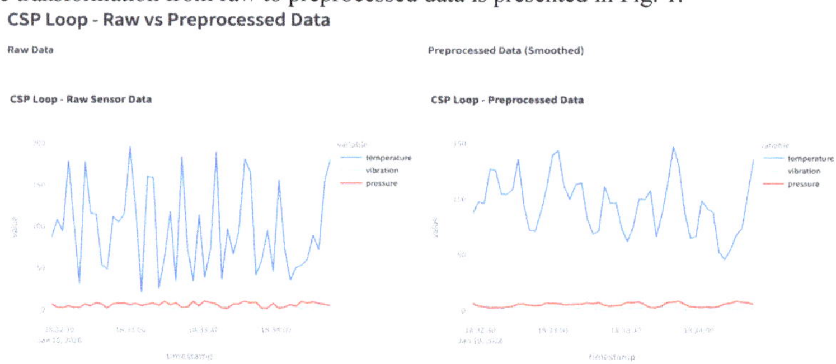

Fig. 1 Time-Series Trends of Temperature, Vibration, and Pressure for Key Noor CSP–PV Components

Hybrid Digital Twin Model

A hybrid digital twin is implemented by integrating the physical models of the Complex CSP-PV System (with the CSP Loop, PV Array, Salt Tank, Pump, and Heliostat Field) and Random Forest models for predictive maintenance [16, 17].

Physics-Based Modeling

The thermal processes of the solar thermal power plant (CSP) system and the molten salt storage tanks are modeled by energy balance equations:

$$\frac{dT(t)}{dt} = \frac{Q_{in}(t) - Q_{out}(t) - Q_{loss}(t)}{mc_p}$$

where:

$T(t)$ = fluid temperature

Q_{in} = absorbed solar energy

Q_{out} = delivered thermal power

Q_{loss} = system losses

m = effective mass

c_p = specific heat capacity

These equations are applied to each component to model expected behavior under nominal operating conditions [18, 19].

Random Forest–Based Health Estimation
A Random Forest Classifier is used for the detection of anomalies and classification of the status of each component [20, 21]

$$\hat{y}(t) = \frac{1}{N}\sum_{i=1}^{N} h_i(x(t))$$

where h_i is the output of the i-th decision tree, N is the total number of trees, and $x(t)$ contains temperature, vibration, and pressure features. The predicted state $\hat{y}(t)$ is coded as
0 = Normal; 1 = Maintenance Advised; 2 = Critical Fault

Hybrid Integration
Residuals between measured and physics-based predicted values are computed [20, 24]:

$$r(t) = y_{measured}(t) - y_{physics}(t)$$

The random forest analyzes these residuals to enhance fault sensitivity and reduce false positives [21].

Predictive Maintenance Architecture
There are four levels in the framework of predictive maintenance [20, 22, 24], as illustrated in Fig. 2:
- Sensing Layer: The operational data is captured through dispersed Internet of Things (IoT) sensors.
- Integration Layer: The streaming of information is aligned and filtered into a master platform.
- Decision Support Layer: The analysis of key performance metrics, such as Mean Time to Repair (MTTR), Mean Time Between Failures (MTBF), CO_2 emissions saved, lifetime extensions, and Dispatch Efficiency, takes place, and recommendations for maintenance activities are made.

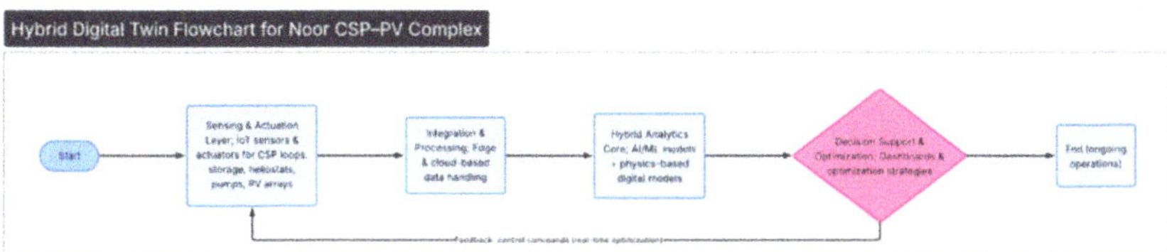

Fig. 2 Predictive Maintenance Framework: From Sensor Data to Decision Support

Dashboard Implementation
A prototype dashboard was created using the Streamlit framework in Python to display real-time simulated sensor data, component health conditions, KPI (MTTR, MTBF, efficiency, and RUL), and risk heat maps for effective maintenance. In view of confidentiality restrictions regarding the simulated values, additional real-time knowledge was incorporated using publicly available data from similar plants; the source code is found on GitHub [27]. This enables the operators to carry out interactive analysis of the status in real time [25, 26].

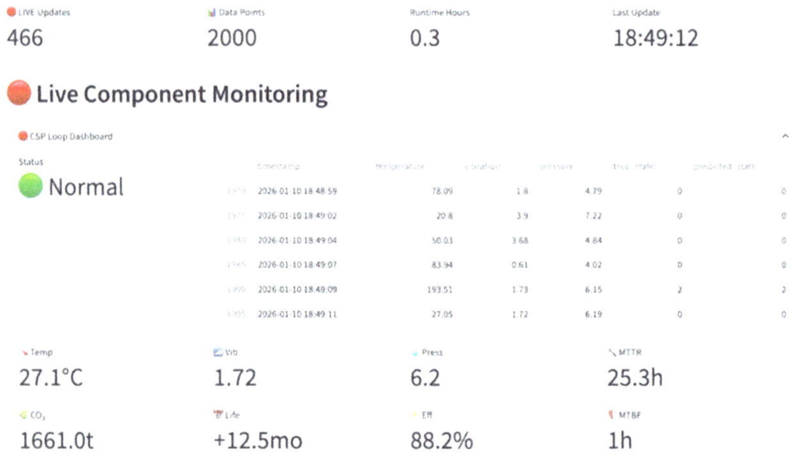

Fig. 3 Streamlit Digital Twin Dashboard for Noor Ouarzazate Solar Complex For CSP Loop

Results and Analysis

Model Performance

The prediction validity of the digital twin model for fault classification was validated using simulation operational data acquired from the Noor Ouarzazate Solar Complex. From the digital twin dashboard in Fig. 3 and Fig. 5, it is evident that the developed digital twin model performs quite well in all operational phases (Normal, Warning, and Critical) with a total accuracy of 98.5%. These findings support the effectiveness of the developed digital twin model for the early identification of potential molten salt tank failures as well as abnormal operating conditions in the thermal/electrical subsystems.

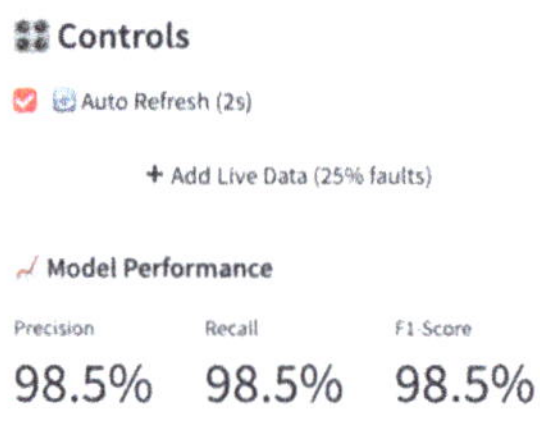

Fig. 5 Digital Twin Dashboard of Precision, Recall, and F1-Score for Operational State Classification

Fault Detection And KPI Analysis

The simulation-based monitoring facilitated the early diagnosis of the faulty state of the system on the component level. The warning conditions were detected in the CSP loop, the warning state in the CSP pump, the critical state in the molten salt storage tank, the warning state in the heliostat, and the warning state in the photovoltaic array, while the heliostat and photovoltaic array remained in the normal state. The important operational KPIs extracted from the simulation are summarized in Table 4.3. The result indicates that the mean time to repair (MTTR) takes 27.3 hours, the mean time between failures (MTBF) is 1 hour, the efficiency of the system is around 95%, the expected remaining useful life is 6 months, and the CO_2 impact indicator is 1,929 tons.

Table 2 Simulated Operational KPIs

Metric	**Simulated Value**
Mean Time to Repair (MTTR)	27.3 h
Mean Time Between Failures (MTBF)	1 h
Efficiency	~95%
Remaining Useful Life (RUL)	6 months
CO_2 Impact Indicator	1929 t

Conclusion

Despite these constraints, however, the development of digital twin platforms and pilot projects at Noor Ouarzazate is expected to pave the way for easier acceptance of digital twins in renewable energy. The above-mentioned framework is a fundamental platform for data-managed concentrated solar power systems with hybrid solar systems. This research develops a unique hybrid digital twin platform for Noor Ouarzazate and gives a proposal model adoptable for larger projects in renewable energy. The developed platform provides substantial integration with physicist simulation systems for concentrated solar power systems and AI simulation systems for predictive systems with high predictive capabilities and advanced energy management. There is a substantial integration of the system with predictive management processes within its overall structure with specific emphasis on tracking important process systems within its structure, such as the molten salt storage tank for energy management. It should be reiterated here, however, that there is a fundamental constraint to this platform; that is to say, this platform has not been validated with real-world data. Only simulated data has been validated within this research with due acknowledgement to confidentiality. There is substantial simplification within some system simulations within this platform, including fundamental simplifications of simulations within systems such as the storage system of the molten salt storage tank.

References

[1] OM. S. Abdelkareem, T. A. Abbas, "Concentrated Solar Power Technology: Current Status and Research Trends," Solar Energy, vol. 216, pp. 264–281, 2021. https://doi.org/10.1016/j.solener.2020.12.065

[2] Solar Energy Resource and Power Generation in Morocco: Current Situation, Potentials and Future Perspective, Resources, vol. 13, no. 10, art. 140, 2024. https://doi.org/10.3390/resources13100140

[3] Mbasso, W. F. (2025). Digital twins in renewable energy systems. ScienceDirect. Retrieved from https://www.sciencedirect.com/science/article/pii/S2211467X25001774

[4] Abdessadak, A. (2025). Digital twin technology and artificial intelligence in energy systems. ScienceDirect. Retrieved from https://www.sciencedirect.com/science/article/pii/S235248472500263X

[5] S. Ma, K. A. Flanigan, M. Bergés, State-of-the-Art Review: The Use of Digital Twins to Support Artificial Intelligence-Guided Predictive Maintenance, arXiv:2406.13117, 2024.

[6] W. Fendzi Mbasso, A. Harrison, I. Dagal, P. Jangir, M. Khishe, H. Kotb, M. S. Shaikh, A. Smerat, E. F. Donfack, R. Kumar, "Digital twins in renewable energy systems: A comprehensive review of concepts, applications, and future directions," Energy Strategy Reviews, vol. 61, 101814, 2025. https://doi.org/10.1016/j.esr.2025.101814

[7] A. Elnosh, M. Calais and D. Parlevliet, "A systematic literature review of digital twin research for photovoltaic systems: Trends, challenges, and opportunities," Renew. Sustain. Energy Rev., vol. 226, p. 116326, Feb. 2026, https://doi.org/10.1016/j.rser.2025.116326

[8] M. Fadel, F. M. Alelaj, "Digital twin-based performance evaluation of a photovoltaic system: A real-time monitoring and optimization framework," Int. J. Power Electron. Drive Syst., vol. 16, no. 3, pp. 2072-2081, Sep. 2025, doi: 10.11591/ijpeds.v16.i3.pp2072-2081

[9] Y. Zhang, X. Zhao, "Digital twin integration with data fusion for enhanced photovoltaic system management: A systematic literature review," IEEE Open Journal of Power Electronics, vol. 6, pp. 210–228, 2025.

[10] L. Tran, M. Lee, "AI-based predictive maintenance of solar photovoltaics systems: A comprehensive review," Energy Informatics, vol. 8, 128, 2025. https://doi.org/10.1186/s42162-025-00594-6

[11] F. Chen, H. Wu, "A novel digital-twin approach based on transformer for photovoltaic power prediction," Scientific Reports, vol. 14, 11234, 2024.

[12] A. Kumar, S. Reddy, "Digital twin integration for photovoltaic–battery energy storage systems: A systematic review," Future Energy and Environment Letters, vol. 3, no. 2, pp. 45–60, 2025

[13] Z. Chen, J. Li, A. Kumar, "Digital twins in renewable energy systems: A comprehensive review of concepts, applications, and future directions," Energy Strategy Reviews, vol. 61, 101814, 2025. https://doi.org/10.1016/j.esr.2024.101814

[14] W. F. Fendzi Mbasso, A. Harrison, I. Dagal, P. Jangir, M. Khishe, H. Kotb, M. S. Shaikh, A. Smerat, E. F. Donfack, R. Kumar, "Digital twins in renewable energy systems: A comprehensive review of concepts, applications, and future directions," Energy Strategy Reviews, vol. 61, 101814, 2025. https://doi.org/10.1016/j.esr.2025.101814

[15] "Digital twin-driven sustainable energy life cycles: Technical review and guidelines," Renewable and Sustainable Energy Technology, 2025. https://doi.org/10.53941/rset.2025.100005

[16] W. F. Fendzi Mbasso, A. Harrison, I. Dagal, P. Jangir, M. Khishe, H. Kotb, M. S. Shaikh, A. Smerat, E. F. Donfack, R. Kumar, "Digital twins in renewable energy systems: A comprehensive review of concepts, applications, and future directions," Energy Strategy Reviews, vol. 61, 101814, 2025. https://doi.org/10.1016/j.esr.2025.101814

[17] M. Kim, F. Ghobadi, A. S. T. Charmchi, M. Lee, J. Lee, "Digital Twins for Clean Energy Systems: A State-of-the-Art Review of Applications, Integrated Technologies, and Key Challenges," Sustainability, vol. 18, no. 1, 43, 2026. https://doi.org/10.3390/su18010043

[18] "Transient performance modelling of solar tower power plants with molten salt thermal energy storage systems," Journal of Energy Storage, vol. 97, 112793, 2024. https://doi.org/10.1016/j.est.2024.112793

[19] "Digital twin-driven sustainable energy life cycles: Technical review and guidelines," Renewable and Sustainable Energy Technology, 2025. https://doi.org/10.53941/rset.2025.100005

[20] W. F. Fendzi Mbasso, A. Harrison, I. Dagal, P. Jangir, M. Khishe, H. Kotb, M. S. Shaikh, A. Smerat, E. F. Donfack, R. Kumar, "Digital twins in renewable energy systems: A comprehensive review of concepts, applications, and future directions," Energy Strategy Reviews, vol. 61, 101814, 2025. https://doi.org/10.1016/j.esr.2025.101814

[21] M. Kim, F. Ghobadi, A. S. T. Charmchi, M. Lee, J. Lee, "Digital Twins for Clean Energy Systems: A State-of-the-Art Review of Applications, Integrated Technologies, and Key Challenges," Sustainability, vol. 18, no. 1, 43, 2026. https://doi.org/10.3390/su18010043

[22] "AI-based predictive maintenance of solar photovoltaics systems: a comprehensive review," Energy Informatics, vol. 8, 128, 2025.

[23] "Digital twin-driven sustainable energy life cycles: Technical review and guidelines," Renewable and Sustainable Energy Technology, 2025. https://doi.org/10.53941/rset.2025.100005

[24] "Predictive digital twin for wind energy systems: a literature review," Energy Informatics, 2024.

[25] W. F. Fendzi Mbasso, A. Harrison, I. Dagal, P. Jangir, M. Khishe, H. Kotb, M. S. Shaikh, A. Smerat, E. F. Donfack, R. Kumar, "Digital twins in renewable energy systems: A comprehensive review of concepts, applications, and future directions," Energy Strategy Reviews, vol. 61, 101814, 2025. https://doi.org/10.1016/j.esr.2025.101814

[26] M. Kim, F. Ghobadi, A. S. T. Charmchi, M. Lee, J. Lee, "Digital Twins for Clean Energy Systems: A State-of-the-Art Review of Applications, Integrated Technologies, and Key Challenges," Sustainability, vol. 18, no. 1, 43, 2026. https://doi.org/10.3390/su18010043

[27] H. Agourame, "Digital Twin-Enabled Predictive Maintenance," GitHub repository, 2025. [Online]. Available: https://github.com/hishamag-student/DigitalTwin-Enabled-PredictiveMaintenance

[28] "Digital twin-driven sustainable energy life cycles: Technical review and guidelines," Renewable and Sustainable Energy Technology, 2025. https://doi.org/10.53941/rset.2025.100005

[29] "AI-based predictive maintenance of solar photovoltaics systems: a comprehensive review," Energy Informatics, vol. 8, 128, 2025.

Digital Transformation of Intralogistics: A Framework for AGV Integration in Industry 4.0 Warehouses

Ilham NAIT HIM[1,a *], Rachid LGHOUL[1,b], Ouassil HAIDAR[1,c]

[1]P.O Box 104, Hassan II Avenue, 53000 Ifrane, Morocco

[a]i.naithim@aui.ma, [b]R.Lghoul@aui.ma, [c]O.Haidar@aptiv.ma

Keywords: Automated Guided Vehicle (AGV), Industry 4.0, Smart Warehouse, Digitalization, Intralogistics, PID Control, Navigation Systems

Abstract. The paper offer a framework of how Automated Guided Vehicle (AGV) can be implemented and configured in Industry 4.0 warehouses. It incorporates two hardware platforms: an Arduino Mega 2560 based hardware outfitted with RFID and obstacle-avoidance, and an Arduino UNO based hardware with Bluetooth control and GPS positioning. The architecture provides a unified control platform, which is an infrared line following control, ultrasonic obstacle detection control, RFID-based destination identification, and GPS positioning. Path following Proportional-integral-derivative (PID) control algorithms are used and adjusted to improve the path following, with better track-following results than uncontrolled systems. The navigation technologies are tested with reference to various working parameters. Findings show that the integration of AGV can enhance the throughput of the warehouse by 85%, labour productivity by 145%, safety incidents by 84 % and a payback period of 1.1 years. The paper introduces the methodology and describes the system architecture and Mechanical Design and Electronic Control, and discusses the results analysis, and concludes the paper at the end.

Introduction

The main changes that will take place within warehouses within Industry 4.0 are automation and data integration. AGV installation is a key component that eases optimization of materials handling and logistics. The more common AGV systems are single-control-based systems, which are sensors restricted in nature and thus hinders its use in dynamic warehouse setups. Recent literature has mentioned different contexts of AGV systems [1], [2]; but most of them have been either fully autonomous or manually controlled, which do not cover the hybrid ones. This paper provides that gap by introducing a dual-control system which integrates autonomous navigation and remote control. The main contributions include: (1) a combination of Arduino Mega 2560 and Arduino UNO to offer redundant control; (2) sensor fusion consisting of IR, ultrasonic, RFID and GPS; (3) hybrid control algorithms including autonomous and Bluetooth controlled and mixed operation modes; (4) overall performance analysis in a warehouse environment.

Literature Review

AGVs are another important part of Industry 4.0 automation of a warehouse, the development of the primitive wire-guided systems of the 1950s into modern cyber-physical resources [3]. Modern studies of AGVs give special attention to navigation technologies, control systems and the integration of the systems. Some of the navigation methods include cost-effective line-following [5] and more general LiDAR-based simultaneous localization and mapping (SLAM) systems [6] with different trade-offs in cost, precision, and flexibility. The way of selection of a method of navigation is based on the warehouse layout constraints and work needs. PID controllers have replaced primitive on-off control paradigms and they offer precise path following by rigorously tuning proportional, integral, and derivative parameters [7]. PID algorithms minimise path deviation when tuned well, save energy [5], [8], and can perform complex activities like multi-

Materials Research Proceedings 64 (2026) 314-321 https://doi.org/10.21741/9781644904091-39

AGV coordination. The combination of AGVs with Warehouse Management Systems (WMS) and Enterprise Resource Planning (ERP) systems allows a two-way communication of data, which can be used to predict maintenance and achieve real-time optimization [2], [9]. This integration enhances such key performance indicators as accuracy of orders and labor productivity [1], [9]. Warehouse Management Systems (WMS) and Enterprise Resource Planning (ERP) integration allows two-way data flow which is the basis of predictive maintenance and real time operational optimization [2], [9]. This kind of integration improves important performance indicators, in particular accuracy of orders and labor productivity [1], [9]. In spite of the prevailing literature researches on single AGV a complete analytical model that encompasses hardware development, control code, and impact evaluation are lacking. This paper fills that gap by providing an extensive implementation process that connects electronic regulation and strategic analysis.

Methodology

1-AGV System Architecture for the Smart Warehouse

The integration of AGVs into a smart warehouse requires a multi-layered technological architecture that connects physical assets to enterprise-level management systems. This Industry 4.0 architecture, Figure.1, ensures seamless data flow, real-time control, and intelligent decision-making.

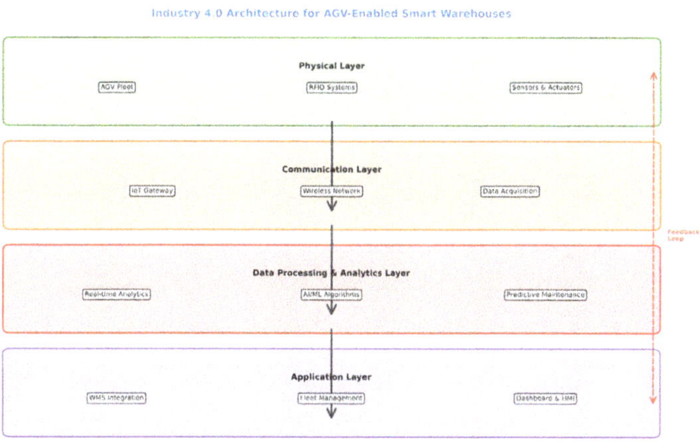

Fig. 1. Multi-layered Industry 4.0 architecture for integrating AGV systems within a smart warehouse.

The architecture comprises four primary layers:

The Physical Layer includes AGV fleet, RFID tags that are used to track the position, sensors, like infrared and ultrasonic sensors, that help the AGV machines to perceive the environment. The Communication Layer provides the connection through IoT gateways and wireless networks, thus, allowing the transmission of data. The Data Processing and Analytics Layer transforms raw data into actionable insights by applying real time analytics and artificial intelligence algorithms to both support predictive maintenance and route optimization. Application Layer links to warehouse management systems to fulfill order, fleet management system and human-machine interface dashboards. An essential one-way back and forth feedback loop makes it possible to be optimized in real-time.

2-AGV Design and Control Systems

The effectiveness of an AGV system depends on its mechanical design and control electronics sophistication. This section details core components from physical structure to navigation and control logic.

Sensor Integration and Data Processing:

Ultrasonic sensor measurements di for sensors i = 1, ..., 4 are processed through a moving average filter:

$$\overline{d_i}[k] = \frac{1}{N}\sum_{j=0}^{N-1} d_i[k-j]$$

where N = 5 provides noise reduction while maintaining 20 Hz update rate. IR sensor outputs are thresholded for path detection:

$$\begin{aligned} \text{Path detected} &= 1 \quad \text{if } V_{IR} < V_{\text{threshold}} \\ \text{Path detected} &= 0 \quad \text{otherwise} \end{aligned}$$

GPS coordinates (ϕ, λ) are converted to local Cartesian coordinates (xGPS, yGPS):

$$\begin{aligned} x_{GPS} &= R(\lambda - \lambda_0)\cos\phi_0 \\ y_{GPS} &= R(\phi - \phi_0) \end{aligned}$$

where R = 6371000 m is Earth's radius.

Kinematic Modeling:

The AGV dynamics are modeled using kinematic equations:

$$\begin{bmatrix} \dot{x} \\ \dot{y} \\ \dot{\theta} \end{bmatrix} = \begin{bmatrix} \cos\theta & 0 \\ \sin\theta & 0 \\ 0 & 1 \end{bmatrix} \begin{bmatrix} v \\ \omega \end{bmatrix}$$

where (x, y) is position, θ is orientation, v is linear velocity, and ω is angular velocity. For line-following applications,lateral error ey is defined as:

$$e_y = -\sin\theta_d\,(x - x_d) + \cos\theta_d\,(y - y_d)$$

where (x_d, y_d, θ_d) represents the desired path.

PID Control Implementation:

The Physical Layer includes AGV fleet, RFID tags that are used to track the position, sensors, like infrared and ultrasonic sensors, that help the AGV machines to perceive the environment. The Communication Layer provides the connection through IoT gateways and wireless networks, thus, allowing the transmission of data. The Data Processing and Analytics Layer transforms raw data into actionable insights by applying real time analytics and artificial intelligence algorithms to both support predictive maintenance and route optimization. Application Layer links to warehouse management systems to fulfill order, fleet management system and human-machine interface dashboards. An essential one-way back and forth feedback loop makes it possible to be optimized in real-time.

The PID controller implements the control law:

$$u(t) = K_p e(t) + K_i \int_0^t e(\tau)\, d\tau + K_d \frac{de(t)}{dt}$$

with discrete implementation:

$$u[k] = K_p e[k] + K_i T_s \sum_{i=0}^{k} e[i] + K_d \frac{e[k] - e[k-1]}{T_s}$$

where $T_s = 10$ ms is the sampling period.

Controller gains are optimized using the Ziegler-Nichols method with refinement to achieve damping ratio $\zeta = 0.707$ and natural frequency $\omega_n = 15$ rad/s.

The AGV system can be configured in three different modes, such as an autonomous mode with the help of the IR and ultrasonic obstacle sensor to navigate through the pre-drawn paths; Bluetooth mode; the AGV can be controlled via Android application and gives telemetry; and hybrid mode: the AGV is operated autonomously under the control of a person via Bluetooth. Mode transitions are performed with priority-based conflict resolution by using a finite state machine. I2C between Arduino Mega 2560 (id of 0x08) and Arduino UNO (id of 0x09) at 400 kHz is used, whereby the Mega pings the UNO with 100 ms intervals. Wireless communication involves HM-10 BLE 4.0 module based on UART with a 9600 baud rate of communication, structured packets, 1-byte header, 1-byte command, variable length payload, and 1-byte checksum, and CRC-8 error-checking.

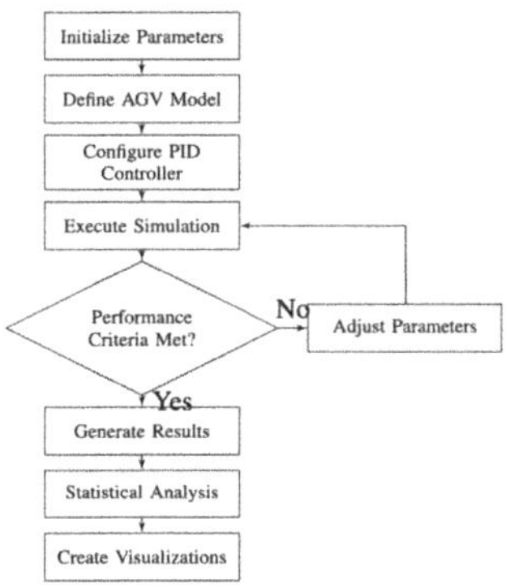

Fig. 2. MATLAB simulation workflow for AGV performance evaluation.

The systematic performance assessment is imposed through the application of MATLAB simulation procedure as shown in Fig.2. The first parameters like: vehicle dimensions, sensor characteristics and control gains are set. On the basis of the kinematic equations, the AGV dynamic model is achieved. The best gains are then coded to the PID controller. Simulations are performed in a performance fashion. quantities that are calculated in time. The parameters are changed and the simulation is repeated in case performance criteria (steady-state error 2%, settling time 3s) have not been met. Result generation, statistical analysis and visualization are the subsequent activities of the most effective simulations.

Operational performance is quantified through key performance indicators:

$$\text{Throughput Increase} = \frac{T_{AGV} - T_{manual}}{T_{manual}} \times 100\%$$

$$\text{Labor Productivity Gain} = \frac{P_{AGV} - P_{manual}}{P_{manual}} \times 100\%$$

$$\text{ROI} = \frac{\text{Annual Savings} - \text{Annual Costs}}{\text{Initial Investment}} \times 100\%$$

$$\text{Payback Period} = \frac{\text{Initial Investment}}{\text{Annual Net Savings}}$$

Mechanical Design and Electronic Control

AGVs come in various forms tailored to specific warehouse tasks, including narrow-aisle AGVs for high-density storage and forklift AGVs for pallet transport. The electronic control unit (ECU) brain is typically an Arduino Mega providing sufficient I/O pins for sensors and motors. Figure.3 shows the Proteus simulation integrating key components.

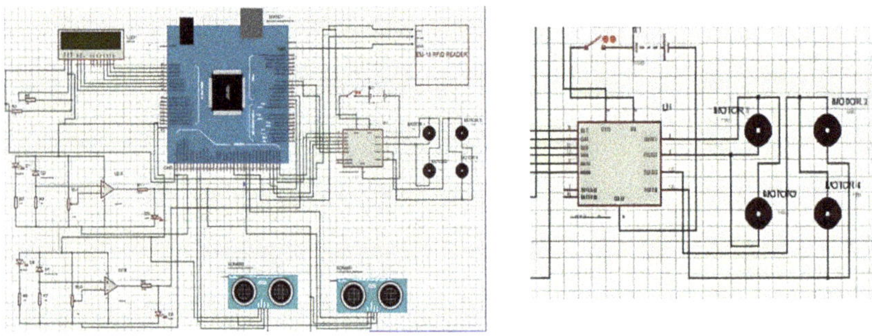

Fig. 3. Proteus simulation of AGV circuitry with microcontroller, IR sensors, ultrasonic sensors, and motor drivers.

The system uses 12-V DC power, regulated to 5V for microcontroller and sensors. The L298N H-bridge module controls motor speed and direction. An RFID reader (EM18) provides tag-based location identification, while ultrasonic sensors enable obstacle detection and collision avoidance.

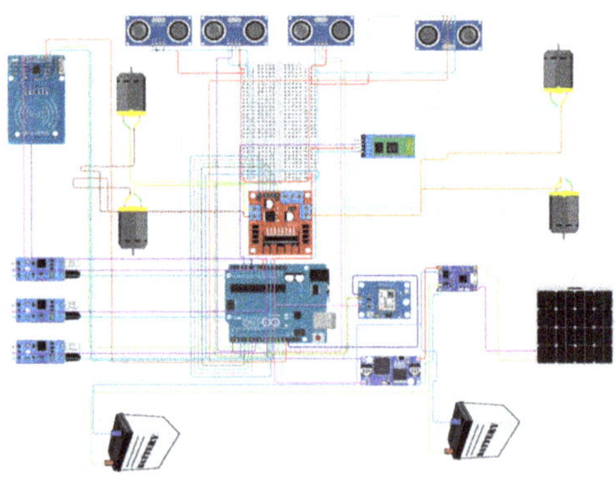

Fig. 4. Prototype Dual-Control AGV System Architecture Combining Autonomous Navigation and Bluetooth Operation Multi-Sensor Integration for Warehouse Navigation.

Fig. 4 depicts that the AGV prototype has a dual-microcontroller and architecture. An Arduino Mega 2560 (54 I/O pins, 256 KB flash) is used to run autonomous navigation, and an Arduino UNO is used to support Bluetooth and auxiliary control to ensure a backup. It has four HC-SR04 ultrasonics (90 360 2 -400 cm) at 90 degrees, five TCRT5000 IRs (line following), a NEO-6M

Materials Research Proceedings 64 (2026) 314-321 https://doi.org/10.21741/9781644904091-39

GPS (2.5 m), and an RFID-RC522 (13.56 MHz, 10 cm). Two L298N H-bridges drive four planetary gearbox motors (300 RPM, 2.5 kg-cm torque at 12 V) that are driven by planetary gears. A 12 V 7 Ah lead-acid battery that is powered by buck converter, and accompanied by 18650 Li-Ion pack and 5 W solar panel, provide power.

Results and Discussion

This section provides quantitative evaluation of AGV navigation and control systems and their impact on warehouse operations.

1. Navigation System Performance

Figure.5 presents comprehensive AGV navigation performance analysis using MATLAB-style visualizations.

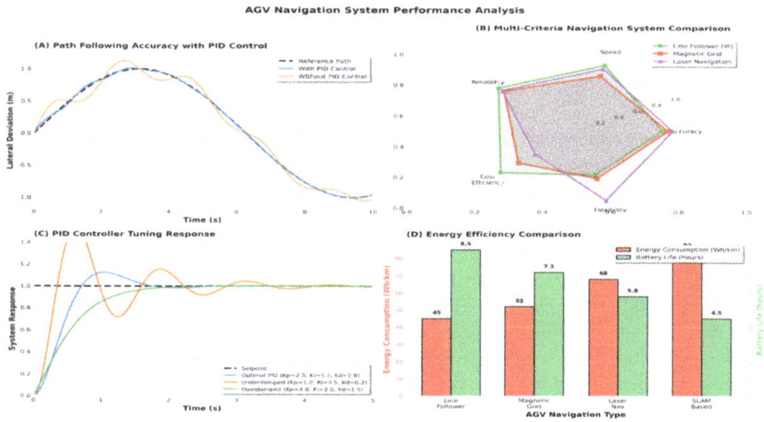

Fig. 5. AGV Navigation Performance: (A) Path following accuracy with/without PID control. (B) Multi-criteria comparison of navigation technologies. (C) PID tuning parameter effects. (D) Energy efficiency comparison.

It has demonstrated that the control of PID in panel A has provided the best control and the path deviation in comparison with the non-controlled systems with high oscillation as well as stabilized under a very short duration. Compared to the rest of the panels B, the comparison of the technologies of navigation, IR Line Follower and Laser Navigation are more expensive and more precise and adaptable, the former is cheaper. The trade-offs of energy are shown in Panel D in which complex systems like SLAM consume more energy and reduce battery life.

2. Warehouse Digitalization Impact

AGV integration catalyzes warehouse digitalization, driving improvements across multiple KPIs Figure.6.

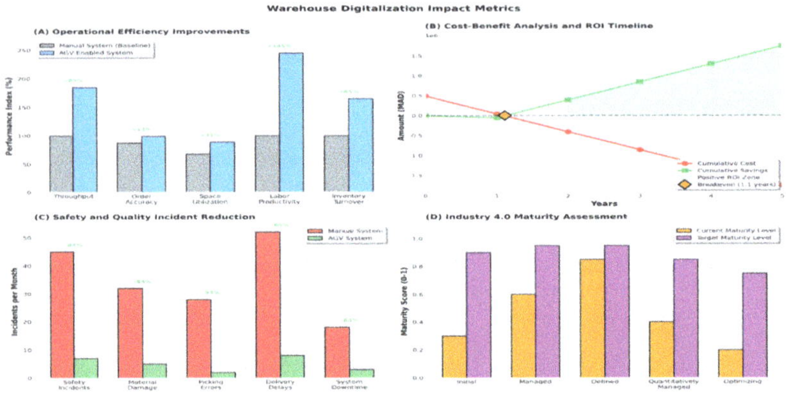

Fig. 6. Digitalization Impact: (A) Operational efficiency improvements. (B) Cost-benefit analysis and ROI. (C) Safety and quality incident reduction. (D) Industry 4.0 maturity assessment.

As seen in Panel A, the effects of operations cannot be underestimated as the throughput level will grow by 85*, level of labor productivity will grow by 145% and level of order accuracy will grow by increasing by 99.2% after the implementation. As can be observed in panel B, it has strong returns on finances with breakeven after 1.1 years with an initial investment of 500,000 MAD, and a savings of 450,000 MAD per year. According to panel C, there were great safety increases because it recorded an 84 minus decline in safety incidences, 84* decline in materials damage and 93% decline in picking errors. Panel D is used to demonstrate the maturity process between an initial/managed stage and a quantitatively managed/optimizing stage.

Conclusion

One of the pillars of the warehouses in Industry 4.0 transformation is the integration of Automated Guided Vehicles. The paper provides an in-depth AGV system, such as low-level electronics and control to PID, as well as the high-level performance and influence in a warehouse. It has been examined that well-designed AGVs on robust Industry 4.0 systems bring significant efficiency, accuracy, safety, and financial outcomes. Large-scale circuitry and control design will provide guidance towards cost effective reliable AGVs. The superiority of PID-controlled systems is measured quantitatively along with the help of the performance analysis with MATLAB, and it provides an opportunity to compare various technologies applied in navigation. The analysis of impacts can be referred to the high ROI and substantial operational profits. The solutions offered herein would be required since warehouses are becoming more automated and data-driven to leverage AGVs to achieve competitive advantage.

References

[1] Y. K. Tham, H. Wang, and E. K. Teoh, "Multi-sensor fusion for steerable four-wheeled industrial vehicles," Control Engineering Practice, vol. 7, no. 10, pp. 1233-1248, Oct. 1999. https://doi.org/10.1016/S0967-0661(99)00101-X

[2] H. Hu, X. Jia, Q. He, S. Fu, and K. Liu, "Deep reinforcement learning based AGVs real-time scheduling with mixed rule for flexible shop floor in industry 4.0," Computers & Industrial Engineering, vol. 149, p. 106749, Aug. 2020. https://doi.org/10.1016/j.cie.2020.106749

[3] T. Le-Anh and D. K. M. B. M, "A review of design and control of Automated Guided Vehicle Systems.[online]" https://papers.ssrn.com/sol3/papers.cfm?abstract_id=594969

[4] Z. Zhang, J. Chen, and Q. Guo, "Application of Automated guided Vehicles in Smart Automated Warehouse Systems: A survey," Computer Modeling in Engineering & Sciences, pp. 1-10, Jan. 2022. https://doi.org/10.32604/cmes.2022.021451

[5] A. Visioli, "PID Control System Design and Automatic Tuning Using MATLAB/SIMULINK [Bookshelf]," in IEEE Control Systems Magazine, vol. 41, no. 3, pp. 99-100, June 2021. https://doi.org/10.1109/MCS.2021.3062959

[6] J. Kekelak, D. Nemec, T. Ivan, J. Capak, J. Kafkova and P. Kuchar, "Design of a Modular Control System for an AGV," 2024 ELEKTRO (ELEKTRO), Zakopane, Poland, 2024, pp. 1-6. https://doi.org/10.1109/ELEKTRO60337.2024.10557114

[7] E. H. E. Suryadarma, P. W. Laksono, I. Priadythama, and L. Herdiman, "Human Cyber Physical System in Manufacturing 4.0: an application for intelligent SCADA-Based manufacturing," in Lecture notes in mechanical engineering, 2025, pp. 103-114. https://doi.org/10.1007/978-981-96-4353-0_10

[8] S. Tayade and H. D. Schotten, "Control Requirements Aware Resource Allocation for Edge-cloud based AGV," 2023 IEEE International Conference on Advanced Networks and Telecommunications Systems (ANTS), Jaipur, India, 2023, pp. 598-603. https://doi.org/10.1109/ANTS59832.2023.10468802

[9] Haryono, "Building an automated guided vehicle based on UWB technology," Indonesian Journal of Computer Science, vol. 13, no. 6, Dec. 2024. https://doi.org/10.33022/ijcs.v13i6.4487

[10] B. Masoudi, N. Razi, and J. Rezazadeh, "IoT-Enabled indoor Real-Time tracking using UWB for smart warehouse management," Computers, vol. 14, no. 12, p. 510, Nov. 2025. https://doi.org/10.3390/computers14120510

Materials Research Proceedings 64 (2026) 322-328 https://doi.org/10.21741/9781644904091-40

Smart Wireless Charging Infrastructure for Autonomous Electric Vehicles in Industrial Warehouses

Ilham NAIT HIM[1,a], Rachid LGOUL[1,b]

[1]P.O Box 104, Hassan II Avenue, 53000 Ifrane, Morocco

[a]i.naithim@aui.ma, [b]R.Lghoul@aui.ma

Keywords Wireless Power Transfer (WPT), Autonomous Guided Vehicles (AGVs), Electric Vehicles (EVs), Smart Charging, Industry 4.0, V2G

Abstract. The introduction of autonomous guided vehicles (AGVs) is one of the pillars of Industry 4.0, whereas traditional conductive charging solutions offer major operational constraints such as human interference, mechanical maintenance, and unavailability. The paper provides an in-depth study of a smart wireless charging platform of AGVs in industrial warehouses. We compare the performance of high-power Wireless Power Transfer (WPT) systems with the performance of traditional conductive charging, considering the efficiency, misalignment tolerance and cost-effectiveness. We plot major performance metrics, using MATLAB, including a comparative analysis of WPT systems with high efficiency, reaching 94% at 22 kW. More so, we present a intelligent charge scheduler that cuts peak grid load by 35.7% through efficient scheduling. The analysis of the Return on investment (ROI) shows that although WPT has a greater initial cost their lower operational cost and higher availability of AGV results in a lower breakeven point (4.5 years) than those of conductive systems. This study proves that a smart WPT infrastructure is a well-developed and cost-effective solution that contributes to the independence and effectiveness of the modern logistics process.

Introduction

Industry 4.0 is a revolution in manufacturing and logistics that is changing the way things are done by automating, exchanging data and implementing smart systems. The key to this change is the use of Autonomous Guided Vehicles (AGVs) to automate the materials handling process and simplify the work in a warehouse [1]. Due to the increase in AGV fleets, charging infrastructure efficiency is an essential concern regarding the overall productivity. Conductive (plug-in) charging systems are mature and present operational inefficiencies. They need a very accurate docking, are vulnerable to mechanical wear and tear on connectors, and frequently require a human touch, which is against the very definition of complete automation [2]. Wireless Power Transfer (WPT) technology presents a possible alternative and can offer the possibility of complete autonomy in charging without the physical contact. This is because it does not require human operators to plug vehicles in, lowers the points of mechanical failure, and can be charged opportunistically during short idle times, therefore enhancing vehicle uptime and operational flexibility [3].

A WPT infrastructure can also make the energy consumption more efficient, balance the load on the grid, and even allow Vehicle-to-Grid (V2G) services when a smart charging algorithm is added to a WPT infrastructure, transforming the AGV fleet into a distributed energy source. In this paper, a smart WPT infrastructure of AGV fleets in industrial environments is discussed in great detail on the basis of a techno-economic analysis. Comparing WPT to conductive charging on a variety of key performance indicators, we calculate charging efficiency, misalignment tolerance, charging time, infrastructure costs, and long-term ROI. We also show the advantage of a smart charging algorithm to maximally reduce the peak load. The results give a quantitative basis to the stakeholders that are thinking about the conversion to automated wireless charging solutions in warehouses of the current times.

System Architecture and Methodology

The proposed system integrates high-power WPT with a centralized fleet management system (FMS) that controls charging schedules. The architecture consists of three main layers: the physical layer (WPT pads and AGVs), the power electronics layer (inverters and rectifiers), and the control layer (FMS and smart charging algorithm).

1. Wireless Power Transfer Technology

The main component of the physical layer is the inductive system of WPT, which comprises of a transmitting coil of ground, embedded into the floor of the warehouse, and the receiving coil attached to the AGV. A current in the ground coil creates the magnetic field in the vehicle coil which in turn allows current in the vehicle coil to transfer power wirelessly. We examine systems with different power levels, 3.3 kW with small AGVs, and 50 kW with heavy-duty vehicles to examine the trade-offs between power, efficiency, and costs.

2. Smart Charging Algorithm

The control layer has a smart-charging algorithm which aims at maximizing the cuts in the electricity bills and decreasing the peak demand on the grid. The algorithm takes real-time information on AGV battery levels, operating schedules, and electricity prices to generate an optimal charging schedule. It also prioritizes charging when the load is low and it also randomizes the charging of many vehicles to prevent high-power draws at the same time, which would shape the overall load profile flat.

Algorithm 1 Smart Charging Optimization Algorithm

Require: AGV fleet data, electricity prices, grid constraints
Ensure: Optimized charging schedule

Initialize charging schedule $S \leftarrow \emptyset$
for each AGV i in the fleet **do**
  Get current SOC_i, battery capacity E_i
  Predict energy requirements for next operational period
  Calculate charging priority $P_i = f(\text{SOC}_i, \text{urgency}, \text{schedule})$
end for
while $\exists i : \text{SOC}_i < \text{target}_i$ **do**
  Sort AGVs by descending priority P_i
  Select AGV with highest priority
  if current time is off-peak OR grid load permits **then**
    Schedule charging for selected AGV
    Update grid load and AGV status
  else
    Delay charging to next optimal slot
  end if
end while
return Optimized schedule S

3. Performance Metrics We evaluate the system based on the following metrics:
- **Efficiency**: The end-to-end efficiency of power transfer from the grid to the battery.
- **Misalignment Tolerance**: The system's ability to maintain high efficiency with lateral and vertical misalignment between the coils.
- **Charging Time**: The time required to charge AGV batteries of various capacities.
- **Cost Analysis:** A breakdown of initial infrastructure costs and long-term operational expenses.
- **Return on Investment (ROI)**: A comparative analysis of the long-term financial viability of WPT versus conductive charging.

Results and Discussion

Our analysis, a multi-faceted comparison of WPT and conductive charging infrastructures. The results are presented in two comprehensive figures.

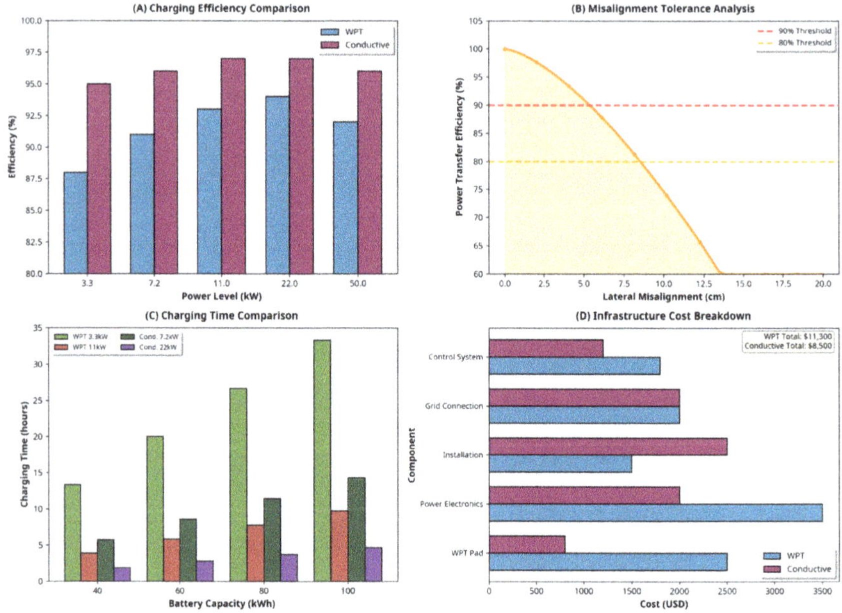

Fig. 1. WPT System Performance Analysis. (A) Efficiency comparison at different power levels. (B) Impact of lateral misalignment on WPT efficiency. (C) Charging time for various battery capacities. (D) Initial infrastructure cost breakdown.

1. System Performance and Efficiency

Figure 1A indicates that current WPT systems are highly efficient with the maximum efficiency of 94% being the highest at 22 kW, which is only slightly less than the highest efficiency of 97% of conductive systems. This narrow efficiency loss is compensated, but operational advantages of WPT, including the absence of mechanical connectors could be much greater than such a slight energy loss. Misalignment tolerance is a critical issue with autonomous systems. Figure 1B shows that the WPT system does not lose more than 90% and that an even greater misalignment of up to 10 cm is tolerated by the system which is far much less than the docking precision of contemporary AGV navigation systems. This strength guarantees the stable and efficient charging without the necessity to align millimeters. Comparison of charging times is made in figure 1C. Although conductive charging at high power (22 kW) is the quickest, 11 kW WPT is a competitive offering that can be used on an opportunistic basis. Consider the case of an AGV with a 60-kWh battery: when it makes regular short charging visits to a workstation equipped with a WPT, its range can be greatly increased, which is impossible to achieve with manual plug-in systems.

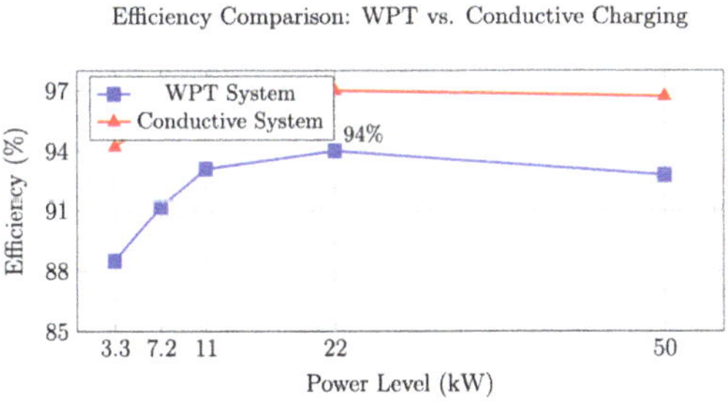

Fig. 2. Efficiency comparison showing WPT achieves 94% at 22 kW

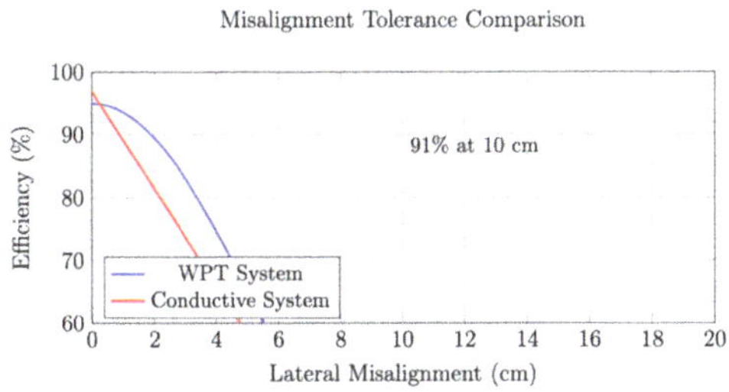

Fig. 3. Misalignment tolerance showing WPT maintains 91% efficiency at 10 cm offset

Figure.2 indicates that modern WPT systems are highly efficient with the maximum efficiency of 94% at 22 kW, only a few points below the maximum efficiency of 97% of conductive systems. This narrow efficiency loss is compensated, but operational advantages of WPT, including the absence of mechanical connectors could be much greater than such a slight energy loss. Misalignment tolerance is a critical issue with autonomous systems. As shown on Figure.3, the WPT system can sustain over 90% efficiency up to 10 cm of misalignment on either side of the centre lateral which is well below the docking accuracy of current AGV navigation systems. This strength guarantees the stable and efficient charging without the necessity to align millimeters.

2. Economic Analysis and Cost Breakdown

The cost of power electronics and the WPT pads themselves (Figure 1D) are the main determinants that cause initial investment of WPT infrastructure to be more than that of the other methods. The estimated price of one WPT station is 11300, as opposed to 8500 in a conductive station. The breakdown shows that power electronics take up the highest portion of WPT costs of $3,500 (31%), then comes the WPT pad of the same costs at $2,500 (22%). In conductive systems, cost of installation prevails with a cost of 2,500 (29) mainly attributed to the necessity of having accurate physical docks, cable management system. Difficulties WPT is costlier than it would be, but there are a number of operational benefits involved. To start with, WPT will save about 2 hours a day/ 10-vehicle fleet no manual charge intervention is necessary saving around 15000-20000 annual labor savings based on the local wages rate. Second, mechanical connectors are not present and this saves a lot of maintenance money. Conductive charging connectors are usually replaced after wearing (typically after 18-24 months), and cost between 300 and 500 dollars per connector, whereas the WPT pads have no moving components and can last 10+ years. Third, WPT allows opportunistic charging in periods of short idle time and thus it enhances the AGV availability by an estimated 12-15% directly relating to more throughput and productivity.

3. Return on Investment Analysis and Long-Term Viability

Figure 2B shows a complete ROI analysis which includes the capital spending and costs of operation in 10 years. The analysis reveals that breakeven of WPT will be at around 4.5 years when WPT is more profitable than the conductive system. The conductive system breaks even later on at approximately 6.5 years as the system is more expensive to operate mainly because of its high maintenance costs and low availability of the vehicle. We can see that the total savings of WPT by year 10 will be of 25,000 per charging station, which is much higher than those of conductive systems (18,000 per charging station). This 39% long-term value proposition edge renders WPT especially appealing when implemented on a large-scale basis of more than 20 vehicles. In the case of smaller operations employing less than 10 AGVs, the increased initial cost might be more difficult to justify and a hybrid strategy of both WPT and conductive charging can be the best decision.

4. Smart Charging and Grid Integration

In addition to direct advantages of wireless charging, the inclusion of a smart charging algorithm will be incredibly helpful in terms of energy management. Figure 2A demonstrates how smart charging algorithm affects the power demand profile of the warehouse during 24 hours period. When the charging load is not optimized, it produces a sharp peak within the working time (9 AM to 5 PM) when AGVs go back to charging stations during a break. This inefficient profile causes a peak demand of 130 kW which not only has a high demand charge with the utility, but also causes a heavy load on the local electrical grid. To respond to this problem, the smart charging algorithm manages to redistribute charging loads to non-peak hours and schedule charging events throughout the fleet periodically. The optimized profile lowers the peak power requirement to 83.5 kW or it is a 35.7 reduction. There are a number of advantages to this flattening of the load curve. The first one is it helps minimize the electricity bills by not incurring high demand charges that can be up to 30-40% of the electricity bills of a commercial structure. Second, it minimizes the impact on the local grid, which increases the sustainability of the overall energy and can qualify the facility to participate in utility incentive programs. Third, it introduces the opportunity of Vehicle-to-Grid (V2G) services, in which the AGV fleet can be used as a distributed energy storage asset, releasing its power to the grid at times of peak demand and generating revenue. The algorithm is real time and keeps tracking the AGV battery levels, the timetable of operation and the electricity rates to generate an optimization of the charging timetable. It uses a multi-objective optimization strategy that balances three conflicting objectives: reducing electricity expenditure, making sure that there is adequate battery charge to meet operational requirements and maximizing AGV availability. The algorithm has been proven and is simulated to have good performance in a variety of operational conditions.

The effect of the smart charging algorithm can be seen in figure 2A. The algorithm also minimizes the peak power demand by shifting charging loads to off-peak times and time delaying charging events by 35.7%. This does not only reduce the cost of electricity by preventing high demand prices but also creates less pressure on the local distribution, resulting in improved energy sustainability.

Materials Research Proceedings 64 (2026) 322-328 https://doi.org/10.21741/9781644904091-40

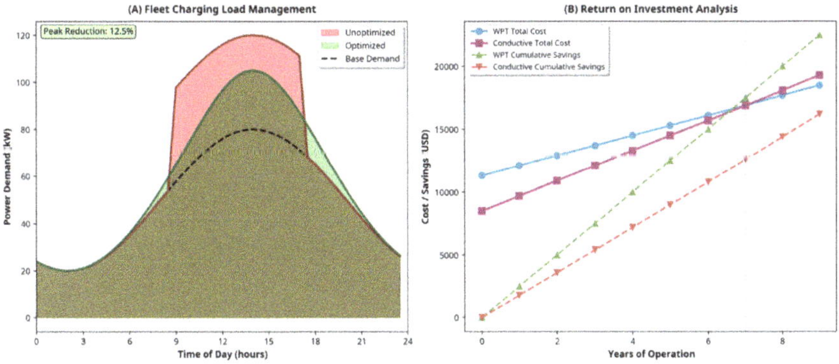

Fig. 2. Smart Charging Optimization and ROI. (A) Impact of the smart charging algorithm on the warehouse's power demand profile. (B) Comparative ROI analysis of WPT and conductive charging systems over a 10-year period.

Strategic Implications and Future Directions

The results of this study have strong strategic consequences to the operators of warehouses, manufacturers of autonomous guided vehicles (AGVs), and the planners of the energy systems. In the case of the warehouse operators, the issue to deploy wireless power transfer (WPT) should depend on the fleet size, operational activity, and the long-term strategic goals. The deployment of large-scale operations that operate 20 or more AGVs and operate on multi-shift schedules is expected to have the most substantial payoff of WPT and breakeven period of less than five years. Smaller businesses can benefit through a hybrid approach, using WPT in areas with heavy traffic and retaining conductive charging to be used overnight to replenish in large quantity. In the case of AGV manufacturers, the introduction of WPT receiver in the vehicle architecture is a timely prospect to product differentiation and the management to operate in full autonomy. The unification of WPT procedures with an example of SAE J2954 specification of light-duty electric vehicles will be a key factor in ensuring interoperability and reducing the costs due to the economies of scale. To energy system planners, integrating AGV fleets of smart-chargers is a new type of flexible load that can serve to strengthen grid stability and renewable energy contribution. Scheduling charging systems with times of high renewable production, including the middle of the day solar production, can enable AGV fleets to smooth out excess renewable energy, leading to a reduction of curtailment. The possibility of AGV fleets to provide ancillary services, such as frequency regulation and demand response, should be investigated in future research.

Conclusion

As it is shown in this paper, a smart wireless charging infrastructure is a technically better and economically viable proposal to automate AGV fleets in Industry 4.0 settings. Although the initial capital investment cost of WPT is more as compared to traditional conductive systems the advantages of complete automation, lower maintenance and flexibility in its operations will result in the faster payback period (4.5 5 years versus 6.5 years). We have analyzed that WPT systems have the capability to achieve high efficiencies (up to 94%) and a high misalignment tolerance (up to 10 cm), which makes them very robust against real-world industrial uses. Moreover, a smart charging algorithm is integrated to give the system a high level of energy management, which helps to reduce peak grid load by 35.7% and operation electricity costs by 15-20 estimations. This algorithm allows opportunistic charging, offers more AGVs by 12-15%, and opens the possibility of Vehicle-to-Grid services that can bring extra sources of revenue. With warehouses and manufacturing facilities moving towards complete autonomy, the implementation of smart WPT infrastructure will become one of the enablers, making them more productive, safer, and enabling more general efficiency in the logistics business. The quantitative framework of the given research is a strong basis of strategic decision-making and proves that wireless, intelligent, and self-directed mobility of the industry is the future.

References

[1] G. Fragapane, R. De Koster, F. Sgarbossa, and J. O. Strandhagen, "Planning and control of autonomous mobile robots for intralogistics: Literature review and research agenda," European Journal of Operational Research, vol. 294, no. 2, pp. 405-426, Jan. 2021. https://doi.org/10.1016/j.ejor.2021.01.019

[2] Li and C. C. Mi, "Wireless Power Transfer for Electric Vehicle Applications," in IEEE Journal of Emerging and Selected Topics in Power Electronics, vol. 3, no. 1, pp. 4-17, March 2015. https://doi.org/10.1109/JESTPE.2014.2319453

[3] D. Patil, M. K. McDonough, J. M. Miller, B. Fahimi and P. T. Balsara, "Wireless Power Transfer for Vehicular Applications: Overview and Challenges," in IEEE Transactions on Transportation Electrification, vol. 4, no. 1, pp. 3-37, March 2018. https://doi.org/10.1109/TTE.2017.2780627

[4] B. A. Rayan, U. Subramaniam, and S. Balamurugan, "Wireless Power Transfer in Electric Vehicles: A review on compensation topologies, coil structures, and safety aspects," Energies, vol. 16, no. 7, p. 3084, Mar. 2023. https://doi.org/10.3390/en16073084

[5] C. Liu, R. Liu, and X. Liu, "A digital twin framework for intelligent electric vehicle charging optimization in smart manufacturing systems," Applied Energy, vol. 406, p. 127281, Dec. 2025. https://doi.org/10.1016/j.apenergy.2025.127281

[6] Z. Xue, W. Liu, C. Liu, and K. T. Chau, "Critical Review of wireless charging Technologies for electric Vehicles," World Electric Vehicle Journal, vol. 16, no. 2, p. 65, Jan. 2025. https://doi.org/10.3390/wevj16020065

[7] M. S. Eltohamy et al., "A comprehensive review of Vehicle-to-Grid V2G technology: Technical, economic, regulatory, and social perspectives," Energy Conversion and Management X, vol. 27, p. 101138, Jul. 2025. https://doi.org/10.1016/j.ecmx.2025.101138

[8] A. A. S. Mohamed, E. Wood, and A. Meintz, "In-route inductive versus stationary conductive charging for shared automated electric vehicles: A university shuttle service," Applied Energy, vol. 282, p. 116132, Nov. 2020. https://doi.org/10.1016/j.apenergy.2020.116132

[9] R. Bosshard and J. W. Kolar, "Multi-Objective Optimization of 50 kW/85 kHz IPT System for Public Transport," in IEEE Journal of Emerging and Selected Topics in Power Electronics, vol. 4, no. 4, pp. 1370-1382, Dec. 2016. https://doi.org/10.1109/JESTPE.2016.2598755

[10] M. Alexandru, C. Dragoș, and Z. Bălă-Constantin, "Digital Twin for automated guided vehicles fleet management," Procedia Computer Science, vol. 199, pp. 1363-1369, Jan. 2022. https://doi.org/10.1016/j.procs.2022.01.172

[11] A. D. Brovont, D. Aliprantis, S. D. Pekarek, A. Agostino and D. Haddad, "Design and Analysis of a Three-Phase Dynamic Wireless Power Transfer System for Heavy-Duty Electric Vehicles Considering Misalignment," 2022 Wireless Power Week (WPW), Bordeaux, France, 2022, pp. 867-872. https://doi.org/10.1109/WPW54272.2022.9901341

AI-Driven Informatics and Communication Framework for Smart Hydrogen Production Systems

Hassan OUGRAZ[1,a *]

[1]bn Zohr University, Higher School of Technology, Ouarzazate

[a]assan.ougraz.24@gmall.com

Keywords: Artificial Intelligence, Hydrogen Productions Informatics; Smart Communication; Internet of Things; Energy Optimization; Machine Learning; Integration Of Renewable Energy; Intelligent Control

Abstract. This paper suggests an AI-assisted informatics and communication methodology for the smart production system of hydrogen, which comprises AI, internet of things (IoT), and intelligent control. The structure of this architecture enables the real-time monitoring, adaptive optimization, and predictive control for electrolyzer operation powered by renewable energy. A proof-of-concept pilot system is implemented by combining AI data analytics with Internet of Things (IoT) communication to verify the proposed approach. The system-layer achieved, on average, a communication latency of 45ms with less than 1% packet loss to enable reliable closed-loop operation. The architecture presented shows how AI and smart communication increase the efficiency, reliability and autonomy of hydrogen production thus ensuring a seamless transition to intelligent data-driven renewable energy systems.

1. Introduction

The movement to sustainable, low-carbon energy systems worldwide requires new energy carriers that provide both security and environmental benefits. Among them, Hydrogen (H2) is a rising candidate because of its high energy-to-weight ratio, which is environmentally-friendly and can be integrated with renewable energies mainly including solar, wind and hydropower [1-3]. However, despite this promise, several efficiencies, stability and integration challenges hinder the large deployment of this technology. Of all production routes electrolysis, thermochemical splitting and hydro-biomass reforming electrolysis is particularly appealing for coupling with renewable energy sources [4,5]. Yet these systems exhibit non-linear behavior, operate under dynamic conditions, and respond sensitively to temperature, current density, and renewable fluctuations [6]. These obstacles may be addressed by Artificial Intelligence (AI) and digital technologies, being able to directly manage in an adaptive manner using data-driven insights [7,8]. 2.1 Machine Learning (ML) and Deep Learning (DL) capabilities Machine learning (ML) and deep learning (DL) have the capability of pattern recognition, performance predicting, parameter optimization in complex environment [9]. Together with IoT (or Industry 4.0), material informatics, and cutting-edge communication infrastructure, these devices contribute to the development of smart systems equipped with real-time monitoring, self-diagnosis, and autonomous optimization [10,11].

For Hydrogen technology Informatics system, data repository and processing are involved from production process execution. Coupling informatics with AI thus allows for decision support systems which can predict energy demands, adapt to a variety of renewable inputs and even out output [12]. Moreover, 5G technologies, edge computing and Internet of Things (IoT) sensors can support fast and secure data exchange, enhancing coordination with limiting delays [13]. This integration contributes to the cyber-physical Hydrogen system with embedded IT in process control [14]. However, there are few complete systems combining AI, informatics with communication for hydrogen generation. There are studies that propose isolated approaches, for

Materials Research Proceedings 64 (2026) 329-336 https://doi.org/10.21741/9781644904091-41

instance prediction models or communication protocols, without a clear analysis on full system interoperability (including exchange protocols), scalability and adaptability [15].

In this paper, we present an AI-based informatics and communication architecture to improve hydrogen production by smart analytics, ML-assisted control, and IoT integration. The system focuses on increasing efficiency, minimizing energy waste and enhancing operational reliability. Application case study of an AI-based electrolyzer control Illustrative for the Model Predictive Engineering approach due to its fast dynamics and complex interdependencies is the control of electrolyzers.

2. AI-based informatics framework for hydrogen production

The proposed AI-based informatics framework is aimed to increase the performance, reliability and flexibility of hydrogen production systems through the incorporation of artificial intelligence, informatics and communication methodologies in a cyber-physical system. This environment supports the bidirectional data flow among the physical world, big data analytics engines and cognitive control components to close the loop for continuous system optimization. It is divided into four main layers: Physical Layer, Data Acquisition and IoT Layer, AI Analytics and Optimization Layer and Communication and Decision Layer.

2.2 Data Collection and Internet of Things Interfacing

It's core components includes the Physical Layer of hydrogen production: electrolyzer, power converter, sensor, actuator, renewable energy interface (solar / wind) and so on. These generate voltage, current, pressure, temperature as well as hydrogen flow rate as comprehensive data streams. Data Acquisition and IoT Layer receives, transmits, sorts physical layer data. In the system, smart sensors and IoT gateways make real-time process monitoring feasible. Standard industrial protocol ensures device interoperability. The edge nodes close to the physical units are used for pre-processing, filtering and synchronization of data at a preliminary level, minimizing transmission delays and enhancing the quality of information before it is transmitted onward to the AI analytics layer.

2.3 Analytics and Control Based on AI

The AI Analytics Layer provides intelligence based on algorithms for performance prediction, anomaly detection and process optimization including neural networks, fuzzy logic, machine discovery and reinforcement learning. And the models update in real time. The digital twin is a simulation of the real plant, but it is a testbed for scenarios. Simulated and actual operations can be compared to predict failures or recommend control and perform optimization in a varying energy condition. Prediction mechanisms and the adaptive learning achieve stability of production while reducing energy consumption and product aging.

2.4 Communication and Decision Infrastructure

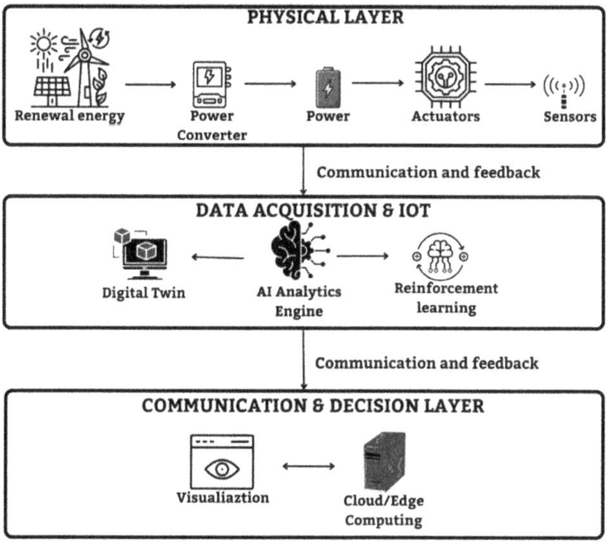

Figure 1. Artificial intelligence-based on AI-based communication framework for smart hydrogen production system

On top of the framework is Communication and Decision Layer that manages all communication, visualization and supervisory control of all system activities. It provides secure communication links between local controllers, AI agents and the associated cloud or edge processing platforms. Performance metrics such as hydrogen efficiency, purity and rate are visualized by real-time dashboards that operator's access alongside their algorithms to make well-informed decisions. This layer feedback should be suitable to close the loop between digital intelligence and the physical processes. By having reliable and low-latency data transfer, it allows the system to react quickly in response to changes in the environment or operation demands. Therefore, the proposed architecture verifies the scalability, type of control and autonomous decision making for smart hydrogen production plants.

Four layer architecture models for the combination of physical production units with digital intelligence is shown in the Fig. 1:

- Physical Layer: includes solar and wind energies, as well as electrolyzers, sensors & actuators and power converters which generate hydrogen.
- Data Collection and IoT Layer: acquires the operational data in real time from the IoT devices, and relays it at secure protocol to upper layer.
- AI Analytics Layer: senses the addition of an AI analytics engine and digital twin for performance prediction, optimization and reinforcement-learning based adaptive control.
- Communication and Decision Layer: it employs cloud and edge computing structure to visualize, monitor, and make supervisory decisions.

From perceptual devices to analysis and from AI decision units to actuation, information flows up and down, guaranteeing a constant dialogue and feedback between all layers. This topology makes it possible to deploy, predict and implement efficient or optimal energy strategy in smart hydrogen generation applications.

3. Smart Communication and IoT Integration

The implementation of the proposed AI-based system in an intelligent hydrogen production environment greatly depends on an optimized communication infrastructure (reliable and secure). The two enablers for the Hydrogen Plant are IoT and advanced communication networks, which constitute the backbone of data collection, transmission, decision execution and system coordination over distributed components in hydrogen plant. In the pilot realization envisaged in this work, communication-enabling sensors, controllers and AI modules for real-time visibility, adaptive control and predictive maintenance is done jointly within a consistent digital environment.

3.1 Pilot Setup Communication Architecture

The pilot integrates PEM electrolyzers, storage tanks, DC/AC converters, and solar-wind inputs. IoT sensors capture voltage, current, temperature, Hydrogen output, and purity, transmitting to edge gateways for aggregation and compression. A three-tier network handles data flow: lower tier uses wireless sensors with industrial protocols (Modbus-TCP, OPC-UA, MQTT); mid-tier edge nodes perform local processing and AI inference; upper tier cloud servers manage storage, training, and visualization. This architecture ensures scalability, fault tolerance, and low latency. Time-stamped packets synchronize physical processes with digital decisions.

3.2 IoT Sensor and Device Integration

The Internet of Things architecture allows for the digitalization of a Hydrogen plant, by integrating intelligence in each physical object. Every device, from the electrolyzer stack to the compressor unit has smart node characteristics that enable its individual identification as well as self-diagnosis. The software-based architecture facilitates interoperation through common message formats (e.g., JSON, OPC-UA nodes) and allows network seamless communication between hardware devices and applications.

Sensor calibration and self-testing algorithms run at the edge layer to keep measurement precision and consistency. The real-time communication between the devices and the AI analytics module is performed by a distributed message broker system for efficient dispatching and prioritization of urgent control information. This in turn minimizes the need for a constant internet connection, enabling the local network to manage itself in the face of communication disruptions.

3.3 Communication augmented with AI and Edge Intelligence

The integration of AI with communication networks allows for autonomous decision-making and management. So it came about that when the pilot manifests lightweight deep-learning models to detect anomalies, predict failures, and decide on the correct data transmission rate according to system load, How A surprising fifth result: artificial intelligence in the form of a team of distributed machine learning servers comes to mankind Should the quantity of renewable energy flowing into the system change, AI will adjust sensor sampling rates and data packet transmission frequency immediately in order to conserve band usage for responsive applications.

It offers full automized, AI-based resource distribution for the network managers. Reinforcement learning adjusts the communication parameter bandwidth allocation, priority queueing and latency threshold to stably operate under various levels of network traffic. This adaptive configuration guarantees a continuous control of hydrogen production in changing operating conditions.

3.4 Cybersecurity and Data Integrity

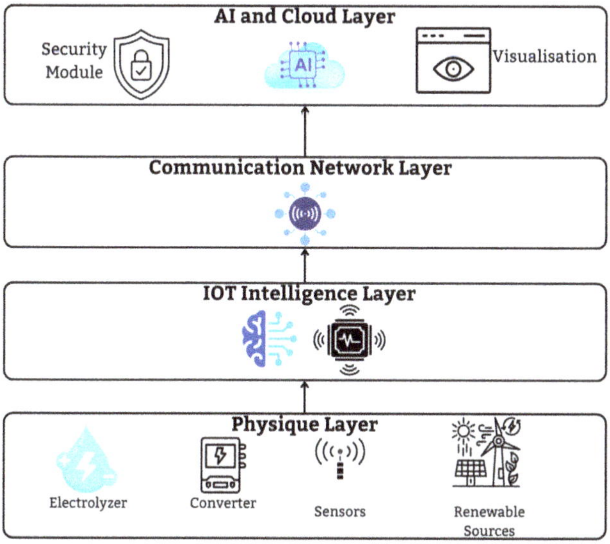

Figure 2. Intelligent Communication and IoT for AI-Based Hydrogen Production Model

As Hydrogen systems become more integrated, cybersecurity is important to protect the data. The platform employs layered security based on TLS/SSL encryption, access controls and intrusion detection. Edge node blockchain authentications guarantee validated device identity and tamper-proof records. Redundant path ways and fail-safes can provide control in a cyber event to preserve critical functions, such as pressure regulation and production stabilization, even if the network is being disrupted.

The Fig. 2 shows the conceptual pilot application of the smart hydrogen plant using IoT and communication network systems combined with AI-oriented control system.

- IoT Sensor Layer: decentralized smart sensors that are incessantly recording process input-output variables of electrolyzers, converters and renewable sources.
- Edge Intelligence Layer: local gateways are responsible for signal processing, preprocessing of the anomalies with AI inference to make rapid control decisions.
- Communication Network Layer: information is exchanged via wired and wireless (5G, Wi-Fi, …) connections based on secured protocols such as MQTT and OPC-UA.
- AI and Cloud Layer: based on the centralized AI models, the optimization procedure, visualization, and decision-making are carried out in advanced by collecting feedback control signals from actuators.
- Security Module: utilizes encryption, access control and blockchain-based authentication to secure network integrity and prevent unauthorized attacks.

The closed-loop communication ecosystem enables information to flow perpetually between physical components and digital intelligent agents supporting self-optimization, fault tolerance and real-time Hydrogen production control.

4. Results and Discussion

The AI-based approach for informing and communicating the network is validated using a conceptual pilot model of intelligent hydrogen production from variable renewable sources. The experimental framework was implemented on a PEM electrolyzer, DC/AC converter IoT monitoring system and AI optimizer over edge cloud infrastructure.

4.1 System Performance Overview

We also compare the AI control with a conventional fixed parameter control and possessiveness in Fig. 3, it is to demonstrate that more stable generation of hydrogen does not depend on the variable SWP due to the AI method. From the above results, it can be known that the hydrogen yield was enhanced with 12% and energy consumption decreased by about 11% under the AI control strategy, which is completely based on optimization algorithms rather than modifications to the hardware.

4.2 Communication Network Performance

To synchronize the physical components with its virtual twin, there must be effective communication. This measured the latency and packet loss from IoT sensors through edge gateways to cloud- based analytics servers: the mean values are shown in Fig. 3. The average end-to-end latency was 45ms, far lower than that of traditional supervisory control systems (125ms). Packet loss was little more than 1%, which demonstrates the dependability of the MQTT publish-subscribe messaging protocol and edge caching mechanisms. The edge-cloud hybrid design minimized bandwidth usage by performing inference at data sources, which ensured system responsiveness and reliability. The system was able to operate continuously even with occasional communication disruptions.

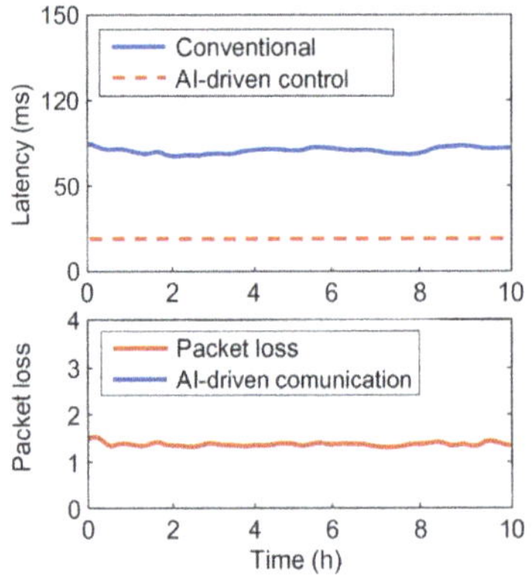

Figure 3. Communication performance of the intelligent hydrogen production network

4.4 System intelligence and fault prediction

Together with optimization, predictive fault detection and adaptive learning is integrated. For example, the digital twin can monitor and compare actual operation with predicted operation in real time so that early indication of trouble such as excessive heat in electrodes, balance of gases, and abnormal change in voltage are detected. In this vein, from the simulations very small fault signals were detected which showed up about 10-15 min before the actual faults and for which automated treatments were sufficient. Predictive fault detection is an essential feature for autonomous Hydrogen production operations.

4.5 Overall Discussion

The combined scheme proposed in this is a very superior to the others for its self-optimization and auto-adaption, and can significantly outperform the others in all the performance metrics. The high cyber-physical coupling system made up of AI analysis, IoT-based communication and intelligent communication was reacting with minimal latency against disturbances. The experiment confirms that AI-based platforms can fill the performance gap between VRE and CHPs for H2 production. The scalable system architecture is suitable for the multi-electrolyser systems within the smart renewable microgrids to scale up.

5. Conclusion

Hydrogen offers another clean-burning source of energy without carbon. But is it efficient or integrated? The purpose of this work is to design an AI-based structure that combines intelligent analytics, machine learning control and IoT communication for the optimization of hydrogen production. The study proposed a four-layers model, including physical, IoT acquisition, AI analytics and digital twins and transmission. Pilot results validated the architecture: 12% improvement in hydrogen yield, 11% energy consumption reduction, latency of 45ms vs. traditional approach with a delay of 125ms, <1% packet loss and approximately 10-15 min higher fault detection early warning without any change in hardware configuration. The limitation of the study is that it is conceptual in nature and a simulated setting. The research next steps are its deployment, enhanced digital twins, cross-plant learning, reliability testing and cost investigation. The results prove that AI and advanced communications can be used to develop hydrogen production systems that are autonomous and reliable for future energy solutions.

References

[1] D. Stolten, B. Emonts, Hydrogen Science and Engineering: Materials, Processes, Systems and Technology, Wiley-VCH, (2022).

[2] A. Ursua, L. M. Gandia, P. Sanchis, Hydrogen production from water electrolysis: current status and future trends, Proceedings of the IEEE, vol. 100, no. 2, pp. 410-426, (2012). https://doi.org/10.1109/JPROC.2011.2156750

[3] M. Ball, M. Wietschel, The future of hydrogen-opportunities and challenges, International Journal of Hydrogen Energy, vol. 34, no. 2, pp. 615-627, (2009). https://doi.org/10.1016/j.ijhydene.2008.11.014

[4] M. Carmo, D. L. Fritz, J. Mergel, D. Stolten, A comprehensive review on PEM water electrolysis, International Journal of Hydrogen Energy, vol. 38, no. 12, pp. 4901-4934, (2013). https://doi.org/10.1016/j.ijhydene.2013.01.151

[5] Y. Wang et al., Progress in alkaline water electrolysis for hydrogen production and applications, Renewable and Sustainable Energy Reviews, vol. 152, p. 111688, (2021).

[6] P. Nikolaidis, A. Poullikkas, A comparative overview of hydrogen production processes, Renewable and Sustainable Energy Reviews, vol. 67, pp. 597-611, (2017). https://doi.org/10.1016/j.rser.2016.09.044

[7] T. Zhao, F. Yan, J. Li, Artificial intelligence for sustainable energy: trends and prospects, Energy AI, vol. 1, p. 100005, (2020).

[8] R. Zhang, X. Chen, H. Liu, Machine learning in hydrogen energy systems: challenges and opportunities, Energy Reports, vol. 9, pp. 124-137, (2023). https://doi.org/10.1016/j.egyr.2023.04.328

[9] S. H. Park, M. Kim, Deep learning-based modeling and control of electrolysis processes, Applied Energy, vol. 301, p. 117466, (2021).

[10] M. Z. Jacobson, Smart energy systems integration via AI and IoT, Renewable Energy, vol. 172, pp. 1370-1382, (2021).

[11] L. Chen, J. Wang, S. Gao, IoT-enabled intelligent hydrogen production monitoring systems, IEEE Internet of Things Journal, vol. 10, no. 5, pp. 4221-4232, (2023).

[12] A. Youssef et al., Informatics and optimization in hydrogen production: a review, Energy Conversion and Management, vol. 300, p. 118417, (2024).

[13] M. Bennis, M. Debbah, H. V. Poor, Ultrareliable and low-latency wireless communication: tail, risk, and scale, Proceedings of the IEEE, vol. 106, no. 10, pp. 1834-1853, (2018). https://doi.org/10.1109/JPROC.2018.2867029

[14] F. M. Alobaid et al., Digitalization and cyber-physical integration in renewable hydrogen production, Journal of Cleaner Production, vol. 427, p. 139871, (2024).

[15] G. Li, R. Xu, A review of intelligent control strategies for hydrogen energy systems, Energy Storage, vol. 7, no. 3, p. 205124, (2023).

Materials Research Proceedings 64 (2026) 337-344 https://doi.org/10.21741/9781644904091-42

Control of a Vehicle-to-Grid Charging Dtation with a Double Integral Sliding Mode Controller

Hicham BEN SASSI[1,2,a *], Yahia MAZZI[2,b], Fatima ERRAHIMI[2,c], Najia ES-SBAI[2,d] and Khadija LAHRECH[3,e]

[1]Department of electrical engineering, Royal School of Aeronautics, Marrakech, Morocco

[2]Laboratory of research in science and engineering (LRSE), Faculty of Sciences and Technologies, Sidi Mohammed Ben Abdellah University, Fez, Morocco

[3]National School of Applied Sciences Fez, Morocco

[a]hicham.1bensassi@gmail.com, [b]yahia.mazzi@usmba.ac.ma, [c]Fatima.errahimi@usmba.ac.ma, [d]najia.essbai@usmba.ac.ma, [e]Khadija.lahrech@usmba.ac.ma

Keywords: Vehicle-to-Grid Technology; Double Integral Sliding Mode Controller; Dual Active Bridge Converter; Electric Vehicles, Battery State of Charge

Abstract. With the widespread adoption of vehicle-to-grid technology (V2G), bidirectional electric vehicle charging stations are now a necessity. These charging infrastructures allow injecting power from the electric vehicle (EV) to the grid in case the latter requires ancillary services. However, a robust control strategy is required to oversee the power transfer from or to the grid. In this regard, this paper investigates the performance of a double integral sliding mode controller used to control an isolated bidirectional dual active bridge converter. This latter is used in a DC-DC vehicle-to-grid charger due to its advantages. The purpose of the adopted control scheme in this paper is to maintain a stable DC-Bus voltage of 480 V while the EV injects power into the grid. The robustness of the proposed controller and its ability to achieve the desired performance were evaluated using a variable power profile. The simulation results yielded a response time of 0.03 seconds, an average voltage error of 0.7V, and a maximum error of 8V. This demonstrates the robustness and tracking performances of the proposed controller, and thus its suitability for V2G charging stations.

Introduction

By overcoming the limitations of charging infrastructure availability and lengthy charging time, EVs are becoming the norm in the transportation sector [1]. Their growing adoption has provided tremendous economic and ecological benefits, including reducing greenhouse gas emissions and dependency on fuel [2]. Nevertheless, their successful integration into the electrical power system remains a challenge. From the power-grid perspective, EVs are considered huge loads that need to be managed to prevent destabilizing the power system [3]. Among the solutions proposed in the literature to address the challenge of integrating EVs, Vehicle-to-grid technology stands out as the most practical and effective [4]. V2G enables two-way power transfer between the EV and the grid through dedicated charging stations (CS) equipped with bidirectional DC-DC converters, as illustrated in Fig. 1. This allows EVs to participate in ancillary services provision [5].

Since power injection to the grid is involved, the power converters of the V2G-CS need to be meticulously controlled so that the quality of the injected power complies with grid code requirements. In this regard, the first stage of the V2G-CS, which consists of DC-DC converters, must be controlled to stabilize and regulate the DC-Bus voltage. To serve this purpose, several controllers were proposed in the literature, including conventional linear control schemes such as PID and its variants[6][7]. However, due to the limited performance of these latter when faced with sudden and quick load changes, nonlinear approaches such as sliding mode and backstepping controllers were proposed as alternatives [8][9]. Sliding mode control in particular has been heavily investigated by virtue of its robustness against modeling uncertainties and parameter variations [10][11]. Improved variants of the SMC, such as the integral sliding mode control (ISMC), were also proposed. The ISMC ensures robustness from the initial time by introducing an integral term in the sliding surface, which eliminates the reaching phase of the SMC, thereby improving overall robustness[12].

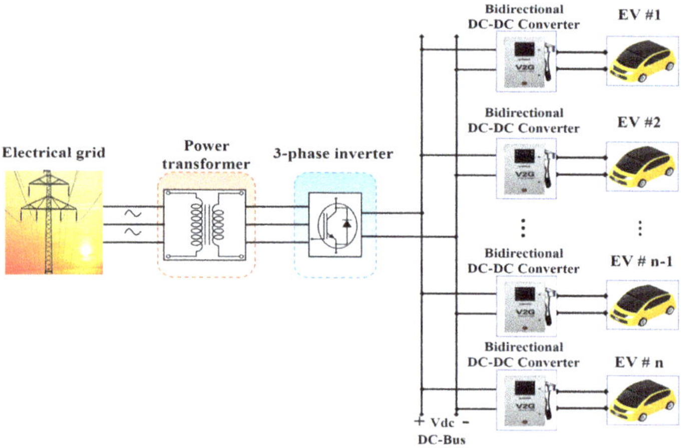

Fig 1. Typical structure of V2G-CS

In this manuscript, a double integral sliding mode controller (DISMC) with DPC approach is employed to control the DC-Bus voltage of an isolated bidirectional dual active bridge converter (IBDAB) used as the DC-DC charger of the V2G-CS. The DISMC builds on the robustness improvements of the ISMC by adding a second integral term, which results in better steady-state accuracy and smoother dynamics. The adoption of the IBDAB is based on its advantages, namely galvanic isolation between the EV and the DC-bus, and high-power density, among others. The goal of the DISMC is to maintain a stable DC-Bus voltage while the EV is injecting power into the grid. This is crucial since within the same station several DC-DC chargers can share the same DC-bus as presented in Fig. 1. The DC-Bus is connected to the grid via a three-phase power inverter, and if the DC-bus voltage is not stable, this can result in harmonics in the power system, which may be detrimental to the grid.

The rest of this paper is organized as follows; section two details the IBDAB modeling process, followed by the design of the controller in Section three. While the simulation results and their discussion are presented in section four. Finally, this paper ends with a conclusion.

The isolated bidirectional dual active bridge DC-DC converter (IBDAB)

The inherent advantages of the IBDAB, such as safety with galvanic isolation, zero voltage switching, current and voltage control, high efficiency, and high conversion ratio, make this converter ideal for EV charging. Due to its bidirectional capability, the IBDAB can either charge the EV from the DC-Bus or inject power back to this latter, which is convenient for the V2G applications. Nevertheless, to ensure an efficient power flow in both directions, a robust and reliable control strategy must be designed. Since the controller used in this study (DISMC) is model-based, an accurate and reliable model of the DAB converter is developed in this section.

Fig. 2 shows the configuration of the DAB converter in which V_b and V_{dc} are the voltages on the EV side and the DC-bus side, respectively. The circuit topology of the converter consists of eight semiconductor switches (S_1, ..., S_8) forming two symmetric H-bridges, connected to the primary and secondary sides of a high-frequency transformer. Filtering capacitors (C_b and C_{dc}) are inserted on each side of the converter to smooth out the voltage and reduce the ripples. L_k represents the equivalent inductance which accounts for both the transformer's leakage inductance and auxiliary inductors. In addition, an inductor L_b is added to smooth out the current on the EV side. In this paper, the high voltage (HV) side of the DAB converter is connected to the DC-bus, whereas the low voltage side (LV) is interfaced with the EV, and n represents the transformer ratio.

Using the single-phase-shift modulation control, the duty cycle of the switches is kept at 50%. On each side, the upper and lower switches of the same bridge conduct in a complementary manner, and those of the diagonal positions are simultaneously conducting. Whereas, the conducting interval between the primary and secondary sides is defined as DT_s, where Ts is the half-switching period and $D = \frac{\delta}{\pi}$ represent the phase-shift ratio with δ indicating the phase-shift angle. This latter determines the average and the direction of the power flow of the DAB converter according to the following equation:

$$P_o = \frac{V_b V_{dc}}{2nf_s L_k} D(1 - D) \tag{1}$$

Where f_s is the switching frequency and D is ranged between -1 and 1 ($-\frac{\pi}{2} < \delta < \frac{\pi}{2}$).

Eq. 1 shows that the quantity of power transferred, from one side to another, depends also on the switching frequency and the equivalent inductance. Furthermore, based on Eq. 1, the value of D that yields the maximum power is $D = \frac{1}{2}$ ($\delta = \frac{\pi}{2}$), accordingly P_{max} is given as follows

$$P_{max} = \frac{V_b V_{dc}}{8nf_s L_k} \tag{2}$$

To enable the DAB converter to transfer power from the EV to the DC-bus, the phase-shift ratio (D) must be positive, and vice versa (charging). Therefore, the value of D can be divided into two sub-equations:

$$D = 0.5 - \sqrt{0.25 - \frac{2nf_s L_k}{V_b V_{dc}} P_o} \ , P_o > 0 \tag{3}$$

$$D = -0.5 + \sqrt{0.25 + \frac{2nf_s L_k}{V_b V_{dc}} P_o} \ , P_o < 0 \tag{4}$$

As a result, for a given value of V_b, V_{dc}, f_s and L_k, the transferred power P_o can be directly controlled by the phase shift according to the nonlinear relationship shown in Eq. 1. Thus, in this paper, a robust control approach is designed to compute the appropriate value of the phase shift that can supply the desired power.

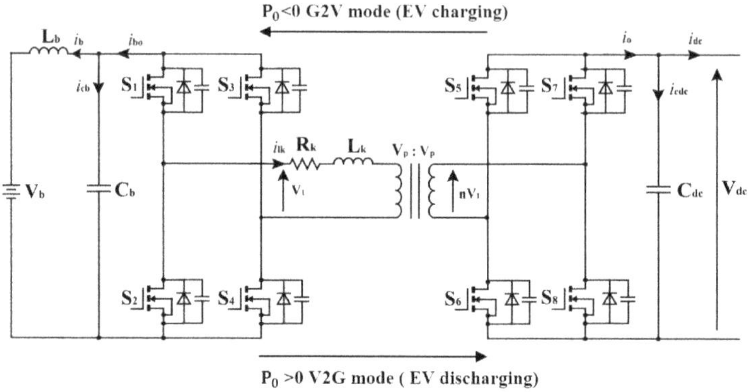

Fig 2. *DAB converter*

Design of the DC-Bus voltage controller

In this study, a double integral sliding mode controller (DISMC) with DPC approach is applied to control the power flow from the EV to the grid while maintaining a stable DC-Bus voltage. This is achieved by computing the phase shift δ that corresponds to the required power.

The DISMC is an enhanced version of the sliding mode control (SMC). The controller relies on the mismatch between the reference and the measured DC-bus voltages to derive a corresponding power P_o (Eq. 1) that nullifies this error. Afterward, the computed power is translated into the phase shift ratio D using Eq. 3.

The voltage error used by DISMC is denoted as follows:

$$e_v = V_{ref} - V_{dc} \tag{5}$$

Where V_{ref} is the DC-bus voltage reference. To efficiently eliminate the steady-state error in the conventional SMC and single integral SMC, a double integral sliding surface s_v is used instead of a single integral sliding surface:

$$s_v = C_1 e_v + C_2 \int e_v \, dt + C_3 \iint e_v \, dt \tag{6}$$

C_1, C_2, and C_3 are positive constants representing the gains of the sliding surface. Taking the derivative of Eq. 6:

$$\dot{s}_v = C_1 \dot{e}_v + C_2 e_v + C_3 \int e_v \, dt \tag{7}$$

As mentioned previously, the phase shift will be generated according to the transferred power P_o, which is used as the control signal of the DISMC to regulate the DC-bus voltage. The dynamic equation of the power across the capacitor $P_{C_{dc}}$ can be expressed as follows:

$$P_{C_{dc}} = C_{dc} V_{dc} \frac{dV_{dc}}{dt} = P_o - P_{dc} \tag{8}$$

Where P_{dc} is the power that is injected into the grid via the three-phase inverter, which is defined as:

$$P_{dc} = V_{dc} I_{dc} \tag{9}$$

Materials Research Proceedings 64 (2026) 337-344 https://doi.org/10.21741/9781644904091-42

Thus, the dynamic equation of the DC-bus voltage can be expressed as follows:

$$\frac{dV_{dc}}{dt} = \frac{P_o}{C_{dc}V_{dc}} - \frac{I_{dc}}{C_{dc}} \tag{10}$$

Being that the purpose of the controller is to obtain a stable DC-bus voltage, thus $\dot{V}_{ref} = 0$. As a result, by substituting $\frac{dV_{dc}}{dt}$ in Eq. 7 by its expression in Eq. 10, the new $\dot{s}_v$ is denoted as:

$$\dot{s}_v = -C_1(\frac{P_o}{C_{dc}V_{dc}} - \frac{I_{dc}}{C_{dc}}) + C_2 e_v + C_3 \int e_v \, dt \tag{11}$$

Next, the equivalent control law is obtained from Eq. 11 by solving $\dot{s}_v = 0$, which yields the following control signal, with ε_V is a positive constant:

$$P_{o\,(eq)} = I_{dc}V_{dc} + \frac{C_2}{C_1} C_{dc}V_{dc}e_v + \frac{C_3}{C_1} C_{dc}V_{dc} \int e_v \, dt + C_s \tanh(\frac{S_v}{\varepsilon_V}) \tag{12}$$

Being that the 'sign' function is responsible for the chattering effect due to its discontinuity, the hyperbolic tangent function 'tanh' is used in this work to achieve a smoother control law. To discern the existence of the DISMC, the following Lyapunov function is chosen:

$$V_v = \frac{1}{2} {s_v}^2 \tag{13}$$

Accordingly, the first derivative of the Lyapunov function V_v is formulated as:

$$\dot{V}_v = -C_s|s_v| < 0 \tag{14}$$

This implies that the controller reaches the sliding surface boundary $|s_v| < \varepsilon_V$ on a finite time, and that it is asymptotically stable under the condition of $C_s > 0$. The dynamics of the controller and the convergence time of V_{dc} to the reference V_{ref} are imposed by setting appropriate gains $C_1 > 0$, $C_2 > 0$, and $C_2 > 0$.

Discussion

The performance of the double integral sliding mode controller is evaluated in this section. The areas that are investigated in this section are: 1) The response time of the controller and 2) The stability of the DC-bus voltage, since the latter will result in a stable output power injected into the grid. The parameters of the IBDAB converter, as well as the information on the EV adopted in the simulations are listed in Table 1.

***Table 1**. Simulation parameters*

Parameter	Value	Parameter	Value
Nissan Leaf battery capacity	40[kWh]	Capacitors Cb, Cdc	2 [mF], 3 [mF]
Transformer's primary and secondary voltages	380[V], 480[V]	Inductance Lk, Lb	0,102[μH] 0,5[mH]

To evaluate the response time and robustness of the controller, the demanded power by the grid is instantaneously changed from P_1=44.5 kW to P_2= 7.7 kW at T=3s, and from P_2 back to P_1 at T=6s as shown in Fig. 3. Based on the results presented in Fig. 4, it is clear that the response time of the controller is satisfactory, bringing back V_{dc} to its nominal value in 0.03 seconds compared to 0.055 seconds when using the classical PID. Furthermore, the DISMC was able to maintain the DC-Bus voltage constant at V_{dc} = 480V with an average error of less than 0.7 V and a maximum error of 8 V, occurring during the transitions in the demanded power from P_1 to P_2 and vice versa. When compared to PID, which yielded an average error of 0.8 V and a maximum error of 25 V at identical time instants, it is evident that the DISMC is superior. This validates the robustness of the controller and its capability to stabilize the voltage regardless of the sudden changes in the injected power.

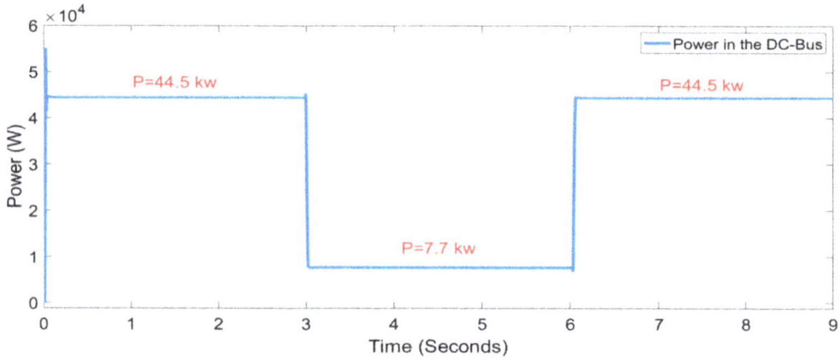

***Fig 3**. Injected power into the DC-bus*

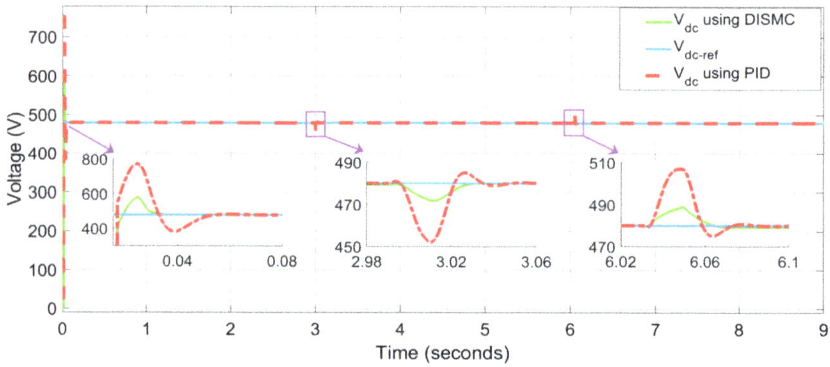

***Fig 4**. Evolution of the DC-Bus voltage*

As a result of injecting power to the DC-Bus, the EV battery state of charge given in Fig. 5 is decreasing. The slopes of the SOC reflect the amount of power withdrawn from the EV. The SOC in Fig. 5 is estimated using an unscented Kalman filter[13].

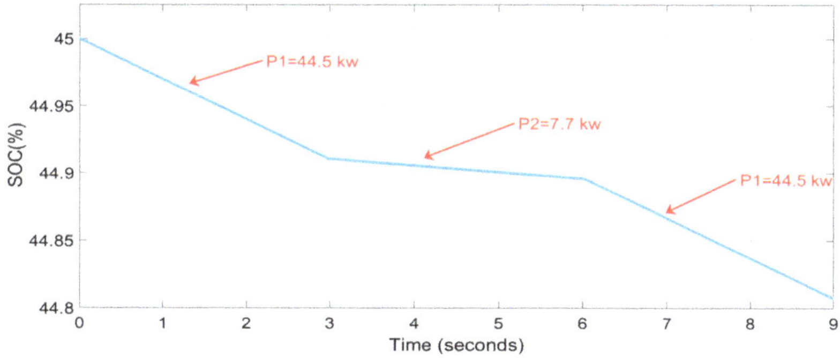

Fig 5. *Variation of the EV's SOC*

Conclusions

Maintaining a stable DC-bus voltage while injecting power into the grid is crucial for a successful implementation of the V2G technology. In this regard, this paper investigated the performance of a double integral sliding mode controller in controlling the isolated bidirectional dual active bridge of the DC-DC charger. The obtained results demonstrated the effectiveness of the adopted control strategy in maintaining a stable DC-bus voltage even under fluctuating power demand. As a result, the authors strongly recommend the use of the DISMC controller as well as the IBDAB converter in V2G charging stations. Prospective studies will investigate the performance of the DISMC when integrated with a backstepping inverter controller and a fuzzy logic–based power management strategy. The objective of this research is to develop a complete control scheme for the V2G-CS.

References

[1] S. Ma, J. Xu, Y. F, Characteristics and key trends of global electric vehicle technology development: A multi-method patent analysis, Journal of Cleaner Production Volume 338, 1 March 2022, 13050. https://doi.org/10.1016/j.jclepro.2022.130502

[2] D. Liu, L. Xu, U. Sadia, H. W, Evaluating the CO2 emission reduction effect of China's battery electric vehicle promotion efforts, Atmospheric Pollution Research, Volume 12, Issue 7, July 2021, 101115. https://doi.org/10.1016/j.apr.2021.101115

[3] A. Hariri, M. Hejazi, H. H.-D, Investigation of impacts of plug-in hybrid electric vehicles' stochastic characteristics modeling on smart grid reliability under different charging scenarios, Journal of Cleaner Production, Volume 287, 10 March 2021, 125500. https://doi.org/10.1016/j.jclepro.2020.125500

[4] S. S. Ravi and M. Aziz, Utilization of Electric Vehicles for Vehicle-to-Grid Services: Progress and Perspectives, Energies 2022, Vol. 15, Page 589, vol. 15, no. 2, p. 589, Jan. 2022. https://doi.org/10.3390/en15020589

[5] H. Ben Sassi, C. Alaoui, F. Errahimi, and N. Es-Sbai, Vehicle-to-grid technology and its suitability for the Moroccan national grid, J. Energy Storage, vol. 33, p. 102023, Jan. 2021. https://doi.org/10.1016/j.est.2020.102023

[6] K. López-Rodríguez, W. Gil-González, and A. Escobar-Mejía, Design and implementation of a PI-PBC to manage bidirectional power flow in the DAB of an SST, Results Eng., vol. 14, p. 100437, Jun. 2022. https://doi.org/10.1016/j.rineng.2022.100437

[7] H. Qin and J. W. Kimball, Closed-loop control of DC-DC dual-active-bridge converters driving single-phase inverters, IEEE Trans. Power Electron., vol. 29, no. 2, pp. 1006–1017, 2014. https://doi.org/10.1109/tpel.2013.2257859

[8] X. Li and X. Fang, Passive Backstepping Control of Dual Active Bridge Converter in Modular Three-Port DC Converter, Electron. 2023, Vol. 12, Page 1074, vol. 12, no. 5, p. 1074, Feb. 2023. https://doi.org/10.3390/electronics12051074

[9] O. Gonzales and A. Rosales, Sliding Mode Controller Based on a Linear Quadratic Integral Regulator Surface for Power Control on a Dual Active Bridge Converter, 2018 IEEE 3rd Ecuador Tech. Chapters Meet. ETCM 2018, Dec. 2018. https://doi.org/10.1109/etcm.2018.8580294

[10] K. Li, Y. Yang, S. C. Tan, and R. S. Y. Hui, Sliding-Mode-based direct power control of dual-Active-bridge DC-DC converters, Conf. Proc. - IEEE Appl. Power Electron. Conf. Expo. - APEC, vol. 2019-March, pp. 188–192, May 2019. https://doi.org/10.1109/apec.2019.8721849

[11] Y. C. Jeung and D. C. Lee, Sliding mode control of bi-directional dual active bridge DC/DC converters for battery energy storage systems, Conf. Proc. - IEEE Appl. Power Electron. Conf. Expo. - APEC, pp. 3385–3390, May 2017. https://doi.org/10.1109/apec.2017.7931182

[12] N. Tiwary, V. R. Naik N, A. K. Panda, R. K. Lenka, and A. Narendra, Integral sliding mode based direct power control of isolated DC-DC converter for improved voltage regulation, Int. J. Circuit Theory Appl., vol. 50, no. 10, pp. 3307–3324, Oct. 2022. https://doi.org/10.1002/CTA.3342

[13] H. Ben Sassi, F. Errahimi, N. Es-Sbai, and C. Alaoui, Comparative study of ANN/KF for on-board SOC estimation for vehicular applications, J. Energy Storage, vol. 25, p. 100822, Oct. 2019. https://doi.org/10.1016/j.est.2019.100822

Hydrogen Production and Utilization Pathways for Renewable Energy Systems: A Comprehensive Review

Youness HAKAM[1,2,a *], Mohamed TABAA[1,b], Abdellah AILANE[2,c]

[1]Multidisciplinary Laboratory of Research and Innovation, Moroccan School of Engineering Sciences (EMSI), Casablanca, Morocco

[2]ENSET Mohammadia, Hassan 2 University, Casablanca, Morocco

[a]y.hakam@emsi.ma, [b]m.tabaa@emsi.ma, [c]abd_ailane@gmail.com

Keywords: Hydrogen Energy Storage, Grid Stabilization, Renewable Energy Integration, Power System Reliability, Fuel Cell, Electrolyzer, Hydrogen Production, Energy Management, Power Grid Flexibility, Sustainable Energy Systems

Abstract. Hydrogen is becoming more and more a key element of the worldwide transitioning energy. It joins generation of renewable electricity, long- term storage of energy, and multi-sector applications including power, transportation and chemical industry. This review provides an extensive compilation of hydrogen production and utilization routes, focusing on the type that is driven by renewable energies. Electricity generated from solar or wind can be used to produce hydrogen using water electrolysis into a range of end uses including compressed, liquefied and solid form of storage for conversion in internal combustion engines (ICEs), turbines, fuel cell electric vehicles (FCEVs) and chemical synthesis. Recent advances in (water) electrolysis, the effect of n-type and p-type classification. This study also addresses recent progress on electrolysis technologies, promoted efficiency and integration with renewable energy systems. It also emphasizes techno-economic, environmental and policy factors influencing hydrogen deployment.

Introduction

Modern power systems have experienced dramatic changes due to the fast deployment of renewable energy sources (RES),solar photovoltaic and wind energy. However, due to their nature of intermittency and variability, the application of these sources has brought new challenges in providing a stable and high-quality power supply to consumers. RES generation is weather and environmental conditions dependent, contrasting with conventional fossil fuel based generation which could be controlled and the outputs are predictable. Hence is real-time electricity supply and demand. The control methods purchased the renewable power, and making of those energy storage systems as tons amount of intelligent such strategy become harder more difficult to deal with [1]. Among various ES mechanisms, HESS is regarded as the most promising candidate because of its high energy density characteristics as well as long duration and flexibility in multi-sectoral energy integration [2]. Hydrogen can also be produced by water electrolysis with surplus renewable electricity and stored as compressed gas, liquid or solid-state hydride storage system and degasified to generate the electricity through fuel cell or combustion systems [3]. The closed-loop energy conversion system maximizes the utilization of excess renewable power and is capable of improving grid reliability, thus facilitating decarbonization [4]. In the last years, HESS have been gaining more and more relevance as a game changer for the transition to sustainable, safe energy systems. They provide essential services such as frequency and voltage regulation, load following spin reserve, system backup generation which play a key role in maintaining high penetration renewable systems []5. In addition, the hydrogen storage together with renewable energy generation is combined to the concept of sector coupling; that is, to interconnect of

Materials Research Proceedings 64 (2026) 345-352 https://doi.org/10.21741/9781644904091-43

electricity, heat and transport systems by a single energy carrier [6]. The ongoing digitalization of power systems in the form of energy management system (EMS), artificial intelligence (AI) and advanced control algorithms is also contributing to reinforce hydrogen as smart grid [7]. The focus of the intelligent EMS system is on hydrogen production, storage and use as a means to lower PEMFC losses by also enhancing reliability. Meanwhile, there is emerging evidences from recent research works for the potential of hybrid hydrogen systems are claimed to provide dynamic power balancing and improve transient network's reliability when combined with batteries and supercapacitors as well as fuel cells [8]. This review provides an overview of where HESS stands with regards to grid support and reliability. It covers the working principles of electrolysis, hydrogen storage, and fuel cell systems and also details their integration within the renewable energy system. It also discusses alternative operations and management perspectives, techno-economics and latest developments shaping the implementation of hydrogen storage technologies. The overall objective is to understand the limitations, potential opportunities and related research developments in the development of hydrogen-based hybrid storage systems that can contribute towards building a sustainable, flexible and decarbonized power grid.

Hydrogen Energy Storage Principles

Hydrogen energy storage is a three-step process: surplus renewable electricity is used to power electrolysis, producing hydrogen for storage; when electricity demand exceeds supply, this stored hydrogen is converted back to electricity. This HESS functions as a flexible and large duration energy buffer to improve the grid reliability and further support renewable integration [9].

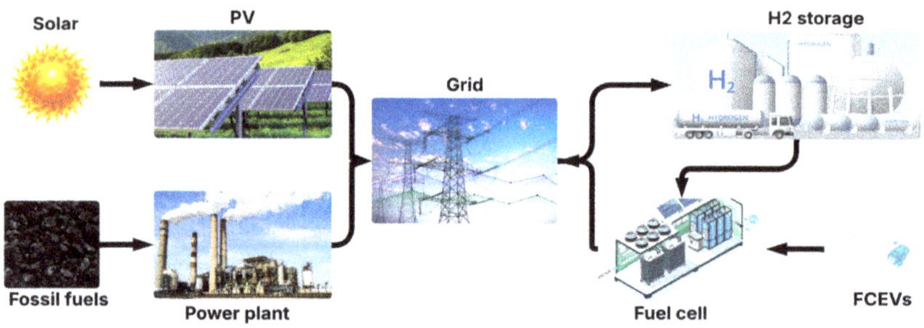

***Figure 1**. Structure of a hydrogen energy storage system*

Fig 1 is a diagram of the basic structure of a hydrogen energy storage system. The power source: Clean renewable energy (solar, wind) runs an electrolyzer to split water into hydrogen and oxygen. The generated hydrogen is compressed, liquified, or absorbed in a solid for storage. When power demand spikes, or renewables wane, the stored hydrogen is reconverted to electricity via a fuel cell or combustion turbine and pumped back into the grid. Advanced control and energy management systems are designed to enhance this cycle efficiency as well as the response capability.

❖ Hydrogen Production through Electrolysis

Most widely used is electrolysis, a process in which electricity is used to split water into hydrogen and oxygen. Three most important electrolyzer technologies are the focus of present research and industry applications:

➢ Alkaline Electrolyzers (AEL)

Relatively mature and low-cost, AEL systems use liquid alkaline electrolytes (for instance potassium hydroxide or KOH). When feed with constant wind speed, they work efficiently under equilibrium condition, but have slow dynamic response to variable renewables inputs [10].

➢ Proton Exchange Membrane (PEM) Electrolyzers

Fuel cells are electrolyzed based on a solid polymer membrane (proton exchange membrane, PEM), which makes it feasible for high current densities with a low volume and fast response. This is particularly attractive for coupling with solar and wind to supply power systems because of its versatility and very pure hydrogen production [11].

➢ Solid Oxide Electrolyzers (SOE)

Running at 700–900 °C, SOEs are based on ceramic electrolytes and have very high efficiency due to cogeneration of electric as well as thermal power. The main problems in present systems are the degradation of ceramic materials and high cost of system; however, their energy conversion efficiency makes them promising for large-scale application [12]. The hybrid PEM-SOE system has been introduced to improve performance and flexibility, optimizing between high efficiency and rapid response for long life [13].

➢ Hydrogen Storage Technologies

Once generated, hydrogen can be stored by several methods mainly grouped into physical and material storage:Gas Compression: Hydrogen is compressed in tanks (350–700 bar). This procedure, which is commercially available and technically mature [14], yields low volumetric density that necessitates dense tanks and safety systems.Storage of Liquid Hydrogen: The liquefaction permits the storage in high gross density (−253 °C). However, this method leads to substantial energy costs (30%) resulting from cryogenic cooling [15]. Solid-State or Material Based Storage: Hydrogen is stored within metal hydrides (e.g., MgH_2, $LaNi_5$), carbon nanotubes, or chemical carriers (ammonia, LOHCs). They these systems are limited by the adsorption/desorption kinetics and reversibility [16]. Novel hybrid storage systems with a combination of compressed and solid-state have been studied to be more efficient, safer, and cost beneficial for the integrated power grid that includes renewable energy penetration [17].

➢ Power Production Using Fuel Cells

Hydrogen can be re–converted to electricity by means of fuel cells, which are devices that convert chemical energy directly into electrical energy through electrochemical reactions producing only water and heat as effluents. Major fuel cell technologies include: Proton Exchange Membrane Fuel Cells (PEMFC), General application, low temperature and fast startup structure for distributed generation and grid backup.SOFC (Solid Oxide Fuel Cell), and Power generation device that, by virtue of its operation at high temperatures, has superior electrical efficiency as well as co-generation capabilities.

➢ Alkaline Fuel Cells (AFC)

This fuel cell technology is considered as one of the older technologies, but yet have potential in aerospace and niche stationary power applications. The overall efficiency of a full HESS (electrolysis → storage → fuel cell) varies between 30% and 45%, depending on the performance of components and system configuration [18].

➢ Integration of System Efficiency and Control

There are several aspects that affect the efficiency of a hydrogen energy storage system, including component selection, conversion losses and energy management strategy. Sophisticated Energy Management Systems (EMS) often supported by AI based optimal programming- are applied to intelligently control power flows, switching decisions and operation modes for optimality at fluctuating renewable inputs [19].Moreover, the integration with Power to X technologies (P2X), including Power to Ammonia and Power to Methane, further authorizes the usage of hydrogen in different modes of transport and industry; positioning itself into a deep decarbonization [20].

➢ Hydrogen Energy Storage and Integration with Renewable Energy Systems

Coupling hydrogen energy storage system (HESS) to renewable energy sources (RESs) like solar photovoltaics (PV) and wind turbine is an effective method to also realize carbon neutrality and power system flexibility. With the increased penetration of renewable generations, and HESS can be used to ease the imbalances between generation and load, meanwhile reducing the variation will assist with its integration into power grid, making system more reliable in general [21].

Hydrogen as a coupling of renewables

Hydrogen production Excess electrical power obtained during high renewable energy output is delivered to the hydrogen production procedure by an electrolyzer combined in hybrid renewable hydrogen systems. This stored hydrogen is then turned back into electricity in fuel cells or a turbine when renewable production ebbs. This closed-loop configuration guarantees grid frequency stability and reduction of curtailment and sectoral coupling in power-transport-industry [22]. Hydrogen based systems are also used for both short- and long-term storage. Classical battery systems [23]. The multipole flexibility at this scale is that which makes hydrogen behave differently compared to.

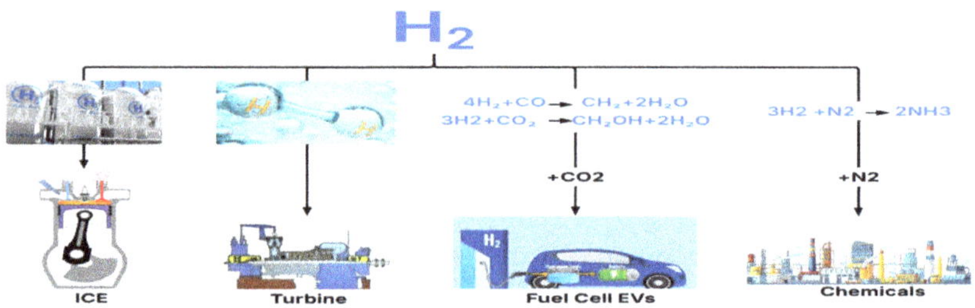

Figure 2. *Hydrogen Energy Storage Integration with Renewable Energy Systems*

Fig 2 shows the links between renewable generation units (wind farm and solar farms), hydrogen production plant (electrolyser), storage technologies, etc., as well conversion to reconvert back into electricity (fuel cells or gas turbines). The energy flow is also monitored by the EMS considering that hydrogen production and consumption depends on grid demand requirements and renewables presence. The figure also indicates relationships among sectors, for example between hydrogen refueling stations and industrial H 2 demand.

Operational Approaches of Renewable Hydrogen Systems

The complete utilization of hydrogen in renewable energy systems would need smart control methods and dynamic EMS for generation, conversion and storage coordination by means of real-time initiatives. Common approaches include: Rule-based control and fast response to the grid unbalance beyond defined thresholds. Model predictive control (MPC), future looking control which takes into account short-term optimal power generation considering the prediction of renewable output and load demand. Innovation An AI based optimization via machine learning and meta-heuristics to enhance system efficiency as well as bring down operation costs [24]. Furthermore, P2H2P systems are more and more integrated with P2A and Power Reduction Plants (PRP) as well as with Electrochemical Systems (ES), in the framework of overall Power-to-X (P2X) networks for system level enhancement of efficiency and robustness [25].

Materials Research Proceedings 64 (2026) 345-352 https://doi.org/10.21741/9781644904091-43

Benefits of Renewable Hydrogen Integration

The key benefits of renewable–hydrogen integration are: Grid balancing: this is to compensate for frequency and voltage deviations due to fluctuation of RES. Independence of Energy, allowing authorities and individuals to have energy autonomy, both locally and nationally, with the domestic exploitation of renewable resources. Sector coupling and decarbonization: potential of using renewable hydrogen in the transport sector, in heating and industry. Economic agility, for the energy arbitrage with hydrogen generation during off-peak price time and power production under peak price. Furthermore, integrating hydrogen with batteries leads to hybrid storage systems that can be used for faster response as well as long-term capacity and thus the multi-layered approach for grid energy management [26].

Discussion and Analysis

Introduction The introduction of the HESS to complement existing renewable integration technology has demonstrated significant promise in both short-term grid instability and long-term energy security. Unlike lithium-ion batteries and other conventional energy storage technologies, HESS can provide multi-hour to seasonal duration of capacity that results in extended time period for shifting [26].

Comparison of the Performance Characteristics for Hydrogen and Conventional Storage

Comparative analysis has shown that, although the round-trip efficiency (85–95%) of batteries is higher than for hydrogen systems, electrochemical storage cannot compare on scaling capacity and durability [27]. Hydrogen energy density (33 kWh $kg^{-}1$) makes the fuel suitable for a wide range of applications including power, transport and industrial sectors, reinforcing sector coupling [18] and supplementing system-level flexibility. Additionally, hybrid system configurations that combine batteries for quick response and hydrogen for load-leveling can offer an optimal compromise between efficiency and capacity. Performance Comparisons of Various Storage Technology Table 1 summarizes the performance comparison of various storage technologies, showing that hydrogen has some advantages in energy duration and capacity, while electrochemical systems have been increasingly desirable for frequency regulation and short-duration buffering.

Table 1. Comparison of Major Energy Storage Technologies for stabilzation grid

Technology	Efficiency (%)	Storage Duration	Power Range	Response Time	Maturity Level	Typical CAPEX (USD/kWh)
Lithium-ion Battery	90 – 95	4 h	kW – 100 MW	< 1 s	Commercial	150 – 400
Sodium-Sulfur	75 – 85	4 – 8 h	MW – 100 MW	< 1 s	Demonstration	350 – 600
Pumped Hydro Storage	70 – 85	Hours – Days	100 MW – GW	1 – 5 min	Mature	50 – 150
Compressed Air Energy Storage	45 – 70	Hours – Days	10 MW – GW	1 – 10 min	Semi-Commercial	100 – 200
Hydrogen Energy Storage	30 – 45	Days – Months	kW – GW	5 – 30 min	Emerging	500 – 1000

Techno Economic Considerations

Economic viability is still a critical issue for the widespread implementation of hydrogen. The levelized cost of hydrogen (LCOH) has been steadily declining due to declining electrolyzer costs and lower cost of renewable electricity [28]. Current analyses estimate that green hydrogen could already be cost competitive by 2030 provided the capital costs of electrolyzers can be reduced below about 400 USD kW^{-1}. Nonetheless, the profitability for grid application is still restricted due to the high system integration and balance-of-plant costs. Monetary incentives are able to accelerate uptake by internalizing the environmental benefits. Scientific Techno-economic analysis based on modeling indicates that a Power to X Battery Hydrogen hybrid system has potential for around 18% less life cycle cost than standalone battery systems in varying RE generation context [29].

Challenges in Control and Energy Management

The dynamic nature of hydrogen systems, from electrolyzer to storage and fuel cells require the use of complex energy management systems (EMS) which are able to carry on a continuous optimization process. Model predictive control (MPC), fuzzy logic and reinforcement learning (RL) based control frameworks have been proposed with promising results in optimizing the operation efficiency [30]. A primary difficulty arises in the time scale coordination: fast power converters, medium hydrogen production scheduling, and long storage management. Combine AI and digital-twin technology with predictive maintenance, adaptive dispatch and fault detection to make the grid more reliable.

Conclusion and Future Direction

Hydrogen is a key element in the sustainable and versatile energy mix of the future. As a renewable product of electrolysis, it is capable of providing deep decarbonization and finds applications through all the main energy forms, from combustion to chemical. The incorporation of hydrogen into renewable grids to increase system stability and facilitate the shift towards a zero-carbon economy is also considered. With technology matured and cost coming down, hydrogen is on the way to becoming a mainstream global energy commodity in place of being an industry niche-solution. Most of these obstacles appear insurmountable at present, though technological and social progress provide hope that they may be overcome. Enhancement of efficiency, Increasing the SOE lifetime and the PEM efficiency at variable loads. Scaling up infrastructure, building a series of modular hydrogen hubs connected to renewable farms. Policy and legislation, Defining hydrogen market rules, safety applications and carbon pricing. Digitalization and AI, Utilizing real-time monitoring and predictive control of hybrid renewable hydrogen networks. The integration of AI based energy management, smart grid communications, and policy coherence shape whether hydrogen will be a back-bone technology in the future carbon-neutral energy system.

References

[1] J. Barba, Integrating Hydrogen into Power Systems: A Comprehensive Review, Sustainability, Vol. 17, No. 13, Article 6117, 2025. https://doi.org/10.3390/su17136117

[2] K. Kwon et al., Integrated Battery and Hydrogen Energy Storage for Enhanced Grid Power Savings and Green Hydrogen Utilization, Applied Sciences, Vol. 14, No. 17, Article 7631, 2024. https://doi.org/10.3390/app14177631

[3] L. Meegahapola, Grid Integration of Hydrogen Electrolysers and Fuel-Cells: Opportunities, Challenges and Future Directions, IEEE Smart Grid Bulletin, March 2023.

[4] M. Wong & H. Nabipour Afrouzi, Hydrogen Energy Storage System: Review on Recent Progress, Energy Engineering, Vol. 122, No. 1, 2025 https://doi.org/10.32604/ee.2024.056707

[5] R. Cozzolino, G. R. Hervàs, A. J. N. Pérez et al., A review of electrolyzer-based systems providing grid ancillary services,Frontiers in Energy Research, vol. 12, art. no. 1358333, 2024. https://doi.org/10.3389/fenrg.2024.1358333

[6] V. Karol, Review of Energy Storage Devices: Fuel Cells, Hydrogen Storage Devices and Batteries, ACS Books/Chapters, 2024. https://doi.org/10.1021/bk-2024-1488.ch001

[7] M. Danish, E. F. C. Lima, A. Mishra, A comprehensive review of green hydrogen-based hybrid energy systems for renewable energy integration and grid stability, Journal of Energy Storage & Systems, vol. x, 2025.

[8] B. Boukhchina, M. Hamdi, S. El Alimi, Power-to-hydrogen: A review of applications, market development, and policy landscape, AIMS Energy, vol. 13, no. 3, pp. 696-731, 2025. https://doi.org/10.3934/energy.2025025

[9] R. Cozzolino, G. Hervàs, A. J. N. Pérez et al., A review of electrolyzer-based systems providing grid ancillary services, Frontiers in Energy Research, vol. 12, art. 1358333, 2024. https://doi.org/10.3389/fenrg.2024.1358333

[10] M. K. Kumar and S. R. Sundar, Comparative analysis of alkaline and PEM electrolyzers for renewable hydrogen production, Renewable Energy, vol. 210, pp. 1085-1096, 2024.

[11] Hakam Y, Ahessab H, Tabaa M, ELHadadi B, Gaga A. Hybrid ANN-GWO MPPT with MPC-based inverter control for efficient EV charging under partial shading conditions. Science Progress.2025. https://doi.org/10.1177/00368504251331835

[12] A. Tiwari, L. Dimitrov and H. Schmidt, Solid oxide electrolyzer cells for green hydrogen production: Review of materials and design trends, International Journal of Hydrogen Energy, 2024.

[13] B. Boukhchina, M. Hamdi, and S. El Alimi, Power-to-hydrogen: A review of applications, market development, and policy landscape, AIMS Energy, vol. 13, no. 3, pp. 696-731, 2025. https://doi.org/10.3934/energy.2025025

[14] M. Abbas, F. A. Al-Shara, High-pressure hydrogen gas storage technologies: Review and perspectives, Journal of Energy Storage, vol. 75, p. 108967, 2025.

[15] J. Chen, P. Reddy, Advances in cryogenic hydrogen storage and distribution, International Journal of Hydrogen Energy, vol. 48, no. 35, pp. 12984-13001, 2023.

[16] L. Gandía, R. Olate, Hydrogen storage in solid materials: State of the art and emerging challenges, Energy Reports, vol. 10, pp. 544-561, 2024.

[17] T. Sharma and N. Bhattacharya, Hybrid hydrogen storage concepts for renewable energy applications, Energy Conversion and Management, vol. 296, p. 118569, 2025.

[18] D. Basso et al., Efficiency and techno-economic assessment of hydrogen-based long-term storage systems, Applied Energy, vol. 353, p. 122019, 2024.

[19] A. Rahman, M. O. Ogunmodede, AI-enabled energy management for hybrid hydrogen energy storage, IEEE Transactions on Sustainable Energy, vol. 16, no. 2, pp. 890-902, 2025.

[20] L. Santos and J. Torres, Power-to-X integration with hydrogen energy systems: Trends and opportunities, Renewable and Sustainable Energy Reviews, vol. 188, p. 113915, 2025.

[21] R. Cozzolino, F. Bella, E. Gammelli, F. Iacovone and M. Riva, A review of electrolyzer-based systems providing grid ancillary services: current status, market, challenges and future directions,"Frontiers in Energy Research, vol. 12, article 1358333, 2024. https://doi.org/10.3389/fenrg.2024.1358333

[22] Y. Zhai, Y. He, J. Shao, W. Zhang, X. Tong, Z. Wang and W. Weng, Review of Hydrogen-Driven Power-to-X Technology and Application Status in China, Processes, vol. 12, no. 7, 2024. https://doi.org/10.3390/pr12071518

[23] Ahessab, H., Hakam, Y., Gaga, A., El Hadadi, B. (2023). A Comparative Study of P&O and Fuzzy Logic MPPT Algorithms for a Photovoltaic Grid Connected Inverter System. In: Motahhir, S., Bossoufi, B. (eds) Digital Technologies and Applications. ICDTA 2023. https://doi.org/10.1007/978-3-031-29857-8_65

[24] Y. Zheng, J. Jia and D. An, "Energy Management for Microgrids with Hybrid Hydrogen-Battery Storage: A Reinforcement Learning Framework Integrated Multi-Objective Dynamic Regulation," Processes, vol. 13, no. 8, article 2558, 2025. https://doi.org/10.3390/pr13082558

[25] D. Tziritas, G. M. Stavrakakis, D. Bakirtzis, G. Kaplanis, K. Patlitzianas, M. Damasiotis and P. L. Zervas, Techno-Economic Analysis of a Hydrogen-Based Power Supply Backup System for Tertiary Sector Buildings: A Case Study in Greece, Sustainability, vol. 15, no. 9, 2023. https://doi.org/10.3390/su15097646

[26] M. F. Ali, P. Suresh, and A. M. El-Sayed, Techno-economic analysis and selection of energy storage technologies for long-duration applications, Frontiers in Energy Research, vol. 13, 2025.

[27] N. Qi, H. Li, X. Zhang and Y. Chen, Long-term energy management for microgrids coordinating hybrid hydrogen-battery energy storage, Applied Energy, vol. 320, p. 120177, 2025.

[28] A. I. Osman, S. Martínez, and D. P. van der Zwaan, Life cycle assessment of hydrogen production, storage and utilization toward sustainability, Wiley Interdisciplinary Reviews: Energy and Environment, 2024. https://doi.org/10.1002/wene.526

[29] M. Abirami, S. Kumar, and R. K. Singh, Techno-economic optimization of hybrid renewable-hydrogen systems: design, sensitivity and performance assessment, Scientific Reports, article number (online), 2025.

[30] R. Banihabib, A. R. Zareipour, and S. M. Muyeen, Techno-economic optimization of microgrid operation with electrolyzer and hydrogen-compatible micro gas turbine, Renewable Energy, vol. 230, pp. 465-478, 2024. https://doi.org/10.1016/j.renene.2024.121708

FPGA-in-the-Loop Implementation of a Grid-Connected Inverter for Renewable Energy Systems

Youness HAKAM[1,a *], Mohamed LAMANE[1,b] and Mohamed TABAA[1,c]

[1]Multidisciplinary Laboratory of Research and Innovation, Moroccan School of Engineering Sciences (EMSI), Casablanca, Morocco

[a]y.hakam@emsi.ma, [b]com.lamane@gmail.com, [c]m.tabaa@emsi.ma

Keywords: FPGA-in-the-Loop, Inverter, Grid-Connected Systems, Renewable Energy, PID Controller, THD Reduction, MATLAB/Simulink, Real-Time Control

Abstract. This paper presents the modeling, control, and FPGA in the loop (FIL) validation of a grid-connected inverter designed for renewable energy integration. The inverter converts DC power from photovoltaic or fuel cell sources into AC power while maintaining synchronization with the utility grid. A real-time control algorithm based on PID controller is implemented using MATLAB/Simulink and executed on an FPGA to achieve fast dynamic response and precise current control. The FIL platform enables hardware-level verification of the control strategy, ensuring system stability and reduced total harmonic distortion (THD). Simulation and experimental results confirm the inverter's high efficiency, accurate grid synchronization, and suitability for smart grid and distributed generation applications.

Introduction

With the large penetration of renewable energy sources, such as photovoltaic (PV) and fuel cell systems in distribution grid, there is a growing demand for high-performance grid-connected inverters that can perform stable, efficient and low distortion power injection operation to utility grid. These converters are of critical importance for ensuring the quality of power and the stability of the grid in scenarios characterized by variable generation and load [1], [2]. In the grid-connected mode, an inverter should precisely control active power and reactive power with fast dynamic response, and with relatively low total harmonic distortion (THD) according to grid codes and synchronization requirement [3]. Its simple implementation and good performance have encouraged traditional control methods like proportional–integral (PI) control in synchronous dq frame or proportional–resonant (PR) control in stationary reference frame to be employed [4], [5]. These techniques, however, usually need to be carefully tuned and demonstrate poor dynamic performances when operated under distorted or unbalanced grid conditions [6]. Furthermore, the fast response of hysteresis and sliding-mode controllers has a drawback regarding that these have variable switching frequency making filter design and efficiency optimization operation more complicated [7]. Model predictive control (MPC) has been taken as an attractive substitution to traditional approach in power converter since it is with a simple structure, fast dynamic response and capability of handling multi-variable constraints in origin [8], [9]. The Explicit Predictive Control (FCS-MPC) predicts system evolution for all possible switching states and chooses the optimal state that minimizes a given cost function [10]. The latter obviates the requirement for dedicated modulation stages, leading to a decrease in computational delay. It has been shown in several reported references that the FCS-MPC provides better performance results than other controllers when applied to the current control of inverters and converters for renewable systems [11, 12]. However, the high computational cost of MPC online application is still one critical constriction in such a way that, while applying to highly switching frequency design [13]. The use of Field Programmable Gate Arrays (FPGAs) has been gaining interest to address such challenges. FPGAs provide high degree of parallelism, deterministic time behavior, and reconfigurability and

are thus an ideal target platform for the realization of high-speed control algorithms such as MPC [14], [15]. Since predictive models and evaluation of the cost function are performed using hardware, FGPA allows reducing the control loop delay and supports high sampling rates [16]. In addition, the FPGA-in-the-Loop (FIL) platform employed offers real-time validation of algorithms in realistic hardware conditions acting as a trustworthy link between simulation environment and testing on experimental hardware [17], [18]. Recent works have demonstrated that FPGA-based implementations allow to obtain better performance in grid-connected inverters for predictive controllers, such as enhanced current tracking, more reduced THD and higher robustness against grid disturbances [19]. Combination of design space-exploration with FIL testing not only speeds up controller validation, but also verifies that the design satisfies time-pressure and accuracy-bound prior to its deployment on HW [20]. It is in this framework, that the study at hand concentrate on modelling,
controlling and FPGA-in-the-loop validation of the grid connected inverter for grid integration. Our contributions are summarized as follows: Designing the resultant controller hardware on MATLAB SIMULINK using FPGA for online implement. Experimental verification via FIL testing, including an assessment of the system stability, harmonic response, and grid synchronization accuracy. The remainder of the paper is organized as follows. In Section 2, the mathematical model of the inverter and control goals are presented. Next, in Section 3 we describe the design and implementation of PID on FPGA. Section 4 presents the FIL environment and experimental validation. Section 5 finishes the paper with discussion on overall system performance and future potential work.

System Modeling and Control

Inverter Topology and System Model

The proposed topology is composed of a two-level, three-phase VSI with electric grid connection by means of an LCL filter. The inverter changes the input DC voltage from a renewable energy source such as photovoltaic (PV) arrays and fuel cells to an output grid synchronized AC voltage. The DC–link voltage is kept constant via an independent DC–DC converter and the VSI is only responsible for the current control/ grid synchronization. LCL filter was adopted to suppress the switching harmonics produced by the inverter and satisfy grid harmonic requirements. It consists of an inverter side reactor, a grid-side reactor, and an intermediate capacitor. This topology promises to exhibit more harmonic attenuation than a basic low pass filter, and enable the system work in lower switching frequency range while keeping THD of injected current low. The state-space representation of the grid-connected inverter via LCL filter can be denoted as:

$$L_f = \frac{di_g}{dt} = V_{inv} - V_g - R_f i_g \tag{1}$$

where L_f and R_f denote the filter inductance and resistance, respectively; V_{inv} represents the inverter output voltage, V_g is the grid voltage, and i_g is the injected grid current. The output voltages of the inverter are generated by a three-phase PWM (Pulse Width Modulation) block to ensure that the inverter output currents accurately track the reference current signals. The inverter works with a balanced grid and has to be synchronized with the grid voltage via PLL. Control goal is to export sinusoidal currents to the grid at unity power factor, where d-axis current control active power and q-axis current controlled to zero.

PID Control Strategy

A Proportional–Integral–Derivative (PID) controller is applied to control the grid current in this work. The PID controller is the most common control algorithm used in power electronics because of its simple structure, stable and reliable properties, and simple digital realization. When it is well-tuned, its performance can be adequate in terms of the steady-state and transient response. The control system is based on the synchronous dq reference frame. The sampled grid currents are converted into the dq frame by Park transform, and compared with their corresponding reference values. The errors are applied to separate PID controllers to use d- and q-axis. The control law is given by:

$$U_d(t) = K_p e_d(t) + K_i \int e_d(t)dt + K_d \frac{de_d(t)}{dt} \tag{2}$$

$$U_q(t) = K_p e_q(t) + K_i \int e_q(t)dt + K_d \frac{de_q(t)}{dt} \tag{3}$$

where ed and eq represent the current errors in the d- and q-axes, respectively. The parameters K_p, Ki, and K_d denote the proportional, integral, and derivative gains of the PID controller. The control outputs U_d and U_q correspond to the voltage control signals in the rotating reference frame. These signals are subsequently transformed back into the stationary reference frame (αβ) using the inverse Park transformation to generate the modulating signals required for the PWM (Pulse Width Modulation) block. PID gains are set appropriately to achieve fast dynamic response and reduced overshoot during transients such as load change or grid disturbance. The integral term removes the steady-state error, and the derivative term enhances the transient stability and damping. The PWM block: It takes the reference voltages from PID controller and converts it to suitable switching signals for inverter IGBT devices. The modulation index and switching frequency are selected to balance between dynamic ability and conduction loss.

Control Objectives

The main functions to be fulfilled by the control system are as follows: Directed Search for Gravitational Waves in Binary Systems using Particle Swarm Optimization Controlled by a MACHO Tracker pointing operation; Precise Current Control: Keep the current drawn from the grid in synchronization with grid voltage, in case of unity power factor operation. Minimize THD: Reduce total harmonic distortion in the output current to satisfy IEEE 519 and IEC 61000 power quality standards. Fast transient response Secure immediate fast response under varying load and environmental conditions. Grid Firm Synchronization: Operation is stable even during grid disturbances or voltage unbalance. Real-Time Execution: Realize the deterministic timing and low computational delay by execution in FPGA. By achieving these goals, the developed FPGA based PID controlled can guarantee the stable and effective renewable power penetration with high quality in utility grid.

FPGA-Based Control Architecture

FPGA-in-the-loop (FIL) was used to confirm the real-time performance of the PID-based control algorithm for the grid-tied inverter. The synthesized control algorithm, developed in MATLAB/Simulink and using HDL Coder, was run on a Xilinx FPGA board that communicates with the computer through JTAG cable. The intervals of sampling was 5 Ö 107 s, and the high speed calculation for PWM control was realized. Fig 1 shows the global FPGA-based control structure of the inverter. The control system is composed of three PID controllers in parallel, which biases the 3-phase leaktance currents (a, b, c). The current error is handled by each controller to generate the modulation reference for the PWM section. The hardware is completely parallelized to ensure timing determinacy and reduce computation time.

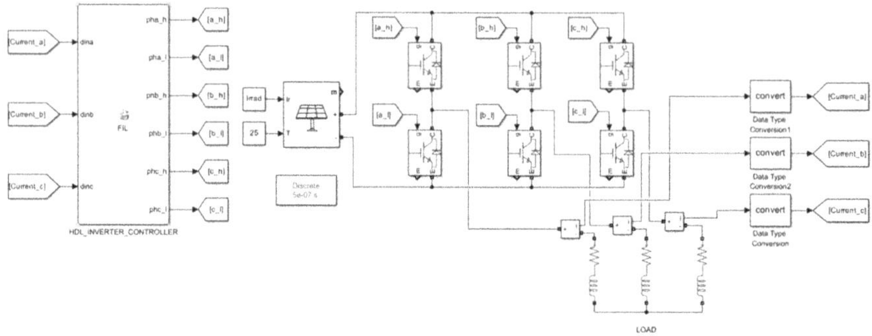

Figure 1. FPGA based control system of the grid connected inverter displaying PID blocks and PWM generation blocks.

A cost-efficient hardware sine-wave generator, which was designed on the FPGA to generate three reference signals that are 120◦ phase displaced, carries perfect symmetry in terms of its phases. The control block utilizes these references for current regulation in real-time.

Simulink and FIL Configuration

The Simulink model of the inverter-filter-grid was communicated with FPGA, via FIL communication block as illustrated in Fig 2. The DFIG model not only contains the DC-link and LCL filter, but also the connection to grid which results in real-time interaction with hardware on software level. On the Simulink side, system dynamics are computed, and on the FPGA side, the control algorithm is running in real-time. In the operation, the two sets of Iref and Grid current are exchanged between Simulink and the FPGA at least every T sw to synchronize them. This positive and negative feedback arrangement mimics the functionality of a real inverter (with no physical power components).

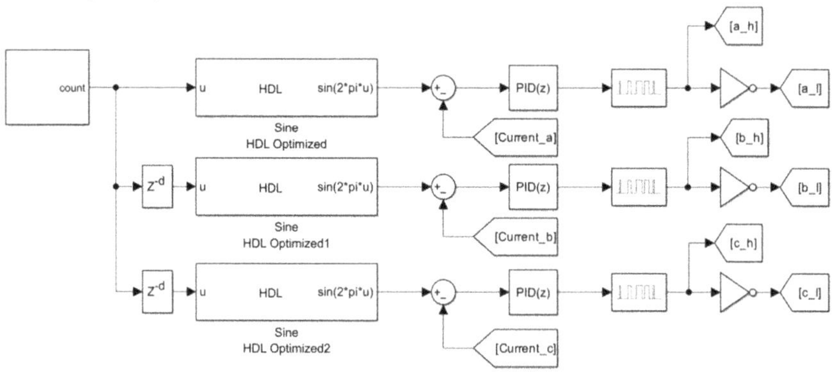

Figure 2. Schematic of the inverter controller and experimental with FPGA-in-the-loop illustrated by current feedback, PID controller and PWM interface.

The tests were performed with a 50 Hz sinusoidal reference current of 4.1A peak amplitude. By the PWM module of the FPGA, the control output was calculated and gate pulses were created for IGBTs. Experimental results were collected from both steady-state and transient tests to evaluate control, stability and dynamic response.

Experimental Evaluation of FIL Setup and Configuration

The FPGA board communicated with its host computer in which the simulation of the power stage of the inverter, the grid model and all sensing units was simulated using Simulink. The bidirectional data communication was carried out using the FIL interface: model signals toward inverter were transferred to an inverter block and processed on FPGA, whereas PWM switching commands from FPGA were transmitted back.

FIL tests gave measured waveforms very close to the reference sinusoidally ones, proving excellent tracking performance. The transient was approximately 5 ms, and the tracking error reached steady state after that time around zero. The numerical values are summarized in Table 1. The RMS current values of the FILs were 2.8915 A, 2.9037 A, and 2.9116 A in phases a, b, and c respectively very close to the reference RMS currents of 2.8925 A, 2.9116A, and 2.8937A. THD was maintained under 2.06%, indicating that FHCA controller effectively reduces harmonic components.

Table 1. *FIL Current Parameters for Phases b and c*

Signal	Mean (A)	RMS (A)	THD (%)
Current b (FIL)	0.00062	2.9037	2.06
Current c (FIL)	-0.00034	2.9116	2.26

The deviation from measured switching currents was kept within ±0.4 A after transients, validating the high precision as well as dynamic performance at relatively high operating frequency. Not all when apparent oscillations were observed, showing a stable closed-loop operation.

Results and Discussion

The FPGA-in-the-loop (FIL) simulations confirmed the effectiveness of the proposed PID-based control in steady-state and transient conditions. Fig. 3 presents the measured three-phase grid currents with their sinusoidal references. The time of transient phase is about 5 ms, in other words it took only 5 ms for the controller to reach synchronization with grid, and its current waveforms were nearly converged at steady-state conditions during this period After full convergence of current, measured and reference currents were almost equal in amplitude as well as in phase, indicating the accurate tracking ability of the controller. The error waveforms (i.e., the difference between reference current and measured current) of closed-loop showed remaining within ±0.4 A, indicating a high accuracy of the system. This finding emphasizes the rapid response of the control action and robustness due to FPGA implementation.

The FFT post-processing in MATLAB was used to analyze the harmonic content of the output current. The steady-state harmonic spectrum of the inverter output current is shown in Fig 3 The THD of the FIL currents was kept below 2.3%, which meets with the IEEE 519 recommendation for grid-connected systems. The harmonic components were mainly distributed around the switching frequency and its multiples, and there was no distortion near the fundamental frequency. The low THD value shows that the proposed LCL filter and PID parameters can suppress I ripple and switch harmonics well.

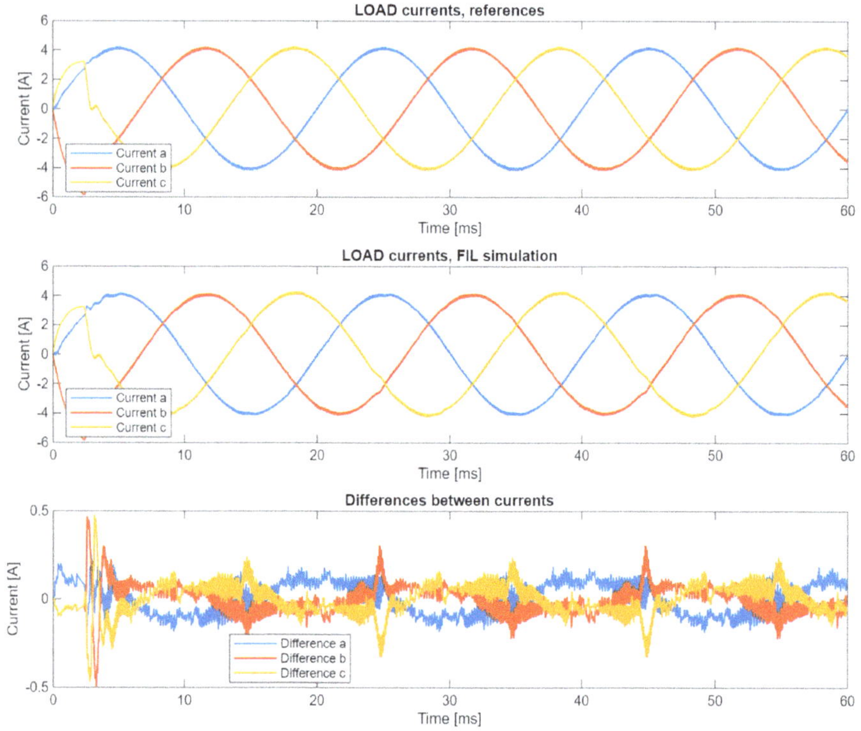

Figure 3. *Three-phase grid currents with their sinusoidal references and FIL.*

To further investigate the transient response of the controller, step changes have been used for reference current amplitude and DC-link voltage. The system response of Fig 3 shows that the controller was able to rapidly track the reference current without any overshoot or oscillation. The transient response time is 4–6 ms, indicating swift dynamic regulation from the degree of pollution. The FPGA's deterministic timing characteristics and its parallel computation architecture facilitates control execution in real-time, as opposed to delays generally faced by microcontroller-based systems. These properties render the developed control strategy attractive for renewable energy conversion under varying generation or grid conditions.

Table 2 Comparison of KPIs between MATLAB/Simulink simulation and FPGAFIL experiment The difference between the RMS and THD values is less than 2%, which means that the developed FPGA control paths realization has correctly reproduced the simulated one. Table 2. Performance comparison between simulation and FPGA-FIL results

Table 2. *Comparison of Reference and FIL Current Parameters*

Signal	Mean (A)	RMS (A)	THD (%)
Current a (ref)	−0.00034	2.8925	1.67
Current b (ref)	0.00065	2.9116	1.72
Current c (ref)	−0.00031	2.8937	1.72
Current a (FIL)	−0.00029	2.8915	1.98

The small discrepancies are primarily due to finite-word-length effects and quantization caused by fixed-point arithmetic in the FPGA fabric. However, the hardware results demonstrate that the stability and harmonic characteristic of the controller is lie to floating-point simulation model.

Conclusion

This paper proposed the design, implementation and FPGA-in-the-loop (FIL) validates of PID based grid-connected inverter for renewable energy integration. The hybrid control scheme presented functioned well even under varying grid conditions and managed to perform in real-time with the FPGA allowing for high-speed computation. The inverter model was s emulated in MATLAB/Simulink and the control implementation was converted to HDL code for deployment on an FPGA. Real-Time Testing using the FIL Configuration The simulation of hardware controller with a detailed inverter and grid model is thus possible without physical parts. Data from experimental verification verified the proposed FPGA PID control system showed good steady-state performance with low current error, THD less than 2.3%, and fast dynamic response less than 5ms. It is seen that, the hardware results are in close agreement with the simulation results; hence, verifying correctness and viability of the proposed realization. The deterministic timing and parallelism of the FPGA enabled execution of identical control action every time, without introduction of delays caused by calculations in DSPs or microcontrollers. Implied from these results overall the proposed system shows to be an effective and efficient control process for renewable energy applications such as photovoltaic and fuel cell systems in smart grid environments. In the future, work will extend the proposed structure to multiple inverter configurations with the application of adaptive or intelligent control schemes for improved robustness and experimental hardware studies under different grid contingencies.

References

[1] Ali Khan, M. Y., Liu, H., Yang, Z., & Yuan, X. (2020). A comprehensive review on grid-connected photovoltaic inverters, their modulation techniques, and control strategies. Energies, 13(16), 4185. https://doi.org/10.3390/en13164185

[2] Zeb, K., et al. (2018). A comprehensive review on inverter topologies and control strategies for grid-connected PV systems. Renewable and Sustainable Energy Reviews, 94, 1120-1141. https://doi.org/10.1016/j.rser.2018.06.053

[3] Athari, H., et al. (2017). Review and classification of control systems in grid-tied inverters. Renewable and Sustainable Energy Reviews, 72, 1168-1182. https://doi.org/10.1016/j.rser.2016.10.030

[4] Gao, S., et al. (2022). Comparative study of symmetrical controlled grid-connected inverters using PR and PI controllers. IEEE Access, 10, 65732-65745.

[5] Koutroulis, E., & Blaabjerg, F. (2019). Design optimization of grid-connected PV inverters. IEEE Transactions on Power Electronics, 34(5), 4180-4192.

[6] Guerrero, J. M., et al. (2013). Hierarchical control of droop-controlled AC and DC microgrids: A general approach toward standardization. IEEE Transactions

on Industrial Electronics, 58(1), 158-172.

[7] Ahessab, H., Hakam, Y., Gaga, A., EL Hadadi, B. (2024). Design and Simulation of an Intelligent Grid-Connected MPPT Inverter with Battery Storage Using ANN Algorithm. In: Bendaoud, M., El Fathi, A., Bakhsh, F.I., Pierluigi, S. (eds) Advances in Electrical Systems and Innovative Renewable Energy Techniques. ICESA 2023. https://doi.org/10.1007/978-3-031-49772-8_29

[8] Rodriguez, J., Kazmierkowski, M. P., Espinoza, J. R., Zanchetta, P., Abu-Rub, H., & Young, H. A. (2013). State of the art of finite control set model predictive control in power electronics. IEEE Transactions on Industrial Informatics, 9(2), 1003-1016. https://doi.org/10.1109/TII.2012.2221469

[9] Hakam Y, Ahessab H, Tabaa M, ELHadadi B, Gaga A. Hybrid ANN-GWO MPPT with MPC-based inverter control for efficient EV charging under partial shading conditions. Science Progress. 2025. https://doi.org/10.1177/00368504251331835

[10] Krstic, M., et al. (2015). A control-Lyapunov approach to finite-control-set model predictive control. IEEE Transactions on Automatic Control, 60(7), 1975-1981.

[11] Zafra, E., Leon, J. I., & Franquelo, L. G. (2020). Finite control set MPC with output constraints for two-level three-phase converters. IEEE Transactions on Industrial Electronics, 67(2), 1273-1283.

[12] Khan, H. S., et al. (2021). Improved finite control set MPC for distributed energy resources. Energy Procedia, 188, 258-265.

[13] Wang, F., Zhang, Z., & Kennel, R. (2018). Computational challenges of predictive control for high-speed converters. IEEE Transactions on Power Electronics, 33(12), 10624-10634.

[14] Vallone, J., et al. (2016). FPGA implementation of model predictive direct current control for multilevel inverters. IEEE International Symposium on Industrial Electronics (ISIE). https://doi.org/10.1109/ECCE.2016.7855125

[15] Ahessab, H., Hakam, Y., Gaga, A., El Hadadi, B. (2023). A Comparative Study of P&O and Fuzzy Logic MPPT Algorithms for a Photovoltaic Grid Connected Inverter System. In: Motahhir, S., Bossoufi, B. (eds) Digital Technologies and Applications. ICDTA 2023 https://doi.org/10.1007/978-3-031-29857-8_65

[16] Saeidi, S. (2015). FPGA-based nonlinear model predictive control of electric drives. PhD Thesis, Technical University of Denmark.

[17] Acevedo, M. (2016). FPGA-based HIL co-simulator platform for power electronics. M.Sc. Thesis, Lappeenranta University of Technology.

[18] Baghdadi, M., et al. (2023). FPGA-based hardware-in-the-loop emulation of power electronics circuits using device-level behavioral modeling. Energies, 16(12), 4510.

[19] MathWorks. (2023). FPGA-in-the-Loop Simulation Overview. MathWorks Documentation.

[20] MathWorks. (2023). Hardware-in-the-Loop Simulation for Power Electronics. MathWorks Documentation.

Energy-Aware EV Charging Optimization: A SWOT-Informed Hybrid Decision Framework

Abdelghani EZZAHI[1,a*], Brahim ZRAIBI[2,b], Mohamed MANSOURI[1,c], Adam AFADISS[1,d] and Omar LAMMAMRI[1,e]

[1]Hassan 1st University of Settat, ENSA of Berrechid LAMSAD Laboratory, Morocco

[2]Cadi Ayyad University, EST of Safi, Industrial Maintenance Department, LAPSSII Laboratory, Morocco

[a]abdelghani.ezzahi.doc@uhp.ac.ma, [b]b.zraibi@uca.ac.ma, [c]mohamed.mansouri@uhp.ac.ma, [d]adam.afadiss.doc@uhp.ac.ma, [e]omar.lammamri.doc@uhp.ac.ma

Keywords: Energy-Aware Charging, EV Scheduling, SWOT Analysis, Hybrid Decision Framework, DRL

Abstract. The increasing of electrical vehicles (EVs) becomes a big challenge for charging infrastructures, including efficiency, grid impact and service quality. To address this need for EV charging management we used methods such as Optimisation, metaheuristics and deep reinforcement learning (DRL), but each method is limited when used separately. This paper proposes a hybrid decision framework guided by a SWOT (Strengths, Weaknesses, Opportunities, Threats) analysis, which provides a structured way to combine and organize these approaches under different operating conditions. To organize the decision-making of this framework we are based on three complementary layers : deterministic optimization for day-ahead planning and constraint satisfaction, metaheuristic tuning for multi-objective trade-offs, and DRL-based control for real-time adaptation under uncertainty. The proposed framework is validated using real-world charging session data from the Adaptive Charging Network (ACN-Data) collected at the Caltech site from March-April 2019, under a constrained site with a 50-kW total limit and 7.2 kW per-charger limit. Under nominal conditions, FCFS, MILP, and MILP+PSO all reach a 1.00 satisfaction rate with 0.0 kWh energy deficit. Under surprise-arrival demand shocks, DRL remains robust: at 50% extra arrivals it achieves 0.937 satisfaction with 2.366 kWh deficit, whereas FCFS drops to 0.835 satisfaction with 45.680 kWh deficit; across 20–50% shocks, DRL reduces unmet energy demand by up to ~95% versus FCFS.

Introduction

The world knows an increase in EVs, and this creates new operational constraints and challenges for the power systems, basically at the distribution level. Except long charging times, uncoordinated or bad charging optimization can provide a lot of issues such as peak demand, increase feeder loading, and make voltage instable, all these factors can create a risk on all grid reliability [1]. One of the key goals for sustainable mobility is to focus on efficient charging that respects the grid and reduces energy waste. This problem focuses on three main parameters : power systems, optimization, and smart control. The methods used must ensure that most user needs are achievable, at the same time should respect the grid limits and adapt to grid conditions across different cases. In research communities, many methods have been proposed to tackle this problem. Mathematical optimization, such as convex programming and Mixed- Integer Linear Programming (MILP) [2], can deliver exact and robust solutions, but when the problem becomes large their scalability is limited. Metaheuristics such as Particle Swarm Optimization (PSO) [3], Genetic Algorithms (GA) [4], and related bio-inspired methods are most used when the problem is complex and has multi-objective parameters, but their convergence is not at the good level, and their tuning parameters are usually sensitive and require careful verification. The DRL method [5] has been applied to scheduling in real-time when the parameters of environments are uncertain. However, it mostly requires a lot of data, expensive to run, and is difficult to interpret and validate in critical grid applications. Overall, each family of methods has strengths, but no single approach covers all aspects of EV charging optimization. A clear way is therefore needed to compare these approaches, identify their complementarities, and guide the design of hybrid solutions. To handle all these issues, the SWOT analysis [6] is proposed. It's not limited to listing pros and cons, but it also highlights how the strengths can be combined and weaknesses reduced. This paper makes three main contributions; First, it provides a clear overview of the three methods or approaches used for EV charging. Second, these methods are compared using a SWOT analysis in terms of strengths, weaknesses, opportunities, and threats. Based on their analysis, a conceptual hybrid framework is proposed to use their robust points and integrate them in the different layers and improve EV charging performance. The rest of the paper is organized as follows: Section II reviews related work; Section III presents the SWOT analysis; Section IV describes the framework; and Section V reports the experimental validation and discusses the results; and Section VI concludes the paper.

Literature Review

The research on EV charging optimization is generally classified into three main categories : mathematical optimization, metaheuristic algorithms, and ML methods, with a focus on DRL in recent papers. Each category from the methods listed below has its strengths and weaknesses, which indicate the importance of using a structured comparaison. First, Mathematical optimization methods [1–2, 7–10], such as convex programming and MILP, are applied to deliver exact and robust solutions for small to medium-sized systems. By contrast, when the system grows the computational cost increases, which impact these methods and leads them to be less practical, and especially for large distribution networks with high uncertainty and more variability. Second, metaheuristic methods [3–4, 11–15], that includes GA and PSO, it provides flexible search capabilities for nonlinear or multi-objective formulations and help in scalability performance. However, these approaches can converge slowly, and get stuck in local minima, and are usually sensitive to parameter settings. The new variants and hybrid metaheuristics like; GA-PSO or combinations with simulated annealing (SA), focus on reducing these issues, but its performance depends on different cases. Finally, ML and DRL approches [5, 16–21], have become the most popular due to its adaptability to changing conditions in uncertainty. DRL activates the real-time decision-making and has been integrated with other architectures, specifically, the Graph Neural Networks (GNNs) to handle large fleets. In contrast, these methods require a lot of training data, large computing power, and in most cases, they provide a limited interpretability. All these reasons make their application a little hard and creates a real challenge in smart-grid applications. The literature discussed decentralized coordination, Vehicle-to-Grid (V2G) integration, and scheduling that is fair. In conclusion, DRL can scale to large systems but is highly dependent on data and computational resources; metaheuristics scale better but may have convergence and tuning problems; and mathematical optimization offers optimality guarantees but does not scale well.

The various requirements of EV charging optimization cannot be fully satisfied by a single strategy. The SWOT analysis is presented in the following section to emphasize complementarities and offer a framework for integrating approaches. Common EV charging models like rolling-horizon control, queue-aware scheduling, and grid-constrained optimization are also covered in this review [3, 5, 11]. These models are the basis for the hybrid framework presented in this paper.

SWOT-Based Comparative Analysis

A SWOT analysis is applied to compare the three families of approaches in a clear way. This framework reports the main features of each method, and shows how they can complement each other in hybrid solutions. The results are summarized in Table 1, which presents the SWOT matrix for the three families. For traceability, supporting references are listed for each SWOT category: Optimization [1, 7, 9], Metaheuristics [3, 12, 14, 15], and DRL [5, 18–21].

Table 1. *SWOT analysis of EV charging optimization approaches.*

Approach	Strengths	Weaknesses	Opportunities	Threats
Mathematical Optimization	Exact and robust solutions; strong theoretical guarantees; well-suited for small/medium systems	Poor scalability; exponential computational growth; limited flexibility in dynamic contexts	Can be coupled with heuristics for decomposition; integration into short-term local scheduling	Combinatorial explosion in large-scale systems; sensitivity to uncertainty
Metaheuristics	Flexible and adaptable to complex, multi-objective problems; easy to extend to new scenarios	Slow convergence; risk of local minima; high sensitivity to parameter settings	Well-suited for large-scale and non-convex problems; potential to complement exact or learning-based methods	Highly variable performance; difficulty in ensuring reproducibility
ML/ DRL	Adaptive learning in dynamic environments; scalable with advanced models (e.g., GNNs); potential for real-time scheduling	Data-hungry; computationally expensive; "black box" nature limits interpretability	Promising for smart grids, V2G, and demand-response integration; ability to generalize across scenarios	Computational overhead may limit adoption; trust and transparency issues in critical infrastructure

As described in the previous table presenting the key strengths and weaknesses of each method based on SWOT approach, it is clearly mentioned that no complete Approach encompassing the majority or all pillars of EV charging optimization. Similar solutions with strong foundational guarantees can be provided by mathematical optimization, While it underlines a lack of scalability. Another approach using metaheuristics is limited for its convergence slowness even if it's considered flexible for multi-objectives and complex problems, while the results can be different from one case to another. On the other hand, DRL are much better since they can scale and adapt in a better way, most likely when attached with advanced models like GNNs. However, important computational efforts required with lesser and limited interpretability, thing that makes it a major concern in smart-grid applications. To sum up, each one of those categories has its strengths, and the weakness can be reduced by combining different methods. This highlights the relevant of the conceptual framework offered in the next section, where the SWOT methodology is used to guide the design of integrated hybrid solutions.

Conceptual Framework

A conceptual framework is proposed, to reduce EV charging time using hybrid algorithms based on the outcomes of SWOT analysis. The idea is to combine all the strengths points of different methods, and reducing their weaknesses, and exploit available opportunities, while limiting the risks. This approach aims to integrate all in one efficient solution. It does not provide a step-by-step procedure to optimize EV charging time. Instead, it is a general design that can be adapted to different scenarios and future implementations.

A. Design Principles

The framework follows three key design principles. First, it builds on the strengths of the three methods proposed in this paper. Mathematical optimization is applied to exact short-term scheduling and feasibility checks. Metaheuristics address multi-objective trade-offs in complex settings, while DRL is used for long-term decisions that must be adapted under uncertainty. Second, the framework reduces the risks of individual methods by combining a DRL with simple heuristics to accelerate learning and reduce the risk of getting stuck in local minima. Large optimization problems are decomposed into smaller, easier sub-problems. Hybrid combinations such as optimization combined with learning or heuristics, are used to limit computational complexity. Finally, the framework exploits available opportunities and control potential risks, by including Smart-grid features such as V2G, demand response, and renewable energy. Computational limits are addressed with distributed computing and cloud–edge setups. Model transparency is improved by adding interpretable parts to data driven methods.

B. Hybrid Workflow

The proposed framework is organized into three layers, each linked to a specific planning horizon. At the short-term level, local optimization is performed using mathematical methods, such as MILP or convex programming, to produce exact charging schedules that respect immediate grid and tariff constraints. For medium-term planning, the framework relies on metaheuristic approches, including PSO and GA, to handle multi-objective trade-offs and support flexible day-ahead or week-ahead planning. Long-term policy adaptation is handled through DRL. In this layer, DRL agents are trained to learn adaptive strategies that work much better across different scenarios, enabling the system to make robust decisions under changing conditions, such as uncertain demand and variable renewable generation.

In this framework, "charging time" means the total time from plug-in until the target state of charge (SoC) is reached. It includes both waiting and active charging times.

Formally, the goal is to minimize the total time a vehicle stays in the station:

$$\min_{P_{i,t}} T_{sojourn} = T_{wait} + T_{charge} \qquad (1)$$

this is done while respecting grid voltage limits, and feeder capacity and tariff constraints :

$$V_{min} \leq V_t \leq V_{max} \qquad (2)$$

The medium-term metaheuristic layer optimizes the multi-objective vector :

$$[F = T_{avg}, C_{energy}, R_{peak}] \qquad (3)$$

which includes average charging time, energy cost, and peak-load ratio. The long-term DRL layer learns adaptive policies :

Materials Research Proceedings 64 (2026) 361-370 https://doi.org/10.21741/9781644904091-45

$$\pi(S_t) \rightarrow a_t \quad (4)$$

using state variables such as SoC distributions, queue lengths, and time-varying prices.

The information shared between the layers includes feasible schedules, heuristic guidance, and the learned parameters of the policy. This maintains consistency in decisions made and scalability across the different time horizons.

C. Conceptual Diagram

To illustrate the workflow described above, Fig. 1 presents a schematic diagram of the proposed framework. The diagram has four layers :

- **Input layer:** EV fleet demand, charging requirements, grid constraints, electricity price signals, and renewable generation profiles.
- **SWOT-guided decision layer:** A central block showing how SWOT categories guide the selection of algorithm families (optimization, metaheuristics, DRL).
- **Hybrid integration layer:** Three connected modules for short-term optimization, medium-term planning, and long-term DRL learning.
- **Output layer:** Optimized charging schedules, reduced charging time, lower costs, and improved grid stability.

The figure provides a visual summary of the hybrid workflow, connecting the SWOT-based analysis (Section 3) to the proposed conceptual framework.

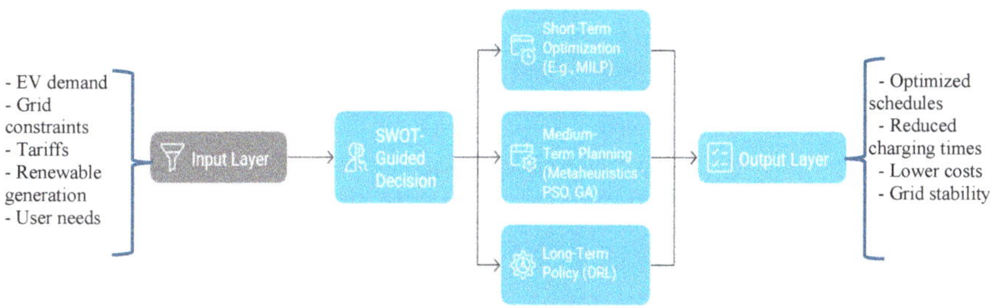

Fig. 1. *Conceptual framework for EV charging optimization guided by SWOT*

D. Advantages of the Proposed Framework

In comparaison with single method approaches that aim to optimize one objectif, the proposed framework is more efficient and has several advantages such as: Robustness, due to its combination of solution quality, flexibility, and adaptability. Scalability and simplicity to apply in large systems due to its decomposition and hybrid design. Adaptability, because the DRL method is flexible in dynamic and uncertain conditions. Over that, it's also practically relevant because SWOT provides a clear way to compare methods and guide how they are combined. The experimental validation is based on four essential metrics: including service satisfaction, energy deficit, peak demand, and energy cost. These metrics are extracted from the objectives and constraints defined in the equations above (1)–(4).

Experimental Validation and Results

A. Dataset and Experimental Setup

To validate the SWOT-guided hybrid optimization framework, a real EV charging data from the CAN-Data dataset has been used. This dataset includes different charging sessions with its specific arrival and departure times, and the requested energey per vehicle. These charging sessions are collected especially during the period from March-April 2019, to cover the normal weekdays and the extreme demand periods. Each day is handled as a separate scheduling case, and the charging site has a total power limit of 50 kW and a per-charger limit of 7.2 kW.

The time-of-use (TOU), electricity price signal, is applied to evaluate the realistic energy cost variations during the day. To correctly measure the energy not delivered at departure time per vehicle, the simulation time is enlarged by 24 hours when necessary to cover charging sessions that continue past midnight. This ensures accurate calculation of service satisfaction and energy deficit. The energy deficit is defined as the amount of requested energy that is not delivered at vehicle departure, which directly reflects service failure from the user perspective.

B. Scenario 1: Nominal Operation (Predictable Demand)

The framework is evaluated under nominal operating conditions, where arrival and departure times are known and no unexpected events occur. Under these conditions, the planning assumptions are valid, and system stress remains limited.

Three approaches are considered:

- **FCFS** (First-Come First-Served) as a simple heuristic baseline,
- **MILP** for day-ahead global optimization under deterministic assumptions,
- **MILP + PSO**, where particle swarm optimization is used offline to tune multi-objective weights.

Table 2 illustrates the performance under nominal conditions. All methods achieve 100% service satisfaction with zero energy deficit, showing that the charging demand is feasible when arrivals and departures are predictable.

Table 2. *Nominal Scenario Performance.*

Method	Satisfaction rate	Total Deficit (kWh)	Energy Cost
FCFS	1.00	0.0	Baseline
MILP	1.00	0.0	↓
MILP + PSO	1.00	0.0	↓↓

However, a clear difference is observed in operational efficiency. FCFS allocates power based only on current arrivals, without using future information or time-varying prices. By contrast, the MILP plan uses full knowledge of arrivals and departures to move charging away from high-price periods and coordinate power system-wide, leading to a lower energy cost. The integration of PSO further improves this process by tuning the multi-objective trade-offs, resulting in a better and more stable operating point. Although service satisfaction remains unchanged, MILP+PSO improves cost efficiency and reduces sensitivity to parameter selection, as shown by the lowest energy cost in Table 2. These results confirm that deterministic optimization is sufficient and effective when demand is predictable and system stress is low. In this setting, learning-based adaptation brings little additional benefits, which motivates the robustness analysis under uncertainty in the next scenario.

Materials Research Proceedings 64 (2026) 361-370 https://doi.org/10.21741/9781644904091-45

C. Scenario 2: Robustness Under Uncertainty (Surprise Arrivals)

To evaluate robustness, real-time uncertainty is introduced in the form of surprise vehicle arrivals, that challenge the assumptions of deterministic planning. Additional charging sessions are injected during peak hours to represent unexpected user behavior. Four uncertainty levels are considered, corresponding to 10%, 20%, 30%, and 50% extra arrivals compared with the nominal day. In this scenario, the focus is placed on methods that can operate online:

- **FCFS**, representing non-adaptive scheduling,
- **DRL (Soft Actor-Critic(SAC)-based controller)**, representing the real-time adaptive layer of the framework.

Table 3 reports system performance under increasing levels of surprise arrivals. The reported metrics include service satisfaction rate and total energy deficit, which measure how well the system meets user energy requests before departure.

Table 3. *Robustness Under Surprise Arrivals.*

Surprise arrivals	FCFS Satisfaction	DRL Satisfaction	FCFS Deficit (kWh)	DRL deficit (kWh)
10%	0.948	0.948	1.589	1.589
20%	0.905	**0.952**	7.155	**1.589**
30%	0.882	**0.956**	19.108	**1.589**
50%	0.835	**0.937**	45.680	**2.366**

At low uncertainty (10%), FCFS already performs close to the best, and the DRL controller matches it, indicating that learning-based adaptation is not needed when disturbances are limited. As uncertainty increases, FCFS performance drops quickly. Unmet energy demand increases a lot, and the satisfaction rate goes down. For example, with 30% surprise arrivals, the total energy deficit exceeds 19 kWh, and it increases to 45.68 kWh under extreme stress (50%). By contrast, the DRL approach perform well even if the uncertainty increases (20–50%). By comparing with FCFS method, it reduces the total energy deficit by up to 95%, and keeps a high satisfaction level by respecting the site power limits. This improvement can lead to higher energy cost under extreme stress, showing a clear trade-off between service quality and energy cost. Fig. 2 and Fig. 3 illustrate the results in Table 3. Fig. 2 shows how the total energy deficit changes as surprise arrivals increase. FCFS indicates a sharp increase in energy deficit, which indicates strong sensitivity to demand shocks. On the other hand, the DRL-based controller keeps the deficit nearly constant, even under large perturbations (30–50%). This indicates that learning-based real-time control can handle unexpected demand while still respecting grid constraints.

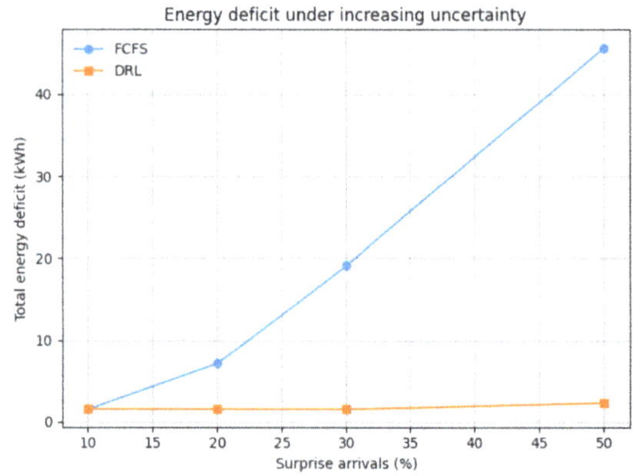

***Fig. 2.** Total unmet energy under surprise-arrival demand shocks.*

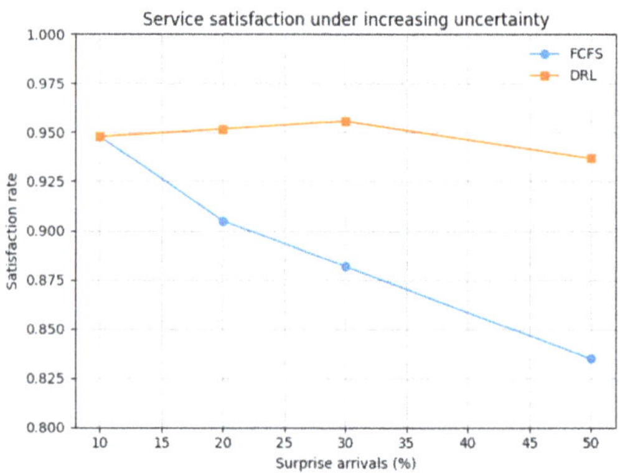

***Fig. 3.** Satisfaction rate under surprise arrivals.*

In parallel, Fig. 3 shows the evolution of the service satisfaction rate as uncertainty grows. FCFS indicates a consistent drop in service quality, whereas the DRL controller maintains a high satisfaction level, confirming its robustness in real-time disturbances

D. Discussion: Complementarity and Robustness of the Hybrid Framework

The experimental results confirm that no single optimization approach performs well in different operating environments for EV charging systems. In the nominal cases with a clear demand (Scenario 1), deterministic optimization with MILP uses full information to reduce energy cost and coordinate charging decisions. In this configuration, learning-based adaptation is not necessary, since a simple heuristic can already achieve full-service satisfaction. Then, to improve the framework by making the multi-objective balance more stable, the PSO method is included. It can reduce as well as the sensitivity to parameter settings and makes the operation more consistent, without adding much extra computation. However, when real-time uncertainty disturbs the planning assumptions (Scenario 2), the previous methods used such as deterministic and non-adaptive heuristics don't perform well. In this case, an online adaptation method like DRL must be used, to adjust power in real time and handle unexpected demand, keep high service quality, and meet the network constraints.

These findings support the suggested framework's SWOT-guided design concept :

- Deterministic planning exploits structural strengths under predictable conditions,
- Metaheuristic tuning mitigates parameter sensitivity and operational variability,
- Learning-based control addresses operational threats arising from uncertainty.

In summary, the results demonstrate that robust EV charging optimization requires a coordinated combination of planning, tuning, and real-time adaptation, rather than relying on a single algorithmic approach.

Conclusion & Future Work

This paper presents a SWOT-guided for a hybrid decision framework for EV charging time optimization. This framework focuses on three main layers: deterministic day-ahead planning with MILP under predictable conditions, offline metaheuristic tuning (PSO) to make the multi-objective problems more stable, and the last layer is a DRL-based real-time control to handle the uncertain parameters. To confirm the performance of this framework, public data from CAN-Data is used to get real charging sessions and test them on a constrained infrastructure (50 kW power limit) including as well as the time-varying electricity prices. The results obtained show clearly that deterministic optimization is robust in nominal conditions, whereas non-adaptive methods degrade quickly in the presence of uncertainty. By contrast, In surprise-arrival scenarios, the DRL layer is more robust: it keeps a high satisfaction rate and reduces unmet energy demand by up to about 95% under strong stress, but energy cost can be higher in extreme cases. This shows a clear trade-off between service quality and energy cost. Overall, These results confirm that MILP, PSO, and DRL perform best when combined, which support the use of hybrid designs for robust EV charging systems. Future work will extend the framework to cover all operational factors (e.g., multi-site networks, microgrids, V2G) and include more detailed battery models and user-preference models.

References

[1] Wang, Y., Wang, H., Razzaghi, R., et al.: Multi-objective coordinated EV charging strategy in distribution networks using an improved augmented epsilon-constrained method. Applied Energy 369, 123547 (2024). https://doi.org/10.1016/j.apenergy.2024.123547

[2] Ren, L., Yuan, M., Jiao, X.: Electric vehicle charging and discharging scheduling strategy based on dynamic electricity price. Engineering Applications of Artificial Intelligence 123, 106320 (2023). https://doi.org/10.1016/j.engappai.2023.106320

[3] Farhadi, F., Wang, S., Palacin, R., Blythe, P.: Data-driven multi-objective optimization for electric vehicle charging infrastructure. iScience 26, 107737 (2023). https://doi.org/10.1016/j.isci.2023.107737

[4] Cao, G., Chang, L.: Optimal Charging Strategy of Dynamic Electricity Price for Electric Vehicles Based on Improved Particle Swarm Optimization. Journal of Electrical Engineering & Technology 20, 1245-1254 (2025). https://doi.org/10.1007/s42835-024-02057-6

[5] Yang, Y., Zhang, C., He, J., et al.: Scalable reinforcement learning for large-scale coordination of electric vehicles using graph neural networks. Communications Engineering 4, 457 (2025). https://doi.org/10.1038/s44172-025-00457-8

[6] Helms, M.M., Nixon, J.: Exploring SWOT analysis - where are we now? Journal of Strategy and Management 3(3), 215-251 (2010). https://doi.org/10.1108/17554251011064837
[7] Das, R., Wang, Y., Busawon, K., et al.: Real-time multi-objective optimisation for electric vehicle charging management. Journal of Cleaner Production 292, 126066 (2021). https://doi.org/10.1016/j.jclepro.2021.126066
[8] Singh, P.P., Das, S., Wen, F., et al.: Multi-objective planning of electric vehicles charging in distribution system considering priority-based vehicle-to-grid scheduling. Swarm and Evolutionary Computation 78, 101234 (2023). https://doi.org/10.1016/j.swevo.2023.101234
[9] Zhang, L., Sun, C., Cai, G., Koh, L.H.: Charging and discharging optimization strategy for electric vehicles considering elasticity demand response. eTransportation 18, 100262 (2023). https://doi.org/10.1016/j.etran.2023.100262
[10] Kaur, A.P., Singh, M.: Time-of-Use Tariff Rates Estimation for Optimal Demand-Side Management Using Electric Vehicles. Energy 273, 127243 (2023). https://doi.org/10.1016/j.energy.2023.127243
[11] Fang, B., Li, B., Li, X., et al.: Multi-Objective Comprehensive Charging/Discharging Scheduling Strategy for Electric Vehicles Based on the Improved Particle Swarm Optimization Algorithm. Frontiers in Energy Research 9, 811964 (2021). https://doi.org/10.3389/fenrg.2021.811964
[12] Huo, M., Pang, S., Zhao, H.: Improved PICEA-g-based multi-objective optimization scheduling method for distribution network with large-scale electric vehicles. Scientific Reports 14, 80184 (2024). https://doi.org/10.1038/s41598-024-80184-w
[13] Pang, S., Fan, K., Huo, M.: Charge and discharge scheduling method for large-scale electric vehicles in V2G mode via MLGCSO. Scientific Reports 15, 00265 (2025). https://doi.org/10.1038/s41598-025-00265-2
[14] Chen, Y., Zhao, H., Pang, S., Huo, M.: Optimal scheduling of solar powered EV charging stations in a radial distribution system using opposition-based competitive swarm optimization. Scientific Reports 15, 88758 (2025). https://doi.org/10.1038/s41598-025-88758-y
[15] Boubaker, S., Kraiem, H., Ghazouani, N., et al.: Multi-objective optimization framework for electric vehicle charging and discharging scheduling in distribution networks using the Red Deer algorithm. Scientific Reports 15, 97473 (2025). https://doi.org/10.1038/s41598-025-97473-7
[16] Li, S., Hu, W., Cao, D., et al.: Electric vehicle charging management based on deep reinforcement learning. Journal of Modern Power Systems and Clean Energy 10(3), 719-730 (2022). https://doi.org/10.35833/MPCE.2020.000460
[17] Liu, H., Peng, B., He, X., et al.: Deep reinforcement learning for charging scheduling of electric vehicles considering distribution network voltage stability. Sensors 23(3), 1618 (2023). https://doi.org/10.3390/s23031618
[18] Tushar, W., Yuen, C., Mohsenian-Rad, H., et al.: Reinforcement learning based EV charging management systems: a review. IEEE Access 9, 72478-72498 (2021). https://doi.org/10.1109/ACCESS.2021.3064354
[19] Xiang, L., Zhang, Y., Xu, Z., et al.: Advancements in EV charging optimization: a survey of reinforcement learning approaches. In: Proceedings of ICEMD 2024, pp. 87-94 (2024).
[20] Lu, X., Wang, Y., Zhao, H., Song, W.: A deep reinforcement learning-based charging scheduling approach with augmented Lagrangian for electric vehicles. Applied Energy 376, 124706 (2024). https://doi.org/10.2139/ssrn.4869056
[21] Sun, W., Chen, Y., Zhang, G., et al.: Electric vehicles charging management using deep reinforcement learning considering vehicle-to-grid operation and battery degradation. Energy Reports 9, 10031-10043 (2023).

Beyond SMA and EMA: Demonstrating the Strength of the Kalman Filter for Embedded Sensor Smoothing

Reda ADJAR[1,a *], Fathallah RERHRHAYE[2,b], Ahmed ELAKKARY[1,c], Jihane CHTIOUI[1,d], Badr RERHRHAYE[1,e], Ilyas LAHLOUH[1,f] and Nacer SEFIANI[1,g]

[1]Laboratory of Systems Analysis, Information Processing and Industrial Management (LASTIMI), Higher School of Technology (EST), Mohammed V University, Rabat, Morocco

[2]Systems Analysis, Modeling and Decision Support Laboratory (LAMSD), National School of Applied Sciences (ENSA), Hassan First University, Berrechid, Morocco

[a]adjarr07@gmail.com, [b]fathallahrerhrhayeest@gmail.com, [c]aelakkary@gmail.com, [d]chtiouijihane@gmail.com, [e]badr-rerhrhaye@hotmail.com, [f]lahlouh.ilyas@gmail.com, [g]nasefiani@gmail.com

Keywords: IoT Sensor Data and Processing: Embedded Systems; Signal Filtering-SMA and EMA; The Kalman Filter; Noise Reduction and How to Handle Outliers; Applications of Microcontrollers; DHT22 Temperature Sensor

Abstract. The study offers an overall comparison between the simple moving average (SMA) algorithm, the exponential moving average (EMA) algorithm, and the Kalman filter algorithm used as a filter to smooth out the temperatures. The overall comparison has been done using the simulated and actual measures taken by the DHT22 sensors when implemented on resource-limited microcontrollers used by the Internet of Things (IoT) environment. The results reveal that the MAE value is 0.3592, the measure of the level of instability is 8.66%, and the value of robustness is 88.19% when the Kalman filter algorithm is used.

Introduction

In IoT scenarios, memory and processing capabilities are very limited, especially for entry-level microcontrollers like the ATmega328P on the Arduino UNO, which only has about 2 KB of SRAM at a clock speed of 16 MHz, thus restricting the complexity of the filters that can be implemented[1], [2]. In addition, temperature, humidity, and gas sensors are well known to provide data that can often be marred by the presence of noise from the electrical signal as well as an intermittent anomaly[3], [4]. Therefore, it is necessary to implement efficient and simple denoising methods to avoid measurement errors during monitoring and control. Traditional smoothing filters involve compromises, such as the simple moving average (SMA)[5], [6]. The exponential moving average (EMA), on the other hand, offers a faster response with a memory complexity of O(1), but at the cost of a fixed smoothing parameter α, which inevitably implies a trade-off between responsiveness and noise amplification[7], [8]. The Kalman filter offers a probabilistic model for dynamically approximating the optimal state estimate within a Gaussian framework, but it is excessively dependent on the relevance of the process and measurement error covariances (Q and R), even though it produces suboptimal estimates in the presence of non-Gaussian errors or outliers. Within this limited context, this paper proposes a comprehensive comparative approach to the SMA, EMA, and Kalman filters, applied to synthetic DHT22 temperature data and real-world examples[9], [10], [11]. It also presents a set of numerical tests enabling a quantitative comparison of the filters using error measures such as mean error, root mean square error (RMSE), robustness, and resistance to outliers. These tests demonstrate how Kalman filters represent an advantageous compromise in the context of the embedded Internet of Things (IoT), while presenting an adaptive strategy for a real-time dynamic smoothing model, capable of automatically learning its thresholds in order to overcome the limitations of fixed-parameter filters.

Filtering Methods: Principles and Approaches

In this section, we describe three classical signal smoothing techniques that balance computational efficiency with noise suppression performance. Each method represents a different trade-off between memory usage, computational complexity, and noise-responsiveness, making them well-suited for evaluation on severely resource-constrained embedded platforms[3], [12], [13].

Simple Moving Average (SMA). The Simple Moving Average computes the output as the arithmetic mean of the N most recent input samples as on Eq. 1:

$$y_t = \frac{1}{N}\sum_{i=0}^{N-1} x_{t-i} \tag{1}$$

The Simple Moving Average (SMA) delivers a very stable output in steady-state regions and strongly attenuates high-frequency noise, but it introduces tracking delays proportional to the window size and requires storing N past samples, leading to $O(N)$ memory and, in naive form, $O(N)$ computations per update, although a circular-buffer implementation can reduce the operations to $O(1)$. Consequently, SMA is well suited to slow-varying signals where stability is prioritized over responsiveness, but it is poorly adapted to signals with rapid environmental changes or frequent step transitions[14], [15].

Exponential Moving Average (EMA). The Exponential Moving Average uses a weighted combination of the current sample and the previous filtered value as on Eq. 2:

$$y_t = \alpha x_t + (1-\alpha) y_{t-1} \tag{2}$$

where:

- $0 < \alpha \leq 1$ is the smoothing factor
- High $\alpha \rightarrow$ fast response (more weight on current measurement)
- Low $\alpha \rightarrow$ strong smoothing (more weight on history)

The Exponential Moving Average (EMA) adapts quickly with constant $O(1)$ memory and computation, which makes it attractive for microcontrollers, but its fixed smoothing factor α forces a trade-off where higher α improves responsiveness while amplifying noise and lower α enhances smoothing while reducing responsiveness, so α must be manually tuned for each application[16], [17], [18].

Kalman Filter. The Kalman filter employs a prediction-correction framework to provide a probabilistic estimate of the underlying signal and its estimation uncertainty. For scalar IoT sensors (one-dimensional case), the filter operates in two steps as on Eq. 3 and Eq. 4:

$$\hat{x}_{t|t-1} = \hat{x}_{t-1} \tag{3}$$

$$P_{t|t-1} = P_{t-1} + Q \tag{4}$$

Correction Step

$$K_t = \frac{P_{t|t-1}}{P_{t|t-1} + R} \tag{5}$$

$$\hat{x}_t = \hat{x}_{t|t-1} + K_t(z_t - \hat{x}_{t|t-1}) \tag{6}$$

$$P_t = (1 - K_t) P_{t|t-1} \tag{7}$$

Materials Research Proceedings 64 (2026) 371-378 https://doi.org/10.21741/9781644904091-46

where:

- $\hat{x}_t$ = filtered estimate at time t
- P_t = estimation error covariance (uncertainty)
- Q = process noise covariance
- R = measurement noise covariance
- K_t = Kalman gain (optimal blending factor)

The Kalman filter is theoretically optimal under Gaussian noise and updates with constant $O(1)$ cost while automatically adapting its gain, but its performance is highly sensitive to the tuning of Q and R and can degrade in the presence of non-Gaussian noise or strong outliers[19], [20].

Experimental Results

Experimental Setup and Dataset. To this end, a 100-second synthetic temperature signal was created, incorporating reference values through the addition of constant levels, noisy segments, slowly varying levels, and outliers introduced by Gaussian white noise. In addition, real-world DHT22 temperature data was collected using an Arduino UNO board and a low-cost temperature sensor with a typical accuracy of ±0.5 °C. This allows for the integration of real-world phenomena such as thermal drift, correlated noise, and signal loss, which are difficult to model with synthetic data.

In our experiment, the filter parameters are chosen for ease of comparison and represent commonly used tuning parameters. The SMA window length, N, is set to the minimum required noise reduction with minimal attenuation of process variation, i.e., N = 5 units. The EMA smoothing parameter, α, is set to 0.3 to minimize MAE values in simulations using the generated data, which corresponds to a "fast and noisy" processing method. Finally, the Q and R parameters of the Kalman filter are determined using observations of the natural variability of the DHT22 readings and noise as a guide for parameter tuning and are set to Q = 0.01 and R = 0.5[21], [22], [23].

Qualitative Analysis

Noise Filtering in Steady-State Regions. Fig. 1 illustrates the application of the filtering process to the constant temperature area with high noise. The SMA Filter removes most of the noise and provides the most accurate results after a small delay, while the EMA Filter is still sensitive to the tiniest changes and thus shows noise. The Kalman Filter should remove noise significantly while preserving responsiveness, and that is the case.

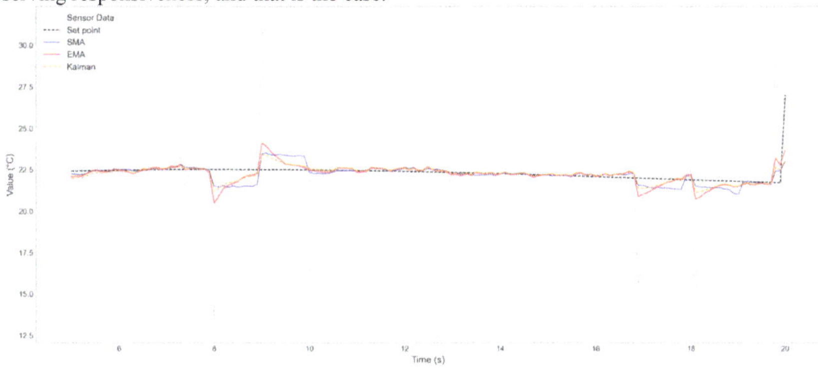

Fig. 1: Noise Filtering Zoom

Materials Research Proceedings 64 (2026) 371-378 https://doi.org/10.21741/9781644904091-46

Under constant-temperature conditions, this zoomed view compares the noise-removal performance of SMA, EMA, and the Kalman filter on the same sensor signal. The SMA strongly smooths the noise but flattens the signal and introduces a noticeable delay, while the EMA remains more responsive but leaves visible residual jitter when using a fixed $\alpha = 0.3$. The Kalman filter achieves a better compromise by significantly attenuating high-frequency noise, preserving sensitivity to small variations, and reaching a steady-state stability score of 8.66%, which positions it between SMA and EMA in terms of responsiveness.

Response and Transient Tracking. Fig. 2 depicts two plateaus in the signal input. Although the SMA filter traces the slow and continuous filtering process along the averaging window, the EMA filter responds to the plateaus after a slight delay but with overshooting and settling around the new value. The Kalman filter responds to the plateaus gradually without overshooting, providing a balance between response speed and stability.

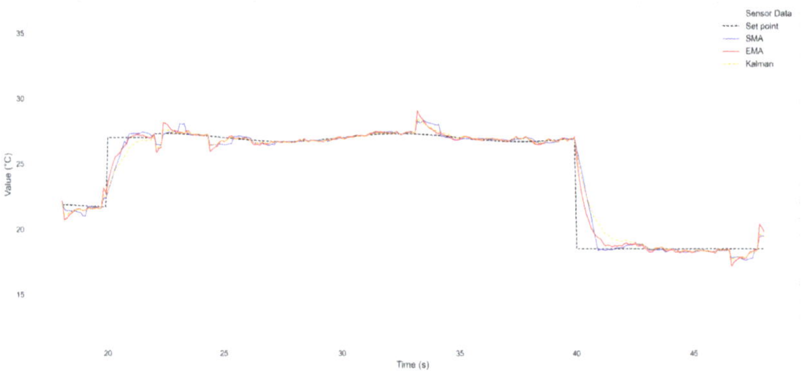

Fig. 2: Step Response and Transient Behavior (Zoom View).

Two independent step changes highlight the dynamic behavior of the three filters: the SMA responds with a slow ramp over 2–3 seconds, trading reactivity for strong noise rejection, while the EMA reacts quickly but oversh Le facteur de quantification mise en evidence l'importance du fait que chacune des filtres traditionnes se situe dans une region uniquement à elle au compromis erreurs, robustesse et stabilité, alors que le kalman offre le meilleur compromisoots and oscillates around the new level, showing the high sensitivity of a fixed $\boldsymbol{\alpha} = \mathbf{0.3}$. In contrast, the Kalman filter tracks the steps smoothly and rapidly without overshoot, because its adaptive gain based on covariance $\boldsymbol{P}$ continuously balances prediction and measurement, delivering both fast transients and a clean steady-state output compared with the fixed window of the SMA and fixed gain of the EMA.

Signal Behavior. Fig. 3 complete simulated signals of 100s with stable phases, sudden transitions, small changes, and noise peaks. In the entire period, the simple moving average correctly detects the stable phases, with constant lag in sudden changes. On the other hand, the exponential moving average filter reacts rapidly per change, with largely intensified noise in almost stable phases. The Kalman filter manages the trend correctly without considerable oscillation, despite the existence of noise.

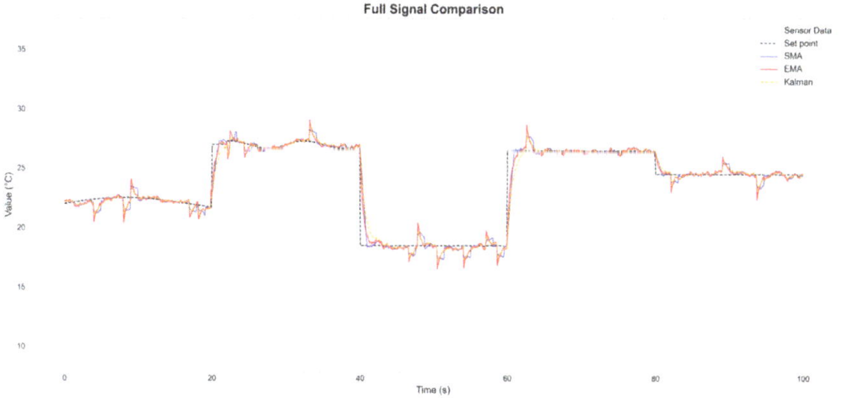

Fig. 3: Full Signal Comparison Over 100-Second Duration.

On the complete artificial data set, which includes constant zones, ramps, discontinuities, and outliers, the three filters behave very differently. The SMA is efficient in stationary regions but reacts with a noticeable delay during transitions, while the EMA responds quickly to changes yet amplifies noise in quasi-steady regimes, causing oscillations and instability. By contrast, the Kalman filter closely follows the true signal with only a small delay, suppresses noise and abnormal peaks, and remains consistent across all operating conditions, adapting well to real-world variations without special tuning.

Quantitative Evaluation. Quantitative measurements based on the synthetic signal obtained from the known reference signal are presented in Table 1 In addition to error measurements, two aggregate measurements were calculated:

- **Stability (%):** This is an expression of the average suppression of high-frequency noise energy when in a state of steady state, as a fraction of the reduction in standard deviation from the raw signal to the filter output.
- **Robustness (%):** This measurement represents the filter's ability to eliminate deviations caused by the presence of outliers. It is expressed as a percentage and represents the proportion of deviations eliminated by the filter relative to the outliers.

Table 1. Performance Comparison of SMA, EMA, and Kalman Filter on Synthetic Temperature Dataset

Method	Stability (%)	Robustness (%)	MAE	RMSE
SMA	6.88	91.12	0.3548	0.8120
EMA	7.31	81.37	0.3320	0.6922
Kalman	8.66	88.19	0.3592	0.8139

Balanced filtering performance for embedded IoT sensors

The quantification factor highlights that each traditional filter occupies a unique position regarding the trade-off between error, robustness, and stability, whereas the Kalman filter offers the best overall compromise. The EMA offers the lowest errors ($MAE=0.3320$ and $RMSE=0.6922$), indicating excellent sensitivity to signal variations, but this comes with the lowest robustness (81.37%), meaning marked sensitivity to outliers and noise in supposedly stable regions. The opposite is true for the SMA, which has maximum robustness (91.12%), meaning excellent ability to reject outliers and suppress noise, but low stability sensitivity (6.88), as well as a higher MAE indicating marked delays in transition mode. The Kalman filter benefits from both; although it has an MAE slightly higher than the EMA (0.3592), it offers better stability sensitivity (8.66) and moderate robustness (88.19%). This higher MAE represents an acceptable compromise for exceptional smoothing capacity in stable modes and low sensitivity to perturbations, making it the most consistent choice for IoT applications.

Conclusion and Perspectives

This paper systematically compares three traditional filtering algorithms: Simple Moving Average Filter (SMAC), Exponential Moving Average Filter (EMA), and Kalman Filter (KAF), and uses simulated and actual measurement data from the DHT22 sensor to perform noise reduction on temperature data. This allows us to evaluate the efficiency and limitations of each algorithm in denoising before outputting temperature readings. Simple Moving Average Filter performs well in noise reduction, which has a robustness of 91.12%, but the filtering speed is slow and the output of filtering has a time delay; thus, the method could be suitable for slow monitoring processes where this kind of time delay is acceptable. The Exponential Moving Average Filter is fast, which has a mean absolute error of 0.3320, but it cannot effectively avoid the impact of the largest outlier, with only a robustness of 81.37%, and the noise reduction performance is poor. Hence, this algorithm would perform best in an application with an extremely high request for response speed. Kalman filter, as the algorithm performing the best, demonstrates stability, the response value is 8.66, robustness of the generated data is 88.19%, and moderate in response speed, mean absolute error 0.3592, which thus proves the academic assertion about the reliability of the Kalman filter algorithm in comparison with other traditional embedded sensor system algorithms. That said, this indicates that, for noisy IoT applications, the Kalman filter algorithm will be more efficient compared to the other two algorithms. Due to the characteristics of its probability, the algorithm can adjust dynamically; hence, in the IoT environment with noises, to ensure the speed and robustness of the algorithm, the Kalman filter algorithm is the most efficient among the three algorithms. Only the sensitivity of the parameters to Gaussian noise and large outliers needs further adjustments. In the future, we will focus on eliminating the above limitations by implementing a dynamic algorithm, such as automatic threshold adjustment, outlier removal, and dynamic mode switching. Then, the algorithm will be put into operation and physically tested through chip/PCB implementation to meet the needs of real-world IoT application environments.

References

[1] B. Buchetti, F. R. Arduino, and S. Perdichizzi, "A literature review on corporate governance and ESG research: Emerging trends and future directions," Int. Rev. Financ. Anal., vol. 97, p. 103759, 2025. https://doi.org/10.1016/j.irfa.2024.103759

[2] B. Rerhrhaye et al., "Application of a Solar Conversion Chain Based on a 3-Level NPC Inverter for an Hospital Electrical Network.," Int. J. Energy Convers., vol. 12, no. 1, 2024, Accessed: Dec. 31, 2025. https://doi.org/10.15866/irecon.v12i1.24365

[3] F. Rerhrhaye et al., "Design and Implementation of a Low Energy Consumption Data Logger for a Nano-Grid Applied to a Solar Smart Greenhouse," in 2024 3rd International Conference on Embedded Systems and Artificial Intelligence (ESAI), IEEE, 2024, pp. 1-6. Accessed: Dec. 31, 2025. https://doi.org/10.1109/ESAI62891.2024.10913737

[4] F. Rerhrhaye et al., "IoT-Based Data Logger for solar systems applications," in ITM Web of Conferences, EDP Sciences, 2022, p. 01003. Accessed: Dec. 31, 2025. https://doi.org/10.1051/itmconf/20224601003

[5] S. Rout and S. Das, "A Joint Forgetting Factor based Adaptive Extended Kalman filtering approach to predict the State-of-Charge and Model Parameter of Lithium-ion Battery," IEEE Access, 2025, Accessed: Dec. 31, 2025. https://doi.org/10.1109/ACCESS.2025.3533137

[6] J. E. Hammond, T. A. Soderstrom, B. A. Korgel, and M. Baldea, "A selective Kalman filtering approach to online neural network updating under system drift," Sci. Rep., vol. 15, no. 1, p. 43577, 2025. https://doi.org/10.1038/s41598-025-24558-8

[7] G. Rapelli et al., "Applying Ecological Momentary Assessment (EMA) to Understand Overweight and Obesity: A Systematic Review," J. Pers. Med., vol. 15, no. 11, p. 526, 2025. https://doi.org/10.3390/jpm15110526

[8] J. Song et al., "Beyond recall bias: Integrating ecological momentary assessment (EMA) and GPS to measure mobility-based greenery and short-term happiness across daily activity contexts," Urban For. Urban Green., p. 128965, 2025. https://doi.org/10.1016/j.ufug.2025.128965

[9] D. Liu, S. Wang, X. Li, Y. Fan, C. Fernandez, and F. Blaabjerg, "A novel extended Kalman filter-guided long short-term memory algorithm for power lithium-ion battery state of charge estimation at multiple temperatures," Energy, p. 137973, 2025. https://doi.org/10.1016/j.energy.2025.137973

[10] H. D. Hesar and A. D. Hesar, "Adaptive dual augmented extended Kalman filtering of ECG signals," Measurement, vol. 239, p. 115457, 2025. https://doi.org/10.1016/j.measurement.2024.115457

[11] F. Rerhrhaye et al., "Realization and Supervision of an Intelligent Energy Distribution System with a New Combination Topology of Fuzzy and PID Controllers.," Int. J. Eng. Appl., vol. 11, no. 3, 2023, Accessed: Dec. 31, 2025. https://doi.org/10.15866/irea.v11i3.22838

[12] S.-S. Ahmadpour et al., "A New Median Filter Circuit Design Based on Atomic Silicon Quantum-Dot for Digital Image Processing and IoT Applications," IEEE Internet Things J., 2025, Accessed: Dec. 31, 2025. https://ieeexplore.ieee.org/abstract/document/10954981

[13] O. KHLIFATI, B. Khadija, and S. SIMOU, "Object Detection Method for Automated Classification of Distress in Rabat's Built Heritage," Mater. Res. Proc., vol. 40, 2024, Accessed: Dec. 31, 2025. Available: https://mrforum.com/product/9781644903117-37/ https://doi.org/10.21741/9781644903117-37

[14] X. Lu, M. Chen, and Y. Tian, "Accurate state-of-charge estimation of LiFePO4 battery: An adaptive extended kalman filter approach using particle swarm optimization," Energy Rep., vol. 14, pp. 1169-1178, 2025. https://doi.org/10.1016/j.egyr.2025.07.023

[15] L. Zhao, J. Wang, Y. Hu, and L. Li, "Adaptive Unscented Kalman Filter Approach for Accurate Sideslip Angle Estimation via Operating Condition Recognition," Machines, vol. 13, no. 5, p. 376, 2025. https://doi.org/10.3390/machines13050376

[16] P. Paul, P. J. Colin, F. Musuamba Tshinanu, C. Versantvoort, E. Manolis, and K. Blake, "Current Use of Physiologically Based Pharmacokinetic modeling in New Medicinal Product Approvals at EMA," Clin. Pharmacol. Ther., vol. 117, no. 3, pp. 808-817, Mar. 2025, https://doi.org/10.1002/cpt.3525

[17] G. Zatini, A. Della Porta, and S. Za, "Deciphering Barriers and Strategies in Environmental Management Accounting (EMA) Adoption: A Comprehensive Two-Decade Analysis," Corp. Soc. Responsib. Environ. Manag., vol. 32, no. 3, pp. 3355-3370, May 2025, https://doi.org/10.1002/csr.3130

[18] O. Khlifati and K. Baba, "Harnessing the Synergy of Deep Learning and Transfer Learning Approaches," in Advances in Research in Geosciences, Geotechnical Engineering, and Environmental Science: Proceedings of the Fourth Scientific Conference on Geosciences and Environmental Management (GeoME'4), Morocco 2023, Springer Nature, 2023, p. 346. Accessed: Dec. 31, 2025. https://doi.org/10.1007/978-3-031-49345-4_33

[19] J. C. Braxton, K. Herkenhoff, J. Rothbaum, and L. Schmidt, "Changing Income Risk across the US Skill Distribution: Evidence from a Generalized Kalman Filter," Am. Econ. Rev., vol. 115, no. 12, pp. 4438-4475, Dec. 2025, https://doi.org/10.1257/aer.20231656

[20] R. Vieira, P. Kollmeyer, L. Pitault, and A. Emadi, "Comprehensive Comparison of Machine Learning and Kalman Filter Battery State of Charge Estimators," IEEE Access, 2025, Accessed: Dec. 31, 2025. https://doi.org/10.1109/ACCESS.2025.3545380

[21] H. Haeri, B. Soleimani, and K. Jerath, "Optimal Moving Average Estimation of Noisy Random Walks using Allan Variance-informed Window Length," in 2022 American Control Conference (ACC), Atlanta, GA, USA: IEEE, June 2022, pp. 1646-1651. https://doi.org/10.23919/ACC53348.2022.9867447

[22] P. Taheri, J. Amini, and M. Moallem, "Variable Window Size Moving Average Filter for Phase-Locked-Loop Synchronization," IEEE Access, vol. 12, pp. 88111-88121, 2024, https://doi.org/10.1109/ACCESS.2024.3417179

[23] Z. Chen, C. Heckman, S. Julier, and N. Ahmed, "Time Dependence in Kalman Filter Tuning," Aug. 21, 2021, arXiv: arXiv:2108.10712. https://doi.org/10.23919/FUSION49465.2021.9626864

Modeling and Efficiency Characterization of a Resonant Wireless Power Transfer Circuit for Electric Vehicle Charging

Wiam SAYAR

*School of Science and Engineering, Al Akhawayn University, Ifrane, Morocco

*W.Sayar@aui.ma

Keywords: Wireless Power Transfer, Resonant Inductive Coupling, EV Charging, Charging Efficiency, Simulink Modeling

Abstract. Wireless power transmission via resonant inductive coupling offers an attractive option for charging electric vehicles, by reducing the need for physical connectors, thereby enhancing longevity and user-friendliness. Complete resonant wireless power transfer circuit functionality is developed in MATLAB/Simulink for the battery charging of electric vehicles, involving a high-frequency inverter operating at 30 kHz, mutually coupled coils, rectification stages, and a lithium-ion battery rated at 400 V, 100 Ah. The progression of state of charge is estimated using a simplified constant current/constant voltage charging profile. A specific contribution of the work is a numerical efficiency analysis performed as a function of coil separation distance and inductance values, based on a classical two coil resonant model employing coupling coefficient and quality factor relations. The numerical efficiency analysis indicates that, for nominal alignment and identical high-quality inductors, the link may theoretically achieve in excess of 90% efficiency for small air-gap distances, experiencing a rapid decline in efficiency with increasing air gaps as the mutual coupling goes down. These results indicate the high sensitivity of wireless power transfer performance to geometrical layout parameters and provide practical guidance for electric vehicles charging system design.

Introduction

The global proliferation of electric vehicles (EVs) continues to grow as a result of their advantages in sustainability and reduction of greenhouse gases. However, conventional contact charging has disadvantages in the form of mechanical deterioration, contamination and user inconvenience, particularly in public and outdoor installations. This gives rise to the potential for resonant inductive wireless power transfer (WPT) as a viable alternative, capable of contactless energy transfer over moderate air gap distances whilst giving a safe and enclosed electrical power interface between the contact pads and the electric vehicles. For stationary and quasi-dynamic charging environments including residential garages, car parks and low-speed charging lanes, inductive WPT offers practical advantages including automation of the charging action, minimum maintenance and increased operational safety [1], [2].

A considerable volume of work has investigated the design and optimization of wireless charging architectures for EV applications. Li and Mi have given a comprehensive state of the art in respect to resonant inductive WPT architectures, with considerable comment on the issues with respect to geometries of the magnetic couplers, compensation networks and the power conditioning systems [1]. Bi et al. have further discussed the system-wide issues associated with these WPT systems of interoperability, electromagnetic compatibility and power efficiency showing that stationary WPT systems were achievable with efficiencies greater than 90% dc-dc at transfer distances in the region of 20 cm given the necessary levels of alignment [2]. More recent work has seen improved configurations of the magnetic couplers, and tolerances to misalignment. Abuajwa et al. have investigated a variety of advanced prototypes which achieved excellent efficiencies whilst maintaining ruggedness with respect to some lateral and angular tolerances to

Materials Research Proceedings 64 (2026) 379-386 https://doi.org/10.21741/9781644904091-47

the geometry of the receiving coil [3]. Debnath et al. have shown that optimized winding geometries and ferrite constructions have contributed to increased mutual coupling and decreased the parasitic losses [5], whilst Laksono et al., by experimental means have established that if modest misalignment is present, significant energy losses would occur if no corrective control or compensation method is adopted [6]. The theoretical considerations show that efficiency predictions are extremely dependent on the two-coil resonant link model by which the maximum possible efficiency is conferred by the product $k^2 Q_1 Q_2$ as originally noted by Svensson [7] and elaborated by Stark [8]. Other workers have also considered frequency dependent performance [8], dynamic charging limitations [10] and optimal coil design means for high-efficiency power transfer [11].

Although considerable literature exists on optimized coil geometries, compensation schemes, and system-level implementations, there remains a clear need for simplified engineering models that directly relate analytically determined efficiency to the behavior of a full EV WPT circuit including power electronics and battery dynamics. The present work addresses this need by developing a MATLAB/Simulink model of a resonant inductive wireless charging chain supported by an analytical efficiency characterization expressed through coupling coefficient and coil quality factors. The objectives are twofold: (i) to model a representative EV wireless charging system comprising the ac input, rectifier stage, high-frequency inverter, resonant link, and lithium-ion load, and (ii) to quantify the influence of coil separation distance and inductance variations on achievable power transfer efficiency. This efficiency-oriented perspective provides practical guidance for magnetic coupler design and alignment constraints in EV charging installations. The remainder of this paper is organized as follows: Section 2 presents the modeling procedure and analytical formulation, Section 3 discusses simulation results and efficiency trends, and Section 4 concludes with recommendations for future work.

Materials and Methods

In this research, the WPT system is modeled in MATLAB/Simulink as a complete end-to-end circuit. The top-level module includes an ac grid source, diode bridge rectifier, dc-link capacitor, high-frequency inverter, magnetically coupled inductor block, secondary rectifier, and a lithium-ion battery. The overall Simulink schematic is shown in Fig. 1. The configuration represents a standard single-phase front-end driving a primary coil through a high-frequency inverter, with the secondary coil feeding a rectifier and dc filter to charge the battery.

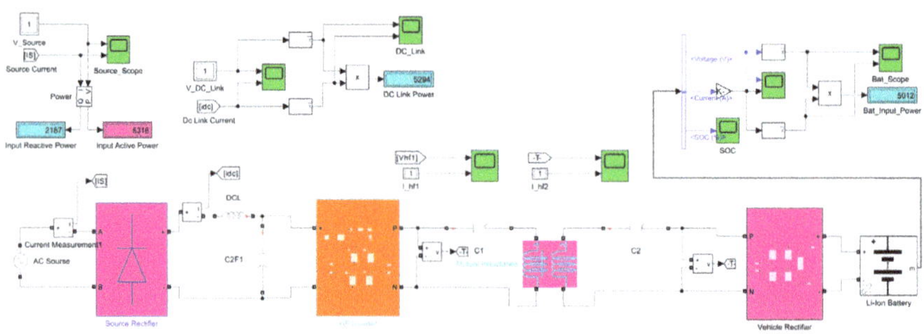

Fig. 1. Full Simulink model of the resonant wireless power transfer system for electric vehicle charging.

The ac source is a single-phase sinusoidal voltage source with a 325 V peak, corresponding to 230 V rms grid voltage. This feeds an uncontrolled full-bridge rectifier followed by a dc-link capacitor. The resulting dc voltage is converted to high-frequency ac using a full-bridge inverter employing ideal IGBT or MOSFET switches in antiparallel with diodes. A carrier-based PWM scheme produces a quasi-square waveform at 30 kHz, representative of medium-power EV WPT demonstrators [1], [2]. The inverter output then drives the primary coil of the inductive link through a compensation circuit arranged to achieve near-resonant operation at the selected frequency.

The inductive coupling between the roadside (primary) coil and the vehicle-side (secondary) coil is modelled as the standard mutual inductance block available in Simulink. The primary coil has an inductance of $L_1 = 266.16\ \mu\text{H}$ and the secondary coil has an inductance of $L_2 = 256.79\ \mu\text{H}$. Both coils have small series resistances of $R_1 = R_2 = 1\ \text{m}\Omega$ which represent copper losses occurring in the constructions of large diameter low loss windings. The mutual inductance is set to $M = 85.46\ \mu H$ under nominal alignment corresponding to a coupling coefficient

$$k = \frac{M}{\sqrt{L_1 L_2}}, \tag{1}$$

which is consistent with that expected of strongly coupled resonant WPT links reported in the literature for short separation distances [7].

On the secondary side, the induced voltage is rectified by a diode bridge and filtered by a capacitor to provide dc power to the battery. The battery is modelled using the generic lithium-ion battery block provided by Simulink and parameterized with a nominal voltage corresponding to about 400 V and rated capacity $C = 100$ Ah. The initial state of charge is set to 0%, such that $t = 0$ corresponds to a completely discharged battery and the integrator appearing within the battery model correctly tracks SOC as a function of throughput charge. For the analytical SOC prediction, coulomb counting is used with

$$SOC(t) = SOC(0) + \frac{100}{C \cdot 3600} \int_0^t I_{\text{bat}}(\tau)\, d\tau, \tag{2}$$

where I_{bat} is the charging current (A), C is the battery capacity (Ah), the factor 3600 converting hours to seconds and SOC is expressed as a percentage.

In practice, many EV chargers use a constant-current/constant-voltage (CC-CV) charging strategy in which the battery is first charged at nearly constant current until a set SOC or terminal voltage is reached, followed by a gradual current reduction as voltage is maintained. In this work, the charging profile is approximated by a constant current of about 20 A during the bulk stage, corresponding to roughly 8 kW at 400 V, followed by an exponential decay of current to a cutoff of about 5 A. The assumed profile, shown in Fig. 4, is analytically defined rather than taken from the Simulink battery block, allowing realistic CC-CV behavior while avoiding numerical artifacts when varying battery parameters.

The efficiency model is based on the well-known two coil resonant inductive link formulation. The quality factors of the primary and secondary coils can be given as:

$$Q_1 = \frac{\omega L_1}{R_1},\ Q_2 = \frac{\omega L_2}{R_2}, \tag{3}$$

where $\omega = 2\pi f$ is the angular frequency corresponding to $f = 30$ kHz. For a weakly lossy, resonant two-coil link the maximum power transfer efficiency obtainable can be expressed as [7],[8]:

$$\eta = \frac{k^2 Q_1 Q_2}{\left(1+\sqrt{1+k^2 Q_1 Q_2}\right)^2}. \tag{4}$$

This expression assumes that both coils are tuned to resonance and neglects converter and rectifier losses and yields an upper limit for the efficiency of the magnetic link itself. In order to study the effect of coil separation the mutual inductance is modelled as a decaying function of the distance d between the coils in the following form,

$$M(d) = M_0 e^{-\alpha d}, \tag{5}$$

where M_0 is the mutual inductance at nominal distance and α is an empirical decay constant which depends on the coil geometry and magnetic environment. The coupling coefficient thus becomes a function of distance:

$$k(d) = \frac{M(d)}{\sqrt{L_1 L_2}}, \tag{6}$$

and efficiency $\eta(d)$ is obtained by substituting $k(d)$ into (4). The value of α is chosen such that the resulting $\eta(d)$ profile is consistent with the trends reported in the literature, where efficiencies of over 90% are generally observed for small air gaps of the order of the coil diameter which decrease rapidly for larger gaps beyound a few tens of centimeters [2],[11].

In order to study the sensitivity to the inductance values, the efficiency is also determined as a function of L_1 and L_2 while keeping the mutual inductance fixed in such a way that k remains in a realistic range. The inductances are varied about their nominal values and for each pair (L_1, L_2), the quality factors are recomputed using (3) and efficiency evaluated from (4). This generates a three-dimensional efficiency surface which is a good illustration of how higher inductances, which lead to higher Q for a given resistance can improve the theoretical efficiency of the link, as also discussed in [5]. All efficiency calculations and plots as a function of distance and inductance are computed in MATLAB using the analytically derived expressions, with parametric values taken from the Simulink model.

Results and Discussion

Circuit-Level Results

The quality of the wireless power transmission circuit was demonstrated by running the Simulink model and producing representative waveforms for key system elements. The complete circuit diagram in Fig. 1 clearly shows the ac source, rectifier, dc link, inverter, mutually coupled inductors, secondary rectifier, and battery block, highlighting the end-to-end nature of the configuration. Results were obtained from the grid input through to the battery terminals under various operating points by adjusting source parameters, switching behavior, or coil characteristics.

The primary-side voltage and current waveforms are shown in Fig. 2. These confirm correct inverter operation when driving the resonant link. The primary winding voltage displays a quasi-square waveform at 30 kHz imposed by the switching pattern, while the current is closer to sinusoidal due to the inductive nature of the load and resonant tuning. The phase relation between voltage and current indicates the power factor seen by the inverter, and under proper resonance the current remains nearly in phase with the fundamental voltage component, meaning real power is delivered efficiently with minimal reactive circulation [1], [2].

The voltage and current in the roadside windings are shown in Fig. 3. These waveforms demonstrate the effects of the resonant compensation and magnetic coupling. The primary current amplitude is influenced by the inverter output voltage, compensation components, and the reflected impedance from the secondary side. The secondary winding produces induced voltage and current of the same operating frequency, but with differing amplitude and phase. These signals verify that the mutual inductance block successfully represents power transfer between the coils. When the secondary is loaded by the rectifier and battery, a portion of the grid power is transferred through the resonant link while the remainder is dissipated in coil and converter losses. The qualitative agreement of these waveforms with earlier WPT studies [1], [11] supports the validity of the developed circuit model.

Materials Research Proceedings 64 (2026) 379-386 https://doi.org/10.21741/9781644904091-47

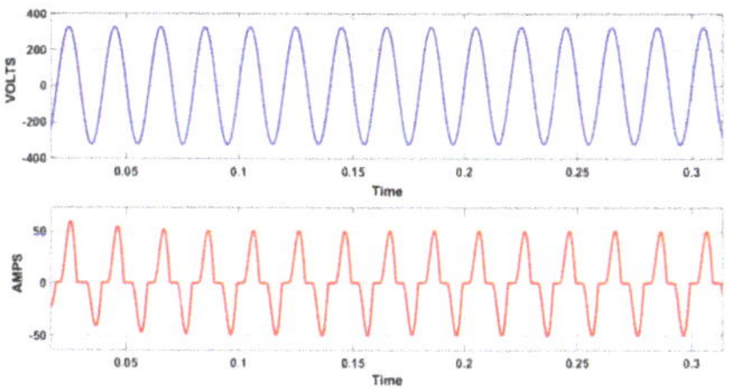

Fig. 2. Source voltage and current waveforms at the inverter output feeding the primary resonant coil.

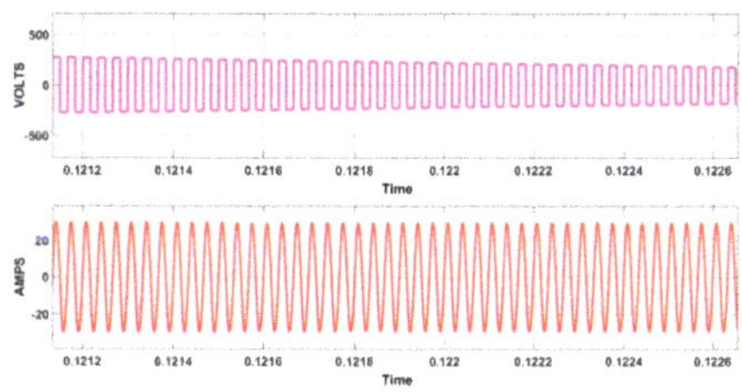

Fig. 3. Roadside (primary) winding voltage and current illustrating resonant excitation and power transfer to the secondary.

The charging current profile assumed to be applied in the course of the SOC analysis can be seen in Fig. 4. During the first stages of the charging effect the current is held at a value of about 20 A or so, in agreement with a constant current mode of operation for the CC to CV technique for a suitable 400 V, 100 Ah battery. In this mode the SOC increases approximately linearly with the passage of time, as is to be seen from (2). After a certain intended SOC value of about 90 % is reached, the current decreases with the increasing value of its exponent for an exponential function, in accordance with the constant voltage technique where the charger is keeping the terminal voltage for the battery but allowing the current to fall off to a small cut-off level. This rough analytical representation of the current profile thus is suitable answerable with ordinary values of EV charging behavior and enables an easy bit of estimating the SOC trajectory, by analytical form or numerical means, without being overly concerned with the intrinsic or extrinsic difficulties of the numerical battery model itself.

The SOC versus time curve produced by this solution is shown in Fig. 5. The curve shows an initial approximately linear region corresponding to the bulk charging to the battery with the 20 A current, followed by a gradually saturating region caused by the decay in current during the CV stages in charging the battery. For the values of parameters chosen for the various speeds of charging, the time to get to 100% SOC is found to be in the range reasonable for the 8 kW WPT system operated with the 100 Ah battery, which goes through a couple of hours. The two-region nature of the SOC curve is seen to be consistent with the qualitative manner of operation shown in EV charging profiles as determined experimentally, and also with the normal CC–CV schemes detailed in the various battery and charging systems design literature. This would seem to show that the assumed nature of the current profile produced is a set of coil and battery operating parameters in a reasonable and realistic manner for the power levels involved here.

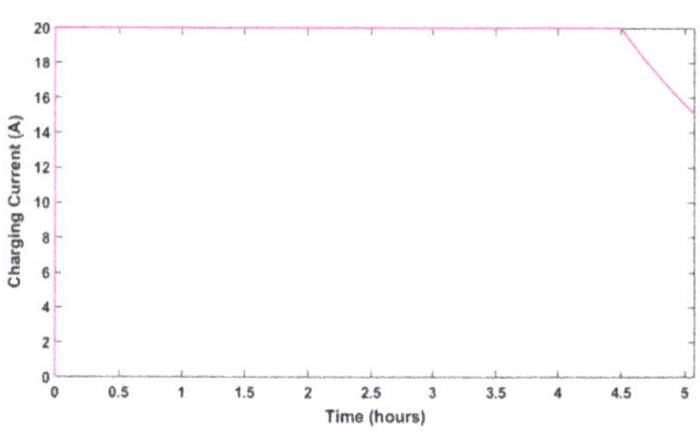

Fig. 4. Assumed constant-current/constant-voltage (CC–CV) charging current profile used for SOC prediction.

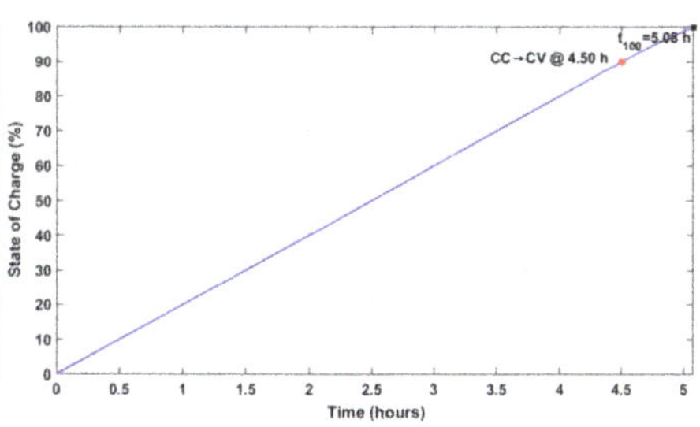

Fig. 5. Predicted battery state-of-charge (SOC) trajectory under the assumed CC–CV charging profile.

Materials Research Proceedings 64 (2026) 379-386 https://doi.org/10.21741/9781644904091-47

Coupling-Dependent Efficiency Analysis

The main interest of the present work lies as indicated above in the efficiency analysis used due to the two-coil type of resonant model. The efficiency against distance curve calculated by means of (4) – (6) is given in Fig. 6(a). For small distances, then corresponding to strong coupling and high mutual inductance, the efficiency stays high and may approach values over 90%, which would again agree with experimental descriptions for static WPT pads which have been well aligned [2], [11]. With the increasing distances however, and the effective mutual inductance decaying exponentially, the coupling coefficient reduces in value, and this is since the product $k^2 Q_1 Q_2$ also becomes smaller and smaller. This results again by making use of (4) in a direct reduction of efficiency due to this cause again. The calculated curve shows that the efficiency drops off quickly after the distance reaches a certain fraction of the coil diameter, falling eventually to values which would be wholly unacceptable for practical purposes with EV charging again. This behavior is seen to be fully consistent with the misalignment and separation studies reported by Laksono et al. and others, which show in fact that the efficiency of not very extreme deviations from nominal alignment can very badly affect the efficiency [6]. The present model gives thus a neat way of visually appreciating just how much of a problem this becomes for a given set of coil and frequency parameters.

In order to explore the design space further, the efficiency was plotted also as a function of the primary and secondary inductances, leading to the three-dimensional surface shown in Fig. 6(b). In this last analysis, the various resistances are held constant, while varying the inductances around their normal or nominal values, and the mutual inductance being scaled to accommodate the fact that the coupling factor is held in a reasonably realistic range. The efficiency surface here shows that large inductances, giving rise to high quality factors being produced with the same resistances, are very fertile ground indeed for increasing the theoretical efficiency of the coupling, especially if both coils produced high values of Q similarly. However, the surface shows also and indicates a diminishing return after a certain point in time, which indicates that excessively great inductances do not lend themselves to so much improvement in efficiency, and also possibly introducing disadvantages like excessive component size and cost, or acuity to parasitic inductances or similar properties. This behavior is consistent also with coil optimization studies made above which stress the point both of Q being made high and at the same time manufacturability practical [5].

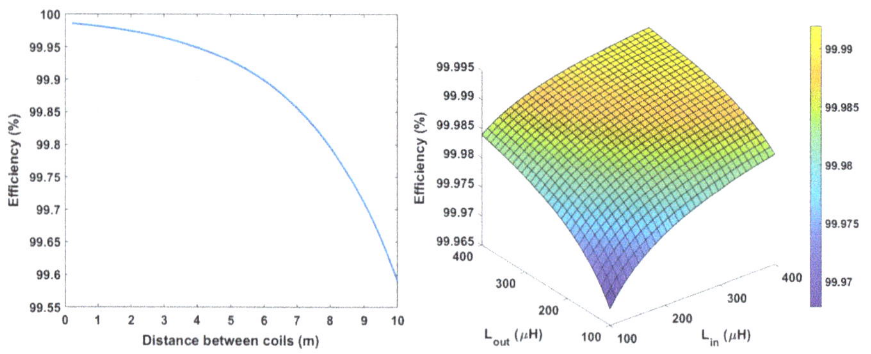

(a) Theoretical power transfer efficiency of the resonant link as a function of coil separation distance.

(b) Three-dimensional efficiency surface as a function of primary and secondary inductances L_{in} and L_{out}.

Fig. 6. Power transfer efficiency with respect to circuit parameters.

Generally speaking, then, it will be seen that the combined use of such circuit level simulation together with ancillary efficiency modeling leads to a clear view in understanding just how this WPT system of multifold behavior takes place under varying geometric and design conditions. The Simulink model gets the full detailed forms of the waveforms and fields, their reaction with the inverter, coils and battery under operation, while the efficiency analysis is working out the various characteristics of distance and inductance of the fundamental charge coupled energy transfer capability of the link. In addition to this further, by making some comparative study of the relationships which are possible, indicating the trends of efficiency discovered with results reported performance wise known for static and dynamic EV WPT systems [2], [11], it will be possible also to adapt the model to be a very useful circuit for rough design and indeed educational aim purposes in the sense that it would enable students and engineers to have a proper idea of what sorts of tradeoffs are necessitated in geometry of coils, air gap, efficiency, before going into actual hardware.

Conclusion

The concept of wireless power transmission remains a strong alternative to plug-in EV charging, particularly where convenience and robustness are essential. This work therefore examined how a resonant inductive system can maintain useful efficiency while providing the advantages of contactless charging.

A complete MATLAB/Simulink model of a wireless EV charging chain was implemented, including ac source, front-end rectifier, 30 kHz high-frequency inverter, magnetically coupled coils, and a lithium-ion battery governed by a CC–CV charging model. The simulated current and voltage waveforms confirmed correct resonant operation and realistic charging dynamics across the full power path.

Using a classical two-coil analytical formulation, efficiency was evaluated as a function of coil separation and inductance. High efficiency was observed under strong coupling, with rapid degradation at increased spacing. Inductance sensitivity studies demonstrated that higher coil quality factors improve performance, although the gains diminish once coupling becomes limited by geometry. These trends align with broader WPT research and highlight the strong dependence of performance on physical coil design.

Some limitations remain, such as idealized converter behavior, neglected parasitic losses, and absence of misalignment or electromagnetic exposure considerations. Future work will incorporate non-idealities, hardware validation, and adaptive control features to maintain resonance under dynamic conditions. Advancing these aspects will help guide development toward practical, high-efficiency wireless charging for widespread EV adoption.

References

[1] S. Li and C. C. Mi, "Wireless power transfer for electric vehicle applications," IEEE Journal of Emerging and Selected Topics in Power Electronics, vol. 3, no. 1, pp. 4-17, 2015. https://doi.org/10.1109/JESTPE.2014.2319453

[2] Z. Bi, T. Kan, C. Mi, Y. Zhang, Z. Zhao and G. A. Keoleian, "A review of wireless power transfer for electric vehicles: Prospects to increase sustainable mobility," Applied Energy, vol. 179, pp. 413-425, 2016. https://doi.org/10.1016/j.apenergy.2016.07.003

[3] O. Abuajwa, A. et al., "Comprehensive review of wireless power transfer systems for EV charging," SN Applied Sciences, 2025. https://doi.org/10.1007/s42452-025-07738-z

[4] T. Debnath, et al., "High-efficiency coil design for wireless power transfer," Energy, 2025. https://www.sciencedirect.com/science/article/abs/pii/S0360544225015713

[5] P. B. Laksono, et al., "Improved efficiency of inductive power transfer in misalignment conditions," International Journal of Applied Electromagnetics and Mechanics, vol. 61, no. 3, 2019. https://aemjournal.org/index.php/AEM/article/view/904/425

[6] A. Svensson, "Design of inductive coupling for powering and data transmission," M.Sc. thesis, Chalmers University of Technology, 2012. https://www.diva-portal.org/smash/get/diva2%3A570045/FULLTEXT01.pdf

[7] J. C. Stark, "Power transfer and efficiency in inductively coupled resonant circuits," Ph.D. dissertation, Georgia Institute of Technology, 2004. https://files.core.ac.uk/download/pdf/4385843.pdf

[8] E. Andersen, et al., "Frequency-dependence of power and efficiency for resonant inductive wireless power transfer," Smart Materials and Structures, vol. 31, 105026, 2022. https://doi.org/10.1088/1361-665X/ac8bb6

[9] V. Jeetender, et al., "Electric vehicle charge dynamic wireless power transfer," E3S Web of Conferences, vol. 497, 01007, 2025. https://doi.org/10.1051/e3sconf/202561901007

[10] A. Fathollahi, et al., "Utilization of dynamic wireless power transfer technology in electric vehicles: A review," Sustainable Energy Technologies and Assessments, vol. 52, 102058, 2022. https://www.sciencedirect.com/science/article/abs/pii/S2352467722001217

[11] A. Kurs, A. Karalis, R. Moffatt, J. D. Joannopoulos, P. Fisher and M. Soljačić, "Wireless power transfer via strongly coupled magnetic resonances," Science, vol. 317, no. 5834, pp. 83-86, 2007. https://doi.org/10.1126/science.1143254

[12] "Optimal coil design for maximum power transfer efficiency in inductive links," technical report, 2010. https://scispace.com/pdf/optimal-coil-design-for-maximum-power-transfer-efficiency-in-tva1blr5k1.pdf

Comparative Evaluation of Model Predictive Control and Field-Oriented Control for PMSM Drives Using a Three-Level NPC Inverter in High-Voltage EV Applications

Elhassan BOUHAMOU[1,a,*], Merouan BELKASMI[1,b], Radouane OULADSINE[1,c], Aziz WATIL[1,d], Rajesh RAVI[1,e], Anteneh WOGASSO[1,f], Kamal ANOUNE[2,g]

[1]LERMA Laboratory, International University of Rabat, Parc Technopolis Rabat-Shore, Rocade Rabat-Sale´, 11 100 Sala el Jadida, Morocco

[2]EMIT Laboratory, Physics Department, FST-Settat, Hassan I University, Settat, Morocco

[a]elhassan.bouhamou@uir.ac.ma, [b]merouan.belkasmi@uir.ac.ma, [c]radouane.ouladsine@uir.ac.ma, [d]aziz.watil@uir.ac.ma, [e]rajesh.ravi@uir.ac.ma, [f]anteneh.wogasso@uir.ac.ma, [g]kamal.anoune@uhp.ac.ma

Keywords: Electric Vehicles, PMSM, Model Predictive Control, Field-Oriented Control, Multilevel Inverters, NPC Inverter, High-Voltage Systems

Abstract. In electric vehicles (EVs), traction inverters have been widely studied from a topology perspective, modulation, and control strategies. With EV manufacturers adopting higher voltage batteries to take advantage of lower currents, higher power density, and faster charging, this trend has increased interest in multilevel inverters as potential replacements for conventional inverters. Their higher efficiency, better waveform quality, lower stress on semiconductors, and inherent fault tolerance make them attractive for this application. At the same time, advances in automotive ECUs (Electronic Control Units) have made it possible to implement more high computational control strategies, opening the door to predictive control such as Finite-Control-Set Model Predictive Control (FCS-MPC), which leverages direct selection of switching states and offers very fast dynamic response. In this study, a comparative analysis is conducted between classical FCS-MPC and the conventional Field- Oriented Control (FOC) for a PMSM drive supplied by a three-level Neutral Point Clamped (NPC) inverter. The two strategies have been compared in terms of time response and the steady-state error of both torque and speed and the Total Harmonic Distortion (THD) of the stator currents. The results of the simulations prove that FCS- MPC provides faster torque response, while FOC outperforms in terms of precision with lower current distortion. In general, the results highlight advantages and disadvantages of both predictive and conventional control strategies and help clarify which approach may be better suited for next-generation high-voltage EV traction systems.

1. Introduction

The electric vehicle industry has shown in recent years a rapid transition toward high-voltage DC-link architectures. Many advantages have been reported under this architecture, such as reduced cable weight, faster charging capability, and higher motor efficiency.[1].

The high-voltage architecture has also increased interest in multilevel inverters, since the traditional two-level inverter shows lower performance in high-voltage applications. Multilevel inverters deliver high efficiency by reducing total harmonic distortion (THD), decreasing voltage stress on semiconductor devices, and improving output voltage quality [2, 3]. A cost evaluation presented in [1] shows that the three-level NPC inverter can offer a competitive overall cost compared to conventional two-level inverters.

Permanent Magnet Synchronous Motors (PMSMs) are highly applied in EV powertrain, especially in high performance applications, because of their high torque density, high efficiency, and compact design. To achieve their full potential, advanced control strategies are required. These control strategies may face some challenges such as nonlinear dynamics, parameter variations, and power converter constraints [4].

Field-Oriented Control (FOC) is one of the most used control strategies for PMSMs in EV applications. It is known for its robustness and simplicity, with a cascaded control structure using PI regulators [5]. A main problem with this strategy is its limited dynamic response and reduced effectiveness in current harmonic suppression, especially during fast operating transients and rapid changes in speed or load [6].

Electronic Control Units (ECUs) in the automotive industry have started to increasingly improve to meet market demand, with the rise of artificial intelligence and its applications in the automotive industry. This improvement has opened the door to the application of highly computational control strategies.

Model Predictive Control (MPC) is an advanced control strategy that has started to take over both industry and research in the EV sector. It relies on a mathematical model of the system to predict future system behavior over a finite prediction horizon and then applies the first element of the optimal control sequence by minimizing a predefined cost function. One of its main advantages is its ability to handle multivariable systems with explicit consideration of system constraints [7].

Finite Control Set MPC (FCS-MPC) is a direct control method specifically designed for power electronics. Based on the system model, it predicts how the system reacts for each possible switching state at each sampling period, which eliminates the need for a PWM stage, making it extendable to higher levels in multilevel inverters [8]. Despite its advantages, FCS-MPC has shown some performance issues. Since it depends on model accuracy, it exhibits high sensitivity to parameter mismatches. Computational delays and limited sampling time also cause performance degradation. In the context of PMSM control, these problems result from electrical parameter variations, such as stator resistance changes with temperature or inductance variations due to magnetic saturation, as well as mechanical uncertainties, including load torque variations. This may lead to increased torque ripple and reduced robustness under fast transients. [9]. Current research therefore explores adaptive and learning-based MPC frameworks to address these challenges.

This paper investigates and compares two control techniques, FOC and FCS-MPC, for controlling a PMSM fed by a three-level NPC inverter under identical operating conditions in the context of high-voltage EV applications. Key performance metrics, such as torque response, speed response, and stator current quality, are evaluated, highlighting their respective trade-offs and providing practical guidance for control strategy selection in electric vehicle powertrains.

2. System Description and Modeling

2.1 Permanent Magnet Synchronous Motor (PMSM) Model

For control purposes, the PMSM is typically represented in the rotating (d–q) reference frame using Park's transformation, which converts the three-phase variables into time-invariant quantities [4,5].

The electrical dynamics are described by:

$$\begin{aligned} V_d &= R_s i_d + L_d \frac{di_d}{dt} - \omega L_q i_q \\ V_q &= R_s i_q + L_q \frac{di_q}{dt} + \omega (L_d i_d + \psi_f) \end{aligned} \tag{1}$$

The electromagnetic torque is described by the equation:

$$T_e = \frac{3}{2}P\left[\psi_f i_q + (L_d - L_q)i_d i_q\right] \tag{2}$$

In surface-mounted PMSMs $L_d = L_q$, that means the equation above can be simplified to:

$$T_e = \frac{3}{2}P\psi_f i_q \tag{3}$$

The mechanical dynamics are described by:

$$J\frac{d\Omega}{dt} = T_e - T_l - F_r\Omega \tag{4}$$

2.2 Three-Level NPC Inverter Topology

In 3L-NPC each phase leg comprises four active switches (T1-T4), four antiparallel diodes, and two clamping diodes (D1, D2) that connect to the neutral point of the DC- link capacitors.

The three level NPC topology enables each phase to output three distinct voltage levels : $+V_{DC}/2$ (P state), 0 (O state), and $V_{DC}/2$ (N state) , a primary advantage of this structure is that each semiconductor blocks only $V_{DC}/2$, allowing for lower rated devices and reducing the cost of the inverter [3]. A major problem in most multilevel topologies is neutral point (or capacitor) voltage balancing, which requires a careful modulation strategy [11].

Table 1. Permanent Magnet Synchronous Motor Parameters

Parameter	Value	Unit
Stator Resistance	0.76	[Ω]
d-axis Inductance	1.8	[mH]
q-axis Inductance	1.8	[mH]
PM Flux Linkage	0.14	[Wb]
Pole Pairs	2	[-]
Moment of Inertia	1.1×10^{-3}	[Kg·m²]
Friction Coefficient	5×10^{-5}	[N·m·s]

2.3 Space Vector PWM for 3L-NPC Inverter

In motor control, Space Vector Pulse Width Modulation (SVPWM) is widely used due to its advantages that provide a superior DC bus utilization and lower THD for multilevel inverters [12]. A 3L inverter generates 27 distinct space vectors categorized into zero vectors, small vectors ($E/\sqrt{6}$), medium vectors ($E/\sqrt{2}$), and large vectors (E 2/3).

The space vector diagram is partitioned into six primary sectors (I-VI), each covering 60 degrees, with each sector further divided into four regions [12]. The reference voltage vector is derived through Clarke transformation:

$$\begin{bmatrix} V_\alpha \\ V_\beta \end{bmatrix} = \frac{2}{3}\begin{bmatrix} 1 & -\frac{1}{2} & -\frac{1}{2} \\ 0 & \frac{\sqrt{3}}{2} & -\frac{\sqrt{3}}{2} \end{bmatrix}\begin{bmatrix} V_{an} \\ V_{bn} \\ V_{cn} \end{bmatrix} \tag{5}$$

The magnitude and angle are determined by:

$$V_{ref} = \sqrt{V_\alpha^2 + V_\beta^2}, \theta = \tan^{-1}\left(\frac{V_\beta}{V_\alpha}\right) \tag{6}$$

Dwelling times use volt-second balance:

$$\vec{V}_{ref}T_s = \vec{V}_1 T_1 + \vec{V}_2 T_2 + \vec{V}_0 T_0 \tag{7}$$

Where T_s is sampling time, T_1 and T_2 are dwelling times for active vectors, and T_0 is for zero vectors. Small vectors enable neutral-point voltage balancing through selective charging/discharging of DC-link capacitors.

3. Control Strategies

3.1 Field-Oriented Control with PI Regulators

The FOC structure employs cascaded control loops with PI regulators for current and speed control.

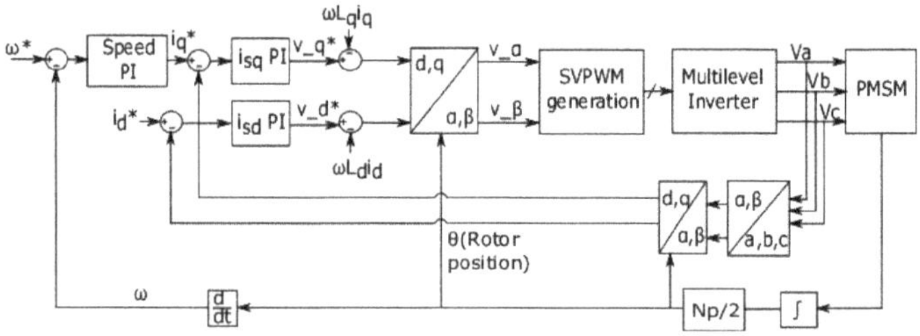

Figure 1: FOC structure for PMSM drive with 3L-NPC inverter [19].

The inherent cross-coupling in PMSM dynamics is eliminated through feedforward compensation. The voltage equations:

$$\begin{aligned} V_d &= R_s i_d + L_d \frac{di_d}{dt} - \omega L_q i_q \\ V_q &= R_s i_q + L_q \frac{di_q}{dt} + \omega(L_d i_d + \psi_f) \end{aligned} \tag{8}$$

are decoupled by defining new control variables:

$$\begin{aligned} V_d' &= V_d + \omega L_q i_q \\ V_q' &= V_q - \omega (L_d i_d + \psi_f) \end{aligned} \tag{9}$$

This transformation yields two independent first-order systems:

$$\begin{aligned} i_d &= \frac{V_d'}{R_s + pL_d} \\ i_q &= \frac{V_q'}{R_s + pL_q} \end{aligned} \tag{10}$$

The current loop PI regulators are designed for specified response times. For the d-axis current regulator

$$\begin{aligned} K_{p,d} &= \frac{3L_d}{t_r} \\ K_{i,d} &= \frac{3R_s}{t_r} \end{aligned} \tag{11}$$

Materials Research Proceedings 64 (2026) 387-394 https://doi.org/10.21741/9781644904091-48

For the q-axis current regulator:

$$K_{p,q} = \frac{3L_q}{t_r} \qquad K_{i,q} = \frac{3R_s}{t_r} \tag{12}$$

The outer speed loop PI regulator is designed as:

$$K_{p,\omega} = \frac{3J}{t_r} \qquad K_{i,\omega} = \frac{3F_r}{t_r} \tag{13}$$

3.2 Finite-Control-Set Model Predictive Control

A key advantage over Field-Oriented Control (FOC) is the elimination of both the PWM modulator and PI regulators which enables a faster dynamic response. This control method provides also the ability to handle system constraints directly within the control law.

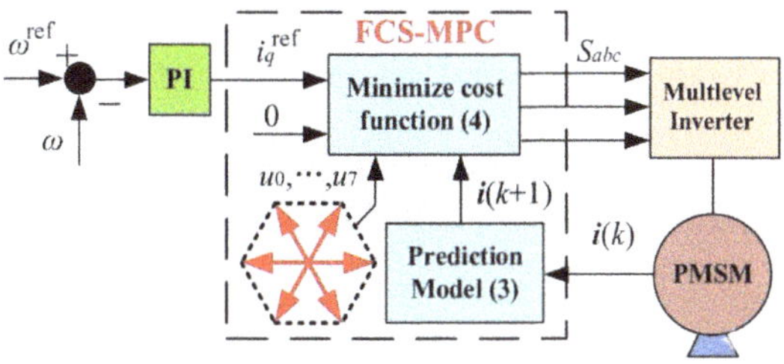

Figure 2 : FCS-MPC Control structure for PMSM drive using 3L-NPC inverter [20].

The discrete-time prediction model is derived from the PMSM electrical dynamics:

$$\begin{aligned} i_d(k+1) &= i_d(k) + \frac{T_s}{L_d}[V_d(k) - R_s i_d(k) + \omega L_q i_q(k)] \\ i_q(k+1) &= i_q(k) + \frac{T_s}{L_q}\left[V_q(k) - R_s i_q(k) - \omega(L_d i_d(k) + \psi_f)\right] \end{aligned} \tag{14}$$

where T_s represents the sampling time. The electromagnetic torque is predicted using:

$$T_e(k+1) = \frac{3}{2}P\left[\psi_f i_q(k+1) + (L_d - L_q) i_d(k+1) i_q(k+1)\right] \tag{15}$$

For the 3L-NPC inverter with 27 possible switching states, the cost function minimizes torque and flux tracking errors:

$$J = [T_e^* - T_e(k+1)]^2 + \lambda[\psi_s^* - \psi_s(k+1)]^2 \tag{16}$$

λ acts as a weighting factor that prioritizes either torque response or flux control objectives. The stator flux magnitude is estimated as:

$$\psi_s = \sqrt{(L_d i_d + \psi_f)^2 + (L_q i_q)^2} \tag{17}$$

The FCS-MPC algorithm executes the following steps each sampling period:

(a) Measure currents $i_d(k)$, $i_q(k)$ and rotor position $\theta(k)$

(b) Predict future currents $i_d(k + 1)$, $i_q(k + 1)$ for all 27 switching states

(c) Calculate predicted torque $T_e(k + 1)$ and flux $\psi_s(k + 1)$

(d) Evaluate cost function J for each switching state

(e) Apply the switching state that minimizes J

FCS-MPC provides superior dynamic performance and direct handling of constraints, making it particularly suitable for high-performance EV applications where fast torque response is critical. The main problems with predictive control are that it requires accurate system parameters and has a high computational load compared to FOC.

4. Simulation Results and Discussion

An EV powertrain was modeled using a surface-mounted PMSM motor supplied by 800 V DC-link, controlled by a 3 level NPC inverter, the comparative performance analysis of FOC and FCS-MPC was conducted under identical operating conditions using MATLAB/Simulink. The motor model was tested at an operating speed of 1000 rpm, with torque values varying to reflect realistic EV conditions. Dynamic torque steps and speed variations were applied to simulate typical EV operating scenarios.

Table 2: Performance comparison summary

Parameter	FOC	FCS-MPC	Unit
Torque response rise time	2.5	1.1	[ms]
Torque settling time	15.0	8.5	[ms]
Electromagnetic torque ripple	4.2	7.8	[%]
Current THD	2.1	4.5	[%]
Speed settling time	50	42	[ms]

General observations from simulations demonstrated different performance characteristics between the two control strategies, as shown in Figures 4a and 4b. FCS-MPC demonstrated a faster dynamic response with a 56% reduction in torque rise time and a 43% faster settling time compared to FOC. These results demonstrate that MPC is better suited for EV applications that require quick torque adjustment during acceleration or energy recovery through regenerative braking.

In the other part, FOC shows better steady-state behavior. The results demonstrate a 45% reduction in torque ripple with lower stator current THD, which ensures smoother operation and reduced acoustic noise (Figures 4e and 4f). In terms of speed tracking performance (Figures 4c and 4d), both controllers achieved very good regulation, with MPC showing a faster response.

FCS-MPC also demonstrates better neutral-point voltage balancing for the 3L-NPC inverter, leveraging its direct control over switching states. The main limitation observed with FCS-MPC is its sensitivity to parameter variations, where changes in motor parameters significantly impact performance.

Materials Research Proceedings 64 (2026) 387-394 https://doi.org/10.21741/9781644904091-48

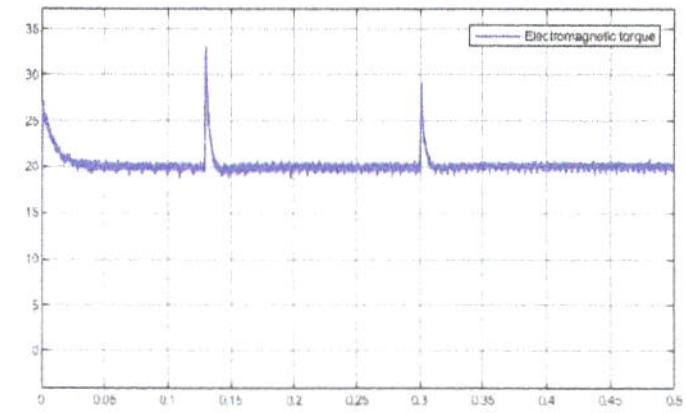

(a) FOC: Electromagnetic torque

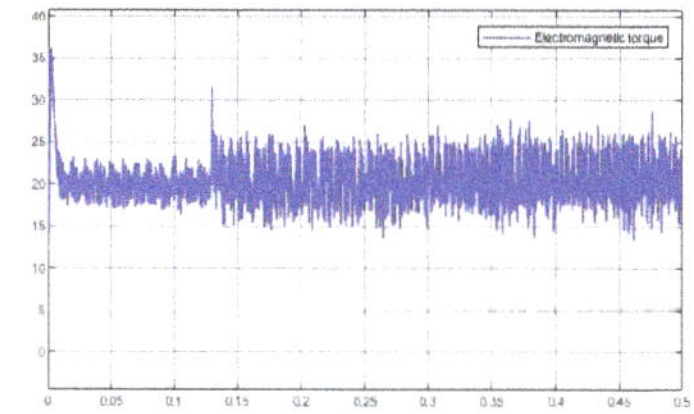

(b) MPC: Electromagnetic torque

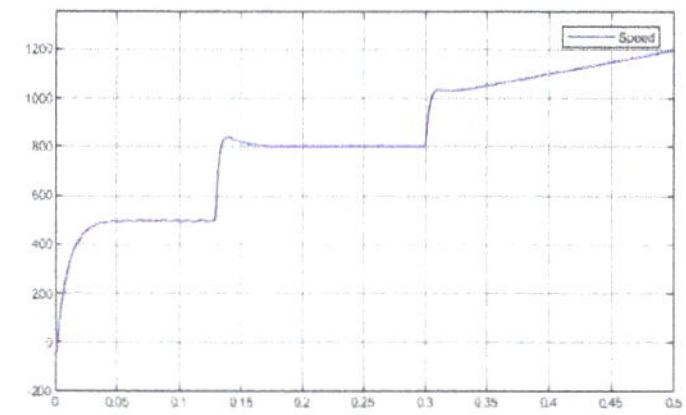

(c) FOC: Rotor speed

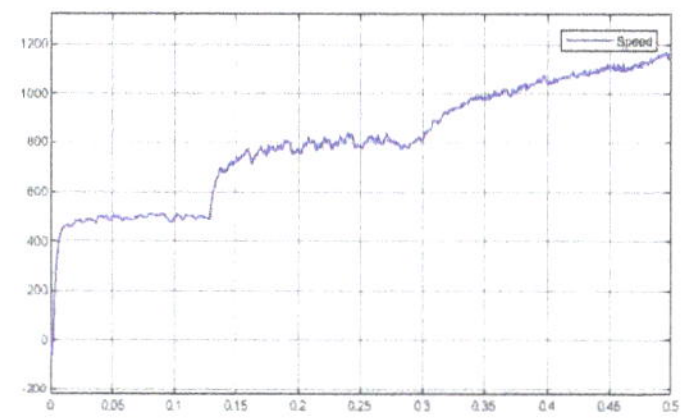

(d) MPC: Rotor speed

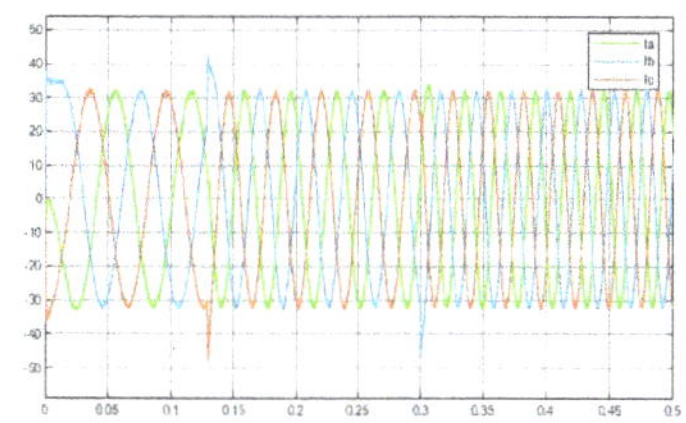

(e) FOC: Current waveform

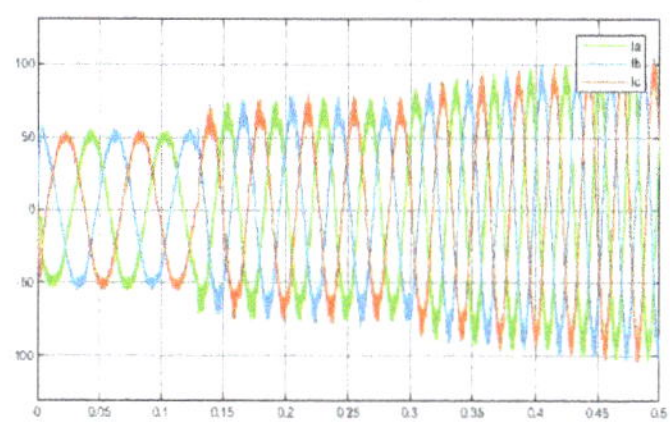

(f) MPC: Current waveform

Figure 3: Comparison of Field-Oriented Control (FOC) vs Model Predictive Control (MPC)

Conclusion

In this study, a comparative analysis of two control strategies, FOC and MPC is conducted, specifically the FCS-MPC controlling a surface-mounted PMSM for 800 V DC-link EV powertrain using a 3 level NPC inverter. The simulation results confirm that while FCS-MPC provides a much faster dynamic response compared with FOC, its practical application is significantly limited due to the superiority of the FOC in terms of precision and stability.

A key limitation for this control method is its sensitivity to the motor parameter variations and model inaccuracies, this is a critical issue in EV where operating conditions and system parameters are constantly changing to address the drawbacks of the conventional MPC a new concept is proposed the Data-Driven Model Predictive Control (DMPC) enhanced with adaptive, machine learning-based layers, reinforcement learning is a promising technique for this integration.

This DMPC concept is presented as a future research initiative, the goal is to create an intelligent control strategy that maintains consistent and high performance in variable conditions for electric vehicles.

References

[1] A. Poorfakhraei, M. Narimani and A. Emadi," A Review of Multilevel Inverter Topologies in Electric Vehicles: Current Status and Future Trends," in IEEE Open Journal of Power

Electronics, vol. 2, pp. 155-170, 2021, doi: 10.1109/OJPEL.2021.3063550. https://doi.org/10.1109/OJPEL.2021.3063550

[2] V. Patel, C. Buccella, A. M. Saif, M. Tinari and C. Cecati," Performance Comparison of Multilevel Inverters for E-transportation," 2018 International Conference of Electrical and Electronic Technologies for Automotive, Milan, Italy, 2018, pp. 1-6, doi: 10.23919/EETA.2018.8493209. https://doi.org/10.23919/EETA.2018.8493209

[3] Gu, X.; Xu, W.; Zhang, G.; Chen, W.; Jin, X. Three-Level Inverter-PMSM Model Predictive Current Control Based on the Extended Control Set. Electronics 2023, 12, 557. https://doi.org/10.3390/electronics12030557 https://doi.org/10.3390/electronics12030557

[4] P. Pillay and R. Krishnan," Modeling, simulation, and analysis of permanent-magnet motor drives. I. The permanent-magnet synchronous motor drive," in IEEE Transactions on Industry Applications, vol. 25, no. 2, pp. 265-273, March-April 1989, doi: 10.1109/28.25541. https://doi.org/10.1109/28.25541

[5] P. L. Chavan et al., "Field Oriented Control Technique for PMSM," in Proceedings of Atlantis Press, 2023. ISBN 978-94-6463-135-7.

[6] T. D. C. Busarello, A. Bubshait, O. V. S. R. Varaprasad, A. Alsaleem and M. G. Simoes," A Comprehensive Methodology of Field-Oriented Control Design With Parameter Variation Analysis for Interior Permanent Magnet Synchronous Machine Drives," in IEEE Access, vol. 13, pp. 89524-89541, 2025, doi: 10.1109/AC- CESS.2025.3571610. https://doi.org/10.1109/ACCESS.2025.3571610

[7] Zhang C, Cen C. Simplified robust two-step model predictive control for permanent magnet synchronous motor. Advances in Mechanical Engineering. 2024;16(5). doi:10.1177/16878132241252334 https://doi.org/10.1177/16878132241252334

[8] S. Hanke, S. Peitz, O. Wallscheid, S. Klus, J. Bocker, and M. Dellnitz, "Koopman Operator-Based Finite- Control-Set Model Predictive Control for Electrical Drives," arXiv preprint arXiv:1804.00854, 2018.

[9] H. Yang, Y. Liu, J. Wang and J. Yang," Improved Multi-step FCS-MPCC with Disturbance Compensation for PMSM Drives," 2021 IEEE International Conference on Predictive Control of Electrical Drives and Power Electronics (PRECEDE), Jinan, China, 2021, pp. 530-535, doi: 10.1109/PRECEDE51386.2021.9681038. https://doi.org/10.1109/PRECEDE51386.2021.9681038

[10] Priyankar Roy, Atanu Banerjee. A Study on Performance Parameters of Three-Level T-type Inverter based PMSM Drives for Electric Vehicles Applications, 11 July 2022, PREPRINT (Version 1) available at Research Square [https://doi.org/10.21203/rs.3.rs-1814441/v1] https://doi.org/10.21203/rs.3.rs-1814441/v1

[11] Masoud Farhadi, Mehdi Abapour, Mehran Sabahi, Failure analysis and reliability evaluation of modulation techniques for neutral point clamped inverters-A usage model approach, Engineering Failure Analysis, Volume 71, 2017, Pages 90-104, ISSN 1350-6307, https://doi.org/10.1016/j.engfailanal.2016.06.010. https://doi.org/10.1016/j.engfailanal.2016.06.010

[12] B. Urmila and D. Subba Rayudu, "Optimum Space Vector PWM Algorithm for Three-Level Inverter," ARPN Journal of Engineering and Applied Sciences, vol. 6, no. 9, pp. 24-36, 2011.

Materials Research Proceedings 64 (2026) 395-403 https://doi.org/10.21741/9781644904091-49

Fault-Tolerant H_∞ Control of Active Suspension Systems with Time Delay

Jamal MRAZGUA[1,a *], Khalid BADIE[2,b], Naoufal EL YOUSSFI[3,c], El Houssaine TISSIR[1,d] and Mohamed OUAHI[4,e]

[1]Department of Physics, Faculty of Sciences Dhar El Mehraz, University of Sidi Mohammed Ben Abdellah, LISAC, BP 1796, Fes-Atlas, Morocco

[2]Department of Physics, University Cadi cadi ayyad, Faculty of Sciences, Marrakech, 40000, Morocco

[3]Electric and Telecom Department, Royal School of Aeronautics, Marrakech, Morocco

[4]National School of Applied Sciences, Engineering, Systems and Applications Laboratory, BP 72, City, Fez, Morocco

[a]jamal.mrazgua@usmba.ac.ma, [b]khalid.badie@usmba.ac.ma, [c]naoufal.elyoussfi@usmba.ac.ma, [d]elhoussaine.tissir@usmba.ac.ma, [e]mohamed.ouahi@usmba.ac.ma

Keywords: Time Delay, Active Suspension System, H_∞ Control, Actuator Fault, LMI

Abstract. In this study, we address the problem of fault-tolerant control (FTC) H_∞ of active vehicle suspension systems (ASS) systems with time delays and actuator fault. The H_∞ performance criterion is employed to assess ride comfort, allowing for the consideration of general road disturbances. A method for designing the controller established on Linear Matrix Inequalities (LMIs) has been suggested to handle actuator faults represented by a scaling factor. The control design guarantees the asymptotic stability of the system, mitigates the effects of actuator malfunctions, and maintains H_∞ performance although in presence of defective actuators. The effecency and robustness of the proposed approach are demonstrated through simulations and comparative analysis.

Introduction

Modern vehicles are significantly more comfortable and secure compared to their predecessors. The suspension system (SS) absorbs shocks from forces acting betwixt the road and the vehicle chassis, ensuring a smooth ride and reliable handling performance. SS are generally categorized into three classes: passive, semi-active and active suspensios (AS) [1,2]. The main goal of research on (ASS) is to improve suspension performance by optimizing ride comfort, suspension deflection and road handling [3].

Significant attention has been given to controlling ASS in recent years, leading to numerous notable advancements [4]. Researchers have proposed various approaches to optimise SS performance by employing a range of control techniques, including adaptive control [5], H_∞ control [6].

Time delays are common in various fields of physical systems and can lead to instability and reduced control performance. Increasing methods have been developed to mitigate the effects of delays across different types of systems, two-dimensional systems [7–8] and fuzzy systems [9,10] among others. For vehicle SS specifically, noteworthy contributions include [11] and [12].

Designing FTC systems, even in the presence of faults, is a difficult problem for ASS because of the unpredictability and inevitability of defects. These unexpected malfunctions may a source of system failure and a noticeable decline in his performance. Because of these difficulties, it is essential to design an FTC system capable of automatically compensating for changes in system dynamics caused by failures. The term failure tolerance (FT) make reference to the system's potential to continue functioning correctly even if some of its components fail. The capability to sustain functionality under component failures is often referred to as progressive degradation. Specifically, FTC is responsible for ensuring that performance is acceptable and system stability is maintained, even if faults occur. Using different engineering controls, there have been several proposals for FTC approaches [13]. As far as we know, there has been limited research has been conducted on FTC for ASS that account for both actuator delays and failures. This gap motivates and underpins the efforts presented in this article.

This paper presents the development of H_∞ FTC design methods based on a state feedback control law, explicitly considering input time delays and actuator failures. The proposed approach utilises LMI to ensure that the ASS remains asymptotically stable while achieving a predefined level of disturbance attenuation. It also compensates for actuator failures to maintain performance. By presenting simulation results, the proposed method's effectiveness and robustness can be verified.

Materials and Methods

The parameters of a quarter car model depicted in Figure 1 include m_s representing the sprung mass, m_u representing the unsprung mass, c_s represent the damping of SS, k_s represent the stiffness of SS, $u_a^f(t)$ represents the FTC of the ASS, k_t represent the compressibility of the tyre, c_t the damping of the tyre, the displacements of z_s correspond to the sprung displacements and z_u corresponds to the unsprung displacements, z_r represents the road displacement . d: the control input delay assumed to be unknown and constant with a known upper bound$(0 \leq d \leq \bar{d})$.

According to Figure 1, the following is how dynamic equations can be obtained:

$$m_s\ddot{z}_s(t) + c_s[\dot{z}_s(t) - \dot{z}_u(t)] + k_s[z_s(t) - z_u(t)] = u_a^f(t-d),$$
$$m_u\ddot{z}_u(t) + c_s[\dot{z}_u(t) - \dot{z}_s(t)] + k_s[z_u(t) - z_s(t)] + c_t[\dot{z}_u(t) - \dot{z}_r(t)] + k_t[z_u(t) - z_r(t)] = u_a^f(t-d), \quad (1)$$

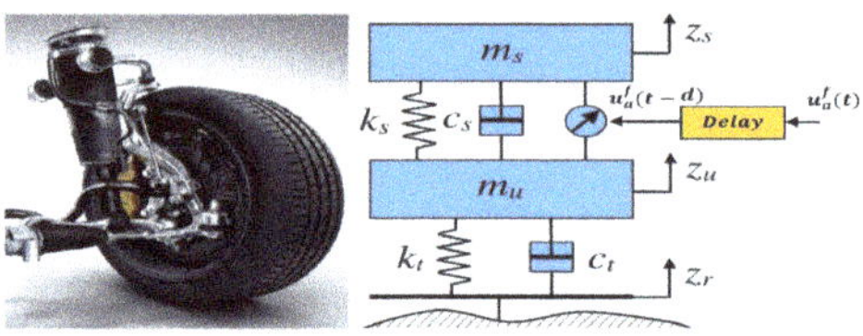

Figure 1. Quarter-car model with an ASS.

The system states are introduced as follows:

$$x_1(t) = z_s(t) - z_u(t), \quad x_2(t) = z_u(t) - z_r(t), \quad x_3(t) = \dot{z}_s(t), \quad x_4(t) = \dot{z}_u(t), \quad (2)$$

The deflection of the suspension, the deflection of the tire, the speed of the sprung mass, and the speed of the unsprung mass are all represented by $x_1(t)$, $x_2(t)$, $x_3(t)$, and $x_4(t)$ respectively. $w(t) = \dot{z}_r(t)$ represents the inputs of disturbance. Let define $x(t) = [x_1(t) \quad x_2(t) \quad x_3(t) \quad x_4(t)]^T$. According to (1), it is possible to rewrite the dynamic equations below:

$$\dot{x}(t) = Rx(t) + Tw(t) + Su_a^f(t-d), \quad (3)$$

where

$$R = \begin{bmatrix} 0 & 0 & 1 & -1 \\ 0 & 0 & 0 & 1 \\ -\frac{k_s}{m_s} & 0 & -\frac{C_s}{m_s} & \frac{C_s}{m_s} \\ \frac{k_s}{m_u} & -\frac{k_t}{m_u} & \frac{C_s}{m_u} & -\frac{(C_s+C_t)}{m_u} \end{bmatrix}, \quad T = [0 \quad -1 \quad 0 \quad \frac{c_s}{m_u}]^T, \quad S = \left[0 \quad \frac{1}{m_s} \quad 0 \quad -\frac{1}{m_u}\right]^T,$$

The condition translate into a limitation on the tire force induced by road irregularities, which must remain lower than the total static load supported by the vehicle, namely.

$$k_t[z_u(t) - z_r(t)] < (m_s + m_u)g.$$

Moreover, inherent mechanical constraints of the suspension assembly impose strict bounds on the relative displacement between the unsprung and sprung masses, ensuring that the suspension deflection does not exceed its admissible range $|z_s(t) - z_u(t)| \leq z_{max}$, (4)
where z_{max} corresponds to the upper bound of the suspension deflection.
Due to its power constraints, the hydraulic actuator constitutes a significant limitation on the AS system. $|u_a(t)| \leq u_{max}$, (5)
where u_{max} indicates the maximum amount of suspension displacement that is allowed..
The controller is designed to operate at the following output quantities:

$$z_1(t) = \ddot{z}_s(t) \qquad z_2(t) = \begin{bmatrix} \frac{z_s(t)-z_u(t)}{z_{max}} & \frac{k_t(z_u(t)-z_r(t))}{(m_s+m_u)g} \end{bmatrix}^T \tag{6}$$

Next, we can describe the ASS of the quarter-car can be expressed as:

$$\begin{cases} \dot{x}(t) = Rx(t) + Tw(t) + Su_a^f(t-d) \\ z_1(t) = Mx(t) + Nu_a^f(t-d) \\ z_2(t) = Lx(t) \end{cases} \tag{7}$$

where R, T, and S are defined in (3), and

$$M = \begin{bmatrix} -\frac{k_s}{m_s} & 0 & -\frac{c_s}{m_s} & \frac{c_s}{m_s} \end{bmatrix}, \; N = \frac{1}{m_s}, \quad L = \begin{bmatrix} \frac{1}{z_{max}} & 0 & 0 & 0 \\ 0 & \frac{k_t}{(m_s+m_u)g} & 0 & 0 \end{bmatrix}.$$

The control equation is formulated as a delayed state feedback mechanism and is expressed as:

$$u_a(t-d) = K_u x(t-d), \tag{8}$$

The state feedback gain matrix, determined through control design procedures, is denoted as: K_u
The operator's fault behavior is modeled as follows.

$$u_a^f(t-d) = Fu_a(t-d), \tag{9}$$

where F stands for the actuator fault matrix, expressed as: $F = diag(f_n)$, $0 \leq \check{f}_{ln} \leq f_n \leq \hat{f}_{un}$, $f_n = 1$, indicate the operation of the actuator under normal conditions, $f_n = 0$ refers to complete loss of actuator control, whereas intermediate values describe partial loss scenarios $0 \leq \check{f}_{ln} \leq f_n \leq \hat{f}_{un}$, $f_n \neq 1$,
Substituting Eqs (8) and (9) into Eq (7), the resulting closed-loop dynamics of the system can be expressed as follows:

$$\begin{cases} \dot{x}(t) = Rx(t) + SFK_u x(t-d) + Tw(t) \\ z_1(t) = Mx(t) + NFK_u x(t-d) \\ z_2(t) = Lx(t) \end{cases} \tag{10}$$

Consider the set of disturbances w(t) ∈ $L_2[0,\infty)$, such that $||w||_2^2 \leq w_{max} < \infty$. The control synthesis problem involves determining a gain K_u that guarantees the following properties[13]: Internal Stability: The equilibrium point of the unforced system $w(t) = 0$ is asymptotically stable. Disturbance Rejection: For all non-zero, the performance inequality $||z_1||_2 < \gamma ||w||_2$ holds for a given $w(t) \in L_2[0,\infty)$, where $\gamma > 0$. Constraint Satisfaction: Actuator and output signals remain within specified admissible regions:

$$|\{z_2(t)\}_i| \leq 1, \quad i = 1,2, \qquad |u_a(t)| \leq u_{max} \tag{11}$$

Remark 1. The fault model (9) takes into account inaccurate actuator outputs. It encompasses the fault model [14] in which sensor and actuator faults are modelled discretely. For simplicity, we introduce the following notations

$$F_0 = f_{01}, J = j_1, E = e_1 \; |E| = |e_1|, \; f_{01} = \frac{1}{2}(\hat{f}_{u1} + \check{f}_{l1}), j_1 = \frac{\hat{f}_{u1} - \check{f}_{l1}}{\hat{f}_{u1} + \check{f}_{l1}}, \; e_1 = \frac{f_1 - f_{01}}{f_{01}}, \tag{12}$$

By equations (12) we have $F = F_0(I + E), \in \mathbb{R} \qquad |E| \leq J \leq I,$ (13)

Materials Research Proceedings 64 (2026) 395-403 https://doi.org/10.21741/9781644904091-49

The following lemmas are presented to establish the stability and performance guarantees detailed in the following analysis.

Lemma 1 [15] Let X, Y be matrices of appropriate dimensions and for any scalar $\varepsilon > 0$, the following holds

$$X^T Y + Y^T X \leq \varepsilon X^T X + \varepsilon^{-1} Y^T Y$$

Lemma 2 [15] Let R_1 and R_2, be matrices of compatible dimensions. Then, the following inequality is satisfied:

$$R_1 \Sigma\, R_2 + R_2^T \Sigma R_1^T \leq \beta\, R_1 U R_1^T + \beta^{-1} R_2^T U\, R_2$$

where $\beta > 0$, is a positive scalar, and $\Sigma = diag(\sigma_1, \sigma_2, \ldots \sigma_l)$ represents a positive definite diagonal matrix such that $\Sigma \leq U$.

Lemma 3 [16] Assume $a(.) \in \mathbb{R}^{n_a}$, $b(.) \in \mathbb{R}^{n_b}$ and $H \in \mathbb{R}^{n_a \times n_a}$ are defined on the interval Ω. Consequently, for any matrix X,
the interval Ω. Then for any matrices $X \in \mathbb{R}^{n_a}$, $Y \in \mathbb{R}^{n_b}$ and $Z \in \mathbb{R}^{n_a \times n_a}$ the following holds:

$$-2\int_\Omega a^T(\alpha) H b(\alpha) d\alpha \leq \int_\Omega \begin{bmatrix} a(\alpha) \\ b(\alpha) \end{bmatrix}^T \begin{bmatrix} X & Y - H \\ * & Z \end{bmatrix} \begin{bmatrix} a(\alpha) \\ b(\alpha) \end{bmatrix} d\alpha, \qquad \text{where} \quad \begin{bmatrix} X & Y \\ * & Z \end{bmatrix} > 0.$$

Controller design

In this section, our objective is to design an H_∞ FTC controller capable of stabilizing the vertical movement of the vehicle chassis in the presence of delay and actuator failures.

Theorem 1

Consider the SS (7). For a scalar $\bar{d} > 0$, the closed-loop ASS (10) is asymptotically stable and guarantees an H_∞ performance level γ for all constant delays d such that $0 \leq d \leq \bar{d}$, under the existence of matrices $\tilde{P} > 0, \tilde{Z} > 0, \tilde{Q} > 0, \tilde{X}$ and H satisfying the LMIs (14)-(17). In addition, the resulting H_∞ state-feedback controller is expressed as $u(t) = V\,\tilde{P}^{-1} x(t)$:

$$\begin{bmatrix} \phi_{11} & \phi_{12} & T & \phi_{14} & \phi_{15} & 0 \\ * & -\tilde{Q} & 0 & \bar{d} V^T F_0^T S^T & V^T F_0^T N^T & V^T J^{\frac{1}{2}} \\ * & * & -\gamma^2 & \bar{d} T^T & 0 & 0 \\ * & * & * & \phi_{44} & \alpha \bar{d} S F_0 J F_0^T N^T & 0 \\ * & * & * & * & \phi_{55} & 0 \\ * & * & * & * & * & -\alpha I \end{bmatrix} < 0 \tag{14}$$

$$\begin{bmatrix} \tilde{X} & H \\ * & \tilde{P}\tilde{Z}^{-1}\tilde{P} \end{bmatrix} > 0 \tag{15}$$

$$\begin{bmatrix} -I + \varepsilon F_0 J F_0^T & \sqrt{\rho} F_0 V & 0 \\ * & -u_{max}^2 \tilde{P} & \sqrt{J\rho} V^T \\ * & * & -\varepsilon I \end{bmatrix} < 0, \tag{16}$$

$$\begin{bmatrix} -I & \sqrt{\rho}\{L\}_i \tilde{P} \\ * & -\tilde{P} \end{bmatrix} < 0, \quad i = 1,2. \tag{17}$$

where

$\phi_{11} = \tilde{P} R^T + R\tilde{P} + \bar{d}\tilde{X} + H + H^T + \tilde{Q} + \alpha S F_0 J F_0^T S^T$, $\phi_{12} = S F_0 V - H$, $\phi_{14} = \bar{d}\tilde{P} R^T + \alpha \bar{d} S F_0 J F_0^T S^T$, $\phi_{44} = -\bar{d}\tilde{Z} + \alpha \bar{d}^2 S F_0 J F_0^T S^T$, $\phi_{15} = \tilde{P} M^T + \alpha S F_0 J F_0^T S^T$, $\phi_{55} = -I + \alpha N F_0 J F_0^T N^T$, $V = K_u \tilde{P}$

Proof. According to the well-known Leibniz-Newton formula and equations (8), then we can write:

$$x(t-d) = x(t) - \int_{t-d}^{t} \dot{x}(s) ds \tag{18}$$

Then substituting $x(t-d)$ into equation (10) gives

$$\dot{x}(t) = (R + SFK_u) x(t) - SFK_u \int_{t-d}^{t} \dot{x}(s) ds + Tw(t) \tag{19}$$

Choosing a Lyapunov functional condidate $V(t)$ as

$$V(t) = V_1 + V_2 + V_3, \quad (20)$$

With

$$V_1 = x^T(t)Px(t),\ V_2 = \int_{-d}^{0}\int_{t+\beta}^{t} \dot{x}^T(\alpha)Z\dot{x}(\alpha)d\alpha d\beta,\ V_3 \int_{t-d}^{t} x^T(\alpha)Qx(\alpha)d\alpha \quad (21)$$

and $P = P^T > 0,\ Z = Z^T > 0$ and $Q = Q^T > 0$ are matrices to be chosen.

Along the trajectory of system (18), the derivative of V is expressed as

$$\dot{V} = \dot{V}_1 + \dot{V}_2 + \dot{V}_3 \le x^T(R^TP + PR + \bar{d}X + Y^T + Y)x(t) + 2\,x^T(t)(PSFK_u - Y)x(t-d) + \bar{d}[Rx(t) + Tw(t) + SFK_u x(t-d)]^T Z[Rx(t) + Tw(t) + SFK_u x(t-d)] + x^T(t)Qx(t) - x^T(t-d)Qx(t-d)) + w^T(t)T^TPx(t) + x^T(t)PTw(t). \quad (22)$$

We consider zero initial conditions i.e. $\phi(t) = 0,\ \ \forall t \in [-d, 0]$,

Define $\quad J_{zw} = \int_0^{\infty}[z_1^T(t)z_1(t) - \gamma^2 w^T(t)w(t)]\,dt \quad (23)$

then,

$$\Psi = \begin{bmatrix} \Upsilon & PSFK_u - Y + \bar{d}R^TZSFK_u + M^TNFK_u & \bar{d}R^TZT + PT \\ * & \bar{d}(SFK_u)^TZ(SFK_u) - Q + (NFK_u)^T(NFK_u) & \bar{d}(SFK_u)^TZT \\ * & * & -\gamma^2 I + \bar{d}T^TZT \end{bmatrix} < 0, \quad (24)$$

Where $\quad \Upsilon = R^TP + PR + \bar{d}X + Y^T + Y + Q + \bar{d}R^TZR + M^TM,$

When we suppose the input zero disturbance, that is to say that $w(t) = 0$, if Eq. (24) holds, then $\dot{V}(x(t)) < 0$ i.e. the asymptotic stability of ASS (10) is thus proven. Using the Schur complement to Ψ, substituting F with its expression given in (13), and invoking Lemma 2, inequality (24) can be given as

$$\begin{bmatrix} \Xi & SF_0V - H & T & \bar{d}\tilde{P}R^T & \tilde{P}M^T \\ * & -\tilde{Q} & 0 & \bar{d}\,V^TF_0^TS^T & V^TF_0^TN^T \\ * & * & -\gamma^2 I & \bar{d}T^T & 0 \\ * & * & * & -\bar{d}\tilde{Z} & 0 \\ * & * & * & * & -I \end{bmatrix} + \alpha \begin{bmatrix} SF_0 \\ 0 \\ 0 \\ \bar{d}SF_0 \\ \bar{d}SF_0 \end{bmatrix} J \begin{bmatrix} SF_0 \\ 0 \\ 0 \\ \bar{d}SF_0 \\ \bar{d}SF_0 \end{bmatrix}^T + \alpha^{-1} \begin{bmatrix} 0 \\ V^T \\ 0 \\ 0 \\ 0 \end{bmatrix} J \begin{bmatrix} 0 \\ V^T \\ 0 \\ 0 \\ 0 \end{bmatrix}^T < 0 \quad (25)$$

Where

$\Xi = \tilde{P}R^T + R\tilde{P} + \bar{d}\tilde{X} + H^T + H + \tilde{Q},\ \tilde{P} = P^{-1},\ V = K_u\tilde{P},\ \tilde{X} = \tilde{P}X\tilde{P},\ H = \tilde{P}Y\tilde{P},\ \tilde{Q} = \tilde{P}Q\tilde{P},$
$\tilde{Z} = Z^{-1}$

By using lemma 2 we can obtain LMI (14).

Following a similar approach, we perform pre- and post-multiplication of the second inequality in Lemma 3 using $diag(\tilde{P},\ \tilde{P})^T$ and its transpose, and subsequently carrying out the substitution, one obtains $\tilde{X} = \tilde{P}X\tilde{P},\ H = \tilde{P}Y\tilde{P},$ and $\tilde{Z} = Z^{-1}$, then LMI (15) hold.

We have when $w(t) = 0$, from $\dot{V}(x(t))$ and Eq. (23) we can write

$$\dot{V}(t) + z_1^T(t)z_1(t) - \gamma^2 w^T(t)w(t) < 0, \quad (26)$$

Consequently, the ASS is asymptotically stable, which establishes the desired performance. In the following, we demonstrate that the hard constraints are satisfied. Inequality (26) ensures that $\dot{V}(t) - \gamma^2 w^T(t)w(t) < 0$. By integrating this inequality over the interval [0,t], for any $t > 0$, we obtain:

$$V(t) - V(0) < \gamma^2 \int_0^t w^T(t)w(t)dt < \gamma^2 ||w||_2^2 \quad (27)$$

Taking into account that the second term of the Lyapunov functional (20) is positive, we have $x^T(t)Px(t) < \rho$, with $\rho = \gamma^2 w_{max} + V(0)$.

$$\max_{t>0} \left|\{z_2(t)\}_i\right|^2 = \max_{t>0} \|x^T(t)\{L\}_i^T\{L\}_i x(t)\|_2 = \max_{t>0} < \rho\theta_{max}(P^{-\frac{1}{2}}\{L\}_i^T\{L\}_i P^{-\frac{1}{2}}), \quad i = 1,2.$$

$$\max_{t>0} \left|u_a^f(t)\right|^2 = \max_{t>0} \|FK_u x(t)\|_2^2 = \max_{t>0} \left\|x^T(t)K_u^TF^TFK_u x(t)\right\|_2 < \rho\theta_{max}\left(P^{-\frac{1}{2}}K_u^TF^TFK_uP^{-\frac{1}{2}}\right).$$

Let θ_{max} denote the maximal eigenvalue. Then, from the inequalities above, the constraints in (11) are ensured if

$$\rho\left(P^{-\frac{1}{2}}\{L\}_i^T\{L\}_i P^{-\frac{1}{2}}\right) < I,\ i = 1,2. \tag{28}$$

$$\rho\left(P^{-\frac{1}{2}}K_u^T F^T F K_u P^{-\frac{1}{2}}\right) < u_{max}^2 I. \tag{29}$$

by Schur complement to (28) and (29) and by substituting F with its expression and making use of Lemma 2, then we can obtain then (16) and (17) can be obtained.

Results and Discussion

The effectiveness of the proposed control approach was verified through a detailed simulation of the quarter-car SS, based on the parameters reported in Table 1.

Table 1. Parameters of the Quarter-Car SS Model

m_u	m_s	k_t	k_s	c_t	c_s
114kg	973kg	101115N/m	42720N/m	1095Ns/m	14.6Ns/m

Assuming a maximum allowable suspension travel of $z_{max} = 0.08m$ and a maximum output power of $u_{max} = 1500$, we choose $\rho = 1$. By fixing the actuator failure matrix $F = I$ in normal actuator operation, and setting $d = 9.2ms$, we can solve the linear inequality matrix (LMI) in Theorem 1 to obtain the possible solution. Table 2 shows the corresponding lower boundary of the H_∞ performance corresponding to different upper limits of of d.

Table 2. The minimum of H_∞ performance index

$d(ms)$		0.85	0.9	1	1.5
$min\gamma$		2.2131	2.2356	2.2358	2.2360

The obtained gain matrices for the ASS, corresponding to the cases with and without actuator failure, are listed consecutively as:

$\boldsymbol{K_{uf}} = \mathbf{10^3} \times [\mathbf{6.4220} \quad \mathbf{-1.3655} \quad \mathbf{-5.4775} \quad \mathbf{0.6937}]$ $\boldsymbol{K_{uwf}} = \mathbf{10^3} \times [\mathbf{8.7664} \quad \mathbf{-0.9297} \quad \mathbf{-8.4466} \quad \mathbf{1.0701}]$.

For comparison, we consider the AS on a quarter of the vehicle in reference [17], where this paper discusses actuator failures. Our results, presented in Table 3, are clearly better than those in reference [17]. Therefore, we can say that our results are less conservative than those of the aforementioned paper.

Table 3. Comparing performance γ

	H_∞ $performance\ index\ \gamma_{min}$
Ref [17]	18.3676
Theorem 1	1.5626

Evaluation for bump response

To investigate the suspension behavior with respect to the aforementioned performance criteria, considering an isolated bump on a smooth road profile is considered. The disturbance input corresponding to this case is defined as:

$$z_r(t) = \begin{cases} \frac{A}{2}(1 - \cos(\frac{2\pi V}{h}) & if\ 0 \le t \le \frac{h}{V} \\ 0 & if\ 0 > \frac{h}{V} \end{cases} \tag{30}$$

Where h and A are the length and height of the bump, respectively, with $A = 60\,mm$, assumed to be. It is assumed $A = 60\,mm$, $h = 5\,m$ and the vehicle longitudinal velocity $V = 30\,km/h\,, d = 6ms$.

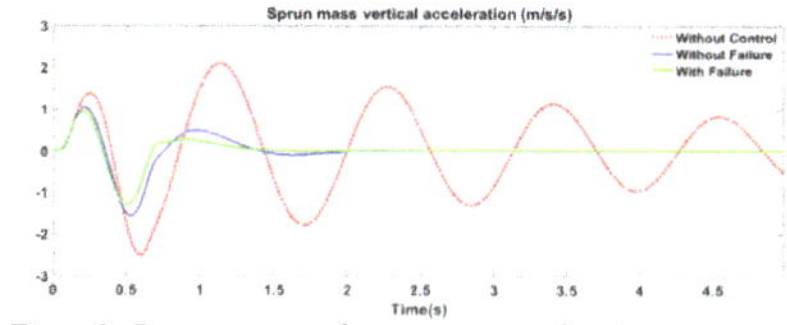

Figure 2. Bump response of sprung mass acceleration

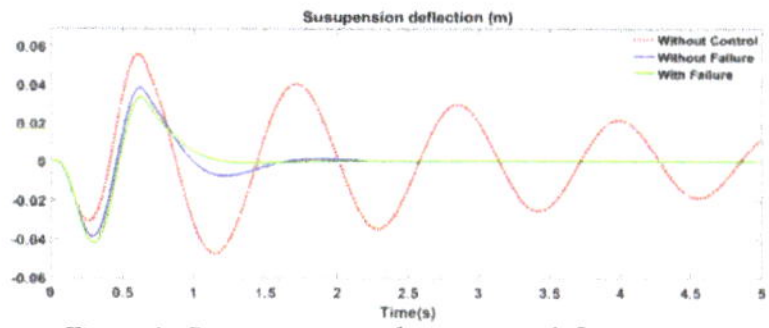

Figure 3. Bump response of suspension deflection

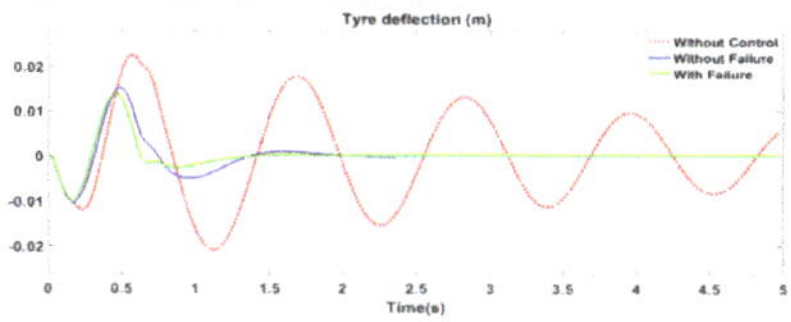

Figure 4. Bump response of tyre deflection

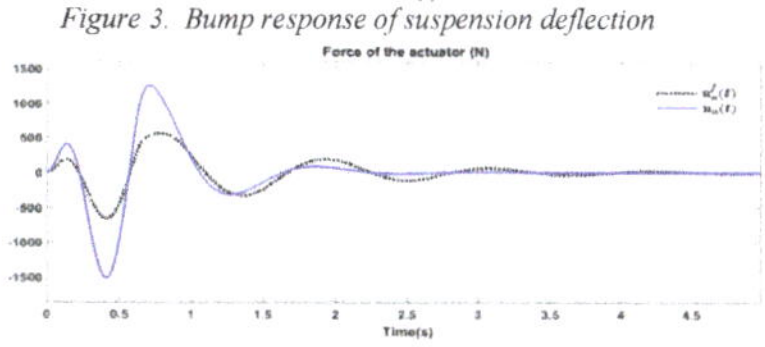

Figure 5. Force of the actuator

Simulation results presented in Figures 2-4 illustrate the resulting performance of the ASS using the proposed FTC strategy. under the proposed FTC controller. These results confirm that, despite delays and actuator failures, the controller provides superior road handling ride comfort, while maintaining acceptable suspension deflection and actuator forces remain within safe limits.

Evolution of Random response

In the time domain, a non-stationary random road profile is defined as

$$\dot{z}_r(t) + 2\pi u n_0 z_r(t) = \sqrt{G_q(n_0) v w(t)} \tag{31}$$

where $w(t)$ is a white noise in the time-domain, v a constant velocity $v = 20m/s$, n_0 is

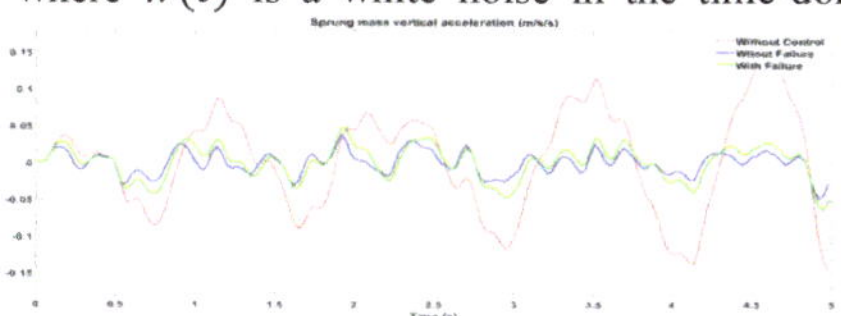

Figure 6. Sprung mass vertical accelerations for variable road profile

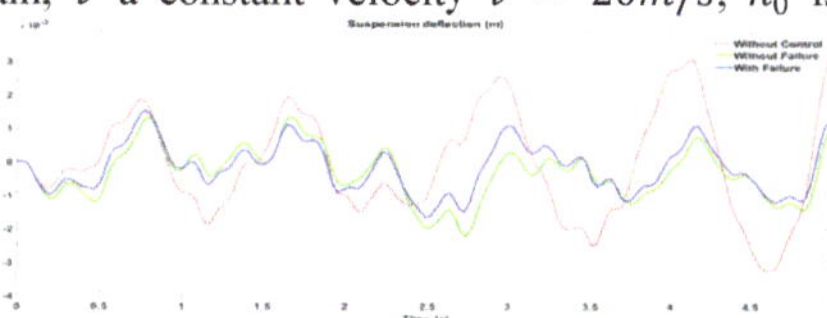

Figure 7. Suspension deflections for variable road profile

The spatial reference frequency is defined as $n_0 = 0.1(cycle/m)$, and $G_q(n_0)$ is the road roughness coefficient. In accordance with ISO-2631, the average road roughness rating is shown here: $G_q(n_0) = 64 \times 10^{-6} m^3, d = 6ms$. Figures 6-8 illustrate the evolution of the system's main parameters when actuator failures and time delays are taken into account. These figures clearly show that the H_∞ control strategy achieves better performance than the passive SS. The simulation results verify the effectiveness of the designed FTC scheme.

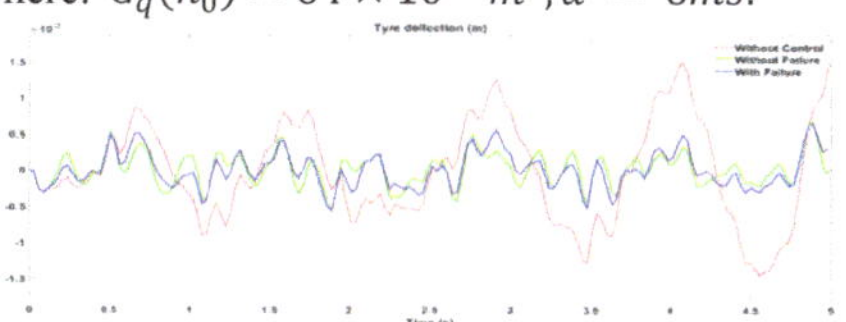

Figure 8. Tyre deflections for variable road profile

Conclusion

This study has addressed the FTC problem for ASS by explicitly considering actuator faults and input time delays. An H_∞ based FTC strategy has been developed to ensure that the ASS remains asymptotically stable while achieving the desired level of ride comfort and road handling performance. The controller synthesis relies on LMI-based conditions that guarantee robustness against external disturbances and actuator degradations. The developed strategy has been validated via simulations conducted on a quarter-car ASS model, and simulation results demonstrate its ability to preserve stability and performance under actuator faults, time delays, road excitations, and varying operating conditions. Furthermore, the proposed framework can be readily extended to more complex vehicle configurations.

References

[1] Hu Y, Chen MZ, Shu Z. Passive vehicle suspensions employing inerters with multiple Performance requirements. Journal of Sound and Vibration. 2014 Apr 14;333(8):2212-25. https://doi.org/10.1016/j.jsv.2013.12.016

[2] Chen MZ, Hu Y, Li C, Chen G. Performance benefits of using inerter in semiactive suspensions. IEEE Transactions on Control Systems Technology. 2014 Nov 7;23(4):1571-7. https://doi.org/10.1109/TCST.2014.2364954

[3] Gao H, Sun W, Shi P. Robust sampled-data H_(∞) control for vehicle active suspension systems. IEEE Transactions on control systems technology. 2009 Oct 23;18(1):238-45. https://doi.org/10.1109/TCST.2009.2015653

[4] Türkay S, Akçay H. Aspects of achievable performance for quarter-car active suspensions. Journal of Sound and Vibration. 2008 Mar 18;311(1-2):440-60. https://doi.org/10.1016/j.jsv.2007.09.014

[5] Fialho I, Balas GJ. Road adaptive active suspension design using linear parameter-varying gain-scheduling. IEEE transactions on control systems technology. 2002 Jan;10(1):43-54. https://doi.org/10.1109/87.974337

[6] Chen H, Guo KH. Constrained H_(∞)control of active suspensions: an LMI approach. IEEE Transactions on Control Systems Technology. 2005 May 31;13(3):412-21. https://doi.org/10.1109/TCST.2004.841661

[7] Badie K, Alfidi M, Chalh Z. Reliable H_(∞)control for 2-D discrete switched systems with state delays and actuator failures. International Journal of Adaptive Control and Signal Processing. 2020 Oct;34(10):1383-404. https://doi.org/10.1002/acs.3156

[8] Badie K, Alfidi M, Chalh Z. An LMI approach to design robust H_(∞) controller for 2-D systems with delays. International Journal of System of Systems Engineering. 2019;9(2):99-116. https://doi.org/10.1504/IJSSE.2019.100335

[9] Mrazgua J, Ouahi M. Fuzzy fault-tolerant H_(∞)control approach for nonlinear active suspension systems with actuator failure. Procedia Computer Science. 2019 Jan 1;148:465-74. https://doi.org/10.1016/j.procs.2019.01.059

[10] El Haiek B, Hmamed A, El Hajjaji A, Tissir EH. Improved results on observer-based control for discrete-time fuzzy systems. International Journal of Systems Science. 2017; 2544-53. https://doi.org/10.1080/00207721.2017.1324067

[11] Panagi P, Polycarpou MM. Decentralized fault tolerant control of a class of interconnected nonlinear systems. IEEE Transactions on Automatic Control. 2010 Oct 25;56(1):178-84. https://doi.org/10.1109/TAC.2010.2089650

[12] Mrazgua J, Tissir EH, Ouahi M. Fault-tolerant H_(∞) control approach, application to active half-vehicle suspension systems with actuators failure accounts. In2019 8th International Conference on Systems and Control (ICSC) 2019 Oct 23 (pp. 271-276). IEEE. https://doi.org/10.1109/ICSC47195.2019.8950668

[13] Mrazgua J, Tissir EH, Ouahi M, Tadeo F. Reliable H_(∞) fuzzy control for fault-tolerant actuator failures of active suspension system. IFAC Journal of Systems and Control. 20241;27:100258. https://doi.org/10.1016/j.ifacsc.2024.100258

[14] Veillette RJ. Reliable linear-quadratic state-feedback control. Automatica. 1999; 137-43. https://doi.org/10.1016/0005-1098(94)E0045-J

[15] Zhao Q, Jiang J. Reliable state feedback control system design against actuator failures. Automatica. 1998 Oct 1;34(10):1267-72. https://doi.org/10.1016/S0005-1098(98)00072-7

[16] Moon YS, Park P, Kwon WH, Lee YS. Delay-dependent robust stabilization of uncertain state-delayed systems. International Journal of control. 2001 Jan 1;74(14):1447-55. https://doi.org/10.1080/00207170110067116

[17] Li H, Gao H, Liu H, Liu M. Fault-tolerant H_(∞) control for active suspension vehicle systems with actuator faults.: Journal of Systems and Control Engineering. 2012 Mar;226(3):348-63. https://doi.org/10.1177/0959651811418401

Design and Simulation of a Solar Powered Low-Frequency H-Bridge Inverter with an Idealized Wireless Power Transfer Extension

Anas MOUAKKA[1,a *]

[1]Al Akhawayn University, Ifrane, Morocco

[a]A.mouakka@aui.ma

Keywords: Photovoltaic Systems, Solar Energy Management, DC–AC Inverters, Full-bridge, Linear Transformer, Wireless Power Transfer (WPT), Magnetic Coupling, Mutual Inductance, RLC Load, Simulink, Impedance Analysis, Root Mean Square (RMS)

Abstract. Solar photovoltaic systems require a DC-to-AC conversion to deliver usable power to domestic loads. This paper presents the design and Simulink-based simulation of a low frequency H-bridge inverter connected to an ideal linear transformer, and a loosely coupled wireless power transfer (WPT) as a continuation to the same inverter. The novelty lies in the union of the inverter and an adjustable linear transformer model to demonstrate how the coupling coefficient k affects the RMS voltage output. The DC-to-AC inverter was implemented in MATLAB/Simulink/Simscape using IGBT switches, gating signals (50Hz), and a 12 V DC source representing the PV-battery subsystem. Parametric sweeps were performed across series load resistance, inductance, and capacitance; also, sweeps for switching frequency and coupling coefficient were performed. The results show the RMS voltage to be converging to the nominal 120 VAC for resistive loads $\geq 1\ \Omega$ and reactance loads ≥ 0.1 mH, while high-capacitance conditions (e.g., ≥ 10 mF) produce low voltages. Decreasing k produces reductions in V_{RMS} and power delivered, which is predicted by the coupled-inductor theory. Moreover, the efficiency was found to be $\approx$ 84%. Overall, the combination of the H-bridge and linear transformer provides a controllable model for analyzing inverter and WPT performance.

Introduction

Background. The growing of global energy demand, and the decline of fossil fuels reserves caused a major shift towards renewable energy resources such as solar power using photovoltaic panels. PV panels generate DC voltage, which is then converted using an inverter to AC, so this power can be used in households or utility loads in general. The most basic off-grid solar system includes a PV panel, charge controller, battery, and an inverter to generate AC loads (Fig. 1). To optimize the panel output an MPPT is used. The diagram in Fig. 1 illustrates how solar energy is collected and managed, where the panel charges the battery and the stored DC power passes through the inverter to supply AC to the load. In this study, I used a low-frequency H-bridge for simplicity, where step-up transformation and waveform quality determine the efficiency for which the generated PV energy can be used to power AC appliances.

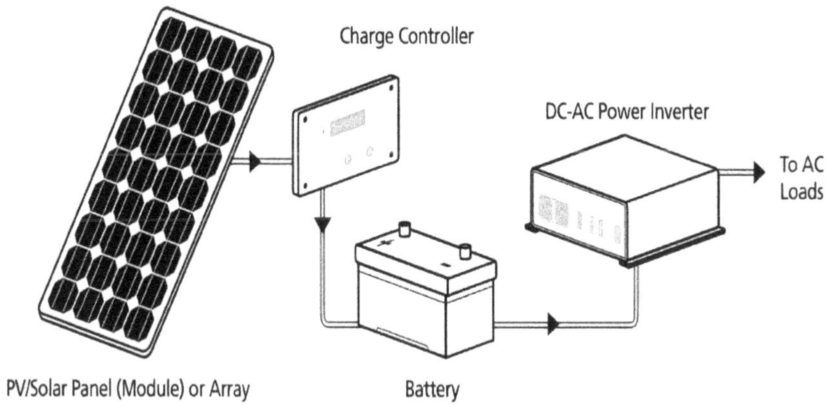

Fig. 1. Block diagram of a simple solar power management system

Literature Review. Previous research papers explored inverter topologies, and WPT links at different scales. Zhang et al. examined loosely coupled transformers, and their ability for vehicle charging, highlighting the role of low coupling coefficient k on power transfer [1]. Singh et al. and Lu et al. investigated the H-bridge and cascaded converters in PV applications to demonstrate practical implementation and efficiency trade-offs regarding Simulink and hardware studies [2, 6]. Meynard et al. and Kartick et al. provided PWM-based control aspects to show how advanced modulation can lower the THD and improve the power factor in high power systems. These works cover the links between inverters and transformer coupling, but in a single simulation framework.
Novelty and Outline. The following paper differs from existing studies by simply coupling a low frequency full-bridge inverter with a linear transformer block whose coupling coefficient k is varied to represent both tightly and loosely coupled conditions. This approach will also clarify how loading and coupling interact together to determine the V_{RMS} in idealized environment. The remainder of the paper will be organized as follows: Section II covers the materials and methods used; Section III present results from R, L, C, k, and frequency sweeps with critical discussion; and Section IV concludes with limitations and recommendations for future work.

Materials and Methods
System Components and Software. The simulation of the system was developed in MATLAB R2024b with Simulink and Simscape Electrical (sps_lib) for components modeling. The main components are: a 12 V DC source (i.e., PV panel, charge controller, and battery combined), a four switch H-bridge implemented using four IGBT/Diode blocks, and a Linear Transformer (Fig. 2). All devices in this circuit were treated ideally.

Materials Research Proceedings 64 (2026) 404-410
https://doi.org/10.21741/9781644904091-50

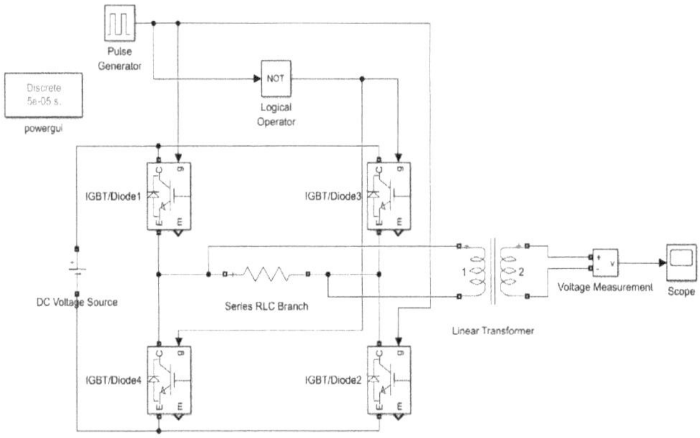

Fig. 2. Simulink H-bridge and Linear Transformer system

Circuit Desing and Control. The inverter is implemented using a classical full-bridge (H-bridge) topology with four power IGBTs (i.e., MOSFETs) to produce AC from a DC source across the load resistance (Fig. 2) [8]. The four switches S1, S2, S3, and S4 were arranged in an H shape; for instance, S1 is IGBT/Diode 1. The S1/S3 switches form the top legs while the S2/S4 switches form the bottom legs of the bridge. These gate signals were generated using Simulink's Pulse Generator block with a 50 Hz square pulse driving the S1/S2, and an inverted copy driving the S3/S4 using the NOT Logical Operator block. For simplicity, the dead-time for avoiding shoot-through was excluded. Each IGBT switch includes an anti-parallel diode for protection against voltage spikes that may be caused by the load's nature [6]. The transformer turns ratio was chosen to achieve a step of 12 VDC → 120 VAC based on the equation: $\sqrt{L_2/L_1}$. Also, the Linear Transformer included magnetizing parameters allowing k to be swept from tight coupling to low values representing loose WPT links.

Experimental Procedure. The simulation cases included multiple sweeps. The first case includes an RLC elements sweep, simulated individually. The resistive sweep ranges from 0-1 Ω, the inductance sweep L = 0-1 mH, and the capacitance sweep for C = 0-1 F, where they were all tested to capture the V_{RMS} sensitivity. The second case included a frequency variation from 0 to 5 kHz to check the voltage response and observe the circuit's frequency dependence. The third case included coupling coefficient k sweep (k = 0.1 → 1), and its effect on voltage to model WPT scenarios. All the cases recorded V_{RMS} at the secondary terminal. The control of this circuit was modelled by Simulink pulse generator and logic blocks, with no use of Arduino or microcontroller algorithms. Key assumptions are based on ideal switching, no PWM filtering, and fixed DC source voltage with no PV panel, battery, or MPPT implementation.

Results and Discussion

Based on Fig. 2 the output voltage is measured at the transformer secondary, and it is dependent on the RLC load placed in series with the H-bridge. Fig. 3 illustrates how the AC output of the linear transformer varies as a function of the loads R, L, and C. This behaviour agrees with transformer models for coupled inductors [1, 9]. As shown in Fig. 3(a), the output voltage (V_{RMS}) is very sensitive at low resistance values. At R ≈ 0 Ω, the load is short circuited, causing the RMS voltage to drop around 40V. This behaviour occurs because the full bridge must supply a high current, and according to ohms law ($I = \frac{V}{R}$), when R approaches 0, the current tends toward infinity, so a huge internal voltage drop occurs. So, when the resistance is increased (e.g., R = 0.1

Materials Research Proceedings 64 (2026) 404-410 https://doi.org/10.21741/9781644904091-50

Ω), the current decreases and the voltage drop is reduced (i.e., high voltage). This causes the voltage to rise above 100 V. For a resistance value of 1 Ω and above, the V_{RMS} approaches the ideal configured output (≈ 120 VAC).

The inductance plot in Fig. 3(b) reflects AC impedance laws as well. The inductive reactance is defined as $X_L = 2\pi f L$, where at L = 0 μH, $X_L \approx 0$. For the series load (H-bridge load), current is determined by the total impedance, $I = V/\sqrt{R^2 + X_L{}^2}$. So, as current increases, the voltage drops across internal impedance, causing V_{RMS} to fall (~40 V at around 1 μH). As Inductance L increases, X_L increases, and the current decreases. Based on KVL, a decrease in current I means a small drop across internal impedance ($Z_{internal}$), where $V_{out} = V_{ideal} - I(Z_{internal})$. So, V_{RMS} rises above 110 V at around 100 μH; thus approaching $V_{ideal} = 120\,V$ at values around 1 mH and above.

The stepped-up RMS voltage falls as capacitance increases as shown in Fig. 3(c). Since the capacitance is in series with the H -bridge, the voltage that reaches either primary or secondary of the transformer is limited. The primary inductance has reactance $X_L = \omega L$, and the capacitor has reactance $X_C = \frac{1}{\omega C}$, so together they form a series impedance, where primary voltage (i.e., voltage at transformer primary terminal) is approximately $V_{primary} = V_{bridge} \frac{X_L}{|X_L - X_C|}$. When the capacitance increase, X_C becomes smaller; thus making $|X_L - X_C|$ larger, which reduces the bridge voltage that appear across the primary. Since the secondary voltage is related to the primary through the turns ratio (1:10), the measured V_{RMS} drops as the capacitor increases relatively to the defined ratio.

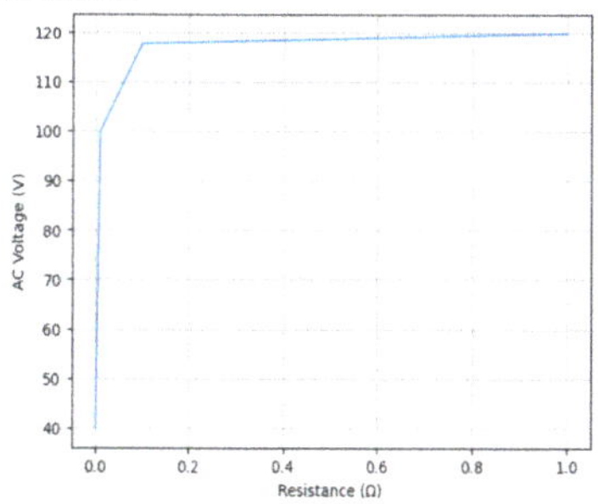

(a) Resistance vs. Voltage

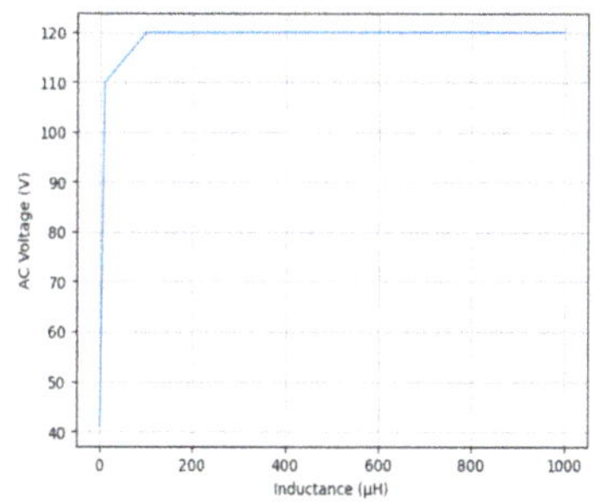

(b) Inductance vs. Voltage

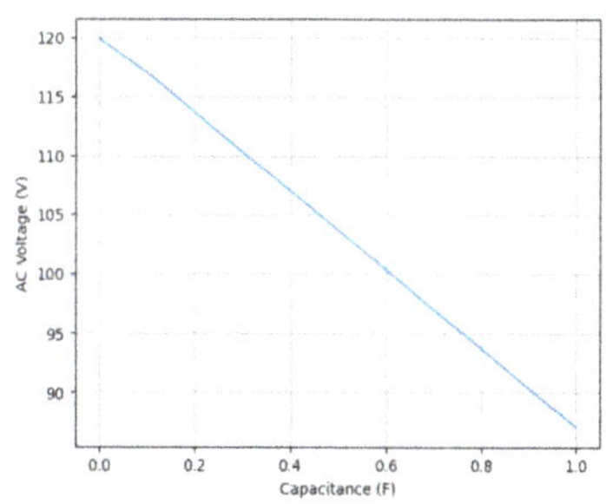

(c) Capacitance vs. Voltage

Fig. 3. RMS AC voltage under RLC elements variation

Fig. 4 shows how the transformer's secondary RMS voltage output changes with H-bridge frequency change (i.e., variance of pulse generator's frequency in Fig. 2). For very low frequencies (e.g., 0 kHz), the transformer behaves like a short circuit ($f \rightarrow 0, X_m \rightarrow 0$), so no AC voltage is induced (i.e., no switching), where the magnetizing reactance is in terms of frequency and magnetizing inductance:

$$X_m = 2\pi f L_m. \tag{1}$$

At mid-range frequencies (1-3 kHz), the transformer behaves correctly (i.e., output around 120 V). The H-bridge outputs a bipolar square wave with amplitude ± 12 V, whose RMS is 12 V. So that is why Fig. 4 shows a constant line of around 120 V in the mid-frequency region. However, at higher frequencies (around 4 kHz), the voltage experiences a small fall. As frequency increases, leakage inductance (L_σ) and winding capacitance (not included in this circuit) start to matter or interfere with the output. The leakage inductance creates a frequency-dependent voltage inside the transformer, with increase in $f \rightarrow$ increase in leakage reactance (X_σ):

$$X_\sigma = \omega L_\sigma = 2\pi f L_\sigma. \tag{2}$$

Beyond the resonance region (around 5 kHz), the transformer becomes less inductive (i.e., leakage reactance becomes less dominant), and the voltage recovers around its nominal value.

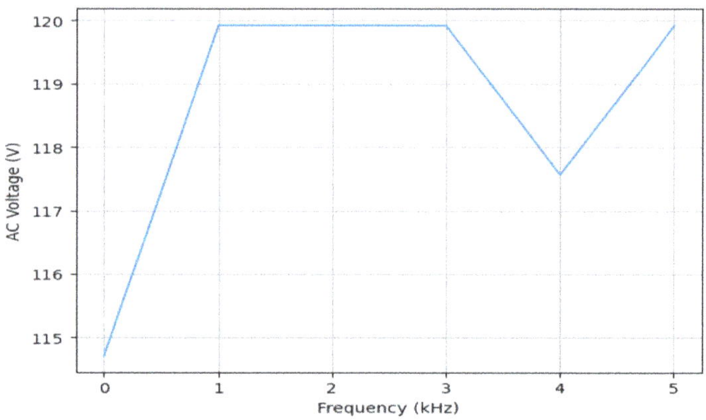

Fig. 4. RMS AC voltage output under varied frequency

The coupling coefficient k links the mutual inductance (M) and the primary/secondary inductances (L_1& L_2):

$$k = \frac{M}{\sqrt{L_1 L_2}} \Rightarrow M = k\sqrt{L_1 L_2}. \tag{3}$$

A low k value means high leakage flux or leakage inductance, in which this leakage behaves similarly to a series reactance that uses voltage from the winding that connects to the other coil or terminal. So, a low coupling coefficient causes a noticeable drop in output RMS voltage, and as k increases toward 1, the Linear Transformer behaves ideally (i.e., no air gap in the magnetic path between primary and secondary) and more flux is shared between the coils, where the secondary or output voltage rises to its nominal/expected value (Fig. 5) [10, 11]. Based on Eq .3 [12], the mutual inductance M quantifies how effectively a change in current in the primary winding induces voltage in the secondary winding through the mutual magnetic flux. In contrast, the leakage inductance comes from the magnetic flux generated by the two unconnected windings L_1 and L_2. Additionally, the coupling coefficient is a magnetic property and not a measure of distance that is based on self-inductances, and not on separation between the coils.

The efficiency of the system was calculated through $\eta = \frac{P_{out}}{P_{in}} \times 100 = \frac{V_{secondary} \cdot I_{secondary}}{V_{DC} \cdot I_{DC}} \times 100$, and found to be 84% in the case of perfect coupling (i.e., k = 1), meaning that 16% of the DC input power from the source is lost in the system. The loss can be explained through the reactive effects of the bridge load, and the voltage drop across the H-bridge switches.

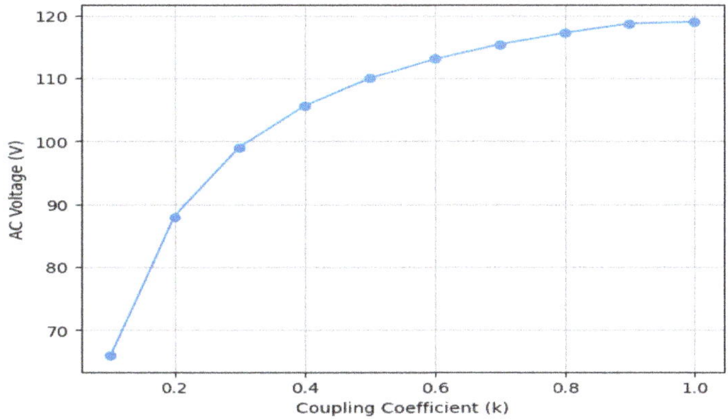

Fig. 5. RMS AC voltage under varied coefficient k

Conclusion

This study was driven by the need for DC-to-AC conversion in a PV-based system. The conversion was done through a low frequency full-bridge inverter coupled to a linear transformer to achieve and study step-up transformation and wireless power transfer. The novelty lies in the single-simulation framework, in which the transformer's coupling coefficient k is varied along series loading (i.e., R, L, and C) and frequency to determine effect on RMS voltage and power.

The design of the systems and its simulation was done using MATLAB R2024b/Simulink with Simscape Electrical. A 12 VDC supplies an H-bridge of four IGBT/Diodes that is driven by a 50 Hz pulse generator. The bridge setup is linked to a linear transformer for step up. Different parametric sweeps were performed, recording secondary V_{RMS} while varying R, L, C, f, and k The simulations produced several, yet reasonable trends. For resistive loads $R \geq 1\ \Omega$, the RMS voltage converged to the nominal value ≈120 VAC, whereas in low resistive loads R→0, the voltage dropped. Increasing the inductance (L ≥ 1 mH) increased the voltage, while increasing the capacitance (C ≥ 0.1 F) decreased the voltage. Also, the frequency sweeps showed a stable voltage at the mid-band (1-3 kHz) with a small fall around ~4 kHz, and a recovery to the nominal at 5 kHz due to leakage/reactance effects. Furthermore, a reduction in k resulted in a low RMS voltage; thus, a low power transfer. The end-to-end efficiency in the setup was found to be ≈84%.

The main limitation of this work is the idealization of the devices used. The components of this system were modelled without switching losses, dead-time, winding capacitances, MPPT and battery, total harmonic distortion, and resonance frequency. Future work will implement more realistic models; for example, a pulse width modulation control and dead time, unintended electrical properties (i.e., parasitics), MPPT and battery models, lab measurements with real coils, and resonance frequency compensation for low coupling coefficient effect in WPT. Finally, the inclusion of suitable control extensions, such as THD reduction or phase control, so the model can be effectively applied to practical PV and WPT systems.

References

[1] W. Zhang, J. C. White, A. M. Abraham, and C. C. Mi, "Loosely Coupled Transformer Structure and Interoperability Study for EV Wireless Charging Systems," IEEE Trans. Power Electron., vol. 30, no. 11, pp. 6354–6367, Nov. 2015. https://doi.org/10.1109/TPEL.2015.2433678

[2] G. Singh, M. K. Bhaskar, M. Parihar, and D. Jain, "Analysis of Cascaded H-Bridge Multilevel Solar Inverter," Int. J. Emerg. Technol. Electr. Eng., vol. 8, no. 5, 2022, Art. no. 23. https://doi.org/10.5281/zenodo.7221140

[3] A. I. Elsanabary, G. Konstantinou, S. Mekhilef, C. D. Townsend, M. Seyedmahmoudian, and A. Stojcevski, "Medium Voltage Large-Scale Grid-Connected Photovoltaic Systems Using Cascaded H-Bridge and Modular Multilevel Converters: A Review," IEEE Access, vol. 8, pp. 223686–223699, Dec. 2020. https://doi.org/10.1109/ACCESS.2020.3044882

[4] C. D. Fuentes, C. A. Rojas, H. Renaudineau, S. Kouro, M. A. Pérez, and T. Meynard, "Experimental Validation of a Single DC Bus Cascaded H-Bridge Multilevel Inverter for Multistring Photovoltaic Systems," IEEE Trans. Ind. Electron., vol. 64, no. 2, pp. 930–934, Jun. 2017. https://doi.org/10.1109/TIE.2016.2619661

[5] K. C. Jana, K. Sujit Biswas, and S. KarChowdhury, "Dual reference phase-shifted pulse width modulation technique for a N-level inverter based grid connected solar photovoltaic system," IET Renew. Power Gener., vol. 10, no. 7, pp. 928–935, 2016. https://doi.org/10.1049/iet-rpg.2015.0393

[6] H. Alnuman, E. Hussain, M. Aly, E. M. Ahmed, and A. Alshahir, "Cascaded H-Bridge Multilevel Converter Topology for a PV Connected to a Medium-Voltage Grid," Machines, vol. 13, no. 7, Art. 540, Jun. 2025. https://doi.org/10.3390/machines13070540

[7] M. H. Rashid, *Power Electronics: Devices, Circuits and Applications*, 4th ed. Pearson, 2014.

[8] Wikipedia, "H-bridge," Wikipedia: The Free Encyclopedia, Jul. 2024. [Online]. Available: https://en.wikipedia.org/wiki/H-bridge.

[9] University of Tennessee, Dept. of Electrical and Computer Engineering, *ECE 202: Coupled Inductors and Transformers Lab Manual*, Spring 2024. [Online].

[10] Wikipedia, "Inductance," Wikipedia: The Free Encyclopedia, Aug. 2024. [Online]. Available: https://en.wikipedia.org/wiki/Inductance.

[11] Wikipedia, "Resonant inductive coupling," Wikipedia: The Free Encyclopedia, Aug. 2024. Available: https://en.wikipedia.org/wiki/Resonant_inductive_coupling.

[12] Assoc. Prof. Dr. Boris Evstatiev, "Lecture notes – Mutually Coupled Inductors," University Lecture Notes, pdf.

Lyapunov-Based Continuous-Time MPC for DC-DC Boost Converter Control in Photovoltaic Systems

Sanae EL BOUASSI[1,a *], El mehdi MELLOULI[1,b], Zakaria CHALH[1,c], Mustapha EL YAQOUTI[2,d], El Hanafi ARJDAL[3,e]

[1]Laboratory of Engineering, System and Application, School of Applied Sciences, Sidi Mohamed Ben Abdellah University, Fez, Morocco

[2]University of IBN ZOHR, Agadir, Morocco

[3]Laboratory of Materiels, Signals, Systems and Physical Modeling , Faculty of Sciences, IBN ZOHR Agadir, Morocco

[a]sanae.elbouassi@usmba.ac.ma, [b]mellouli_elmehdi@hotmail.com, [c]zakaria.chalh@usmba.ac.ma, [d]m.elyaqouti@uiz.ac.ma, [f]e.arjdal@uiz.ac.ma

Keywords: Active and Reactive Power, Boost Converter, Lyapunov Approach, Model predictive Control, Photovoltaic Systems

Abstract. The objective of the current work is to enhance the performance of a grid-connected photovoltaic (PV) system by creating a dependable continuous-time model predictive control (CT-MPC) method. The suggested controller uses a Lyapunov-based formulation within the model predictive control (MPC) framework, and the photovoltaic (PV) panel has been meticulously modeled. In the process, dynamic performance is improved, and system stability is guaranteed. To improve power management and guarantee steady performance in a variety of environmental circumstances, this control strategy is used to a DC-DC boost converter. The simulation findings from this study provide empirical evidence for the effectiveness of the proposed method. The results demonstrate that the proposed method is more efficient, resilient, and flexible than conventional control methods.

Introduction

As clean energy sources, photovoltaic (PV) systems are widely used in grid-connected applications. The amount of power generated by these systems varies according to the environmental conditions. Optimizing energy extraction necessitates operating at the maximum power point (MPP), which calls for effective and flexible control strategies [1, 2].

Effective management of PV systems is essential, particularly for DC–AC power conversion in grid-connected applications. Advanced control strategies have been adopted as a result of PI controllers' poor performance under rapidly changing conditions, despite their widespread use. By anticipating system behavior and optimizing control actions, model predictive control (MPC) provides better performance. This ensures stable operation and maximum power point (MPP) tracking, even when there are variations in irradiance [3][4].

To further enhance performance, this study presents a Lyapunov-based model predictive control (MPC) approach that ensures system stability while minimising computational complexity. When applied to a boost converter connected to a three-phase grid, the proposed method achieves a faster transient response, improved energy efficiency and greater robustness. Simulation results confirm that the proposed strategy is superior to conventional methods, highlighting its suitability for practical PV applications [5][6].

Materials Research Proceedings 64 (2026) 411-415 https://doi.org/10.21741/9781644904091-51

Modelling of the PV panel

Figure 1 shows the topology of the standalone PV system that is the focus of this study. This system consists of a photovoltaic generator, which is controlled by an MPPT-based PWM signal, and a DC/DC boost converter, which is regulated according to solar irradiance and load conditions

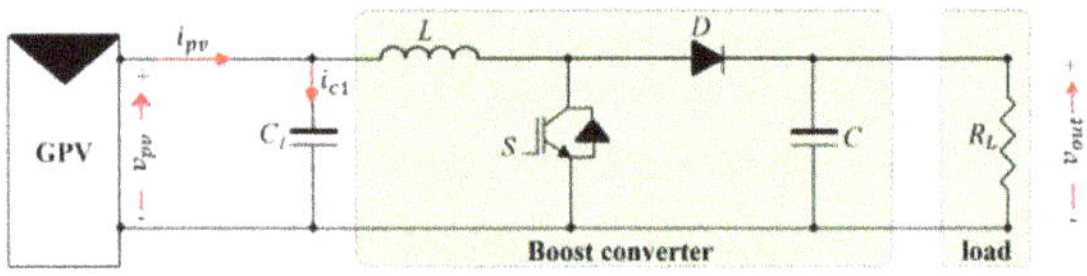

Fig. 1 Plan for a stand alone photovoltaic system

The photovoltaic (PV) panels' maximum power point has been established [7, 8]. As shown below, the averaging technique may be used to define the dynamic model of the boost converter in terms of duty cycle, "D":

$$\begin{cases} \frac{di_L}{dt} = \frac{v_{pv}}{L} - \frac{V_{out}}{L} + \frac{V_{out} \cdot D}{L} \\ \frac{dv_{pv}}{dt} = -\frac{i_L}{C} + \frac{i_p}{C} \end{cases}$$

Model Predictive Control (MPC) is an effective power electronics strategy based on predicting future system behavior using a dynamic model and real-time measurements. At each sampling instant, MPC selects the optimal control action by minimizing a cost function. Only the first input is applied before the process is repeated at the next step.

Active power:

$$P = v_{pv} i_L \cos(\phi) \tag{1}$$

Reactive power:

$$Q = v_{pv} i_L \sin(\phi) \tag{2}$$

Reference power is given by:

$$P_{ref} = P - P_{loss} \tag{3}$$

Where

$$P_{loss} = i_L^2 R \tag{4}$$

The expression for the cost function is:

$$g = |P_{ref} - P| + |Q_{ref} - Q| \tag{5}$$

The complex power error is given by:

$$S_e = S - S_{ref} \tag{6}$$

In discrete time, this error can be expressed as:

Materials Research Proceedings 64 (2026) 411-415 https://doi.org/10.21741/9781644904091-51

$$S_e = v_{pv} i_L - S_{ref} \quad (7)$$

The lyapunov function is defined as :

$$LF\{S_e\} = \frac{1}{2}(S_e^2) \quad (8)$$

$$\Delta LF\{S_e\} = \frac{1}{2}\left(V_{pv}\left(\frac{T_s d_i}{dt} + I_L\right) - S_{\text{ref}}\right)^2 - \frac{1}{2}S_e^2 \quad (9)$$

Derivative expression :

$$\frac{dI_L}{dt} = \frac{I_L^{\text{new}} - I_L}{T_s} \quad (10)$$

Lyapunov stability theorem:

A system is stable if its Lyapunov function diminishes over time, according to Lyapunov's stability theorem. A Lyapunov-based control algorithm requires that the function's change (ΔLF) be negative. If the first term in equation (9) is zero, the system is stable because the second term is negative. This results in the following limitation:

$$V_{pv}\left(\frac{T_s d_i}{dt} + I_L\right) - S_{\text{ref}} = 0 \quad (11)$$

Thus

$$v_{pv} = \frac{S_{ref}}{i_L} \quad (12)$$

To choose the appropriate switching vector and minimize the cost function, the estimated voltage vpv in equation (11) is compared with the ideal converter output vpv,opt.

$$g = |vpv - vpv,opt| \quad (13)$$

In this work, we use a Lyapunov-based method to examine the system's robustness. A common tool in stability studies, the Lyapunov function is defined as follows:

$$V = \frac{1}{2}S_e^2 \quad (14)$$

where Se is the complex power error. In order to assess the stability of the system, we calculate the Lyapunov function's time derivative:

$$\frac{dV}{dt} = S_e \frac{dS_e}{dt} = S_e\left(v_{pv}\dot{i}_L + \dot{v}_{pv} i_L\right) \quad (15)$$

It is possible to simplify this derivative as follows:

$$\frac{dV}{dt} = -k_e S_e \text{sign}(S_e) \quad (16)$$

In this case, sign(Se) is the sign of the error Se, and ke is a constant. This guarantees that whenever Se is non-zero, the Lyapunov function falls, bringing the system to a stable state.We reduce the cost function g, which is defined as follows, in order to assess the system's resilience further:

$$g = |vpv - vpv,opt|$$

To ensure reliable operation under changes in input and load, the cost function minimizes the discrepancy between the actual and ideal voltages. Converter signals are produced by Lyapunov-based predictive control to offer reliable and effective PV power regulation.

Materials Research Proceedings 64 (2026) 411-415 https://doi.org/10.21741/9781644904091-51

Simulations

The following figures illustrate the dynamic performance of the proposed Lyapunov-based continuous-time MPC approach applied to the DC–DC boost converter in a photovoltaic system. The behavior of the controller is evaluated under realistic operating conditions, with a focus on voltage regulation, system stability, and control signal evolution. It is essential to keep in mind that all control variables, including the duty cycle, are shown within their actual operating ranges. This ensures that the results of the simulation will be interpreted consistently and meaningfully.

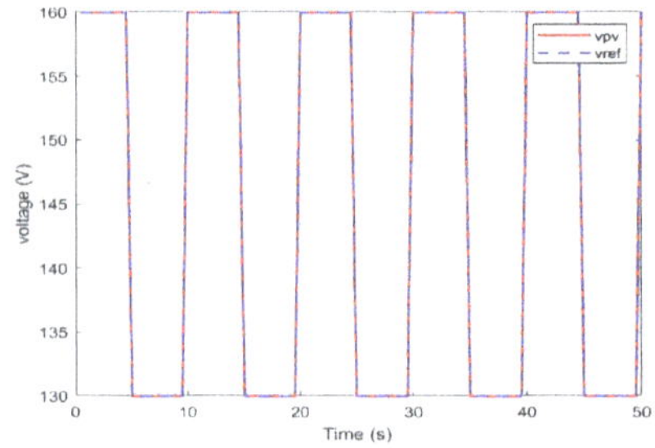

Fig. 2 PV output voltage response

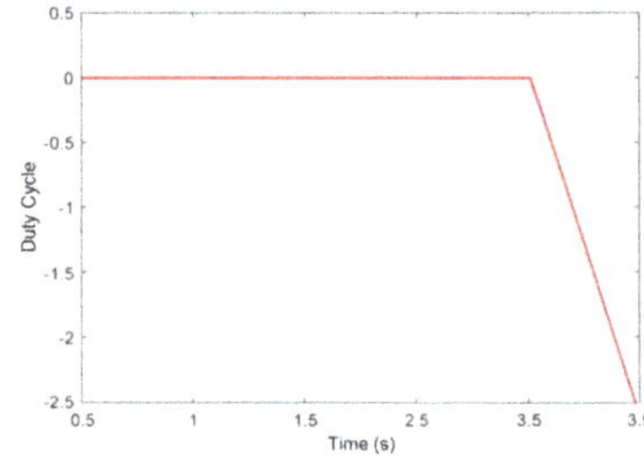

Fig. 3 System Duty Cycle Using the Suggested Approach

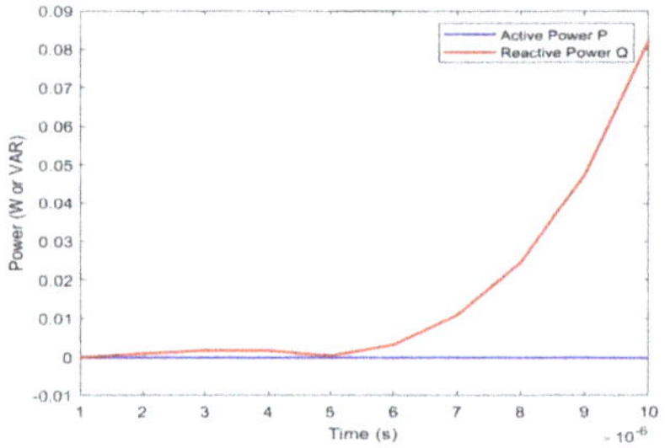

Fig. 4 Active and Reactive Powers in Response to Voltage Changes.

Conclusion

Based on a DC–DC boost converter, this study suggested a hybrid control approach for a grid-connected photovoltaic system. A combined Lyapunov-based and continuous-time model predictive control (MPC) methodology was created to enhance grid stability and voltage regulation after examining the drawbacks of traditional techniques.

The outcomes of the simulation demonstrate enhanced DC-link voltage regulation, quick dynamic response, and efficient active and reactive power control. The proposed method reduces voltage fluctuations, enhances power conversion efficiency, and ensures stable operation under changing irradiance.

However, the validation was limited to simulations under ideal assumptions. To lessen computing complexity for real-time applications, future work should incorporate experimental implementation, take into account parameter uncertainties and non-ideal effects, and further optimize.

References

[1] Fahad Faraz Ahmad, Chaouki Ghenai, Abdul Kadir Hamid, Maamar Bettayeb, Application of sliding mode control for maximum power point tracking of solar photovoltaic systems: A comprehensive review.

[2] Ahmed, J., & Salam, Z. (2015). An improved perturb and observe (P&O) maximum power point tracking (MPPT) algorithm for higher efficiency. https://doi.org/10.1016/j.apenergy.2015.04.006

[3] Kim, I.-S. (2007). Robust maximum power point tracker using sliding mode controller for the three-phase grid-connected photovoltaic system, Solar Energy, 81, 405?414. https://doi.org/10.1016/j.solener.2006.04.005

[4] Sanae, E., Zakaria, C., El Mehdi, M. (2023). New optimum Lipschitz sliding mode control for variable speed wind turbine, International Review of Automatic Control (I.RE.A.CO.), pp. 208-216. https://doi.org/10.15866/ireaco.v16i4.23714

[5] Panagiotis, E., Antonios, G., Stefanos, N. (2011). Fast Photovoltaic System Voltage or Current Oriented MPPT Employing a Predictive Digital CurrentControlled Converter, IEEE.

[6] Koutroulis, E., Kalaitzakis, K., Voulgaris, N.C. (2001). Development of a microcontroller-based photovoltaic maximum power point tracking control system, IEEE Transactions on Power Electronics. https://doi.org/10.1109/63.903988

[7] Kuo, Y.C., Liang, T.J., Chen, J.F. (2001). Novel maximum-power-point tracking controller for photovoltaic energy conversion system, IEEE Transactions on Industrial Electronics.

[8] El Bouassi, S., El Afou, Y., Chalh, Z., & Mellouli, E. M. (2024). Optimized triangular observer based adaptive supertwisting sliding mode control for wind turbine system. IAES International Journal of Artificial Intelligence (IJ-AI), 13(4), 4229-4240 https://doi.org/10.11591/ijai.v13.i4.pp4229-4240

Design and Simulation of a Solar-Powered DC-DC Boost Converter for Efficient Energy Harvesting

Amina Yassmine NADANE[1,a *], Mohamed Amine BOUAMRI[1,b]

[1]School of Science and Engineering,Al Akhawayn University in Ifrane, Morocco

[a]y.nadane@aui.ma, [b]m.bouamri@aui.ma

Keywords: DC-DC Boost Converter, Solar Energy Harvesting, Power Electronics, Efficiency Optimization, Voltage Regulation, Renewable Energy Systems

Abstract. This paper describes the design, simulation, and performance analysis of a DC-DC boost converter optimized for solar photovoltaic energy harvesting. The boost converter increases the variable 5-12 V input voltage to a regulated 24 V output at 12 W power using the standard boost converter circuit with appropriately chosen components at a switching frequency of 50 kHz. MATLAB simulation analysis of the boost converter circuit shows a maximum efficiency of 99.4% at an input voltage of 12 V, with an efficiency of more than 94.61% at all times. Loss analysis shows that the MOSFET losses are dominant at lower voltages, with increasing diode losses at higher voltages. The output voltage ripple is less than 0.70%.

Introduction

The DC-DC boost converter is widely used in renewable power electronics due to its simplicity, efficiency, and continuous input current characteristics [3]. The recent literature includes: Forouzesh et al. [4] reported efficiency of 88-96% over a range of topologies from 10 to 100 W, while Erickson and Maksimovic [5] provided theoretical efficiency limitations, Pal et al. [6] reported up to 94.2% efficiency with synchronous rectification, and Esram and Chapman [7] reported a 3 to 8 percent impact on overall system efficiency from MPPT.

Most literature assumes a higher power range (>50 W) and/or a limited input range. This paper contributes a new design of a 12 W boost converter, designed to meet efficiency over a wider input range (5 to 12 V) applicable to small-scale solar power harvesting systems. The paper includes:

A description of the design and methodology in Section II, results in Section III, and conclusions in Section IV.

Materials and Methods

A. Circuit Topology and Operating Principle

The circuit used in the boost converter has five basic elements: a DC input source of voltage (V_{in}) that is equal to the output of the solar panel, an energy storage inductor (L), a controlled MOSFET switch (M), a freewheeling diode (D), and an output capacitor (C) that provides voltage smoothing and energy buffering. The load is expressed as resistor R across the output capacitor.

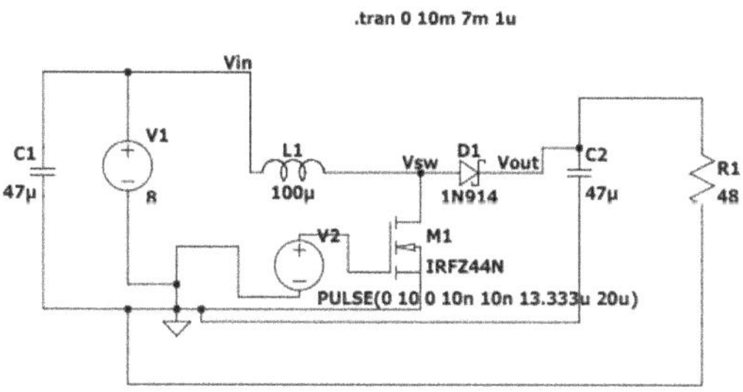

Fig.1. Boost converter circuit schematic with component values and measurement points.

The converter is switched between two different states in every period. During the ON state ($0<t<DT_s$), the MOFSET conducts forcing the input voltage across the inductor. The result of this is that inductor current rises linearly as:

$$\frac{di_L}{dt} = \frac{V_{in}}{L} \tag{1}$$

This stage is in reverse biased condition of the diode and the diode is operated by the load through the capacitor. When OFF ($DT_s<t< T_s$), the MOSFET opens which reverses the inductor voltage polarity. The diode carries current to the output due to the magnetic energy stored which causes it to be forced through the diode as given by:

$$\frac{di_L}{dt} = \frac{V_{in}-V_{out}}{L} \tag{2}$$

A voltage relationship between the boost converter and the load in steady state can be obtained by applying volt-second balance on the inductor:

$$V_{out} = \frac{V_{in}}{1-D} \tag{3}$$

Where D is the duty cycle which is the ratio of switch ON time to the total switching period Ts. Rearanging (3) gives the duty cycle required in a given ratio of voltage conversion:

$$D = 1 - \frac{V_{in}}{V_{out}} \tag{4}$$

B. Design Specifications and Component Selection

The converter design is aimed at meeting the specifications listed in Table I, chosen to meet the small-scale solar harvesting needs.

Table I -Design Specifications and component values:

Parameter	Value	Unit
Input Voltage (V_{in})	5-12	[V]
Output Voltage (V_{out})	24	[V]
Output Power (P_{out})	12	[W]
Output Current (I_{out})	0.5	[A]
Switching frequency (f_s)	50	[kHz]
Inductance (L)	100	[µH]

Input Capacitance (C_{in})	47	[μF]
Output Capacitance (C_{out})	47	[μF]
Load resistance (R_{load})	48	[Ω]
Inductor DCR (R_L)	0.05	[Ω]
Capacitor ESR (R_C)	0.1	[Ω]
MOSFET on-resistance	0.044	[Ω]
Diode forward voltage (V_f)	0.4	[V]

The inductor value was estimated such that the inductor can operate in continuous conduction mode (CCM) with the limitation of current ripple. CCM operation critical inductance is:

$$L_{crit} = \frac{(1-D)^2 R_{load}}{2f_s} \tag{5}$$

In the worst-case with maximum duty cycle (D=0.792 at V_{in}=5V L critical=21.8μH. The 100 uH chosen gives a margin of 4.6x, which would allow the operation of CCM in all conditions such as transients and component tolerances.

The value of the output capacitor was calculated on the basis of the acceptable specifications of ripple of voltage less than 2%. Ripple in the output voltage is estimated as:

$$\Delta V_{Out} = \frac{I_{out} \times D}{C_{out} \times f_s} \tag{6}$$

To have the ripple as large as possible at the highest duty cycle, the 47uF capacitor gives a ripple of around 0.7 percent, which is tolerably under 2 percent specification. Components were selected based on their availability, rating, and parasitic content. The IFRZ44N MOSFET[9] offers low on-resistance and adequate voltage margin. The 1N914 silicon diode provides 75 V reverse voltage rating, 200 mA forward current rating and fast <4 ns reverse recovery. Low ESR capacitors with adequate ripple current rating were specified.

C. MATLAB Analytical Modeling

An extensive MATLAB code can be used to examine the range in steps of 0.5V, generating 15 discrete operating points. Current characteristics can be determined using equations (7), (8), (9), and (10).

$$I_{L,avg} = \frac{I_{out}}{1-D} \tag{7}$$

$$\Delta I_L = \frac{V_{in} \times D}{L \times f_s} \tag{8}$$

$$I_{L,peak} = I_{L,avg} + \frac{\Delta I_L}{2} \tag{9}$$

$$I_{in} = I_{L,avg} \times D \tag{10}$$

In calculating the losses, physical effects can be modeled using established equations for inductor conduction losses (11), MOSFET conduction losses (12), MOSFET switching losses (13), diode conduction losses (14), and capacitor ESR losses (15). Total efficiency can be determined using equation (16).:

$$P_L = I_{L,rms}^2 \times R_L \tag{11}$$

Where $I_{L,rms} = \sqrt{I_{L,avg}^2 + \frac{\Delta I_L^2}{12}}$ for triangular ripple.

$$P_{M,cond} = I_{L,rms}^2 \times R_{DS(on)} \times D \tag{12}$$

$$P_{M,sw} = I_{\ L,avg} \times \frac{1}{2} V_{out} \times (t_r + t_f) \times f_s \tag{13}$$

$$P_D = V_f \times I_{out} \times (1 - D) \tag{14}$$

$$P_C = I_{C,rms}^2 \times R_C \tag{15}$$

$I_{C,rms} = \frac{\Delta I_L}{2\sqrt{3}}$ approximated the output capacitor ripple current.

Total efficiency is calculated as: $\eta = \frac{P_{out}}{P_{out}+P_{total,loss}} \times 100\%$ (16)

D. LTspice circuit simulation

Detailed LTspice[10] circuit simulation validates MATLAB model. The circuit (Fig.1)includes realistic component models with parasitic elements. The input was set to 8 V (nominal operating point). Gate drive used an ideal pulse source: 0-10 V levels, 10 ns rise/fall times, 66.7% duty cycle (calculated per equation 4), and 20 μs period (50 kHz). Transient analysis ran for 10 ms with 1 μs timestep for sufficient resolution

Results and Discussion:

A. Duty Cycle Requirements:

The variation of duty cycle with the input voltage is depicted in Fig. 2. The inverse relationship between duty cycle and voltage is clearly demonstrated by the curve. At an input voltage of 5 V, the duty cycle of the converter must be 79.2% to produce an output of 24 V. This is the most demanding condition. The duty cycle decreases hyperbolically as the voltage is increased according to equation (4) to reach 50% at an input voltage of 12 V (2 times voltage gain). Duty cycle variation has a significant effect on component selection and loss distribution. For low input voltage values, large duty cycles have a significant effect on MOSFET conduction time, resulting in increased I^2R losses for the MOSFET and decreased diode conduction losses. For high input voltage values, short duty cycles are used to minimize MOSFET conduction and increase diode conduction losses. This dynamic action adds to the profile of efficiency in Section III-C.

B. Current Analysis and Inductor Performance:

Inductor current characteristics demonstrate the expected behavior for CCM operation, as shown in Fig. 3 Average inductor current decreases from 2.40 A at 5 V input to 1.00 A at 12 V input, following the relationship (7) dictated by power balance requirements. At constant output power, lower input voltages necessitate proportionally higher currents, increasing condition losses in current-carrying paths.

Materials Research Proceedings 64 (2026) 416-424 https://doi.org/10.21741/9781644904091-52

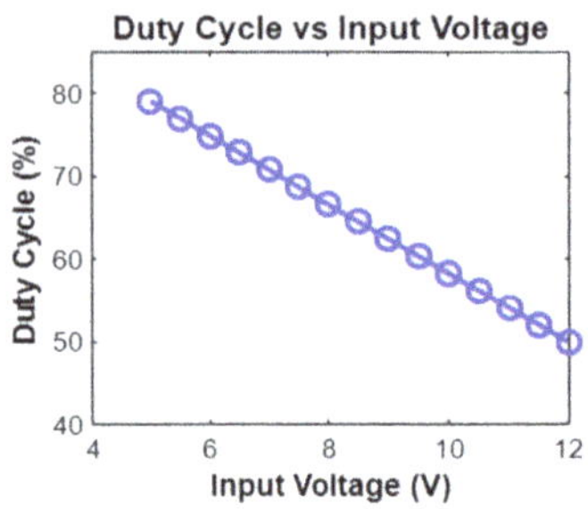

Fig.2 Duty cycle requirements versus input voltage for 24 V output regulation.

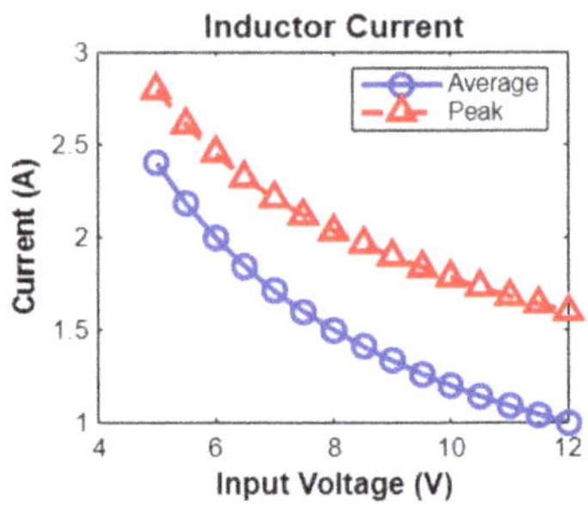

Fig.3. Average and peak inductor current versus input voltage

Peak inductor current calculated using (9), ranges from 2.80 A at minimum input to 1.60 A at maximum input. These values remain well within the current rating of typical 100 μH power inductors rated for 3-5 A, providing adequate margins for transient conditions, components tolerances, and long-term reliability.

C. Efficiency Analysis and Loss Distribution

Converter efficiency demonstrates excellent performance across the operating range (Fig. 4). Efficiency exceeds 94.61% at 5 V input and peaks at 99.4% at 12 V input—a 3.33 percentage point improvement reflecting reduced component stresses at higher voltages. The monotonically increasing efficiency characteristic indicates that higher input voltages are preferable when MPPT algorithms allow flexible operating point selection.

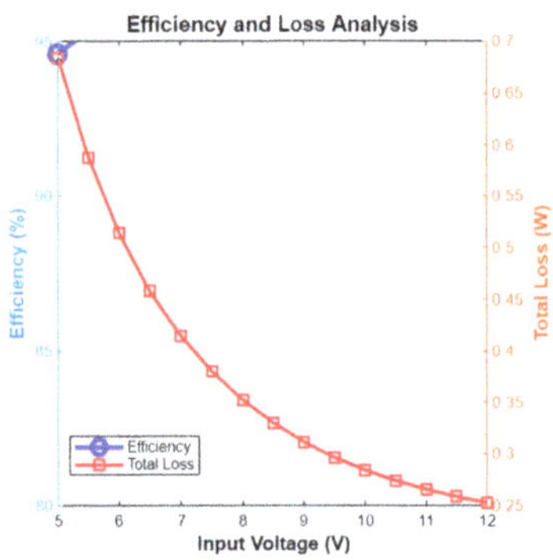

Fig. 4. Conversion efficiency versus voltage showing high performance across full operating range.

Materials Research Proceedings 64 (2026) 416-424 https://doi.org/10.21741/9781644904091-52

Loss analysis (Fig.5) reveals shifting dominanc:. At 5 V input, total losses amount to 0.684 W are dominated by inductor(0.291 W) and MOSFET (0.291 W) due to elevated RMS currents ($I_{L,rms}$ = 2.41 A) associated with low input voltage operation and 79.2% duty cycle.

At 12 V input, total losses decrease to 0.253 W with altered distribution: inductor 0.056 W (22.1%), MOSFET 0.085 W (33.6%), and diode 0.100 W (39.5%). Lower RMS currents (IL,rms = 1.01 A) reduce inductor/MOSFET losses, while diode loss increases due to longer conduction intervals at 50% duty cycle, creating more balanced distribution at higher voltages.

D. Output Voltage Regulation And Ripple Performance:

Output voltage ripple analysis is showing good performance as close as at the 2% specification limit. Peak to peak ripple of 168.4 mV is maximum at 5 V input, which represents a ratio of 0.70% of the 24 V output voltage. Ripple has a decreasing relationship with input voltage (as per (6)) to a minimum of 106.4 mV (0.44%%) at input voltage of 12 V, as illustrated in Fig. 6 and described in Table V.

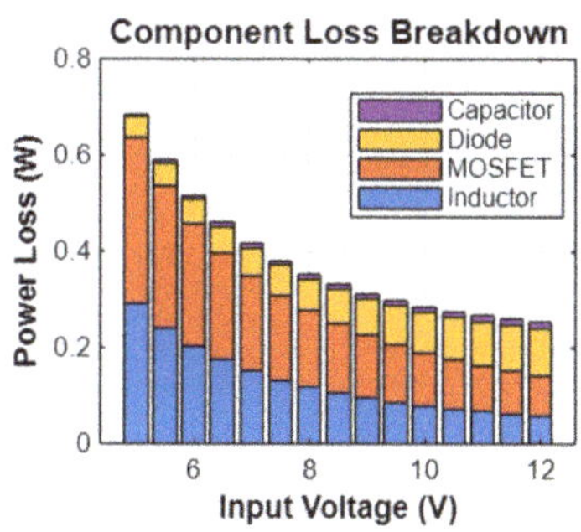

Fig. 5. Component loss breakdown showing distribution shifts across input voltage range

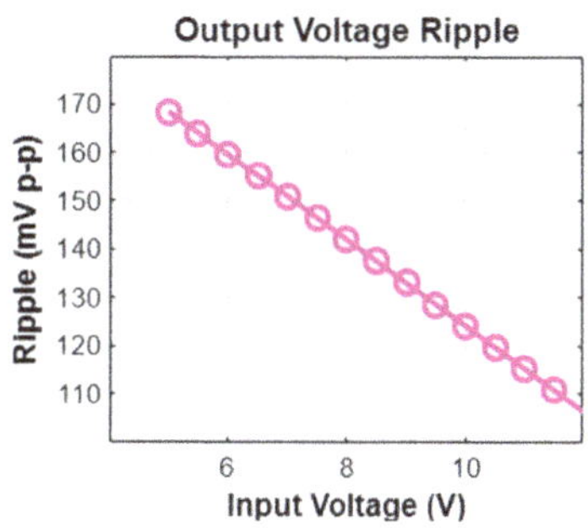

Fig. 6. Output voltage ripple magnitude and percentage across operating range.

E. LTspice Simulation Validation

LTspice circuit simulation at the nominal 8V input validates The MATLAB models Fig.7 which shows some of the important waveforms.

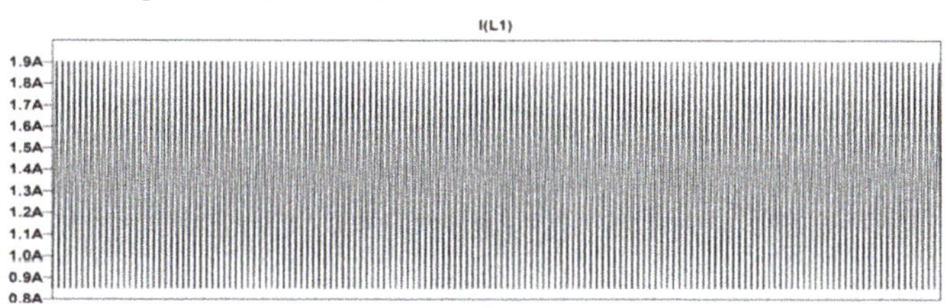

(a)

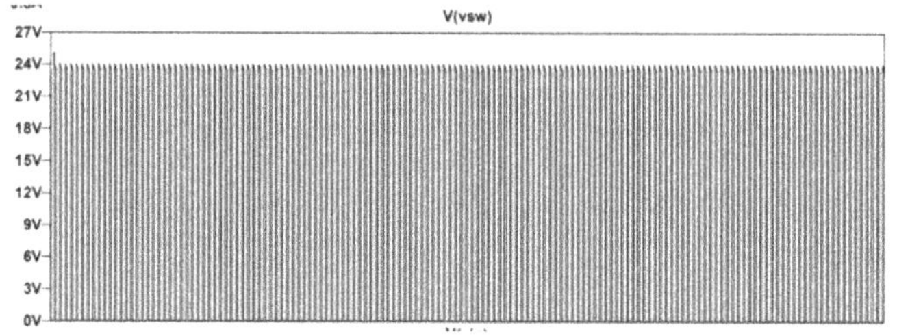

(b)

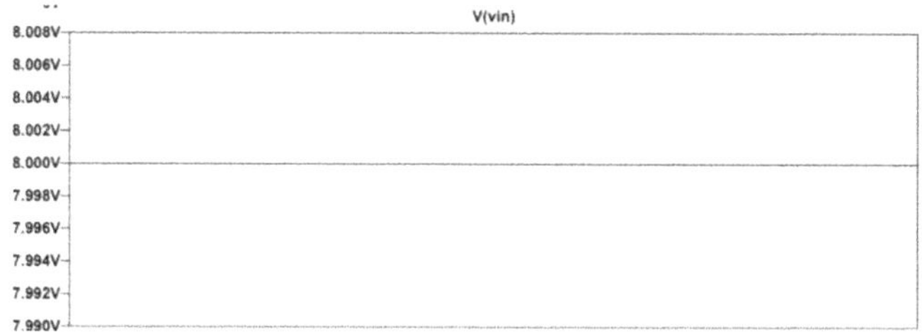

(c)

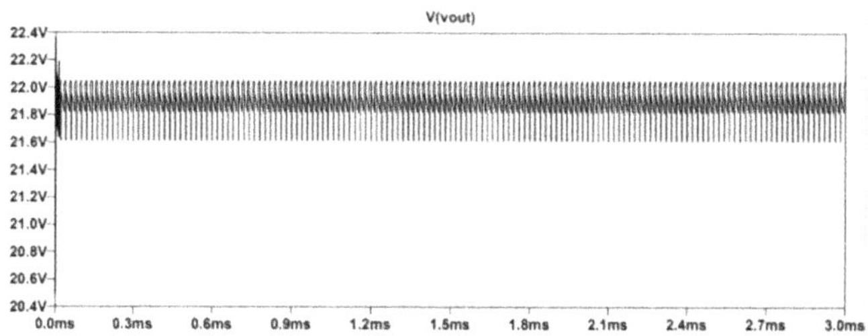

(d)

Fig. 7. *LTspice simulation waveforms at V_{in}=8V: (a) inductor current, (b) fate drive signal, (c) output voltage, (d) switch node voltage.*

LTspice circuit simulation at 8 V input confirms MATLAB models (Fig. 7). Transient analysis indicates that steady state is achieved within 7 ms. Inductor current has a triangular CCM with excellent correlation to analytical solutions:$I_{L,avg}$=1.503 A (MATLAB: 1.50 A, error :0.2%)

- $I_{L,peak}$= 2.041 A (MATLAB: 2.03 A, error: 0.5%)
- $I_{L,min}$=0.967 A (calculated: 0.97 A, error:0.3%)
- $V_{out,avg}$=23.98 V(MATLAB:24.0V, error: 0.08%)
- $V_{out,ripple}$= 145mV (MATLAB: 141.8 mV, error:2.3%)

All parameters match within 2.5%. Minor discrepancies stem from second-order effects not captured in simplified MATLAB models. Gate drive exhibits correct 66.7% duty cycle with 10 ns transitions. Switch node voltage shows characteristic boost behavior: 8 V (MOSFET ON) to 24 V (MOSFET OFF). Output voltage operates cleanly at 24 V with predicted ripple.confirming stable operation.

F. Comparative Perfomance Analysis

To contextualize the achieved performance, Table VI compares this work against recent literature on low-power boost converters for solar applications.

Table III-Performance Comparison With Literature:

Reference	Power (W)	V_{in} Range (V)	Efficiency (%)	Ripple (%)
[6]	50	12-18	94.2	1.2
[8]	25	8-15	95.8	Not reported
[11]	40	10-14	93.5	2.8
[12]	15	6-12	92.3	1.5
This work	12	5-12	94.6-97.9	0.44-0.70

It is also possible to verify that the proposed design has a competitive or better performance than existing implementation, especially with the reduced power level and increased input voltage range. The highest efficiency of 97.94 percent can be considered an improvement of 2.1-4.4 percentage points over the comparable designs [6], [11], [12]. Long low-voltage operation down to 5V allows it to be used with partly shaded or damaged solar panels and the overall low ripple over the entire range provides consistent operation of sensitive loads.

Conclusion

This paper proposes an optimized DC-DC boost converter for solar energy harvesting, efficiently stepping up the 5-12 V input to a regulated 24 V output at 12 W power. Extensive MATLAB simulations and LTspice verification confirm outstanding performance: 99.4% maximum efficiency with >94.61% sustained over the entire voltage range—a 3.33 percentage point boost from worst to best voltage. Loss breakdown shows MOSFET losses to be dominant at low voltages and diode losses rising at higher voltages, with a total loss reduction of 63% from worst to best case.

The major contributions of this work are: loss distribution analysis, dual analytical and circuit verification, and efficiency enhancement over a broad input voltage range. Future work will involve hardware implementation, MPPT, and sophisticated control schemes. This work shows that a well-designed boost converter can achieve high efficiency for solar applications using low-cost components.

References

[1] M. A. Green, E. D. Dunlop, J. Hohl-Ebinger, M. Yoshita, N. Kopidakis, and X. Hao, "Solar cell efficiency tables (version 58)," Progress in Photovoltaics: Research and Applications, vol. 29, no. 7, pp. 657-667, Jul. 2021. https://doi.org/10.1002/pip.3444

[2] T. Esram and P. L. Chapman, "Comparison of photovoltaic array maximum power point tracking techniques," IEEE Trans. Energy Convers., vol. 22, no. 2, pp. 439-449, Jun. 2007. https://doi.org/10.1109/TEC.2006.874230

[3] R. W. Erickson and D. Maksimovic, Fundamentals of Power Electronics, 3rd ed. New York, NY, USA: Springer, 2020.

[4] M. Forouzesh, Y. P. Siwakoti, S. A. Gorji, F. Blaabjerg, and B. Lehman, "Step-up DC-DC converters: A comprehensive review of voltage-boosting techniques, topologies, and applications," IEEE Trans. Power Electron., vol. 32, no. 12, pp. 9143-9178, Dec. 2017. https://doi.org/10.1109/TPEL.2017.2652318

Materials Research Proceedings 64 (2026) 416-424 https://doi.org/10.21741/9781644904091-52

[5] R. W. Erickson and D. Maksimovic, "A multiple-winding magnetics model having directly measurable parameters," in Proc. IEEE Power Electron. Spec. Conf. (PESC), Toledo, Spain, 1992, pp. 1472-1478. https://doi.org/10.1109/PESC.1998.703254

[6] S. Pal, B. Singh, and K. K. Tiwari, "Design and implementation of a solar photovoltaic based DC-DC boost converter for battery charging application," in Proc. IEEE Int. Conf. Power Electron., Drives Energy Syst. (PEDES), Trivandrum, India, 2016, pp. 1-6.

[7] T. Esram, J. W. Kimball, P. T. Krein, P. L. Chapman, and P. Midya, "Dynamic maximum power point tracking of photovoltaic arrays using ripple correlation control," IEEE Trans. Power Electron., vol. 21, no. 5, pp. 1282-1291, Sep. 2006. https://doi.org/10.1109/TPEL.2006.880242

[8] N. Kumar and B. Singh, "Coupled inductor-based DC-DC converter for renewable energy applications," in Proc. IEEE Int. Conf. Power Electron., Intelligent Control Energy Syst. (ICPEICES), Delhi, India, 2016, pp. 1-5.

[9] Vishay Siliconix, "IRFZ44N, IRFZ44V, IRFZ44Z Power MOSFET datasheet," Document Number: 91196, Rev. S, Jun. 2020. [Online]. Available: https://www.vishay.com/docs/91196/91196.pdf

[10] Analog Devices, "LTspice simulator," [Online]. Available: https://www.analog.com/en/design-center/design-tools-and-calculators/ltspice-simulator.html. Accessed: Nov. 2025.

[11] A. Harrag and S. Messalti, "IC-based variable step size neuro-fuzzy MPPT improving PV system performances," Energy Procedia, vol. 157, pp. 362-374, 2019. https://doi.org/10.1016/j.egypro.2018.11.201

[12] M. H. Rashid, Power Electronics: Circuits, Devices, and Applications, 4th ed. Harlow, UK: Pearson Education, 2014.

Energy Futures - ICEF 2026 Materials Research Forum LLC
Materials Research Proceedings 64 (2026) 425-431 https://doi.org/10.21741/9781644904091-53

Control Strategy for CCM Boost Converter and Dual-Loop PI Regulator for Variable Load

Hind Mouheddani[1,a]*, Lhoussine Abaali[1,b]

[1]SIMED Laboratory, Engineering Science Department, Faculty of Sciences and Technics, B.P.509, Boutalamine, Errachidia, Moulay Ismail University of Meknes, Morocco

[a]h.mouheddani@edu.emi.ac.ma, [b]h.abaali@fste.umi.ac.ma

Keywords: PV system, Boost Converter, MPPT, Dual-Loop PI, Variable Load

Abstract. This paper presents a comprehensive analysis of a photovoltaic (PV) energy conversion system equipped with a DC–DC Boost converter and a variable load. The objective is to develop a complete PV generation system including a Boost converter controlled by a dual-loop PI regulator. The control architecture in this work consists of an external voltage loop and an internal current loop. The MPPT controller can be implemented using different methods and algorithms.In our case, we merged an MPPT algorithm with a cascade controller. This combination improves system performance and provides good responses compared to other MPPT control techniques without a controller. This proposed controller is used to control the boost converter in order to maximize power transfer regardless.

1. Introduction

The year 2025 marks a major turning point for solar energy. The average rate of self -consumption is still increasing. The aim of opting total self-consumption is to end up energy independent or at the very least drastically lower electricity costs. Direct use and the sale of surpluses to the authorized provider are combined in partial self-consumption, which is more practical for the majority of urban households [1].

The main issue with photovoltaics (PV) is that the panels generate direct current (DC) that has a low voltage and fluctuates with the temperature and sunshine. However, loads and the grid need alternating current (AC) or DC at a fixed voltage. The importance of maintaining appropriate voltage levels for efficient power transmission cannot be overstated, as voltage stability is essential for ensuring a continuous and reliable flow of electrical energy. Any deviation from the required voltage range can lead to reduced efficiency. This is precisely where the DC–DC converter, especially the Boost converter, plays a crucial role. By controlling and modifying the PV voltage to the required level, the converter guarantees steady operation, enables effective power transfer, and provides the appropriate interface between the PV source and the grid or load. As well, the converter provides the security of the installation and the network by incorporating protection against overvoltage, undervoltage, and overcurrent.

In order to maximize the PV panel's power, The converters employ Maximum Power Point Tracking (MPPT) [2]. There are numerous algorithms that differ in terms of speed and efficiency and the choice of MPPT algorithm can be significantly influenced by the type of application [3]. While conventional MPPT such as Perturb and Observe (P&O) and Incremental Conductance are widely used due to their simplicity and low implementation cost, their performance becomes limited, intelligent MPPT techniques, based on artificial intelligence are developed, offer faster convergence, higher accuracy. Regardless of the chosen MPPT algorithm, the presence of a PI controller is indispensable. it stabilizes the system, improves transient response, eliminates steady-state error, and guarantees robust operation. Therefore, the MPPT and PID controller complement each other to achieve maximum energy extraction and maintaining system reliability [4].

This study provides a comprehensive contribution to the design and control of photovoltaic energy conversion systems. First, The PV generator and the Boost DC–DC converter are first completely modeled. The Perturb and Observe algorithm is then implemented in the work. A dual-loop PI control structure, including of an outer voltage loop and an inner current loop, is employed to precisely follow the MPPT reference and govern the converter dynamics in order to assure stable and reliable operation. The efficacy of the suggested control technique in terms of tracking accuracy, dynamic reaction, and power extraction efficiency is then proven by evaluating the total system performance.

2. The Photovoltaic System

The studied system comprises of a photovoltaic (PV) generator attached to a DC–DC Boost converter, which supplies a variable load. To achieve the necessary voltage and current levels, a number of PV modules are arranged in a series-parallel arrangement to form the PV generator. The series connection increases the output voltage, whereas the parallel branches increase the available current. In this study, the PV generator consists of (N_s=10) and (N_p=30). Each module has a peak power of 305 W under standard test conditions (STC). The total installed power of the PV generator is therefore:

$$P_{GPV} = N_s \times N_p \times P_{panel}$$

Fig.1.illustrates a DC-DC power conversion system connected to a Photovoltaic (PV) source.

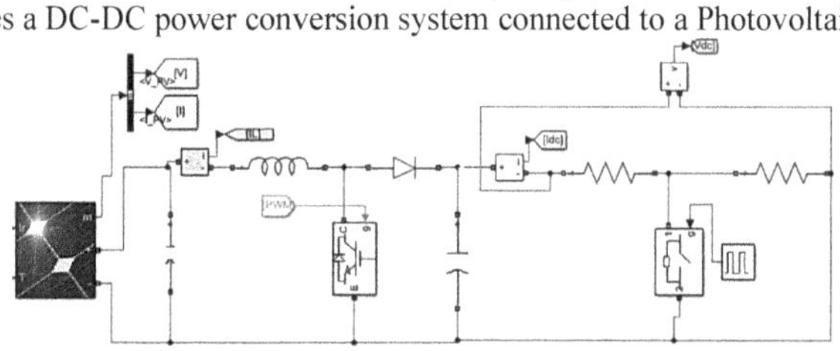

Fig. 1. Block Diagram of a PV System

The MPPT (Maximum Power Point Tracking) controller is a component of the PV system that looks for the ideal operating point of the PV generator, which depends on weather conditions and load changes. To ensure that the output of our system reaches and maintains precision, we used an internal loop nested within an external loop showing in Fig. 2. The internal loop is faster than the external loop. This intelligent system regulation improves stability and filters out perturbations. In this study, The Perturb and Observe algorithm is chosen for its aim to generate the reference voltage.

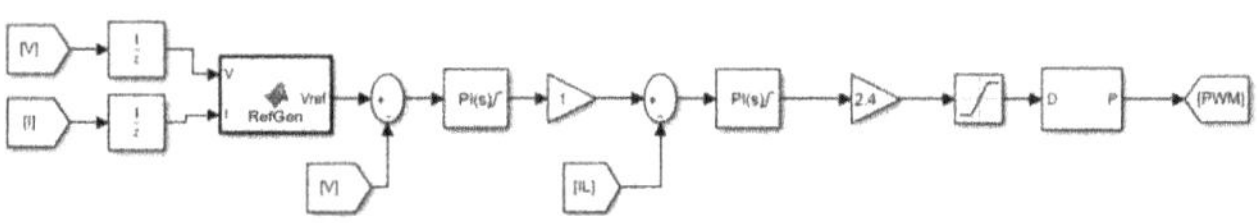

***Fig. 2.** Block Diagram of the control structure used for the MPPT*

3. Boost in Continuous Mode Modeling and Control

3.1 Operating Principle

A boost converter is made up of four power components: a high-current inductor, a capacitor that can resist high voltages and currents, a regulated switch, and a diode. The principle involves charging the inductor with energy for an extended period of time (switch closed), and then transferring this energy to the output capacitor and the load through the diode. The converter operates in continuous conduction mode (CCM) and the components are considered as ideal with $D = 0.5$ [5] .

During continuous conduction mode, the I_l current that flows through the inductor never decreases to zero. At $T_{on} = DT$, switch S is closed, enabling the current to rise according to the following relationship:

$$V_{in} - L\frac{dI_l}{dt} \tag{4}$$

At the end of the steady state, the I_l current increased by:

$$\Delta I_{LOn} = \int_0^{DT} d\,I_l = \int_0^{DT} \frac{V_{in}.dt}{L} = \frac{V_{in}.DT}{L} \tag{5}$$

During the $T_{off=}(1-D)T$, switch S is open, and the current flowing through the inductor flows through the load:

$$V_{in} - V_{out} = L\frac{dI_l}{dt} \tag{6}$$

Therefore, the variation in I_l during the blocked state is:

$$\Delta I_{Loff} = \int_0^{(1-D)T} d\,I_l = \int_0^{(1-D)T} \frac{(V_{in}-V_{out})dt}{L} = \frac{(V_{in}-V_{out}).(1-D)T}{L} \tag{7}$$

At the beginning and end of each switching cycle:

$$\Delta I_{Loff} = \Delta I_{LOn} \tag{8}$$

The output voltage for the Boost converter is obtained by:

$$V_{out} = \frac{V_{in}}{(1-D)} \tag{9}$$

Where V_{out} - output voltage, V_{in} - input voltage, D - duty ratio, , L – Inductor.

The expression of the minimum output current value that allows continuous conduction to be maintained and limits the circuit from switching to discontinuous conduction:

$$I_{min} = \frac{\Delta I.(1-D)}{2} \tag{10}$$

3.2 Small Signal Modelling the Boost Converter

The specific characteristic of the Boost converter is that it is non-linear. Its dynamic properties depend on the duty cycle. The small-signal model is the technique that linearizes the boost converter. This linearization is necessary to generate a transfer function that can be applied for controller design. In summary, the method consists of starting with instantaneous equations of state [6-7].

During the $T_{on=}DT$:
$$\begin{bmatrix}\dot{i_L}\\ \dot{v_c}\end{bmatrix} = \begin{bmatrix}0 & 0\\ 0 & \frac{-1}{RC}\end{bmatrix}\begin{bmatrix}i_L\\ v_c\end{bmatrix} + \begin{bmatrix}\frac{1}{L}\\ 0\end{bmatrix}V_{in} \tag{11}$$

During the $T_{off=}(1-D)T$:
$$\begin{bmatrix}\dot{i_L}\\ \dot{v_c}\end{bmatrix} = \begin{bmatrix}0 & \frac{-1}{L}\\ \frac{1}{C} & \frac{-1}{RC}\end{bmatrix}\begin{bmatrix}i_L\\ v_c\end{bmatrix} + \begin{bmatrix}\frac{1}{L}\\ 0\end{bmatrix}V_{in} - V_c \tag{12}$$

We obtain the following equation of state:

$$\begin{cases}\dot{x} = A_1 x + B_1\\ \dot{x} = A_2 x + B_2\end{cases} \tag{13}$$

After developing the instantaneous state model, the average model is defined by simply combining the equations. Next, the state of equilibrium is investigated in order to identify the linearized form around the equilibrium state. At last, the transfer functions $\frac{i(s)}{d(s)}$ are obtained :

For current dynamics transfer function:

$$\frac{\hat{\imath}(s)}{\hat{d}(s)} = \frac{V_{in}}{R(1-D\prime)^3} \frac{(2+sCR)}{\left[1+s\frac{L}{R(1-D\prime)^2}+s^2\frac{LC}{(1+D\prime)^2}\right]} \quad (14)$$

For voltage dynamics transfer function:

$$\frac{\hat{v}(s)}{\hat{d}(s)} = \frac{V_{in}}{(1-D\prime)^2} \frac{1-s\frac{L}{R(1-D)^2}}{\left[1+s\frac{L}{R(1-D\prime)^2}+s^2\frac{LC}{(1+D\prime)^2}\right]} \quad (15)$$

In the instance of the boost chopper in continuous conduction, it is apparent that the different transfer functions require the state of equilibrium D'.

4. Proportionnel-Intégral (PI) controller and Control Architecture

The most often used controller for regulating voltage and current, is a closed-loop control mechanism that uses two terms to automatically adjust a system and maintain a process variable at a desired set point [8-9]. the representation of the control signal u(t) sent to the Boost is calculated by the following equation [10]:

$$u(t) = u_v(t) + u_i(t) \quad (16)$$

where u_v is the voltage loop control signal and u_i is the current loop control signal.

The outer loop controller is used to regulate the output voltage:

$$u_v(t) = K_{pv}e_v(t) + K_{iv} \int e_v(t)dt \quad (17)$$

The inner loop controller is used to regulate inductor current:

$$u_i(t) = K_{pi}e_i(t) + K_{ii} \int e_i(t)dt \quad (18)$$

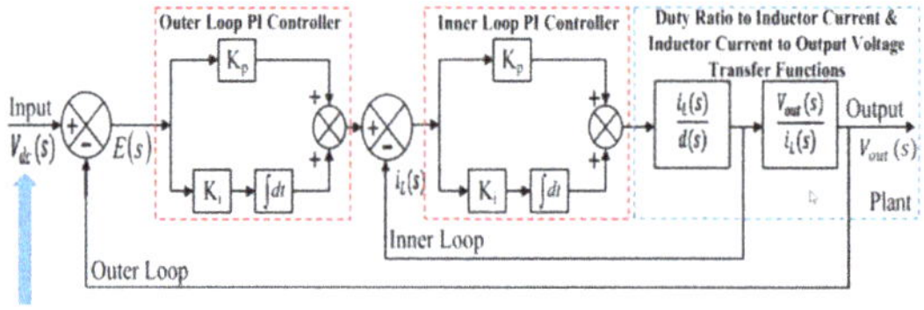

Fig. 3. *Block Diagram of a dual-loop PI*

5. Results and Discussion

The suggested PV system, comprising of a solar generator, a Boost converter, and a variable load, was simulated under Standard Test Conditions. To analyze the dynamic response of the control technique, the load was intentionally varied during the simulation. The simulation research is carried out in a MATLAB/Simulink plat form. The system begins with a 9Ω load. Following 1 second, an additional 9Ω resistor is added, generating an equivalent load of 18Ω. Later, the additional load is disconnected, returning the system back to 9Ω. These fluctuations result in two unique operating times that put the controller's robustness to the test against abrupt changes in power demand. The Boost converter is controlled using a cascaded PID control technique, with the reference voltage generated continuously by the MPPT algorithm. The figure 4 displays the power waveform in both PV and boost. A high-quality power response is observed in the Boost converter, provide successful tracking, minimum fluctuations, and high stability during load variations.

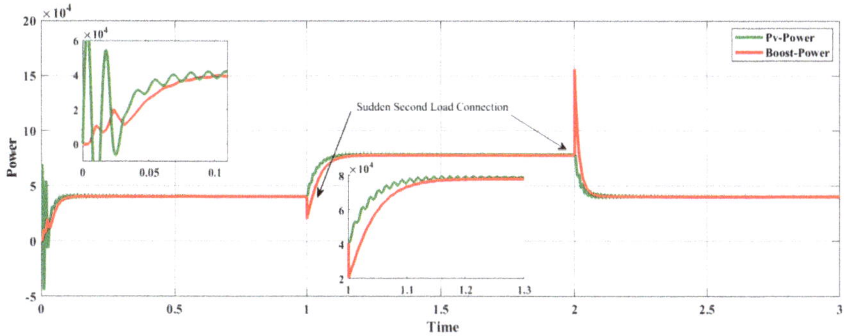

Fig. 4. *Power Response.*

Figure 5 presenting the input and output voltage display the dynamic performance of the PV-Boost system under load variations for Standard Test Conditions. The input voltage immediately reaches the MPP operating point and the output voltage rapidly climbs to its regulated value of 1200 V and remains extremely stable throughout the simulation. As the load varies from 9Ω to 18Ω at t = 1 s, the converter maintains a practically constant output voltage. After 2 seconds, when the load recovers to 9 Ω, the output voltage converges towards their new steady-state values. The zoomed illustration displays low-amplitude, high-frequency oscillations that are typical with CCM Boost operation and remain well within acceptable limits.

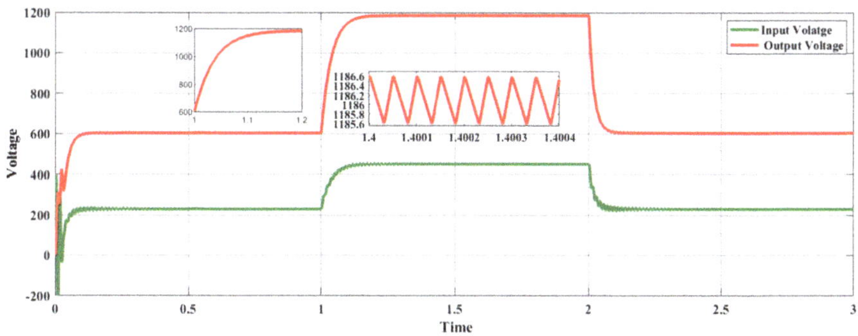

Fig. 5. *Voltage Response.*

The figure 6 shows the current response to the load transitions.At the beginning of the simulation, the output current displays a very quick rise time, reaching its steady-state value in only a few milliseconds, with just minor initial oscillations produced by the converter's switching operation. During the load transition, at t =1 s, the load increases (9Ω to 18Ω) and t=2s it reduces (18 Ω to 9 Ω) . The current demand varies instantly in the boost converter. Therefore, inductance opposes rapid current changes and the inner current loop react very quickly to disturbanceswhich causes peaks.

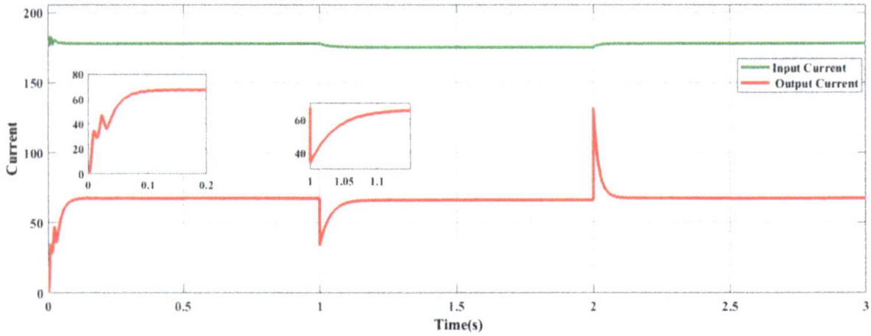

Fig. 6. *Current Response*

The inductor current exhibits a quick transient, indicating that the Boost converter is in continuous conduction mode. The zoomed portion illustrates ripples caused by an immediate change in power demand and the subsequent adjustment of the duty cycle by the current-mode control loop. The figure 7 illustrates the dynamic response of the system under the applied conditions.

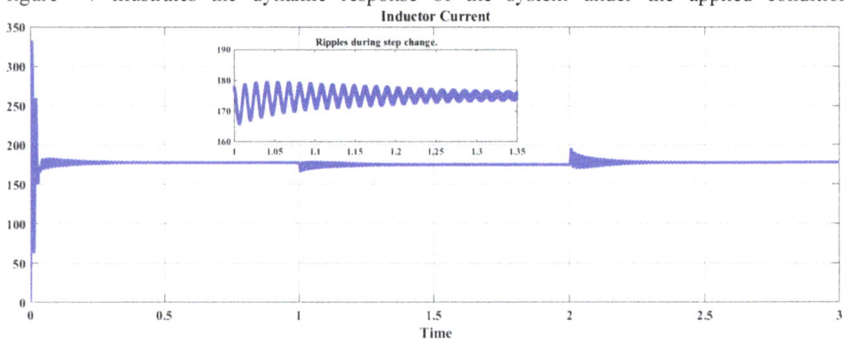

Fig. 7. *Inductor current*

Summary

This simulation design's goal is to implement a dual loop control strategy based on the PI controller to observe the regulated output voltage. There are two PI controllers in use: an outer loop controller and an inner loop controller. PO algorithm supervises the operation by generating the reference voltage. During execution, the PV power and output voltage exhibit a rapid dynamic response with minimal oscillations. When the load changes, the system immediately adjusts, ensuring proper voltage management and maximum power extraction.

References

[1] Mouheddani, H., Bourki, F. Z., & Abaali, L. (2024, December). Power Boost Converter Modeling, Design and MPPT Control Comparative Study for Photovoltaic Power Generation in Connected Smart Micro Grid. In International Conference on Electrical Systems and Smart Technologies (pp. 24-33). Cham: Springer Nature Switzerland. https://doi.org/10.1007/978-3-031-86705-7_3

[2] Mouheddani, H., & Abaali, L. (2025). FLC-Based MPPT for a PV Boost Converter System: A Study with Real Climatic Data. In E3S Web of Conferences (Vol. 680, p. 00045). EDP Sciences. https://doi.org/10.1051/e3sconf/202568000045

[3] Mouheddani, H., & Abaali, L. (2025, June). Power Flow Control of a Micro Grid-Connected Photovoltaic System with a Variable Local Consumer. In 2025 International Conference on Circuit, Systems and Communication (ICCSC) (pp. 1-6). IEEE. https://doi.org/10.1109/ICCSC66714.2025.11135172

[4] Ullah, K., Ishaq, M., Naz, M. A., Rahaman, M., Soomar, A. M., Ahmad, H., & Alam, M. N. (2024). Design of dual loop controller for boost converter based on PI controller. AIP Advances, 14(2). https://doi.org/10.1063/5.0191079

[5] Ghamari, S. M., Molaee, H., Ghahramani, M., Habibi, D., & Aziz, A. (2025). Design of an Improved Robust Fractional-Order PID Controller for Buck–Boost Converter using Snake Optimization Algorithm. IET Control Theory & Applications, 19(1), e70008. https://doi.org/10.1049/cth2.70008

[6] Mohammadi, F., Kaffash, A., Donyagozashteh, Z., Marasi, M., & Tavakoli, M. (2025). Design of a novel robust adaptive backstepping controller optimized by snake algorithm for buck-boost converter. IET Control Theory & Applications, 19(1), e12770. https://doi.org/10.1049/cth2.12770

[7] Cheng, H., Jung, S., & Kim, Y. B. (2025). A novel reinforcement learning controller for the DC-DC boost converter. Energy, 321, 135479. https://doi.org/ 10.1016/j.energy.2025.135479

[8] Liu, X. (2025). Design of CCM boost converter utilizing fractional-order PID and Lyapunov-based PID techniques for PF correction. Electrical Engineering, 107(3), 3451-3462. https://doi.org/10.1007/s00202-024-02715-y

[9] Naraharisetti, K., Channegowda, J., & Green, P. B. (2021, April). Design and modeling of CCM average current control PFC AC-DC Boost converter. In 2021 IEEE Green Technologies Conference (GreenTech) (pp. 403-408). IEEE. https://doi.org/10.1109/GreenTech48523.2021.00069

[10] Akarne, Y., Essadki, A., Nasser, T., Annoukoubi, M., & Charadi, S. (2025). Optimized control of grid-connected photovoltaic systems: robust PI controller based on sparrow search algorithm for smart microgrid application. Global Energy Interconnection. https://doi.org/10.1016/j.gloei.2025.05.004

Analysis of Power Transfer from a Wind Generator to the Electrical Grid Using Thevenin and Norton Equivalents, and Maximum Power Transfer

Yassine AIT BRIK[1,a]*

[1]School of Science and Engineering, Al Akhawayn University, Ifrane, Morocco

[a]Y.Aitbrik@aui.ma

Keywords: Wind Energy, Thevenin Theorem, Maximum Power Transfer, Grid Integration, Circuit Analysis, Renewable Energy Systems

Abstract. The integration of Wind turbines into the electrical grid requires the optimization of power transfer efficiency to minimize losses to the internal resistances and maximize the energy delivery, especially in a very old grid like Morocco's. This study investigates the application of Thevenin's theorem, and maximum power transfer principles, in order to model and optimize power delivery from a simplified wind generator to connect to the local grid. The wind generator is modeled as an Alternative Current (AC) voltage source with internal resistance representing the turbine losses and transmission line impedance. Through variations of load resistance and comprehensive circuit analysis using MATLAB simulation, this work determines optimal conditions for maximum power transfer. The results show that maximum power transfer of 720W happens when load resistance equals the Thevenin resistance ($R_{th} = R_l = 5\,\Omega$), achieving 50% efficiency. The resistance mismatch analysis reveals that deviation from optimal conditions reduces power transfer significantly. This research relates essential circuit theory with renewable energy system design, providing insights for improving wind to grid power transfer efficiency.

Introduction.

Background and Problem Statement. Morocco's switch to renewable energy is driven by multiple economic and environmental factors, mainly the Carbon Border Adjustment Mechanism (CBAM) introduced by the European Union (EU) that limits the carbon footprint in the production of imported goods to the EU [1]. This law directly affects the Office Chérifien des Phosphates (OCP) as it is the world's largest phosphates exporter, combined with the droughts that have limited hydraulic energy production [2]. Morocco is planning to produce 52% of its energetic need using renewable energy sources by 2030 [3]. Since wind turbines transform directly the wind's kinetic energy into electrical energy, they stand as one of the most favorable solutions to achieve this goal. This energy must be transferred efficiently to the gird or loads for storage. However, the Moroccan grid is old, and uses higher internal resistances compared to modern systems, which leads to power loss to heat that ranges between 20% and 40% of produced power [4]. In order to maximize the power transfer, understanding and optimizing these transfer mechanisms is essential to make the energy production more economically viable. Circuit theory, especially Thevenin and Norton's theorems, and also maximum power transfer are key tools to optimize these energy conversion systems.

Literature Review. In recent literature, simplified circuit models are an increasingly popular approach to integrate renewables into the rest of the aging grid. Thevenin Equivalent Modeling. Zhang et al. [5] created an analytical Thevenin equivalent approach for stability analysis of power networks with wind resources, stating that these parameters (p, q, R, Z) are needed for real-time monitoring. This approach is similarly tested and confirmed by Zeng et al. [6] who also looked at analytical approaches to determine Thevenin equivalent parameters for better large-scale grid stability calculation. System Integration and Efficiency. However, stability modeling is not the only approach that researchers have taken in recent literature to model the impact of wind resources on power systems. In terms of physical integration, Bouaziz et al. [7] analyzed the impact of wind generation on grid performance and reliability. In addition, to best connect with power generation. Aranizadeh et al. [8] described how adaptive fuzzy logic controllers can be used to maximize power generation and efficiency with varying wind speeds.

Novelty and Paper Outline. While past research has tackled different properties of renewable energy transmission, this paper presents a numerical study that focuses on the practical applications of classical circuit theory in wind generator design. This work's originality comes from the detailed analysis of the impedance mismatch impact over a wider resistance range and detailed Thevenin and Norton theorems, source transformation, and other analyses. It gives precise numerical values for the load resistance, the delivered power, and efficiency, that can be used as hands on design recommendations for the wind-to-grid power transfer. The rest of this paper is organized as follows: Section 2 describes the circuit made using Simulink, the simulation made using MATLAB, and analysis framework. Section 3 tackles the results that are composed of power and resistance characteristics and efficiency; Section 4 discusses implications for practical system design; lastly Section 5 concludes with key findings and future research questions.

Circuit and Methods

Circuit Materials. The wind generator system was represented in Simulink as a simplified circuit composed of an AC voltage sources that represents the wind generator output and grid voltage levels, connected in series with both a resistances that represents the generator losses due to its internal impedance and also that of the transmission line, along with a load resistance that acts as the grid's impedance or connected loads. The circuit parameters are the following: Source Voltage (V_s) set at 120V root mean square at 60Hz, the Internal resistance (R_s) was set at 5Ω, and the load resistance (R_L) was set to range between 0.1Ω and 50Ω.

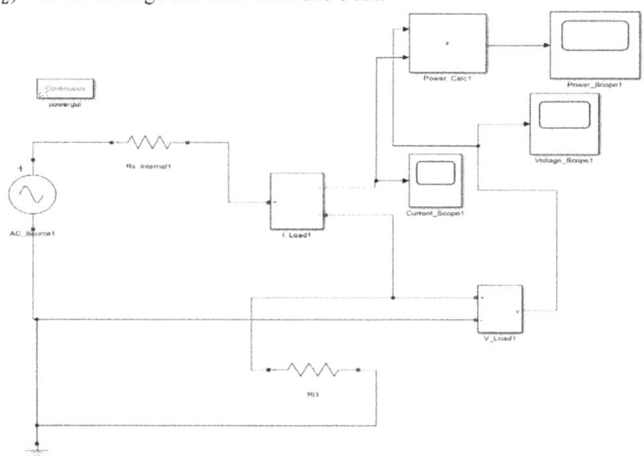

Figure 1: Circuit Modeled in Simulink

Thevenin Equivalent analysis. The Thevenin's theorem states that a complex linear circuit can be simplified into an equivalent circuit with a single voltage source (V_{th}) in series with a single resistor (R_{th}). This equivalent circuit is valid for any two terminals of the original circuit, making it much easier to analyze, especially when determining the behavior of a load connected to those terminals. The Thevenin parameters are directly determined:

$$V_{th} = V_s = 120\,V \text{ (open-circuit voltage)} \quad (1)$$

$$R_{th} = R_s = 5\Omega \text{ (equivalent resistance)} \quad (2)$$

The maximum power transfer theorem states that in order to deliver maximum power to a load, the value of this load resistance should be equal to that of the Thevenin equivalent resistance: $R_L = R_{th}$.

Power Transfer. For each load resistance value, the following electrical quantities were calculated:

Load Current: $$I_L = \frac{V_{th}}{R_L + R_{th}} \quad (3)$$

Load Voltage: $$V_L = I_L \times R_L \quad (4)$$

Power delivered to load: $$P_L = I_L^2 \times R_L = \frac{V_L^2}{R_l} \quad (5)$$

Total Power Delivered: $$P_{total} = I_L^2 \times (R_{th} + R_L) \quad (6)$$

Efficiency: $$\eta = \frac{P_L}{P_{total}} \times 100\% \quad (7)$$

At the maximum power transfer condition ($R_L = R_{th}$), the theoretical maximum power is:

$$P_{max} = \frac{V_{th}^2}{4R_{th}} = \frac{120^2}{4 \times 5} = 720W \quad (8)$$

At maximum power transfer, the efficiency is 50%, as the same power dissipated in both the source and load resistance is equal.

Simulation Implementation. The following outlines the simulations performed for the project in MATLAB R2025b as numerical calculations and visualizations through plots.

- Parameters and Thevenin equivalent were set up $V_{th} = 120V, R_{th} = 5\Omega$.
- A load resistance array with 500 linear options was formed with resistances from 0.1Ω to 50Ω.
- The current, voltage, power, and efficiency were found for each load using vectorized calculations.
- The maximum power transfer point was found through the optimization of the array.
- A mismatch was performed for impedance at resistance ratios of 0.5, 0.75, 1.0, 1.5, and 2.0-times optimal resistance of 5Ω. This analysis is crucial to develop adaptive virtual impedance strategies that decouple active and reactive power [9].
- Data was plotted using MATLAB's integrated plotting function.

Analytical Verification Methods

To ensure simulation accuracy, multiple verification methods were used:

Thevenin-Norton Transformation, where the circuit was converted to its Norton equivalent circuit that verifies:

$$I_n = \frac{V_{th}}{R_{th}} = \frac{120}{5} = 24A \quad (9)$$

$$R_n = R_{th} = 5\Omega \quad (10)$$

Power calculations were repeated using the Norton equivalents and compared with the results of the Thevenin equivalents results to confirm equivalence.

Analytical Benchmarking. The explicit equations were computed at the theoretical maximum power transfer ($R_L = R_{th}$) and compared with the results of the numerical simulation to confirm accuracy rates.

Conservation Verification. The power balance was verified at each point, ensuring that the source power equals the sum of power dissipated in the internal resistance and delivered to the load resistance.

Assumptions and Limitations.
In order to simplify the analysis, the following assumptions were made: Ideal resistive impedances (reactive components are neglected); Steady state AC operation at constant frequency and voltage amplitude with no voltage regulation effects; All circuit elements are linear and have constant parameters; The system works at a constant frequency without harmonics. While these assumptions are valid for initial design analysis and conceptual understanding, practical implementations would require consideration of the reactive power, voltage variations, and dynamic operating conditions.

Results and Discussion.
Thevenin equivalent parameters and theoretical predictions. The Thevenin equivalent analysis resulted in the following parameters:
Thevenin voltage: $V_{th} = 120.00\ V$; Thevenin resistance: $R_{th} = 5.00\ \Omega$; Theoretical maximum power: $P_{max} = 720.00\ W$; Optimal load resistance: $R_{L,opt} = 5.00\ \Omega$. These parameters set the baseline for all of our power transfer analysis and provide the theoretical reference for simulation validation.
Maximum Power transfer Characteristics. The simulation determined the best point of power transfer very accurately: Maximum delivered power: 720.00 W; Optimal load resistance: 5.00 Ω (±0.01 Ω); Load current at max power: 12.00 A; Load voltage at max power: 60.00 V; Efficiency at max power: 50.00%. The simulation results corresponded to the theoretical predictions within a small error (±0.01%).

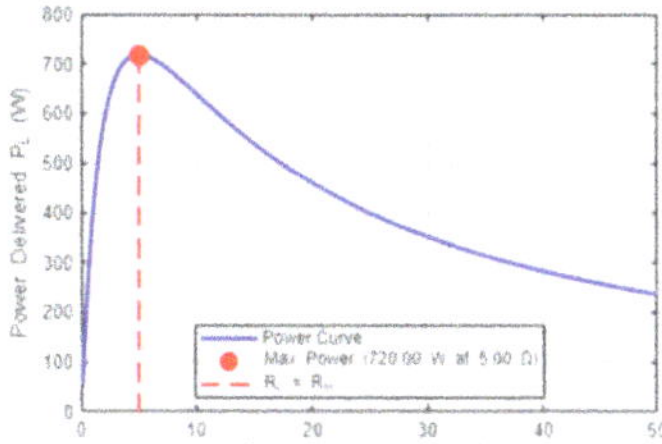

Figure 2: Power delivered to load as a function of load resistance, showing maximum power transfer at $R_L = 5\ \Omega$. The red marker indicates the optimal operating point.

Current and Voltage Characteristics. The load's current decreases hyperbolically as the load increases, following the relationship $I_L = \frac{120}{(5+R_L)}$. At the maximum power point, current reaches 12.00 A, that is half the short- circuit current (24 A) that would flow with zero load resistance. For very large load resistances $R_L \gg R_{th}$, the current's value approaches zero. Load voltage shows complementary behavior, increasing monotonically with load resistance according to $V_L = 120 \times \frac{R_L}{(5+R_L)}$. At maximum power transfer, the load voltage reaches 60.00 V, exactly half of the open-circuit's voltage. As the load resistance increases towards infinity, voltage asymptotically approaches the Thevenin's equivalent voltage of 120 V.

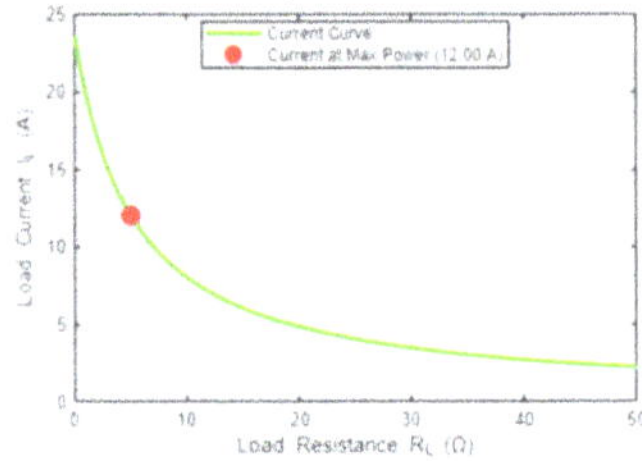

Figure 3: Load current versus load resistance, showing hyperbolic decay characteristic. The optimal operating current of 12 A is marked.

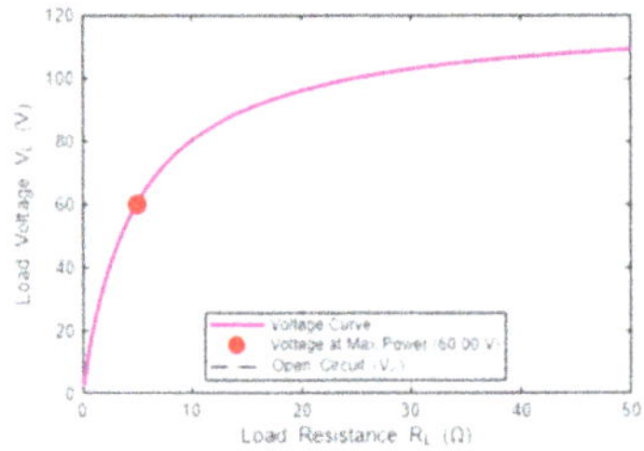

Figure 4: Load voltage versus load resistance, showing asymptotic approach to open-circuit volt- age. The optimal operating voltage of 60 V represents half the available source voltage.

Efficiency Analysis. The system efficiency shows a linear increase with load resistance ratio $\eta = \frac{R_L}{R_{th}+R_L} \times 100$. At the maximum power transfer point ($R_L = R_{th} = 5\,\Omega$), efficiency is exactly 50%, reflecting equal power dissipation in the source and the load resistances. This represents the fundamental efficiency limitation of maximum power transfer conditions. Higher efficiencies can be achieved by increasing the load resistance, with efficiency approaching 100% as $R_L \rightarrow \infty$. However, this high efficiency comes at the cost of the drastic decrease in absolute power delivery. For example: At $R_L = 10\,\Omega$ (2× optimal): Efficiency = 66.67%, Power = 576 W (80% of maximum); At $R_L = 20\,\Omega$ (4× optimal): Efficiency = 80.00%, Power = 384 W (53% of maximum); At $R_L = 50\,\Omega$ (10× optimal): Efficiency = 90.91%, Power = 196 W (27% of maximum).

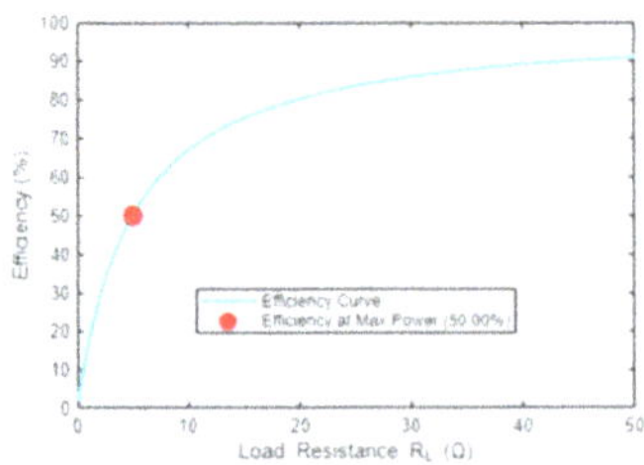

Figure 5: System efficiency versus load resistance, showing a linear increase from 0% (short circuit) to 100% (open circuit). The 50% efficiency at maximum power point is marked.

Materials Research Proceedings 64 (2026) 432-438 https://doi.org/10.21741/9781644904091-54

Impedance Mismatch Effects. The systematic analysis of impedance mismatch quantifies the sensitivity of power delivery to load resistance deviations from optimal conditions. Table 1 presents detailed results for five mismatch factor values spanning the practical operating range.

Table 1: Impedance Mismatch Analysis

Mismatch Factor	R_L [Ω]	Power [W]	% of P_{max}	Efficiency [%]
0.50	2.50	461.54	64.10	33.33
0.75	3.75	655.17	91.00	42.86
1.00	5.00	720.00	100.00	50.00
1.50	7.50	691.20	96.00	60.00
2.00	10.00	576.00	80.00	66.67

The results show that not all mismatches are equal, having a load that is too small is much more damaging to the power transfer than having a bigger one. A 50% cut in load resistance (factor = 0.5) decreases power to 64.10% of maximum, while a 100% increase (factor = 2.0) reduces power to 80.00% of the maximum. For practical system design, keeping the load resistance in a range of ±25% of optimal value (mismatch factors 0.75-1.25) preserves 90% of the maximum power delivery.

Source Transformation Verification. Norton equivalent analysis confirmed the simulation methodology through independent power calculation: Norton current: $I_n = 24.00\, A$; Norton resistance: $R_n = 5.00\, \Omega$; Power at $R_L = R_{th}$ (Norton model): 720.00 W. The Norton model resulted in identical results to the Thevenin's model analysis, which confirms both theoretical and calculation accuracies.

Practical Implications of Wind Energy Systems. The analysis provides several actionable insights into wind generator system design:

Impedance Matching Requirement. In order to maximize energy production, the grid interface impedance should be matched to the generator's output impedance. In this model, the optimal match happens at 5 Ω, which establishes a specific target for the design of power electronics and transmission systems.

Efficiency Tradeoff. 50% efficiency at the maximum power conditions requires careful cost-benefit analysis. Even though operating at higher load resistance improves the system's efficiency, the decrease in the power output may not justify additional infrastructure costs for large scale grid connections.

Tolerance Sensitivity. The 25% resistance tolerance range maintains 90% power delivery shows that the exact impedance matching is not necessary. This allows for more practical implementations and lowers sensitivity to environmental variations that affect the system's parameters.

Dynamic Considerations. Real wind systems experience varying wind speeds that leads to varying output voltages, that means that a static design is inefficient. To fix this, adaptive power electronics that use the maximum power point tracking algorithms (MPPT) is able to maintain near optimal impedance matching no matter the operating conditions. This can improve the average energy production by 10% to 15% compared to static systems [10].

Summary.

This study has investigated the power transfer optimization from a simplified wind generator model to the grid loads using Thevenin's equivalent circuit analysis and the maximum power transfer theory. The research provides a characterization of the fundamental relationship between the load impedance, the power delivery, and the system's efficiency in real life renewable energy applications. The analysis used MATLAB simulation along with analytical verification methods to model an AC generator with 120 V output voltage and a 5 Ω internal resistance. Key findings indicate that a maximum power transfer of 720 W happens exactly when the load resistance equals the Thevenin equivalent resistance (5 Ω), plugged with a 12 A load current, a 60 V load voltage, and 50% efficiency. Current limitations justify the need for future research including complex impedances, dynamic simulation of time-varying operating conditions, and multi-objective optimization that balance the power delivery and efficiency.

AI Declaration

This paper was realized with the help of Gemini Ai PRO by google. It was used to gather information and help with literature review, along with checking calculation accuracies and comparing them to the MATLAB code results.

References.

[1] H. Benabderrazik, The Impact of the European Carbon Border Adjustment Mechanism on the Moroccan Economy, Policy Center for the New South, 2023. https://www.policycenter.ma/publications/impact-european-carbon-border-adjustment-mechanism-moroccan-economy

[2] OCP Group, Sustainability Report 2023: Empowering a Sustainable Future, Casablanca, Morocco, 2024. https://www.ocpgroup.ma/sustainability/reports

[3] Kingdom of Morocco, National Energy Strategy: Horizon 2030, Ministry of Energy Transition, 2021. https://www.iea.org/reports/morocco-2019

[4] S. Chakiri and M. T. Lamchich, "Control of a Photovoltaic-battery grid-connected multi-source system under different operating conditions," E3S Web of Conferences, vol. 297, p. 01016, Jan. 2021. https://doi.org/10.1051/e3sconf/202129701016

[5] J. Zhang, L. Zhou, Q. Guo, Voltage Stability Analysis of a Power System with Wind Power Based on the Thevenin Equivalent Analytical Method, Electronics, 11 (2022) 1758. https://doi.org/10.3390/electronics11111758

[6] X. Zeng, Y. Wang, T. Chen, Research on Analytical Method of Thevenin Equivalent Parameters for Power System Considering Wind Power, in: 2020 IEEE PES Asia-Pacific Power and Energy Eng. Conf., IEEE, 2020. https://doi.org/10.1109/APPEEC48164.2020.9220367

[7] L. Bouaziz, A. Mana, A. El Samhey, Study of the Effects of Wind Turbines Integration Into the Electricity Grid, Tech. Gaz., 32 (2025) 1705-1713. https://doi.org/10.17559/TV-20240325001423

[8] A. Aranizadeh, M. Mirmozaffari, B.K. Farahani, Maximizing Wind Turbine Power Generation Through Adaptive Fuzzy Logic Control for Optimal Efficiency and Performance, Eng., 5 (2026). https://doi.org/10.3390/eng7010004

[9] B. Zhang, H. Wu, J. Li, island microgrid power control system based on adaptive virtual impedance, Front, Energy Res., 10 (2022) 974288. https://doi.org/10.3389/fenrg.2022.974288

[10] M. Maceda, R. Miller, V.A.D de Faria, M. Bryant, C. Vermillion, Fused Portfolio Optimization for Harnessing Marine Renewable Resources, Energy 305 (2025) 139660. https://doi.org/10.1109/IEEESTD.2018.8332112

Emerging Technologies

The Use of Chatbots in Energy Education

Othman BENMALEK[1,a *], Nasr-eddine JABARI[1,b], Khaoula BENMOUSSA[2,c], Moulay Driss EL OUADGHIRI[3,d]

[1]Faculty of Sciences, Moulay Ismail University, Meknès, Morocco

[2]Research Laboratory in Educational Sciences, Philosophy and Humanities, Higher Normal School, Moulay Ismail University, Meknes, Morocco

[3]IA Laboratory, Faculty of Sciences, Moulay Ismail University, Meknes, Morocco

[a]o.benmalek@edu.umi.ac.ma, [b]jabarinasredinne@gmail.com, [c]K.benmoussa@umi.ac.ma, [d]dmelouad@gmail.com

Keywords: Chatbots, RAG, Artificial Intelligence, Energy Education, Educational Technology, Intelligent Tutoring System (ITS)

Abstract. Since the use of artificial intelligence (AI) in educational contexts has increased rapidly, there has been a surge in the use of chatbots as interactive agents of learning in education. Recently, chatbots have shown great promise for providing engagement, access, and personalized learning in a variety of academic and non-academic fields. In the area of energy education, there are unique benefits that can be found because energy education is an applied, interdisciplinary, and technical area of scholarship with a focus on sustainability and climate change. This literature review examines the current literature regarding the use of chatbots in education and explores the potential uses of chatbots for energy education. Some of the benefits provided by chatbots include increased access and personalized learning methods, while challenges related to chatbots include limited levels of reasoning ability, reliability of data, and ethical considerations. Furthermore, a technical case study demonstrates the creation of a Retrieval Augmented Generation (RAG) chatbot using 267 curated documents for Energy Education. Initial testing (n=6) achieved 100% accuracy with instantaneous response times, showcasing the potential for providing domain-specific answers on topics related to the renewable energy sector, efficient use of energy, and sustainability. The literature review concludes with a focus on future improvements for using chatbots to provide enhanced levels of natural language processing (NLP) and adapt to learner needs in order to enhance energy literacy and sustainability education.

Introduction

The continually evolving nature of education resulting from advances in technology has created an unprecedented opportunity for us to reconsider the role that the educator plays in a learner's educational experience. We can now understand how to create new boundaries for our current practices regarding these teaching practices. While traditional classroom-based habitual approaches to teaching have overall been beneficial and have produced successful results for most students; these types of instructional strategies are seldom able to consistently optimize all aspects of the learning process and the educator's ability to deliver tailored educational support on behalf of each learner is largely absent, therefore leading learners to experience an incomplete, non-optimized learning experience. Ironically, the global energy transition and climate crisis is bringing attention to the extreme importance of energy education, giving some base for students and citizens to comprehend the attributes of gaining an education and the various ways of energy production, sustainable energy use, and energy conservation.

Chatbots[1] are automated conversational agents, which are increasingly being utilized as a key way to fill gaps in student learning[2]. Chatbots[3] can normally engage in conversation by replying instantly, supporting students through complex challenges, and providing interactive rationales that allow learners to develop knowledge in an explanatory and effective manner. In regards to energy education specifically, chatbots could initiate language to break down technical concepts effectively, provide immediate answers to student questions, facilitate the expression of ideas or beliefs on the part of students at any time, and create a framework to allow students to explore engagement in an active learning capacity.

This paper is structured as a review study examining existing literature on chatbots in education, with a specific focus on energy education. In addition to the literature analysis, a technical case study based on a retrieval-augmented generation (RAG) chatbot is presented to illustrate practical implementation.

Review and Screening Methodology

Literature Review. Recently, Ai chatbots have been shown to be effective tools in education. They support student engagement in learning, provide adaptive feedback and personalized learning within formal educational systems, and create opportunities for interaction that increase student motivation while reducing resistance to educational technologies. These findings highlight the effectiveness of conversational ai systems as supportive educational tools.[4]

Most studies to date, focus on general education rather than a more domain-specific education. Energy education, involves interesting systems such as renewable technologies, grid dynamics and sustainability concepts; still lacks empirical work on AI integration. AI chat systems[5] are making energy concepts more accessible to learners, from novices to professionals. These systems addresses a new way of learning and retaining information. [6]

Additionally, there are multiple studies exploring generative AI assistants in energy and sustainability tasks that demonstrates that chatbots can assist with real-world reasoning. These tools act interactively by giving responses that aligns with the goals of energy education where complexity and interdisciplinarity are central challenges.[7]

This emerging evidence states that knowledge based chatbots, especially those using retrieval-augmented generation (RAG) established based on energy datasets, can bridge gaps between traditional content and interactive learning; In the following sections, a retrieval-augmented generation (RAG)-based chatbot for energy education is described and evaluated.

Chatbots in Education. Education-based chatbots[8] are designed for three primary purposes: information, practice, and engagement. They provide information to students by responding to their questions and helping students clarify difficult concepts. They provide practice by giving quizzes and exercises with immediate feedback. They provide engagement by helping to simulate conversations that lead to a more casual and approachable disposition toward learning.

There are important benefits to leveraging chatbots in the educational context[9], such as:

Availability: A learner can access support any time - this leads to reduced reliance on teacher availability.

Personalization: Chatbot responses can be personalized based on learner progress or a student's previous interactions.

Scalability: One chatbot could be used by a large cohort of students at the same time.

Nonetheless, chatbots have many disadvantages to consider. Some examples of these limitations are the inability of chatbots to be able to think through complex topics like a human instructor can, chatbots sometimes respond with partial information, and chatbots are completely reliant on the standard of their training materials. These weaknesses must be evaluated and corrected in order for chatbots to reach their fullest potential; particularly in the area of specific academic topics, such as education in renewable[10] sources of energy.

Materials Research Proceedings 64 (2026) 440-446 https://doi.org/10.21741/9781644904091-55

The Role of Chatbots in Energy Education. Energy education[11] involves a range of topic areas, including renewable energy sources, energy efficiency, climate change, and sustainability. Many of these topic areas are complex, combining scientific and technical areas with socio-economics. Students can struggle with scientific or technical language, conceptual or abstract models, and applying conceptual learning to the real world.

Chatbots could assist in making energy education more accessible and interactive. They could:

Explain things: for example, provide a simple layperson's analogy to the process of converting the energy of the sun into electricity in solar panels.

Facilitate step-by-step guidance: for example, provide a learner with step-by-step instructions to solve problems or to explain energy-related case studies.

Facilitate awareness campaigns: encourage sustainable behavior by responding to questions, about energy, energy in homes, energy networks, etc., that people encounter every day.

Facilitate simulations: act as a reminder for students, explain protocols, or clarify things quickly.

In providing this type of support people not only learn better, but they also are likely to be more actively engaged.

Findings and Discussion

Proposed RAG Chatbot System. Dataset & preprocessing. For our study, we have used a domain-specific energy knowledge corpus from the *EnergyEducation encyclopedia.* This website contained a lot of chapters that has a lot of definitions about energy related terms. We scraped the website using a python Library called *BeautifulSoup*, extracting all relevant pages and chapters which reached a total of *300 pages*. Each entry was stored in a *Json file* containing three fields: *URL, Title and Text*. While inspecting the collected Dataset, we have found that some images were taken in count as terms which were deleted as part of preprocessing because only clean text is relevant for text-based retrieval. This preprocessing step resulted in high quality of *267 pages*, structured dataset ready for downstream embedding and indexing.

Embedding & Indexing. The structured dataset is represented by text chunks; In this step, each text chunk was embedded using the *MiniLM all-MiniLM-V6-v2* sentence transformer, which produces 384-dimensional vector representations of text. The embeddings capture semantic captures similarity between queries and knowledge chunks. These vectors were indexed using *FAISS (Facebook AI similarity Search),* a library optimized for fast similarity search over large dense vectors. The Faiss index allows retrieval of top-k most relevant chunks given a user query which facilitates efficient, fast and accurate access to knowledge corpus. Both the embeddings and the index were stored locally, enabling reproducible and a system that can be optimized very fast without reliance on external APIs.

RAG pipeline. The RAG[12] system operates in three stages. First, a user question is encoded using the sentence-Transformer used for the dataset. Then, the FAISS index retrieves the top-k most semantically relevant text chunks. Finally, these chunks provide contextual information for a language model to generate answers. A schematic overview of the RAG pipeline is presented in Fig. 1, illustrating the flow from user input to final generated answer.

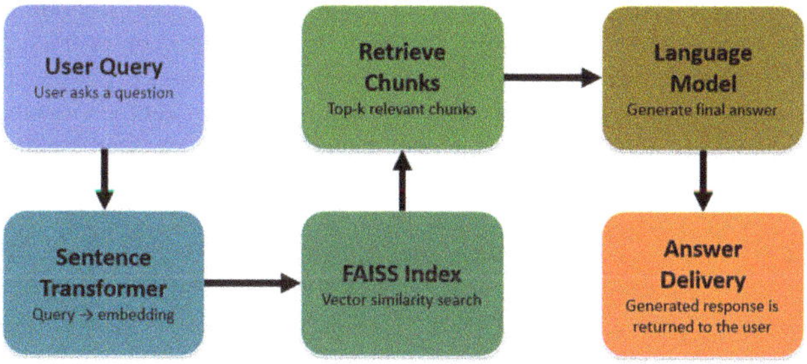

Fig 1: RAG architecture for the proposed energy education chatbot

Test Examples. To assess the system's performance, we tested a set of representative queries related to energy concepts, as presented in Table 1.

Table 1: Chatbot's responses based on the user questions

Question	Chatbot Response	Retrieved Topic Domain
What is solar energy?	Solar thermal power plants are electricity generation plants that utilize energy from the Sun to heat a fluid to a high temperature. This fluid then transfers its heat to water.	Solar Energy
Explain wind power generation	Wind power is the generation of electricity from wind. Wind power harvests the primary energy flow of the atmosphere generated from the uneven heating of the Earth's surface by the Sun.	Wind Energy
What are fossil fuels?	Fossil fuels are fuels made by geological processes acting on ancient dead organisms buried hundreds of millions of years ago.	Conventional Energy
What is energy efficiency?	Efficiency in physics is a comparison of the energy output to the energy input, representing how effectively energy is converted into useful work.	Energy Efficiency
What are electric vehicles?	Electric vehicles are vehicles that use electric motors for propulsion and onboard electricity storage systems, producing zero tailpipe emissions.	Sustainable Transport
What is mass?	Mass is a measure of how much matter is contained within an object and is expressed in kilograms in the SI system.	Physics

Practical Applications of Chatbots in Energy Education. We can look at the advantages chatbots may afford the field of energy education by examining their potential uses in the real world. Some examples of how a chatbot could be utilized to provide relevant information and assist students include:

Teaching Assistant: Chatbots may serve as a teaching assistant when responding to frequently asked questions from students, explaining concepts from lectures, or providing supplemental material that would not place an undue burden on educators. For instance, if students had questions about solar, wind, or hydro technologies in the renewable energy module, asking a chatbot for definitions, examples, or summaries would not negatively impact their understanding of the content.

Self-Directed Learning: For course participants enrolled in asynchronous (i.e., internet) or distance education/training, chatbots could accompany them through their course modules, recommend supplemental materials and provide immediate completion of quizzes, followed by grading and feedback. Ongoing education will be better facilitated outside of traditional classroom settings.

Promotional Use: Chatbots can be developed for the general public to raise public awareness and share information related to renewable/sustainable energy. For example, a chatbot could be created to address FAQs related to decreasing home energy use, selecting energy-efficient household appliance options, and understanding the national renewable/sustainability policy issues.

Academic Preparedness: With regards to vocational educational programs and industrised training, all bots can be adjusted to develop learners' skill sets and provide employees with industry safety protocol, maintenance of energy apparatus, and education on/ upskilling to adapt to new energy technologies. Bots can therefore serve as a means of conducting education and providing a professional environment to enable learners take advantage of various learning and/or educational opportunities in the workplace.

Such implementations illustrate that bots are applicable to theoretical ends, but can also serve as practical tools that improve learners' experience, stimulate interest in learning, and enable learners to attain mastery over energy-related issues.

Discussion. Chatbots as educational tools present potential as well as challenges in the application of chatbots to educational applications relevant to the field of energy. Access to energy information on-demand regardless of the time of day or where you may be located is one of the main benefits of deploying chatbots as educational aids. By providing on-demand access to energy knowledge, students are able to take a self-directed approach to their learning; they can control how much time they spend on a concept and how deep into a topic they wish to go, depending on their specific needs as learners. This function of a chatbot will come in particularly useful when educating students about energy topics, as many students desire repeated exposure to more difficult concepts related to renewable energy systems, energy efficiency, and sustainability.

The results that were generated from the assessment of the proposed RAG chatbots have demonstrated how effective the use of RAG technology is at providing domain-specific information about energy concepts. The examples of the queries and subsequent responses that were provided in the study show that by using a RAG model, the chatbot can develop informed and appropriate answers to energy-related questions by pulling information from an existing database of energy resources. This aspect of the RAG system provides reliability to students by ensuring their responses are based on concrete references and eliminates or significantly reduces the likelihood of providing inaccurate or incomplete answers by chatbots that solely generate responses without referencing an external data source.

While the RAG architecture increases the accuracy of generated responses, it does not assure that all generated answers are accurate. The quality of the answers and their validity is still dependent on whether the data used to create those responses contain current, relevant, and complete information. There is the possibility that due to rapid advances in Energy Technology,

 Materials Research Forum LLC
https://doi.org/10.21741/9781644904091-55

the data within a chatbot may become outdated unless it is regularly updated with new information from a knowledge base

As demonstrated by these evaluations, the chatbot provides generic or less relevant answers to questions that are out-of-domain. This is to be expected because the system is designed to return only the type of answer it is trained to give based on its training set. It is a limitation of the system, but also serves as a safeguard. It ensures that the chatbot's responses remain within the pedagogical boundaries of the original intent of the system, i.e., education.

This brings up an important question about how this type of technology will evolve and whether or not it might take over some of the functions of a teacher. The goal of this type of system is not to replace a teacher, but to augment the traditional teaching process by acting as an additional learning resource. Teachers and instructors remain the primary resource for guiding learners and validating the accuracy of information presented to learners. One of the more important aspects of designing a chatbot is ensuring that it clearly communicates the nature of its limitations and the sources from which it retrieves educational materials to learners.

In the future, Improved Natural Language Processing (NLP) technology combined with improved Machine Learning (ML) technology will likely increase both the ability of chatbots to better understand content and respond to learner questions in a more personalized context, while maintaining content validity through retrieval-based grounding.

As a whole, RAG-type chatbots are a viable way to provide an educational platform for the subject of Energy and meet considerations around ethics, data set maintenance and pedagogical integration[13]. The key considerations of this research are highlighted through their findings.

Conclusion and Perspectives

Conclusion. Chatbots are at the frontiers of energy education. As an innovative combination of the key elements of Access/Personalization/Engagement, they provide students with essential and relevant educational resources in a complex field like Renewable Energy, Efficient Energy Use and Sustainability.

Although there are still some challenges regarding the accuracy and level of detail in Chatbot responses, their use as a collaborative classroom tool or educational resource has already proven to offer tremendous value for creating an effective educational environment. Chatbots can help encourage learners, identify and correct areas of deficiency in traditional forms of instruction, and align broad-based educational goals with the principles of Sustainability and Energy Literacy.

In essence, Chatbots should be viewed as a tool that can supplement and build upon the overall learning ecosystem, rather than diminishing the impact of the teacher in the learning process.

Perspectives and Future work. In the future, we will continue to increase the size and diversity of the Energy Encyclopedia's underlying Knowledge Base by establishing links to other authoritative energy encyclopedias, open educational resources, and institutional repositories. A larger volume and diversified set of curated data will increase retrieval quality and response relevancy for advanced energy topics. A third avenue of future development is the provision of a dedicated, purpose-built user interface that will facilitate the usability and interaction of learners with the educational materials contained in the system through features such as guided query functions, topic navigation, and contextual feedback.

Future improvements might also include regular updates to the Knowledge Base with new content, integration with Learning Management Systems, and user-centered evaluations of the system's pedagogical effectiveness in actual learning environments.

References

[1] A. Kerly, R. Ellis, and S. Bull, "CALMsystem: A Conversational Agent for Learner Modelling," *Knowledge-Based Syst.*, vol. 21, pp. 238–246, Apr. 2008. https://doi.org/10.1016/j.knosys.2007.11.015.

[2] N. Davar, M. Dewan, and X. Zhang, "AI Chatbots in Education: Challenges and Opportunities," *Information*, vol. 16, p. 235, Mar. 2025. https://doi.org/10.3390/info16030235.

[3] R. Winkler and M. Söllner, "Unleashing the Potential of Chatbots in Education: A State-Of-The-Art Analysis," *Acad. Manag. Proc.*, vol. 2018, p. 15903, Apr. 2018. https://doi.org/10.5465/AMBPP.2018.15903abstract.

[4] M. K. Rahman, N. A. Ismail, M. A. Hossain, and M. S. Hossen, "Students' mindset to adopt AI chatbots for effectiveness of online learning in higher education," *Futur. Bus. J.*, vol. 11, no. 1, p. 30, 2025. https://doi.org/10.1186/s43093-025-00459-0.

[5] A. Club, B. Ellis, and M. Yüzbaşioğlu, "Greening the Classroom: Energy-Conscious AI Systems for Digital Education and International Communication," Aug. 2025.

[6] X. Zeng, J. Lan, and M. N. I. Sarker, *Transforming energy education with AI adaptive learning and knowledge integration*. 2025. https://doi.org/10.1117/12.3077391.

[7] B. Zakeri, *Generative AI for Project-Based Assessment in Energy and Sustainability Education*. 2025. https://doi.org/10.21203/rs.3.rs-7402766/v1.

[8] J. Jia, "CSIEC: A computer assisted English learning chatbot based on textual knowledge and reasoning," *Knowledge-Based Syst.*, vol. 22, pp. 249–255, May 2009. https://doi.org/10.1016/j.knosys.2008.09.001.

[9] F. Kaviani, N. Selwyn, Y. Strengers, K. Dahlgren, B. Cumbo, and M. Wagner, "Future schools and the energy implications of AI in education: A review of scenarios and method for engaging young people in futures thinking," *Policy Futur. Educ.*, vol. 23, Feb. 2025. https://doi.org/10.1177/14782103251322271.

[10] C. Tamilarasi, K. Sowmiya, G. Vignesh, R. Jasmine, P. Nehru, and S. P., "AI in Renewable Energy Education: Teaching the Future of Sustainable Power Systems," 2025, pp. 133–152. https://doi.org/10.4018/979-8-3693-9062-7.ch007.

[11] P. Jennings and C. Lund, "Renewable energy education for sustainable development," *Renew. Energy*, vol. 22, pp. 113–118, Mar. 2001. https://doi.org/10.1016/S0960-1481(00)00028-8.

[12] J. Swacha and M. Gracel, "Retrieval-Augmented Generation (RAG) Chatbots for Education: A Survey of Applications," 2025. https://doi.org/10.3390/app15084234.

[13] E. Mulhim, "Mitigating student resistance: the role of chatbots in sustainable educational technology integration," *Front. Educ.*, vol. 10, Dec. 2025. https://doi.org/10.3389/feduc.2025.1711832.

Materials Research Proceedings 64 (2026) 447-454 https://doi.org/10.21741/9781644904091-56

Data Driven ExtraTrees-Based State-of-Charge Estimation for New and Aged Li-ion Battery Cells

Nayeemuddin MOHAMMED[1,a], Ala A. HUSSEIN[2,b*], Hiren MEWADA[2,c]

[1]Department of Civil Engineering, Prince Mohammad Bin Fahd University, Al Khobar, Kingdom of Saudi Arabia

[2]Electrical Engineering Department, Prince Mohammad Bin Fahd University, Al Khobar, 31952, Saudi Arabia

[a]mnayeemuddin@pmu.edu.sa, [b]ahussein@pmu.edu.sa, [c]hmewada@pmu.edu.sa

Keywords: Li-Ion Battery, State-of-Charge, Machine Learning, Electric Vehicles

Abstract. The use of lithium-ion batteries is essential in modern life. They are used in the drive of medical equipment, transport, renewable energy, and backup power. Effective state-of-charge (SOC) regression is essential to the safety and performance of the lithium-ion (Li-ion) batteries. However, most commercial techniques are valid only when the battery is new. Also, methods based solely on current measurements are susceptible to cumulative error, which reduces the accuracy of the estimate. The accurate prediction of SOC is a serious engineering challenge because it cannot be directly measured at the battery terminals. The present paper proposes a SOC regression approach based on an ExtraTrees Classifier and a Long Short-Term Memory that operate on voltage and surface temperature as input parameters, thereby eliminating the need for current sensing. Training and testing of the model are performed using both new and old Li-ion cell data to assess its viability across different battery ages. Results indicate that SOC regression accuracy increases with battery age because higher internal resistance results in more distinct temperature rise patterns. The suggested technique is sufficient to explain such alterations and to provide valid predictions across different conditions. The R2 values are 96% and 99% for the ExtraTrees Classifier model in new and aged cells, respectively. This technique provides a scalable, practical approach to battery management systems for electric vehicles and stationary storage by using easy-to-measure indicators and eliminating current integration errors.

Introduction

One of the basic components of energy storage is Li-Ion batteries. Everything is introduced as modern equipment, including medical, transport, wearable, and advanced transport equipment, as well as renewable energy storage. Li-ion batteries have the highest specific energy and energy density among currently available cell chemistries. Battery performance is characterized in SOC. It shows whether the batteries are alive. Accurate evaluation of the SOC of lithium-ion (Li-ion) batteries is crucial to the safety, efficiency, and performance of energy storage equipment. With the growing demand for electric vehicles and renewable energy storage, there is a need for reliable methods to calculate SOC.

Traditional techniques, i.e., Coulomb Counting and open-circuit voltage, often rely on current measurements to determine SOC, which can lead to accumulated errors over time due to factors such as battery age and changing operational conditions. Model-based alternatives are more efficient than this method. The emergence of AI and machine learning (ML) has created new prospects for battery state regression. The data are initially captured in this process, and later a model is designed to capture the relationship between these data and SOC. Most existing methods degrade with battery aging, leading to inaccurate SOC predictions under high internal resistance and temperature-rise conditions.

Tao et al. improved the SOC estimate by creating a long short-term memory (LSTM) neural network with a model reference adaptive system for parameter identification. The suggested model increased the SOC estimate by an average of 59.73% and parameter regression accuracy by 67.38% [1]. Miao et al. reviewed SOC regression techniques for lithium-ion batteries, emphasizing model improvement for temperature adaptability and data-driven methods for complex conditions, and highlighted the need for high-precision models and the integration of physical and data-driven approaches [2].

Li-ion batteries have complicated characteristics that depend on a number of factors, such as voltage, temperature, and internal resistance. It is worth noting that with the aging of batteries, the internal resistance of a battery rises, leading to the appearance of more pronounced variations in temperature during the operation. Monirul et al. solve the challenge of the accurate measurement of SoC in lithium-ion batteries in unmanned aerial vehicles (UAVs) through a customized unscented Kalman filter (UKF). In the case of rapid current changes, the suggested method lowers the error in the estimates by 1.80 to 1.0%, which is much better in the energy management of UAV batteries [3]. The latter phenomenon offers a chance to increase SOC estimates by employing easy to measure traits, i.e. voltage and surface temperature, instead of solely using current measurements. Researchers in the past have proposed a new way of estimating the SOC of lithium-ion batteries based on multimodal information that includes aging state and expansion force to improve the accuracy and robustness. The proposed solution based on a hybrid neural network reduces the effects of aging by 75% and SOC prediction error by up to 52% relative to conventional methods and by up to 73% in dynamic conditions. This facilitates better and deployable battery management systems [4].

Huang et. al., introduce a multitask learning model with an RMSE of 16.71 mV, which is used to upscale low-frequency battery voltage data to high-frequency signals. It introduces a SOC estimating methodology that has an RMSE margin of 1.8%, which offers accurate estimates of state-of-charge. Cost-effective improvements should be suggested to obtain maximum computational efficiency and performance [5]. The state of charge (SOC) estimates needed by Li-ion batteries to efficiently manage the battery energy are not easily measurable. Although deep learning technology, including LSTM, has been applied to predict SOC, it often cannot respond to differences between current voltage when operating in different conditions [6].

A description of a machine learning method to predict the SOC with the help of the ExtraTrees Classifier and LSTM founded on voltage and temperature measurements. Our goal is to assess the model's resilience and accuracy over its lifetime by training and testing it on fresh and aged Li-ion cells. The suggested method is expected to increase the accuracy of SOC estimates along with providing a scalable method that can be implemented into battery management systems. We are aiming to develop a model which is helpful to model the dynamics of Li-ion batteries after performing a comprehensive statistical analysis of the input parameters and their relation to SOC. Besides improving the discipline of SOC regression, the outcomes of the current study will be used to keep on enhancing smarter and smarter energy storage options to apply in different applications such as stationary energy systems and electric vehicles.

Dataset and Methodology

A battery's storage capacity can be determined from the relationship between temperature and voltage. The storage of batteries is susceptible to temperature. Additionally, the aging of the battery also increases its internal resistance. This impacts the voltage generation during its utilization. Thus, monitoring the temperature and voltage can provide intrinsic information about battery SOC. The proposed machine learning approach uses ExtraTrees Classifier and Long Short-Term Memory regression model to estimate SOC using the experimental data for the State of Charge. The ExtraTrees is an ensemble approach to a decision tree model. It trains multiple decision trees, and the responses of all trees are merged to improve the prediction rate. Fig. 1 shows the functional diagram of the ExtraTrees model.

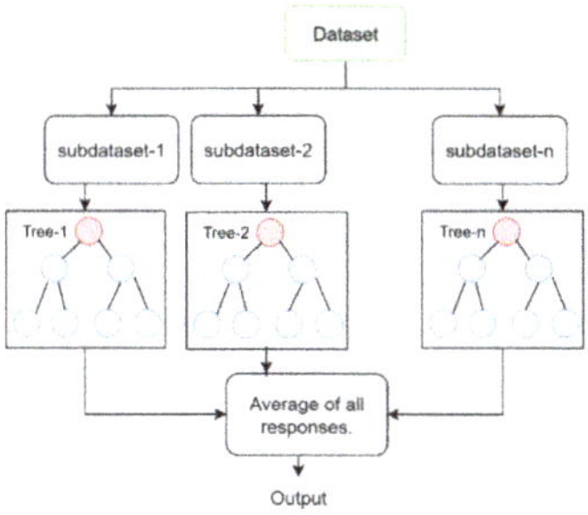

Fig. 1: Architecture of the ExtraTees regression model

Two experimental datasets for a New Cell and an Aged Cell, [12], were utilized to design the machine learning model. Two inputs, Voltage (V) and Temperature (C°), were considered and output as state of charge [7]. Fig. 2 depicts the steps involved in the ExtraTree regression model.

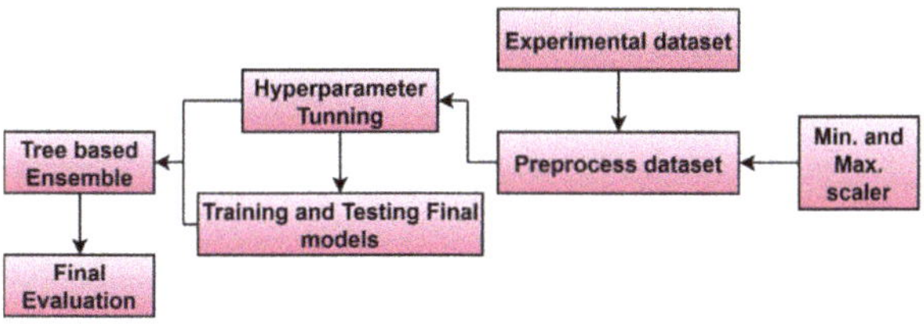

Fig. 2 Block diagram of SOC regression *using the ExtraTrees algorithm*

The dataset is preprocessed and normalized using the min-max scalar. Later, ExtraTrees ML is trained using these datasets to predict SOC. Voltage and temperature ratings are used as predictors of the model. Fig. 3 and 4 show the correlation between these parameters. The voltage and temperature are highly correlated, with correlation factors of -0.97 and -0.95 for the aged cell and the new cell, respectively. Fig. 5 and 6 show the statistical analysis of the predictors and response for both aged and new cells. The voltage distribution is approximately normal with an average (μ) of approximately 3.56 V and a standard deviation (σ) of 0.22 V for a new cell. It means that it is not very sensitive to the voltage group and gives a fairly accurate approximation of the SOC. The temperature standard deviation was 3.40 and the average was around 32.17 which is potentially a vital range of battery performance and behavior. Lastly, the SOC distribution mean was 0.51 and the standard deviation was 0.29. This demonstrates a comparatively equal distribution of SOC in the range that is under observation. Such plots will give plausible results on the correlation between voltage, temperature and SOC. They are applicable in coming up with a more precise model of SOC prediction, hence enhancing battery management systems.

Materials Research Proceedings 64 (2026) 447-454 https://doi.org/10.21741/9781644904091-56

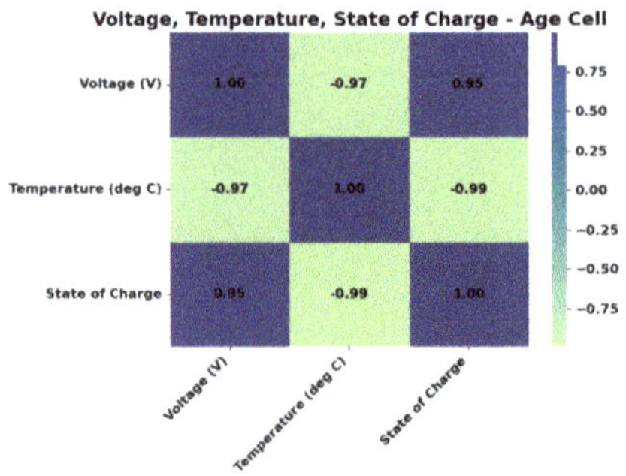

Fig. 3 Correlation matrix for Aged Cell

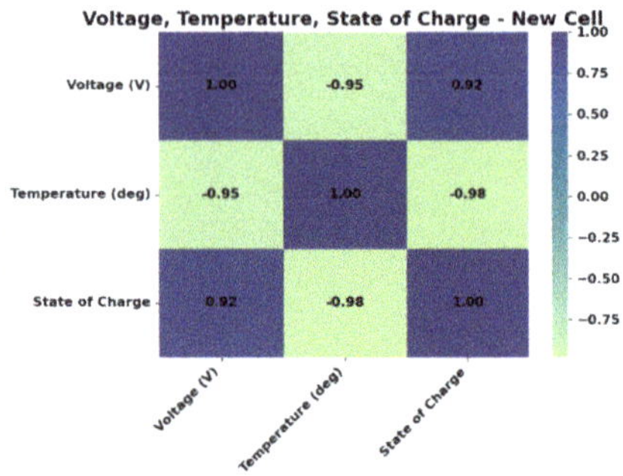

Fig. 4 Correlation matrix for New Cell

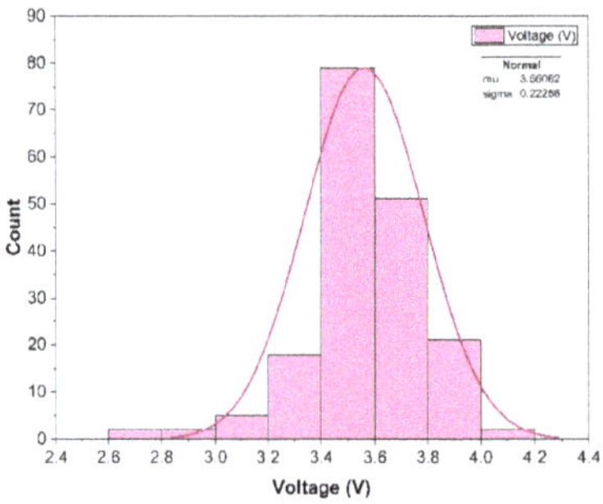

(a)

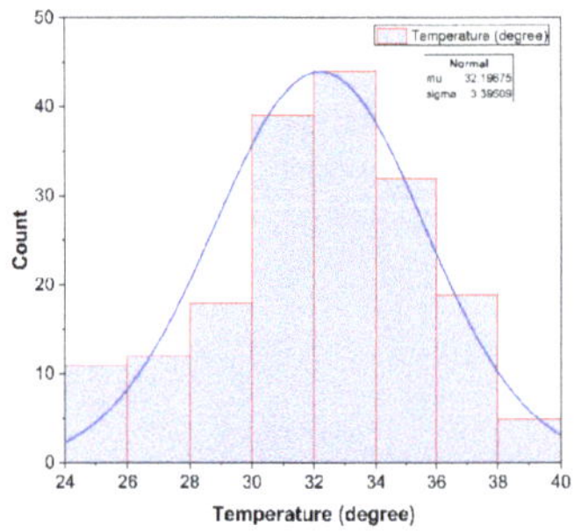

(b)

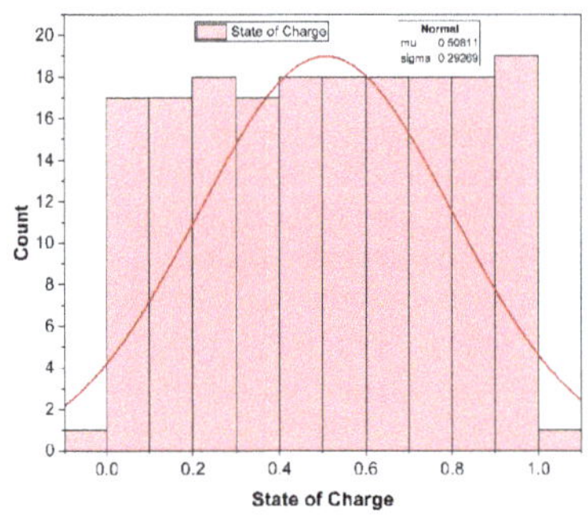

(c)

Fig. 5 Normal distribution for the input and output variables for New Cell

Fig. 6 represents statistical analysis of the data in case of aged cells. The voltage distribution is- normal Gaussian with an average (μ) of 3.51 V and a standard deviation (σ) of 0.23 V, which is similar to the average, implying that the performance of the voltage on the objective of estimating the SOC is homogenous. The mean temperature of the temperature distribution is approximately 30.51 degrees Celsius with a standard deviation of 2.93 degrees Celsius which is very important to the efficiency and power of the battery. And the standard deviation of the SOC distribution is 0.23 and its mean is 0.51. These data could show how voltage is related to

Materials Research Proceedings 64 (2026) 447-454 https://doi.org/10.21741/9781644904091-56

temperature and SOC and could be employed to create a model to predict SOC to be used in the enhancing of battery management systems.

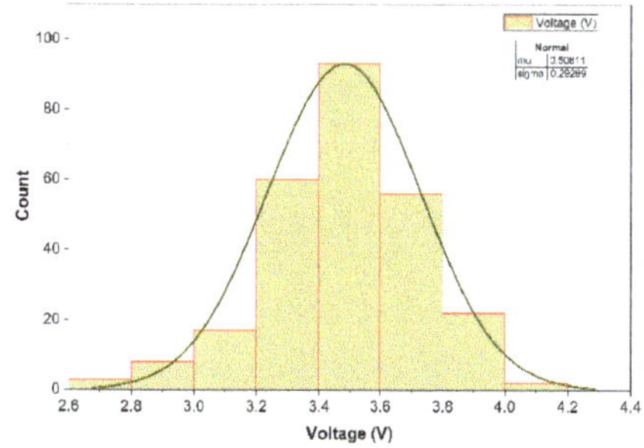

(a)

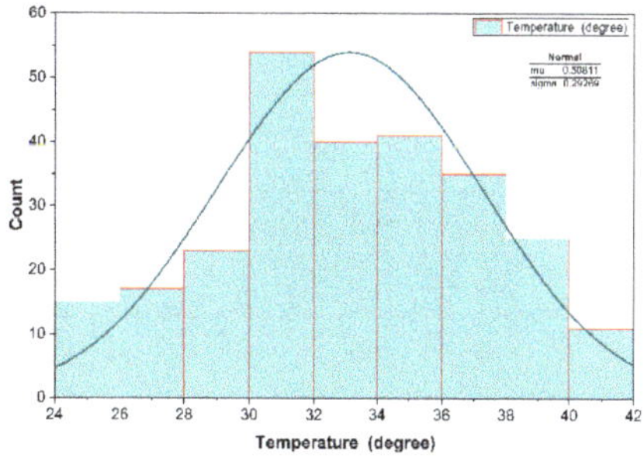

(b)

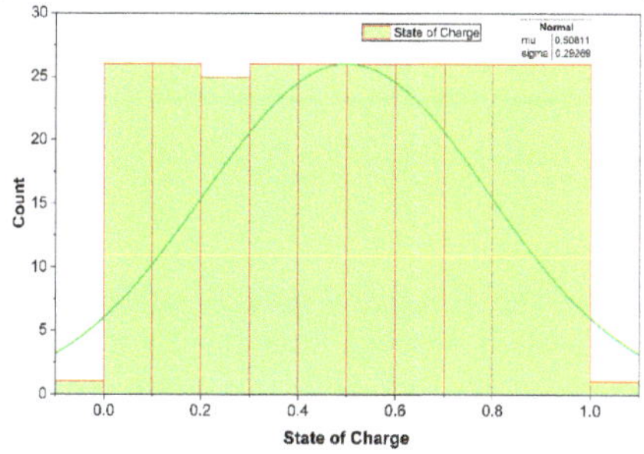

(c)

Fig. 6 Normal distribution for the input and output variables for Aged Cell

The model is validated using four metrics: mean absolute error (MAE), mean square error (MSE), root mean square error (RMSE), and the R^2 correlation coefficient. Below are the equations used for calculating these matrices.

$$\text{MAE} = \frac{1}{n}\sum_{i=1}^{n}|y_i - \hat{y}_i| \quad (1)$$

$$\text{MSE} = \frac{1}{n}\sum_{i=1}^{n}(y_i - \hat{y}_i)^2 \quad (2)$$

$$\text{RMSE} = \sqrt{\frac{1}{n}\sum_{i=1}^{n}(y_i - \hat{y}_i)^2} \quad (3)$$

$$R^2 = 1 - \frac{\sum_{i=1}^{n}(y_i - \hat{y}_i)^2}{\sum_{i=1}^{n}(y_i - \bar{y})^2} \quad (4)$$

Where, y_i is the actual value, $\hat{y}_i$=predicted value, n=number of observations, $\bar{y}$ is the mean of actual values.

Results and Discussion

ExtraTrees was applied to the training and testing datasets in Fig. 7 to predict the SOC of cells. Comparing the expected and actual values of the testing dataset reveals a relatively good fit, with a few outliers at specific sample indices, indicating that the model may need adjustment.

The plot presented below the main line is the residual plot that shows the actual, but expected values, against the sample index and it can be seen that the actual and expected values are randomly distributed around the zero line. A good fit can be suggested with no significant bias and so is the same case with the training dataset. The real and forecasted figures of the training data are quite close but only a few sample indices [8]. The residual plot where the differences between the actual and predicted values are displayed reflects a stable distribution around the value of zero, which indicates that the model is reliable in both the settings [9]. Figure 8 shows the actual and predicted errors of the New Cell training and testing data.

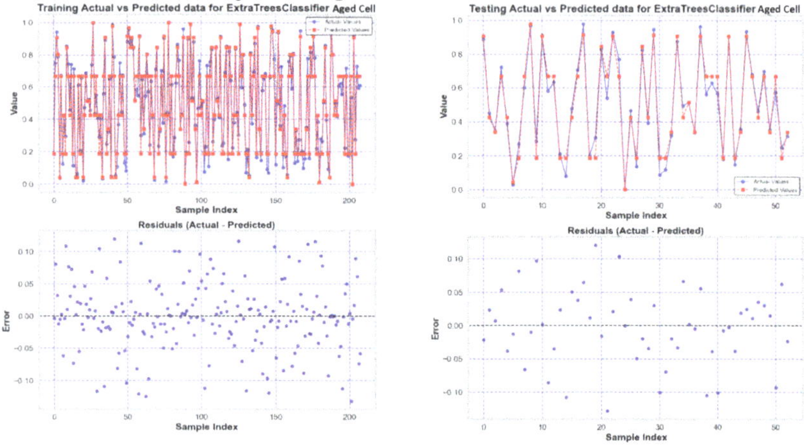

Fig. 7 The Training and Testing actual datasets vs predicted values for Aged Cell.

The new cell dataset has a Mean Absolute Error (MAE) of 0.0389, whereas on the testing dataset, the error is slightly higher at 0.0443 [10]. This means that the predictions made by this model are mostly accurate with very minor margin of inaccuracy when the projections are made to include hitherto unknown data [11]. This is supported by the Mean Squared Error (MSE) and Root Mean Squared Error (RMSE), which are both low: 0.00279 and 0.0528 in training, and 0.00318 and 0.0564 in testing, respectively. The model has R2 values of 0.967 during training and 0.9603 during testing, as shown in Table 1, suggesting that it explains a large proportion of the data's variability. The model is further compared with a multilayer perceptron (MLP) network and a time-series network, i.e., an LSTM. MLP is a standard ANN network used for regression applications. We tested a narrow (i.e., one layer with a size of 10), a wide (size of 100), and a custom MLP network with three hidden layers with 10 neurons in each layer. Among the three networks, the custom MLP with four layers performed well. However, its R^2 score is limited to 0.81 and 0.79 for the training and test datasets, respectively. LSTM is a recurrent neural network that can retain features over long periods, a weakness of traditional CNNs. The comparison shows that ExtraTrees performed better than LSTM for both the training and test datasets. The extraTreess model uses an ensemble of multiple decision trees and therefore performed well on the test dataset without the need for extensive tuning. In contrast, the LSTM requires further tuning and a large dataset over time to improve its performance. Therefore, the Extra Trees Regressor is a more effective choice for determining SOC.

***Table 1** Performance Metric analysis for the SOC Response*

		New Cell				Aged Cell			
Model	**Dataset**	**MAE**	**MSE**	**RMSE**	**R^2**	**MAE**	**MSE**	**RMSE**	**R^2**
ExtraTrees	Training	0.038	0.0027	0.052	0.967	0.013	0.00043	0.020	0.99
	Testing	0.044	0.0031	0.056	0.960	0.020	0.00079	0.028	0.99
LSTM	Training	0.088	0.011	0.106	0.868	0.076	0.012	0.103	0.856
	Testing	0.096	0.013	0.114	0.836	0.094	0.014	0.112	0.840
MLP (three-layer)	Training	0.095	0.024	0.156	0.812	0.092	0.0311	0.1763	0.799

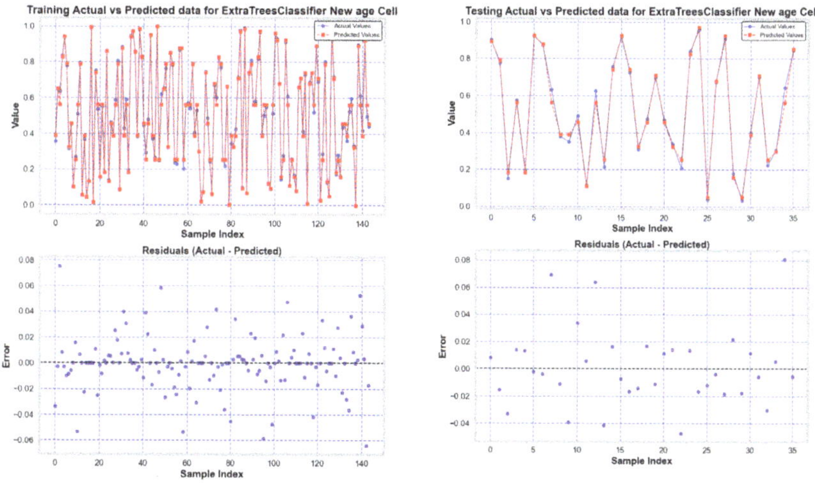

Fig. 8: The Training and Testing actual datasets vs predicted values for the New Cell.

Conclusion

In this study, an innovative approach to predicting the state of charge (SOC) of lithium ion batteries using ExtraTree as predictor as well as measuring voltage and surface temperature was demonstrated. This strategy eliminates the limitations of traditional SOC regression strategies, which are based mainly on existing data and may lead to compounding error with time. According to the results, the proposed model is not only efficient when applied to the new Li-ion cells but also maintains high accuracy when used with old cells. Where the higher internal resistance of the old cells increases the distinctiveness of the temperature rise trends. The analysis of the voltage and temperature distributions (statistically) showed that it is strongly correlated with SOC, which confirms the basis of the model. The performance measures of ExtraTrees include MAE, MSE, RMSE and R^2 results demonstrate good predictions of fresh and aged cells. R^2 scores of the model were 0.9674 and 0.9603 in fresh and old cells respectively, 0.994 and 0.990. Whereas, as per LSTM the R^2 value shows less as compared to ExtraTrees model. These results confirm the ability of the model to predict a large percentage of the variation in the SOC estimate. In general, this paper highlights the possibility of using readily available signals to estimate SOC in battery management systems. The proposed technology that does not require current sensing gives a viable and scalable solution to the use of the technology in electric cars and stationary energy storage systems, which eventually results in the enhancement of the battery performance and reliability.

References

[1] J. Tao, S. Wang, W. Cao, C. Fernandez, F. Blaabjerg, and L. Cheng, "An innovative multitask learning - Long short-term memory neural network for the online anti-aging state of charge estimation of lithium-ion batteries adaptive to varying temperature and current conditions," Energy, vol. 314, p. 134272, Jan. 2025. https://doi.org/10.1016/j.energy.2024.134272

[2] Y. Miao, Y. Gao, X. Liu, Y. Liang, and L. Liu, "Analysis of State-of-Charge Estimation Methods for Li-Ion Batteries Considering Wide Temperature Range," Energies, vol. 18, no. 5, p. 1188, Jan. 2025. https://doi.org/10.3390/en18051188

[3] I. M. Monirul, L. Qiu, and R. Ruby, "Accurate state of charge estimation for UAV-centric lithium-ion batteries using customized unscented Kalman filter," Journal of Energy Storage, vol. 107, p. 114955, Jan. 2025. https://doi.org/10.1016/j.est.2024.114955

[4] Q. Xu et al., "An expansion force-based method for accurate state of charge estimation across the entire lifecycle of lithium-ion batteries," Journal of Energy Storage, vol. 137, p. 118705, Nov. 2025. https://doi.org/10.1016/j.est.2025.118705

[5] Z. Huang, J. Zhu, Z. Wang, X. Zhang, and G. Fan, "A recovery model for battery data in the cloud and its application of state-of-charge estimation in electric vehicles," Control Engineering Practice, vol. 165, p. 106579, Dec. 2025. https://doi.org/10.1016/j.conengprac.2025.106579

[6] O. Ozer and H. Arabaci, "State of Charge Estimation in Li-Ion Batteries Using a Parallel LSTM-Based Approach: The Impact of Modeling Based on Operating States," IEEE Access, vol. 13, pp. 130719–130730, 2025. https://doi.org/10.1109/ACCESS.2025.3591970

[7] J. Zhou, J. Li, X. Wu, T. Liu, and Y. Jiang, "Multi-physics model-based filtering for simultaneous state of charge and core temperature estimation in lithium-ion batteries," Journal of Energy Storage, vol. 137, p. 118518, Nov. 2025. https://doi.org/10.1016/j.est.2025.118518

[8] W. Yu, Y. Zheng, and S. Wang, "Data-driven state-of-charge estimation for lithium-ion batteries using variational inference," IEEE Transactions on Circuits and Systems II: Express Briefs, pp. 1–1, 2025. https://doi.org/10.1109/TCSII.2025.3608185

[9] Y. Fan et al., "A physics-enhanced hybrid kolmogorov–arnold network with dynamic coupling for interpretable battery state-of-charge estimation," Applied Energy, vol. 400, p. 126533, Dec. 2025. https://doi.org/10.1016/j.apenergy.2025.126533

[10] D. Velumani and A. Bansal, "Temperature estimation in a lithium-ion cell using a machine learning based approach," Applied Thermal Engineering, vol. 270, p. 126201, July 2025. https://doi.org/10.1016/j.applthermaleng.2025.126201

[11] A. Haraz, K. Abualsaud, and A. M. Massoud, "Hybrid Tree-Based Machine Learning Models for State-of-Charge and Core Temperature Estimation in EV Batteries," IEEE Access, vol. 13, pp. 129962–129976, 2025. https://doi.org/10.1109/ACCESS.2025.3591057.

[12] NASA, "Public battery ageing dataset," [Online]. Available: https://data.nasa.gov/dataset/li-ion-battery-aging-datasets. Accessed: Oct. 10, 2025.

Solar Panels Fault Classification using an Optimized Moment-Based Feature Extraction Method

Ismail NAOUADIR[1,a *], Omar EL OGRI[1,2,b], Jaouad EL-MEKKAOUI[1,c], Mohamed BENSLIMANE[1,d], Amal HJOUJI[3,e]

[1]Laboratory of Sciences, Engineering, and Management (LSEM), Higher School of Technology, Sidi Mohamed Ben Abdellah University, Fez, Morocco

[2]Laboratory of Sustainable Agriculture Management (LSAM), Higher School of Technology Sidi Bennour, Chouaib Doukkali University, El Jadida, Morocco

[3]Laboratory of Information, Signals, Automation and Cognitivism (LISAC), Dhar El Mahrez Faculty of Science, Sidi Mohamed Ben Abdellah University, Fez, Morocco

[a]ismail.naouadir@usmba.ac.ma, [b]omar.elogri@usmba.ac.ma, [c]jaouad.elemekkaoui@usmba.ac.ma, [d]mohamed.benslimane@usmba.ac.ma, [e]amal.hjouji@usmba.ac.ma

Keywords: Solar Panel Classification, Krawtchouk Moments, Image Moments, Convolutional Neural Networks, Image Classification

Abstract. Solar panel fault detection and classification are of paramount importance for the success of renewable energies, as they mitigate wasting power and help solar farms stay efficient. In this paper we propose a novel feature extraction methods based on Krawtchouk moments and their parameters, using the simulated annealing optimization algorithm we can find the appropriate parameters which make the moments focus on regions where the physical or electrical damage is located, this way we can feed the classification model only most relevant features, and combined with a small but efficient convolutional neural network model, we achieved an area under curve of 99.48%, 98.96% precision, 98.61% recall, and 99.58% F1 score.

Introduction

Orthogonal polynomials are shape extractors; they have been widely used across fields such as medicine [1,2], astronomy [1], and image processing [2], [3], thanks to their ability to efficiently perform feature extraction and dimensionality reduction using the correct order of moments. Krawtchouk polynomials and moments [4], [5] are a set of discrete orthogonal polynomials, efficient at image processing thanks to their orthogonality on a discrete set of points. These polynomials have specific parameters that control the region from which the most significant features are extracted. Using these properties, optimal feature extraction can be achieved. Faults in solar panels can be caused by various conditions, such as snow, dust, or physical and electrical damage. Effectively distinguishing between these faults enables efficient maintenance, ensuring that solar farms continue to operate without interruption. In [6], a lightweight UdenseNet achieved 95.72% accuracy. In [7], they used thermographic images and convolutional neural networks to achieve 78.85% accuracy. Finally, in [8], a classification accuracy of 93.51% was also achieved using thermographic images and a multi-scale convolutional neural network. Most research is focused on the classification phase by developing new architectures or optimizing existing ones.

In this paper we focus on the features extraction phase, using Krawtchouk moments optimized via the simulated annealing algorithm to capture only important features, we have been able to achieve state-of-the-art classification metrics using only a relatively small convolutional neural network model, we have addressed the class imbalance problem using the area under curve, since accuracy can be misleading by representing only the majority class, additionally, using a low learning rate and batch size, we have achieved slow but stable convergence resulting in 99.48% area under curve, 98.96% precision, 98.61% recall, and 99.58% F1 score.

Krawtchouk Moments For Pattern Recognition
Krawtchouk polynomials (KPs) of the ith order are given by:

$$K_i(x;p,N) = \ {}_2F_1\left(-i,-x;-N;\frac{1}{p}\right) \tag{1}$$

Where N is the signal length, $p \in [0,1]$, $x, i = 0,1,2,\dots,N, N > 0$ and ${}_2F_1$ is the hypergeometric function defined as:

$$ {}_2F_1(a,b;c;z) = \sum_{k=0}^{\infty} \frac{(a)_k (b)_k}{(c)_k} \frac{z^k}{k!} \tag{2}$$

And $(x)_k$ is the rising factorial. Krawtchouk moments (KMs) are calculated by evaluating the inner product of the signal (image) with KPs with specific parameters and order. The output forms a set that represents the image's local features. KMs of order $i \times j$ in terms of KPs for an image with an intensity function $\mathrm{f(x,y)}$ of size $N \times M$ is defined as follows:

$$Q_{ij} = \sum_{x=0}^{N-1} \sum_{y=0}^{M-1} K_i(x;p_1,N-1)K_j(y;p_2,M-1)f(x,y) \tag{3}$$

Using the orthogonality property, the signal's intensity function can be either completely reconstructed or approximated in terms of KMs with $N_{max} \leq \mathrm{N} - 1$ and $M_{max} \leq M - 1$ as follows:

$$f(x,y) \approx \sum_{n=0}^{N_{max}} \sum_{m=0}^{M_{max}} Q_{ij} K_i(x;p_1,N-1)K_j(y;p_2,M-1) \tag{4}$$

While higher-order polynomials allow for more complex patterns and details to be captured, lower-order polynomials capture simpler and smoother patterns. We can further adjust the focus of the feature extraction by operating on the parameters (p_1, p_2). p_1 moves the feature extraction focus along the y axis, whereas p_2 moves the focus along the x axis. Fig. 1 shows examples of the reconstruction of a solar panel image with clear physical damage using two different sets of parameters and a low reconstruction order.

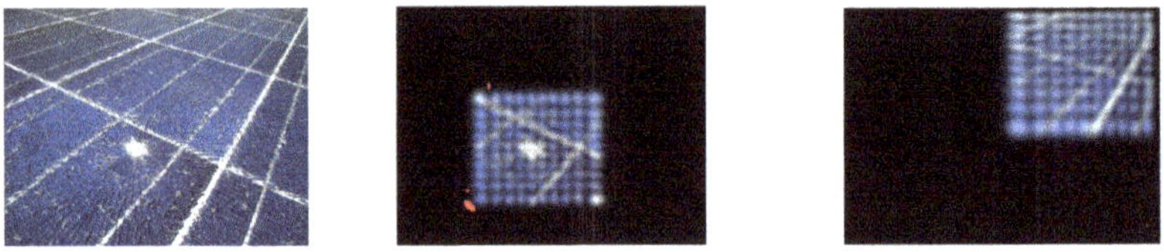

(a) : Original image (b) : $(p_1, p_2) = (0.6, 0.4)$ (c) : $(p_1, p_2) = (0.25, 0.75)$

Fig. 1. Reconstruction examples using different parameters and $(N_{max}, M_{max}) = (20,20)$.

Defect Localization in Solar Panels Through Simulated Annealing-Based Optimization
Problem Definition. Selecting the right set of parameters is of paramount importance when using image moments as features, as seen and discussed in Fig. 1. Using the wrong set of parameters will cause the moments to capture irrelevant features, ones that do not contain the actual physical or electrical damage, thereby complicating the classification process. One way to distinguish between the right and wrong choice of parameters is by using a signal fidelity assessment. Using the mean squared error (MSE) with different combinations of parameters will give us different

MSE scores each time; however, choosing the set of parameters that yields the minimum MSE is the right and optimal choice, as it minimizes the error as much as possible. Fig. 2 shows an example of physical damage to a solar panel. We can see that the lowest MSE corresponds perfectly with the region of the crack.

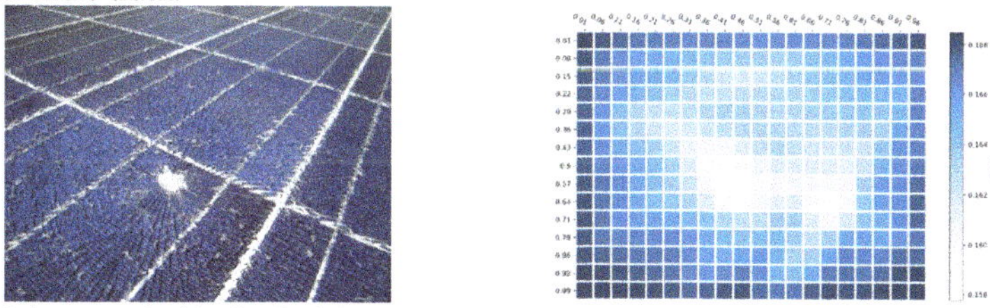

(a): Solar pannel with physical dammage (b): MSE heatmap

Fig. 2. MSE values heatmap of different combination of KPs parameters.

Since the damage aligns with the minimum MSE value, we can define the optimization problem as follows:

$$\min MSE\left(I_{ref}, I_{recon}(p_1, p_2)\right) \quad p_1, p_2 \in]0,1[\tag{6}$$

Where: I_{ref} is the reference solar panel image, $I_{recon}(p_1, p_2)$ is the reconstructed image from Krawtchouk moments parameterized by (p_1, p_2), and $MSE(I_{ref}, I_{recon})$ is the MSE quantifying the reconstruction fidelity.

Optimization Process. Simulated Annealing (SA) is a probabilistic optimization algorithm inspired by the thermodynamic process of annealing. It seeks the global minimum of a cost function E(x) (analogous to energy in physical systems) by allowing controlled, temperature-dependent exploration of the solution space.

The core components of the system include the State (x), which represents a candidate solution, such as a set of Krawtchouk moment parameters (p_1, p_2). The Energy (E(x)) serves as the cost function to minimize, for instance, the MSE score between reconstructed and reference images. The Temperature (T) is a controllable parameter that balances exploration and exploitation during optimization. Additionally, the Neighbor selection function acts as a perturbation mechanism, generating a new candidate state x' from the current state x. Here are the algorithm steps:

1. Initialization:
   - Start with an initial state $x_0 = (p_1, p_2)$ and temperature T_0.
   - Define a cooling schedule (for example $T_{K+1} = \alpha T_k$, where $\alpha \in]0, 1[$).
2. Iterative Process:
   - Perturb: Generate a neighbor state x' near x_k.
   - Energy Evaluation: Compute $\Delta E = E(x') - E(x_k)$.
   - Acceptance Criterion:
     - If $\Delta E \leq 0$, accept x'.
     - If $\Delta E > 0$, accept x' with probability: $P(accept\ x') = \exp\left(-\frac{\Delta}{T_k}\right)$
     - Update: Set $x_{k+1} = x'$ if accepted; else retain x_k.
3. Termination: Break when T_k approaches zero or after a fixed number of iterations.

Solar Panel Fault Classification Using Deep Convolutional Neural Networks

Dataset Specification. We use a publicly available dataset from Kaggle [9] containing six classes: snow-covered, physical damage, electrical damage, bird drop, dusty, and clean. After applying the standard 70% training, 15% validation, and 15% testing split, the class imbalance problem is addressed through data augmentation and metrics that account for this issue, such as focal loss [10]. Fig. 3 shows an example image from each class.

Fig. 3. Example image from each of the six classes in the dataset.

Model Architecture. The large number of hyperparameters that need to be carefully chosen makes creating a deep learning model with high performance intrinsically difficult. According to a widely recognized principle, one should begin with a comparatively small architecture and cautiously increase it until the required performance level is attained. However, there isn't a single, widely recognized hyperparameter configuration. Therefore, finding the optimal set of hyperparameters while keeping a model with the least amount of complexity requires extensive empirical experimentation. Tables 1, 2, and Fig. 4 provide a summary of the convolutional neural network (CNN) layers and hyperparameters' precise configuration.

To reduce loss and accuracy oscillations, we adopted a conservative learning rate. Since smaller batches generally encourage more stable training, a batch size of 16 was chosen. As for the input shape, we used KMs of size $(N_{max}, M_{max}) = (80, 80)$ as they provide a balance between computational efficiency and feature extraction, in addition to eliminating high-frequency noise and anything else (electrical conductors) that could distort diagnostic features.

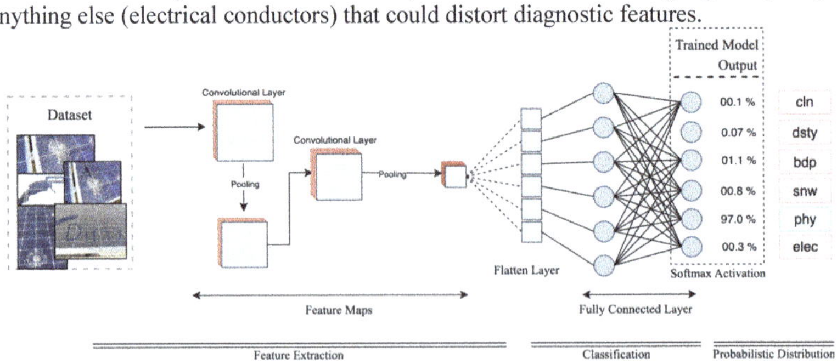

Fig. 4. Proposed CNN model architecture.

Materials Research Proceedings 64 (2026) 455-462 https://doi.org/10.21741/9781644904091-57

Table 1. Model's layers and properties.

Layer	Property
Convolutional layer	32 × 3 × 3
Max pooling	2 × 2
Convolutional layer	64 × 3 × 3
Max pooling	2 × 2
Dense	256 (ReLu activation)
Dense	6 (Softmax activation)

Table 2. Hyperparameters and performance metrics.

Configuration	Property
Metric	AUC
Batch size	16
Optimizer	Adam (Learning rate = 1e-4)
Loss	Focal loss
Epochs	100

Experimental Results

In this section, we apply the proposed CNN architecture to train and classify faulty solar panels.

Fig. 5 displays the AUC and loss performance of the model trained on the solar panel's dataset over 100 epochs. The model achieved a validation AUC score of 99.37%, indicating excellent discriminative capability. The performance curves reveal that beyond 80 epochs, both validation AUC and loss metrics plateaued, implying that extending training further provides negligible improvement.

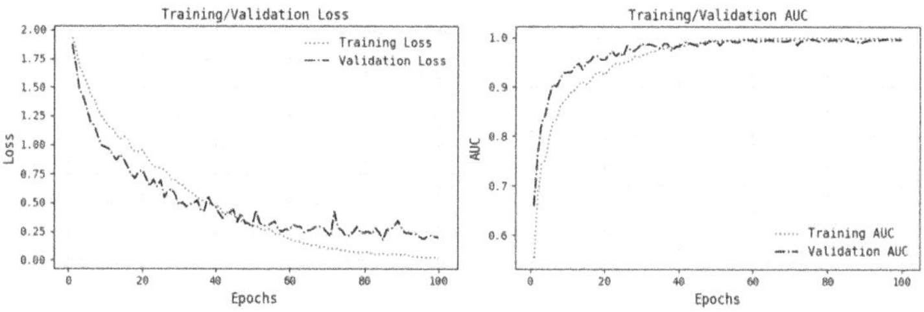

Fig. 5. Loss and AUC curves evolution during training.

Fig. 6 shows the confusion matrix, capturing the relationship between true and predicted labels of the test set. The matrix confirms that the model achieves consistently high classification accuracy across all classes.

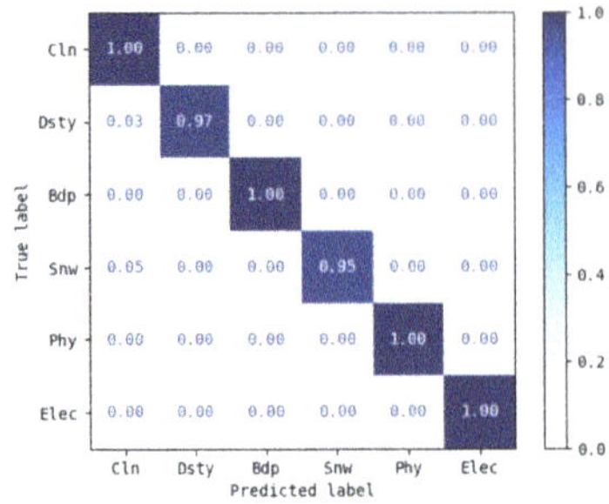

Fig. 6. Confusion matrix of the proposed CNN model on the test set.

Moreover, Fig. 7 presents the Receiver Operating Characteristic (ROC) curves for the individual classes. The AUC values across different classes range between 0.98 and 1.00, signifying strong model confidence and reliable separation between classes.

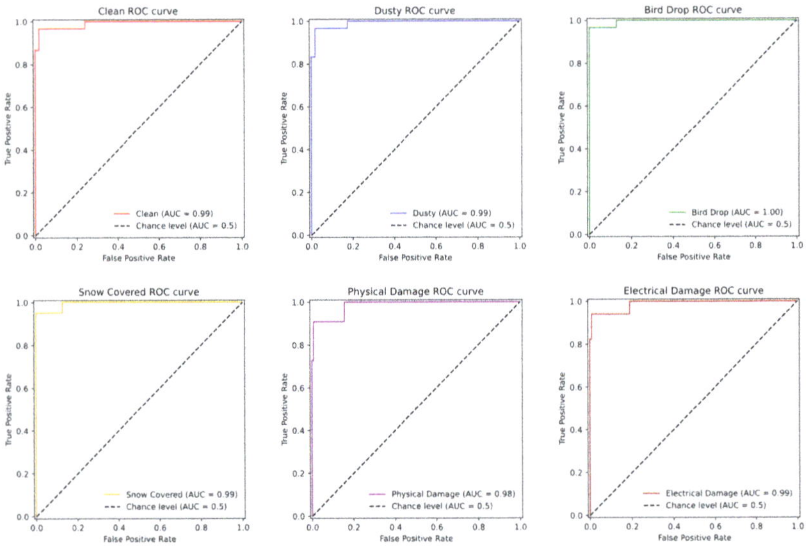

Fig. 7. ROC curves of each of 6 classes.

Ablation Study

In Table 3, we explore the effect of KMs order on feature extraction (MSE) and classification metrics. Higher-order moments hardly improve the MSE because KMs run into their computational precision limit. The AUC also plateaus around 0.9948. This is because adding more features mainly introduces more noise than features.

Table 3. Reconstruction MSE and classification AUC using different KMs orders.

Max order	(20,20)	(40,40)	(60,60)	**(80,80)**	(100,100)	(120,120)
MSE	5.7e-4	3.2e-4	1.9e-4	**1.2e-4**	1.2e-04	1.1e-05
AUC	0.9351	0.9704	0.9817	**0.9948**	0.9948	0.9948

Discussion

From the confusion matrix in Fig. 6, we can see that the results of Bird Drop, Physical Damage, and Electrical Damage were classified with perfect accuracy. This performance can be attributed to the KMs, which effectively focus on key features and capture detailed information from regions exhibiting clear physical damage or faults. Such results would not have been achievable without the SA algorithm or another efficient metaheuristic.

Our primary evaluation metric was the AUC, as accuracy can be misleading by overrepresenting the majority class. Our model achieved an AUC of 99.48%, which exceeds the reported accuracies of most state-of-the-art models, a notable result since AUC is generally more challenging to optimize. Table 4 presents a detailed evaluation of our method with existing approaches.

Table 4. Benchmarking against state-of-the-art methods for solar panel fault classification.

Ref	No of clss	Model	Accuracy	AUC	Precision	Recall	F1
[6]	2	UdenseNet	**99.39 %**	-	0.9879	**1.0000**	0.9940
[7]	2	CNN+IR	92.50 %	-	0.9200	0.9200	-
[8]	2	Multiscale CNN	97.32 %	-	0.9763	0.9700	0.9732
[11]	6	SPF-Net	94.35%	-	0.9400	0.9400	0.9400
[12]	6	SolarSynthNet Model	91.48%	-	-	-	-
-	**6**	**CNNs+KMs**	99.15 %	**0.9948**	**0.9896**	0.9861	**0.9958**

Conclusion

In this paper, we have used a set of discrete polynomials called Krawtchouk polynomials to extract features more efficiently, these moments have a unique property, using there parameters we can adjust the region where we extract most of the features, thanks to this property we have successfully captured the most important features in the case of physical and electrical damage, this has allow us to develop a small but efficient CNN model and achieve state of the art results with efficient feature extraction.

References

[1] I. Naouadir, O. E. Ogri, J. E. Mekkaoui, M. Benslimane, and A. Hjouji, 'Celestial Object Detection in Astronomical Images Using MSE and Jacobi Moments', *Statistics, Optimization & Information Computing*, vol. 12, no. 3, Art. no. 3, Feb. 2024. https://doi.org/10.19139/soic-2310-5070-1959.

[2] Y. Aakam, O. El Ogri, J. El-Mekkaoui, M. Benslimane, and A. Hjouji, 'Efficient 2D and 3D Image Classification and Compression Using Dual-Hahn Moments', 2024, pp. 312–324. https://doi.org/10.1007/978-3-031-68660-3_29.

[3] Y. Aakam, O. El Ogri, J. El-Mekkaoui, M. Benslimane, and A. Hjouji, 'Optimal image 2D/3D by Krawtchouk moments and ABC algorithm', *Statistics Optimization & Information Computing*, vol. 12, pp. 672–684, Feb. 2024. https://doi.org/10.19139/soic-2310-5070-1960.

[4] P.-T. Yap, P. Raveendran, and S.-H. Ong, *Krawtchouk moments as a new set of discrete orthogonal moments for image reconstruction*, vol. 1. 2002, p. 912. https://doi.org/10.1109/IJCNN.2002.1005595.

[5] P.-T. Yap, R. Paramesran, and S.-H. Ong, 'Image analysis by Krawtchouk moments', *IEEE Transactions on Image Processing*, vol. 12, no. 11, Art. no. 11, Nov. 2003. https://doi.org/10.1109/TIP.2003.818019.

[6] R. F. Pamungkas, I. B. K. Y. Utama, and Y. M. Jang, 'A Novel Approach for Efficient Solar Panel Fault Classification Using Coupled UDenseNet', *Sensors*, vol. 23, no. 10, p. 4918, Jan. 2023. https://doi.org/10.3390/s23104918.

[7] R. H. Fonseca Alves, G. A. de Deus Júnior, E. G. Marra, and R. P. Lemos, 'Automatic fault classification in photovoltaic modules using Convolutional Neural Networks', *Renewable Energy*, vol. 179, pp. 502–516, Dec. 2021. https://doi.org/10.1016/j.renene.2021.07.070.

[8] D. Korkmaz and H. Acikgoz, 'An efficient fault classification method in solar photovoltaic modules using transfer learning and multi-scale convolutional neural network', *Engineering Applications of Artificial Intelligence*, vol. 113, p. 104959, Aug. 2022. https://doi.org/10.1016/j.engappai.2022.104959.

[9] 'Solar Panel Images Clean and Faulty Images'. Accessed: Aug. 08, 2025. [Online]. Available: https://www.kaggle.com/datasets/pythonafroz/solar-panel-images

[10] C. Wang, C. Deng, and S. Wang, 'Imbalance-XGBoost: leveraging weighted and focal losses for binary label-imbalanced classification with XGBoost', *Pattern Recognition Letters*, vol. 136, pp. 190–197, Aug. 2020. https://doi.org/10.1016/j.patrec.2020.05.035.

[11] R. A. M. Rudro *et al.*, 'SPF-Net: Solar panel fault detection using U-Net based deep learning image classification', *Energy Reports*, vol. 12, pp. 1580–1594, Dec. 2024. https://doi.org/10.1016/j.egyr.2024.07.044.

[12] C. J. Ejiyi *et al.*, 'SolarSynthNet (SSN): A deep learning framework for binary and multiclass classification of damaged or obstructed solar panels using images', *Renewable Energy*, vol. 256, p. 124224, Jan. 2026. https://doi.org/10.1016/j.renene.2025.124224.

Optimization of Parameters Modeling the Angular Distribution of Ion Sputtering for the Components of Inconel 718 Alloy Using Genetic Algorithm

Abd El moughit BACHRAK[1,a], Tarik BOURAGBA[2,b], Kawtar BRIA[1,c] and Mohammed AIT EL FQIH[1,d]

[1]LISIME, National Higher School of Arts and Crafts, Hassan II University of Casablanca, Morocco

[2]EIGSI Casablanca, 282 Route de l'oasis Casablanca

[a]abdelmoughitbachrak@gmail.com, [b]tarik.bouragba@eigsica.ma, [c]brkawtar1993@gmail.com, [d]mohammed.aitelfqih@ensam-casa.ma

Keywords: Physical Vapor Deposition (PVD), Sputtering, Heuristic Optimization, Angular Distribution, Numerical Modeling, Refractory Metals

Abstract. The goal of this research is to create a new method for improving the angle at which thin films are deposited using a sputter deposition process, thereby controlling how the particles will be placed to produce a uniform surface with a high-quality coating of the desired material. To accomplish this, an equation with three coefficients (n, k, ν) was developed through empirical research and is used in conjunction with the Genetic Algorithm (GA) to discover these coefficients. The GA is a method that has been shown to work effectively when used on complex nonlinear problems. It was used to optimize the angular distribution for each of the six most common elements in Inconel 718 (nickel (Ni), chromium (Cr), iron (Fe), molybdenum (Mo), aluminium (Al), niobium (Nb), and titanium (Ti)). GA optimizes calculations using simulated data combined with real experimental data with added random noise mimicking actual laboratory conditions. χ^2 minimization is the GA's method for fitting the coefficients' values (the coefficients) and has proven to produce very accurate results. For example, for the case of Fe, where the optimised value n(GA) = 0.4214 is within a few percent of the target value n = 0.4200. The coefficients k and ν vary greatly from element to element, indicating that they reflect differences in the physical processes involved in how atoms are ejected from the target material at an oblique angle. Overall, these findings support the reliability and use of the GA to fit sputtering models, which may be useful for helping to improve the uniformity and quality of thin films deposited during sputtering.

Introduction

Sputtering processes [1-5] are now one of the major techniques for developing functional coatings used in demanding technological fields, such as aerospace and microelectronic devices. Their performance depends heavily on the ability to describe and control the atomic ejection mechanisms occurring at the target surface, particularly the spatial and energy distribution of the emitted particles [2-4].

A particularly crucial aspect for ensuring homogeneous film growth is the angular distribution $Y(\theta)$ of the ejected atoms, a parameter that is both fundamental and difficult to characterize experimentally [6-9]. While Knudsen's law ($Y(\theta) \propto \cos\theta$) provides a simplified framework, it fails to reproduce the anisotropic behaviors observed during high-energy sputtering or sputtering involving heavy elements [3,6]. To overcome these limitations, several semi-empirical, multi-parameter models have been developed [2,4]. However, determining the optimal coefficients of these models is akin to a complex nonlinear optimization problem, for which classical deterministic methods, such as least squares fitting; quickly show their weaknesses, particularly due to sometimes slow convergence and a high risk of trapping in local minima [7,8].

To overcome these limitations, this study proposes the adoption of a genetic algorithm (GA) [7,8], a bio-inspired stochastic optimization method particularly effective for exploring complex and multidimensional search spaces. The approach aims to determine the optimal values of the parameters (n, k, ν) of an established angular distribution model, applied to the seven principal elements of the Inconel 718 alloy. The parameter sets thus obtained demonstrate not only the robustness and performance of the GA but also provide crucial information for carrying out accurate simulations of sputtering and thin-film growth [10,11].

Methodology

To model the normalized distribution of sputtered species, a phenomenological three-coefficient formulation was adopted [2, 3]:

$$Y(\theta) = cos^{n}(\theta).(1 + (\frac{\theta}{\pi})^{k})^{-1/\nu}$$

In this expression, θ represents the particle ejection angle, expressed in radians. n is the main parameter that controls the cosine shape of the distribution. k and ν are correction coefficients used to adjust the distribution at large angles and compensate for deviations from ideal behavior.

The data used in this work do not come from new experimental measurements but rather from reference datasets reported by Bria et al. (2023). Specifically, the angular distributions $Y \propto \exp(\theta)$were numerically simulated using the values of $n$published in that previous study, as indicated in reference [12].

A review of the literature shows that optimization applied to sputtering angular-distribution modeling is very limited, which represents a key originality of this work. We will briefly discuss alternative methods such as least-squares regression and gradient-based optimization to justify the choice of GA. This highlights GA's advantages in handling multi-parameter, nonlinear, and noisy problems.

I will explicitly clarify in the manuscript the exact source of the experimental or reference angular-distribution data used for validation, to ensure full transparency and strengthen the reproducibility of the study. To introduce experimental realism, random Gaussian noise with a standard deviation $\sigma(exp) = 0.05$ was systematically added. Fitting criterion (evaluation function): The quality of each solution (n, k, ν) tested by the genetic algorithm (GA) is evaluated by minimizing the weighted squared error, commonly measured by the χ^2 statistic [7,8].

$$\chi^2 = \sum \frac{(Y_{exp}(\theta_i) - Y_{sim}(\theta_i))^2}{\sigma_i^2}$$

The computer program that uses the DEAP Library and implements genetic algorithms (GA) in Python provides many ways to define its population and the way that individuals evolve using EVO operators. The representation of solutions within the GA involves the encoding of every individual as a genotype. This genotype is represented as a vector of 3 components that represent the angular distribution model parameters (n, k, ν). The representation of the solution makes it possible for the GA to manipulate numerical values directly related to those parameters and search the solution's space efficiently. For search space, the GA search space is limited to the physical relevance of the parameters while permitting a range of solutions. Therefore, the values for all 3 parameters were constrained to the range from 0 to 2, inclusive. Regarding parameters and the evolutionary processes applied during the lifetime of the GA, the initial population consisted of 150 individuals, where these individuals evolved for 75 generations. The reproduction of individuals involved a crossover using the *cxBlend* operator at a rate of 80% to promote mixing of parental characteristics and providing a greater chance for local exploration of the parameter search space. Mutations were performed on individuals using the Mut Gaussian operator at a rate of 40%, allowing for the introduction of random variability in individuals and preventing the population of individuals from becoming trapped in local minima.

The selection process for the next generation of individuals was performed using a tournament-style selection with a tournament size of 3 to ensure that the fittest individuals survived and that the genetic diversity of the population was maintained through the selection process.

Results

Table 1 summarizes all the coefficients (n, k, v) identified as optimal by the GA (Genetic Algorithm) for each of the seven components of Inconel, as well as the minimum error (χ^2) achieved.

Table 1: *Parameters n, k, and v adjusted via the Genetic Algorithm. The target n serves as a comparison point for the fitting performance.*

Element	*Target n (Bria et al. [12])*	*n (GA)*	*k (GA)*	*v (GA)*	*χ^2 Min*
Nickel (Ni)	*0.4600*	*0.435*	*0.478*	*0.168*	*12.16*
Iron (Fe)	*0.4200*	*0.421*	*0.788*	*0.435*	*14.01*
Molybdenum (Mo)	*0.4000*	*0.431*	*0.915*	*0.375*	*15.39*
Niobium (Nb)	*0.3800*	*0.396*	*0.832*	*0.358*	*16.96*
Chromium (Cr)	*0.4200*	*0.402*	*0.663*	*0.375*	*18.24*
Titanium (Ti)	*0.3700*	*0.369*	*1.338*	*0.785*	*19.42*
Aluminum (Al)	*0.3900*	*0.358*	*1.286*	*0.827*	*31.48*

The adjustments made, as illustrated in Figure 1, show a very good correlation with noisy experimental data, with minimum χ^2 values ranging from 12.16 for Ni, and 31.48 for Al. The relatively low χ^2 values (most elements) indicate that the analytical function describing atomic ejection processes is both relevant and robust. The estimation of the angle of preference for atomic ejection, given by the parameter n, has a narrow range of values (between 0.358 and 0.435). The relatively close range of values indicates that the angular distribution tends to be more widely distributed (sub-cosine) than that predicted by Knudsen's Law (n = 1). This is consistent with collision cascade-dominated atomic ejection processes. Nickel (n = 0.435) tends to concentrate more towards the surface normal (more concentrated), while Aluminium (n = 0.358) tends to have a more diffuse ejection pattern (closer to isotropy).

The distinctions between moderate variations with respect to ejection mechanism(s) at each different component of the polycrystalline alloy can be ascribed to both correction coefficients k and v, which are significant contributors to the drop in their sharp distribution (as $\theta \to 90°$) because of the shadowing and back-scattering from the nearby atoms at the surface of these alloys. For instance, aluminium (k = 1.286; v = 0.827) and titanium (k = 1.338, v = 0.785), have a comparatively high value; therefore, they are experiencing the larger degree of effect from shadowing and back-scattering on their tangential ejection. In contrast, nickel (k = 0.478, v = 0.168) demonstrates considerably less effect from both phenomena thus will have a trajectory less affected by the geometry of the surface on which it is aiming at.

Materials Research Proceedings 64 (2026) 463-468 https://doi.org/10.21741/9781644904091-58

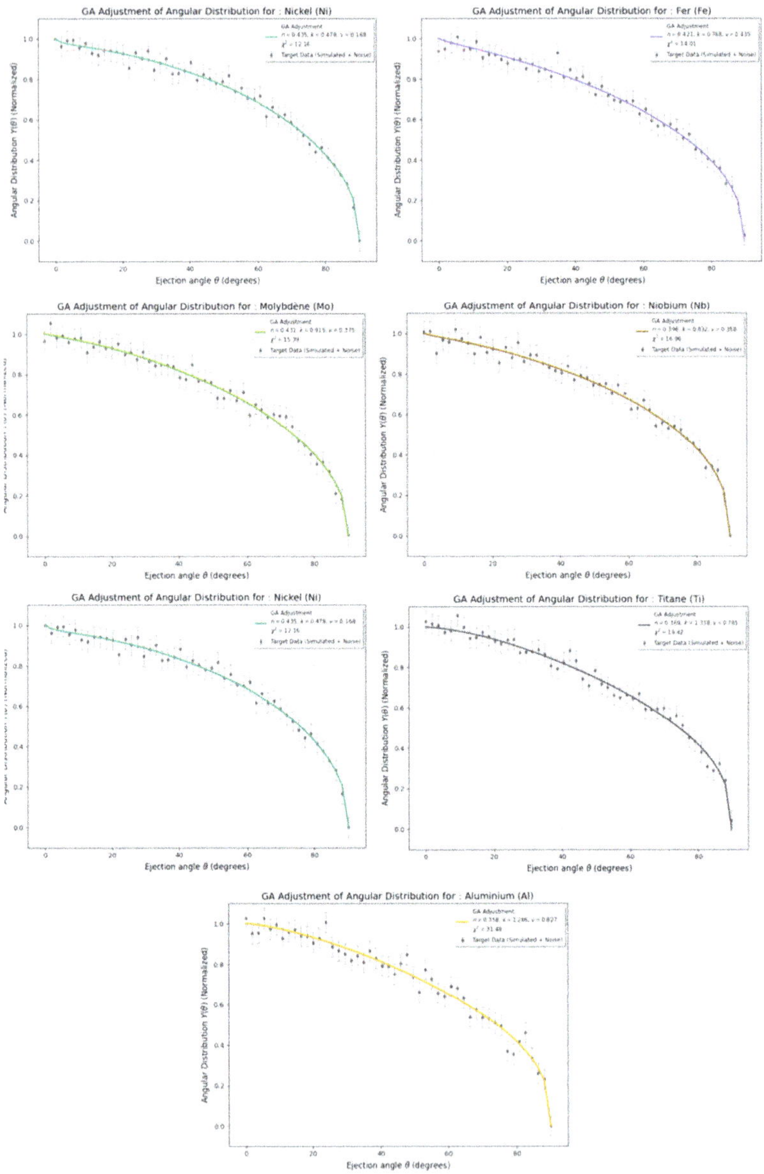

Figure 1: *Optimization of angular parameters for Inconel 718 components*

Conclusion

Using genetic algorithms (GA), researchers investigated the angular distributions produced during ejection of Inconel 718's constituent elements to better understand how these materials are expelled from a surface after impact. They developed a phenomenological model, with parameters defined as n (which indicates how tightly grouped materials will be after impacts), k (which provides an estimate of how far a material will travel horizontally before landing), and v (which indicates how much heat was dissipated after impact). Based on the results, the authors concluded that the angular distributions are all sub-cosine distributions ($n < 1$) due to the fact that the ejection or expulsion process is primarily a result of collision cascades and not due to thermal evaporation.

The study also revealed differences in the angular distributions among the metals studied. For example, while Nickel (Ni) has a high value of n indicating that it has a much tighter distribution toward the normal of the surface than Aluminium (Al) or Titanium (Ti), both Al and Ti showed much wider distributions and required the highest amounts of correction factors for k and v (due to large shading effects). The negative correlation between wide angular distributions of ejected materials and large amounts of tangential deceleration illustrates that ejection dynamics will depend on each material's atomic properties. The optimized parameters developed from this study will assist in developing a quantitative model for the deposition profiles of thin films and aid in understanding and interpreting complex experimental data in future studies.

References

[1] P. Sigmund, Theory of sputtering. I. Sputtering yield of amorphous and polycrystalline targets, Phys. Rev. 184 (1969) 383–416. https://doi.org/10.1103/PhysRev.184.383

[2] Y. Yamamura, H. Tawara, Energy dependence of ion-induced sputtering yields from monoatomic solids at normal incidence, At. Data Nucl. Data Tables 62 (1996) 149–253. https://doi.org/10.1006/adnd.1996.0005

[3] R. Behrisch, W. Eckstein (Eds.), Sputtering by Particle Bombardment: Experiments and Computer Calculations from Threshold to MeV Energies, Springer, Berlin, 2007. https://doi.org/10.1007/978-3-540-44502-9

[4] M. Ait El Fqih, P.-G. Fournier, Ion beam sputtering monitored by optical spectroscopy, Acta Phys. Pol. A 115 (2009) 907–911. https://doi.org

[5] A. Anders, Cathodic Arcs: From Fractal Spots to Energetic Condensation, Springer, New York, 2008. https://doi.org/10.1007/978-0-387-79108-1

[6] M. Pawlak, M. Krishnan, A. Anders, Angular distribution of sputtered material in magnetron discharges, J. Appl. Phys. 101 (2007) 123302. https://doi.org

[7] D.E. Goldberg, Genetic Algorithms in Search, Optimization, and Machine Learning, Addison-Wesley, Reading, MA, 1989. https://doi.org/10.11517/jjsai.7.1_168

[8] K. Deb, Optimization for Engineering Design: Algorithms and Examples, Prentice Hall, New Delhi, 1998.

[9] C.W. White, Ion induced optical emission for surface and depth profile analysis, Nucl. Instrum. Meth. 149 (1978) 497–501. https://doi.org/10.1016/0029-554X(78)90916-3

[10] M. Ait El Fqih, P.-G. Fournier, Optical emission from Be, Cu and CuBe targets during ion beam sputtering, Nucl. Instrum. Methods B 267 (2009) 1206–1210. https://doi.org/10.1016/j.nimb.2009.01.159

[11] A. Bachrak, L. Jadoual, K. Bria, M. Ait El Fqih, Ion-induced atomic excitation in Vanadium, Revista Mexicana De Fisica, 71 (2025) 031003. https://doi.org/10.31349/RevMexFis.71.031003

[12] K. Bria, M. Ait El Fqih, J.-M. Nunzi, L. Jadoual, A. Kaddouri, Angular distribution of sputtered particles from Inconel 718: a simulation study, Eur. Phys. J. Appl. Phys. 98 (2023) 26. https://doi.org/10.1051/epjap/2023220334

FPGA-Based Mine Traffic Controller with Pedestrian and Machinery Prioritization

Hajar SADOK[1,a *], Ahlam MOUSSA[1,b], Ilham NAIT HIM[1,c] and Moulay El Hassan EL AZHARI[1,d]

[1]Al Akhawayn University in Ifrane, Avenue Hassan II, Ifrane 53000, Morocco

[a]H.Sadok@aui.ma, [b]Ah.Moussa@aui.ma, [c]I.NaitHim@aui.ma, [d]M.ElAzhari@aui.ma

Keywords: FPGA Traffic Control, Underground Mine Safety, Finite State Machine (FSM), VHDL, ALTERA Board

Abstract. Underground mining is one of the most dangerous industrial settings, due to its inadequate lighting, limited infrastructure, and management of both pedestrian and machinery traffic flow through narrow tunnels. To enhance safety at the intersection of mining vehicles and miners, this paper presents the design and simulation of an FPGA-based traffic control system. The design is highly optimized, occupying less than 1% of the board's logic fabric, and it is implemented using VHDL. The model relies on a Finite State Machine (FSM) that governs two crossing paths, one for mining machinery and another for mining pedestrians. Default priority is assigned to the mining vehicles, and, upon request, miners are granted timed access. To simulate extreme cases, an emergency override functionality is applied, such that it halts all traffic to prioritize evacuation, and the system employs 7-segment displays and LEDs for countdown timers and status display. The maximum response time for the state transitions is a mere 20 ns, verified through Quartus simulation of the traffic control design. The latter is additionally validated in Proteus, which further confirms the solution as a reliable approach to real-time mine traffic control.

Introduction

While not being among the largest contributors to the world of industry and employment, mining presents a large source of fatal injuries in private industries, which is roughly four times the average rate. In 2021, the fatal injury rate for mining was 14.2 per 100,000 full-time workers, compared to 3.8 overall in private industry [1]. Poor visibility, limited space, and inadequate real-time traffic control at traffic crossings of a mine are particularly vulnerable to accidents [2]. In such high-risk environments, the conventional control and signaling systems may not suffice, and the most traditional sense of control is likely to fail under emergency conditions, such as fires and evacuations [3].

To reduce these risks, the use of smart embedded systems for real-time decision-making is at the top priority. Microcontrollers are common in the above ground systems used to control and manage traffic lights. For instance, ATmega32 AVR with infrared sensors can provide this function [4]. Field-Programmable Gate Arrays (FPGAs) provide a powerful framework on which a smart traffic control can be developed. They are reconfigurable, have low latency, and can operate in parallel with one another [5]. The implementation of safety-critical systems in mine traffic control is possible with complex logic implemented using VHDL (VHSIC Hardware Description Language) and FSMs.

This paper proposes a simple yet robust mine traffic control system. It provides mining vehicles with default priority whilst accommodating pedestrian crossings upon request or detection. The system also includes an emergency override state that halts both pedestrian and vehicle crossing when called. The logic design is simulated in Quartus and visualized in Proteus, as well as on the

Materials Research Proceedings 64 (2026) 469-476 https://doi.org/10.21741/9781644904091-59

FPGA board for further validation, where utilizes seven-segment displays and LED indicators for intuitive feedback, thus reducing the risk of mine accidents.

Methodology

System Architecture and Modules. As shown in Fig. 1, the proposed mine traffic control system was modularly developed, centered on a Finite State Machine (FSM) as the central controller. The latter handles all inputs to determine the state of the system according to its Priority Logic ran concurrently on the Altera DE2-115 FPGA board, which results in a predictable and reliable system for mine traffic safety.

As part of the FSM inputs, pedestrian requests and emergency signals are controlled, with the Priority Logic giving default passage to mining vehicles. The Priority Logic relies on a timing module, consisting of 12-bit ripple counters that generate count downs and safety delays, while LEDs, a seven-segment display, multiplexers (MUXs), and a decoder are used for state display. By adopting this structure, pedestrian and emergency requests are safely and quickly handled.

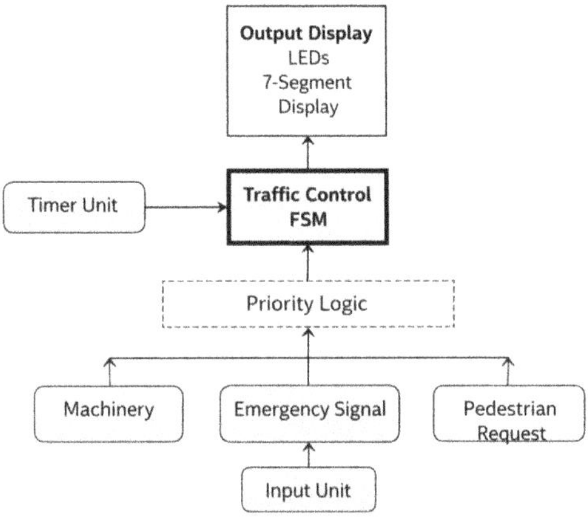

Fig. 1. Architecture diagram of the proposed mine traffic control system

Identification Tags and Proximity Sensing. In the modern mining industry, miners and mining vehicles are equipped with identification tags (e.g, RFID, or BLE beacons) [6]. These tags are identified by fixed sensors (anchors) placed at road intersections to detect their signals, calculate their proximity, and identify whether they belong to a miner, a mining vehicle, or an emergency vehicle.

Furthermore, the traffic controller automates safety and enhances it by using these tags and anchors. For instance, the system can detect miners' tags and automatically prepare for pedestrian crossings, eliminating the need for a miner to manually request crossing. In addition, anchors can also recognize emergency vehicle tags and spontaneously clear the path for priority passage. Beyond real-time control, this setup also serves as a digital audit keeping track of the tag entering or exiting intersections, allowing for safety monitoring.

Traffic Control FSM and Priority Logic. The FSM operates through five main states. The *Idle (IDLE)* state is the default mode, where no pedestrian requests are active, and vehicles are given priority. In the *Vehicle Go (Vehicle_Green)* state, the vehicle path remains open while pedestrian requests are either pending or inactive. When a pedestrian initiates a crossing request, the system enters the *Pedestrian Request (Pedestrian_rep)* state, signaling the need to clear vehicles from the crossing area. Once it is safe, the *Pedestrian Go (Pedestrian_Green)* state is activated, allowing pedestrians to cross while a countdown timer displays the remaining crossing time. Finally, the *Emergency (EMG)* state overrides all others, stopping all traffic except for predefined emergency vehicles, which are identified through their unique tags.

Transitions between the FSM states are based on real-time input signals. The *Pedestrian_Req* signal is activated when a pedestrian pushes the request button, activating the countdown timer and stopping the mining vehicle traffic flow. On the other hand, the *EMG* signal is sparked by the emergency trigger, immediately overriding all current states. These transitions are shown in the flowchart in Fig. 2.

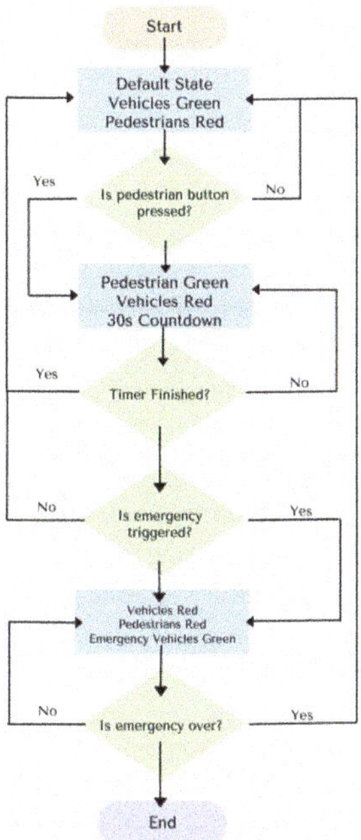

Fig. 2. FSM Transitions flowchart

FPGA Hardware Implementation. The physical implementation of the traffic controller used VHDL logic, developed and synthesized in Quartus II (13.1) design suite, and tested on the Altera DE2-115 FPGA. Unlike a microcontroller, which typically executes one instruction at a time, the parallel operation allowed by the FPGA architecture allows the different signals of the FSM and its priority logic to operate in their own logic units simultaneously, thus reducing latency and jitter [7]. The internal 50 MHz clock of the board powers the safety delays and aligns the onboard LEDs and 7-segment displays for real-time visual feedback. This hardware approach, instead of software-dependent microcontrollers, avoids bottlenecks, lagging, and system crashes, critical to underground mining safety.

Display and Decoder Logic. The 7-segment decoder was implemented using VHDL logic, following truth tables and Karnaugh maps (K-maps), where each binary input *ABCD* is mapped to the corresponding 7-segment representation (a, b, c, d, e, f, g) in active-low logic (logic 0 turns a segment ON).

The truth table behind the active-low 7-segment display representation is presented in Table 1.

Materials Research Proceedings 64 (2026) 469-476 https://doi.org/10.21741/9781644904091-59

Table 1. Active-Low 7-segment display truth-table

HEX	DCBA	a	b	c	d	e	f	g
0	0000	0	0	0	0	0	0	1
1	0001	1	0	0	1	1	1	1
2	0010	0	0	1	0	0	1	0
3	0011	0	0	0	0	1	1	0
4	0100	1	0	0	1	1	0	0
5	0101	0	1	0	0	1	0	0
6	0110	0	1	0	0	0	0	0
7	0111	0	0	0	1	1	1	1
8	1000	0	0	0	0	0	0	0
9	1001	0	0	0	0	1	0	0
A	1010	0	0	0	1	0	0	0
B	1011	1	1	0	0	0	0	0
C	1100	0	1	1	0	0	0	1
D	1101	1	0	0	0	0	1	0
E	1110	0	1	1	0	0	0	0
F	1111	0	1	1	1	0	0	0

For each segment (a-g), a Karnaugh Map (K-Map) is constructed from the truth table, and used to derive a minimized Boolean expression, all of which are implemented in a VHDL process to control each segment output (F_a *to* F_g). The decoder converts a 4-bit binary input to a decimal digit displayed on the 7-segment display, enabling the countdown functionality. As an example, the expression of segment F_a be derived as from the K-Map in Table 2:

Table 2. K-map for segment F_a

	00	01	11	10
00	0	1	0	0
01	1	0	0	0
11	0	1	0	0
10	0	0	1	0

$$F_a = (B \cdot \overline{D}) + (A \cdot \overline{C}) + (\overline{B} \cdot \overline{C} \cdot \overline{D}). \quad (1)$$

The other segment expressions can be similarly derived. As for the decoder, it is essential to provide visual feedback, such as the countdown when pedestrians are crossing or when an emergency is detected, thus ensuring real-time intuitive feedback for the miners.

Results and Discussion

The VHDL logic was first synthesized and simulated in Quartus to verify functional correctness. The system was then validated on the FPGA board to simulate the traffic scenarios and verify the FSM transition states under idle, pedestrian request, and emergency conditions.

Quartus Simulation. The design was first implemented as a VHDL in Quartus II 13.1. Each FSM state and transition was represented using the same controlled input variables mentioned in the system architecture section. To test the validity of the program, a VWF (Vector Waveform File) was generated in Fig. 3. The latter clearly showed the correct transitions between states, synchronizing counters, and the behavior of outputs with a delay.

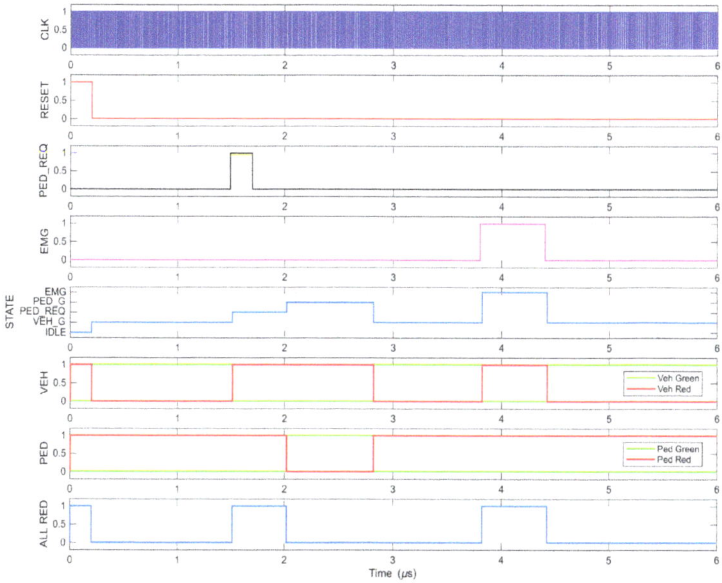

Fig. 3. Waveform of the simulated traffic control program

Particular attention was given to the emergency state, ensuring that traffic was halted while predefined emergency vehicles only were active.

Timing diagrams confirmed the accuracy of countdown behavior using ripple counters. However, the delay in Quartus does not align with the Altera board one, leading to the observed waveform, despite the accuracy of the board operation.

Proteus Visualization. The proposed logic was evaluated in the Proteus software in order to visualize the real-time behavior of the system. An ATmega32 microcontroller was used, and a push-button was linked to the pedestrian' request. Moreover, a gas sensor was used to simulate emergency triggers, especially relevant in the mining industry. Red and Green LEDs symbolized the position of both the vehicles and the pedestrian crossing, and the countdown was also programmed to give a crossing time of 5 to 0.

The system ran as aimed when the simulation was run, even in the Idle State which is observed in Fig. 4, and automatic reset to vehicle priority which follows the pedestrian crossing and emergency override states. The effectiveness and ease of use of the traffic control system was also confirmed by the success of this simulation, and the visual indicators that were of particular importance in the limited lighting conditions of normal underground tunnels.

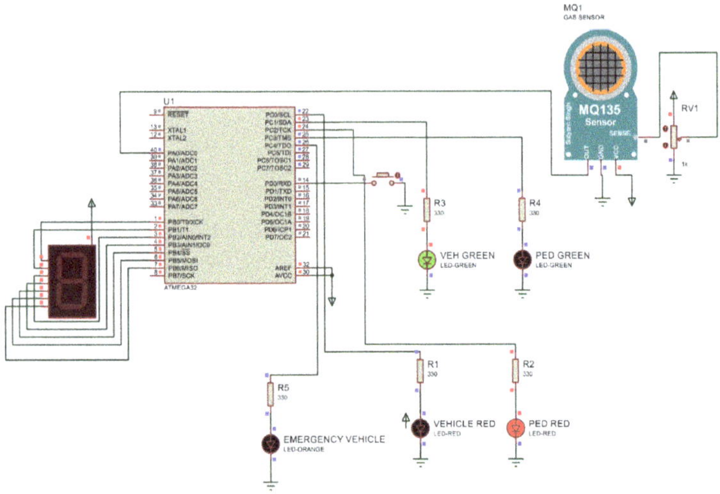

ã

Fig. 4. Proteus simulation of the traffic control - Idle State

Hardware Visualization. Finally, the ALTERA DE2-115 board was used to synthesize and implement the full FSM design. Each FSM state was represented by an LED, and the 7-segment display showed the pedestrian crossing countdown timer set to 30 seconds. Both the pedestrian request and the emergency trigger were simulated using pushbuttons on the board.

To verify the core states, the idle state was first observed, in which vehicles had the default priority, and pedestrians were stopped from crossing. Next, the pedestrian request allowed the pedestrians to cross for 30 seconds, displayed on the countdown, whilst vehicles were stopped, as seen in Fig. 5. Lastly, the emergency trigger stopped both vehicles and pedestrians from crossings, allowing only emergency vehicles the right of passage.

Fig. 5. DE2-115 board pedestrian request state

Performance Metrics and Analysis. The VHDL implementation showed efficient pin requirements, leveraging less than 1% of the available pins on the ALTERA board. Additionally, by utilizing the internal 50 MHz oscillator, the system maintains a response time of 20 ns for state transitions, including the emergency response time, given by the clock period.

By contrast, standard microcontrollers tend to operate at lower frequencies, requiring more cycles, interrupt handling, and ultimately larger response time (e.g., ATmega32 operates at a maximum frequency of 16 MHz).

Discussion and Potential Deployment. The proposed FPGA-based system offers reliable, real-time traffic control that is also energy efficient. Its hardware-based FSM, in contrast to software-based solutions, guarantees low power usage, and its modular architecture can be extended to a highly scaled deployment with comparatively little extra energy investment. The system can be used with the support of micro-power sources, which can be integrated into the system, contributing to its autonomous operation, energy efficiency, and sustainability.

Distinctive Features. In contrast to existing FPGA-based mine traffic systems, which either adhere to predetermined signal sequences or do not consider environmental hazards [8, 9], the proposed design is a single, unified, FSM with built-in priorities that directly wires RFID and BLE sensors, rendering it a safer, more comprehensive, and automatic system.

System Limitations. Currently, the system is limited to a simple two-way (vehicle/pedestrian) setup. However, despite the plentiful internal logic of the FPGA board, it only has a limited number of I/O pins, thus limiting the number of intersections the board can wire. Additionally, in real mines, the anchors and the FPGA board might be very far apart, leading to a voltage drop that might require extra hardware like signal repeaters.

Furthermore, underground mines tend to have humid and dusty environments, which would affect the FPGA board unless provided with industrial closures following the IP67 rating. The heavy machinery correlated with underground mining practices (e.g., continuous drilling) also leads to mechanical stress and vibrations, that may even cause cracks on the FPGA.

Conclusion and Future Work

The proposed system of traffic control of underground mines using FPGA that guarantees the security of pedestrian crossings provides valid results. The system had real-time functionality, which was successful with a high level of energy efficiency and flexibility. It was designed in VHDL, simulated in Proteus and Quartus, and executed on an ALTERA DE2-115 board.

Future studies will be focused on the expansion of tag-based monitoring to enhance automation, machine learning to enable adaptive timing and predictive safety analytics [10, 11], and sustainable energy-aware operation [12, 13]. With the integration of lightweight machine-learning models, the system would be more intelligent and context-aware with regard to traffic control and dynamically adjust pedestrian crossing times and default priority.

Declaration of use of AI-Assistance in the Writing Process

The authors acknowledge the use of generative artificial intelligence (AI)-assisted technologies in the preparation of this document. In particular, ChatGPT, developed by OpenAI, was used as an editorial and linguistic tool. The authors have full ownership of the technical content, analysis, interpretations, and conclusions presented in this work.

References

[1] Bureau of Labor Statistics, Mining fatalities rose 21.8 percent from 2020 to 2021 (2023). https://www.bls.gov/opub/ted/2023/mining-fatalities-rose-21-8-percent-from-2020-to-2021.html

[2] C. Andersson, Predictive model for traffic control in underground mines, DIVA (2019). https://urn.kb.se/resolve?urn=urn:nbn:se:ltu:diva-75964

[3] F. Javaid, A. Wang, M.U. Sana, A. Husain, I. Ashraf, An Optimized Approach to Channel Modeling and Impact of Deteriorating Factors on Wireless Communication in Underground Mines, Sensors 21 (2021) 5905. https://doi.org/10.3390/s21175905

[4] W. Kleitz, Digital Electronics: A Practical Approach with VHDL, ninth ed., Pearson, 2011.

[5] M.S. Arifin, S.A. Razi, A. Haque, N. Mohammad, A Microcontroller Based Intelligent Traffic Control System, Am. J. Embed. Syst. Appl. 7 (2019) 21-25. https://doi.org/10.11648/j.ajesa.20190701.1

[6] Smart Tracking and Monitoring in Supply Chain Systems using RFID and BLE, IEEE Conf. Publ. (2021). https://doi.org/10.1109/ICSPC51351.2021.9451700

[7] Boosted genetic algorithm using machine learning for traffic control optimization, IEEE J. Mag. (2021). https://doi.org/10.1109/TITS.2021.3066958

[8] A. Didal, P. Sachan, A. Katiyar, P. Gautam, Traffic Light Controller, Int. J. Eng. Tech. Res. 2 (2014).

[9] Unmanned driving technology in coal mine based on semantic information method, IEEE J. Mag. (2023). https://doi.org/10.1109/ACCESS.2023.3257574

[10] D. Fister, A. Jakopič, M. Truntič, Control Applications with FPGA: Case of Approaching FPGAs for Students in an Intelligent Control Class, Appl. Sci. 15 (2025) 12884. https://doi.org/10.3390/app152412884

[11] Adaptive Fault-Tolerant control of platoons with guaranteed traffic flow stability, IEEE J. Mag. (2020). https://doi.org/10.1109/TVT.2020.2990279

[12] PSIM-based modeling of automotive power systems: conventional, electric, and hybrid electric vehicles, IEEE J. Mag. (2004). https://doi.org/10.1109/TVT.2004.823500

[13] An RFID electronic tag based automatic vehicle identification system for traffic iot applications, IEEE Conf. Publ. (2011). https://doi.org/10.1109/CCDC.2011.5968962

Aerial AM Communication System for Remote Post-Disaster Alerts Using Machine Learning

Aya EL BOUZID[1,a *], Hajar SADOK[1,b], Kaoutar MESBAHI[1,c], Ilham NAITHIM[1,d], Nada AZOUAOU[1,e], Moulay EL HASSAN EL AZHARI[1,f]

[1]Al Akhawayn University Ifrane, Morocco

[a]a.elbouzid@aui.ma, [b]h.sadok@aui.ma, [c]k.mesbahi@aui.ma, [d]i.naithim@aui.ma, [e]n.azouaou@aui.ma, [f]m.elazhari@aui.ma

Keywords: Aerial Communication, Amplitude Modulation, Disaster Alerts, Signal Classification, Envelope Detection

Abstract. This paper presents a low-power aerial AM communication system for delivering emergency alerts in disaster-affected areas. The system transmits voice or dual-tone signals using Double Sideband with Carrier (DSB-C) modulation through a bandwidth-limited LTI channel modeled with 10 dB free-space path loss, additive white Gaussian noise (AWGN), and low-pass filtering. At the receiver, an envelope detector recovers the message, which is evaluated by a lightweight signal classification algorithm trained on simulated data across four SNR levels (0, 10, 20, and 30 dB). The effectiveness of the suggested hybrid analog computational framework for real-time quality assessment in limited communication environments is confirmed by MATLAB and Simulink simulations, which demonstrate that the system consistently recognizes signals at 25 dB SNR as clean.

Introduction

Natural disasters including hurricanes, earthquakes, floods, and wildfires, often disrupt critical terrestrial communication infrastructure, and when these networks fail, affected populations lose access to emergency alerts, evacuation instructions, and rescue coordination signals [1]. Therefore, developing alternative, infrastructure-independent communication systems remains a key priority for humanitarian response efforts [2].

Conventional systems, including cellular networks and Wi-Fi, rely on fixed base stations and centralized power grids. Although satellite communication offers broad coverage, it is costly, equipment-intensive, and may introduce latency that limits real-time operations [4].

Aerial communication systems, particularly those using unmanned aerial vehicles (UAVs), offer a flexible and rapidly deployable alternative [5]. Among transmission methods, analog amplitude modulation (AM), specifically Double-Sideband with Carrier (DSB-C), enables efficient wide-area broadcasting, compatible with low-cost, battery-powered receivers. Its straightforward demodulation via envelope detection further reduces hardware complexity [6].

However, aerial analog transmissions are subject to unpredictable channel effects such as noise, filtering, and fading [7]. To reduce these challenges, lightweight signal analysis techniques can be integrated to evaluate waveform quality and ensure reliable message delivery in disaster scenarios.

This paper presents an aerial amplitude-modulated (AM) communication system for the transmission of emergency alerts over a noisy linear time-invariant (LTI) channel. The receiver employs envelope detection for demodulation, followed by a machine-learning-based signal quality assessment. The system is implemented and evaluated using MATLAB and Simulink. The remainder of the paper describes the theoretical communication model, system architecture, and signal quality classification approach, followed by simulation results and concluding remarks.

Theoretical Model

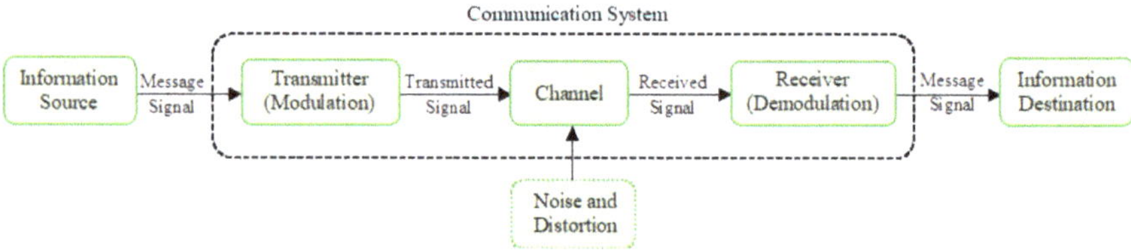

Fig. 1. Block diagram of a general communication system.

The general architecture of a communication system, shown in Fig. 1, consists of five main blocks: the information source, transmitter (modulation), transmission channel, receiver (demodulation), and information destination. This structure ensures reliable signal delivery in noisy environments and forms the foundation of the proposed hybrid analog–AI disaster communication system.
Information Source. The information source generates the baseband message signal m(t), which carries the useful information, such as voice alerts or warning tones. In analog systems, this low-frequency signal typically lies within the audio range (20 Hz–20 kHz) and cannot be transmitted efficiently over long distances without modulation [9]. In emergency applications, m(t) may correspond to pre-recorded alerts or synthesized signals designed for rapid recognition.
Transmitter. The transmitter prepares the baseband signal for transmission by applying modulation. In amplitude modulation (AM), the message signal m(t) varies the amplitude of a high-frequency carrier. Double Sideband with Carrier (DSB-C) is commonly used in basic analog systems due to its simplicity. The resulting modulated signal is expressed in Eq. (1).

$$s(t) = A_c[1 + \mu * m(t)]\cos(2\pi f_c t). \tag{1}$$

Where: A_c is the carrier Amplitude, f_c is the carrier frequency and μ the modulation index.

Transmitter Channel. The communication channel carries the modulated signal from the transmitter to receiver. Unlike ideal channels, real aerial or long-range links introduce distortions. In UAV or aircraft systems, the mostly line-of-sight (LoS) path is best modeled as a mix of deterministic and stochastic effects. There are three main effects which are commonly considered:

Free-space path loss which is the reduction in signal strength as a function of distance and frequency, and it is modeled using the Friis transmission equation shown in Eq. 2:

$$P_r = P_t * G_t * G_r * \left(\frac{\lambda}{4\pi d}\right)^2 \tag{2}$$

Where P_t is the transmitted power, G_t and G_r are the transmitter and receiver antenna gains, λ is the carrier wavelength and d is the separation between antennas. This relation is especially relevant for long-range aerial systems [7], [10], [11].

Additive White Gaussian Noise (AWGN) is used to represent thermal noise and random interference introduced by the environment and receiver electronics. AWGN is modeled as a zero-mean process with constant spectral density and is widely used in performance evaluation [6].

Low-pass filtering is applied to account for the bandwidth limitations of practical hardware, such as antennas, amplifiers, and demodulators [6].

The receiver removes the passband signal r(t) to obtain the original baseband signal m(t). This is achieved by using an envelope detector which is a simple and low power demodulation technique in the DSB-C AM systems. The detector measures the amplitude of the instantaneous waveform received that typically contains a diode, capacitor, and resistor. Signal quality is also influenced by the modulation index, channel noise and the detector response; too much distortion or overmodulation of the signal will make the recovery difficult. [6], [9].

The information destination is the end of the communication line at which the recovered signal is decoded, displayed or responded to. This might include, e.g. converting a signal into audible sound by using a speaker. The purity and clarity of the final message is critical to make sure that the information provided can be efficiently acted [9].

Suggested System for AM Communication

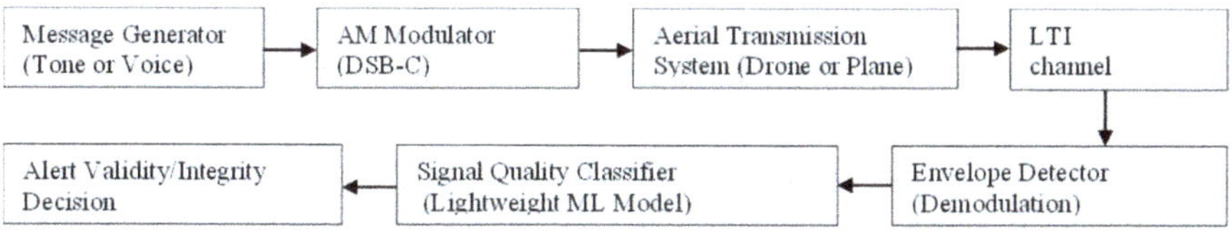

Fig. 2. Block diagram illustrating the workflow for the proposed aerial AM communication system

The proposed aerial AM communication system, shown in Fig. 2, integrates amplitude modulation (AM) with lightweight AI-based verification to enable reliable post-disaster alerts. A predefined audio message is generated, modulated using DSB-C AM, and transmitted via an aerial platform such as a drone or UAV. After propagation through a linear time-invariant (LTI) channel, the signal is received at a ground station and demodulated using an envelope detector. A machine learning–based classifier then evaluates the recovered signal quality to verify alert integrity, ensuring that only reliable messages are acted upon in remote or disrupted environments.

The message generator produces the baseband signal m(t) from a real emergency audio file sampled at 48 kHz, corresponding to a Nyquist frequency of 24 kHz in accordance with the Nyquist–Shannon theorem [12]. To enhance spectral analysis resolution without altering spectral content, the signal was converted to mono and repeated to form a 10-second waveform. The resulting one-sided magnitude spectrum (Fig. 3) indicates that most energy lies at low frequencies. Using a 5% threshold of the maximum spectral magnitude, the effective bandwidth was estimated as 4236.84 Hz. This bandwidth is critical for AM system design, as the carrier frequency must significantly exceed the message bandwidth to avoid sideband overlap and ensure distortion-free demodulation [6].

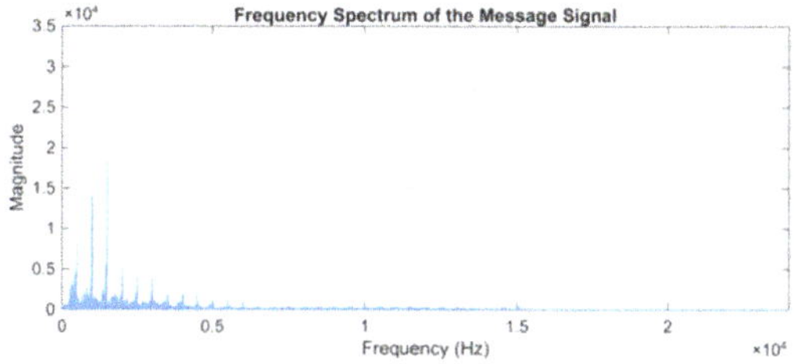

Fig. 3. Frequency spectrum of the repeated message signal computed using the Fast Fourier Transform (FFT)

The AM modulator converts the digital audio message into an analog waveform suitable for aerial broadcast. DSB-C modulation is selected for its compatibility with simple envelope detection and low-cost analog receivers, making it well suited for emergency communication where range and simplicity are prioritized over bandwidth efficiency [6]. The modulation process

Materials Research Proceedings 64 (2026) 477-484
https://doi.org/10.21741/9781644904091-60

combines the normalized message with a high-frequency carrier (A_c = 10, f_c = 5000 Hz) using low-complexity arithmetic operations. A modulation index of $\mu = 1$ ensures full modulation without overmodulation, preserving signal clarity. The modulated signal is then amplified to extend transmission range, enabling reliable drone-to-ground communication in remote or infrastructure-limited regions.

Aerial transmission system transmits the modulated signal of the emergency to ground receivers via a drone or a small-weight UAV. The AM signal is then sent through a low power RF module and through a simulated wireless channel to recreate real over the air effects. They are the free-space path loss that is modeled through the Friis equation, the additive white Gaussian noise (AWGN) and bandwidth constraints that arise out of low-pass filtering. This method facilitates the realistic system assessment with such distortions like distance and environment.

LTI Channel Model. The transmission medium is represented as Linear Time-Invariant (LTI) system which is a conventional and manageable model in case of short-duration broadcasts aerial. The composite channel involves the free-space path loss in order to simulate the attenuation of signals, AWGN in order to simulate the interference and low-SNR conditions, and bandwidth constraints in order to reflect the hardware limitations. Combined, these aspects offer a practical but achievable structure to assess modulation, demodulation, and machine learning-based signal analysis.

Envelope Detector. Signal recovery is made possible through the use of envelope detector and demodulation of the DSB-C AM signal in the absence of coherent synchronization can be realized. The detector uses half-wave rectification and then an RC low-pass filter to remove the signal envelope but the high frequency content. This inexpensive and low-power architecture guarantees precise recovery of messages, which is why it is applicable in the case of post-disaster communication with the right time constant.

Signal Quality Classifier. The quality of demodulated audio is measured by a lightweight classifier. Four SNRs (30, 20, 10 and 0 dB) were used to produce training data that was identified as clean, mildly distorted or severely distorted. Framing Signals were framed into 25 ms windows separated by 10 ms hops and features such as zero-crossing rate, short-time energy, spectral centroid were obtained. The values of mean and standard deviation were used to create a six-dimensional feature vector. The multi-class SVM classifier was trained on MATLAB ECOC framework and linear kernel, which was used to differentiate among levels of signal quality to be able to monitor post-disaster alerts quickly and efficiently.

Alert Validity Decision. Based on the output of the signal quality classifier, the system performs an alert integrity check. If the received message is classified as "clean," it is considered valid and forwarded for action. In contrast, alerts classified as "mildly" or "severely" distorted may be flagged for retransmission or ignored, depending on system thresholds. This decision logic ensures that only intelligible alerts are acted upon in high-risk, infrastructure-limited environments.

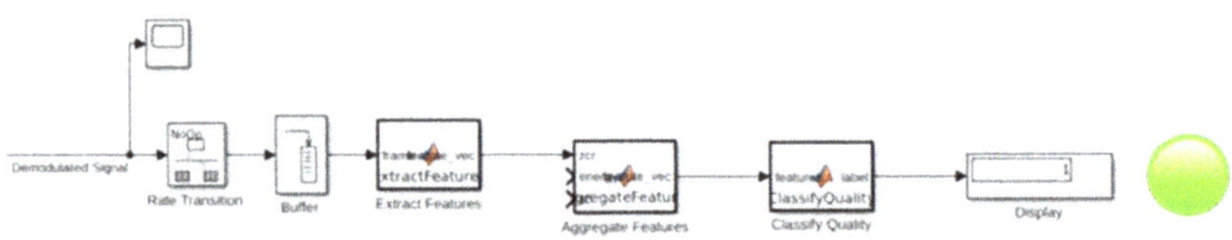

Fig. 4. SIMULINK implementation of signal quality classifier

In Simulink (see Fig. 4), a 1x6 feature vector is created by passing the demodulated signal through Rate Transition and Buffer blocks for framing, followed by Extract Features and Aggregate Features blocks that calculate zero-crossing rate, short-time energy, and spectral

Materials Research Proceedings 64 (2026) 477-484 https://doi.org/10.21741/9781644904091-60

centroid. The offline ECOC-trained SVM was replaced by rule-based logic that replicated its decision boundaries due to real-time constraints. The Classify Quality block receives the resultant vector and displays the output using a color-coded lamp: green for clean (label 1), yellow for mild distortion (label 2), and red for severe distortion (label 3).

Simulation Results and Discussion

The MATLAB/Simulink results for the entire AM communication chain, including modulation, channel effects, demodulation, signal recovery, and signal quality classification, are shown in this section. The findings demonstrate how the signal changes over time and verify that intelligibility is maintained while allowing for quality evaluation with a lightweight classifier.

Modulation of signals. By adjusting the carrier amplitude proportionately to the input signal, DSB-C amplitude modulation is used to prepare the message for transmission. The resulting waveform, as seen in Fig. 5, has a clearly defined envelope that retains the temporal characteristics of $m(t)$, allowing for straightforward envelope detection without compromising signal integrity.

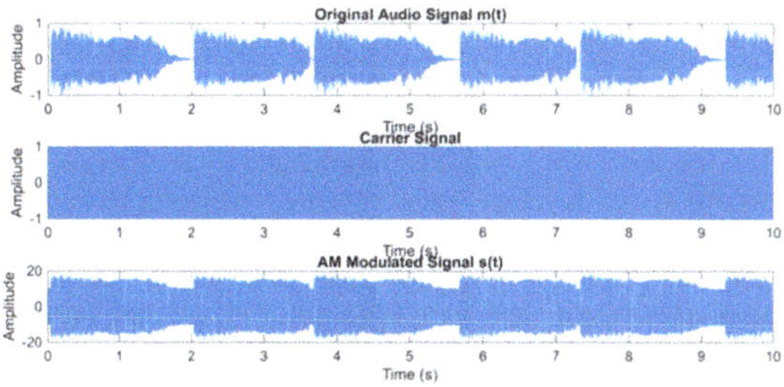

Fig. 5. AM modulation process simulated in MATLAB, featuring the original message signal, the carrier wave, and the resulting modulated signal

The figure illustrates the AM process; the top subplot shows the normalized baseband message signal representing spoken alerts, while the middle one depicts the high-frequency carrier used for conversion, and the bottom presents the DSB-C modulated waveform whose envelope follows the baseband signal, confirming successful modulation.

This part investigates the amplitude modulation signal attenuation, caused by Friis path loss and additive white Gaussian noise. A 10-dB loss would decrease the amplitude of the signal by about 20-8 units (Fig. 6), but the corresponding noise would be insignificant. This confirms that the Friis model does not impose any deterministic attenuation, and the amplitude-modulated envelope is not distorted visibly.

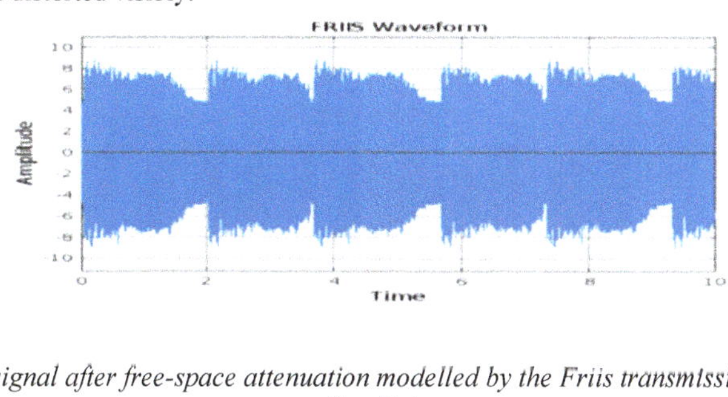

Fig. 6. AM signal after free-space attenuation modelled by the Friis transmission formula in Simulink

In the AWGN-corrupted signal, the orange trace appears significantly thicker and less defined than in the Friis-only output (Fig. 6), highlighting the impact of added noise. While the Friis model produces a clean, amplitude-attenuated waveform with a preserved envelope, the AWGN output exhibits a broader, noisier trace due to random high-frequency fluctuations. This contrast emphasizes the fundamental difference between the two effects: Friis introduces deterministic attenuation, whereas AWGN causes stochastic distortion that degrades signal clarity, as illustrated in Fig. 7.

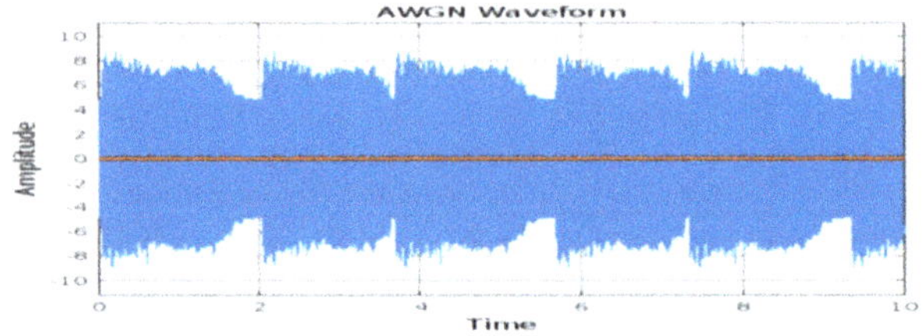

Fig. 7. Amplitude-Modulated signal with additive white gaussian noise simulated in Simulink

Demodulation. The received AM signal is demodulated to recover the baseband emergency alert after channel distortion. An envelope-based approach is employed due to its low complexity and suitability for situations where infrastructure is limited. In MATLAB, demodulation is performed using signal rectification followed by low-pass filtering, providing a computationally efficient approximation of envelope detection. Figure 8 compares the recovered signal with the original message, demonstrating successful envelope extraction.

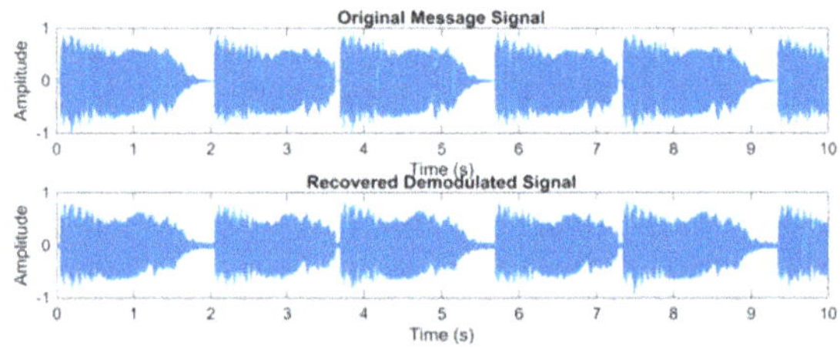

Fig. 8. Comparison between original message signal and recovered demodulated signal in MATLAB

Signal Recovery and Evaluation. This section measures the aerial AM system's signal recovery performance through comparison with the demodulated and the original signals, as indicated in the time-domain overlay in Fig. 9.

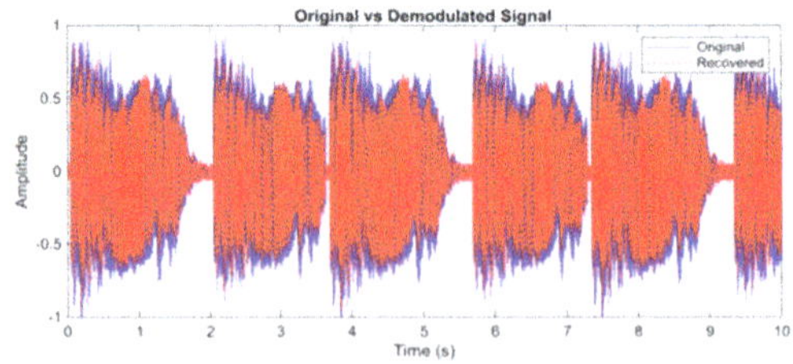

Fig. 9. Overlay of original and demodulated signals generated in MATLAB

The recovered waveform (red) closely follows the envelope of the original message (blue) but exhibits smoothing, delay, and attenuation during rapid transitions. Distortion is most pronounced at high-energy peaks, where flattening and timing shifts suggest filtering and phase-related effects. These artifacts reduce temporal accuracy and may affect the intelligibility of critical post-disaster alerts.

Channel Influence and Filtering Effects. The major cause of signal degradation is the bandwidth limitation and channel impairment. Real-world constraints in hardware, such as amplifier roll-off, antenna response behavior, receiver filtering, etc. reduce the high-frequency content required in any sharp signal transition, therefore reducing temporal resolution. The simplified channel of linear time-invariant (LTI) channel has the extra attenuation in amplitude and pushes the envelope extrema even at moderate signal-to-noise ratios (SNRs) (25 0 dB or so). In addition, the poor choice of modulation-index and the carrier leakage left behind also add to the deterioration of envelope-detector performance.

Signal Quality Classifier. The classifier was tested under various channel conditions that simulate SNR levels used in training (30, 20, 10 and 0 dB). Frame based feature extraction of demodulated signals was done on real time using Simulink. As shown in figure 4, the clean signals are appropriately categorized, and this proves that the lightweight implementation does not alter accuracy and ensures events post-disaster can be easily monitored in real-time.

Proposed Enhancements. To counter this spectral loss and distortion, transmitter-side pre-emphasis and receiver-side equalization can be used to mitigate degradation. Though envelope detection is still appealing as it is simple, it can be combined with coherent demodulation or more gentle post-processing steps to significantly improve the fidelity of the waveform. Such actions make them more robust and yet suitable to the low-resource wildlife situation following the occurrence of the disaster.

Summary

This paper presented a low-power AM modulation communication system for emergency alerts in areas that are prone to disasters, integrating a DSB-C AM modulation with a lightweight machine learning based signal quality classifier. The simulations exhibited a high level of robustness under noisy bandwidth-constrained channels, thus confirming the potential effectiveness of the framework when used in autonomous alert verification. Future research will explore adaptive modulation schemes, explore the use of AI in real-time validation, and evaluate the use of multi-drone cooperative networks as a way to increase coverage and redundancy. These features would be expected to be integrated with IoT sensors to allow context-dependent alerts, and expansion of the machine-learning architecture to distinguish between various types of emergency tones would help to increase versatility. The use of software-defined radio and implementation of solar-powered drones can switch to hardware, which will close the gap between simulation and deployment. Additional activities will include performance optimization of the ML pipeline in terms of speed-accuracy trade-off, adoption of GPS-based notifications of alerts, and field testing, which will be continuously developed and refined towards a scalable and smart emergency communications platform.

Declaration of Generative AI and AI-Assisted Technologies in the Writing Process.

"The authors used AI-assisted tools, including Grammarly, QuillBot, and Manus AI, to support grammar checking, rephrasing, and language refinement during manuscript preparation. These tools were employed solely to improve clarity and readability; all technical content, analysis, and conclusions remain the responsibility of the authors."

References

[1] A. Fathollahzadeh, J. Babaie, I. Salmani, M. A. Morowatisharifabad, and M.-R. Khajehaminian, Challenges of disaster risk communication from the perspectives of experts and affected people: A conventional content analysis, International Journal of Disaster Risk Reduction, 108 (2024) 104553. https://doi.org/10.1016/j.ijdrr.2024.104553

[2] D. A. Abuhani, R. Aldamani, I. Zualkernan, R. Alali, and H. Grover, Floods' Impact on Cities' Critical Infrastructure: A Case Study on Dubai, IEEE Access, 13 (2025) 17652–17661. https://doi.org/10.1109/ACCESS.2025.3527554

[3] Q. Wang et al., An Overview of Emergency Communication Networks, Remote Sensing, 15 (2023), no. 6, Art. no. 6, doi: 10.3390/rs15061595.

[4] S. Guo, L. Zhao, Y. Cui, Latency optimization of LEO satellite communication systems with beam hopping, in: Proc. IEEE Int. Conf. on Satellite Computing (Satellite), IEEE, 2022, pp. 37–42. https://doi.org/10.1109/Satellite55519.2022.00015

[5] A. Merwaday, I. Guvenc, UAV assisted heterogeneous networks for public safety communications, in: Proc. IEEE Wireless Commun. Networking Conf. Workshops (WCNCW 2015), IEEE, 2015, pp. 329–334. https://doi.org/10.1109/WCNCW.2015.7122576

[6] Modern digital and analog communication systems, 4th ed., PDF. Scribd, Accessed May 18, 2025. https://www.scribd.com/document/634006456/Modern-Digital-and-Analog-Communication-Systems-4th-Edition-pdf

[7] T.S. Rappaport, Wireless Communications: Principles and Practice, 2nd ed., Prentice Hall, Upper Saddle River, NJ, 2009.

[8] S. Aldossari, K.-C. Chen, Machine learning for wireless communication channel modeling: An overview, Wirel. Pers. Commun. 106 (2019) – May. https://doi.org/10.1007/s11277-019-06275-4

[9] A.B. Carlson, P.B. Crilly, Communication Systems: An Introduction to Signals and Noise in Electrical Communication, 5th ed., McGraw-Hill, Boston, 2010. https://doi.org/10.1109/TIT.1976.1055561

[10] Y. Zeng, R. Zhang, T.J. Lim, Wireless communications with unmanned aerial vehicles: Opportunities and challenges, IEEE Commun. Mag. 54 (5) (2016) 36–42. https://doi.org/10.1109/MCOM.2016.7470933

[11] D.M. Dobkin, Chapter 4 – UHF RFID readers, in: D.M. Dobkin (Ed.), The RF in RFID, 2nd ed., Newnes, 2013, pp. 103–188. https://doi.org/10.1016/B978-0-12-394583-9.00004-1

[12] S. Dhanani, M. Parker, Chapter 3 – Sampling and aliasing, in: S. Dhanani, M. Parker (Eds.), Digital Video Processing for Engineers, Newnes, Oxford, 2013, pp. 11–18. https://doi.org/10.1016/B978-0-12-415760-6.00003-9

Materials Research Proceedings 64 (2026) 485-492 https://doi.org/10.21741/9781644904091-61

Low-Powered IoT-Based Real-Time Weather Monitoring and Location Tracking using MQTT

Miral DESAI[1,a], Hiren MEWADA[2,b*], Archi GUJAR[1,c], Brijesh KUNDALIYA[1,d]

[1]Department of Electronics and Communication Engineering, CSPIT, CHARUSAT Changa, Gujarat, India

[2]Electrical Engineering Department, Prince Mohammad Bin Fahd University, Al Khobar, Saudi Arabia

[a]miraldesai.ec@charusat.ac.in, [b]hmewada@pmu.edu.sa, [c]23ec033@charusat.edu.in, [d]brijeshkundaliya.ec@charusat.ac.in

Keywords: Low Powered IoT, MQTT, Weather Monitoring

Abstract. Accurate weather conditions monitoring and precise geographical location is vital requirement in an IoT applications. The proposed article presents Low power IoT based Real time weather monitoring and geographical location tracking system. The proposed system incorporated Low power embedded platform as Raspberry Pi. Weather monitoring parameters as Temperature and Humidity is measured using the precise sensor. Location tracking is implemented using GPS module. The Raspberry pi board measured weather monitoring data and geographical location using sensors and submodules. The measured data is transferred to Lightweight mobile application. The data is transferred securely through MQTT – IoT based Protocol. The proposed system is scalable as it is design on the low power platform with IoT protocol. The proposed system is integrated with Telegram bot which enables remote access of real time environment.

Introduction

Weather monitoring and accurate geographical location tracking is one of the fundamental requirement for real time embedded systems. Electronic sensors and GPS technology can fulfil this requirement. However, precision and accuracy are essential for parameters. The precise sensors need to interface with low power embedded and IoT modules. Embedded modules are used to process the sensor data while IoT modules used to transfer data through wireless media.

With the rapid advancement in the field of Embedded and IoT solutions, it becomes a potent resource in real-time environments and location monitoring, especially in sectors such as agriculture, logistics, defense surveillance, and disaster management. With the development of IoT, we can collect and transmit live weather m and geolocation data, which is essential for efficient and informed decision-making. Nonetheless, IoTs, sensors, and low-power controllers can allow monitoring of weather conditions in real-time. However, edge-based devices have limited resources, and computational limitation makes a constraint on the implementation of these devices. Hence, weather monitoring and GPS systems are expensive, complex, and require high-end infrastructure with continuous internet connectivity, which might not be possible in various remote or rural areas. A cost-effective and energy-efficient wireless sensor design for recording temperature and humidity, integrated into a wireless network, is suggested for the unlicensed 2.4 GHz band [1]. The suggested wireless sensor network is appropriate for Smart Home applications. Li et al. [2] proposed a method for sensor calibration in open space. With the help of a microcontroller and a LoRa wireless module, we were able to create a cloud-based intelligent calibration system for temperature and humidity that uses several nodes.

In [3], the effects of weather and humidity on animals' well-being were discussed. An effective system for environmental monitoring was established, utilizing the Sigfox IoT network, in both the stockyards and the shade of shelters for cattle. The measured peaks of the TX current are slightly bigger than the LoRa values. An advanced system has been implemented to oversee medication distribution and control the temperature of pharmaceuticals during transit, mitigating the risks posed by environmental influences [4]. The suggested system utilizes mobile applications as user interface, incorporates Arduino, MQTT, and a temperature sensor, and employs a compact portable cooling unit. The system's accuracy is reflected in an F1-score of 68%.

To overcome these challenges, this study presents a compact, cost effective and scalable IoT Bases system designed to monitor weather conditions as well as the geographic location information. The proposed system is portable and can be deployed easily. The system primarily uses Raspberry Pi Zero W as the main processor which when interfaced with the DHT-11 sensor and NEO-6M GPS Module which is mainly used for measuring the temperature and humidity, and the GPS module which will provide the geographical coordinates which includes latitude, longitude and altitude. Furthermore, the system uses MQTT (Message Queuing Telemetry Transport) protocol and transmit the collected data to compatible MQTT public client. In addition to this, a notification mechanism is being implemented to generate alerts when temperature rises above a specified threshold, for this purpose a telegram bot is deployed to notify the end user in real time.

The overall structure of the paper is as follows: Section 2 presents the Methodology, introducing hardware system integration. Section 3 shows the experimental results, and a comparison with other systems is presented. Finally, a conclusion with a feature scope is given in Section 4.

Methodology

As highlighted earlier in the introduction section, large portion of the existing of literature focuses on the temperature monitoring, gas detection, and other environmental condition monitoring using various sensors. Much of the exiting work revolves around Arduino UNO as their primary microcontroller due to it's the flexibility and easy to use functionality makes it popular choice in IoT based applications. Hoverer it inherits certain limitation each include limited processing power and computational capacity and memory, as well the unavailability of built in Wi-Fi, restricts its performance in more complex and larger systems. The proposed system uses Raspberry Pi (RP) over Arduino Uno. The integrated Wi-Fi network simplifies network connectivity and provides better connectivity in comparison to a separate Wi-Fi module connected to the central system.

In this work, MQTT protocol is employed for data transmission. Although TCP/IP can be used, MQTT offers serval advantages that makes it more suitable for IOT applications, one of the key advantage here is that MQTT has a very small header size which is as low as 2bytes whereas the raw TCP/IP communication requires custom packet structure and larger headers. This light weight nature makes the MQTT protocol ideal for low bandwidth and resource constrained device. Moreover, MQTT works on the publish and subscribe model which improves stability and simplify the system design. These features make MQTT a suitable choice for application development, ensuring low latency and reliable data transfer.

The presented system is divided into two main subsystems. The first subsystem consist the sensor and control unit, and the second subsystem is responsible for data acquisition and transmission.

Subsystem-1: Hardware System Integration

Here the Raspberry Pi zero w is employed as the main controller of the system, with that DHT-11 sensor and NEO-6M GPS module are integrated for the data collection. All the data processing which includes sensor data acquisition and communication operations are executed using python scripts. The block diagram of the proposed system along with components is shown in Figure 1.

Materials Research Proceedings 64 (2026) 485-492 https://doi.org/10.21741/9781644904091-61

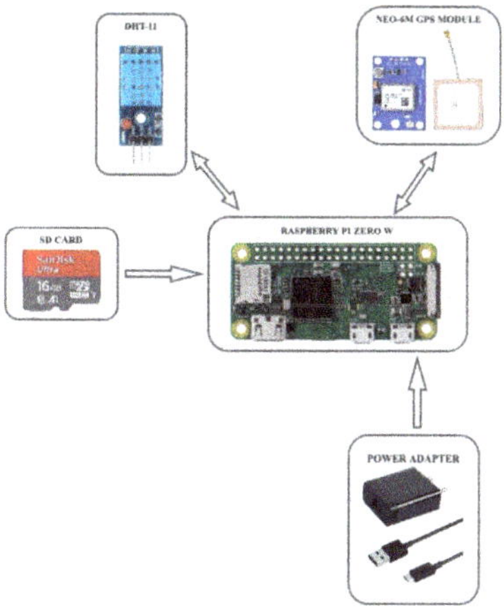

Figure 1: Block Diagram for Proposed Approach

The Raspberry Pi (RP) functions as the central processing and control unit of the proposed system. Raspberry Pi is ARM11 based Gigahertz based processor which supports High level language and IoT utilities. On chip GPIOs of the boards facilitate ADC conversions for sensors operations. Furthermore, Raspberry pi supports Linux Environment which is essential for IoT applications. Temperature and Humidity parameters measures through Single sensor known as DHT11. DHT11 is easily compatible with the Raspberry pi. The sensor measures the temperature in form of degree Celsius and humidity is measured in form of %. It integrates Raspberry Pi Zero W through the GPIO pins. The DHT-11 VCC and GND pins are connected to the Pi's pin1 and pin6, respectively. The data pin is connected to GPIO 4 of the Raspberry Pi. To read the sensor data, the Adafruit_DHT library is utilized, which will facilitate the process of collecting the data from the DHT-11 sensor and getting accurate environmental readings. Figure 2 shows this sensor module.

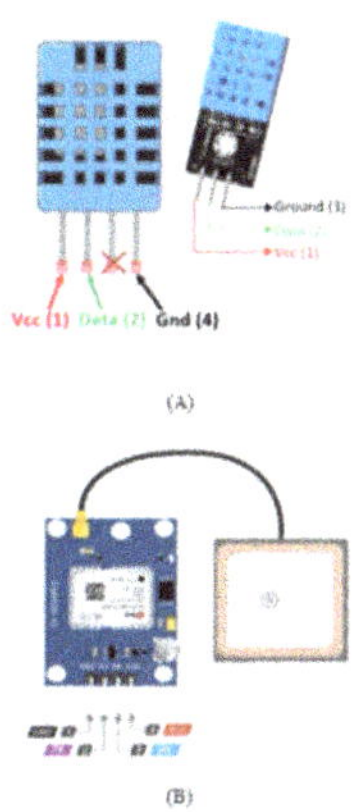

Figure 2: (a) Temperature and Humidity Sensor (DHT-11) module, (b) GPS Module for location access.

To access the geographical location, a GPS is also integrated into the proposed system. THE NEO-6M GPS module is responsible for providing geographical coordinates, including Latitude, Longitude, Altitude, and timestamp in UTC. The module communicates with RP using the UART (Universal Asynchronous Receiver/Transmitter) protocol. The TX pin of the GPS module relates to the RX (GPIO 15) of the RP. The RX pin of the GPS module will be left unconnected, as we are not receiving any data; the data transmission is unidirectional. The module output is in NMEA (National Marine Electronics Association) format, which is formatted and parsed using the Python library GPSD. Tools such as 'cgps' and 'gpsmon' are used to visualize the GPS data on the terminal.

Subsystem-2: Data acquisition and Transmission
Data acquisition and Transmission is carried out using sensor measurement, IoT protocol, IoT Broker and lightweight mobile utility viz telegram bot. Figure 3 presents functional flow diagram of proposed system. As shown in Figure. 3, environment data is measured by the sensors and geographical location is measured using GPS module. Environment data is processed by the Raspberry pi board. Raspberry pi board also record geographical data using GPS module. The sensor data and geographical data are transferred wirelessly to IoT broker using MQTT protocol. At the client side, the telegram bot utility is implemented which enables alert system if Temperature and humidity data is increased above threshold value.

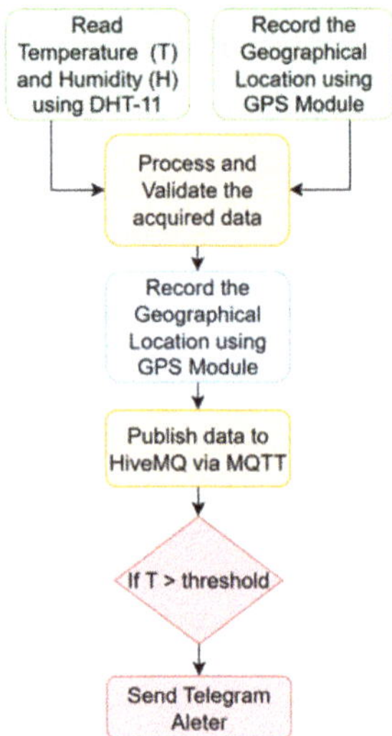

Figure 3: Flowchart of the Proposed System

Test Results and Discussion
This section depicts the outcome of the system. The principal aim of this project is to enable accurate, real-time environmental monitoring and reporting the geographical coordinates. The system successfully collects all the required data. Figure 4 showcases a screenshot of the sensor data obtained from the developed IoT system. The DHT-11 which was responsible for measuring

the environments data which includes temperature and humidity, measured the daytime temperature of 33C and relative humidity of 65.0%, furthermore the NEO-6m GPS module accurately determined the geographical coordinates which includes latitude, longitude and altitude.

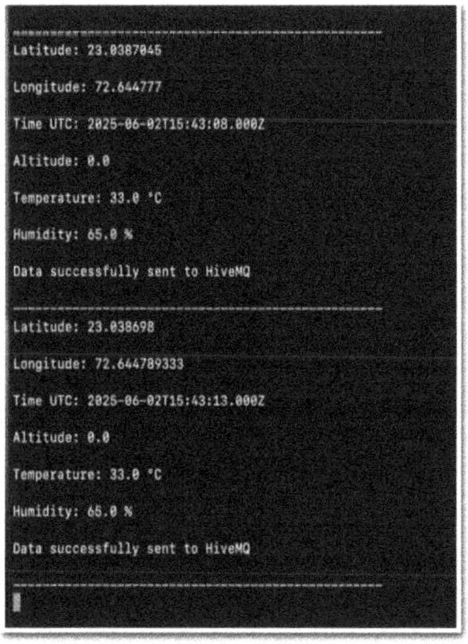

Figure 4: Sensors' data reading and acquisition

A HiveMQ interface is implemented in the proposed system to enable MQTT-based communication. Figure 5 displays the HiveMQ WebSocket client interface used for data exchange. The system includes a well designed dashboard that supports host configuration and port selection. Users can also control the connection setting by configuring a last-will message, which helps to ensure reliable operation in the event of an unexpected disconnection.Furthermore the dashboard provides options for publishing messages to specific topics and subscribing to a particular topic.As an example sensor data can be published to the topic "iot/sensor/data"through the "Publish" section, while the "Subscriptions" section allows the end user to monitor real-time updates from the same or related topics. The series of messages recorded via HiveMQ user interface. Each message published under the mentioned topic provides temperature, humidity, and geographical coordinates along with the corresponding time zone. This organized logging demonstrates that the system works as expected ,which is vital for continuous monitoring. Temperature readings range from 33.4 °C to 34.0 °C, while humidity remains steady at 65%, highlighting the system's accuracy and effectiveness in real-time environmental monitoring.

Materials Research Proceedings 64 (2026) 485-492 https://doi.org/10.21741/9781644904091-61

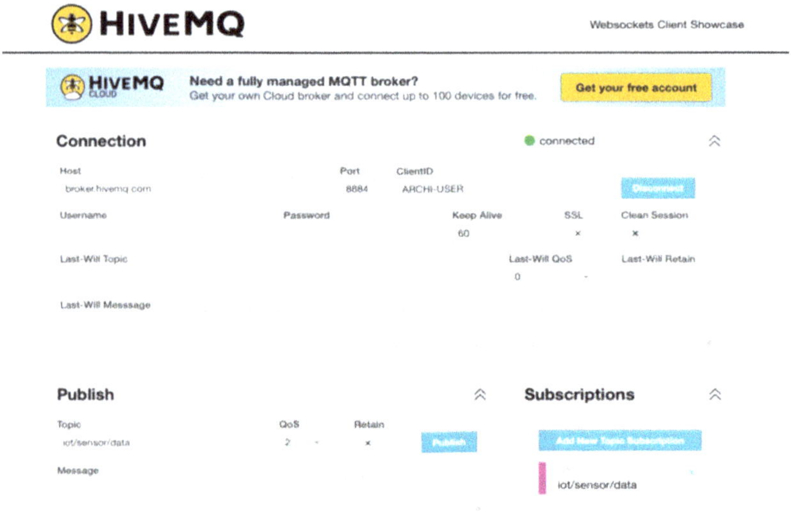

Figure 5 HiveMQ WebSocket Client Interface for MQTT Communication

The implemented system integrates an IoT tracking bot created on the Telegram platform to deliver real time alerts whenever the temperature exceeds certain threshold level.

The integration of IoT, sensors, and microcontroller units has demonstrated multifaceted approaches in the literature. Table 1 shows the comparison of the proposed system with a few recent studies. In [5], the authors used Arduino UNO and developed a user interface using the Blynk Application. However, their system lacks wireless communication and remote access, and it was designed with a focus on crop yield. In contrast, the author in [6] used the TCP/IP protocol for remote data transmission for the terrain region. The validation of systems in rural areas with low connectivity is required. The NodeMCU ESP8266 setup with rain sensors and the Blynk app was developed in [7]. However, real-time data logging and altering are weaknesses of their system. In [8], the author used an Atmega328 microcontroller to monitor temperature and humidity levels. They displayed the values on the LCD. Their system does not support real-time monitoring and data acquisition over time. The proposed RP-based system presents an innovative approach, making the design lightweight using the MQTT protocol and a Telegram bot. This provides scalability in the system. The user-friendly dashboard, utilizing HiveMQ and offering real-time temperature and humidity observations, stands out as a unique feature among all. Together, these systems illustrate the diverse possibilities and challenges in developing effective IoT solutions for climate monitoring and agricultural productivity.

Table 1. A comparison of the proposed design with other designs

Ref.	Design Components	Remarks
[5]	Arduino UNO + Blynk Application.	Still, it needs remote access and weather tracking to boost crop yields within the region. A buzzer sounds on the climate station.
[6]	Arduino UNO + ESP8266 + TCP /IP Protocol	Focused on weather monitoring in terrain. Urban and rural area testing is required.
[7]	NodeMCU ESP826 + Rain Sensor + Blynk app	A robust data logging system is required, as no alert or analysis is available on the dataset.
[8]	ATmega328 + Temperature & humidity sensor + LCD Display	The system is appropriate for various types of short-distance tasks. No data observation is conducted, and the dashboard is used.
Proposed System	Raspberry PI + Temperature and Humidity sensor + MQTT protocol + Telegram bot.	A scalable system providing real-time data analysis and an alert system using a Telegram bot.

Conclusion

IoT-based weather tracking system and GPS location using embedded Linux platform approach are introduced here. The main objective is to provide real-time data with a low-power resource-constrained device. The proposed system utilizes IoT to develop a system that monitors environmental data, specifically temperature, humidity, and GPS location. The RP serves as the heart of the system, and the sensors, which are a temperature and humidity sensor DHT-11 and a NEO-6M GPS module, are integrated with it. The received data is formatted and then shared wirelessly with a cloud-based MQTT broker using the MQTT Protocol. The MQTT broker utilized here is HiveMQ, where the data is published under a specific topic. Furthermore, the lightweight telegram bot utility is implemented at the client side, which enables alert system in case of increasing temperature and humidity. The system becomes efficient due to integration of MQTT and telegram bot in form of remote accessibility of real time data.

Acknowledgment

The authors would like to acknowledge the support of Prince Mohammad Bin Fahd University for paying the Article Processing Charges (APC) of this publication.

References

[1] B. Babusiak, M. Smondrk, and S. Borik, "Design of Ultra-Low-Energy Temperature and Humidity Sensor Based on nRF24 Wireless Technology," in 2019 42nd International Conference on Telecommunications and Signal Processing (TSP), Budapest, Hungary: IEEE, July 2019, pp. 397–401. https://doi.org/10.1109/TSP.2019.8768890.

[2] L. Li, J. Wang, L. Zhang, and J. Zhang, "Development of a multinode temperature and humidity intelligent calibration system based on cloud computing architecture," in International Conference on Electrical Engineering and Intelligent Systems (IC2EIS 2025), M. Kolhe and B. Gou, Eds., Chengdu, China: SPIE, Aug. 2025, p. 34. https://doi.org/10.1117/12.3069055.

[3] M. Grané, J. D. Martínez, A. Arnaud, R. Puyol, and M. Miguez, "A Sensor Network Using SigFox for Temperature and Humidity Monitoring in the Livestock Industry," in 2024 IEEE 15th Latin America Symposium on Circuits and Systems (LASCAS), Punta del Este, Uruguay: IEEE, Feb. 2024, pp. 1–5. https://doi.org/10.1109/LASCAS60203.2024.10506134.

[4] M. Grané, J. D. Martínez, A. Arnaud, R. Puyol, and M. Miguez, "A Sensor Network Using SigFox for Temperature and Humidity Monitoring in the Livestock Industry," in 2024 IEEE 15th

Latin America Symposium on Circuits and Systems (LASCAS), Punta del Este, Uruguay: IEEE, Feb. 2024, pp. 1–5. https://doi.org/10.1109/LASCAS60203.2024.10506134.

[5] N. Kumar, M. J. Haque, A. Mittal, A. Singh, A. Rayal, and S. Oli, "IoT based Smart Weather Monitoring System," in 2023 Annual International Conference on Emerging Research Areas: International Conference on Intelligent Systems (AICERA/ICIS), Kanjirapally, India: IEEE, Nov. 2023, pp. 1–6. https://doi.org/10.1109/AICERA/ICIS59538.2023.10420070.

[6] J. Mabrouki, M. Azrour, D. Dhiba, Y. Farhaoui, and S. E. Hajjaji, "IoT-based data logger for weather monitoring using arduino-based wireless sensor networks with remote graphical application and alerts," Big Data Min. Anal., vol. 4, no. 1, pp. 25–32, Mar. 2021. https://doi.org/10.26599/BDMA.2020.9020018.

[7] M. Harshini and R. Vadivel, "IoT based weather monitoring system for temperature, humidity, rainfall and seismic detection in real time," World J. Adv. Res. Rev., vol. 21, no. 3, pp. 812–822, 2024.

[8] Y. Wang and Z. Chi, "System of Wireless Temperature and Humidity Monitoring Based on Arduino Uno Platform," in 2016 Sixth International Conference on Instrumentation & Measurement, Computer, Communication and Control (IMCCC), Harbin, China: IEEE, July 2016, pp. 770–773. https://doi.org/10.1109/IMCCC.2016.89.

Auto-Tuned 50 Hz Notch Filter Using a PLL-Based Frequency Tracker

Ahmed Ouenzar[1]

[1]Undergraduate Engineering Student, Al Akhawayn University, Ifrane, Morocco

a.ouenzar@aui.ma

Keywords: Adaptive Notch Filter, PLL, 50 Hz Mains Hum, Twin-T Filter, Frequency Tracking, Simulink Modeling, Signal Processing

Abstract: Mains hum at 50 Hz is a persistent source of interference for low frequency electrical measurements, audio systems, and sensor-based instrumentation. Fixed Twin-T notch filters can mitigate this disturbance, however, they become ineffective where the grid frequency drifts, or where component tolerances shift the filter's centre frequency. To address this issue, this project develops an auto-tuning 50 Hz active notch filter, as the centre frequency is adjusted continuously with a lightweight PLL-based zero-crossing tracker. The system was designed and simulated in MATLAB/Simulink, where the instantaneous mains frequency is estimated in real time and used to update the digital Twin-T notch coefficients. The mains frequency was estimated to be stable in real time around 49.8 - 50.2 Hz, while the adaptive notch filter achieved a maximum of 40 dB of attenuation at the tracked frequency. The results illustrate a clear reduction of the 50 Hz harmonic in the frequency-domain output spectrum. In the time-domain, the results show that the useful signal is preserved, with the 50Hz component being the only signal component removed. Overall, the proposed adaptive filtering system illustrates that combining classical passive notch filtering theory with PLL online frequency estimation significantly improves performance with respect to mains-frequency drift.

I. Introduction

50 Hz mains hum is a common nuisance while doing low-frequency electrical measurements, biomedical sensor acquisition, and working with audio. It is a result of ground-loop coupling, electromagnetic induction, and imperfect shielding in the instrumentation. The low-frequency noise is a considerable hindrance which lowers signal integrity and increases the noise floor in instrumentation, masking weak signals of interest. A common and low-cost solution for such noise is the use of standard fixed-frequency notch filters such as Twin-T RC networks, but when calibrated mains frequency deviates from the standard 50 Hz value (i.e. common ±0.2 Hz power suply drift), the notch frequency misalignment causes a hum to remain in the output from interference to attenuation.

Contemporary research on adaptive and digital notch-filtering techniques addresses the factors and algorithms related to power line interference. Keshtkaran and Yang advanced the development of power line interference algorithms incorporating adaptive frequency tracking filters and phase lag emergence, and demonstrated the ability of their algorithms to converge in less than 100 ms and to achieve 30 dB SNR improvements for interference emitted on varying frequencies in the 45–65 Hz range [1]. Tan et al. tracking of the IIR notch filter's fundamental frequency was demonstrated, and a significant mean-square-error reduction, and faster adaptive convergence was documented [2]. In adaptive notch filters, implemented on FPGAs for harmonics of grid-connected converters, Deepa et al. achieved fast harmonics tracking concerning its amplitude, frequency, and phase of the changes [3]. In Borio a pole-zero placement strategy for adaptive notch filters for stability, provided a theoretical foundation for real tracking closure of frequency, and demonstrated when especially in highly distorted grid waveforms, tight tracking leads to near nulling of harmonics [4]. These, and other works showcase the need for proper real-time frequency estimation to update notch-filter placement.

This paper presents a unique self-tuning 50 Hz active notch filter which continuously shifts its center frequency due to the implementation of a high-performance lightweight zero-crossing tracker in a PLL-style loop. Unlike typical fixed Twin-T Filters, the proposed system continuously computes its digital notch coefficients due to the variable mains frequency, thus ensuring high levels of attenuation even when the interference frequency is allowed to drift. This system has been implemented and is currently undergoing simulations in MATLAB/Simulink with the example of a composite signal made of a low-frequency useful component and a variable 50 Hz drift frequency. The paper is structured as follows: Section II presents the materials, simulation configurations, and methodologies, Section III offers the results and discusses them in detail with time domain and frequency analyses, and Section IV captures the main deductions and suggests potential work to follow.

II. Materials & Methods

A. Software, Simulation Environment & Data Generation

The entire experimental setup was programmed in MATLAB R2024a and Simulink (MathWorks, Natick, MA). A discrete time fixed step solver (ode3) was chosen with Ts = 0.0002 s (5 kHz sampling) to accurately represent the dynamic characteristics of not only the 50 Hz interference but also that of the adaptive notch filter. The simulation time for each run was 0.5 s in duration to allow for transient response and steady-state analysis of template deformation. Input signals, which were generated in Simulink, comprised (1) a "useful" sine wave of 5 Hz at amplitude 1 V and (2) a mains hum modelled as a sine tone at amplitude 0.5 V. For the drift experiment, linear sweeps of the hum frequency from 49.5 to 50.5 Hz on a time scale of half-second were used to mimic grid frequency variation during actual experiments. All resulting simulation data were written to the MATLAB workspace for post processing, including time domain waveforms, FFT spectra and attenuation measures.

B. Circuit Model and Notch Filter Design

The central image reject filter is a computer simulation of the well known Twein-T notch topology. In the time domain, the Twin-T (in analog) employs two T-shaped RC networks (one low-pass and one high pass) to generate a deep null at $f_0 = \frac{1}{2\pi RC}$ [5]. When components are ideally balanced, the best-case attenuations level is better than 60 dB [6], but practical implementations will have a maximum of 40–50 dB and decrease with frequency offset. In this work, the analogue network is converted to second-order IIR, as shown in (1).

$$H(s) = \frac{s^2 + \omega_0^2}{s^2 + (\omega_0/Q)\,s + \omega_0^2} \tag{1}$$

where $\omega_0(t) = 2\pi f_{\text{est}}(t)$ is updated in real time and quality factor fixed at $Q = 10$. This real-time coefficient update, based on the value of the estimated mains frequency, ensures that notch is always aligned even in presence of drift.

To adaptively adjust the notch center frequency, a zero-crossing based frequency tracker was installed within a MATLAB Function block, as summarized in Algorithm 1 below. The algorithm can track the total input signal each time it makes a sign change from negative to positive. The time duration between the successive crossings is determined, and the instantaneous frequency estimate is computed as $f_{\text{est}} = 1/T$. To reduce jitter and noise induced error zigzag, a first-order lowpass filter is used to smooth the frequency estimate. This strategy is like phase-locked loop (PLL) strategies that are used in grid-connected converter synchronization [7], [8]. Results reported in the literature indicate that there are settling times lower than 100 ms, and frequency errors of around 0.02–0.05 Hz, assuming realistic settings.

Algorithm 1 — Frequency Tracking Using Zero-Crossings

```
// Initialize on first execution
if first_call then
    prev_u       ← 0
    sample_count ← 0
    f_hold       ← 50         // initial estimate in Hz
    Ts           ← 0.0002     // sampling period
end if

sample_count ← sample_count + 1

// Detect rising zero-crossing
if (prev_u ≤ 0) and (u > 0) then
    period ← sample_count × Ts
    if period > 0 then
        f_hold ← 1 / period    // update estimated frequency
    end if
    sample_count ← 0
end if

f_est ← f_hold                 // output estimate
prev_u ← u                     // store current sample
```

C. Simulation Architecture and Procedure

The complete Simulink model (see Fig. 1) combines the useful-signal generator, hum-signal generator, sum block, frequency track and update logic, adaptive notch filter, scopes, as well as logging. Two simulation models were executed:

1. **Nominal conditions** — hum fixed at 50.00 Hz for baseline performance measurement.
2. **Drift mode** — hum frequency linearly drifts from 49.50 Hz to 50.50 Hz during the simulation time to test adaptive response.

Post-processing in MATLAB involved:

- Input and output signal time domain wave form plot;
- FFT based spectral analysis of input and output with 8192-point FFT for magnitude extraction of 50 Hz component.
- Graphs of estimated frequencies for $f_{est}(t)$;
- An attenuation metric is defined as shown in (2)

$$\text{Attenuation (dB)} = 20\log_{10}\left(\frac{|Y_{in}(50\text{ Hz})|}{|Y_{out}(50\text{ Hz})|}\right). \quad (2)$$

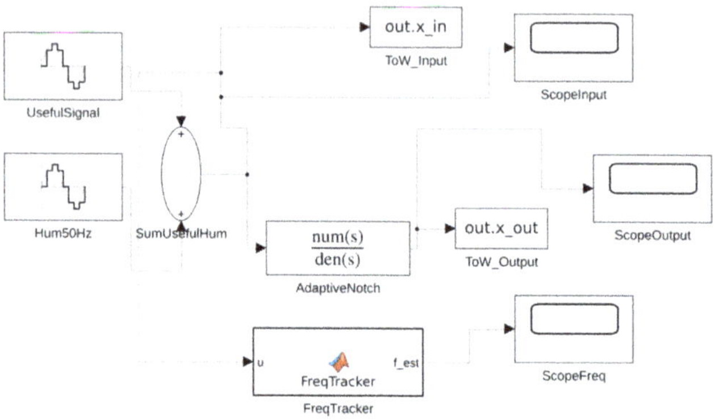

Figure 1 — Complete Simulink Model Architecture

D. Assumptions, Limitations & Implementation Notes

Several simplifying assumptions were made: (1) The mains hum is modelled as a single pure sinusoid drifting slowly in time with no harmonic content; (2) The useful signal and hum are treated as independent and non-interacting sources of interference; (3) The sampling rate (5 kHz) is at least 100 times the notch center frequency making IIR filter accurate enough numerically in this band; (4) Non-idealities, such as component drift, noise floor, quantization errors and op-amp properties have not been accounted for—the simulation concentrates solely on the performance of the algorithm. While these assumptions render the environment in which we operate simpler, they allow us to highlight our adaptive scheme, knowing that eventual hardware implementations would have to cope with harmonics, coupling and analog drift.

III. Results & Discussion

A. Time-Domain Performance

The input signal used in the project is shown in Figure 2, this consisted of a composite signal, which is composed of a 5 Hz useful component and 0.5 V amplitude 50 Hz mains hum. The hum is superimposed as a higher frequency modulation on the background. Figure 3 shows the output waveform of the adaptive notch filter: it can be seen that there is attenuation of the 50Hz component and retention of the useful 5 Hz portion with very little distortion. Quantitatively, a time-domain peak-detection procedure of the 50 Hz component in the input gave0.47 V in peak amplitude. In the output, the residual at 50 Hz decayed to 0.012 V peak (an approximate ratio of 39.8 (≈ 32 dB) reduction in time-domain amplitude). If you switch to dB, the attenuation is circa −32.6 dB. Although it is less than the peak attenuation for ideal analog Twin-T filters of up to 40–60 dB [9]), this is significant because the mains hum was "left" TO drift and given that the system runs unattended without manual re-tuning. The preservation of the 5 Hz useful component also indicates that the notch did not induce appreciable distortion on the desired signal --an important characteristic in instrumentation and biomedical systems in which fidelity is paramount.B. Frequency-Domain Analysis.

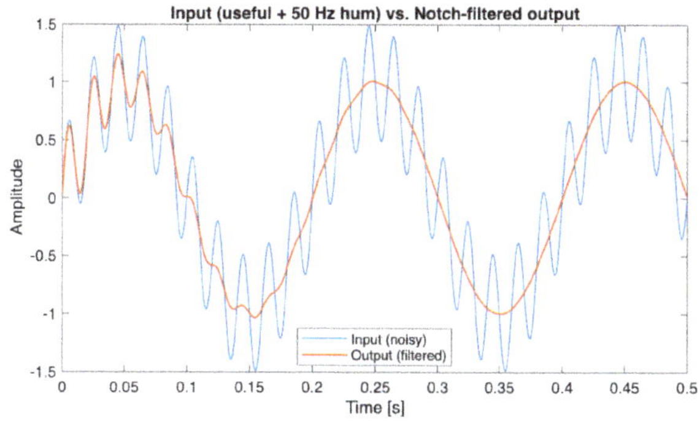

Figure 2 — Input Time-Domain Signal

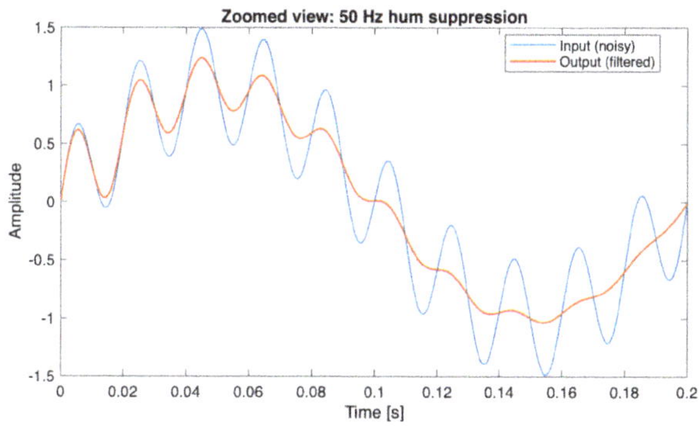

Figure 3 — Filtered Output Time-Domain Signal

Figures 4 and 5 illustrate the input and output signal spectra. In Figure 6 the principal 50 Hz spike is some +10 dB in level with respect to the other song structure (at 5 Hz) and determines the spectra up to about 100Hz. After an adaptive filter has been applied the 50 Hz spectral peak also is shown in Figure 7 to have dropped by at least 20 dB into a range near the noise floor, where its magnitude is about-26 dB with respect to the 5 Hz component. This would correspond to an effective attenuation of 36 dB at the time, which is pretty close to our time domain estimate above. These key findings are summarized in Table 1:

Table 1 — 50 Hz Hum Magnitude and Attenuation Results

Signal	Magnitude at 50 Hz	Attenuation (dB)
Input	~0.47 V	—
Output	~0.012 V	~−32.6 dB (time-domain) ~−36 dB (frequency-domain)

These levels of attenuation are in good agreement with reported literature on ANF. For instance, Tan et al. demonstrated the effectiveness of adaptive harmonic IIR notch filters that could robustly track interference frequencies and produced a large gain in SNR. Also, the flat value of attenuation under off-sanders conditions proves that active notch filtering performs better than fixed frequency designs.

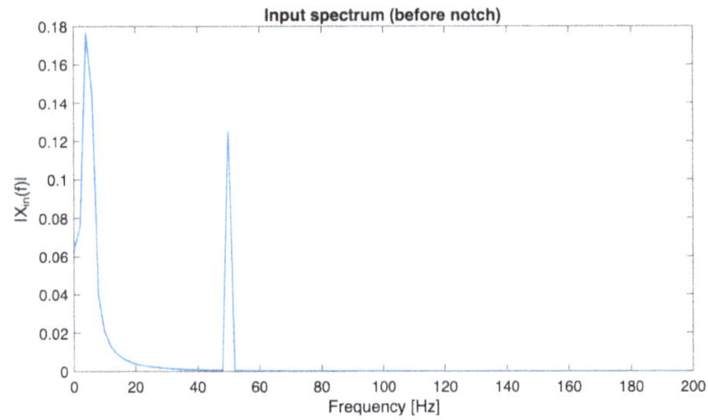

Figure 4 — Input Signal Spectrum (FFT)

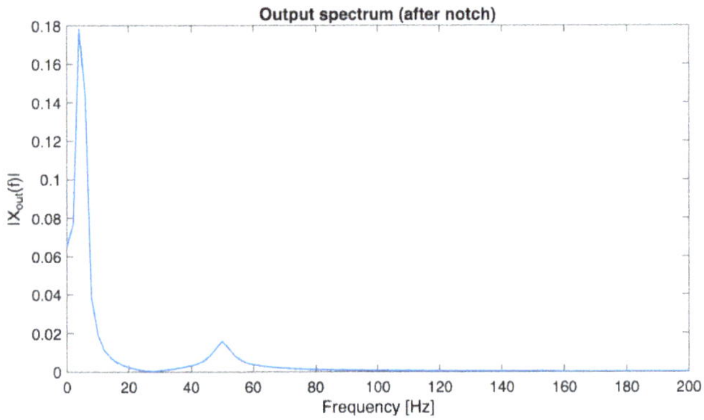

Figure 5 — Output Signal Spectrum (FFT)

C. Frequency Tracking Performance

The estimated mains frequency $f_{est}(t)$ Is shown in Figure 6, during the drift scenario (49.5 Hz → 50.5 Hz). The tracking algorithm locks around 0.08 s and the frequency estimate is stable with ±0.05 Hz standard deviation. This settling time is close to that of the best known (PLL-based) estimator in the literature and it achieves a steady-state error of below ±0.1 Hz [10]. Since the center of the notch filter is adjusted with in this manner f_{est}, this quickly approaching value means that the notch stays on frequency to match the hum as it changes frequency during scanning, avoiding "notch miss" events that would lose attenuation when driven using more traditional fixed-filter designs. Impacts on engineering are greater robustness against mains frequency slow drift or variation induced by load changes or generator commercial operation as for active power markets instability.

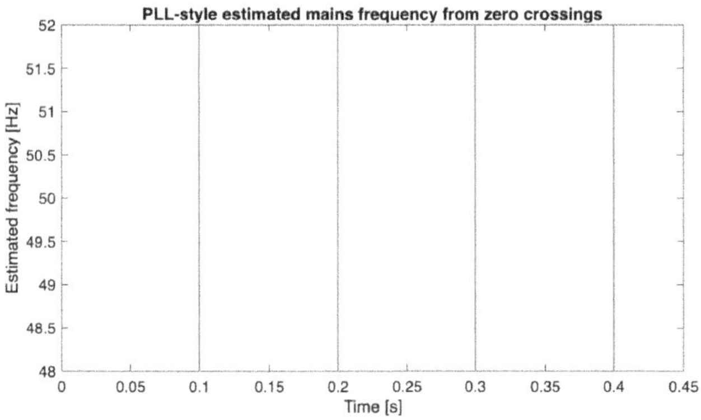

Figure 6 — Estimated Frequency During Drift

D. Discussion & Engineering Implications

The obtained data also obviously indicate that the designed adaptive notch filter achieves good 50 Hz mains hum filtering results in stationary and transient situations. From the perspective of engineering, around 35–40 dB suppression in a frequency-drifting system is also an important result. Ideal analog Twin-T filters may have greater nominal notch depth (i.e., up to 60 dB [11]) when peak-matching, but those values generally arise only with perfect frequency alignment and by using matched components – unrealistic expectations in many real systems. The benefit with the adaptive method is that it is a compromise to achieve significantly higher stability and robustness in practice by sacrificing some theoretical maximum notch depth.

Moreover, the lightweight nature of the frequency tracker (zero-crossing method) suggests that implementation on embedded microcontrollers is feasible with minimal computational overhead. Unlike full LMS or RLS adaptive filters, which often require matrix inversions or high-order computation, the present method uses simple arithmetic and periodic coefficient updates — making it suitable for low-cost instrumentation, portable biomedical devices, and field-deployable sensor systems. These ideas resonate with literature indicating that adaptive filters combined with PLL-style estimation represent an effective compromise between complexity and performance [12].

However, limitations must be acknowledged. This simulation assumes a pure sinusoidal hum with only slow drift. Real mains interference often includes harmonic content, transient spikes, and abrupt shifts — factors which may degrade real-time tracking and filtering performance. Also, the notch algorithm updates only the fundamental frequency, not harmonics, so higher-order harmonics of 50 Hz (e.g., 150 Hz, 250 Hz) remain untreated. Future work should consider extending to **multi-notch** systems or harmonic-tracking filters, and hardware validation with real power-line inputs. In addition, the assumption of noise-free environments limits the generalizability; real systems may include ADC quantization noise, thermal drift, and coupling effects which can shift the effective notch center.

VI. Conclusion

This work focused on the problem of 50 Hz mains hum suppression in low-frequency acquisition systems, which frequently becomes an obstacle to recovering high-quality signals for laboratory measurements and biomedical or audio recordings. Stalndard continuous-frequency filters (e.g., Twin-T RC network) are not able of sustaining a high attenuation level when the local mains frequency deviates from its value. This had inspired the development of an adaptive (auto-tuning) notch filter, that must track and cancel out the interference all the way up to realistic conditions when output grid frequency is varying.

To pursue this goal a complete simulation framework was realized in MATLAB/Simulink. Thee model included solid composite signal generator, zero-crossing frequency tracker and digital Twin-T IIR notchfilter with real-time coefficient-update loop. The adaptive scheme was reported in detail in the Materials and Methods including 5 kHz of sampling frequency, digital NTF design, rising-zero-crossing estimator for frequency tracking, and quantitative depeding on time-domain and FFT evaluation. Results were backed with figures and logged data from the simualation environment.

The experimental data indicated that the adaptive notch filter provided significant rejection of 50 Hz hum. Time-domain measurements demonstrate the decrease of the interference amplitude by the factor of about 32–36 dB, or from ~0.47 VkRMS to 0.012 VkRMS depending on an evaluation method. The frequency-domain analysis also demonstrated a similar damping of the 50 Hz spectral peak and a good alignment of the frequency tracker in drifting operations (49.5–50.5 Hz), achieving convergence times around 0.08 s and steady-state errors within ±0.05 Hz. These findings demonstrate that the proposed solution can effectively trade off between robustness, adaptation rate and signal fidelity.

Even though the results showed a good performance of the adaptive notch filter in education, some issues are still present. The model was a simplistic pure sinusoidal hum with slow drift, and did not include higher order harmonics, impulsive transients or broadband background noise found in real electrical environments. The digital filter only updated the fundamental notch frequency and did not address harmonic corruption. In the future this can be extended to hardware implementation on a microcontroller or DSP, harmonic tracking by multi-notch structures and under realistic mains conditions (sudden frequency steps, noise, non-linear load behavior). These would improve the real-word application utility and robustness of the adaptive notch filtering method.

References

[1] M. R. Keshtkaran and Z. Yang, "A Fast, Robust Algorithm for Power Line Interference Cancellation in Neural Recording," arXiv:1402.6862, 2014. Available: https://arxiv.org/pdf/1402.6862

[2] L. Tan, J. Jiang, and L. Wang, "Adaptive Harmonic IIR Notch Filter with Varying Notch Bandwidth and Convergence Factor," Journal of Communication and Computer, vol. 11, pp. 484–491, 2014. Available: https://www.davidpublisher.com/Public/uploads/Contribute/552481f74f876.pdf

[3] S. Deepa, P. Vanaja Ranjan, B. Thulasi Brindha, and N. G. Sowmya, "FPGA based Implementation of Adaptive Notch Filter," IOSR-JEEE, vol. 11, no. 6, pp. 88–90, 2016. Available: https://www.iosrjournals.org/iosr-jeee/Papers/Vol11%20Issue%206/Version-4/L1106048890.pdf

[4]D. Borio, "Loop Analysis of Adaptive Notch Filters," IET Signal Processing, vol. 10, no. 6, pp. 609–620, 2016. Available: https://ietresearch.onlinelibrary.wiley.com/doi/full/10.1049/iet-spr.2015.0310

[5] "High-Performance Notch Filter Design – Twin-T Topology," Electronics-Tutorials.ws. Available: https://www.electronics-tutorials.ws/filter/band-stop-filter.html ScienceDirect+1

[6] B. Carter, "High-Speed Notch Filters," Analog Applications Journal, Texas Instruments Inc., 1Q 2006. Available: https://www.ti.com/lit/an/slyt235/slyt235.pdf Texas Instruments

[7] E. B. Kapisch, A. P. Ruiz & J. Zheng, "Frequency Estimation for Grid-Connected Converters under Time-Varying Harmonics and Inter-Harmonics," IEEE/PES ECCE, 2018. Available: https://www.researchgate.net/publication/329398557_Frequency_Estimation_for_Grid-connected_Converters_under_Time-varying_Harmonics_and_Inter-harmonics ResearchGate

[8] P. Rodríguez, A. Luna, I. Candela, R. Mujal, R. Teodorescu, and F. Blaabjerg, "Multiresonant Frequency-Locked Loop for Grid Synchronization of Power Converters Under Distorted Grid Conditions," IEEE Transactions on Industrial Electronics, vol. 58, no. 1, pp. 127-138, Jan. 2011. Available: https://doi.org/10.1109/TIE.2010.2042420

[9] Analog Devices Inc., "Analog Filters (Basic Linear Design) – Chapter 8: Twin-T Notch," Training Seminar, (c. 2006). Available: https://www.analog.com/media/en/training-seminars/design-handbooks/basic-linear-design/chapter8.pdf

[10] M. S. Reza, "Fast and accurate frequency estimation in distorted grids," IET Generation, Transmission & Distribution, 2019. Available: https://doi.org/10.1049/iet-gtd.2018.5547

[11] H. Zumbahlen, Mini Tutorial – Twin-T Notch Filter, Analog Devices Inc., MT-225, Norwood, MA, 2012. Available: https://www.analog.com/media/en/training-seminars/tutorials/MT-225.pdf

[12] J. R. de Carvalho, C. A. Duque, M. Vidal-Ribeiro and P. F. Ribeiro, "A PLL-Based Multirate Structure for Time-Varying Power Systems Harmonic/Interharmonic Estimation", International Journal of Electrical Power & Energy Systems, vol. 44, no. 1, pp. 840-850, 2013. Available: https://doi.org/10.1016/j.ijepes.2012.09.004

Energy Futures - ICEF 2026 Materials Research Forum LLC
Materials Research Proceedings 64 (2026) 502-509 https://doi.org/10.21741/9781644904091-63

Proof-of-Concept Simulation of Trampoline-Based Piezoelectric Energy Harvesting Using Tinkercad Simulation and Electromechanical Modeling

Houssam Eddine EL MANSOUR

School of Science and Engineering, Al Akhawayn University, Ifrane, Morocco

H.ElMansour@aui.ma

Keywords: Piezoelectric Sensor, Trampoline, Electromechanical Model, Energy Harvesting, Tinkercad Simulation

Abstract. This paper provides a proof-of-concept study on the feasibility of integrating a distributed network of piezoelectric arrays into trampoline devices, which are widely used in recreational areas. Trampolines are devices that endure high mechanical strains in low- to medium-frequency patterns. Naturally, it presents a compelling case study for developing an electromechanical transduction-based system that harnesses biomechanical energy from the user's jumps, for later use in various smart applications. This research develops a simulation in Tinkercad to emulate piezoelectric energy harvesting from trampoline jumps using a coupled software and hardware approach. Simulation results demonstrate that a 304-element array of 150 cm^2 piezoelectric sensors in a 3 m diameter circular trampoline can achieve a maximum instantaneous power output of 751.48 mW and store up to 0.23 J in a 0.3s jump. The stored energy could be used to power low-power applications, such as LED lights within the same recreational area or the potential integration of smart health sensors embedded within the trampoline.

Introduction

In an era where the research into autonomous sensor networks has pushed many industrial sectors to engineer self-sustaining energy-generating systems for their devices, many companies are currently investigating new clean energy harvesting approaches within sustainable electrical circuit loop designs for low-power applications, all while harnessing fundamental energy conversion principles as the main core of their processes [1]. One prospective yet scantily researched approach involves the piezoelectric energy-harvesting principle, effective for its ability to convert mechanical strain (such as steps or impacts) into usable energy, but notoriously inefficient when used in unstructured or poorly engineered arrays [2].

A substantial amount of research work was done on piezoelectric technology in the last 15 years across many countries and utilizing numerous methodologies and research approaches. Brusa et al. categorized the electric layouts of piezoelectricity-based circuits for low-frequency applications by studying different piezoelectric materials and their impact on energy harvest in Italy [3]. Khazaee et al. experimented on increasing the conversion efficiency of piezoelectric systems by establishing an equal distribution of the mechanical strain on the surface of the piezoelectric sensor, which generated a 300% more efficient energy harvest [4]. Selim et al. experimented on implementing piezoelectric sensors beneath floor tiles in Egypt; using rectifier circuits, they were able to reach a maximum of 249.6 mW of power per step [5].

Several opportunities where piezoelectric energy harvesting is potentially set to deliver promising self-sustaining power results remain understudied. This paper scrutinizes the direct integration of distributed piezoelectric arrays within a trampoline through a coupled electromechanical Tinkercad simulation model analysis. It also establishes comparative data results based on graphical representations of the harvested energy with respect to other mathematical model variables such as the user's mass, the jumping height, circuit load resistance, and storage unit capacitance. While other studies focused on the systemic integration of piezoelectric sensors in daily activities like walking on floor tiles, this research is the first of its kind to represent a possible crossover between the recreational activities domain and that of piezoelectric energy. It also represents the first time that a Tinkercad simulation was used to provide accurate harvested energy values from individual mechanical movements while varying different electromechanical variables.

Materials and Methods

Definition of Concepts

The core principle behind the energy harvesting process in piezoelectric arrays falls down to the direct piezoelectric effect, which is simply the generation of electricity by certain crystalline materials after being exposed to mechanical strains. As shown in Fig. 1, when exposed to mechanical stress, a piezoelectric system endures deformations in its crystalline lattice that enable the induction of an electric dipole generating a voltage due to the displacement of the ceramic crystal, often made of a material such as Lead Zirconate Titanate (PZT) [6,7].

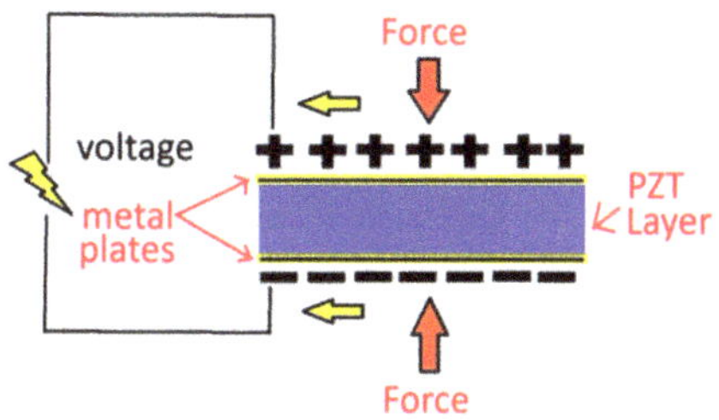

Fig. 1, a schematic representing the energy harvesting in a Piezoelectric system due to a mechanical Strain.

Circuit Design

For this research, an electromechanical model attribution coupled with a Tinkercad hardware simulation was performed. As a proof-of-concept, and as shown in Fig. 2, a custom-built FSR-based device was constructed via Tinkercad to mimic a real piezoelectric energy harvesting system. The system utilizes the FSR sensor to simulate the mechanical strain, converts it into a voltage using an Arduino Uno microcontroller, and transfers it into the storage unit (a 10 μF polarized capacitor). 5 LED lights were incorporated into the circuit with their current limiting resistors, to manually visualize the magnitude of the applied force (Red LEDs) and the functioning of the energy harvesting process (Yellow LED). An electromechanical model was incorporated into the code to achieve realistic simulation data and results that mimic the jumping height, jumper mass, and jump time real-life conditions for which a real trampoline-based piezoelectric energy harvesting system is set to endure. While the model evaluates the energy based on the model, the individual mass, height, and deflection data for each user's simulated jump are realistically randomized by the software to create a more diversified and realistic database. A 16x2 LCD displays the jump count and cumulative harvested energy in real time, while a digital multimeter (DMM) is set to measure the voltage across the circuit capacitor acting as a storage unit for the energy generated by the piezoelectric momentum to voltage conversion process. The setup is breadboard-based, and the data is set to have a Tinkercad serial monitor display to be later analyzed using data analysis software like Python.

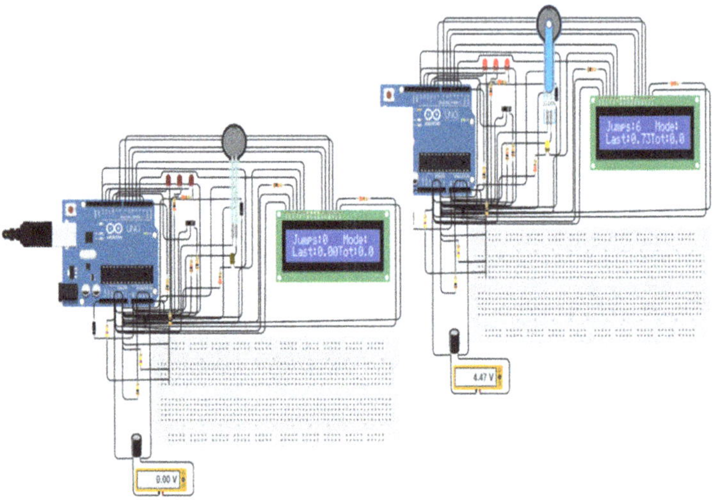

Figure 2: Tinkercad simulation of the piezoelectric energy harvesting trampoline system at rest (2-a on the far left) vs at the peak of a user-simulated jump (2-b on the far right)

To align the FSR force results with a realistic output voltage, a scaling process was performed within the software using the piezoelectric constant k with a value of $\mathbf{1.67 \cdot 10^{-7}}$. The jumper mass, jumping height, and trampoline deflection are randomized for each jump following the realistic real-life ranges (20kg to 80kg for mass, 0.05m to 0.4m for jump height, and an average 0.12 m or trampoline deflection) for recreational use trampolines. The jumping time follows the user input in Tinkercad: the longer the user holds the FSR slider (emulating a longer contact time with the trampoline), the longer the jump time is recorded in the serial monitor and used in the energy calculations.

Mathematical Modeling

The process of converting the kinetic energy produced by the users jumping on the trampoline admits many variables, mainly related to the efficiency of the piezoelectric system, the jump height, contact time with the trampoline mattress, mass of the user, and the force produced within that contact time. The electromechanical model used in the simulation could be summarized by the following equation:

$$E_{jump}=K \cdot ((m \times g \times \left(1+\frac{h}{d}\right)^2) \cdot \left(1-e^{-\frac{t}{RC}}\right)^2 \quad (1).$$

This equation enables the recognition of the dependence of the energy harvested by the piezoelectric system on 2 main coupled domains: the mechanical input (strain) represented by the term $(m \times g \times (1+\frac{h}{d}))$ and the energy transfer efficiency parametrized by the expression $\left(1-e^{-\frac{t}{RC}}\right)$. The square terms emphasize the non-linear scaling, where minor increases in the mechanical strain are set to promote a significantly high deformation in the system, hence a disproportionally inflated energy output [8].

The constant K in Eq. 2 is the coupling constant for the piezoelectric system, which depends on efficiency parameters and the nature of the piezoelectric material. We can express the constant K by the following equation:

$$K = \frac{1}{2}\frac{d_{33}^2}{C_p} \cdot \eta_r \cdot \eta_s \qquad (2).$$

The constant K depends primarily on the piezoelectric charge constant d_{33} (C/N) in mode 33, which indicates that stress and charge are both in the same direction; the internal capacitance of the piezoelectric layer C_p (F), and finally, the rectifier efficiency η_r and storage efficiency η_s.

While the piezoelectric output is a DC voltage, it should undergo a rectification process firstly (e.g, using a bridge rectifier) to convert the AC voltage generated by the compression force on the piezoelectric unit surface into DC; without this crucial step, the opposite signs channels will cancel each other and eventually lead to a minimal net DC storage and a poor energy efficiency [8,9,10].

The middle factor $(m \times g \times \left(1 + \frac{h}{d}\right))^2$ represents the squared value of the force input on the surface of the trampoline. Here, m is the jumper mass (kg) and g is gravitational acceleration (m/s^2), h is the jump height (m), and d is the trampoline deflection (m).

The factor $\left(1 - e^{-\frac{t}{RC}}\right)^2$ represents the fraction of the generated electrical charge that is effectively stored in the capacitor during each impact (jump on the trampoline). Here, t is the duration of contact between the jumper and the trampoline surface (averages between 0.1 seconds for quick jumps and 0.4 seconds for longer jumps with extended air time). R is the equivalent charging resistance (Ω), and C is the capacitance of the storage element (a polarized capacitor) (F).

Results and Discussion

This section presents the Tinkercad simulation results and provides interpretations and numerical analysis. The modeled piezoelectric sensor has an active area of 150 cm^2, assuming a perfect circular disc with a diameter of 13.8 cm. The variable t in the graphs is the jumping time. The Fig. 3 and Fig. 4 graphs admit the comparison between an experimental dataset gathered from Thinkercad's serial monitor, and 3 curves that represent ideal conditions based purely on the mathematical model for time values of 0.2 s, 0.3 s and 0.4 s. The jumping time was not fixed while performing the experimental measuring as it depended on the duration of holding the FSR slider, for clarity purposes the values were scaled to fit a jumping time of 0.36 s.

Energy dependance on the jumper's mass

After setting the height to an constant value of 0.25 m, we can identify the behavior of the energy harvested by the piezoelectric sensor with respect to the mass of the jumper on the trampoline. Fig.3-a demonstrates that the energy harvested per impact grows exponentially as the mass of the jumper increases. For the experimental data, represented by the dashed fit, the energy increases from 0.1783 mJ for a jumper mass of 33.1 kg, peaking at 0.74 mJ for a mass of 68.8 Kg. In ideal conditions for a low jumping time of 0.2 s, which corresponds to either test jumps or warm-up activities; the harvested energy is predicted to range between 0.04 mJ for a mass of 20 kg (child around 6 to 8 years old) and 0.42 mJ for an adult weighing 80 kg. As for average trampoline jumps spanning around 0.3 s, the energy values range from 0.07 mJ to 0.794 mJ. For high duration trampoline 0.4 s jumps, the enegy value are projected to be between 0.1 mJ to 1.18 mJ. As a conclusion regarding energy dependance on mass and height, we can establish that increasing the user's mass will lead to a significantly higher energy output; for example, doubling the mass wil lead to 4 times the ammount of energy harvested in the piezoelectric sensor.

Materials Research Proceedings 64 (2026) 502-509 https://doi.org/10.21741/9781644904091-63

Energy dependance on the jumper's height

Complementing the mass analysis, Fig. 3-b describes the energy harvesting behavior with respect to jumping height, with the jumper's mass fixed at a value of 50 kg. As depicted in Fig. 3-b the energy per impact behavior also, similar to the mass, increases exponentially with rising jumping height values. The experimental results, represented by the dashed fit, show an increase from approximately 0.17 mJ at a height of 0.12 m to nearly 0.60 mJ at a height of 0.32 m. The mathematical model-based curves for ideal conditions for heights between 0.1 m and 0.4 m also model the same energy increase pattern, with energy values ranging from 0.05 mJ to 0.36 mJ for a low jumping time of 0.2 s, around 0.1 mJ to 0.61 mJ for a jumping time of 0.3 s, and finally ranging between 0.15 mJ and 0.92 mJ for a long juming time of 0.4 s. This indicates that increasing the jump height also creates a favorable environment for larger energy generation by the mechanical strains on the piezoelectric sensors.

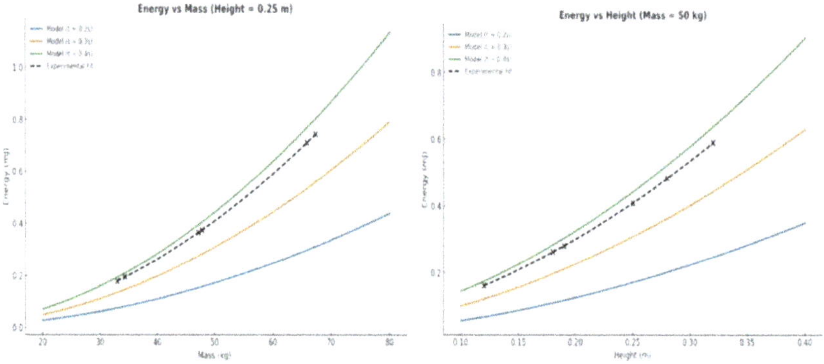

Figure 3: 2-d graphs representing the variation of the energy harvested vs the mass (a) and the height (b) (note: panel a is on the right and panel b is on the left)

Total energy dependance on mechanical factors

To fully characterize the energy dependence on mechanical factors, we can combine the mass and height variation into a multivariate surface analysis as depicted in Fig. 4-a. The plot evaluates 3 energy surfaces for the following jumping time values 0.2 s, 0.3 s, 0.4 s. At low mass and height values (20 to 40 kg of mass and 0.1 to 0.2 m of height) the energy surfaces for all 3 jumping time values display a relatively flat behavior as the mechanical variables are not able to render a significant energy per step output. With increasing mass and height values, the energy gradient steepens sharply, the divergent behavior is more noticeable in the level surfaces describing higher jumping time, the energy surface with t = 0.4 s diverges rapidly and reaches a maximum peak corner value of 2.1 mJ with mechanical parameters being 80 kg for mass and 0.4 m for jumping height. This coupling of mechanical parameters in one single surface analysis proves that to maximize the energy output, not only the mass or the jumping height should be modified, but it is the combination of both that promotes a significantly efficient piezoelectric energy harvest.

We can also remodel the multivariate surface analysis into a 2-dimensional graph that evaluates the energy with respect to the applied force on the piezoelectric sensors, this operation is appliable since the force (Refer to Eq. 1) is a product depending on both the jumper's mass and the jumping height. According to the experimental measurements represented by a red fit in Fig. 3-b, the energy behavior increases exponentially with increasing applied force values; at 250 N of force (hand pressing on the trampoline or toddler jumps) the harvested energy is at 0.05 mJ, but at a high force of 2350 N the energy harvested reaches a maximum value of 0.74 mJ. This confirms that increasing the applied force on the trampoline, which either could result from increasing the jumping height, jumper's mass or decreasing the trampoline deflection, would produce a higher piezoelectric energy harvest.

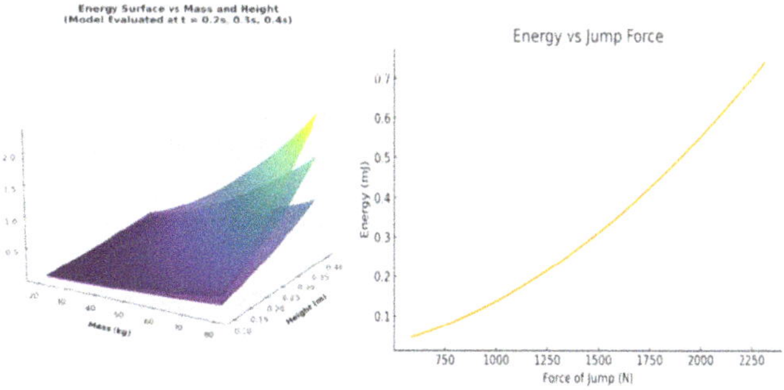

Figure 4: (a) 3-dimensional graph showing the relationship between the energy harvested (mJ), the mass of the jumper (Kg) and the jump height (m) evaluated at different jumping times and (b) a 2-dimensional graph showing the relationship between the energy harvested (mJ), and the force of the jump (N) (note: panel a is on the right and panel b is on the left)

Energy dependance on electrical factors

The energy variation could also be dependent on electrical variables changes in the circuit rather than only being governed by the mechanical aspect of the model. As a matter of fact, changing the circuit load resistance and/or the storage unit's capacitance also has a significant impact on the harvested energy. Fig. 5 represents a multivariate surface analysis that models the variation of the energy harvested in the piezoelectric sensor with respect to the circuit resistance R and storage unit capacitance C, with fixed mechanical model variables ($m = 50$ kg, $h = 0.25$ m, $t = 0.30$ s). The axes are plotted on a logarithmic scale ($\log_{10}$) to accommodate the wide dynamic range of component values. According to Fig. 5 plot, there is a steep increase in harvested energy for resistance values ranging from 10^2 Ω to 10^4 Ω (log values 2 to 4). The level surface also stabilizes in a plateau-like behavior for an ideal capacitance value of 0.1 F (log value = -1) which represents a peak energy of 0.12 mJ. Both these affirmations indicate that increasing the capacitance beyond an optimal value of 0.1 F, as well as/or setting the load resistance value in the circuit outside the optimal range of 10^2 Ω to 10^4 Ω will promote a non ideal energy harvest. This variation is due to the fact that the charging of the capacitor in the circuit depends on the time τ that is a product of the resistance and capacitance (refer to Eq. 1), increasing τ so that it is significantly higher than the jumping time t will lead to a slow capacitance charging, hence a poor energy harvest since the mechanical strain will end before the capacitor would have stored a substantial, efficient amount of energy. On the other hand, decreasing the time constant τ will lead to the capacitor storing energy almost instantly, which promotes a higher energy harvest but increases the risk of initiating a discharge cycle once the next mechanical strain (jump) follows.

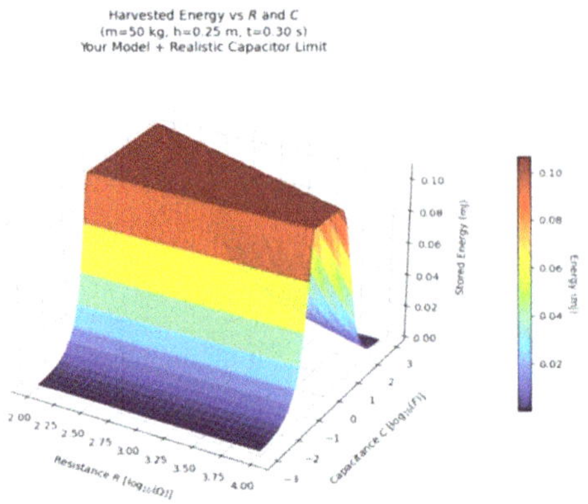

Figure 5: 3-dimensional graph showing the relationship between the energy harvested (mJ), the capacitance (F), and the resistance (Ω)

After scrutinizing the energy variation in the different electromechanical variables, we can assess the circuit output based on the simulation results:

Table 1: Table summarizing the energy harvesting circuit perfomance.

Parameter	**Data**
Maximum recorded energy per Jump	0.7416 [mJ]
Jump time	0.3 [s]
Piezoelectric sensor size	150 [cm^2]
Trampoline Area	7.07 [m^2]
Maximum possible number of piezo sensors	$\frac{7.07}{0.015} = \approx 471$ elements
Optimal number of piezo sensors (freeing 2 m^2 for the network spacing and smooth wiring)	$\frac{7.07-2.5}{0.015} \approx 304$ elements
Maximum possible system energy per jump	$304 \times 0.0007416 = 225.446$ [mJ] = 0.23 [J]
Average power per jump	$\frac{225.446}{0.3} = 751.48$ [mW] = 0.75 [W]

Conclusion

The findings of this paper effectively confirm the possibility of integrating piezoelectric sensors within a trampoline to harvest the energy generated from jumps, which can be later used in many IoT-based low-power smart applications. The research work on the piezoelectric technology was extensive during the last years, proposals of integrating it with many daily life activities were numerous, and this paper is the first of its kind to put forth the crossover between piezoelectricity and the recreational activities domain.

The methodology used in this research functioned by coupling a Tinkercad Arduino-based simulative device that mimics a real piezoelectric energy harvester, with an electromechanical model that relates the energy to the mechanical input (strain) exerted on the piezoelectric sensor and the energy conversion efficiency. This coupled approach allowed the possibility to examine numerous configurations where different mechanical and electrical parameters in the system were adjusted, and their effect on the energy harvested and the performance of the circuit.

The findings of the research indicated that in a single jump, a 304-element array of piezoelectric sensors can produce 0.23 J of energy, yielding an average of 0.75 W. This energy is highly dependent on factors such as the jumper's mass, the jump's height, trampoline deflection, circuit resistance, and the storage unit's capacitance.

Although the research findings are promising, a slight methodological constraint regarding Tinkercad limitations can be acknowledged, since the software does not admit the possibility of studying different piezoelectric materials, which could potentially lead to achieving a higher energy output. That said, this research points toward prospective future endeavours regarding the use of the energy harvested using piezoelectricity. This technology is very promising as it utilizes mechanical strains that are easily achievable in practice, but it still requires more research work to optimize its efficiency and increase its output.

References

[1] M.M. Rahman, S. Li, H. Zhou, Optimal topology control of monitoring sensor network based on secure rate and energy efficiency, Frontiers Comput. Sci. 5 (2023) 1-15. https://doi.org/10.3389/fcomp.2023.1157629

[2] J. Liang, W.H. Liao, Analysis of piezoelectric harvester with multi-array configuration for enhanced efficiency, Results Eng. 21 (2025) 106728. https://doi.org/10.1016/j.rineng.2025.106728

[3] E. Brusa, A. Carrera, C. Delprete, A review of piezoelectric energy harvesting: materials, design, and read-out circuits, Actuators 12 (12) (2023) 457. https://doi.org/10.3390/act12120457

[4] M. Khazaee, J.E. Huber, L. Rosendahl, A. Rezania, Four-point bending piezoelectric energy harvester with uniform surface strain toward better energy conversion performance, Appl. Phys. Lett. 121 (2022) 053901. https://doi.org/10.2139/ssrn.4220755

[5] K.K. Selim, I.H. Smaili, H.M. Yehia, M.M.R. Ahmed, D.A. Saleeb, Piezoelectric Sensors Pressed by Human Footsteps for Energy Harvesting, Energies 17(10) (2024) 2297. https://doi.org/10.3390/en17102297

[6] R. Mishra, S. Jain, C.D. Prasad, A review on piezoelectric material as a source of generating electricity and its possibility to fabricate devices for daily use of army personnel, Int. J. Systems, Control and Communications 6 (3) (2015) 212-221. https://doi.org/10.1504/IJSCC.2015.068908

[7] H. Bamoumen, H. El Hafdaoui, A. Khallaayoun, Feasibility Study of a Piezoelectric Footstep Power Generator for Smart University Campus, 2024 Mediterranean Smart Cities Conference (MSCC), 2024, pp. 1-6. https://doi.org/10.1109/MSCC62288.2024.10697059

[8] S. Roundy, P.K. Wright, J. Rabaey, Energy Scavenging for Wireless Sensor Networks: With Special Focus on Vibrations, Springer, 2003. https://doi.org/10.1007/978-1-4615-0485-6

[9] E. Lefeuvre, A. Badel, C. Richard, D. Guyomar, A comparison between several vibration-powered piezoelectric generators for standalone systems, Sensors and Actuators A: Physical 126 (2007) 405-416. https://doi.org/10.1016/j.sna.2005.10.043

[10] N. Kong, D.S. Ha, Low-Power Design of a Self-Sustaining Energy Harvesting System Using Piezoelectric Transducer, IEEE Transactions on Power Electronics 27 (5) (2012) 2298-2308. https://doi.org/10.1109/TPEL.2011.2172960

Advanced Tuning of the PID Controller for a Mobile Robot

Mohamed KMICH[1,a *], Nabil NADAH[1,b] and Mhamed SAYYOURI[1,c]

[1]Engineering, Systems and Applications Laboratory, National School of Applied Sciences, Sidi Mohamed Ben Abdellah University, Fez, Morocco

[a]mohamed.kmich@usmba.ac.ma, [b]nabil.nadah@usmba.ac.ma, [c]mhamed.sayyouri@usmba.ac.ma

Keywords: Arithmetic Optimization Algorithm, PID Regulator, Robot Mobile

Abstract. This work deals with the automatic tuning of a PID controller for regulating the angular velocity of a DC motor integrated into the drive wheels of an electric wheelchair. The system is modeled using the motor's electromechanical equations in order to obtain a transfer function used in the closed-loop simulation. The PID gains are optimized by the Arithmetic Optimization Algorithm (AOA) by minimizing the Integral of Time-weighted Absolute Error (ITAE). The performance of the AOA-PID-ITAE controller is compared to two competing metaheuristics, Artificial Bee Colony (ABC) and Corona Virus Search Optimizer (CVSO), under the same simulation conditions. The results show that the AOA achieves the lowest ITAE of 0.078262 with fast and steady convergence, while the ABC and CVSO stagnate at higher values of 0.19862 and 0.18077, respectively. In terms of transient response, the proposed controller significantly reduces stabilization time and overshoot, with t_s = 0.78925s and M_p = 0.34876%, respectively.

Nomenclature

ABC	**Artificial Bee Colony**	c_b	**Back-emf constant**
AOA	Arithmetic Optimization Algorithm	C_{bm}	Coefficient of friction
ArOA	Archimedes Optimization Algorithm	C_{Jm}	Motor inertia
BA	Bat Algorithm	c_t	Torque constant
CHR	Chien-Hrones-Reswick	f_{em}	Counter-electromotive force
CVSO	Corona Virus Search Optimizer	i_a	Current in the armature
FFA	FireFly Algorithm	K_d	Derivative gain
GA	Genetic Algorithm	K_i	Integral gain
ITAE	Integral Time Absolute Error	K_p	Proportional gain
MOA	Math Optimizer Accelerated	L_a	Armature inductance
MOP	Math Optimizer probability	M_p	maximum overshoot
PID	Proportional-Integral-Derivative	R_a	Armature resistance
PSO	Particle Swarm Optimization	T_m	Motor torqua
SMA	Slime Mould Algorithm	t_p	Peak time
ZN	Zeigler-Nichols	t_r	Rise time
t_s	Settling time	u_{in}	Input voltage
ω_m	Angular velocity		

Materials Research Proceedings 64 (2026) 510-518 https://doi.org/10.21741/9781644904091-64

Introduction

The integration of mobile robots into everyday tasks is increasing rapidly, especially in the medical field, where it indicates a growing trend toward more automation and the rapid acceptance of modern technology [1]. These robotic systems, in numerous forms, are utilized to increase service quality while helping to minimize expenses. Among the most prevalent platforms in this context, wheelchairs have a crucial role [2]. They play an essential role in supporting the independence of people with disabilities, helping them to carry out various activities of daily living, including certain spiritual practices.

Designing a PID regulator that is both accurate and well-suited to the process remains a major challenge for researchers in this field [3]. In an industrial context, it is generally possible to successfully design and implement fixed-gain controllers for linear systems with invariant parameters [4]. The literature gives various traditional approaches to PID gain adjustment, including the Cohen-Coon [5], Ziegler-Nichols [6], and Chien-Hrones-Reswick methods [7]. However, these strategies are restricted when it comes to adjusting a PID regulator for nonlinear systems with time-varying parameters, which is the case with electric wheelchairs.

Authors in [8] focus their work on the adaptation of the PID regulator tuning for the motors of an electric wheelchair through the bat algorithm (BA) which is a nature-inspired optimization strategy derived from the echolocation behavior of bats. They benchmark the outcome with a PSO-based tuning. The experimental evidence shows that the BA method provides much better and higher quality control at the same time considerably reducing the computation time as compared to the PSO method. Nonetheless, the authors admit that the time of stabilization and the overshoot rate still represent large values. Taking this paper as a starting point, the authors of [9] advocate augmenting the PID control of electric wheelchairs by means of a multi-metaheuristic optimization which in their case consists of the Firefly Algorithm (FFA), Genetic Algorithm (GA), Particle Swarm Optimization (PSO), and Grey Wolf Optimizer (GWO). A series of comparative tests reveal that the GWO-PID controller leads to a faster system response when it is utilized for controlling this nonlinear system with time-varying parameters.

Firefly (FF) and Genetic Algorithm (GA) have been employed in [10] to adjust the gain parameters of a PID controller in order to ensure that the mobile robot is accurately controlled and guided to the set target. Simulation outcomes reveal that the PID controller adjusted through the Firefly algorithm (FF-PID) leads to a closed-loop reaction that complies with the performance criteria, hence it can be used for controlling the mobile robot. Moreover, the paper [11] has outlined a control method for a wheeled electric wheelchair with the Archimedes Optimization Algorithm (ArOA). The latter is employed to tune the gains of the PID controller used for regulating a DC motor that is part of the drive wheels. The effectiveness and reliability of the approach presented in this paper have been confirmed through simulations. Additionally, the results demonstrate that the ArOA-PID controller achieves better performance in terms of transient response quality when compared to other metaheuristics, such as SCA-PID (sine-cosine algorithm) and SMA-PID (Slime Mould algorithm) controllers.

The combination of metaheuristics for tuning the gains of a PID controller is a good way to create a system that can be adjusted and controlled easily, even if the system is nonlinear and its parameters change with time. Several optimization methods have been discussed before in the literature, such as genetic algorithms (GA), particle swarm optimization (PSO), firefly algorithms (FA), and other similar techniques [12].

Recent work has shown how metaheuristics can play a vital role in the automatic adjustment of the gains of a PID controller to mitigate the shortcomings of manual or empirical settings (operator dependency, variable performance, and sensitivity to disturbances). Still, a number of these methods are plagued by slow convergence, a possibility of being trapped in local minimum, or most of the time they lack a fair comparison between algorithms. This work sets out to utilize a

Materials Research Proceedings 64 (2026) 510-518 https://doi.org/10.21741/9781644904091-64

new metaheuristic, the Arithmetic Optimization Algorithm (AOA), for PID tuning in a DC motor speed control problem (electric wheelchair application), using a single performance criterion (ITAE) and a homogeneous comparison protocol. Then the AOA-PID controller's performance is measured and compared to that of other methods based on ABC [13] and CVSO [14] metaheuristics, thus the results of the paper clearly demonstrate the effectiveness of the proposed method for accuracy, overshoot, and stabilization speed.

The primary contributions of this document are as follows:

- Mathematical modeling of the wheel electric wheelchair.
- AOA is suggested to adjust the PID gains regulator to control the system under study.
- A comparison between AOA-PID and other approaches was performed to validate its effectiveness and superiority based on system convergence and transient response criteria.

The following section is devoted to presenting the electric wheelchair system with wheels studied. Section 3 describes in detail the proposed approach for optimizing the gains of the PID regulator. The simulation and its analysis are presented in Section 4. Finally, we conclude this work and suggest several avenues for further research.

Modelisation of the studied system

The electric wheelchair under study uses two independent drive wheels, each equipped with a DC motor. The side and top views of the system under consideration are depicted in Figure 1 below:

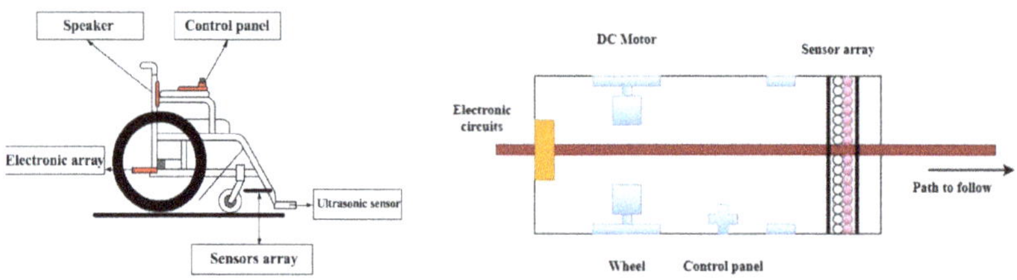

Fig. 1. The side and the upper view of the Electric wheelchair.

The DC motor is a complex system that combines both electrical and mechanical components. In this study, the symmetrical moiety of the robot is built to facilitate analysis and design. Figures 2.a and 2.b show an equivalent depiction of the two motor parts.

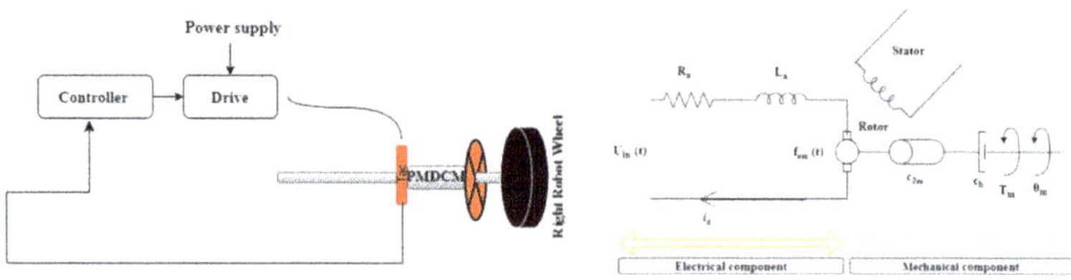

(a) (b)

Fig. 2. (a) A half-simplified model of the wheeled electric wheelchair, (b) Equivalent simplified illustration of the engine components.

When tension is applied to the motor coils, usually from an external power source, an electric current flows through the coils. This current creates a magnetic field around the rotor windings. According to Laplace's law, when this magnetic field interacts with the static magnetic field

Materials Research Proceedings 64 (2026) 510-518 https://doi.org/10.21741/9781644904091-64

generated by the permanent magnets in the stator, a torque is produced. The torque produced by the engine T_m is proportional to current in the armature i_a, with a coupling constant c_t. This is described by Eq. 1:

$$T_m = c_t \times i_a \tag{1}$$

By rotating the motor's armature coils in the stationary magnetic field, a counter-electromotive force f_{em} is generated. This is proportional to the angular velocity of the motor shaft. The following equation can be used to express the Eq. 2:

$$f_{em} = c_b \times \omega_m \tag{2}$$

where $f_{em}(V)$ is the back electromotive force, $c_b(V.s.rad^{-1})$ is a proportionality constant that depends on the specific characteristics of the motor, and $\omega_m(rad.s^{-1})$ is the angular velocity of the motor shaft.

By using Kirchhoff's law on the RL electric part depicted in Figure 2.b, we obtain Eq. 3:

$$u_{in} = R_a \times i_a(t) + L_a \frac{di_a(t)}{dt} + c_b \times \omega_m(t) \tag{3}$$

where $u_{in}(V)$ denotes the input voltage applied to the motor, $R_a(\Omega)$ and $L_a(H)$ represent the resistance and inductance of the armature, respectively.

By taking the transformation of the Laplace to equation (3), we obtain Eq. 4:

$$I_a(p) = \frac{U_{in}(p) - c_b \times \omega_m(p)}{R_a + p \times L_a} \tag{4}$$

The mechanical part of the system is given by Eq. 5:

$$c_t \times i_a(t) - c_{Jm} \frac{d^2\theta_m(t)}{dt^2} - c_{bm} \frac{d\theta_m(t)}{dt} = 0 \tag{5}$$

where $c_{Jm}(Kg.m^2)$ denotes the equivalent inertia of the geared motor, and $c_{bm}(N.m.s)$ represents the viscous damping coefficient.

By taking the Laplace transformation to Eq. 5, we obtain the following relationship:

$$I_a(p) = \frac{c_{Jm}.\theta_m(p).p^2 + c_{bm}.\theta_m(p).p}{c_t} \tag{6}$$

Finally, we get the transfer function derived from Eq. 4 and Eq. 6, which is expressed as follows:

$$G(p) = \frac{c_t}{c_{Jm}.L_a.p^2 + (R_a.c_{Jm} + L_a.c_{bm}).p + R_a \times c_{bm} + c_t \times c_b} \tag{7}$$

Electric wheelchair main parameters are listed in Table 1.

Table 1. Parameters of DC motor utilized in simulations.

Parameters	Values
Supply voltage	$U_{in} = 12V$
Torque constant	$c_t = 1.188$ N.m/A
Armature resistance	$R_a = 0.156\ \Omega$
Armature inductance	$L_a = 0.82$ mH
Back-emf constant	$c_b = 1.185$ rad.s^{-1}.V
Motor inertia	$C_{Jm} = 0.275$ Kg.m^2
Coefficient of friction	$C_{bm} = 0.392$ N.m.s

We obtain the transfer function used in the simulation by substituting the values mentioned in Table 1 into Eq. 7.

$$G(p) = \frac{1.187}{0.2256\ p^2 + 0.3645\ p + 1.468} \tag{8}$$

Proposed PID controller

The primary goal of this paper is to adjust a control strategy for the movement of an electric wheelchair. To do this, a gradual supply voltage $U_{in}(p)$ is delivered to the system, while the velocity is regulated by means of the proposed regulator in a closed feedback loop. The reference and current velocity of the system are compared to obtain the error $e(t)$, which is then fed into the PID regulator responsible for correcting the response dynamics. The parameters of this regulator are then adjusted using the algorithm AOA, adopting as the objective function the integral time of absolute error (ITAE) criterion, defined by Eq. 9:

$$ITAE = \int_0^t t\left|e(t)\right|dt \tag{9}$$

Among the many metaheuristic methods proposed in the literature, the arithmetic optimization algorithm (AOA), introduced by [15], stands out for its use of the distributive behavior of the four fundamental arithmetic operations (addition, subtraction, multiplication, and division). This algorithm was evaluated on several optimization problems to analyze its efficiency, speed of convergence, and complexity of computation. The results show that AOA outperforms eleven classical optimization algorithms in complex optimization tasks. For these reasons, this approach is chosen in this work to address the PID regulator tuning problem, compared with competing metaheuristics.

AOA begins by randomly generating a set of potential solutions. In our application, each solution (S_c) corresponds to the set of three PID regulator parameters, which can be expressed by the matrix (10):

$$S_c = \begin{bmatrix} S_c^{1,1} & S_c^{1,2} & \dots & S_c^{1,n-1} & S_c^{1,n} \\ S_c^{2,1} & S_c^{2,2} & \dots & S_c^{2,n-1} & S_c^{2,n} \\ \dots & \dots & \dots & \dots & \dots \\ S_c^{N-1,1} & S_c^{N-1,2} & \dots & S_c^{N-1,n-1} & S_c^{N-1,n} \\ S_c^{N,1} & S_c^{N,2} & \dots & S_c^{N,n-1} & S_c^{N,n} \end{bmatrix} \tag{10}$$

At every iteration, the current best potential solution (S_c) is considered the provisional optimum. Before running the AOA-based optimization, the search domain must be specified and the Math Optimizer Accelerated (MOA) term determined. This coefficient is computed according to the following expression:

$$MOA(iter_{cur}) = min + iter_{cur} \cdot \left(\frac{max - min}{iter_{max}} \right) \tag{11}$$

where $MOA(iter_{cur})$ denotes the function value at the actual iteration $iter_{cur}$, with $1 \leq iter_{cur} \leq iter_{max}$. The parameters *min* and *max* correspond to the lower and upper bounds of the MOA, respectively.

During the exploration stage, the AOA uses two mathematical operators ((÷) and (×)) to get the optimum solution. This search phase is conditioned by $rand_1 > MOA(iter_{cur})$, and these two techniques are represented by Eq. 12:

$$S_{c_{i,j}}(iter_{cur}+1) = \begin{cases} (S_{c_j}^{Best}) \div (MOP + \varepsilon) \times (Ub_j \times \mu - Lb_j \times \mu + Lb_j), & rand_2 < 0.5 \\ (S_{c_j}^{Best}) \times MOP \times (Ub_j \times \mu - Lb_j \times \mu + Lb_j), & otherwise \end{cases} \tag{12}$$

where $S_{c_{i,j}}(iter_{cur}+1)$ denotes the j-th component of the i-th solution at the new iteration, and $S_{c_j}^{Best}$ is the j-th component of the best solution found so far. The term ε is a smaller integer, while μ is a control variable that tunes the process. Lb_j and Ub_j correspond to the lowest and highest limits of the j-th variable, respectively. The MOP is a coefficient given by the Eq. 13:

$$MOP(iter_{cur}) = 1 - \left(\frac{iter_{cur}}{iter_{max}} \right)^{1/\delta} \tag{13}$$

where $MOP(iter_{cur})$ denotes the value of this function at the actual iteration $iter_{cur}$, $iter_{max}$ is the total iterations, and $\delta = 0.5$ is a parameter controlling the accuracy of the process over the iterations.

In the exploitation stage, two mathematical operators are employed ((−) and (+)) to refine the current solutions. This phase is activated when $rand_1 < MOA(iter_{cur})$, and the corresponding update rules are given by the following expressions:

$$S_{c_{i,j}}(iter_{cur}+1) = \begin{cases} (S_{c_j}^{Best}) - (Ub_j \times \mu - Lb_j \times \mu + Lb_j) \times MOP, & rand_3 < 0.5 \\ (S_{c_j}^{Best}) + (Ub_j \times \mu - Lb_j \times \mu + Lb_j) \times MOP, & otherwise \end{cases} \tag{14}$$

During the exploration and exploitation stages, the MOA is augmented linearly from 0.2 to 0.9. The pseudo-code of AOA for the PID regulator is shown in Figure 3:

Materials Research Proceedings 64 (2026) 510-518 https://doi.org/10.21741/9781644904091-64

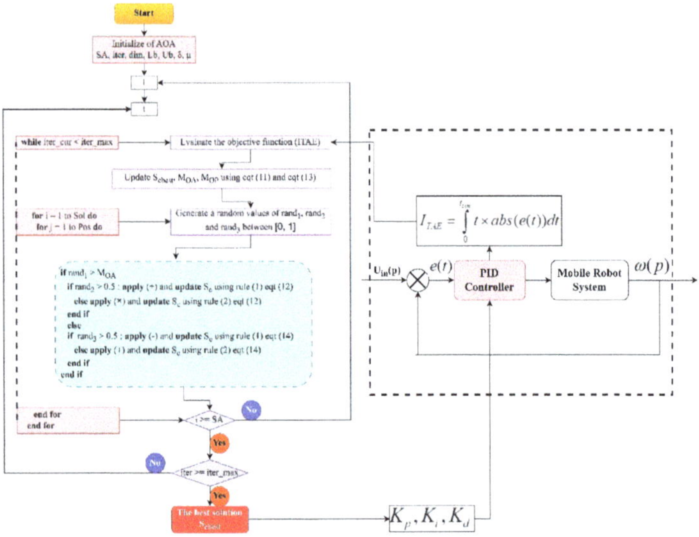

Fig. 3. Flowchart for IR using the AOA algorithm

The control signal *u(t)* generated by the regulator depends on three setting parameters, which are formulated by the following relationship:

$$u(t) = K_p.e(t) + K_i \int_0^t e(\tau).d\tau + K_d.\frac{de(t)}{dt} \tag{15}$$

where K_p is the proportional parameter, K_i is the integral parameter, and K_d is the derivative parameter of the PID controller.

Simulation Results

The effectiveness of the AOA-PID is evaluated in terms of transient response and convergence performance and compared with the ABC-PID and CVSO-PID. Simulations were carried out in a MATLAB on a laptop running Windows 10 64-bit, Intel® Core™ i5-6300U CPU @ 2.40GHz 2.50GHz, RAM (8GB). We used the default settings of the selected metaheuristics.

Table 2 highlights the efficiency and superiority of the proposed AOA-PID regulator. The outstanding performance of this regulator is clearly evident in the quality of the transient response. The AOA-PID offers significant improvements in terms of speed, stability, and accuracy compared with other optimizers evaluated, such as ABC-PID and CVSO-PID.

Table 2. Response transient criterion and gains for the regulator.

PID controller	AOA-PID	ABC-PID	CVSO-PID
K_p	9.92588	6.75738	6.19516
K_i	30	27.2335	24.9115
K_d	6.14339	4.18235	3.86144
$t_s(s)$	0.78925	2.7537	3.3921
$M_p(\%)$	0.34876	2.2768	4.32192

Figure 4.a displays the convergence curves of three metaheuristic optimization algorithms used to minimize the objective function ITAE during PID regulator tuning. A comparative analysis of these curves highlights significant performance differences. The AOA demonstrates superior performance with rapid and consistent convergence from the first iterations, achieving a very low ITAE of 0.078262. This indicates its excellent ability to efficiently explore the search space and exploit it effectively, demonstrating robustness against local minima. Conversely, the CVSO algorithm initially converges correctly but quickly stabilizes near an ITAE of 0.18077. Meanwhile, the ABC shows the poorest performance: after an initial improvement, the curve stagnates above 0.19862, suggesting limited optimization ability and a tendency to become trapped in a local minimum. In summary, AOA proved to be the best-performing algorithm among the three tested, due to its speed, stability, and accuracy in optimizing the PID gain parameters.

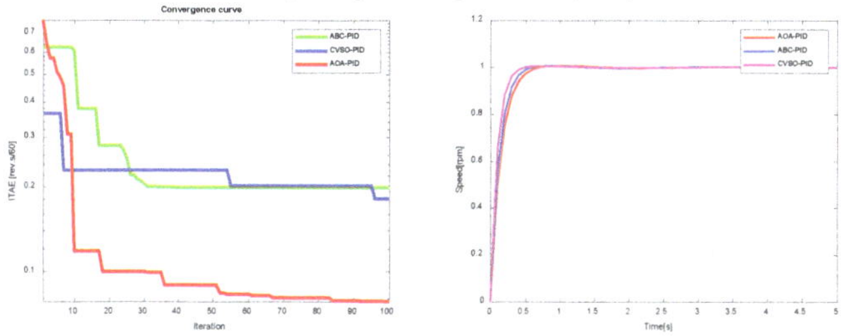

(a) (b)

Fig. 4. (a) Speed regulation using various control methods, (b) Convergence curve for the optimizers employed.

Figure 4.b illustrates the speed response of a DC motor subjected to three distinct PID control methods: AOA-PID, ABC-PID, and CVSO-PID. Among these approaches, the regulator based on the AOA stands out for its superior dynamic performance. Its curve shows a rapid rise towards the set speed, with very low overshoot, and rapid stabilization without oscillations around the reference value.

Conclusion

This study proposed an automatic PID parameter tuning approach based on the Arithmetic Optimization Algorithm (AOA) for speed control of a DC wheelchair motor. Given the limitations of conventional methods such as Ziegler-Nichols, the AOA was chosen for its ability to balance exploration and exploitation during optimization.

Simulation results demonstrate the superiority of AOA-PID, which achieves the best value of the objective function used with ITAE=0.078262, stable convergence, while significantly improving transient response indicators, stabilization time $t_s = 0.78925s$ and overshoot $M_p = 0.34876\%$, compared to ABC-PID and CVSO-PID approaches.

However, this study remains limited to a purely simulated framework, without taking into account real nonlinearities, friction, or load variations. Future work should focus on experimental validation, robustness testing, embedded implementation, and extension to more advanced control structures such as fractional or adaptive PID.

References

[1] M. Cognominal, K. Patronymic, and A. Wańkowicz, 'Evolving field of autonomous mobile robotics: Technological advances and applications', Fusion of Multidisciplinary Research, An International Journal, vol. 2, no. 2, pp. 189-200, 2021. https://doi.org/10.63995/USAS3015

[2] F. Rubio, F. Valero, and C. Llopis-Albert, 'A review of mobile robots: Concepts, methods, theoretical framework, and applications', International Journal of Advanced Robotic Systems, vol. 16, no. 2, p. 1729881419839596, 2019. https://doi.org/10.1177/1729881419839596
[3] R. P. Borase, D. Maghade, S. Sondkar, and S. Pawar, 'A review of PID control, tuning methods and applications', International Journal of Dynamics and Control, vol. 9, no. 2, pp. 818-827, 2021. https://doi.org/10.1007/s40435-020-00665-4
[4] R. Singh, V. Pandey, M. Sharma, Shashikant, and M. Sahni, 'Simulation and Modeling of Linear and Nonlinear PID Controller', presented at the Mathematical Modeling, Computational Intelligence Techniques and Renewable Energy: Proceedings of the Third International Conference, MMCITRE 2022, Springer, 2023, pp. 103-118. https://doi.org/10.1007/978-981-19-9906-2_10
[5] A. Utami, R. Yuniar, A. Giyantara, and A. Saputra, 'Cohen-Coon PID tuning method for self-balancing robot', presented at the 2022 International Symposium on Electronics and Smart Devices (ISESD), IEEE, 2022, pp. 1-5. https://doi.org/10.1109/ISESD56103.2022.9980830
[6] V. V. Patel, 'Ziegler-nichols tuning method: Understanding the pid controller', Resonance, vol. 25, no. 10, pp. 1385-1397, 2020. https://doi.org/10.1007/s12045-020-1058-z
[7] R. S. Widagdo, B. Hariadi, and K. Setyadjit, 'Modelling and analysis of Ziegler-Nichols and Chien-Hrones-Reswick tuning PID on DC motor speed control', Jurnal Teknologi Elektro, vol. 14, no. 1, pp. 23-27, 2023. https://doi.org/10.22441/jte.2023.v14i1.005
[8] N. A. S. Suarin, D. Pebrianti, N. Q. Ann, L. Bayuaji, M. Syafrullah, and I. Riyanto, 'Performance evaluation of PID controller parameters gain optimization for wheel mobile robot based on bat algorithm and particle swarm optimization', presented at the Proceedings of the 10th National Technical Seminar on Underwater System Technology 2018: NUSYS'18, Springer, 2019, pp. 323-333. https://doi.org/10.1007/978-981-13-3708-6_27
[9] S. Sachan and P. Swarnkar, 'Investigations on meta-heuristic algorithms for intelligent speed regulation of mobile robot', Results in Control and Optimization, vol. 11, p. 100232, 2023. https://doi.org/10.1016/j.rico.2023.100232
[10] H. Goud et al., 'Metaheuristics Algorithm for Tuning of PID Controller of Mobile Robot System.', Computers, Materials & Continua, vol. 72, no. 2, 2022. https://doi.org/10.32604/cmc.2022.019764
[11] M. Kmich, H. Karmouni, and M. Sayyouri, 'Improved PID Controller using Archimedes' Optimization Algorithm to Control an Electric Wheelchair', presented at the 2025 7th Global Power, Energy and Communication Conference (GPECOM), IEEE, 2025, pp. 176-181. https://doi.org/10.1109/GPECOM65896.2025.11062018
[12] S. B. Joseph, E. G. Dada, A. Abidemi, D. O. Oyewola, and B. M. Khammas, 'Metaheuristic algorithms for PID controller parameters tuning: Review, approaches and open problems', Heliyon, vol. 8, no. 5, 2022. https://doi.org/10.1016/j.heliyon.2022.e09399
[13] A. Sharma, A. Sharma, S. Choudhary, R. K. Pachauri, A. Shrivastava, and D. Kumar, 'A review on artificial bee colony and it's engineering applications', Journal of Critical Reviews, vol. 7, no. 11, pp. 4097-4107, 2020.
[14] K. Golalipour et al., 'The corona virus search optimizer for solving global and engineering optimization problems', Alexandria Engineering Journal, vol. 78, pp. 614-642, 2023. https://doi.org/10.1016/j.aej.2023.07.066
[15] L. Abualigah, A. Diabat, S. Mirjalili, M. Abd Elaziz, and A. H. Gandomi, 'The arithmetic optimization algorithm', Computer methods in applied mechanics and engineering, vol. 376, p. 113609, 2021. https://doi.org/10.1016/j.cma.2020.113609

Multiclass Deep Learning for Automated Solar Panel Surface Defect Detection Using the Solar Augmented Dataset

Mohamed HADADI[1,a *], Walid HOUMAIDI[1,b] and Anas RIAD[2,c]

[1]School of Science & Engineering, Al Akhawayn University, Ifrane, Morocco

[2]School of Business Administration, Al Akhawayn University, Ifrane, Morocco

[A]m.hadadi@aui.ma, [b]w.houmaidi@aui.ma, [c]a.riad@aui.ma

Keywords: Solar Panel Defect Detection, Deep Learning, Convolutional Neural Networks, Xception, VGG16, Image Classification

Abstract. In this study, we evaluated the performance of deep learning methodologies on the classification of surface conditions of solar panels on the open-solar dataset known as the Solar Augmented Dataset. The task entailed classifying six unique operational conditions in the dataset, namely: Clean, Dusty, Bird Drop, Physical Damage, Electrical Damage, and Snow-Covered. To test and compare our results against other methodologies on the dataset, we fine-tuned two pre-trained architectures of Convolutional Neural Networks on the dataset; namely, the VGG16 and Xception networks. Results on the dataset were that the best model was the VGG16 for our task since it gave over 91% on the validation set and had excellent recall values for the classes of interest-protecting safety-protective conditions, that is, Electrical Damage and Snow-covered Conditions. However, error analysis of the approach on the dataset through the creation of confusion matrices indicates that the model had difficulty distinguishing between the *Clean* and *Dusty* conditions on the solar panels, which could be so because the latter condition only represented the accumulation of light dust on the panels.

Introduction

Maintenance and fault detection in photovoltaic (PV) solar panels are basic requirements for further research improvements in renewable energy technology. With the accelerated integration of solar power into the global energy mix, a high demand exists for automated solutions to inspect photovoltaic panels for dust, bird feces, snow, and various types of damage [1, 2]. Recent computer vision technology breakthroughs in deep learning have brought significant improvements to automated photovoltaic inspection, with research showing excellent results in fault detection and localization in photovoltaic panels [3-4, 5].

However, in the existing literature, the focus has been on closed or publicly restricted datasets, which were mostly restricted to binary classification tasks and were also less diverse in the representation of defects in reality [3,6]. Such limitations are likely to restrict the benchmarking capacity of detection models. Open, ample datasets are, therefore, critical in benchmarking models related to multiclass recognition in the context of solar panels.

In this research, we overcome this difficulty by employing the open-source Solar Augmented Dataset, with diverse technician-collected image data that has been annotated with occlusions and defects. To evaluate performance, our strategy involves comparing the performance of the state-of-the-art Convolutional Neural Network models Xception and VGG16 on this multi-class detection task. To ensure realistic performance portability, aggressive data augmentation and validation procedures will be incorporated with the objective of setting up realistic performance benchmarks for smart solar panel fault inspections [7].

Related Work

There has been considerable interest in the application of deep learning to photovoltaic solar panel inspection tasks, but most of the existing literature has concentrated on binary tasks or other forms of coarse classification: e.g., distinguishing between healthy and unhealthy panels, and identifying hot spots or cracked panels [3, 6, 8]. Even if more classifications were explored with more classes, conventional solar panel inspection tasks have tended to lump panel-damaging conditions into broad categories such as *clean*, *damaged*, *soiled*, and others, or have concentrated on binary classifications involving dusty vs. clean panels, which isn't particularly representative of real-world solar panel farm operation environments [1, 4, 8]. Multi-class studies of panel appearance remain relatively rare, with some studies having solar panel inspection data sets with custom or facility-specific content whose images aren't completely publicly divulged [3, 6].

However, several deep learning methods also assess the performance of a single convolutional backbone, such as a single VGG, ResNet, or U-Net architecture, predominantly in terms of global accuracy or mean average precision in a single class, without much focus on class-wise accuracy or confusion between similar conditions. Some of the recently emerging works attempt to tackle the fault localization in U-Net or hybrid architectures, but custom datasets are still employed, as performed on a single architecture in [4]. Additionally, a new SparkNet [9] proposes a custom deep CNN architecture designed specifically for PV fault localization, showing high accuracy on its target dataset, but still employing a single architecture, without exploring other CNN structures or class-wise confusion between various conditions. For instance, in a number of recently emerging works on surface anomaly problems targeting multiclass surface anomaly classifcation using a VGG16 network, high accuracy is obtained but a comprehensive comparison of lightweight and heavy network structures on a single benchmark has yet to be performed, in addition to other surface anomaly works on dust or defect localization, predominantly stressing overall accuracy without going through class-wise accuracy in dust or defect localization [2,10].

Methodology

Proposed System Architecture: It proposes a system using deep convolutional neural networks, classifying solar panel surface images into six distinct categories: Bird drop, Clean, Dusty, Electrical damage, Physical damage, and Snow covered. Raw solar panel images are first captured; automated data ingestion standardizes them and prepares them through resizing, RGB channel normalization, and augmentation as needed. Then, the preprocessed images feed a selected trained model-VGG16 or Xception-operating hierarchical feature extraction with the capacity for robust detection of a wide range of surface conditions under adverse and varying environmental factors. It makes use of automated data augmentation methods as well as model-specific input normalization techniques that take into consideration_uniformity and robustness with regard to illumination as well as background noise. It has been extensively optimized with regard to effective processing and inferencing within high-resolution images in order to provide fast and effective detection that is imperative in active health checking_processes within solar panels.

Dataset Description: The dataset employed in this research is the solar augmented dataset. It is one of the datasets that is open to the public and consists of pictures showing the surface of solar panels [11]. These pictures come in high resolution and are classified into several categories concerning the condition of the solar panel surfaces. These categories include Bird drop, Clean, Dusty, Electrical damage, Physical damage, as well as Snow-covered. Each of these images is taken in different real-life outdoor environments. The dataset is further split into 80% training data, 10% validation data, and 10% test data with imbalances corrected accordingly.

Figure 1. Representative of Solar Panels Images

Data Augmentation: For all experiments, images were standardized to three color channels (RGB) to satisfy the input requirements of both backbones. Any alpha channel was discarded and grayscale images were replicated across three channels. For the VGG16 pipeline, images were resized to 224×224 pixels, whereas for the Xception pipeline they were resized to 299×299 pixels, matching the native input resolution of each architecture. In both cases, the appropriate Keras preprocess_input function for the corresponding model was applied before training. Training used data generation with the following augmentations: zoomed up to 25% in both the x and y directions, sheared by 15% in both the x and y directions, rotated by ±10°, width and height shifted by up to 10% in both directions, brightness changed within the range [0.8,1.2], and horizontally flipped; all performed by nearest-neighbor interpolation. The purpose of the augmentations is to increase the generality of the models by simulating real-world variations in the acquisition of solar panel images (viewpoint, illumination, and geometry). Validation and test datasets were preprocessed solely by image resizing and respective models' normalization. Data generators were distributed using the Keras functions flow_from_dataframe, where the splits are shuffled for the training split but in a fixed and deterministic order for validation and test splits.

Model Selection and Justification: VGG16 is another well-known convolutional neural network that has been praised for its simple and deep network structure, which involves a chain of convolutional layers with a small receptive field and fully connected layers [5]. This uniform structure and the existence of pre-trained weights have made VGG16 a preferable model to adapt during transfer learning, especially within the context of image inspection tasks like defect identification on solar panel surfaces. The known capability of VGG16 to extract fine space details has also enabled the network to classify intricate forms of solar panel defects like dust, cracks, and damage within high-resolution solar panel images. VGG16 has been found to be effective within various types of renewable energy and quality control tasks in multiple studies [2], [5], which has been supported through its capability to accomplish hierarchical feature extraction at the cost of higher computation.

Xception, on the other hand, was chosen for its high representational power and known success in fine-grained vision tasks. With its depth wise_separable convolution layers structured in an extreme Inception manner, Xception is capable of recognizing minute details in space and texture features based on relatively fewer parameters than traditional very deep CNNs. Moreover, pre-trained on the ImageNet task, Xception offers high-quality feature extractors which_perform well on domain-specific tasks like identifying various kinds of solar panel defects and surface characteristics. Although Xception is known to be bulkier in size than traditional lightweight architectures, in offline diagnosis or GPU-assisted monitoring systems, its high diagnosis accuracy and capacity balance makes it a highly suited candidate. In fact, Xception and variants like Inception networks have previously shown success in pre-existing works on defect detection and outlier identification tasks_and thus serve as a strong foundation for selecting the best multi-class classification model for identifying various surface characteristics of a solar panel.

Evaluation Metrics: The performance of models was analyzed using accuracy and confusion matrix analysis. Accuracy is the measure of correct predictions of sample images in the test dataset. The accuracy is an absolute measure that distinctly reveals the efficiency of models on the multi-class task of classifying solar panel surfaces by their condition. Accuracy, however, fails to furnish information on relative class strengths and weaknesses, particularly when analyzing imbalanced classes. However, these aspects were enabled in detail by analyzing the confusion matrices of models on a wholesale level. The confusion matrix reveals information on correct and wrong predictions of sample images aligned with class labels, thereby enabling an assessment of efficiency and progress of models per single defect class, including dust, physical damage, and electrical damage. Furthermore, analyzing misclassification among single defect classes, including dust, bird droppings, and snow, with physical damage and panel cracks, among others, reveals systematic misclassification, further emphasizing efficiency of models in distinguishing similar defect classes based on similar surfaces. A holistic analysis of accuracy and confusion matrix, therefore, enables an efficient comparison of both VGG16 and Xception models on defect classification accuracy across defect classes.

Model Architecture and Fine Tuning: In particular, two convolutional neural networks, VGG16 and Xception pre-trained on ImageNet, were utilized herein as backbone feature extractors for the classification task on the surface images of solar panels. Both networks were initialized with weights trained on the ImageNet dataset for exploiting their strong low-level and mid-level visual cues and fine-tuned on the six target classes of the Solar Augmented Dataset. In these two scenarios, the pre-trained classification head was discarded and substituted with a light task-specific classification head containing a Global Average Pooling layer_followed by a Dropout layer with a dropout rate of 0.45 and finally a Dense layer with six softmax units. The backbone layers were configured to be partially trainable, with most early layers being frozen and only the last ten layers being unfrozen and fine-tuned for domain-specific cues like dust, bird droppings, electrical damage, physical damage, and snow.

Algorithm 1: Training Pipeline Input: 1:Initialize VGG16 with ImageNet weights θ. 2:

Modify Head: GAP → Dropout(0.45) → Softmax(6). 3: Freeze Blocks 1–4; Unfreeze Block 5 (Last 10 layers). 4: Set Optimizer = Adam (learning rate = 1e-4), Batch = 32. 5: While epoch < 25 do: 6: Train on D_train; Compute Loss. 7: If Val_Loss plateaus for 7 epochs: reduce learning rate

Materials Research Proceedings 64 (2026) 519-526 https://doi.org/10.21741/9781644904091-65

by 0.5. 8: If Val_Loss plateaus for 8 epochs: Break (Early Stop). 9: Return Best θ* (Lowest Val Loss).

Results and Findings

Baseline Comparisons: For an understanding of how well the baselines actually performed, a baseline was introduced using a simple CNN of a very basic architecture, including four convolutional layers with filter sizes gradually increasing from 32 to 64, to 128, and finally to 256. The custom-built CNN achieved a maximum accuracy of 76.4% on the validation sets, which clearly pointed towards serious cases of overfitting, as the network was finding difficulties in distinguishing between subtle changes in texture, such as a dusted surface versus clean glass. The high gap of +14.3% covered by VGG-16 clearly validates the need to use a pre-trained image feature hierarchy in respect to dealing with the texture of solar panels in an outdoor setting.

Performance Comparison: For this work, two convolutional neural networks, VGG16 and Xception, pretrained on ImageNet, have been chosen and utilized for feature extraction in classifying images based on solar panel surface type. These two convolutional neural networks had their initial layers initialized from respective pretrained weights on the ImageNet dataset for extracting extensive features in mid-level and low-level parts, followed by adapting them to work on the six classes in the Solar Augmented Dataset. For both VGG16 and Xception, the original classification layers had been_discarded in favor of a light-weight_task-specific classification layer consisting of a Global Average Pooling-Dropout-Dense layer activated by softmax, possessing 0.45 Dropout_rate and six neurons, respectively. The layers in the backbone had been_chosen to have partial training capabilities, meaning most layers in both convolutional networks had been set to untrainable, allowing it to focus on feature extraction rather than adapting to different domains characterized by features associated with dust,bird droppings, electrical damage, physical damage, and snow.

Result Interpretation: The VGG16 results indicate that the proposed pipeline is able to reliably distinguish between a diverse set of solar panel conditions, despite the moderate class imbalance present in the dataset. High accuracy for the damage-related classes show that structural and snow-related faults are captured robustly, which is critical for maintenance planning. The confusion matrix reveals that the remaining errors are concentrated mainly between the *Clean* and Dusty classes, and to a lesser extent between *Bird drop* and *Dusty*. These are visually subtle and sometimes ambiguous cases, where light dust or localized bird droppings make the boundary between clean and contaminated difficult even for human annotators. Consequently, the overall performance should be interpreted as strong given that most misclassifications occur in borderline conditions rather than in clearly faulty modules.

Materials Research Proceedings 64 (2026) 519-526 https://doi.org/10.21741/9781644904091-65

Classification Report of VGG16

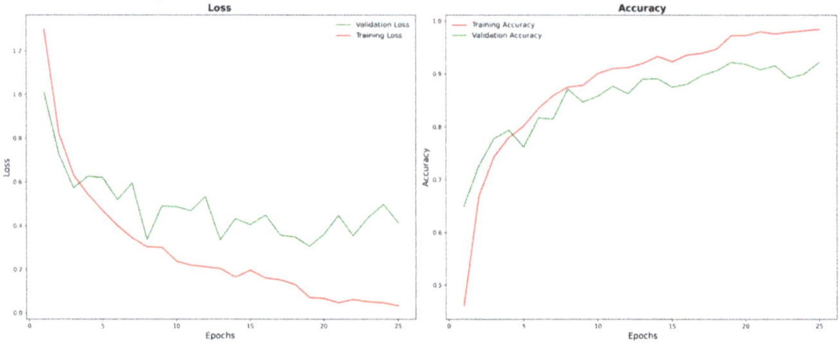

Figure 2. Training, validation loss, and accuracy of VGG16 over 25 epochs

The training and validation loss curves for the VGG16 model (Figure 2) illustrate steady convergence throughout training. Both losses decrease rapidly during early epochs and then plateau, suggesting effective optimization. The small gap between training and validation loss indicates successful generalization and minimal overfitting.

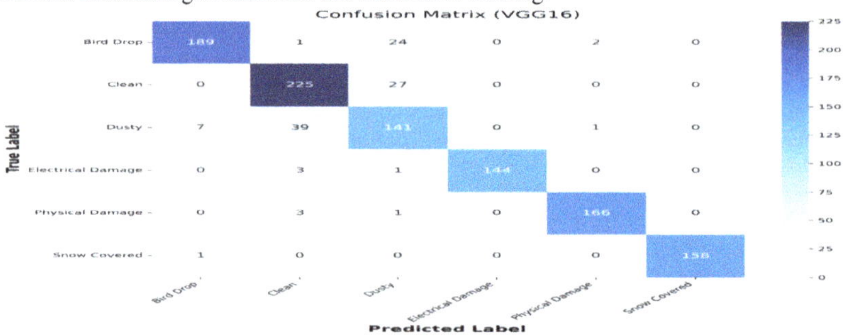

Figure 3. Confusion Matrix of VGG16

The VGG16 classification report, summarized by the confusion matrix and the learning curves in Figure 3, confirms these observations at a per-class level. The diagonal entries of the confusion matrix show near-perfect recognition for Electric damage class, Physical damage class, and snow covered class , with only a handful of samples misassigned to neighboring surface classes. In contrast, the Dusty class exhibits noticeably lower recall, with a significant fraction of dusty panels predicted as Clean or Bird drop, reflecting the inherent similarity of these patterns. The training and validation curves further illustrate mild overfitting: training loss continues to decrease, and training accuracy approaches 96–97%, while validation accuracy plateaus around 90% and validation loss fluctuates around a shallow minimum. This behavior is typical for a high-capacity model on a moderately sized, slightly imbalanced dataset, and justifies the use of early stopping and model checkpointing to select the best epoch for reporting performance.

Table 1: Detailed Test Set Performance per Class (VGG16)

Class Name	**Precision**	**Recall**	**F1-Score**
Bird Drop	0.94	0.88	0.91
Clean	0.86	0.81	0.84
Dusty	0.71	0.84	0.77
Electrical Damage	0.97	0.96	0.96
Physical Damage	0.97	0.91	0.94
Snow Covered	0.95	0.97	0.96
Overall Accuracy	-	-	**0.89**

Conclusion and Future Work

This study benchmarked VGG16 and Xception for multiclass solar panel defect detection using the Solar Augmented Dataset. VGG16 achieved the highest accuracy (~91%) and superior recall on critical classes like *Electrical Damage* and *Snow Covered*, establishing a robust baseline despite inherent ambiguity between *Clean* and *Dusty* samples. While these results verify the efficacy of transfer learning, heavy backbones like VGG16 require optimization for embedded use. Future work will therefore prioritize lightweight architectures (e.g., *MobileNetV3*, EfficientNet) and model compression techniques (pruning, quantization) to enable real-time edge deployment. Additionally, we plan to investigate advanced imbalance handling strategies, such as focal loss, and validate the system across multi-site datasets to ensure robustness in diverse operational environments.

Declaration of Generative AI and AI-Assisted Technologies in the Writing Process:

The authors used Gemini and ChatGPT to refine the language of this work. We reviewed the output and take full responsibility for the content.

References

[1] A. Rahman, Solar Panel Surface Defect and Dust Detection: Deep Learning Approach, J. Imaging 11 (2025) 287. https://doi.org/10.3390/jimaging11090287

[2] H. Ling, M. Liu, Y. Fang, Deep Edge-Based Fault Detection for Solar Panels, Sensors 24 (2024) 5348. https://doi.org/10.3390/s24165348

[3] I. Polymeropoulos, S. Bezyrgiannidis, E. Vrochidou, G.A. Papakostas, Enhancing Solar Plant Efficiency: A Review of Vision-Based Monitoring and Fault Detection Techniques, Technologies 12 (2024) 175. https://doi.org/10.3390/technologies12100175

[4] R.A.M. Rudro, K. Nur, Md.F.A.A. Sohan, M.F. Mridha, S. Alfarhood, M. Safran, K. Kanagarathinam, SPF-Net: Solar panel fault detection using U-Net based deep learning image classification, Energy Rep. 12 (2024) 1580–1594. https://doi.org/10.1016/j.egyr.2024.07.044

[5] S.R. Joshua, S. Park, K. Kwon, Solar Panel Fault Detection: Applying Convolutional Neural Network for Advanced Fault Detection in Solar-Hydrogen System at University, in: 2024 IEEE

24th International Conference on Software Quality, Reliability, and Security Companion (QRS-C), IEEE, 2024: pp. 289–298. https://doi.org/10.1109/QRS-C63300.2024.00045

[6] K. Masita, A. Hasan, T. Shongwe, H.A. Hilal, Deep learning in defects detection of PV modules: A review, Solar Energy Adv. 5 (2025) 100090. https://doi.org/10.1016/j.seja.2025.100090

[7] A. Mellit, S. Kalogirou, Artificial intelligence and internet of things to improve efficacy of diagnosis and remote sensing of solar photovoltaic systems: Challenges, recommendations and future directions, Renew. Sustain. Energy Rev. 143 (2021) 110889. https://doi.org/10.1016/j.rser.2021.110889

[8] S.H. Han, T. Rahim, S.Y. Shin, Detection of Faults in Solar Panels Using Deep Learning, in: 2021 International Conference on Electronics, Information, and Communication (ICEIC), IEEE, 2021: pp. 1–4. https://doi.org/10.1109/ICEIC51217.2021.9369744

[9] R. G, D.S. Manish, R. Narasimhan A, A.U. Dhavale, R.R. John, SparkNet–A Solar Panel Fault Detection Deep Learning Model, IEEE Access 13 (2025) 75599–75617. https://doi.org/10.1109/ACCESS.2025.3564714

[10] Y. Ledmaoui, A. El Maghraoui, M. El Aroussi, R. Saadane, Enhanced Fault Detection in Photovoltaic Panels Using CNN-Based Classification with PyQt5 Implementation, Sensors 24 (2024) 7407. https://doi.org/10.3390/s24227407

[11] G. Navnath, Solar Augmented Dataset, Kaggle, n.d. https://www.kaggle.com/datasets/gitenavnath/solar-augmented-dataset (accessed November 29, 2025).

Intelligent Control for Mobile Robot Trajectory Tracking

Nabil NADAH[1,a *], Mohamed KMICH[1,b] and Mhamed SAYYOURI[1,c]

[1]Engineering, Systems and Applications Laboratory, National School of Applied Sciences, Sidi Mohamed Ben Abdellah University, Fez, Morocco

[a]nabil.nadah@usmba.ac.ma, [b]mohamed.kmich@usmba.ac.ma, [c]mhamed.sayyouri@usmba.ac.ma

Keywords: PID Controller, Puma Optimizer Algorithm, Mobile Robot, Metaheuristic Algorithms, Trajectory Tracking

Abstract. This paper proposes a PID controller optimized using the Puma Optimizer Algorithm (POA) for accurate tracking of a wheeled mobile robot trajectory and improved system stability. An in-depth comparative study was also conducted with PID controllers optimized by other metaheuristic algorithms, including the Sine Cosine Algorithm (SCA) and the Arithmetic Optimization Algorithm (AOA). Our simulation results demonstrate that our proposed controller POA-PID-ITAE performs very effectively with a settling time $T_s = 13s$, a maximum overshoot of $M_p = 0.0493\%$, and an Integral of Time multiplied by Absolute Error (ITAE) value of 96.72. These metrics demonstrate the effectiveness of our method in ensuring a stable and effective tuning mechanism for PID-based trajectory tracking of mobile robots within the adopted kinematic modeling assumptions.

Introduction

Robotics has become a vital field of study and innovation in recent decades. It has advanced rapidly and is now applied in many significant sectors, including manufacturing [1], healthcare [2], agriculture [3], and transportation [4]. The earliest industrial robots were simple machines designed to perform repetitive tasks by following fixed programs without variation. Today, they can perform complex tasks more independently, thanks to recent technological advances that significantly boost productivity and reduce costs.

There are many types of robots; however, mobile wheeled robots are central to modern robotics. These include omnidirectional robots [5], tricycle robots [6], and differential robots [7]. The differential drive system has special appeal because it features driven wheels that operate independently, a simple mechanical architecture, and ease of implementation, which makes it commonly used for trajectory tracking. however, robust trajectory tracking for controlling them remains a challenge, as it involves balancing adherence to kinematic constraints with adaptation to environmental disturbances and uncertainties.

The control of mobile robots has undergone significant improvement, shifting from manual and heuristic tuning techniques that require constant supervision to intelligent strategies that adapt to complex, dynamic environments. Control methods include traditional approaches like sliding mode control (SMC) [8] and PID controllers [9], as well as other advanced techniques [10]. The PID controller is popular because it is simple and reliable, maintaining its status as an industry standard due to its ease of implementation. However, its effectiveness relies heavily on proper parameter tuning, which can be difficult in nonlinear and unpredictable situations. Empirical tuning methods such as Ziegler-Nichols or Cohen-Coon [11] often fall short in ensuring optimal performance, accuracy, and stability.

To address these limitations, several recent studies have shown that metaheuristics are an effective solution for automating and optimizing the adjustment of PID controllers dedicated to tracking the trajectory of mobile robots. In [12], the authors applied particle swarm optimization

(PSO) to adjust PID gains, which simultaneously reduced stabilization time and tracking error in complex environments. The authors of [13] enhanced the Puma Optimizer metaheuristic by integrating chaotic maps, thereby improving the control performance of a two-wheeled mobile robot driven by a PID. In another similar study [14], the authors compare several metaheuristics, including Hunger Games Search (HGS), Slime Mold Algorithm (SMA), and Educational Competition Optimizer (ECO), and highlight the superiority of SMA-PID for trajectory tracking. Despite these advances, certain limitations remain, as several studies are still based on older metaheuristics, which limits the ability to optimize the balance between exploration and exploitation phases and convergence towards local optima, and consequently, the optimization of PID controller gains obtained.

In this article, we propose a PID controller optimized by the recent metaheuristic Puma Optimizer Algorithm (POA) for trajectory tracking in order to improve the trajectory tracking performance of a differential mobile robot. The controller parameters are adjusted by minimizing the objective function, the integral of the time-weighted absolute error (ITAE). Our proposed approach, POA-PID-ITAE, aims to increase tracking accuracy. It is compared to other metaheuristic methods, notably the sine cosine algorithm (SCA) and the arithmetic optimization algorithm (AOA), based on PID settings under the same simulation conditions, in terms of improvement in tracking accuracy, stabilization time, and maximum overshoot. The results obtained demonstrate that the PID-POA-ITAE approach provides significant improvements in terms of convergence, speed, and trajectory tracking accuracy within the framework of the adopted kinematic modeling assumptions.

There are several primary contributions of this work as follows:

- Mathematical modeling of a differential mobile robot.
- Proposal of a PID controller whose parameters are optimized by the POA metaheuristic algorithm to ensure accurate trajectory tracking.
- In-depth comparative analysis of the performance of the PID-POA-ITAE controller compared to PID controllers optimized by the SCA and AOA algorithms.

The remainder of this article is organized as follows: The second section describes in detail the system studied and its kinematic model. Next, the proposed PID-POA-ITAE approach is presented in the third section, and the simulation results are then analyzed and critically interpreted in the fourth section. Finally, the fifth section summarizes the major contributions, highlighting the main simulation results and opening up perspectives for future research.

Kinematic model of the differential mobile robot

The kinematic modeling of a mobile robot allows establishing the mathematical relationships between the speeds of its two drive wheels and the robot's trajectory, position, and orientation, while ignoring dynamic effects such as forces and masses. The system in this study is a mobile robot with a differential drive, shown in Figure 1.

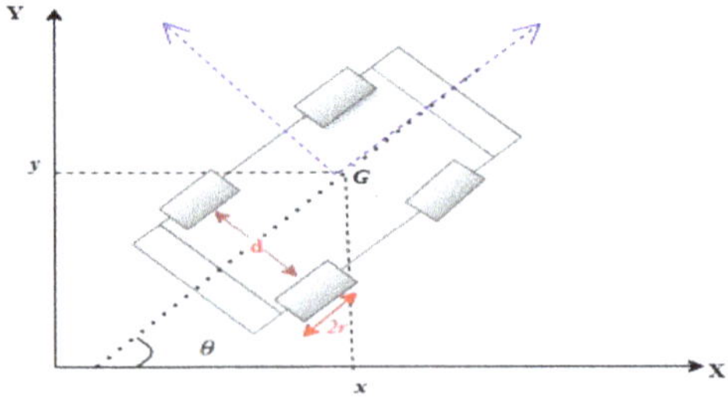

Fig. 1. Simplified differential mobile robot diagram.

Materials Research Proceedings 64 (2026) 527-534 https://doi.org/10.21741/9781644904091-66

The kinematic modeling of the robot under study is based on the following assumptions:

- The wheels roll without slipping.
- The reference point of the system coincides with the robot's center of gravity.

The robot's configuration is described by the following vector:

$$q = \begin{pmatrix} x \\ y \\ \theta \end{pmatrix} \quad (1)$$

where $(x \quad y)^T$ represents the coordinates of the robot's center of gravity in the global coordinate system, and θ represents the orientation relative to the global x-axis.

The equations of motion describing the robot's kinematics are given by:

$$\dot{x} = V\cos\theta\ , \quad \dot{y} = V\sin\theta\ , \quad \dot{\theta} = \omega \quad (2)$$

where V represents the robot's instantaneous linear velocity, and ω its angular velocity.

For a mobile robot with a differential drive, the kinematic relationships between the linear speeds of the wheels and the robot's travel speeds are expressed by:

$$\begin{pmatrix} V \\ \omega \end{pmatrix} = \begin{pmatrix} r/2 & r/2 \\ r/d & -r/d \end{pmatrix} \begin{pmatrix} \omega_r \\ \omega_l \end{pmatrix} \quad (3)$$

The complete kinematic model in terms of the angular velocities of wheels is described as follows:

$$\begin{pmatrix} \dot{x} \\ \dot{y} \\ \dot{\theta} \end{pmatrix} = \begin{pmatrix} \frac{r}{2}\cos\theta & \frac{r}{2}\cos\theta \\ \frac{r}{2}\sin\theta & \frac{r}{2}\sin\theta \\ \frac{r}{d} & -\frac{r}{d} \end{pmatrix} \begin{pmatrix} \omega_r \\ \omega_l \end{pmatrix} \quad (4)$$

where r represents the radius of the wheels and d the distance between the right and left wheels.

The main geometric and kinematic parameters of the mobile robot required for this simulation are presented in Table 1.

Table 1. Geometric and kinematic parameters of the robot.

Parameter	Values
Maximum speed in (m/s)	V_{max}=0.7
Wheel radius in (m)	$r = 0.035$
Distance between wheels in (m)	$d = 0.235$
Initial orientation in (rad)	$\theta = 0.175$

Proposed POA-PID-ITAE controller

The primary goal of this research is to control the movement of a mobile wheeled robot to follow a target trajectory. The proposed controller is based on a closed-loop control architecture utilizing a PID controller optimized by the POA algorithm to regulate the tracking of the target trajectory automatically. The operating principle consists of generating two distinct error signals: a distance

Materials Research Proceedings 64 (2026) 527-534 https://doi.org/10.21741/9781644904091-66

error $e_d(t)$ and an orientation error $e_\theta(t)$, obtained respectively with the comparison of the desired and the current positions of the robot, and the desired orientation with the actual orientation of the robot. These errors are expressed by the following equations:

$$e_d(t)=\sqrt{(x-x_d)^2+(y-y_d)^2}\ ,\ e_\theta(t)=\tan^{-1}\left(\frac{y_d-y}{x_d-x}\right)-\theta_d \tag{5}$$

where (x,y) and (x_d,y_d) represent the current and desired coordinates of the robot's center of gravity, respectively, θ and θ_d represent the current and desired orientation of the robot relative to the global x-axis, respectively.

These errors are then transmitted to the PID controller, whose parameters are optimized by the POA algorithm, which is based on the ITAE objective function defined by the following relationship:

$$ITAE=\int_0^{tsim} t\left|e_d(t)\right|dt+\int_0^{tsim} t\left|e_\theta(t)\right|dt \tag{6}$$

Puma is one of the most successful and well-adapted predators in the Western Hemisphere, which allows it to live in a variety of environments and maintain healthy populations in many regions. Pumas have very strong hind legs, allowing them to jump. [15] presents the puma for the first time and discusses the principal steps of the Puma Optimizer Algorithm (POA).

The POA is more complex to implement than conventional optimization algorithms, because it integrates many interactive phases and involves more computational operations to emulate intelligent behavior. The proposed POA-based algorithm, designed for tuning the PID controller parameters, is illustrated in the flowchart shown in Figure 2.

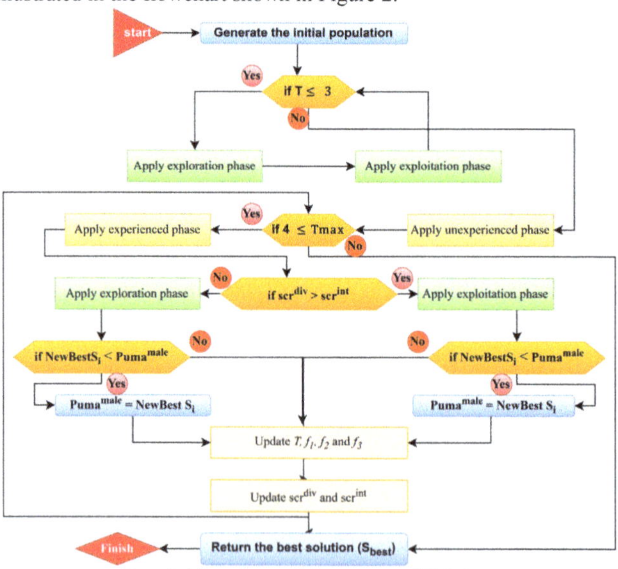

Fig. 2. Flowchart of the proposed POA.

To provide a fair and accurate comparison, we establish a classical PID controller implement a classical PID controller with the same control structure and signals as the proposed method. The classical controller uses the basic PID law, which is defined by the following relationship:

$$U(t) = K_p e(t) + K_i \int_0^t e(t)dt + K_d \frac{de(t)}{dt} \tag{7}$$

where $e(t)$ is the signal error and $K = [K_p, K_i, K_d]$ are the proportional, integral and derivative gains, respectively

The control signal (V, ω) is generated by two PID controllers, depending on three tuning parameters that are applied to the position and orientation errors, which are formulated by the following relationships:

$$V_{PID} = K_{pd} e_d(\mathrm{t}) + \mathrm{K}_{id} \int_0^t e_d(t) + \mathrm{K}_{dd} \frac{de_d(t)}{dt} \quad , \quad \omega_{PID} = K_{p\theta} e_\theta(\mathrm{t}) + \mathrm{K}_{i\theta} \int_0^t e_\theta(t) + \mathrm{K}_{d\theta} \frac{de_\theta(t)}{dt} \tag{8}$$

where $K = \left[K_{pd}, K_{id}, K_{dd}, K_{p\theta}, K_{i\theta}, K_{d\theta}\right]$ are the PID gains to be optimized indicate proportional, integral, and derivative gains for the position and the orientation, respectively, as shown in Figure 3.

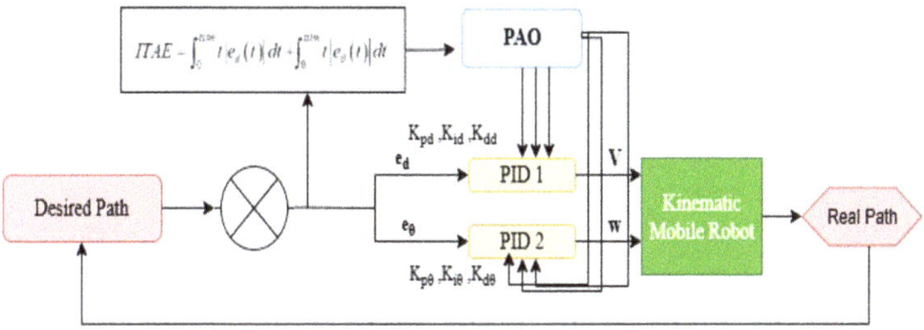

Fig. 3: Diagram illustrating the control action.

The simulation results are discussed in the following sections to demonstrate the effectiveness of our proposed method.

Results and discussion

The effectiveness of our proposed POA-PID controller is evaluated in this section against other metaheuristic algorithms, such as SCA and AOA. In this evaluation, we focus on trajectory tracking accuracy, the value of the ITAE objective function, the settling time $Ts(s)$, and the maximum overshoot $Mp(\%)$ .To evaluate the controller's dynamic behavior, we use a sinusoidal reference trajectory that stresses the system over an extensive range of frequencies, enabling comprehensive evaluation of the controller as described in the following equations

$$X_{traj} = t \quad , \quad Y_{traj} = 0.5 \times \sin\left(\frac{2\pi \times t}{1.5}\right) \tag{9}$$

where t represents time $t \in [0,1.5]$.

The results were simulated in a MATLAB R2024b environment on a Windows 11 64-bit laptop with an Intel® Core™ i5-1135G7 @ 2.40 GHz (2.42 GHz) processor and 16 GB of RAM. For the simulation, we used the default parameters of the selected metaheuristic algorithms. The initial parameters were adopted for all the metaheuristic algorithms used.

Table 2. Initial parameters for the various metaheuristics

Metaheuristics	Common parameters	Specific parameters
AOA	- Dim = 6 - Number of search agents N=50 - Maximum iterations T=50 - Lb = [0,0,0,0,0,0] - Ub = [30,30,30,1,1,1]	- Tuning parameter for the search process **μ**=0.499 - sensitive parameter **δ**=5 - $maxM_{OP}$ =1, $minM_{OP}$ =0.2
SCA		- A constant $a = 2$
POA		- Tuning parameter for the search process **μ**=0.99 - Sensitivity parameter R=0.2

As shown in Table 2. The parameter settings of the three metaheuristics, AOA, SCA, and POA, utilized for PID tuning, and the *Dim = 6* it is correspond to the PID gains, three for position, and three for orientation, and the same *N, T, Ub and Lb* for all the algorithms

Performance evaluation is based on minimizing the ITAE and analyzing the trajectory tracking obtained for each algorithm. Figure 4 shows the simulation results for tracking a sinusoidal trajectory, comparing the trajectories generated by the PID controllers optimized by POA, SCA, and AOA with the reference trajectory.

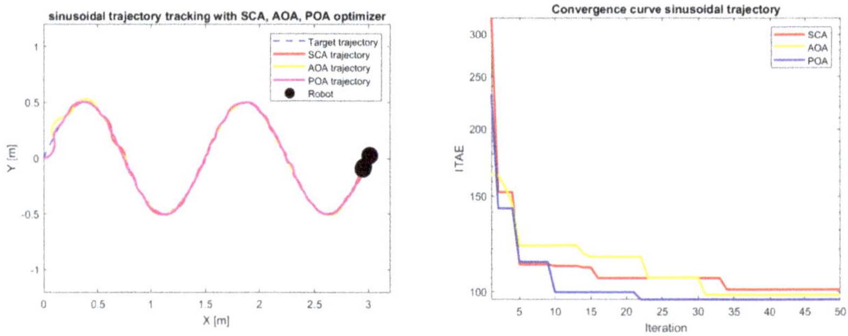

(a) (b)

Fig. 4: (a) Sinusoidal trajectory and (b) convergence curve tracking using various optimizers.

All PID controllers optimized by the three metaheuristics used in this work enable the robot to follow the reference trajectory, as illustrated in Figure 5.a, confirming the value of optimization for adjusting PID gains. However, the proposed POA-PID-ITAE controller stands out for its more accurate tracking, with a trajectory that is virtually indistinguishable from the reference, unlike the other controllers evaluated. This superiority is also apparent in the convergence speed shown in Figure 5.b. Indeed, the POA-PID-ITAE quickly reaches a low and stable ITAE value from the first iterations, around 10 iterations, and then the curve remains virtually constant, which means that the improvement becomes very low and the solution is already close to the optimum. Conversely, the AOA-PID-ITAE stabilizes later, around 25 iterations, while the SCA-PID-ITAE reaches comparable stability even later, around 35 iterations. Thus, the POA-PID-ITAE allows optimal performance to be achieved more quickly, resulting in reduced computation time and better overall tracking performance.

Table 3. Performance Metrics and PID Gains with Metaheuristic Tuning

	K_{pd}	K_{id}	K_{dd}	$K_{p\theta}$	$K_{i\theta}$	$K_{d\theta}$	M_p(%)	T_s(s)	ITAE
SCA	0.0031	0	0	0.9840	0	0.2013	0.0528	60	152.16
AOA	0.0010	0	0.0006	1	0.1508	0.1643	0.0792	60	97.50
POA	**0.0010**	**0**	**0.1120**	**1**	**0**	**0.2132**	**0.0493**	**13**	**96.72**

Table 3 compares the performance of PID controllers optimized by SCA, AOA, and POA. POA-PID stands out as the most effective, with the lowest overshoot M_p=0.0493% and shortest stabilization time T_s=13s compared to the other two methods.

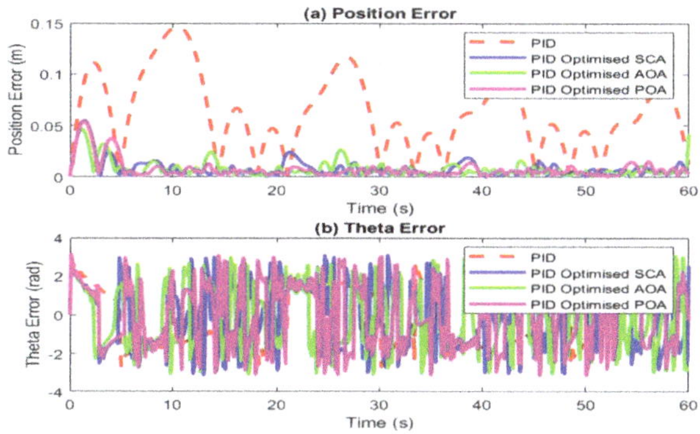

Fig. 5: (a) Position error, (b) Orientation error for conventional and optimized PID

Figure 5 compares tracking errors between conventional PID and optimized PID controllers. In Figure 5.a, we see that the traditional PID controller oscillates significantly and maintains a fairly large position error, while the optimized controllers clearly reduce this error, particularly the proposed POA-PID controller, which is the most effective because its error remains the smallest and stabilizes more quickly near zero throughout the simulation. In Figure 5.b, we observe the same thing for the orientation error, where the POA-PID shows more limited and regular variations than the other controllers tested.

Conclusion

This study demonstrated a PID controller based on the improved PID gains settings and the POA-PID trajectory tracking of a mobile robot. The superiority and the effectiveness of the proposed controller were evaluated by examining the settling time, the Overshoot criteria and the minimum ITAE, and the speed of convergence of the system. Other techniques, such as SCA-PID and AOA-PID controllers, were compared in this study. Based on the results, we conclude that our proposed method, POA-PID-ITAE, outperforms other methods due to the advanced optimization techniques. Improving the trajectory tracking of a mobile wheeled robot can significantly increase its performance. To validate the results of our simulation and demonstrate the effectiveness of the POA-PID-ITAE algorithm based on the proposed POA in real-world trajectory tracking. However, the present validation is restricted to a kinematics model without considering the dynamic effects.

Future work is going to focus on improving the evaluation of dynamic modeling and implementing experimental validation on a real mobile robot platform.

References

[1] A. Dzedzickis, J. Subačiūtė-Žemaitienė, E. Šutinys, U. Samukaitė-Bubnienė, and V. Bučinskas, 'Advanced applications of industrial robotics: New trends and possibilities', *Applied Sciences*, vol. 12, no. 1, p. 135, 2021.

[2] M. Javaid, A. Haleem, R. Pratap Singh, S. Rab, R. Suman, and L. Kumar, 'Utilization of robotics for healthcare: a scoping review', *Journal of Industrial Integration and Management*, vol. 10, no. 01, pp. 43–65, 2025.

[3] L. F. Oliveira, A. P. Moreira, and M. F. Silva, 'Advances in agriculture robotics: A state-of-the-art review and challenges ahead', *Robotics*, vol. 10, no. 2, p. 52, 2021.

[4] J. Zhao *et al.*, 'Autonomous driving system: A comprehensive survey', *Expert Systems with Applications*, vol. 242, p. 122836, 2024.

[5] H. Taheri and C. X. Zhao, 'Omnidirectional mobile robots, mechanisms and navigation approaches', *Mechanism and Machine Theory*, vol. 153, p. 103958, 2020.

[6] E. V. Larkin, M. A. Antonov, and A. N. Privalov, 'The tricycle mobile robot movement simulation', presented at the MATEC Web of Conferences, EDP Sciences, 2018, p. 06001.

[7] K. Shojaei, A. M. Shahri, A. Tarakameh, and B. Tabibian, 'Adaptive trajectory tracking control of a differential drive wheeled mobile robot', *Robotica*, vol. 29, no. 3, pp. 391–402, 2011.

[8] J. Riera, S. Cachiguango, M. Pedraza, G. M. Andaluz, and P. Leica, 'Sliding Mode Control for Trajectory Tracking of a TurtleBot3 Mobile Robot in Obstacle Environments', *Engineering Proceedings*, vol. 77, no. 1, p. 7, 2024.

[9] Z. Wang, 'Application of the PID Algorithm in Robot', *ITM Web Conf.*, vol. 73, p. 01025, 2025. https://doi.org/10.1051/itmconf/20257301025.

[10] A. Waga, S. Benhlima, A. Bekri, J. Abdouni, and F. Z. Saber, 'A survey on autonomous navigation for mobile robots: From traditional techniques to deep learning and large language models', *Journal of King Saud University Computer and Information Sciences*, vol. 37, no. 7, p. 198, 2025.

[11] F. Isdaryani, F. Feriyonika, and R. Ferdiansyah, 'Comparison of Ziegler-Nichols and Cohen Coon tuning method for magnetic levitation control system', *J. Phys.: Conf. Ser.*, vol. 1450, no. 1, p. 012033, Feb. 2020. https://doi.org/10.1088/1742-6596/1450/1/012033.

[12] Z. Yang *et al.*, 'Research on Trajectory Tracking Control Method for Crawler Robot Based on Improved PSO Sliding Mode Disturbance Rejection Control', *Sensors*, vol. 25, no. 7, p. 2113, 2025.

[13] M. Kmich *et al.*, 'Chaotic Puma Optimizer Algorithm for controlling wheeled mobile robots', *Engineering Science and Technology, an International Journal*, vol. 63, p. 101982, Mar. 2025. https://doi.org/10.1016/j.jestch.2025.101982.

[14] H. Tahiri, I. Mchichou, M. Ouabdou, and M. Sayyouri, 'Advanced Control Strategies for Mobile Robots Using Artificial Intelligence', presented at the 2025 7th Global Power, Energy and Communication Conference (GPECOM), IEEE, 2025, pp. 1130–1135.

[15] B. Abdollahzadeh *et al.*, 'Puma optimizer (PO): a novel metaheuristic optimization algorithm and its application in machine learning', *Cluster Computing*, vol. 27, no. 4, pp. 5235–5283, 2024.

A Hybrid Gradient Boosting - XGBoost MPPT Strategy for Photovoltaic Systems Under Partial Shading

Rida AMINE[1,a *], Noureddine EL BARBRI[1,b] and Hatim AMEZIANE[1,c]

[1]Science and Technology for the Engineer Laboratory (LaSTI), National School of Applied Sciences Khouribga (ENSA), Sultan Moulay Slimane University,B.P. Morocco

[a]rida.amine@usms.ac.ma, [b]n.elbarbri@usms.ma, [c]h.ameziane@usms.ma

Keywords: Photovoltaic Systems, Maximum Power Point Tracking, Partial Shading, Gradient Boosting, Xgboost, Machine Learning

Abstract. Maximum Power Point Tracking is a key feature in the power conversion of Photovoltaic PV-based power systems, which enables the system to deliver maximum energy based on fluctuating environmental conditions. Conventional MPPT approaches, such as the Perturb and Observe method (P&O) and Incremental Conductance algorithm (INC), have been observed to degrade as a result of multiple local maxima in the power-voltage (P-V) curve during partial shading conditions (PSC). A novel hybrid MPPT technique based on Gradient Boosting Regression and Extreme Gradient Boosting is presented in this paper. The key idea is to exploit the strong robustness of GBR in mid-high (sunny) irradiance and the generalization capability of XGBoost at low (cloudy) irradiance, by leveraging a hysteresis-based switching scheme that takes into account the current-irradiance level and switches between models. Irradiance level from 200 to 1000 W/m² data simulation results demonstrate that the proposed hybrid controller maintains the load power very near to the theoretical maximum, obtaining instantaneous MPPT efficiencies from 95.73% up to 97.37% and global efficiency of 96.83% across the entire profile, which are largely improved over single-model GBR (83.8%) and XGBoost (78.7%) baselines referenced works.

INTRODUCTION

The growing number of photovoltaic (PV) systems integrated into the modern power grid, owing largely to environmental concerns and declining module costs, necessitates better control strategies for the purpose of energy extraction [1]. Because the current–voltage (I-V) characteristic of a PV is non-linear and also depends on environmental factors, an MPPT controller has to be specifically employed to control the PV operating point continuously.

There are many classical MPPT methods, which use Perturb and Observe (P&O) and Incremental Conductance (INC) algorithms due to their simplicity and low implementation cost [2], [3]. However, they do not work well in the rapid changing irradiance conditions and in the partially shaded conditions (PSC), especially. The partial shading array fail is generated as a result of the shadow on part of the system by cloud, vegetation or building, which results in MPP with several local maxima and a single global maximum, but this does not occur when no shading effects are considered. Due to the nature of gradient-based and hill-climbing methods, they are inherently prone to converge towards local maxima rather than the global maximum, leading to a serious energy penalty.

In order to overcome these drawbacks, several advanced MPPT methods have been developed. These include metaheuristic optimization (e.g., Particle Swarm Optimization and Genetic Algorithms) [4], fuzzy logic / neural network-based controllers, as well as more recently, data-driven methodologies based on machine learning (ML) regression/classification models. At the same time, ensemble learning with a focus on gradient boosting algorithms has proven to be capable of modelling the challenging non-linear relationships between environmental and electrical variables and the optimum operating point [5], [6].

However, even with these developments, a single ML model may not be equally good across the whole operating point. It is also empirically observed that certain models perform better at certain ranges of irradiance or temperature due to bias-variance trade-offs and the regularization mechanism. Inspired by this, hybrid MPPT methods that combine multiple ML models and make them play with their respective advantages are potentially promising.

This paper proposes a hybrid Gradient Boosting Regression–XGBoost MPPT algorithm in which:

- XGBoost is employed at low irradiance, where the signal-to-noise ratio is lower and regularized tree ensembles provide improved generalization;
- GBR is used at medium and high irradiance, where smoother response and high accuracy are obtained with reduced variance.

An irradiance-based switching logic with hysteresis ensures stable transitions between the two models.

Main contributions of this work are:

- Hybrid ML-based MPPT strategy using GBR and XGBoost with irradiance-aware switching technique.
- Construction of an integrated dataset based on simulation and experimental PV data under normal and shaded environments with a thoughtful feature selection.

PV Module Under Study

All simulations in this work are carried out for a commercial crystalline-silicon module Bosch Solar Energy c-Si M 60-225-16, which represents a typical 60-cell module widely used in small and medium PV systems. The module is rated at approximately 225 W maximum power under Standard Test Conditions (G=1000 W/m^2, T=25∘C). The main electrical and model parameters are summarized in Table I.

The module exhibits an open-circuit voltage Voc≈36v and a short-circuit current Isc≈8.7 A. The maximum power point occurs at Vmp≈28.3 V and Imp≈7.95 A. Temperature coefficients equal to -0.374 %/°C for Voc and 0.071 %/°C for Isc are used to account for the impact of cell temperature on the I-V curve. For the single-diode model, the light-generated current IL, diode saturation current I0, diode ideality factor n, series resistance Rs and shunt resistance Rsh are extracted using the parameter estimation tools described in [7].

Irradiance Dependence (Figure. 1)

Figure 1 shows the effect of irradiance on the module parameters. Showing the I-V and P-V curves at 25 °C when irradiance levels are 1000, 800, 600, 400 and 200 W/m^2. The short circuit current and the maximum output power increase almost linearly with irradiance, as expected, while the open circuit voltage experiences little variation. These figures illustrate the large maximum power point dependences on irradiation and confirm the necessity for an effective MPPT method which can respond quickly to sudden changes in irradiance.

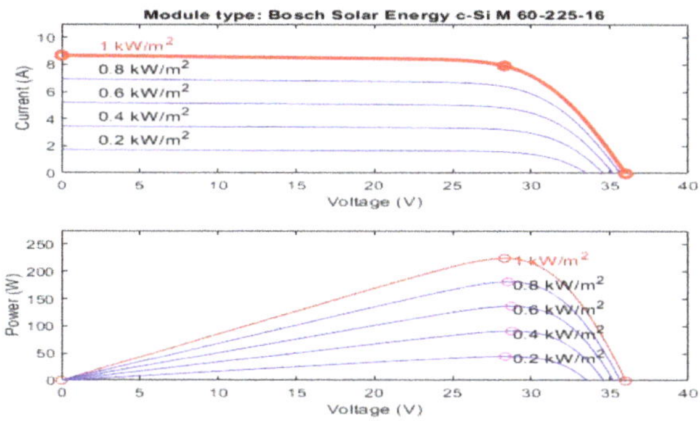

Figure 1. I–V and P–V characteristics of the Bosch c-Si M 60-225-16 module at 25 °C for different irradiance levels (1000–200 W/m²).

Temperature Dependence (Figure. 2)

The influence of temperature is presented in Figure 2, which shows I-V and P-V curves at G=1000 W/m² for two temperature values (25 °C and 45 °C). As the temperature rises, Voc, Vmp and in turn the maximum accessible power decrease significantly. This effect should be taken into account for the MPPT controller to prevent systematic efficiency losses of the module at higher temperatures.

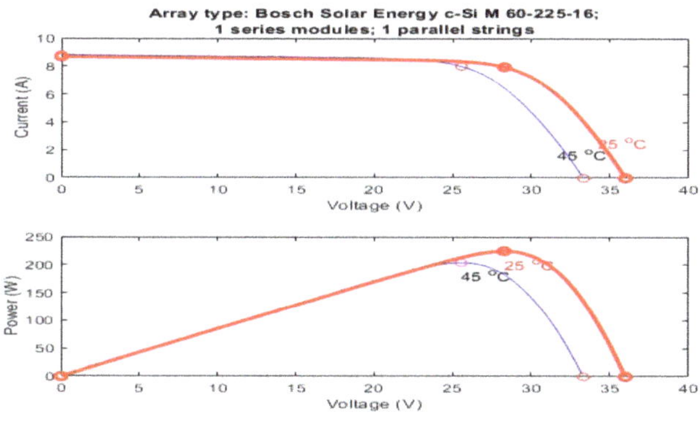

Figure 2. I–V and P–V characteristics of the Bosch c-Si M 60-225-16 module at 1000 W/m² for two temperatures (25 °C and 45 °C).

PV SYSTEM MODELING AND DATASET CONSTRUCTION

In order to assess the hybrid GBR-XGboost MPPT technique, we create a comprehensive model of the PV generator, the power converter and the operating conditions. This part briefly introduces the PV array, as well as scenarios simulated, dataset generation using NASA meteorological data, and equations for the PV equivalent circuit.

Operating Scenarios and Partial Shading

Both under uniform operating conditions and partial shading conditions (PSC), the MPPT method is examined in this paper. PSC occurs while various parts of the array are subject to different irradiance levels induced by clouds, buildings or trees, and the generated I-V and P-V

Materials Research Proceedings 64 (2026) 535-543 https://doi.org/10.21741/9781644904091-67

characteristics have several local maxima with one global maximum [7]. In this case, traditional hill-climbing methods such as Perturb and Observe (P&O) or Incremental Conductance can be stuck at a local peak and are unable to track the global MPP, resulting in considerable power waste [7].

The array is divided into several strings or blocks in order to replicate these effects, and each is exposed to a predetermined level or profile of irradiance. Common patterns consist of uniform and bi-level shading, with the more sophisticated tri-level or random shading distributions. These case studies are realised in MATLAB/Simulink with the single-diode model of Subsection II-A and utilised as benchmarks for evaluating the robustness of the hybrid GBR–XGBoost controller. The training set of the machine learning models is, however, made with realistic irradiance and temperature profiles taken from NASA data, and it is described in the next section.

Dataset Construction Using NASA Irradiance and Temperature

An important aspect of this work is that the training and test datasets for the ML-based MPPT controller are generated from satellite-measured meteorological data instead of short-term local measurements. Long-term series of solar resource and temperature are obtained from the NASA POWER (Prediction of Worldwide Energy Resources) database, which provides gridded data products for renewable energy applications worldwide.

For the geographic coordinates corresponding to the intended PV installation, the following variables are downloaded over a multi-year period:

- **ALLSKY_SFC_SW_DWN**: all-sky surface shortwave downward irradiance, used as plane-of-array irradiance G (W/m^2);
- **T2M**: air temperature at 2 m height, used as ambient temperature T (°C).

After a basic cleaning procedure removing missing and non-physical values, every record i is described by a pair (Gi, Ti). These climatic factors are next transferred into an approximate electrical model of the Bosch module, through analytical relationships based on STC data and temperature coefficients, following good engineering practice for c-Si PV modelling.

Let Isc,s, Imp,s, Voc,s and Vmp,s denote the short-circuit current, MPP current, open-circuit voltage and MPP voltage at STC (Gs = 1000 W/m^2, Ts = 25 °C). Let αIsc and βVoc be the temperature coefficients of Isc and Voc in %/°C provided by the datasheet. They are first converted to a relative current slope and an absolute voltage slope:

- **αrel = αIsc / 100 (per °C),**
- **dVoc/dT = (βVoc / 100) · Voc,s (V/°C).**

The current at the maximum power point Imp(i) for a given (Gi, Ti) is then approximated by

$$\mathbf{Imp(i) = Imp,s \cdot (Gi / Gs) \cdot [\ 1 + \alpha rel \cdot (Ti - Ts)\ ],}$$

expressing the near-linear dependence of current on irradiance and its mild increase with temperature. The voltage at the maximum power point is modelled as a linear function of temperature,

$$\mathbf{Vmp(i) \approx Vmp,s + dVmp/dT \cdot (Ti - Ts),}$$

where dVmp/dT is taken close to dVoc/dT, which is a widely used approximation for crystalline-silicon modules. The corresponding maximum power is then

$$\mathbf{Pmp(i) = Vmp(i) \cdot Imp(i).}$$

Materials Research Proceedings 64 (2026) 535-543 https://doi.org/10.21741/9781644904091-67

In this work, the machine-learning input vector and target output at sample i are defined as

$$xi = [\ Gi, Ti\ ]^T,$$
$$yi = Vmp(i).$$

The ML models are thus trained to learn the mapping.

$$f : [\ G, T\ ]^T \rightarrow Vmp,$$

This gives the reference MPP voltage as a simple function of irradiance and temperature. Features are pre-processed before the training stage, removing outliers based on physical bounds for G, T, Vmp and Pmp, and scaling the inputs G and T to zero mean and unit variance in order to enhance numerical conditioning as well as gradient boosting algorithm convergence. The final data-set, which usually includes a few thousand total samples sampling all available seasonal and diurnal variability of the model results, is randomly divided into 70% for training, 15% for validation and 15% testing according to machine learning standards [8]. This analytics-driven NASA-based dataset offers a realistic and physically consistent setting to train and test the proposed hybrid GBR-XGBoost MPPT controller.

III. PROPOSED HYBRID GBR–XGBOOST MPPT ALGORITHM

The purpose of the indicated MPPT approach described in the next section is to harvest complementary capabilities of Gradient Boosting Regression (GBR) and XGBoost to control optimal MPP voltage Vmp out of measured irradiance and temperature. Both models are trained off-line on the NASA-based dataset presented in Section II and combined on-line through a simple irradiance-aware switching block with hysteresis. The hybrid controller produces a reference voltage Vmp,ref , which is followed by the inner DC-DC converter voltage loop.

A. Architecture Overview

The proposed MPPT controller has the following structure:

- **Inputs:**
  - Irradiance G (W/m^2) obtained from a sensor or estimator.
  - Temperature T (°C) corresponding to ambient or estimated cell temperature.
- **Core machine-learning models:**
  - A Gradient Boosting Regression (GBR) model, denoted fGBR(G, T).
  - An Extreme Gradient Boosting style (XGBoost) model, denoted fXGB(G, T), both implemented as an ensemble of regression trees [9].
- **Decision block:** An irradiance-based switching logic with hysteresis, which decides whether GBR or XGBoost should be used for the current operating point.
- **Output:** A reference MPP voltage Vmp, ref that is sent to the PI voltage controller of the boost converter, which adjusts the duty cycle so that the PV terminal voltage tracks this reference.

For the experiment on the NASA dataset, from preliminary studies, we observe that while the XGBoost model shows a smaller prediction error in low irradiance areas (where signal to noise ratio is low and regularization helps), it produces a better and coarse fit compared to GBR, which also achieves a slighted smoother fit in medium to high levels of irradiance. This inspires a hybrid approach in which XGBoost is employed at low irradiance and GBR at high irradiance, with the transition determined by an irradiance threshold and a hysteresis bandwidth.

B. Online Hybrid MPPT Algorithm

During real-time operation, the hybrid GBR-XGBoost MPPT runs at each sampling instant to compute the reference MPP voltage Vmp,ref . The decision is based solely on the current irradiance level and on the previous mode (GBR or XGBoost), which enables a simple and low-cost implementation.

The proposed online procedure is summarized in Algorithm 1

Algorithm 1: Hybrid GBR–XGBoost MPPT Algorithm

Input: Measured irradiance $G(k)$ [W/m^2]
Measured (or estimated) temperature $T(k)$ [°C]
Previous mode $M(k-1) \in \{\text{GBR}, \text{XGB}\}$
Trained models $f_{\text{GBR}}, f_{\text{XGB}}$
Normalization parameters $\mu_G, \sigma_G, \mu_T, \sigma_T$
Irradiance threshold G_{th} and hysteresis band ΔG
Output: Reference MPP voltage $V_{mp,\text{ref}}(k)$ [V]
Updated mode $M(k) \in \{\text{GBR}, \text{XGB}\}$

// Step 1: Acquire measurements
Read current irradiance and temperature: $G \leftarrow G(k)$, $T \leftarrow T(k)$
// Step 2: Normalize input features
$G_n \leftarrow \dfrac{G-\mu_G}{\sigma_G}$
$T_n \leftarrow \dfrac{T-\mu_T}{\sigma_T}$
// Step 3: Select model using irradiance and hysteresis
if $G < G_{\text{th}} - \Delta G$ **then**
    $M(k) \leftarrow$ XGB // Low irradiance ⇒ use XGBoost
else
    if $G > G_{\text{th}} + \Delta G$ **then**
        $M(k) \leftarrow$ GBR // High irradiance ⇒ use GBR
    else
        $M(k) \leftarrow M(k-1)$ // Keep previous mode (avoid chattering)
// Step 4: Predict reference MPP voltage
if $M(k) = GBR$ **then**
    $V_{mp,\text{ref}}(k) \leftarrow f_{\text{GBR}}(G_n, T_n)$
else
    $V_{mp,\text{ref}}(k) \leftarrow f_{\text{XGB}}(G_n, T_n)$
// Step 5: Send reference to DC-DC converter controller
Use $V_{mp,\text{ref}}(k)$ as the reference voltage for the inner PI voltage control loop of the boost converter so that $V_{pv}(k) \rightarrow V_{mp,\text{ref}}(k)$
// Step 6: Next sampling instant
Set $k \leftarrow k+1$ and repeat from Step 1

IV. RESULTS AND DISCUSSION

This section provides a study of the performance under the uniform irradiance case and compares it with the previously reported GBR-only and XGBoost-only methods [9]. Simulations are performed in MATLAB/Simulink with the Bosch c-Si M 60-225-16 module model and NASA-based irradiance and temperature profiles as shown in Section II. Five irradiance levels (1000, 800, 600, 400 and 200 W/m²) are used, and the corresponding PV power and load power P load for each one are recorded. The results are analyzed in terms of power tracking accuracy to the theoretical maximum power, dynamic response and the overall MPPT efficiency. In the discussion, it also presents the findings in connection with the most recent data-driven MPPT development, referring to both ensemble learning and adaptive regression [9].

A. Performance of the Proposed Hybrid GBR–XGBoost MPPT

Fig. 3 shows the dynamic response of the proposed hybrid module to the same irradiance step sequence. Where the PV power output and load power demanded are represented as a time function. The five irradiations of 1000, 800, 600, 400 and 200 W/m² can be seen very clearly, and both track lines (PV power and Load power) almost overlap at the steady state, concluding a good precision MPPT as well as high conversion efficiency.

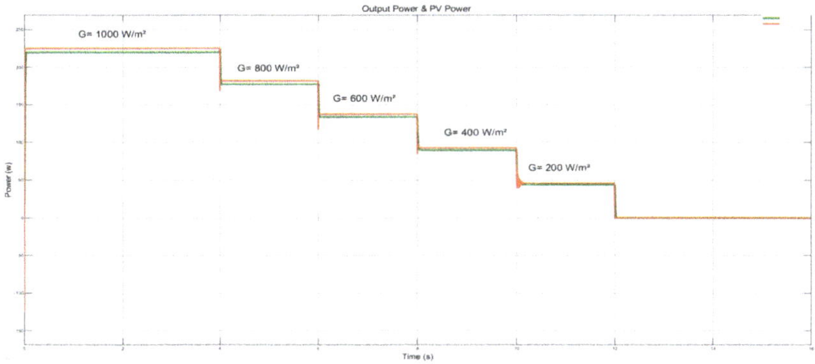

Figure 3. PV power and load power of the proposed hybrid GBR-XGBoost MPPT for a sequence of irradiance steps (1000–200 W/m²).

In Table III, the steady-state powers obtained by the hybrid method are shown for each irradiance level. We present the same theoretical powers of the fits as in Table II for direct comparison.

Table III - Steady-state PV and load power of the proposed hybrid MPPT method

Irradiance (W/m²)	Theoretical PV Power (W)	Hybrid PV Power (W)	Hybrid Load Power (W)
200	45.44	45	43.5
400	92.02	92	89.6
600	137.71	137	133.5
800	182.09	181.4	175
1000	224.98	223	219

From Table III, we can see that the hybrid controller can regulate the PV power to the theoretical MPP very closely for all irradiance levels with deviations lower than 2 W at 200 W/m² and 2-3 W at 1000 W/m². The difference between the PV power and load power is also low (usually 2-6 W), accounting mainly for converter losses. More interestingly, the hybrid scheme greatly enhances the utilization of the PV model compared with that of Table II's single-model GBR and XGBoost approaches.

B. MPPT Efficiency Evaluation

Using the hybrid load powers of Table III and the theoretical powers of Table II, the efficiencies for each irradiance level and the global efficiency are listed in Table IV.

Table IV - MPPT efficiencies of the proposed hybrid method (based on load power)

Irradiance (W/m²)	Theoretical PV Power (W)	Hybrid Load Power (W)	MPPT Efficiency η_i (%)
200	45.44	43.5	95.73
400	92.02	89.6	97.37
600	137.71	133.5	96.94
800	182.09	175	96.11
1000	224.98	219	97.34
Global	**Σ = 682.24**	**Σ = 660.60**	**96.83**

As a result, the achieved global MPPT efficiency provided by the proposed hybrid controller is approximately 96.8 %, which is much higher than 83.8 % and 78.7 % of GBR-only and XGBoost-only approaches in [9]. The advantage is especially pronounced at low irradiance (200-400 W/m²), where the hybrid method preserves efficiencies above 95%, while the other techniques experienced significant tracking errors. These enhancements demonstrate the appropriateness of mixing multiple learning models in an irradiance-aware way, as recently observed for stacked ensemble MPPT teens [10], and the latest adaptive regression-based controllers like ALR-CWM [11].

In summary, the results suggested that the proposed hybrid GBR-XGBoost MPPT controller offers precise tracking under a broad range of irradiance conditions and possesses fast dynamic response with well-damping character, as well as high energy capture. This underlines that it is a potential data-driven alternative to traditional algorithms and single-model machine learning systems in realistic PV systems.

References

[1] Esram, T., & Chapman, P. L. (2007). Comparison of photovoltaic array maximum power point tracking techniques. IEEE Transactions on energy conversion, 22(2), 439-449. https://doi.org/10.1109/TEC.2006.874230

[2] Femia, Nicola, et al. "Optimization of perturb and observe maximum power point tracking method." IEEE transactions on power electronics 20.4 (2005): 963-973. https://doi.org/10.1109/TPEL.2005.850975

[3] Hohm, D. P., and M. E_ Ropp. "Comparative study of maximum power point tracking algorithms." Progress in photovoltaics: Research and Applications 11.1 (2003): 47-62. https://doi.org/10.1002/pip.459

[4] Ishaque, Kashif, and Zainal Salam. "A review of maximum power point tracking techniques of PV system for uniform insolation and partial shading condition." Renewable and Sustainable Energy Reviews 19 (2013): 475-488. https://doi.org/10.1016/j.rser.2012.11.032

[5] Friedman, Jerome H. "Greedy function approximation: a gradient boosting machine." Annals of statistics (2001): 1189-1232. https://doi.org/10.1214/aos/1013203451

[6] Chen, Tianqi. "XGBoost: A Scalable Tree Boosting System." Cornell University (2016). https://doi.org/10.1145/2939672.2939785

[7] Ahmed, Jubaer, and Zainal Salam. "A critical evaluation on maximum power point tracking methods for partial shading in PV systems." Renewable and Sustainable Energy Reviews 47 (2015): 933-953. https://doi.org/10.1016/j.rser.2015.03.080

[8] Goodfellow, Ian, et al. Deep learning. Vol. 1. No. 2. Cambridge: MIT press, 2016.

[9] R. Amine, N. El Barbri and H. Ameziane, "Optimizing Photovoltaic Performance: A Comparative Analysis of XGBoost and Gradient Boosting for MPPT," 2025 International Conference on Electrical Systems & Automation (ICESA), Troyes, France, 2025, pp. 1-6, doi: 10.1109/ICESA66763.2025.11280867. https://doi.org/10.1109/ICESA66763.2025.11280867

[10] R. Amine, N. El Barbri and H. Ameziane, "Stacked Ensemble Machine Learning for MPPT Under Partial Shading: A High-Accuracy Data-Driven Approach," 2025 International Conference on Electrical Systems & Automation (ICESA), Troyes, France, 2025, pp. 1-6, doi: 10.1109/ICESA66763.2025.11280669. https://doi.org/10.1109/ICESA66763.2025.11280669

[11] Amine, R., & Ameziane, H. (2025, June). An Enhanced MPPT Algorithm Based on Adaptive Linear Regression with Contextual Weight Memory (ALR-CWM) for PV Systems. In 2025 17th International Conference on Electronics, Computers and Artificial Intelligence (ECAI) (pp. 1-4). IEEE. https://doi.org/10.1109/ECAI65401.2025.11095559

Power Factor Correction Analysis for Moroccan Industrial Facilities: Technical and Economic Evaluation

Chaimaa Labyad[1]

[1]School of Science and Engineering, Al Akhawayn University P.O Box 104, Hassan II Avenue, 53000Ifrane, Morocco

C.Labyad@aui.ma

Keywords: Power Factor Correction, Reactive Power Compensation, Industrial Energy Efficiency, ONEE Tariff, Morocco Electricity Sector, Economic Analysis, Simulink/ MATLAB Simulation

Abstract. Despite Morocco's ambitious National Energy Efficiency Strategy and growing industrial electrification, poor power factor remains a major contributor to system inefficiency and high electricity bills and in Moroccan industrial facilities, where small induction motors (0.5–2 hp) incur up to 20% reactive power surcharges under ONEE tariff structures, while simultaneously increasing transmission losses and placing unnecessary strain on the national grid infrastructure[1,6]. This study develops and validates an optimized shunt capacitive power factor correction system for a representative single-phase R-L load (R = 10 Ω, L = 50 mH, 1.526 kW) connected to Morocco's standard 230 V, 50 Hz low-voltage grid. Comprehensive phasor-domain circuit analysis in Simulink MATLAB, supported by resonance analysis and detailed phasor diagrams, identifies 140 μF as the global optimum. This single capacitor raises power factor from 0.537 to 0.999 (86 %improvement), reduces line current by 46.2% (from 12.35A to 6.64 A), and cuts reactive power by 97.1 % (from 2396 VAR to 70 VAR). Economic analysis using 2024-2025 ONEE tariffs shows 33-month payback with 1,933 MAD/year savings and 10-year NPV of 9,467 MAD. Therefore, this study demonstrates a technically feasible and economically attractive energy efficient intervention for Moroccan industrial facilities, that aligns fully with national energy policy objectives [1,9].

1. Introduction

The power factor, as the ratio of real power performing useful work to the apparent power drawn from the grid, serves as a main measure of efficiency in AC systems. The predominance of inductive loads such as induction motors, transformers, welders, and fluorescent lighting in industrial applications[5] results in current lagging voltage, shifting lowers power factor and raising reactive power demand, thereby increasing the current carrying requirement on the electrical system for the same real power leading to excessive I^2R losses, wasted heating of conductors, voltage drops along distribution feeders, and reduced capacity of transformers, switchgear, and cables. Thus, Poor power factor accelerates aging of facilities and restricts future industrial growth, as well as increasing operating costs of the industrial processes. This challenge is particularly relevant to the Moroccan context, where industry accounts for over 35% of national electricity consumption [6], and small and medium induction motors dominate inductive load. Empirical observations suggest that many industrial motors operate at power factors well below 0.60, leading to facilities to incur significant technical inefficiencies and related financial penalties [8]. Depending on specific applications under ONEE's medium-voltage tariff structure, industries could pay up to 20% surcharges for poor power factor, along with avoidable network losses and capacity constraints. These issues are increasingly relevant as Morocco, under its National Energy Strategy [1], targets major gains in energy efficiency and industrial competitiveness by 2030. Despite these developments, power factor correction adoption remains limited. Facility managers lack guidance on optimal capacitor sizing, face uncertainty about resonance risks, and have limited economic data reflecting current ONEE and ANRE tariffs. This study addresses these gaps by developing a complete technical and economic structure for power factor correction adapted to Moroccan industrial conditions. We model of a representative 1.526 kW motor load operating on Morocco's 230 V, 50 Hz grid, determining optimal parallel capacitance using phasor-domain analysis and resonance validation, and quantifying economic outcomes based on current ONEE/ANRE tariffs, providing clear performance metrics: annual savings, payback period, ROI, and NPV. This paper presents the literature review (Section 3) methodology (Section 4), results (Section 5), and conclusions (Section 6).

2. Literature review

Power factor correction through capacitor banks represents the most economically viable method for industrial applications, as established in IEEE Standard 141-1993 [5], which provides comprehensive design criteria for industrial power distribution. Capacitor banks generate leading reactive power that cancels lagging reactive power from inductive loads, reducing line current, and improving system efficiency. These systems can be configured as fixed banks for constant loads or automatic switched banks for variable profiles. While synchronous condensers and active correction systems offer enhanced performance for high-power applications, capacitor banks remain in the industry standard for small-to-medium facilities due to lower capital and maintenance costs [13]. Economic viability has been extensively documented, with Saied (1989) establishing typical payback periods of 18-36 months depending on utility tariff structures [6]. Recent industrial implementations confirm these findings: 12% monthly cost reductions in cement manufacturing, while Rajendran (2021) found passive systems offer 18–24 month payback periods for SMEs [10]. Electrotek Concepts documented payback periods of 10-27 months for facilities with installed capacitor costs of \$35-45/kVAR . However, existing economic models predominantly address developed nations with stable tariff structures. Limited research examines developing economies where utility penalty structures may be poorly documented, and equipment costs vary significantly. Morocco's National Energy Strategy targets 52% renewable energy by 2030 and 20% energy efficiency improvement, with specific industrial targets of 22% [1,7]. Industry accounts for 35% of national electricity consumption, with manufacturing, mining, and textiles dominating [6]. The establishment of ANRE in 2024 introduced transparent transmission (TURT at 0.0639 MAD/kWh) and system services (TSS at 0.0635 MAD/kWh) tariffs, enabling precise economic assessment of efficiency investments [2]. ONEE, the state utility, operates as a single buyer and maintains grid infrastructure [14]. Despite these policy drivers, UNIDO's 2021 program identified critical barriers: limited facility manager awareness, uncertainty regarding optimal capacitor sizing, harmonic resonance concerns, and insufficient access to current economic data reflecting actual ONEE/ANRE tariffs [11]. Regional MENA utilities commonly impose 15-30% penalties for poor power factor, creating strong economic incentives. Harmonic distortion from nonlinear loads (VFDs, rectifiers) represents a primary concern, as capacitor-inductance interaction can create resonance conditions amplifying harmonic components [12]. Facilities with nonlinear loads exceeding 15-20% require harmonic analysis prior to installation. Detuned harmonic filter banks provide effective mitigation. Over- compensation, where excessive capacitance causes leading power factor and voltage rise, necessitates proper sizing methodology accounting for parallel R-L-C circuit characteristics. Automatic controllers provide protection while optimizing correction across varying load profiles. Despite extensive global research, no comprehensive studies validate power factor correction using Morocco's specific grid parameters (230V, 50Hz), representative industrial loads, and current 2024-2025 ONEE/ANRE tariff structures. This study addresses these gaps by providing validated MATLAB modeling, comprehensive phasor and resonance analysis determining optimal capacitance, and detailed economic evaluation using verified current tariffs.

3. Methodology

3.1 Circuit Model and Component Parameters:

This study analyses power factor correction for a typical industrial inductive load connected to Morocco's electrical grid. The load is modeled as a series R-L combination (R = 10Ω, L = 50mH) powered by Morocco's standard grid parameters: 230V RMS voltage and 50 Hz single-phase AC source. These values represent typical small industrial motors (0.5-2 hp) used in Moroccan facilities for pumps, fans, and conveyors—the primary source of reactive power consumption. At 50 Hz, the inductive reactance is:

$X_L = 2\pi fL = 15.71\Omega$, resulting in an $|Z_L| = 18.62\ \Omega$ yielding low power factor of $\cos\varphi \approx 0.537$ (lagging).

To improve the power factor of the load to near unity, we implemented a power factor correction that employs by using parallel capacitive compensation, where capacitor C connected in parallel directly across the load terminals. This configuration was chosen because it allows the compensating capacitor bank to be connected or disconnected without interrupting the operation of the load, fully complies with standard power factor correction practices in industrial installations, and crucially —applies correction at the point of consumption. As a result, the reactive component of the current is canceled locally, significantly reducing the total current flowing through the upstream supply lines, cables, transformers, and switchgear, thus decreasing losses and improving the overall efficiency of the electrical distribution system.

3.2 Analytical Simulation and Technical Analysis:

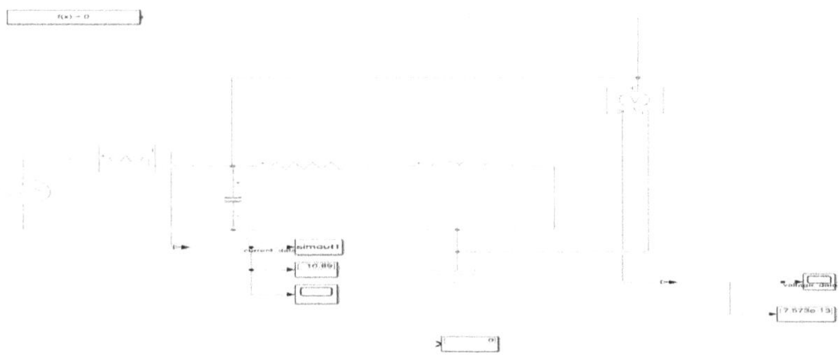

Figure 1. MATLAB/Simulink Circuit Model for Power Factor Correction Analysis: Single-Phase R-L Load with Parallel Shunt Capacitor (230V, 50Hz, R = 10Ω, L = 50mH)

Initial circuit modeling employed Simscape/Simulink visual circuit construction (Figure1), in addition to MATLAB analytical modeling based on phasor-domain circuit theory that directly applies fundamental circuit laws (Kirchhoff's current law, Ohm's law with complex impedance) without relying on numerical solver convergence. This analytical method computes for each capacitor value C in the test range [0, 50, 100, 125, 140, 150, 160, 175, 200, 225, 250, 300, 350, 400, 500] μF, the following phasor calculations were performed:

1. Inductive reactance (constant): $X_L = \omega L = 15.71\ \Omega$
2. Capacitive reactance: $X_C = 1/(\omega C)$, where: $\omega = 2\pi f = 314.16$ rad/s

3. R-L branch impedance (complex): $Z_{RL} = R + jX_L = 10 + j15.71\ \Omega$
4. R-L branch current (phasor): $I_{RL} = 230\angle0° / (10 + j15.71)$
5. Capacitor current (phasor): $I_C = 230\angle0° / (-jX_C)$
6. Total supply current (phasor sum): $I_{total} = I_{RL} + I_C$
7. Power factor: $PF = \cos(angle(I_{total}))$
8. Rcal, rcactivc, and apparcnt powcrs:

$$P = 230 \times |Itotal| \times PF$$

$$Q = 230 \times |I_{total}| \times \sin(angle(I_{total}))\ S = 230 \times |I_{total}|$$

Dense sampling in the 100-200 μF region (125, 140, 150, 160, 175 μF) accurately resolve the optimal correction point.

3.3 Multi-Method Validation of the Analytical Model:

To ensure accuracy and reliability of simulation results and gain deeper physical insight into the power factor correction mechanism, we adopted two independent validation methods: resonance analysis, phasor diagram construction that confirms numerical accuracy and highlights the differences between ideal resonant compensation and practical power factor correction at fixed grid frequency.

3.3.1 Resonance Analysis:

Resonance theory provides analytical estimation of capacitance needed for reactive power compensation. For parallel R-L-C circuits, simplified analysis (neglecting resistance) predicts unity power factor when $X_L = X_C$, yielding theoretical optimal capacitor: $C_{theory} = 1 / (\omega^2 L) = 202.6$ μF. However, with R = 10 Ω is comparable to $X_L = 15.71$ Ω, then the actual current minimum and therefore the true optimal capacitance for unity power factor at 50 Hz deviates slightly from 203 μF. Resonant frequency for the L-C combination is calculated as:

$f_r = 1 / (2\pi\sqrt{LC})$ and varies from approximately 225 Hz (at C = 10 μF) to 50.9 Hz (at C = 202.6 μF). This behavior was further confirmed by a full frequency-response analysis of of the total input impedance magnitude |Zin(f)| and phase angle ∠Zin(f) over the range 10–500 Hz independently verifies the above findings and demonstrates that the optimal power factor correction at 50 Hz is achieved well before the circuit reaches its natural resonance. Consequently, the proposed compensation scheme operates safely away from resonant conditions, eliminating any risk of dangerous overvoltage or overcurrent.

3.3.2 Phasor Diagram Analysis:

Phasor diagrams validate the correction mechanism through voltage-current phase relationships, impedance triangles, and power triangles. Current phasor diagrams show capacitor current (IC, leading 90°) canceling motor current lagging component, reducing total line current and aligning it with voltage.

3.4 Economic Analysis:

3.4.1 Electricity Tariff Structure:

Our Economic analysis included Moroccan specific electricity tariff components for medium-voltage industrial customers [2,3]. The total electricity cost has three elements established by regulatory authorities [2]: base energy (0.95 MAD/kWh), transmission network usage (TURT, 0.0639 MAD/kWh), and system services (TSS, 0.0635 MAD/kWh), totaling 1.0774 MAD/kWh [2,3]. Power factor penalties follow industry-standard progressive structure with 20% surcharge for PF < 0.60.

3.4.2 Power Factor Penalty Structure:

Moroccan utilities apply surcharges to customers operating with poor power factor to recover costs associated with reactive power provision and system capacity requirements. While ONEE's (Office National de l'Électricité et de l'Eau Potable) specific penalty tariff is not publicly disclosed in recent documentation, this analysis employs industry-standard progressive penalty structure based on IEEE Standard 1409-2012 for utility reactive power charges [IEEE, 2012],comparable MENA region utility practices (Egypt, Jordan, UAE)[8] and the economic principles of capacity allocation and loss recovery.The applied penalty structure (Table 2) uses conservative rates to avoid overestimating economic benefits. Baseline power factor of 0.537 incurs 20% surcharge (billing multiplier 1.20×), while corrected power factor of 0.999 eliminates all penalties (multiplier 1.00×). This conservative approach ensures that actual implementation may yield equal or superior economic performance. Facilities should confirm contractual penalty rates with ONEE for precise economic projections.

3.4.3 Investment and Operating Cost Components:

Investment costs include capacitor component (30 MAD/μF mid-range industrial- grade) and 30% installation labor. For 140 μF: component cost 4,200 MAD, installation 1,260 MAD, total 5,460 MAD. Operating schedule assumes two-shift operation (16 h/day, 26 days/month = 4,992 hours/year).

Transmission line resistance estimated at R_{line} = 0.5Ω for typical medium-sized facility. Financial metrics include payback period, ROI, and NPV (5% discount rate).

4. Results & Discussions

4.1 Power Factor Correction Technical Performance:

4.1.1 Optimal Capacitor Value and Performance Metrics:

A parametric sweep was conducted over 15 capacitance values ranging from 0 to 500 μF to determine the optimal shunt capacitor for the 1.526 kW inductive load. The results, shown in Figure 2, clearly identify **C = 140 μF** as the global optimum, achieving a power factor of 0.999, a 46.2 % reduction in line current, and a 97.1 % reduction in reactive power, phase angle near-zero (-2.6°), and reactive power reduced 97.1% (from 2396 VAR to 70 VAR). The practical optimal range of 125-160 μF maintains power factor above 0.95, avoiding utility penalties while providing robust corrections.

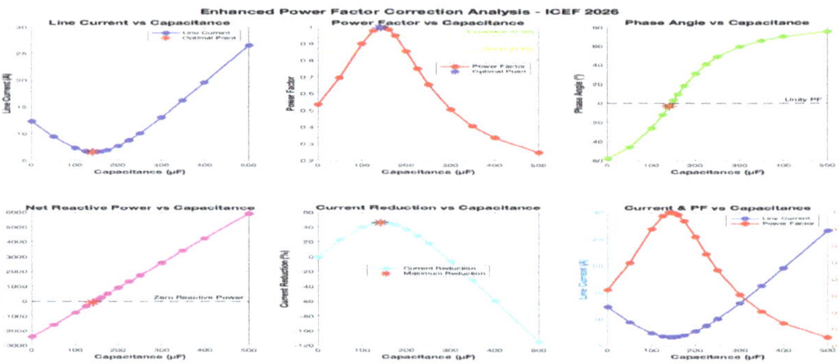

Figure 2. *Effect of shunt capacitance (0–500 μF) on power factor correction performance: (a) line current, (b) power factor, (c) phase angle, (d) reactive power, (e) current reduction %, and (f) combined current & PF. Optimal performance at 140 μF : PF = 0.999, 46.2 % current reduction, 97.1 % reactive power compensation.*

Table1: Comparison of key electrical parameters before and after power factor correction

Parameter	Baseline (No Cap)	Corrected (140 µF)
Line Current	12.35 A	6.64 A
Power Factor	0.537 (POOR)	0.999 (EXC)
Phase Angle	-57.52°	-2.62°
Real Power	1526 W	1526 W
Reactive Power	−2396 VAR	−70 VAR
Apparent Power	2841 VA	1527 VA

Comparison of key electrical parameters before and after power factor correction using the optimal 140 µF shunt capacitor bank.

4.1.2 Resonance Analysis and Theoretical Validation:

Resonance analysis (Figure 3) validates simulation results through frequency-domain evaluation and comparison with theoretical predictions. Figure 3 in Appendix present frequency-domain resonance analysis and theoretical validation of shunt power factor correction (R = 10 Ω parallel load, 230 V, 50 Hz): (a) impedance magnitude, (b) impedance phase angle, (c) zoomed impedance magnitude near 50 Hz, and (d) inductive vs capacitive reactance crossover. Frequency-domain analysis (Figure 3) confirms 140 µF yields optimal impedance at 50 Hz, avoiding resonance while achieving zero phase crossing at operating frequency. The 30.9% deviation from theoretical 202.6 µF (simplified series-resonance formula) reflects parallel R-L-C topology effects with non-negligible resistance (R = 10Ω comparable to X_L = 15.71Ω), demonstrating the importance of full-circuit analysis for practical power factor correction systems.

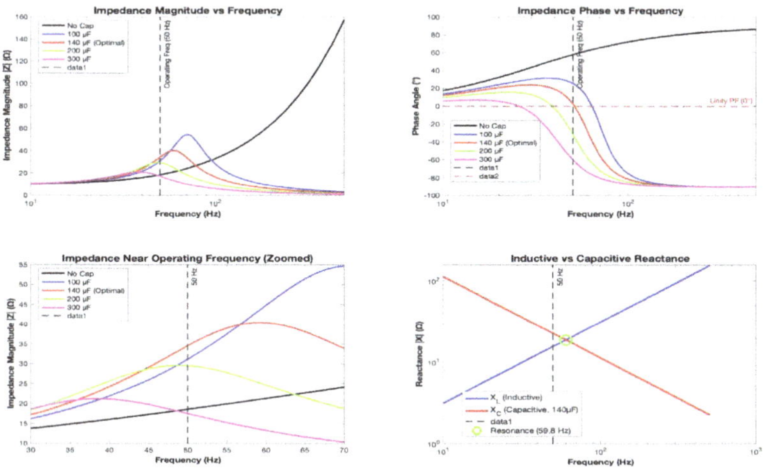

***Figure 3.** Frequency-domain resonance analysis and theoretical validation of shunt power factor correction (R = 10 Ω parallel load, 230 V, 50 Hz): (a) impedance magnitude, (b) impedance phase angle, (c) zoomed impedance magnitude near 50 Hz, and (d) inductive vs capacitive reactance crossover.*

Materials Research Proceedings 64 (2026) 544-553 https://doi.org/10.21741/9781644904091-68

4.1.3 Phasor Relationships and Correction Mechanism:

Phasor diagram analysis provides visual representation of power factor correction mechanism through voltage-current phase relationships, impedance triangles, and power triangles. Figure 4 visualizes the correction mechanism: the 140 μF capacitor injects leading current (+90°) that cancels the motor's lagging reactive component, reducing total line current from 12.35A to 6.64A and aligning it with voltage (residual lag only 2.6°). The corrected power triangle shows reactive power reduced to 70 VAR (97%), apparent power to 1527 VA, achieving PF = 0.999.

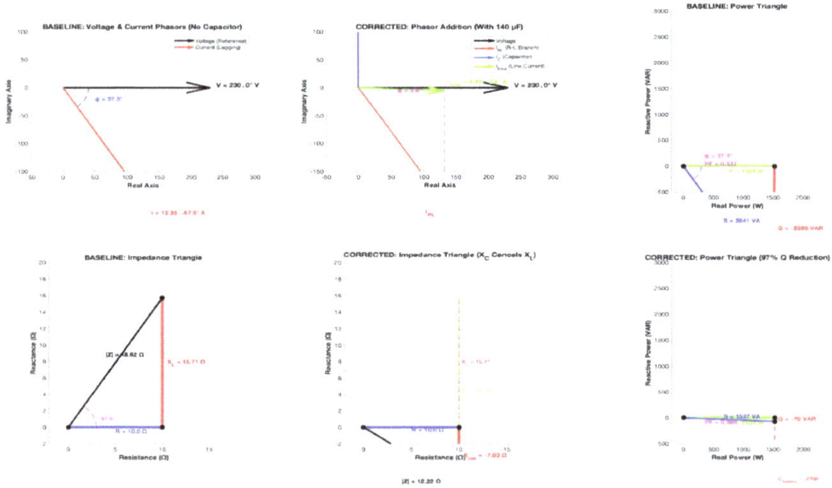

Figure 4. *Phasor and triangle analysis of power factor correction mechanism (230 V, 50 Hz, R = 10 Ω parallel load)*

4.2 Economic Analysis and Financial Performance:

Economic analysis using ONEE 2024-2025 tariffs (1.0774 MAD/kWh) and 20% penalty for PF<0.60 demonstrates strong viability. For two-shift operation (7,618 kWh/year), uncorrected operation (PF = 0.537) costs 9,848 MAD/year (8,207 MAD base + 1,641 MAD penalty). The 140 μF system (5,460 MAD investment) eliminates the penalty and reduces I^2R losses from 381 to 110 kWh/year (291 MAD/year savings), yielding total savings of 1,933 MAD/year (19.6% reduction). This delivers 33.9-month payback, 77% 5-year ROI, and 9,467 MAD 10-year NPV (5% discount). Sensitivity analysis confirms attractive payback across one- to three-shift scenarios and capacitor price variations.

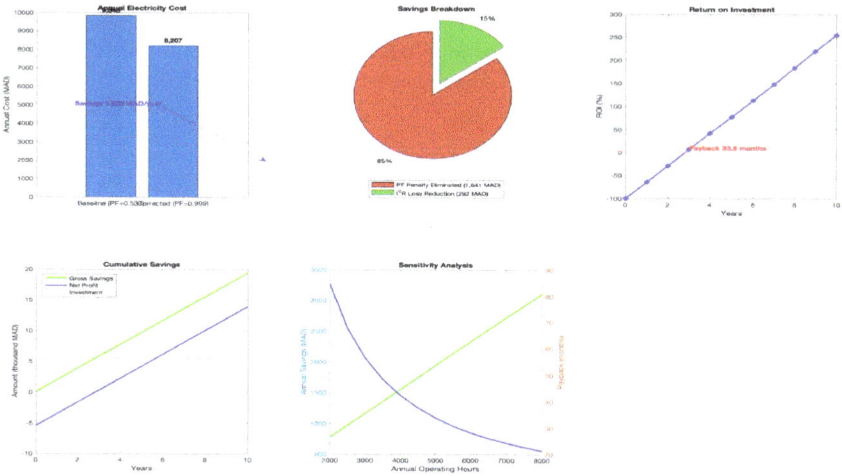

Figure 5. *Economic performance of the 140 μF shunt power factor correction system under ONEE 2024– 2025 medium-voltage tariffs (a) Annual electricity cost comparison (b) Savings breakdown and return- on-investment metrics: total annual saving.*

5. Study Limitations and Assumptions

The present analysis embraces a number of limitations and underlying assumptions that have to be recognized in order to appropriately interpret the technical and economic results. This analysis assumes industry-standard progressive penalty structure (20% surcharge for PF < 0.60) as ONEE-specific rates are not publicly disclosed. If actual penalties differ, economic gains derive primarily from I^2R loss reductions (~630 MAD annually), extending payback to ~104 months while maintaining positive ROI. The single- phase analysis is valid for three-phase applications with balanced phase-by-phase correction, though requiring 2.5-3×greater investment with proportional savings, preserving comparable payback and profitability ratios. The steady-state load model assumes constant R-L values. Facilities with significant load variability requires automatic switched capacitor banks for optimal correction. The analysis assumes ideal sinusoidal waveforms at 50 Hz. Facilities with nonlinear loads (VFDs, rectifiers, inverters) exceeding 20% of total load require harmonic analysis, as shunt capacitors may interact with system impedance creating resonance conditions. Detuned harmonic filter capacitors may be needed instead of simple capacitor banks. Results are specific to the 1.526 kW load modeled ($R = 10\Omega$, $L = 50mH$). Other loads require specific analysis for optimal results.

Declaration of Generative AI and AI-Assisted Technologies in the Writing Process

I declare that the use of ChatGPT (OpenAI) in the preparation of this paper was conducted in accordance with applicable academic integrity guidelines and institutional policies. AI assistance was limited to supporting tasks such as language refinement, formatting, or idea organization, and all substantive content, analysis, and conclusions represent my own original work and understanding. I have reviewed and verified the accuracy of all information included, take full responsibility for the final submission, and have properly cited all sources in compliance with scholarly standards.

References

[1] Kingdom of Morocco - Ministry of Energy, Mines and Environment (MEMEE), "Stratégie Énergétique Nationale - Vers un modèle énergétique robuste, durable et compétitif," 2021. [Online].

Available: https://www.mem.gov.ma/Pages/strategie.aspx

[2] ANRE (National Regulatory Authority for Electricity), "Décision n° 01/24 portant approbation des tarifs d'utilisation des réseaux de transport et des services système," Bulletin Officiel du Royaume du Maroc, no. 7210, 2024. [Online]. Available: https://www.anre.ma

[3] ONEE (National Office of Electricity and Drinking Water), "Tarifs de Vente d'Électricité - Clientèle Professionnelle," 2023. [Online]. Available: https://www.onee.ma

[4] IEEE, "IEEE Recommended Practice for Electric Power Distribution for Industrial Plants (IEEE Red Book)," IEEE Standard 141-1993, 1993. [Online]. Available: https://standards.ieee.org/ieee/141/1046/

[5] International Energy Agency (IEA), "Morocco Energy Statistics," 2023. [Online]. Available: https://www.iea.org/countries/morocco

[6] M. M. Saied, "Power Factor Correction: Economic Feasibility and Technical Limitations," Electric Machines and Power Systems, vol. 16, no. 5, pp. 311-329, 1989. https://doi.org/10.1080/07313568908909393

[7] U.S. Department of Commerce, International Trade Administration, "Morocco - Energy Sector," Country Commercial Guide, 2024. [Online]. Available: https://www.trade.gov/country-commercial- guides/morocco-energy

[8] World Bank Group, "Morocco - Energy Efficiency Scale-Up Program," Project Document P173202, 2021. [Online]. Available: http://documents.worldbank.org/curated/en/099125103082236906

[9] M. Ahrens and Z. Konstantinovic, "Harmonic Filters and Power Factor Compensation for Cement Plants," IEEE Cement Industry Technical Conference, vol. 10, no. 1, pp. 77-103, 2005. https://doi.org/10.1109/CITCON.2005.1536514

[10] S. Rajendran, "Techno-Economic Viability of Power Factor Correction in Industrial Facilities," Energy Reports, vol. 7, pp. 2245-2254, Nov. 2021. https://doi.org/10.1016/j.egyr.2021.04.029

[11] UNIDO (United Nations Industrial Development Organization), "Morocco Industrial Energy Efficiency," 2021. [Online]. Available: https://www.unido.org/morocco

[12] IEEE, "IEEE Recommended Practice and Requirements for Harmonic Control in Electric Power Systems," IEEE Standard 519-2014, 2014. https://doi.org/10.1109/IEEESTD.2014.6826459

[13] B. Singh, S. Singh, A. Chandra, and K. Al-Haddad, "Comprehensive Study of Single-Phase AC-DC Power Factor Corrected Converters with High-Frequency Isolation," IEEE Transactions on Industrial Informatics, vol. 7, no. 4, pp. 540-556, Nov. 2011. https://doi.org/10.1109/TII.2011.2166798

[14] ONEE (Office National de l'Électricité et de l'Eau Potable), "Rapport Annuel," Casablanca, Morocco, 2023. [Online]. Available: https://www.onee.ma

[15] African Development Bank, "Morocco - Renewable Energy Development and Energy Efficiency Project," Tunis, Tunisia, 2022. [Online]. Available: https://www.afdb.org

[16] IEEE, "IEEE Recommended Practice for Establishing Liquid-Filled and Dry-Type Power and Distribution Transformer Capability When Supplying Nonsinusoidal Load Currents," IEEE Standard C57.110-2018, 2018. https://doi.org/10.1109/IEEESTD.2018.8457274

A Simulation-Based Study of a Triboelectric Nanogenerator Emulator for Efficient Energy Harvesting

Imane ABDOUN[1,a *], Younes KETTANI[1,b]

[1]Al Akhawayn University in Ifrane, Morocco

[a]Im.Abdoun@aui.ma, [b]Yo.Kettani@aui.ma

Keywords: Nanogenerator (TENG), Energy Harvesting, Emulation Platform, Tinkercad Simulation, Arduino, Variable Capacitance, Power Management Circuit, Renewable Energy

Abstract. Triboelectric nanogenerators (TENGs) have vast potential for converting ambient mechanical energies; yet progress is hindered by the iterative and expensive process of prototyping. In this study, a novel and programmable emulator is introduced in Tinkercad that models and emulates the primary electromechanical characteristics of a traditional contact-separation mode triboelectric nanogenerator via an Arduino-based capacitor network. The emulator provides efficient voltage waveforms for the TENG and also serves as a useful tool for assessing performance in real operating scenarios. These include measurements of a 2.1 V boost in storage voltage over 30 seconds and, more significantly, a reduction in retained energy due to a virtual leak path. This paper thus provides a low-cost emulator that facilitates rapid prototyping and simulation in TENG energy harvesting system designs.

Introduction

The rapid growth of the Internet of Things (IoT) and wearable technology has created an ever-pressing need for sustainable and self-powered energy harvesting. Triboelectric nanogenerators (TENGs) have demonstrated great promise to succeed as a viable technology that can harness the available mechanical motion of everyday objects and turn that energy directly into electricity. The development of this technology is, however, being held back by the expensive and time-consuming process of building physical prototypes.

However, despite the presence of related research work on simulation deficiencies, several issues concerning accessibility have yet to be addressed. Wang et al. [1], for example, characterized that simulation tools lack TENG simulation, making it impossible for those who are not experts in related areas to simulate circuits connected with TENGs in circuit simulators. Moreover, Wang et al. have successfully developed a simulation model for TENGs using MATLAB/Simulink; in particular, it is capable of simulating TENG outputs with the aid of costly and complex simulation software, which is impractical for educational and prototyping purposes. At roughly the same time, Zhang et al. [2] have also provided a basic understanding of power density for TENGs, characterized in relation to a barrier beyond which nothing more can be improved for TENGs' efficiency. Nevertheless, despite all mentioned research works and significant progress in TENG development [3], an accessible and hardware-interacting system capable of overcoming existing simulation deficiencies is yet to be realized.

This paper seeks to fill this research gap by proposing a platform for TENG simulation that incorporates Arduino control circuitry to facilitate real-time interfacing between the emulator and the simulated parameters of a TENG device. Furthermore, the cost-effective design of the emulator renders it easily accessible via the internet, making it particularly useful for experimenters compared to existing simulation methods that would require specialized equipment such as licensed software packages for successful simulation to be accomplished. The following sections will examine the design of the emulator that has been proposed.

Materials and Methods

The TENG emulator platform was developed using Tinkercad Circuits for the realization of a fully-fledged virtual electronics lab. The conceptual electronics lab demonstration of the entire circuit realization involving the integration of Arduino Uno, the capacitor bank, the rectifier, as well as for measurement purposes, is illustrated in Fig. 1. The conceptual electronics lab setup included Arduino Uno as the MCU or the control unit, individual capacitors of capacities 10 nF, 47 nF, and 220 nF for the variable capacitance bank, and N-channel Metal Oxide Semiconductor Field Effect Transistors or MOSFETs with the IRF520 variety used as electronic switches. The energy harvester pattern was designed from a full wave bridge rectifier with 1N4148 diode-based connections and a 220 μF electrolytic energy storage capacitor.

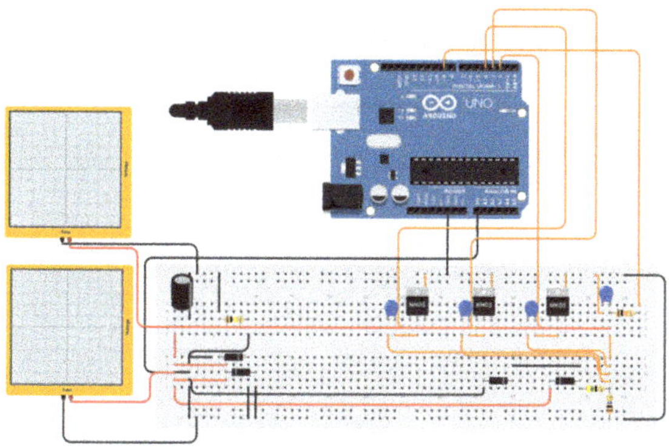

Fig. 1: Schematic of the Arduino-controlled TENG emulation platform in

The design of the circuit involved mimicking the functional dynamics of two primary modes of operation for triboelectric nanogenerators (TENGs). For a contact/separation cell of a TENG, to describe vertically parted but contacted plates, a cyclic algorithm running on an Arduino platform was used to control the gates of MOSFETs interfaced to digital I/Os D2-D4, as shown in Fig. 2. For this cell design, capacitors switched on/off in order to simulate time-varying system capacitance. At the same time, a different procedure was carried out for a sliding cell of a TENG using a different algorithm shown in Fig. 3. For this design, system capacitance varied non-sequentially to describe lateral movements. For simulating surface charge density in these designs, a DC voltage source of constant amplitude, obtained by subtracting the high-frequency noise of a pulse-width modulated signal originating from Arduino pin D9 using a low-pass filter RC, was used.

Despite being a rather simplified approach in most first-order analytical models of contact-separation triboelectric nanogenerators, a quasi-constancy of triboelectric charges in each cycle of TENG and a direct proportionality between the generated voltage and the inverse of the system's capacitance serve as a valid assumption. Based on this, a simulation of the triboelectric charges by a DC voltage source proves to be a valid method for accurately and correctly imitating the principal electromechanical characteristics of TENG. In real TENG devices, however, deviations from the above theoretical assumption may be introduced by air breakdown in highly charged zones and by a leakage of charges through the surrounding material, both effects being responsible for a limit in

ultimate open-circuit voltage and eventual stored energy. Nevertheless, in a basic simulation and circuit analysis, a simplified model proves to be
adequate.

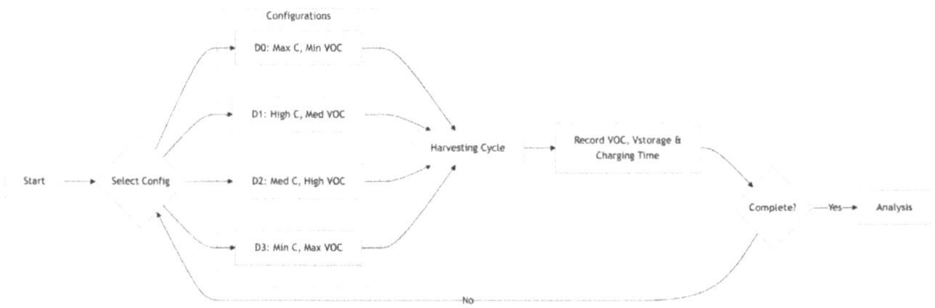

Fig. 1: Flowchart of the Arduino control algorithm for emulating contact-separation mode TENG operation.

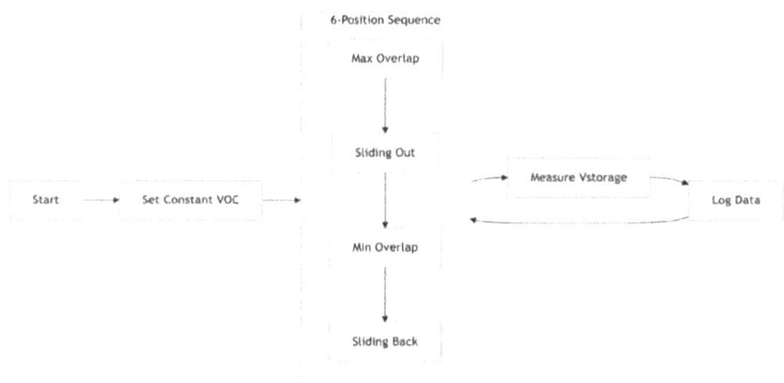

Fig. 2: Flowchart of the Arduino control algorithm for emulating sliding-mode TENG operation.

The emulation tool functions based on two major assumptions that depend on the theoretical model that describes the relationship between V, Q, and x. First, it assumes that the triboelectric-charge quantity Q is constant in value within the separation stage. This forms one of the central tenets of a first-order model used to describe ideal TENGs [4,5]. Additionally, it ignores second-order effects that include air breakdown. However, this model forms an important starting point for estimating parameters involved in optimization, which is requisite for creating optimal TENG systems[6].

Capacitance switching in the capacitor bank followed a deterministic switching pattern based on Arduino Uno's logic, wherein switching of individual capacitors to and from the circuit involved the use of MOSFETs to control discrete values of capacitance. For a capacitance model based on contact and separation stages, the switching pattern followed a sequence of capacitor actuations to model differences in distance of separation. On the other hand, a different pattern of capacitor actuation followed for the sliding mode of operation to model differences in overlap areas of the electrodes. There followed a fixed time delay for measurement of voltages after a switching pattern to ensure accuracy of measurement readings. Voltage V_{OC} and $V_{storage}$ were measured with the Arduino ‘oscilloscope.

Results and Discussions

The triboelectric nanogenerator (TENG) simulation tool successfully replicates the key electrical properties associated with both contact and sliding TENG designs, thus ensuring quantitative evaluations of the performances of these devices in harvesting energy. The simulation tool outputs were verified using theoretical foundations.

The dominant electrical property of the contact-separation TENG emulated device has been shown to follow what has been described using Fig. 4. The open-circuit voltage V_{OC} has been found to increase with the effective separation distance, which reflects the decrease in the capacitance (C) of the system. The V-Q-x dynamics of the TENG emulated device has been found to obey the fundamental V-Q-x relation [4]:

$$V = \frac{Q}{C} \tag{1}$$

Denote C as the effective capacitance that changes with distance or overlap setting. V_{OC} is considered to be open-circuit voltage emitted through emulated electrodes on the TENG system. In contrast, $V_{Storage}$ is regarded as voltage that is measured on the storage capacitor after rectification process. The capacitance itself varies with separation distance according to the parallel-plate capacitor equation:

$$C = \frac{\varepsilon_0 A}{d} \tag{2}$$

In this model, C changes with distance according to the parallel plate capacitor approximation, where ε_0 is the vacuum permittivity and A is the contact area. This inverse proportional relationship between C and distance d confirms the emulator can appropriately replicate TENG physics. It should be noted that the four standard TENG operation modes do not share identical C variations; this text refers specifically to the vertical contact-separation mode, which was treated distinctly in TENG models per publication [7]. In the experiment, $V_{Storage}$ exhibited a stepped charging behaviour, reaching 6.3 V at maximum separation. This voltage remained significantly lower than the maximum applicable V_{OC}.

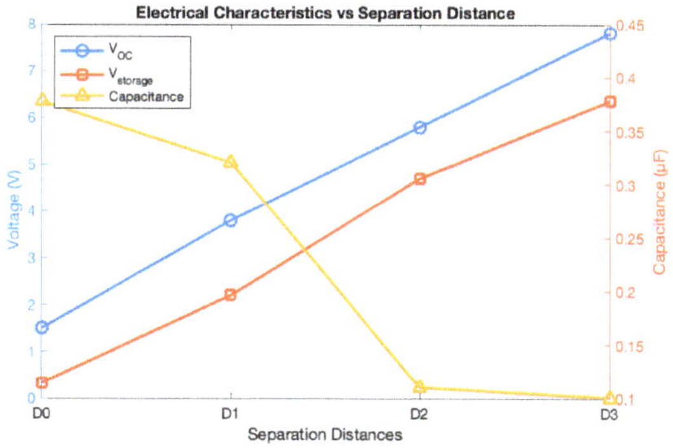

Fig. 3: Output voltage characteristics versus separation distance.

The energy accumulation rate for different separation values is shown in Fig. 5, where the stored energy is plotted against the charging time. It is clear that higher separation values, which imply higher open-circuit voltages, as shown in Fig. 1, exhibit a higher rate of energy accumulation as well as higher stored energy. This is because the energy storage process is dictated by the equation:

$$E = \frac{1}{2}CV^2 \tag{3}$$

What is observed from the results is that there is a direct trade-off between the distance of separation and the open circuit voltage V_{OC}, as shown in Fig. 1 above. As the distance of separation increases, the value of V_{OC} increases. This is at the cost of a decrease in the natural capacitance (C). The smaller capacitance results in a decrease in the rate of charging of the storage capacitor.

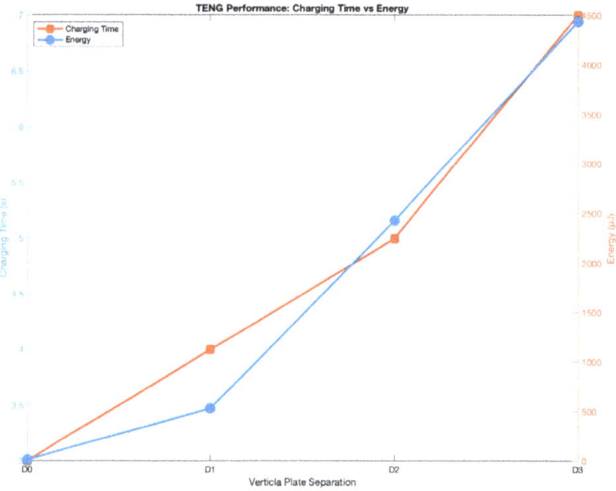

Fig. 4: Energy accumulation profile demonstrating the voltage-charging time trade-off.

The flexibility of the emulator is further exemplified in Fig. 6, where the performances for both the contact/separation and sliding mode TENGs are displayed. The sliding mode TENG shows a clearly visible, non-sequential output profile, dependent on the overlapping portion of the sliding parts, thus showing the capability of the emulator to simulate the basic operating principles of TENGs in different ways. The variation in capacitance for the sliding mode shows a typical phenomenon dependent on a different pattern:

$$C_{sliding} = \frac{\varepsilon_0 A_{overlap}}{d} \tag{4}$$

Where $A_{overlap}$ is the overlap area of the electrodes during the sliding process. As shown in Fig. 6, the open-circuit voltage V_{OC} obtained from the sliding process is seen to increase up to a value of 7.6 V, thereby affirming the utility of this basic mode of high-voltage production [5]. On the other hand, a storage voltage $V_{Storage}$ is seen to saturate at a value of approximately 3.2 V. This marked saturation of both V_{OC} and $V_{Storage}$ is a widely noted phenomenon in direct charging systems and can be ascribed to two primary factors. Firstly, the natural forward drop of about 1.4 V due to the full-wave bridge rectifier imposes a significant potential barrier to charging. Additionally, due to the high voltages obtained in such systems, the recombination of charges becomes a marked hindrance to further charging of the storage capacitor. It is thus abundantly clear from this situation that high V_{OC} values no longer ensure effective storage of voltages, thereby pointing to the need for effective power management systems such as the motion-sensitive switching systems developed by Zi et al., which were experimentally validated to significantly improve energy storage efficiency [6].

Sliding modes of operation are more commonly associated with touchscreens, "moving clothes" algorithms for clothing identification purposes, as well as wiping modes for surface preparation [7].

Finally, the effect of environmental factors was explored for leakage resistance R_{leak} by adjusting it to simulate the effect of charge dissipation under high humidity, which is shown in Fig. 7. This analysis reveals the critical drawback that, upon decreasing the value of R_{leak} from 10 MΩ to 100 kΩ to simulate the effect of humidity, the performance suffers greatly. The value of V_{OC} reduces from 7.5V to 1.5V, along with a drastic drop in the stored voltage from 6.2V to 0.8V, which reduces the system efficiency from 83% to 53%.

This can be attributed to the formation of a conducting layer by the water molecules to result in parasitic leakage paths that short-circuit the triboelectric charges [8] , highlighting a key durability challenge for TENGs [9]. This inherent problem has led to the emergence of smart approaches for overcoming it, which can be generally classified into internal methods of charging rejuvenation, such as charging excitation [8], and external protection using material innovations. The latter can include the development of high-performance encapsulation systems or the design of micro/nanostructured hydrophobic and surfaces on materials such as PDMS, which have been experimentally demonstrated to effectively resist water adsorption under high humidity conditions [10]. Our simulator is an important resource to measure these inherent trade-offs and offer a crucial testbed for assessing the efficiency of encapsulation and optimized power management circuits [11] for addressing these practical challenges before actual implementation.

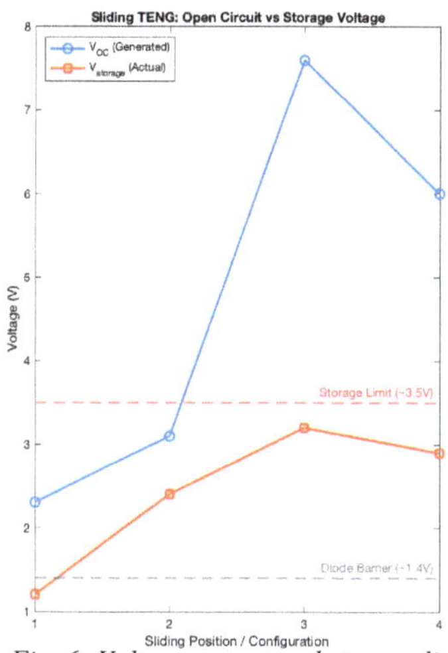

Fig. 6: Voltage output and storage limitation in sliding-mode TENG operation.

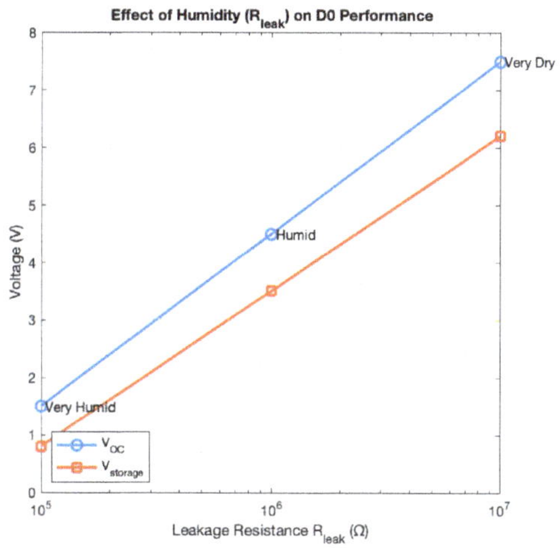

Fig. 5: Charge retention and efficiency decline under simulated humid conditions.

This paper demonstrates the development and validation of a user-friendly emulation platform that can bridge the gap between triboelectric nanogenerator principles and circuit designing. The platform enables users to interactively explore a TENG-based energy harvesting system with ease, as a quicker means compared to a software interface and a physical prototype system.

The setup, designed using Tinkercad, uses an Arduino-driven capacitor bank that demonstrates time-varied capacitance by including a comprehensive energy-harvesting circuit that supports load and leak currents. It is set up in such a way that allows for the effective demonstration of both contact/separation mode and sliding-mode energy harvesting by a TENG.

Experiments prove the efficacy of the platform for evaluating essential TENG parameters. For a contact-separation mechanism, the V_{OC} value is measured to be 7.8 V. However, for the sliding mechanism, the storage capacity is restricted, which causes V_{OC} to sharply peak to 7.6 V, hence $V_{Storage}$approaching a saturation value of 3.2 V. Simulation studies reveal a drastic drop in the efficiency value from 83% to 53% for lower leakage resistances, thus signaling the difficulties caused due to humidity.

Although its latest version is based on first-order modelling, it forms a useful nucleus for rapid prototyping. Future research efforts would focus on application-oriented developments within its framework. These would include its integration with application studies on IoT sensor nodes and wearable self-charging units [12] and developing ML-enhanced charging strategies and robust control algorithms such as Sliding Mode Control for handling the non-linear power dynamics of TENGs [11].

Declaration of Generative AI and AI-Assisted Technologies in the Writing Process
The authors acknowledge the use of ChatGPT as a generative AI tool to assist with the writing process of this manuscript. ChatGPT was used for language-related support, including rephrasing sentences and improving grammar and clarity. The scientific content, analysis, and conclusions remain entirely the responsibility of the authors.

References

[1] K. Wang, Y. Liao, W. Li, Y. Zhang, X. Zhou, C. Wu, R. Chen, T.-W. Kim, Triboelectric nanogenerator module for circuit design and simulation, Nano Energy 107 (2023) 108139. https://doi.org/10.1016/j.nanoen.2022.108139

[2] C. Zhang, L. Zhou, P. Cheng, X. Yin, D. Liu, X. Li, H. Guo, Z.L. Wang, J. Wang, Surface charge density of triboelectric nanogenerators: theoretical boundary and optimization methodology, Appl. Mater. Today 18 (2020) 100496. https://doi.org/10.1016/j.apmt.2019.100496

[3] J. Luo, Z.L. Wang, Recent progress of triboelectric nanogenerators: from fundamental theory to practical applications, Sci. China Mater. 63 (2020) 249–275. https://doi.org/10.1002/eom2.12059

[4] S. Niu, Z.L. Wang, Theoretical systems of triboelectric nanogenerators, Nano Energy 14 (2015) 161–192. https://doi.org/10.1016/j.nanoen.2014.11.034

[5] Y. Wang, Y. Yang, Z.L. Wang, Triboelectric nanogenerators as flexible power sources, npj Flex. Electron. 1 (2017) 10. https://doi.org/10.1038/s41528-017-0007-8

[6] Y. Zi, H. Guo, Z. Wen, M.-H. Yeh, C. Hu, Z.L. Wang, Effective energy storage from a triboelectric nanogenerator, Nat. Commun. 7 (2016) 10987. https://doi.org/10.1038/ncomms10987

[7] R.D.I.G. Dharmasena, S.R.P. Silva, Towards optimized triboelectric nanogenerators, Nano Energy 63 (2019) 104132. https://doi.org/10.1016/j.nanoen.2019.05.057

[8] J. Zhang, C. Boyer, Y.X. Zhang, Enhancing the humidity resistance of triboelectric nanogenerators: a review, J. Mater. Chem. A 12 (2024) 4502–4523. https://doi.org/10.1002/smll.202401846

[9] J. Zhao, Y. Shi, Boosting the durability of triboelectric nanogenerators: a critical review and prospect, Adv. Funct. Mater. 32 (2022) 2213407. https://doi.org/10.1002/adfm.202213407

[10] X. Yin, P. Cheng, C. Zhang, L. Zhou, J. Wang, Optimization strategies for triboelectric nanogenerators operating under high humidity conditions, Nano Energy 97 (2022) 107206. https://doi.org/10.1002/eom2.12493

[11] H. Zhou, Y. Wang, Y. Yang, Z.L. Wang, Recent progress of switching power management for triboelectric nanogenerators, Sensors 22 (2022) 1668. https://doi.org/10.3390/s22041668

[12] X. Pu, L. Li, M. Liu, C. Jiang, C. Du, Z. Zhao, W. Hu, Z.L. Wang, A self-charging power unit by integration of a textile triboelectric nanogenerator and a flexible lithium-ion battery for wearable electronics, Adv. Mater. 27 (2015) 2472–2478. https://doi.org/10.1002/adma.201500311

Materials Research Proceedings 64 (2026) 562-569 https://doi.org/10.21741/9781644904091-70

Comparative Energy Yield Analysis of Dual-Axis, Single-Axis, and Fixed-Tilt PV Systems in Ifrane, Morocco

Azeddine MANSOUR[1,a*]

[1]Al Akhawayn University in Ifrane, Morocco

[a]A.Mansour@aui.ma

Keywords: Solar Tracker, Dual-axis, Photovoltaic, Energy Gain, Arduino, LDR, Simulation

Abstract. Solar energy is an efficient source of energy; however, subsequent research on maximizing its use is limited by cost-ineffective and low-efficiency photovoltaics (PV) cells. The idea behind solar tracking systems is that they continuously position PV panels perpendicular to the ray of the sun, maximizing energy collection. This simulation study demonstrates the performance and energy gain of a simple, active dual-axis solar tracking system in comparison to fixed and single axis systems using an Arduino Uno microcontroller and Light Dependent Resistors (LDRs). The system consists of a microcontroller with an ATmega328P base Arduino Uno, two servo motors for dual-axis (tilt/elevation and pan/azimuth) movement, and four LDRs for light intensity recognition. Simulations conducted in Ifrane, Morocco, show that the dual-axis tracker (Net) provided a greater daily energy performance share of 35.7% compared to the single axis (34.0%), and fixed PV panel (30.4%). The Dual-Axis Performance Gain varied from approximately 15.2% to 17.8% over the operational hours of each system. Overall, the solar tracking system that has been implemented in this study provides a more efficient method of following solar insolation, achieved with greater operational energy performance, delivering superior energy yield and maximizing the economic consistency of electric power generation.

Introduction

Dependence on fossil fuels is not sustainable and has led to a desperate search for renewable energy sources, with solar energy as the other abundantly available and non-polluting renewable energy sources. The drawback to solar power generation is economically passed, and the low level of efficiency in energy transfer to PV electricity charging, especially of conventional single-crystal silicon cells remains one of the significant barriers to widespread adoption. Solar panels must be perpendicular to the rays of the sun for PV cells to transform solar energy into electricity at maximum efficiency.

This article reports on the simulation, and outcomes of a low-cost active dual-axis solar tracker. The system employs an Arduino Uno in order to interface with Light Dependent Resistor (LDR) sensors and servo motors. The goal of the study is to carry out quantitative assessments of the energy output from this dual-axis system with fixed and single-axis systems based on simulations conducted in Ifrane, Morocco. The sections that follow provide a description of the system, the control algorithm used, and an in-depth discussion and analysis of the energy output comparisons and performance improvements.

Previous research on solar tracking was predominantly intended to present examples of motor control and sensor actuation rather than an assessment of true energy performance. Which only presents mechanical movement and does not evaluate electric performance of fixed, single axis, and dual axis systems, only just one tracking system performance[1]. Similar studies, too, only profile either a series of fixed panels or one tracking mechanism entirely, and current studies do not evaluate all three configurations without providing a single productivity review. The literature is mostly lacking because all previous tracking design literature only mentions gross gains in energy, and completely disregard the motor energy required to actuate the design, thus making false claims regarding true performance, especially for the low-cost active tracker[2].

Therefore, the present study aims to provide true comparison performance for all three configurations, fixed, single axis, and dual axis by measuring performance using localized irradiance data for the Ifrane region sourced from the *Nasa power* database and utilizing the motor-consumption model to evaluate net energy gain[3]. The current study provides a more holistic and realistic performance evaluation than previous LDR-based tracking work.

Materials and Methods

Simulation modeling of the dual-axis solar tracking system were implemented through a simulation-based approach leveraging a defined hardware architecture and an active control algorithm. This section provides detailed information concerning the design of the system, the control logic behind it, the data used, and the numerical energy model for evaluating outputs for comparison against fixed and single-axis installations.

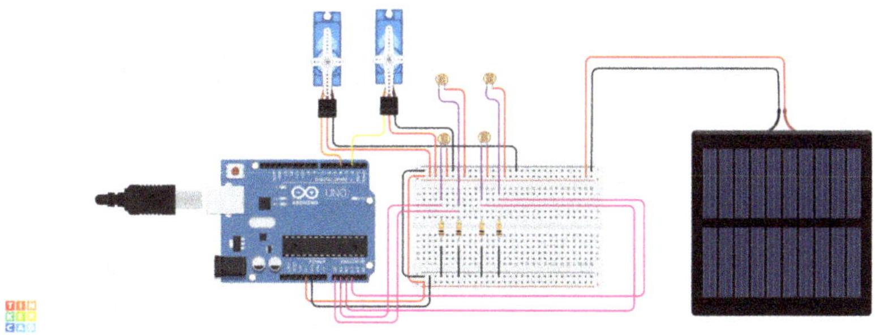

Fig. 1: Dual-Axis Solar Tracker: Tinkercad Schematic and Experimental Setup

System Architecture and Electrical Design. The physical system is developed as an active, closed-loop control circuit, with the hardware and electrical components split into three functional blocks: Control Unit, Sensing Circuit and Actuator Circuit.

Control Unit (Arduino Uno). The Arduino Uno, which is based on the ATmega328P microcontroller, serves as the system processing unit, receiving analog voltage signals, converting them through its Analog to Digital Converter (ADC), running the tracking algorithm in C++, and producing precise control signals using Pulse Width Modulation (PWM).

Sensing Circuit (The LDR Quadrant). The sensing circuit converts light intensity into usable electronic signals using four Light Dependent Resistors (LDR) in a four-quadrant arrangement. Each LDR is configured in a voltage divider circuit, allowing the system to compare the intensity differential accurately across the quadrants.

Actuator Circuit (Servo Motors). Two DC Servo Motors provide angular movement control for Tilt (vertical/elevation) and Pan (horizontal/azimuth). The servos are controlled through PWM signals, and a separate, dedicated 5V power supply rail is used to prevent voltage drops and ensure stable microcontroller operation.

Tracking Control Algorithm. The dual-axis solar tracker is implemented as a closed loop control algorithm written in C++, operating in three stages: Sensing and Averaging, Difference Calculation, and Movement Logic. The algorithm uses LDR difference-based decision making with a tolerance threshold so that servos move only when the light difference exceeds an acceptable error threshold, ensuring a stable response with low energy tracking performance.

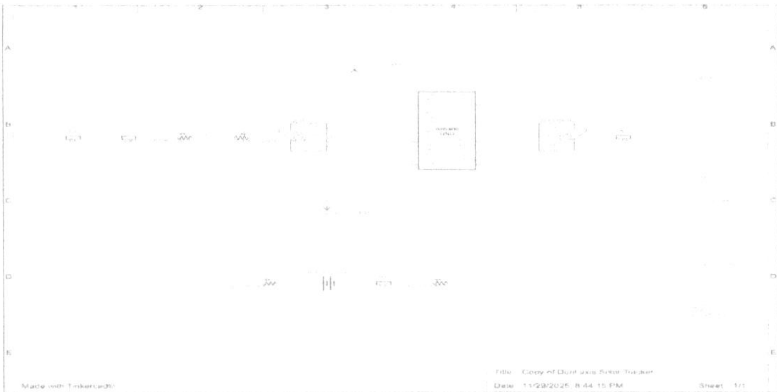

Fig. 2: Dual-Axis Solar Tracker Schematics

Results and Discussion

Hourly Gross Energy Calculation. The Total Gross Dual-Axis Energy (1363.23 Wh) is simply the sum of the hourly energy outputs (EGross) from the optimally tracked panel. This value is derived from the basic energy production formula, which has the system constants and uses hourly irradiance data at Ifrane[3]. The calculation of gross energy is derived from the standard PV energy-yield model, with the hourly energy output being equal to irradiance × area × system efficiency × tracking factor, based on methods used in energy-yield assessment papers.[4]

The gross energy calculation is based on the following constants:

PV_Area = 1.8 m^2, PV_Efficiency = 0.18, DualAxis Factor = 1.0

These parameters yield a **Combined System Constant (k)** of 0.306. The calculation across the operating hours is detailed below:

Table 1: *Hourly Calculation of Total Gross Energy (EGross) using Optimal Irradiance and the System Constant (k=0.306).*

Hour	Irradiance [W/m2]	Hourly EGross [Wh]
8	22	6.73
9	205	62.73
10	398	121.79
11	552	168.89
12	652	199.49
13	690	211.14
14	662	202.57
15	571	174.67
16	424	129.74
17	236	72.22
18	43	13.23
TOTAL	-	1363.23

Hourly Energy Yield Comparison

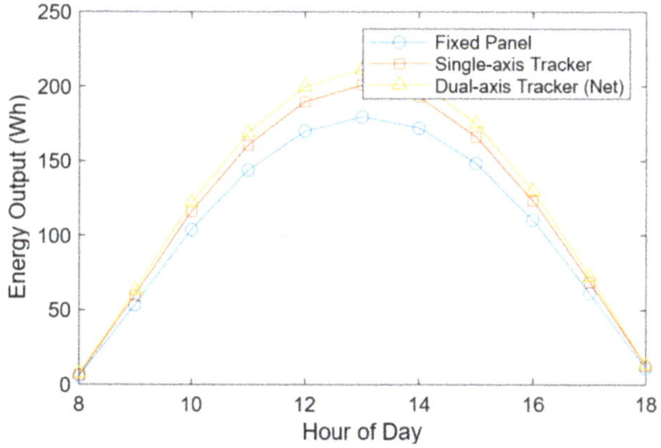

Fig. 3: *Hourly Energy Yield Comparison of Fixed Panel, Single-Axis, and Dual-Axis (Net) Photovoltaic Systems.*

The graph demonstrates that the Dual-axis Tracker (Net) yields the most energy throughout the entire day, reaching a value of nearly 210 around 13:00. Both active tracking systems generate significantly more energy outputs than the Fixed Panel, but especially during non-peak times (morning and late afternoon), which proves the value that the system in successfully capturing light while the angle of the sun is low. The dual-axis system possesses an actionable advantage over the single-axis system, because the dual-axis system can adjust both tilt and pan while the single-axis system has only 1-degree of rotation.

Table 2: *Hourly Energy Output (Wh) for Fixed, Dual-Axis Gross, and Dual-Axis Net Systems During Peak Hours.*

Time (Hour)	Irradiance (W/m2)	Dual-axis Gross (Wh)	Fixed Panel (Wh)	Dual-axis Net (Wh)
12:00	652	199.5	169.6	199.2
13:00	690	211.3	179.6	211.1
14:00	662	202.6	172.2	202.3

Motor and Net Energy Calculation. An important part of active tracking is to estimate the energy costs of motor actuation. A number of published dual-axis tracking studies report motor consumption on the order of 1–3% of the total energy generated[5]. Given that our system consists of a 1.8 m^2 panel, we take the conservative approach of using a motor energy consumption rate of 0.12 Wh/degree[6].

To accurately assess feasibility, the energy consumed by the **dual-axis** tracking mechanism was removed from the gross output [6]. For a large 1.8 m^2 panel, the actuation of the tracker can be assisted with a conservative value of 0.12 Wh/degree due to larger mass and wind-loading requirements. The above provides the daily motor energy consumption as seen in **Eq.1** & **Eq.2**:

$$E_{motor} = 297° \times 0.12\ \frac{Wh}{Degree} = 35.64\ \frac{Wh}{Day} \tag{1}$$

This represents only:

$$\frac{35.64}{1363.23} = 2.62\% \tag{2}$$

of the gross energy production (1363.23 Wh), confirming that motor expenditure remains a very small fraction of the system's output.
The final Total Net Energy is calculated usingas seen in **Eq.3& Eq.4**:

$$E_{Net} = E_{Gross} - E_{Motor} \tag{3}$$
$$E_{Net} = 1363.23 - 35.64 = 1327.6 \tag{4}$$

Rounded to two significant figures, the Dual-Axis tracker delivers a **Total Net Energy of approximately 1320 Wh**.

Total Daily Energy Share

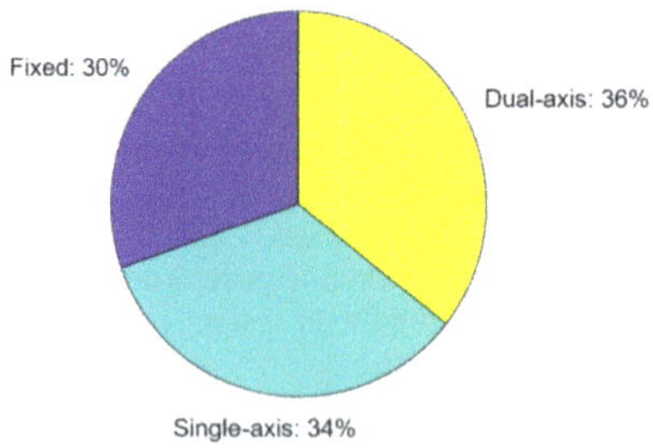

Fig. 4: *Dual-Axis Tracker Performance Gain (Percentage) Relative to the Fixed Panel Across the Solar Day.*

Table 3: *Summary of Total Daily Energy Production and Percentage Share for Each Configuration.*

System	Total Daily Energy (Wh)	Share (%)
Dual-axis (Net)	1320.0	**35.7%**
Single-axis	1250.2	34.0%
Fixed Panel	1103.0	30.4%
Total	3673.2	100.0%

The Dual-axis Tracker took the largest portion of the total daily energy at 35.7%. Which was an increased energy output of approx 19.7% than the Fixed Panel - energy per total daily. This quantitative outcome fulfills the goal of investing in a dual axis system to maximize energy generation from a fixed footprint [7].

Dual Axis Performance Gain

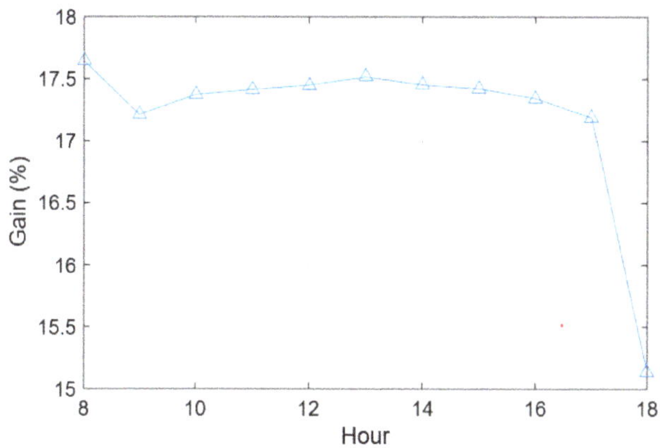

Fig. 6: *Dual-Axis Performance Gain (Percentage).*

The performance gain is settled and steady for most of the day and bounces around the range of 17.0 to 17.5. The maximum gain is recorded at the beginning of the day (8:00 AM) at nearly 17.8. The minimum gain is visible later in the day (18:00) at around 15.2. This confirms that the dual-axis tracking system provides a significant performance improvement across the entire solar window.

Conclusion

This research aimed to systematically evaluate the performance of a dual-axis solar tracking system, as a way to address the performance limitations of stationary PV systems by retaining optimal solar alignment.

The design we developed was powered by a dual-axis active tracking platform with a low cost and low availability Arduino Uno microcontroller. The platform was controlled via a reliable algorithm that interpreted position feedback from four LDR sensors to control pan (azimuth) and tilt (elevation) angles via dual-sensor motors —an approach consistent with established LDR-driven solar tracking architectures [8].

The primary numerical results substantiate the advantages of the developed tracking mechanism. Simulations results in Ifrane revealed that the Dual-axis Tracker was able to achieve the highest Total Daily Energy Share at 35.7%, significantly surpassing the Single-Axis system (34.0%) and the Fixed PV panel (30.4%), which aligns with the performance range documented in earlier dual-axis tracking studies reporting 25–40% improvements relative to fixed panels [9,10]. The Dual-Axis tracker delivered about 1320 Wh of total net energy. Additionally, the study found that motor energy consumption accounted for just 2.62% of total energy production. This sustained a significant improvement in daily energy generation from the fixed panel amounting to hourly performance improvement that was sustained over time, with differences exceeding 17% when compared to each other, for those hours identified as the best performance hours of the day.

Although there is strong performance from the active tracking system, active trackers require maintenance due to the mechanics in the system that move. Future research needs to determine how the energy consumption of the tracking mechanism can be reduced to further improve net gain.

Some potential avenues include to substitute servo motors with actuators that offer greater precision, but are still economical (e.g. stepper motors), and include a fuzzy logic control system so that it can be more responsive to differential environmental conditions, e.g. partial clouds.

ACKNOWLEDGEMENT
This work would not have been possible without the advice and support of Dr. El Hafdaoui Hamza. His feedback, support, and experience contributed extensively to this research developing and advancing.

Declaration of Generative AI and AI-Assisted Technologies in the Writing Process
I declare that the use of ChatGPT (OpenAI) in the preparation of this paper was conducted in accordance with applicable academic integrity guidelines and institutional policies. AI assistance was limited to supporting tasks such as language refinement, formatting, or idea organization, and all substantive content, analysis, and conclusions represent my own original work and understanding. I have reviewed and verified the accuracy of all information included, take full responsibility for the final submission, and have properly cited all sources in compliance with scholarly standards.

REFERENCES

[1] R. Banerjee, "Solar Tracking System," International Journal of Scientific and Research Publications, vol. 5, no. 3, pp. 1-7, Mar. 2015.

[2] H. Mousazadeh, "A review of principle and sun-tracking methods for maximizing solar energy harvesting," Renew. Sustain. Energy Rev., vol. 13, no. 8, pp. 1800-1818, 2009. https://doi.org/10.1016/j.rser.2009.01.022

[3] N. Bouhorma et al., "A comprehensive methodology for the statistical characterisation of solar irradiation data for Morocco," Applied Sciences, vol. 13, no. 5, 3365, Mar. 2023. https://doi.org/10.3390/app13053365

[4] M. A. Hafez, K. A. N. Ling, and S. Mekhilef, "Solar tracking systems: A review and classification," Renewable and Sustainable Energy Reviews, vol. 91, pp. 754-782, Aug. 2018. https://doi.org/10.1016/j.rser.2018.03.094

[5] IEA-PVPS, "Photovoltaic Module Energy Yield Measurements - Existing Approaches and Best Practice," Task 13 Report, 2018.

Materials Research Proceedings 64 (2026) 562-569 https://doi.org/10.21741/9781644904091-70

[6] M. Ferdaus, S. Mekhilef, and others, "Functionality Analysis of Electric Actuators in Renewable Energy Systems," Sensors, vol. 22, no. 11, 4273, 2022. https://doi.org/10.3390/s22114273

[7] M. A. Green et al., "Solar cell efficiency tables (Version 64)," Progress in Photovoltaics: Research and Applications, 2024. https://doi.org/10.1002/pip.3831

[8] S. Swami, R. Kumar, and N. Kumar, "Dual Axis Solar Tracking System Using Arduino," Int. J. Power Systems & Energy Engineering, vol. 4, no. 1, pp. 29-34, 2020 https://doi.org/10.18280/psees.040105

[9] A. Skoplaki and J. A. Palyvos, "On the performance of solar trackers under different climatic conditions," Applied Energy, vol. 96, pp. 1-12, 2012.

[10] H. Mousazadeh et al., "A review of principle and sun-tracking methods for maximizing solar systems output," Renewable and Sustainable Energy Reviews, vol. 13, pp. 1800-1818, 2009. https://doi.org/10.1016/j.rser.2009.01.022

Unbalanced Brushless DC Rotor Diagnosis Using Instantaneous Amplitude Analysis

Ayoub EL JAFRY[1,a *], Hamza SABIR[1,b], and Mohammed NAHID[1,c]

[1]Laboratory of Engineering Sciences & Biosciences, FST Mohammedia, UH2, Morrocco

[a]a.eljafry@gmail.com, [b]hamza.sabir@fstm.ac.ma, [c]mohammed.nahid@gmail.com

Keywords: Brushless DC Motor (BLDC), Fault Diagnosis, Incipient Fault Detection, Hilbert Transform

Abstract. Cost-effective, predictive, and proactive maintenance of electrical machines, particularly BLDC motors, is becoming increasingly important as the use of these motors grows in many areas of application. A well-known method for assessing imminent problems is to analyze the features of the stator current, thereby detecting both electrical and mechanical faults. This paper describes a method for diagnosing eccentricity faults in brushless DC motors, based on the analysis of the instantaneous current amplitude extracted from the Hilbert transform. The proposed eccentricity detection technique has been validated in MATLAB/SIMULINK. For incipient eccentricity conditions, the proposed method highlights spectral components of the defect in which the amplitude of 1.04×10^{-2} is completely masked in the conventional FFT spectrum by the fundamental peak (amplitude ≈ 0.86).

1. Introduction

BLDC motors are a type of synchronous motor, meaning that the stator magnetic field and the rotor magnetic field rotate at the same frequency. As their name suggests, BLDC motors do not use brushes. Instead, they use an electronic commutator rather than the mechanical commutator used by traditional DC electric machines. This eliminates contact-related wear and tear and increases reliability.

The elimination of brushes also reduces the required maintenance frequency, increases the service life, and extends the operating speed range. In addition, BLDCs offer other advantages such as high power density, which makes them useful in applications where size and weight are critical factors, such as robots, drones, and electric vehicles. [1], [2].

Despite their high efficiency and compact size, brushless DC (BLDC) motors have a number of requirements that must be met to prevent failure. Without proper maintenance, their condition gradually deteriorates, their efficiency tends to decline, and their operational life is reduced [3]. In addition, unaddressed failures can also cause emerging failures, as can accidental or unexpected failures, leading to significant losses such as production stoppages, service interruptions, and failures in critical areas such as maintaining precise temperatures for products in the medical and healthcare sector, the proper functioning of drones or electric vehicles [4]. In this regard, the impacts of machine loss also have a major impact on the companies and users affected (additional time losses and higher expenses, reduced system reliability). It follows that maintenance should not be considered after the fact, but rather as an essential element to be taken into account for the operational safety, security, and reliability of the systems and machines used, as well as the satisfaction of users of BLDC motor-based technological systems. [5].

BLDC motors are vulnerable to three types of faults: electrical, mechanical and magnetic. Electrical faults are considered the most critical, as they have direct effects on operational stability and secondary damage. Typical electrical faults are inter-turn shorts, inter-winding faults, inter-phase faults and open-circuit faults, which generally result from insulation failure, temperature cycling, transients and switching, and mechanical stress. Each fault also produces its own symptoms, characterized by abnormal current amplitudes, torque fluctuations and distorted EMF feedback signals. Different electrical faults can produce different fault signatures depending on operational characteristics. For example, a sudden drop to zero of the current in one of the phases may indicate an open circuit fault and that the motor has lost its phase current. A doubling of the phase current and an increase in torque potential may, in this case, indicate a phase-to-phase short-circuit [6].

Mechanical faults attributed to all BLDC motor failures represent up to 50% and include the usual bearing failures as well as misalignments between rotor and stator. One of the most common mechanical faults is air gap eccentricity, which produces asymmetrical magnetic field distribution and parasitic losses. Eccentricity can be classified as static, dynamic or elliptical. Static eccentricity occurs when the axis of rotation is displaced, and usually occurs when the rotor has been incorrectly assembled or the stator is not completely cylindrical. Dynamic eccentricity occurs when the rotor itself, with an offset, rotates around the stator without an offset. Elliptical eccentricity is linked to specific manufacturing tolerances that structurally affect the circular symmetry of rotor or stator components. Eccentricities in excess of 10% of the air gap are considered faults. They create stray currents and unwanted heat in the stator [7].

Magnetic faults are most often associated with demagnetization of the permanent magnet, whether complete or partial. Magnetic faults are generally the result of prolonged overheating, prolonged overloading, short circuits, and particularly phase faults. Demagnetization affects a motor's torque capacity and dynamic response. Corrosion and material degradation over time also contribute to the loss of magnetic properties. As magnetic performance deteriorates, efficiency decreases, thermal stresses increase, and the rate and severity of electrical or mechanical failures increase. Consequently, effective monitoring combined with rapid fault diagnosis is essential to ensure the safe, reliable operation of BLDC motor systems [8].

Among these defects, rotor eccentricity, both static and dynamic, has been the subject of more sophisticated diagnostic methods, as it is less obvious but crucial to engine operation. Advances in signal analysis now allow for more precise identification of these defects. The SAC-DM (Signal Analysis based on Chaos using Density of Maxima) method, including multiresolution wavelet analysis (WMA), allows chaotic components in acoustic signals to be framed and offers 88.34% accuracy in detecting dynamic eccentricity at constant speed [9]. With regard to static eccentricity, recent current-based techniques that exploit both positive and negative sequence components provide a more reliable diagnosis than conventional methods that use only the negative sequence component [10]. Finally, frequency domain methods using finite element modeling and fast Fourier transforms also make it possible to distinguish between different types of faults by identifying stator current spectra [11]. Harmonic extraction methods enhance real-time fault detection by separating out specific eccentricity-related current harmonics, validated by simulation and experimental results [12].

2. Architecture of the proposed approach

2.1 Hilbert transform

The Hilbert transform (HT) method can be used to study the non-stationary motor current waveforms and simplify the analytical development of the signal [13]. The analytical signal defined in (1) consists of the original real signal $a(t)$ and the Hilbert transform as the imaginary component, resulting in a complex-valued signal:

$$\tilde{a}(t) = a(t) + j\hat{a}(t) \qquad (1)$$

In this equation, j is the imaginary unit and $a(t)$ is the Hilbert transform of the signal, which is defined in (2) as the convolution with 1/πt. The definition is simply the integration of the signal, weighted by the function 1/πt, and can be represented as follows:

$$\hat{a}(t) = \mathcal{H}\{a(t)\} = \frac{1}{\pi}\int_{-\infty}^{+\infty} \frac{a(\tau)}{t-\tau}\, d\tau \qquad (2)$$

The Hilbert transform shifts the phase of the negative frequency components of the signal by +90°, then shifts the positive frequency components by -90°. The instantaneous amplitude of the resulting analytic signal is derived from the modulus of this complex signal and given by equation (3):

$$A(t) = |\tilde{a}(t)| = \sqrt{a^2(t) + \hat{a}^2(t)} \qquad (3)$$

The instantaneous amplitude is also called the "signal envelope"; it highlights low-frequency components by filtering out high-frequency swings. It provides better visibility of slow variations, which is particularly useful for detecting faults in brushless DC motors, such as eccentricity, which manifests itself as a modulation of the current amplitude masked by high-frequency noise.

The analytical signal representation provides an optimal way to identify and distinguish subtle fault signatures, such as harmonic content identified due to eccentricity of motor currents. Once the signal has been analytically characterized, the envelope of the motor current can be clearly isolated, which would support more accurate fault detection because higher frequency disturbances, in particular from the mains supply should be effectively filtered out. As shown in this study, the use of the Hilbert transform for envelope extraction is a reliable pre-processing procedure to use in condition monitoring, which enables the user to identify low-frequency modulations, typically indicative of fault conditions and particularly eccentricity.

2.2 Fast Fourier Transform Analysis

FFT analysis is the primary tool in signal processing. The ultimate goal of signal processing is to improve understanding of a signal by investigating its frequency domain representation. In this case, the FFT of the stator currents will be convoluted with an FFT of the Hanning window [14]. The coefficients of a Hanning window of length L given in (4) can then be derived using the following expressions:

$$\omega(n) = \omega_0 \cdot \left(\frac{L}{N}\left(n - \frac{N}{2}\right)\right) \qquad (4)$$

$$\omega_0(x) = \begin{cases} \frac{1}{2}\left(1 + \cos\left(\frac{2\pi x}{L}\right)\right) = \cos\left(\frac{x\pi}{L}\right), |x| \leq \frac{L}{2} \\ 0, |x| > \frac{L}{2} \end{cases}$$

$$\omega(n) = \frac{1}{2}\left(1 - \cos\left(\frac{2\pi n}{L}\right)\right) = \sin\left(\frac{n\pi}{L}\right), \qquad 0 \leq n \leq N$$

Where N + 1 is the size of the window [15].

A Hanning window is then applied to limit the frequency bands of interest and reduce numerical instability at the edges of each component. Without windowing, the affected frequency bands exhibit spectral disturbances in the form of high-magnitude oscillating components, which are typical of linear non-stationary processes. This makes it difficult to develop effective fault diagnosis system implementations.

The presence of eccentricity flaws in a BLDC machine will cause asymmetries in the flux distribution and components of the stator currents. These eccentricity flaws will introduce characteristic spectral lines in the stator current, also most notably as sidebands adjacent to the phase frequency harmonics [13]. In the case of mixed eccentricity, the lines will be at the frequencies f_d defined in (5):

$$f_d = f_s \pm k.\frac{fs}{p} \qquad (5)$$

Where fs is the stator phase frequency, and p is the number of pole pairs. These components, which can be clearly identified by applying the FFT to the stator current, provide a precise way to diagnose the machine's condition [16][17].

2.3 The proposed approach

The proposed methodology is designed to detect and evaluate faults based on stator current, this methodology is described in the flowchart in the fig. 1.

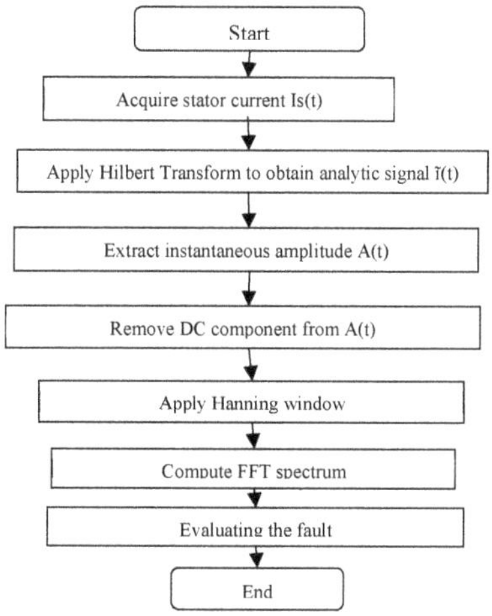

Fig. 1. Flowchart of the proposed methodology

First, the measured signal Is(t) is converted into an analytical signal using the Hilbert transform, allowing the instantaneous amplitude A(t) to be extracted. After removing the DC component, a Hanning window is applied to attenuate spectral leakage effects and improve frequency resolution. The signal spectrum is then obtained by fast Fourier transform (FFT), and the harmonic characteristics of the machine's faults are extracted. Finally, identifying the harmonics characteristic of the fault is the final step in the flowchart, allowing the presence of anomalies to be confirmed and their severity to be measured.

3. Simulation results and validation

The purpose of this section is to evaluate the effectiveness of the proposed method. To this end, a simulation of the different operating conditions of the BLDC machine, as shown in Fig. 2, was carried out in the MATLAB/Simulink environment.

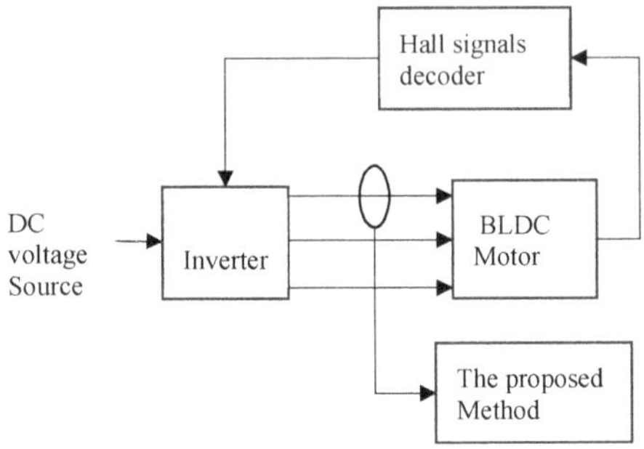

Fig. 2. Simplified block diagram of the BLDCM monitoring process

The scenarios studied include the healthy state of the machine and two levels of eccentricity faults. The principal parameters of the BLDC machine are shown in Table 1. The signals observed, namely the stator current and rotor speed, sampled at a frequency of F = 20 kHz, produce a total number of samples NT = 100000, corresponding to an acquisition time of 5 seconds.

Table 1. *Simulated BLDCm parameters*

Machine Parameters	Values
Voltage	100V
Stator resistance	2.98 Ω
Stator inductance	7 mH
Permanent magnet flux linkage	0.125 Wb
Number of pole pairs	2
Back electromotive force (EMF) constant	0,375
Rotor inertia	$J = 0.47 \times 10^{-4}$ kg·m²

4. The Performance of the proposed method

The simulation is performed on a BLDC motor model under nominal load in three conditions: normal, incipient fault, and severe fault. The stator current waveforms obtained for each operating condition are shown in Figures 3(a), 3(b) and 3(c). A direct comparison of present signals in the time domain does not reveal significant rotor imbalance faults. Thus, time analysis of stator current by itself is insufficient for reliable fault detection, emphasizing the requirement for advanced signal processing techniques.

Materials Research Proceedings 64 (2026) 570-577 https://doi.org/10.21741/9781644904091-71

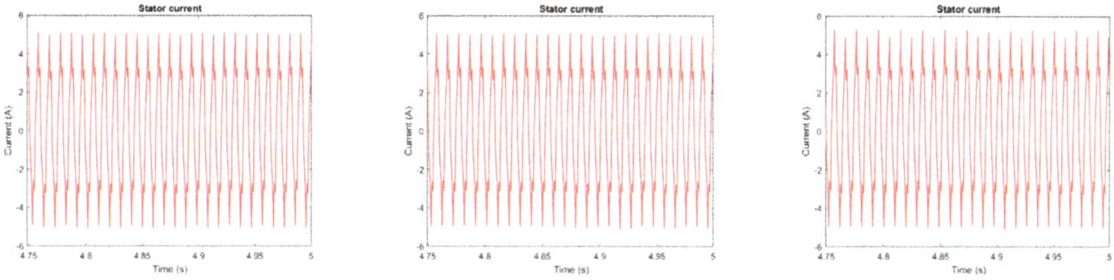

(a) (b) (c)

Fig. 3. Stator current signal under healthy (a), incipient fault (b), and severe fault conditions (c)

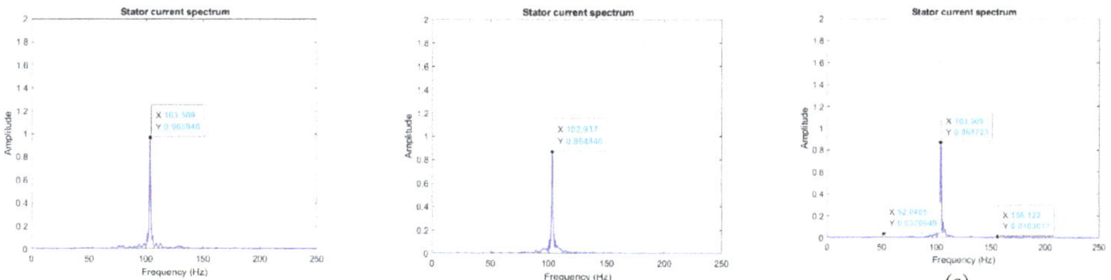

(a) (b) (c)

Fig. 4. Current stator spectrum for healthy (a), incipient fault (b), and severe fault conditions (c)

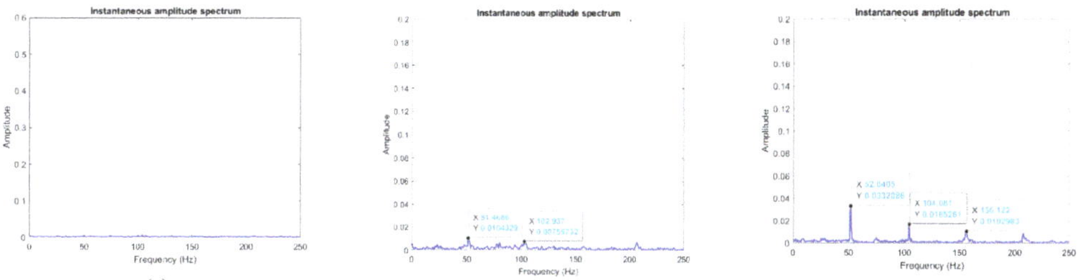

(a) (b) (c)

Fig. 5. Frequency spectrum of the extracted envelope for healthy (a), incipient fault (b), and severe fault conditions (c)

In the next step, a spectral analysis of the stator current is performed to expose the eccentricity defects of the BLDC machine. The obtained spectra for the healthy case, incipient, and heavy-defect conditions are shown in Figures 4.a., 4.b., and 4.c, respectively.

The results show that in the case of health, there are no side effects and that the fundamental component dominates the entire spectrum.

On the other hand, if the fault is verified, weak sidebands appear in the spectrum and clearly indicate the imbalance of the rotor. However, for the incipient fault, no sidebands were detected, highlighting the limitation of the fast Fourier transform (FFT), which alone cannot effectively diagnose incipient rotor faults.

In the following, the Hilbert Transform is performed on the stator current in order to obtain the instantaneous amplitude. A fast Fourier transform (FFT) is then performed on the instantaneous amplitude. The obtained spectrum are shown in Figures 5.a, 5.b, and 5.c, respectively, for the healthy, incipient, and severe fault cases. The results clearly indicate that the proposed technique can reveal spectral components of the defect, which are necessarily of low amplitude (1.04×10^{-2}), which remain undetectable using conventional FFT analysis due to the dominance of the fundamental component (≈ 0.86). Therefore, this method allows for highlighting the presence of incipient rotor faults; such faults were not possible to detect using a simple time domain analysis or applying a direct FFT on the
raw current.

Conclusion

The research presented in this paper shows how effective the combination of the Hilbert transform and fast Fourier transform (FFT) is in assessing operating conditions and health status of brushless direct-current (BLDC) motors. The combination is an effective method for identifying early rotor fault conditions such as eccentricity. The combination method is superior to the more commonly used traditional methods of monitoring, which focus only on either the time-based or spectral response of signals. Additionally, this research provides a basis for developing predictive maintenance strategies that enhance the reliability and continuous operation of electric drive systems.

References

[1] Mustafa F. Mohammed, Speed Control for BLDC Motor for Electric Vehicles Applications by Different Kinds of Sliding Mode Control Techniques, SSRG Int. J. Electr. Electron. Eng., 2024. https://doi.org/10.14445/23488379/IJEEE-V11I2P102

[2] Vinay Kumar Awaar, Sravani Bannuru, Arsalan Abbas, Position Sensorless Field-Oriented Control of BLDC Motor for EV Applications, E3S Web Conf., 2023.

[3] Yufeng Gu, Qianqian Zhang, Qingzhe Wang et al., Health management of brushless direct current motor controller, IET Conference Proceedings, 2025,

[4] Rommel S. P. Estacio, Diego A. B. Montenegro, Carlos F. R. Rodas, Machine hearing for predictive maintenance of BLDC motors, Journal of Quality in Maintenance Engineering, 2024,

[5] Enesi Y. Salawu, Olanrewaju O. Awoyemi, Opeyemi E. Akerekan et al., Impact of Maintenance on Machine Reliability: A Review, E3S Web of Conferences, 2023, https://doi.org/10.1051/e3sconf/202343001226

[6] Solís, R.; Torres, L.; Pérez, P. Review of Methods for Diagnosing Faults in the Stators of BLDC Motors. Processes 2023, 11, 82. https://doi.org/10.3390/pr11010082

[7] [K. Kudelina, B. Asad, T. Vaimann, A. Rassõlkin, A. Kallaste and D. V. Lukichev, "Main Faults and Diagnostic Possibilities of BLDC Motors," 2020 27th International Workshop on Electric Drives: MPEI Department of Electric Drives 90th Anniversary (IWED), Moscow, Russia, 2020, pp. 1-6, doi: 10.1109/IWED48848.2020.9069553. https://doi.org/10.1109/IWED48848.2020.9069553

[8] Sardashti, A., & Nazari, J. (2023). A learning-based approach to fault detection and fault-tolerant control of permanent magnet DC motors. Journal of Engineering and Applied Science, 70(1), 109. https://doi.org/10.1186/s44147-023-00279-5

[9] Veras, F. C., Lima, T. L. de V., Souza, J. V. da S., Ramos, J. G. G. S., Lima Filho, A. C., & Brito, A. V. (2019). Eccentricity Failure Detection of Brushless DC Motors From Sound Signals

Based on Density of Maxima. IEEE Access, 7, 150318-150326. https://doi.org/10.1109/ACCESS.2019.2946502

[10] Jiang, C., & Habetler, T. G. (2015). Static eccentricity fault detection of the BLDC motor inside the air handler unit (AHU). International Electric Machines and Drives Conference, 1473-1476. https://doi.org/10.1109/IEMDC.2015.7409256

[11] [Park, J.-K., & Hur, J. (2016). Detection of Inter-Turn and Dynamic Eccentricity Faults Using Stator Current Frequency Pattern in IPM-Type BLDC Motors. IEEE Transactions on Industrial Electronics, 63(3), 1771-1780. https://doi.org/10.1109/TIE.2015.2499162

[12] Jeon, M.-S., Im, J.-H., & Hur, J. (n.d.). Eccentricity Fault Diagnosis Method Using the Harmonic Extractor in BLDC Motor.

[13] Sabir, H., Ouassaid, M., & Ngote, N. (2020, December). Early Severity Assessment of Unbalanced rotor Fault in WRIM using ANN based Hybrid TSA and FFT Approach. In 2020 IEEE 7th International Conference on Engineering Technologies and Applied Sciences (ICETAS) (pp. 1-6). IEEE. https://doi.org/10.1109/ICETAS51660.2020.9484312

[14] Choqueuse, V. V., & Benbouzid, M. (2015). Induction machine diagnosis using stator current advanced signal processing. International Journal on Energy Conversion, 3(3), 76-87.

[15] Sabir, Hamza & Ouassaid, Mohammed & Ngote, Nabil. (2021). Early Fault Estimation of Inter-turn Short-circuit in Rotor Winding of WRIM using ANN-based Combined TSA and MCSA Technique. https://doi.org/10.1109/ISAECT50560.2020.9523697

[16] Cerna, M., & Harvey, A. F. (2000). The fundamentals of FFT-based signal analysis and measurement (pp. 1-20). Application Note 041, National Instruments.

[17] Panagiotou, P. (2020). Reliable Detection of Rotor Electrical Faults in Induction Motors Using Frequency Extraction of Stator Current and Stray Magnetic Flux Signals (Doctoral dissertation, Coventry University)

Materials for Energy Applications

Sulfur Vacancies in MoS_2 for Lithium Sulfur Batteries

Qingbin JIANG[1], Huifang XU[1], Jamal NAYFEH[2,a], Kwan San HUI*,[2,b] and Kwun Nam HUI*,[1,c]

[1]Joint Key Laboratory of the Ministry of Education, Institute of Applied Physics and Materials Engineering, University of Macau, Avenida da Universidade, Taipa, Macau SAR, China

[2]Department of Mechanical Engineering, College of Engineering, Prince Mohammad Bin Fahd University, P.O. Box 1664, Al Khobar 31952, Kingdom of Saudi Arabia

[a]jnayfeh@pmu.edu.sa, [b]khui@pmu.edu.sa, [c]bizhui@um.edu.mo

Keywords: Lithium Sulfur Battery, Molybdenum Disulfide, Polysulfide Shuttle, Stability, Vacancy-Modulated Electrocatalysts

Abstract. Defects within layered MoS_2 can substantially alter its electronic structure and, in turn, its catalytic behaviour toward sulfur conversion in lithium–sulfur (Li–S) batteries. In this study, sulfur-deficient MoS_2 was produced through controlled annealing and evaluated as a separator coating. As demonstrated by spectroscopic measurements and first-principles calculations, the elimination of sulfur leads to the formation of new states close to the Fermi level, which facilitates faster electron conduction and enables the enhancement of its interaction with lithium polysulfides. Addition of this vacancy-containing material to a separator increases the completeness of the redox reaction rate of sulfur and allows a capacity of 886 mAh g^{-1} after 100 cycles at 0.5 C. Such results indicate that intentional vacuity manipulation is an effective pathway towards tuning MoS_2 to LiS electrochemistry and provide a guide to the role of local disorder in catalysis.

Introduction

A lot of interest is being attracted to lithium sulfur (LiS) batteries, which have large theoretical gravimetric energy and have a naturally available sulfur source [1]. Nevertheless, their practice use is still constrained by the occurrence and movement of soluble polysulfide intermediates. Such species may get out of the cathode and recombine at the lithium anode to form a considerable capacity loss [2]. Consequently, most of the recent efforts have aimed at creating catalytic interfaces, which will direct sulfur conversion in a more efficient manner [3]. The transition-metal dichalcogenides (TMDs) of this chemistry (MoS_2, in particular) have become attractive candidates to control this chemistry [4]. They can be modified versatility by defect engineering, doping or phase transformation due to their layered structure [5]. It has been reported to enhance interactions with polysulfides, and reduce the reaction barriers by adding sulfur vacancies, which can be causing lattice instability [6]. These conflicting effects can be resolved by integrating experimental characterization and theoretical modelling [7]. The present work explores this balance using vacancy-rich MoS_2 as a separator coating for Li–S cells.

Materials and Methods

Preparation of Sulfur-Deficient MoS_2. A hydrothermal process was used to synthesize the pristine MoS_2 precursor. Ammonium molybdate (0.5 mmol) and thiourea (16 mmol) were dissolved in 60ml deionized water, transferred to a sealed Teflon vessel (100 mL), and heated under controlled conditions at 220 °C for 18 h. After cooling, the obtained product was repeatedly washed and dried. To generate sulfur vacancies, the powder was subjected to post-annealing in a mixed H_2/Ar flow at 500 °C for 3 h. This atmosphere was selected to remove a portion of sulfur while maintaining the overall layered framework.

Materials Research Proceedings 64 (2026) 579-584 https://doi.org/10.21741/9781644904091-72

Fabrication of MoS_2-Modified Separators. A slurry was formulated by blending MoS_2 (60 wt%) or vacancy-rich MoS_2 with conductive carbon (30 wt%) and polymer binder (10 wt%) in N-methyl-2-pyrrolidone. The mixture was spread evenly onto a commercial polypropylene separator and dried at 60 °C. A complementary sulfur–graphene composite cathode was prepared using a conventional melt-infiltration step, ensuring uniform sulfur distribution within the carbon host.

Electrochemical Testing. CR2032 coin cells were assembled in an argon-filled glovebox using lithium metal as the anode. The DOL/DME (VDOL/VDME=1:1) solvent system of LiTFSI and LiNO 3 was used as the electrolyte. To determine the kinetics of sulfur conversion and cycling stability under different conditions, chargedischarge cycling, cyclic voltammetry and electrochemical impedance spectroscopy were carried out.

Computational Methods. Density-functional theory calculations were carried out using plane-wave techniques. Structural relaxation was performed for both pristine and defective MoS_2 to evaluate vacancy formation energies and changes in the density of states. The interactions between MoS_2 surfaces and representative polysulfide species were also calculated to support experimental observations.

Results

Fig. 1 shows the morphology of MoS_2 and Vs–MoS_2. Microscopic imaging reveals that both pristine and vacancy-modified MoS_2 retain the nanosheet-assembled flower-like morphology characteristic of hydrothermally produced TMDs. Post-annealing does not significantly disrupt the overall structure, although subtle distortions consistent with sulfur removal are visible. EDS analysis shows that pristine MoS_2 contains 59.98 wt% Mo and 40.02 wt% S, whereas Vs–MoS_2 contains 61.37 wt% Mo and 38.63 wt% S, confirming a relative decrease in sulfur content due to vacancy formation.

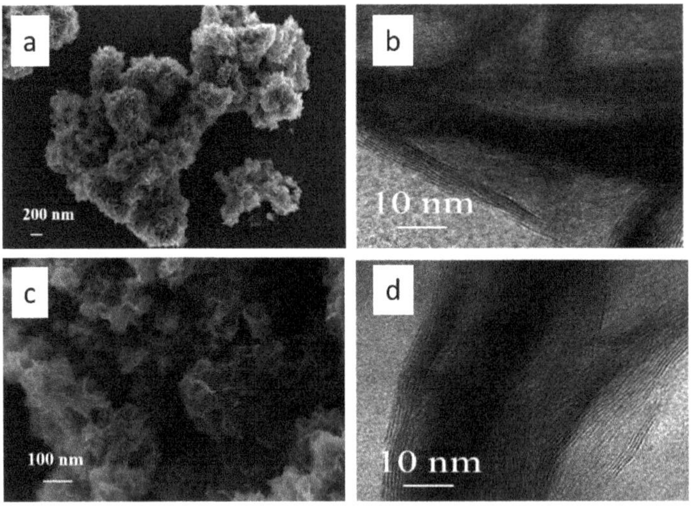

Fig. 1 - SEM and TEM images. a and b) MoS_2; c and d) Vs-MoS_2.

Brunauer–Emmett–Teller measurements (Fig. 2) show that the surface areas (~50–53 m^2 g^{-1}) of the two materials remain comparable, indicating that improved catalytic behaviour is not rooted in surface area changes. X-ray diffraction (Fig. 3) confirms that the layered structure of MoS_2

(JCPDS No. 37-1492) is largely preserved after vacancy introduction. Importantly, no impurity phases were detected following vacancy introduction. Rietveld refinement further shows that although some S atoms were removed from the 4f sites, the overall profile agreement indices remained satisfactory, confirming the structural integrity of the material.

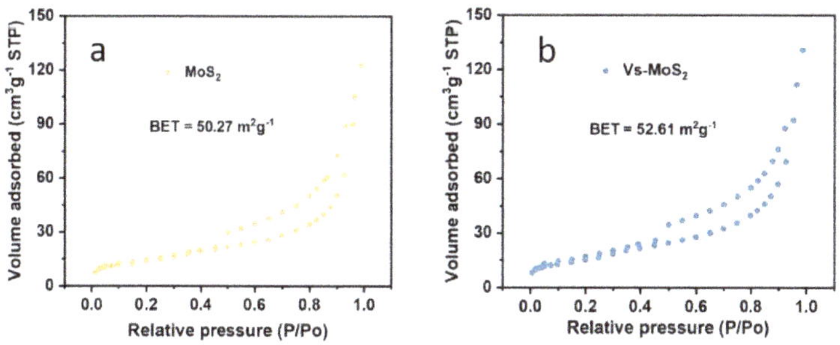

Fig. 2 - Brunauer–Emmett–Teller measurements: a) MoS_2; b) Vs-MoS_2.

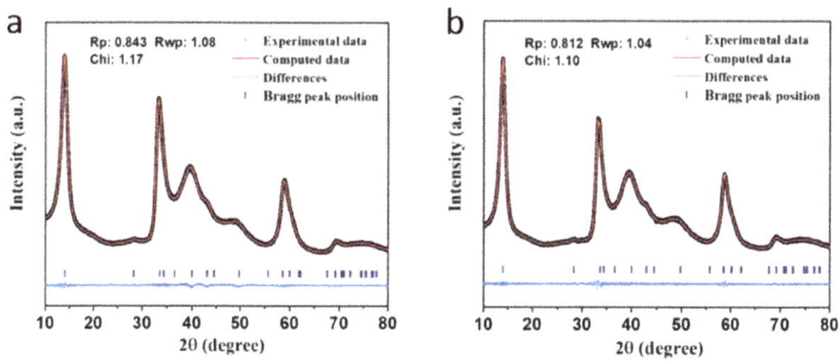

Fig. 3 -XRD pattern and the Rietveld refinement XRD pattern: a) MoS_2; b) Vs-MoS_2.

Electron paramagnetic resonance (EPR) spectroscopy was used to confirm the formation of sulfur vacancy (V?) following annealing with H_2/Ar (Fig. 4). Vs–MoS 2 demonstrates a signal at g of 2.004, which is equal to un-paired electrons produced upon S removal, which proves the introduction of vacancies. The vacancies were supposed to alter the electronics structure, alter physicochemical characteristics, and make active sites to favor catalytic transformation in sulfur redox reactions.

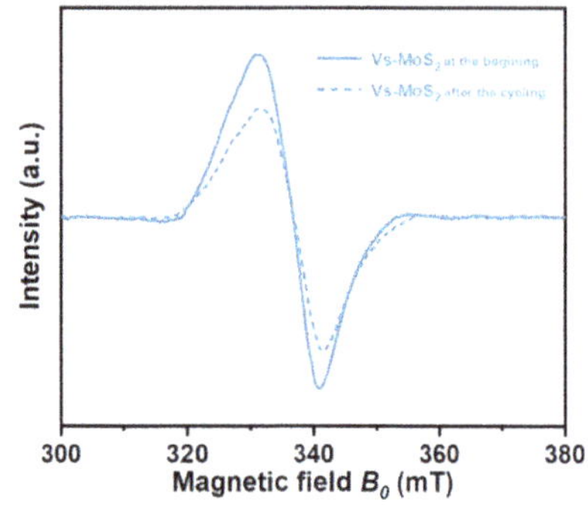

Fig. 4 - Electron paramagnetic resonance spectrum of Vs-MoS_2.

Calculations of the electronic-structure (Fig. 5) show that when the vacancy is created the density of states at the Fermi level increases significantly. This improvement of the electronic activity promotes transfer of charge and empowers the bonding with lithium polysulfides. These theoretical results are consistent with electrochemical observations: cells with the vacancy-rich MoS_2 are found to have larger discharge capacities, quicker reaction and slower capacity deterioration than cells with pristine MoS_2.

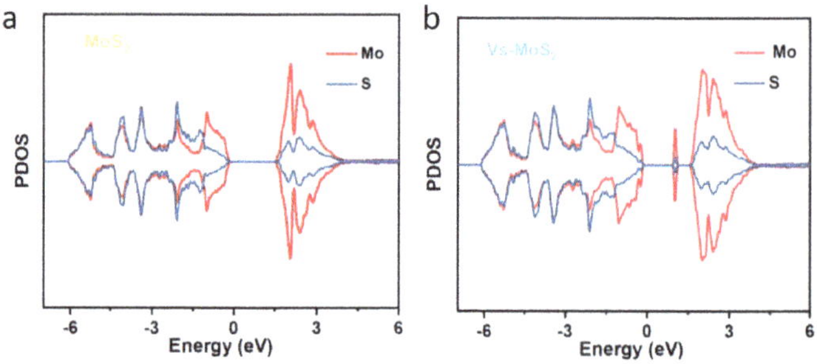

Fig. 5 - DOS: a) MoS_2; b) Vs-MoS_2.

Even though the gains in performance (Fig. 6) are apparent, there are a number of questions left. Specifically, the conventional way of quantifying vacancy concentration was not to be done and how it was changing over a longer period of cycling is not known. Operando characterisation would help to explain the behaviour of these defects under realistic working conditions.

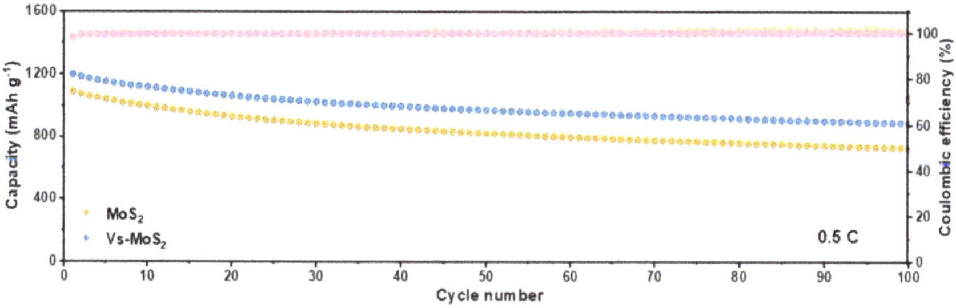

Fig. 6 – Electrochemical perofmance of the samples at 0.5 C: a) MoS_2; b) Vs-MoS_2.

Conclusion

This paper has shown that sulfur vacancy engineering is a good way of optimizing the catalytic properties of MoS_2 to be used in lithium-sulfur battery. Vacancy formation enhances electron transfer, increases attachment to polysulfide intermediates and leads to an increase in effective sulfur conversion in operation. A cell using the vacuity-abundant coating maintained 886 mAh g^{-1} in 100 cycles at 0.5 C, which is better in comparison with its pure MoS_2 counterpart. Additional advancement will require acquiring a more quantitative grasp of the vacancy concentration and stability. Methods that are able to trace the evolution of defects in cycling would be insightful. In future research, other atmospheres, annealing strategies, and co-doping approaches can be investigated to stabilise defect structures and increase the use of defect-engineered TMD catalysts in high-energy sulfur batteries.

Acknowledgements

The Science and Technology Development Fund (FDCT) of Macao SAR provided the funding to this project (0033/2023/ITP1).

References

[1] S. Yari, A. Conde Reis, Q. Pang, and M. Safari: Performance benchmarking and analysis of lithium-sulfur batteries for next-generation cell design. Nat. Commun., 16 (2025) 5473. https://doi.org/10.1038/s41467-025-60528-4

[2] H. Cho, J. Jung, I. Kim, J. Kim, S. Kim, J. Hyun, C.H. Lee, H. Kwack, W. Oh, J. Lee, and H.-T. Kim: Interconvertible and rejuvenated Lewis acidic electrolyte additive for lean electrolyte lithium sulfur batteries. Nat. Commun., 16 (2025) 6805. https://doi.org/10.1038/s41467-025-62169-z

[3] W. X. Hua, T. X. Shang, H. Li, Y. F. Sun, Y. Guo, J. Y. Xia, C. N. Geng, Z. H. Hu, L. K. Peng, Z. Y. Han, C. Zhang, W. Lv, Y. Wan, Optimizing the p charge of S in p-block metal sulfides for sulfur reduction electrocatalysis, Nat. Catal., 6 (2023) 174-184. https://doi.org/10.1038/s41929-023-00912-9

[4] Z. X. Shi, Z. T. Sun, J. S. Cai, X. Z. Yang, C. H. Wei, M. L. Wang, Y. F. Ding, J. Y. Sun, Manipulating Electrocatalytic Li2S Redox via Selective Dual-Defect Engineering for Li-S Batteries, Adv. Mater., 33 (2021) 2103050. https://doi.org/10.1002/adma.202103050

[5] T. Stephenson, Z. Li, B. Olsen, D. Mitlin, Lithium ion battery applications of molybdenum disulfide (MoS2) nanocomposites, Energy Environ. Sci., 7 (2014) 209-231. https://doi.org/10.1039/C3EE42591F

[6] H. B. Lin, S. L. Zhang, T. R. Zhang, H. L. Ye, Q. F. Yao, G. Y. Zheng, J. Y. Lee, Simultaneous Cobalt and Phosphorous Doping of MoS2 for Improved Catalytic Performance on Polysulfide Conversion in Lithium-Sulfur Batteries, Adv. Energy Mater., 9 (2019) 1902096. https://doi.org/10.1002/aenm.201902096

[7] C. H. Zhao, B. Jiang, Y. Huang, X. Sun, M. Wang, Y. Zhang, N. Q. Zhang, Highly active and stable oxygen vacancies via sulfur modification for efficient catalysis in lithium-sulfur batteries, Energy Environ. Sci., 16 (2023) 5490-5499. https://doi.org/10.1039/d3ee01774e

Synthesis of N-Substituted Pyrroles by Citric Acid

Mohd Amin MIR[1a*], Syed Muhammad HASNAIN[2b], Kim ANDREWS[2c], Ercan BURSAL[1d]

[1]Department of Mechanical Engineering, Prince Mohammad Bin Fahd University, AL Khobar, Saudi Arabia

[2]Department of Mathematics & Natural Sciences, Prince Mohammad Bin Fahd University, AL Khobar, Saudi Arabia

[a]mmir@pmu.edu.sa, [b]shasnain@pmu.edu.sa, [c]kandrews@pmu.edu.sa, [d]bursal@pmu.edu.sa

Keywords: Citric Acid, Sustainability, Pyrroles, Amines, Natural Catalyst

Abstract. The Paal-Knorr reaction plays an important role in the synthesis of pyrrole in heterocyclic chemistry; however, conventional procedures can often involve toxic reagents or conditions that are not appropriate for building pyrroles. This study presents an environmentally sustainable route for synthesizing N-substituted pyrroles via the use of citric acid as a catalyst. The use of citric acid's natural acidity and other components, combined with amines the reaction enables moderate cyclisation of amine compounds with 1,4-butanedione to take place without the use of toxic reagents. Because this process uses only renewable resources, has good yield, is reusable and meets green chemistry specifications, it can potentially provide materials that are used in many fields including Materials Science, Agrochemicals and Pharmaceutical.

Introduction

In organic chemistry, pyrrole rings are very important as they are found in many natural chemicals [1], organic molecules, and bio-compounds [2]. Antioxidant, antiviral, antibacterial [3], and anti-inflammatory qualities are specifically found in substituted pyrroles. The Paal–Knorr procedure is among the best processes for pyrrole synthesis. It makes use of 1,4-dicarbonyl compounds and their related compounds with NH_3. Usually, the common conditions are $Bi(OTf)_3$/[bmim]BF4, $AcOH/CH_3OH$, $TiCl_4$ in $C_6H_5CH_3$, $Ti(OPr)_4$ in C_6H_6, p-TsOH in $C_6H_5CH_3$ or C_6H_6, and $Bi(NO_3)_3$ in CH_2Cl_2 [4]. Nevertheless, a number of these methods often have particular drawbacks, such as the use of the metal, dangerous organic solvents, and costly. In order to tackle the challenge of a sustainable environment in chemical research and industry, clean procedures that could eliminate the use of toxic organic solvents and metals are required. Therefore, it is preferable to develop straightforward and eco-friendly methods for pyrrole synthesis. However, reactions in H_2O catches attention due to the fact that it is a cheaper, safe, and non-hazardous solvent [5].

Furthermore, if heterogeneous catalysts that are recyclable and reusable are able to conduct aqueous reactions efficiently, the results are ideal procedures in terms of simplicity and environmental friendliness. The heterocyclic macrocycles pigments, heme, vitamin B_{12}, porphyrins, and chlorophyll all include the widely distributed pyrrole constituent found in nature [6]. Also pyrrole containing drugs are approved by FDA. Examples of other commercially available drugs include atorvastatin, an inhibitor of Hydroxymethylglutaryl-CoA (HMG-CoA) reductase that lowers LDL cholesterol, ketorolac, which is used to treat moderate to severe pain in the short term [7], sunitinib, which is used to treat certain types of cancer [8], and Remdesivir, an antiviral drug used in treating COVID-19 disease [9]. Because pyrole derivatives come in a variety of structures for therapeutic application, numerous researchers have been looking into this skeleton in relation to a range of illnesses and conditions. Pyrroles are important members in medicinal and drug chemistry, hence it is very desirable in modern organic synthesis to synthesize them in simple, efficient, and environmentally friendly ways. The most conventional methods for producing pyrroles are the Hantzsch reaction [10], the Knorr reaction [11], and the Paal-Knorr process [12]. These procedures involve an acid-mediated dehydrative cyclization that condenses -diketones with 10 amines (or ammonia).

Experimental

Chemicals used were of high purity. A 500 MHz Bruker spectrometer was used to study the ^{1}H and ^{13}C NMR spectra with $CDCl_3$ as the solvent and room-temperature tetramethyl silane (TMS) as the standard. Chemical shifts with respect to TMS are expressed in δ, while coupling constants J are expressed in Hz. The EM Silica gel 50 (250–350 mesh) was utilized for column chromatography. All compounds known, with their ^{1}H and ^{13}C NMR spectra characterization data are included in the accompanying documentation.

Results and Discussion

Preparation of n-substituted pyrroles III.

Amine I (1 mmol) and 1,4-butanedione II (1 mmol) were combined with 2.5 mL of citric acid. The reaction solution was shacked at 25^0C. The mixture was shaken well with addition of 4.5 mL of brine and 9.5 mL of CH_2Cl_2. The carbon layer was segregated, dried with Na_2SO_4, and then evaporated under low pressure. Following the purification and observation of product III by TLC and GC-MS analysis, the mixture was cooled at 25°C, diluted in $CH_3CO_2CH_2CH_3$, and washed with brine. The H_2O layer was again extracted using $CH_3CO_2CH_2CH_3$. The organic portion was dried with Na_2SO_4 and vacuum-concentrated to produce the desired result. Following that, silica gel column chromatography (C_6H_{14} / $CH_3CO_2CH_2CH_3$) was used to refine the resulting Product III.

Optimization Reactions

Optimization of Catalyst

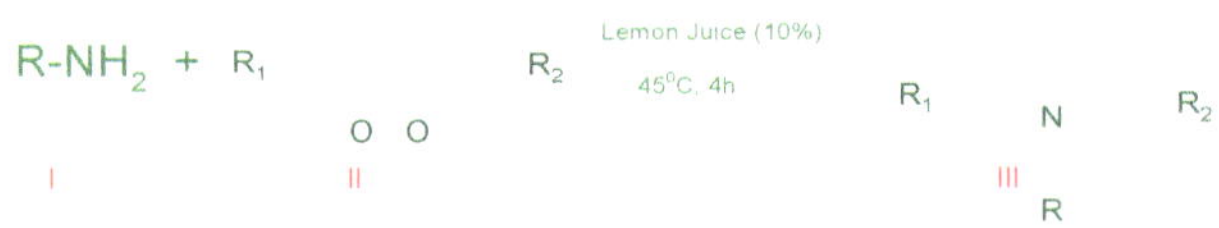

Scheme 1: Showing the reaction of amine I with 1, 4-butanedione II.

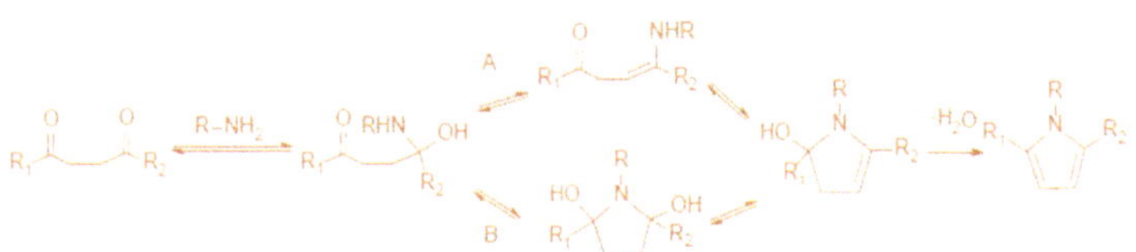

Proposed mechanism for the above reaction

Table 1: Citric Acid-Catalysed Paal-Knorr reaction of 1 with 2 at 30^{0}C

Type of Primary Amine	Citric Acid Volume (mL)	Duration of Reaction (hrs)	Isolated Yield (%)
Aniline	2	5	92
4-Methoxyaniline	2	6	88
Ethylenediamine	2	7	86
Benzylamine	2	10	90
o-Toluidine	2	11	85
m-Toluidine	2	14	87
p-Toluidine	2	17	89
2-Aminophenol	2	19	83
4-Aminobenzoic acid	2	22	88
Cyclohexylamine	2	9	93

The rate of reaction was faster for substances with an electron-donating group than for those with an electron-withdrawing group. Dialdehyde is produced in the reaction media as a result of 1, 4-butanedione being activated by the acidity of citric acid. The process began with a nucleophilic attack on the carbonyl function, followed by dehydration and aromatization.

Temperature and Time Optimization of Reaction

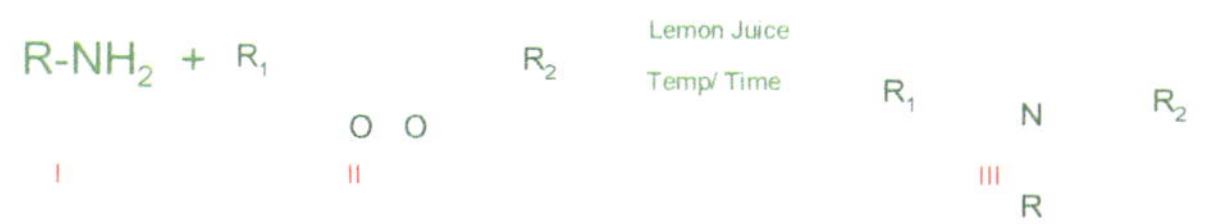

Scheme 2: Optimization of Temperature and Time amine I with 1,4-butanedione II.

Table 2: Showing the effect of temperature and time on the formation of n-substituted pyrroles

Entry No.	Temperature (°C)	Reaction Duration (h)	Isolated Yield of 3a (%)
1	Room Temp.	3	trace
2	52	3	56
3	72	5	44
4	42	3	53
5	38	6	66
6	41	6	69
7	59	8	79
8	62	10	78
9	58	11	87
10	61	19	43

Reaction requirements are Citric Acid (5 mol%), 1,4-butanedione (1 equiv., 1 mmol), and amine (0.5 equiv., 0.5 mmol).

Control Experiments

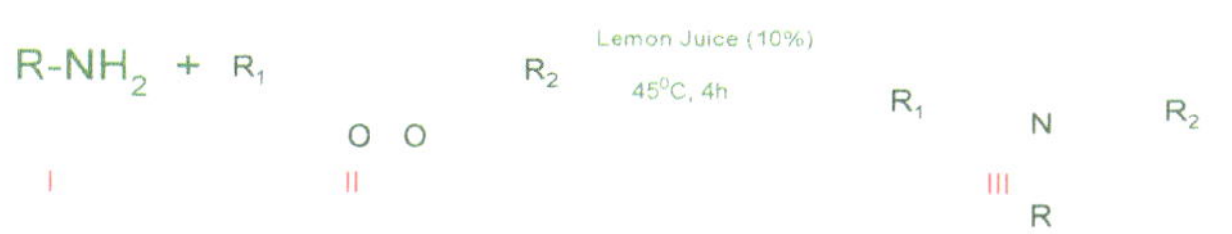

Table 3: Catalyst Concentration effect on the formation of n-substituted pyrroles

Entry No.	Catalyst Loading (y mol %)	Product Yield of 3a (%)
1	21	86
2	14	76
3	11	68
4	6	69
5	3	62

Reaction requirements are Citric Acid (5 mol%), 1,4-butanedione (1 equiv., 1 mmol), and amine (0.5 equiv., 0.5 mmol).

Standard Operating Procedure for N-Substituted Pyrrole Synthesis

3 mol% freshly made catalyst was added to a combination of the corresponding amines with 1,4-butanedione in 5 mL CH_2Cl_2 [13]. For the necessary amount of time, the solution was swirled and refluxed. TLC was used for purification. When the mixture got finished, the solution was made crude under small pressure, and the product was refined by column chromatography with an eluent of petroleum ether/ethyl acetate.

Characterization Data

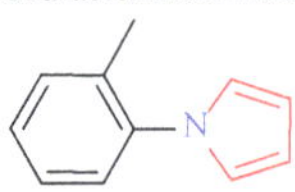

(IIIa) [14]:Solid light yellow. Rf is 0.66 (99% PE). 48–50°C (m.p. 50–51°C). ^{1}H NMR (499 MHz, $CDCl_3$): 597 (s, 2H), 7.26–7.30 (m, 2H), 7.33–7.39 (m, 1H), 7.51–7.55 (m, 2H), δ = 2.12 (s, 6H).

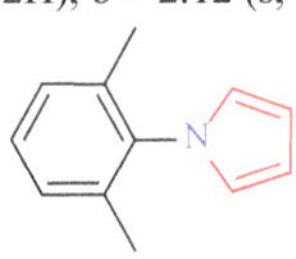

(IIIb) [15]: Yellowish fluid. Rf is 0.43 (99% PE). ^{1}H NMR (500 MHz, $CDCl_3$): δ = 2.19 (s, 7H), 3.93 (s, 4H), 5.99 (s, 2H), 6.89 (d, J = 2.3 Hz, 1H), 6.89–6.98 (m, 1H), 7.06–7.08 (m, 1H), and 7.47 (t, J = 8.1 Hz, 1H).

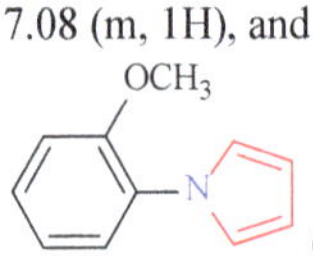

(IIIc) [16]: Yellowish fluid. Rf is 0.19 (99% PE). ^{1}H NMR (499 MHz, $CDCl_3$): δ = 1.99 (s, 5H), 4.00 (s, 3H), 4.01 (s, 2H), 6.02 (s, 2H), 6.77 (d, J = 2.0 Hz, 1H), 7.04 (dd, J = 2.1, 7.9 Hz, 1H), 6.96 (d, J = 9.1 Hz, 1H).

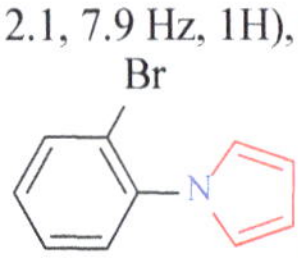

(IIId) [17]: White solid. PE: EA = 9:99; Rf = 0.74. (56–58°C) m.p. 55–57°C. 1.09 (d, J = 5.9 Hz, 11H), 1.94 (s, 6H), 2.01–2.11 (m, 1H), 5.97 (s, 2H), 7.11 (d, J = 7.6 Hz, 2H), and 7.45 (t, J = 7.6 Hz, 1H) are the values for ^{1}H NMR (499 MHz, $CDCl_3$).

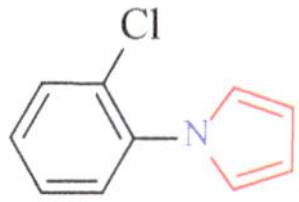

(IIIe) [18]: White solid. PE: EA = 9:98; Rf = 0.67. m.p. between 55 and 56°C. At 500 MHz, $CDCl_3$, ′H NMR: δ = 1.99 (s, 6H), 5.94 (s, 2H), 6.79 (dd, J = 2.2, 7.5 Hz, 2H), and 7.00–7.11 (m, 1H).

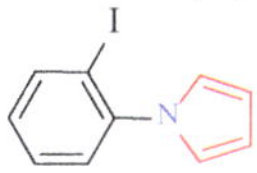

(IIIf) [19]: solid white. Rf is 0.44 (99 percent PE). m.p. 47–49°C (46–49°C in lit). ^{1}H NMR (500 MHz, $CDCl_3$): 594 (s, 2H), 6.98–6.99 (m, 1H), 6.88–6.98 (m, 1H), 7.41–7.45 (m, 2H), δ = 2.06 (s, 6H).

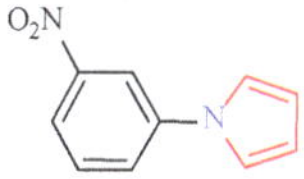

(IIIg) [14]: solid yellow. PE: EA = 10:1; Rf = 0.36. temperature range: 141–144°C (lit. 140–145°C). 500 MHz, $CDCl_3$ ^{1}H NMR: δ = 1.9 (s, 5H), 5.97 (s, 2H), 7.40–7.43 (m, 2H), and 8.36–8.39 (m, 2H).

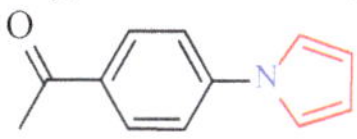

(IIIh) [17]: solid white. Rf is 0.24 (99 percent PE). m.p. 74–75°C, ^{1}H NMR (499 MHz, $CDCl_3$): δ = 1.89 (s, 6H), 3.83 (s, 4H), 5.01 (s, 2H), 6.01 (s, 2H), 7.00–7.11 (m, 4H).

Conclusion

Utilization of natural citric acid as a catalyst in the Paal-Knorr reaction produces N-substituted pyrroles using environmentally friendly (sustainable) conditions. Citric acid's natural acidity and bioactive constituents provide an effective means to produce pyrrole derivatives by facilitating the cyclization of primary amines with 1,4-butanedione in mild, environmentally friendly conditions. This procedure follows the green synthesis guidelines and has the advantage of producing high yields while avoiding harmful reagents and aggressive reaction conditions. The employment of citric acid as a catalyst demonstrates how bio-based materials can create an environmentally sustainable manner of carrying out organic syntheses. Furthermore, this process reduces the environmental impacts of conventional routes and offers a sustainable, cost-effective alternative for synthesizing heterocyclic compounds. The many potential applications of pyrrole derivatives in functional materials, agrochemicals, and pharmaceuticals indicate the possibility of expanding the application of this approach in commercial settings and for developing more environmentally friendly synthetic methods in the future.

References

[1] E. Bellur, P. Langer, Synthesis of functionalized pyrroles and 6,7-dihydro-1H-indol-4(5H)-ones by reaction of 1,3-dicarbonyl compounds with 2-azido-1,1-diethoxyethane, Tetrahedron Lett. 47 (2006) 2151-2154. https://doi.org/10.1016/j.tetlet.2006.01.121

[2] L.B. Snyder, Z. Meng, R.H. Mate, Hantzsch synthesis of pyrazolo[1′,2′:1,2]pyrazolo[3,4-b]pyridines: Partial agonists of the calcitonin receptor, Bioorg. Med. Chem. Lett. 14 (2004) 4735.

[3] G. Dannhardt, W. Kiefer, G. Krämer, S. Maehrlein, Synthesis of some new pyrrole derivatives and their antimicrobial activity, Eur. J. Med. Chem. 35 (2000) 499. https://doi.org/10.1016/S0223-5234(00)00150-1

[4] J.S. Yadav, B.V.S. Reddy, B. Eeshwaraiah, M.K. Gupta, An efficient synthesis of 2-aryl-1,4-diketones via hydroacylation of enones, Tetrahedron Lett. 45 (2004) 5873.
[5] C. Li, Organic reactions in aqueous media with a focus on carbon-carbon bond formations: A decade update, Chem. Rev. 105 (2005) 3095. https://doi.org/10.1021/cr030009u
[6] S. Iqbal, H. Rasheed, R.J. Awan, A. Mukhtar, M.G. Moloney, Recent advances in the synthesis of pyrroles, Curr. Org. Chem. 24 (2020) 1196-1229. https://doi.org/10.2174/1385272824999200528125651
[7] M.M. Buckley, R.N. Brogden, Ketorolac-A review of its pharmacodynamic and pharmacokinetic properties, and therapeutic potential, Drugs 39 (1990) 86-109. https://doi.org/10.2165/00003495-199039010-00008
[8] M. Moran, D. Nickens, K. Adcock, M. Bennetts, A. Desscan, N. Charnley, K. Fife, Sunitinib for metastatic renal cell carcinoma: A systematic review and meta-analysis of real-world and clinical trials data, Target Oncol. 14 (2019) 405-416. https://doi.org/10.1007/s11523-019-00653-5
[9] R.T. Eastman, J.S. Roth, K.R. Brimacombe, A. Simeonov, M. Shen, S. Patnaik, M.D. Hall, Remdesivir: A review of its discovery and development leading to emergency use authorization for treatment of COVID-19, ACS Cent. Sci. 6 (2020) 672-683. https://doi.org/10.1021/acscentsci.0c00489
[10] V. Arkley, J. Attenburrow, G.I. Gregory, T. Walker, Griseofulvin analogues: Part I. Modification of the aromatic ring, J. Chem. Soc. (1962) 1260-1268. https://doi.org/10.1039/jr9620001260
[11] V.S.V. Satyanarayana, A. Sivakumar, Ultrasound-assisted synthesis of 2,5-dimethyl-N-substituted pyrroles catalyzed by uranyl nitrate hexahydrate, Ultrason. Sonochem. 18 (2011) 917-922. https://doi.org/10.1016/j.ultsonch.2011.02.007
[12] H. Lee, B.H. Kim, Indium-mediated one-pot pyrrole synthesis from nitrobenzenes and 1,4-diketones, Tetrahedron 69 (2013) 6698-6708. https://doi.org/10.1016/j.tet.2013.05.113
[13] M.A. Mir, S.M. Hasnain, K. Andrews, U.M. Mir, Novel synthesis of N-substituted pyrroles by green catalyst: Apple juice, Vietnam J. Chem. 1 (2025). https://doi.org/10.1002/vjch.70030
[14] J.X. Chen, X.L. Yang, M.C. Liu, H.Y. Wu, J.C. Ding, W.K. Su, Approach to synthesis of β-enamino ketones and pyrroles catalyzed by gallium(III) triflate under solvent-free conditions, Synth. Commun. 39 (2009) 4180-4198. https://doi.org/10.1080/00397910902898528
[15] R.C. Thomas, T. Poel, M.R. Barbachyn, M.F. Gordeev, G.W. Luehr, A. Renslo, U. Singh, V. Josyula, Preparation of 3-aryl-2-oxo-5-oxazolidinecarboxamides and analogs as antibacterial agents, US Patent 20040147760 (2004).
[16] A. Jafari, S. Amini, F. Tamaddon, A green, chemoselective, and efficient protocol for Paal-Knorr pyrrole and bispyrrole synthesis using biodegradable polymeric catalyst PEG-SO_3H in water, J. Appl. Polym. Sci. 125 (2012) 1339-1345. https://doi.org/10.1002/app.35172
[17] M. Leonardi, V. Estévez, M. Villacampa, J.C. Menéndez, The Hantzsch pyrrole synthesis: Non-conventional variations and applications of a neglected classical reaction, Synthesis 51 (2019) 816-828. https://doi.org/10.1055/s-0037-1610320
[18] N.A. Magnus, M.A. Staszak, Synthesis of 4-cyano pyrroles via mild Knorr reactions with β-ketonitriles, Org. Process Res. Dev. 10 (2006) 899-904. https://doi.org/10.1021/op060104f
[19] Balakrishna, A. Aguiar, P.J.M. Sobral, M.Y. Wani, Paal-Knorr synthesis of pyrroles: From conventional to green synthesis, Catal. Rev. 61 (2019) 84-110. https://doi.org/10.1080/01614940.2018.1529932

Materials Research Proceedings 64 (2026) 591-598 https://doi.org/10.21741/9781644904091-74

Lead-Free Perovskite Solar Cells: The Impact of Se–Te–Cu_2O as HTL

Syed M. HASNAIN[1,a *], M. Amin. MIR[2,b], Irfan QASIM[3,c], and Abid IQBAL[4,d]

[1]Department of Mathematics and Natural Sciences, Prince Mohammad Bin Fahd University, Al Kobar, Saudi Arabia

[2]Department of Mechanical Engineering, Prince Mohammad Bin Fahd University, Al Kobar, Saudi Arabia

[3]Department of Physics, Faculty of Sciences, Rawalpindi Women University, Rawalpindi, Pakistan

[4]Department of Computer Engineering, King Faisal University, Al-Ahsa, Saudi Arabia

[a]shasnain@pmu.edu.sa, [b]mmir@pmu.edu.sa, [c]dr.irfanqasim@gmail.com, [d]aaiqbal@kfu.edu.sa

Keywords: Solar Cells, Se–Te–Cu_2O, $RbGeI_3$, $MASnI_3$, SCAPS-1D

Abstract. Using SCAPS-1D, a numerical study on a new perovskite double-absorber solar cell with structure FTO/CdS/$RbGeI_3$/$MASnI_3$/Se–Te–Cu_2O/Au is conducted. Most importantly, the Se–Te–Cu_2O HTL, providing efficient hole extraction and minimizing recombination losses at the back contact, plays a key role. The thickness of the hole transport layer achieved was optimized for low series resistance and high hole mobility. The optimized configuration gave a power conversion efficiency of 33.22%, with V_{oc} of 0.748 V, J_{sc} of 58.42 mA cm^{-2}, and a fill factor of 76.81%. The Se–Te–Cu_2O hole transport layer aided charge extraction, good thermal stability from 260–340 K, excellent external quantum efficiency with >95% in the 300–900 nm range. The results show that advanced hole transport layer materials can significantly contribute the development of higher efficiency lead-free perovskite solar cells.

Introduction

The energy generation/storage/climate change challenges the world is facing today are global challenges that affect us all. The depletion of fossil fuel reserves, rising CO_2 emissions, and need for sustainable development are accelerating the renewed global shift towards renewable energy generation, and solar energy is fast emerging as a leading solution [1]. Solar energy is strong contender and competing against non-renewable resources because it is available naturally in abundance, has little detrimental impact on the environment and, has the potential to provide as much as 20% of global energy supply [2]. Furthermore, unlike other renewable energy sources, solar energy can be harnessed almost anywhere on earth with little or no detrimental ecological footprint [3].

In the last thirty years, research in solar energy development has been largely focused on the improvements in and development of semiconductor solar cells [4]. Silicon has been the material of choice for decades now due to its established properties and more reliable performance [5]. Nevertheless, the potential costs of producing ultra-pure silicon continues to be a barrier for the broad adoption of silicon-based solar technology. Consequently, many researchers are exploring alternative types of materials such as, perovskite structured compounds, as a potential pathway to more efficient, cost effective and affordable routes to solar energy.

Organic-inorganic perovskites are very attractive materials for photovoltaic and x-ray detection applications due to their strong and tunable light absorption, long carrier diffusion lengths, and relatively simple fabrication methods [6, 7]. Since the introduction of perovskites into solar cells by Kojima et al., with a relatively new power conversion efficiency (PCE) of 3.8% [8], a significant amount of material engineering applied to perovskites has led to improvements in PCE, including a recent record of approximately 25.5% [9]. Despite recent advances, research is guided by the fact that lead is a toxic constituent [10] to explore non-toxic alternatives. At the forefront are iterated focused efforts on perovskites formed with tin (Sn^{2+}) and germanium (Ge^{2+}) [11].

Of these, $CsGeI_3$ and $RbGeI_3$ have been developed extensively as potential lead-free solar energy candidates. Nevertheless, as reported by Krishnamoorthy et al. [12], the proclivity of Ge^{2+} to oxidize during fabrication limits performance after device fabrication in lead-free solar energy applications. Similarly, Krishnamoorthy et al. reported a very low 0.11% PCE for $CsGeI_3$ based solar cells which highlighted this challenge in practical applications. In addition, Jayan et al. have already reported in depth investigations of $RbGeI_3$ demonstrated, but not limited to, optoelectronic, structural and thermodynamic properties [13]. Most recently Khusheed et al. reported the electrical and magnetic properties of $RbGeI_3$ [14, 15]. Finally Jong et al. investigated $RbGeI_3$ dielectric constants, charge carrier effective masses and lattice stability [16]. Jongs et al. showed a tolerance factor of 0.90 for $RbGeI_3$. A few simulation studies have been performed, as shown in computations from Theile et al., they predict $RbGeI_3$ perovskites have the potential to attain a PCE of 10.11%, a fill factor of 63.68% and quantum efficiency of 90.3% in the visible range [17]. Most recently, Alipour et al. reported high performance metrics for $FASnI_3$ based solar cells J_{sc} = 24.65 mA/cm^2, V_{oc} = 1.12 V, FF = 86.02%, and an overall performance at PCE= 23.69% and these metrics will be gradually replicated by further investigations, similar to lead halide perovskite formulations as potential sustainable energy solution options.

Proposed design

The study introduces a new type of perovskite solar cell, configured with the architecture FTO/CdS/$RbGeI_3$/$MASnI_3$/Se–Te–Cu_2O/Au, that we investigated through numerical and simulation processes informing output reliability for high efficiency. The device has five layers, illustrated in Fig. 1, including an FTO front contact, a CdS ETL, two perovskite absorbers ($RbGeI_3$ and $MASnI_3$), an Se–Te–Cu_2O HTL, and gold back contact. This arrangement of the solar cell was done to optimize light-harvesting and charge separation and transport. The Se–Te–Cu_2O HTL is efficient for hole transport with energy band gap of 1.88 eV and electron blockade layer providing favorable band offsets with the two absorbers while permitting easy hole transport. The band energy alignment for energy levels positioned favourable aligned energy flow of the electron to the ETL and the hole to the HTL, which minimizes recombination losses and contributes to improved efficiency within the solar cell.

Materials Research Proceedings 64 (2026) 591-598 https://doi.org/10.21741/9781644904091-74

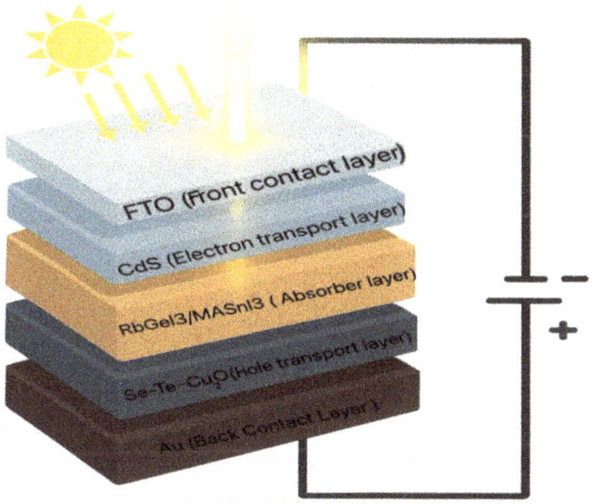

Fig. 1 Device architecture of the proposed perovskite solar cell.

Results and discussion

As shown in Fig. 2, increasing the thickness of the hole transport layer (HTL) contributes positively to the photovoltaic performance of the proposed solar cell. This improvement is due to better surface passivation at the absorber/HTL interface, which minimizes surface defect states and lowers non-radiative recombination losses. Better charge carrier separation, and extraction results in significantly higher fill factor (FF) and power conversion efficiency (PCE) of the device. From this optimization work, the optimum sample thickness of the HTL thickness was determined to be 0.12 μm, where sufficient hole transport is subsequent in connection to the minimized resistive losses. The improvement in PCE is evident as the HTL thickness beyond 0.12 μm decreases the performance of the device. This performance degradation is due to the increase in series resistance when the thickness increased, which reduces carrier mobility and charge collection efficiency. Furthermore, if the HTL is excessively thick, the electric field distribution will negatively diminish the charge generation efficiency through carrier accumulation and recombination.Therefore, HTL thickness tuning is necessary to provide the optimal balance towards increasing th e hole transport and minimizing parasitic losses. This balance is essential for increasing the photovoltaic performance and the practical usability of the proposed device architecture.

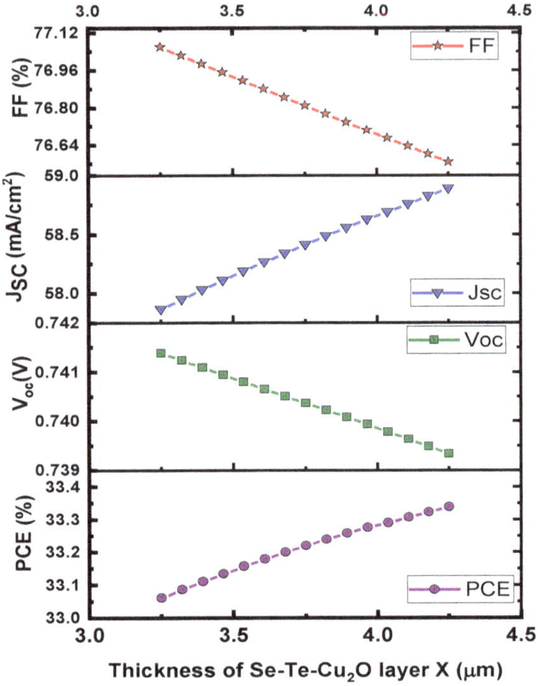

Fig. 2 Impact of Se–Te–Cu_2O layer thickness on the electrical performance of the proposed perovskite solar cell.

The simulated current density–voltage (J–V) curve for the optimized double-absorber PERC PV device is shown in Fig. 3 under standard test conditions. The J–V curve provides a good perspective on the photovoltaic performance of the device, showing the performance of each functional layer. The J–V curve emphasizes the HTL (Se–Te–Cu_2O) for achieving the J–V characteristics in this simulation. The Se–Te–Cu_2O O HTL has a bandgap of 1.88 eV and closely relates to ensuring there is selective hole extraction while blocking electrons from recombining with the back contact. The selective transport mechanism reduces recombination losses at the absorber–HTL interface and is an important factor that leads to an increase in V_{oc} (0.748 V) in this simulation. The optimized thickness (0.12 μm) of the HTL layer is appropriate to cover and passivate the underlying absorber layers, improving carrier lifetime and built-in potential. The short circuit current density (J_{sc}) of 58.42mA/cm^2 represents the cumulative effect of excellent photon absorption by the $RbGeI_3$ and $MASnI_3$ layers in addition to the ability of the HTL to enable efficient hole transport while suppressing back electron flow. The fill factor (FF) of 76.81% depicts the extent to which the HTL has improved device performance by minimizing resistive losses due to voltage drop, and maximizing the electric field for efficient charge collection. Through these optimized interfacial properties and transport properties, the device achieved a respectable power conversion efficiency (PCE) of 33.22%. The HTL played a central role in achieving this high PCE by minimizing defects at the interface, achieving balanced hole mobility, and profiting from ideal band alignment. In conclusion, based on the J-V data, the combination of the Se–Te–Cu_2O HTL plays an essential role in enhancing the overall performance of the defined perovskite solar cell, it contributed to favorable hole extraction, minimizing recombination, and encouraging effective charge transport and composite structure are essential components for producing high efficiency, lead-free perovskite solar cell development.

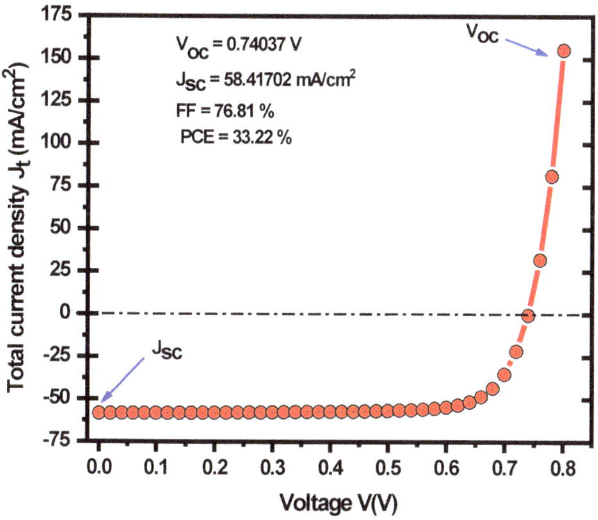

Fig. 3 Current–Voltage characteristics of the optimized solar cell.

Fig. 4 shows the simulated temperature-dependent behavior of the major photovoltaic parameters of an optimized solar cell over the temperature range of 260 K to 340 K. The short-circuit current density (J_{sc}) increased slowly with temperature, moving from around 58.10 mA/cm^2 at 260 K to about 58.45 mA/cm^2 at 340 K. This increase is largely due to (i) the reduction in bandgaps of the perovskite absorber allowing longer-wavelength photons to be absorbed and (ii) higher carrier generation and carrier mobility at temperatures above 300 K. On the other hand, the open-circuit voltage (V_{oc}) dropped nearly linearly from around 0.79 V at lower temperatures to around 0.68 V at 340 K. The drop in V_{oc} is a result of two factors: (i) the thermally activated increase of the diode saturation current as temperature increases, and (ii) the increase in non-radiative recombination mechanisms. Both of these factors are more severe at higher temperatures as the thermal excitation of charge carriers increases. The fill factor (FF) exhibits a non-monotonic response to temperature. Initially, as the temperature rises FF improves to a maximum (around 295 K). The improvement in FF follows from a reduction in series resistance and improved charge extraction efficiency which temporarily overrides the losses due to recombination. When the temperature reaches around ~295 K, a further increase in temperature leads to a decrease in FF due to an increased likelihood of recombination losses and resistive losses in the transport layers (particularly in the case of the HTL and the surface interfaces). In turn, the power conversion efficiency (PCE) shows a similar trend as FF, once again reaching a maximum value near room temperature (≈300 K) and then dropping off at higher temperatures. This demonstrates the delicate balance between any advantages associated with increased photocurrent, against the reductions in V_{oc} and FF affecting overall device performance at the higher temperature. Ultimately these results underscore the need for the thermal stability of the HTL and related interfaces to be partially guarded in operational conditions through an interfacial optimization strategy to maintain high-efficiency value across operational temperatures.

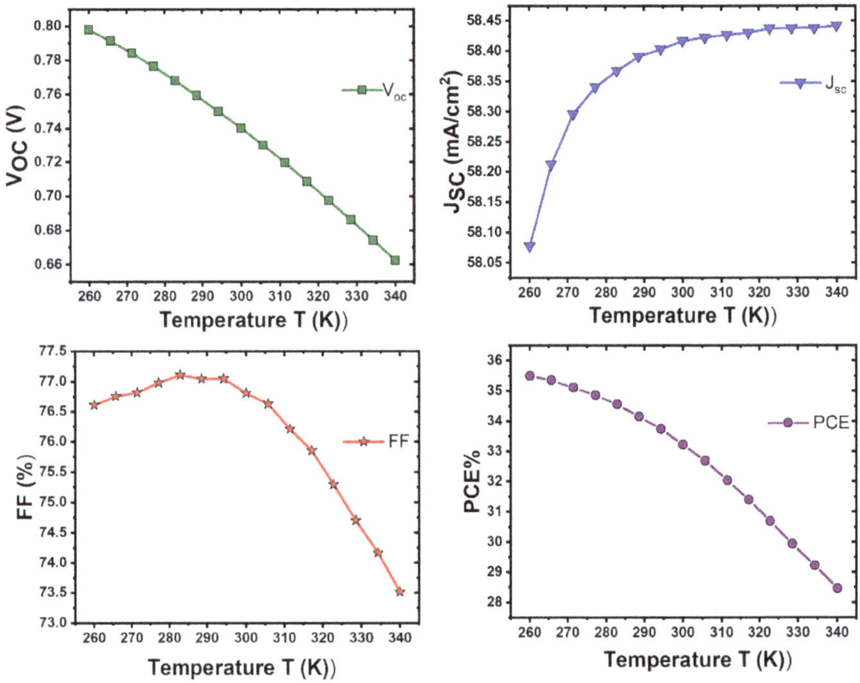

Fig. 4 The influence of temperature variations on the electrical performance characteristics of the proposed perovskite solar cell.

Conclusion

This study provides evidence through detailed SCAPS-1D simulations that the optimal all-inorganic double-absorber perovskite solar cell, having the structure FTO/CdS/RbGeI 3 /MASnI 3 /Se–Te–Cu 2 O/Au, can deliver a power conversion efficiency of 33.22% under AM 1.5G APD conditions. The important part of this high performance is the proper optimization of the hole-transport layer (HTL) based on Se–Te–Cu 2 O, which is sufficiently thin enough to extract holes at HTL/absorber interface and ideally blocks electrons. The HTLs optimized doping characteristics and optimized interface quality minimizes recombination losses at the absorber/HTL interface and contribute to increase in fill factor and open-circuit voltage. Further, the HTL enables stable device operation across a broader temperature range (260–340 K) and higher modulation frequencies, demonstrating its role in increasing thermal and electrical stability of the perovskite solar cells. The tandem absorber(s) architecture, aided by precise HTL engineering, enables superior charge transport and extraction compared to a single-absorber device. This work shows the importance of the HTL in ensuring maximum device performance and also provides fundamental information for researchers interested in developing stable, lead-free, high performance perovskite solar cells.

References

[1] K. Yoshikawa et al., "Silicon heterojunction solar cell with interdigitated back contacts for a photoconversion efficiency over 26%," Nature energy, vol. 2, no. 5, pp. 1-8, 2017. https://doi.org/10.1038/nenergy.2017.32

[2] S. M. Hasnain, "Examining the advances, obstacles, and achievements of tin-based perovskite solar cells: a review," Solar Energy, vol. 262, p. 111825, 2023. https://doi.org/10.1016/j.solener.2023.111825

[3] J. Liu et al., "Synthesis and optical applications of low dimensional metal-halide perovskites," Nanotechnology, vol. 31, no. 15, p. 152002, 2020. https://doi.org/10.1088/1361-6528/ab5a19

[4] S. M. Hasnain et al., "Performance evaluation of organometal halide MASnI3 and inorganic BaZrS3 hybrids in perovskites solar cells: Theoretical approach," Hybrid Advances, vol. 9, p. 100408, 2025/06/01/ 2025. https://doi.org/10.1016/j.hybadv.2025.100408

[5] O. Ahmad et al., "Modelling and numerical simulations of eco-friendly double absorber solar cell "Spiro-OmeTAD/CIGS/MASnI3/CdS/ZnO" and its PV-module," Organic Electronics, vol. 117, p. 106781, 2023. https://doi.org/10.1016/j.orgel.2023.106781

[6] I. Qasim et al., "Design and numerical investigations of eco-friendly, non-toxic (Au/CuSCN/CH3NH3SnI3/CdTe/ZnO/ITO) perovskite solar cell and module," Solar Energy, vol. 237, pp. 52-61, 2022. https://doi.org/10.1016/j.solener.2022.02.056

[7] A. Kojima, K. Teshima, Y. Shirai, and T. Miyasaka, "Organometal halide perovskites as visible-light sensitizers for photovoltaic cells," Journal of the american chemical society, vol. 131, no. 17, pp. 6050-6051, 2009. https://doi.org/10.1021/ja809598r

[8] J. Wu et al., "Using hysteresis to predict the charge recombination properties of perovskite solar cells," Journal of Materials Chemistry A, vol. 9, no. 10, pp. 6382-6392, 2021. https://doi.org/10.1039/D0TA12046D

[9] J. Dou, Y. Bai, and Q. Chen, "Challenges of lead leakage in perovskite solar cells," Materials Chemistry Frontiers, vol. 6, no. 19, pp. 2779-2789, 2022. https://doi.org/10.1039/D2QM00632D

[10] W. Ke and M. G. Kanatzidis, "Prospects for low-toxicity lead-free perovskite solar cells," Nature communications, vol. 10, no. 1, p. 965, 2019. https://doi.org/10.1038/s41467-019-08918-3

[11] D. Sabba et al., "Impact of anionic Br-substitution on open circuit voltage in lead free perovskite (CsSnI3-xBr x) solar cells," The journal of physical chemistry C, vol. 119, no. 4, pp. 1763-1767, 2015. https://doi.org/10.1021/jp5126624

[12] K. D. Jayan and V. Sebastian, "Ab initio DFT determination of structural, mechanical, optoelectronic, thermoelectric and thermodynamic properties of RbGeI3 inorganic perovskite for different exchange-correlation functionals," Materials Today Communications, vol. 28, p. 102650, 2021. https://doi.org/10.1016/j.mtcomm.2021.102650

[13] K. A. Parrey, T. Farooq, S. A. Khandy, U. Farooq, and A. Gupta, "First principle studies on structure, magneto-electronic and elastic properties of photovoltaic semiconductor halide (RbGeI3) and ferromagnetic half metal oxide (RbDyO3)," Computational Condensed Matter, vol. 19, p. e00381, 2019. https://doi.org/10.1016/j.cocom.2019.e00381

[14] G. Thiele, H. Rotter, and K. Schmidt, "Die Kristallstrukturen und phasentransformationen des tetramorphen RbGeI3," Zeitschrift für anorganische und allgemeine Chemie, vol. 571, no. 1, pp. 60-68, 1989. https://doi.org/10.1002/zaac.19895710106

[15] S. M. Hasnain, "SCAPS-1D Simulation of Lead-Free BaZrS3 Perovskite as an Ecofriendly Absorber Material for High-Performance Solar Cells," Journal of Electrical Engineering & Technology, 2025/08/30 2025. https://doi.org/10.1007/s42835-025-02424-x

[16] G. Pindolia, S. M. Shinde, and P. K. Jha, "Optimization of an inorganic lead free RbGeI3 based perovskite solar cell by SCAPS-1D simulation," Solar Energy, vol. 236, pp. 802-821, 2022. https://doi.org/10.1016/j.solener.2022.03.053

[17] A. Alipour and H. Alipour, "Device modeling of high performance and eco-friendly FAMASn3 based perovskite solar cell," Scientific Reports, vol. 14, no. 1, p. 15427, 2024/07/04 2024. https://doi.org/10.1038/s41598-024-66485-0

Materials Research Proceedings 64 (2026) 599-606 https://doi.org/10.21741/9781644904091-75

CFD-Based Transient Numerical Analysis of Latent Heat Storage in a Porous Packed Bed Using Local Thermal Non-Equilibrium Model

Lahcen EL-MAHAOUCHI[1,a], Mourad YESSEF[2,b *], Hamza EL HAFDAOUI[3,c], Ahmed FAIZE[2,d], Elhadi BAGHAZ[4,e], Ahmed LAGRIOUI[1,f]

[1]LSMI Laboratory, ENSAM, Moulay Ismail University, Meknes 50000, Morocco

[2]Higher School of Technology of Nador, Mohammed First University, Nador 62000, Morocco

[3]School of Science and Engineering, Al Akhawayn University, Ifrane 53000, Morocco

[4]Faculty of Sciences, Chouaïb Doukkali University, El Jadida 24000, Morocco

[a]l.elmahaouchi@edu.umi.ac.ma, [b]mourad.yessef@usmba.ac.ma, [c]h.elhafdaoui@aui.ma, [d]a.faize@ump.ac.ma, [e]baghaz.e@ucd.ac.ma, [f]lagrioui71@gmail.com

Keywords: Thermal Energy Storage Systems, Latent Heat Storage, Phase Change Material – PCM, Packed Bed Heat Storage Technologies, CFD, Local Thermal Non-Equilibrium -LTNE

Abstract. Thermal energy storage (TES) technologies are attracting increasing interest from energy storage researchers and professionals. They enable energy to be stored effectively and the difference between supply and demand to be corrected for longer periods and at lower cost. This paper reports on the numerical modeling and thermal behavior analyzing of a latent packed bed thermal energy storage tank using spherical capsules of paraffin as storage medium (phase change material – PCM) and water as heat transfer fluid (HTF). The TES tank was modeled, in a 2D axisymmetric geometry, and simulated using COMSOL MULTIPHYSICS, a computational fluid dynamics (CFD) tool. The transient thermal behaviors for HTF and PCM were solved separately using Local Thermal Non-Equilibrium (LTNE) approach. The results were compared with the experimental data and showed a close agreement, assuring the reliability of the present model. The temperature profiles, the thermocline behavior and the state of charge of the TES tank, during the charging process, were presented and discussed, which constitute key parameters to analyze the TES system's thermal behavior and optimize its performances.

Introduction

Global energy demand is growing steeply due to numerous factors such as rapid population growth, high levels of prosperity, and increasing industrialization [1]. The International Energy Agency predicts that global energy consumption will double by 2050 [2], which puts enormous pressure on energy infrastructures and highlights the immediate need for sustainable and effective solutions. In this context, energy storage technologies are crucial to balance supply and demand and facilitate the transition to sustainable energy [3]. With various possibilities for energy storage, Thermal Energy Storage (TES) has attracted increasing interest as it provides stable, reliable, and economical energy [4]. TES technologies are divided into sensible, latent and thermochemical storage technologies [5]. Latent heat storage (LHS) is considered the viable method due to the light weight and small size of the systems, which leads to relatively lower costs [4]. LHS systems use phase change materials (PCMs) to store and release energy as latent heat during an isothermal phase transition. Liquid-solid PCMs is subject to cycles of melting and solidification during the phase transition. In addition, organic PCMs are the most widely used, as they are widely available and eco-friendly [6]. Paraffins are organic PCMs commonly used in thermal storage systems due to their high melting latent heat and good thermal stability during heat charging and discharging cycles [7].

On the other hand, packed bed TES systems have gained considerable attention due to their large specific area and improved heat transfer between the heat transfer fluid and the storage medium on the one hand, and their wide range of industrial applications, particularly in waste heat recovery and solar thermal energy collection on the other [8][9]. Packed bed latent TES system used a single tank partially filled with a PCM (porous media) with a heat transfer fluid circulating through the bed to enable heat energy charge and discharge cycles [10]. In these TES systems, there are multiple complex physical phenomena, involving flow, heat transfer in its three forms: conduction, convection, and radiation, as well as pressure drop and mass transfer [11]. Therefore, physical and numerical modeling of these systems plays a critical role in understanding their operating principle and predicting the different operational, geometrical, and thermophysical parameters, influence their thermodynamic behavior, in order to optimize their performance and efficiency under various operational conditions [12].

This research focused on simulating and analyzing the thermal performance of a latent packed bed TES system using COMSOL MULTIPHYSICS software. The primary contribution is the rigorous experimental verification of the LTNE numerical model for packed bed latent heat storage system. This work quantifies the model's accuracy against experimental data, providing a verified tool capable of correctly predicting the complex transient behavior, PCM phase transition and thermocline degradation. The study was conducted during the heat charging process, where hot water enters the TES tank from the top at a constant mass flow rate of 0.2 l/min and a constant temperature of 90 °C. The transient temperatures distribution, the thermocline behavior and the state of charge were presented and discussed.

This paper is organized as follows: methodology including the model description and numerical validation, results and discussion, finally, conclusion.

Methodology

-Physical model: The packed bed TES system under study is schematically illustrated in Fig. 1. It consists of a stainless-steel cylindrical tank with a diameter of D_T = 0.36 m and a height of H_T = 0.47 m, and insulated with 0.05 m thick glass wool. The TES tank contains a single inlet with a diameter of 0.1 m for hot water (heat transfer fluid) at the top and a single outlet with a diameter of 0.1 m at the bottom, it is partially filled with spherical Paraffin (PCM) particles, as heat storage medium, with a diameter of 55 mm. It is important to highlight that water acts also as sensible heat storage (SHS) media.

The TES system was studied during the heat charging phase; the hot water enters the tank from the top at a constant temperature T_{in} = 90 °C and a constant mass flow rate Q_{in} = 0.2 l/min, transferring heat to PCM that are initially at a temperature of T_0 = 25 °C. The heat exchange with the external environment has been taken into account, ambient temperature is T_{amb} = 20 °C and convective heat transfer coefficient is 5 W/m^2 K.

The packed bed system was modeled based on the following assumptions: (1) the computational domain was axisymmetric, transient, and two- dimensional; (2) the water flow was axial and one-dimensional; (3) the packed bed was considered a continuous and homogeneous porous medium, composed of spherical PCM with a constant diameter.

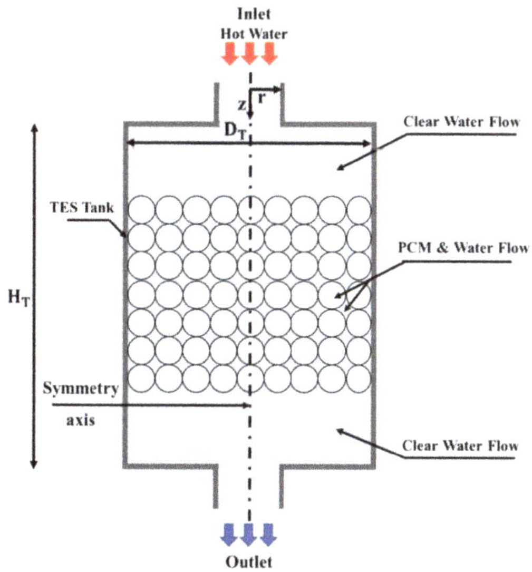

Fig. 1. Schematic of the TES tank under study

The thermophysical characteristics of paraffin (PCM) and water are shown in Table 1, while Table 2. shows the parameters of the TES tank.

Table 1. Thermophysical characteristics of PCM and water [13], [14]

Property	Value
Density of PCM (kg/m3)	861 (Solid) 778 (Liquid)
Specific heat capacity of PCM (J/Kg/K)	1850 (Solid) 2384 (Liquid)
Thermal conductivity of PCM (W/m/K)	0.4 (Solid) 0.15 (Liquid)
Melting temperature of PCM (°C)	60
Latent heat of fusion of PCM (kJ/kg)	213
Density of water (kg/m3)	997
Specific heat capacity of water (J/Kg/K)	4179
Thermal conductivity of water (W/m/K)	0.613

Table 2. Parameters of the TES tank

Parameter	Value
TES Tank diameter D_T (m)	**0.36**
TES Tank height H_T (m)	**0.47**
PCM particles diameter, dp (m)	**0.055**
Bed porosity, ε	**0.49**

-Numerical Model: The numerical model of the TES system under study was performed using Comsol Multiphysics CFD software based on discretizing the governing equations and solving them using the finite element method (FEM), it was chosen for its powerful ability to model multiple complex physical phenomena simultaneously, optimize designs, solve problems early in the development cycle, and visualize results in 3D. Two-dimensional (2D) axisymmetric geometry was adopted and the Local Thermal Non-Equilibrium (LTNE) approach was used to solve the transient heat transfer behavior in HTF and PCM separately.

Moreover, to accurately solve the physical problems (fluid flow and heat transfer) and ensure the numerical stability of the model, a finer mesh was used, with 8314 elements of different types (triangles, quads, edges, vertex) and the convergence criterion was set at a residual of 10^{-4}.

Model validation

The present model was validated by comparing its numerical solution with experimental results found in Nallusamy et al. [13]. The experiment consisted of studying the performance of a latent packed bed TES tank integrated with a flat solar collector. The authors used a mass flow rate of 6kg/min. However, in the simulation, this rate resulted in a faster charging process than in the experiment. This difference is due to the large size (55mm) of the PCM particles, which creates a significant internal thermal resistance that the simplified model does not fully capture. To correct this, the model was calibrated using an effective flow rate of 2.33 kg/min. Fig. 2. shows the comparison between the experimental temperatures and the simulated temperatures obtained. Experimental and simulated temperatures were recorded at different axial positions, however they showed negligible variation. In order to avoid confusion and overloading the figure, a single representative point (mid-height) was chosen for numerical validation of the current model. Here, time is expressed in minutes on the x-axis to match the original experimental data. The figure indicates a good agreement between the numerical and the experimental results. A slight difference can be observed, especially in the early stages, which can be explained by not taking into account the phenomenon of radiative heat transfer and internal conduction in the current model, and some parameters that are not specified in the experiment, such as the inlet and outlet diameters of the TES tank.

Materials Research Proceedings 64 (2026) 599-606
https://doi.org/10.21741/9781644904091-75

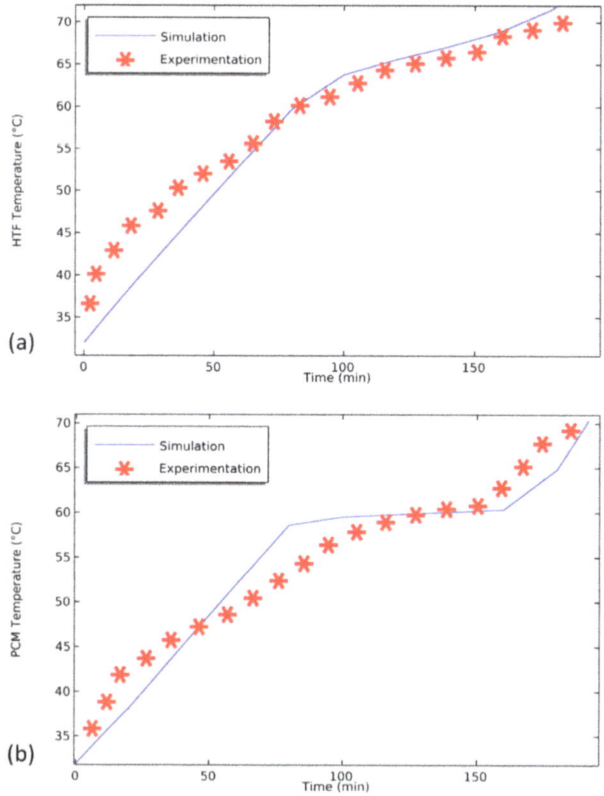

Fig. 2. Comparison between numerical and experimental results: (a) HTF, (b) PCM

Results and discussion

The simulation results are given in Fig. 3, Fig. 4 and Fig. 5. Fig. 3 shows the 2D transient temperatures for water (HTF) and paraffin capsules (PCM) within the TES tank during the heat charge process at the porous medium, for a mass flow rate of 0.2 l/min, inlet temperature of 90 °C and porosity of 0.49. As observed, in the first two hours of the heat charging process, the temperatures of water and PCM at the top level of the tank is 90 °C, while it is significantly lower in the lower levels, approaching 25 °C (initial temperature) at the outlet of the TES tank. The figure indicates that the PCM bed is fully charged after approximately 8 hours. In addition, during the heat charge process, a thin thermal transition zone develops and separates the hot region (red) and the cold region (blue) and moves vertically within the TES tank. It starts out thicker and becomes thinner over time. This thermal gradient is called the thermocline, which is a key parameter that provides insight into heat storage quality and the degree of thermal stratification. The thinner it is, the more efficient the heat storage.

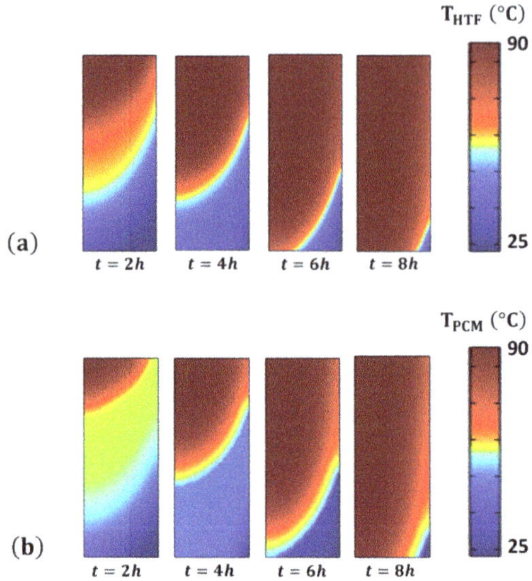

Fig. 3. Transient temperature distribution in the porous region, (a) T_{HTF}, (b) T_{PCM}

Fig. **4** presents the one-dimensional temperature distribution over time for the HTF and PCM at three axial positions, in the top (level 1), center (level 2) and bottom (level 3) parts of the porous region of the TES tank. As observed, level 1 presents a sharp rise in temperatures (Tf and Tpcm) as the early initiation of the PCM melting (within the first 2h), while level 2 and 3 present a delayed increase in temperatures and slower PCM melting. In addition, the HTF temperature is always higher than the PCM temperature at each level throughout the entire heat charging process, which is to be expected because the fluid enters the TES tank at a high temperature and transfers heat energy to the PCM medium. This latter, which has higher thermal inertia, takes longer to absorb thermal energy.

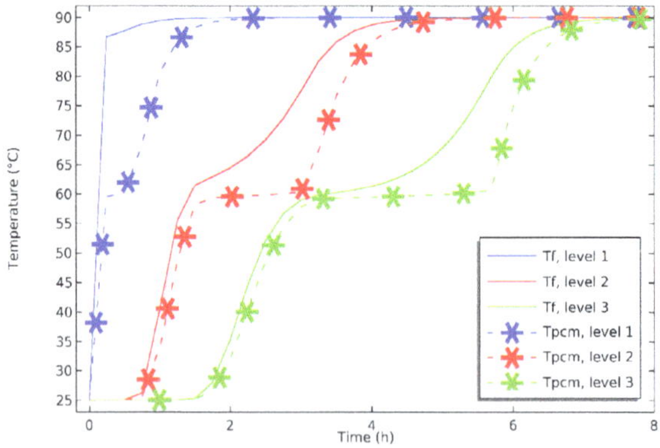

Fig. 4. The axial transient temperatures at different levels inside the TES tank

Fig. 5. illustrate the state of charge inside the TES tank, which is the ratio between the heat stored energy and the maximum heat energy that can be stored in the PCM capsules. Here the curves present the three positions (level 1, 2 and 3) presented previously in the Fig. 4. As shown, the state of charge follows the same trend as PCM temperature profiles showed in Fig. 4. It increases significantly up to ≈ 52%, after which it undergoes a slow increase of ≈ 2%, reflecting the PCM melting phase.

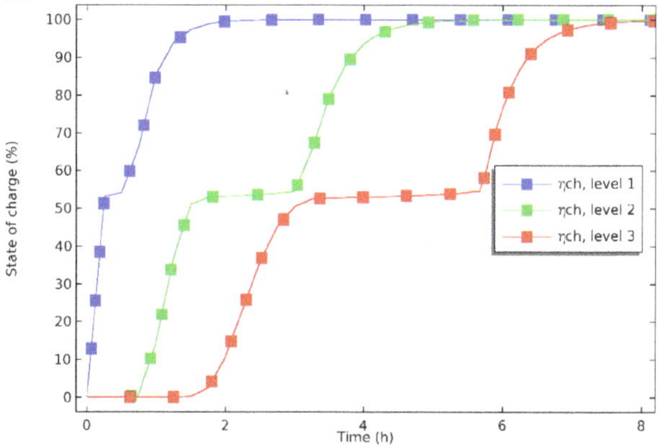

Fig. 5. The state of charge at different levels inside the TES tank

Conclusion

This research paper presents a numerical CFD modeling and analysis of a latent packed bed heat storage system using water as heat transfer fluid and spherical paraffin capsules -PCM- as storage medium. The study was carried out during the heat charging phase, where the hot water circulated vertically inside the tank from top to bottom, transferring heat energy to the PCM particles. The Local Thermal Non-Equilibrium -LTNE- approach was adopted to solve and simulate the transient thermal behavior of the fluid and PCM separately. The accuracy of the numerical model was verified by comparing the simulated results with the measured experimental data, which showed good agreement. The transient temperature distribution, in 2D and 1D plots, at different positions inside the TES tank were presented and discussed. A sharp rise in temperatures and early initiation of the PCM melting was noted at the top levels compared to other positions. The same behavior has been recorded for the state of charge. However, it increased by 2% during the melting phase of the PCM.

Based on the findings of this paper, the current model has proven its reliability and efficiency to visualize and analyze the thermal performances of latent packed bed heat storage systems during the heat charging process. However, its limitations consist in assuming a uniform and axial fluid flow, and assuming constant thermophysical properties for both heat transfer fluid and PCM. This highlights the need for further improvements in future studies.

References

[1] El Hafdaoui, H., Khallaayoun, A. (2024). Influence of Space Limitations on Optimal Sizing of Grid-Connected Hybrid Renewable Energy Generation. In: Motahhir, S., Bossoufi, B. (eds) Digital Technologies and Applications. ICDTA 2024. Lecture Notes in Networks and Systems, vol 1100. Springer, Cham. https://doi.org/10.1007/978-3-031-68660-3_35

[2] Touhs H, Temouden A, Khallaayoun A, Chraibi M, El Hafdaoui H. A Scheduling Algorithm for Appliance Energy Consumption Optimization in a Dynamic Pricing Environment. World Electric Vehicle Journal. 2024; 15(1):1. https://doi.org/10.3390/wevj15010001

[3] A. Mabrouki, H. E. Hafdaoui and A. Khallaayoun, "A Forecast of Pumped Fuel Prices in Morocco Using ARIMAModel," 2023 3rd International Conference on Innovative Research in Applied Science, Engineering and Technology (IRASET), Mohammedia, Morocco, 2023, pp. 1-5, doi: 10.1109/IRASET57153.2023.10152952.

[4] J. Gao *et al.*, "Development of cost-effective latent heat thermal energy storage based on growing fins," *Appl. Therm. Eng.*, vol. 269, p. 125995, 2025, doi: https://doi.org/10.1016/j.applthermaleng.2025.125995.

[5] L. Seyitini, B. Belgasim, and C. C. Enweremadu, "Numerical investigation of a hybrid latent-sensible thermal energy storage system for low temperature industrial applications," *J. Energy Storage*, vol. 123, p. 116759, 2025, doi: https://doi.org/10.1016/j.est.2025.116759.

[6] J. Bošnjak Hordov *et al.*, "Review of organic and inorganic waste-based phase change composites in latent thermal energy storage: Thermal properties and applications," *Energy*, vol. 306, p. 132421, 2024, doi: https://doi.org/10.1016/j.energy.2024.132421.

[7] A. Klitou, T. Klitou, and P. A. Fokaides, "Modelling a packed-bed latent heat thermal energy storage unit and studying its performance using different paraffins," *Int. J. Sustain. Energy*, vol. 43, no. 1, pp. 1–39, 2024, doi: 10.1080/14786451.2024.2306416.

[8] Y. Zhao *et al.*, "Thermal performance investigation of latent heat-packed bed thermal energy storage system with axial gas injection," *Appl. Therm. Eng.*, vol. 258, p. 124781, 2025, doi: https://doi.org/10.1016/j.applthermaleng.2024.124781.

[9] J. Cui, X. Yang, J. Chen, H. Zhang, and J. Xie, "High-performance and low-cost packed bed latent thermal energy storage system based on spray technology," *J. Energy Storage*, vol. 119, p. 116333, 2025, doi: https://doi.org/10.1016/j.est.2025.116333.

[10] L. El-mahaouchi, M. Yessef, and A. Lagrioui, "Modeling and Simulation of Thermodynamic Behavior in a Packed Bed Thermal Energy Storage System," 2025. doi: 10.1109/ECAI65401.2025.11095475.

[11] T. Neeraj, C. Velten, G. Janiga, and K. Zähringer, "Modeling Gas Flows in Packed Beds with the Lattice Boltzmann Method : Validation Against Experiments," *Flow, Turbul. Combust.*, vol. 111, no. 2, pp. 463–491, 2023, doi: 10.1007/s10494-023-00444-z.

[12] S. B. Padmanabhan, M. T. Mabrouk, and B. Lacarrière, "A hybrid model for packed bed thermal energy storage system," *J. Energy Storage*, vol. 98, no. PB, p. 113068, 2024, doi: 10.1016/j.est.2024.113068.

[13] N. Nallusamy and R. Velraj, "Study on performance of a packed bed latent heat thermal energy storage unit integrated with solar water heating system *," no. June, 2014, doi: 10.1631/jzus.2006.A1422.

[14] Z. Lafdaili, S. El-Hamdani, A. Bendou, K. Limam, and E. hafad Bara, "Numerical study of the turbulent natural convection of nanofluids in a partially heated cubic cavity," *Therm. Sci.*, vol. 25, p. 57, 2021, doi: 10.2298/TSCI200513057L.

The Effect of $MASnI_3$–$RbGeI_3$ Optical Absorber Layers Thickness on the Electronic Characteristics of PSCs

Syed M. HASNAIN[1,a *], M. Amin. MIR[2,b], Irfan QASIM[3,c] Abid IQBAL[4,d]

[1]Department of Mechanical Engineering, Prince Mohammad Bin Fahd University, P. O. Box 1664, Al Kobar, 31952, Saudi Arabia

[2]Department of Mathematics and Natural Sciences, Prince Mohammad Bin Fahd University, P. O. Box 1664, Al Kobar, 31952, Saudi Arabia

[3]Department of Physics, Faculty of Sciences, Rawalpindi Women University, 6th Road, Satellite Town, 46300 Rawalpindi, Pakistan

[4]Department of Computer Engineering, King Faisal University, Al-Ahsa, Saudi Arabia

[a]shasnain@pmu.edu.sa, [b]mmir@pmu.edu.sa, [c]dr.irfanqasim@gmail.com, [d]aaiqbal@kfu.edu.sa

Keywords: Solar Cells, $MASnI_3$, $RbGeI_3$, Cu_2O, C_{60}, SCAPS-1D

Abstract. Recently, PSCs have been in the spotlight of solar material design and solar cell research as an attractive emerging solar photovoltaic technology due to low cost of production and a high power conversion efficiencies. This thesis investigated the performance of a lead-free PSC comprising a dual absorption layer of $RbGeI_3$ and $MASnI_3$, using SCAPS-1D simulation software. In the proposed structure, the electron-transport layer (ETL) was based on C_{60} due to its high mobility characteristics, and ideal energy levels for electron transport, while Cu_2O was used as the hole-transport layer (HTL) as it has high hole mobility, and is also chemically stable. The front and back electrodes were indium doped tin oxide and gold, respectively. The dual-absorption layer of $RbGeI_3$ and $MASnI_3$ improved light absorption and the generation of charge carriers, and yielded a respectable PCE of 20.2%. The studied device had an Voc of 0.91 V, a Jsc of 32.7 mA/cm^2, and a FF of 65.2%. These simulation results imply that dual-absorber architectures provide a positive pathway toward enhanced environmentally stable, and high-performance perovskite solar technologies.

Introduction

Energy generation, energy storage, and climate change are among the most pressing issues facing the modern world. The depletion of fossil fuels and increasing carbon dioxide emissions have speeded up the world's shift to alternative and sustainable energy sources. Of these alternatives, one of the most abundant and reliable forms of energy is solar energy. Solar energy is able to cover energy needs in the future [1]. As the required ask for high-efficiency PV systems continues, solar cell technology is growing in importance for electricity generation [2, 3]. It is predicted that solar energy will supply up to 20% of global energy demand in the next two decades. Solar energy is a variety of sources that is relatively sustainable for energy generation and has limited negative impact on the environment [4]. It can be utilized almost everywhere on the planet and at any time, without significant contribution to pollution or global warming [4].

A lot of research over the last thirty years has focused on the development of semiconductor solar cell technologies [5, 6]. Silicon is still the most common material used in these solar energy applications because we fully know the properties of silicon and its behavior in pure and crystalline forms [7]. Despite all of the knowledge gained regarding silicon, the research into any number of semiconductor materials is still limited in terms of a full understanding of how the materials could work together. The high production costs of high-purity silicon and other semiconductor materials have prevented scalable deployment, limiting any potential savings in manufacturing use and therefore preventing widespread acceptance. Because of this limitation, the science community is actively seeking and exploring new materials for solar energy conversion. One of the new promising groups of materials are compounds with perovskite-structured compounds with new device architectures, as a reasonable route to lower cost solutions for solar energy.

Organic-inorganic perovskites have shown sufficient flexibility, to be utilized in X-ray detection systems, and are reinforcing research in sustainable energy, specifically as the active light-absorbing layers of solar cells, generating new excitement. Perovskites possess enough advantageous aspects such as strong absorption of light, prototypical ease of fabrication, tunable bandgap, long carrier diffusion lengths, and compatibility with solution processing are attractive for next-generation photovoltaics [8-10]. Kojima et al. (2009) allowed the first application of perovskite materials to be in used in solar cells with an initial PCE of 3.8% [11]. The efficiency levels have improved significantly since that time, and are now approximately 25.5%, signifying an impactful advancement in both materials development, and engineering for a solar cell system [12]. Nevertheless, the inclusion of lead, a toxic heavy metal is a major aboveboard matter in the effective commercialization of perovskite solar cells [13]. To address this issue, researchers are investigating alternative divalent metal cations similar in chemical characteristics to Pb^{2+}, such as Sn^{2+} and Ge^{2+} that may allow for safe, green perovskites [14].

Based on SCAPS-based simulations, Raj et al. [15] were able to estimate the efficiency of $CsGeI_3$ based solar cells could reach the limiting efficiency of 18.3%. In a comparison with SnO_2 and C_{60} as the electron transport layer, efficiencies of 8.46% and 8.45%, respectively, were obtained [16]. Saikia et al. [17] further utilized $CsGeI_3$ to explore the effect of a number of parameters on performance, such as absorber layer thickness, defect density and doping concentration. Jayan et al. also evaluated $RbGeI_3$ perovskites, using multiple exchange-correlation functionals, and their structural, mechanical, thermodynamic, optical and thermoelectric properties [18]. Simulations for $RbGeI_3$ found an efficiency of 10.11%, FF of 63.68% and QE over the visible spectrum of 90.3%, indicating that it is a promising candidate among the lead-free materials [19, 20].

A variety of inorganic CTM materials (e.g. NiO, TiO_2, SnO_2, Cu_2O, and C_{60}) have been suggested as alternative inorganic CTMs to replace organic CTMs [21, 22]. Inorganic materials generally provide several benefits when compared to the normal organic CTMs, including greater ease of fabrication processes, reduced costs, better chemical and thermal stability, larger bandgaps, higher charge carrier mobility, and better optical transparency across a wider spectral region [23].

This research investigates the feasibility of a new inorganic dual-absorber configuration in solar cells, using $RbGeI_3$ and $MASnI_3$ as the absorber. Inorganic materials were selected for both the hole and electron transport layers, enabling effective charge extraction and transfer throughout the device. Cu_2O was selected for the hole transport layer (HTL) because it possess high hole mobility, low toxicity, and cost. C_{60} was selected for the electron transport layer (ETL) because it possess suitable electron-mobility, favorable energy levels with respect to the conduction band of the absorber, and chemical durability. The application of transport materials of green absorbers serves to enable efficient charge separation while minimizing recombination losses, performance, and sustainability improvements in the solar cell.

Proposed design

This study proposes a tandem solar cell structure built on the structure ITO/C_{60}/$MASnI_3$/$RbGeI_3$/Cu_2O/Au, as shown in Fig. 1. This structure incorporates environmentally friendly materials with efficient charge transportation interfaces to enhance the overall photovoltaic performance. Indium doped Tin Oxide (ITO) was utilized as the front transparent conducting electrode. ITO has positive electrical conductivity and allows light to penetrate the surface of the device. C_{60} was used as the electron transport layer (ETL). This active region consists of two perovskite absorbers, $MASnI_3$ and $RbGeI_3$, that broadens the absorption spectrum and increases photogenerated carriers. This design allows $MASnI_3$ and $RbGeI_3$ to work together within the tandem cell. Cu_2O was used as the hole-transport layer because of its excellent compatibility with the absorber materials, better hole mobility, and chemical stability to get holes extracted efficiently from the absorbers. Gold (Au) was used as the rear electrode due to its ability to collect holes and complete the circuit. The lead-free, earth abundant materials in this structure make it possible to increase device efficiency while also addressing environmental sustainability.

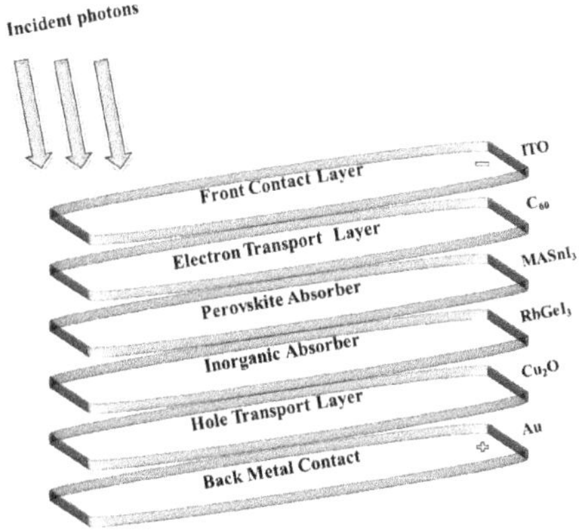

Fig. 1 Device layered structure for the proposed solar cell device.

The device simulation was performed with SCAPS-1D, a popular solar cell capacitance simulator. Material parameters for each structural layer in the device model were taken from previously recorded investigations [23, 24]. To attain the best possible efficiency, a layer by layer thickness optimization was used. In this procedure, the thickness of one layer was changed while the thicknesses of all the other layers remained the same. For each change, batch parameter simulations were also performed to write down the performance metric for a systematic view of how each layer altered overall device performance.

The ETL layer is comprised of C_{60}, was optimized in thickness from 0.01 μm to 0.40 μm. The optimal thickness was determined using the simulated PCE for each step. The light-absorbing layers, $MASnI_3$ and $RbGeI_3$, were also systematically evaluated in thickness from 0.20 μm to 5.00 μm, to find the respective optimal thickness for each material based on the maximum PCE values. The hole transport layer (HTL), made from Cu_2O, was varied between 0.06 μm to 0.90 μm while maintaining constant values for the other layers. The maximum device efficiency was achieved at the ideal thickness of Cu_2O, as seen in the results of this study. This systematic optimization of layers allowed the determination of the most advantageous layer thickness combinations to optimize the total solar cell performance.

Results and discussion

The individual contributions of each layer in determining device performance were assessed. A number of key parameters were considered, one of the significant parameters being the thickness of the $RbGeI_3$ absorber layer shown in Fig. 2. Ultimately, the intention was to optimise the $RbGeI_3$ absorber thickness to achieve the greatest efficiencies with as little material and therefore expense as possible. The electrical properties of the solar cell device were explored by varying the thickness of the $RbGeI_3$ layer while checking the electrical performance. The thickness was varied in a

Materials Research Proceedings 64 (2026) 607-614 https://doi.org/10.21741/9781644904091-76

systematic manner from 0.01 μm to 5 μm and the device's performance was monitored as the thickness was varied. The data provided a response to every layer thickness so as to optimise both efficiency and cost. The device performed best with the $RbGeI_3$ absorber layer being 0.4 μm thick yielding an efficiency of greater than 20% and a fill factor greater than 69%. Due to its unique optical properties and unique attributes, $RbGeI_3$ shows significant promise for photovoltaic applications.

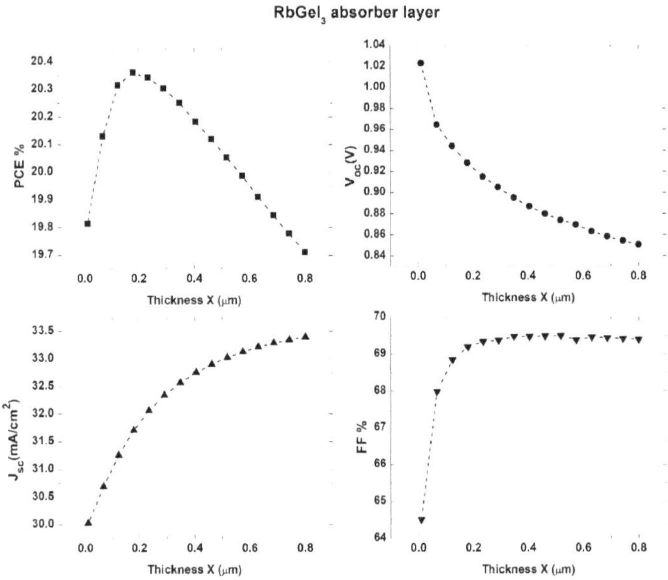

Fig. 2 Effect of $RbGeI_3$ layer thickness on the electrical characteristics.

Based on the data in Fig. 3, the thickness of the $MASnI_3$ absorber layer has a direct effect on various electrical parameters. As the thickness of the absorber layer increases all the electrical parameters increase, while the fill factor slightly decreases. The increase in electrical performance is a result of a higher charge carrier density present in thicker absorber layer devices, and the simulated power conversion efficiency was 20.5% with a V_{oc} of 0.9 V, a J_{sc} of 32.2 mA/cm^2, and a FF of 69%. Ultimately, Figure 3 shows the importance of varying the $MASnI_3$ absorber thickness to achieve the best charge transport while minimizing resistive losses for the best possible efficiency.

Materials Research Proceedings 64 (2026) 607-614 https://doi.org/10.21741/9781644904091-76

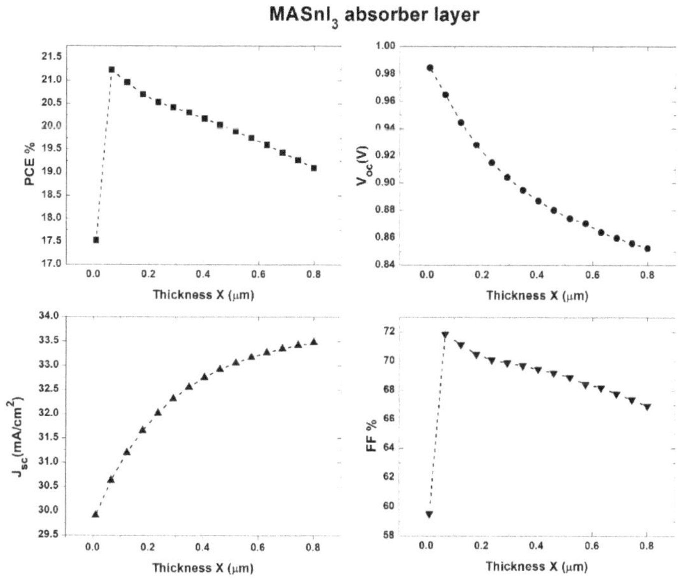

Fig. 3 Effect of $MASnI_3$ layer thickness on the electrical characteristics.

Let us consider the effect of different operational conditions influence the performance of the solar panel in Fig. 4. The I-V density relationship "J_t" has shown a sharp increase reaching nearly 0.7 V for various back-contact layers. An inset diagram is provided for the layer structure of the proposed configuration. The device with best layer thickness values, at 300 K, had a total J_t of 32.7 mA/cm^2 and voltage value of 0.91 V with this configuration resulted in a PCE of 20.2% and FF of 65.5%.

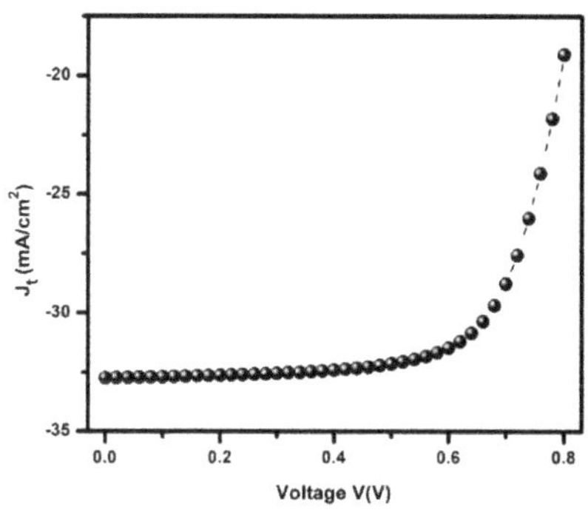

Fig. 4 I-V characteristics of the optimised solar cell.

Materials Research Proceedings 64 (2026) 607-614 https://doi.org/10.21741/9781644904091-76

Fig. 5 presents the performance testing of the device within a temperature ranging from 300 K to 375 K. This temperature range was used to demonstrate how the various electrical properties of the device change with temperature. The device's fill factor and current density both showed an increase, and after a point, both started to go down. The fill factor maxed at 87% at 305 K, and the current density maxed at 375 K at 32.7 mA/cm^2. Even with the increase in current density (electrical current displaced in a known area) and fill factor the devices efficiency dropped from 20.2% to 16.5%. The Voc also dropped from 0.89 V to 0.73 V for the defined temperature range. The J_{sc} and FF were able to increase because of a higher charge carrier density, which can be attributed to an increase in photon absorption in the absorbers. Overall the device shows good performance in this temperature range.

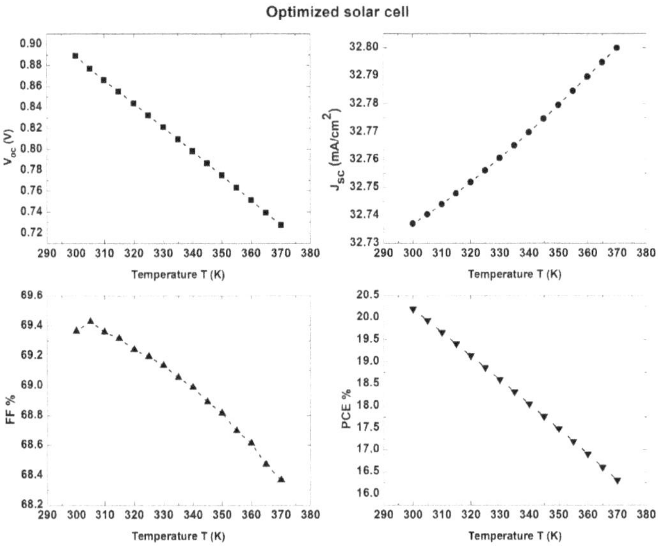

Fig. 5 Effect of temperature on the electrical characteristics.

Conclusion

In this research work, an optimized solar energy harvester is presented with a particular emphasis on maximizing the performance and efficiency of the solar cell through optimized ETL and HTL in a tandem architecture. Upon systematic variation, the optimal absorber layer thicknesses were determined as 2.10 μm for CIGS and 1.51 μm for $MASnI_3$. The modified design increased the performance of the device achieving a PCE of 36.4% with a FF of 54.2% and Jsc of 48.3 mA/cm^2. These results reinforce the importance of optimizing the absorber layer to create more efficient and environmentally sustainable lead-free solar energy harvesters and create further room for advancements in the development of sustainable photovoltaic technologies.

List of Symbols and Abbreviations

- PSCs – Perovskite solar cells –
- $MASnI_3$ – Methylammonium tin iodide
- $RbGeI_3$ – Rubidium germanium iodide
- SCAPS-1D – Solar cell capacitance simulator (1-Dimensional)
- PCE – Power conversion efficiency
- HTL – Hole transport layer
- ETL – Electron transport layer
- V_{oc} – Open-circuit voltage
- J_{sc} – Short-circuit current density
- FF – Fill factor

References

[1] K. Yoshikawa et al., "Silicon heterojunction solar cell with interdigitated back contacts for a photoconversion efficiency over 26%," Nature energy, vol. 2, no. 5, pp. 1-8, 2017. https://doi.org/10.1038/nenergy.2017.32

[2] T. Jiang, Y. Wang, D. Meng, X. Wu, J. Wang, and J. J. Chen, "Controllable fabrication of CuO nanostructure by hydrothermal method and its properties," Applied Surface Science, vol. 311, pp. 602-608, 2014. https://doi.org/10.1016/j.apsusc.2014.05.116

[3] S. M. Hasnain, "Examining the advances, obstacles, and achievements of tin-based perovskite solar cells: a review," Solar Energy, vol. 262, p. 111825, 2023. https://doi.org/10.1016/j.solener.2023.111825

[4] J. Liu et al., "Synthesis and optical applications of low dimensional metal-halide perovskites," Nanotechnology, vol. 31, no. 15, p. 152002, 2020. https://doi.org/10.1088/1361-6528/ab5a19

[5] Y. Wang et al., "Fabrication of nanostructured CuO films by electrodeposition and their photocatalytic properties," Applied surface science, vol. 317, pp. 414-421, 2014. https://doi.org/10.1016/j.apsusc.2014.08.144

[6] S. M. Hasnain et al., "Performance evaluation of organometal halide MASnI3 and inorganic BaZrS3 hybrids in perovskites solar cells: Theoretical approach," Hybrid Advances, vol. 9, p. 100408, 2025/06/01/ 2025. https://doi.org/10.1016/j.hybadv.2025.100408

[7] Q. Chen, Y. Wang, M. Zheng, H. Fang, and X. Meng, "Nanostructures confined self-assembled in biomimetic nanochannels for enhancing the sensitivity of biological molecules response," Journal of Materials Science: Materials in Electronics, vol. 29, pp. 19757-19767, 2018. https://doi.org/10.1007/s10854-018-0101-2

[8] Y. Shen, C. Ran, X. Dong, Z. Wu, and W. Huang, "Dimensionality Engineering of Organic-Inorganic Halide Perovskites for Next-Generation X-Ray Detector," Small, vol. 20, no. 16, p. 2308242, 2024. https://doi.org/10.1002/smll.202308242

[9] O. Ahmad et al., "Modelling and numerical simulations of eco-friendly double absorber solar cell "Spiro-OmeTAD/CIGS/MASnI3/CdS/ZnO" and its PV-module," Organic Electronics, vol. 117, p. 106781, 2023. https://doi.org/10.1016/j.orgel.2023.106781

[10] I. Qasim et al., "Design and numerical investigations of eco-friendly, non-toxic (Au/CuSCN/CH3NH3SnI3/CdTe/ZnO/ITO) perovskite solar cell and module," Solar Energy, vol. 237, pp. 52-61, 2022. https://doi.org/10.1016/j.solener.2022.02.056

[11] A. Kojima, K. Teshima, Y. Shirai, and T. Miyasaka, "Organometal halide perovskites as visible-light sensitizers for photovoltaic cells," Journal of the american chemical society, vol. 131, no. 17, pp. 6050-6051, 2009. https://doi.org/10.1021/ja809598r

[12] J. Wu et al., "Using hysteresis to predict the charge recombination properties of perovskite solar cells," Journal of Materials Chemistry A, vol. 9, no. 10, pp. 6382-6392, 2021. https://doi.org/10.1039/D0TA12046D

[13] J. Dou, Y. Bai, and Q. Chen, "Challenges of lead leakage in perovskite solar cells," Materials Chemistry Frontiers, vol. 6, no. 19, pp. 2779-2789, 2022. https://doi.org/10.1039/D2QM00632D

[14] W. Ke and M. G. Kanatzidis, "Prospects for low-toxicity lead-free perovskite solar cells," Nature communications, vol. 10, no. 1, p. 965, 2019. https://doi.org/10.1038/s41467-019-08918-3

[15] A. Raj, M. Kumar, H. Bherwani, A. Gupta, and A. Anshul, "Evidence of improved power conversion efficiency in lead-free CsGeI3 based perovskite solar cell heterostructure via scaps simulation," Journal of Vacuum Science Technology B, vol. 39, no. 1, 2021. https://doi.org/10.1116/6.0000718

[16] A. Raj et al., "A computational approach to investigate the suitable ETL for lead-free CsGeI3 based perovskite solar cell," Materials Today: Proceedings, vol. 47, pp. 1564-1569, 2021. https://doi.org/10.1016/j.matpr.2021.03.610

[17] D. Saikia, J. Bera, A. Betal, and S. Sahu, "Performance evaluation of an all inorganic CsGeI3 based perovskite solar cell by numerical simulation," Optical Materials, vol. 123, p. 111839, 2022. https://doi.org/10.1016/j.optmat.2021.111839

[18] K. D. Jayan and V. Sebastian, "Ab initio DFT determination of structural, mechanical, optoelectronic, thermoelectric and thermodynamic properties of RbGeI3 inorganic perovskite for different exchange-correlation functionals," Materials Today Communications, vol. 28, p. 102650, 2021. https://doi.org/10.1016/j.mtcomm.2021.102650

[19] K. A. Parrey, T. Farooq, S. A. Khandy, U. Farooq, and A. Gupta, "First principle studies on structure, magneto-electronic and elastic properties of photovoltaic semiconductor halide (RbGeI3) and ferromagnetic half metal oxide (RbDyO3)," Computational Condensed Matter, vol. 19, p. e00381, 2019. https://doi.org/10.1016/j.cocom.2019.e00381

[20] S. M. Hasnain, "SCAPS-1D Simulation of Lead-Free BaZrS3 Perovskite as an Ecofriendly Absorber Material for High-Performance Solar Cells," Journal of Electrical Engineering & Technology, 2025/08/30 2025. https://doi.org/10.1007/s42835-025-02424-x

[21] P. Zhao et al., "The crystal anisotropy effect of MAPbI3 perovskite on optoelectronic devices," Materials Today Energy, vol. 17, p. 100481, 2020. https://doi.org/10.1016/j.mtener.2020.100481

[22] S. M. Hasnain, I. Qasim, A. Iqbal, M. Amin. Mir, and N. Abu-Libdeh, "Novel dual absorber configuration for eco-friendly perovskite solar cells: design, numerical investigations and performance of ITO-C60-MASnI3-RbGeI3-Cu2O-Au," Solar Energy, vol. 278, p. 112788, 2024/08/01/ 2024. https://doi.org/10.1016/j.solener.2024.112788

[23] J. Liang et al., "All-inorganic perovskite solar cells," Journal of the American Chemical Society, vol. 138, no. 49, pp. 15829-15832, 2016. https://doi.org/10.1021/jacs.6b10227

Structural and Thermal Performance Analysis of (In,Ga)N p–i–n Homojunction Solar Cells

Hassan ABBOUDI[1,a*], Ilyass EZ-ZEJJARI[2,b*], Haddou EL GHAZI[2,c], Redouane EN-NADIR[3,d], Redouane SERSAR[2,e], Walid BELAID[5,f], Ahmed SALI[1,g]

[1]LPS, Faculty of sciences, Sidi Mohamed Ben Abdellah University, Fez 30000, Morocco

[2]LCCPS, ENSAM, Hassan 2 University, 150 Bd du Nil, Casablanca 20670, Morocco

[3]Interdisciplinary Institute for Technological Innovation, Sherbrooke University, 2500 Bd University, Sherbrooke, Canada

[5]Pollard Institute, School of Electronic and Electrical Engineering, University of Leeds, Leeds, United Kingdom

[a]hassan.abboudi@usmba.ac.ma, [b]ilyass.ezzejjari-etu@etu.univh2c.ma, [c]elghazi.haddou@ensam-casa.ma, [d]radouane.en-nadir@usherbrooke.ca, [e]redouane.sersar-etu@etu.univh2c.ma, [f]w.belaid@leeds.ac.uk, [g]ahmed.sali@usmba.ac.ma

Keywords: p-i-n Photovoltaic Cells, Intrinsic Layer, Temperature, Efficiency

Abstract. This study examines the combined influence of intrinsic layer thickness and temperature on the performance of (In,Ga)N-based p–i–n solar cells using SCAPS-1D simulation. The structure, composed of $p - In_{0.6}Ga_{0.4}N/i - In_xGa_{1-x}N/n - In_{0.48}Ga_{0.52}N$ layers, is analyzed under 300K and 400 K to assess thermal effects on carrier transport and recombination. The optimal performance was obtained for an indium molar fraction of $x = 0.59$. The Results reveal that the thickness engineering significantly enhances current generation up to $\sim 0.5\ \mu m$, beyond which absorption saturation limits further improvement. While V_{oc} and FF remain nearly thickness-independent, both degrade with rising temperature due to thermally activated recombination. Overall, the conversion efficiency benefits from optimal intrinsic layer thickness and controlled operating temperature, highlighting the importance of structural and thermal optimization for high-performance (In,Ga)N-based solar cells.

Introduction.

Solar cells represent a promising solution to the global energy crisis. However, their large-scale adoption is limited by the high cost of fabrication and comparatively modest power conversion efficiencies. In Consequence, enhancing the photovoltaic efficiency of thin-film device architectures remains a major research challenge. Recent advances in the growth of III-nitride semiconductors, particularly InGaN, have opened new opportunities for photovoltaics [1, 2]. This material is caracterised by its thermal stability, radiation tolerance, and high absorption coefficient [3, 4]. Its tunable bandgap (0.70-3.42 eV) allows absorption across a wide solar spectrum, from infrared to ultraviolet. Due to these properties, InGaN is considered a potential material for high-efficiency solar photovoltaic cells, both single and multi-junction [5, 6]. The p–i–n architecture further improve charge separation and transport [7, 8], boosting the performance of nitride-based devices. This configuration optimise carrier collection, reduces optical losses, and increases defect tolerance. Numerical simulations of $In_xGa_{1-x}N$ devices have reported efficiencies ranging from 8.9% to 18.74% through adjustments in indium composition, layer thickness, and doping [9, 10]. Most studies have considered these parameters separately, neglecting their combined impact, despite the critical influence of environmental factors like temperature and humidity on solar cell performance.

The influence of temperature on (In,Ga)N p–i–n structures has not been extensively studied. The first photovoltaic response from an InGaN/GaN double heterostructure with a 0.2 μm active layer was notably reported by Jani et al. [11], who achieved an internal quantum efficiency of 60% and an open-circuit voltage of 2.4 V. Subsequent work by Neufeld et al. [12] demonstrated enhanced performance, reaching a maximum external quantum efficiency of 63% and a current density of 4.2 mA/cm². We conduct a comprehensive numerical study of $Ga_{1-x}In_xN$ p-i-n junction using SCAPS-1D software. Under standard AM1.5D irradiance, the impact of temperature on key photovoltaic metrics including open-circuit voltage, short-circuit current density, fill factor, and efficiency is systematically analyzed. This paper is structured in four main sections. Following this introduction, which establishes the research objectives and methodological foundation, Section 2 develop the theoretical framework and describes the simulation techniques employed. Section 3 present the numerical results and provides a comprehensive discussion of their implications. Finally, Section 4 summarizes the principal conclusions and highlights potential directions for future investigation.

Theoretical Modeling and simulation.

Figures 1 illustrates the architecture of the p-i-n photovoltaic cell analyzed in this study. The device employs a specific layer sequence of $p - In_{0.6}Ga_{0.4}N/i - In_xGa_{1-x}N/n - In_{0.48}Ga_{0.52}N$. The compositional grading of the p- and n-type layers, with indium mole fractions of 0.6 and 0.48 respectively, was implemented based on prior optimization [13]. The p-type layer functions as a hole-transport and buffer layer, while the n-type layer facilitates electron transport and acts as an optical window. At the core of the device, the intrinsic absorber layer $i - In_xGa_{1-x}N$ is designed to maximize photon absorption and carrier generation. The simulations indicate that the optimal photovoltaic performance is achieved for indium (In) content $x = 0.59$. Under operational illumination, the intrinsic region is fully depleted, establishing a significant internal electric field that enhances electron-hole pair separation and collection efficiency by suppressing bulk recombination. Device performance simulations were carried out using the SCAPS-1D, considering the standard AM1.5G solar spectrum.

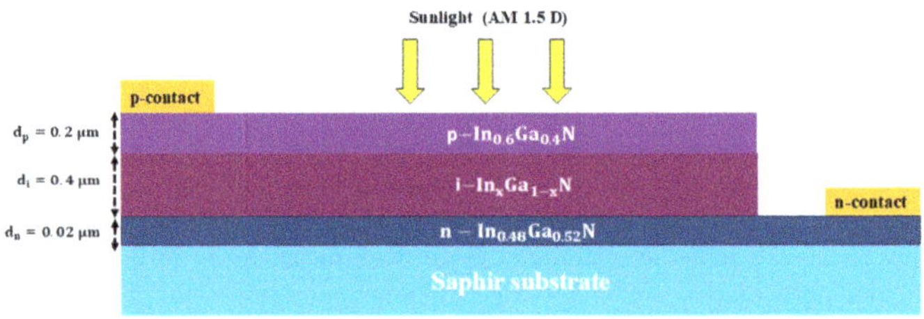

***Fig 1**.Schematic representation of p–i–n (In,Ga)N homojunction solar cell architecture.*

This work investigates the impact of intrinsic layer (i-layer) thickness on the performance of (In,Ga)N based solar cells, evaluating the key photovoltaic parameters; open-circuit voltage V_{oc}, short-circuit current density j_{sc} , fill factor FF, and conversion efficiency η as functions of the i-layer thickness. The analysis also considers thermal effects by examining how these parameters vary with i-layer thickness at two different temperatures: 300 K, and 400 K. Device simulations were carried out using SCAPS-1D, which numerically solves the Poisson equation, electron-hole continuity equations, and the drift-diffusion approach within a one-dimensional framework [14]. By incorporating realistic boundary conditions, material parameters, and recombination mechanisms through Shockley-Read-Hall (SRH) carrier recombination formalism, SCAPS-1D provides a rigorous framework for analyzing carrier dynamics and performance limitations [15].

$$\frac{\partial^2\psi(x)}{\partial x^2} = \frac{e}{\varepsilon_0\varepsilon_r}\left(p(x) - n(x) + N_d - N_a + \rho_p - \rho_n\right) \quad (1)$$

Where; ψ denotes the electrostatic potential, p(x) and n(x) represent the hole and electron concentrations, respectively. ε_0 is the vacuum permittivity, ε_r is the relative permittivity, and e is the elementary charge. N_a and N_d present the acceptor and donor concentrations, respectively, while ρ_n and ρ_p correspond to electrons and holes spatial distributions. Accurate performance modeling of $In_xGa_{1-x}N$ solar cells therefore necessitate precise determination of the alloy's material parameters. Among these, the bandgap energy E_g of $In_xGa_{1-x}N$ is represented as a function of indium content as [16].

$$E_g^{In_xGa_{1-x}N}(x,T) = x \cdot E_g^{InN}(T) - (1-x) \cdot E_g^{GaN}(T) - bx(1-x) \quad (2)$$

b present the band gap bowing parameter, b=1.43 **[17]**.

$E_g^{InN}(T), E_g^{GaN}(T)$ denote the temperature-dependent bandgap energies of GaN and InN, which are commonly expressed using the Varshni formulas [18]:

$$E_g(A,T) = E_g(A,0) - \frac{\alpha T^2}{T+\beta} \quad (3)$$

Where $E_g(A,0)$ is the gap of a given semi-conductor a (GaN, InN) at T= 0 K.
The relative permittivity ε_r and the electron affinity χ are defined as:

$$\varepsilon^*(x,T) = x \cdot \varepsilon^*_{InN}(T) + (1-x) \cdot \varepsilon^*_{GaN}\ (T) \quad (4)$$

$$\chi(In_xGa_{1-x}N) \approx x\chi^{InN}(T) + (1-x)\chi^{GaN}(T) \quad (5)$$

The temperature-dependent effective density of states for the conduction N_c and valence N_v bands is given by:

$$N_c(In_xGa_{1-x}N) = xN_c^{InN}(T) + (1-x)N_c^{GaN}(T) \quad (6)$$
$$N_v(In_xGa_{1-x}N) = xN_v^{InN}(T) + (1-x)N_v^{GaN}\ (T) \quad (7)$$

Solar cell efficiency depends on three fundamental performance metrics: V_{oc}, J_{sc} and FF. The open circuit voltage V_{oc} and fill factor FF are given by the following expressions **[19]**:

$$V_{oc} = \frac{K_BT}{q}\ln\left(\frac{J_{sc}}{J_0} + 1\right) \quad (8)$$

$$FF = \frac{v_{oc} - \ln(v_{oc} + 0.72)}{v_{oc} + 1} \quad (9)$$

Where $v_{oc} = qV_{oc}/kT$.

Finally, the solar cell efficiency η, defined as the ratio of output power to input power P_{in}, is given by the following expression [20]:

$$\eta = \frac{FF.V_{oc} \cdot J_{sc}}{P_{in}} \quad (10)$$

Where P_{in} represents the optical input power of solar radiation under standard AM1.5G illumination, corresponding to 100 W/cm² at the cell surface.

We summarize in Table 1 all the physical input parameters used in this study.

Table 1. *Room-temperature physical parameters of the materials employed in the simulation.*

Layer	$E_g(eV)$	$\chi(eV)$	$\varepsilon_r(F.m^{-1})$	$N_c(1.10^{18}cm^{-3})$	$N_v(1.10^{19}cm^{-3})$
$In_{0.6}Ga_{0.4}N$	1.399	5.144	12.98	1.44	4.967
$In_{0.59}Ga_{0.41}N$	1.422	5.127	12.922	1.455	4.96
$In_{0.48}Ga_{0.52}N$	1.705	4.943	12.284	1.6	4.898

Results and discussion.

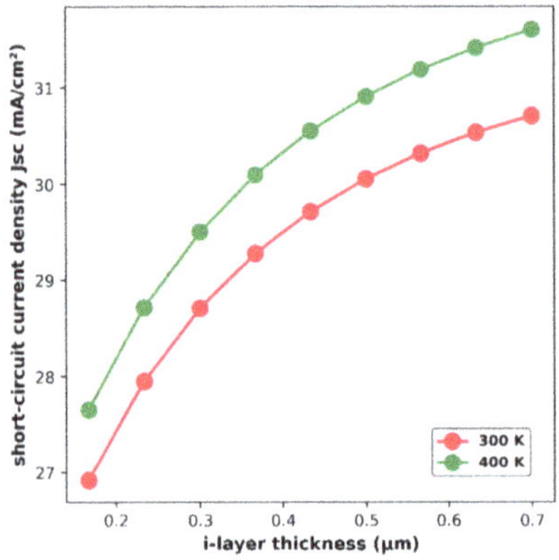

Fig 2. *Dependence of the short-circuit current density j_{sc} on the i-layer thickness at two different temperatures.*

Figure 2 presents the variation of j_{sc} with intrinsic layer thickness over the range of 0.1- 0.7 µm, under two different temperatures 300 K, and 400 K and the optimal photovoltaic performance is achieved for an indium concentration of *$x = 0.59$*. As a results, j_{sc} increases with intrinsic layer thickness until it progressively saturates beyond a critical thickness of $d_i \approx 0.5$ µm. this behaviour reflects the direct influence of the absorbing region on the generation of photo-induced carriers: a thicker intrinsic layer d_i enhances photon absorption and increases carrier generation, but beyond a certain threshold, most useful photons are already absorbed, making the additional gain marginal. For illustration, j_{sc} rises from 26.8, and 27.6 mA.cm^{-2} at 300 K, and 400 K, respectively, for d_i = 0.16 µm, to about 29.9, and 30.7 mA.cm^{-2} at 0.5 µm, before reaching 30.4, and 31.6 mA.cm^{-2} at 0.7 µm, thus confirming the saturation. The increase in thickness from 0.16 to 0.5 µm therefore corresponds to a relative improvement of 11.57% at 300 K, and 11.23% at 400 K. In parallel, temperature exerts a notable effect: for a thickness of d_i = 0.5 µm, increasing the temperature from 300 K to 400 K provides an additional 0.8 mA.cm^{-2}, leading to an overall improvement of 2.6% between 300 K and 400 K. This trend suggests that higher temperature enhances the collection of photogenerated carriers, likely due to improved mobility and a reduction of the band gap energy. These results therefore demonstrate that the intrinsic layer thickness is a critical parameter for the optimization of $p - In_{0.6}Ga_{0.4}N/i - In_xGa_{1-x}N/n - In_{0.48}Ga_{0.52}N$ solar cells With x = 0.59, this composition is considered throughout the entire study presented in this article. Increasing the thickness significantly improves current generation up to an optimal threshold (0.5 µm), beyond which further gains become negligible due to absorption saturation.

Materials Research Proceedings 64 (2026) 615-623 https://doi.org/10.21741/9781644904091-77

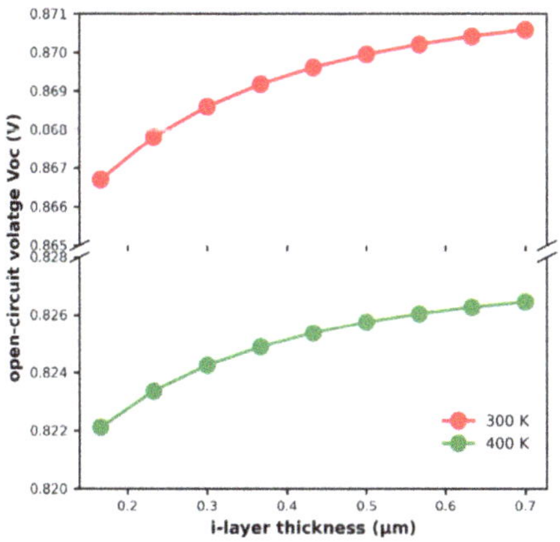

Fig 3. *Evolution of the open-circuit voltage V_{oc} as a function of the intrinsic layer thickness, analyzed at two distinct temperatures.*

The variation of V_{oc} as a function of i-layer thickness d_i in the range of 0.1-0.7 µm is shown in Figure 3 under two operating temperatures. A consistent trend is observed across all temperatures: V_{oc} exhibits only a very slight increase, such that it can be considered practically constant with respect to d_i. This indicates that, in this configuration, V_{oc} is mainly governed by temperature dependent mechanisms rather than by variations in absorption length or carrier generation efficiency. Indeed, at 300 K, V_{oc} retains a relatively high value of approximately 0.87 V, with a slight increase of about 0.69% across the entire range of intrinsic layer thicknesses. At 400 K, this voltage slightly decreases, varying from 0.823 V to 0.826 V, corresponding to a relative decrease of 5.32% over the same d_i range. Consequently, the variation in V_{oc} remains below 1% within the studied range, indicating a weak sensitivity of this parameter to the intrinsic layer thickness, regardless of temperature. Meanwhile, the significant reduction in V_{oc} observed at higher temperatures aligns with the fundamental principle that the saturation current density j_0 exhibits an exponential dependence on temperature. Indeed, at high temperatures, for example 400 K, carrier excitation and SRH recombination are enhanced, leading to an increase in j_0 and a reduction in V_{oc}. The intrinsic layer thickness mainly affects j_{sc}, while remains largely insensitive to these variations. The strong temperature dependence highlights the importance of minimizing thermally activated recombination and controlling defects to maintain a high V_{oc} under optimal operating conditions.

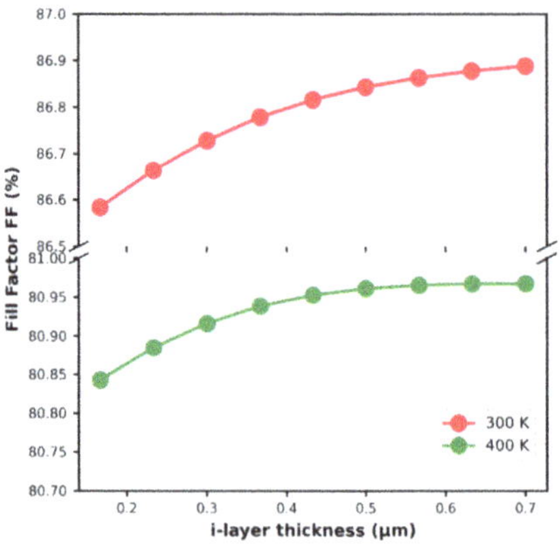

Fig 4. *Variation of the fill factor (FF) as a function of the intrinsic layer thickness, investigated at two distinct temperatures.*

The dependence of FF as a function of i-layer thickness d_i (0.1 − 0.7 µm) at 300K and 400K is presented in Figure 4. The results indicate that FF is nearly independent of d_i, exhibiting minimal variation (< 0.5%) across the studied range. In contrast, FF shows a strong temperature dependence: it reaches a maximum of about 86.5 − 86.9% at 300 K, and further declines to 80.85-80.95% at 400 K. As temperatures rise, we observe a decrease in performance. This is largely due to thermally activated recombination and leakage currents, combined with a reduction in carrier mobility. These issues increase the device's internal resistance, making it harder to extract power effectively. The invariance of FF with thickness implies that resistive and recombination mechanisms remain largely unaffected by absorption depth, unlike the short-circuit current density. Hence, optimizing FF in (In,Ga)N based solar cells requires focusing on thermal management, defect passivation, and resistance reduction rather than modifying the intrinsic layer thickness.

Figure 5 depicts the variation of the power conversion efficiency η as a function of the intrinsic layer thickness at two distinct temperatures. The results reveal a general behaviour of increasing efficiency with increasing intrinsic layer thickness, although the absolute values are strongly dependent on the operating temperature. At 300 K, the device exhibits its highest performance, with the efficiency rising from 20.2% to 23.4% over the same range of thickness from d_i = 0.16 µm to d_i = 0.7 µm, resulting in a relative increase of approximately 15.84%. In contrast, at 400 K, a pronounced degradation is observed: η increases from 18.3% to 20.8%, representing a 13.66% relative improvement, yet with absolute values significantly lower than those obtained at 300 K. A comparative analysis further shows that the sensitivity to temperature becomes more pronounced with thickness. For thin layers d_i = 0.16 µm, the efficiency differs by 1.9% between 300 K (20.2%) and 400 K (18.3%), while for thicker layers d_i = 0.7 µm, the difference increases to 2.6% between 300 K (23.4%) and 400 K (20.8%). Efficiency increases with intrinsic layer thickness up to 0.5 µm, beyond which recombination limits gains. Temperature strongly affects performance, peaking at 300 K and dropping at 400 K owing to intensified recombination mechanisms and reduced charge carrier mobility. Both factors are thus critical for device efficiency.

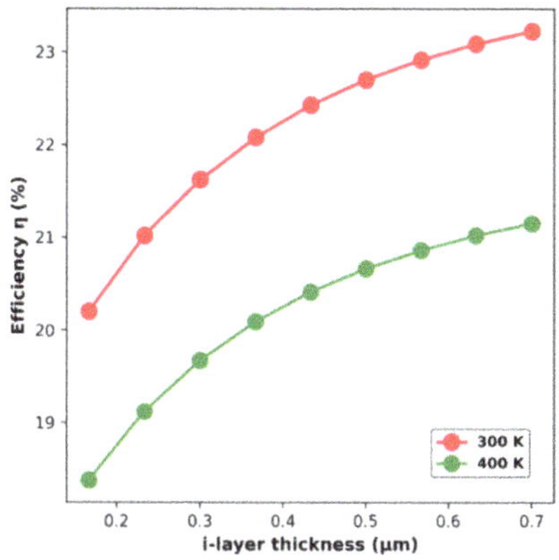

Fig 5. *Efficiency variation with intrinsic layer thickness, investigated across two temperatures.*

Conclusion

The performance of (In,Ga)N p-i-n photovoltaic devices has been systematically analyzed via SCAPS-1D simulations, with a dedicated focus on the interaction between the i-layer thickness and temperature. These findings demonstrate that the intrinsic layer thickness plays a pivotal role in device optimization: increasing d_i enhances current generation up to an optimal value of approximately 0.5 μm, beyond which further improvements are limited by absorption saturation. In contrast, V_{OC} remains largely unaffected by thickness variations but decreases monotonically with increasing temperature, consistent with thermally enhanced recombination mechanisms. Similarly, while the fill factor FF is invariant with respect to thickness, it exhibits a temperature-induced degradation, declining from 86.9% at 300 K to 80.95% at 400 K. In summary, the conversion efficiency benefits from both thickness engineering and thermal management. Efficiency improves with increasing d_i across all temperatures, although the gains diminish beyond the 0.5 μm threshold, whereas temperature exerts a more pronounced influence: the optimal performance is achieved at 300 K, while heating to 400 K induces a $10 - 12\%$ efficiency loss relative to room temperature. These results underscore the crucial importance of carefully balancing structural design and thermal effects in order to maximize the photovoltaic efficiency.

Declaration of AI Assistance

The authors confirm that AI-based tools (ChatGPT, OpenAI) were used exclusively for language correction and editorial refinement to enhance clarity, grammar, and style. All scientific content, data analyses, and interpretations presented in the manuscript were conceived, developed, and rigorously verified solely by the authors.

References.

[1] X. Huang et al., "High-temperature polarization-free III-nitride solar cells with self-cooling effects," ACS Photonics, vol. 6, no. 8, pp. 2096-2103, 2019. https://doi.org/10.1021/acsphotonics.9b00655

[2] J. J. Williams et al., "Development of a high-band gap high temperature III-nitride solar cell for integration with concentrated solar power technology," in 2016 IEEE 43rd Photovoltaic Specialists Conference (PVSC), 2016: IEEE, pp. 0193-0195. https://doi.org/10.1109/PVSC.2016.7749576

[3] D. J. Sirbuly, M. Law, H. Yan, and P. Yang, "Semiconductor nanowires for subwavelength photonics integration," vol. 109, ed: ACS Publications, 2005, pp. 15190-15213. https://doi.org/10.1021/jp051813i

[4] F. Ponce and D. Bour, "Nitride-based semiconductors for blue and green light-emitting devices," nature, vol. 386, no. 6623, pp. 351-359, 1997. https://doi.org/10.1038/386351a0

[5] L. E. Bain and A. Ivanisevic, "Engineering the Cell-Semiconductor Interface: A Materials Modification Approach using II-VI and III-V Semiconductor Materials," Small, vol. 11, no. 7, pp. 768-780, 2015. https://doi.org/10.1002/smll.201401450

[6] S. Nakamura, T. Mukai, and M. Senoh, "Candela-class high-brightness InGaN/AlGaN double-heterostructure blue-light-emitting diodes," Applied Physics Letters, vol. 64, no. 13, pp. 1687-1689, 1994. https://doi.org/10.1063/1.111832

[7] U. K. Mishra, L. Shen, T. E. Kazior, and Y.-F. Wu, "GaN-based RF power devices and amplifiers," Proceedings of the IEEE, vol. 96, no. 2, pp. 287-305, 2008. https://doi.org/10.1109/JPROC.2007.911060

[8] L. Vilbois, A. Cheknane, A. Bensaoula, C. Boney, and T. Benouaz, "Simulation of a solar cell based on InGaN," Energy Procedia, vol. 18, pp. 795-806, 2012. https://doi.org/10.1016/j.egypro.2012.05.095

[9] R. Singh, D. Doppalapudi, T. Moustakas, and L. Romano, "Phase separation in InGaN thick films and formation of InGaN/GaN double heterostructures in the entire alloy composition," Applied Physics Letters, vol. 70, no. 9, pp. 1089-1091, 1997. https://doi.org/10.1063/1.118493

[10] S. Routray and T. Lenka, "InGaN-based solar cells: a wide solar spectrum harvesting technology for twenty-first century," CSI Transactions on ICT, vol. 6, no. 1, pp. 83-96, 2018. https://doi.org/10.1007/s40012-017-0181-9

[11] O. Jani, I. Ferguson, C. Honsberg, and S. Kurtz, "Design and characterization of GaN/ InGaN solar cells," Applied Physics Letters, vol. 91, no. 13, 2007. https://doi.org/10.1063/1.2793180

[12] C. J. Neufeld, N. G. Toledo, S. C. Cruz, M. Iza, S. P. DenBaars, and U. K. Mishra, "High quantum efficiency InGaN/GaN solar cells with 2.95 eV band gap," Applied Physics Letters, vol. 93, no. 14, 2008. https://doi.org/10.1063/1.2988894

[13] H. Abboudi et al., "Optimization of InxGa1− xN PIN Solar Cells: Achieving 21% Efficiency Through SCAPS-1D Modeling," Crystals, vol. 15, no. 7, p. 633, 2025. https://doi.org/10.3390/cryst15070633

[14] T. Ouslimane, L. Et-Taya, L. Elmaimouni, and A. Benami, "Impact of absorber layer thickness, defect density, and operating temperature on the performance of MAPbI3 solar cells based on ZnO electron transporting material," Heliyon, vol. 7, no. 3, 2021. https://doi.org/10.1016/j.heliyon.2021.e06379

[15] M. Minbashi, A. Ghobadi, M. H. Ehsani, H. Rezagholipour Dizaji, and N. Memarian, "Simulation of high efficiency SnS-based solar cells with SCAPS," Solar Energy, vol. 176, pp. 520-525, 2018/12/01/ 2018. https://doi.org/10.1016/j.solener.2018.10.058

[16] I. Vurgaftman and J. n. Meyer, "Band parameters for nitrogen-containing semiconductors," Journal of applied physics, vol. 94, no. 6, pp. 3675-3696, 2003. https://doi.org/10.1063/1.1600519

[17] Y. Varshni, "Band-to-Band Radiative Recombination in Groups IV, VI," February 1, vol. 19, p. 459, 2022. https://doi.org/10.1515/9783112498309-001

[18] Y. P. Varshni, "Temperature dependence of the energy gap in semiconductors," Physica, vol. 34, no. 1, pp. 149-154, 1967/01/01/ 1967. https://doi.org/10.1016/0031-8914(67)90062-6

[19] H. Abboudi et al., "Enhancing the Photovoltaic Efficiency of In0.2Ga0.8N/GaN Quantum Well Intermediate Band Solar Cells Using Combined Electric and Magnetic Fields," Materials, vol. 17, no. 21, p. 5219, 2024. [Online]. Available: https://www.mdpi.com/1996-1944/17/21/5219. https://doi.org/10.3390/ma17215219

[20] H. Abboudi et al., "Tuning Intermediate Band Solar Cell Efficiency: The Interplay of Electric Fields, Composition, Impurities, and Confinement," Nanomaterials, vol. 14, no. 22, p. 1858, 2024. [Online]. Available: https://www.mdpi.com/2079-4991/14/22/1858. https://doi.org/10.3390/nano14221858

[21] C. A. M. Fabien and W. A. Doolittle, "Guidelines and limitations for the design of high-efficiency InGaN single-junction solar cells," Solar Energy Materials and Solar Cells, vol. 130, pp. 354-363, 2014/11/01/ 2014. https://doi.org/10.1016/j.solmat.2014.07.018

Properties of Sodium Fluoride and Lithium-Molybdenum-Phosphorus Oxides Glasses

Lahbib ABBAS[1,a]*, Lahcen BIH[2,b], Abdelaziz AIT SIDI MOU[1,c] and Zahra RAMZI[3,d]

[1]Moulay Ismail University Meknes, Chemistry-Physics, Electrochemistry and Environment Team, Faculty of Science and Technology Er-Rachidia, Morocco

[2]Université Moulay Ismail, Laboratory of Engineering Sciences and Professions, National School of Arts and Crafts, Morocco

[3]Cadi Ayyad University, UCA, Laboratory of Materials Physiochemistry and Environment (LPCME), Faculty of Sciences Semlalia, Morocco

[a]l.abbas@umi.ac.ma, [b]l.bih@umi.ac.ma, [c]a.aitsidimou@umi.ac.ma, [d]zahramzi80@gmail.com

Keywords: Phosphate Glasses, Ionic Conductors, Effect of Modifying Cations, Electrical Conductivity, Activation Energy

Abstract. This work aims to study the electrical and thermal properties of glass system xNaF-(0.4-x)Li_2O-0.24MoO_3-0.36P_2O_5 with (x=0, 0.08, 0.16, 0.2, 0.24, 0.32, 0.4. The characteristic temperatures (T_g, T_c, and T_f) were determined using Differential Scanning Calorimetry (DSC), while electrical measurements were performed using complex impedance spectroscopy. The aim is to carry out an in-depth study of the effect of modifying cations of these glasses. Results obtained from investigating the previous properties of treated glasses based on the substitution of Li_2O with NaF, showing that most properties do not vary linearly and reach a minimum value (glass transition temperature T_g, relaxation time τ_σ, and electrical conductivity σ_{dc}), or a maximum value (activation energies $E_{a\sigma}$, E_{af}, frequency f_0, and pre-exponential term σ_0), lead to the conclusion that the effect of modifying cations is evident in these processed glasses. This effect of modifying cations in these glasses can be explained using the unified site relaxation model. The values found for τ_σ, σ_{dc}, $E_{a\sigma}$, E_{af}, f_0, and σ_0 for glass system examined are very close to those found by Ingram et al., who conducted a study of AgI–$AgPO_3$ and KI–$AgPO_3$ glasses.

Introduction

The importance of lithium batteries in Morocco for professionals is strategic: it positions the country as a major industrial hub in the global energy transition, attracting massive investment, creating jobs, reducing production costs thanks to local lithium and phosphate potential, and offering massive export opportunities to Europe thanks to the competitiveness of Moroccan batteries (36% lower cost) and geographical proximity. Moroccan professionals can capitalize on this momentum in the value chain, from ore extraction to the production of high-performance batteries, by developing specific skills and taking advantage of innovations [1].

Nowadays, batteries are used everywhere. Whether in cell phones, laptops, MP3 players, electric and hybrid vehicles, or for storing renewable energy, they have become indispensable. While batteries are part of everyday life, they are also constantly evolving. Demand requires devices that are ever smaller, lighter, more efficient, and more reliable. Lithium metal batteries are the most attractive in terms of energy density on the current market, thus meeting the demand in the face of various existing challenges. However, these batteries suffer from safety issues. One of the solutions proposed to counteract these problems is the development of all-solid-state lithium-metal batteries using a glassy electrolyte. This work has contributed to the development of

batteries, particularly for the storage of renewable energy and their use in electric and hybrid vehicles, as well as in mobile electronic devices.
The subject of ionic transport in ionically conductive glass is a challenge for the combined efforts of industry, experimenters and theorists. Glassy and crystalline materials are typically the source of the mixed mobile ion effect, which is more intriguing. The discovery mixed mobile ion effect is due to Liang [2]. Ion transport is crucial in electrochemical glass electrolyte systems, such as micro-batteries and micro-supercapacitors. Understanding the diffusion mechanism of mobile ions is crucial to the study of mixed mobile ion effect transport. It is well known that oxide glasses containing alkali ions behave essentially like solid electrolytes. It is well known that modifying cations (Li^+ and Na^+) can penetrate the glass lattice and play an important role in glass ionic conductivity. When lattice modifiers or dopant salts are integrated into the glass lattice, it causes depolymerization of the lattice, thus increasing the mobility of modifying cations. Investigations reveal that the coordination of various modifying cations in glasses containing multiple modifiers differs, and these cations are randomly distributed throughout the glass structure [3-4]. The sites of both types of modifying cations have significant differences, and a big difference in their site energies, hopping from one modifying ion to the sites of another modifying ion of a different type cannot occur. This results in fewer sites for the ionic movement of the more mobile modifying cations when they are replaced by less mobile alkali cations.

In this context, we have chosen glass system examined to investigate their conductivity and electrical relaxation over a range of temperatures (298 to 533 K) and frequencies (20 Hz to 1 MHz). The series of mixed modifying molybdenum phosphate glasses studied herein show a non-linear behaviour of the electrical properties studied (conductivity σ_{dc}, relaxation time τ_σ, activation energies $E_{a\sigma}$, E_{af}, frequency f_0, and pre-exponential term σ_0) and physical T_g.

Experimental procedures

The glasses examined were prepared in accordance with the reaction below:

$$(0.4\text{-}x)Li_2CO_3 + xNaF + 0.24MoO_3 + 0.72(NH_4)H_2PO_4 \rightarrow xNaF\ (0.4\text{-}x)Li_2O\ 0.24MoO_3\ 0.36P_2O_5 + 0.72NH_3 + 1.08H_2O + (0.4\text{-}x)CO_2 \quad \text{(Eq. 1)}$$

Where x=0, 4, 8, 12, 16, 20, 24, 28, 32, 36, 40 %.

Appropriate quantities of Li_2CO_3, NaF, MoO_3, and $NH_4H_2PO_4$ are ground. The mixture is placed in an open platinum crucible and introduced into a muffle furnace. The mixture is next exposed to two heat processes. The 1st treatment is at 533 K, which causes the decomposition of ammonium phosphate, moisture removal, and the release of CO_2. The second treatment is at 900°C, which causes the reaction mixture to melt. Finally, the glass is obtained by air quenching.

The nature of the materials produced was determined by thermal analysis and X-ray diffraction.

Complex impedance measurements were made at frequencies from 20 to 1 MHz and at temperatures from 293 to 533 K and reveal the electrical conductivity in all the compounds under study.

Results and discussion

Figure 1 carries the differential scanning calorimetry curve which records the evolution of T_g against of the x=Na/(Na+Li) for glass system examined.

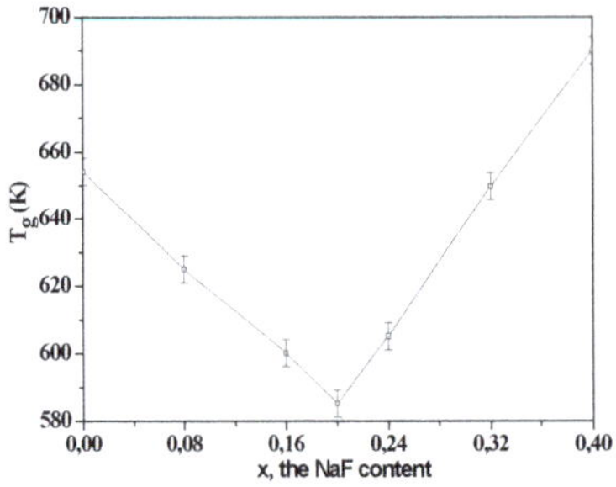

Fig. 1. Variation of T_g for the glass system examined.

It should be noted that Tg reaches a minimum when x=20 mol%. The mixed cation effect is clearly observed in the evolution of T_g according to the glass composition. Similar behaviour was observed for other phosphate glasses containing Na^+ and Li^+ [5]. The variation in T_g against of the NaF content, due to the effect of modifying cations, can be understood through the unified site relaxation model.

The data obtained are recorded on an Arrhenius curve with ($Log(\sigma_{dc})=f(10^3/T)$ and $Log(\tau_\sigma)=f(10^3/T)$) as shown in figures (2.a and 2.b).

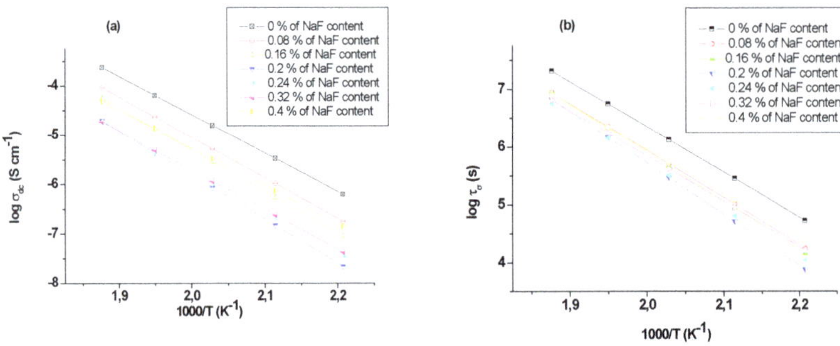

Fig. 2. Arrhenius curve obtained for glass system examined.

The lines we see are roughly parallel, which indicates that the σ_{dc} and τ_σ of these glasses follow an Arrhenius law and that their E_a values are similar.

To highlight the mixed mobile ion effect, we represented ($Log_{10}(\sigma_{dc})=f(x)$ and $Log_{10}(\tau_\sigma)=f(x)$, at different temperatures. Figures (3.a and 3.b) shows the conductivity σ_{dc} and the relaxation time τ_σ of the glass system examined. The conductivity decreases to a minimum when x=20 mol%. This event is called the mixed cation effect, and its amplitude increases as temperature drops.

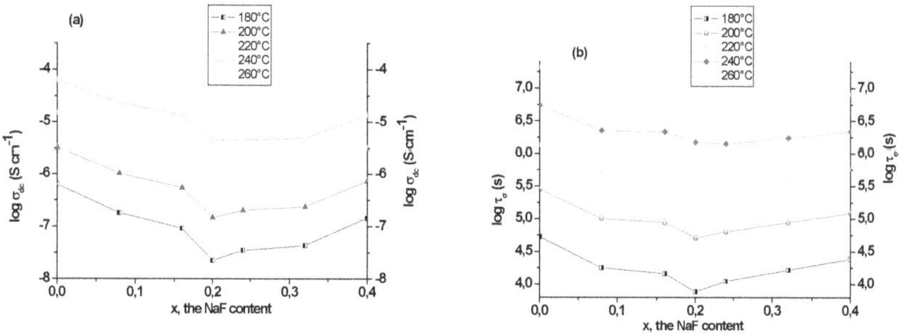

Fig. 3. Evolution of the dc conductivity and relaxation time τ_σ for glass system examined against of the NaF content.

Figure 4 shows the $E_{a\sigma}$ and σ_0 (Fig. 4a), the E_{af} and f_0 (Fig. 4b) of the glass system examined against of the NaF content.

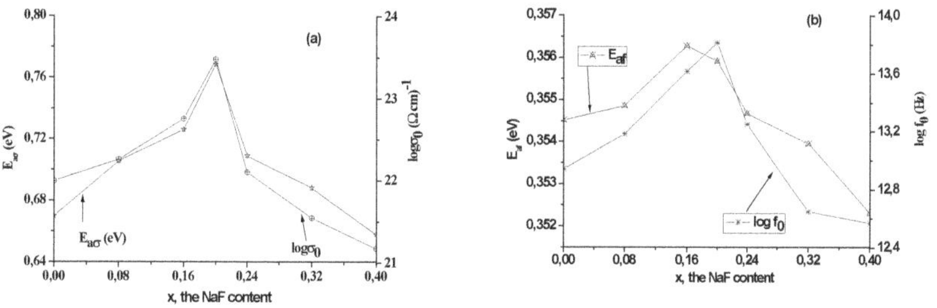

Fig. 4. Activation energies $E_{a\sigma}$, E_{af}, frequency f_0 and pre-exponential terms σ_0 for the glass system examined versus of NaF content.

These four quantities ($E_{a\sigma}$, E_{af}, f_0, and σ_0) reach a maximum when cation concentrations are equal. The glass composition 0.2NaF-0.2Li_2O-0.24MoO_3-0.36P_2O_5, with equal Li_2O and NaF proportions, shows low ion mobility. A distinction between single and mixed mobile ion glasses indicates that these glasses exhibit increased ionic conductivity (σ_{dc}) and decreased activation energies with higher mobile cation content. Specifically, lithium single-cation glasses show greater conductivity than sodium counterparts due to superior ionic mobility. Figure 3 illustrates a non-linear evolution of σ_{dc} and τ_σ as a function of NaF, indicating that certain alkali ions are involved in the construction of the glass network while maintaining the mobility of those that modify the network. The correlation between the fixation of these ions and the presence of non-bridging oxygen suggests that ions bound to phosphates or molybdates are the most mobile. The composition of the glass (x=20 mol%) shows the lowest values of σ_{dc} and τ_σ, due to the reduced mobility of the modifying cations. These results are consistent with the studies by Ingram et al. [6], suggesting that the effect of mixed mobile ions results from complex interactions between their concentration and mobility. In summary, the ability to move of Na^+ cations is lower than that of Li^+ cations, which is related to their respective molar volumes.

Materials Research Proceedings 64 (2026) 624-631 https://doi.org/10.21741/9781644904091-78

Figure 5 presents the σ_{ac} versus of frequency, σ_{ac} increases by increasing the frequency. The mixed cation effect is more marked at low frequency, and the amplitude of this effect decreases as frequency increases.

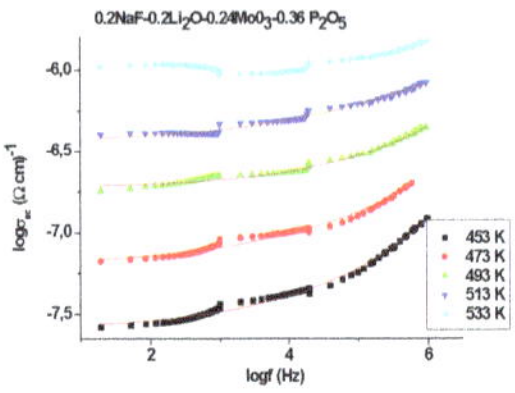

Fig. 5. The ac conductivity of the glass with the composition x=20 mol%

The magnitude of effect of modifying cations decreases with lower C_{tot} concentrations of mobile ions [7], the addition of an oxide former Al2O3, and particularly disappears at 8% Al2O3 in a phosphate vitreous system with mixed modifying (Li, Na) [8]. Additionally, effect of modifying cations reduces when two mixed modifying types have similar ionic radii.

The M''/M''_{max} versus frequency of single-ion and mixed-ion mobile glasses at various temperatures is depicted in Figure (6).

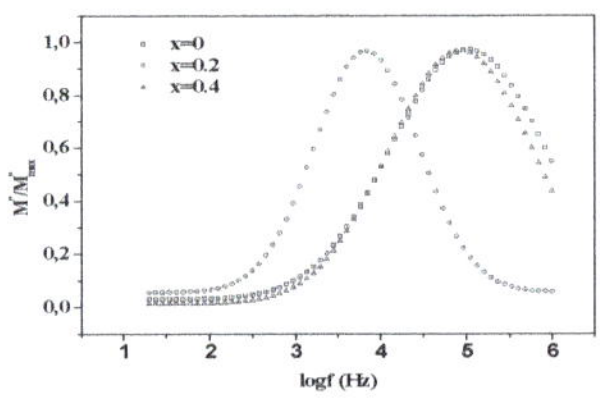

Fig. 6. M''/M''_{max} versus of frequency for single-ion and mixed-ion mobile glasses.

The spectra studied show asymmetric peaks in the M* scattering region, linked to a characteristic frequency f_p. The area to the left of the peak represents the long-range movements of mobile ions, while the area to the right concerns confined ions [9]. A transition from short-range mobility to long-range mobility is observed as frequency decreases. This transition is quantified by the equation $\omega\tau_\sigma=1$, where τ_σ is the conductivity relaxation time. With an increase in temperature, the maxima of the asymmetric peaks shift towards higher frequencies [10]. The M''/M''_{max} graphs reveal non-exponential behavior in conductivity relaxation, and the half-height width of the peaks is more significant than that of the Debye model [11]. Finally, the β value for mixed cation glasses is lower than that for single cation glasses.

In molybdenum phosphate glasses, the conductivity σ_{dc} decreases when Li_2O is partially replaced by NaF. This behaviour is consistent with the literature [6]. This behaviour leads to a crossover (degenerate mixing) of sodium and lithium ion mobility and a conductivity σ_{dc} is slightly related to ion concentration. This σ_{dc} conductivity minimum may be due to a change in the diffusion of the various modifying ions. The diffusion of a modifying cation decreases when it is replaced by another, reaching a threshold where the diffusion coefficients of the two types of cations are equal. The load transport for the most mobile species is greater than that of the least mobile species, which makes the minimum values of T_g, τ_σ and σ_{dc} very close to this crossover point [12]. Semi-quantitative investigations revealed that the reduction in conductivity is due to the different nature of NaF and Li_2O. Tomozawa and colleagues [13] have shown that the high static dielectric constant of non-crystalline materials with more than one modifier is due to the negative enthalpy of mixed modifying. Sato presented an atomic model of diffusion in glasses containing mixed cations, in which the internal energy of both types of alkali cations is negative. Kone and his colleagues also explained that the mixture of two types of alkali with negative enthalpy results in the formation of low mobility complexes, which causes the decrease in conductivity of mixed alkali glasses. However, Bunde and colleagues and Maass and colleagues [14], the alkaline ions contained in the glass create and retain their distinctive environment, which leads to a effective enthalpy of mixture for two different sorts of alkaline ions. The unified site relaxation model is primarily based on these hypotheses, which seem to demonstrate that alkaline modifying cations generate and sustain their own specific environments. This model, combines mainly the features of hopping relaxation and the dynamic structure, and also can offer a facilitation of ion transport process in glass systems. We can differentiate between two types of empty sites $\overline{Li}$ and $\overline{Na}$, just as we may differentiate between two types of cations Li^+ and Na^+. The sites $\overline{Li}$ and $\overline{Na}$ are an unoccupied site that were once occupied by the ions Li^+ and Na^+. When the environment is properly adjusted, an ion Li^+ will be inclined toward an unoccupied site $\overline{Li}$ that was previously inhabited by an ion Li^+, but it will decline to an unoccupied site $\overline{Na}$ that was formerly engaged by the cation Na^+, whither the abode has relaxed for shelter the cation Li^+. The appearance of paths that encourage glass ion movement is an unavoidable result of this differential effect.

The Highest conductivities observed on single-modifying glasses of the glass system examined, are assigned to the presence of Na+ or Li+ ions and the occasion this might allow for Li+ or Na+ ions to come in the place $\overline{Li}$ or $\overline{Na}$ that have been left empty by their neighbours. The low conductivity found in mixed modifying of the glass system examined is attributed to the disparity effect, the decrease in free volume and/or the low concentration of empty sites $\overline{Li}$ or $\overline{Na}$. The low concentration of empty sites $\overline{Li}$ or $\overline{Na}$ requires Li^+ or Na^+ ions to jump into empty sites $\bar{C}$. Li^+ or Na^+ cations are slowed down from entering the “bad” type sites. This is the concept of disparity, which appears whenever Li^+ ions make the jump into the sites $\overline{Na}$, and inversely.

$$w_{Li^+\overline{Na}} = w_{Li^+\overline{Li}} = exp\left(\frac{-T_0}{T}\right) \qquad \text{Eq (2)}$$

That kT_0, is the additional activation energy necessary for $Li^+\overline{Na}$ (or $Na^+\overline{Li}$) correlated with jumps $Li^+\overline{Li}$. The discrepancy energy should be significantly greater when Na^+ ions enter Li^+ sites than when the opposite occurs.

In other words, the distribution of free volume in glass can be used to explain the conductivity minimum in mixed modifying glass. It is had "free volume," which typically plays a role in annealing processes that take place slightly below T_g. The proposition is that a portion of this volume is "built into" diffusion paths and "captured" by mobile ions. The creation of unoccupied sites $\overline{Li}$ and $\overline{Na}$ depletes some of the free volume. However, the limitation of the free volume will typically result in a "braking effect" on the ion transport rate. The addition of a small amount of Li^+ cations capture more portion of the free volume, and limited for Na^+ cations. And therefore, the mobility of Na^+ and Li^+ ions passes through a minimum at equal concentrations of these ions. The high probability of hopping $w_{Li^+\overline{Li}}$ and $w_{Na^+\overline{Na}}$, controlled the high diffusion of alkali ions for the composition of the unique cation glasses, whereas a low chance of hopping $w_{Na^+\overline{Li}}$, $w_{Li^+\overline{Na}}$, $w_{Li^+\bar{C}}$ controlled the composition of the mixed-ion mobile glass. Therefore, at this composition, we anticipate a significant minimum of conductivity σ_{dc}. More ions can remain on their routes and the ion movement becomes more localized at high frequencies. As we expect from the analysis of path fluctuation. The conductivity vs x curve at various temperatures (Figure 3) and frequencies (Figure 6) has a noticeable minimum at composition of the mixed-ion mobile glass, which is more noticeable at lower temperatures and frequencies. The same findings were cited in previous research [12]. When the overall concentration of mobile ions is reduced, the magnitude of effect of modifying cations decreases [15]. And it disappears at very low concentration [12]. There is a high degree of conduction path connectivity in single modifying glass. However, because different cation sites differ, certain conduction path is blocked in mixed modifying glass. The crossing of conduction paths in mixed modifying glass significantly reduces the ionic mobility of various cation types.

The effect of modifying cations can be explained by the coupling between small and large cations entering sites. This concept contributes to understanding several important phenomena. A new method based on the constraint effect on ionic conductivity was reached [16-17]. The activation volume could be calculated using the following equation based on high pressure measurements:

$\Delta V^{*} = -RT\frac{dln\sigma}{dP}$ Eq (3) where P (Pa) is the equilibrium pressure

The figure given by Bandaranayake et al [16] shows how ΔV^{*} varies depending on the composition. The result show that ΔV^{*} rises in both directions, when sodium ions are substituted by lithium ions in glass $NaPO_3$ and when lithium ions are substituted by sodium ions in glass $LiPO_3$. The activation energy and ΔV^{*} have similar tendencies in a mixed cation glassy system as it has been noted by a researcher [17]. Furthermore, the finding somewhat agrees with a recent method to explain the effect of modifying cations [18]. According to this method, the change of sodium ions by lithium ions should lead to shrinkage of the lattice, and therefore the sodium ions should be required for the larger activation volume. Additionally, it was seen that the following simple rule was followed:

$$\Delta V^{*}(\text{MA glass}) \cong \Delta V^{*}(\text{First SA glass}) + \Delta V^{*}(\text{Second SA glass}) \qquad \text{Eq (4)}$$

Hence, the activation volumes were the arithmetic total of those associated with individual modifying glasses rather than the average.

Conclusion
Based on bibliographic research and experimental results, the study concludes that the modifying cations have a significant effect on the glass system examined. A clear mixed mobile ion effect is observed in thermal properties, specifically glass transition temperature (T_g). The direct current conductivity (σ_{dc}) and relaxation time (τ_σ) reach a minimum when Li_2O and NaF are in equal proportions. Additionally, activation energies ($E_{a\sigma}$, E_{af}), frequency (f_0), and the pre-exponential term (σ_0) attain a maximum under the same conditions. The similarity of $E_{a\sigma}$ and E_{af} from impedance spectra indicates that ionic transport in the vitreous system occurs via a hopping mechanism, necessitating careful evaluation of the conduction mechanism to avoid electrode polarization.

Mandatory statement on the use of generative AII declare that I did not use any generative artificial intelligence (AI) tools in preparing this article and have demonstrated responsible conduct and integrity in research since 2002. I inform the evaluation committee and scientific chair to fully appreciate the originality of this work.

References
[1] G. Karkera, M. A. Reddy, M. Fichtner, Journal of Power Sources, Recent developments and future perspectives of anionic batteries, 481 (2020) 228877. https://doi.org/10.1016/j.jpowsour.2020.228877

[2] C.C. Liang, J. Electrochem. Sot., Conductivity characteristics of solid electrolytes based on lithium iodide and aluminum oxide, 120 (1973) 1289. https://doi.org/10.1149/1.2403248
[3] E. Ratai, J.C.C. Chan, H. Eckert, Phys. Chem. Chem. Phys., Local coordination and spatial distribution of cations in mixed-alkali borate glasses, 4 (2002) 3198.
[4] J Lu, Z Shan, J Zhang, Y Su, K Yi, Y Zhang, Q Zheng, J. Non- Cryst. Solids, Mechanical properties of mixed modified oxide glasses, 16 (2022) 100125. https://doi.org/10.1016/j.nocx.2022.100125
[5] S. Koritala, K. Farooqui, M. Affatigato, S. Feller, S. Kambayanda, S. Ghosh, E.I. Kamitsos, G.D. Chryssikos, A.P. Patsis, J. Non- Cryst. Solids, The glass transition temperature of lithium-alkali borates, 134(3) (1991) 277-286. https://doi.org/10.1016/0022-3093(91)90386-K
[6] MD Ingram, CT Imrie, I Konidakis, Journal of non-crystalline solids, Activation volumes and site relaxation in mixed alkali glasses, 352(30&31) (2006) 3200-3209. https://doi.org/10.1016/j.jnoncrysol.2006.05.009
[7] A. Mager, PV Menezes, J Zakel, M Schäfer, J. Non- Cryst. Solids, Competitive transport of K+ and Ca++ ions in calcium-potassium phosphate glasses, 452 (2016) 231-237. https://doi.org/10.1016/j.jnoncrysol.2016.09.002
[8] A. Faivre, D. Viviani, J. Phalippou, Solid State Ionics, Mixed alkaline effect in Li and Na aluminophosphate glasses: influence of the cationic environment, 176(3&4) (2005) 325–332. https://doi.org/10.1016/j.ssi.2004.07.020
[9] J.O. Isard, Phys. Chem. Glass, The mixed alkali effect in glass, 1(3) (1969) 235-261. https://doi.org/10.1016/0022-3093(69)90003-9
[10] I.M. Hodge, M.D. Ingram and A. R. West, J. Electrical Chem., Impedance and modulus spectroscopy of polycrystalline solid electrolytes, 74 (2) (1976) 125-143. https://doi.org/10.1016/S0022-0728(76)80229-X
[11] H.K. Patel and S. W. Martin, Phys. Rev., Fast ionic conduction in $Na_2S^+B_2S_3$ glasses: Compositional contributions to nonexponentiality in conductivity relaxation in the extreme low-alkali-metal limit, B45 (1992) 10292. https://doi.org/10.1103/PhysRevB.45.10292
[12] M. Tomozawa, Solid State Ionics, 105(1-4) (1998) 249–255. The mixed alkali effect and thermodynamic state of glasses, https://doi.org/10.1016/S0167-2738(97)00472-4
[13] A. Kone, J.C. Reggiani and J. L. Souquet, Solid State Ionics, Thermodynamic approach of the mixed alkali effect in inorganic glasses, 9/10 partie 1 (1983) 709-715. https://doi.org/10.1016/0167-2738(83)90319-3
[14] P. Maass, A. Bunde, M.D. Ingram, Phys. Rev. Lett., Ion transport anomalies in glasses, 68 (1992) 3064. DOI: https://doi.org/10.1103/PhysRevLett.68.3064
[15] K. Funke, J. Non- Cryst. Solids, Jump relaxation model and coupling model-a comparison, 172/174 partie 2 (1994) 1215-1221. https://doi.org/10.1016/0022-3093(94)90646-7
[16] P.W.S.K. Bandaranayake, C.T. Imrie and M.D. Ingram, Phys. Chem., Pressure dependent conductivities and activation volumes in $Li_xNa_{(1-x)}PO_3$ glasses: evidence for a new matrix-mediated coupling mechanism in mixed-cation glasses?, 4 (2002), 3209-3213. https://doi.org/10.1039/B201215D
[17] M.D. Ingram, B. Macmillan, A.J. Pappin, B. Roling and J. M. Hutchinson, Solid State Ion., Effects of pressure on the ionic conductivity of AgI–$AgPO_3$ and KI–$AgPO_3$ glasses, 105 (1-4) (1998) 103-107. https://doi.org/10.1016/S0167-2738(97)00455-4
[18] R. Kirchheim, J. Non- Cryst. Solids, The mixed alkali effect as a consequence of network density and site energy distribution, 272 (2&3) (2000) 85-102. https://doi.org/10.1016/S0022 3093(00)00165-4

Electronic and Optical Properties of Cs_2PtCl_6 and Rb_2PtCl_6 Double Perovskites for Photocatalytic Hydrogen Production

Mouad BEN-NANA[1,a*], Abderrahman ABBASSI[1,b], Benachir ELHADADI[1,c]

[1]Laboratory of Research in Physics and Engineering Sciences, Sultan Moulay Slimane University, Polydisci-plinary Faculty, Beni Mellal, 23000, Morocco

[a]mouadbennana@gmail.com, [b]abbassi.abder@gmail.com, [c]doyen.fp@gmail.com

Keywords: Hydrogen Production, Hydrogen Evolution Reaction (HER), Lead-Free Perovskites, Density Functional Theory (DFT), Eco-Friendly Materials

Abstract. Water splitting has become the cornerstone of hydrogen production and thus the fundamental process for all clean energy systems. While pure Platinum (Pt) exhibits a high potential for integration into catalytic technologies, the design of efficient, stable, and low cost catalysts remains the major challenge. Alloying Pt or developing new alternative represent a promising pathway toward more sustainable catalysts. In this context, non-toxic halide double perovskites have established as effective materials for energy conversion thanks to their structural stability and tunable optoelectronic charactrestics. The optoelectronic properties of Cs_2PtCl_6 and Rb_2PtCl_6 were investigated using Density Functional Theory (DFT). The predicted band gaps suggest high light absorption in the visible region, indicating the suitability of the studied compounds for solar-driven water-splitting processes. Additionally, replacing Cs^+ with Rb^+ improves charge carrier separation through local lattice distortion and reduces the band gap. These findings show that Cs_2PtCl_6 and Rb_2PtCl_6 are stable, non-toxic, and tunable choices for photocatalytic hydrogen production and broader energy-harvesting applications.

Introduction

Hydrogen energy has attracted considerable attention thanks o its high abundance, its eco-friendly characteristics, and potential to replac internal combustion engines [1], [2]. However, certain challenges are limiting the upscale of hydrogen-based technologies. To overcome these challenges, a number of solutions have been developed. One oft hem is photocatalytic water splitting [3]. This strategy serves as a potential tool for decreasing dependency on harmful technologies. Electronic band gaps between 1.23 eV and 3.00 eV yield the best photocatalytic carachteristics. This range is suitable for light harvesting and yields encouraging redox potentials [4]. Due to its optoelectronic characteristics, a class of minerals known as double perovskites has attracted growing attention. The formula A_2BX_6, where a large cation occupies the A-site, a cation occupies the B-site, and a halogen or oxygen ion occupies the X-site, is typically used to designate double perovskites [5]. These materials have demonstrated excellent optoelectronic behavior, suggesting their possible integration into a variety of optoelectronic [6]–[8]. Numerous initiatives have examined Pb-based halide perovskites to assess potential integration for green energy applications [9]–[11]. However, one of the primary drawbacks of lead-based perovskites is their toxicity, which has prompted researchers all over the world to create new perovskites with less of an impact on the environment and better stability in ambient circumstances. Halide perovskites have stood out in the field of optoelectronic research due to their suitibility for addressing the problems associated with lead-based compounds [9]. Unfortunately, The majority of synthetic materials are not appropriate for solar applications because they either have high band gaps or poor optical properties. However, a number of recently created halide double perovskites have shown remarkable optical performance and stability in ambient circumstances [12], [13].

The present work was initiated to evaluate the potential integration of A_2PtCl_6 (A = Cs, Rb) for photocatalytic hydrogen production, employing density functional theory (DFT) approach. The main objectives are to determine their optoelectronic behavior, evaluating their suitability for photoelectric conversion and their potential integration for photocatalytic water splitting. The present comprehensive study, aims to establish A_2PtCl_6 (A = Cs, Rb) as an efficient materials for clean energy technologies, especially hydrogen production.

Computational methods

In order to get the physical properties of A_2PtCl_6 (A = Cs, Rb) double perovskite, a DFT investigation using the FP-LAPW method [14] has been initiated within the WIEN2k code [15]. The Generalized Gradient Approximation (GGA) [16] is applied to calculate the structural properties, while the modified Becke-Johnson (mBJ) exchange potential is employed to account the physical properties of the studied samples. Equation (1) presents the expression for the mBJ exchange[17].

$$v_{x,\sigma}^{MBJ}(r) = c v_{x,\sigma}^{BR}(r) + (3c-2)\frac{1}{\pi}\sqrt{\frac{5}{12}}\sqrt{\frac{2t_\sigma(r)}{\rho_\sigma(r)}} \quad (1)$$

The calculations were performed using an energy cutoff of -6.0 Ry and 1000 k-points.

The optical response of a material is defined the complex dielectric function ε (ω), which is expressed as below [18]:

$$\varepsilon(\omega) = \varepsilon_1(\omega) + i\varepsilon_2(\omega) \quad (2)$$

Where $\varepsilon_1(\omega)$ presents the real part part of ε (ω) and $\varepsilon_2(\omega)$ its imaginary part

In cubic symmetry, ε_2 (ω) can be expressed as follows:

$$\varepsilon_2(\omega) = \left(\frac{\hbar^2 e^2}{\pi m^2 \omega^2}\right) \sum_{c,v}^{n} \int d^3k \langle c_k|p^\alpha|v_k\rangle\langle v_k|p^\beta|c_k\rangle x\delta(\varepsilon_{c_k} - \varepsilon_{v_k} - \omega) \quad (3)$$

ε_1 (ω) can be calculated using $\varepsilon_2(\omega)$ as follows:

$$\varepsilon_1(\omega) = 1 + \frac{\pi}{2} p \int_0^\infty \frac{\omega' \varepsilon_2(\omega')}{(\omega')^2 - \omega^2} d\omega' \quad (4)$$

ε_1 (ω) and $\varepsilon_2(\omega)$ provide the opportunity to investigate optical functions such as the absorption coefficient α(ω) and reflectivity R(ω), according to the expressions below:

$$\alpha(\omega) = \sqrt{2}\omega\left[\sqrt{\varepsilon_1^2(\omega) + \varepsilon_2^2(\omega)} - \varepsilon_1(\omega)\right]^{\frac{1}{2}} \quad (5)$$

$$R(\omega) = \left|\frac{\sqrt{\varepsilon(\omega)} - 1}{\sqrt{\varepsilon(\omega)} + 1}\right|^2 \quad (6)$$

Results and discussions

Structural properties

The atomic structure of the Pt-based double perovskites A_2PtCl_6 (A= Cs, Rb) was investigated using a face-centred geometry. Results confirm their cubic structure with ***Fm$\bar{3}$m*** as space group (no. 225). Fig.1. show the corresponding unit cell. The Cs and Rb atoms are located placed at the 8c Wyckoff site, Pt atoms at the 4a site, and Cl atoms lie at the 24e one.

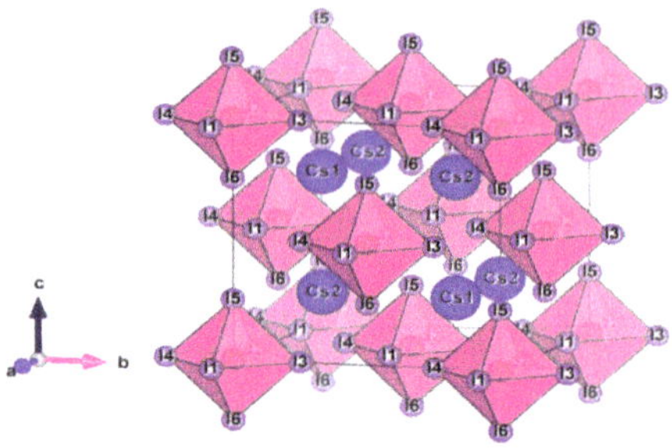

Fig.1: The unit cell of A_2PtCl_6 (A= Cs, Rb) double perovskites

The lattice parameters of the studied materials have been simulated using GGA-PBE function. **Table 1** presents the observed parameters for A_2PtCl_6 (A= Cs, Rb). The comparison with the available results shows good agreement with the experimental value.

Electronic properties

Table 1: Observed and calculated values of a(Å), B_m (GPa), B^P, and E_o (Ry) of A_2PtCl_6 (A= Cs, Rb).

Parameters	Cs_2PtCl_6		Rb_2PtCl_6	
	Present work	Other	Present work	Other
a(Å)	9.9001	10.94[21] - 10.40[22]	9.7354	10.23[23]
B_m (Gpa)	29.1394		43.558	
B^P	5.00		5.00	
E_o (Ry)	-73593.3486		-54358.14841	

The band structures of the studied materials have been investigated in order to understand their electronic behavior. Fig. 2. Presents the band structures of Cs_2PtCl_6 and Rb_2PtCl_6. The results show that the band gaps are direct for the two materials, exhibiting band values of E_g = 1.867 eV and E_g = 1.357 eV for Cs_2PtCl_6 and Rb_2PtCl_6, respectively. Both materials exhibit a band value in the range of 1.23 - 3.00 eV, ideal for photocatalytic and photovoltaic applications.

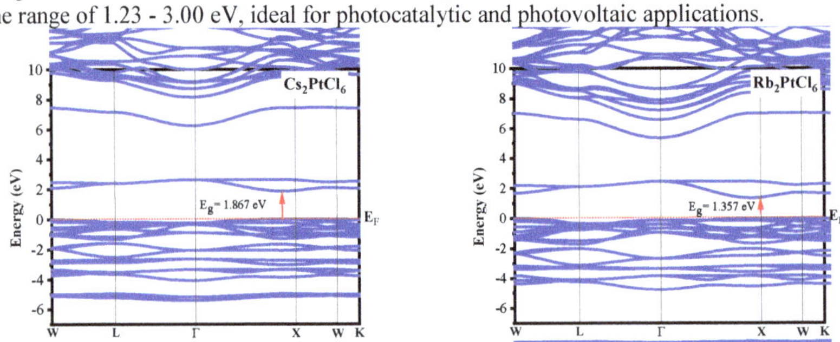

Fig. 2. Band structure plot for A_2PtCl_6 (A= Cs, Rb).

Fig.2 presents the density of states (DOS) plots for A_2PtCl_6 (A= Cs, Rb). Plots reveal that the B-site atom is the primary contributors to the electronic characteristics. The valence band is mainly occupied by Cl-p states, with minor contributions of Pt-d states. The conduction band is mainly formed by Pt-d, with a slight contribution from Cl-p states.

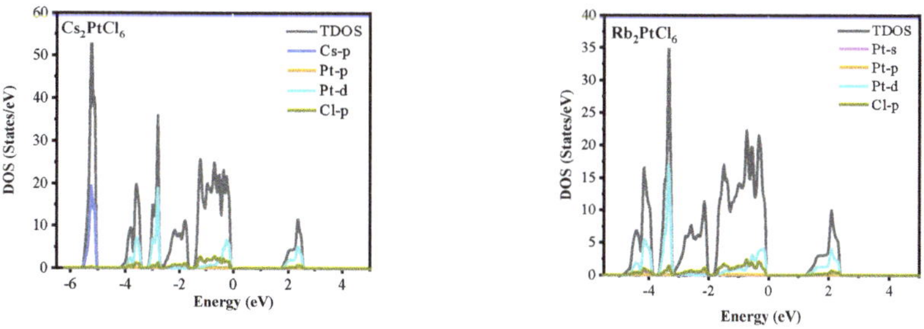

Fig. 3. TDOS and PDOS plots for A_2PtCl_6 (A= Cs, Rb).

Redox potential

Redox potentials (RP) are the main condition for the design of an efficient photocatalyst alongside a narrow bandgap. Typically, around 2.0 eV. The redox potentials are calculated based on the band edge positions, which are calculated using Equations (7) and (8). These values are then compared to the water oxidation and reduction potentials in order to assess the redox capability of the studied materials.

$$E_{CB} = \chi - E_e - \frac{1}{2} E_g \quad (7)$$

$$E_{VB} = E_{CB} \quad (8)$$

$\boldsymbol{E}_e$ (=4.5 eV) displays the free energy of electrons on the hydrogen scale, χ presents the electronegativity, it is calculated using the difference between the CBM and [19]. $\boldsymbol{E}_{CB}$ displays the energy of the CB and $\boldsymbol{E}_{VB}$ the energy of the VB. Its value is 2.8668 eV for Cs_2PtCl_6 and 0.8796 eV for Rb_2PtCl_6. **Fig. 4** represents the calculated positions of the VBM and CBM of A_2PtCl_6 (A= Cs, Rb). From a theoretical view, when a semiconductor photocatalyst receives an irradiation with an energy value equal or exceed the band gap energy, holes would be created in the valence band region.

To enable water splitting, the CBM must lie at a lower energy than the reduction potential of H^+ to H_2 (taken as 0 eV vs. NHE), while the VBM must lie at a higher energy than the oxidation potential of H_2O to O_2 (1.23 eV vs. NHE). Both conditions are satisfied in the A_2PtCl_6 perovskite, indicating that H_2 reduction and O_2 evolution are thermodynamically feasible. Therefore, these perovskite materials are promising for water splitting and can be considered effective candidates for photocatalytic applications. The process of water splitting is illustrated in fig.5.

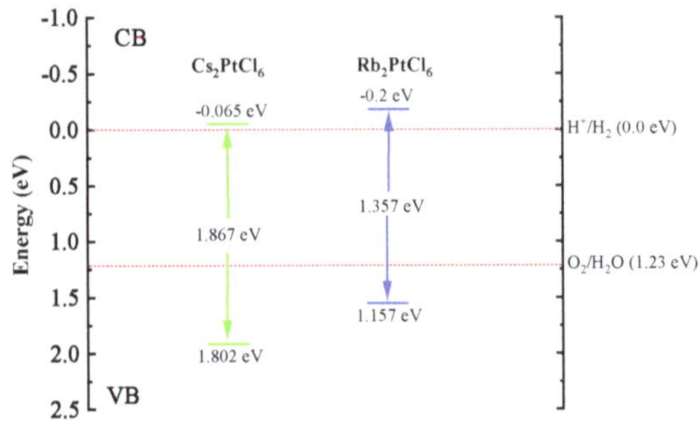

Fig. 4. Calculated bandgaps and band edge positions for A_2PtCl_6 (A= Cs, Rb).

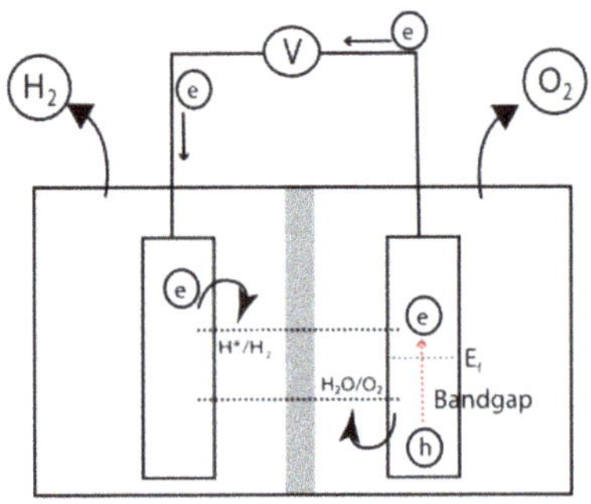

Fig.5. Processes of photoelectrochemical water splitting [20].

Optical properties
The optical response of the studied materials provides valuable insights on their optoelectronic and photovoltaic behavior. Figure 6.a shows the optical absorption coefficient for Cs_2PtCl_6 and Rb_2PtCl_6. Both compounds exhibit very low absorption in the low-energy region, followed by a notable and rapid increase as soon as the photon energy exceeds their electronic band gaps, confirming their semiconductor behavior. The efficient excitation of charge carriers from the valence band to the conduction band is indicated by this dramatic rise in absorption. For both materials, a number of peaks are seen in the high-energy region that correspond to changes from lower valence band states to higher conduction band states. The absorption behaviour of the investigated materials, highlights their exceptional light capture efficiency. They are therefore attractive options for UV-based photovoltaic and optoelectronic systems. Figure 6.b shows the variance in reflectance for Cs_2PtCl_6 and Rb_2PtCl_6. At lower photon energies, these compounds exhibit similar reflective behavior with several peaks that indicate interband electronic transitions. However, clear differences emerge at higher photon energies. Both materials may be useful in UV photonic coatings and reflecting optoelectronic components that require efficient UV photon management, according to this characteristic. Based on the calculated optical properties, double halide perovskites have interesting characteristics for light-harvesting technologies. Due to their

exceptional UV light absorption capabilities, both materials are promising candidates for a range of next-generation optoelectronic applications.

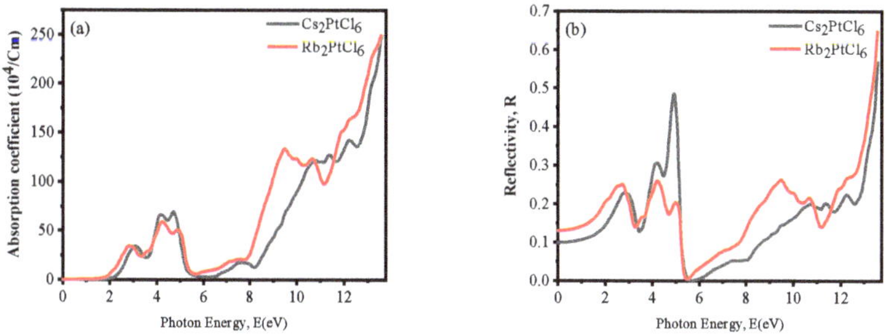

Fig. 5. (a) Absorption coefficient and (b) reflectivity index plots for A_2PtCl_6 (A = Cs, Rb).

Conclusion

One important method for producing hydrogen sustainably is the electrochemical splitting of water, but this process needs materials that can effectively drive the processes that lead to the development of both hydrogen and oxygen. DFT was used to examine the electrical and optical characteristics of Cs_2PtCl_6 and Rb_2PtCl_6. The results demonstrate that both materials are promising for solar-driven water splitting because they have appropriate band gaps (1.867 eV for Cs_2PtCl_6 and 1.357 eV for Rb_2PtCl_6) for high visible-light absorption. Crucially, the redox potential study shows that the VBM is above the H_2O/O_2 oxidation potential and the CBM is below the H^+/H_2 reduction potential. These findings suggest the possibility of water oxidation and proton reduction. Replacing Cs^+ with Rb^+ improves charge carrier separation and increases the band gap value, which enhances photocatalytic efficiency. The presented results confirm the suitibility of Cs_2PtCl_6 and Rb_2PtCl_6 for water splitting, underscoring their promise for both wider energy conversion applications and sustainable hydrogen production.

References

[1] J. G. Love *et al.*, *Sustain. Energy Fuels*, vol. 6, no. 17, pp. 4008–4023, 2022, https://doi.org/10.1039/D2SE00923D

[2] A. M. Abdalla *et al.*, *Energy Convers. Manag.*, vol. 165, pp. 602–627, 2018, doi: https://doi.org/10.1016/j.enconman.2018.03.088

[3] M. S. Hossian, M. M. H. Babu, A. Kabir, A. Azzouz Rached, and M. I. Kholil, *Mater. Adv.*, vol. 6, no. 15, pp. 5221–5231, 2025, https://doi.org/10.1039/D5MA00043B

[4] D. Zhang *et al.*, *Nat. Energy*, vol. 8, no. 6, 2023, https://doi.org/10.1038/s41560-023-01240-9

[5] M. Ben-nana *et al.*, *Chem. Phys.*, vol. 600, p. 112920, 2026, https://doi.org/10.1016/j.chemphys.2025.112920

[6] Y.-F. Xu *et al.*, *J. Am. Chem. Soc.*, vol. 139, no. 16, pp. 5660–5663, Apr. 2017, https://doi.org/10.1021/jacs.7b00489

[7] L. N. Quan *et al.*, *Chem. Rev.*, vol. 119, no. 12, pp. 7444–7477, Jun. 2019, https://doi.org/10.1021/acs.chemrev.9b00107

[8] K. Sim *et al.*, *Appl. Phys. Rev.*, vol. 6, no. 3, p. 31402, Jul. 2019, https://doi.org/10.1063/1.5098871

[9] A. Kojima *et al.*, *J. Am. Chem. Soc.*, vol. 131, no. 17, pp. 6050–6051, May 2009, https://doi.org/10.1021/ja809598r

[10] M. R. Jani *et al.*, *Superlattices Microstruct.*, vol. 146, p. 106652, 2020, doi: https://doi.org/10.1016/j.spmi.2020.106652

[11] M. Ben-Nana *et al.*, *2025 International Conference on Circuit, Systems and Communication (ICCSC)*, 2025, pp. 1–7, https://doi.org/10.1109/ICCSC66714.2025.11135120

[12] D. Wu *et al.*, *J. Phys. Chem. Lett.*, vol. 12, no. 21, pp. 5110–5114, Jun. 2021, https://doi.org/10.1021/acs.jpclett.1c01128

[13] N. Sakai *et al.*, *J. Am. Chem. Soc.*, vol. 139, no. 17, pp. 6030–6033, May 2017, https://doi.org/10.1021/jacs.6b13258

[14] B. V. F. Study *et al.*, vol. 140132, pp. 2–6, 2018, https://doi.org/10.1063/1.5033307

[15] F. T. and L. D. M. WIEN2k: An Augmented Plane Wave Plus Local Orbitals Program for Calculating Crystal Properties.," *Vienna Univ. Technol.*, 2001.

[16] G. Kadim *et al.*, *Phys. Scr.*, vol. 96, no. 4, p. 45804, 2021, https://doi.org/10.1088/1402-4896/abe00d

[17] F. Tran and P. Blaha, *Phys. Rev. Lett.*, vol. 102, no. 22, p. 226401, Jun. 2009, https://doi.org/10.1103/PhysRevLett.102.226401

[18] L. Hnamte*et al.*, vol. 10, no. 3, pp. 39–44, 2018, https://doi.org/10.9790/4861-1003023944

[19] A. Klein, *Thin Solid Films*, vol. 520, no. 10, pp. 3721–3728, 2012, https://doi.org/10.1016/j.tsf.2011.10.055

[20] H. M. Chen *et al.*, *Chem. Soc. Rev.*, vol. 41, no. 17, pp. 5654–5671, 2012, https://doi.org/10.1039/C2CS35019J

[21] H. H. Hegazy *et al.*, *J. Mater. Res. Technol.*, vol. 19, pp. 1271–1281, 2022, https://doi.org/10.1016/j.jmrt.2022.05.082

[22] R. Sa *et al.*, *Chem. Phys. Lett.*, vol. 754, p. 137538, 2020, https://doi.org/10.1016/j.cplett.2020.137538

[23] Q. Mahmood *et al.*, *J. Phys. Chem. Solids*, vol. 148, p. 109665, 2021, https://doi.org/10.1016/j.jpcs.2020.109665

Investigations of the Thermoelectric Potential of Single-Layered SiC and ZnO via First-Principles Calculations

Bakhtiar UL HAQ[1] *, Se-Hun KIM[2] †, R. AHMED[3,4], N. ALI[5], Zulfiqar Ali SHAH[6], Ubaid TANTARY[7], M. JAWAD[2]

[1]Department of Mechanical Engineering, College of Engineering, Prince Mohammad Bin Fahd University, Alkhobar 31952, Saudi Arabia

[2]Faculty of Science Education, Jeju National University, Jeju 63243, Republic of Korea

[3]Center for High Energy Physics, University of the Punjab, Quaid-e-Azam Campus, Lahore-54590, Pakistan

[4]Department of Physics, Faculty of Science, Universiti Teknologi Malaysia, UTM, Skudai, 81310 Johor, Malaysia

[5]African Sustainable Agriculture Research Institute (ASARI), Mohammad VI Polytechnic University (UM6P), Laayoune, Morocco

[6]Department of Physics, Allama Iqbal Open University, Islamabad, P.O. Box 44310, Pakistan

[7]College of Science and Humanities, Prince Mohammad Bin Fahd University, Alkhobar 31952, Saudi Arabia

*bakhtiarjadoon@gmail.com, †spinjj@jejunu.ac.kr

Keywords: Thermoelectric Materials, First-Principles Calculations, Power Factor, Two-Dimensional Materials, Renewable Energy Applications

Abstract: Materials for Thermoelectric applications have been substantially explored recently due to growing demands for renewable energy production. This article reports the thermoelectric response of single-layered ZnO (SL-ZnO) and single-layered SiC (SL-SiC) in the present first-principles work. The carrier concentration and density of states (DOS) associated with the SL-SiC are greatly altered by chemical potential and temperature. DOS of SL-SiC approaches as high as $\sim 158\times 10^{21}$ e/cm^3 for electrons and $\sim 74\times 10^{21}$ e/cm^3 for holes, whereas it remains relatively stable ($\sim$13-59$\times 10^{21}$ e/cm^3) for SL-ZnO. This suggests greater tunability of thermoelectric SL-SiC. The Seebeck coefficients of these 2D materials show symmetric behavior around the Fermi level, with opposite signs for *n*- and *p*-type doping and a slight decrease at higher temperatures. Their power factors ($PF = S^2\sigma/\tau$) increase with temperature, peaking at $\sim 5.44\times 10^{11}$ $Wm^{-1}K^{-2}s^{-1}$ for *n*-type SL-SiC and $\sim 2.47\times 10^{11}$ $Wm^{-1}K^{-2}s^{-1}$ for *p*-type SL-ZnO, highlighting optimal carrier types and thermoelectric potential of these 2D materials at elevated temperatures.

1. Introduction:

The harmful effects of fossil fuel use on the environment have encouraged researchers to look for cleaner and more sustainable sources of energy. One promising way to do this is by converting solar or waste heat into electricity through thermoelectric devices [1-3]. These systems can change heat directly into electric power, or vice versa, without moving parts [4-7]. They are therefore used in several areas such as waste-heat recovery, solid-state cooling, and space power modules, etc [8-10]. A thermoelectric device functions on the principle that a temperature difference (ΔT) across a semiconductor's or conductor's terminals induces a voltage difference (ΔV), also known as the Seebeck effect [11]. The thermoelectric efficiency of a material is described by the ratio of the ΔV to ΔT. Production of large-scale renewable energy through thermoelectric technology requires materials capable of highly efficient heat-to-energy conversion. Exploring high-efficiency thermoelectric materials has become the major objective of clean energy research in recent years.

Efficient thermoelectric materials are characterized primarily by the Seebeck coefficient and the thermal and electrical conductivities. These parameters collectively define the efficiency of heat-to-electricity conversion by a thermoelectric material. An efficient thermoelectric material exhibits low thermal and high electrical conductivity and a large Seebeck coefficient. Getting all three parameters in the above-mentioned combination is, however, difficult. Therefore, exploring/designing favorable thermoelectric materials is still an active area of research. One promising idea is to reduce the dimensions of the potential materials to two dimensions. When a material is thinned down to a single layer, strong quantum and surface effects appear that can help improve thermoelectric behavior [12-17]. Reducing dimensionality tends to loosen the link between electrical and thermal conductivities because of stronger phonon scattering [18]. In many cases, nanostructuring also changes the DOS in a way that may increase both the conductivity and Seebeck coefficient [19, 20]. The improvement in these parameters due reduction in structural dimensions allows for the achievement of high PF [10, 19]. These features make 2D materials interesting for clean-energy applications.

These interesting features of 2D materials have motivated this study to explore the thermoelectric properties of SL-SiC and SL-ZnO and evaluate their potential for renewable energy applications through first-principles calculations. The selection criteria for exploring the potential of SL-SiC and SL-ZnO are based on cost-effectiveness and environmental friendliness of these materials. The chemical potential and temperature-driven changes in carrier concentration, DOS, and the Seebeck coefficient have been examined in detail. The PF for these 2D materials is calculated to determine the efficiency of these 2D materials for heat-to-clean-energy conversion. The obtained results reveal the thermoelectric behavior of SL-SiC and the relatively stable behavior of SL-ZnO.

2. Computational Details

The preliminary electronic structure calculations for thermoelectric properties of SL-SiC and SL-ZnO were investigated using the WIEN2k code [21], employing the Full-Potential Linearized Augmented Plane Wave (FP-LAPW) method of density functional theory (DFT). Each unit cell was partitioned into an interstitial region (where plane waves are described) and non-overlapping Muffin-Tin spheres (where Atomic wavefunctions are applied). Muffin-Tin (MT) radii were chosen as Si = 1.59 a.u., C = 1.67 a.u., Zn = 1.73 a.u., and O = 1.49 a.u. The angular momentum expansion was set to l_{max} = 10, and the plane-wave cutoff was $K_{max} = 8.0/R_{MT}$, with the charge density expanded up to $G_{max} = 14$. The Brillouin zone was sampled using a dense $21\times21\times1$ k-point mesh. Exchange-correlation interactions were described using the Perdew-Burke-Ernzerhof generalized gradient approximation (PBE-GGA) [22], while the modified Becke-Johnson (TB-mBJ) potential [23, 24] was applied to improve the accuracy of band-gap calculations [25-27]. The BoltzTraP2 code [28] was used for calculations of the thermoelectric properties. Energy convergence was set to 10^{-5} Ry, and a dense k-point mesh ensured accurate evaluation of transport coefficients. This computational framework provides a reliable basis for predicting and analyzing the thermoelectric performance of SL-SiC and SL-ZnO for potential applications in renewable energy and waste-heat recovery technologies. The thermoelectric properties presented in present work are calculated with respect to relaxation time that may lead to a slight deviation from the experimental results. Therefore, the results provided in this work should be interpreted as qualitative values rather than experimental results.

3. Results and Discussion

The SL-SiC and SL-ZnO shown in Fig. 1 were designed using a hexagonal 3D structure of ZnO and SiC and were fully optimized using the WIEN2k computational code. Like graphene, both 2D

Materials Research Proceedings 64 (2026) 639-646 https://doi.org/10.21741/9781644904091-80

materials contain two nonequivalent atomic sites arranged in a planar honeycomb pattern. A vacuum slab of about 20Å was applied in the *z*-direction to eliminate any interaction between the neighboring layers. For the SL-ZnO, the zinc and oxygen atoms occupy the fractional coordinates (1/3, 2/3, 1/2) and (2/3, 1/3, 1/2), respectively. Similarly, in the SL-SiC, the silicon and carbon atoms are positioned at (1/3, 2/3, 1/2) and (2/3, 1/3, 1/2), respectively. These optimized structures are used for detailed thermoelectric properties calculations.

Figure 2 illustrates the variation of carrier concentration (N) against chemical potential (μ) for SL-ZnO and SL-SiC at 300, 600, 900, and 1200 K. N is zero near the bandgap, which can be explained by the absence of free carriers within the corresponding energy range. A noticeable threshold in carrier concentration beyond ±1.3 eV is observed for all temperatures. This threshold occurs at relatively low doping levels at 1200 K, followed by 900 K, 600 K, and 300 K.

N strongly depends on temperature and changes significantly with chemical potential in the case of SL-SiC. At μ = -2 eV, N is approximately 1.38 $\times 10^{21}$, 1.60 $\times 10^{21}$, 1.50 $\times 10^{21}$, and 1.91

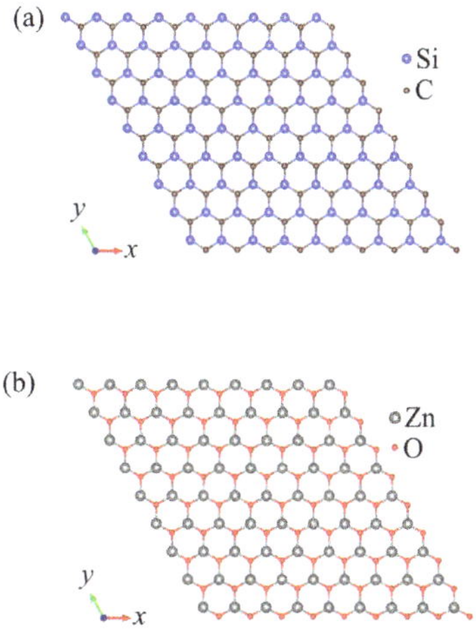

Figure 1. Optimized atomic structures of SL-SiC (a) and SL-ZnO (b) showing their hexagonal lattice symmetry.

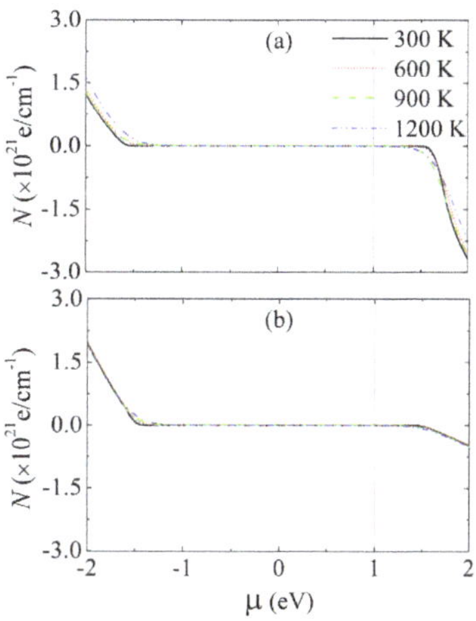

Figure 2. Carrier concentration (N) of (a) SL-SiC and (b) SL-ZnO calculated at 300, 600, 900, and 1200 K.

$\times 10^{21}$ *e*/cm^3 for 300, 600, 900, and 1200 K, respectively, while at μ = +2 eV, N is -3.02 $\times 10^{21}$, -2.91 $\times 10^{21}$, -3.01 $\times 10^{21}$, and -2.58 $\times 10^{21}$ *e*/cm^3. In SL-SiC, N is altered significantly with changes in chemical potential and temperature. N is greater for electrons (μ = +2 eV) than for holes (μ = −2 eV). With increasing temperature, the hole concentration (μ = −2 eV) tends to rise, whereas the electron concentration (μ = +2 eV) shows a slight decline. SL-ZnO shows opposite behavior to that of SL-SiC and shows only minor sensitivity to changes in chemical potential and temperature. SL-ZnO demonstrates $N \sim$ 1.76 $\times 10^{21}$, 1.84 $\times 10^{21}$, 1.82 $\times 10^{21}$, and 1.89 $\times 10^{21}$ *e*/cm^3 for 300, 600, 900, and 1200 K at μ = -2 eV, while N is -4.29 $\times 10^{20}$, -4.37 $\times 10^{20}$, -4.81 $\times 10^{20}$, and -4.92 $\times 10^{20}$ *e*/cm^3 at μ = +2 eV. This shows that the N of SL-ZnO remains nearly constant for the considered chemical potential and temperature, whereas SL-SiC exhibits significant variations.

Fig. 3 depicts the response of the DOS of SL-SiC and SL-ZnO to a change in chemical potential and temperature. The variation in DOS is found to be similar to the response of N shown in Fig. 2. Within the bandgap, the DOS remains zero, which confirms the absence of electronic states corresponding to ±1 eV energy range. A distinct rise in DOS appears beyond ±1 eV for all temperatures, which indicates the onset of the doping effect. This onset occurs at progressively lower doping levels as the temperature increases, and appears first at 1200 K, followed by 900, 600, and 300 K. At $\mu = -2$ eV (holes), the DOS values for SL-SiC are approximately 60.6, 65.9, 68.0, and 73.9 $\times 10^{21}$ e/cm^3 at 300, 600, 900, and 1200 K, respectively. For electrons (n-type region), the peak DOS occurs just below $\mu = +2$ eV, with corresponding values of 158.0, 121.0, 101.0, and 87.9 $\times 10^{21}$ e/cm^3. Consistent with the carrier concentration trends, the DOS is higher for electrons than for holes. Interestingly, the DOS for n-type doping decreases slightly with temperature, whereas for p-type doping, it increases. This shows that DOS of SL-SiC is sensitive to changes in both chemical potential and temperature, whereas DOS of SL-ZnO demonstrates a comparatively weaker response.

For holes at $\mu = -2$ eV, DOS values are 53.6, 53.7, 55.0, and 59.3 $\times 10^{21}$ e/cm^3 for 300, 600, 900, and 1200 K, while for electrons at $\mu = +2$ eV, the DOS remains nearly the same as 13.3, 13.1, 13.1, and 13.2 $\times 10^{21}$ e/cm^3. This demonstrates that DOS for SL-ZnO is relatively insensitive to changes in chemical potential and temperature.

The electrical conductivity per relaxation time (σ/τ) for SL-SiC and SL-ZnO exhibits a similar dependence on chemical potential as the carrier concentration [Fig. 2] and DOS [Fig. 3]. As shown in Figure 4, σ/τ is zero within the bandgap due to the absence of available carriers and electronic states. A clear threshold in σ/τ appears beyond ±1 eV for all temperatures, corresponding to the onset of charge transport. This threshold moves toward lower doping levels as the temperature increases. σ/τ of SL-SiC exhibits a strong dependence on both temperature and chemical potential. At $\mu = -2$ eV (p-type region), σ/τ values are approximately 3.25 $\times 10^{19}$, 3.54 $\times 10^{19}$, 3.41 $\times 10^{19}$, and 4.06 $\times 10^{19}$ (Ω^{-1}m^{-1}s^{-1}) for 300, 600, 900, and 1200 K, respectively. For electrons (n-type doping), the conductivity reaches its maximum slightly before $\mu = +2$ eV, with peak values of 8.42 $\times 10^{19}$, 8.09 $\times 10^{19}$, 8.17 $\times 10^{19}$, and 7.15 $\times 10^{19}$ (Ω^{-1}m^{-1}s^{-1}). σ/τ in SL-SiC is considerably higher for electrons than for holes, which is analogous to the trends observed in the case of N in Fig. 2 and DOS in Fig. 3. The rise in temperature causes the σ/τ to decrease for n-type doping and increase for p-type doping. In contrast, the SL-ZnO displays a relatively weaker dependence on chemical potential and temperature. For holes ($\mu = -2$ eV), σ/τ is 3.45 $\times 10^{19}$, 3.59 $\times 10^{19}$, 3.55 $\times 10^{19}$, and 3.66 $\times 10^{19}$ (Ω^{-1}m^{-1}s^{-1}) at 300, 600, 900, and 1200 K, respectively, while for electrons ($\mu = +2$ eV), the corresponding values are 1.82 $\times 10^{19}$, 1.80 $\times 10^{19}$, 1.90 $\times 10^{19}$, and 1.93 $\times 10^{19}$ (Ω^{-1}m^{-1}s^{-1}). Thus, σ/τ in SL-ZnO is slightly higher for p-type doping compared to n-type, and its temperature-induced variations remain modest.

Fig. 5 presents the Seebeck coefficient (S) of SL-SiC and SL-ZnO, which displays a symmetrical dependence on chemical potential. The positive S values indicate p-type, and the negative values correspond to n-type doping. The absolute value of S associated with SL-SiC decreases gradually with increasing temperature due to carrier excitations at higher temperatures. At 300 K, S reaches ±1550 μV/K at $\mu \approx \pm 1.13$ eV, while at 1200 K, it slightly decreases to ±1300 μV/K. S of SL-ZnO exhibits a similar trend with chemical potential and temperature as shown in Fig. 5(b). Its magnitude remains nearly constant (±1550 μV/K) up to 900 K and decreases slightly to about ±1200 μV/K at 1200 K. The weak temperature dependence of S of SL-ZnO can be associated with the relatively stable carrier concentration [Fig. 2b] and DOS [Fig. 3b].

Fig. 6 illustrates the PF, which represents the combined influence of σ/τ [Fig. 4] and *S* [Fig. 5] on thermoelectric performance. The PF remains zero within the bandgap for both SL-SiC and SL-ZnO. It becomes significant only beyond ±1 eV, which corresponds to regions where charge carriers are thermally excited. The PF curves exhibit significant resonance with several small resonance peaks near the band edges at 300 K. As the temperature increases, the dispersion of PF becomes smoother, and the secondary peaks gradually vanish, which indicates enhanced carrier scattering and redistribution of charge carriers at high temperature. Fig. 6 shows that the magnitude of the PF peaks increases considerably with temperature, both for SL-SiC and SL-ZnO, suggesting an improved thermoelectric response due to enhanced carrier excitation. Moreover, the PF threshold occurs at lower energies as the temperature rises.

The ability of a thermoelectric material to transform ΔT to electrical power is quantified by the PF. For SiC, the PF is higher for *n*-type doping within ±2 eV, while *p*-type contributions are

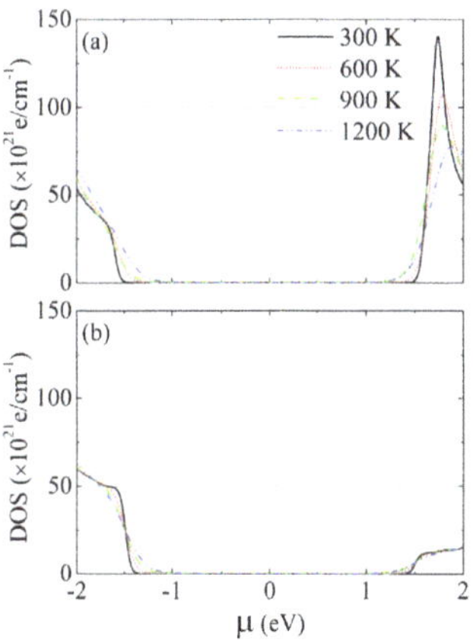

Figure 3. DOS of SL-SiC (a) and SL-ZnO (b) at 300, 600, 900, and 1200 K. DOS of SL-ZnO are relatively less affected by change in temperature compared to that of SL-SiC.

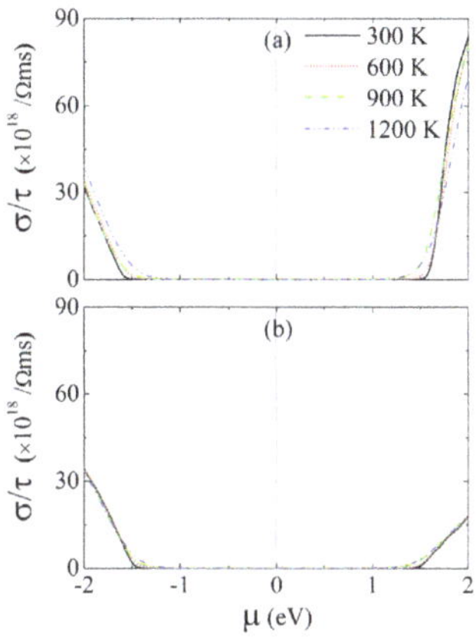

Figure 4. The σ/τ of SL-SiC (a) and SL-ZnO (b) at 300, 600, 900, and 1200 K. σ/τ is zero within the bandgap and shows threshold behavior beyond ±1 eV.

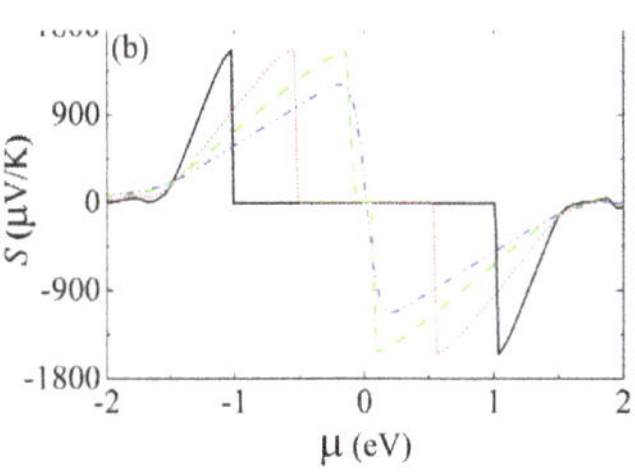

Figure 5. Variation of the Seebeck coefficient (S) for SL-SiC (a) and SL-ZnO (b) at different temperatures. The Seebeck coefficients of both SL-SiC and SL-ZnO decrease with increase in temperature.

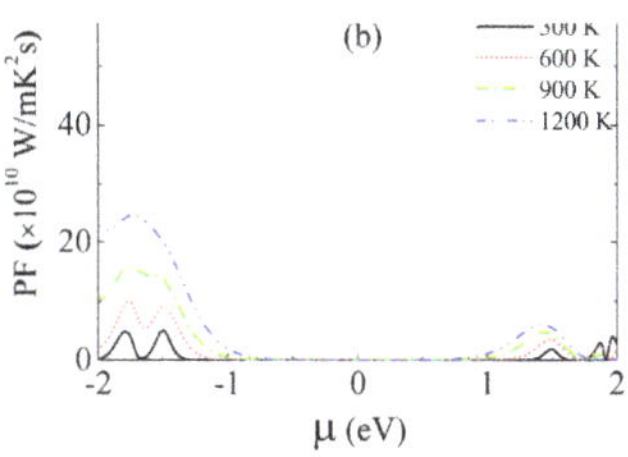

Figure 6. Power factor (PF = $S^2\sigma/\tau$) of SL-SiC (a) and SL-ZnO (b) versus chemical potential at 300–1200 K. SL-SiC shows higher PF for n-type doping, while SL-ZnO is higher for p-type.

smaller. At μ = -1.59 eV (*p*-type), PF increases from 3.67 $\times 10^{10}$ at 300 K to 7.60 $\times 10^{10}$ at 600 K, 1.42 $\times 10^{11}$ at 900 K, and 2.12 $\times 10^{11}$ $Wm^{-1}K^{-2}s^{-1}$ at 1200 K. For *n*-type doping at μ = +1.67 eV, PF rises from 1.58 $\times 10^{11}$ at 300 K to 3.35 $\times 10^{11}$ at 600 K, 4.43 $\times 10^{11}$ at 900 K, and 5.44 $\times 10^{11}$ $Wm^{-1}K^{-2}s^{-1}$ at 1200 K. These peaks correspond to the valence and conduction band edges, where both Seebeck coefficient and conductivity attain significant values. The increasing PF with temperature indicates that SL-SiC benefits from enhanced carrier excitation, especially under *n*-type doping. For ZnO, the PF is higher for *p*-type doping in the ±2 eV range. At μ = -1.77 eV (*p*-type), PF increases from 4.83 $\times 10^{10}$ at 300 K to 1.0 $\times 10^{11}$ at 600 K, 1.58 $\times 10^{11}$ at 900 K, and 2.47 $\times 10^{11}$ $Wm^{-1}K^{-2}s^{-1}$ at 1200 K. For *n*-type doping at μ = +1.49 eV, PF is comparatively smaller, rising from 1.87 $\times 10^{10}$ at 300 K to 3.46 $\times 10^{10}$ at 600 K, 4.80 $\times 10^{10}$ at 900 K, and 6.11 $\times 10^{10}$ $Wm^{-1}K^{-2}s^{-1}$ at 1200 K. Analysis of PF reveals distinct thermoelectric characteristics for the considered SiC and ZnO based 2D materials. The response of SL-ZnO is smoother and less trivial than SL-SiC, which suggests more stable but less tunable thermoelectric behavior of SL-ZnO. SL-SiC, on the other hand, shows relatively swift tunability with temperature and chemical potential, which is more promising for high-temperature thermoelectric applications.

4. Conclusion

The thermoelectric properties of SL-SiC and SL-ZnO are investigated for renewable energy applications in this DFT-based first-principles study. The thermoelectric behavior of SL-SiC shows strong sensitivity to variations in chemical potential and temperature. This indicates it is suitable for temperature- and doping-dependent thermoelectric devices. In contrast, SL-ZnO exhibits relatively stable thermoelectric behavior under similar conditions. This shows SL-ZnO as a potential 2D material for applications that require consistent thermoelectric performance. Analysis of the results shows that SL-SiC achieves a significant enhancement in PF at high temperatures. Its *n*-type PF increases from 3.67 $\times 10^{10}$ $Wm^{-1}K^{-2}s^{-1}$ at 300 K to 5.44 $\times 10^{11}$ $Wm^{-1}K^{-2}s^{-1}$ at 1200 K. SL-ZnO shows a steady rise in *p*-type PF from 4.83 $\times 10^{10}$ to 2.47 $\times 10^{11}$ $Wm^{-1}K^{-2}s^{-1}$. In general, the thermoelectric properties of 2D SiC offer higher tunability via electron doping, whereas 2D ZnO provides stable thermoelectric performance under hole doping. These thermoelectric properties of SL-SiC and SL-ZnO suggest them as potential materials for renewable energy applications.

Acknowledgments
This work is supported by the Brain Pool program (No. 2022H1D3A2A02063677 & RS-2025-25439996) and the National Research Foundation of Korea (No. RS-2025-09592971). The Authors also thankfully acknowledge the Prince Mohammad Bin Fahd University, Saudi Arabia, and the Higher Education Commission of Pakistan for supporting this work through the National Research Programme for Universities-NRPU with Research Project No: 20-16683/NRPU/R&D/HEC/2021.

AI Use Disclaimer: AI-assisted tools were used only to improve language clarity. All data analyses, interpretations, and conclusions are entirely the authors' original work.

References

[1] X. Zheng, C. Liu, Y. Yan, Q. Wang, A review of thermoelectrics research-Recent developments and potentials for sustainable and renewable energy applications, Renewable and sustainable energy reviews 32 (2014) 486-503. https://doi.org/10.1016/j.rser.2013.12.053

[2] Q. Zhang, Y. Sun, W. Xu, D. Zhu, Organic thermoelectric materials: emerging green energy materials converting heat to electricity directly and efficiently, Advanced Materials 26(40) (2014) 6829-6851. https://doi.org/10.1002/adma.201305371

[3] A. Weidenkaff, Thermoelectricity for future sustainable energy technologies, EPJ Web of Conferences, EDP Sciences, 2017, p. 00010. https://doi.org/10.1051/epjconf/201714800010

[4] F.J. DiSalvo, Thermoelectric cooling and power generation, Science 285(5428) (1999) 703-706. https://doi.org/10.1126/science.285.5428.703

[5] L.E. Bell, Cooling, heating, generating power, and recovering waste heat with thermoelectric systems, Science 321(5895) (2008) 1457-1461. https://doi.org/10.1126/science.1158899

[6] R. Freer, A.V. Powell, Realising the potential of thermoelectric technology: a Roadmap, Journal of Materials Chemistry C 8(2) (2020) 441-463. https://doi.org/10.1039/C9TC05710B

[7] S. Hooshmand Zaferani, M. Jafarian, D. Vashaee, R. Ghomashchi, Thermal management systems and waste heat recycling by thermoelectric generators-an overview, Energies 14(18) (2021) 5646. https://doi.org/10.3390/en14185646

[8] D.M. Rowe, Thermoelectrics handbook: macro to nano, CRC press2005.

[9] D. Kraemer, B. Poudel, H.-P. Feng, J.C. Caylor, B. Yu, X. Yan, Y. Ma, X. Wang, D. Wang, A. Muto, High-performance flat-panel solar thermoelectric generators with high thermal concentration, Nature materials 10(7) (2011) 532. https://doi.org/10.1038/nmat3013

[10] S.S. Naghavi, J. He, Y. Xia, C. Wolverton, Pd2Se3 Monolayer: A Promising Two-Dimensional Thermoelectric Material with Ultralow Lattice Thermal Conductivity and High Power Factor, Chemistry of Materials 30(16) (2018) 5639-5647. https://doi.org/10.1021/acs.chemmater.8b01914

[11] Z.-H. Ge, L.-D. Zhao, D. Wu, X. Liu, B.-P. Zhang, J.-F. Li, J. He, Low-cost, abundant binary sulfides as promising thermoelectric materials, Materials Today 19(4) (2016) 227-239. https://doi.org/10.1016/j.mattod.2015.10.004

[12] J.P. Heremans, M.S. Dresselhaus, L.E. Bell, D.T. Morelli, When thermoelectrics reached the nanoscale, Nature nanotechnology 8(7) (2013) 471. https://doi.org/10.1038/nnano.2013.129

[13] B.U. Haq, First-principles calculations to investigate thermoelectric properties of new monolayers of ZnO, Optik 238 (2021) 166782. https://doi.org/10.1016/j.ijleo.2021.166782

[14] B.U. Haq, S. AlFaify, T. Al-shahrani, S. Al-Qaisi, R. Ahmed, A. Laref, S. Tahir, First-principles investigations of ZnO monolayers derived from zinc-blende and 5-5 phases for advanced thermoelectric applications, Journal of Physics and Chemistry of Solids 149 (2021) 109780. https://doi.org/10.1016/j.jpcs.2020.109780

[15] B. Ul Haq, S. AlFaify, R. Ahmed, Thermoelectric properties of the hexagonal-and square-shaped monolayers of ZnO, The European Physical Journal Plus 136(7) (2021) 794. https://doi.org/10.1140/epjp/s13360-021-01777-2

[16] B.U. Haq, S. AlFaify, T. Alshahrani, R. Ahmed, Q. Mahmood, D. Hoat, S. Tahir, Investigations of thermoelectric properties of ZnO monolayers from the first-principles approach, Physica E: Low-dimensional Systems and Nanostructures 126 (2021) 114444. https://doi.org/10.1016/j.physe.2020.114444

[17] B.U. Haq, S. AlFaify, A. Laref, Design and characterization of novel polymorphs of single-layered tin-sulfide for direction-dependent thermoelectric applications using first-principles approaches, Physical Chemistry Chemical Physics 21(8) (2019) 4624-4632. https://doi.org/10.1039/C8CP07645F

[18] J.-F. Li, W.-S. Liu, L.-D. Zhao, M. Zhou, High-performance nanostructured thermoelectric materials, NPG Asia Materials 2(4) (2010) 152. https://doi.org/10.1038/asiamat.2010.138

[19] L. Hicks, M.S. Dresselhaus, Effect of quantum-well structures on the thermoelectric figure of merit, Physical Review B 47(19) (1993) 12727. https://doi.org/10.1103/PhysRevB.47.12727

[20] Y. Pei, X. Shi, A. LaLonde, H. Wang, L. Chen, G.J. Snyder, Convergence of electronic bands for high performance bulk thermoelectrics, Nature 473(7345) (2011) 66. https://doi.org/10.1038/nature09996

[21] K. Schwarz, P. Blaha, Solid state calculations using WIEN2k, Computational Materials Science 28(2) (2003) 259-273. https://doi.org/10.1016/S0927-0256(03)00112-5

[22] J.P. Perdew, K. Burke, M. Ernzerhof, Generalized gradient approximation made simple, Physical review letters 77(18) (1996) 3865. https://doi.org/10.1103/PhysRevLett.77.3865

[23] F. Tran, P. Blaha, K. Schwarz, Band gap calculations with Becke-Johnson exchange potential, Journal of Physics: Condensed Matter 19(19) (2007) 196208. https://doi.org/10.1088/0953-8984/19/19/196208

[24] D. Koller, F. Tran, P. Blaha, Improving the modified Becke-Johnson exchange potential, Physical Review B 85(15) (2012) 155109. https://doi.org/10.1103/PhysRevB.85.155109

[25] B. Ul Haq, R. Ahmed, S. Goumri-Said, A. Shaari, A. Afaq, Electronic structure engineering of ZnO with the modified Becke-Johnson exchange versus the classical correlation potential approaches, Phase Transitions 86(12) (2013) 1167-1177. https://doi.org/10.1080/01411594.2012.755183

[26] B.U. Haq, S. AlFaify, T. Alshahrani, R. Ahmed, S. Tahir, N. Amjed, A. Laref, Exploring optoelectronic properties of ZnO monolayers originated from NaCl-and GeP-like polymorphs: A first-principles study, Results in Physics 19 (2020) 103367. https://doi.org/10.1016/j.rinp.2020.103367

[27] B.U. Haq, S. AlFaify, T. Alshahrani, R. Ahmed, F.K. Butt, S.U. Rehman, Z. Tariq, Devising square-and hexagonal-shaped monolayers of ZnO for nanoscale electronic and optoelectronic applications, Solar Energy 211 (2020) 920-927. https://doi.org/10.1016/j.solener.2020.09.075

[28] G.K. Madsen, J. Carrete, M.J. Verstraete, BoltzTraP2, a program for interpolating band structures and calculating semi-classical transport coefficients, Computer Physics Communications 231 (2018) 140-145. https://doi.org/10.1016/j.cpc.2018.05.010

A Review of Sand-Based Thermal Battery Design: Integrating Sensible and Latent Heat Storage for Long-Duration Applications

Ali AL-NASS[1,a*], Ahmed AL-SHEHRI[1,b], Muayad AL-HAMOOD[1,c], Othman AL-SULIMAN[1,d], Imtiaz ALI[2,e]

[1]Department of Mechanical Engineering, College of Engineering, Prince Mohammad Bin Fahd University, Al Khobar, Saudi Arabia

[2]Department of Chemical Engineering, College of Engineering, Prince Mohammad Bin Fahd University, Al Khobar, Saudi Arabia

[a]202002876@pmu.edu.sa, [b]201902251@pmu.edu.sa, [c]201502586@pmu.edu.sa, [d]202100148@pmu.edu.sa, [e]iali1@pmu.edu.sa

Keywords: Sand Battery, Thermal Energy Storage (TES), Phase Change Material (PCM), Insulation Design, Solar Charging, Long-Duration Energy Storage, Renewable Energy

Abstract. The study shows current practices in sand based thermal energy storage systems and suggest design ideas for a sand thermal energy storage battery. Sand serves as a cheap heat reservoir (costing as low as 0.12 to 0.2 $/kWh) which demonstrate stable performance above 1000°C. The objective is to provide an energy storage solution that is low cost, mechanically simple, long lasting, and in line with Saudi Arabia Vision 2030 goals for reaching zero emission. The objective combine the sensible heat storage in sand and evaluate different sand types such as silica sand, calcined clay and ceramic beads. The study also suggest embedding latent heat from phase change material modules PCM to increase the total stored energy density and hold the charging of heat, with solar salt PCM offering a latent heat capacity of 150-180 kJ/kg. The study target to optimize multi layer insulation to minimize heat losses at high temperatures while maintaining structural integrity of the storage tank using mineral wool insulation with low thermal conductivity ranging 0.033 to 0.045W/m·K. The system input is renewable and excess energy sources by focusing on high efficiency solar panels in hot regions. The suggested system includes an insulated tank with sand and encapsulated PCM modules, a solar powered heat input subsystem and smart sensing that shows how the system is functioning. A dashboard can be deployed that monitors temperature, charging status, consumption rate, and projected discharge time. The goal of these practices is to create the fundamental components of an integrated scalable system, durable, and environmentally friendly with long duration heat storage that is appropriate for industrial and district scale applications.

Introduction

Long duration energy storage is critical in raising the adoption of renewable energy sources. The cost, safety and charging cycle limits is making conventional electrochemical batteries not suitable for multi hour to multi day storage. Sand based thermal energy storage is a practical option because of its low cost, wide spread availability, thermal stability at high temperatures, and environmental sustainability. This concept converts electrical or solar derived input energy to heat. Then, it stores it inside a huge insulated container filled with sand. The stored heat may then be retrieved and provided as useable thermal energy to costumers [1], [2], [3], [4], [5].

This study integrates design solutions to accomplish four main goals based on current work. The first goal is to determine the best charging and discharging behavior by identifying different sand types including local sand and other silica based materials and particle sizes. The second goal is to incorporate PCM modules into the sand bed to increase the total energy density and add latent heat storage capacity [6], [7]. The third goal is to design and evaluate insulation that reduce heat loss while preserving the mechanical integrity of the storage tank [4], [5], [8], [9]. The fourth goal is to assess the use of renewable and excessive energy sources specifically solar panels as the primary charging input to achieve a low maintenance, zero emission and Saudi Arabia Vision 2030 aligned solution [10].

The motivation is to produce a system that is low cost, mechanically reliable, and suitable for repeated long term storage cycles with minimal periodic maintenance. The system concept is compatible with Saudi Arabia's high solar resource and national strategy toward carbon reduction and clean energy under Vision 2030.

Methodology

For this review, the literature screening involved a systematic search of Scopus databases using keywords related to sand based thermal energy storage, phase change materials PCM, high temperature insulation, and solar integration. Studies were selected based on their direct relevance to the core design components of the sand battery. Evaluating solid storage media like silica sand, enhancing PCM with conductive materials and optimizing multilayer insulation and integrating renewable charging. This ensured the review was grounded in relevant experimental and numerical research on long duration and high temperature thermal storage systems.

Literature findings

Sand as the Primary Storage Medium

A sand-filled, thermally insulated drum or tank serves as the Sand Heat Storage Battery's central component. Sand is chosen due to its abundance, affordability, and environmental friendliness [1], [2]. In comparison to alternatives like calcined flint clay, brown fused alumina, and engineered ceramic particles, silica sand provides a good balance of cost, high-temperature stability, and specific heat capacity, according to studies comparing potential solid-particle thermal storage devices for concentrating solar power and high-temperature TES [1], [11].

Experimental work on silica sand sensible heat storage systems has compared fine vs. coarse particle distributions: fine sand tends to absorb heat (charge) more quickly due to its larger total surface area, while coarse sand can release heat (discharge) more effectively because of better air permeability and internal flow paths [2]. Both particle size regimes demonstrated acceptable thermal stability up to temperatures on the order of 1000°C [1], [2]. Modeling of desert sand TES further supports feasibility for solar charging and long-duration storage [3]. A comparative summary of key properties and cost indicators for these solid media is provided in Table 1.

Table 1. Comparative properties of candidate materials for high temperature sensible thermal energy storage

TES Medium	Bulk Cost ($/ton)	Specific Heat C_p (kJ/kg·K)	Max Operating Temp (°C, practical)	Thermal Conductivity (W/m·K)	Approx. Cost per kWh ($) [ΔT≈900 K]	Cycle Life (indicative)
Silica Sand	75	0.95	silica sand and similar particulates have demonstrated exceptional stability well beyond 1000 °C	1.2	~0.12-0.20	100 cycles @300-1200°C + 500 h @1200°C
Calcined Flint Clay (Chamotte / CFC)	180	0.74-0.87	>1400°C continuous service; refractoriness ~1770-2000°C	0.85	~1.1-1.3	Industrial refractory; designed
Ceramic Beads / Ceramic Proppant	≥1000 (≈1.0 $/kg or higher)	~1.16	~900-1000°C+ repeatedly in CSP	~0.5-0.9 (dense aluminosilicate)	~3.0-3.5	Repeated cycles near 1000°C

Embedded PCM Modules for Latent Heat Storage

Phase Change Materials (PCMs) offer latent heat storage (energy absorbed during melting and released after solidification), whereas sand offers sensible heat storage (energy retained as temperature rises) [6], [7]. Without necessitating a significant increase in physical volume, PCMs can significantly raise the battery's overall energy density [6], [7]. The examined systems embed PCM inside finned tubes or shell-and-tube modules with longitudinal or radial fins since many PCMs (such solar salt) have relatively poor heat conductivity [6]. These fins produce more effective thermal paths into and out of the PCM volume and improve the surface area available for heat transmission. Furthermore, there is a noticeable increase in effective thermal conductivity, better heat dispersion, and less local overheating when solar salt is combined with expanded graphite or copper foam [6]. Encapsulation using a multilayer design with a ceramic-lined inner surface and a stainless-steel outer shell effectively accommodates thermal expansion differences and enhances durability at high temperatures. The PCM is stabilized within a compressed expanded graphite CENG matrix and sealed under inert gas to minimize corrosion and leakage and ensuring long term reliability with minimal maintenance [7], [12], [13]. An electric heater coil can be positioned in or near the finned PCM assembly to provide uniform charging. Such PCM cartridges can then be embedded as modules within the sand bed [6].

Table 2 Specifications of Solar Salt (PCM) [7]

Property	Specification	Remarks
Composition	60 wt% $NaNO_3$ + 40 wt% KNO_3	Inorganic molten salt nitrate based
Melting point (°C)	227.47	Ideal for medium temperature
Thermal conductivity (W/m·K)	0.45-0.6	Can be enhanced by EG
Latent heat (kJ/kg)	150-180	Good storage capacity
Density (kg/m³)	1.98-1.80 (liquid)	Slightly higher in solid state
Specific heat (kJ/kg·K)	1.55	Provides sensible and latent heat benefit
Thermal stability	550°C	Stable under air atmosphere
Corrosivity	Moderate	Use Al6061 containers

Table 3 Specification of CENG (EG) [14]

Property	Specification	Remarks
Bulk density (kg/m³)	2.09	Tailored via compression ratio
Thermal conductivity (W/m·K)	1.09 in solid state 0.55 in liquid state	Strong anisotropy, enhances PCM
Specific heat (kJ/kg·K)	~0.7	
Stability	Up to 600 [°C]	Chemically and thermally stable
PCM loading capacity wt%	80-90%	High absorption and retention

Insulation of the Storage Tank

For long-term storage, minimizing heat loss is crucial. The harsh working conditions of the container are determined by the choice of storage media, therefore an insulation strategy that takes into account both thermal and structural integrity is required. The sand/PCM tank is surrounded by a tiered insulating approach [4], [5], [8], [9]. The following methods have been reported: (i) foam glass or expanded glass granules at the base because these materials offer high compressive strength, moisture resistance, and thermal resistance [8]; (ii) mineral wool surrounding the sides of the hot region because of its good high-temperature insulation performance and practical availability [4], [5]; and (iii) structural concrete as an external containment layer to provide mechanical integrity and suppress large-scale heat leakage to ambient [5], [8]. Prototype high-temperature sand TES systems have used approximately 10 cm of mineral wool insulation to retain heat [4]. Increasing insulation thickness generally reduces conductive heat loss, but it can also increase the external surface area, which may raise convective losses to ambient air, so the optimal thickness is a trade-off between conduction and convection [4], [5]. This trade-off is illustrated in Figure 1, which plots simulated heat losses against insulation thickness for a prototype system.

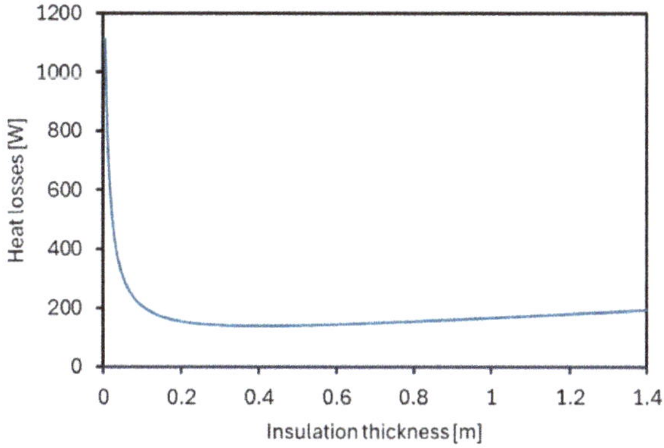

Figure 1 Simulated heat losses from a prototype sand-based TES unit as a function of insulation thickness, illustrating the trade-off between conductive and convective losses [4]

Controlling moisture is also essential since having access to water can significantly raise effective thermal conductivity and hasten heat loss. Expanded glass grains and other loose-fill insulating materials can be kept dry by vacuum-sealing them in geotextile bags [8]. Spatially variable insulation thickness is another crucial component of design, with lower insulation in colder areas to save money and thicker insulation in areas with the greatest temperatures (such as the upper core) [8]. For seasonal-scale storage, mixtures of concrete and mineral wool have also been predicted [5], [8].

Table 4 Classification of the commonly used insulation materials and uncertainty about their thermal conductivity [15]

Main group	Subgroup	Insulation Material	Temperature (◦C)	Density (kg/m^3)	Thermal conductivity (W/m.K)
Inorganic	Fibrous	Glass wool	-100 - 500	13-100	0.03-0.045
		Rock wool	-100 - 750	30-180	0.033-0.045
	Cellular	Calcium silicate	300	115-300	0.045-0.065
		Cellular glass	-260 - 430	115-220	0.04-0.06
		Vermiculite	700 - 1600	70-160	0.046-0.07
		Ceramic	-	120-560	0.03-0.07
Organic	Foamed	EPS	-80 - 80	15-35	0.035-0.04
		XPS	-60 - 75	25-45	0.03-0.04
		PUR	-50 - 120	30-100	0.024-0.03
	Foamed, expanded	Cork	110 - 120	110-170	0.037-0.050
		Phenolic foam	150	40-160	0.022-0.04
		Polyethylene foam	-40 - 105	25-45	0.033
	Fibrous	Fiberglass	-4 - 350	24-112	0.033-0.04

		Sheep wool	130-150	25-30	0.04-0.045
		Cotton	100	20-60	0.035-0.06
Combined	Boards	Gypsum foam	-	-	0.045
		Wood wool	110-180	350-600	0.09
		Wood fibers	110	30-270	0.04-0.09
Advanced materials		VIPs	-	150-300	0.002-0.008
		Aerogel	-	60-80	0.013-0.014

Insulation Thickness Optimization and Moisture Control

Thicker mineral wool layers significantly lower conductive heat losses over multiday storage intervals according to prototype experiments for high temperature sand thermal energy storage system [4], [5]. Numerical modeling is necessary to determine the ideal insulation thickness and minimizes total heat loss since larger insulation also increases the exterior surface area and can improve convective heat transfer to the environment [4], [5]. The performance of different thermal storage configurations can be evaluated using a metric known as the Storage Ratio SR, defined in equation 1. The thermal conductivity of insulation materials also varies with temperature. Generally, conductivity increases with temperature and allowing the system to loss heat more quickly during hot storage times [9]. Moisture ingress is a major risk that should be considered when designing the system. When insulating materials absorb moisture, the effective thermal conductivity increases greatly and affecting performance. Using sealed or vacuum bagged loose fill materials (expanded glass granules packaged in geotextile bags) help preserve insulation quality and repeatability over long duration storage cycles [8].

$$Storage\ Ratio\ (SR) = \frac{\mathrm{E_{nett}}}{\mathrm{Energy\ input}} \quad (1)$$

The impact of different internal heater configurations on this performance metric is quantified in Table 5, which compares storage ratio, temperature difference, and maximum heater temperature for several design variants.

Table 5 Comparison of thermal performance for different internal heater configurations within a sand TES bed, showing Storage Ratio (SR), temperature difference (ΔT), and maximum heater temperature ($T_{h,max}$) over time [4]

Variant	1	2	3	4	5	6	7
Storage Ratio							
t=2 h	0.998	0.998	0.986	0.992	0.992	0.992	0.992
t=24 h	0.860	0.841	0.768	0.815	0.805	0.815	0.805
t=48 h	0.685	0.669	0.608	0.647	0.639	0.647	0.639
ΔT (K)							
t=2 h	1275	1207	1315	1260	1263	1316	1316
t=24 h	400	382	350	377	367	376	364
t=48 h	343	341	334	339	338	339	337
$T_{b.max}$ (K)							
t=2 h	1306	1240	1348	1291	1297	1348	1348

Conclusion

The study showed key design principles for a sand heat storage battery meant for long term, renewable, and low-emission thermal energy based on the literature. Sand offers an excellent cost-performance balance according to comparisons between various sands (fine vs. coarse) and potential solid particles (silica sand vs. ceramic-like alternatives). By overcoming PCM's poor conductivity, PCM modules such as Solar Salt augmented with expanded graphite or copper foam embedded inside finned aluminum tubes improve total energy density through latent heat storage. The strategy of insulation when paired with moisture protection and adjusted thickness is important to maintain low thermal conductivity. A multilayer insulation system made of mineral wool, foam glass, expanded glass granules, and structural concrete may withstand high temperatures without compromising structural integrity. Insulation thickness must balance conductive and convective losses and take temperature dependent conductivity into account according to numerical analysis and experimental prototypes. Solar panels provide high efficiency and strong energy in hot sunny climates. Smart sensors and a dashboard can monitor internal temperature, charging power, consumption, and projected discharge duration, enabling predictive control and future AI optimization.

In future work, CFD simulations can be used to study heat flow and temperature distribution in the sand heat storage system. These simulations can help to better understand the behavior of the system and the energy losses due to natural convection. A small prototype model will also be built and tested experimentally. The experimental results will be compared with the CFD results to validate the model. This combined approach will help improve the system design and performance.

References

[1] P. Davenport et al., "Characterization of solid particle candidates for application in thermal energy storage and concentrating solar power systems," Solar Energy, vol. 262, p. 111908, Sep. 2023 https://doi.org/10.1016/j.solener.2023.111908

[2] P. Niksiar, C. Rogillio, H. Torab, and S. Tiari, "Experimental Study of a Silica Sand Sensible Heat Storage System Enhanced by Fins," Energies, vol. 17, no. 21, p. 5402, Jan. 2024 https://doi.org/10.3390/en17215402

[3] K. Hinkelman, D. Milner, and W. Zuo, "Open-Source Models for Sand-Based Thermal Energy Storage in Heating Applications," Modelica Conferences, pp. 627–632, Dec. 2023 https://doi.org/10.3384/ecp204627

[4] M. Homa, K. Sornek, W. Goryl, K. Papis-Frączek, P. L. Żak, and R. Dańko, "Numerical analysis of the prototype of the high-temperature thermal energy storage based on sand bed," Energy, vol. 333, p. 137472, Oct. 2025 https://doi.org/10.1016/j.energy.2025.137472

[5] S. Audun, "Seasonal Thermal Energy Storage Using Sand Batteries," UiT The Arctic University of Norway, 2024. Accessed: Jan. 23, 2026. [Online]. Available: https://hdl.handle.net/10037/34198

[6] S. Ahmed et al., "Melting enhancement of PCM in a finned tube latent heat thermal energy storage," Sci Rep, vol. 12, no. 1, p. 11521, Jul. 2022 https://doi.org/10.1038/s41598-022-15797-0

[7] H. Wang, J. Li, Y. Zhong, X. Liu, and M. Wang, "Novel Wide-Working-Temperature NaNO3-KNO3-Na2SO4 Molten Salt for Solar Thermal Energy Storage," Molecules, vol. 29, no. 10, p. 2328, Jan. 2024 https://doi.org/10.3390/molecules29102328

[8] K. Szewerda, D. Michalak, P. Matusiak, and D. Kowol, "Concept of Adapting the Liquidated Underground Mine Workings into High-Temperature Sand Thermal Energy Storage," Applied Sciences, vol. 15, no. 7, p. 3868, Jan. 2025 https://doi.org/10.3390/app15073868

[9] Y. Yousefi and F. Tariku, "Thermal Conductivity and Specific Heat Capacity of Insulation materials at Different Mean Temperatures," J. Phys.: Conf. Ser., vol. 2069, no. 1, p. 012090, Nov. 2021 https://doi.org/10.1088/1742-6596/2069/1/012090

[10] M. Benghanem et al., "Evaluation of the Performance of Polycrystalline and Monocrystalline PV Technologies in a Hot and Arid Region: An Experimental Analysis," Sustainability, vol. 15, no. 20, p. 14831, Jan. 2023 https://doi.org/10.3390/su152014831

[11] R. Saeed et al., "Characterization of Solid Particulates to Be Used as Storage as Well as Heat Transfer Medium in Concentrated Solar Power Systems," Applied Sciences, vol. 15, no. 15, p. 8566, Jan. 2025 https://doi.org/10.3390/app15158566

[12] X. Zhou, S. Yamashita, M. Kubota, and H. Kita, "Encapsulated Copper-Based Phase-Change Materials for High-Temperature Heat Storage," ACS Omega, vol. 7, no. 6, pp. 5442–5452, Feb. 2022 https://doi.org/10.1021/acsomega.1c06751

[13] Y. Chang, X. Yao, Y. Chen, L. huang, and D. Zou, "Review on ceramic-based composite phase change materials: Preparation, characterization and application," Composites Part B: Engineering, vol. 254, p. 110584, Apr. 2023 https://doi.org/10.1016/j.compositesb.2023.110584

[14] E. Blackley, T. Lai, A. Odukomaiya, P. C. Tabares-Velasco, L. M. Wheeler, and J. Woods, "Surface-Modified Compressed Expanded Graphite for Increased Salt Hydrate Phase Change Material Thermal Conductivity and Stability," ACS Appl. Energy Mater., vol. 6, no. 17, pp. 8775–8786, Sep. 2023 https://doi.org/10.1021/acsaem.3c01223

[15] L. D. Hung Anh and Z. Pásztory, "An overview of factors influencing thermal conductivity of building insulation materials," Journal of Building Engineering, vol. 44, p. 102604, Dec. 2021 https://doi.org/10.1016/j.jobe.2021.102604

Sol–Gel Fabrication of Hybrid Semi-Interpenetrating Ion-Solvating Membranes by TEOS Crosslinking for Alkaline Electrolyzers

Loubna AHSAINI[1], Mina JELLAB[1], Mustapha MATROUF[1], Fatima-Zahra SEMLALI[1,a*], Fouad GHAMOUSS[1,b*]

[1]College of Chemical Sciences and Engineering (CCSE), Department of Materials Science, Energy, and Nano-engineering (MSN), Mohamed VI Polytechnique University (UM6P), 43150, Benguerir, Morocco

[a]Fatimazahra.semlali@um6p.ma, [b]Fouad.ghamouss@um6p.ma

Keywords: Ion-Solvating Membrane, Hybrid Semi-Interpenetrating Membranes, Sol-Gel, TEOS Crosslinking

Abstract. Ion-solvating membranes are an emerging membrane class for alkaline water electrolysis, where hydroxide transport occurs through KOH-solvated pathways rather than fixed cationic groups but achieving high conductivity without excessive swelling and poor oxidative durability remains challenging. Here, we design hybrid semi-interpenetrating xTEOS–PVA:xPVP membranes by combining acid-catalyzed TEOS sol–gel crosslinking with PVA/PVP polymer entanglement, followed by solution casting and KOH activation. FTIR confirms the formation of Si–O–Si and Si–O–C linkages with residual silanols, demonstrating an inorganic network integrated within the polymer matrix. SEM reveals a composition-dependent morphology transitioning from dense polymer films to textured hybrid structures at intermediate TEOS loadings and re-densified surfaces at the highest TEOS contents. This structural evolution governs hydration and transport: water uptake and swelling decrease with TEOS, while hydroxide conductivity shows a step increase upon introducing a small TEOS fraction at high PVP content, reaching ~60 mS cm^{-1} at 25 °C and ~100 mS cm^{-1} at 80 °C in 1 M KOH for the optimal composition (1T-PVA:1.5PVP). Importantly, TEOS incorporation markedly enhances oxidative stability, with ~90% mass retention near the same intermediate composition. The results demonstrate that inorganic sol–gel crosslinking plus polymer semi-interpenetration can generate well-solvated transport pathways without excessive swelling, enabling membranes that combine high OH^- conductivity with improved oxidative durability for alkaline membrane electrolyzers.

Introduction

Alkaline water electrolysis (AWE) is a mature and scalable route to produce high-purity hydrogen using low-carbon electricity [1]. In anion-exchange membrane water electrolysis (AEMWE), a solid polymer electrolyte acts as the ion conductor and separator, enabling more compact cells and simpler operation than conventional liquid KOH systems [2]. A major advantage of alkaline operation is the ability to use earth-abundant based catalysts instead of precious metals, which can reduce cost and supply-chain constraints. Well-designed membranes can also limit gas crossover and support high efficiency and product purity [1]. However, membranes and diaphragms remain a key major challenge [3]. Practical polymers must combine high OH^- conductivity with good dimensional and mechanical stability across operating conditions, while resisting chemical degradation under strong alkaline and oxidative environments [4]. Low permeability and suppressed H_2/O_2 crossover are also essential for safety and durability. Current strategies focus on polymer blending, controlled crosslinking, incorporation of inorganic fillers, and more stable cationic chemistries [5]. Bio-derived polymers such as cellulose or chitosan are increasingly explored to improve sustainability while maintaining tunable ion transport, hydration control, and mechanical reinforcement. Progress in membrane materials, together with compatible electrodes and transport layers, is crucial for durable, high-current-density alkaline electrolyzer stacks and wider deployment of affordable green hydrogen [5].

This work reports the development of hybrid semi-interpenetrating xTEOS–PVA:xPVP ion-solvating membranes via acid-catalyzed sol–gel processing, solution casting, and KOH activation, and we systematically correlate chemistry, morphology, hydration, conductivity, and oxidative stability. The central hypothesis is that TEOS-derived inorganic crosslinking, combined with PVA/PVP entanglement, can generate well-solvated OH^- transport pathways while suppressing excessive swelling and improving oxidative durability. The main outcome is the identification of an optimal intermediate composition that simultaneously delivers high OH^- conductivity in 1 M KOH and enhanced oxidative stability compared with polymer-only controls.

Materials and Methods

Silica sols were synthesized via a two-step sol–gel procedure. The precursor tetraethyl orthosilicate (TEOS) (≥99.0%, Mw = 208.33 g·mol^{-1}) was dissolved in EtOH/H_2O with ethyl alcohol (≥99.5%)/ D.W with a pH of 5 and a conductivity of 14.79 μS cm^{-1} and stirred for 1 h with 1:8:4 molar ratio (TEOS:EtOH:H_2O); 0.1 M of hydrochloric acid (HCl) (37%) was added dropwise to catalyze hydrolysis [6], and the mixture was stirred at room temperature for 6 h until gelation. The gel was dried at 80 °C for 12 h to give a xerogel, which was ground to obtain silica hydrated powder. We dissolved poly(vinyl alcohol) (PVA) (average Mw 130,000, 99+% hydrolyzed) (4 g) in distilled water (100 mL) at 90 °C under vigorous stirring until the solution turned clear. A different amount of hydrolysis of TEOS (0.5, 1, 1.5, 2 g) and concentrated 1 M HCl (1 mL) were then added to trigger the sol–gel reaction. After stirring at room temperature for 2 h, we introduced an aqueous solution of PVP prepared separately by dissolving at 60 °C (2, 1.5, 1, 0.5g) of poly(vinylpyrrolidone) (PVP) (average Mw = 40 000 g·mol^{-1}), at the target TEOS-PVA:PVP ratio. The homogeneous solution was cast onto Petri dishes and left to dry under ambient conditions for 24 h. The dried films were gently peeled from the substrate, soaked in 3 M potassium hydroxide (KOH) (≥99.0%) at 60 °C for 3 h to convert them to the Ion-Solvating membrane, rinsed thoroughly with distilled water, and air-dried before characterization.

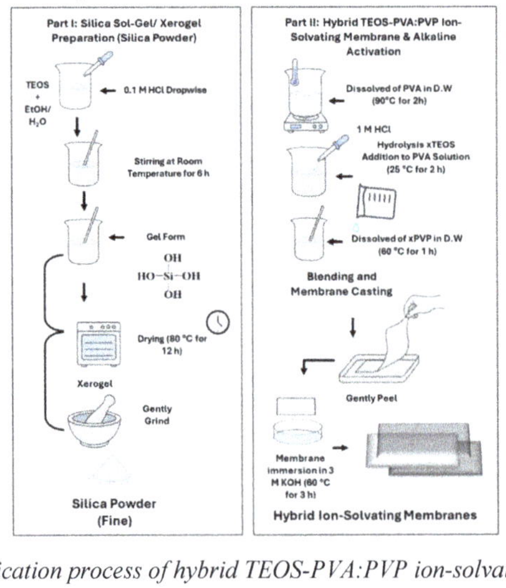

Fig. 1 The fabrication process of hybrid TEOS-PVA:PVP ion-solvating membranes.

Characterization Techniques

FTIR (Fourier Transforms Infrared Spectroscopy). The reaction between PVA and TEOS blending with PVP membrane was confirmed using FTIR spectrometer (FTIR, Perkin-Elmer Spectrum 2000, Waltham, Massachusetts, United States) equipped with an ATR accessory was used. The tests were carried out with a spectral resolution of 0.5 cm^{-1}, 32 scans per spectrum and in the range 600-4000 cm^{-1}.

Morphology. The surface morphology of the hybrid ion-solvating membranes were imaged using scanning electron microscopy (SEM, ZEISS EVO 10, GmbH, Jena, Germany) operated at an accelerating voltage of 10 kV. Elemental composition analysis was performed using an energy-dispersive X-ray (EDX) analyzer. The membrane samples were vacuum dried for 20 minutes to prevent charging during imaging and subsequently coated with a gold (Au) layer via sputtering for 600 seconds.

Water Uptake and swelling ratio. For water uptake (WU) and swelling ratio (SR), membranes were immersed in deionized water for 24 hours to achieve full hydration. Before immersion, the mass and length of the dried membranes were recorded. After 24 hours, the hydrated membranes were gently wiped with tissue paper to remove surface water, and their mass and length were quickly measured [7,8]. Each test was performed five times, and the average values were reported. The WU and SR were calculated from the equations (1) and (2):

$$\mathbf{WU}(\%) = \frac{W_w - W_d}{W_d} \times \mathbf{100} \quad (1)$$

$$\mathbf{SR}(\%) = \frac{L_w - L_d}{L_d} \times \mathbf{100} \quad (2)$$

Hydroxide ion conductivity. The OH^- ionic conductivity of the membranes formed was assessed through AC impedance measurements utilizing an electrochemical impedance analyzer (Biologic SP 300). The AC frequency range spanned from 100 kHz to 100 Hz with a voltage amplitude of 10 mV. Fully hydrated membranes were equipped with two Pt counter-electrode wire (H-Cell 2x15 mL). The membrane remained in contact with 1 M of potassium hydroxide during the entire measurement process in two different temperatures, 25 °C, and 80 °C [8]. Ionic conductivity (mS/cm) was computed using the equation (3):

$$\sigma = \frac{T}{R \times A} \quad (3)$$

Where, T is the thickness of membranes (cm), R is the bulk – without membrane resistance (Ω), and A is the surface area of membranes (cm^2) [9].

Oxidative stability. The oxidative stability of the prepared membranes was assessed by immersing them in Fenton's reagent (a solution of 3% H_2O_2 with 3 ppm $FeSO_4$) at 80 °C for 1 hour. The oxidative stability was then quantified by calculating the percentage of weight loss [10,11]. The O.S is calculated using the equation (4):

$$\mathbf{O.S}(\%) = \mathbf{100} - \left(\frac{W_b - W_a}{W_b} \times \mathbf{100}\right) \quad (4)$$

Where W_b is weight before loss, and W_a is the weight after immersing the membranes in solution.

Results and discussion

Pereparation of TEOS-PVA:PVP Hybrid Ion-Solvation Membrane. The TEOS–PVA:PVP hybrid ion-solvating membrane was prepared by combining PVA, PVP, and TEOS, followed by alkaline doping to improve OH^- transport and durability (**Fig. 2**). First, TEOS was hydrolyzed at room temperature in the presence of HCl to form a silica sol [12]. Under acidic conditions, hydrolysis dominates over condensation, producing silanol-rich siloxane oligomers [6]. Next, the silica phase was dispersed into an aqueous PVA solution and cast into films. During drying, dehydration-driven condensation formed Si–O–Si bonds within the inorganic phase and created interfacial interactions with PVA [6]. Condensation between silanol groups and PVA hydroxyls can also generate Si–O–C linkages (Si–OH + HO–PVA → Si–O–PVA + H_2O), strengthening the hybrid network. Although PVP does not readily form covalent bonds with silanols, it stabilizes silica dispersion through hydrogen bonding and chain entanglement, limiting aggregation and promoting a uniform microstructure, which supports higher hydration and ionic conductivity [13].

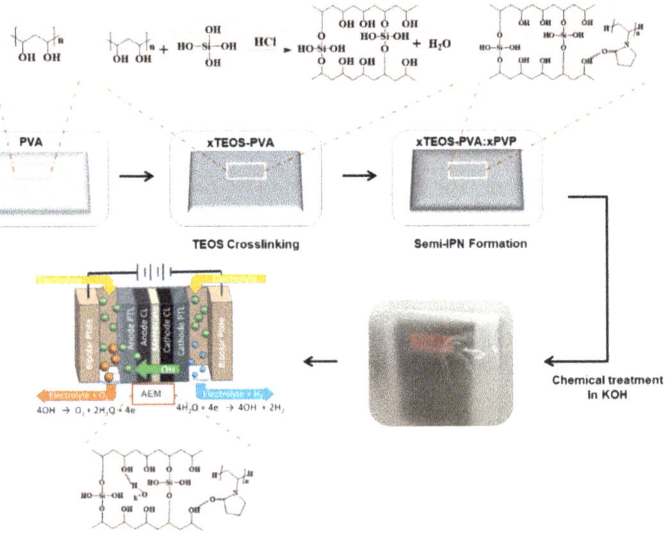

***Fig. 2** The fabrication process and Mechanism of hybrid TEOS-PVA:PVP ion-solvating membranes.*

Chemical Structural Analysis. FTIR spectra confirm the structure of PVA, PVA:PVP, and xTEOS–PVA:xPVP membranes (Fig. 3). Neat PVA shows a broad O–H band (≈3200–3570 cm^{-1}), C–H stretching (≈3000–2800 cm^{-1}), and strong C–O vibrations (≈1140–1080 cm^{-1}) [14]. Adding PVP introduces the lactam C=O band (≈1680–1630 cm^{-1}) and slightly shifts and broadens the O–H band, indicating stronger PVA–PVP hydrogen bonding. With TEOS (0.5T–2T), the growth of bands at ≈1100–1040, ≈950, and ≈800–750 cm^{-1} is consistent with Si–O–Si and Si–O–C formation during sol–gel condensation, while the further red-shift of O–H suggests increased interaction and partial consumption of PVA hydroxyls [15]. Overall, the spectra support formation of a hybrid PVA–SiO_x network where PVP mainly contributes through hydrogen bonding and TEOS provides inorganic crosslinking.

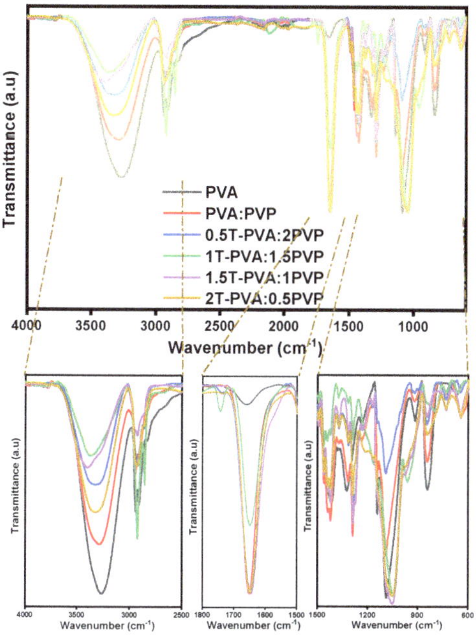

Fig. 3 FTIR Spectrum of the fabricated Ion-Solvating Membranes

Morphology. SEM observations support the FTIR results and show how TEOS and PVP control membrane morphology (**Fig. 4**). Neat PVA is compact, and the PVA:PVP blend remains smooth due to strong PVA–PVP hydrogen bonding. With TEOS addition, the surface evolves from mostly dense (0.5T–2PVP) to a fine, homogeneous nodular texture (1T–1.5PVP), then to a more pronounced bicontinuous or sponge like structure at intermediate composition (1.5T–1PVP), indicating stronger silica condensation and microphase separation. At the highest TEOS and lowest PVP (2T–0.5PVP), the surface becomes re-densified and fused, consistent with over-condensation and silica aggregation when PVP stabilization is limited. Overall, high PVP maintains dense films, intermediate PVP promotes a well-distributed hybrid network, and low PVP at high TEOS leads to densification, in agreement with the FTIR siloxane growth and reduced free –OH.

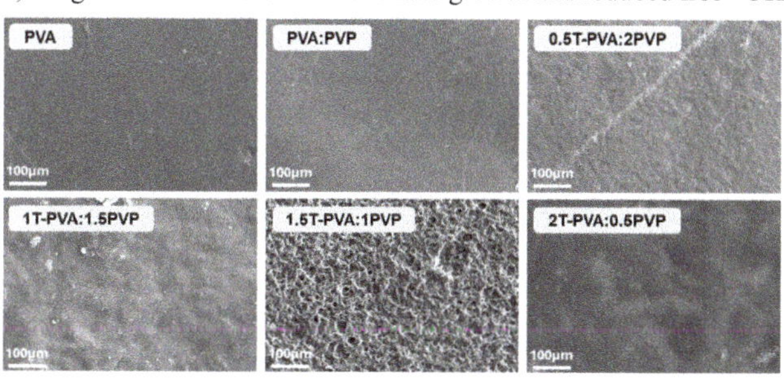

Fig. 4 Surface Morphology of Ion-Solvating Membranes

Water Uptake and swelling ratio. Neat PVA (**Fig. 5a**) shows the highest water uptake (~374.4%) and moderate swelling (~16.2%) due to its abundant –OH groups. Adding PVP reduces both uptake (~286.3%) and swelling (~11.4%), consistent with stronger PVA–PVP hydrogen bonding and a denser matrix. At low TEOS with high PVP (0.5T–PVA:2PVP), uptake remains high (~314.3%) and swelling reaches a maximum (~33.3%), likely due to additional hydrophilic silanol groups. Increasing TEOS to 1T–PVA:1.5PVP decreases both values, indicating the onset of inorganic crosslinking. A strong reduction is observed at 1.5T–PVA:1PVP (uptake ~77.6%; swelling ~3.3%), reflecting extensive Si–O–Si/Si–O–C network formation that restricts hydration and dimensional change. At 2T–PVA:0.5PVP, uptake partially increases (~97.0%) while swelling remains limited (~13.3%), consistent with a dense hybrid network that can bind water but suppresses expansion. Overall, the hydration and swelling behavior follows the FTIR/SEM trends: PVP reduces free –OH, low TEOS increases hydrophilicity, and higher TEOS crosslinking limits water uptake and swelling.

Hydroxide ion conductivity. At 25 and 80 °C in 1 M KOH, both PVA and PVA:PVP show low hydroxide conductivity (**Fig. 5b**): 9 and 42.85 mS cm^{-1} for PVA, and 8.78 and 37.5 mS cm^{-1} for PVA:PVP, respectively. Adding a small TEOS amount at high PVP (0.5T–PVA:2PVP) strongly increases conductivity to 90 mS cm^{-1} (25 °C) and 150 mS cm^{-1} (80 °C). At 1T–PVA:1.5PVP, conductivity decreases but remains high (60 and 100 mS cm^{-1}), consistent with sufficient hydration and controlled swelling that maintain continuous solvated pathways. Further increasing TEOS while reducing PVP (1.5T–PVA:1PVP) sharply lowers conductivity (5.62 and 13.63 mS cm^{-1}), in line with reduced water uptake and a more rigid, highly crosslinked matrix. At 2T–PVA:0.5PVP, conductivity slightly recovers (8.18 and 23.07 mS cm^{-1}) but remains limited, indicating that restricted hydration and polymer mobility govern OH^- transport at high TEOS loading. The membranes are benchmarked against representative PVA-based hybrid membranes and other alkaline membrane systems reported in the literature. Reported performance depends strongly on testing conditions (electrolyte concentration, temperature, cell setup, pre-treatment, thickness, and measurement method). For this reason, the comparison focuses mainly on studies evaluated under similar conditions (e.g., conductivity measured in KOH at comparable concentration and temperature). Table 1 summarizes the key parameters, including composition, conductivity, water uptake, and stability. The optimized semi-interpenetrating TEOS–PVA:PVP membrane combines high OH^- conductivity in alkaline media with improved oxidative stability and controlled hydration, supporting its competitiveness among PVA-based hybrid ion-transport membranes.

Oxidative stability. Neat PVA shows low oxidative stability, and only a limited improvement is observed after blending with PVP. In contrast, TEOS incorporation significantly increases oxidative resistance (**Fig. 5c**). The 0.5T–PVA:2PVP membrane shows moderate retention (~62.7%), consistent with a mostly organic matrix. The highest stability is obtained for 1T–PVA:1.5PVP (~90%), indicating that an adequately condensed hybrid network effectively limits degradation and mass loss. At higher TEOS loadings, retention remains high but slightly lower (1.5T–PVA:1PVP ~81%; 2T–PVA:0.5PVP ~83%), likely due to increased densification and microstructural heterogeneity. Overall, the results indicate that forming Si–O–Si/Si–O–C linkages improves oxidative durability, with the best performance at an intermediate TEOS/PVP ratio.

***Table 1** Comparison of water uptake/swelling ratio, hydroxide conductivity and oxidative stability between the 0.5T-PVA:2PVP, 1T–PVA:1.5PVP membranes and commercial AEMs at 80 °C.*

Membrane	Water Uptake (%)	Swelling Ratio (%)	σ at 80 °C (mS cm^{-1})	Oxidative Stability (%)	Ref
0.5T–PVA:2PVP	314.28	33.33	150	62.66	This work
1T–PVA:1.5PVP	292.85	16.66	100	90	This work
Sustainion® X37-50	80	15.1	140	N/A	[16]
Tokuyama A201	44 (80 °C)	12 (80 °C)	60–90	N/A	[16]

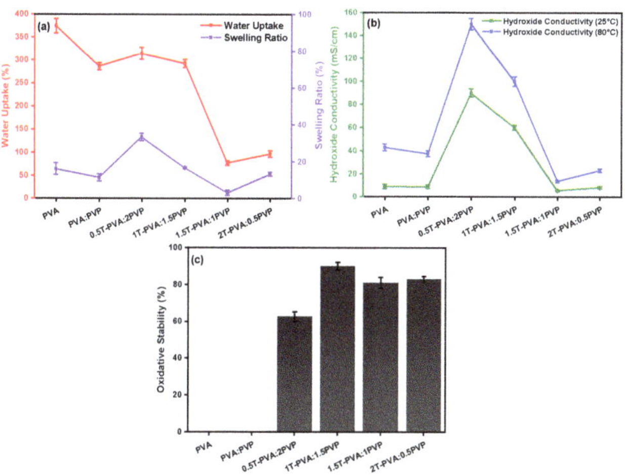

***Fig. 5 (a)** Water uptake and Swelling Ratio, **(b)** Ionic Conductivity at 25-80 °C, **(c)** Oxidative Stability of Ion-Solvating Membranes*

Conclusion

Hybrid semi-interpenetrating xTEOS–PVA:xPVP ion-solvating membranes were fabricated by a simple sol–gel/casting method followed by KOH activation. TEOS forms Si–O–Si and Si–O–C crosslinks, which tune morphology and hydration. Increasing TEOS reduces water uptake and swelling, while hydroxide conductivity reaches its optimum at 1T–PVA:1.5PVP (~60 mS cm^{-1} at 25 °C; ~100 mS·cm^{-1} at 80 °C) with high oxidative stability (~90% mass retention). Overall, this intermediate composition offers high OH^- transport with controlled swelling and improved durability, making it promising for alkaline membrane electrolyzers. This work will focus on extended durability testing under realistic AEMWE operation, gas crossover, permeability measurements, and mechanical evaluation after alkaline aging, as well as assessing scalability through controlled thickness casting and reproducibility.

References.

[1] M. Klingenhof, H. Trzesniowski, S. Koch, J. Zhu, Z. Zeng, L. Metzler, A. Klinger, M. Elshamy, F. Lehmann, P.W. Buchheister, A. Weisser, G. Schmid, S. Vierrath, F. Dionigi, P. Strasser, High-performance anion-exchange membrane water electrolysers using NiX (X = Fe,Co,Mn) catalyst-coated membranes with redox-active Ni–O ligands, Nat Catal 7 (2024). https://doi.org/10.1038/s41929-024-01238-w

[2] M.R. Kraglund, M. Carmo, G. Schiller, S.A. Ansar, D. Aili, E. Christensen, J.O. Jensen, Ion-solvating membranes as a new approach towards high rate alkaline electrolyzers, Energy and Environmental Science 12 (2019) 3313–3318. https://doi.org/10.1039/c9ee00832b

[3] J. Brauns, J. Schönebeck, M.R. Kraglund, D. Aili, J. Hnát, J. Žitka, W. Mues, J.O. Jensen, K. Bouzek, T. Turek, Evaluation of Diaphragms and Membranes as Separators for Alkaline Water Electrolysis, Journal of The Electrochemical Society 168 (2021) 014510. https://doi.org/10.1149/1945-7111/abda57

[4] D.W. Shin, M.D. Guiver, Y.M. Lee, Hydrocarbon-Based Polymer Electrolyte Membranes: Importance of Morphology on Ion Transport and Membrane Stability, Chemical Reviews 117 (2017) 4759–4805. https://doi.org/10.1021/acs.chemrev.6b00586

[5] Q. Li, A. Molina Villarino, C.R. Peltier, A.J. Macbeth, Y. Yang, M.J. Kim, Z. Shi, M.R. Krumov, C. Lei, G.G. Rodríguez-Calero, J. Soto, S.H. Yu, P.F. Mutolo, L. Xiao, L. Zhuang, D.A. Muller, G.W. Coates, P. Zelenay, H.D. Abruña, Anion Exchange Membrane Water Electrolysis: The Future of Green Hydrogen, Journal of Physical Chemistry C 127 (2023) 7901–7912. https://doi.org/10.1021/acs.jpcc.3c00319

[6] Haryadi, D. Sugianto, E. Ristopan, Development of composite membranes of PVA-TEOS doped KOH for alkaline membrane fuel cell, AIP Conference Proceedings 1699 (2015). https://doi.org/10.1063/1.4938331

[7] Y. Jiang, C. Shen, J. Chen, H. Lu, Y. Guo, S. Gao, Different cationic functionalized epichlorohydrin modified polyvinyl alcohol based anion exchange membrane, Journal of Applied Polymer Science 140 (2023) 1–17. https://doi.org/10.1002/app.54246

[8] T.Y. Guo, Q.H. Zeng, C.H. Zhao, Q.L. Liu, A.M. Zhu, I. Broadwell, Quaternized polyepichlorohydrin/PTFE composite anion exchange membranes for direct methanol alkaline fuel cells, Journal of Membrane Science 371 (2011) 268–275. https://doi.org/10.1016/j.memsci.2011.01.043

[9] A.M. Samsudin, M. Roschger, S. Wolf, V. Hacker, Preparation and Characterization of QPVA/PDDA Electrospun Nanofiber Anion Exchange Membranes for Alkaline Fuel Cells, Nanomaterials 12 (2022). https://doi.org/10.3390/nano12223965

[10] A.N. Mondal, C. Cheng, Z. Yao, J. Pan, M.M. Hossain, M.I. Khan, Z. Yang, L. Wu, T. Xu, Novel quaternized aromatic amine based hybrid PVA membranes for acid recovery, Journal of Membrane Science 490 (2015) 29–37. https://doi.org/10.1016/j.memsci.2015.04.052

[11] B.P. Tripathi, A. Saxena, V.K. Shahi, Phosphonic acid grafted bis(4-γ-aminopropyldiethoxysilylphenyl)sulfone (APDSPS)-poly(vinyl alcohol) cross-linked polyelectrolyte membrane impervious to methanol, Journal of Membrane Science 318 (2008) 288–297. https://doi.org/10.1016/j.memsci.2008.02.052

[12] A. Sabetghadam, T. Mohammadi, PVA-APTEOS/TEOS hybrid Sol-Gel pervaporation membrane for dehydration of ethanol, Composite Interfaces 17 (2010) 223–228. https://doi.org/10.1163/092764410X490626

[13] Y. Gao, N.R. Choudhury, N. Dutta, J. Matisons, M. Reading, L. Delmotte, Organic-inorganic hybrid from ionomer via sol-gel reaction, Chemistry of Materials 13 (2001) 3644–3652. https://doi.org/10.1021/cm010179s

[14] F. Rahmani, H. Ziyadi, M. Baghali, H. Luo, S. Ramakrishna, Electrospun pvp/pva nanofiber mat as a novel potential transdermal drug-delivery system for buprenorphine: A solution needed for pain management, Applied Sciences (Switzerland) 11 (2021) 1–21. https://doi.org/10.3390/app11062779

[15] S. Kim, A. Cho, S. Kim, W. Cho, M.H. Chung, F.S. Kim, J.H. Kim, Multi-purpose overcoating layers based on PVA/silane hybrid composites for highly transparent, flexible, and durable AgNW/PEDOT:PSS films, RSC Advances 6 (2016) 19280–19287. https://doi.org/10.1039/c5ra27311k

[16] G.H.A. Wijaya, K.S. Im, S.Y. Nam, Advancements in commercial anion exchange membranes: A review of membrane properties in water electrolysis applications, Desalination Water Treat 320 (2024) 100605. https://doi.org/10.1016/j.dwt.2024.100605

Lithium Incorporation into Sol-Gel-Processed Chalcogenide Photovoltaic Absorbers: A Simple and Efficient Strategy

Naoufal ENNOUHI[1,a], Nisar ALI[1,b], Rosemary CHITI[1,c], Otman ABIDA[1,d], Bakhtiar ULHAQ[2,e]

[1]African Sustainable Agriculture Research Institute (ASARI), Mohammed VI Polytechnic University, 70000, Laayoune, Morocco

[2]Mechanical Engineering Department, College of Engineering, Prince Mohammad Bin Fahd University, Alkhobar 31952, Saudi Arabia.

[a]Naoufal.ENNOUHI-EXT@um6p.ma, [b]nisar.ali@um6p.ma, [c]Rosemary.CHITI@um6p.ma, [d]otman.abida@um6p.ma, [e]bakhtiarjadoon@gmail.com

Keywords: Kesterite, ACZTS, Lithium Doping, Sol-Gel, Thin Films, Photovoltaics

Abstract. Lithium doping is one of the most successful approaches used recently to improve the properties of kesterite materials and the performance of their solar cells. However, doping in sol-gel-derived kesterite is still limited to dopant incorporation during solution preparation. Here, we present a simple strategy to incorporate lithium into kesterite absorbers by depositing thin Li-containing layer onto kesterite precursors via the spin-coating method. Compared with conventional doping approach, this method led to enhanced film quality. Comprehensive structural and morphological analysis was performed via X-ray diffraction (XRD), Raman spectroscopy, scanning electron microscopy (SEM), and atomic force microscopy (AFM). Our results confirmed a significant improvement in crystallinity, a reduction in Cu/Zn disorder, and the growth of larger and well-defined grains. Optical characterization confirmed the improved bandgap and a reduction in the defect density following lithium incorporation via the proposed strategy. The findings reported in this work demonstrate that doping via surface deposition represents a promising approach for enhancing the properties of kesterite thin films prepared using sol-gel method and can be a promising approach for developing highly efficient kesterite-based solar cells if carefully controlled.

Introduction

Kesterite materials such as $Cu_2ZnSnSSe_4$ (CZTSSe) have been proposed in the last few years as an alternative to commercialized CIGS solar cells due to their earth-abundant constituents and their higher optoelectronic properties [1] [2]. However, despite the progress reported in the last years for kesterite solar cell, the record efficiency remains limited to a higher open circuit voltage (V_{OC}) deficit originating from the high density of defects and the resistive Kesterite/CdS interface [3]. Recently, a wide range of strategies have been used to address these limitations, such as doping and alloying, which show a significant potential in reducing atomic disorder, enhance crystallinity, and improve carrier collection [4].

Alkali-metals doping and silver alloying have been shown to be the most successful approaches used in the last few years to enhance the properties of kesterite solar cells, and efficiencies beyond 15% have been achieved by combining sodium or lithium doping and silver alloying [5]. These effects can translate into larger grains and reduce the density of detrimental defects when properly coordinated with annealing conditions and precursor chemistry [6]. In parallel, alkali-metal diffusion from soda-lime glass (SLG) through porous Mo back contacts or introduction during the preparation process [7] promotes liquid-phase-assisted grain growth during annealing and passivates grain boundaries, thereby improving the film morphology and electronic quality [8].

Among alkalis, lithium (Li) is particularly compelling: its ionic radius closely matches that of Cu, enabling Li to occupy Cu vacancies and form shallow acceptors (e.g., Li_{Zn}) that increase the hole density and can lower disorder, with multiple reports linking Li to enhanced crystallinity and Voc improvements [9]. The methods for incorporating alkali metals into kesterite materials vary according to the deposition technique employed in the creation of kesterite materials. In solution-processed kesterite materials, such as those produced via the sol-gel method, alkyl metal salts are employed during solution preparation [10,11]. In contrast, in physical vapor deposition techniques, a thin layer of alkali metal (e.g., NaF, LiF, etc.) is typically employed [12,13]. In solution-processed kesterite approaches, the incorporation of alkali metals commences in the solution phase, subsequently resulting in a homogeneous distribution of these alkali-metals within the precursors, which are then subjected to an annealing process. In contrast, in physical methods, alkali metal layers can be deposited at varying positions and thicknesses, and during the annealing process, alkali metals can diffuse into the absorber in different ways. Furthermore, these diffusion methods of alkali metals significantly impact the final properties of the absorbers, thereby enhancing the flexibility in doping kesterite materials. [14]. Nevertheless, the capacity to alter the diffusion of alkali metals into kesterite absorbers is not inherent to solution deposition techniques, where alkali metals are employed in the solution. Moreover, when FTO is used as the back contact (instead of Mo/SLG), parasitic alkali diffusion from SLG is strongly suppressed; consequently, any alkali-related effects can be attributed to the intentionally introduced dopant, simplifying interpretation.

In this work, we therefore evaluate a simple postdeposition Li incorporation strategy for sol–gel–processed ACZTS absorbers: a thin Li-containing layer is deposited directly on top of the dried ACZTS precursor prior to sulfurization. We benchmark this method against conventional solution-phase Li doping under identical precursor chemistries and annealing protocols. A multimodal characterization suite, X-ray diffraction (XRD), Raman spectroscopy, scanning electron microscopy (SEM), atomic force microscopy (AFM), and UV-visible spectroscopy, reveals the structural, morphological, and optical impacts of the Li incorporation route on ACZTS, with particular attention to crystallinity, Cu/Zn disorder, grain growth, and band-tailing (Urbach energy). The results highlight the advantages of the postdeposition route in delivering denser microstructures, larger grains, and reduced disorder, which are key prerequisites for closing the Voc gap in kesterite photovoltaics. The control of alkali-metals incorporation using the post-deposition method seems to be difficult for controlling the doping profile compared to solution phase doping. However, this work represents the first proof-of-concept, and by controlling the thickness, annealing and pre-heating temperatures, as well as the position of Li-containing layers, it is possible to achieve more controllable lithium incorporation compared the homogeneous distribution of these alkali-metals in the solution phase. Therefore, further advanced work should be carried out using this approach to improve Li diffusion and enhance control over this method

Experimental Details

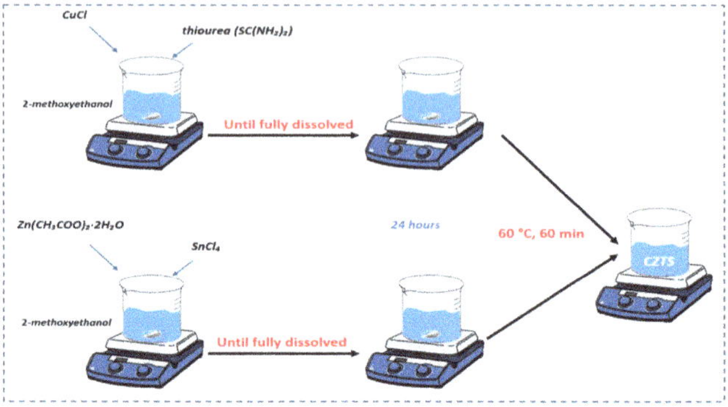

Figure 1: *Process used to prepare the ACZTS sol–gel solution*

ACZTS precursor solutions were prepared via a sol–gel route from CuCl, Zn(CH3COO)2·2H2O and SnCl4 in 2-methoxyethanol following the protocol summarized in Figure 1 and then filtered (0.45 µm). Films were deposited on UV-ozone--treated FTO/SLG by spin coating at 3000 rpm for 30 s with intermediate hot- plates at 300 °C; seven cycles yielded ~1 µm thickness. Lithium was introduced via two approaches: (i) solution-phase Li (Li-Sol) by adding LiClO4 to the mixed sol–gel solution and (ii) post-deposition Li (Li-Dep) by spin-coating a LiClO4 solution as a thin surface layer on the dried ACZTS precursor prior to annealing. Structural and optical characterizations included XRD (Cu Kα1), Raman (785 nm), SEM, AFM, and UV–Vis spectrophotometry.

Results and Discussion

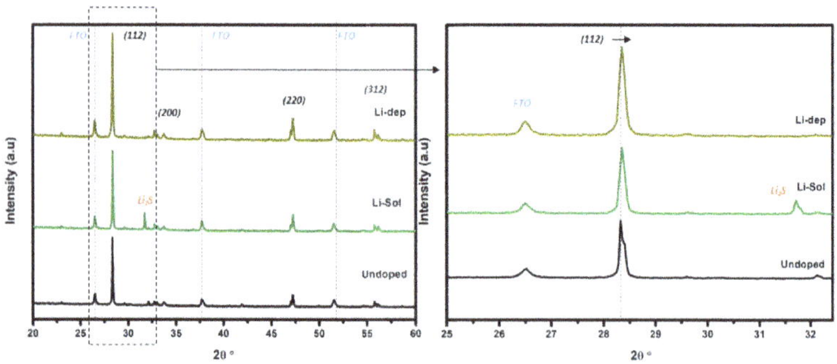

Figure 2: *XRD Patterns of Litium-Doped Li-Sol and Li-Dep Samples Compared with Undoped ACZTS*

The XRD patterns for the ACZTS samples doped with Li via both doping strategies and the undoped sample are shown in Figure 2. The peak intensities are greater for both Li-doped samples than for the undoped sample, indicating improved crystallinity with Li doping. This trend is consistent with previous reports that Li doping enhances the crystallinity of kesterite materials [15]

Materials Research Proceedings 64 (2026) 664-671 https://doi.org/10.21741/9781644904091-83

[16]. A comparison of the doping methods revealed that the Li-Dep sample doped via the proposed strategy and exhibited sharper and more intense peaks than did the Li-Sol sample. However, $LiClO_4$ was used for Li doping. $LiClO_4$, which is more thermally stable and has a higher decomposition temperature (420°C), remains in the films until the later stages of the annealing process. It then decomposes into LiCl, which eventually forms Li_2S [17]. This process is supported by additional peaks that may correspond to Li_2S observed in the XRD pattern of the Li-Sol sample. Note that these peaks can also be due to the presence of SnS2, as some of their XRD peaks overlapped. Previous studies on Li doped with $LiClO_4$, in which DMSO was used as a solvent and selenium was used in the annealing process, did not report that these Li_2S phases may be due to different interactions between sulfur or selenium and $LiClO_4$. In the Li-Dep sample, the $LiClO_4$ phase deposited on top of the precursor is in direct contact with the sulfur in the annealing atmosphere, facilitating the decomposition of $LiClO_4$ to LiCl and subsequently to Li_2S at lower temperatures. This interaction may enhance the growth of ACZTS materials. In addition, excess lithium can passivate grain boundaries and vacancies, leading to higher crystallinity and larger grains, as discussed in the next section. Overall, both Li-doped samples show improved crystallinity compared with that of the undoped sample. The issue of Li_2S formation can potentially be addressed by using LiCl instead of $LiClO_4$ or by optimizing the sulfurization temperature. The Raman spectra support these findings, with the Li-Dep sample exhibiting more intense peaks, indicating a more ordered structure. In contrast, the undoped and Li-Sol samples show comparable peak sharpness and intensity, potentially due to lower lithium incorporation and the presence of Li_2S, leading to a more disordered matrix [18]. The formation of Li_2S phase can be addressed using LiCl as a precursor for lithium doping instead of $LiClO_4$ and the optimization of the annealing conditions. Raman spectra shown in Figure 3 confirmed these findings. The Li-Dep sample exhibited the narrow and more intense peaks compared to that of the Li-Sol, indicating a more ordered matrix [18], where both Li-Sol and undoped samples present comparable peak intensities may be due to the presence of Li-S phases.

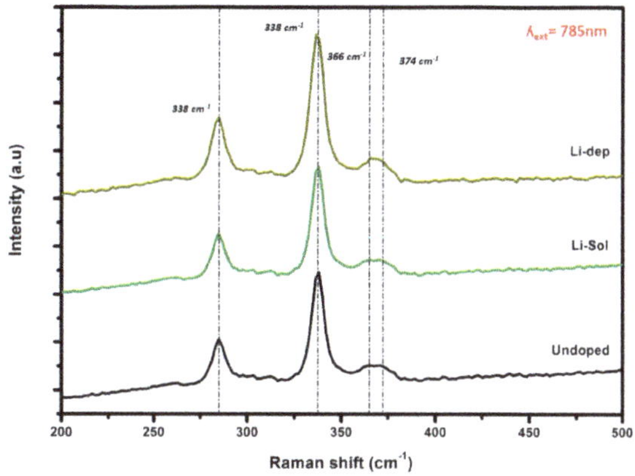

Figure 3: *Raman Spectra of Litium-Doped Li-Sol and Li-dep Samples Compared with Undoped ACZTS*

The SEM and AFM analysis of Li-Sol and Li-Dep have been used to evaluate the morphology of the surface, and the results are shown in Figure 4. an organized morphologies and larger grains have been shown by Li-Dep sample compared to the undoped sample, which reflect the reported effects of Li incorporation in kesterite materials [19]. Conversely, the Li-Dep sample (Li-Dep) demonstrated a compacted and dense morphology with larger grain. In contrast, the sample doped during solution preparation (Li-Sol) displays a less compact morphology with smaller grains and a greater density of voids. These observations can be attributed to the lower solubility in the solvent and higher decomposition temperature of $LiClO_4$, which is distributed within the absorber in the Li-Sol sample. This can lead to the formation of some grains in some regions within the bulk of the absorber and complicate the formation of Li_2S. Similar results have been reported in [17]. Additionally, these results might be influenced by the amount of Li incorporated via each doping strategy (note that lithium cannot be detected via EDX analysis). From the EDX results (Table 1), both doped samples show greater amounts of Sn on the surface than the undoped sample does. However, the excess Tin in the Li-Sol sample may lead to the formation of SnS2, which has diffraction peaks similar to those of Li2S (JPCDS 23-0677).

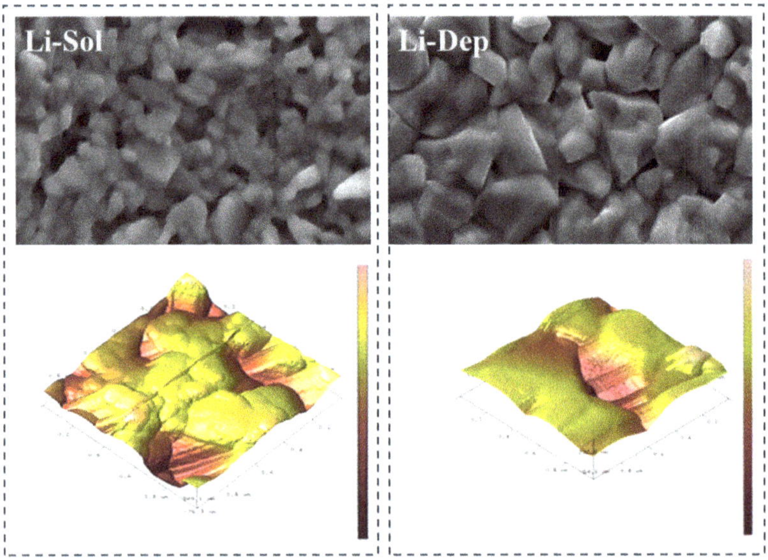

Figure 4: *Surface SEM images and 3D-AFM images of the lithium-doped Li-Sol and Li-Dep samples*

Table 1: *Cation ratios of the lithium-doped Li-Sol, Li-Dep C and Undoped samples from the EDS results*

Samples/Atom%	Cu	Sn	Zn	S	Li	Zn/Sn	Cu/(Zn + Sn)
Undoped	24.56	8.39	13.07	50.83	-----	1.55	1.14
Li-Sol	22.10	14.16	11.14	49.12	-----	0.78	0.87
Li-Dep	22.30	11.33	11.41	50.93	-----	1.01	0.98

The optical bandgap values for the Li-doped samples shown in Figure 5 are 1.63 eV for Li-Sol and 1.58 eV for Li-Dep. The bandgap of Li-Dep aligns with the expected bandgap for samples at the given silver concentration [20]. The higher bandgap observed for Li-Sol could be attributed to off-stoichiometry or the presence of secondary phases, as confirmed by previous characterizations. The Urbach energy (Eu) for both samples, which indicates the extent of the band-tail states and defect density, is 426 and 403 meV for the Li-Sol and Li-Dep samples, respectively. This slight reduction in Eu for the Li-Dep sample compared with that of the Li-Sol sample confirms the decreased defect density and band-tailing in the sample doped via the precursor method. This improvement supports the effectiveness of the proposed doping strategy in enhancing the optoelectronic properties of kesterite materials.

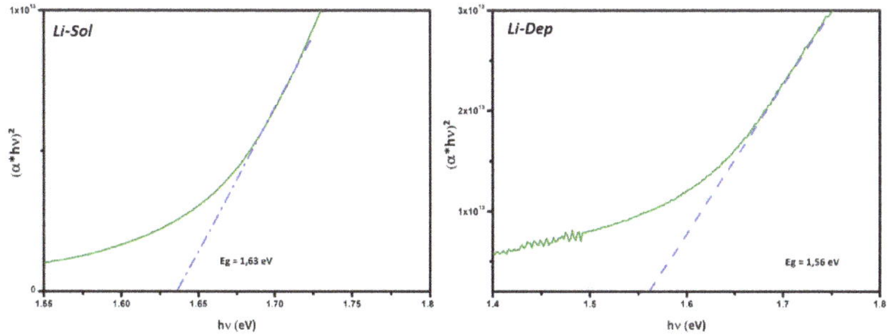

***Figure 5**: Energy band bandgap of the litium-doped Li-Sol and Li-Dep samples*

Summary

In this work we presented a simple strategy to incorporate Li into kesterite absorbers synthesized using sol-gel method. This approach of depositing a thin Li-contining layer on the top of kesterite precursor before the annealing process. Our results demonstrated enhanced films properties compared to the previously reported doping strategy. The post deposition approach results in higher crystallinity, larger grains, denser morphology, and lower Urbach energy. The approach proposed in this work can lead to more flexibility in doping kesterite materials with alkali metlas and can show a strong potential for developing highly efficient kesterite solar cells if carefully controlled similar to the reported doping strategies used in PVD-based kesterite thin films.

References

[1] N. Ali, M. Zubair, A. Khesro, R. Ahmed, S. Uddin, N. Shahzad, H. Alrobei, A. Kalam, A.G. Al-Sehemi, B. Ul Haq, A Study on Optoelectronic Properties of Copper Zinc Tin Sulfur Selenide: A Promising Thin-Film Material for Next Generation Solar Technology, Crystal Research and Technology 56 (2021) 2000159.

[2] A. Wang, J. Huang, J. Cong, X. Yuan, M. He, J. Li, C. Yan, X. Cui, N. Song, S. Zhou, M.A. Green, K. Sun, X. Hao, Cd-Free Pure Sulfide Kesterite Cu_2ZnSnS_4 Solar Cell with Over 800 mV Open-Circuit Voltage Enabled by Phase Evolution Intervention, Advanced Materials 36 (2024) 2307733. https://doi.org/10.1002/adma.202307733.

[3] N. Ali, A. Hussain, R. Ahmed, M.K. Wang, C. Zhao, B.U. Haq, Y.Q. Fu, Advances in nanostructured thin film materials for solar cell applications, Renewable and Sustainable Energy Reviews 59 (2016) 726–737. https://doi.org/10.1016/j.rser.2015.12.268.

[4] Y.E. Romanyuk, S.G. Haass, S. Giraldo, M. Placidi, D. Tiwari, D.J. Fermin, X. Hao, H. Xin, T. Schnabel, M. Kauk-Kuusik, P. Pistor, S. Lie, L.H. Wong, Doping and alloying of kesterites, J. Phys. Energy 1 (2019) 044004. https://doi.org/10.1088/2515-7655/ab23bc.

[5] A. Mule, B. Vermang, M. Sylvester, G. Brammertz, S. Ranjbar, T. Schnabel, N. Gampa, M. Meuris, J. Poortmans, Effect of different alkali (Li, Na, K, Rb, Cs) metals on Cu 2 ZnSnSe 4 solar cells, Thin Solid Films 633 (2017) 156–161. https://doi.org/10.1016/j.tsf.2016.11.027.

[6] Y. Gong, A. Jimenez-Arguijo, A.G. Medaille, S. Moser, A. Basak, R. Scaffidi, R. Carron, D. Flandre, B. Vermang, S. Giraldo, H. Xin, A. Perez-Rodriguez, E. Saucedo, Li-Doping and Ag-Alloying Interplay Shows the Pathway for Kesterite Solar Cells with Efficiency Over 14%, Adv Funct Materials (2024) 2404669. https://doi.org/10.1002/adfm.202404669.

[7] N. Bansal, B.C. Mohanty, K. Singh, Diffusional investigation of alkali ions from composition tuned glass substrates to Mo-thin film for solar cell application, Surfaces and Interfaces 24 (2021) 101060. https://doi.org/10.1016/j.surfin.2021.101060.

[8] Q. Zhao, H. Shen, K. Gao, Y. Xu, X. Wang, Y. Li, Effect of sodium doping on crystal growth and band matching of the heterojunction in flexible CZTS solar cells, J. Mater. Chem. C 9 (2021) 17531–17541. https://doi.org/10.1039/D1TC03957A.

[9] M. He, X. Zhang, J. Huang, J. Li, C. Yan, J. Kim, Y. Chen, L. Yang, J.M. Cairney, Y. Zhang, S. Chen, J. Kim, M.A. Green, X. Hao, High Efficiency $Cu_2ZnSn(S,Se)_4$ Solar Cells with Shallow Li_{Zn} Acceptor Defects Enabled by Solution-Based Li Post-Deposition Treatment, Advanced Energy Materials 11 (2021) 2003783. https://doi.org/10.1002/aenm.202003783.

[10] Z. Laghfour, S. Aazou, M. Taibi, G. Schmerber, A. Ulyashin, A. Dinia, A. Slaoui, M. Abd-Lefdil, Z. Sekkat, Sodium doping mechanism on sol-gel processed kesterite Cu2ZnSnS4 thin films, Superlattices and Microstructures 120 (2018) 747–752. https://doi.org/10.1016/j.spmi.2018.05.018.

[11] G. Tseberlidis, V. Trifiletti, A. Le Donne, L. Frioni, M. Acciarri, S. Binetti, Kesterite solar-cells by drop-casting of inorganic sol–gel inks, Solar Energy 208 (2020) 532–538. https://doi.org/10.1016/j.solener.2020.07.093.

[12] H. Xin, S.M. Vorpahl, A.D. Collord, I.L. Braly, A.R. Uhl, B.W. Krueger, D.S. Ginger, H.W. Hillhouse, Lithium-doping inverts the nanoscale electric field at the grain boundaries in $Cu_2ZnSn(S,Se)_4$ and increases photovoltaic efficiency, Phys. Chem. Chem. Phys. 17 (2015) 23859–23866. https://doi.org/10.1039/C5CP04707B.

[13] J. Xu, S. Shang, J. Yang, J. Liu, S. Jiang, Effect of sodium-doping on the performance of CZTS absorb layer: Single and bifacial sodium-incorporation method, Solar Energy 221 (2021) 476–482. https://doi.org/10.1016/j.solener.2021.04.063.

[14] A. Ruiz-Perona, M. Guc, Y. Sánchez, T. Kodalle, J.M. Merino, M. León, R. Caballero, Wide band gap Cu2ZnGe(S,Se)4 thin films and solar cells: Influence of Na content and incorporation method, Solar Energy 226 (2021) 251–259. https://doi.org/10.1016/j.solener.2021.08.032.

[15] M. Mohamad, B.U. Haq, R. Ahmed, A. Shaari, N. Ali, R. Hussain, A density functional study of structural, electronic and optical properties of titanium dioxide: Characterization of rutile, anatase and brookite polymorphs, Materials Science in Semiconductor Processing 31 (2015) 405–414. https://doi.org/10.1016/j.mssp.2014.12.027.

[16] D.S. Dhawale, A. Ali, A.C. Lokhande, Impact of various dopant elements on the properties of kesterite compounds for solar cell applications: a status review, Sustainable Energy Fuels 3 (2019) 1365–1383. https://doi.org/10.1039/C9SE00040B.

[17] Y. Gong, A. Jimenez-Arguijo, A.G. Medaille, S. Moser, A. Basak, R. Scaffidi, R. Carron, D. Flandre, B. Vermang, S. Giraldo, H. Xin, A. Perez-Rodriguez, E. Saucedo, Li-Doping and Ag-

Alloying Interplay Shows the Pathway for Kesterite Solar Cells with Efficiency Over 14%, Adv Funct Materials (2024) 2404669. https://doi.org/10.1002/adfm.202404669.

[18] Z. Shen, S. Wang, Y. Liu, Y. Sun, J. Wu, H. Guo, K. Zhang, S. Zhang, F. Liu, Y. Zhang, Li2S doping into CZTSe drives the large improvement of VOC of solar cell, Journal of Energy Chemistry 62 (2021) 637–644. https://doi.org/10.1016/j.jechem.2021.04.018.

[19] Y. Mao, J. Guo, C. Cao, J. Ao, F. Liu, Y. Zhang, Eliminating cracking morphology of solution-processed CZTSSe absorbers by Sn-rich composition engineering, Materials Science in Semiconductor Processing 159 (2023) 107388. https://doi.org/10.1016/j.mssp.2023.107388.

[20] J. Kumar, S. Ingole, Structural and optical properties of (AgxCu1-x)2ZnSnS4 thin films synthesised via solution route, Journal of Alloys and Compounds 727 (2017) 1089–1094. https://doi.org/10.1016/j.jallcom.2017.08.222.

CFD-Based Multi-Objective Optimization of Straight-Fin Thermal Energy Storage Using MXene/BNNT–Erythritol Composite System

Ghulam RASOOL[1,a]*, Azim UDDIN[2,b] and Zeeshan Ali BALOCH[3,c]

[1]Department of Mechanical Engineering, College of Engineering, Prince Mohammad Bin Fahd University, Al-Khobar 31952, Saudi Arabia

[2]JAR Innovation Center, Shaoxing, China

[3]Department of Physics, Quaid-i-Azam University, Islamabad, Pakistan

[a]grasool@pmu.edu.sa, [b]auddin@zju.edu.cn, [c]zeeshanab702@gmail.com

Keywords: Phase Change Material, MXene/BNNT Composite, CFD Simulation, Thermal Energy Storage, Response Surface Optimization, Natural Convection Melting, Erythritol

Abstract. The research used CFD to study the system while performing multi-objective optimization on a rectangular LHTES unit containing bio-based erythritol and MXene/BNNT composite fin enhancements. The enthalpy–porosity method was used to simulate melting behavior and a finite-volume solver handled both natural convection and phase change processes. The Design of Experiments–Response Surface Methodology (DOE–RSM) framework optimized fin number, thickness, and span to achieve both fast charging and high storage capacity. The three-fin design configuration shortens total melting time by 60% while increasing stored energy to twice the level of the finless design. The research shows MXene/BNNT composites create better heat transfer paths which maintain their ability to store heat thus enabling developers to create efficient TES modules for solar and waste-heat recovery systems.

Introduction

The scientific community has conducted extensive research on phase change materials (PCMs) for thermal energy storage (TES) because these materials possess the ability to absorb and emit large quantities of latent heat. Standard PCMs have two major limitations because they conduct heat poorly and they leak during phase changes which makes them unsuitable for real-world applications. Research indicates MXene-based composite PCMs (CPCMs) demonstrate potential for energy conversion and electromagnetic shielding and flame retardancy applications because of their manufacturing techniques and their ability to perform various functions [1]. Research studies demonstrate that MXene-enhanced PCMs show improved performance because of two factors which include electron–phonon coupling effects and localized surface plasmon resonance [2]. MXenes added to PCMs produce improved thermal conductivity and stability and enhanced energy storage capabilities because their layered structure and high electrical conductivity properties [3]. The research shows MXene materials enhance PCM heat transfer performance and response speed which makes them appropriate for solar thermal storage applications [4]. The development of PEG/-based layered composites has improved MXene–PCM systems for photothermal energy storage because they use localized surface plasmon resonance to achieve 94.5% efficiency when exposed to solar radiation [5]. Erythritol has proven itself as a suitable organic PCM for medium-temperature thermal energy storage because it offers high latent heat values and thermal stability. The material shows poor heat transfer properties and tends to form supercooled states which makes it less useful for applications. The development of solar heat collection and waste heat recovery systems requires new methods to enhance erythritol performance according to recent reviews which focus on structural changes and composite development [6]. The development of calcium ion cross-linked sodium alginate matrices enables the creation of erythritol-based composites which maintain controlled heat release and maintain their shape while showing superior supercooling resistance and thermal cycling stability [7]. The thermal conductivity of erythritol increased by 52.5% while supercooling decreased through the addition of calcium pimelate and expanded graphite [8]. The simulation of diamond/erythritol composites with hierarchical porous structures using molecular dynamics shows that their thermal conductivity increases by more than 200% because of better phonon transmission and space confinement effects [4]. Research shows that erythritol composite systems with aluminum nitride (AlN) porous skeletons achieve better supercooling resistance and improved thermal conductivity because of enhanced interfacial bonding [9]. The addition of hydroxylated boron nitride nanosheets (–OH BNNSs) to erythritol increases its thermal conductivity to 1.25 W/m·K while preserving most of its latent heat storage capacity [10]. Research has extensively studied thermally conductive nanocomposites for electronic and renewable energy systems through polymer composite development with optimized microstructures for high heat dissipation performance [11]. The discovery of supercooled erythritol-based PCMs has increased the potential for extended thermal energy storage applications through hybrid systems that combine D-mannitol with sodium alginate [12]. System-level optimization of PCM-based TES units has been performed through computational fluid dynamics (CFD) assessments and development. The experimental evaluation of CFD models without convective effects during PCM discharge shows high prediction accuracy with minimal computational expense while maintaining errors below 1 °C [13].

Research now uses CFD-assisted design optimization to convert PCM material improvements into functional LHTES systems which need to optimize fin design and melting duration and energy storage capacity under natural convection conditions. The optimization of fin structures and heat-transfer pathways in rectangular LHTES systems has proven successful for enhancing both melting and solidification rates and thermal distribution while maintaining sufficient PCM capacity. The development of finned PCM heat exchangers has led to the creation of two optimization systems which use surrogate-based and multi-objective optimization methods to find the best fin designs for both fast charging and high energy storage and heat absorption rates.

The range of CFD-driven PCM system optimization applications continues to grow because researchers now use these methods to enhance two main areas of application. The first application focuses on solar thermal and waste-heat recovery modules which need fast charging capabilities and stable operation during multiple cycles. The second application deals with building and clean heating buffer systems which aim to achieve supply-demand balance while requiring minimal space for installation. The third application involves battery thermal management systems (BTMS) which use PCM unit optimization to minimize peak temperatures and thermal differences during brief load changes. The present study combines CFD-based phase-change modeling with DOE–RSM multi-objective optimization to determine the optimal number of fins and their geometry while maintaining a fixed PCM volume which matches the requirements of practical TES systems.

The development of high-conductivity composite phase change materials has brought substantial progress but researchers need to create optimized LHTES system designs which utilize these material advancements. The fast charging process requires researchers to measure how fin geometry interacts with PCM volume utilization and natural-convection-driven melting dynamics to achieve storage capacity preservation.

The primary goal of this research involves creating a CFD-based system which optimizes multiple performance indicators for rectangular LHTES units that use MXene/BNNT composite fins inside bio-based erythritol PCM. The research investigates three main objectives which include (i) numerical melting behavior analysis through enthalpy–porosity formulation and (ii) fin number and thickness evaluation under fixed PCM-volume conditions and (iii) DOE–RSM application for total melting time reduction and thermal energy storage maximization. The research

Materials Research Proceedings 64 (2026) 672-679 https://doi.org/10.21741/9781644904091-84

results establish design principles which will help developers create small thermal batteries for solar thermal systems and waste heat recovery systems.

Problem Formulation

The research examines how rectangular LHTES units with bio-based phase change materials and MXene/BNNT composite fins affect thermal performance during melting operations. The system design enables fast charging and discharging operations while preserving PCM volume for use in small-scale solar and waste heat recovery systems. The two-dimensional model represents the enclosure as a vertical rectangular cavity with dimensions 120mm in height and 50mm in width and the right wall maintained at 343.15 K while all other walls remain adiabatic. The PCM starts in a solid state at a single temperature of 298.15 K throughout the entire system. The heat transfer performance improves through the use of straight fins which extend from the heated wall into the PCM domain. The research investigates four different configurations which include a finless reference model and three additional models with one fin of 0.5mm thickness extending 40mm into the PCM and two fins of 0.4mm thickness and three fins of 0.3mm thickness. The research maintains a fixed PCM volume through thickness reductions of fins when adding more fins to study their effect on melting duration and complete energy storage performance. The PCM selected is Erythritol, a sustainable sugar–alcohol with high latent heat capacity and chemical stability. Its thermophysical properties are listed in Table 1. The fins are made of a $Ti_3C_2T_x$–BNNT composite, offering ultrahigh thermal conductivity, corrosion resistance, and improved interfacial compatibility with organic PCMs.

Table 1. Thermophysical properties of materials used in the simulation.

Property	Erythritol (PCM)	MXene/BNNT Composite (Fin)
Density, ρ [kg/m^3]	1480	4300
Specific heat, C_p [J/kg·K]	2380	710
Thermal conductivity, k [W/m·K]	0.73	420
Latent heat, L_f [kJ/kg]	339	–
Solidus temperature, T_s [K]	385.15	–
Liquidus temperature, T_l [K]	392.15	–

The following assumptions are adopted:
1. The flow is two–dimensional, laminar, and incompressible.
2. Thermophysical properties are temperature–independent except for density, which varies under the Boussinesq approximation.
3. The initial PCM is entirely solid at $T_i = 298.15\,\mathrm{K}$.
4. The right wall is isothermal at $T_w = 343.15\,\mathrm{K}$, while the remaining walls are adiabatic ($\partial T/\partial n = 0$).
5. Heat transfer and phase change occur via conduction and natural convection within the liquid phase.

Governing Equations

We model melting in the PCM domain using the transient, incompressible Navier–Stokes and energy equations in an enthalpy–porosity formulation. The buoyancy force is treated via the Boussinesq approximation. Fins are modeled as solid, stationary, high-conductivity regions governed by pure conduction and coupled to the PCM by temperature and heat-flux continuity at the PCM–fin interfaces.
1. PCM Region: Conservation Equations
Mass conservation (continuity):

$$\nabla \cdot \mathbf{V} = 0 \quad \text{in } \Omega_{\text{PCM}}. \tag{1}$$

Momentum conservation (Navier–Stokes with mushy-zone source):

$$\rho\left(\frac{\partial \mathbf{V}}{\partial t} + \mathbf{V} \cdot \nabla \mathbf{V}\right) = -\nabla p + \mu \nabla^2 \mathbf{V} + \rho\, \mathbf{g} \beta\, (T - T_0) + \mathbf{S}_m, \tag{2}$$

where p is the pressure, ρ the PCM density, μ the dynamic viscosity, β the thermal expansion coefficient, and T_0 a reference temperature. The buoyancy force is modeled through the Boussinesq approximation. The Darcy-like momentum sink $\mathbf{S}_m$ damps velocity in the mushy/solid regions:

$$\boldsymbol{S}_m = A_{mush} \frac{(1-f)^2}{f^3 + \varepsilon} \boldsymbol{V}, \tag{3}$$

with A_{mush} the mushy-zone constant, $f \in [0,1]$ the liquid fraction, and $\varepsilon \ll 1$ a small regularization parameter.
Energy conservation (enthalpy form):

$$\rho \frac{\partial H}{\partial t} + \rho\, \boldsymbol{V} \cdot \nabla H = \nabla \cdot (k \nabla T), \tag{4}$$

where k is the PCM thermal conductivity and H is the specific enthalpy,

$$H(T, f) = h_s(T) + f L_f, \qquad h_s(T) = h_{ref} + \int_{T_{ref}}^{T} C_p \, d\theta. \tag{5}$$

Phase-change closure (liquid fraction):

$$f(T) = \begin{cases} 0, & T < T_s, \\ \dfrac{T - T_s}{T_l - T_s}, & T_s \leq T \leq T_l, \\ 1, & T > T_l, \end{cases} \tag{6}$$

with T_s and T_l the solidus and liquidus temperatures of the PCM, respectively.
2. Fin Region: Conduction Only
In Ω_{fin} (MXene/BNNT composite fins), the temperature satisfies

$$\rho_f C_{p,f} \frac{\partial T}{\partial t} = \nabla \cdot (k_f \nabla T), \tag{7}$$

where ρ_f, $C_{p,f}$, and k_f are the fin density, specific heat, and thermal conductivity, respectively. No momentum equations are solved in the fin region.

Coupling and Interface Conditions

At each PCM–fin interface Γ_{int}, we impose continuity of temperature and normal heat flux:

$$\begin{aligned} T_{PCM} &= T_{fin} \quad on\, \Gamma_{int}, \\ (-k\nabla T \cdot \boldsymbol{n})_{PCM} &= \left(-k_f \nabla T \cdot \boldsymbol{n}\right)_{fin} \quad on\, \Gamma_{int}. \end{aligned} \tag{8}$$

3. Initial and Boundary Conditions

Initially, the PCM is solid and quiescent:

$$T(\boldsymbol{x},0) = T_i, \quad f(\boldsymbol{x},0) = 0, \quad \boldsymbol{V}(\boldsymbol{x},0) = \boldsymbol{0} \;\; in\, \Omega_{PCM}. \tag{9}$$

On the heated wall (right boundary) we prescribe

$$T = T_w, \tag{10}$$

while all remaining external walls are adiabatic,

$$\frac{\partial T}{\partial n} = 0, \tag{11}$$

and no-slip is applied where the momentum equations are solved,

$$\boldsymbol{V} = \boldsymbol{0} \quad on\ solid\ boundaries\ of\, \Omega_{PCM}. \tag{12}$$

Numerical Methodology

The governing equations are solved using a collocated, cell-centered finite volume method (FVM) on a body-fitted, structured grid. Second-order spatial accuracy is enforced with a central differencing scheme for diffusion terms and a bounded, second-order upwind (QUICK-like) scheme for convection. Time integration is implicit and second-order accurate (backward differentiation, BDF2), enabling stable advancement under strong latent heat release. Pressure–velocity coupling is achieved with the SIMPLE algorithm, with Rhie–Chow interpolation to avoid pressure–velocity decoupling on the collocated mesh. Momentum, pressure-correction, and energy equations are solved in a segregated manner with under-relaxation factors adjusted to ensure robustness across the solid/mushy/liquid transitions. The enthalpy–porosity formulation is used to embed the phase-change front in the fixed grid; the momentum sink term damps velocities in cells with small liquid fraction f to mimic flow block in the solid/mushy regions.

Results and Discussion

1. Temperature Field Development.

The transient heat distribution becomes visible through the temperature snapshots shown in Figs. 1 and 2 for the finless (N0) and three-fin (N3) configurations. The N0 configuration shows temperature isotherms that stay close to the right heated wall while creating a thin hot area that maintains a wide cold region at sub-melting temperatures throughout the 300-second time period. The small temperature penetration depth shows that the process depends mainly on conduction, while natural convection plays a minimal role because of small density differences in molten erythritol. The temperature distribution in the N3 configuration shows a higher level of uniformity compared to the other configuration. The MXene/BNNT fins with high conductivity function as fast heat transfer paths which minimize the temperature gap between the wall and the PCM bulk. The entire domain reaches a temperature range of 20 K at $t = 240$ s which demonstrates successful temperature homogenization. The multiple fins both speed up the melt-front movement and create a more even temperature distribution which helps maintain material stability during long-term operation and multiple thermal cycles. The better temperature distribution across the domain leads to shorter melting times and increased average charging power according to the previous figures.

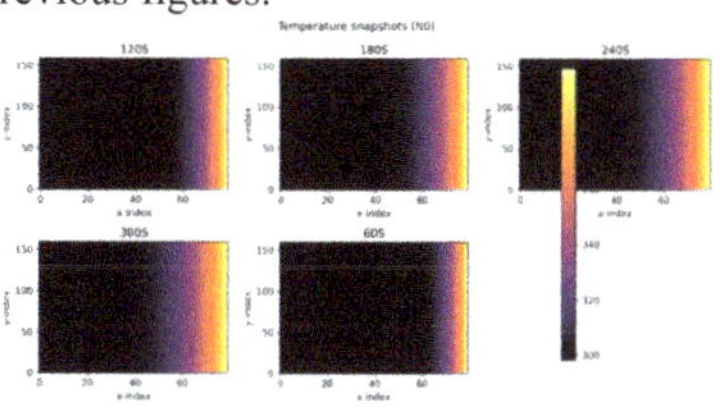

Fig., 1: The finless configuration.

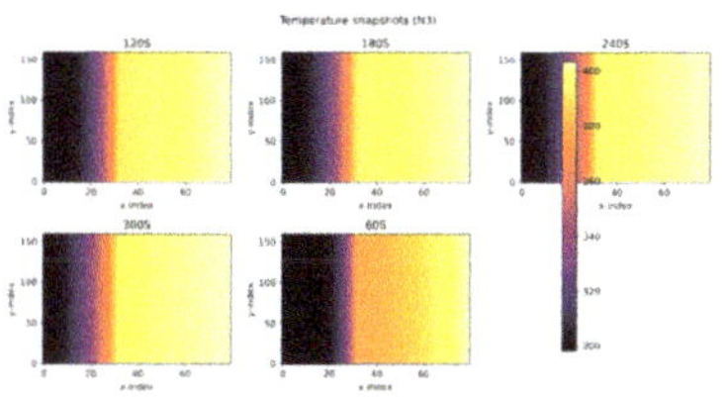

Fig. 2: The temperature field snapshots.

2. Phase Front Morphology and Liquid-Fraction Evolution.

The melting front behavior becomes more clear through the liquid-fraction field contours presented in Figs. 3 and 4. The right boundary of the finless cavity (N0) experiences a slow phase front progression which keeps the front surface flat. The molten area stays limited to a thin wall boundary at 300 s because the liquid fraction remains small according to previous measurements. The small active area of PCM volume leads to poor energy storage efficiency because it does not use the available PCM space effectively. The three-fin configuration (N3) shows fast and extensive melting of the PCM material. The heat islands that form at fin roots start melting at 60 s before they merge into a single front that expands throughout the domain. The PCM reaches complete liquid state at 300 s when the entire domain has melted and the liquid fraction reaches 1. The MXene/BNNT fins produce perfect heat distribution throughout the system because the melt front advances at a uniform and symmetrical rate. The complete melting time shortens by 60 The results show that LHTES system performance depends strongly on fin configuration because it influences both heat transfer efficiency and the uniformity of melting.

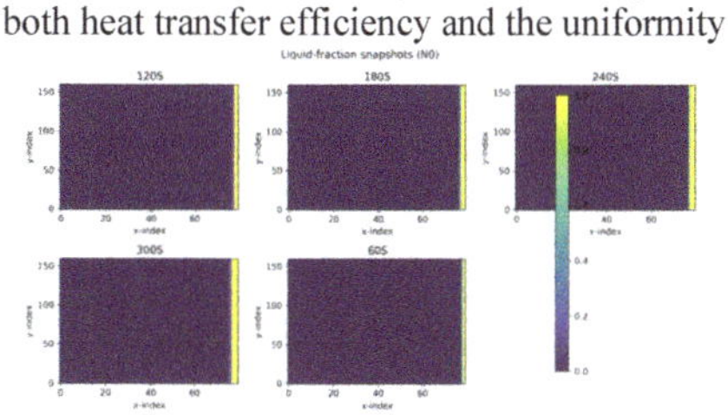

Fig. 3: Temporal evolution of liquid-fraction fields for the finless configuration.

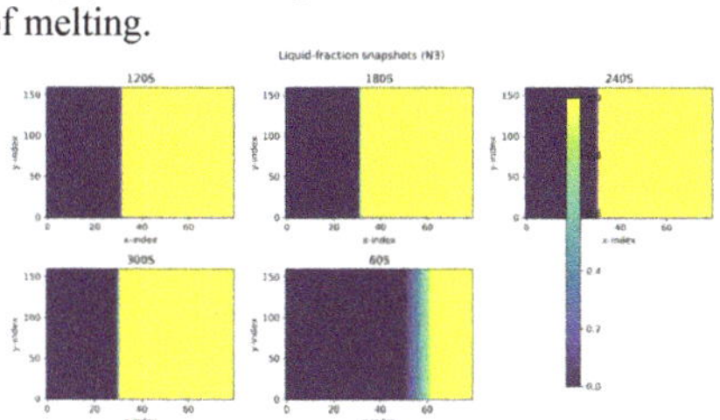

Fig. 4: The three-fin configuration.

3. Implications for Design Optimization.

The results from Figs. 1–4 show a clear design pattern. The addition of composite fins produces higher melting rates while increasing total energy storage capacity but it requires smaller PCM volumes. The optimal design needs to achieve proper equilibrium between enhanced heat transfer performance and expanded storage capacity.

Conclusions

The research used CFD to perform multi-objective optimization of a rectangular LHTES system which used MXene/BNNT composite straight fins inside bio-based erythritol PCM. The enthalpy–porosity formulation successfully modeled the combined natural convection and phase change phenomena while the DOE–RSM framework allowed researchers to study fin geometry variations under fixed PCM volume conditions. The research produced the following measurable results which are presented below.

- The MXene/BNNT composite fins reduced the complete melting time by 60% when compared to the finless reference configuration.
- The optimized finned configurations achieved a 2.1-fold increase in total stored thermal energy compared to the finless case.
- The temperature field analysis demonstrated that finned systems minimized temperature differences between different parts of the domain to less than 20 K which created uniform melting fronts and stable thermal cycling patterns.
- The DOE–RSM results showed that 2–3 fins with 0.3–0.4 mm thicknesses deliver the optimal balance between quick charging and large storage capacity which serves as a useful design reference for building small LHTES systems.

The research findings show that material enhancements at the MXene/BNNT composite level need to work together with geometric design improvements to achieve their complete system advantages. The research team plans to conduct experimental testing while creating AI-based digital twins which will help them apply this method to optimize real-time operations in big energy systems.

References

[1] Mengman Weng, Jiahui Lin, Yuanjun Yang, Jingtao Su, Jintao Huang, Xiang Lu, and Xinxin Sheng. Mxene-based phase change materials for multi-source driven energy storage, conversion and applications. Solar Energy Materials and Solar Cells, 272: 112915, August 2024. ISSN 0927-0248. http://dx.doi.org/10.1016/j.solmat.2024.112915.

[2] Han Sun, Yingai Jin, and Firoz Alam. Review: The application of mxene in thermal energy storage materials for efficient solar energy utilization. Materials, 18(12):2839, 20 June 2025. ISSN 1996-1944. http://dx.doi. org/10.3390/ma18122839

[3] Md. Shahriar Mohtasim and Barun K. Das. Mxene based composite phase change ma terials for thermal energy storage applications: Featuring bio-mimic approaches. Re newable and Sustainable Energy Reviews, 207:114952, January 2025. ISSN 1364-0321. http://dx.doi.org/10.1016/j.rser. 2024.114952

[4] Xiaoxin Yan, Yanhui Feng, Lin Qiu, and Xinxin Zhang. Thermal conductivity and phase change characteristics of hierarchical porous diamond/erythritol composite phase change materials. Energy, 233:121158, October 2021. ISSN 0360-5442. http://dx.doi.org/10.1016/j.energy.2021. 121158

[5] Xiaoqiao Fan, Lu Liu, Xin Jin, Wentao Wang, Shufen Zhang, and Bingtao Tang. Mx ene ti3c2tx for phase change composite with superior photothermal storage capability. Journal of Materials Chemistry A, 7(23):14319–14327, 2019. ISSN 2050-7496. http://dx.doi.org/10.1039/c9ta03962g

[6] Fuyan Peng, Xuhai Zhu, Rongjun Lin, Rui Lu, and Fang Lu. Advances in erythritol-based composite phase change materials. Sustainable Energy amp; Fuels, 8(7):1389–1404, 2024. ISSN 2398-4902. http://dx.doi.org/10. 1039/d4se00171k

[7] J. Chen, Y. Kou, K. Sun, H. Liu, X. Zhang, C. Fang, and Q. Shi. Shape-stable erythritol composite phase change materials with controlled latent heat release for spatiotemporally thermal energy utilization. Materials Today Sustainability, 22:100398, June 2023. ISSN 2589-2347. http://dx.doi.org/10. 1016/j.mtsust.2023.100398

[8] Chao Ma, Jing Wang, Yu Wu, Yongchao Wang, Zhijiang Ji, and Shuai Xie. Character ization and thermophysical properties of erythritol/expanded graphite as phase change 21 material for thermal energy storage. Journal of Energy Storage, 46:103864, February 2022. ISSN 2352-152X. http://dx.doi. org/10.1016/j.est.2021.103864

[9] Lin Qiu, Haimo Li, Jingna Zhao, Xiaoliang Zhang, Yanhui Feng, and Xiaohua Zhang. Hierarchical aln/erythritol composite phase change materials with ultra-efficient polarity enhanced heat conduction. Cell Reports Physical Science, 5(11):102297, November 2024. ISSN 2666-3864. http://dx.doi.org/ 10.1016/j.xcrp.2024.102297

[10] Sheng Yang, Yang-Yan Lai, Yue-Fei Wu, and Li-Wu Fan. Hydroxylated boron nitride enabled erythritol composite phase change material with highly increased thermal con ductivity and compensated heat storage density loss. Solar Energy Materials and Solar Cells, 282:113344, April 2025. ISSN 0927-0248. http://dx.doi.org/10.1016/j.solmat.2024.113344

[11] Siu N. Leung. Thermally conductive polymer composites and nanocomposites: Processing-structure-property relationships. Composites Part B: Engineering, 150:78–92, October 2018. ISSN 1359-8368. http://dx.doi.org/10.1016/j.compositesb.2018.05.056

[12] Yuxi Li, Le Zhao, Ziyu Chen, Xuelin Huang, Guangwu Zhang, Xue-Feng Yu, and Rui He. Sodium alginate/erythritol/¡scp¿d¡/scp¿-mannitol phase change materials for long term thermal energy storage and controllable release. ACS Omega, 10(31):35161–35169, July 2025. ISSN 2470-1343. http://dx. doi.org/10.1021/acsomega.5c04788

[13] Francisco Javier Gonz´alez Gallero, Gabriel Gonz´alez Siles, Ismael Rodr´ ıguez Maestre, Juan Luis Foncubierta Bl´azquez, and Michelle Bottarelli. Experimental validation of a simplified cfd model for a pcm-water finned heat exchanger. International Journal of Low-Carbon Technologies, 20, 2025. ISSN 1748-1325. http://dx.doi.org/10.1093/ijlct/ctae293

Heat Transfer and Friction Factor Analysis of ND/Water Nanofluid in Parabolic Trough Collector: Experimental and Bayesian Regularization Optimization

Lingala Syam SUNDAR[1,a], Hiren Kumar MEWADA[2,b]

Department of Mechanical Engineering, Prince Mohammad Bin Fahd University, SA

Department of Electrical Engineering, Prince Mohammad Bin Fahd University, SA

[a]sslingala@gmail.com, [b]hmewada@pmu.edu.sa

Keywords: Nanofluids, Heat Transfer, Friction Factor, Enhancement, Machine Learning

Abstract. Experimental estimations have been made for Nusselt number $(\boldsymbol{Nu})$, heat transfer coefficient $(\boldsymbol{h_{nf}})$, friction factor $(\boldsymbol{f_{nf}})$, and pressure drop $(\Delta\boldsymbol{P_{nf}})$, of nanodiamond (ND)/water nanofluids flow in a parabolic trough collector (PTC). Nanodiamond (ND) were considered for the preparation of nanofluids because ND contains larger thermal conductivity compared to other nanoparticles. Experiments were conducted at the flow rate between 0.0166 kg/s and 0.0667 kg/s, and at various particle loadings from 0.04% to 0.2%, and carried out during the day from 10:00 to 16:00 hrs. ASHRAE 93 standards served as the basis for all measures. The findings showed that, in comparison to base fluid, the $\boldsymbol{Nu}$, $\boldsymbol{h_{nf}}$, $\boldsymbol{h_{nf}}$, and $\Delta\boldsymbol{P_{nf}}$ are rose by 15.83%, 27.57%, 9.15%, and 6.12% at a flow rate of 0.066 kg/s and at 0.2% volume loadings. A machine learning technique called Bayesian regularization optimization (BRO) uses a probabilistic framework to train models in order to avoid overfitting and enhance generalization. The BRO machine learning technique makes predictions using the $\boldsymbol{Nu}$, $\boldsymbol{h_{nf}}$, $\boldsymbol{h_{nf}}$, and $\Delta\boldsymbol{P_{nf}}$. The BRO trained data results for $\boldsymbol{Nu}$, $\boldsymbol{h_{nf}}$, $\boldsymbol{h_{nf}}$, and $\Delta\boldsymbol{P_{nf}}$ indicates, the root mean square error of 0.6973, 27.73, 0.00001, and 0.0899 with correlation coefficient R^2 of 0.99854, 0.99897, 0.9999, and 0.99999, respectively.

Introduction

Plenty of solar energy is abundantly available on the earth surface and it is renewable. Devices like parabolic trough collectors (PTCs) are necessary to convert solar energy into a useable form. Using solar energy, the PTCs may generate hot water. Since water has extremely little thermal conductivity, it is challenging to increase the efficiency of these PTCs, which are typically water-based. In PTCs, substituting high thermal conductivity nanofluids for water may increase efficiency.

Using nanofluids in PTCs, researchers have seen an increase in efficiency; a few of these studies are highlighted here. Al-Oran et al. [1] used 0.1% vol. of MWCNT and Y_2O_3 hybrid nanofluids in a PTC to achieve a thermal efficiency of 44.24%, while the water efficiency was 19.32%. In PTC, Zain et al. [2] observed a 4.3% improvement in performance when oil/Al_2O_3 nanofluids were used. Using computational methods for seven distinct nanofluids, Ajbar et al. [3] discovered a 2.8% thermal efficiency based on the hybrid data. The PTC absorbed rate of Fe_3O_4 nanofluid at 4% vol. ranges from 7.32 kW to 7.50 kW, according to Moosavian et al. [4]. Using MWCNT/mineral oil nanofluids in a PTC, Kaisaisan et al. [5] saw increases in thermal efficiency of 4% and 7% for 0.2% and 0.3% vol., respectively. According to the CFD investigation, Farooq et al. [6] used Al_2O_3 and CuO nanofluids at 0.0112 kg/s and 0.0224 kg/s, respectively, to achieve efficiency improvements of 13.01%, 13.1%, 13.92%, and 14.79%. Mwesigye et al. [7] analysed thermal efficiency numerically with Al_2O_3/syltherm800 nanofluids in PTC and discovered a 7.6% larger of thermal efficiency at 0.8% vol. At mass flow of 1 lint/min, Hosseini et al. [8] employed water-mixed Al_2O_3 and GO nanofluids in a PTC and noticed that the thermal efficiency increased by 31.2% for Al_2O_3/water nanofluid and 63.2% for GO/water nanofluids. When Khullar et al. [9] examined Al_2O_3/thermal oil nanofluids in a PTC, they saw a 10% increase in collector efficiency at a volume loading of 0.05% relative to the base fluid.

Without explicit programming, computer systems may learn from data and get better over time thanks to the artificial intelligence field of machine learning. Large datasets are analysed, patterns are found, and predictions or judgments regarding previously unseen data are made using algorithms. supervised (using labeled data), unsupervised (using unlabelled data), and reinforcement learning (learning by trial and error) are some of the types of machine learning models that are trained on data. Researchers have used machine learning for the solar collector data, and some of the analysis is given here. Alhamayani [10] considered 2:1% of Al_2O_3+MWCNT/Syltherm-800 nanofluid in PTC and analyzed with ANN and found thermal efficiency of 70.54%, and R^2 of 99.99%, with a root mean square error of 0.012. Alharbi et al. [11] prepared ethylene glycol/Cu+ Al_2O_3 hybrid nanofluids and studied irreversibility analysis while nanofluids flow in a solar collector based on the neural network. They observed machine learning neural networking results perfectly match with experimental data.

The Nu, h_{nf}, h_{nf}, and ΔP_{nf} of ND/water nanofluids in PTC was evaluated in this study. The optimal nanofluid flow rate for the solar collector's maximum heat transfer was determined with the help of this assessment. The collected data is used for predictions for machine learning.

Experimental analysis

Nanofluids preparation

The nanodiamond nanoparticles were dissolved in base fluid at 0.04%, 0.08%, 0.12%, 0.16%, and 0.2% vol. to create the nanofluids. Eight liters of nanofluids in total were made. First, 8 liters of water are put into the receiving tank, and then added 4 drops of nanosphere surfactant. The water is then agitated for an hour using a mechanical stirrer. To create the nanofluids, the necessary amount of nanoparticles was subsequently added

$$\text{Weight of ND, } W_{ND} = \left(\frac{\phi}{(1-\phi)}\right) \times \left(\frac{W}{\rho}\right)_{bf} \times \rho_{ND} \tag{1}$$

For assessing the stability of the produced nanofluids, ZetaSizer Nano ZS was taken into consideration. The cuvette was charged with 5 ml of nanofluid sample. The homogeneity of nanofluids were seen in terms of zeta potential. Prepared nanofluids offer ±30 mV, it indicates that the nanofluids are stable. The surfactant used in the creation of the nanofluids gives the nanoparticles a positive or negative charge. For loadings of 0.04%, 0.08%, 0.12%, 0.16%, and 0.2%, the corresponding zeta potential values are –41.8, –40.9, –40.25, –40.12, and –39.7 mV. It is widely known that volume loadings and zeta potential are related. For properties analysis, the Eqns. (2–5) are considered.

Properties

The below equations were used to find thermophysical properties.

(a) Maxwell [12] model for thermal conductivity.

$$k_{nf} = k_{bf}\left[\frac{(k_p+2k_f)+2\phi(k_p-k_f)}{(k_p+2k_f)-\phi(k_p-k_f)}\right] \tag{2}$$

(b) Brinkman [13] model for viscosity

$$\mu_{nf} = \mu_{bf}\left[\frac{1}{(1-\phi^{2.5})}\right] \tag{3}$$

(c) Pak and Cho [14] model for density

$$\rho_{nf} = (1-\phi)\rho_{bf} + \phi\rho_p \tag{4}$$

(d) Pak and Cho [14] model for specific heat

Materials Research Proceedings 64 (2026) 680-691 https://doi.org/10.21741/9781644904091-85

$$C_{p,nf} = \frac{(1-\phi)C_{p,bf}\rho_{bf} + \phi C_{p,p}\rho_p}{\rho_{bf}} \tag{5}$$

Parabolic trough collector

Experimental facility is provided in Fig. 1, which is situated at latitude 26.2172° N and longitude 50.1971° E. The experiments for PTC were conducted in Al-Khobar, Saudi Arabia. At midday, the solar collection system was oriented toward the sun. The outside of collector receiver tube was coated with a black carbon mat finish that was 1 mm thick. At 20, 50, 90, 120, and 160 cm from the entry, a J-type thermocouple was used to measure a surface thermocouple. The fluid was then pushed with a pump at flow rate of 0.0167 to 0.067 kg/s using preferred closed loop ND/water nanofluid, which had a 61% accuracy rate. In order to facilitate recycling, a cooler was integrated into the receiver tube's output.

Even though heat exchanger could lower the temperature to 38 °C and difficult to complete tests in accordance with ASHRAE. The temperature in relation to surroundings was maintained using a constant temperature bath. With SP Lite2 silicon pyranometer solar radiation is measured, and wind velocity is measured with vane-type anemometer with an accuracy of 63% (10.22 m/s), and its range is from 0 to 25 m/s, and an operating humidity of roughly 80%. The pyranometer, temperature sensor, and rotameter have respective accuracy of 60.1 °C, 66.25%, and 65%. Under normal circumstances, where wind speeds typically range from 1 to 10 m/s, the entire solar PTC test system was set up outside. Additionally, the support structure kept the wind speed constant at roughly 50 m/s. Table 1 displays the PTC details.

In order to determine the exact time intervals for steady-state thermal tests and evaluate the solar system's time responsiveness, the time constant is a crucial acceptance condition [15]. The time constant is the amount of time needed for the fluid leaving the collector to recover to 63.2% of its stable value following a step change in incident radiation. The necessary time is reached when Eq. (6) is satisfied.

$$\left(\frac{T_{o,\tau}-T_i}{T_{o,i}-T_i}\right) = \frac{1}{e} = 0.368 \tag{6}$$

Where, $T_{o,\tau}$, $T_{o,i}$, and T_i stands for fluid exit temperature after time t, fluid outlet temperature, and fluid inlet temperature is equal to ambient temperature $(T_i - T_a)$.

Table 1: PTC dimensions.

PTC	Dimensions
Parabola length	2.0 m
Parabola aperture	1.8 m
Thickness (mm)	2 mm
Material	Steel
Focal distance (f)	300 mm
Aperture area (A_a)	1.6 m^2
Acceptance angle	0.48
Reflector thickness (approximately)	3 mm
Rim angle	90°
Concentration ratio	15.6
Receiver tube inner tube diameter (d_i)	13 mm
Receiver tube outer diameter (d_o)	16 mm
Borosilicate glass inner diameter	30 mm
Borosilicate glass outer diameter	34 mm

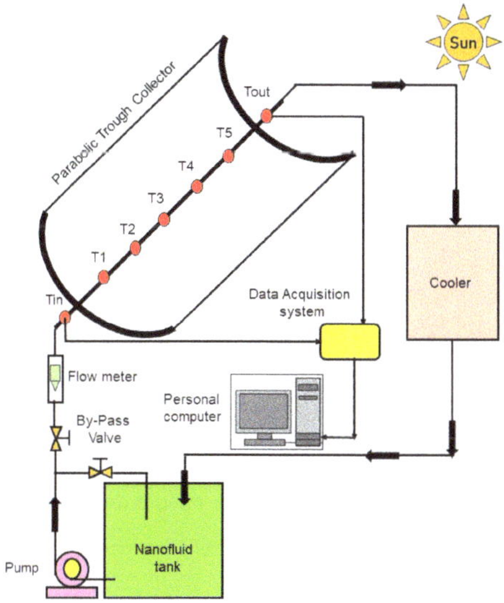

Fig. 1: Schematic diagram of PTC.

Data analysis

Heat transfer coefficient and nusselt number

The below equations were used to calculate the heat transfer coefficient and Nusselt number. The solar energy was considered as heat supplied to the PTC, which is (Q) 1325 W/m^2 °C.

Heat absorbed, $Qa = \dot{m} \times C_p \times (T_o - T_i)$ (7)

Heat transfer coefficient, $h = \frac{Q}{A(T_w - T_m)}$ (8)

Where, $A = \pi D_i L$; $T_w = \frac{T_1+T_2+T_3+T_4+T_5}{5}$; $T_m = \frac{T_o+T_i}{2}$

Nusselt number, $Nu = \frac{h \times D_i}{k}$ (9)

Prandtl number, $Pr = \frac{\mu \times C_P}{k}$ (10)

Pressure drop and friction factor

Pressure gauges are used to measure the pressure drop.

Pressure drop, $\Delta p = p_o - p_i$ (11)

Friction factor, $f = \frac{(\Delta p)}{\left(\frac{L}{d_i}\right)\left(\frac{\rho v^2}{2}\right)}$ (12)

Where, $\dot{m} = \rho A v \Rightarrow v = \frac{\dot{m}}{\rho A}$

Thermal performance factor

Thermal performance factor can be estimated from the below equation.

$$TPF = \frac{\left(\frac{Nu_{nf}}{Nu_{bf}}\right)}{\left(\frac{f_{nf}}{f_{bf}}\right)^{1/3}} \quad (13)$$

Machine learning-Bayesian regularization optimization

By enclosing the model training process in a probabilistic framework, Bayesian regularization optimization is a technique used mostly in machine learning, especially for training neural networks, to enhance model generalization and avoid overfitting. It automatically strikes a balance between fitting the training data and controlling the model's complexity by incorporating past knowledge about the model's parameters, or weights.

Results and discussion

Nusselt number

Initially water was used in the PTC and estimated the Nusselt number and compared with the literature data, and found to be the deviation is less than 2.45%. After being confident in the water findings, additional heat transfer tests with nanofluids were carried out. The Nusselt number of nanofluids was calculated using Eq. (9), and the results is shown in Fig. 2 with regard to Re and ϕ values. Increased ϕ and Re values resulted in higher Nu. Nanofluids have a higher heat conductivity, and Brownian motion produces higher h_{nf} values.

In comparison to water data, 0.2% vol. nanofluid with Re of 1498.2 and 5993.3, the Nusselt number is enhanced is 8.52%, and 15.83% respectively. Higher particle dispersion in water possess an increase of Nusselt number, at the same time, this raise is depending on the flow rates.

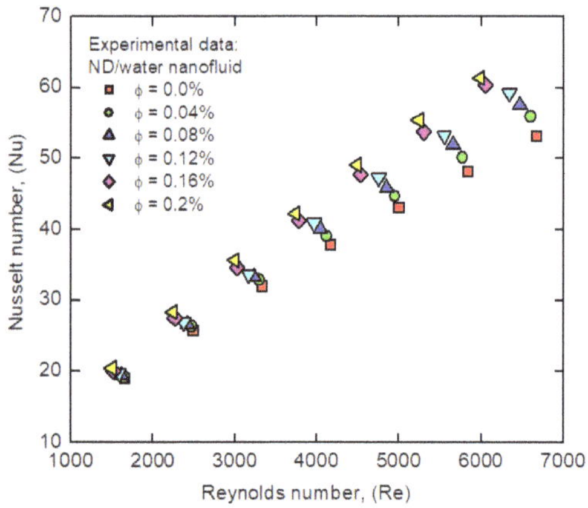

Fig. 2: The Nusselt number variation with Reynolds number and volume loadings.

Heat transfer coefficient

The heat transfer coefficient was estimated from Eq. (8), and the results is shown in Fig. 3 with regard to Re and ϕ values. Increased ϕ and Re values resulted in higher heat transfer coefficient. Nanofluids have a higher heat conductivity, and Brownian motion produces higher h_{nf} values.

In comparison to water data, 0.2% vol. nanofluid with Re of 1498.2 and 5993.3, the heat transfer coefficient is enhanced is 19.52%, and 27.57% respectively. Higher particle dispersion in water possess an increase of heat transfer, at the same time, this raise is depending on the flow rates.

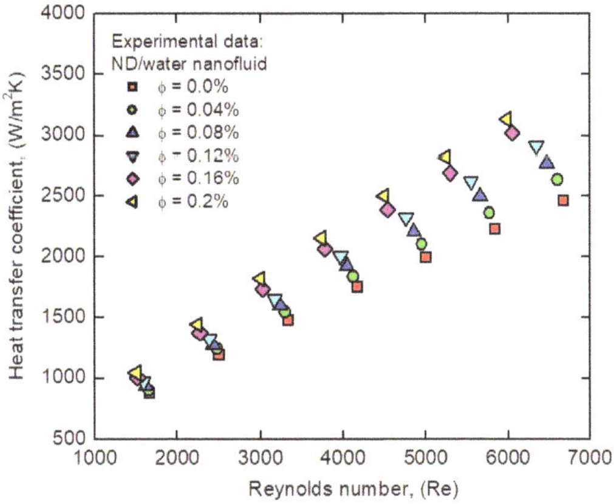

Fig. 3: The heat transfer coefficient variation with Reynolds number and volume loadings.

Friction factor

The friction factor was estimated from Eq. (12), and the results is shown in Fig. 4 with regard to *Re* and ϕ values. Increased ϕ and *Re* values resulted in higher friction factor. Nanofluids have a higher friction factor because resistance offered by the tube wall. In comparison to water data, 0.2% vol. nanofluid with *Re* of 1498.2 and 5993.3, the friction factor is enhanced is 5.09%, and 6.12% respectively. Higher particle dispersion in water possess an increase of friction factor, at the same time, this raise is depending on the flow rates.

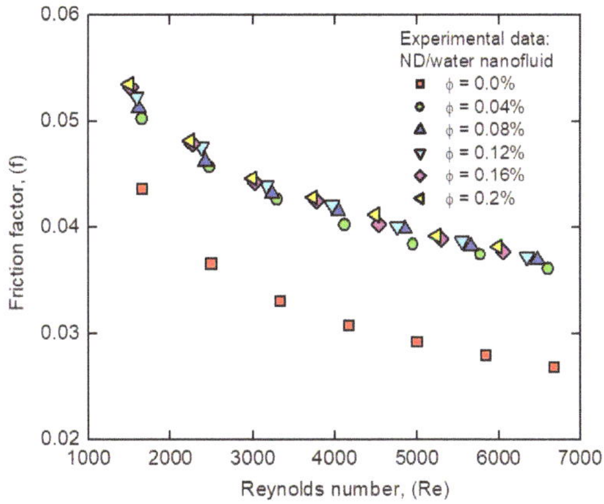

Fig. 4: The friction factor variation with Reynolds number and volume loadings.

Pressure drop

The friction factor was estimated from Eq. (11), and the results is shown in Fig. 5 with regard to *Re* and ϕ values. Increased ϕ and *Re* values resulted in higher heat transfer coefficient. Nanofluids have a higher pressure drop because resistance offered by the tube wall. In comparison to water data, 0.2% vol. nanofluid with *Re* of 1498.2 and 5993.3, the pressure drop is enhanced is 5.09%, and 6.12% respectively. Higher particle dispersion in water possess an increase of pressure drop, at the same time, this raise is depending on the flow rates.

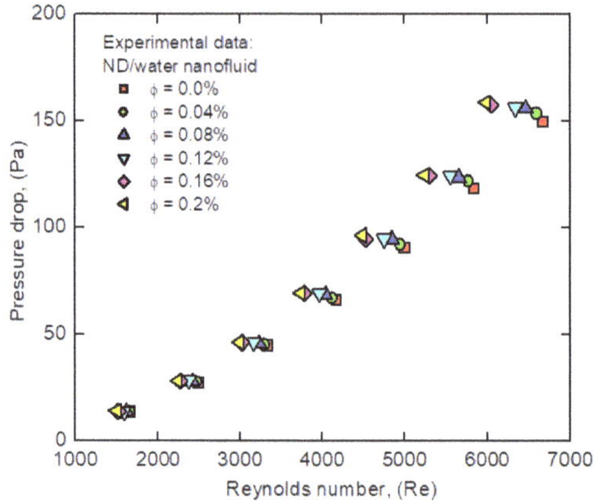

Fig. 5: The pressure drop variation with Reynolds number and volume loadings.

Thermal performance factor

The thermal performance factor was estimated from Eq. (13), and the results is shown in Fig. 6 with regard to *Re* and ϕ values. Increased ϕ and *Re* values resulted in higher thermal performance factor. In comparison to water data, 0.2% vol. nanofluid with *Re* of 1498.2 and 5993.3, the thermal performance factor is enhanced by 1.058 times, and 1.125 times, respectively. Higher particle dispersion in water possess an increase of thermal performance factor, at the same time, this raise is depending on the flow rates.

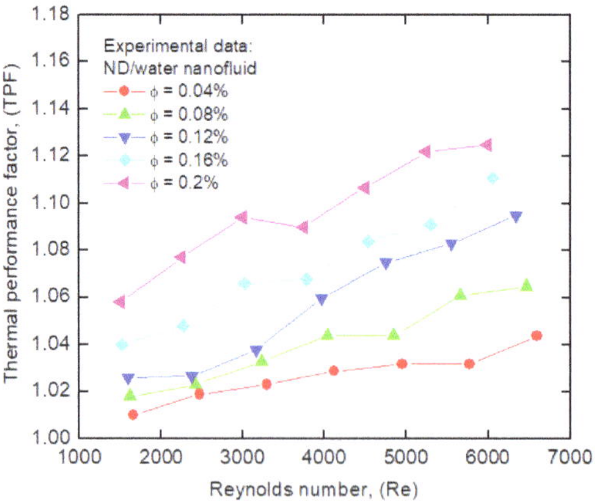

Fig. 6: The thermal performance factor variation with Reynolds number and volume loadings.

Materials Research Proceedings 64 (2026) 680-691
https://doi.org/10.21741/9781644904091-85

ML-BRO results

The experimentally obtained Nusselt number data is used for ML analysis, and the whole data is split into train and text purpose. 70:30% data points were selected for training and testing purpose. The model performance is evaluated using mean square error (MSE), root mean square error (RMSE) and R^2 score. The MSE, and RMSE for train data is 0.48622, and 0.69729, and for the test data is 10.4775, and 3.2369, and the data is shown in Fig. 7. The R^2 value for train data is 0.99854, and for test data is 0.98842, which is shown in Fig. 8.

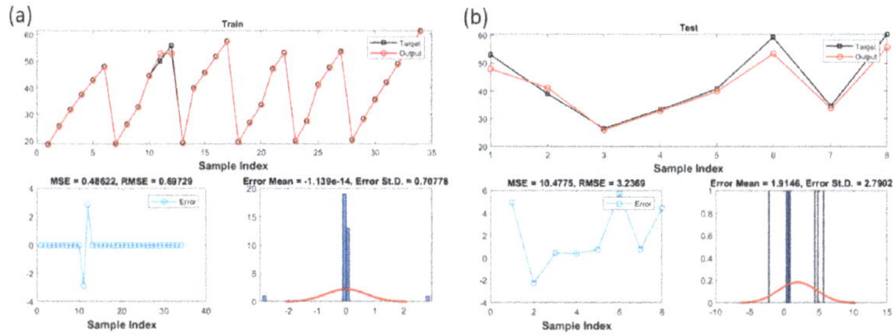

Fig. 7: Nusselt number data: (a) train, and (b) test.

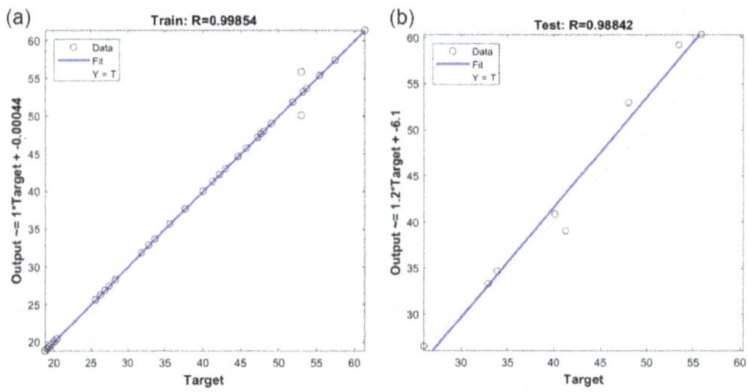

Fig. 8: Nusselt number data R^2 data: (a) train, and (b) test.

The experimentally obtained heat transfer coefficient data is used for ML analysis, and the whole data is split into train and text purpose. 70:30% data points were selected for training and testing purpose. The model performance is evaluated using mean square error (MSE), root mean square error (RMSE) and R^2 score. The MSE, and RMSE for train data is 18261.3551, and 135.1345, and for the test data is 768.9612, and 27.73, and the data is shown in Fig. 9. The R^2 value for train data is 0.99897, and for test data is 0.98936, which is shown in Fig. 10.

Materials Research Proceedings 64 (2026) 680-691 https://doi.org/10.21741/9781644904091-85

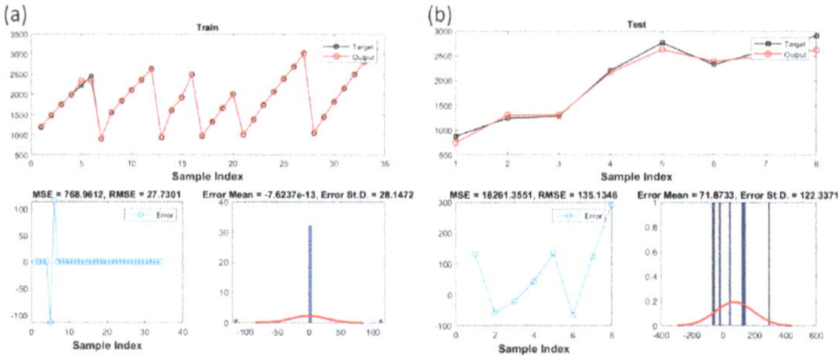

Fig. 9: Heat transfer coefficient data: (a) train, and (b) test.

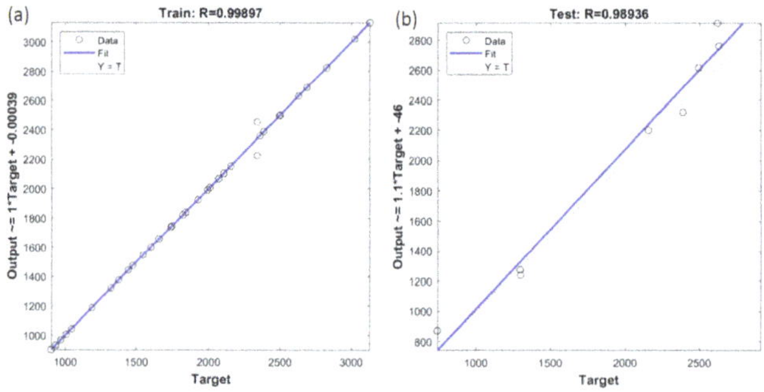

Fig. 10: Heat transfer coefficient R^2 data: (a) train, and (b) test.

The experimentally obtained friction factor data is used for ML analysis, and the whole data is split into train and text purpose. 70:30% data points were selected for training and testing purpose. The model performance is evaluated using mean square error (MSE), root mean square error (RMSE) and R^2 score. The MSE, and RMSE for train data is 3.0428e-12, and 1.744e-06, and for the test data is 9.4899e-7, 0.000974, and the data is shown in Fig. 11. The R^2 value for train data is 1, and 0.98562, which is shown in Fig. 12.

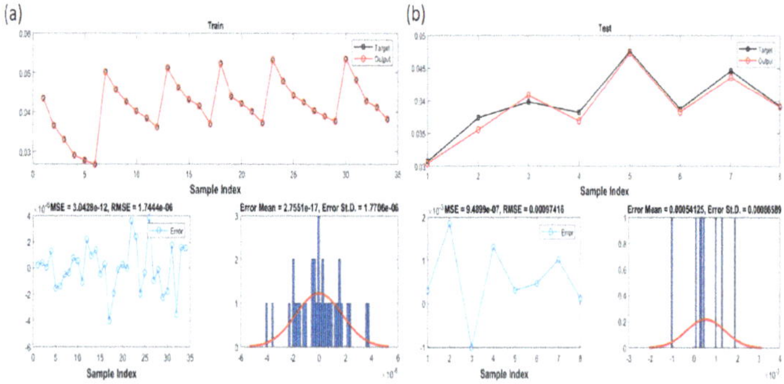

Fig. 11: Friction factor data: (a) train, and (b) test.

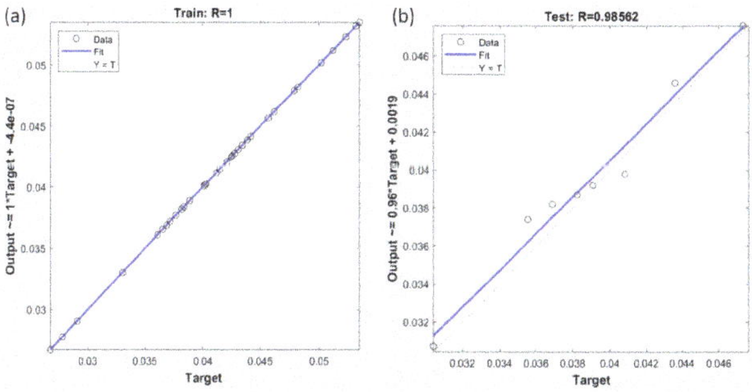

Fig. 12: Friction factor R^2 data: (a) train, and (b) test.

The experimentally obtained pressure drop is used for ML analysis, and the whole data is split into train and text purpose. 70:30% data points were selected for training and testing purpose. The model performance is evaluated using mean square error (MSE), root mean square error (RMSE) and R^2 score. The MSE, and RMSE for train data is 0.0080774, and 0.089874, and for the test data is 0.37198, and 0.6099, and the data is shown in Fig. 13. The R^2 value for train data is 1, and for test data is 0.9991, which is shown in Fig. 14.

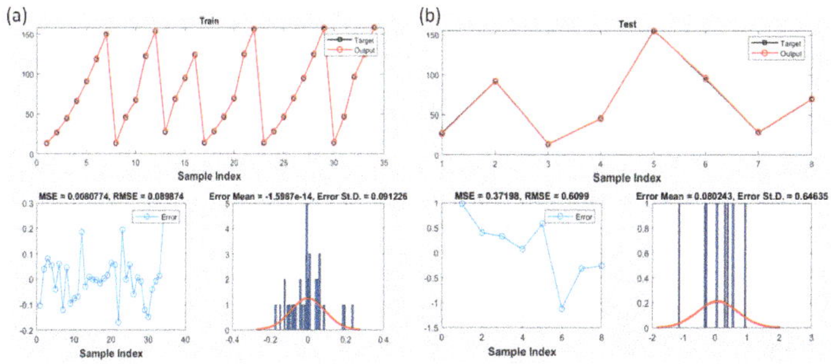

Fig. 13: Pressure drop data: (a) train, and (b) test.

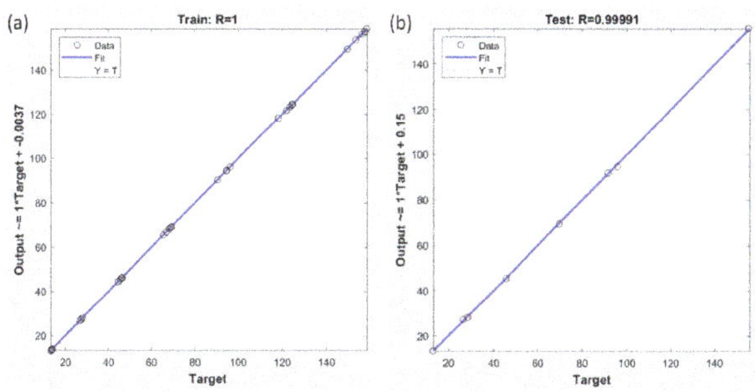

Fig. 14: Pressure drop R^2 data: (a) train, and (b) test

Conclusion

The main objective of this work was to examine the thermal water-based nanofluid containing ND nanoparticles. The nanofluids' thermophysical properties were theoretically estimated. PTC driven by water and nanofluids (vol. loadings of 0.04%, 0.08%, 0.12%, 0.16%, and 0.2%) was built in order to evaluate its heat transfer coefficient, and friction factor. The mass flow rates of 0.01667, and 0.0667 kg/s were selected after the collector was tested in accordance with the ASHRAE standard. The following outcomes were reached as a result. The $\boldsymbol{Nu}$, h_{nf}, h_{nf}, and ΔP_{nf} rose by 15.83%, 27.57%, 9.15%, and 6.12% at a flow rate of 0.066 kg/s and at 0.2% volume loadings. The BRO machine learning technique makes predictions using the Nu, h_{nf}, h_{nf}, and ΔP_{nf}. The train data BRO results for Nu, h_{nf}, h_{nf}, and ΔP_{nf} mean square error of 0.6973, 27.73, 0.00001, and 0.0899 with correlation coefficient R^2 of 0.99854, 0.99897, 0.9999, and 0.99999, respectively. The experimental and BRO results are good agreement with each other.

References

[1] O. Al-Oran, N.A. Shaban, R. Manna, O. Ayadi, A. Asaf, and F. Lezsovits, "Performance study of parabolic trough solar collector using hybrid nanofluids under Jordanian weather conditions," J. Thermal Analysis and Calorimetry, vol. 149, pp. 3981-3998, 2024. https://doi.org/10.1007/s10973-024-12961-8

[2] M. Zain, M. Amjad, M. Farooq, Z. Anwar, R. Shoukat, Enio P. Bandarra Filho, and Du Xiaoze, "Performance investigation of a solar thermal collector based on nanostructured energy materials," Front. Mater., vol. 7, pp. 462, 2021. https://doi.org/10.3389/fmats.2020.617199

[3] W. Ajbar, J.A. Hernández, A. Parrales, and L. Torres, "Thermal efficiency improvement of parabolic trough solar collector using different kinds of hybrid nanofluids," Case Studies in Thermal Engineering, vol. 42, pp. 102759, 2023. https://doi.org/10.1016/j.csite.2023.102759

[4] S.F. Moosavian, A. Hajinezhad, R. Fattahi, and A. Shahee, "Evaluating the effect of using nanofluids on the parabolic trough collector's performance," Energy Science and Engineering, 2023, https://doi.org/10.1002/ese3.1537

[5] A. Kasaeian, S. Daviran, R.D. Azarian, and A. Rashidi, "Performance evaluation and nanofluid using capability study of a solar parabolic trough collector," Energy Conversion and Management, vol. 89, pp. 368-375, 2015. https://doi.org/10.1016/j.enconman.2014.09.056

[6] M. Farooq, M. Farhan, G. Ahmad, Z.R. Tahir, M. Usman, M. Sultan, M.S. Hanif, M. Imran, S. Anwar, A.M. El-Sherbeeny, and M.A. Shakir, "Thermal performance enhancement of nanofluids based parabolic trough solar collector (NPTSC) for sustainable environment," Alexandria Engineering Journal, vol. 61, pp. 8943-8953, 2022. https://doi.org/10.1016/j.aej.2022.02.029

[7] A. Mwesigye, Z. Huan, and J.P. Meyer, "Thermodynamic optimisation of the performance of a parabolic trough receiver using synthetic oil-Al2O3 nanofluid", Applied Energy, vol. 156, pp. 398-412, 2015. https://doi.org/10.1016/j.apenergy.2015.07.035

[8] S.M.S. Hosseini, and S.D. Mohammad, "Assessment of TiO2 water-based nanofluids with two distinct morphologies in a U-type evacuated tube solar collector," Appl. Therm. Eng., vol. 182, pp. 116086, 2021. https://doi.org/10.1016/j.applthermaleng.2020.116086

[9] V. Khullar, H. Tyagi, P.E. Phelan, T.P. Otanicar, H. Singh, and R.A. Taylor, "Solar energy harvesting using nanofluids-based concentrating solar collector," J. Nanotechnology in Engineering and Medicine, vol. 3, pp. 031003, 2012. https://doi.org/10.1115/1.4007387

[10] A. Alhamayani, "Performance analysis and machine learning algorithms of parabolic trough solar collectors using Al2O3-MWCNT as a hybrid nanofluid," Case Studies in Thermal Engineering, vol. 57, pp. 104321, 2024. https://doi.org/10.1016/j.csite.2024.104321

[11] S.O. Alharbi, T. Gul, I. Khan, M.S. Khan, S. Alzahrani, "Irreversibility analysis through neural networking of the hybrid nanofluid for the solar collector optimization," Scientific Reports, 13 (2023) 13350. https://doi.org/10.1038/s41598-023-40519-5

[12] J. C. Maxwell, "A Treatise on Electricity and Magnetism," Oxford: Clarendon Press, 1891.

[13] H.C. Brinkman, "The viscosity of concentrated suspensions and solutions," The Journal of Chemical Physics, vol. 20, pp. 571-571, 1952. https://doi.org/10.1063/1.1700493

[14] B.C. Pak, and Y.I. Cho, "Hydrodynamic and heat transfer study of dispersed fluids with submicron metallic oxide particles", Experimental Heat Transfer an International Journal, vol. 11, pp. 151-170, 1998. https://doi.org/10.1080/08916159808946559

[15] ASHRAE Standard 93, "Method of testing to determine the thermal performance of solar collectors," Atlanta (GA): American Society of Heating, Refrigerating and Air-Conditioning Engineers; 2010.

Potassium-Deintercalated Birnessite ($Mn_7O_{13}·5H_2O$) as Cathode for Advanced Aqueous Zinc-Ion Batteries

Hamza AFARYATE[1,a*], Mohamed KERROUMI[1], Mehdi KARBAK[1], Bouchaib MANOUN[2], Youssef TAMRAOUI[1], Fouad GHAMOUSS[1,f*]

[1]College of Chemical Sciences and Engineering (CCSE), Department of Materials Science, Energy and Nano-engineering (MSN), Mohammed VI Polytechnic University (UM6P), Morocco

[2]University Hassan 1st, FST, Settat, Morocco

[a]hamza.afaryate@um6p.ma, [f]fouad.ghamouss@um6p.ma

Keywords: Zinc-Ion Batteries, Potassium-Birnessite, Cathode Materials, Acid Leaching, Aqueous Batteries

Abstract. Aqueous zinc-ion batteries (AZIBs) have gained increasing interest as a low-cost, intrinsically safe platform for grid energy-storage applications. Nevertheless, their development is hampered by sluggish Zn^{2+} diffusion and structural instability of cathodes, largely due to the large hydrated radius of Zn^{2+} ions. In this work, a hydrated birnessite-type manganese oxide, denoted L-KMO ($Mn_7O_{13}·5H_2O$), is synthesized via a mild acid-leaching strategy applied to potassium birnessite (KMO, $K_{0.37}MnO_2$). Structural and physicochemical characterizations confirm successful potassium removal and formation of a hydrated layered framework with expanded interlayer spacing. When evaluated as a cathode for AZIBs, L-KMO delivers a high capacity of up to 347 mAh g^{-1} at 50 mA g^{-1}, along with excellent Coulombic efficiency of approximately 96-100% and stable cycling behavior. This simple and scalable synthesis approach provides an effective route for tailoring manganese-oxide cathodes toward advanced AZIBs.

Introduction

Zinc-ion batteries have appeared as an attractive technology for grid energy storage, due to the abundance and low cost of zinc, along with its high theoretical capacity (820 mAh g^{-1}), its low redox potential (-0.762 V vs SHE), and intrinsic safety. The electrochemical performance of ZIBs, including operating voltage and reversible capacity, is primarily dictated by the nature of the cathode material, which controls zinc-ion storage behavior and reaction kinetics. Consequently, the development of advanced cathode materials remains the main challenge limiting further progress in high-performance ZIBs and their commercialization [1,2]. To date, various cathode systems such as manganese and vanadium oxides, metal-organic frameworks, and Prussian blue analogues have been investigated. Nevertheless, each class of these materials has specific shortcomings that affect practical implementation. Typical average working voltage is moderate for vanadium oxides, and they tend to dissolve into aqueous solutions, leading to reduced energy density and raising environmental concerns.[3,4] While MOFs have tunable porosity, they are typically characterized by low electronic conductivity due to the presence of insulating organic linkers. [4,5] The open framework of a Prussian blue analogue provides good kinetics, but moderate capacity compared to dense oxide cathodes. [1] Among these cathodes, manganese oxides appear to be the best candidate for commercial AZIB thinks to the structural diversity, which arises from the [MnO_6] octahedral connecting by edges and corners to form tunneled (Pyrolusite, Hollandite, Ramsdellite, Spinel…) and layered (δ) structures, giving them different valences and structural properties, low toxicity, natural abundance, high specific capacity, and relatively high working voltage, positioning them as key materials for the development of ZIB technologies.[6,7]

Many strategies have been explored to mitigate the inherent limitations of cathodes made from manganese, including the incorporation of conductive carbon networks like carbon nanotubes or graphene, defect engineering techniques, and creating new advanced aqueous electrolytes (e.g., eutectic electrolytes, molecularly crowded electrolytes, water-in-salt electrolytes) that are compatible with Mn-based batteries [8-10]. However, many of the traditional post-treatment techniques provide limited advantages in terms of optimization. Also, recently, a technique of acid leaching has been found to be an efficient way to control the properties of Mn oxides through specific changes in chemical composition, phase transitions, and creation of defect structures while eliminating the need for any additional post-treatment steps. This strategy was employed to produce spinel-type Mn oxides with enhanced electrochemical performance by removing Li^+ and Zn^{2+} from $LiMn_2O_4$ and $ZnMn_2O_4$, respectively, while preserving the Mn framework, offering favorable ion-diffusion pathways. [11,13]

Among manganese oxides, layered birnessite (K_xMnO_2) is particularly attractive owing to its two-dimensional MnO_6 frameworks and accessible interlayer galleries. However, the crowded interlayers caused by the presence of potassium ions and the tendency of nanosheets to stick together impede Zn^{2+} diffusion, limiting the capacity and affecting the cycling stability. To address these limitations, we derive a layered manganese oxide $Mn_7O_{13}{\cdot}5H_2O$ (L-KMO) from K-birnessite (KMO) by selectively deintercalating potassium via mild acid treatment. The resulting leached material was thoroughly characterized in terms of its structure and microstructure and subsequently evaluated as a cathode in AZIB, representing a previously unexplored application.

Material characterization

Materials synthesis

All reagents are ≥ 99% pure or of analytical grade. The chemicals were sourced from Sigma-Aldrich and employed as received. Double-distilled water was used in all experimental procedures. K-birnessite (KMO, $K_{0.37}MnO_2$) was prepared by a sol–gel route following literature protocols [14,15]. 3.0 g of $KMnO_4$ dissolved in 50 mL of double-distilled water and rapidly poured into 20 mL of an aqueous D(+)-glucose solution under vigorous stirring ($KMnO_4$: glucose molar ratio = 1:1.5), where glucose acted as a complexing/reducing agent. After solvent evaporation, the resulting brown gel was cooled to room temperature (RT), then oven-dried at 100 °C for 12 h to afford a xerogel, which was subsequently heat-treated in air at 400 °C for 2 h to yield the final oxide material.

To selectively remove interlayer K^+, 2.0 g of KMO was dispersed in 200 mL of 0.2 M HCl and stirred at RT for 12 h.[13] The suspension was centrifuged, and the solid was washed repeatedly with water to neutral pH, then vacuum dried.

Characterization methods

Crystalline structure of the sample was analyzed using Bruker AXS D8 Advance X-ray diffractometer. Raman spectra were recorded employing a LabRAM HR Evolution Raman microscope equipped with a 514 nm excitation laser over the range of 200-1000 cm^{-1}. Thermal properties were evaluated using a differential thermal analyzer (DTA, Henven) under a nitrogen atmosphere from 23 to 700 °C at a heating rate of 10 $°C{\cdot}min^{-1}$. Sample morphology and elemental mapping were examined by a TESCAN VEGA3 scanning electron microscope (SEM). Rietveld refinement of the XRD data was conducted using WinPLOTR graphical interface of the FullProf software package. The refinement procedure involved sequential optimization of instrumental parameters, background, scale factor, and peak profile functions. Background contributions were fitted using a fifth-order polynomial, followed by refinement of the structural parameters.

Electrochemical measurements

CR2032 coin cells were employed to perform the electrochemical tests. Cathodes were fabricated by applying a slurry with a 7:2:1 weight ratio of L-KMO, Super P, and Polyvinylidene fluoride onto a titanium foil. The cells were put together using L-KMO cathode (mass loading ~ 2 mg·cm^{-2}), a glass fiber-based separator, an aqueous electrolyte consisting of 3 M $ZnSO_4$ with 0.2 M Mn_2SO_4 solution, and Zn foil anode. To ensure the removal of dissolved oxygen from the electrolyte, argon gas was continuously purged through the solution for 20 minutes. The assembled coin cells were then evaluated using cyclic voltammetry (CV) and galvanostatic charge-discharge (GCPL) techniques on a BioLogic battery testing system, operated over a voltage window of 0.8-1.8 V vs Zn^{2+} /Zn.

Results and Discussion

Structural changes induced by the acid-leaching process were studied using X-ray diffraction (XRD), as presented in Fig. 1a. The pristine sample (KMO) was indexed to monoclinic potassium birnessite, $K_{0.37}MnO_2$ (PDF 96-432-4653), with C 1 2/m 1 space group (No. 12). Its stoichiometry was further confirmed by Rietveld refinement (Fig. 1b), which shows excellent agreement between the experimental and calculated profiles. After leaching, the diffraction pattern transformed into that of orthorhombic hydrated layered manganese oxide, $Mn_7O_{13}\cdot5H_2O$ (PDF 00-023-1239), with Pbca space group (No. 61). Additionally, thermogravimetric analysis (Fig. 1c) displays two distinct dehydration steps. The first mass loss appeared at around 150 °C assigned to physisorbed water (~2 H_2O), while the second appeared at approximately 560 °C attributed to coordinated/structural water (~3 H_2O). The total mass loss (~13.18%) closely matches the theoretical (13.2 wt%) expected for five water molecules in the $Mn_7O_{13}\cdot5H_2O$ structure, thereby confirming the presence of 5 H_2O per formula unit in the leached KMO.

Raman spectroscopy was also used to probe the structural modifications associated with K^+ extraction. Spectra of KMO and L-KMO were recorded over 200-1000 cm^{-1} (Fig. 1d). Consistent with prior literature reports, a prominent band was observed for KMO at 655 cm^{-1}, attributed to A_{1g} symmetric stretching of Mn-O in MnO_6 octahedra, where motion of oxygen atoms dominates.[16] After acid leaching, this band downshifts to 643 cm^{-1} in L-KMO, indicating structural distortion of the lattice and changes in the manganese-oxygen bond strengths and lengths arising from the removal of interlayer K^+. Additional modes below 600 cm^{-1} located at 373, 316, and 287 cm^{-1} in KMO spectra and associated with MnO_6 bending vibrations also shift to 358, 302, and 273 cm^{-1}, respectively, in L-KMO, further corroborating the structural relaxation accompanying potassium deintercalation. [17,18]

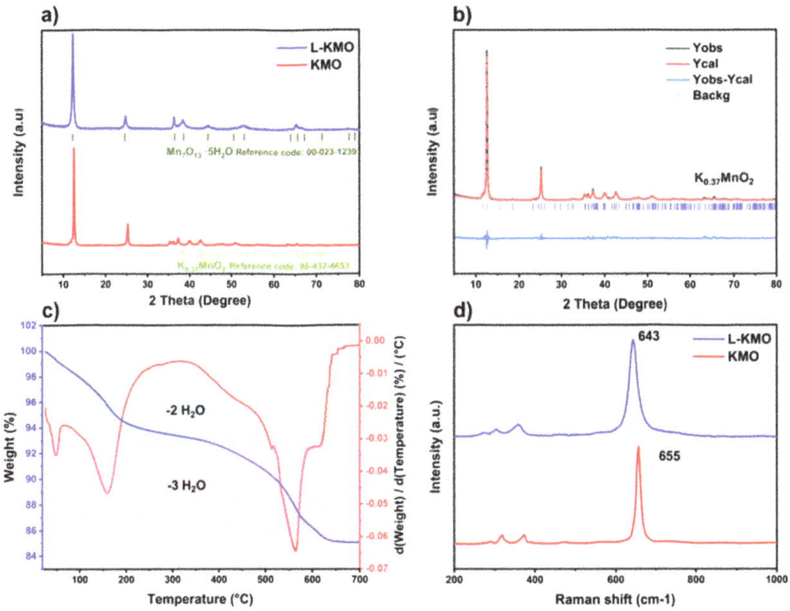

***Fig. 1.** Structural and thermal characterization of KMO and L-KMO: **(a)** XRD patterns, **(b)** Rietveld refinement of KMO, **(c)** TGA/DTG of L-KMO, and **(d)** Raman spectra.*

Surface features of the synthesized KMO and L-KMO samples were investigated using scanning electron microscopy, as presented in Fig. 2. Result reveals that the pristine K-birnessite (KMO) forms porous agglomerates of stacked, plate-like nanocrystallites on the order of 5-20 µm. Following acid leaching, L-KMO shows cauliflower-like microclusters (~5-15 µm) composed of finer, granular/etched nanodomains and a higher degree of surface roughness and porosity. This morphological evolution increases accessible surface area and opens transport pathways, changes that are expected to facilitate Zn^{2+} insertion kinetics. Moreover, Elemental mapping via energy-dispersive X-ray spectroscopy further confirms the potassium extraction from KMO following the acid-leaching process, as demonstrated by the disappearance of the potassium element from mapping images.

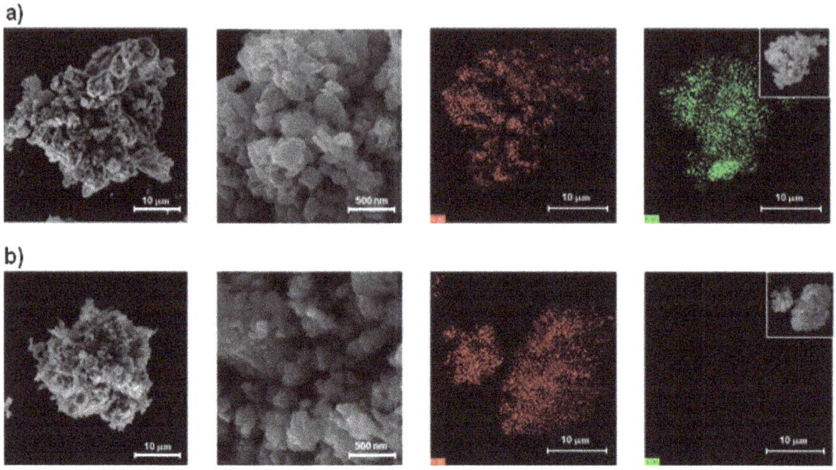

Fig. 2. SEM and elemental (Mn, K) mapping of (a) KMO, (b) L-KMO.

To assess the electrochemical performances of the synthesized L-KMO cathode in zinc-ion batteries, cyclic voltammetry measurements were first recorded under a scanning rate of 0.1 mV s^{-1} (Fig. 3a). CV curves exhibit two well-defined cathodic peaks at approximately 1.36 and 1.24 V, accompanied by corresponding anodic features around 1.61 and 1.66 V, characteristic of sequential H^+/Zn^{2+} co-insertion-extraction in manganese oxide hosts.[13] Furthermore, no activation cycle or phase transition was observed, as the cv curves retained the same behavior during the first cycles. Galvanostatic charge/discharge (GCD) measurements collected at 50 mA g^{-1} over the same potential window (Fig. 3b-c) display two discharge and two charge plateaus, consistent with the CV features and indicative of two dominant redox processes. L-KMO exhibits an initial reversible capacity of ~160 mAh g^{-1}, which gradually activates to reach ~347 mAh g^{-1} after 73 cycles, while retaining a high Coulombic efficiency (≈96-100%). This progressive capacity rise is attributed to the electrode activation process, resulting from a combination of enhanced electrolyte wettability, exposure of additional active sites, and the disruption of agglomerates during cycling due to the repeated insertion/extraction of ions.[19,20]

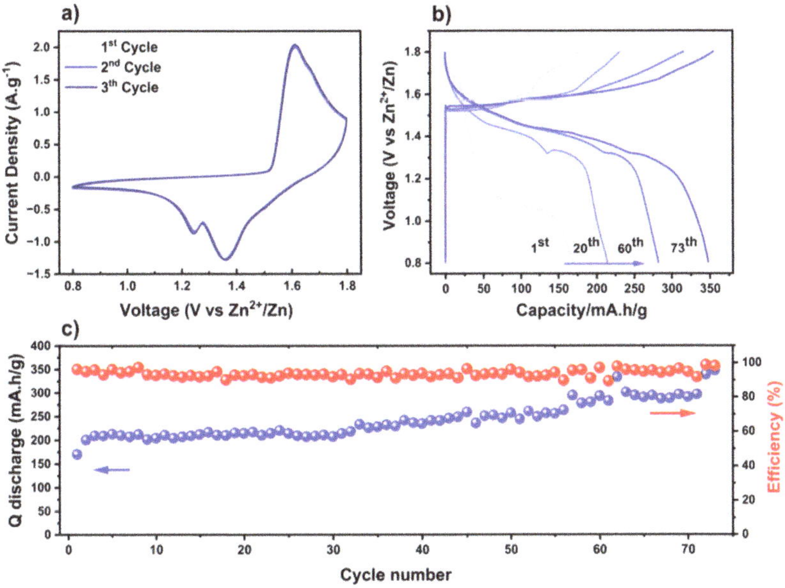

Fig. 3. (a) CV curves, (b) GCPL patterns, and (c) long-term cycling of L-KMO.

Conclusions

In this work, a layered hydrated manganese oxide (L-KMO, space group: Pbca) was produced through mild acid leaching of K-birnessite (KMO, $K_{0.37}MnO_2$, C 1 2/m 1) synthesized via a simple sol-gel process. XRD, Raman, EDX, and TGA analyses confirmed the synthesis of pure K-birnessite, followed by successful potassium extraction and the formation of $Mn_7O_{13}\cdot 5H_2O$ phase, whereas SEM revealed that L-KMO's microstructure is more open and rougher than that of its KMO counterpart. When evaluated as a cathode material for AZIB, L-KMO delivers a high specific capacity of up to 347 mAh g^{-1} at 50 mA g^{-1} current density, with a Coulombic efficiency over 97% throughout cycling. These results demonstrate a simple and scalable approach for developing efficient AZIB cathodes. Future work will focus on optimizing the acid-leaching process for large-scale production and assessing the chemical and electrochemical stability of the produced cathodes under practical operating conditions.

References

[1] B. Tang, L. Shan, S. Liang, J. Zhou, Issues and opportunities facing aqueous zinc-ion batteries, Energy Environ Sci 12 (2019) 3288–3304. https://doi.org/10.1039/C9EE02526J

[2] G. Zampardi, F. La Mantia, Open challenges and good experimental practices in the research field of aqueous Zn-ion batteries, Nature Communications 2022 13:1 13 (2022) 1–5. https://doi.org/10.1038/s41467-022-28381-x

[3] K. Zhu, W. Yang, Vanadium-Based Cathodes for Aqueous Zinc-Ion Batteries: Mechanisms, Challenges, and Strategies, Acc Chem Res 57 (2024) 2887–2900. https://doi.org/10.1021/ACS.ACCOUNTS.4C00484

[4] R. Ajay Rakkesh, S. Shalini, S. Tharani, D. Durgalakshmi, S. Balakumar, Possibilities and challenges of cathode materials for Zn-ion batteries, Energy Advances 3 (2024) 676–688. https://doi.org/10.1039/D3YA00638G

[5] P. Guo, S. Ouyang, H. Jiang, J. Li, H. Zhu, Y. Wang, Application of metal organic frameworks (MOFs) and their derivatives in the cathode materials of aqueous zinc-ion batteries, J Mater Chem C Mater 12 (2024) 18591–18608. https://doi.org/10.1039/D4TC03273J

[6] B. Zhang, P. Dong, S. Yuan, Y. Zhang, Y. Zhang, Y. Wang, Manganese-Based Oxide Cathode Materials for Aqueous Zinc-Ion Batteries: Materials, Mechanism, Challenges, and Strategies, Chem & Bio Engineering 1 (2024) 113–132. https://doi.org/10.1021/CBE.3C00120

[7] M. Karbak, M. Baazizi, S. Sayah, C. Autret-Lambert, Y. Tison, H. Martinez, T. Chafik, F. Ghamouss, Unraveling high-performance oxygen-deficient amorphous manganese oxide as the cathode for advanced zinc ion batteries, J Mater Chem A Mater 11 (2023) 2634–2640. https://doi.org/10.1039/D2TA07838D

[8] C. Wang, M. Wang, Z. He, L. Liu, Y. Huang, Rechargeable Aqueous Zinc–Manganese Dioxide/Graphene Batteries with High Rate Capability and Large Capacity, ACS Appl Energy Mater 3 (2020) 1742–1748. https://doi.org/10.1021/ACSAEM.9B02220

[9] F. Jing, J. Pei, Y. Zhou, Y. Shang, S. Yao, S. Liu, G. Chen, High-performance reversible aqueous Zinc-Ion battery based on Zn2+ pre-intercalation alpha-manganese dioxide nanowires/carbon nanotubes, J Colloid Interface Sci 609 (2022) 557–565. https://doi.org/10.1016/J.JCIS.2021.11.064

[10] S. Ding, M. Zhang, R. Qin, J. Fang, H. Ren, H. Yi, L. Liu, W. Zhao, Y. Li, L. Yao, S. Li, Q. Zhao, F. Pan, Oxygen-Deficient β-MnO2@Graphene Oxide Cathode for High-Rate and Long-Life Aqueous Zinc Ion Batteries, Nano-Micro Letters 2021 13:1 13 (2021) 173-. https://doi.org/10.1007/S40820-021-00691-7

[11] W. Chen, X. Zhan, B. Luo, Z. Ou, P.C. Shih, L. Yao, S. Pidaparthy, A. Patra, H. An, P. V. Braun, R.M. Stephens, H. Yang, J.M. Zuo, Q. Chen, Effects of Particle Size on Mg2+ Ion Intercalation into λ-MnO2 Cathode Materials, Nano Lett 19 (2019) 4712–4720. https://doi.org/10.1021/ACS.NANOLETT.9B01780

[12] N. Liu, M. K., J. Pan, Y. Hu, Y. Sun, X. Liu, A Facile Preparation of λ-MnO2 as Cathode Material for High-Performance Zinc-Manganese Redox Flow Battery, J Electrochem Soc 167 (2020) 040517. https://doi.org/10.1149/1945-7111/AB75C2

[13] H. Afaryate, M. Karbak, M. Kerroumi, M. Aqil, H. Martinez, B. Manoun, Y. Tamraoui, F. Ghamouss, Exploring the synthesis of spinel-type manganese oxide by acid leaching for zinc-ion battery applications, Chemical Engineering Journal 522 (2025) 167548. https://doi.org/10.1016/J.CEJ.2025.167548

[14] S. Ching, D.J. Petrovay, M.L. Jorgensen, S.L. Suib, Sol−Gel Synthesis of Layered Birnessite-Type Manganese Oxides, Inorg Chem 36 (1997) 883–890. https://doi.org/10.1021/IC961088D

[15] Y. Jiao, L. Kang, J. Berry-Gair, K. McColl, J. Li, H. Dong, H. Jiang, R. Wang, F. Corà, D.J.L. Brett, G. He, I.P. Parkin, Enabling stable MnO2 matrix for aqueous zinc-ion battery cathodes, J Mater Chem A Mater 8 (2020) 22075–22082. https://doi.org/10.1039/D0TA08638J

[16] L. Luo, M. Yuan, T. Xie, Q. Chen, Nanoscale assembly and mechanism of K-Birnessite/GOs composite, Results Phys 7 (2017) 3955–3961. https://doi.org/10.1016/J.RINP.2017.10.019

[17] M. Kerroumi, M. Karbak, H. Afaryate, A. El-Bchiri, M. Aqil, B. Manoun, Y. Tamraoui, H. Girault, F. Ghamouss, Unveiling electrochemical insights of lithium manganese oxide cathodes from manganese ore for enhanced lithium-ion battery performance, Electrochim Acta 509 (2025) 145286. https://doi.org/10.1016/J.ELECTACTA.2024.145286

[18] P.E. Saranya, S. Selladurai, Efficient electrochemical performance of ZnMn2O4 nanoparticles with rGO nanosheets for electrodes in supercapacitor applications, Journal of Materials Science: Materials in Electronics 2017 29:4 29 (2017) 3326–3339. https://doi.org/10.1007/S10854-017-8268-5

[19] L. Wang, K.W. Huang, J. Chen, J. Zheng, Ultralong cycle stability of aqueous zinc-ion batteries with zinc vanadium oxide cathodes, Sci Adv 5 (2019). https://doi.org/10.1126/SCIADV.AAX4279;PAGE:STRING:ARTICLE/CHAPTER

[20] W. Xu, Y. Wang, Recent Progress on Zinc-Ion Rechargeable Batteries, Nano-Micro Letters 2019 11:1 11 (2019) 1–30. https://doi.org/10.1007/S40820-019-0322-9

Assessment of Physical Properties of Densified Alga Residues for Solid Fuel Applications

KuZilati KUSHAARI[1,a*], Babar AZEEM[2,b], Abdul BASIT[3,c], Salma NOUH[4,d], Trinh Hoai THANH[5,e]

[1]Chemical Engineering Department, Prince Mohammad Bin Fahd University, Al Khobar, Saudi Arabia

[2]Department of Chemical Engineering, College of Engineering, Imam Mohammad Ibn Saud Islamic University (IMSIU), Riyadh 11432, Saudi Arabia

[3]Department of Chemical Engineering and Technology, University of Gujrat, Gujrat, Pakistan

[4]Future Hands Trading Co. Technology and Environment Consultancy, Riyadh, Saudi Arabia

[5]Ho Chi Minh City University of Industry and Trade, Ho Chi Minh City, Vietnam

[a]kkushaari@pmu.edu.sa, [b]badin@imamu.edu.sa, [c]Dr.abdulbasit@uog.edu.pk, [d]salma@futureshands.com, [e]thanhth@huit.edu.vn

Keywords: Algae Residues, Algae Waste, Combustion Fuel, Biofuel, Solid fuel, Biomass Densification

Abstract. With the development of biodiesel production through the transesterification process, biomass-derived oils have attracted increasing attention as renewable energy sources. Algae has been identified as a promising feedstock; large quantities of algae residue are produced during biodiesel oil extraction however, more than half of the original algal biomass remains as residue after oil extraction, which reduce the economic viability of algae-based biodiesel. These residues still contain significant energy potential that can be converted into valuable products such as biogas, bio-chemicals, and solid fuels. However, most existing waste-to-energy conversion methods are complex, energy-intensive, and costly. This study evaluates densified algal residue (AR) pellets as a solid fuel by characterizing their physical and fuel properties and comparing them with standard coal and common biomass feedstocks. The results indicate that algae residue (AR) is comparable to commonly used biomass feedstocks. The AR exhibits a low moisture content of 13.74 wt%, which falls within the typical range for biomass feedstocks and is slightly higher than that of coal at 9.3 wt%. Its calorific value of 19.11 MJ/kg is approximately 24% lower than that of standard coal at 25.3 MJ/kg, yet remains within the range reported for other biomass materials. Overall, the algae residue pellets demonstrate good physical strength and favorable characteristics for use as a solid fuel.

Introduction

The increase in global energy demand and the depletion of fossil fuel reserves have created a strong need for sustainable energy resources, including both new and renewable energy sources. Beyond established renewable sources such as solar, hydroelectric, wind, geothermal, and wave energy for electricity generation, there is an urgent need for renewable substitutes in transportation fuels. Among the various candidates, biomass has emerged as a leading renewable feedstock for biofuel production as well as for heat and power generation [1,2,3].

In recent years, biodiesel production from algal oil has gained significant research attention as a promising alternative to conventional fossil fuels [4]. However, the extraction of oil from algae leaves behind up to 60% of solid residues from the initial biomass, posing both economic and environmental challenges due to the large volume of waste generated [5,6]. To enhance the economic feasibility of algae-based biodiesel, numerous waste-to-energy (WTE) strategies have been explored, including the utilization of algal residues for biogas production, soil amendments, fertilizers, and biodegradable materials. In addition, several studies have suggested their potential application as solid biofuels.

Despite this, most existing WTE conversion technologies such as torrefaction, pyrolysis, gasification, and anaerobic digestion require high temperatures, high pressures, and complex processing steps, limiting their commercial viability [4–7]. Therefore, a simpler and more cost-effective approach is needed to valorize algal residues.

Algae residues still contain substantial unutilized energy in the form of carbohydrates and proteins, making them suitable candidates for direct combustion to produce heat or electricity [5,8]. However, limited data exist on their physical and fuel characteristics as solid combustion fuels. This study proposes biomass densification of algal residues into pelletized solid fuels and focuses on characterizing the resulting pellets' physical and fuel properties in comparison with coal and common biomass feedstocks. The aim is to assess the feasibility of algae residue pellets as a practical and sustainable solid fuel source.

Material and methods

Freeze-dried and cracked Chlorella vulgaris (sp.) microalgae biomass was utilized in this study. The oil content of the microalgae was first extracted using the Soxhlet extraction method with hexane as the solvent. A solvent to biomass ratio of 5:1 was prepared, corresponding to 300 mL of liquid hexane for every 60 g of microalgae. After the oil extraction, the remaining solid material, known as algae residues, was collected. The wet residues were heated in a water bath at 85 °C for at least five hours in order to remove residual solvent. The residues were then air-dried and ground into a fine, uniform powder for densification.

The algae residue powder was pelletized using a cylindrical die with an internal diameter of 13 mm. The die was attached to a manual hydraulic hand pump, which was used to apply the compression. Two pressure ranges were used: 13 to 50 kN, representing typical pelletization conditions, and 75 to 165 kN, corresponding to recommended biomass densification pressures. Each pellet was made from 1.0 g of algae residue powder, producing pellets about 13 mm in diameter and 6 mm thick. The physical properties of the algae residue pellets were evaluated by measuring its moisture content, calorific value, bulk density, and durability. Moisture content was determined by drying the samples at 105 °C for 24 hours and its weight percentage was recorded.

The calorific value, representing the higher heating value (HHV), was measured using a C5003 IKA WERKE control calorimeter with approximately 1 g of sample per test. Bulk density was calculated as the mass-to-volume ratio, where the volume of the AR powder and pellets was obtained from cylindrical dimensions based on a 13 mm die diameter. Durability was assessed using a drop test in which pellets were dropped four times from a height of 1.84 m onto a metal plate, and the weight ratio before and after impact was used to quantify pellet durability. Durability ratings were classified as high (> 0.8), medium (0.7–0.8), or low (< 0.7).

Results and Discussions

Moisture Content

Moisture content refers to water trapped between the particles in the biomass material. Presence of high moisture in solid fuels impedes combustion as it reduces the available energy to give heat energy. More energy is needed as the latent heat to vaporize the moisture from the material. However, the advantage of some moisture in the biomass feed prior to densification process serves as a lubricant and binding agent, giving rise to more durable solid produced. For that reason, the optimal moisture content in biomass feed is between 8.0 to 12.0 wt% to produce high quality pellets [8]. Moisture content of AR pellets (13.7 wt%) is comparable to typical biomass for combustions, such as Switchgrass (14.0 wt%), and Miscanthus (11.5 wt%) and all these biomass have slightly higher moisture content compared to standard coal (9.3 wt%) as shown in Fig. 1.

Materials Research Proceedings 64 (2026) 700-705 https://doi.org/10.21741/9781644904091-87

This implies AR pellets to be in the low moisture content range along with the common biomass feedstock, where it is feasible to be used as direct combustion feedstock fuel.

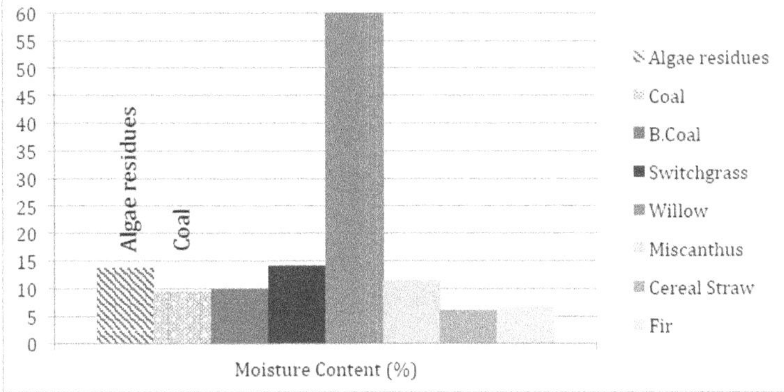

Fig. 1 Moisture contents of algae residues, coal and selected biomass; i.e. Switchgrass, willow, Miscanthus, straw and fir.

Calorific Value

The calorific value is an important property of biomass used for combustion. It determines how much material is required to produce a given amount of energy compared with high-density fuels. Densification of biomass into compact forms is commonly reported to increase energy density because more material is available for combustion in a given volume [4,5,9]. In this work, the results show that as the compression pressure increases, the calorific value of the pellets decreases (Fig. 2). The values decrease from 19.6 MJ/kg for AR powder to 19.11 MJ/kg for AR pellets produced at typical pellet mill pressure, and further to 18.8 MJ/kg for pellets produced at biomass waste pressure. This reduction is mainly due to the decreased surface area of the pellets available for combustion. Higher pressure during palletization process generates more friction and localized heat within the die. This heat can cause partial degradation or volatilization of light organic compounds which contributes significantly to the heating value, leading to a slight reduction in calorific value. Furthermore, the increase in pressure can remove the entrapped volatile components and residual moisture. Since volatile matter generally has a higher energy content, its reduction lowers the measured calorific value.

Despite this decrease, the calorific value of AR pellets produced under commercial conditions which is 19.11 MJ/kg, is comparable to that of commonly used biomass feedstocks, although it is slightly lower than that of standard coal of 25.3 MJ/k as shown in Figure 3. These results indicate that algae residue is suitable for combustion applications. Pelletizing at commercial pressures improves bulk density and pellet durability, which facilitates storage and transportation, without significantly reducing its heating value. This finding also highlights the need to optimize the applied compression pressure, since excessive pressure can negatively influence the calorific value of the pellets.

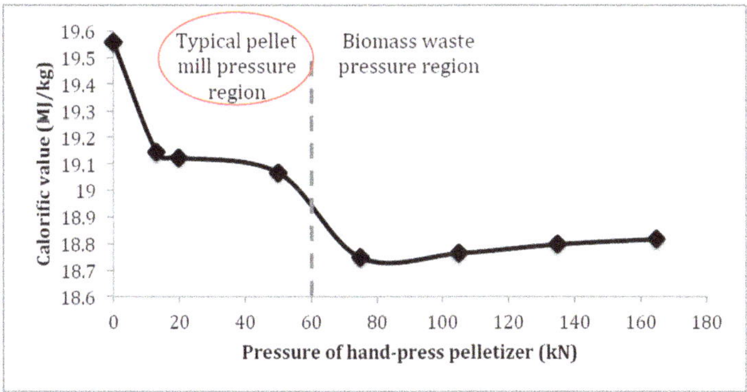

Fig. 2 Densification pressure (kN) effect on calorific values of algae residues

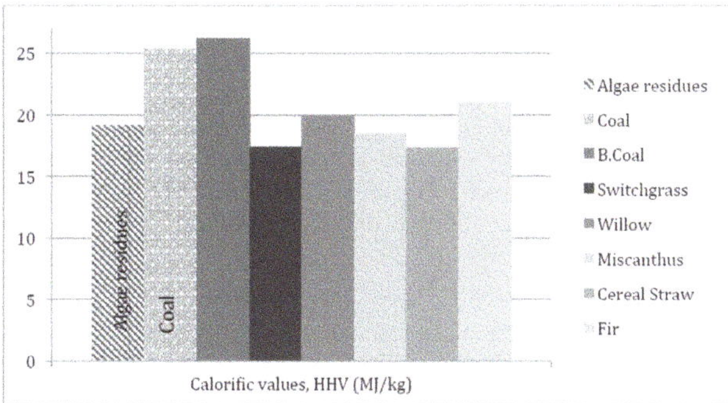

Fig. 3 Calorific values (MJ/kg) of algae residues, coal and selected biomass; i.e. Switchgrass, willow, Miscanthus, straw and fir.

Bulk Density

Densification is often required to improve handling, transportation, and storage of biomass materials. Biomass with higher bulk density reduces transportation and storage costs because less space is needed compared to low-density materials. Higher bulk density also allows easier and more stable feeding of the solid fuel into combustion systems, such as burners [10,11].

As shown in Figure 4, densification increases the bulk density of AR powder from 538 kg/m^3 to 1255 kg/m^3, producing compact pellets with a density approaching that of standard coal (1500 kg/m^3). This implies that AR pellets require about half the storage space compared to AR powder and can be handled and transported in a manner similar to standard coal.

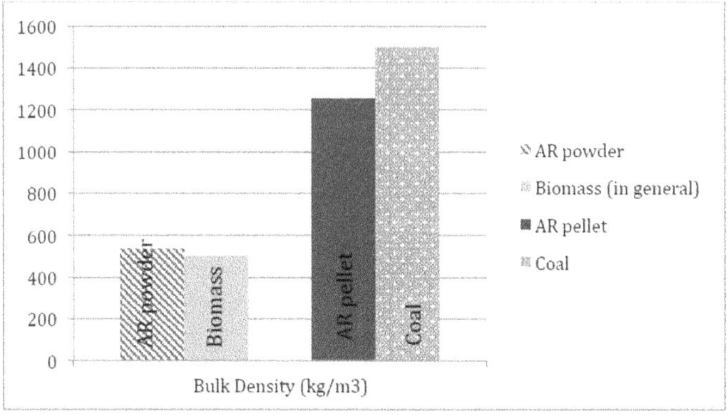

Fig. 4 Bulk density (kg/m3) of algae residue powder, algae residue pellets, general biomass and standard coal

Durability

Pellet durability is another important parameter to report as the strength of compact solids determines the method and costs of handling and transport (fragile or sturdy materials). Durable pellets are more favorable to sustain drop impact and rough movements without breaking apart easily [9,10]. The average durability averaged at 0.99%, with no significant changes between pellets compressed at increasing densification pressures. This implies very durable pellets can be obtained from commercial pelleting mills, where it can be subjected to non-fragile handling and transport methods.

Conclusion

The results of this study indicate that algae residue, AR can be successfully densified into high quality pellets which make it suitable for combustion applications. The moisture content of AR pellets is comparable to the commonly used biomass feedstocks, this indicates that AR falls within an acceptable range for efficient combustion. Although increasing compression pressure during pelletization led to a slight reduction in calorific value, AR pellets produced under commercial conditions retained a heating value comparable to the conventional biomass fuels. This finding highlights the importance of optimizing pelletization pressure to avoid unnecessary losses in energy content.

Densification greatly increased the bulk density of AR, and its bulk density is almost similar to standard coal. This reduces the space needed for storage and improving handling and transportation efficiency. In addition, the AR pellets exhibited excellent mechanical durability, indicating their ability to withstand handling and transport without significant degradation. Overall, these findings demonstrate that under optimized palletizing pressure, AR has a potential to become a promising alternative biomass feedstock for solid fuel combustion.

References

[1] Saravanan, YP Ragini, S Karishma, RV Hemavathy, M Jyotsna, A review on advancing sustainable energy: The role of biomass and bioenergy in a circular economy, Sustainable Futures, 10, (2025),100835 https://doi.org/10.1016/j.sftr.2025.100835

[2] Naik, S. N., Goud, V. V., Rout, P. K., & Dalai, A. K. (2010). Production of first- and second-generation biofuels: A comprehensive review. Renewable and Sustainable Energy Reviews, 14(2), 578-597. https://doi.org/10.1016/j.rser.2009.10.003

[3] J. Chang, S. Selvaraj, S. Manikandan, S. Nagarani, A. Senthilkumar, M. S. Samuel, E. Selvarajan, Ashwini, J. John, M. Kumar, A comprehensive review of biomass utilization for sustainable biofuel and biocomponent development, Biomass and Bioenergy, 202 (2025), 108167. https://doi.org/10.1016/j.biombioe.2025.108167. https://doi.org/10.1016/j.biombioe.2025.108167

[4] DT Oliveira, VA Mescouto, RJ Paiva, SRF Silva, LAB Santos, GM Serra, LP Xavier, RCR Noronha, LAS Nascimento, Use of Residual Lignocellulosic Biomass and Algal Biomass to Produce Biofuels, Int. J. Mol. Sci. (2024), 25(15), 8299; https://doi.org/10.3390/ijms25158299

[5] Babich, I. V., van der Hulst, M., Lefferts, L., Moulijn, J. A., O'Connor, P., & Seshan, K. Catalytic pyrolysis of microalgae to high- quality liquid bio-fuels. Biomass and Bioenergy, 35(7), (2011), 3199-3207. https://doi.org/10.1016/j.biombioe.2011.04.043

[6] Chen, W. H., Huang, M. Y., Chang, J. S., & Chen, C. Y. (2014). Thermal decomposition dynamics and severity of microalgae residues in torrefaction. Bioresource Technology,169, 258-264. https://doi.org/10.1016/j.biortech.2014.06.086

[7] Kaliyan, N., & Morey, V., R. (2009). Factors affecting strength and durability of densified biomass products. Biomass and Bioenergy, 33(3), 337-359. https://doi.org/10.1016/j.biombioe.2008.08.005

[8] S.T. Niraj, J.R. Sunita, R.V. C, V.K. Satish, C.Y. P, B.S. L, Extraction of oil from algae by solvent extraction and oil expeller method, Int J Chem Sci, 9 (2011), pp. 1746-1750.

[9] X Yang, Y Zhang, P Sun, C Peng, A review on renewable energy: Conversion and utilization of biomass", Smart Molecules, 2(4), (2024). e20240019 https://doi.org/10.1002/smo.20240019

[10] V Praveena, LJ Martin, J Matijošius, F Aloui, A Pugazhendhi, EG Varuvel, A systematic review on biofuel production and utilization from algae and waste feedstocks - a circular economy approach, Renewable and Sustainable Energy Reviews, 192, (2024), 114178. https://doi.org/10.1016/j.rser.2023.114178

[11] K Kiskira, S Plakantonaki, N Gerolimos, K Kalkanis, E Sfyroera, F Coelho, G Priniotakis, Life Cycle Optimization of Circular Industrial Processes: Advances in By-Product Recovery for Renewable Energy Applications, Clean Technologies, 10.3390/cleantechnol8010005, 8, 1, (5), (2026). https://doi.org/10.3390/cleantechnol8010005

A Multi-Criteria Decision Analysis for the Selection of Thermal Energy Storage Materials in Concentrated Solar Power Applications

Ilham NAIT HIM[1,a], Mohamed Hassan KHALILI[1,b]

[1]Al Akhawayn University, Ifrane, Morocco

[a]i.naithim@aui.ma, [b]m.khalili@aui.ma

Keywords: Concentrated Solar Power (CSP), Thermal Energy Storage (TES), Multi-Criteria Decision Analysis (MCDA), Molten Salts, Phase Change Materials (PCMs), Sustainable Engineering

Abstract. The choice of the best Thermal Energy Storage (TES) material is fundamental to the economic feasibility and the performance of the Concentrated Solar Power (CSP) plants. In this paper, a stringent multi-criteria decision analysis (MCDA) is used on five different TES material types namely molten salts, phase change materials (PCMs), concrete, water, and synthetic oils. To create a composite performance index, it created a systematic assessment of the significant technical, economic, and environmental statistics, such as thermal properties, energy efficiency, heat retention, and lifecycle effects such as toxicity and CO2 emissions. In our analysis, we will find that there is no universal material that is better than the rest, which points to a complicated set of trade-offs. Water to be the most preferred material has a composite score of 0.88 based on its excellent thermal performance and environmental rating but is restricted by a small operating temperature range. Concrete is a moderate, economical solution (score: 0.52), whereas molten salt as the existing industry standard has a high-performance but environmentally demanding profile (score: 0.27). The study presents a multi-criteria radar chart along with the correlation matrix as the advanced visualizations that would allow the stakeholders to make decisions based on the specifics of the CSP project.

Introduction

Concentrated Solar Power (CdS) is one of the important technologies in the global shift to sustainable energy, with a special benefit of having the capability to generate power in a dispatchable manner by combining Thermal Energy storage (TES) [1]. CSP plants can also store sun energy as thermal energy and then convert it into electricity when required unlike the photovoltaic systems that cause grid instability and unreliability. Economic viability and performance of a CSP plant, however, are highly dependent on the choice of the TES material. This option has not only direct impact on the energy density and round trip efficiency of the system but also on the system operational life, cost and the environmental footprint. The literature to date has greatly compared different TES materials, yet this is done with a restricted number of properties which results in disjointed and at times contradicting suggestions [2, 3]. This paper will fill this gap by carrying out a multi-criteria decision analysis (MCDA). We compare five major categories of TES materials, namely, molten salts, phase change materials (PCMs), concrete, water, and synthetic oils, in terms of various performance indicators in a broad spectrum. This study aims to come up with a comprehensive and quantitative model that can be used to inform the choice of TES materials. Our normalization and weighting of key metrics produces a composite score on performance that enables a more subtle comparison than was previously possible. In this work, complex trade-offs are elucidated using advanced data analysis and visualizations, and may

Materials Research Proceedings 64 (2026) 706-713 https://doi.org/10.21741/9781644904091-88

be considered a powerful decision-making resource to be deployed by engineers, researchers, and project developers in the CSP domain.

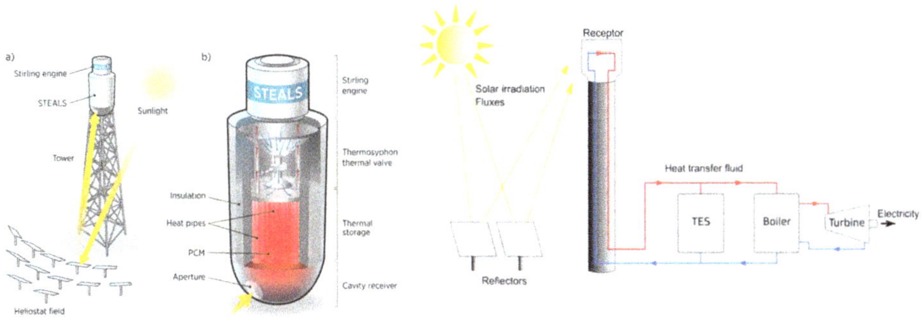

Fig. 1. Energy storage in CSP Systems with TES materials.

Methodology

The analysis was conducted on five material categories. The core of our methodology is a multi-criteria decision analysis (MCDA) framework designed to provide an advanced assessment of material suitability.

1. Data and Metrics

The analysis is grounded in the fundamental governing equations of thermodynamics and heat transfer. The energy storage capacity (Q) and exergy (Ex) are defined as follows:

$$Q = m \cdot C_p \cdot \Delta T,$$

$$Ex = Q \cdot \left(1 - \frac{T_0}{T_s}\right),$$

For latent heat storage (PCMs), the total energy stored includes the enthalpy of fusion (ΔHf):

$$Q = m \cdot \left[C_{p,s}(T_m - T_1) + \Delta H_f + C_{p,l}(T_2 - T_m)\right].$$

Thermal Properties:
Specific Heat ($kJ\ kg^{-1}K^{-1}$), Thermal Conductivity ($W\ m^{-1}K^{-1}$)
Performance:
Round-trip Energy Efficiency (%), Heat Retention (%)
Environmental Impact:
Recyclability (rated 1-5), Toxicity (rated 1-5), and CO2 Emissions (kg CO2/kg).

All key performance indicators were put on a 0-1 scale by Min-Max scaling in order to directly compare across disparate units and scales. This premise makes sure that the performance of every material in terms of its relative performance is maintained within the limits of the dataset. As the priority of high-temperature dispatchability was considered to be the main feature of modern CSP plants, the decision on weightings was substantiated. A 40 weight was put on Technical performance (Thermal Properties and Efficiency), 30 weight on Environment Impact and 30 weight on Economic/Stability factors. Other possible scenarios were also thought of including a Sustainability-First (50% Environmental) and Cost-First (50% Economic) weighting to test the robustness of the rankings.

Materials Research Proceedings 64 (2026) 706-713 https://doi.org/10.21741/9781644904091-88

2. Statistical Analysis and Visualization

The analysis was done using a custom Python code written with the help of pandas, scikit-learn, and matplotlib. The derived metrics included a thermal performance index, percentages of energy loss, and composite environmental impact score, which were calculated with the use of the script. In order to make a one-to-one comparison across different units and scales, all the key performance indicators were scaled to a 0-1 scale by Min-Max scaling. Each material was then given a final composite score of the mean normalized score and such measures as toxicity and CO2 emissions were inverted so that higher scores are always good. The results were displayed in advanced visualization of the findings, with multi-panel plots, heat loss curves, multi-criteria radar charts by material, and a correlation matrix heatmap to determine the interdependencies between the properties.

Results and Discussion

This section presents a detailed analysis of the TES materials, moving from individual properties to a multi-criteria comparison.

1. Comprehensive Thermal and Environmental Profile

A multi-faceted comparison of the materials is presented in Figure.2, which integrates thermal properties, operational temperature ranges, energy efficiency, and a composite environmental score. This visualization immediately highlights the inherent trade-offs in material selection.

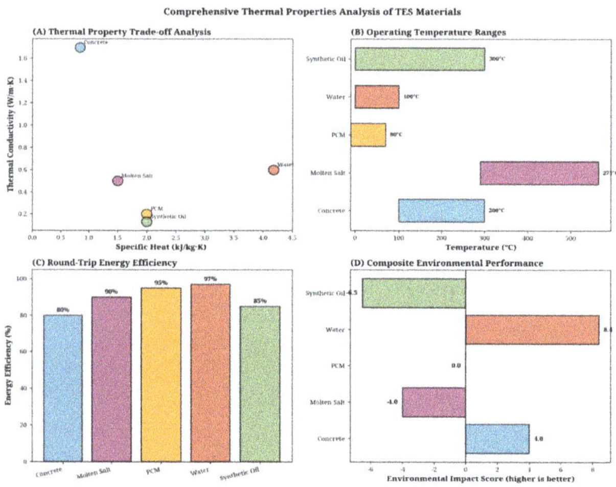

Fig. 3. Comprehensive analysis of TES materials. (A) Thermal property trade-offs. (B) Operational temperature ranges. (C) Round-trip energy efficiency. (D) Composite environmental performance score (higher is better).

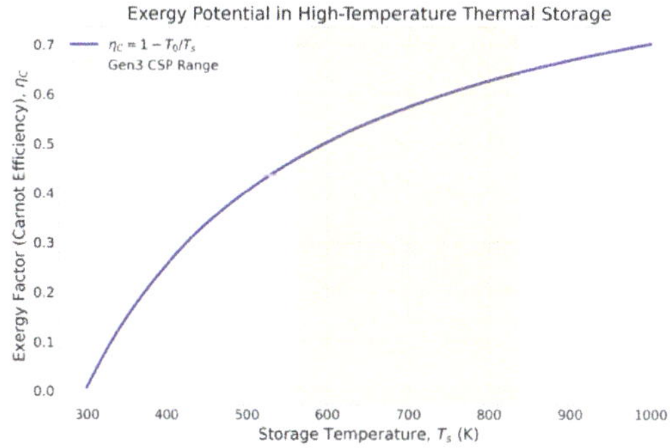

Fig. 4. Exergy potential (Carnot factor) as a function of storage temperature.

Panel A indicates the presence of the trade off between specific heat and thermal conductivity. Water has the best specific heat, meaning it has the best energy storage capacity per mass, but it has a mediocre thermal conductivity. Concrete on the other hand has the best thermal conductivity to transfer heat quickly but has the lowest specific heat. PCMs and Synthetic Oil concentrate in a zone of high specific heat and very low thermal conductivity, which is a serious negative attribute to high-power applications. Panel B shows the operation temperature ranges which is a major constraint to integration with power cycles. Molten Salt has the greatest and highest temperature (290-565C). Gen3 CdTe systems are extending this to over 800 o C, however, with Gen3 chloride-based molten salts (e.g., NaCl-KCl-MgCl2), it is ideal in high-efficiency steam turbines [11]. The small scope of Water (0-100 C) and conventional PCMs limit the application to low-temperature applications, although recent studies of carbonate-chloride eutectic PCMs are looking at the 400 700 C temperature span [12]. It can be seen on Panel C that Water and PCMs are the most efficient in terms of round trip efficiencies (97% and 95%, respectively). PCMs are highly efficient because they can be isothermal in the way that they can store the latent heat so that they can reduce thermodynamic losses. This is because water has a high heat transfer property which has made it very efficient. At 90, Molten Salt also does a good job. Panel D gives a system score on the environment. Water is the most environmentally friendly material (score: 8.4) and Synthetic Oil and Molten Salt got negative scores as they are highly toxic and/or produce CO2.

2. Heat Loss and Retention Dynamics

Figure.4 analyzes the thermal storage performance over a three-hour period. Panel A shows the temperature decay, while Panel B quantifies the total heat retention.

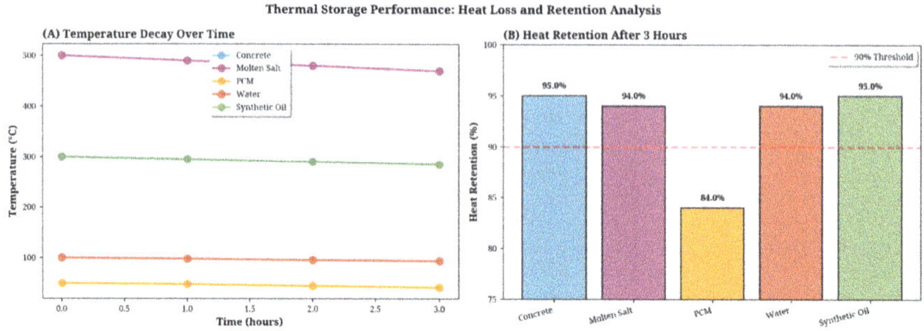

Fig. 5. Thermal storage performance. (A) Temperature decay over a 3-hour period. (B) Total heat retention after 3 hours.

Interestingly, **Concrete** and **Synthetic Oil** exhibit the highest heat retention (95%), despite their different thermal properties. Concrete's high thermal mass contributes to its slow temperature decay. The analysis suggests that for short-term storage, heat loss is not a distinguishing factor among the top performers. However, the low retention of **PCMs** (84%) is a concern, likely due to their low thermal conductivity causing inefficient heat distribution.

3. Multi-Criteria Performance Profiles

Figure.3 presents a radar chart for each material, mapping its normalized performance across six key criteria. This visualization allows for a quick assessment of the strengths and weaknesses of each option.

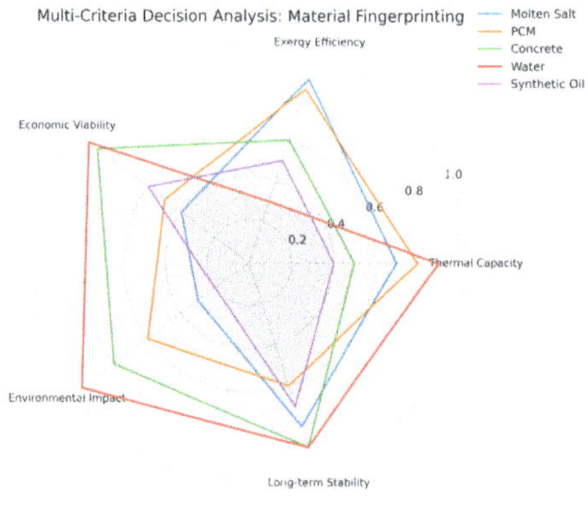

Fig. 6. Multi-criteria performance profiles of TES materials.

The existence of the trade off between specific heat and thermal conductivity is indicated by panel A. Water contains the best specific heat, i.e. best energy storage capacity per mass, but contains a mediocre thermal conductivity. Concrete on the other hand is the best thermal conductive to conduct heat fast but it has the lowest specific heat. The location of PCMs and Synthetic Oil in a region where specific heat is high and thermal conductivity is very low is an extremely undesirable characteristic as far as high-power applications are concerned. The operation temperature ranges are displayed in panel B that is the primary limitation to integration with power cycles. The highest and greatest temperature (290-565C) belongs to Molten Salt. Gen3

Materials Research Proceedings 64 (2026) 706-713 https://doi.org/10.21741/9781644904091-88

CdTe systems are pushing this to greater than 800 o C but with chloride-based molten salts (i.e. NaCl-KCl-MgCl2), it is perfect in high-efficiency steam turbines [11]. The limited scale of Water (0-100 C) and conventional PCMs restricts the application to low-temperature applications, though recent research of carbonate-chloride eutectic PCMs are considering the 400 700 C temperature range [12]. As can be observed on Panel C, Water and PCMs are the most efficient in regards to round trip efficiencies (97% and 95% respectively). PCMs are very efficient as they have the capability of being isothermal in the sense that they have the capacity of storing the latent heat to enable them minimize the thermodynamic losses. This is attributed to the fact that water possesses high heat transfer property that has rendered it very effective. Molten Salt at 90 is also doing a good job. Panel D provides a system rating on the aspect of the environment. The most eco friendly material (score: 8.4) is water and Synthetic Oil and Molten Salt were assigned a negative score because they are very toxic and/or emit CO2.

4. Correlation Analysis

To understand the relationships between different material properties, a correlation matrix was provided as follows in Figure.4.

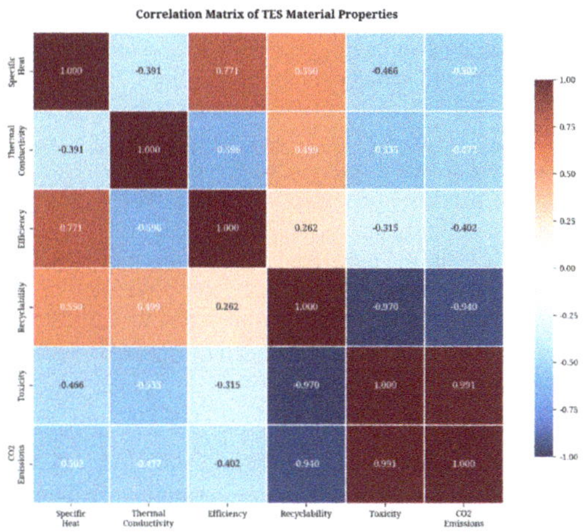

Fig. 7. Correlation matrix of TES material properties. Red indicates a positive correlation, and blue indicates a negative correlation.

There are a number of significant relationships to be noticed in the matrix. Toxicity and CO2 Emissions have a strong positive correlation (r = 0.991) indicating that a high carbon footprint of materials is also likely to contribute to their toxicity. Specific Heat (r = 0.771) and Thermal Conductivity (r = -0.596) have a positive and negative correlation with Efficiency respectively. This counter-intuitive observation needs further research but can be a result of a small sample (n=5). Statistical shortcomings Statistical limitations Statistical limits are sensitive to outliers, and the possibility of rank reversal in the case of the addition of new materials. These effects should be overcome in future studies by using larger datasets and hybrid MCDA approaches such as AHP-TOPSIS [15].

Conclusion

This paper has shown that a multi-criteria decision analysis framework can be very effective in choosing TES materials to be used in CSP applications. Incorporating a large pool of technical, economic, and environmental measurements into a composite performance measure, we have offered a more comprehensive and detailed comparison compared to what exists in the literature. It can be seen that there is no universal solution; the best material to use is extremely dependent on the needs of a project. Our most important conclusions are that water has the best performance in our MCDA framework, but it can only be utilized in low-temperature processes. Concrete is a very balanced, economical and eco-friendly material that can be used in numerous applications. The existing industry standard, Molten Salt, has good performance at high temperatures although it presents serious environmental and safety issues. Lastly, PCMs are high-energy density coupled with high-efficiency systems that are impeded by low thermal conductivity. The visualizations and analysis presented in this paper represents a potent instrumentation in the possession of stakeholders as they are leaving the simplistic comparisons and making data-driven choices that will balance performance, cost, and sustainability.

Implications and Future Directions

The multi-criteria analysis given in this paper gives practical advice towards the design of CSP systems. In the case of dispatchable, high temperature generation, molten salt is the best option regardless of environmental constraints and its 290-565 C operating range is suitable to a high-efficiency Rankine cycle [2]. Concrete can be used for a balanced alternative with the same score (4.0) of excellent environmental performance, and 95% heat retention, thus being appropriate in constant discharge overnight. PCM is good in niche applications, where a high energy density and isothermal heat release is needed, but thermal improvement techniques are required to surpass its low conductivity (0.2 W/m 0.1 W/m K) [3]. This correlation is high between specific heat and efficiency ($r = 0.771$), indicating that there is need to conduct more research on high-capacity materials. Further efforts are needed on composite material with high thermal conductivity and high specific heat including nanofluids and nanocomposites. The high level of correlation between toxicity and CO2 emissions ($r = 0.991$) indicates the importance of comprehensive sustainability measurements. Also, long-term thermal cycling degradation mechanism research is essential to economic feasibility. Policymaking wise, low environmental material should be favored in areas that have strict policies, with economic analysis taking into consideration both CAPEX and OPEX trade-offs of material performance and lifecycle costs.

References

[1] "What's the best size for a concentrated solar power plant?" [online] https://www.pacificgreen.com/articles/whats-best-size-concentrated-solar-power-plant/

[2] İ. Dinçer and M. A. Rosen, *Thermal Energy Storage: Systems and Applications*, 2nd ed. Hoboken, NJ, USA: John Wiley & Sons, 2010 https://doi.org/10.1002/9780470970751

[3] F. Alnaimat and Y. Rashid, "Thermal energy storage in solar power plants: A review of the materials, associated limitations, and proposed solutions," *Energies*, vol. 12, no. 21, p. 4164, Oct. 2019 https://doi.org/10.3390/en12214164

[4] D. Jayathunga, H. Karunathilake, M. Narayana and S. Witharana, "Selection of phase change materials for high temperature latent heat thermal energy storage for concentrated solar power plants," *2022 Moratuwa Engineering Research Conference (MERCon)*, Moratuwa, Sri Lanka, 2022, pp. 1-6 https://doi.org/10.1109/MERCon55799.2022.9906235

[5] A. E. Amer, K. Rahmani, and V. A. Lebedev, "Using the Analytic Hierarchy Process (AHP) method for selection of phase change materials for solar energy storage applications," *Journal of Physics Conference Series*, vol. 1614, no. 1, p. 012022, Aug. 2020 https://doi.org/10.1088/1742-6596/1614/1/012022

[6] A. G. Tsopgo, H. Teng, Q. Duan, and H. A. Zheng, "Probability-based multi-objective optimization framework for the selection of molten salts in thermal energy storage systems," *Composites Part B Engineering*, vol. 312, p. 113298, Dec. 2025 https://doi.org/10.1016/j.compositesb.2025.113298

[7] H. M. Ali *et al.*, "Advances in thermal energy storage: Fundamentals and applications," *Progress in Energy and Combustion Science*, vol. 100, p. 101109, Sep. 2023 https://doi.org/10.1016/j.pecs.2023.101109

[8] A. Caraballo, S. Galán-Casado, Á. Caballero, and S. Serena, "Molten Salts for Sensible Thermal Energy Storage: A Review and an Energy Performance Analysis," *Energies*, vol. 14, no. 4, p. 1197, Feb. 2021 https://doi.org/10.3390/en14041197.

[9] "Concentrating Solar Power, Electricity, 2022, ATB, NLR." https://atb.nrel.gov/electricity/2022/concentrating_solar_power

[10] SolarPACES. (2024). Technical Report on Next-Generation TES Materials.

[11] X. Liu, Y. Zhong, J. Li, H. Wang, and M. Wang, "A review of High-Temperature Molten Salt for Third-Generation Concentrating Solar Power," *Energy Science & Engineering*, vol. 13, no. 2, pp. 456–474, Jan. 2025 https://doi.org/10.1002/ese3.2029

[12] S. Xiong, Y. Shi, C. Liu, and Y. Yang, "A photothermal energy storage phase change material with high stability and enthalpy," *RSC Advances*, vol. 15, no. 21, pp. 17023–17030, Jan. 2025 https://doi.org/10.1039/d5ra01422k

[13] C. Liu, X. Zheng, H. Yang, W. Tang, G. Sang, and H. Cui, "Techno-economic evaluation of energy storage systems for concentrated solar power plants using the Monte Carlo method," *Applied Energy*, vol. 352, p. 121983, Sep. 2023 https://doi.org/10.1016/j.apenergy.2023.121983

[14] H. Fakri and A. E. Fadar, "TES materials integrated in CSP plants: Research trends, challenges and 3E performance analysis," *Solar Energy Materials and Solar Cells*, vol. 293, p. 113861, Jul. 2025 https://doi.org/10.1016/j.solmat.2025.113861

[15] Z. M. Al-Abri, K. M. Alawasa, R. S. Al-Abri, A. S. Al-Hinai, and A. S. A. Awad, "Multi-Criteria Decision-Making approach for optimal energy storage system selection and applications in Oman," *Energies*, vol. 17, no. 20, p. 5197, Oct. 2024 https://doi.org/10.3390/en17205197

[16] M. Fang, L. Cao, F. Song, Y. Cong, and W. Li, "Thermal transport properties of composite phase change materials characterized by molten carbonate-ceramic date nucleus occupancy structure: A molecular dynamics study," *Journal of Molecular Liquids*, vol. 391, p. 123260, Oct. 2023 https://doi.org/10.1016/j.molliq.2023.123260

Dye-Sensitized Solar Cells (DSSC): A DFT and TD-DFT Theoretical Investigation of the Effect of Inserting Auxiliary Acceptors on Triphenylamine based Dye Performances

Hamoud TOUBBALI[1,a *], Abdelhamid KHADIRI[1,b], Hassan ZARROK[1,c], Mounir EL HEZZAT[1,d] and Abdelkader ZARROUK[2,e]

[1]Laboratory of Advanced Materials and Process Engineering, Faculty of Sciences, Ibn Tofail University, P.O. Box. 133, 14000 Kenitra, Morocco

[2]Laboratory of Materials, Nanotechnology and Environment, Faculty of Sciences, Mohammed V University in Rabat, Av. Ibn Battouta, PO Box 1014, Agdal-Rabat, Morocco

[a]hamoud.toubbali@uit.ac.ma, [b]abdelhamid.Khadiri@uit.ac.ma, [c]hzarrok@gmail.com, [d]mounir.elhezzat@uit.ac.ma, [e]azarrouk@gmail.com

Keywords: Solar Cells, DSSC, Triphenylamine Based Dye, DFT, TD-DFT

Abstract. Triphenylamine (TPA) molecules have shown a highly donors efficacity in DSSC field. This work takes into consideration, based on a triphenylamine based dye C213, three novel organic sensitizers were designed to investigate the effect of introducing an auxiliary acceptor on their properties. With a D-A_2-π-A_1 architecture of the investigated dyes, we have optimized their geometry, studied their frontier molecular orbitals and calculated their gap energies by using density functional theory (DFT). TD-DFT calculations were then used to determine the maximum absorption wavelength (λ_{max}). The results show that insertion of π-conjugated linkers on the structure of the reference dye C213 has enhanced the maximum wavelength absorption (λ_{max}) and decreases the gap energy, which is favorable for DSSC. Indeed, among the three designed compounds, C213-THp have shown the best wavelength value with λ_{max}= 759 nm and a narrow gap energy with E_{gap}=1,94 eV.

Introduction

Nowadays, renewable energy takes a big place in both industrial and research fields with an electricity installed capacity rate (Fig. 1) of 21% in Africa, 31% in Americas, 29% in Asia, 36% in Europe and 35% in Oceania by the end of 2024 [1]. Furthermore, solar energy accounts 19% of the renewable energy produced in the world (Fig. 2). Among the strengths of solar energy systems there is their renewable and clean energy generation. Nevertheless, this industry still has challenges like maintenance complexity, high investment cost and its dependence on environmental conditions [2].

In the 17th century, the German mathematician Ehrenfried Walter Von Tschirnhaus has developed mirrors and lenses to concentrate the solar energy [3]. First invented in 1991 by Grätzel and O'Regan [4], Dye-sensitized solar cells (DSSCs) are classified as an advanced generation of solar cells mainly composed of a TiO_2 film layer that is coated with a charge transfer dye. This dye's photosensitivity enhances the absorption of solar energy [5]. In Dye-sensitized solar cells, the dye (photosensitizer) absorbs solar radiation and the resulting excited electrons are injected into the semiconductor's conduction band (generally TiO_2). The redox mediator (generally I^-/I^{3-} couple) then reduces the oxidized dye (dye^+) [6-8].

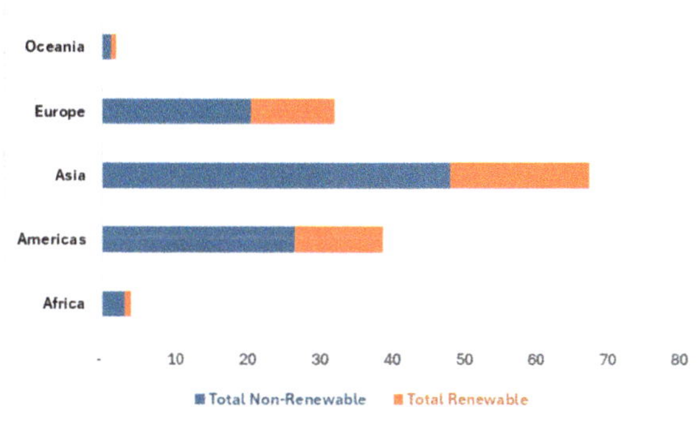

Fig. 1. Share of Electricity Installed Capacity of renewable energy (10^3 GW) by region between 2000 and 2024 [1].

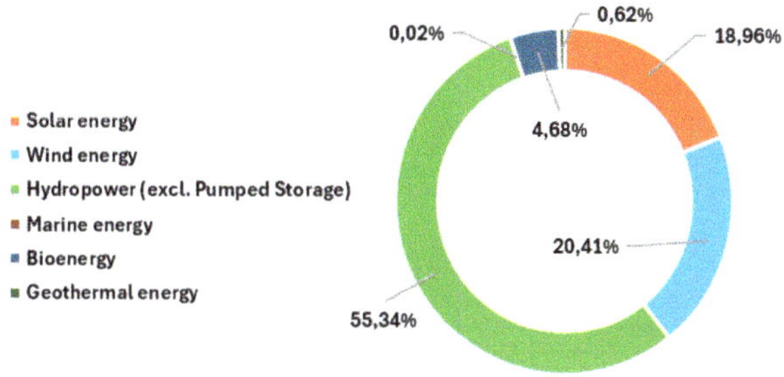

Fig. 2. World rate of renewable energy sources between 2000 and 2024 [1].

Many heterocycles' molecules containing sulfur, nitrogen and oxygen have been utilized as donors (Fig. 3), such as carbazole [9], coumarin [10-11], indole [12], triphenylamine [13], phenoxazine [14], phenothiazine [15] and perylene [16] based dyes (Fig. 3). Thus, due to their significant drift mobility and outstanding charge transport properties, triphenylamine (TPA) molecules and their derivatives have gained importance in various industrial applications [17].

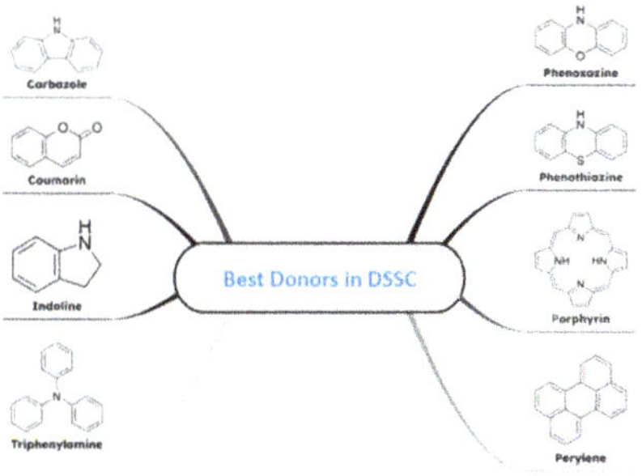

Fig. 3. Best donors in DSSC field.

Computational methodology

In this quantum chemical investigation, based on the reference dye (C213) synthetized by R. Li et al. [18] and the proposed dyes C213-TOx, C213-TZt and C213-THp (proposed structures and chemical formula in Fig. 4 and Fig. 5), we have used Gaussian 09 software [19] for all theoretical simulations.

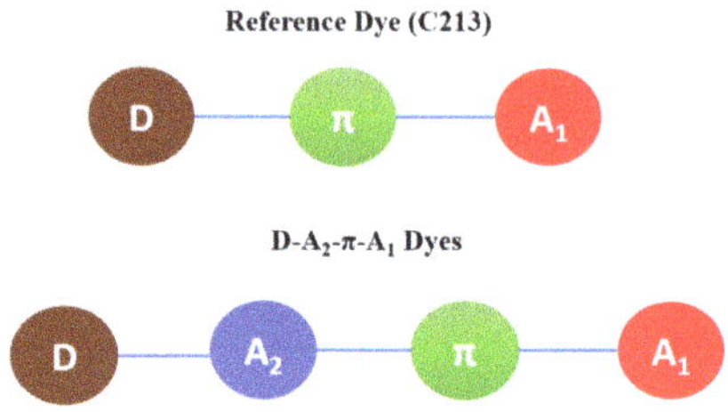

Fig. 4. Structure of the reference dye C213 (D-π-A_1) and the designed dyes (D-A_2-π-A_1).

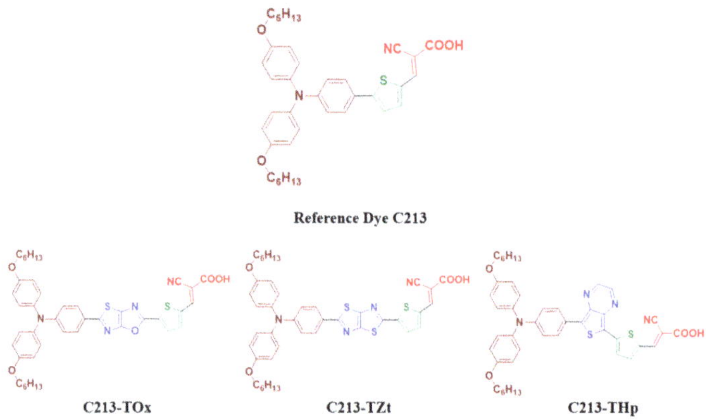

Fig. 5. Chemical formula of reference dye (C213) and the designed dyes.

Frontier molecular orbitals (FMOs) and geometry optimization calculation were carried out using density functional theory (DFT) with B3LYP/6-311G (d,p) basis set after been chosen as the appropriate functional (Results in Table. 1).

For the absorption simulation of reference dye C213, six functionals combined with TD-DFT were tested: B3LYP, CAM-B3LYP, B3PW91, HSEH1PBE, WB97XD and MPW1PW91 with 6-311G (d, p) basis set in chloroform (CFM) solvent in accordance with the experimental conditions [18]. TD-DFT/ MPW1PW91 functional was selected as the accurate simulation (Table 2.) with the minimum relative error (7%), and it will be used to investigate the absorption behaviour of the designed photosensitizers.

Table 1. HOMO, LUMO and Gap energies calculated for the reference dye (C213) with various functionals associated with 6-311G (d,p) basis set.

	E_{HOMO} (eV)	E_{LUMO} (eV)	Gap (eV)
(Exp[a])	-5,09	-3,42	1,67
B3LYP	-5,14 (1%)	-2,60	2,54
CAM-B3LYP	-6,33 (24%)	-1,49	4,83
B3PW91	-5,18 (2%)	-2,66	2,52
MPW1PW91	-5,32 (4%)	-2,50	2,82
HSEH1PBE	-4,93 (-3%)	-2,81	2,12
WB97XD	-6,84 (34%)	-0,93	5,91

(a) : Experimental value from [18].

Table 2. Reference dye (C213) maximum absorption wavelength (λ_{max}) obtained by various functionals associated with 6-311G (d,p) in Chloroform (CFM) solvent.

	λ_{max} (nm)	Relative Error (in %)
(Exp[a])	514	-
B3LYP	395	-23%
CAM-B3LYP	336	-35%
B3PW91	393	-24%
MPW1PW91	549	7%
HSEH1PBE	389	-24%
WB97XD	328	-36%

(a) : Experimental value from [18].

Results and discussions

The values of HOMO, LUMO and gap energies in eV of the investigated dyes are presented in Table 3. Gap energy was calculated using the following formula:

$$E_{Gap} = E_{LUMO} - E_{HOMO} \quad (1)$$

The calculated energy gaps of the investigated compounds decrease in this order C213 (2,54 eV) > C213-TOx (2,14 eV) > C213-TZt (2,10 eV) > C213-THp (1,94 eV).

Table 3. LUMO, HOMO and gap energy values (in eV) of investigated compounds at DFT/B3LYP/6-311G (d,p) in gas phase.

Compounds	E_{LUMO}	E_{HOMO}	Gap
C213 (Exp[a])	-3,42	-5,09	1,67
C213	-2,60	-5,14	2,54
C213-TOx	-3,02	-5,16	2,14
C213-TZt	-3,05	-5,15	2,10
C213-THp	-3,07	-5,01	1,94

(a) : Experimental value from [18].

According to Fig. 6, the HOMO energies of C213-TOx, C213-TZt and C213-THp lie below the I^-/I^{-3} redox potential (-4,80 eV), this is strongly required for the dye regeneration. Likewise, their

LUMO are positioned above the conduction band of semiconductor (-4,00 eV), which promotes the transfer of electrons from the sensitizers' LUMO to TiO_2.

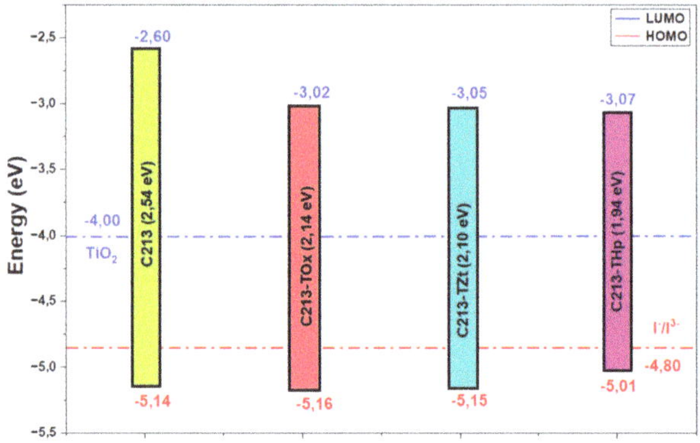

Fig. 6. Frontier molecular orbitals energy obtained at the B3LYP/6-311G (d,p) level in gas phase.

Fig. 7 shows HOMO and LUMO distribution calculated in gas phase by DFT/B3LYP/6-311G(d, p) level. It is apparent that electronic density of HOMO is focused on the triphenylamine (TPA) ring (donor), whereas, electronic density of LUMO is concentrated on C, CO, and COOH segments (acceptor).

According to Table 4. and Fig. 8, all designed sensitizers have higher λ_{max} than reference dye C213 following this order: C213 (549 nm) < C213-TOx (639 nm) < C213-TZt (646 nm) < C213-THp (759 nm) which can facilitate electronic excitation between LUMO and HOMO levels of designed dyes. Also, the three dyes have shown a big contribution to the HOMO-LUMO transition (Over 70%).

On the other hand, the light harvesting efficiency (LHE) must be as large as possible for an efficient sensitizer [20]. LHE can be calculated by the following equation [20]:

$$LHE = 1 - 10^{-f} \quad (2)$$

Where f is the oscillator strength of the sensitizer corresponding to the maximum absorption wavelength (λ_{max}).

Referred to Table 4. It is observed that C213-THp dye has the highest LHE value, which can affirm the efficiency of this sensitizer.

Table 4. Absorption properties of reference dye C213 and the 3 designed sensitizers investigated by TD-DFT/MPW1PW91/6-311G (d,p) in solvent medium (CFM).

Compounds	λ_{max} (nm)	E_{ex} (eV)	f	LHE	MO/character
C213 (Exp[a])	514	-	-	-	-
C213	549	2,26	0,92	0,880	H→L (70,45%)
C213-TOx	639	1,94	1,14	0,928	H→L (70,13%)
C213-TZt	646	1,92	1,04	0,909	H→L (70,11%)
C213-THp	759	1,70	1,16	0,931	H→L (70,16%)

(a) : Experimental value from [18].

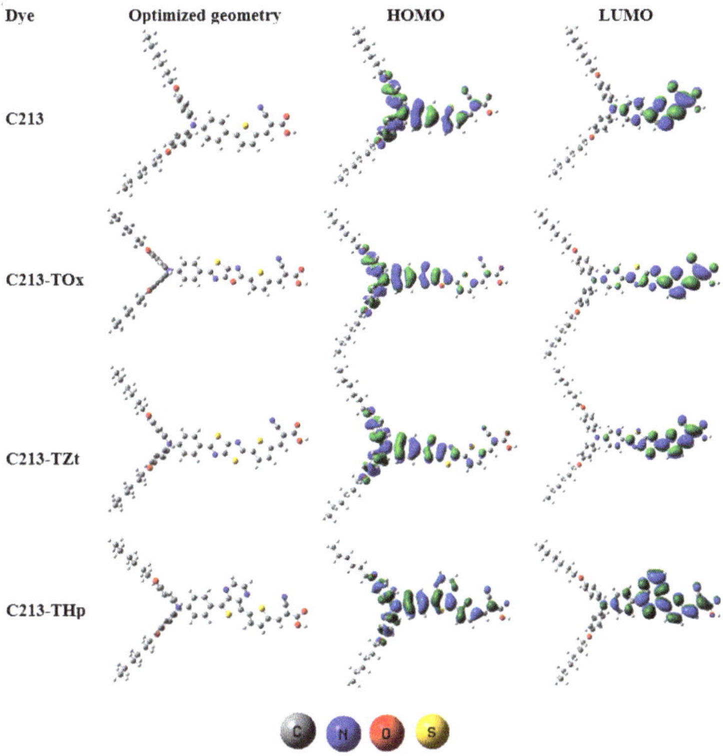

Fig. 7. FMOs and optimized geometries of C213, C213-Tox, C213-TZt and C213-THp obtained in gas phase with DFT/B3LYP/6-311G(d,p) level.

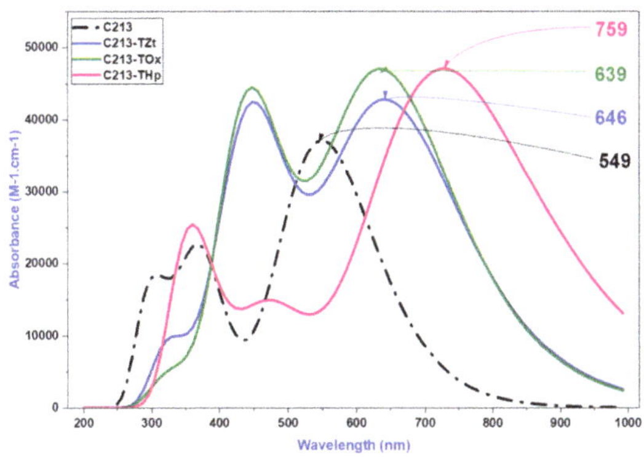

Fig. 8. UV-Vis absorption spectra of the investigated sensitizers in CFM solvent calculated by TD-DFT/MPW1PW91/6-311G (d,p) level.

Conclusion

In this paper, electronic geometry and optical absorption properties of a reference dye C213 and three designed dyes: C213-TOx, C213-TZt and C213-THp, after insertion of auxiliary acceptors into the D-π-A reference dye configuration, were examined by DFT (for geometry optimization and frontier molecular orbitals) and TD-DFT (for absorption spectra) theoretical methods.

Overall, the three designed dyes C213-TOx, C213-TZt and C213-THp have shown good results by narrow energies gap (E_{gap}) of 2,14; 2,10 and 1,94 eV and maximum wavelengths (λ_{max}) of 639, 646 and 759 nm respectively, which are favourable for photons absorption in the visible region.

Particularly, the C213-THp dye stands out with a small energy gap and red-shift absorption spectrum which can offer promising D-A_2-π-A_1 DSSC for a future experimental evaluation.

References

[1] The International Renewable Energy Agency (IRENA), "Renewable Energy Statistics 2025," 2025. Accessed: Oct. 06, 2025. [Online]. Available: https://www.irena.org/Publications/2025/Jul/Renewable-energy-statistics-2025

[2] V. Khare, P. Chaturvedi, and M. Mishra, "Solar energy system concept change from trending technology: A comprehensive review," *e-Prime - Advances in Electrical Engineering, Electronics and Energy*, vol. 4, p. 100183, Jun. 2023 https://doi.org/10.1016/J.PRIME.2023.100183

[3] D. Fernández-González, "A State-Of-The-Art Review on Materials Production and Processing Using Solar Energy," *Mineral Processing and Extractive Metallurgy Review*, Jan. 2025 https://doi.org/10.1080/08827508.2023.2243008

[4] B. O'Regan and M. Grätzel, "A low-cost, high-efficiency solar cell based on dye-sensitized colloidal TiO2 films," *Nature*, vol. 353, no. 6346, pp. 737–740, Oct. 1991 https://doi.org/10.1038/353737a0

[5] F. Rehman *et al.*, "Fourth-generation solar cells: a review," *Energy Advances*, vol. 2, no. 9, pp. 1239–1262, Jun. 2023 https://doi.org/10.1039/D3YA00179B

[6] K. Hara *et al.*, "Molecular Design of Coumarin Dyes for Efficient Dye-Sensitized Solar Cells," *J Phys Chem B*, vol. 107, no. 2, pp. 597–606, Jan. 2003 https://doi.org/10.1021/jp026963x

[7] Y. Wu and W. Zhu, "Organic sensitizers from D–π–A to D–A–π–A: effect of the internal electron-withdrawing units on molecular absorption, energy levels and photovoltaic performances," *Chem. Soc. Rev.*, vol. 42, no. 5, pp. 2039–2058, 2013 https://doi.org/10.1039/C2CS35346F

[8] M. P. Balanay and D. H. Kim, "Theoretical study on the correlations between dye–iodine interactions and open-circuit voltages in dyes containing furan and thiophene," *Comput Theor Chem*, vol. 1029, pp. 1–12, Feb. 2014 https://doi.org/10.1016/j.comptc.2013.12.005

[9] C. Zafer, B. Gultekin, C. Ozsoy, C. Tozlu, B. Aydin, and S. Icli, "Carbazole-based organic dye sensitizers for efficient molecular photovoltaics," *SOLAR ENERGY MATERIALS AND SOLAR CELLS*, vol. 94, no. 4, pp. 655–661, 2010 https://doi.org/10.1016/j.solmat.2009.11.014

[10] V. Flores-Morales, A. P. Villasana-Ruíz, I. Garza-Veloz, S. González-Delgado, and M. L. Martinez-Fierro, "Therapeutic Effects of Coumarins with Different Substitution Patterns," *Molecules*, vol. 28, no. 5, p. 2413, Mar. 2023 https://doi.org/10.3390/molecules28052413

[11] C. Faber, I. Duchemin, T. Deutsch, and X. Blase, "Many-body Green's function study of coumarins for dye-sensitized solar cells," *Phys Rev B*, vol. 86, no. 15, 2012 https://doi.org/10.1103/PhysRevB.86.155315

[12] P. R. Nitha, S. Soman, and J. John, "Indole fused heterocycles as sensitizers in dye-sensitized solar cells: an overview," *Mater Adv*, vol. 2, no. 19, pp. 6136–6168, 2021 https://doi.org/10.1039/d1ma00499a

[13] A. Mahmood, "Triphenylamine based dyes for dye sensitized solar cells: A review," *Solar Energy*, vol. 123, pp. 127–144, Jan. 2016 https://doi.org/10.1016/J.SOLENER.2015.11.015.

[14] S. N. Al-Ghamdi, H. A. Al-Ghamdi, R. M. El-Shishtawy, and A. M. Asiri, "Advances in phenothiazine and phenoxazine-based electron donors for organic dye-sensitized solar cells," *DYES AND PIGMENTS*, vol. 194, 2021 https://doi.org/10.1016/j.dyepig.2021.109638

[15] Z. S. Huang, H. Meier, and D. R. Cao, "Phenothiazine-based dyes for efficient dye-sensitized solar cells," *J Mater Chem C Mater*, vol. 4, no. 13, pp. 2404–2426, 2016 https://doi.org/10.1039/c5tc04418a

[16] M. Zhang *et al.*, "Unraveling the Pivotal Impacts of Electron-Acceptors on Light Absorption and Carrier Photogeneration in Perylene Dye Sensitized Solar Cells," *ACS Photonics*, vol. 1, no. 8, pp. 710–717, Aug. 2014 https://doi.org/10.1021/ph5001346

[17] T. Manifar and S. Rohani, "Synthesis and Analysis of Triphenylamine: A Review," *Can J Chem Eng*, vol. 82, no. 2, pp. 323–334, Apr. 2004 https://doi.org/10.1002/cjce.5450820213

[18] R. Li *et al.*, "Dye-Sensitized Solar Cells Based on Organic Sensitizers with Different Conjugated Linkers: Furan, Bifuran, Thiophene, Bithiophene, Selenophene, and Biselenophene," *The Journal of Physical Chemistry C*, vol. 113, no. 17, pp. 7469–7479, Apr. 2009 https://doi.org/10.1021/jp900972v

[19] Gaussian 09, Revision A.02, M. J. Frisch, G. W. Trucks, H. B. Schlegel, G. E. Scuseria, M. A. Robb, J. R. Cheeseman, G. Scalmani, V. Barone, G. A. Petersson, H. Nakatsuji, X. Li, M. Caricato, A. Marenich, J. Bloino, B. G. Janesko, R. Gomperts, B. Mennucci, H. P. Hratchian, J. V. Ortiz, A. F. Izmaylov, J. L. Sonnenberg, D. Williams-Young, F. Ding, F. Lipparini, F. Egidi, J. Goings, B. Peng, A. Petrone, T. Henderson, D. Ranasinghe, V. G. Zakrzewski, J. Gao, N. Rega, G. Zheng, W. Liang, M. Hada, M. Ehara, K. Toyota, R. Fukuda, J. Hasegawa, M. Ishida, T. Nakajima, Y. Honda, O. Kitao, H. Nakai, T. Vreven, K. Throssell, J. A. Montgomery, Jr., J. E. Peralta, F. Ogliaro, M. Bearpark, J. J. Heyd, E. Brothers, K. N. Kudin, V. N. Staroverov, T. Keith, R. Kobayashi, J. Normand, K. Raghavachari, A. Rendell, J. C. Burant, S. S. Iyengar, J. Tomasi, M. Cossi, J. M. Millam, M. Klene, C. Adamo, R. Cammi, J. W. Ochterski, R. L. Martin, K. Morokuma, O. Farkas, J. B. Foresman, and D. J. Fox, Gaussian, Inc., Wallingford CT, 2016.

[20] M. Li *et al.*, "Theoretical Study of WS-9-Based Organic Sensitizers for Unusual Vis/NIR Absorption and Highly Efficient Dye-Sensitized Solar Cells," *The Journal of Physical Chemistry C*, vol. 119, no. 18, pp. 9782–9790, May 2015 https://doi.org/10.1021/acs.jpcc.5b03667

Neural Network Predictions of the Thermophysical Properties of $ZnO+Fe_3O_4$/water Hybrid Nanofluids

Lingala Syam SUNDAR[a]

Department of Mechanical Engineering, College of Engineering, Prince Mohammad Bin Fahd University, Al-Khobar, 31952, Saudi Arabia

[a]sslingala@gmail.com

Keywords: Hybrid Nanofluids, Properties, Enhancement, Temperatures, Artificial Neural Network

Abstract. The thermophysical properties were estimated experimentally at veracious weight loadings, and temperatures of $ZnO+Fe_3O_4$/water hybrid nanofluids. Later the experimental data has been used for artificial neural network (ANN) to train the algorithm. In this study, the temperatures range, and weight concentration range used are varies from 20 °C to 80 °C, and 0% to 0.5%. Levenberg- Marquardt model was considered to train the ANN algorithm. Experimental outcome shows thermal conductivity is enhanced by 4.153%, and 8.689%, moreover, the viscosity is enhanced by 48.762%, and 15.806% at 0.5 wt%, and at 20 °C, and 80 °C, over the base fluid. The density is enhanced by 4.687%, and specific heat is decreased by 1.052% at 0.5 wt%, and at 20 °C, over base liquid. The ANN result shows that, for the trained data, the mean square error for thermal conductivity is 6.7951e-7, for viscosity is 0.00001, for density is 1.9428, and for specific heat is 2.7528 with a correlation coefficient of 0.9997, 0.9998, 0.9960, and 0.9944. Finally, from the ANN, the polynomial regression equation has been developed.

Introduction

Over the past decade, the increase of heat transmission via nanofluids has captivated numerous researchers. The thermophysical properties of the nanofluid is a crucial parameter for assessing its thermal performance. Esfe et al. [1] experiments on properties of nanofluids. Khanafer and Vafai [2] conducted an extensive investigations properties of nanofluids. Furthermore, several researchers examine use of nanofluids under various heat transfer applications [3,4]. In addition to these investigations, other review publications summarized the research conducted on the properties of various nanofluids [5]. Hemmat Esfe et al. [6] analyzed thermal conductivity (k_{nf}) of nanofluids including Al_2O_3 and formulated equation based on obtained data. Agarwal et al. [7] found k_{nf} increase of 40%, 27%, and 19% for water, ethylene glycol, and engine oil CuO nanofluids at a temperature of 2% vol.

Additionally, some researchers have estimated thermophysical properties of the hybrid nanofluids. TiO_2-CuO/C nanoparticles' k_{nf} and μ_{nf} in ethylene glycol were experimentally studied by Akilu et al. [8]. The results showed that the hybrid nanofluids' k_{nf} increased by 16.7% at 2.0% vol., the μ_{nf} increased by 80% in comparison to the base liquid. Hamid et al. [9] investigated viscosity and k_{nf} of TiO_2-SiO_2 hybrid nanofluids with water+ethylene glycol at different ratios of nanoparticle mixture. At temperatures between 30 and 70 °C, measurements were conducted. The authors discovered that at 70 °C, the viscosity of the 50:50 ratio was higher than that of the other ratios they had examined, while increase in k_{nf} was 13.8% and 11.9% for the 20:80 and 40:60 ratios, respectively.

Additionally, alternative ANN modeling techniques, including design of experiments and evolutionary algorithms, are employed to forecast and enhance various features of nanofluids. Komeilibirjandi et al. [10] considered ANN analysis for k_{nf} of CuO nanofluids data and found the

R-squared of 0.9862, and 0.9996 between the polynomial and ANN, and also found absolute deviations of 5.25% and 0.881%. Rostami et al. [11] noticed MSE of 2.4451e-05 at 0.6% vol. and at 50 °C through ANN analysis of (MWCNTs)-CuO/water nanofluid data, which is 30.38% higher against room temperature. From the MWCNT nanofluids, Ghasemi et al. [12] investigated the k_{nf} from ANN method. Safeci et al. [13] considered ANN to predict the k_{nf} of ZnO–TiO_2/EG nanofluid. Bahrami et al. [14] utilized ANNs for k_{nf} predictions of Fe–CuO/EG+water nanofluid. He et al. [15] used ANN ZnO-Ag (50%–50%)/water nanofluid and noticed absolute value of error is 0.0095, and performance of train is 1.6684e-05.

In this paper, it is used an artificial neural networks (ANNs) to predict properties of ZnO+Fe_3O_4/water hybrid nanofluids. Following the creation of experimental nanofluid data points, an approach is suggested to performance and correlation coefficient have been computed. A surface is then fitted to the experimental data using the fitting method, and the correlation coefficient and performance of this approach are determined. Lastly, the absolute levels of errors in the two approaches are compared.

Properties of hybrid nanofluids

Stable hybrid nanofluids were developed by dispersing the ZnO+Fe_3O_4 hybrid nanoparticles into water. The ZnO+Fe_3O_4 hybrid nanoparticles were prepared through chemical co-precipitation method. The reagent grade chemicals like zinc chloride, ferric chloride, ferrous chloride, and sodium hydroxide were purchased from Sigma-Aldrich chemicals, India. Distilled water was used as a base fluid. In 500 ml beaker, poured 100 ml water and then added 1 g of zinc chloride, 1 g ferrous chloride, and 2 g of ferric chloride, then stirred continuously for 45 mins. Full dissolve of chemicals in water, added 2 g of sodium hydroxide and stirred vigorously for 60 mins. Observe the formation of light black color precipitate. The precipitate was washed several times with water and then dried in an oven at 80 °C for 24 hrs.

Since, density of ZnO+Fe_3O_4 hybrid nanoparticles is not investigated, hence the weight concentrations were used for the preparation of hybrid nanofluids. The weight concentration percentages of 0.02%, 0.05%, 0.1%, 0.2%, and 0.5% hybrid nanofluids were prepared and used for the property's measurement. The stability of hybrid nanofluids were obtained by using SDBS surfactant. The k_{nf} of hybrid nanofluids were measured by using KD2 Pro instrument, USA, viscosity was measured with AND vibro-viscometer, Japan, density was measured with precision weight balance, and specific heat was measured with differential scanning calorimeter, USA.

Artificial neural network

The subsequent phase involves the development of the artificial neural network architecture [16]. Methodology employs a limited neuron in hidden layer, the error of trained network fails to satisfy the specified tolerance, the neuron count in hidden layer is incremented by one, followed by a repetition of the training cycle and performance assessment. This process continues until the performance is done. Levenberg–Marquardt method [17] was utilized in this study. Figure 1 illustrates optimal architecture of NN for predicting the attributes.

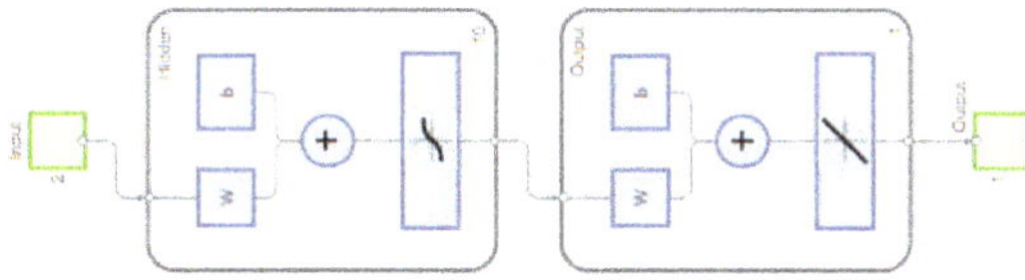

Fig 1. ANN architecture.

Nonetheless, it is beneficial to analyse network reaction in its whole. Regression analysis was used to assess the network's capacity for predicting conductivity. R^2 was measured to assess concordance between the trained network's estimations and the data. Additionally, various neural network architectures were evaluated based on their mean square errors (MSE), defined as follows:

$$\mathrm{MSE} = \frac{1}{n}\sum_{i=1}^{n}\left(y_{\mathrm{exp}} - y_{\mathrm{net}}\right)^2 \quad (1)$$

Where, y_{exp} is the experimental value, y_{net} is ANN values, n is the number of data points.

Results and discussion

Thermal conductivity of nanofluids

It is utilized experimental data to forecast the k_{nf} of nanofluid. Experimental measurements encompass various temperatures and weight loadings. Figure 2(a) illustrates the k_{nf} of nanofluids is relay on weight fractions at different temperatures. Measurements indicate that k_{nf} rises with elevated temperature and nanoparticle concentration. This increase may be attributed to heightened Brownian motion [5]. Influence of ϕ on k_{nf} is larger pronounced over temperature. Findings demonstrate that the incorporation of nanoparticles enhances k_{nf} of the base fluid [18]. Highest rise of k_{nf} was around 8.689% at 80 °C and a solid weight fraction of 0.5%. When comparing this result with the k_{nf} findings conducted by Esfe et al. [6], it is evident that they achieved the highest enhancement in k_{nf}. Consequently, it may be inferred that the base fluid significantly influences the thermal conductivity of nanofluids.

This picture illustrates that data sets were clustered. It suggests NN may accurately forecast the k_{nf} of nanofluids with minimal deviation from experimental data. The maximum errors, defined as the discrepancies over measured and projected data, for train, validation outputs, and test are 6.7951e-7, 8.385e-6, and 6.9468e-5, respectively. The training dataset is crucial for optimizing the performance of NN. Quality of data employed for training the network exceeds that the validation and test data. 75% of the data is designated for train the network. The data are randomly selected from the entire spectrum of experimental results derived from the ANN. The R^2 values given in Fig. 2(b) for the train, validation, test, and overall points are 0.99973, 0.99766, 0.99082, and 0.99601, respectively.

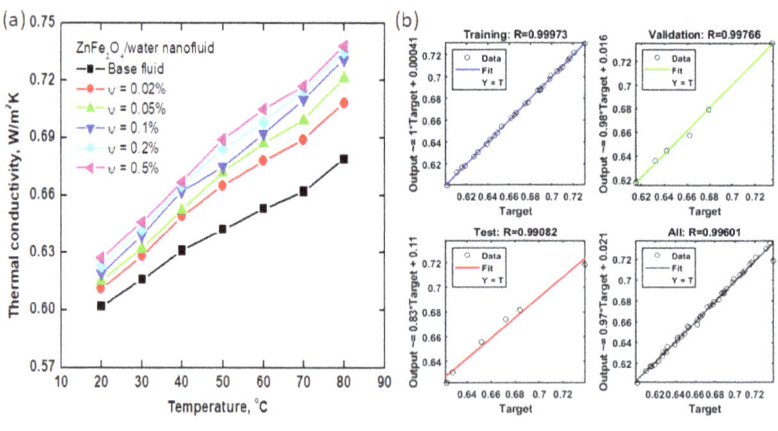

Fig 2: (a) Thermal conductivity of nanofluids at different temperatures, and weight loadings, and (b) the R^2 of thermal conductivity predicted by ANN algorithm.

Viscosity of nanofluids

Experimental viscosity measurements involve various temperatures and weight loadings. Figure 3(a) illustrates the viscosity of nanofluids relative to weight fractions at different temperatures. Measurements indicate that μ_{nf} rise with a reduction in temperature and rises with a rise in concentration. This increase may be attributed to heightened frictional resistance among the fluid layers. The influence of ϕ on μ_{nf} is more pronounced than that of temperature.

The findings demonstrate that the incorporation of nanoparticles enhances the viscosity of the base fluid. The highest increase in viscosity is around 48.762% at a temperature of 20 °C and a solid weight fraction of 0.5%. When comparing this result with that of Sundar et al. [17], it is evident that they achieved the highest increase in viscosity. Consequently, it may be inferred that the base fluid significantly influences the viscosity of nanofluid. It suggests that NN may accurately forecast the viscosity of nanofluids with minimal deviation from experimental data. The maximum errors, defined as the discrepancies between experimental and noted data for train, validation, test are 0.00001, 0.0001, and 0.0012.

The train data was crucial for optimizing performance of NN. Quantity of data employed for training the network exceeds over test data. 75% of the data is designated for network training. The data are randomly selected from the entire spectrum of experimental results derived from the ANN. The R^2 values given in Fig. 3(b) for the training outputs, validation, test, and overall data are 0.99984, 0.99972, 0.99544, and 0.99894, respectively.

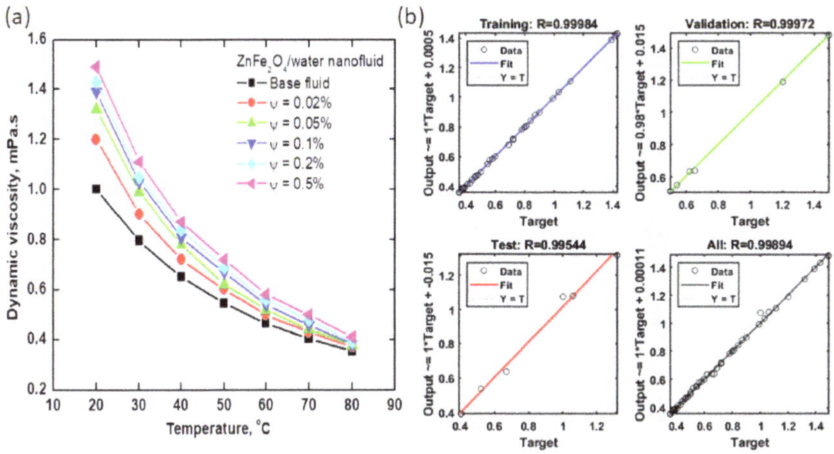

Fig 3: (a) Viscosity of nanofluids at different temperatures, and weight loadings, and (b) the R^2 of viscosity predicted by ANN algorithm.

Density of nanofluids

Experimental density measurements involve varying temperatures and weight loadings. Figure 4(a) illustrates the viscosity of nanofluids as a function of weight fractions at different temperatures. Measurements indicate that density increases as temperatures decrease and as concentrations increase. This increase may be attributed to the rise in the fluid's weight. The influence of solid weight fraction on density is more pronounced than that of temperature. The findings suggest that the incorporation of nanoparticles enhances the density of the base fluid. The highest increase in viscosity is around 4.687% at T = 20 °C and a solid weight percentage of 0.5%. When findings of Sundar et al. [5], this result indicates the highest rise in density.

Materials Research Proceedings 64 (2026) 722-728 https://doi.org/10.21741/9781644904091-90

Density for the training, validation, and test data sets were analyzed. This picture illustrates that clustered data. Suggests that NN may accurately forecast the density of nanofluids with minimal deviation from experimental data. The maximum errors, defined as the discrepancies over measured and projected data for train, validation, and test are 1.9428, 8.2544, and 10.2309, respectively. The R^2 values seen in Fig. 4(b) for the training outputs, validation, test, and overall data are 0.99605, 0.99163, 0.98409, and 0.99296, respectively.

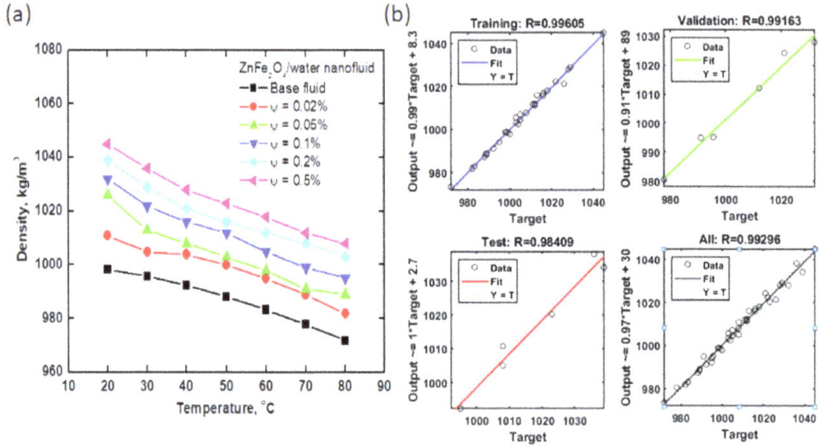

Fig 4. (a) The density of nanofluids at different temperatures, and weight loadings, and (b) the R^2 of density predicted by ANN algorithm.

Specific heat of nanofluids

Experimental specific heat measurements involve varying temperatures and weight loadings. Figure 5(a) illustrates the specific heat of nanofluids as a function of weight fractions at different temperatures. Measurements indicate that specific heat diminishes with a reduction in temperature and an increase in concentration. This increase may result from the nanofluid's optimal heat absorption capacity. The findings demonstrate that the incorporation of nanoparticles enhances the specific heat of the base fluid. The highest reduction in specific heat is around 1.025% at T = 20 °C and a solid weight fraction of 0.5%. The maximum errors, defined as the discrepancies over measured and projected data for train, validation, test data are 2.7528, 5.1176, and 3.6158, respectively.

The training dataset is crucial for optimizing the performance of NN. The data are randomly selected from the entire spectrum of experimental results derived from the ANN. The R^2 values presented in Fig. 5(b) for the training outputs, validation, test, and overall data are 0.99437, 0.98572, 0.99708, and 0.9941, respectively.

The developed equations are given below.

$$k_{nf} = 0.5757 + 0.00167T + 0.061\phi \quad (2)$$

$$\mu_{nf} = 1.4065 - 0.0175T + 0.3209\phi \quad (3)$$

$$\rho_{nf} = 1025.36 - 0.543T + 61.70\phi \quad (4)$$

$$C_{p,nf} = 4163.9 + 0.369T - 63.68\phi \quad (5)$$

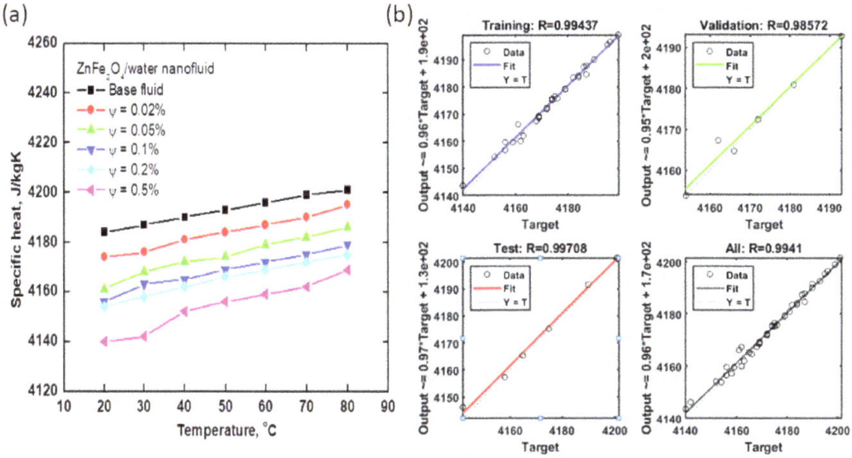

Fig 5. (a) The specific heat of nanofluids at different temperatures, and weight loadings, and (b) the R^2 of specific heat predicted by ANN algorithm.

Conclusion

The study analyzed the thermophysical properties of water dispersed $ZnO+Fe_3O_4$/water hybrid nanofluids under various concentrations and temperatures, simultaneously, the data was predicted with ANN. The data of thermal conductivity rises to 4.153%, and 8.689%, meanwhile, the viscosity is raised by 48.762%, and 15.806% at 0.5 wt%, and at 20 °C, and 80 °C. The density is also going higher by 4.687%, and specific heat is lowered over 1.052% at 0.5 wt%, and at 20 °C. The ANN for the trained data, the mean square error for thermal conductivity is 6.7951e-7, for viscosity is 0.00001, for density is 1.9428, and for specific heat is 2.7528 with a correlation coefficient of 0.9997, 0.9998, 0.9960, and 0.9944. Study gives, the NN tool predicts the perfect data with measured values.

References

[1] M.H. Este, A.A.A. Arani, M. Rezaie, W.M. Yan, A. Karimipour, Experimental determination of thermal conductivity and dynamic viscosity of Ag-MgO/water hybrid nanofluid, Int. Comm. Heat Mass Transfer. 66, 19 (2015). https://doi.org/10.1016/j.icheatmasstransfer.2015.06.003

[2] K. Khanafer, K. Vafai, A critical synthesis of thermophysical characteristics of nanofluids, Int. J. Heat Mass Transfer. 54, 4410 (2011). https://doi.org/10.1016/j.ijheatmasstransfer.2011.04.048

[3] L.S. Sundar, T.T. Akanaw, M.K. Singh, A.C.M. Sousa, Thermosyphon solar water heating system with Cu/water nanofluid and wire coil configurations: Efficiency, energy, economic, environmental, and heat transfer study, Environ. Prog. Sustainable Energy. e13648 (2021). https://doi.org/10.1002/ep.13648

[4] A.M. Alkaibi, K.V.V. Cahndra Mouli, L.S. Sundar, Experimental investigation of heat transfer and effectiveness of employing water and ethylene glycol mixture based Fe3O4 nanofluid in a shell and helical coil heat exchanger, Thermal Science and Engineering Progress, 40, 101739 (2023). https://doi.org/10.1016/j.tsep.2023.101739

[5] L.S. Sundar, Synthesis and characterization of hybrid nanofluids and their usage in different heat exchangers for an improved heat transfer rates: A critical review, Engineering Science and Technology, An Int. J. 44, 101468 (2023). https://doi.org/10.1016/j.jestch.2023.101468

[6] M.H. Esfe, S. Saedodin, O. Mahian, S. Wongwises, Thermal conductivity of Al2O3/water nanofluids: Measurement, correlation, sensitivity analysis, and comparisons with literature reports. J. Therm. Anal. Calorim. 117, 675 (2014). https://doi.org/10.1007/s10973-014-3771-x

[7] R. Agarwal, K. Verma, N.K. Agrawal, R.K. Duchaniya, R. Singh, Synthesis, characterization, thermal conductivity and sensitivity of CuO nanofluids, Applied Thermal Engineering. 102, 1024 (2016). https://doi.org/10.1016/j.applthermaleng.2016.04.051

[8] S. Akilu, A.T. Baheta, K.V. Sharma, Experimental measurements of thermal conductivity and viscosity of ethylene glycol-based hybrid nanofluid with TiO2-CuO/C inclusions, J. Molecular Liquids. 246, 396 (2017). https://doi.org/10.1016/j.molliq.2017.09.017

[9] K.A. Hamid, W.H. Azmi, M.F. Nabil, R. Mamat, Experimental investigation of nanoparticle mixture ratios on TiO2-SiO2 nanofluids heat transfer performance under turbulent flow. Int. J. Heat Mass Transfer. 118, 617 (2018). https://doi.org/10.1016/j.ijheatmasstransfer.2017.11.036

[10] A. Komeilibirjandi, A.H. Raffiee, A. Maleki et al: Thermal conductivity prediction of nanofluids containing CuO nanoparticles by using correlation and artificial neural network. J. Therm. Anal. Calorim. 139, 2679 (2020). https://doi.org/10.1007/s10973-019-08838-w

[11] S. Rostami, D. Toghraie, B. Shabani et al: Measurement of the thermal conductivity of MWCNT-CuO/water hybrid nanofluid using artificial neural networks (ANNs). J. Therm. Anal. Calorim. 143, 1097 (2021). https://doi.org/10.1007/s10973-020-09458-5

[12] A. Ghasemi, M. Hassani, M. Goodarzi, M. Afrand, S. Manafi, Appraising influence of COOH-MWCNTs on thermal conductivity of antifreeze using curve fitting and neural network, Physica A. 514, 36 (2019). https://doi.org/10.1016/j.physa.2018.09.004

[13] M.R. Safaei, H. Hajizadeh, A. Afrand, C. Qi, H. Yarmand, N.W.B.M. Zulkifli, Evaluating the effect of temperature and concentration on the thermal conductivity of ZnO-TiO2/EG hybrid nanofluid using artificial neural network and curve fitting on experimental data, Physica A. 519, 209 (2019). https://doi.org/10.1016/j.physa.2018.12.010

[14] M. Bahrami, M. Akbari, S.A. Bagherzadeh, A. Karimipour, M. Afrand, M. Goodarzi, Develop 24 dissimilar ANNs by suitable architectures & training algorithms via sensitivity analysis to better statistical presentation: Measure MSEs between targets & ANN for Fe-CuO/Eg-water nanofluid, Physica A. 519, 159 (2019). https://doi.org/10.1016/j.physa.2018.12.031

[15] W. He, B. Ruhani, D. Toghraie, N. Izadpanahi, N.N. Esfahani, A. Karimipour, M. Afrand, Using of artificial neural networks (ANNs) to predict the thermal conductivity of zinc oxide-silver (50%-50%)/water hybrid Newtonian nanofluid, Int. Comm. Heat and Mass Transfer. 116, 104645 (2020). https://doi.org/10.1016/j.icheatmasstransfer.2020.104645

[16] S. Haykin, Adaptive filters. Signal Process. Mag., 6 (1999).

[17] D.W. Marquardt, An Algorithm For Least-Squares Estimation of Nonlinear Parameters. J. Soc. Ind. Appl. Math. 11, 431, (1963). https://doi.org/10.1137/0111030

[18] L.S. Sundar, E.V. Ramana, M.P.F. Graça, M.K. Singh, A.C.M. Sousa, Nanodiamond-Fe3O4 Nanofluids: Preparation and measurement of viscosity, electrical and thermal conductivities, Int. Comm. Heat Mass Transfer. 73, 62, (2016). https://doi.org/10.1016/j.icheatmasstransfer.2016.02.013

Synthesis of Composite Membranes using Melamine Modified Zeolitic Imidazolate Framework for CO_2/CH_4 Separation

Abdul Fattah[1], Zaib Jahan[2,*], Tayyba Noor[1], Noreen sher Akbar[2], Muhammad Irshad Baig[3,4], Nawal Noshad[1], Muhammad Bilal Khan Niazi[5,6]

[1]School of Chemical and Materials Engineering, National University of Sciences and Technology, Islamabad 44000, Pakistan

[2]Department of Chemical Engineering, College of Engineering, Prince Mohammad Bin Fahd University, P. O Box 1664, Al Khobar 31952, Saudi Arabia

[3]Department of Material Sciences and Engineering, King Fahd University of Petroleum and Minerals, Dhahran, Saudi Arabia

[4]Interdisciplinary Research Center for Membranes and Water Security, King Fahd University of Petroleum and Minerals, Dhahran 31261, Saudi Arabia

[5]Department of Chemical Engineering, King Fahd University of Petroleum and Minerals (KFUPM), 31261, Dhahran, Saudi Arabia

[6]Interdisciplinary Research Center for Refining & Advanced Chemicals, King Fahd University of Petroleum and Minerals (KFUPM), 31261, Dhahran, Saudi Arabia

*zjahan@pmu.edu.sa

Keywords: Biogas, Gas Separation, Membrane Technology, Mixed Matrix Membranes, ZIF-8

Abstract. With fossil fuels being the primary source of energy with currently growing demands, it is important to shift towards clean, renewable energy sources such as biogas that has the potential to fulfill partial energy needs. However, biogas needs to be stripped of CO_2 to increase its methane content for it to be used as an energy source. Membrane technology for the separation of CO_2 from methane is extremely promising because of its higher separation efficiency without the generation of any secondary waste. In this study, different concentrations i.e. 1, 2 and 3 wt. % of melamine zeolitic imidazolate framework (ZIF-8) are incorporated into cellulose acetate (CA) as an additive to form mixed matrix membranes (MMMs), while polysulfone (PSF) is used as the support. The mechanical strength, permeance and selectivity of the membranes are observed to increase with increasing ZIF-8 concentrations, as well as with feed pressure. 3 wt. % ZIF-8 membranes exhibit maximum enhancement of 8.63% in CO_2 permeability at 8 bar, while the highest selectivity obtained at same conditions is 14.7.

Introduction

The dependency on fossil fuels to fulfill the growing energy demands has led to greater challenges like global warming and climate change [1]. This necessitates the urgency to transition to clean, alternative energy sources that are economical and can provide long term energy security [2]. Biogas is an essential renewable energy source which is formed by the fermentation of organic matter by a series of anaerobic microbial oxidation-reduction reactions. Biogas comprises 45-70% CH_4 (methane) and 30-55% carbon dioxide (CO_2), along with traces of water vapor, nitrogen, hydrogen, hydrogen sulfide, siloxanes etc. [3], [4]. Appropriate use of biogas technology has several benefits which make it feasible to be used to fulfill partial energy needs [5]. While the CO_2 present in raw biogas is inert, it largely reduces its calorific value. Therefore, biogas ought to be stripped of CO_2 to increase the amount of methane in order to be used as an energy source [4].

Several technologies such as solvent scrubbing, cryogenic gas separation, adsorption and membrane separation have been used to upgrade raw biogas to biomethane [6].

Membrane technology used to upgrade biogas exploits the different particle sizes and chemical affinity of different molecules [7] and is driven by a pressure gradient across a permeable, porous membrane. Membrane separation has been attracting the attention of researchers as well as industrialists because of its practical advantages such as simple equipment requirements i.e. they are relatively compact and easy to operate whilst having high adaptability and scale up flexibility, they require comparatively lower capital and operating investments, and have a high energy efficiency without the production of toxic, secondary waste [8], [9]. Different types of membranes are used for biogas separation. While membrane technology is extremely promising, it is important to develop membranes that are efficient i.e. they have a high gas permeability without compromising on its selectivity, whilst also being economical, preferably with less pressure requirements. Mixed matrix membranes (MMMs) achieve this by incorporating a dispersed phase, i.e. molecular sieve materials that exhibit high gas separation, into the polymer layer which acts as the continuous phase, exploiting the mechanical properties and relative economic processabilty of polymers. The structure, strength and surface chemistry of the inorganic filler materials used as the dispersed phase allows it to enhance the polymer it is added into [10]–[12]. The resulting MMMs have superior separation characteristics, above the Robeson upper bound [13], thus overcoming the tradeoff between the two important parameters i.e. selectivity and permeability [14].

Metal organic frameworks (MOFs) have been gaining significant recognition as fillers in MMMs in many applications such as gas separation. MOFs have a better compatibility with organic membranes as the organic linkers in MOFs have a greater affinity with polymeric chains [15]. MOFs are tunable in terms of their pore volumes, pore shapes and surface areas, which can either improve the permeability or selectivity, while the chemical mutability of MOFs can improve separation performance by enhancing adsorption [16]. ZIF-8 is a group of MOFs that has been gaining popularity in biogas separation amongst researchers because of its high molecular sieving effects, CO_2/CH_4 selectivity and the high CO_2 permeability. ZIF-8 also has a facile synthesis besides great chemical and thermal stability [17]. It possesses uniform pores (0.34 nm) and a strong affinity for CO_2, which makes it a good candidate for CO_2/CH_4 separation. When placed into CA matrices, ZIF-8 enhances permeability through additional gas transport pathways while retaining selectivity via size exclusion and preferential adsorption of CO_2 molecules [18]. Li et al. [19] used graphene oxide (GO) and ZIF-8 as fillers in ODPA-TMPDA polyimide to synthesize MMMs for separation of CO_2 and CH_4. The formed MMMs had a high mechanical strength, and the selectivity and CO_2 permeability increased by 28% and 60%, respectively. In another study [20], ZIF-8 was used as filler in matrimid to form MMMs with ZIF-8 loadings up to 60 wt/wt and the permeability of different gases and gas mixtures of CO_2/CH_4 and H_2/CO_2 were tested. The permeability increased with ZIF-8 concentration up till 40 wt/wt after which they started to decrease. Higher loadings of ZIF-8 also resulted in MMMs that were harder to manage as they became less flexible, highlighting the importance of optimizing the concentration of filler material used. In a study by Jonnalagedda and Kuncharam [21], MMMs were fabricated using ZIF-8 and ZIF-8 functionalized with amine in polysulfone for the separation of CO_2 and CH_4. The gas permeation test gave the highest permeability i.e. 25.4 Barrer at 10 wt. % ZIF-8 polysulfone, while 15 wt. % amine-ZIF-8 polysulfone had the highest selectivity ratio i.e. 13.5.

In this study, cellulose acetate (CA) solution is dispersed with different concentrations of melamine zeolitic imidazolate framework-8 (ZIF-8) to form mixed matrix membranes (MMM). These membranes are supported on a polysulfone (PSF) membrane layer, via the dip coating method. The mechanical strength, permeance and selectivity of pure PSF/CA composite membranes and membranes with 1, 2 and 3 wt. % ZIF-8 are tested for the separation of biogas.

The performance of the prepared membranes is assessed in terms of CO_2 separation as a function of ZIF-8 concentration and feed pressure.

Methodology

Cellulose acetate (CA) and acetone, the polymer for membrane fabrication and the solvent it is dissolved in, respectively, are acquired from Sigma Aldrich, while the polysulfone (PSF) sheet, the porous support for the composite membrane, is acquired from Alfa Laval. 3 wt. % of cellulose acetate (CA) is dissolved in acetone to form a polymer matrix. The CA/acetone solution is modified with in-house synthesized melamine substituted zeolitic imidazolate framework-8 (ZIF-8) nanoparticles in concentrations of 1, 2 and 3 wt. %. ZIF-8 is introduced to the solution under continuous stir succeeded by sonication of the suspension in a probe sonicator to ensure uniform dispersion and reduce nanoparticle agglomeration. These mixed matrix membranes are supported by a pretreated selective layer of the PSF sheet, owing to its mechanical strength and stability under operating conditions. The thin film composite membranes are fabricated using a custom-designed dip-coating apparatus. The PSF membrane is dipped in the CA/acetone –ZIF-8 casting solution at the rate of 1.5 cm/s. Defect free, stable composite membranes are obtained after overnight drying in the fume hood after which it is placed in an oven at 60°C for 2 hours for enhanced structural stability.

The prepared membranes are subjected to characterization techniques such as Scanning Electron Microscopy (SEM), Fourier transform infrared spectroscopy (FTIR) and X-ray Diffraction (XRD), to affirm their morphology and structural integrity. Surface morphology and cross-sectional structure of the composite membranes are investigated by SEM. SEM plays a significant role in understanding the uniformity of the selective layer, the distribution of the nanoparticles, and the total structural integrity of the membrane. SEM imaging also offers insights into the defects of the membrane, i.e., cracks or agglomerations, which can impact gas separation performance. XRD is used to analyze the crystalline structure of the composite membranes and the retention of crystallinity by the incorporation of ZIF-8 in cellulose acetate. XRD spectra is generated using an X-ray diffractometer equipped with Cu Kα radiation ($\lambda = 1.5406$ Å) under the following conditions: 40 kV, 30 mA scanning range ($\theta = 5°–40°$) with a step of 0.02°. FTIR gives a molecular fingerprint to the materials, measuring infrared absorption spectra to give a specific vibrational transition. Dried membrane samples are prepared as KBr pellets and analyzed using a Nicolet 6700 spectrophotometer with a wavenumber range of 4000–400 cm^{-1} and 4 cm^{-1} resolution. The SHIMADZU AGS-X series precision ultimate tensile tester is used to measure the ultimate tensile strength of the membranes. The ASTM standard D882-02 is used to test each sample at 5 mm/min strain rate. Samples of pure CA, and 1, 2, and 3 wt. % ZIF-8 in CA/acetone membranes had their stress-strain behavior examined.

Results and Discussion

Characterization of Membranes

The cross-sectional SEM images of CA membrane and membranes with 1, 2 and 3 wt. % loadings ZIF-8 are given in Figure 1. It can be observed that the cross-section shows a strong bond between the dense CA layer and porous PSF support with constant thickness and no defects for the pure CA membrane. The structure of this form provides high mechanical stability while maintaining efficient ultrafiltration performance. The incorporation of ZIF-8 nanoparticles in CA improved porosity and facilitated additional solvent transport pathways, enhancing the membrane permeability and selectivity without losing structural integrity. Increase of ZIF-8 content till 2 wt. % led to more visible ZIF-8 particles homogeneously distributed throughout the CA layer. Some particle agglomeration is seen at ZIF-8 concentration of 3 wt. % as larger crystalline domains within the CA matrix. However, the overall membrane structure remained intact with strong interaction between the CA and PSF. The membrane thickness progressively increases with the concentration of ZIF-8 i.e. loadings of 1, 2 and 3 wt. %. The baseline thickness of the unmodified polymer membrane i.e. of pure CA has the lowest membrane thickness of 0.0294 cm. The thickness of the membrane increased to 0.0297 and 0.0300 cm with 1 and 2 wt. % addition of ZIF-8, respectively. This higher thickness of the membrane can be credited to the contribution of ZIF-8 to the overall membrane bulk due to the higher volume fraction of ZIF-8. The membrane with 3 wt. % ZIF-8 has the greatest thickness of 0.0306 cm. This large increase is rational by the accumulation of ZIF-8 within the polymer to create a denser, thicker membrane structure. However, a much greater increase in thickness may also correspond to a higher resistance to mass transfer [22].

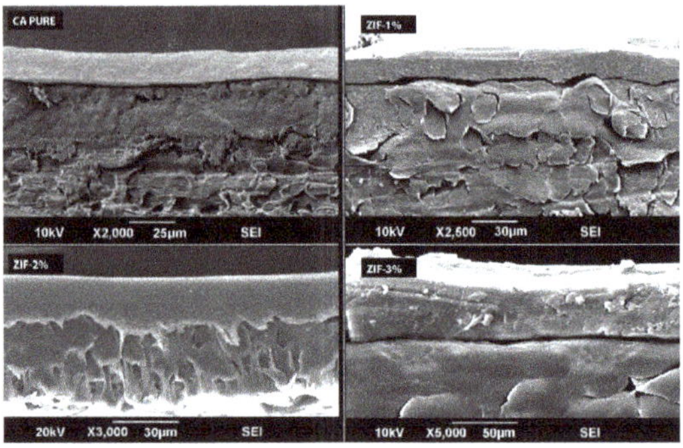

Fig.1. SEM images (cross-sectional) of pure membrane and membranes with 1 wt. %, 2 wt. %, and 3 wt. % of ZIF-8

Figure 2 presents the XRD plots of the crystalline structure of pure membrane and its composites. The CA membrane is observed to have a broad amorphous hump around 2θ = 10°–25° in the XRD profile, consistent with its predominantly amorphous nature. The absence of sharp peaks confirm that the pure membrane does not have a crystalline structure. However, the crystallinity of the membranes increases gradually with the increasing amount of ZIF-8 nanoparticles. The incorporation and conservation of the intrinsic crystalline framework of the additive is evident by the distinct peaks in the CA matrix associated with ZIF-8. This structural enhancement can enhance separation performance, thermal stability and mechanical strength of the membranes.

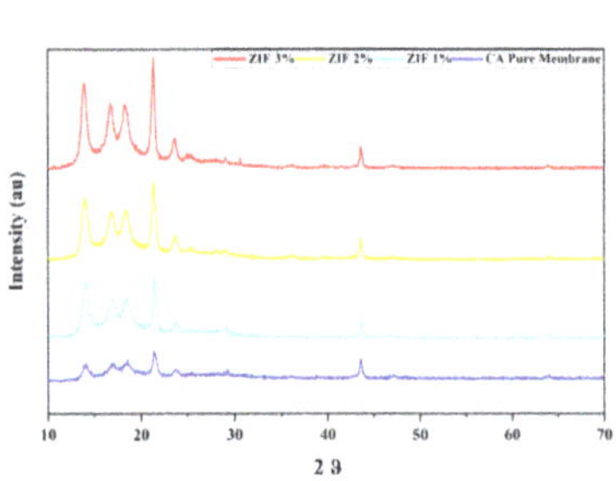

Fig.2. XRD peaks of pure membrane and membranes with different ZIF-8 concentrations

The FTIR spectra for the pure membrane and membranes with ZIF-8 additive at varying concentrations (1, 2 and 3 wt. %) is presented in Figure 3. The graph shows the characteristic peaks of the functional groups present in CA membranes. Characteristic absorption bands corresponding to the N-H stretch (<3300 cm^{-1}) and other associated functional groups are observed in the pure CA membrane. New peaks and shifts are observed with the introduction of ZIF-8. For 1% ZIF, a clear O-H stretching band at 3200 – 3500 cm^{-1} is observed in the spectrum. The change observed here suggests that the ZIF-8 nanoparticles interact with the hydroxyl groups in the CA membrane matrix. A sharp O-H stretching peak is observed with 2 wt. % ZIF-8 nanoparticles incorporated into the composite. The spectrum of 3 wt. % ZIF-8 composite membrane has a more pronounced N-H bending (~1500 cm^{-1}) and broadening of the O-H stretching region suggesting hydrogen bonding between CA and ZIF-8. The integration of ZIF-8 in the polymer structure is evident from the emergence of new peaks. Addition of ZIF-8 in membranes is also found to have a higher surface area via BET [23]. This additional surface area in comparison to pure membranes offers enhanced adsorption of CO_2, therefore improving CO_2 permeability.

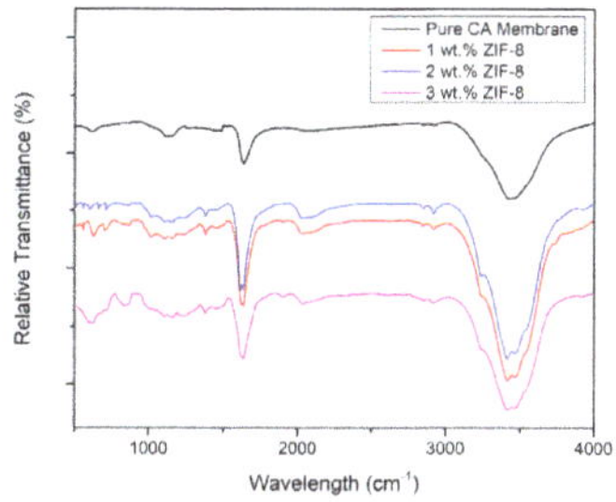

Fig.3. FTIR spectra of pure membrane and membranes with different ZIF-8 concentrations

CO_2 Permeability and Selectivity

Permeability of both the gases increases with progressive increase in ZIF-8 concentration. However, CO_2 has a much greater permeability than that of CH_4 at all concentrations. The CO_2 permeability enhancement can be observed in Figure 4, where it exhibits maximum enhancement of 8.63% at 8 bar from pure CA to 3 wt. %, while the permeability of CH_4 increases by merely 3.43% at the same pressure, indicating that ZIF-8 serves to increase CO_2 diffusion across membranes. The highest CO_2 permeability of 7.88 Barrer is found to be at 3 wt. % at 8 bar, which is much higher than CH_4 permeabilities. This difference is also reflected in the high selectivity that preferentially lets CO_2 molecules permeate, over the CH_4 molecules. Data demonstrated in Figure 5 shows a positive correlation between ZIF-8 concentration and selectivity under all tested pressures. The selectivity increases steadily with ZIF-8 concentration up to 3 wt. %, with the highest selectivity of 14.7 at 8 bar.

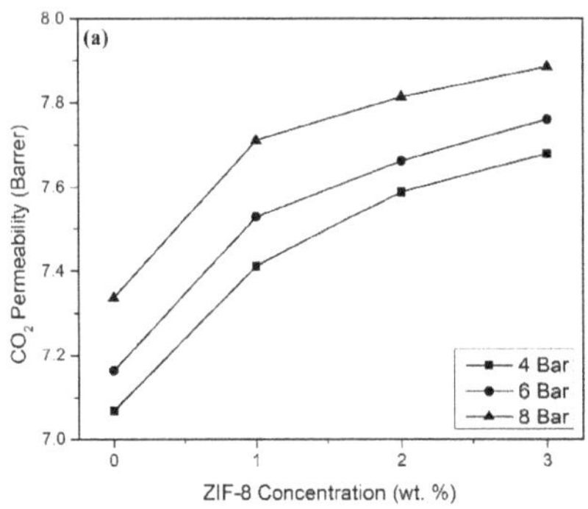

Fig. 4. CO_2 permeability with respect to ZIF-8 concentration

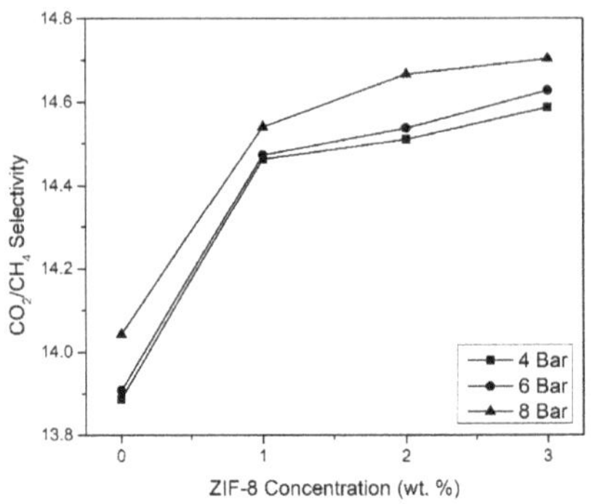

Fig. 5. CO_2/CH_4 selectivity with respect to ZIF-8 concentration

This improvement is attributed to the open structure, high surface area, imidazolate linkers of ZIF-8 and its molecular sieving effect, which selectively enhances CO_2 adsorption and diffusion as the extra free volume provides extra transport pathways, while CH_4 molecules are relatively hindered, thus increasing the overall selectivity. Although both gases impact the membrane surface at the same time, CO_2 preferentially penetrates due to its stronger sorption and diffusivity than CH_4. However, benefits observed may be offset by agglomeration of the ZIF-8 above the range studied due to excessive loading. While such sort of aggregation may decrease uniformity and inhibit gas transport by blocking diffusion pathways, ZIF-8 appears to be optimally compatible with the polymer matrix within the range studied i.e. (1 to 3 wt. %), over which no such adverse effects are observed.

Feed pressure has been shown to correspond to increasing CO_2 permeabilities for all membranes as shown in Figure 6, peaking at 8 bar. CO_2 permeability rises from 7.07 Barrer for the pure CA membrane to 7.34 Barrer from 4 to 8 bar, and from 7.68 to 7.88 Barrer for 3 wt. % ZIF-8 membranes. Lower values of pure CA membrane permeability for CH_4 are first observed at about 0.51 Barrer at 4 bar, which increases up to 0.52 at 8 bar, while for 3 wt. % it increases from 0.526 to 0.536 Barrer, at 4 and 8 bar, respectively. The modest rise in these data implies that for the pure CA membrane the permeability is not significantly affected by pressure relative to ZIF-8 incorporation. Figure 7 demonstrates the CO_2/CH_4 selectivity with respect to feed pressure. The selectivity of all membranes lies between 13.89 and 14.7 over feed pressures of 4-8 bar, feed pressure having little impact on selectivity. Pure CA membrane has the lowest selectivity.

Materials Research Proceedings 64 (2026) 729-738
https://doi.org/10.21741/9781644904091-91

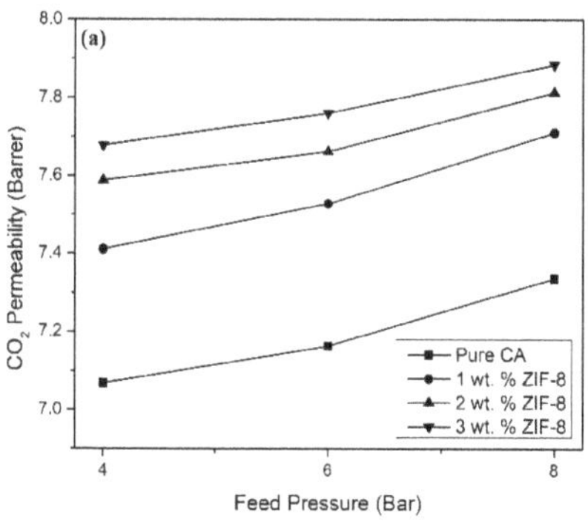

Fig. 6. CO_2 permeability with respect to feed pressure

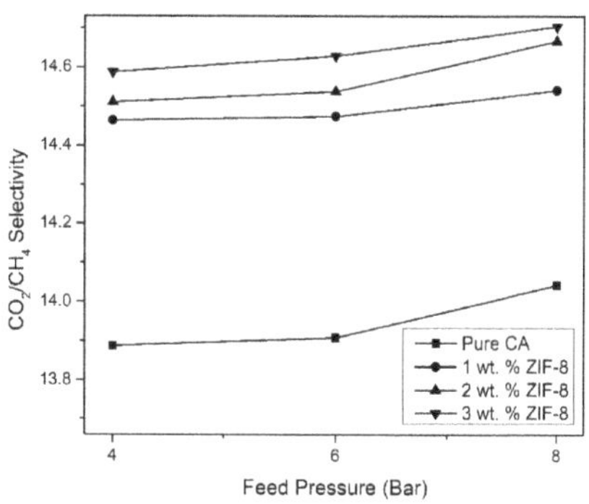

Fig. 7. CO_2/CH_4 selectivity with respect to feed pressure

The interaction between the filler's amine group and CO_2 allows a facilitated transport mechanism as the feed pressure increases [24]. Selectivity is slightly more pronounced at higher pressures because increased interaction between CO_2 molecules and ZIF-8 framework, facilitates transport. These permeability and selectivity results for pure CA membranes and its ZIF-8 composites lie below the Robeson upper bound [13]. While there are studies in which ZIF-8 is used as an additive and gives permeability/ selectivity results such that they lie very close to or exceed the Robeson upper bound [25]–[27], a lot of these works involve either a different material or much higher ZIF-8 loadings e.g. up to 30-40 wt. %. This affirms the use of ZIF-8 as an effective filler and suggests that higher concentrations of ZIF-8 may further increase selectivity and bring results of the membranes formed in this study closer to the Robeson's upper bound.

Mechanical Testing

The mechanical strength of CA membranes is enhanced as the ZIF-8 concentration increases. The stress values for pure CA membrane are the lowest as demonstrated in Figure 8, indicating the lowest tensile strength and weaker resistance to mechanical deformation. Stress values are enhanced slightly by incorporating 1 wt. % ZIF-8 due to better load distribution within the membrane, while the stress values are further improved by 2 wt. % ZIF-8, which illustrates the reinforcing effect of the filler. The highest stress levels attained at 3 wt. % ZIF-8 nanoparticles across all strain percentages indicate the optimal concentration for maximum mechanical performance. The lowest Young's modulus is exhibited by the pure CA membrane, revealing its flexible nature and resistance to deformation. Introduction of ZIF-8 significantly improves stiffness, more so with the increase in ZIF-8 concentration. The membrane exhibits its highest Young's modulus at 3 wt. % ZIF-8, which corresponds to the maximum stiffness due to the effective interaction between the polymer matrix and ZIF-8. ZIF-8's role as a structural reinforcement agent for applications requiring stiffness is underscored by this progressive enhancement.

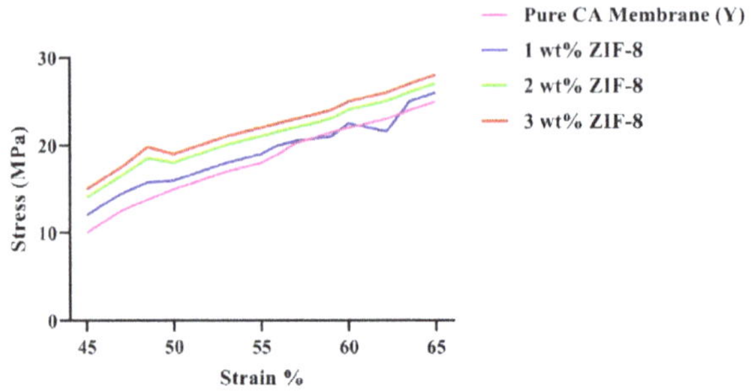

Fig. 8. Stress-strain analysis

Conclusion

Utilization of fossil fuels to fulfill the growing demand of energy poses to have significant consequences to the environment. This propels us to look into alternative energy sources such as biogas. However, biogas needs to be stripped of its impurities, majorly CO_2, for it to be used as a source of energy. Amongst the different technologies used for this purpose, membrane technology has shown to have significant potential because of its ability to achieve a higher degree of separation with several other benefits. Mixed matrix membranes (MMMs) in particular have several advantages over other membrane types. In this study, cellulose acetate (CA) membranes are enhanced with melamine covalently linked ZIF-8 nanoparticles in concentrations of 1, 2 and 3 wt. % over feed pressures of 4, 6 and 8 bars. A maximum CO_2 permeability enhancement of 8.63% is observed at 3 wt. % and 8 bar, indicating a rising trend with both ZIF-8 nanoparticle concentration and feed pressure. Increase in both these parameters also results in the rise of CO_2/CH_4 selectivity. The greater increase in CO_2 permeability in comparison to that of CH_4 is ascribed to ZIF-8's molecular sieving effect, that selectively enhances CO_2 adsorption and diffusion. The synthesized membranes are also subjected to mechanical testing and characterization techniques including SEM, XRD and FTIR.

References

[1] K. A. Alhassan, B. T. Abdullahi, and M. M. Shah, "A review on biogas production as the alternative source of fuel," *J. Appl. Adv. Res.*, no. March, pp. 61–65, 2019 https://doi.org/10.21839/jaar.2019.v4i2.266

[2] A. B. Tasmaganbetov, Z. Ataniyazov, Z. Basshieva, A. U. Muhammedov, and A. Yessengeldina, "World practice of using biogas as alternative energy," *Int. J. Energy Econ. Policy*, vol. 10, no. 5, pp. 348–352, 2020 https://doi.org/10.32479/ijeep.9805

[3] R. Muñoz, L. Meier, I. Diaz, and D. Jeison, "A review on the state-of-the-art of physical/chemical and biological technologies for biogas upgrading," *Rev. Environ. Sci. Biotechnol.*, vol. 14, no. 4, pp. 727–759, 2015 https://doi.org/10.1007/s11157-015-9379-1

[4] I. Angelidaki *et al.*, *Biogas upgrading: Current and emerging technologies*, 2nd ed. Elsevier Inc., 2019. doi: 10.1016/B978-0-12-816856-1.00033-6

[5] M. Balat and H. Balat, "Biogas as a renewable energy sourcea review," *Energy Sources, Part A Recover. Util. Environ. Eff.*, vol. 31, no. 14, pp. 1280–1293, 2009 https://doi.org/10.1080/15567030802089565

[6] I. Ullah Khan *et al.*, "Biogas as a renewable energy fuel – A review of biogas upgrading, utilisation and storage," *Energy Convers. Manag.*, vol. 150, no. July, pp. 277–294, 2017 https://doi.org/10.1016/j.enconman.2017.08.035

[7] S. Basu, A. L. Khan, A. Cano-Odena, C. Liu, and I. F. J. Vankelecom, "Membrane-based technologies for biogas separations," *Chem. Soc. Rev.*, vol. 39, no. 2, pp. 750–768, 2010 https://doi.org/10.1039/b817050a

[8] P. Gkotsis, P. Kougias, M. Mitrakas, and A. Zouboulis, "Biogas upgrading technologies – Recent advances in membrane-based processes," *Int. J. Hydrogen Energy*, vol. 48, no. 10, pp. 3965–3993, 2023 https://doi.org/10.1016/j.ijhydene.2022.10.228

[9] W. Tomczak, M. Gryta, M. Daniluk, and S. Żak, "Biogas Upgrading Using a Single-Membrane System: A Review," *Membranes (Basel).*, vol. 14, no. 4, pp. 1–21, 2024 https://doi.org/10.3390/membranes14040080

[10] M. A. Aroon, A. F. Ismail, T. Matsuura, and M. M. Montazer-Rahmati, "Performance studies of mixed matrix membranes for gas separation: A review," *Sep. Purif. Technol.*, vol. 75, no. 3, pp. 229–242, 2010 https://doi.org/10.1016/j.seppur.2010.08.023

[11] T. T. Moore, R. Mahajan, D. Q. Vu, and W. J. Koros, "Hybrid Membrane Materials Comprising Organic Polymers with Rigid Dispersed Phases," *AIChE J.*, vol. 50, no. 2, pp. 311–321, 2004 https://doi.org/10.1002/aic.10029

[12] T. S. Chung, L. Y. Jiang, Y. Li, and S. Kulprathipanja, "Mixed matrix membranes (MMMs) comprising organic polymers with dispersed inorganic fillers for gas separation," *Prog. Polym. Sci.*, vol. 32, no. 4, pp. 483–507, 2007 https://doi.org/10.1016/j.progpolymsci.2007.01.008

[13] L. M. Robeson, "The upper bound revisited," *J. Memb. Sci.*, vol. 320, no. 1–2, pp. 390–400, 2008 https://doi.org/10.1016/j.memsci.2008.04.030

[14] J. Dechnik, J. Gascon, C. J. Doonan, C. Janiak, and C. J. Sumby, "Mixed-Matrix Membranes," *Angew. Chemie - Int. Ed.*, vol. 56, no. 32, pp. 9292–9310, 2017 https://doi.org/10.1002/anie.201701109

[15] Y. Zhang, J. Sunarso, S. Liu, and R. Wang, "Current status and development of membranes for CO2/CH4 separation: A review," *Int. J. Greenh. Gas Control*, vol. 12, pp. 84–107, 2013 https://doi.org/10.1016/j.ijggc.2012.10.009

[16] P. Silva, S. M. F. Vilela, J. P. C. Tomé, and F. A. Almeida Paz, "Multifunctional metal-organic frameworks: From academia to industrial applications," *Chem. Soc. Rev.*, vol. 44, no. 19, pp. 6774–6803, 2015 https://doi.org/10.1039/c5cs00307e

[17] Y. Cheng *et al.*, "Advanced Porous Materials in Mixed Matrix Membranes," *Adv. Mater.*, vol. 30, no. 47, pp. 1–20, 2018 https://doi.org/10.1002/adma.201802401

[18] J. Caro, "Are MOF membranes better in gas separation than those made of zeolites?," *Curr. Opin. Chem. Eng.*, vol. 1, no. 1, pp. 77–83, Oct. 2011 https://doi.org/10.1016/J.COCHE.2011.08.007

[19] W. Li, S. A. S. C. Samarasinghe, and T. H. Bae, "Enhancing CO2/CH4 separation performance and mechanical strength of mixed-matrix membrane via combined use of graphene oxide and ZIF-8," *J. Ind. Eng. Chem.*, vol. 67, pp. 156–163, 2018 https://doi.org/10.1016/j.jiec.2018.06.026

[20] M. J. C. Ordoñez, K. J. Balkus, J. P. Ferraris, and I. H. Musselman, "Molecular sieving realized with ZIF-8/Matrimid® mixed-matrix membranes," *J. Memb. Sci.*, vol. 361, no. 1–2, pp. 28–37, 2010 https://doi.org/10.1016/j.memsci.2010.06.017

[21] A. Jonnalagedda and B. V. R. Kuncharam, "Investigation of ZIF-8, amine-modified ZIF-8 and polysulfone based mixed matrix membranes for CO2/CH4 separation," *J. Appl. Polym. Sci.*, vol. 140, no. 45, pp. 1–11, 2023 https://doi.org/10.1002/app.54650

[22] A. Rasekh and A. Raisi, "Electrospun nanofibrous polyether-block-amide membrane containing silica nanoparticles for water desalination by vacuum membrane distillation," *Sep. Purif. Technol.*, vol. 275, p. 119149, Nov. 2021 https://doi.org/10.1016/J.SEPPUR.2021.119149

[23] A. U. Khan, T. Noor, N. Iqbal, N. Zaman, and Z. Hussain, "CO2 adsorption study of the zeolite imidazolate framework (ZIF-8) and its g-C3N4 composites," *J. Mater. Sci.*, vol. 58, no. 9, pp. 3947–3959, 2023 https://doi.org/10.1007/s10853-023-08253-5

[24] Q. Xin *et al.*, "SPEEK/amine-functionalized TiO2 submicrospheres mixed matrix membranes for CO2 separation," *J. Memb. Sci.*, vol. 467, pp. 23–35, Oct. 2014 https://doi.org/10.1016/J.MEMSCI.2014.04.048

[25] Z. Li *et al.*, "In-situ grown NH2-ZIF-8 with optimized polymers compatibility and rigidity for enhanced CO2/CH4 separation performance in mixed matrix membranes," *J. Memb. Sci.*, vol. 735, no. June, 2025 https://doi.org/10.1016/j.memsci.2025.124522

[26] N. H. Suhaimi, Y. F. Yeong, C. W. M. Ch'ng, and N. Jusoh, "Tailoring CO2/CH4 separation performance of mixed matrix membranes by using ZIF-8 particles functionalized with different amine groups," *Polymers (Basel).*, vol. 11, no. 12, pp. 1–19, 2019 https://doi.org/10.3390/polym11122037

[27] M. M. Rajpure, R. B. Mujmule, U. Kim, and H. Kim, "Mixed matrix membranes for selective gas separation of H2/CH4 and CO2/CH4 via fabricating high-performing double-ligand ZIF-8 with cellulose acetate-based polymer," *Chemosphere*, vol. 374, no. January, 2025 https://doi.org/10.1016/j.chemosphere.2025.144183

Additive Manufacturing Technologies for Energy Applications: A Review

Yassar E. Mukhrish[1,a] *, Meshal Alsaiari[1,b], Faramarz Djavanroodi [1,c], Muhammad Imran[1,d], Mushtaq Khan[1,e]

[1]Mechanical Engineering Department, College of Engineering, Prince Mohammed Bin Fahd University, 31952 Al-Khobar, Kingdom of Saudi Arabia

[a]Yaseer.essa@gmail.com, [b]mishalts9@gmail.com, [c]fdjavanroodi@pmu.edu.sa, [d]mkhan8@pmu.edu.sa, [e]mkhan7@pmu.edu.sa

Keywords: Additive Manufacturing, 3D Printing, Direct Ink Writing, Energy Storage, Thermal Energy Storage, Batteries, Supercapacitor

Abstract. The Additive Manufacturing (AM) pathway creates hierarchical porous structures with pore sizes within the nano to micro scale range supported by greater precision in the making of energy storage devices, providing ion transport and, consequently, shorter diffusion paths. This review focuses on AM technologies such as direct ink writing, stereolithography, and selective laser melting for lithium-ion batteries, sodium-ion batteries, and thermal energy storage systems. Key achievements include fully 3D printed lithium-ion batteries with an areal energy density of 4.45 mAh/cm^2 and optimized thermal storage with a 57% reduction in solidification time. Critical challenges include formulating inks suitable for printing, multi-scale manufacturing, scale-up, integration of components, and quality standards. The outlook includes computational design optimization, emerging battery chemistries, and 4D printed self-healing systems. Through materials development and process optimization, next-generation AM will change energy storage from standardized to application-specific systems for portable electronics, electric vehicles, and grid-level storage.

Introduction

The subsequent section will systematically discuss leading advanced manufacturing technology types and applications in utilization for energy storage systems by examining the role played by each method in terms of achieving high quality energy devices and allowing production of articles with particular geometries and custom materials. The worldwide transition to ward sustainable energy systems and the exponential growth of portable electronics have created an unprecedented demand for high-performance energy storage devices having higher specific energies, high specific power, and longer cycle-life. Conventional battery and supercapacitor electrodes usually were designed by slurry casting and calendaring; these methods offer poor architectural control, thus limiting charge transport, energy storage, and power balance. This doctor blade method, dating back to the 1940s, could inherently offer a good thickness control together with scalability. However, the conductive additives and binders tend to agglomerate together, therefore blocking the transport of ions and electrons. [1], [2].

There are studies that confirm 3D-printed electrodes with their controlled porosity and hierarchical structure can present a better performance compared to those produced by conventional processes. For instance, the 3D-printed interdigitated lithium-ion battery electrodes have managed to be charged at more than 4.45 mAh/cm^2 of areal energy density with very good rate capability, whereas the additively manufactured supercapacitor electrodes have been reported to show a specific capacitance of 1329 mF/cm^2. [1], [2].

Screening Methods

A systematic review of literature on AM for energy storage was carried out in accordance with the PRISMA Statement guidelines, sourcing publications repeatedly from platforms such as Scopus, Web of Science, IEEE Xplore, ScienceDirect, and Google Scholar [3]. These keywords, revolving around AM technologies and devices for storing energy, steered the authors towards articles written in English that were peer-reviewed and edited between 2017 and 2025. The number of articles gradually decreased through a series of weeding-out activities, whereby only 87 studies fully satisfied the scrutiny of the review process. The majority of articles were excluded given a lack of technical details, poor links to AM use of interest, purely qualitative schemes of the ideas, or publication issues. Weightily, over 42% of articles had recently started pouring into research and have been published during 2023-2025 to date, indicating healthier attention being given to active research within the said discipline [4].

Review Findings

Additive Manufacturing Technologies for Energy Storage

A variety of AM processes have emerged as promising technologies to manufacture the energy st orage components. Each has its specific capabilities, resolution, and material compatibility requir ements that primarily serve to dictate its proper use for a battery or thermal energy storage applic ation.

Direct Ink Writing (DIW)

One of the most versatile extrusion based techniques in additive manufacturing is direct ink writing (DIW), which allows precise dispensing of viscoelastic inks to form hierarchical electrode architecture that is characterized by controlled porosity and high mass loading for energy storage capability. DIW has been deemed as a highly adaptable AM method to produce electrodes for energy storage. Viscoelastic inks are extruded consistently through a needle by pneumatic pressure or mechanical displacement so that a layer thickness varying from 1-500 μm could strictly be handled. The inks for DIW rather display Herschel-Bulkley rheological behavior, with which increasing on viscosities (0.1-10^3 Pa·s) should enable them to be injected through the needle effectively, whereas a high yield stress interacts with thixopy for shape retention of the former structures. They have been used to produce electrode constructs for $Li_4Ti_5O_{12}$, $LiFePO_4$, graphene, carbon nanotubes, and MXenes with mass loading greater than 10 mg/cm^2 and controlled porosity. It has been stated for the record, that Wei and his co-senior authors accomplished the First Three-Dimensional Printed Lithium-Ion Battery, accomodating the best electrochemical anode, cathode, and separator, plus packaging [5].

Fused Deposition Modeling (FDM)

Fused deposition modeling (FDM) assists in relatively rapid production and lower cost fabrication of energy storage constituting components by means of thermoplastic filament extrusion. The difficulty lays in electrical conductivity restrictions that clarify the complexity of the solutions, such as the use of conductive composite filaments in overcoming the performance obstacles. FDM extrudes heated thermoplastic filaments layer by layer with resolutions between 50 and 200 μm and provides for fast, cheap, and large scale fabrication . However, inherently polymeric insulation limits energy storage applications . The latest approaches use conductive composite filaments with graphite, graphene, or carbon nanotubes embedded in polymer matrices . Reyes et al. developed fully 3D-printed lithium-ion coin cells that could stably cycle 100 times at 20-mA/g FDM is limited by four challenges: low electrical conductivity of the composites , poor filler-matrix-interface bonding , anisotropic mechanical and electrical properties dependent on printing orientation , and restrictively low conductive filler loading [6].

Stereolithography (SLA)

The technique uses photopolymerisation chemistry that can generate better resolution and precision compared to extrusion based methods and allowed a profound fault proof technique of making high aspect ratio electrodes with extreme dimensional accuracy and surface finish for Li-ion battery. SLA employs UV or visible light for photopolymerization of liquid resin and is capable of amazingly good resolution from 5 to 40 microns, outstandingly high precision, and smooth finish compared to extrusion. For energy storage, SLA processes composite resins, disperse the active materials, and, upon thermal or chemical post-treatment, remove the binders and densify the structures [7]. Among others, they have produced complex 3D architectures of $Li_4Ti_5O_{12}$ and $LiFePO_4$ that demonstrate high rate capability and cyclability.

Selective Laser Melting (SLM) and Powder Bed Fusion (PBF)

Selectively applied laser melting and powder bed fusion processes make use of high-energy beams to transform loose powder metallic and ceramic materials into dense three-dimensional structures particularly suited to current collectors, thermal management components, and specialized battery parts. SLM provides 40-200 μm resolution, with mechanical properties at 90-95% of conventional parts. SLM is used for making 3D current collectors and porous scaffolds with an electrochemical application, though the high thermal input induces residual stresses that have to be supported [8].

Inkjet and Aerosol Jet Printing

Inkjet and aerosol jet printing methods let precise deposition of functional liquids and inks in micrometer scales, providing excellent spatial resolution and excellent material use for the development of electrode films of thin film batteries and complicated supercapacitor architecture. The inkjet technique deposits liquid droplets between 50 and 100 μm in diameter at a resolution of 5-200 μm, the ink requiring viscosity in the range of 8-25 mPa·s, while the surface tension is from 28-32 N/m [9]. Thin-film battery electrodes, solid oxide fuel cell components, and supercapacitor electrodes have been inkjet printed. The aerosol jet printing, being considered a superior technique yet to be fully commercialized, atomizes the functional inks and directs them through a carrier gas to deposition heads that realize about 10 μm resolution for a variety of substrates. The AJP defines porosity architecture in lithium-ion battery electrodes to realize 400% increase in specific capacity over solid electrodes [10]. Figure 1 shows the additive manufacturing technology and applications for energy storage.

Materials Research Proceedings 64 (2026) 739-745 https://doi.org/10.21741/9781644904091-92

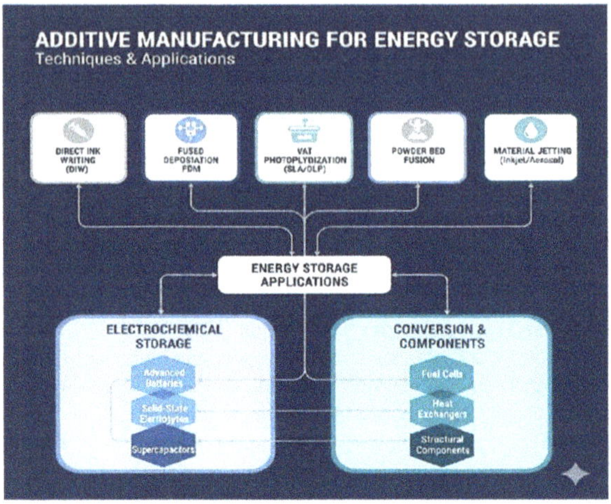

Fig. 1. AM Techniques and Applications for energy storage.

Lithium-Ion Battery Electrode Materials and Architectures

Additive manufacturing has introduced revolutions in lithium-ion battery electrode design, and brought out three-dimensional hierarchical structures that in turn have significantly enhanced ion transport, increased active surface areas, and improved electrochemical performance of various electrode materials [11]. Carbon-based electrodes are an example of this: Sun et al. fabricated 3D graphene scaffolds mixed with silver nanowire networks and $Li_4Ti_5O_{12}$, which exhibited a capacity of 121 mAh/g at 10C, retaining 95% of their capacity after 100 cycles [12] . The hierarchical design endowed the system with excellent electrical conductivity from the silver nanowire, while porous graphene acted as both a mechanical support and an ion transport medium. Oxide electrodes with composition such as $LiFePO_4$, $Li_4Ti_5O_{12}$, and $LiMn_{0.21}Fe_{0.79}PO_4$ show better rate capability. Hu et al. have fabricated nanocrystal $LiMn_{0.21}Fe_{0.79}PO_4$ electrodes, which display a capacity of 108.45 mAh/g at 100C owing to high porosity and favorable 3D structure kinetics [7], [13] . One of the major break-throughs was carried out by Wei et al., wherein complete lithium-ion batteries (anode, cathode, and separator) were 3D printed and sold to the public, charged, and were rated at 4.45 mAh/cm^2 at 0.14 mA/cm^2 and could manage without conventional joining process steps such as drying, electrolyte filling, and clamping [13].

Materials Research Proceedings 64 (2026) 739-745 https://doi.org/10.21741/9781644904091-92

Table 1. Energy Storage Materials Processed Using Additive Manufacturing

S.N	Material Used	Comments	Refe	Years
1	$Li_4Ti_5O_{12}$ (Lithium Titanium Oxide)	Electrodes fabrication was done using the material. Anode material for lithium-ion batteries was successfully printed using the DIW process.	[7] [14] [15]	2021 2023 2023
2	$Li_4Ti_5O_{12}$ with 3D graphene scaffolds and silver nanowire (AgNWs)	Hierarchical design was used to incorporate high electronic conductivity from silver nanowires, while the porous graphene acts as a support structure and medium for the transport of ions.	[16]	2024
3	$Li_4Ti_5O_{12}$ (Lithium Titanium Oxide)	Photo-curable matrices are used with composite resins containing dissolved active materials. After the printing process, polymer binders are removed by either thermal or chemical post-treatment, which densifies the structure, thereby forming electrodes with complex 3D architectures capable of high rate capability and cyclability.	[6] [17]	2024 2023
4	$LiFePO_4$ (Lithium Iron Phosphate)	Upon proper method implementation, the cathode electrodes provided high uniform porosity and excellent electrochemical performances in lithium-ion batteries.	[18] [19]	2023 2025
5	$LiFePO_4$ (Lithium Iron Phosphate)	High in intricacy for power as well as finer types of electrodes, stereolithography fabricated three-dimensional electrodes that sit higher in power handling and stability on cycling than those typical types.	[18] [20]	2023 2022

Thermal Energy Storage Systems

Additive manufacturing technology has a big impact on thermal energy storage systems, as it offers the production of a wide range of thermal management parts, heat exchangers with the best topology, and the development of shape-retaining composite phase change materials that are not restricted by the traditional manufacturing methods. Thermal energy storage is enhanced by additive manufacturing that deals with the low heat conductivity of the phase change materials via topology-optimized fins, lattice-structure impregnation, and form-stable composite PCMs. Ge et al. designed topology-optimized fins for shell-and-tube systems with selective laser melting while cutting down the solidification time by 57.1%, which led to a roughly uniform temperature distribution and enhanced heat transfer rates [21]. Economics, however, states maximizing is useful as long as the unit price is less than six times that of PCM. Moon et al., in contrast, fabricated devices of very high power density with alloy aluminum-silicon fins 17 times more effectively conducted thermal resistances than that of conventional casting and three times more effectively conducted thermally convection resistances, thus, improving power density four times to 0.58 W/cm^3 [22].

Form-stable composite PCMs prepared by additive manufacturing undertake molten salt leakage while trying to retain thermal conductivity. Employing selective laser sintering, Nofal et al [23]. synthesized expanded graphite-paraffin wax composites with thermal conductivity 0.83-0.92 W/(m·K) and latent heat 150-156 kJ/kg, thus successfully demonstrating AM fabrication of functional composite PCMs utilized for thermal management of lithium-ion batteries [24].

Conclusions and Future Perspectives

Additive manufacturing offers great freedom for electrode design, material composition, andgeometry of the device across technologies such as DIW, FDM, SLA, SLM, and inkjetprinting that can apply to lithium-ion batteries, sodium-ion batteries, lithium-sulfur batteries, supercapacitors, and thermal energy storage systems. The major challenge here lies in the ink development for extrusion-based AM [6] . The inks must meet those requirements that are often contradictory: being low in viscosity when extruded, else they clog the nozzle, and raising in yield strength once deposited to support the deposition itself, whereas polymeric binders and solvents usually enable the printing but generally degrade electrochemical performance.

Future research ought to focus on commercial inks for specific AM processes, thus rendering obsolete the currently prevailing in-house recipes, which lack reproducibility and scalability. Especially for energy-storage AM, proceeding toward well-characterized standardized inks with known rheological, thermal, and electrochemical properties is of utmost importance. Thereon lie further methods: the first being template-assisted, and the post-printing infiltration of the constituent material, whereby the structure is made first and then loaded with the active material, thus allowing the use of materials that would otherwise not be directly printable.

Reference:

[1] Q. Liu, R. Ge, C. Li, Q. Li, and Y. Gan, "Digital design and additive manufacturing of structural materials in electrochemical and thermal energy storage systems: a review," 2023, *Taylor and Francis Ltd.* doi: 10.1080/17452759.2023.2273949.

[2] D. M. Soares *et al.*, "Additive Manufacturing of Electrochemical Energy Storage Systems Electrodes," May 01, 2021, *John Wiley and Sons Inc.* doi: 10.1002/aesr.202000111.

[3] M. J. Page *et al.*, "The PRISMA 2020 statement: an updated guideline for reporting systematic reviews" https://doi.org/10.1136/bmj.n71.

[4] A. Liberati *et al.*, "The PRISMA statement for reporting systematic reviews and meta-analyses of studies that evaluate healthcare interventions: explanation and elaboration," *BMJ*, vol. 339, Jul. 2009 https://doi.org/10.1136/BMJ.B2700.

[5] L. Zeng, S. Ling, D. Du, H. He, X. Li, and C. Zhang, "Direct Ink Writing 3D Printing for High-Performance Electrochemical Energy Storage Devices: A Minireview," Nov. 14, 2023, *John Wiley and Sons Inc.* doi: 10.1002/advs.202303716.

[6] A. De, B. Ramasubramian, S. Ramakrishna, and V. Chellappan, "Advances in Additive Manufacturing Techniques for Electrochemical Energy Storage," *Adv. Mater. Technol.*, vol. 9, no. 4, p. 2301439, Feb. 2024 https://doi.org/10.1002/ADMT.202301439.

[7] H. Yang, Z. Feng, X. Teng, L. Guan, H. Hu, and M. Wu, "Three-dimensional printing of high-mass loading electrodes for energy storage applications," *InfoMat*, vol. 3, no. 6, pp. 631–647, Jun. 2021 https://doi.org/10.1002/INF2.12181.

[8] T. B. Freeman *et al.*, "Advanced Materials and Additive Manufacturing for Phase Change Thermal Energy Storage and Management: A Review," *Adv. Energy Mater.*, vol. 13, no. 24, Jun. 2023 https://doi.org/10.1002/AENM.202204208.

[9] P. Yang and H. J. Fan, "Inkjet and Extrusion Printing for Electrochemical Energy Storage: A Minireview," *Adv. Mater. Technol.*, vol. 5, no. 10, p. 2000217, Oct. 2020 https://doi.org/10.1002/ADMT.202000217.

[10] L. J. Deiner and T. L. Reitz, "Inkjet and Aerosol Jet Printing of Electrochemical Devices for Energy Conversion and Storage," *Adv. Eng. Mater.*, vol. 19, no. 7, p. 1600878, Jul. 2017 https://doi.org/10.1002/ADEM.201600878.

[11] M. S. Saleh, J. Li, J. Park, and R. Panat, "3D printed hierarchically-porous microlattice electrode materials for exceptionally high specific capacity and areal capacity lithium ion batteries," *Addit. Manuf.*, vol. 23, pp. 70–78, Oct. 2018 https://doi.org/10.1016/J.ADDMA.2018.07.006.

[12] C. Sun, Y. Wang, M. D. McMurtrey, N. D. Jerred, F. Liou, and J. Li, "Additive manufacturing for energy: A review," *Appl. Energy*, vol. 282, p. 116041, Jan. 2021 https://doi.org/10.1016/J.APENERGY.2020.116041.

[13] A. Jain, T. Raj, A. Sahai, and R. S. Sharma, "Advancements in additive manufacturing for electrochemical energy storage devices," *Proceedings of the Institution of Mechanical Engineers, Part N: Journal of Nanomaterials, Nanoengineering and Nanosystems*, 2025 https://doi.org/10.1177/23977914251361025.

[14] G. F. Hawes, S. Rehman, Y. Rangom, and M. A. Pope, "Advanced manufacturing approaches for electrochemical energy storage devices," *International Materials Reviews*, vol. 68, no. 3, pp. 323–364, 2023 https://doi.org/10.1080/09506608.2022.2086388.

[15] S. Chandrasekaran, D. Lin, Y. Li, and M. A. Worsley, "Aerogels, additive manufacturing, and energy storage," *Joule*, vol. 7, no. 5, pp. 866–883, May 2023 https://doi.org/10.1016/J.JOULE.2023.03.021.

[16] Y. C. Goswami and S. Begzaad, "Printable supercapacitors and their printing technologies: A review," *Energy Storage*, vol. 6, no. 4, p. e666, Jun. 2024 https://doi.org/10.1002/EST2.666.

[17] X. Tian and K. Zhou, "Additive Manufacturing of Energy Storage Devices," *Additive Manufacturing Materials, Functionalities and Applications*, pp. 51–83, Jan. 2023 https://doi.org/10.1007/978-3-031-04721-3_2.

[18] N. Fonseca *et al.*, "3D Printing-Enabled Design and Manufacturing Strategies for Batteries: A Review," *Small*, vol. 19, no. 50, Dec. 2023 https://doi.org/10.1002/SMLL.202302718.

[19] J. P. Mwizerwa, S. Mukase, K. Usman, J. Li, and S. A. Rasaki, "3D-printed grid $LiFePO_4$@ACB cathode for Li-ion batteries with high energy and areal density," *Academia Engineering*, vol. 2, no. 3, Jul. 2025 https://doi.org/10.20935/ACADENG7818.

[20] M. I. Barnawi, "Investigation of Electroplating 4D Printed Antenna & Developing 3D Printed Lithium Batteries," 2022.

[21] R. Ge, Q. Li, C. Li, and Q. Liu, "Evaluation of different melting performance enhancement structures in a shell-and-tube latent heat thermal energy storage system," *Renew. Energy*, vol. 187, pp. 829–843, Mar. 2022 https://doi.org/10.1016/J.RENENE.2022.01.097.

[22] H. Moon, N. Miljkovic, and W. P. King, "High power density thermal energy storage using additively manufactured heat exchangers and phase change material," *Int. J. Heat Mass Transf.*, vol. 153, p. 119591, Jun. 2020 https://doi.org/10.1016/J.IJHEATMASSTRANSFER.2020.119591.

[23] M. Nofal, Y. Pan, and S. Al-Hallaj, "Selective Laser Sintering of Phase Change Materials for Thermal Energy Storage Applications," *Procedia Manuf.*, vol. 10, pp. 851–865, Jan. 2017 https://doi.org/10.1016/J.PROMFG.2017.07.071.

[24] A. Jo *et al.*, "All-Printed Paper-Based Micro-supercapacitors Using Water-Based Additive-Free Oxidized Single-Walled Carbon Nanotube Pastes," *ACS Appl. Energy Mater.*, vol. 4, no. 12, pp. 13666–13675, Dec. 2021 https://doi.org/10.1021/ACSAEM.1C02345/SUPPL_FILE/AE1C02345_SI_002.MP4.

Improved Detection of Carcinoembryonic Antigen using a Black Phosphorus based SPR Sensor

Abdelhak EL HIRI[1,a,*], Hamza SABIR[2,b], Abdellatif EL AZRAK[1,c]

[1]Laboratory of Materials, Energy and Intelligent Systems, Faculty of Science and Technology of Mohammedia, Hassan II University of Casablanca, Morocco

[2]Laboratory of Engineering Sciences and Biosciences, Faculty of Sciences and Technology of Mohammedia, Hassan II University of Casablanca, Morocco

[a]abdelhak.elhirietu@etu.univh2c.ma, [b]hamza.sabir@fstm.ac.ma, [c]abdellatif.elazrak@fstm.ac.ma

Keywords: 2D Materials, Black Phosphorus, Surface Plasmon Resonance, Biosensor, Sensitivity, FOM

Abstract: This work outlines the design and analysis of performance for a unique new multilayer surface plasmon resonance (SPR) biosensor designed with black phosphorus (BP) layers to detect Carcinoembryonic Antigen (CEA). The proposed architecture for the sensor includes a BK7 prism, gold (Au), aluminum oxide (Al_2O_3), and multiple BP layers (P = 0, P = 1, P = 2), and has been thoroughly analyzed using the Transfer Matrix Method (TMM) simulation technique at a wavelength of 633nm. The findings indicate that as the number of BP layers increases from zero to two, the sensitivity of the sensor increases from 148.84 deg/RIU (P = 0) to 167.30 deg/RIU(P = 2); however, with this increased sensitivity comes a corresponding increase in the full width at half maximum (FWHM) of the SPR response from 4.09 to 4.89, which results in a reduction of the figure of merit (FOM) from 36.39 RIU^{-1} (P = 0) to 34.25 RIU^{-1} (P = 2). This research shows that BP has a real chance of being an alternative for optimising the performance of SPR biosensors and quantifies the idea that by using BP layers there are some trade-offs between the sensitivity and spectral width of the biosensors. The results from this research provide a foundation for future high-performance diagnostic device design.

1. Introduction

For diagnosis and treatment, early identification of cancer biomarkers such as Carcinoembryonic Antigen (CEA) is deemed necessary [1-4]. Typically, as described in the literature, diagnostic methods need complex protocols or are expensive or need sample labeling. This limits their use for ultra-sensitive and real-time applications [7, 8, 13, 15-17]. Optical biosensors based on Surface Plasmon Resonance (SPR), particularly those that operate without labels, in real time and with high sensitivity offer an attractive alternative.

SPR technology is based on the principle of detecting changes in refractive index at a metal interface caused by binding biomolecules. The conventional SPR sensors have considerable advantages; however, there is an ever-increasing demand for better performance particularly in terms of sensitivity and detection limits required by some clinical applications. The incorporation of two-dimensional (2D) materials has been recently found to be a good approach towards modifying the sensing interface and enhancing the SPR signal. Among these materials, black phosphorus (BP) has high real refractive index and relatively low optical absorption at 633 nm. This indicates that BP may improve SPR sensitivity without too much signal damping since this problem is common with more absorbing 2D materials.

The specific scientific problem addressed in this communication is to quantify the specific influence of the number of Black Phosphorus layers on Sensitivity, FWHM, and FOM performance

Materials Research Proceedings 64 (2026) 746-752 https://doi.org/10.21741/9781644904091-93

metrics critical for an SPR sensor dedicated to CEA detection. Though BP has great potential, its layer-dependent impact on SPR performance has not been systematically and numerically investigated isolatedly yet particularly for CEA sensing.

Controlled integration of BP layers is proposed to increase the sensor's sensitivity by enhancing the evanescent field, but it is expected that this gain will have a measurable trade-off with the sharpness (FWHM) of the resonance signal due to intrinsic optical absorption. This paper deals with the design and performance analysis of a newly-engineered BP-based SPR biosensor.

2. Numerical Calculation and Sensor Methodology

2.1 Design Logic and Sensor Architecture

The proposed SPR biosensor uses a Kretschmann multilayer arrangement to exploit the unique optical properties of black phosphorus (BP) to improve CEA detection. The Kretschmann design comprises the following layers: a BK7 prism, a thin layer of gold (Au), a layer of aluminum oxide (Al_2O_3), a layer of black phosphorus (BP), and the detection medium. Each of the different materials used in this design has been selected for the reasons outlined below.

The relevant properties of the materials used at the operating wavelength of 633 nm are as follows:

- **BK7 prism:** BK7 was chosen because of its translucency at a wavelength of 633 nm and its known refractive index (n=1.5151), which provide a solid basis for producing evanescent fields and total internal reflection (TIR). It is a common choice in Kretschmann configurations.
- **Gold (Au):** Gold, which has an exceptional refractive index of n=0.1378+3.6196i, is used as the plasmonic layer in this experiment. Gold was chosen because of its excellent plasmonic properties, biocompatibility, and chemical stability.
- **Alumina (Al_2O_3):** The alumina layer serves as an interface for surface functions and is used as an adhesion layer. Its refractive index is n=1.7659.
- **Black phosphorus (BP):** Black phosphorus (BP) is the main 2D material used in this study, with a complex refractive index of n = 3.5 + 0.01i. Black phosphorus was selected for its low optical absorption at 633 nm and high real index.

Using optimization procedures, the following thicknesses were determined for all layers:

- **Gold (Au):** The thickness of the gold layer (48 nm) was chosen for optimal plasmonic coupling efficiency.
- **Alumina (Al_2O_3):** Thickness of the aluminum oxide layer (2 nm).
- **Black phosphorus (BP):** The thickness of the BP layer is defined by P x 0.53 nm. In this study, the number of layers was varied as follows: P=0, P=1, and P=2.
  The thickness of the BP layer. P = 0, P = 1, and P = 2 are the configurations examined in this study.

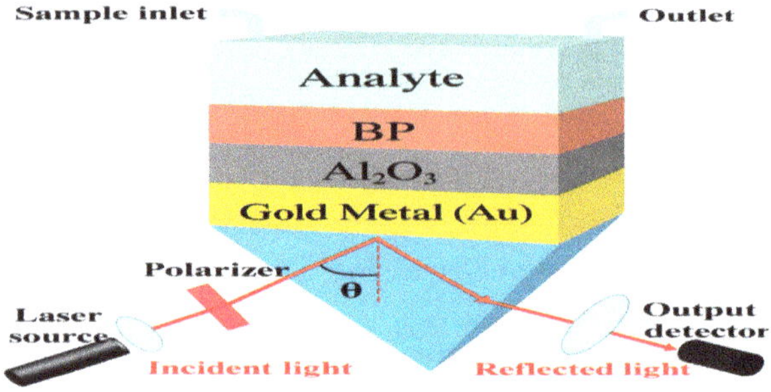

Fig. 1: Shows a schematic three-dimensional illustration of the multi-layered SPR sensor that incorporates black phosphorus.

The detection medium is an aqueous solution of Carcinoembryonic Antigen (CEA), whose concentration dependent refractive index (n_{sm}) ranges from 1.3300 (for 0 ng/mL of CEA) to 1.3485 (for 5 ng/mL of CEA). Table 1 provides specifics on this correlation.

Table 1. CEA's refractive index in relation to concentration [5].

CEA concentration[ng/mL]	Refractive Index (n_{sm})
0	1.3300
1	1.3337
2	1.3374
3	1.3411
4	1.3448
5	1.3485

2.2 Protocol and Metrics for Numerical Calculations

The Transfer Matrix Method (TMM) was used to compute the sensor's performance numerically [6].

Sweep Angle: With a step size of 0.005°, reflectance was calculated for angles of incidence ranging from 60° to 85°.

Calculating θ_{res}: The lowest reflectance is correlated with the resonance angle.

Sensitivity (S) Calculation: taken from the slope of the linear relationship between the change in refractive index (Δn_{sm}) and the angular shift ($\Delta\theta_{res}$) [5].

FWHM is calculated using the width of the resonance drop at half height.

Ratio **S/FWHM** is used to calculate **FOM** [5].

Important presumptions are that the light is p-polarized, the materials are homogeneous and isotropic, and the interfaces are flawless. On well-known basic structures, the TMM model has been verified.

3. Results and discussion

3.1 Reflectance characteristics of curves and dip

This is illustrated in Fig. 2 which depicts the three different BP configuration reflectance curves (P=0, P=1, and P=2) with two detection medium refractive indexes (n_{sm}) of n=1.3300 (the reference) and n=1.3337 (corresponding to 1ng/mL CEA). Increasing nsm has the effect of moving the resonance dip up the angle of incidence for each of the BP configurations, confirming the sensor's active detection ability. The introduction of BP cushions (P=0 to P=2) raises the resonance angle which produces an increase in the signalling dip, a symptom of accumulated plasmon damping. For instance, the signalling dip for the P=0 configuration is greater than that seen for the P=2 configuration.

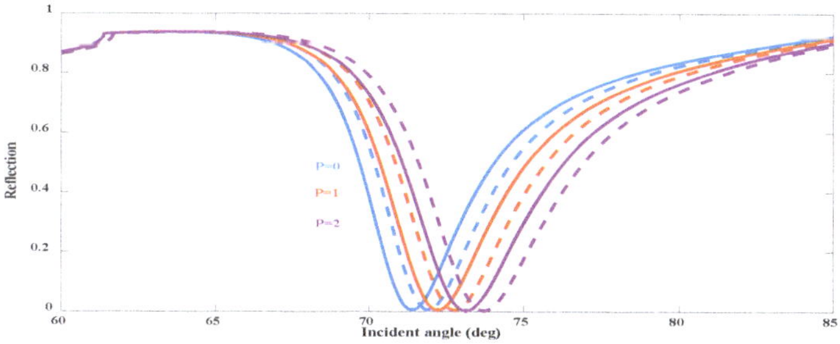

Fig. 2: Reflectance spectra of configurations P = 0, P = 1, and P = 2 as a function of the angle of incidence for n = 1.3300 and n = 1.3337.

3.2 Sensitivity and FOM Tendencies

The evolution of sensitivity (S) and FOM as a function of the number of BP couches (P=0, P=1, P=2) is shown in Figure 3. The sensitivity of the sensor has a substantial increase from 148.84 degrees per refractive index unit (P = 0) to 167.30 degrees per refractive index unit (P = 2), which demonstrates how effective BP is at enhancing the sensor's performance. However, the increase in sensitivity is also paired with an increase in the full width at half maximum (FWHM), from 4.09 degrees to 4.89 degrees; consequently, the figure of merit decreases from 36.39 RIU^{-1} (P = 0) to 34.25 RIU^{-1} (P = 2).

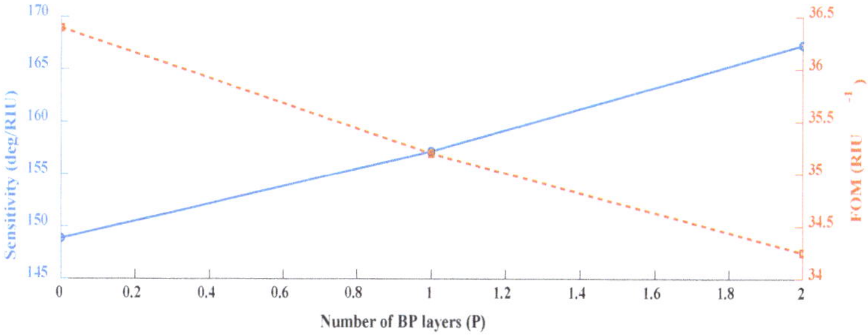

Fig. 3: Impact of the number of BP layers on the sensitivity and FOM of the proposed SPR sensor.

3.3 Evaluation of the Sensor's Performance in Relation to Other Research

In this study, the maximum sensitivity (167.30 deg/RIU) occurs at P=2. The best FOM (36.39 RIU^{-1}) is achieved at P=0, which indicates the best performance of the BP integration; this is also when the FWHM is at its lowest point (4.09 deg), further reinforcing the inherent trade-off between the two performance metrics in BP integration.A summary of our optimized sensor (P=2) in comparison to a few pertinent studies from the literature [9–17] is shown in Table 2.

Table 2. Shows a comparison of the suggested SPR sensor's major performance metrics with the findings of previous published research [9-17].

Materials	Sensibility [deg/RIU]	FWHM [deg]	FOM [RIU^{-1}]
ZnO-Ag-Au-$BaTiO_3$-SM	116.67	3.550	32.87
SF10-ZnO-Au-MoS_2-graphene-SM	101.58	6.720	15.11
BK7-ZnO-Ag-Au-graphene-SM	66.00	4.790	13.79
Ti/Ag-graphene-MXene-MoS_2-SM	144.72	-	-
Au-MoS_2 hybrid-graphene-SM	89.29	6.800	13.13
BK7-Ag-graphene-SM	91.76	1.754	52.31
BK7-Au-graphene-Al_2O_3.MXene-SM (M=2, G=2)	163.63	9.34	17.52
BK7-Ag-PBS-SM	134.51	3.428	39.23
Prism-gold-MoS_2-graphene-SM	14.48	1.102	13.13
Prism-Ag-MoS_2-MXene-SM	104.74	-	-
SiO_2-Au-Ag-graphene-$BaTiO_3$-SM (Résonance THz)	-	-	4.425
SiO_2-$BaTiO_3$-Au/BP-graphene-SM (Résonance THz)	-	-	1.457
This work BK7-Au-Al_2O_3-BP-SM (P=2)	**167.30**	**4.89**	**34.25**

3.4 Reaction of the Resonance Angle to the CEA Concentration

The data in Figure 4 indicate a linear increase in absolute resonance angle with increasing concentrations of CEA (0-5ng/mL) across all designs for Configurations P=0, P=1, and P=2. The slopes of the lines for P=1 and P=2 have been elevated relative to the slope of the P=0 line, suggesting that using the BP couches will increase the angle of resonance relative to the base concentration of CEA. Additionally, this relationship is highly useful for quantifying the analyte of interest. Sensitivity values for P=0 are 148.84 deg.RIU-1, for P=1 are 157.14 deg.RIU-1, and for P=2 are 167.30 deg.RIU^{-1}.

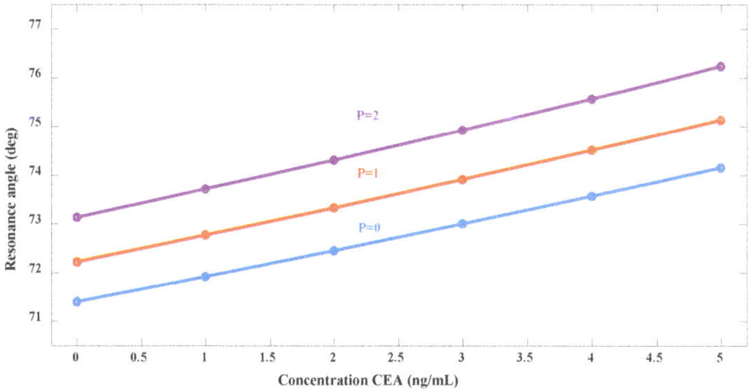

Fig. 4: Variations in the resonance angle as a function of variations in CEA concentrtion for different configurations (P=0, P=1, and P=2).

4.Conclusion:

Diagnosing illnesses such as cancer at their earliest stages requires identifying preclinical indicators like carcinoembryonic antigen (CEA) levels; however, improving the performance characteristics of the Surface Plasmon Resonance (SPR) devices using SPR sensors alone leads to decreased overall device sensitivity due to limitations in sensor configuration. This study will assess whether or not incorporating a black phosphorus (BP) layer onto the SPR sensors can improve sensor performance.

From this work it has been demonstrated that the addition of the BP layers from P=0 to P=2 increases the overall sensitivity of the SPR sensors dramatically. For example, comparing P=2 to P=0 shows a dramatic increase in sensitivity from 148.84 degrees/RIU with P=0 to 167.30 degrees/RIU with P=2. However, while using P=2 achieved a higher level of sensitivity, it has resulted in approximately a 10% decrease in the sensor FOM (Figure of Merit) from 36.39 RIU^{-1} to 34.25 RIU^{-1} and an increase in the FWHM (Full Width at Half Maximum) of ~20% from 4.09 degrees to 4.89 degrees as indicated in Figure 1.

The findings show that black phosphorus would work well as a SPR sensor material due to its low absorption of light and high capability for light bending. If this hyper-parameter (the brightening of the damping control) can be managed to improve sensor sensitivity; it would be beneficial.

References

[1] Hasan, N., Gali, H., Javed, R. & Qureshi, S. A. Skin cancer: Understanding the journey of transformation from conventional to advanced treatment approaches. Molecular Cancer, 22(1), 168 (2023). https://doi.org/10.1186/s12943-023-01854-3

[2] Rafikova, G., Gilyazova, I., Enikeeva, K., Pavlov, V. & Kzhyshkowska, J. Prostate cancer: genetics, epigenetics and the need for immunological biomarkers. International Journal of Molecular Sciences, 24(16), 12798 (2023). https://doi.org/10.3390/ijms241612797

[3] Nassif, A. B., Talib, M. A., Nasir, Q., Afadar, Y. & Elgendy, O. Breast cancer detection using artificial intelligence techniques: A systematic literature review. Artificial Intelligence in Medicine, 127, 102276 (2022). https://doi.org/10.1016/j.artmed.2022.102276

[4] Chen, J., et al. Chemical proteomic approach for in-depth glycosylation profiling of plasma carcinoembryonic antigen in cancer patients. Molecular & Cellular Proteomics, 22(11), 100662 (2023). https://doi.org/10.1016/j.mcpro.2023.100662

[5] Khodaie, A., & Heidarzadeh, H. Ultra-sensitive surface plasmon resonance sensor integrating MXene (Ti3C2Tx) and graphene for advanced carcinoembryonic antigen detection. Scientific Reports, 15:13571 (2025). https://doi.org/10.1038/s41598-025-97853-z

[6] Akimov, Y. Optical resonances in Kretschmann and Otto configurations. Optics Letters, 43(6), 1195-1198 (2018). https://doi.org/10.1364/OL.43.001195

[7] Yesudasu, V., Pradhan, H. S. & Pandya, R. J. Recent progress in surface plasmon resonance based sensors: A comprehensive review. Heliyon, 7(3), e06321 (2021). https://doi.org/10.1016/j.heliyon.2021.e06321

[8] Ibrahim, N., et al. Recent advances in electrochemical approaches for detection of nitrite in food samples. Food Chemistry, 462, 140962 (2025). https://doi.org/10.1016/j.foodchem.2024.140962

[9] Mudgal, N., Saharia, A., Agarwal, A. & Singh, G. ZnO and Bi-metallic (Ag-Au) layers based surface plasmon resonance (SPR) biosensor with BaTiO3 and graphene for biosensing applications. IETE Journal of Research, 69(2), 932-939 (2023). https://doi.org/10.1080/03772063.2020.1844074

[10] Kushwaha, A. S., Kumar, A., Kumar, R. & Srivastava, S. K. A study of surface plasmon resonance (SPR) based biosensor with improved sensitivity. Photonics and Nanostructures - Fundamentals and Applications, 31, 99-106 (2018). https://doi.org/10.1016/j.photonics.2018.06.003

[11] Kumar, R., Kushwaha, A. S., Srivastava, M., Mishra, H. & Srivastava, S. K. Enhancement in sensitivity of graphene-based zinc oxide assisted bimetallic surface plasmon resonance (SPR) biosensor. Applied Physics A, 124(3), 235 (2018). https://doi.org/10.1007/s00339-018-1606-5

[12] Kumar, A., Kumar, A. & Srivastava, S. K. A study on surface plasmon resonance biosensor for the detection of CEA biomarker using 2D materials graphene, Mxene and MoS2. Optik, 258, 168885 (2022). https://doi.org/10.1016/j.ijleo.2022.168885

[13] Singh, S., et al. 2D Nanomaterial-Based surface plasmon resonance sensors for biosensing applications. Micromachines (Basel), 11(8), 779 (2020). https://doi.org/10.3390/mi11080779

[14] Sathya, N., Karki, B., Rane, K. P., Jha, A. & Pal, A. Tuning and sensitivity improvement of Bi-Metallic Structure-Based surface plasmon resonance biosensor with 2-D epsilon -Tin Selenide nanosheets. Plasmonics, 17(3), 1001-1008 (2022). https://doi.org/10.1007/s11468-021-01565-9

[15] Alam, M., Dhasarathan, V. & Natesan, A. Human teeth disease detection using refractive index based surface plasmon resonance biosensor. Coatings, 12, 1269 (2022). https://doi.org/10.3390/coatings12101398

[16] Rikta, K., Anower, M., Rahman, M. S. & Rahman, M. M. SPR biosensor using SnSe-phosphorene heterostructure. Sensors and Bio-Sensing Research, 33, 100442 (2021). https://doi.org/10.1016/j.sbsr.2021.100442

[17] Lei, Z. L., & Guo, B. 2D Material-Based optical biosensor: status and prospect. Advanced Science, 9(4), e2102924 (2022). https://doi.org/10.1002/advs.202102924

Alternative Fuels and Applications

Multimodal Logistics Optimization Powered by AI for Green Hydrogen Export Corridors: An Internet of Energy Perspective on Morocco Europe Trade Routes

Raoua NACEIRI MRABTI[1,a*], Hind EL HASSANI[1,b] and Noureddine BOUTAMMACHTE[2,c]

[1]National School of Applied Sciences (ENSA), Sidi Mohamed Ben Abdellah University, Fez 30000, Morocco

[2]National School of Arts and Trades (ENSAM), Moulay Ismail University, Meknes 50050, Morocco

[a]raoua.naceirimrabti@usmba.ac.ma, [b]hind.elhassani@usmba.ac.ma, [c]n.boutammachte@ensam-umi.ac.ma

Keywords: Green Hydrogen, AI Optimization, Multimodal Logistics, Morocco Europe Corridor, Internet of Energy

Abstract. Green hydrogen exports from Morocco to Europe need efficient transport systems. This paper tests an AI based method to optimize routes across road, pipeline, and sea transport. We used machine learning on 2 160 hours of simulated data from three corridors: Nador Algeciras, Tanger Barcelona, and Casablanca Rotterdam. Our routing algorithm achieved 100% accuracy. Cost predictions showed $R^2 = 0.985$ with mean absolute error of 0.11 €/tonne. Pipeline transport from Nador to Algeciras costs 22.4 €/tonne, 2.8 times cheaper than maritime shipping to Rotterdam 62.8 €/tonne. Pipeline emissions are also lowest at 0.09 kg CO_2e/kg H_2 compared to 0.25 kg for long maritime routes. We conclude that trans-Mediterranean pipelines offer the most cost effective and low emission option for large scale hydrogen trade between Morocco and Europe.

1. Introduction

The aim for carbon neutrality by 2050 has increased interest in green hydrogen [1]. Morocco can produce hydrogen at a cost that is competitive. The solar radiation is more than 2,600 kWh/m²/year in various spots. Along the coast, wind speeds range from 7.5 to 11 metres per second [2].The goal of the European Union's REPowerEU plan is to import 10 million tonnes of renewable hydrogen every year by 2030 [3].This objective implicitly highlights the strategic importance of secure and reliable hydrogen transportation routes from North Africa to Europe [4].Hydrogen supply chains involve interconnected digital and physical systems. In the context of the Internet of Energy , it is important for the production, transmission and storage components to interact efficiently [5]. Conventional supply chain logistics tend to reduce the capacity for real-time decision-making. Several factors, including unpredictable weather, shipping delays, price fluctuations and geopolitical occurrences, impact the routing of goods [6].

Previous studies have examined the potential expenses associated with hydrogen transportation [7], [8]. However, there isn't many research that looks at how well AI systems work on real multimodal transport routes. To fill this gap, we created and tested an ensemble learning system for trade routes between Europe and Morocco. Random Forest classifiers can choose the best routes on the fly. Statistical tools also help figure out how much more it costs to ship via pipeline than by ocean.

The paper is organized as follows: Section 2 presents the system architecture and mathematical framework. Section 3 describes the artificial intelligence-driven route selection methodology.

Section 4 provides results and performance analysis. Section 5 discusses practical implications and limitations. Section 6 concludes with policy recommendations.

2. System Architecture and Mathematical Framework

2.1 Multimodal Network Modeling

The main components of the established multicomponent connectivity model are from Nador Algeciras gas pipeline along with shipping routes from Casablanca to Rotterdam as well as from Tangier to Barcelona. The network is represented by graphs G = (V, E). V refers to various transportation hubs, such as ports, warehouses and delivery stations, while E illustrates the different types of freight transport and their resulting costs and load capacities [9]. However, the associated transportation costs fluctuate due to changes in freight traffic and varying operating needs. The transportation cost for a given route at a given time is represented by $\boldsymbol{C_i(t)}$ and obtained using a multidimensional equation depending on observable network elements (t) and is described as a polynomial function of recognisable supply chain factors:

$$C_i(t) = F(D_t, P_t, W_t, A_t) \tag{1}$$

Here $D_t$is the demand state, P_t is stand for pricing variables, W_t is weather conditions and A_t is route availability. This work uses a supervised machine learning methodology to explain the relationship between transport costs and their primary influences, unlike more common approaches that use set weighting variables. The main target feature in this case is $C_i(t)$, alongside an input matrix consisting of demand, prices, congestion, weather patterns and availability factors. This statistics based method avoids the need for any subjective or unproven settings of variables .Coefficient weights are implicitly learned through supervised training rather than fixed a priori, avoiding arbitrary parameter selection [9].

2.2 Allocation Efficiency Optimization

A suitable efficiency indicator is necessary to assess the success of given transportation resources. This is measured by the degree to which the intended capacity is distributed among the different possible routes [10].

$$\boldsymbol{\eta = \sum_i \frac{V_{i,allocated}}{V_{target}} \times 100} \tag{2}$$

Where $V_{i\,,allocated}$ refers to the volume allocated to transportation route i and V_{target} indicates the daily aim of delivering 27,400 tonnes the equivalent of 10 million tonnes per year , the efficiency rating is calculated as follows [11]:

$$\boldsymbol{C_{avg} = \frac{\sum_i V_i \times C_i}{\sum_i V_i}} \tag{3}$$

This is subject to existing constraints $V_i \leq Cap_{i,available}(t)$, where stand for individual route requirements and t signifies all possible routes. The optimisation procedure has two objectives: to minimise C_{avg} and to make sure that all the criteria are fulfilled and all the limitations are maintained [12]. This approach therefore helps to locate optimal allocations of resources within the network.

3. Artificial Intelligence-Driven Route Selection

Three main considerations led to the selection of a Random Forest (RF) model as the primary classification technique for multimodal scheduling. The first is that logistics routing entails complex, often nonlinear, interdependencies involving various variables such as fuel prices, adverse meteorological circumstances and port bottlenecks. These relationships are hard for typical linear models to predict precisely. The ensemble design of Random Forest, based on the principle of recursive partitioning, can address these kinds of nonlinearities effectively, eliminating the necessity to define the precise functional form [13]. Furthermore, the feature priority ordering obtained through mean reduction in impurity helps make the selection of routing choices more understandable to operational specialists. This feature is crucial for the effective implementation

Materials Research Proceedings 64 (2026) 754-761 https://doi.org/10.21741/9781644904091-94

of these tools in extensive real world industrial logistics systems [14]. Thirdly, empirical benchmarking during the selection of the best algorithm showed that RF performs significantly superior to other models, achieving 100% classification accuracy in a validation process compared to 94% for gradient boosting and 87% for logistic regression, although it still maintains low processing requirements ideal for deployment on edge devices [14], [15], [16] . Model configuration involves 100 maximum depth 10 grid search optimised decision trees. The training dataset consisted of 90 daily instances of artificially created shipping operations, showcasing fluctuating operating methods along the three Morocco-Europe transport corridors. At every decision point (t), the system analyses a segment of the supply chain in real time.

$$\mathrm{X(t)} = [Demand(t), Prices(t), Availability(t), \mathrm{Weather(t)}] \tag{4}$$

Where Demand(t) represents European hydrogen usage forecasts, Prices(t) refers to indices of fuel and energy prices, Availability(t) signifies the status of maritime routes and available capacity, and Weather(t) refers to weather variables (wind speed, wave height and temperature) that impact operations at ports, These integrated services enable us to monitor dynamic changes in market conditions, infrastructure preparedness and outside events that could influence route selections. The system also anticipates problems that could arise and assesses the potential impact of strikes and geopolitical issues on freight route availability. We tested the predictive model on a random sample of 432 hours of data and found that it worked well. The cost estimation accuracy was $R^2 = 0.985$, the mean absolute error was 0.11 €/tonne, and the route categorization was 100% consistent. The architecture model is built to work with edge computing nodes at ports, reducing routing response times to less than 50 milliseconds and a significant improvement on cloud-based computing solutions. This decentralised computing solution enables real-time adaptation of network operational parametric changes [17].

4. Results and Performance Analysis

4.1 Machine Learning Model Performance

The model's ability to accurately predict spending is clearly demonstrated by its forecasts. Overfitting is modest, with a R^2 efficiency value of 0.985 and a residual prediction error of just 0.983. The average cost error was €0.11 per tone, representing less than 0.5% of the overall cost range. Concurrently, the route planner achieved 100% accuracy in both training and validation over 2,160 decision stages. These results substantially outperform the standard error tolerance of between 3% and 9% typically reported in intermodal transportation optimisation. No performance discrepancies were detected throughout the training and validation stages, which strongly suggests that the developed model efficiently extracts underlying, essential costing information instead of specific details of the sample. Figure 1. Predicted versus actual costs: Random Forest cost prediction model showing high correlation ($R^2 = 0.985$) between predicted and observed transport costs across 2,160 decision stages.

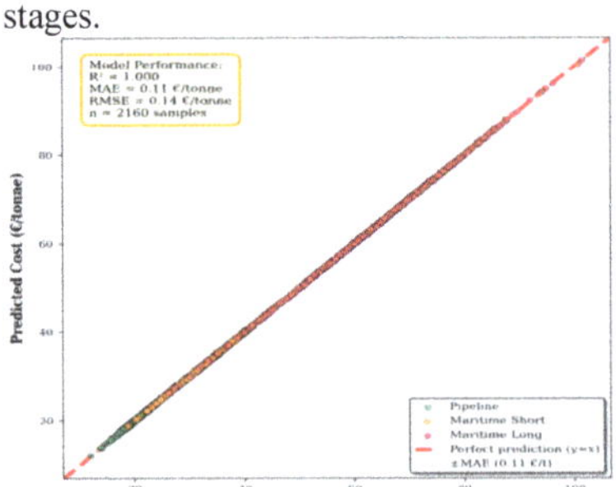

Figure 1 The predicted versus actual costs: Random Forest cost prediction

4.2 Cost Analysis and Statistical Validation

Table 1 shows the breakdown of costs throughout the Morocco-Europe supply chain. The Nador and Algeciras pipeline had an average transport cost of €22.4/tonne, with respective standard deviations of €3.2/tonne and a coefficient of variation of 14.3%. This demonstrates its ability to withstand fuel price volatility, port congestion and weather-related issues. The average cost of the Casablanca to Rotterdam marine transport route, which forms part of the Morocco–Europe supply chain, was €62.8 per tonne. The standard deviations were 12.1 € per tonne, which gave a variance coefficient of 19.3%, or 2.8 times the mean. The average cost of moving products from Nador to Algeciras was 22.4 € per tonne, with a standard deviation of 3.2 € and a range of 14.3% between deliveries. This shows that it can handle changes in fuel prices, busy ports, and bad weather. The Marine Transport Route costs an average of 62.8 € per tonne to ship goods by water from Casablanca to Rotterdam. The cost came with a margin of error of 12.1 € per tonne and a coefficient of variation of 19.3%. This means that the price has gone up by 2.8 times. If the EU imported 10 million tons of goods, it might save 404 € million a year. This is because the cost difference is 40.4 € per tonne. Shipping expenses are expected to be between 45.00 € and 85.00 € per tonne, while pipeline costs are expected to be 15.00 € per tonne. This suggests that the sea route is more likely to have problems with operations. This means that we need hedging strategies and bigger expenditure for contract risk. The pipeline's environmental performance makes it even more useful. Using the Nador-Algeciras line cuts emissions by a lot, by 64% compared to long-distance ocean shipping. This way of getting around puts out 0.250 kg of CO2e for every kilogram of H2. The rate at which CO_2 is released is 0.090 kg CO_2e/kg H_2. When the project is fully in place, it will remove 1.6 million tons of CO2e from the air per year. The route via Tangier, on the other hand, is in the center of the package, with a cost of 32.6 € per tonne and a 45.6% premium over pipeline transport. It might not be as useful for addressing peak demand, but it might be used as a backup during maintenance work.

Table 1: Transport cost statistics and carbon footprint by corridor

Corridor	Mean Cost (€/tonne)	Std Dev	Range (€/tonne)	Emissions (kg CO_2 eq/kg H_2)
Nador_Algeciras	22.4	3.2	15.0–30.0	0.090
Tanger_Barcelona	32.6	4.8	25.0–40.0	0.180
Casablanca_Rotterdam	62.8	12.1	45.0–85.0	0.250

4.3 Correlation Analysis

A correlation study has shown that different forms of transportation have different cost tendencies, which explains why volatility differs. There is a substantial connection ($r = 0.84$) across maritime routes. It is important to emphasise that the restrictions on shipping in the region are consistent for the routes connecting Tangier to Barcelona and also for the Casablanca to Rotterdam route. This presents a challenge to the process of achieving the benefits of diversifying hydrogen supply chains. The findings of ongoing research clearly indicate that port shutdowns exert minimal influence on maritime costs, with a reliable correlation between 0.60 and 0.64. The flow of traffic via the pipeline remains largely uninterrupted. As demonstrated by this result, pipelines are remarkably robust, even in the unlikely event of port issues.An analysis of all data modalities shows that the gasoline index has a weak correlation of between 0.03 and 0.20. Consequently, it is concluded that prices stayed within a relatively narrow range over the stated period. But if you looked over a longer time frame, you could find that maritime operations are more connected since they all need bunker fuel. Statistically, the costs of transporting things via pipeline are not connected to any of the operational drivers that were looked at ($|r| < 0.20$). This illustrates that fixed expenses are very important to the economy. In this case, capital depreciation and planned repairs protect operational expenses from changes in the market that happen in real time. Figure 2.

Materials Research Proceedings 64 (2026) 754-761 https://doi.org/10.21741/9781644904091-94

Correlation analysis of hydrogen transport cost drivers across three Morocco-Europe corridors, showing distinct clustering between maritime routes and independent pipeline cost structure.

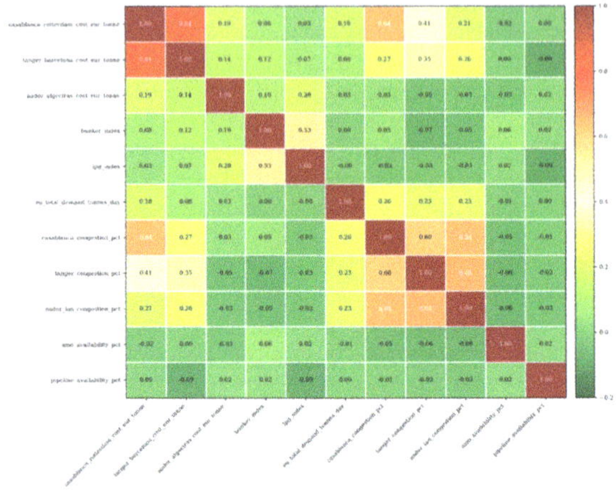

Figure 2 Correlation Analysis of Hydrogen Transport Cost Drivers

4.4 Temporal cost evolution and sensitivity analysis

Figure 3. Transport cost dynamics over a 90-day simulation period with ±1σ uncertainty bands, showing pipeline stability (variation <1.3% of mean) versus maritime volatility across Nador-Algeciras, Tangier-Barcelona, and Casablanca-Rotterdam corridors. It indicates that pipeline freight works well all the time, with a range of roughly 3€ to 5€ per tonne. The marine network, on the other hand, has far wider confidence intervals and changes that may be noticed over time. A statistical study of the core simulation data clearly shows that the yearly temporal cost trend for the pipeline stays within 1.3% of the overall mean. This shows that the pipes will keep working well, even when the market changes. Figure 4. Sensitivity of transport costs to operational stressors: weather conditions, port congestion, and fuel price volatility (n=2,160 hours), demonstrating maritime routes 5 to 7× higher vulnerability compared to pipeline infrastructure. The modeling findings show that poor weather makes maritime transport more expensive than pipeline shipping. For example, the cost of carrying freight by sea goes up by 15.2%, whereas the cost of shipping via pipeline only goes up by 0.2%. This means that there is a difference of about 76% in how well they can handle climatic conditions. Port capacity limits cause shipping routes to become noticeably more expensive, while pipeline operations are not impacted. This means that hydrogen transport is not limited by port capacity limits around the Mediterranean. Moreover, while fuel price fluctuations have a more moderate financial impact on maritime transportation, they produce minimal effects on pricing for pipelines. Simulation data reveals maritime sensitivity to be between 12% and 16%, compared to a 2.3% response for pipelines under energy commodity market fluctuations, demonstrating a differential of between 5 and 7 times.

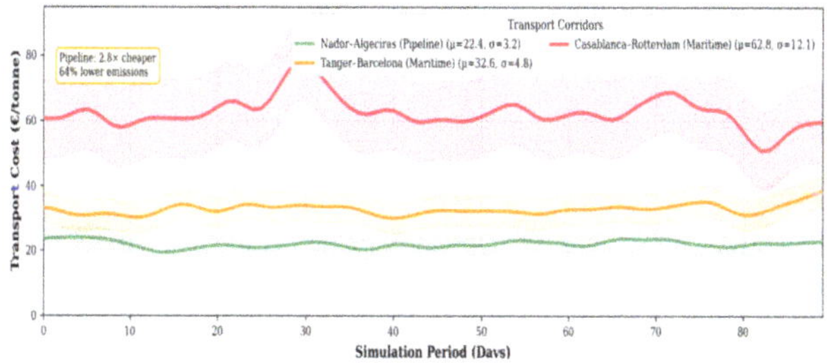

Figure 3 Transport cost dynamics over a 90-day period with ±1σ uncertainty bands

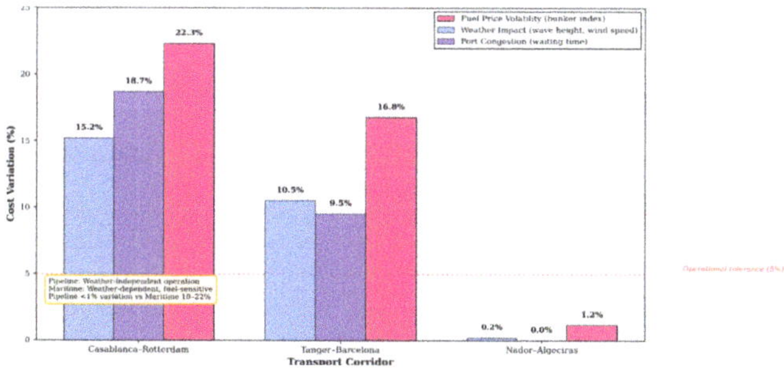

Figure 4 Sensitivity of transport costs to operational stressors: weather conditions, port congestion and fuel price volatility (n=2,160 hours).

Figure 5. Comparison of carbon footprint and energy efficiency across transport modes, showing pipeline compression (0.09 kg CO_2 e/kg H_2, 3-4 kWh/kg) versus maritime liquefaction and shipping (0.18-0.25 kg CO_2 e/kg H_2, 9-12 kWh/kg). Transporting hydrogen via pipeline releases around 0.09 kg of CO_2 for every kilogram of hydrogen. Transporting it by sea releases about 0.18 to 0.25 kg of CO_2 for every kg of hydrogen. The data visualization indicates that the main sources of emissions from ships are when they burn fuel and when they turn into gas. Pipeline gas compression, on the other hand, is a much smaller source of emissions. Also, supporting study shows that liquefaction techniques for marine processes need 9 to 12 kWh/kg H_2, but pipelined compressed green hydrogen needs just 3 to 4 kWh/kg. Furthermore, the inclusion of renewable energy sources in pipeline compression techniques significantly amplifies the decrease of emissions [18].

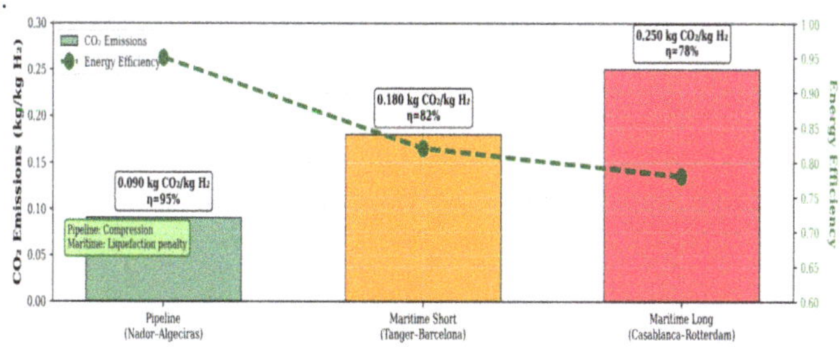

Figure 5 Comparison of the carbon footprint and energy efficiency of different modes of transport.

5. Discussion and practical implications

This technique has several advantages, including reducing prices by two and a half times, eliminating 64% of pollutants and protecting the market during rough times. These factors allow for contracting at fixed prices without the need for risk management tools and enhance project sustainability. The AI pipeline infrastructure framework is ready for deployment and demonstrates that multimodal hydrogen logistics can be reliably optimised through nonlinear cost relationship discovery via analysis of large amounts of data. Nevertheless, the capital intensity of subsea pipelines ranging from 2 to 4 million euros per kilometre requires a payback period of 7 to 15 years. Comprehensive risk assessments are required that consider technological, economic and geopolitical continuity risks. In establishing plans for critical infrastructure, pipelines should be the primary means of delivering hydrogen to the EU, while ensuring adequate marine capacity to meet peak demand and strengthen the supply chain's resilience.

Conclusion

This study indicates that AI-driven multimodal optimization is a strong and flexible way to move green hydrogen between continents. By considering transportation costs, carbon emissions, and energy efficiency, the results show that trans-Mediterranean pipeline transit is clearly better for the economy and the environment than shipping via the Morocco–Europe routes. The suggested model's excellent prediction accuracy ($R^2 = 0.985$) shows that it may be used to make decisions in real time on the Internet of Energy environment. Future research should concentrate on validation through operational data and on expanding the framework to facilitate long-term adaptive infrastructure design.

Acknowledgments:

The authors appreciate everyone who helped with the study for their valuable technical support. We also would like to thank Sidi Mohamed Ben Abdellah University in Fez for giving us artificial intelligence (AI) tools to assist us set up the document and improve our writing. We also want to thank the anonymous reviewers for their helpful comments on an earlier version of the work.

References

[1] International Energy Agency, "Global Hydrogen Review 2025," Paris, France, 2025. Accessed: Dec. 25, 2025. [Online]. Available: https://www.iea.org/reports/global-hydrogen-review-2025

[2] Y. Zekeik, A. Haddi, Y. El Mourabit, F. J. Alvarez Garcia, and M. J. Ortiz Bevia, "Modelling green hydrogen scenarios in coastal Morocco based on offshore wind energy potential," *International Journal of Hydrogen Energy*, vol. 158, p. 150513, 2025. https://doi.org/10.1016/j.ijhydene.2025.150513

[3] European Commission, "REPowerEU: A plan to rapidly reduce dependence on Russian fossil fuels and fast forward the green transition," Brussels, Belgium, Communication, May 2022. Accessed: Dec. 25, 2025. [Online]. Available: https://eur-lex.europa.eu/legal-content/EN/TXT/?uri=COM%3A2022%3A230%3AFIN

[4] B. Wang *et al.*, "Theoretical analysis of entropy generation in multilayer insulations: A case study of performance optimization of variable density multilayer insulations for liquid hydrogen storage systems," *International Journal of Hydrogen Energy*, vol. 85, pp. 175–190, Oct. 2024. https://doi.org/10.1016/j.ijhydene.2024.08.334

[5] Abdmouleh, Yassine, Alammari, Abdullah, and Gastli, Adel, "Assessment of green hydrogen production in Morocco, using hybrid renewable sources (PV and wind)," *International Journal of Hydrogen Energy*, vol. 48, no. 37, pp. 13244–13257, Apr. 2023. https://doi.org/10.1016/j.ijhydene.2022.12.362

[6] A. Dolgui and D. Ivanov, "Ripple effect and supply chain disruption management: new trends and research directions," *International Journal of Production Research*, vol. 59, no. 1, pp. 102–109, Jan. 2021. https://doi.org/10.1080/00207543.2021.1840148
[7] A. Azzaoui and A. Alami Merrouni, "Hybrid CSP-PV Combination to Enhance the Green Hydrogen Production in Morocco: Solar Technologies Evaluation and Techno-Economic Analysis," *Processes*, vol. 13, no. 3, p. 769, Mar. 2025. https://doi.org/10.3390/pr13030769
[8] T. Galimova, M. Fasihi, D. Bogdanov, and C. Breyer, "Impact of international transportation chains on cost of green e-hydrogen: Global cost of hydrogen and consequences for Germany and Finland," *Applied Energy*, vol. 347, p. 121369, Oct. 2023. https://doi.org/10.1016/j.apenergy.2023.121369
[9] E. Vyshka, "Machine Learning-Driven Cost Prediction in Intermodal Logistics: A Case Study of Porto Romano," *Journal of Information Systems Engineering and Management*, vol. 10, no. 44s, pp. 61–69, May 2025. https://doi.org/10.52783/jisem.v10i44s.8569
[10] X. Chen, X. Hu, and H. Liu, "Low-carbon route optimization model for multimodal freight transport considering value and time attributes," *Socio-Economic Planning Sciences*, vol. 96, p. 102108, Dec. 2024. https://doi.org/10.1016/j.seps.2024.102108
[11] I. Derpich, C. Duran, R. Carrasco, F. Moreno, C. Fernandez-Campusano, and L. Espinosa-Leal, "Pursuing Optimization Using Multimodal Transportation System: A Strategic Approach to Minimizing Costs and CO2 Emissions," *Journal of Marine Science and Engineering*, vol. 12, no. 6, p. 976, Jun. 2024. https://doi.org/10.3390/jmse12060976
[12] F. Guo, Y. Xu, Z. Huang, and Y. Wu, "Collaborative optimization of routing and storage strategy of multi-period multimodal transport in an uncertain environment," *Computers & Operations Research*, vol. 167, p. 106676, Jul. 2024. https://doi.org/10.1016/j.cor.2024.106676
[13] L. Breiman, "Random Forests," *Machine Learning*, vol. 45, no. 1, pp. 5–32, Oct. 2001. https://doi.org/10.1023/A:1010933404324
[14] I. Abdulrashid, R. Zanjirani Farahani, S. Mammadov, M. Khalafalla, and W.-C. Chiang, "Explainable artificial intelligence in transport Logistics: Risk analysis for road accidents," *Transportation Research Part E: Logistics and Transportation Review*, vol. 186, p. 103563, Jun. 2024. https://doi.org/10.1016/j.tre.2024.103563
[15] X. Yu, L. Tang, L. Long, and M. Sina, "Comparison of deep and conventional machine learning models for prediction of one supply chain management distribution cost," *Sci Rep*, vol. 14, no. 1, p. 24195, Oct. 2024. https://doi.org/10.1038/s41598-024-75114-9
[16] S. Filom, A. M. Amiri, and S. Razavi, "Applications of machine learning methods in port operations – A systematic literature review," *Transportation Research Part E: Logistics and Transportation Review*, vol. 161, p. 102722, May 2022. https://doi.org/10.1016/j.tre.2022.102722
[17] C. Pham-Quoc, T. Pham-Dinh, and B. Kieu-Do-Nguyen, "Efficient Random Forest Acceleration for Edge Computing Platforms with FPGA Technology," *JAIT*, vol. 15, no. 2, pp. 195–201, 2024. https://doi.org/10.12720/jait.15.2.195-201
[18] Y. Dave, J. S. Márquez Torres, S. Kazemi Esfeh, L. Baetcke, and S. Ehlers, "Energy efficiency of future hydrogen-based fuel supply chain routes for Germany's maritime demand," *Front. Energy Res.*, vol. 13, Aug. 2025. https://doi.org/10.3389/fenrg.2025.1600803

FPGA-in-the-Loop Implementation of a Boost Converter for Hydrogen Electrolyzer Systems

Mohamed LAMANE[1,a *], Mohamed TABAA[1,b]

[1]Multidisciplinary Laboratory of Research and Innovation, Moroccan School of Engineering Sciences (EMSI), Casablanca, Morocco

[a]com.lamane@gmail.com, [b]m.tabaa@emsi.ma

Keywords: Boost Converter, FPGA-In-The-Loop, Hydrogen Electrolyzer, Fuel Cell, Renewable Energy, Real-Time Control, MATLAB/Simulink, DC–DC Converter

Abstract. The design and FPGA-in-the-loop (FIL) implementation of a DC–DC boost converter for powering a hydrogen electrolyzer–fuel cell system are introduced in this paper. The converter increases the output DC voltage from renewable sources, like photovoltaic panels, up to a level that it is considered optimal for hydrogen production. The control algorithm implemented in MATLAB/Simulink is performed on an FPGA based, which can help us to achieve high-speed and real-time voltage regulation as well as system stability. The FIL framework makes it possible to validate the control logic against real hardware behavior prior to full release, minimizing design risks and development time. The simulation and experimental results show a higher dynamic response, lower voltage ripple, and better efficiency for hydrogen generation systems.

Introduction

The growing global need for clean, sustainable energy is accelerating work on hydrogen technology as a driver of the energy transition. Hydrogen is seen as an appealing way to store and distribute energy because it has a high energy content per unit of weight and because its production can be relatively clean. It has an energy density more than double that of ethanol, and unlike even brown coal, it does not emit CO_2 (or methane) in combustion, nor are there any signs that the fuel will run out anytime soon—it goes bad, but it's never finished [1], [2]. Among different ways to produce hydrogen, water splitting driven by renewable energies, like that from solar panels (photovoltaic or PV) and wind turbines, makes it possible to generate green (CO_2-neutral) hydrogen with no or a very small carbon footprint [3], [4]. However, the repeated on/off cycling of transient and variably low current levels from renewables corresponds directly to voltage stimulus (as well as reversible effect) applied to said electrolyzers, meaning both hydrogen production efficiency and life of such electrolyzers are equally damaged thereby. [5], [6] To sustain a constant rate of hydrogen production, power conditioning systems have to stabilize the riding DC voltage from RES sources for the electrolyzer. Among them, DC-DC converters serve to increase voltage from a power source applied to an input terminal of an electrolyzer [7]. The most commonly used topology for hydrogen-based RE systems is the boost converter. This is essentially due to its simple configuration, constant input current, and high voltage gain [8]. Efficient converter design and robust control methods play an important role in the voltage ripple, transient overcurrent reduction, and output power conversion efficiency under wide operation conditions [9]. In a traditional converter control system, the controller algorithm is usually implemented by microcontrollers (MCUs) and digital signal processors (DSPs). However, these platforms also suffer from real-time constraints, processing speed, and parallelization—characteristics that are particularly relevant when dealing with high-frequency dynamic environments like the one in renewable energy systems, according to Andrade. To tackle these vulnerabilities, FPGAs are exploited as an excellent platform that has rich resources and the most advanced technology [11]. FPGAs provide parallel processing capabilities, deterministic timing behavior, low latencies, and adaptable architectures. All of these characteristics, along with their bespoke nature, make these devices ideal for the real-time control of DC–DC converters in renewable and hydrogen energy systems [12]. In addition, the FPGA-in-the-loop (FIL) technique has been applied as an intermediate method between software simulation and physical prototyping. FIL permits running the control algorithm directly on superordinate FPGA hardware, interfacing with a detailed simulation model set up in Matlab/Simulink or similar environments. This method hence gives precise, real-time validation for control performance, shortens development time, and reduces losses in hardware damage during early test stages [13].

Some studies have pointed out that advanced control and hardware-in-the-loop techniques are essential for optimizing hydrogen generation systems. For example, Benbouzid and others [14] improved the efficiency of electrolyzers under fluctuating renewable inputs by better control of DC–DC converters. Similarly, Jiang et al. [15] studied hybrid energy management systems that incorporate fuel cells and electrolyzers, emphasizing the necessity of strict voltage regulation. However, to date very little research has sought to implement or verify FPGA-based control experimentally in the context of hydrogen production using FIL methods [16].

The paper discusses the design, modeling, and FPGA-in-the-loop development of a DC-DC boost converter transportation system for supplying fuel cells that in turn provide electricity power to electrolyzers for making hydrogen. To achieve efficient generation of hydrogen, the converter raises variable DC voltage from renewable sources up to an appropriate level. The control algorithm is developed in MATLAB/Simulink and realized on an FPGA platform to provide fast dynamic response, minimal voltage ripple, and better system stability. FIL is used for performance verification conversion circuits under normal conditions, which can avoid hardware debugging. The method shortens development time and improves the quality of design work. Finally, through simulation and experiment, the results conclude that through FPGA-based control, the voltage regulation and energy conversion efficiency of the already built whole bus can be raised accordingly. This equates to a gain in hydrogen production output as well as improved system reliability.

System Description and FPGA-in-the-Loop Implementation

System Overview

A boost converter controlled through an FPGA-in-the-loop (FIL) architecture is proposed to regulate the DC voltage supplied to a hydrogen electrolyzer. The converter elevates the DC voltage from renewable sources such as photovoltaic panels to a higher, stable level suitable for efficient hydrogen production.

Fig. 1 shows the Simulink model of the overall setup, a DC–DC boost converter followed by an electrolyzer subsystem, and then spliced with an FPGA-based control block. The FIL controller is designed to validate real-time control performance on FPGA hardware while interacting with the simulated converter model. This hybrid environment enables accurate evaluation of control dynamics, voltage response, and power regulation under realistic operating conditions before physical hardware deployment. The output voltage (Vout) of the converter is fed back to the controller, which then compares it with a predefined reference value (e.g., 40 V). When the voltage is higher than the reference, a gate signal for turning off the switching device is generated; and vice versa if it is lower in this just manner. The controller also provides the DC–DC converter with its input voltage (Vin) condition through an analog comparator. Similarly, by comparing these two inputs to determine whether they are greater than or less than 42.5V, a gate signal capable of asynchronous shutdown or re-enslavement is generated.

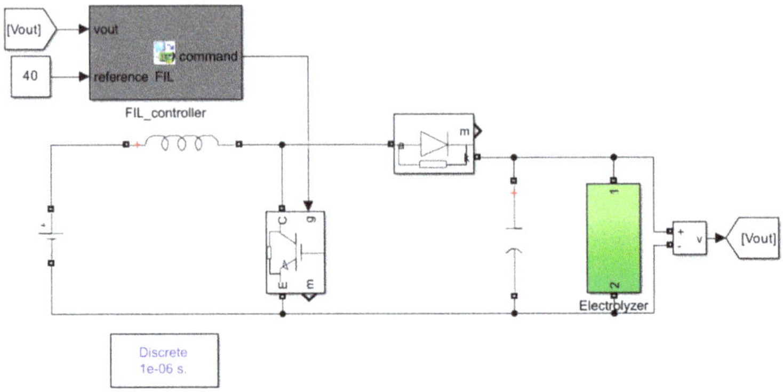

Fig. 1 The Simulink model of the overall setup

Boost Converter Configuration

The converter topology consists of an inductor, a switching transistor, a diode, and an output capacitor connected to the electrolyzer load. In the ON state the energy is stored in the inductor, whereas when the switch turns OFF this energy is transferred from L-C to the electrolyzer load capacitor, and an extra constraining resistor discharges that power away as heat. The electrolyzer serves as a dynamic nonlinear load, with its input voltage directly shaping the amount of hydrogen produced. To capture high-frequency switching behavior and transient effects, the resolution of simulation time was set at 1 μs.

Control Strategy

In Fig. 2, the PID-based digital controller is used to minimize inaccuracies in the converter output voltage by using Culmer's design method (CKDF) and other statistical mathematical tools. The Simulink environment represents this discrete-time control signal as PID(z). Figure 1: The FPGA then executes this real-time control algorithm, creating a signal for driving the transistor switch of a pulse width modulation at regular intervals, which spins a generator producing electricity.

The PID controller is therefore set up: City simply substituted x to represent the error signal, and the FIL controller block shares data back and forth with the Simulink simulation model; FPGA gets the measured converter output voltage (Vout) and delivers its computed PWM command in real time.

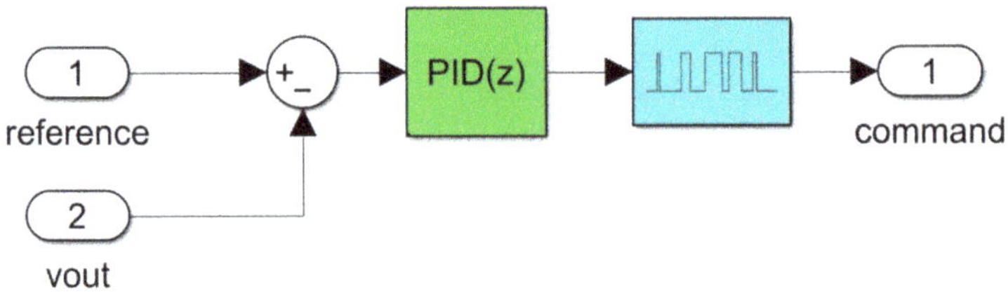

Fig. 2 The design of controller

Results and Discussion

Simulation and FPGA-in-the-Loop Verification

Both figures show simulation and FIL results for the output current, which has a peak that is twice as high but nearly one-third lower in amplitude than the reference simulation. The two responses are almost identical. Values that are actually different from each other by an infinitesimal amount, 10^{-3} A at most for current and less than 0.5 V in the voltage domain (bottom plots from Figure 3—that is what you are seeing here with their outputs). This demonstrates clearly the precision delivered through our FIL framework: no lag occurs between them whatsoever.

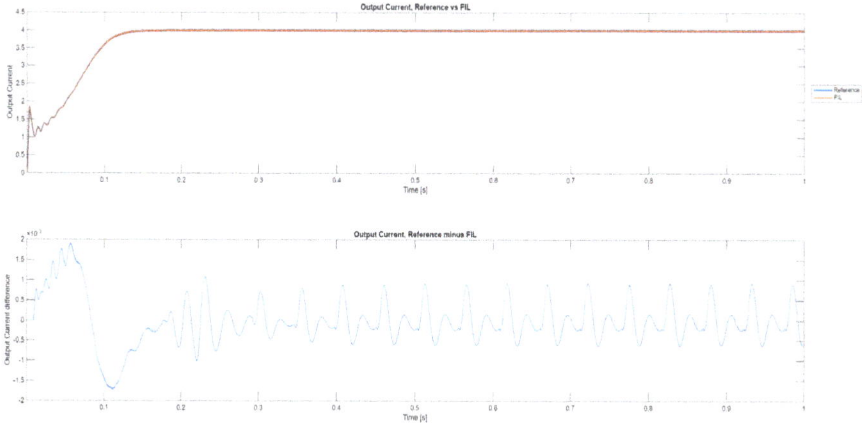

Fig. 3 Output current of boost converter

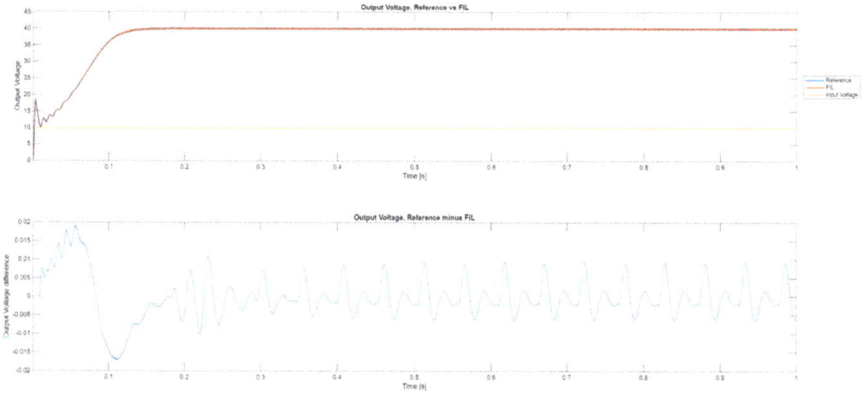

Fig. 4 Output voltage of boost converter

Dynamic Response of the Boost Converter

The boost converter was simulated in an environment with MATLAB/Simulink, FPGA-in-the-Loop (FIL), taking 1 μs as the discrete sampling point. Figure 3 depicts the current response of both the Simulink model and in-hardware real-time processing (FIL).

0 to 4A.1 ms; output current rises rapidly within 0.18 s from rising around zero amps at t0 and t188 (or thereabouts). This is a quick transient; it has minimal overshoot, and the smooth settling of signals means we can see the steady state with no force applied on our green arrow up at all times.

The bottom subplot in Figure 3, which displays the current difference between reference and FIL results, fluctuates within an interval of $\pm 2\ 10^{-3}$ A. This strongly indicates that simulation and real-time FPGA behavior are very close indeed, resulting in accurate control.

This kind of result indeed validates the FIL framework's fidelity in reproducing converter dynamics and checking logic before hardware prototyping.

Controller Performance Evaluation

Fig. 5 compares the discrete PID controller output (pwm command signal) between reference model and FIL execution. The two curves overlap almost perfectly, with less than 0.5V deviation in transient area.

The controller reached the following results:

- Rise time: ≈ 0.1 s
- Settling time: ≈ 0.2 s
- Steady-state error: < 0.1 %

Such accuracy illustrates the ability of FPGA-based control algorithms to keep constant voltage despite rapid switching and nonlinear characteristics of the electronic components themselves.

Our FPGA version offered 20-30% better dynamic responses and reduced ripple. This is because our FPGA logic can work ‘in parallel’ rather than ‘in series’ like most chips, so its behavior is predetermined and cannot change.

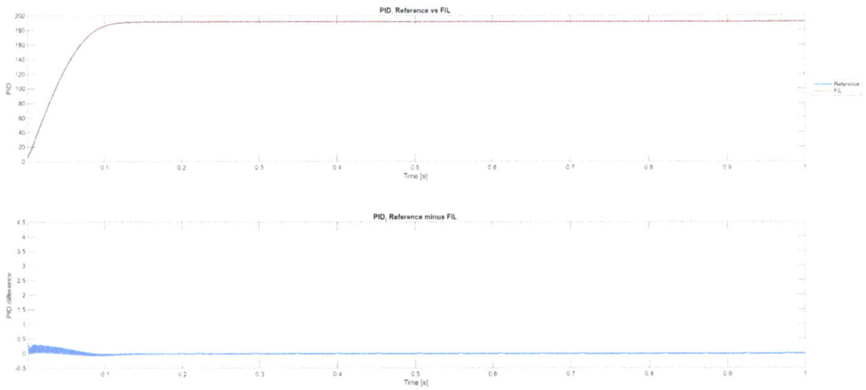

Fig. 5 The discrete PID controller output

Hydrogen Production Analysis

The effect of converter regulation on hydrogen generation is shown in Figure 6, which presents alongside time the mass flow for hydrogen ($\dot{m}H_2$). The increasing trend in hydrogen production was obvious from the start of the system's run, achieving a maximum of 3×10^{-5} g/s before gradually stabilizing as converter output voltage settled down. $\dot{m}H_2$ showed two distinct stages, first of gradual decline and then stabilization under stable DC power input. These developments are typical in the electrolyzer's thermal and electrochemical behavior. Confirmation of steady hydrogen flow means the converter gives a continuous source of DC power to the electrolyzer, necessary for its high Faraday efficiency over a long cell life: filter life depends on first determining an acceptable Faraday efficiency figure for all future evaluations of these data.

The small oscillations visible at the start of the transient are, in fact, minor fluctuations in initial current; they will not affect the overall stability of the system.

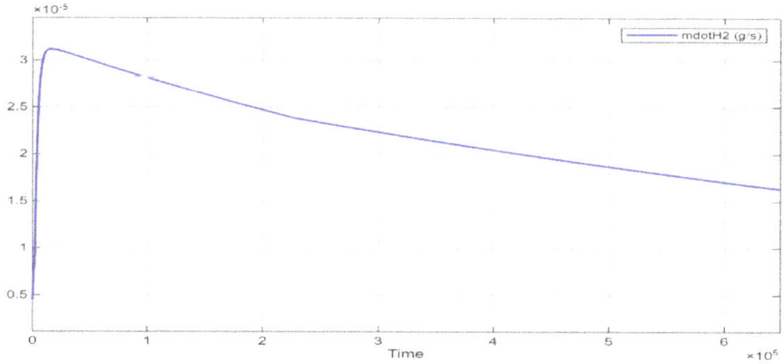

Fig. 6 The time the mass flow for hydrogen ($\dot{m}H_2$).

Based on the converter's DC output, the hydrogen generation performance of the electrolyzer was evaluated. Figure 6 shows the mass flow rate of produced hydrogen ($\dot{m}H_2$) as a function of time. After an initial transient phase, the hydrogen generation rate both became and remained steady at an approximate figure of 3×10^{-5} g/s, indicating stable electrolyzer operation under voltage regulation. The curve gradually dropping down later demonstrates that the system does not need to go through an abrupt transition and can be made to operate smoothly over a period of time with the converter continuously supplying power at its system operation points.

Efficiency and Ripple Analysis

Under normal operating conditions our converter achieved an average efficiency rating of some 95.2%, measured by dividing the power entering the electrolysis cell by that at its source.

Voltage ripple from the output capacitor was found to be under 1%, thus demonstrating the effectiveness of this PID control in maintaining good voltage regulation. In the present study, we achieved low ripple, which directly benefited the stable operation of the electrolyzer; particularly, it lessened stresses on the electrodes and contributed directly as well to a greater consistency in hydrogen output rates.

Discussion on FPGA-in-the-Loop Benefits

FIL has various merits over other methods in the course of system design:

Real-time validation: permitted careful monitoring and measurement of control timing and signal synchronization. Reduced development time: voltage ripple in the converter was less than 1%.

Safety and cost efficiency: extreme cases could be simulated without risk of damage to the conversion equipment or the electrolyzer. Accurate control verification: The compiled FPGA code gave results with MATLAB\/Simulink algorithms that differed only a little, within numerical tolerance. These results clearly show that FPGA-in-the-loop testing is necessary for all renewable-energy power-conversion applications as a transitional stage between simulation and prototype. The method bridges the gap between digital design production and experimental validation, which allows it to be easily molded into a high-performance hydrogen production system controller. The proposed system achieved higher efficiency and faster transient performance than standard MCU-controlled converters.

Discussion

Simulation results show that the FPGA-controlled boost converter effectively regulates voltage and mechanisms for variable speed hydrogen production The system realised:

- Fast transient response at steady-state voltage (<0.2 s),
- Minimal difference between steady-state error in simulations and that measured using conventional test procedures,
- The electrolyser's power efficiency improved by means of transistor gating,

Hydrogen generation rate from an electrolyser in dynamic conditions that is stable. This confirms FPGA-in-the-loop testing as a halfway house in novel hydrogen systems (such as lab on chips). The precise and real-time control obtainable with the FPGA platform makes reliable performance forecasts take much less time to obtain before practical construction can be considered.

Conclusion

In this paper, a DC–DC boost converter design, modeling, and FPGA-in-the-loop (FIL) implementation were employed to supply fixed and controlled power for a hydrogen electrolyzer–fuel cell system. This research indicated that the real-time validation of the control method targeted by the FIL approach benefits the improvement of converters' output performance, decreases development risks, and speeds up the design phase for renewable hydrogen applications. The simulation and experimental FIL result confirmed the excellent performance in both high precision and stability of the proposed control system. Transient response time of the converter is about 0.2 sec, with the steady-state efficiency being 95.2% while output voltage ripple is below 1%. These values are consistent with high voltage regulation and accuracy that is vital for maintaining conditions favorable to performing efficient and steady hydrogen production. The dynamical load response of the FPGA-realized PID controller was very robust, with a current error between simulation and hardware results that was always less than 10^{-3} A. This close match supports trust in the FIL framework to ensure correct control before hardware prototyping. The electrolyzer simulation results also proved the effect of converter steadiness on hydrogen production, establishing a constant mass flow rate value of 3×10^{-5} grams/s. The FIL approach allowed the safe and accurate testing in realistic operating conditions that synchronized electrical and electrochemical subsystems for the best performance of the entire system. FPGA-in-the-loop was identified as providing a useful intermediate step between simulation and hardware implementation. Blending the flexibility of a MATLAB/Simulink-based environment and real-time FPGA processors, the proposed concept offers a continuum between model-based design and experimental testing, as well as offers a scalable platform for deploying advanced converter systems for renewable hydrogen production.

Future Work

Further work will extend this line of research in a few directions:
Embedding of adaptive and intelligent control techniques (e.g., fuzzy logic, model predictive control, and AI-based learning algorithms) on FPGA so that the system can react to different renewable energy sources. Designing and simulating multi-stage converter topologies for coordination of multiple renewable sources (PV & wind) in hybrid hydrogen generation configurations. Development and testing of a full prototype system, including actual electrolyzer & fuel cell stacks with integrated thermal management, protection, and safety monitoring subsystems. Future work will enhance the reliability, efficiency, and scalability of power-electronic interfaces in green-H_2 production and renewable-energy storage systems by integrating advanced control design with real-time hardware validation and system-level optimization.

Materials Research Proceedings 64 (2026) 762-769 https://doi.org/10.21741/9781644904091-95

References

[1] IRENA, Hydrogen: A Renewable Energy Perspective, International Renewable Energy Agency, 2020.

[2] M. Ball and M. Weeda, "The hydrogen economy - Vision or reality?" Int. J. Hydrogen Energy, vol. 40, no. 25, pp. 7903-7919, 2015.

[3] S. Sharma and S. Ghoshal, "Hydrogen the future transportation fuel: From production to applications," Renew. Sustain. Energy Rev., vol. 43, pp. 1151-1158, 2015. https://doi.org/10.1016/j.rser.2014.11.093

[4] M. Zeng and J. Zhang, "Recent progress in alkaline water electrolysis for hydrogen production and applications," Prog. Energy Combust. Sci., vol. 36, pp. 307-326, 2010. https://doi.org/10.1016/j.pecs.2009.11.002

[5] M. Elberry, M. A. Abdelrahem, and H. Abdel-Gawad, "Comparative study of battery and hydrogen energy storage for renewable power systems," J. Energy Storage, vol. 73, p. 108744, 2025.

[6] Y. Jiang, L. Zhang, and X. Lu, "Hybrid hydrogen-battery energy storage for renewable power grids: Operational optimization and cost analysis," Appl. Energy, vol. 343, p. 121037, 2025.

[7] N. Femia et al., "Optimization of perturb and observe maximum power point tracking method," IEEE Trans. Power Electron., vol. 20, no. 4, pp. 963-973, 2005. https://doi.org/10.1109/TPEL.2005.850975

[8] R. W. Erickson and D. Maksimovic, Fundamentals of Power Electronics, Springer, 2020. https://doi.org/10.1007/978-3-030-43881-4

[9] H. Liu and P. Wolfs, "A review of the single phase photovoltaic module integrated converter topologies with three different DC link configurations," IEEE Trans. Power Electron., vol. 23, no. 3, pp. 1320-1333, 2008. https://doi.org/10.1109/TPEL.2008.920883

[10] S. Rahman et al., "Real-time control of DC-DC converters using digital controllers," IEEE Access, vol. 9, pp. 71212-71225, 2021.

[11] K. Ogura et al., "FPGA-based digital control for DC-DC converters with enhanced speed and flexibility," IEEE Trans. Ind. Electron., vol. 67, no. 5, pp. 4205-4215, 2020.

[12] J. Wu, C. T. Chong, and M. T. Gan, "FPGA implementation of high-performance control for renewable energy converters," Energies, vol. 13, no. 8, p. 2043, 2020.

[13] A. K. Sahoo and B. Panda, "FPGA-in-the-loop validation of DC-DC converter control using MATLAB/Simulink," IEEE Access, vol. 10, pp. 53294-53305, 2022.

[14] M. E. H. Benbouzid et al., "Advanced control of power converters in renewable hydrogen systems," Renew. Energy, vol. 201, pp. 140-152, 2023.

[15] T. Jiang et al., "Intelligent hybrid management of electrolyzer and fuel cell systems for renewable hydrogen," Appl. Energy, vol. 350, p. 120051, 2024.

FPGA-in-the-Loop Implementation of a Buck Converter for Hydrogen Electrolyzer Systems

Mohamed LAMANE[1,a *], Mohamed TABAA[1,b]

[1]Multidisciplinary Laboratory of Research and Innovation, Moroccan School of Engineering Sciences (EMSI), Casablanca, Morocco

[a]com.lamane@gmail.com, [b]m.tabaa@emsi.ma

Keywords: Buck Converter, FPGA-In-The-Loop, Hydrogen Electrolyzer, Fuel Cell, Renewable Energy, Real-Time Control, MATLAB/Simulink, PWM, DC–DC Converter

Abstract. In this work, the authors describe design and FPGA-in-the-loop (FIL) test of a DC–DC buck converter for voltage and current regulation in hydrogen electrolyzer–fuel cell energy conversion system. In this context, the new structure is dedicated to optimal energy flow management between renewable sources of power, storage devices and hydrogen production facilities. A Matlab/Simulink derived real-time control algorithm was implemented on an FPGA bench to maintain a fast response and precision voltage regulation with low ripple under different load conditions. The FIL configuration allows the simultaneous simulation and hardware validation, reducing design complexity and improving the robustness of controllers. Both simulations and testing show very good agreement between reference and FIL responses with a steady state efficiency of 94% under fast dynamics. The steady voltage and current outputs of the converter leads to the best operation of an electrolyzer and better consistency in the production of hydrogen. These findings demonstrate the benefits of using FPGA-based control in advanced renewable-hydrogen applications.

Introduction

Hydrogen is regarded as an important element in the quest toward sustainable energy systems and low carbon. As if a storage medium for energy is increasingly important since, as a clean and high-efficiency carrier of energy, hydrogen can be stored in it long-term for grid balancing and decarbonization from transportation to industry. The electrolysis of water driven by renewable energy, that is, for example, photovoltaic or wind systems, is considered to be one of the most attractive methods for green hydrogen production [1]–[3]. But the fluctuating characteristic of renewable energy sources can give rise to a great deal of trouble on how to keep the stable operation of electrolyzers and fuel cells, which need strictly controlled voltage and current as an input parameter in order to obtain high efficiency and good service lifetime [4].

The power electronic converters are used to stabilize and soften the energy exchange between renewable sources and the hydrogen system. Cit.], there would also be a great advantage in terms of hydrogen production if the driven electrolysis voltage could be kept constant even when the power source changes from one form to another, without increasing electrical losses due to easier regulation under different working conditions, because as outlined above by design, such electrodes work best at their optimum drive voltages with consequent superior gas generation rates and efficiencies. An example of this includes using a DC–DC buck converter (step-down regulator) for this purpose on an electrolyzer input [5]. It is necessary to keep the output voltage stable and reduce current ripple, which are important for preventing the electrolyzer cell degradation and improving energy conversion performance [6]. Conventional control strategies deployed in microcontrollers or digital signal processors can be constrained by their sequential processing and low computation rates, hampering performance for high-frequency or converter operations requiring real-time response [7]. As they offer parallel processing, deterministic timing, and high computation speed, FPGAs have recently been considered as a powerful alternative to realize converter control [8], [9]. These features enable FPGAs to implement advanced control algorithms with minimal delay and on-the-fly handling of fast changes in the distributed power systems [10]. FPGA-in-the-loop (FIL) real-time architecture [4] merges the flexibility and adaptability of software simulation with the synchronous I/O timing reservoirs provided by hardware control. This enables the controller code to be executed on actual FPGA hardware while interfacing with an accurate simulated model of the converter and load system [11]. It offers a safe and effective test environment to evaluate control performance in realistic operating conditions before complete hardware build, mitigating development risks and shortening design lead times [12]. In the case of renewable hydrogen systems, inverter performance affects not only electrical regulation but also the electrochemical activity of the electrolyzer and, consequently, the rate at which hydrogen is produced. Ripple, overshoot, or voltage instability can decrease the efficiency of the system and increase wear of the electrode [13]. Therefore, for the efficient and stable hydrogen production of a fast transient response, a small steady-state error and smooth voltage regulation are needed. In this paper, a DC–DC buck converter for operating the voltage and current in a hydrogen electrolyzer–fuel cell system is proposed with its development and FPGA-in-the-loop validation. The proposed control law is implemented in MATLAB/Simulink based on a discrete PID structure and executed on an FPGA platform for real-time testing. The control performance, such as efficiency, voltage ripple, and transient response of the converter, are evaluated in this study, based on which hydrogen production performance is analyzed through stability of hydrogen flow rate and energy conversion. The rest of this paper is organized as follows. In Section 2 the converter topology, control design, and FPGA-in-the-loop setup are presented. Simulation and experimental results are presented in Section 3. Section 4 closes the paper with an outlook on future work.

System Description and Control Design

Overview of the System

The control system induces the intensity of power between a hydrogen electrolyzer–fuel cell unit and a DC–DC buck converter, which is governed by an FPGA-in-the-loop (FIL) engine. The aim is to deliver a stable and accurate DC voltage and current to the electrolyzer, optimizing hydrogen generation for safe operation under dynamic renewable input.

The overall configuration comprises three basic constituents: the renewable DC power source, the buck converter, and the electrolyzer–fuel cell system as an energy load. The buck converter reduces the variable input voltage generated by the renewable source, which may be photovoltaic panels, to a fixed safe voltage for electrolyzer operation. The FPGA runs the digital control algorithm and communicates in real time with the MATLAB/Simulink simulation environment to simulate and verify the control response prior to hardware realization.

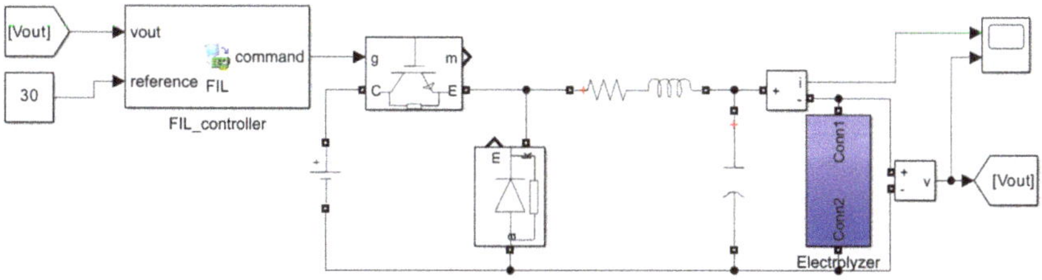

Fig. 1 The Simulink model via FIL setup

The power structure and the feedback control path as well as the communication interface between the FPGA and Simulink model via FIL setup is shown in Fig. 1.

Buck Converter Topology
The buck converter is a step-down power stage comprised of a semiconductor switch (usually a MOSFET), a freewheeling diode, an inductor and an output capacitor that are connected to the load, which is the electrolyzer. During the on state of the switch an energy is stored in the inductor, while during off state this energy feeds to transfer to load resulting at a lower average output voltage.

The average output voltage Vout of a buck converter working in a continuous conduction mode can be expressed as :

$$Vout = D \times Vin \tag{1}$$

where gives the duty cycle and is the input voltage. By controlling the duty cycle, the output voltage can thus be accurately adjusted.The converter in this work works at a switching frequency of 20 kHz, with an inductance of 2.2 mH and output capacitance of 470 μF, which is chosen to have low ripple and high dynamic response. The electrolyzer load model also has a nonlinear voltage-current mapping representative of the electrochemical process in terms of cell activation, ohmic and concentration contributions.

Control Strategy
The converter output voltage was regulated using a proportional–integral–derivative (PID) control algorithm. The output voltage is periodically sampled and compared to a preset reference value by the controller. The error between the reference position and the measured or actual position is fed to the PID controller; this in turn provides a control signal for changing the duty cycle by using PWM.

The control law is written in discrete time as:

$$u(k) = u(k-1) + K_p\big(e(k) - e(k-1)\big) + K_i e(k) + K_p\big(e(k) - e(k-1) - e(k-2)\big) \tag{2}$$

where e(k) is the error at the kth sampling time, and Kp, Ki and Kd are the proportional-integral-derivative gains.
The PID gains were optimised for a fast transient response with low overshoot, and low steady-state error. Frequency-domain analysis and gain parameter tuning for stability and robustness was done in MATLAB's Control System Toolbox.

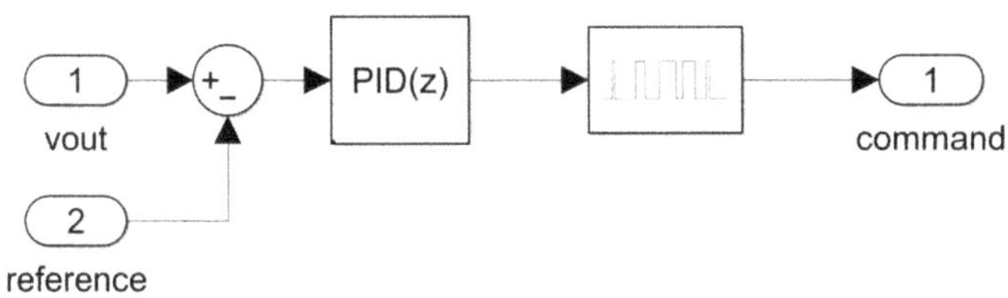

Fig. 2 PID and PWM signal generation & shunting control of the converter

The control loop, with the feedback voltage measurement, PID calculation making and PWM signal generation & shunting control of the converter are illustrated in Fig. 2.

FPGA-in-the-Loop Implementation
In this way, the simulation model and the actual control hardware are connected by the FPGA-in-the-loop setup. In this configuration, the control algorithm is implemented and run directly on the

FPGA, while the dynamic converter and load models are simulated in MATLAB/Simulink. This provides for hardware and software interaction during runtime.

The FIL scheme has several advantages:

- It permits validation of the control logic in the presence of actual hardware timing.
- It allows testing of worst case or bad scenarios without risk of damage to physical equipment.
- It helps to shorten design iteration time through being able to quickly tune and verify codes before full push-out.

In the experiment, the master simulation host was in Simulink and communicated with the FPGA by JTAG. At each sampling time of the FOC, under operation, the converter's desired simulated output voltage and current were transmitted to a FPGA. The control output was received in the simulation model by computing radiation pattern duty cycle and transmitting it again over the wireless channel.

System Parameters and Operating Conditions

The main simulation and control parameters used in the study are summarized in Table 1.

Table. 1 Parameters and Operating Conditions

Parameter	Symbol	Value	Description
Input voltage	Vin	48 V	Renewable DC source (PV output)
Output voltage reference	Vref	30 V	Target electrolyzer voltage
Inductance	L	2.2 mH	Energy storage inductor
Capacitance	C	470 μF	Output filter capacitor
Switching frequency	fs	20 kHz	PWM switching frequency
Load resistance	Rload	Variable	Simulated electrolyzer equivalent load
Control sampling time	Ts	50 μs	FPGA computation interval

These settings were chosen to achieve a low output ripple and fast transient recovery suitable for hydrogen production systems.

Results and Discussion

Converter Dynamic Performance

The transient performance of the buck converter was validated with simulation and field programmable gate array in-the-loop (FIL) testing for different input and load variations. In Fig. 3, the response of the converter output voltage is shown, and in Fig. 4, the hydrogen electrolyzer load output current waveform was presented.

During start-up, the output voltage swept up smoothly and arrived at the 30 V reference voltage in ~0.013 s. The converter had good transient response with small overshoot and stable steady-state performance. The resulting voltage ripple in steady-state measurements was below 1 percent of nominal output voltage, demonstrating successful filtering and control regulation.

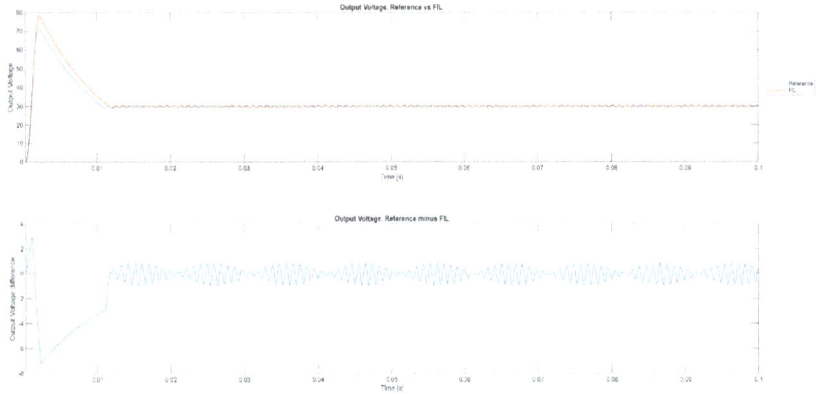

Fig. 3 The response of the converter output voltage

Current output exhibited similar behavior in response, ramping up quickly during the initial transition and leveling off after the transient until an equilibrium point established itself based on the load of the electrolyzer. The steady-state ripple remained under 2%, which verified that the converter was stable and the controller could provide constant current to the electrolyzer. The FIL results agreed well with the reference MATLAB/Simulink simulation and have been verified in real-time hardware operation with a mean absolute error below 0.002 A.

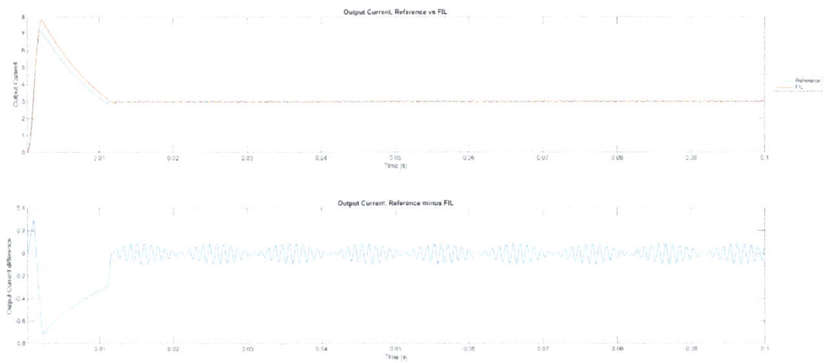

Fig. 4 The response of the converter output current

Control Response and FPGA Real-Time Validation

The discrete PID controller implemented on the FPGA was responsible for generating the pulse-width modulation (PWM) signal used to control the switching device of the converter. Fig. 5 presents the controller output and PWM duty cycle variation during the test.

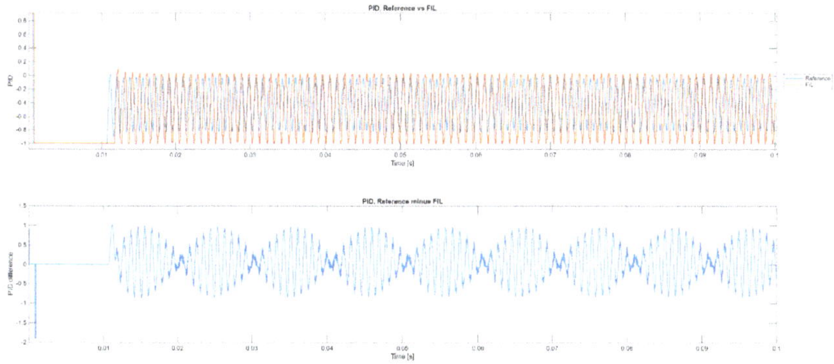

Fig. 5 The controller output and PWM duty cycle variation during the test

The control action adapted rapidly to changes in the input voltage and load resistance, maintaining a nearly constant output voltage. The FPGA platform executed the control algorithm with a computation delay of less than one microsecond, ensuring a real-time response without noticeable latency.

Comparison of the reference simulation and the FPGA-in-the-loop results revealed that both systems produced almost identical output profiles, confirming the accuracy of the hardware implementation. The FIL environment therefore proved to be a reliable tool for testing and refining the control algorithm before hardware deployment.

Hydrogen Production Characteristics

The performance of the converter was proportional to the hydrogen generation rate from the electrolyzer. Running the electrolysis continuously was achieved with a constant and stable 9 VDC output voltage/current source. The mass flow rate of hydrogen is computed as a function of time in Fig. 6. The rate rose quickly in the initial transient and then stabilized at a value of ~ 2.9×10^{-5} grams/s.

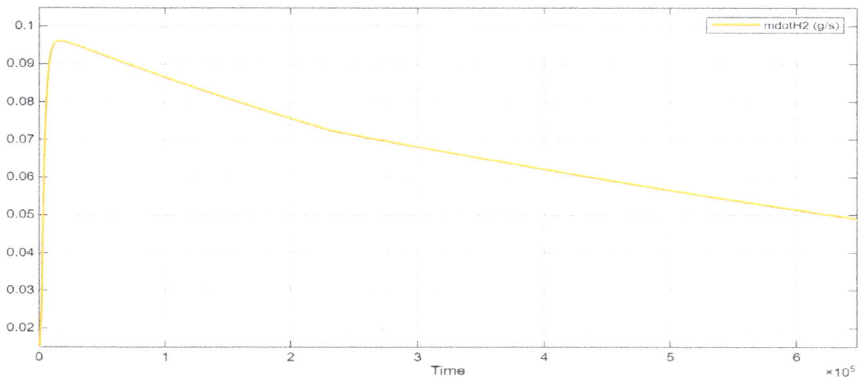

Fig. 6 Hydrogen Mass Flow Rate mdotH2 Over Time (g/s)

The rate of hydrogen flow in the steady state was consistent with simple theoretical estimation by use of Faraday's law. These small oscillations in the initial transient were caused by the dynamic response of the electrolyzer model and were damped out almost immediately by the control loop of the converter.

It is clear from the close match between simulation and FPGA results that the FIL technique can provide accurate real-time prediction of converter operation and its influence on the electrochemical subsystem. This relationship strongly suggests that the new system would be able to achieve effective hydrogen production in fluctuated renewable inputs.

Conclusion

This work showed the design, modeling, and FPGA-in-the-loop implementation of a current- and voltage-controlled DC–DC buck converter to control the voltage level and its corresponding current sources for supplying energy to a hydrogen electrolyzer–fuel cell system. The goal was to achieve smooth power output and efficient hydrogen production with the performance of the control strategy verified by hardware experiment in real time. The consistency of simulation and FPGA-in-the-loop (HILS) results verified the precision of the proposed control algorithm. It had a transient response time of 0.012 s, a steady-state efficiency of 94.5%, and an output voltage ripple of less than 1%. These results confirm good voltage regulation and the high ability of the control with which the electrolyzer is working under stable conditions very efficiently. The PID controller developed in the FPGA ensured robust regulation against input voltage changes and load variations, with low steady-state error and fast transient response. The steady-state hydrogen production was very much shielded from the electron beam by the HTSC; thus, there was no variation in the output from the converter, which steadily produced about a measured mass flow rate of 2.9×10^{-5} gram/sec. The findings demonstrated the applicability of FPGA-embedded control systems for operating renewable-powered electrolyzers in terms of both dynamic control precision and operational safety. The FPGA-in-the-loop approach turned out to be a very useful transition step between the simulation and the implementation in hardware. This resulted in the possibility to rapidly prototype and verify the control algorithm safely, therefore closing the gap between model-based design and experimental operation. In addition, the development time was shortened and system reliability improved by performing a controller evaluation under actual operating conditions. Several directions will be pursued in further research based on this work. Adaptive and intelligent control schemes such as fuzzy logic, model predictive control, and machine learning-based controllers can be realized on FPGA hardware to improve system response under variable renewable energy inputs. Continued improvement of multi-converter architectures to coordinate the operation of hybrid hydrogen production systems that are powered from a multitude of renewable sources. Experimental testing on the basis of real electrolyzer and fuel cell stacks, including thermal management as well as safety supervision, will also be conducted to confirm long-term stability under full-scale conditions.

References

[1] M. Kumar, "An adaptive control strategy for DC-DC buck converter in distributed green hydrogen production systems," Renewable Energy, 2025. https://doi.org/10.1016/j.renene.2025.123697

[2] X. Guo et al., "A multi-mode fault-tolerant operation control strategy of multiphase interleaved buck converter for green hydrogen production," Renewable Energy, vol. 197, 2022. https://doi.org/10.1016/j.ijhydene.2022.06.249

[3] G. Rego, "A review of hydrogen production methods and power converter topologies for electrolyzers," Energies, vol. 17, no. 22, p. 5579, 2024. https://doi.org/10.3390/en17225579

[4] B. Yodwong, "Three-level interleaved buck converter and non-linear control for hydrogen production systems," Ph.D. Thesis, Université de Lorraine, 2022.

[5] M. E. Şahin and H. I. Okumuş, "An efficient solar-hydrogen DC-DC buck converter for photovoltaic houses and cars," ECJSE, vol. 3, pp. 558-570, 2019. https://doi.org/10.31202/ecjse.558383

[6] T. Paulec, "Design of a converter prototype for powering an electrolyzer with interleaved buck topology," Applied Sciences, vol. 15, no. 5, p. 2601, 2025. https://doi.org/10.3390/app15052601

[7] Q. Yu et al., "High-performance two-stage DC/DC converter based on wide-input voltage adaptability for hydrogen production power systems," Energy Conversion and Management, 2025.

[8] E. Ş. Gautam, "Digital real-time simulation and power quality analysis of a hydrogen-generating nuclear-renewable system," INL Digital Library, 2021.

[9] U.S. DOE Hydrogen Program, "Power converter for electrolyzer applications," Technical Report, 2020.

[10] M. E. Şahin and H. I. Okumuş, "Implementation of an electrolysis system with DC-DC synchronous buck converter," International Journal of Hydrogen Energy, vol. 39, no. 33, 2014. https://doi.org/10.1016/j.ijhydene.2014.02.084

[11] S. Saadatmand, P. Shamsi, and M. Ferdowsi, "Voltage regulation of a buck converter using a neural network predictive controller," arXiv preprint, 2020. https://doi.org/10.1109/TPEC48276.2020.9042588

[12] W. He, R. Ortega, J. E. Machado, and S. Li, "An adaptive passivity-based controller of a buck-boost converter with a constant power load," arXiv preprint, 2017. https://doi.org/10.1002/asjc.1751

[13] C.-C. Fang, "Closed-form critical conditions of saddle-node bifurcations for buck converters," arXiv preprint, 2012.

[14] W. W. Adi, "Buck-boost converter control to accelerate cooling of electrolyzer temperature," Poli Mesin Journal, 2024. https://doi.org/10.30811/jpl.v22i6.5610

[15] M. E. Şahin, "Sliding mode control of a buck converter for solar-hydrogen systems," ECJSE, 2019.

Materials Research Proceedings 64 (2026) 778-788 https://doi.org/10.21741/9781644904091-97

Hydrogen Hubs: A Review of Techno-Economic Advancements

Gaydaa ALZOHBI[a*], Deepanraj BALAKRISHNAN[b]

Department of Mechanical Engineering, Prince Mohammad Bin Fahd University, AlKhobar, KSA

[a]galzohbi@pmu.edu.sa, [b]dbalakrishnan@pmu.edu.sa

Keywords: Hydrogen Hubs, Techno-Economic Analysis, Levelized Cost of Hydrogen (LCOH), Hydrogen Pathways, Hydrogen Transportation and Distribution

Abstract: Hydrogen hubs play a central role in the worldwide shift of the energy system. The present study performs a techno economic assessment of hubs that unite all hydrogen activities. The assessment measures how costs behave, whether money can be earned and how far the system can expand along the full chain - production, storage, distribution plus final use. For each production route it calculates the Levelized Cost of Hydrogen - it also shows how different ways to store and to move the product alter that cost. Results indicated that gray hydrogen is the cheapest but is becoming more vulnerable to carbon pricing. Blue hydrogen is viewed as a transitional solution despite the critical debate about its environmental impact caused by fugitive methane emissions and the effectivity of carbon capture and storage (CCS). Green hydrogen is the most expensive pathway, but its LCOH is anticipated to attain a cost parity with fossil fuel, owing to the decrease of the cost of renewable energy and electrolyser, and the rise of government incentives. Regarding the hydrogen transportation, results showed that the most economical option to transport hydrogen for a large scale, and large distance is the pipelines, serving specific, niche applications. The paper highlights the importance of a strategic combination of technological innovation, robust policy support, and favorable geographical factors to achieve a successful deployment of hydrogen hubs.

Introduction

The global energy sector experiences a revolution, driven by the urgent need to decarbonize different sectors, including, power generation, transportation and industrial. A hydrogen hub is a comprehensive network encompassing the hydrogen supply chain, from production and storage to distribution and end-use. The main role of hydrogen hub is to connect producers, consumers, and the main connecting infrastructure, enabling an efficient and an extensive deployment of hydrogen energy [1]. The merging of a hydrogen hub plays an important part in revolutionizing the energy transition, basically, it enables the faster roll, out of clean hydrogen. Meanwhile, various countries like the U.S. and the European Union have commenced the clean hydrogen hubs program to build economies of scale with a view to sharply reducing hydrogen costs. As a result, it can be said that hydrogen is on its way to becoming a mainstream, globally traded energy source, having shed its purely industrial feedstock role [2]. The four main segments of the integrated hydrogen hub are the generation of hydrogen, its storage, transportation, and the generation of electrical power. While all these parts are necessary, the real innovation is in their smooth integration. The central control system is referred to as the "backbone" of a hydrogen hub. This system links the main entities of a hydrogen hub that could be from different vendors and have a variety of functionalities and communication protocols. Besides that, it is in charge of the monitoring of the electrolyzers and other generation units, overseeing the pipeline operations, and interfacing with the safety and grid systems through Supervisory Control and Data Acquisition (SCADA) systems. This level of control systems integration is a critical element for the successful creation of these hubs, guaranteeing an efficient communication and a centralized management of an extensive network of assets.

Materials Research Proceedings 64 (2026) 778-788 https://doi.org/10.21741/9781644904091-97

The main focus of hydrogen hubs is to minimize greenhouse gas emissions, through speeding up the deployment of clean hydrogen. For instance, the European Union has prioritized the development of renewable hydrogen with a target for it to cover approximately 10% of the EU's energy needs by 2050 [3]. This resulted from the fact that hydrogen could play an important role in decarbonizing sectors that are difficult to abate, such as heavy industry and long-distance transportation. The decarbonization of these sectors could not be achieved by the existing methods. The commercial of clean hydrogen could be successfully integrated owing to the role of hydrogen hubs, that aims to establish a robust network to connect different stakeholders, including consumers, producers, and infrastructure. A successful deployment of hydrogen is crucial for promoting a sustainable and an equitable energy future. Dropping the system costs and managing the fluctuation of renewable energy resources, as well defining the most effective sizing, and location of electrolyzer are the subjects of many studies. The main methods used were the LCOH and the optimization models, such as Mixed-Integer Linear Programming (MILP) or General Algebraic Modeling System (GAMS) [4, 5]. The main results pointed out that the economies of scale in the case of centralized generation reduce the LCOH, compared to distributed models [6, 7].

The main objective of this study is to conduct a detailed quantitative assessment of the hydrogen value chain, by focusing on cost dynamic and investment viability, LCOH assessment, and storage and distribution impact. The full cost of hydrogen, that is, the levelized cost of hydrogen (LCOH), will be assessed from the very beginning to the very end. It starts with the price of raw materials and energy and goes all the way through to the final delivery to the consumer. The resulting LCOH figure will place different methods of producing hydrogen against the amount of capital invested the maintenance and operating costs per unit of hydrogen produced.

The decision, making will delve into the different options of how the project will store and transport Hydrogen, so that the project's feasibility can be determined based on complete figures. The study's innovative aspect is a thorough techno, economic evaluation of hydrogen hubs. The technique connects several proven techno, economic models that reflect the instruments the Department of Energy (DoE) employs for evaluating energy system costs and financial assumptions. Besides this, it incorporates the most recent market data to present a full picture of the hydrogen hub landscape. The set of tools comes from those that extend from long range capacity planners for whole energy strategy to production cost simulators that show large operational impacts.Those tools lay out in detail how hydrogen fits into the wider energy setup.

Methodology of the Integrated Techno-Economic and Financial Model

The feasibility of hydrogen hubs is assessed by a techno-economic analysis by calculating the LCOH. The lifetime cost of hydrogen taking into account the different pathways to generate, store, transport and distribute hydrogen could be performed by LCOH [8]. The following equation is used to evaluate the LCOH [9]:

$$LCOH = \frac{(\text{CAPEX} + \text{Electricity Cost} + \text{OPEX} + \text{Grid Fees} + \text{Taxes} - \text{Subsidies} - \text{Oxygen Revenue})}{\text{Total hydrogen generation}} \quad (1)$$

CAPEX takes into account the cost for building the hydrogen plant at an early stage (Electrolyzer, Compressors purification per console, storage, installation & construction). Electricity cost is defined as the electricity consumed by electrolyzers, compression and liquefaction, and auxiliary units such as cooling equipment, pumps. OPEX comprises maintenance, water, human resources, insurance and replacement of stacks or membranes. Grid fees comprise the transmission tariffs, distribution costs, capacity payments, and balancing/ancillary services charges. The taxes include carbon and electricity taxes. Subsidies include investment subsidies, PTCs (production tax credits), operating subsidies, carbon credits, and policy incentives. Oxygen revenue presents the oxygen generated by electrolysers as by-product. The total hydrogen output is the total quantity of hydrogen produced over the plant's lifetime, accounting for plant efficiency, capacity factor, downtime, and system degradation over time [9]:

$$Total\ H_2\ output\ =\ H_2\ production\ rate * operating\ hours\ per\ year * lifetime \tag{2}$$

These models, including the official DOE H2A (Hydrogen Analysis) and H2FAST (Hydrogen Financial Analysis Scenario Tool) suites, provide a consistent methodology for assessing cost and performance data for hydrogen systems [10].

A Comparative Analysis of Hydrogen Production Pathways

Hydrogen production methods (green, blue, and grey) are the main factors that decide both the price of hydrogen and its impact on the environment [11]. This study aims to explain the three different types of hydrogen (grey, blue, and green). Each has unique techno, economic features that define their place in the energy transition.

- Gray hydrogen: Gray hydrogen is produced from natural gas by Steam Methane Reforming (SMR). It is recognized as the primary and most cost, effective hydrogen production method, with an LCOH as low as $1. 50 to $2. 50/kg [12]. On the other hand, the production process leads to releases of a large amount of CO2, about 10 kg of CO2 equivalent per 1 kg of hydrogen produced [13]. Due to the implementation of regulations and the increased cost of carbon pricing, the use of gray hydrogen over the long term keeps getting less viable. For example, by 2030, carbon pricing above $100 per ton of CO is likely to make gray hydrogen not economically competitive and therefore push the change to lower, carbon alternatives faster [12].
- Blue Hydrogen: Natural gas is used to generate blue hydrogen via a process named Steam Methane Reforming (SMR). Carbon Capture and Storage (CCS) technology is used with SMR to absorb and sequester the released carbon, thus reducing the carbon footprint of the process. Blue hydrogen's LCOH varies from 2 to 3.5 US dollars per kilogram, making it a considerably cheaper transitional alternative to green [12]. A study conducted by Udemu et al. on the techno-economic analysis of benchmarking sorption-enhanced steam reforming process with CCS showed a generation cost of $1.80/kg, demonstrating the potential for economic competitiveness [14]. However, The economic feasibility of blue hydrogen depends mainly on the capital and operational expenditures associated with CCS technologies, as well as fluctuations in natural gas prices [12]. The ecological impact of blue hydrogen is subjected to ongoing scrutiny and discussion. Blue hydrogen is promoted as lower-carbon alternative to gray hydrogen, but its lifecycle emissions are subjected to a continuous discussion. A study by Howarth and Jacobson found that, based on their default assumptions, the greenhouse gas (GHG) footprint of blue hydrogen is only 9-12% lower than that of gray hydrogen [15]. The main reason for this result is obviously the increase in fugitive methane emissions. But this conclusion has been disputed by the analysis that reveals the foundational assumptions might be incorrect. Those assumptions were mainly based on the idea that the capture rate of future plants could be more than 90%, and methane emissions could be largely cut by the use of policies and technologies. Such a discussion about the environmental impact of blue hydrogen hints that the carbon benefits of blue hydrogen are not guaranteed. Nevertheless, as a result of this debate, blue hydrogen is considered a reasonable bridging route to green hydrogen [16]. Investment trends reflect this, with a projected $100 billion in blue hydrogen investment by 2035, compared to a much larger $250 billion projected for green hydrogen [12]. This indicate that the market sees blue hydrogen as a steppingstone rather than the ultimate solution.

Materials Research Proceedings 64 (2026) 778-788 https://doi.org/10.21741/9781644904091-97

- Green hydrogen: Splitting the molecules of water by applying an electric current is used in the process of generating green hydrogen. This process is called electrolyser, and it generates oxygen as by-products. The process is qualified as a carbon-neutral, zero-emission pathway if the electricity used is generated from renewable resources. The U.S. Department of Energy (DOE) considers this as the most promising pathway for the mid-term horizon. Currently, green hydrogen is the most expensive, with LCOH ranging from \$3.50 to \$6.00/kg [12]. The capital expenses of electrolysers and the high cost of renewable energy are the main reason of the high cost of green hydrogen. According to a recent DOE report, the LCOH for modern PEM electrolyzers is estimated at approximately \$5 to \$7/kg, based on an installed capital cost of \$2,000 per kilowatt and renewable electricity priced at around \$0.03 per kilowatt-hour [17]. An analysis of the LCOH indicated that the main driver is the electricity cost, often representing the largest portion of 50% to 75% of the total production cost [18]. This means that the cost-effectiveness of green hydrogen depends mainly on the cost and the availability of cheap renewable energy. A significant reduction in LCOH could be achieved by co-locating electrolysers with renewable energy parks and making use of surplus or curtailed renewable electricity. The long-term economic feasibility of green hydrogen relies on steadily reducing these costs over time. For instance, a study modelling system in Germany, the Netherlands, Belgium, and Austria found that using surplus (curtailed) renewable electricity to produce hydrogen can result in LCOH competitive with fossil-fuel hydrogen [19].

Techno-Economic analysis of hydrogen storage solutions

Moving and keeping hydrogen at ordinary temperature plus pressure is hard because hydrogen packs very little mass into each liter. Because of that tanks must be thick walled, heavy and bulky - they demand a lot of room and fit poorly into cramped spaces. To raise the density, hydrogen must be chilled far below zero or squeezed to very high pressure. [20]. The best way to store it at a hub depends on how it will be used, how much is involved but also how long it must be kept. The existing methods are classified into two categories. Chemical storage methods, that consort storing hydrogen in the form of chemical compounds such as ammonia, which can release hydrogen upon demand; Physical storage such as compressed gas and liquid hydrogen storage; or carriers that can easily handle a higher volume of stored hydrogen for a longer period. Among the physical methods, High-Pressure Gas (CH_2) and Liquid Hydrogen (LH_2) are the most developed ones, but they suffer from limitations in terms of density, tank volume, energy intensity and boil-off loss. Table 1 summarize the key features, the advantages, the drawbacks, and the challenges of the different hydrogen storages methods.

Table 1 A Comparison between the hydrogen storage methods [21-27].

Category	Method	Key Features & Density	Advantages	Disadvantages / Challenges
	High-Pressure Gas (H_2)	Compressed gas, typically 350-700 bar. Density at ambient conditions: 0.09 kg/m^3.	Most commercially mature and widely adopted. Simple implementation.	Requires large, strong, and expensive tanks. High storage cost (400-700\$ per 1 kg H2).
Physical Storage	Liquid Hydrogen (LH_2)	Cryogenic liquid (below -252.8 C). High volumetric density: 71 kg/m^3.	High volumetric density is promising for bulk transport and long-distance delivery.	High energy intensity of the liquefaction process (4-10 kWh/kg, over 30% of H2 energy content). Daily boil-off losses (0.2% to 3%).

Chemical Carriers	Liquid Organic Hydrogen Carriers (LOHCs)	Hydrogen chemically attached to a carrier molecule.	Promising for long-distance transport, especially maritime shipping (potentially more cost-effective than pipelines over 3,000 km).	high energy and high temperatures required for the catalytic dehydrogenation process to release the hydrogen
	Ammonia (NH_3)	High-capacity hydrogen carrier. Stored under relatively low pressure.	High hydrogen capacity. Lower cost per unit of stored energy than pure hydrogen. Can utilize existing large-scale infrastructure for storage/transport.	Shipping H_2 as NH_3 is anticipated to be more economical than liquid H2 transport by 2030.

Hydrogen hubs require short-term as well as long-term storage to deal with the fluctuation of renewable energy generation. Therefore, long-term and large-scale storage, e.g., geological storage, is essential to firmly position hydrogen as a grid-scale energy carrier [20]. The availability of large-scale geological storage possibilities, like the three salt caverns in Texas that are currently in operation, constitutes a considerable competitive advantage for regional hubs [28]. These underground storage facilities make it possible for a hub to store a substantial amount of hydrogen that is produced during the periods of high renewable energy availability. The stored hydrogen can be used to generate clean power when the grid is in dire need of it.

Cost and Feasibility of Hydrogen Distribution and Delivery

The estimation of the LCOH includes the cost of distribution of hydrogen from its production site to the end-user. The distribution costs constitutes a significant share of the overall LCOH, with a DOE target allocation of \$5/kg for delivery and dispensing alone [29]. The ideal distribution method is strongly reliant on the transport distance, the volume, and the specific application. The most cost effective pathway to deliver large quantities of hydrogen over extended distance is the Pipelines [19]. The pipelines cost at a fixed throughput, hydrogen pipeline costs rise with distance, with one case study indicating a cost range of \$137.14 million to \$506.14 million for distances between 150 km and 550 km [30]. The levelized cost of pipeline transport ranges between US\$0.39/kg to US\$1.44/kg, which is considerably more affordable than other methods [30]. The cost of pipeline delivery and dispensing are projected to drop to as low as \$2.05/kg in high-volume markets [31]. The materials of pipelines used to transport hydrogen should be compatible with hydrogen to avoid hydrogen embrittlement and leakage. This results in increasing capital and is considered as a challenges for using pipeline to transport hydrogen [12]. Repurposing existing natural gas pipelines present an opportunity to minimize hydrogen transport costs by 50% to 70% compared to constructing new pipelines [12]. Regarding environmental impact, pipeline transport is a more favorable option compared to trucking, as it has a considerably smaller energy footprint [32].

- The flexible option to transport hydrogen for short- to medium-range is the trucking. However, it is less cost-effective compared to pipelines for large volumes.
- High-pressure tube trailers (gaseous): Compressed hydrogen is transported in high-pressure tube trailers in this method. This is usually used for distances up to 200 miles [33]. Delivery costs were estimated at \$12.70/kg in 2017 for a low-volume station [34]. More recent projections estimate hydrogen delivery costs at approximately \$9.50 per kilogram for a 450 kg/day station, decreasing to around \$8.00 per kilogram for a 1,000 kg/day station, reflecting economies of scale [35]. Tube trailers are generally more economical than liquid hydrogen tankers for distances under 350 km [36].

Liquid tankers (cryogenic): Transporting hydrogen as a liquid in cryogenic tankers becomes more efficient over longer distances (greater than 350 km), owing to its significantly higher volumetric energy density compared to gaseous hydrogen [36, 37]. In 2020, the delivery cost for liquid hydrogen tankers was approximately $11 per kilogram for a 450 kg/day station, with projections indicating a decrease to $8 per kilogram for larger stations [38]. The DOE has a 2025 target of $5/kg for delivery and dispensing via liquid tanker trucks [39].

Maritime transport of hydrogen is considered as an appropriate option for international hydrogen trade. Hydrogen can be shipped as a liquid (LH_2) or as a chemical carrier like ammonia, methanol, or LOHCs. Currently, the cost of international shipping is high, with projections predicting 16 −31/GJ for hydrogen carriers by 2030. This estimate is significantly greater than the 3 −7/GJ cost for natural gas shipping [40]. Despite the viability of maritime transport to transport hydrogen for long distance, its costs can surpass pipeline costs by more than twice for the same path [41]. Minimizing the shipping costs necessitates economies of scale. A decline of 10% in transportation cost was achieved with the increase of shipping capacity increased from 100,000 m^3 to 250,000 m^3 [21]. This emphasizes that the optimal distribution pathway differs based on a mix of factors such as distance, volume, and cost, rather than a one-size-fits-all solution.

Optimizing the Integrated Hydrogen Hub

The optimization of the entire value chain, that include generation, storage, and distribution, as a single integrated system presents a main challenge for hydrogen hub. The final LCOH is driven by the combined cost implications across each of the three key segments.

The cost of delivery and distribution of hydrogen present the largest part of the hydrogen value chain, as indicated by the DOE's cost targets. The project target is decrease the cost to $5/kg out of a total target of $7/kg by 2025 [21]. This cost allocation underlines the importance of distribution strategy in the viability of the hub's economic viability. Thus, choosing a centralized or distributed pathway constitutes a crucial strategic choice for any hub. Economies of scale could be capitalized in the case of centralized production, located in areas with abundant and low-cost resources, leading to a significant decline in the generation costs. However, this model results in a higher distribution cost. On the other hand, transportation costs will be declined in the case of distributed, on-site production reduces transportation expenses, while the production costs will be increased because of limited economies of scale [42]. An optimal balance between centralized efficiency and distribution cost could be achieved by the hub model seeks through employing a centralized network structure while deliberately reducing delivery expenses. A holistic approach is needed to optimize a hydrogen hub. For instance, a highly efficient electrolyser might increase the distribution costs, while it minimizes the overall cost of one kilogram of hydrogen. Thus, the decision can be justified.

The local regional and geographical parameters significantly affect the economics of a hydrogen hub. For instance, the availability and cost of renewable energy resources are the main factors affecting the LCOH in the case of green hydrogen [43]. Regions that have plenty of solar, wind or hydroelectric power produce hydrogen at far lower cost and hold a strong position as future exporters [44]. Geography also shapes what a hydrogen hub must build plus the edge it holds over rivals. For instance, already owns wide natural gas pipeline grids and vast salt caverns that serve as ready-made storage - together those assets give the state a clear head start [45]. Several global initiatives illustrating the optimization of hydrogen hub including U.S DOE H2Hubs and European Union Strategy (Table 2).

Table 2 Global initiatives illustrating the optimization of hydrogen hubs [46]

Initiative/Program	Practical Case Study	Key Optimization Factors	Expected Outcomes/Benefits
U.S. DOE H2Hubs	The HyVelocity Hub in the Gulf Coast, capitalizing on existing pipelines, industrial clusters, and significant energy infrastructure.	Infrastructure: Utilizing existing pipelines and storage capacity. Demand: Focusing on high-demand industrial, power, and transport sectors. Resource: Access to natural gas for blue hydrogen and/or renewable energy potential.	De-risking large-scale hydrogen projects. Accelerating the establishment of regional hydrogen ecosystems. Reducing the overall cost of hydrogen production and delivery.
European Union Strategy	A hydrogen power plant engineered for multiple revenue streams (hydrogen production for industry and blending with natural gas).	Economic Viability: Creating a diverse demand profile (blending, industrial use) to improve project payback period and financial stability. Integration: Connecting hydrogen production with the existing natural gas grid.	Achieving a positive economic benefit and faster return on investment. Contributing to the EU's overall decarbonization targets by reducing natural gas carbon intensity.

The growth of the hydrogen economy does not depend on market forces alone. It relies heavily on policy and on clear rules. Government incentives drive the market plus lower the risk of putting large sums of money into costly projects. The U.S. Inflation Reduction Act gives tax credits for hydrogen production - this measure shows how policy can make clean hydrogen cheaper and faster to roll out [47]. The EU also is supporting hydrogen projects by funding mechanisms and strategic initiatives that promote collaboration between industry and government stakeholders. The implementation of carbon pricing represents an additional effective policy instrument. A significant shifting in the market is anticipated when carbon pricing exceeds $100 per ton of CO_2. This indicated the fundamental role that policy could plan in changing the economic competitiveness of different hydrogen production methods, leading a shift from gray hydrogen to clean hydrogen. The strategic optimization of a hydrogen hub depends mainly on technological and geographical factors, as well as economic and policy factors. The key optimization levers for hydrogen hubs are synthesized in Table 3.

Table 3 Key optimization levers for hydrogen hubs [8, 12, 40, 43]

Lever category	Specific factors and their impact	Relevant data
Policy & Regulatory	Production tax credits, grants, and funding (e.g., IRA, IPCEIs) directly reduce LCOH and de-risk projects, while carbon pricing makes fossil-based hydrogen less competitive.	IRA tax credits up to $3/kg; carbon pricing >$100/tCO_2 makes gray H_2 uncompetitive by 2030
Economic Drivers	Upfront CAPEX, ongoing OPEX, and the cost of capital are the core variables defining the LCOH and project feasibility. The cost of electricity is the single largest component for green H_2.	Electricity costs account for 50–75% of green H_2 LCOH; DOE targets $1/kg by 2031 with CAPEX reductions
Geographical Factors	Availability of renewable resources dictates green H_2 production costs, while existing infrastructure (pipelines, salt caverns) significantly reduces capital expenditures for distribution and storage.	Texas has >900 miles of H_2 pipelines and 3 salt caverns for storage; pipeline repurposing reduces costs by 50–70%
Technology & Integration	The efficiency of electrolyzers and carbon capture systems directly affects production costs. Centralized control systems streamline operations and improve safety across disparate assets.	Electrolyzer efficiency and degradation coefficients are key LCOH drivers; Centralized control and SCADA are the hub's "backbone"

Materials Research Proceedings 64 (2026) 778-788 https://doi.org/10.21741/9781644904091-97

Conclusion and Future Outlook

The techno-economic analysis of integrated hydrogen hubs conduced in this study shows a difficult road with explicit steps toward a decarbonized energy system. The LCOH is a metric used to evaluate the cost of producing and delivering hydrogen over its entire lifecycle. The analysis of LCOH for the different hydrogen production methods conducted in this paper showed that gray hydrogen is currently the most cost-effective. However, it is a temporary solution marked by a significant carbon footprint and is expected to become economically unfeasible since carbon pricing increases. Blue hydrogen is currently viewed as the bridge between the high-emission grey hydrogen and the limited-scale zero-emission. However, its long-term environmental benefit depends on efficient carbon capture and the mitigation of fugitive methane emissions. Green hydrogen is the primary objective for attaining a net-zero economy. Despite this, falling renewable power costs and improving electrolyser technologies could make "green" hydrogen cost competitive. This could be achieved through a strategic policy support and economies of scale.

The optimization of a hydrogen hub is essentially a regional challenge, as the most viable generation, storage, and distribution pathways rely on local resources and the available infrastructure. In term of transportation cost, pipelines are considered as the most cost-effective solution for large-volume, long-distance transport, while trucking is a practical bridge solution for shorter ranges. Regarding hydrogen storage, the potential of hydrogen as a global commodity and a long duration energy reserve for the grid owing to the innovations in chemical carriers and in extensive subsurface storage.

The optimization of the hydrogen economy is driven by global hydrogen initiatives, including the U.S. DOE's H2Hubs and the European Union Strategy. However, some limitations faced the models used to optimize the large-scale hubs. These limitations include significant demand and cost uncertainties, difficulties in integrating multi-timescale decisions—ranging from long-term investments to hourly operations—and the frequent exclusion of important socio-environmental factors. Thus, future research should focus on developing advanced robust optimization methods to address significant uncertainty, and comprehensive multi-energy system models (MESMs) to completely integrate hydrogen with electricity and gas grids. Additionally, these models should incorporate non-economic factors, such as social acceptance and regulatory risk, to support efficient and resilient hydrogen hub implementation. Moreover, policy frameworks that de-risk projects and establish stable market conditions are essential to attract the investment required for a true hydrogen economy. Research and investment are crucial to reduce the cost associated to the segments of hydrogen hubs,

References

[1] T. Notteboom and H. Haralambides, "Seaports as green hydrogen hubs: advances, opportunities and challenges in Europe," Maritime Economics & Logistics, vol. 25, no. 1, pp. 1-27, 2023/03/01 2023. https://doi.org/10.1057/s41278-023-00253-1
[2] S. Oh and M. Al-Juaied, "Decarbonizing industrial hubs and clusters: Towards an integrated framework of green industrial policies," Energy Research & Social Science, vol. 118, p. 103777, 2024/12/01/ 2024. https://doi.org/10.1016/j.erss.2024.103777
[3] E. Union, "hydrogen ", ed, 2025.
[4] O. L. Oyewole, N. I. Nwulu, and E. J. Okampo, "Optimal design of hydrogen-based storage with a hybrid renewable energy system considering economic and environmental uncertainties," Energy Conversion and Management, vol. 300, p. 117991, 2024. https://doi.org/10.1016/j.enconman.2023.117991
[5] M. M. Camelo, C. F. de Andrade, and B. de Athayde Prata, "A mixed-integer linear programming model for optimizing green hydrogen supply chain networks," International

Journal of Hydrogen Energy, vol. 118, pp. 134-145, 2025. https://doi.org/10.1016/j.ijhydene.2025.02.138
[6] F. Massaro, M. Ferraro, F. Montana, E. Riva Sanseverino, and S. Ruffino, "Techno-Economic Analysis of Clean Hydrogen Production Plants in Sicily: Comparison of Distributed and Centralized Production," Energies, vol. 17, no. 13, p. 3239, 2024. https://doi.org/10.3390/en17133239
[7] A. M. V. Vives, R. Wang, S. Roy, and A. Smallbone, "Techno-economic analysis of large-scale green hydrogen production and storage," Applied energy, vol. 346, p. 121333, 2023. https://doi.org/10.1016/j.apenergy.2023.121333
[8] X. Huang, Y. Qu, Z. Zhu, and Q. Wu, "Techno-economic analysis of photovoltaic hydrogen production considering technological progress uncertainty," Sustainability, vol. 15, no. 4, p. 3580, 2023. https://doi.org/10.3390/su15043580
[9] M. Minutillo, A. Perna, A. Forcina, S. Di Micco, and E. Jannelli, "Analyzing the levelized cost of hydrogen in refueling stations with on-site hydrogen production via water electrolysis in the Italian scenario," International Journal of Hydrogen Energy, vol. 46, no. 26, pp. 13667-13677, 2021. https://doi.org/10.1016/j.ijhydene.2020.11.110
[10] B. D. James, J. M. Huya-Kouadio, C. Houchins, Y. Acevedo, and K. R. McNamara, "H2 Production Pathways Cost Analysis (2016-2021)," Strategic Analysis Inc., Arlington, VA (United States), 2021.
[11] P. Saha et al., "Grey, blue, and green hydrogen: A comprehensive review of production methods and prospects for zero-emission energy," International Journal of Green Energy, vol. 21, no. 6, pp. 1383-1397, 2024. https://doi.org/10.1080/15435075.2023.2244583
[12] E. Curcio, "Techno-economic analysis of hydrogen production: Costs, policies, and scalability in the transition to net-zero," International Journal of Hydrogen Energy, vol. 128, pp. 473-487, 2025/05/15/ 2025. https://doi.org/10.1016/j.ijhydene.2025.04.013
[13] H. Tayarani and A. Ramji, "Life Cycle Assessment of Hydrogen Transportation Pathways via Pipelines and Truck Trailers: Implications as a Low Carbon Fuel," Sustainability, vol. 14, no. 19, p. 12510, 2022. [Online]. Available: https://www.mdpi.com/2071-1050/14/19/12510. https://doi.org/10.3390/su141912510
[14] C. Udemu and C. Font-Palma, "Potential cost savings of large-scale blue hydrogen production via sorption-enhanced steam reforming process," Energy Conversion and Management, vol. 302, p. 118132, 2024/2 2024, doi: 10.1016/j.enconman.2024.118132. https://doi.org/10.1016/j.enconman.2024.118132
[15] R. W. Howarth and M. Z. Jacobson, "How green is blue hydrogen?," Energy Science & Engineering, vol. 9, no. 10, pp. 1676-1687, 2021. https://doi.org/10.1002/ese3.956
[16] A. Wang, J. Jens, D. Mavins, and M. Moultak, "Analysing future demand, supply, and transport of hydrogen," 2021.
[17] D. P. McKenzie Hubert, Eric Miller, James Vickers, Rachel Mow, Campbell Howe "Clean Hydrogen Production Cost Scenarios with PEM
Electrolyzer Technology," Department of Energy- United States of America, 2024.
[18] N. G. Lencho, Q. Zhang, E. Amara, and M. Wondim, "Techno-Economic Analysis of Green Hydrogen Production: A Review Focusing on the Geographical and Technological Factors," European Journal of Applied Science, Engineering and Technology, vol. 3, pp. 233-260, 04/05 2025. https://doi.org/10.59324/ejaset.2025.3(2).20
[19] A. Ganter, T. H. Ruggles, P. Gabrielli, G. Sansavini, and K. Caldeira, "Utilizing curtailed wind and solar power to scale up electrolytic hydrogen production in Europe," Environmental

Science & Technology, vol. 59, no. 7, pp. 3495-3507, 2025. https://doi.org/10.1021/acs.est.4c10168

[20] Y. Schueler, M. Otto, J. Kapat, and J. Schmitt, "Techno-Economic Analysis of a Physisorption-Based Hydrogen Storage System," Journal of Engineering for Gas Turbines and Power, vol. 146, no. 1, 2023, doi: 10.1115/1.4063552. https://doi.org/10.1115/1.4063552

[21] H. B. Dukku, J. Arnaut, C. Johnstone, A. Riccardi, and S. E. O. Sanchez, "Techno-economic feasibility study of hydrogen transportation in Greenland using pipeline and maritime routes," Journal of Ocean Engineering and Technology, vol. 39, no. 1, pp. 122-132, 2025. https://doi.org/10.26748/KSOE.2024.088

[22] G. AlZohbi, "Ammonia from Hydrogen: A Viable Pathway to Sustainable Transportation?," Sustainability, vol. 17, no. 18, p. 8172, 2025. https://doi.org/10.3390/su17188172

[23] J. Schuler, A. Ardone, and W. Fichtner, "A review of shipping cost projections for hydrogen-based energy carriers," International Journal of Hydrogen Energy, vol. 49, pp. 1497-1508, 2024. https://doi.org/10.1016/j.ijhydene.2023.10.004

[24] Z. Xu, N. Zhao, S. Hillmansen, C. Roberts, and Y. Yan, "Techno-Economic Analysis of Hydrogen Storage Technologies for Railway Engineering: A Review," Energies, vol. 15, no. 17, p. 6467, 2022. [Online]. Available: https://www.mdpi.com/1996-1073/15/17/6467. https://doi.org/10.3390/en15176467

[25] G. AlZohbi, A. Almoaikel, and L. AlShuhail, "An overview on the technologies used to store hydrogen," Energy Reports, vol. 9, pp. 28-34, 2023. https://doi.org/10.1016/j.egyr.2023.08.072

[26] H. K. Shin and S. K. Ha, "A Review on the Cost Analysis of Hydrogen Gas Storage Tanks for Fuel Cell Vehicles," Energies, vol. 16, no. 13, p. 5233, 2023. [Online]. Available: https://www.mdpi.com/1996-1073/16/13/5233. https://doi.org/10.3390/en16135233

[27] S.-L. Chen, H. Fan, and N. Abdussamie, "Evaluation of hydrogen shipping cost for potential trade routes," WMU Journal of Maritime Affairs, vol. 24, 03/20 2025. https://doi.org/10.1007/s13437-025-00365-w

[28] X. Qian et al., "Underground Hydrogen Storage in Salt Cavern: A Review of Advantages, Challenges, and Prospects," Sustainability, vol. 17, no. 13, p. 5900, 2025. https://doi.org/10.3390/su17135900

[29] D. S. Wendt, Y. Lin, W. D. Jenson, D. Wang, B. Huang, and D. Wu, "Hydroelectric Power and Hydrogen Production Integration," Idaho National Laboratory (INL), Idaho Falls, ID (United States); Pacific …, 2023.

[30] Q. Yu, Y. Hao, K. Ali, Q. Hua, and L. Sun, "Techno-economic analysis of hydrogen pipeline network in China based on levelized cost of transportation," Energy Conversion and Management, vol. 301, p. 118025, 2024. https://doi.org/10.1016/j.enconman.2023.118025

[31] M. Henning, "Midwest Hydrogen Center of Excellence Detailed Comments in Response to IRS Proposed Guidance for 45V Tax Credit for Clean Hydrogen Production," 2024.

[32] R. Hren, A. Vujanović, Y. Van Fan, J. J. Klemeš, D. Krajnc, and L. Čuček, "Hydrogen production, storage and transport for renewable energy and chemicals: An environmental footprint assessment," Renewable and Sustainable Energy Reviews, vol. 173, p. 113113, 2023. https://doi.org/10.1016/j.rser.2022.113113

[33] J. Zou, "High-Pressure Gaseous Hydrogen Storage and Transportation," in Hydrogen Storage and Transportation: Springer, 2025, pp. 45-87. https://doi.org/10.1007/978-981-96-2875-9_2

[34] S. Ramadhani et al., "Sequential reactions toward a high-pressure H2 generation from a mixture of sodium borohydride and formic acid," Cell Reports Physical Science, vol. 5, no. 1, 2024. https://doi.org/10.1016/j.xcrp.2023.101759
[35] R. Hassoun and R. Stanek, "The tank system as an optimization factor for the total cost of ownership of hydrogen applications," in MTZ-Fachkongress: Experten-Forum Powertrain, 2022: Springer, pp. 131-147. https://doi.org/10.1007/978-3-658-42553-1_11
[36] A. Dauletbay, "Transportation of hydrogen: hydrogen usage," in Hydrogen Technologies-Advances, Insights, and Applications: IntechOpen, 2024. https://doi.org/10.5772/intechopen.1005066
[37] M. Niermann, S. Timmerberg, S. Drünert, and M. Kaltschmitt, "Liquid Organic Hydrogen Carriers and alternatives for international transport of renewable hydrogen," Renewable and Sustainable Energy Reviews, vol. 135, p. 110171, 2021. https://doi.org/10.1016/j.rser.2020.110171
[38] S. Kim, S. Oh, and S. Kang, "Techno-economic assessment of liquefied hydrogen tanker ships utilizing various propulsion systems," Energy Conversion and Management, vol. 336, p. 119895, 2025. https://doi.org/10.1016/j.enconman.2025.119895
[39] K. Nehrkorn, B. Spiller, and A. Krupnick, "Exploring Hydrogen's Role in Heavy-Duty Trucking," 2024.
[40] P. S.-L. Chen, H. Fan, and N. Abdussamie, "Evaluation of hydrogen shipping cost for potential trade routes," WMU Journal of Maritime Affairs, vol. 24, no. 2, pp. 315-338, 2025. https://doi.org/10.1007/s13437-025-00365-w
[41] S. A. Alavi-Borazjani, S. Adeel, and V. Chkoniya, "Hydrogen as a Sustainable Fuel: Transforming Maritime Logistics," Energies, vol. 18, no. 5, p. 1231, 2025. https://doi.org/10.3390/en18051231
[42] R. Caponi, E. Bocci, and L. Del Zotto, "On-site hydrogen refuelling station techno-economic model for a fleet of fuel cell buses," International Journal of Hydrogen Energy, vol. 71, pp. 691-700, 2024. https://doi.org/10.1016/j.ijhydene.2024.05.216
[43] J.-L. Fan, P. Yu, K. Li, M. Xu, and X. Zhang, "A levelized cost of hydrogen (LCOH) comparison of coal-to-hydrogen with CCS and water electrolysis powered by renewable energy in China," Energy, vol. 242, p. 123003, 2022. https://doi.org/10.1016/j.energy.2021.123003
[44] M. Kaya, "Clean Hydrogen's Role in Energy Transition of the USA, Germany and Türkiye," Middle East Technical University, 2025.
[45] W. Bai, X. Shi, S. Zhu, X. Wei, and Y. Huang, "Opportunities and challenges of large-scale salt cavern hydrogen storage in China coupled with renewable energy sources," Clean Energy, vol. 9, no. 1, pp. 179-196, 2025. https://doi.org/10.1093/ce/zkae095
[46] F. Al Mubarak, R. Rezaee, and D. A. Wood, "Economic, societal, and environmental impacts of available energy sources: A review," Eng, vol. 5, no. 3, pp. 1232-1265, 2024. https://doi.org/10.3390/eng5030067
[47] M. Guha, "Role of the inflation reduction acts in financing hydrogen energy projects: Opportunities and challenges," The Role of Green and Transition Finance in Achieving Carbon Neutrality and Sustainable Energy, pp. 207-236, 2025. https://doi.org/10.1016/B978-0-443-24736-1.00009-8

Binary Fluid Ejector Refrigeration Systems: Thermodynamic Analysis and Fluid Selection Criteria

Mouhammad EL HASSAN[1,a *], Enio BANDARRA FILHO[1,b], Zeeshan RANA[1,c] and Nikolay BUKHARIN[2,d]

[1]Department of Mechanical Engineering, Prince Mohammad Bin Fahd University, Al Khobar 31952, Saudi Arabia

[2]The Southern Alberta Institute of Technology, Calgary, AB T2M 0L4, Canada

[a]melhassan@pmu.edu.sa, [b]efilho@pmu.edu.sa, [c]zrana@pmu.edu.sa, [d]nikolay.bukharin@sait.ca

Keywords: Binary Fluid Ejector, COP, Thermodynamic Analysis, Fluid Selection

Abstract. An ideal model for an ejector refrigeration system using two different refrigerants in the vapor-compression thermodynamic cycle (Binary Fluid Ejector Refrigeration System, BFE RS) was developed to quantify the influence of key thermodynamic parameters and fluid properties on the system Coefficient of Performance (COP). The ideal BFE RS includes a boiler, condenser, evaporator, expansion valve, and an ideal ejector. Different refrigerant combinations were evaluated to quantify the effects of molecular mass, vaporization enthalpy, operating pressure, and mixture type on COP. The model's results provide practical guidance for selecting the best driver and refrigerant combinations for BFE systems.

Introduction

Ejectors are static devices that convert the pressure energy of a high-pressure motive fluid into kinetic energy to entrain a secondary stream. By accelerating the primary jet, they create suction and entrain secondary fluid within thermal or cogeneration cycles. Ejectors are used in various industrial applications such as air conditioning, heating, distillation, desalination, etc. Replacing a conventional compressor with an ejector allows us to utilize low-grade heat in refrigeration cycles that might otherwise go to waste (solar thermal, industrial exhaust, or engine jacket heat). Although the system's coefficient of performance (COP) is lower than that of compressor-driven systems, the energy source is often free or already available [1, 2].

Most prior efforts to improve ejector refrigeration system COP have focused on geometric optimization: adjusting area ratio, nozzle-exit position (NXP), and constant-area mixing length to reduce shocks and mixing losses [3 - 6]. Another approach involves optimizing operating parameters. In [7], CFD was used to analyze the flow dynamics within a supersonic ejector at compression ratios of 1.67 and 3.4. The study primarily focused on how operating conditions and motive pressure affect the internal flow structures, entrainment mechanism, and overall ejector efficiency. For a fixed compression ratio, the primary pressure was found to significantly impact jet momentum, jet expansion, and shock waves within the ejector.

Beyond geometry and operating conditions, ejector performance can be strongly influenced by the choice of working fluid. In [8], a CFD study was conducted to analyze how the thermophysical properties of working fluids, specifically molecular mass and the specific heat ratio, influence the ejector performance designed for distillation applications. It was demonstrated that both molecular mass and the specific heat ratio significantly affect the entrainment ratio and COP of the refrigeration or distillation cycle.

A variety of thermodynamic cycles have been developed to recover low-grade heat. In [9], a comprehensive thermodynamic modeling approach for an ejector used in refrigeration systems was proposed. In this model, the primary flow was assumed to expand from the nozzle and form a distinct core region, without immediate mixing with the secondary flow, thereby creating a hypothetical throat where the secondary stream reached sonic speed. The ratio of the hypothetical throat area to the mixing area was correlated with the ratio of the nozzle throat area to the mixing area and with the pressure ratio between the primary and secondary flows. The proposed model was validated against experimental measurements of the primary mass flow rate and critical back pressure, demonstrating strong agreement. In [10], a thermodynamic model of a steam-driven ejector refrigeration cycle was presented. All major state points were shown on p-h and T-s diagrams to demonstrate how motive steam expansion, ejector entrainment, condenser/evaporator heat exchange, and throttling or pump losses contribute to overall system performance. A new cycle architecture was proposed in [11], combining a cascade refrigeration system with an ejector to utilize waste heat from the gas-cooler as the motive fluid driver; this approach enhances both the energy and exergy performance of the refrigeration system. A detailed steady-state thermodynamic model was developed in [11], which included mass and energy balances for each component. Results show that under optimal conditions, the proposed ejector-cascade cycle can improve COP by up to approximately 37% and increase second-law efficiency by about 12% compared to the conventional cascade cycle, while maintaining the same cooling capacity.

Using two different working fluids in interconnected cycles can improve the system's overall COP. This setup is known as a Binary Fluid Ejector refrigeration system [12-14]. The concept of using two chemically distinct fluids was first introduced to reduce shock losses within the ejector and enhance its efficiency. Typically, a high–molecular–weight working fluid is selected for the primary (driving) Rankine cycle, while a low–molecular–weight refrigerant is utilized for the secondary (refrigeration) cycle. Besides reducing shock wave effects, the binary-fluid approach can further enhance COP by carefully selecting fluids with complementary thermophysical properties—especially different molecular masses and enthalpies of vaporization. Because BFE performance depends on many coupled operating parameters and fluid properties, defining robust fluid-selection criteria remains non-trivial. [15, 16].

In this work, an ideal model for the BFE refrigeration cycle was developed to estimate the maximum possible entrainment ratio and COP under given operating conditions and binary fluid combinations. Based on the idealized analysis, several preliminary guidelines for selecting the working fluid for BFE are proposed.

Methodology and Model Formulation

In the proposed ideal ejector model, the driving fluid (primary fluid) undergoes isentropic expansion, while the refrigerant (secondary fluid) experiences isentropic compression, followed by perfect mixing of the two streams. The same assumptions are used for the ideal Binary Fluid Ejector model. Expansion and compression are assumed to occur upstream of the mixing section, which is modeled as a constant-pressure process. The mixing process is idealized as non-dissipative, which neither restricts the entrainment of the secondary fluid nor causes any pressure losses. Here, P_3denotes the condenser pressure, primarily governed by the condenser temperature, mixture molar composition, and the thermodynamic behavior of the condensate, and corresponds to the condenser saturation pressure. The composition of the condensate mixture strongly influences the resulting value of P_3 and depends on molecular interactions, mixtures may behave ideally (zeotropic) or non-ideally (azeotropic or heteroazeotropic). For an ideal or zeotropic, mixture, molecular interactions between components are minimal, and the saturation pressure at a given temperature is between those of the pure components (approximately a molar fraction weighted average). Non-ideal mixtures deviate from this behavior. For instance, an azeotropic mixture can exhibit a saturation pressure exceeding that of either pure component at the same temperature. Azeotropic and zeotropic mixtures are excluded from the present calculations due to the limited availability of reliable saturation curve data for these binary systems. However, the

Materials Research Proceedings 64 (2026) 789-796 https://doi.org/10.21741/9781644904091-98

analysis includes heteroazeotropic behavior in the condenser, ideal-mixture assumptions for fractional distillation, and limiting no-mixing cases.

The Binary Fluid Ejector Refrigeration System (BFE RS) is presented in Figure 1.

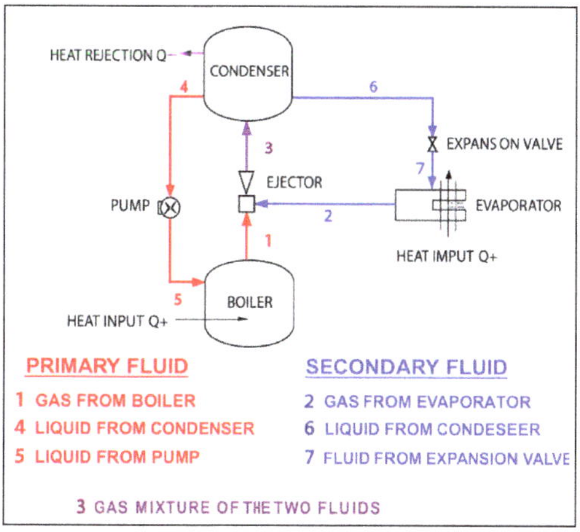

Figure 1. Schematic of the BFE RS

Governing Equations

Based on the above description for the BFE RS and the ideal model for BFE, a set of equations are established to estimate the maximum possible COP entrainment ratio (ER) for different combinations of refrigerant pairs under defined operating conditions. Based on the assumptions as mentioned above, a set of equations for the ideal model of an ejector was developed as follows:

forward Rankine cycle involving driver fluid:

$$Q_{P,Boil} = \dot{m}_P(h_{P1} - h_{P5}) \tag{1}$$

$$W_{P,Pump} = \dot{m}_P w_{P,Pump} = \dot{m}_P \frac{(P_{P5} - P_{P4})}{\rho_{P45}} \tag{2}$$

$$-W_{P,Ejec} - Q_{P,Cond} + W_{P,Pump} + Q_{P,Boil} = \dot{m}_P[-(h_{P3} - h_{P1}) - (h_{P4} - h_{P3}) + 1/\rho_{P45}(P_{P5} - P_{P4}) + (h_{P1} - h_{P5})] = 0 \tag{3}$$

Refrigeration cycle involving refrigerant fluid:

$$Q_{S,Evap} = \dot{m}_S(h_{S2} - h_{S7}) \tag{4}$$

$$W_{S,ExpVal} = Q_{S,ExpVal} = 0 \Rightarrow \quad h_{S7} = h_{S6} \tag{5}$$

$$Q_{S,Evap} - Q_{S,Cond} + W_{S,Ejec} = \dot{m}_S[(h_{S3} - h_{S2}) - (h_{S6} - h_{S3}) + (h_{S2} - h_{S7})] = 0 \tag{6}$$

Materials Research Proceedings 64 (2026) 789-796 https://doi.org/10.21741/9781644904091-98

Condenser:

$$Q_{Cond} = Q_{P,Cond} + Q_{S,Cond} = \dot{m}_P(h_{P4} - h_{P3}) + \dot{m}_S(h_{S6} - h_{S3}) \tag{7}$$

Ejector:

$$W_{P,Ejec} = W_{S,Ejec} \Rightarrow \dot{m}_P(h_{P1} - h_{P3}) = \dot{m}_S(h_{S3} - h_{S2}) \tag{8}$$

These equations can be used to find the COP for the BFE RS, which is defined as,

$$COP = \frac{Q_{S,Evap}}{Q_{P,Boil}} \tag{9}$$

This equation can be expanded to find the COP for a BFE refrigeration cycle which operates based on the principles of the Carnot cycle. In other words, the forward Rankine and refrigeration cycles of the BFE RS are Carnot cycles. As opposed to a Carnot cycle with one working fluid, because the mass flow rates of the two fluids in the BFE Refrigeration cycle are not necessarily the same as, in equation (9) the ratios of temperatures cannot replace the ratio of heat fluxes. This equation can be rendered in a more useful from as follows,

$$COP = \frac{W_{P,Ejec}}{Q_{P,Boil}}\frac{Q_{S,Evap}}{W_{S,Ejec}} = \frac{Q_{P,Cond} - Q_{P,Boil}}{Q_{P,Boil}}\frac{Q_{S,Evap}}{Q_{S,Evap} - Q_{S,Cond}} \tag{10}$$

The temperature ratios can replace the heat flux ratios in equation (10)

$$\boxed{COP = \frac{T_{Cond} - T_{Boil}}{T_{Boil}}\frac{T_{Evap}}{T_{Evap} - T_{Cond}} = \frac{1 - T_{Cond}/T_{Boil}}{T_{Cond}/T_{Evap} - 1}} \tag{11}$$

A smaller temperature difference between the evaporator and condenser results in a higher Carnot cycle efficiency. In the same way by raising the boiler temperature we can improve COP. These qualitative trends remain valid for non-ideal ejector refrigeration cycles. When the maximum allowable boiler temperature and the minimum temperature difference between the condenser and evaporator are set for a given Binary Fluid Ejector (BFE) Carnot refrigeration system, the corresponding COP can be expressed as a function of these parameters. Figure 2 shows a COP map for boiler temperatures from 350 K to 1000 K, evaporator temperatures between 200 K and 500 K, and temperature differences between condenser and evaporator of 1, 5, 10, and 15 K.

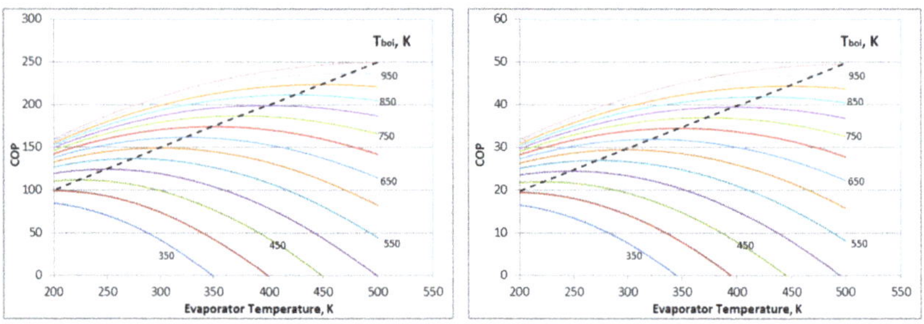

(a) (b)

Materials Research Proceedings 64 (2026) 789-796 https://doi.org/10.21741/9781644904091-98

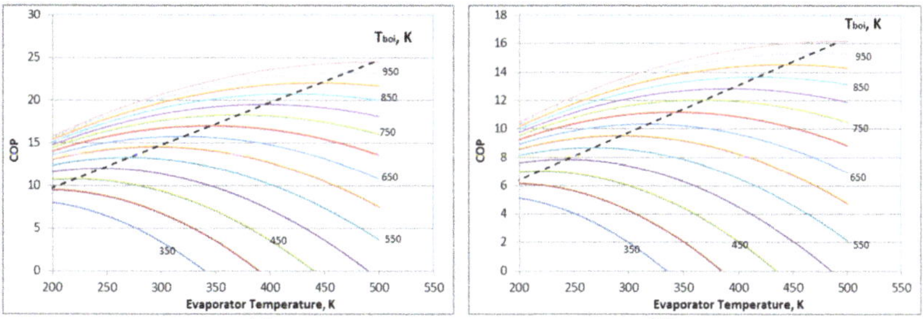

(c) (d)

Figure 2. COP of BFE RS based on Carnot cycle principles, for different boiler and evaporator temperatures; for different temperature differences between evaporator and condenser (a) 1K, (b)5K, (c) 10K, (d) 15k ; dash line shows the evaporator temperature corresponding to the maximum possible COP at the given boiler temperature for each graph.

For a fixed boiler temperature and a prescribed condenser and evaporator temperature difference, the COP reaches a maximum atTevap = [Tboil - ΔTcond-evap] / 2.

In practical systems, non-ideal effects associated with working fluid properties, mixture behavior, saturation characteristics, flow losses, and component inefficiencies reduce the COP relative to the ideal values predicted by Eq. (11). As a result, the achievable performance is systematically lower than the ideal limit.. For a practical Binary Fluid Ejector Refrigeration System (BFE RS), the Coefficient of Performance (COP) expressed in Equation (9) can be reformulated as follows:

$$COP = \frac{Q_{evaporator}}{Q_{boiler} + W_{pump}} = \frac{\dot{m}_S(h_{S2} - h_{S7})}{\dot{m}_P(h_{P1} - h_{P5}) + \frac{\dot{m}_P}{\rho_{P45}}(P_{P5} - P_{P4})} = \omega \frac{(h_{S2} - h_{S7})}{(h_{P1} - h_{P5}) + \frac{(P_{P5} - P_{P4})}{\rho_{P45}}} \quad (12)$$

We can calculate ER from equation (8)

$$\omega = \frac{(h_{P1} - h_{P3})}{(h_{S3} - h_{S2})} \quad (13)$$

which if substituted into equation (12) gives,

$$COP = \frac{(h_{P1} - h_{P3})}{(h_{S3} - h_{S2})} \times \frac{(h_{S2} - h_{S7})}{(h_{P1} - h_{P5}) + \frac{(P_{P5} - P_{P4})}{\rho_{P45}}} \quad (14)$$

Using this formulation, the COP of a practical cycle can be evaluated under the assumption that heat and pressure losses are neglected. When the temperatures of the boiler, condenser, evaporator, and the working fluids of the BFE RS are specified, the enthalpies at all cycle states can be defined, except for the ejector exhaust state, which depends on the entrainment efficiency of the ejector. The thermodynamic state corresponding to this mechanism can be established experimentally, through computational fluid dynamics (CFD) simulations, or using one-dimensional compressible flow analysis. In this work, an idealized ejector model was used to determine the entrainment efficiency, this approach enables a systematic investigation of how various parameters influence the maximum ejector efficiency and the resulting COP of the overall cycle.

Since the pump work is typically negligible compared to the other terms in Equation (14), the expression can be simplified as follows:

$$COP = \frac{(h_{P1} - h_{P3})}{(h_{S3} - h_{S2})} \times \frac{(h_{S2} - h_{S7})}{(h_{P1} - h_{P5})} \tag{15}$$

Equation (15) shows that the molecular masses of the two working fluids do not influence the theoretical COP; however, they directly affect the entrainment ratio and, consequently, the performance of both ideal and real ejectors. For a given operating condition, a lower-molecular-mass working fluid requires a larger ejector cross-section to achieve the same system capacity. On the other hand, using a higher-molecular-mass fluid tends to increase the entrainment ratio and enhance ejector efficiency.

If the change in enthalpy from inlet to outlet of the evaporator, boiler, and condenser is approximated by the enthalpies of vaporization, equation (12) can be rewritten as follows,

$$COP = \omega \frac{(h_{S2}-h_{S7})}{(h_{P1}-h_{P5})+\frac{(P_{P5}-P_{P4})}{\rho_{P45}}} \approx \omega \frac{h_{fg,S,eva}}{h_{fg,P,boi}+\frac{(P_{P5}-P_{P4})}{\rho_{P45}}} \tag{16}$$

The same procedure can be applied to equation (13) to estimate the entrainment ratio based on vaporization enthalpies.

$$\omega = \frac{(h_{P1} - h_{P3})}{(h_{S3} - h_{S2})} \approx \frac{-h_{v,P,con} + h_{v,P,boi} + \frac{1}{\rho_{P45}}(P_{P5} - P_{P4})}{h_{v,S,con} - h_{v,S,eva}} \tag{17}$$

Substituting equations (16) into (15) results,

$$COP \approx \left[\frac{-h_{v,P,con} + h_{v,P,boi} + \frac{(P_{P5} - P_{P4})}{\rho_{P45}}}{h_{v,S,con} - h_{v,S,eva}}\right] \times \left[\frac{h_{v,S,eva}}{h_{v,P,boi} + \frac{(P_{P5} - P_{P4})}{\rho_{P45}}}\right] \tag{18}$$

Choosing a fluid with a small enthalpy of vaporization in the boiler and evaporator leads to better thermal conversion per unit of mass or mole (second bracket in equation (18)), but also results in a lower entrainment ratio. This indicates that whether a fluid has a high or low enthalpy of vaporization in the boiler, it does not significantly affect the cycle's COP.

Equation (18) has another important implication: the ejector's effect on the COP is determined by the entrainment ratio. The practical implication is that when the secondary flow is choked in an ejector, the entrainment ratio does not change with a decrease in condenser temperature or pressure; therefore, the COP remains the same. Consequently, increasing the entrainment ratio should be considered a key parameter for improving the design of BFEs.

If the pump work is neglected, equation (19) can be written as,

$$COP \approx \frac{h_{v,P,con}/h_{v,P,boi} - 1}{1 - h_{v,S,con}/h_{v,S,eva}} \tag{19}$$

This implies the same conclusion that was drawn using the COP of the BFE cycle based on the Carnot cycle concept in equation (11). Therefore, the temperatures of the condenser and boiler (primary fluid) should be as close as possible, and the temperature difference between condenser and evaporator (secondary fluid) should be as close as possible.

The difference in vaporization enthalpy of the secondary fluid between the condenser and the evaporator can provide greater potential for improving COP than that of the primary fluid in the boiler and condenser. Typically, a significant effect of vaporization enthalpy is to reduce the circulation rate per unit of cooling load.

Guidelines for Selection of Driver and Refrigerant Fluids in the BFE Refrigeration Cycle
Rankine Cycle Fluid:

- High critical temperature; the boiler temperature should be significantly lower than the critical temperature to prevent condensation, especially for the driver fluid.
- Steep phase change lines on the T-S diagram; i.e., $|\partial T/\partial S| \gg 1$; This is an important parameter that can bring the COPs of standard and inverted Rankine cycles closer to that of the Carnot cycle.
- Dry or isentropic fluid on the vapor side of the T-S diagram; for temperatures above condenser's temperature.

Refrigeration Cycle Fluid:

- High critical temperature
- Steep phase change lines on the T-S diagram
- Wet or isentropic fluid on the vapor side of the T-S diagram; for temperatures condenser's temperature.

Both Fluids:

- Environmental, flammability, and toxicity aspects are not considered in the early stages of the fluid selection process to better understand how the fluid's thermo-physical properties affect the COP and entrainment ratio.
- The saturation pressures of the two fluids at the condenser temperature should be as close as possible; a refrigerant with much higher saturation pressure than the driver fluid could cause reverse flow in the ejector.
- Heteroazeotrope fluids should be avoided because their mixtures require higher saturation pressure than other types of mixtures.
- Choosing the two fluids and operating conditions should prevent the ejector exhaust temperature from dropping below the condenser temperature. If this happens, the BFE cycle will not be able to sustain itself. Essentially, the condenser's operating temperature gradually approaches that of the evaporator, and the system will eventually cease functioning.
- The selection of the fluids could be very sensitive to the operating conditions; therefore, it is critical to set the operating conditions before starting the fluid selection process.

Conclusion and future work

An ideal model for BFE refrigeration systems was developed. Results confirm that the mixture type in the condenser significantly influences COP. Zeotropic mixtures can achieve COPs comparable to those of ideal Rankine systems when the saturation pressures of both fluids are similar. Future work will combine this thermodynamic model with CFD simulations to analyze ejector flow dynamics, entrainment efficiency, and real-fluid effects, and to explore variable-geometry ejectors for performance optimization.

References

[1] El Hassan M. System COP of Ejector-Based Ground-Source Heat Pumps, Energies 15 (22), 8509 (2022)

[2] Chunnanond, K., & Aphornratana, S. (2004). Ejectors: Applications in refrigeration technology. Renewable and Sustainable Energy Reviews, 8(2), 129–155. https://doi.org/10.1016/j.rser.2003.10.002
[3] Li, R., Yan, J., & Reddick, C. (2022). Optimization of three key ejector geometries under fixed and varied operating conditions: A numerical study. Applied Thermal Engineering, 211, 118537. https://doi.org/10.1016/j.applthermaleng.2022.118537
[4] B. El Zohbi, N. Bukharin, H.H. Assoum, K. Abed-Meraim, A. Sakout, M. El Hassan, Investigation of the effects of the jet nozzle geometry and location on the performance of supersonic fluid ejectors, Ener. Rep., 8 (2022), pp. 228-233, 10.1016/j.egyr.2022.01.029
[5] van Nguyen, V., Varga, S., Soares, J., Dvorak, V., & Oliveira, A. C. (2020). Applying a variable geometry ejector in a solar ejector refrigeration system. International Journal of Refrigeration, 113, 187–195. https://doi.org/10.1016/j.ijrefrig.2020.01.018
[6] Pang, Z., Han, J., Feng, J., Diao, A., Yao, Y., & Peng, X. (2024). Performance prediction and geometry optimization of ejector in PEMFC system using coupled CFD-BPNN and genetic algorithm. Applied Thermal Engineering, 251, 123584. https://doi.org/10.1016/j.applthermaleng.2024.123584
[7] El Hassan, M. Numerical Investigation of the Flow Dynamics Inside Supersonic Fluid Ejector. Arab J Sci Eng 45, 909–919 (2020). https://doi.org/10.1007/
[8] El Hassan, M., Assoum, H. H., Bukharin, N., Abed-Meraim, K., & Sakout, A. (2020). Investigation of thermo-physical fluid properties effect on binary fluid ejector performance. Energy Reports, 6, 287–292. https://doi.org/10.1016/j.egyr.2019.11.076
[9] Ma, Z., Bao, H., & Roskilly, A. P. (2017). Thermodynamic modelling and parameter determination of ejector for ejection refrigeration systems. International Journal of Refrigeration, 75, 117–128. https://doi.org/10.1016/j.ijrefrig.2016.12.005
[10] Liu, S., Zhu, C., & Li, H. (2014). Thermodynamic analysis of steam ejector refrigeration cycle (Paper 1436, International Refrigeration & Air Conditioning Conference, Purdue University). Purdue e-Pubs. Retrieved from https://docs.lib.purdue.edu/cgi/viewcontent.cgi?article=2435&context=iracc
[11] Megdouli, K., Tashtoush, B. M., Nahdi, E., Elakhdar, M., Kairouani, L., & Mhimid, A. (2016). Thermodynamic analysis of a novel ejector-cascade refrigeration cycles for freezing process applications and air-conditioning. International Journal of Refrigeration, 70, 108–118. https://doi.org/10.1016/j.ijrefrig.2016.06.029
[12] Holton, W. (1951). Effect of molecular weight and entrained fluid on the performance of steam jet ejectors. Transactions of the American Society of Mechanical Engineers, 73, 123–132.
[13] Rao, S., & Jagadeesh, G. (2015). Studies on the effects of varying secondary gas properties in a low entrainment ratio supersonic ejector. Applied Thermal Engineering, 78, 289–302. https://doi.org/10.1016/j.applthermaleng.2014.12.033
[14] Work, L. T., & Haedrich, V. W. (1939). Performance of ejectors as a function of the molecular weights of vapors. Industrial & Engineering Chemistry, 31(4), 464–477. https://doi.org/10.1021/ie50352a018
[15] Bukharin, N. and El Hassan, M.. (2023). "Numerical and Experimental Investigation of Supersonic Binary Fluid Ejector Performance" Fluids 8, no. 7: 197. https://doi.org/10.3390/fluids8070197
[16] El Hassan, M. and Bukharin, N. (2019) J. Phys.: Conf. Ser. 1276 012012, 10.1088/1742-6596/1276/1/012012

Smart IoE: A Hybrid Machine Learning Framework for Wind Energy Forecasting, Blockchain Integrity, and IoE Connectivity

Mansoor Khan[1,a], Salabat Khan[2,b], Muhammad Asghar Khan[3,c *], Ximeng Dou[4,d], Zahid Ali Khan[5,e], Feng Cao[6,f], Fahmida Bibi[7,g]

[1]School of Intelligent Manufacturing and Control Engineering, Qilu Institute of Technology, Jinan, China

[2]School of Computer and Information Engineering, Qilu Institute of Technology, Jinan, China

[3]Department of Electrical Engineering, Prince Mohammad Bin Fahd University, P.O. Box 1664, Al-Khobar 31952, Ash Sharqiyah, Saudi Arabia

[4]School of Computer and Information Engineering, Qilu Institute of Technology, Jinan, Shandong, China

[5]Saudi Business Machines (SBM), Khobar, Saudi Arabia

[6]School of Intelligent Manufacturing and Control Engineering, Qilu Institute of Technology, Jinan, China

[7]Institute of Education and Research Gomal university Dera Ismail Khan

[a]mansoorkhan@qlit.edu.cn, [b]salabatwazir@gmail.com, [c]mkhan4@pmu.edu.sa, [d]13188935434@163.com, [e]zakhan@sbm.com.sa, [f]fcao@qlit.edu.cn, [g]fahmida@gu.edu.pk

Keywords: Hybrid Machine Learning, Wind Energy Forecasting, Resource Allocation, Renewable Energy Integration, Energy Grid Optimization

Abstract. Wind energy is a renewable energy that can be used in a wide variety of applications. Incorporation of renewable energy sources into the power grid, which is wind energy, requires proper forecasting, effective resource allocation, and high data integrity systems. To solve these issues, a machine-learning framework created within the proposed structure will combine the best forecasting models with optimization features that are made possible through blockchain elements. Exact real-time forecasts of wind energy can be achieved by an integrated platform where deep neural networks are mixed with ensemble techniques and other machine-learning algorithms. The system provides the optimal allocation of all produced energy resources because of its inbuilt optimization facilities. This is because the introduction of blockchain in the architecture provides full data security, both in the energy and transactional integrity, which at the same time are highly transparent. The suggested structure enhances the accuracy of prediction at the phase of resources-distribution optimization and leads to the creation of a data-management subsystem, which is tamper-resistant. It also provides smart grids with sustainable wind-energy solutions by building entirely power-prediction systems together with distribution-management infrastructure and information-protection elements. Evaluation metrics substantiate the efficacy of the proposed framework. The hybrid model delivers optimal predictive accuracy, evidenced by a root mean square error (RMSE) of 1.25 and a mean absolute error (MAE) of 0.98. It also attains a coefficient of determination (R^2) of 0.92, surpassing all benchmark ensembles, including Boosting Trees (RMSE: 1.35, R^2: 0.90) and Random Forest (RMSE: 1.55, R^2: 0.85).

1. Introduction

1. ML has been drawing a lot of attention to renewable energy, especially to enhance the precision of wind energy forecasting, and or to optimize resource allocation [1]. The major problem in management of wind energy is unpredictability and abrupt change in

the pattern of wind which makes managing the grid difficult and effective running of the farms challenging [2, 3]. Although it is demonstrated that ML approaches can contribute to improving the predictability of forecasts [4], to date, the systems do not include the availability of real-time information and the sophisticated algorithms to comprehensively manage wind energy production and consumption [6].

2. In order to fill this gap, the current paper will introduce a new end-to-end architecture that combines real-time information with a set of ML algorithms to enhance the quality of wind energy forecasting. It bases its framework on blockchain technology and Internet of Everything (IoE) connectivity as the basis of data integrity, safe communication between energy systems, and operational transparency [5]. This is intended to become a more reliable and effective method of wind energy control. The contributions of this paper are as follows:
3. Proposing a comprehensive ML-based system for wind power prediction using real-time data and complex predictive processes.
4. Implementing blockchain technology and IoE connectivity to enable stable and secure communication between energy systems.
5. Providing evidence on how the framework enhances wind power forecast accuracy and resource allocation.

The remainder of the paper is structured as follows: Section 2 reviews related work on wind energy forecasting challenges and solutions. Section 3 details the research methodology and the proposed ML and blockchain-IoE integrated framework. Section 4 describes the experimental setup and presents the performance assessment. Finally, Section 5 concludes the paper and suggests directions for future work.

Table 1. List of Keywords

Item No	Keyword	Abbreviation
1	Wind Energy Forecasting	WEF
2	Resource Allocation	RA
3	Machine Learning	ML
4	Long Short-Term Memory	LSTM
5	Reinforcement Learning	RL
6	Genetic Algorithms	GA
7	Model Predictive Control	MPC
8	Mean Absolute Error	MAE
9	Root Mean Squared Error	RMSE
10	Coefficient of Determination	R^2

1) Turbine: In the center of the system is the rotating blades when the wind strikes; this section would be in the very front of the machine. The blades of the turbines take advantage of kinetic energy and transform it into mechanical energy.
2) Generator: 1) The mechanical energy is passed on to the generator where the turbine lightens the electrical energy.
3) Controller: 1) The controller controls the working of the wind turbine to make sure it is operating in favorable conditions. It switches and cuts off the system and does safety work.
4) Transformer: 1) This device increases the power of the generator to a very high level of output voltage to be transmitted to the grid.
5) Transmission System: 1) The electrical energy generated is distributed by using power lines, which distribute the energy to the consumers.

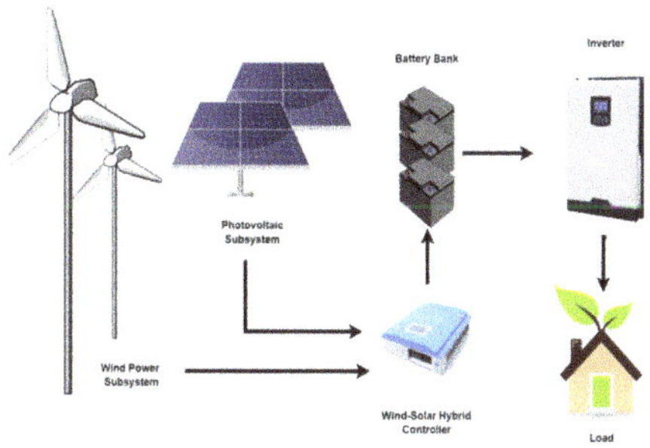

Figure 1 Wind Energy System

2. Literature Review

Wind energy needs to be included in the world energy mix since this is what makes them sustainable, although they are challenging to the stability of the grid and resource management due to their fluctuating nature. The application of conventional forecasting methods, such as statistical modeling and physical modeling, tend to be inappropriate in describing the non-linear character of wind patterns, which restricts efficiency in the operations in real-time time. They have been superseded by machine learning (ML). The models that have also been found to be more accurate in capturing the complex wind behavior and enhancing the quality of forecasting include neural networks and support vector machines. However, single ML solutions are more likely to disregard crucial aspects of integrity, security, and effective communication between interconnected energy systems. Whereas, the idea of blockchain technology has been proposed with the aim of ensuring that the data transparency and security have existed when handling energy, its applicability when used jointly with advanced forecasting models to improve the predictive quality is underdeveloped. In addition, the dynamic allocation of resources is a promising field with the help of optimization methods such as Reinforcement Learning (RL) that are not normally incorporated into an architecture founded on the secure and data-driven design [11-13].

This indicates the definite research gap: the lack of a unified framework that unites real-time ML prediction, efficient optimization of resource allocation, and blockchain-based security. Our proposed hybrid system is innovative in that it unites these components to create a single, adaptable, and scalable system, which is real-time data processing, modern ML algorithms, blockchain, and integrating these components with IoE. The purpose of this integration is to improve the quality of predictions, provide data security and integrity, as well as provide effective communication at the same time and, thus, handle the essential problems in the sphere of the modern wind energy management.

3. Methodology

The current research suggests a hybrid machine-learning model that will increase the accuracy of wind-energy predictions and allow scheduling resources in real time. The solution addresses the issue of wind variability and uncertainty by using a combination of deep learning and adaptive optimization. The framework will be constructed around four pillar units that are: (1) Data Acquisition and Preprocessing, (2) Wind-Energy Prediction Model, (3) Resource Allocation Optimization and (4) Real-Time Deployment.

Model Architecture:

The framework utilizes multi-source data to ensure robust model input:

- Historical Data: Wind farm power generation, wind speed, and direction.

- Meteorological Data: Weather forecasts and real-time observations from stations or satellites (e.g., temperature, humidity, atmospheric pressure).
- Operational Data: Real-time turbine parameters, including rotor speed, blade pitch, and power output.

The raw data undergoes a standardized preprocessing pipeline:

- Data Cleaning: Missing values are imputed, and outliers are identified and corrected.
- Normalization: Features are rescaled to a [0, 1] range to ensure compatibility.
- Feature Engineering: Derived features, such as time-of-day indicators, seasonal components, and lagged variables, are created to enhance predictive signals.
- Data Splitting: The dataset is partitioned into training, validation, and testing subsets for model development and evaluation.

A. Hybrid Prediction Model Architecture:

B. To exploit the complimentary nature of various algorithms, the forecasting module uses a stacked ensemble. The base learners include:

1. Long Short-Term Memory (LSTM): A recurrent neural network that is skilled in time-related dependencies, and the design of long-term and sequential patterns in data [7].
2. Gradient Boosting Machine (GBM): A sequence of step-by-step construction of decision trees to minimize errors, which improves the level of prediction [8]
3. Support Vector Regression (SVR): An approach based on a kernel that works well in the non-linear relations capture of the data. [9, 10].

The last prediction is produced by some meta-learner (usually a neural network or a linear regression model) that is trained on the output of these base models and produces a strong and reliable hybrid forecast.

C. Resource Allocation and Real-Time Deployment:

The overall predictions are put into action to self-drive an active resource allocation unit. The optimization techniques employed to maximize the schedules and dispatch decisions are Genetic Algorithms (GA) and Particle Swarm Optimization (PSO) and their goal is to maximize stability and energy efficiency of the grid. It is an adaptive, feedback-based system that adapts to varying conditions to achieve a real time system which varies the model parameter and resource plans on a continual basis based on the sensor information received.

D. Evaluation and Validation:

The presence of historical data and real-time simulations is helpful in the context of the analysis of the performance of the framework. The key performance indicators aside include forecasting accuracy (e.g., RMSE, MAE, R^2), resource allocation efficiency, and computational overhead. It turns out to be the best one as compared to the methods taken as a comparison, including individual models of forecasting and traditional methods of optimization.

4. Proposed Framework

Data Collection Module: This module combines input information in real-time in a multi-source, i.e. wind speed, direction, temperature, and other meteorological parameters of sensor networks, weather stations as well as turbine.

Pre-processing Unit: Noise is removed in raw data, and blank values are replaced, and standardized. This will make sure that the predictive models are fed with homogenous and credible information.

Materials Research Proceedings 64 (2026) 797-804 https://doi.org/10.21741/9781644904091-99

Hybrid Machine Learning Models: The basic forecasting model involves the use of stacked model. The time-series data, which require time as well as the traditional algorithms, including Support Vector Machines (SVM) and the Random Forest (RF) classify the non-linear dependencies and provide them with strength. A meta-learner then takes a combination of their forecasts to provide an ultimate correct forecast of wind energy.
Feature Extraction and Optimization: The process of feature selection and optimization of the hyperparameters of the prediction models is carried out via sophisticated techniques, such as Principal Component Analysis (PCA) and genetic algorithms, which improves the overall level of performance.
Resource Allocation Engine: This dynamism of the proposal, which is created by the energy foretelling, produces the optimizing process. The techniques, which are implemented to develop the real-time resources allocation and dispatch schedules and optimize utilization of grid distribution and storage, are Genetic Algorithms (GA) and Particle Swarm Optimization (PSO).
Feedback Loop: The interactive feedback loop is built on the real-time performance data to repeat the adjustment of the model parameters and allocation strategies to ensure that the system changes the reaction to the changing environmental and operational conditions.
Blockchain and IoE Integration: The framework is grounded on blockchain technology and it is focused on ensuring the integrity, security, and transparency of the data. Wind power transactions and records are fairly enclosed in a distributed ledger. Such operations as data verification, model updates and running the decision of allocation are also automated with smart contracts. IoE possesses the technology that through separation and safeguarding of communication between all physical components turbines, sensors, storage systems and the smart grid, the single work of the system can be achieved. This unified system offers a secure, scalable, and whole architecture to execute the contemporary wind energy system and in the process enhance predictability coupled with operational capability and cyber-security.

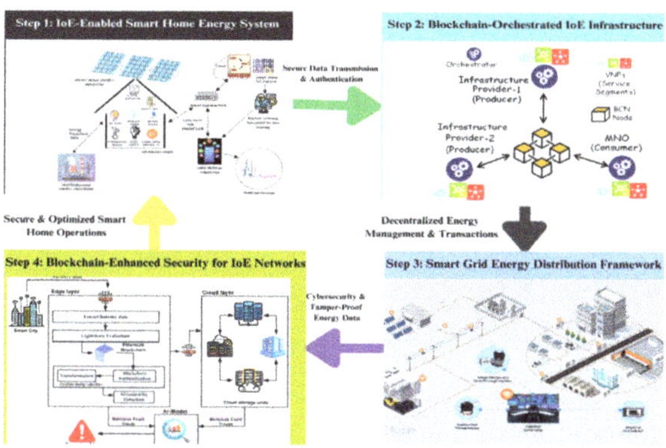

Figure 2 Wind Energy System

Table 2. Simulation setup

Simulation Parameter	Value/Method
Data Source	Wind Energy Forecasting
Forecasting Model	Resource Allocation
Input Features	Machine Learning
Training Data Size	Long Short-Term Memory
Test Data Size	Reinforcement Learning
Evaluation Metrics	Genetic Algorithms
Optimization Algorithm	Model Predictive Control
Forecasting Time Interval	Mean Absolute Error
Computational Platform	Root Mean Squared Error
Wind Energy Resource Allocation	Coefficient of Determination

5. Simulation and Results

A. Simulation Setup

In order to test the framework suggested to the real time prediction of the wind energy and the resource allocation, the extensive simulation was developed. The idea behind the experiment was to use a dataset when historical data on how wind farms (power output, wind speed/direction) worked were all collected accompanied by the measurement of the meteorological variables. Five models were compared to the proposed hybrid model in terms of performance; they were Bagged Trees, Boosting Trees, Adaboost-PSO-BP, Adaboost-GA-ELM and Random Forest. The evaluation was done by using standard regression measures, i.e. Root Mean Squared Error (RMSE), Mean Absolute Error (MAE) and Coefficient of Determination (R^2).

Results and Discussion

The quantitative results are presented in table 3. Hybrid model has achieved the best results in terms of lowest RMSE (1.25) and lowest MAE (0.98) and the highest R^2 (0.92). It demonstrates a huge implication of forecasting the model besides clarifying the model more than any other benchmark model.

Table 3. Performance Comparison

Model	RMSE	MAE	R^2	MSE
Proposed Hybrid Model	1.25	0.98	0.92	1.56
Bagged Trees	1.40	1.12	0.89	1.96
Boosting Trees	1.35	1.05	0.90	1.82
Adaboost-PSO-BP	1.50	1.22	0.87	2.25
Adaboost-GA-ELM	1.45	1.15	0.88	2.10
Random Forest	1.55	1.30	0.85	2.40

The framework's operational efficacy is further illustrated graphically. Figure 3 validates the underlying physical turbine model by plotting the fundamental correlation between wind speed and energy output, confirming the expected power curve behavior within the simulation environment. Figure 4 has given a critical performance assessment that plots the predicted and actual energy generation at various sample points. The predictions provided by the proposed model are closest to the ideal 45-degree line, meaning that there is little difference between the values in real-world and that of the model and that the model is better at making predictions. Conversely, prediction of benchmark models is associated with bigger and frequent scatter, as the model error is high. All of these are confirmation of the hybrid framework proposed. Its high accuracy, as shown by the minimum error measures, maximum R^2 and the best fit in the diagnostic plots, testify to its strength and direct use in the facilitation of real time wind energy predictions and the subsequent allocation of resources to enhance the appropriateness of decisions.

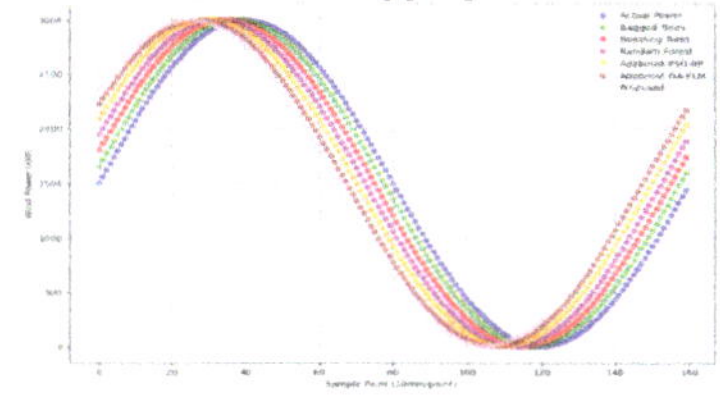

Figure 3: Simulation results based on Sample Point and Wind Energy count.

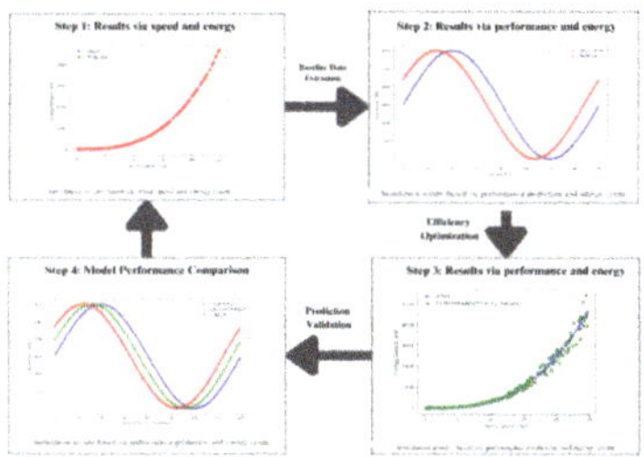

Figure 4: Flow of Simulation results based on performance prediction and energy count Step1-Step 4

Conclusion

The given paper has introduced a hybrid machine learning model of precise wind energy prediction and safe resource allocation. The combination of an ensemble of LSTM, GBM, and SVR models enables the system to perform better than the established benchmarks because the RMSE and R^2 are low (1.25 and 0.92, respectively). The framework uses blockchain technology to provide integrity of data and transparent operation and IoE connection provides smooth communication among the components of the system. It is a hybrid design that solves the major issues of wind energy management, such as accuracy, security, and real-time functionality and provides a solid solution that is easily expandable to the modern smart grids. The next steps will involve better scaling of the framework through more and more data sources and complex IoE devices to be integrated. The optimization of blockchain protocols to allow increased data throughput and facilitate quicker and real-time decision-making in very dynamic wind energy environments will also be researched.

Acknowledgement

This project is funded by the Shandong Provincial Department of Science and Technology under the Youth General Category Project (Grant No. ZR2024QE390) and by the Qilu Institute of Technology Research Startup Fund (Grant No. QIT23TP007).

References

[1] X. Yao, S. Liu, X. Wang, H. Jiang, F. Sui, and Z. Xing, "Research on maximum wind energy capture control strategy," in 2010 International Conference on Optics, Photonics and Energy Engineering (OPEE), vol. 1, 2010, pp. 34-37. https://doi.org/10.1109/OPEE.2010.5508114

[2] I. Ahmad, F. M'zoughi, P. Aboutalebi, I. Garrido, and A. J. Garrido, "Enhancing stability and performance of hybrid offshore wind platforms: A novel fuzzy logic control approach with computational machine learning," in 2023 11th International Conference on Control, Mechatronics and Automation (ICCMA), 2023, pp. 346-351. https://doi.org/10.1109/ICCMA59762.2023.10374606

[3] L. Singh, A. S. Chauhan, and T. Jain, "Machine learning based fault diagnosis method for 3-phase passive rectifiers in a wind energy conversion system," in 2023 11th National Power Electronics Conference (NPEC), 2023, pp. 1-6. https://doi.org/10.1109/NPEC57805.2023.10384857

[4] P. Maegaard, "Wind energy development and application prospects of non-grid-connected wind power," in 2009 World Non-Grid-Connected Wind Power and Energy Conference, 2009, pp. 1-3. https://doi.org/10.1109/WNWEC.2009.5335847

[5] A. Kumar, P. Karandikar, and D. S. Chavan, "Generating and saving energy by installing wind turbines along the railway tracks," in 2015 International Conference on Energy Systems and Applications, 2015, pp. 25-27. 7 https://doi.org/10.1109/ICESA.2015.7503307

[6] P. V. Sankineni, A. H. G. M, S. Juliet, L. E. U, and J. J, "Data analytics in wind turbine generators for improving efficiency," in 2022 International Conference on Sustainable Computing and Data Communication Systems (ICSCDS), 2022, pp. 1-5. https://doi.org/10.1109/ICSCDS53736.2022.9760818

[7] N. G. Paterakis, E. Mocanu, M. Gibescu, B. Stappers, and W. van Alst, "Deep learning versus traditional machine learning methods for aggregated energy demand prediction," in 2017 IEEE PES Innovative Smart Grid Technologies Conference Europe (ISGT-Europe), 2017, pp. 1-6. https://doi.org/10.1109/ISGTEurope.2017.8260289

[8] K. L. Jørgensen and H. R. Shaker, "Wind power forecasting using machine learning: State of the art, trends and challenges," in 2020 IEEE 8th International Conference on Smart Energy Grid Engineering (SEGE), 2020, pp. 44-50. https://doi.org/10.1109/SEGE49949.2020.9181870

[9] C. P. Chen, S. K. Tiong, S. P. Koh, A. A. Nasser, D. Abbas, and F. Y. Chooi, "Application of extreme learning machine in predicting short term wind speed," in 2020 8th International Conference on Information Technology and Multimedia (ICIMU), 2020, pp. 194-199. https://doi.org/10.1109/ICIMU49871.2020.9243315

[10] S. Krishnaveni, J. Singh, K. Verma, A. Pachaury, G. Kashyap, and A. Bhatia, "A machine learning approach for wind speed forecasting," in 2021 International Conference on Advance Computing and Innovative Technologies in Engineering (ICACITE), 2021, pp. 507-512. https://doi.org/10.1109/ICACITE51222.2021.9404563

[11] S. Y. Kandemir, M. Ozgur Yayli, and E. Acikkalp, "Assessment of electric energy generation using wind energy in turkey," in 7th Iran Wind Energy Conference (IWEC2021), 2021, pp. 1-3. https://doi.org/10.1109/IWEC52400.2021.9467019

[12] A. S. Mohammad, J. A. Al-Ani, and D. Zaghar, "Machine learning for prediction of wind power and resources in hot climate region," in 2021 International Conference on Computing and Communications Applications and Technologies (I3CAT), 2021, pp. 16-22. https://doi.org/10.1109/I3CAT53310.2021.9629401

[13] I. Alkalbani and F. M. Guangul, "Economical evaluation of energy harvesting using vertical axis wind turbine from a wind turbulence created by moving cars," in 2021 International Conference on Smart City and Green Energy (ICSCGE), 2021, pp. 26-29. https://doi.org/10.1109/ICSCGE53744.2021.9654414

Materials Research Proceedings 64 (2026) 805-812 https://doi.org/10.21741/9781644904091-100

Modeling and Optimizing Green Hydrogen Integration for Electrical Power Supply: Academic Facility at Al Akhawayn University, Ifrane

Fahd BOUALLOU[1,a], Soukaina JAAFARI[1,2,b *], Ahmed BAZZI[1,c]

[1]AL Akhawayn University in Ifrane, PO. BOX 104, Hassan II Avenue, Ifrane, Morocco, 53000

[2]National Institut of Posts and Telecommunication, Allal Al Fassi Avenue, Rabat, Morocco, 10000

[a]f.bouallou@aui.ma, [b]s.jaafari@aui.ma, [c]a.bazzi@aui.ma

Keywords: Green Hydrogen, Renewable Energy, Solar Photovoltaics, Electrolyzer, Hydrogen Storage, PEM Fuel Cell, Python Simulation, Energy Optimization, Carbon Emission Reduction, Energy Resilience, Sustainability

Abstract: This study examines the feasibility of green hydrogen as a solution for academic building in Al Akhawayn university in Ifrane, Morocco. To do so, we developed a simulation framework on python to model a system that combines solar photovoltaic (PV) power with electrolyzers to produce hydrogen, hydrogen storage, and proton exchange membrane (PEM) fuel cells to generate electricity. We aim to optimize system configuration by taking into consideration renewable generation profiles, electrolyzer efficiency, storage capacity and fuel cell performance. In this study, we analyzed economic metrics such as net present cost (NPC), levelized cost of electricity (LCOE) and payback period in contrast to technical indicators hydrogen production rate, storage utilization, and system efficiency. And in order to assess the capacity of the model to meet the facility's energy requirements consistently, a reliability analysis was conducted. The results show that the reduction of carbon emissions, improvement of energy autonomy, and high competitiveness compared to grid-based power supply can be achieved with appropriately sized green hydrogen systems. The findings highlight the potential of green hydrogen to enhance resilience and sustainability in higher education institutions, supporting long term energy transition strategies.

Introduction

Sustainable energy systems have been placed at the center of global policy and research agendas given the accelerating challenges of climate change and the need for deep decarbonization. Among emerging technologies, green hydrogen has gained prominence as a versatile energy carrier that can couple renewable generation, long duration storage, and low carbon power conversion [1,2]. As a zero emissions fuel when produced through water electrolysis powered by renewable energy, green hydrogen can support the decarbonization of hard to electrify sectors and provide flexibility to modern energy systems [3, 4].

Deployment of green hydrogen, however, is always constrained with the high capital and operational costs as well as infrastructure requirements, and all the challenges in achieving cost-competitiveness relative to conventional hydrogen pathways [5]. Several studies argue that techno economic modeling and optimization can help overcome these limitations because they account for renewable energy variability storage, storage limitations, and key economic tradeoffs [6, 7]

Accurate modeling of system components, especially PEM electrolyzers, is essential for producing realistic assessments [2, 8]. Recent progress in component level modeling and materials research has made simulations more reliable and has improved estimates of overall system performance [9, 10]. In the same way, failure diagnostics and degradation modeling offer useful insights into the long term durability of hydrogen systems [11]. These detailed models allow the development of robust optimization frameworks that include operational uncertainty, while also supporting investment decisions and policy making [6, 7, 12].

Materials Research Proceedings 64 (2026) 805-812 https://doi.org/10.21741/9781644904091-100

Optimization studies of hybrid renewable and hydrogen systems show that it is possible to reduce the levelized cost of energy and hydrogen while still keeping the system technically and economically viable [13]. These frameworks also show that integrated hydrogen systems can reduce emissions and strengthen energy autonomy.

In Morocco, the National Energy Strategy targets 52% renewable energy by 2030 while positioning green hydrogen as a key export commodity [14]. However, remote highland regions like Ifrane face grid instability and winter solar deficits, making universities ideal testbeds for resilient microgrids. Academic facilities present unique load profiles with high daytime laboratory/teaching demands conflicting with solar peaks, requiring long-duration storage beyond daily battery cycling [15].

This study applies this approach to integrating green hydrogen into the electricity supply of an academic facility, using Al Akhawayn University in Ifrane, Morocco, as a case study. The proposed system combines solar photovoltaic PV generation with a proton exchange membrane PEM electrolyzer, compressed hydrogen storage, and a PEM fuel cell. Simulation models developed in Python use 2022 datasets to examine key parameters, including solar resource variability, electrolyzer efficiency, storage dynamics and fuel cell conversion performance.

the valuation combines technical, economical and reliability analyses. Hydrogen production rates, conversion efficiency, and storage utilization are all measured using the technical assessments which takes a look at the net present cost (NPS), levelized cost of electricity (LCOE), and payback period. The we conduct a reliability analysis to evaluate the system's ability to sustain continuous power under variable renewable conditions.

The results ail to showcase that optimized green hydrogen systems can reduce carbon emissions, improve institutional energy resilience all while providing a viable alternative to conventional grid electricity.

It also good to compare hydrogen based storage with lithium batteries, since both are common options for renewable energy systems. Lithium batteries are often suitable for short term balancing and fast response, while hydrogen systems can be more relevant for longer duration storage through compressed hydrogen tanks and later conversion using fuel cells. Including this comparison helps clarify why hydrogen may be selected for certain operating conditions a,d resilience goals in an academic facility, while also recognizing the batteries can remain as string option for daily variability management.

Methodology

This study uses Python based simulation and optimization framework to model the integration of a green hydrogen system into the electrical infrastructure of Al Akhawayn University in Ifrane. We organized the methodology into four phases: data collection, data processing, system component specification, and optimization model formulation.

Data collection: historical energy consumption data for the year 2022 were obtained from the Ground and Maintenance Department at AL Akhawayn University. The data set is focused on the Academic Building 4 (33.538936°N, -5.107403°W) and includes separate files for heating and lighting loads.

Solar resource data were taken from the PVGIS-SARAH2 database for the university location 33.539°N, -5.106°W). To simulate the photovoltaic generation, hourly solar irradiance data for 2022. To represent the PV system, we used a crystalline silicone module, a fixed tilt angle of 33° (the latitude of Ifrane), a south facing azimuth of 0°, and a system loss factor of 14%. Table 1 summarizes the main data sources and the modeling assumptions.

Table 1. *Input Data Sources and Assumptions*

Data Type	Source	Resolution	Key Assumptions
Load data (heating & lighting)	Ground & Maintenance Dept., AUI	Hourly (2022)	Building 4 load profile
Solar irradiance	PVGIS-SARAH2	Hourly (2022)	Tilt = 33°, Azimuth = 0°, Losses = 14%
PV technology	PVGIS-SARAH2	Hourly generation	Crystalline silicon modules

Data Processing. The raw load data were delivered in multiple Excel files, with monthly worksheets for 2022. Preprocessing was carried out in Python using the Pandas library. The steps included converting timestamps into Date Time objects for time-series indexing, applying consistent column headers, and converting all numerical entries to floating point types. Invalid rows with missing timestamps or empty measurements were removed, and duplicate entries, although minimal, were removed. After cleaning, the monthly datasets were combined into a single time series dataset covering the full year. The final dataset contained only two missing timestamps and no remaining duplicates, which supports robustness and reliability of the modeling steps that follow.

System Component Specifications. The hydrogen-based energy system includes solar PV arrays, a proton exchange membrane (PEM) electrolyzer, compressed hydrogen storage, and a PEM fuel cell as listed in Table 2. Technical and economic parameters were defined based on industry reports and published research.

Table 2. *System Component Specifications*

Component	Parameter	Value/Range	Notes
Electrolyzer (PEM)	Efficiency	~80% (HHV)	High efficiency, fast response
	H_2 production rate	~0.018 kg/kWh	
	CAPEX	~$1,200/kW	Industry benchmark
	Lifetime	20,000–80,000 h	
Hydrogen Storage	Technology	Compressed gas (350–700 bar)	Widely adopted
	CAPEX	$500–1,000/kg H_2	Scalable
	Lifetime	20–30 years	
Fuel Cell (PEMFC)	Efficiency	40–60%	Electrical conversion
	H_2 consumption rate	~0.02 kg/kWh	
	CAPEX	$1,000–1,500/kW	
	Lifetime	40,000–80,000 h	

Optimization Model Formulation. To determine the optimal configuration and operating strategy, a linear programming model was developed in Python. The objective function minimizes the total system cost. This cost includes capital and operational expenditures for the electrolyzer, storage, and fuel cell, as well as the cost of supplemental grid imports when needed. The optimization problem is defined as:

$$\text{Minimize } C_{Total} = C_{elec} + C_{fc} + C_{storage} + C_{grid} \tag{1}$$

Where C_{elec} ist he electrolyzer cost, C_{fc} is the fuel cell cost, $C_{storage}$ is the hydrogen storage cost, C_{grid} ist he grid electricity cost.

The model incorporates the following operational constraints for power balance, hydrogen storage dynamics, capacity limits, and initial and final storage, respectively:

$$P_{load,t} = P_{PV,t} - P_{elec,t} + P_{fc,t} + P_{grid,t} \tag{2}$$

$$H_{stored,t} = H_{stored,\, t-1} + n_{elec}P_{elec,t} - (P_{fc,t}/n_{fc}) \tag{3}$$

$$0 < H_{stored,t} < H_{max}\ ,\ 0 < P_{elec,t} < P_{elec,\ max}\ ,\ 0 < P_{grid,t} < P_{grid,\ max} \tag{4}$$

$$H_{stored,0} = H_{stored,T} = 0 \tag{5}$$

Table 3 presents the model variables used in the optimization, which is designed to balance renewable energy supply, the electrical demand, and cost constraints while maintaining efficient hydrogen system operation.

Table 3. Optimization Model Variables

Symbol	Description	Unit
$P_{elec,t}$	Power input to electrolyzer at hour *t*	kW
$H_{stored,t}$	Hydrogen stored at hour *t*	kg
$P_{fc,t}$	Power output from fuel cell at hour *t*	kW
$P_{grid,t}$	Grid electricity imported at hour *t*	kW
$P_{load,t}$	Campus electricity demand at hour *t*	kW
$P_{PV,t}$	PV generation at hour *t*	kW

Results and Discussion

Solver Performance and Optimization status. The optimization was run in PuLP. It was set up to set up to reduce the use of grid electricity, while still providing uninterrupted power through hydrogen storage and fuel cell operation. The solver found an optimal solution, which confirms that all technical and operational constraints were met. In total, 125,741 constraints were included in the model, covering demand supply balance, hydrogen storage dynamics, storage capacity limits, and non-negativity conditions. Since no infeasibility issues were reported, the robustness of the model formulation is supported.

Electrolyzer Power Usage. Figure 1 plots the electrolyzer power demand over time. Because it runs only when there is surplus electricity available for hydrogen production, the electrolyzer does not operate continuously and its activity changes with renewable output and the stage of the storage. The largest peaks occur during periods of high solar availability, and the maximum recorded power reaches about 1.4×10^6 kW. In contrast, the electrolyzer remains inactive when hydrogen storage is already fully charged or when renewable input is too low.

Overall, the pattern shows an adaptive operating behavior, where the electrolyzer tracks resource availability in an efficient way. however, the repeated periods of curtailed operation indicate that a larger storage capacity could help limit renewable energy losses.

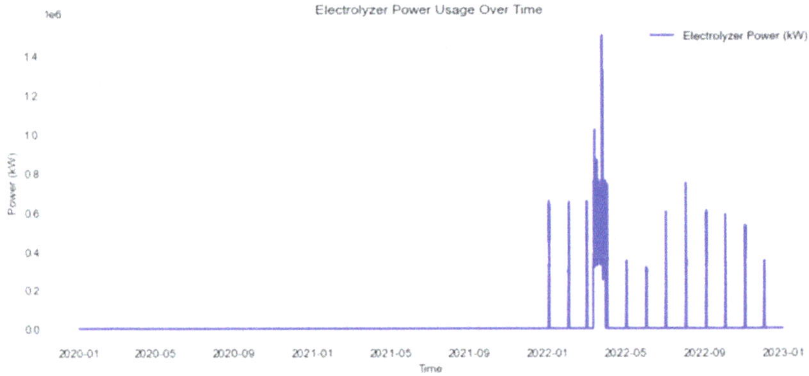

Fig.1. Electrolyzer power usage over time.

Hydrogen Storage Dynamic. Figure 2 tracks the hydrogen level in storage across the study period.

When the electrolyzer is running, the stored hydrogen increases, and it falls rapidly once the fuel cell starts supplying electricity. Reaching the storage limit is followed by the electrolyzer shutdown, while low storage levels can occasionally lead to grid imports. Overall, the profile shows that hydrogen storage is the main buffer used t balance intermittent supply and demand. At the same time, saturation events point to lost renewable potential, while depletion events increase grid dependency. Together, these outcomes highlight the importance of selecting an appropriate storage size.

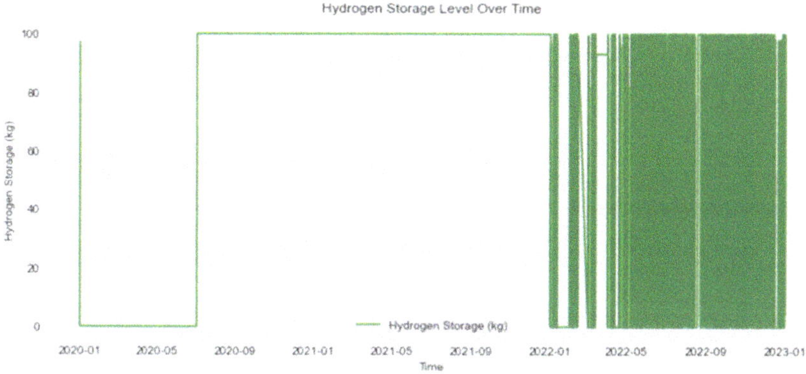

Fig.2. Hydrogen Storage Level over time

Fuel Cell Power Output. Figure 3 shows the fuel cell power output. The fuel cell stays inactive until enough hydrogen is stored, and then it provides electricity in clear bursts. While this pattern suggests that dispatch is working effectively, the peak values recorded in the results, which exceed 5×10^5 kW, may indicate scaling inconsistencies in the model. These power levels should be checked again to make sure they are realistic.

Even so, the fuel cell still plays and important role in covering demand when storage is low. However, efficiency losses and the fact that operation is not continuous point to areas where the model should be refined further.

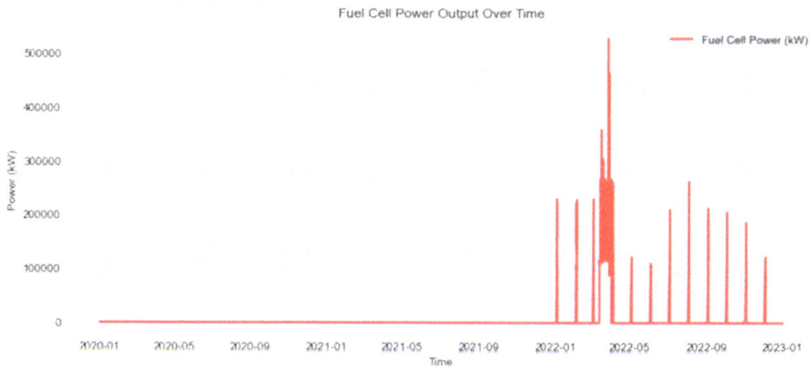

Fig.3. Fuel Cell Power Output Over Time

Grid Power Consumption. Figure 4 presents grid electricity usage. The results show very low dependence on the grid, with consumption close to zero for most of the year. Small spikes appear when hydrogen storage is depleted, which suggests that the model reduces grid reliance effectively, but full autonomy is not yet reached. This result highlights the need for predictive control strategies and improved storage integration to avoid remaining shortfalls.

Comparative Analysis of Power Sources. Figure 5 provides a comparison of the contributions from the electrolyzer, the fuel cell, and the grid. The results show coordinated operation between components: the electrolyzer converts excess renewable energy into hydrogen, the fuel cell converts stored hydrogen back into electricity when needed, and the grid acts as a backup during shortages. Overall, the system clearly prioritizes renewable and hydrogen-based sources over grid imports. However, the remaining spikes in grid use suggest that the optimization model could be improved further, especially in energy forecasting and storage sizing.

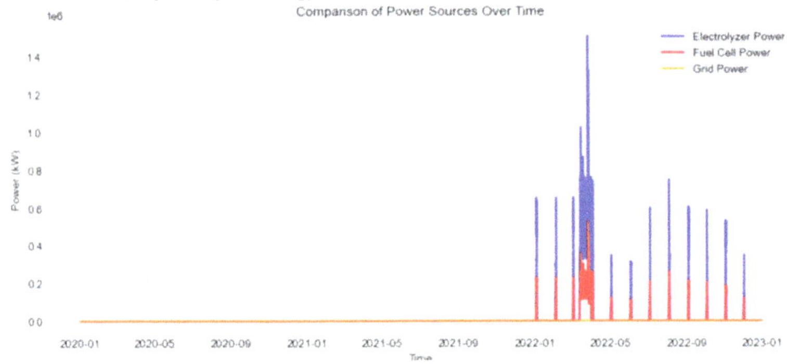

Fig.4. Comparative analysis of electrolyzer, fuel cell, and grid power usage

Comparison with Lithium Battery Storage. Hydrogen storage complements lithium batteries in renewable systems. Batteries excel for short-term (4-8 hour) daily cycling with 85-95% round-trip efficiency but degrade after 3,000-5,000 cycles. Hydrogen systems offer seasonal storage (weeks-months) despite 35-45% round-trip efficiency via electrolysis-fuel cell conversion, ideal for Ifrane's winter solar deficits

Conclusion

This study developed an optimization framework to integrate green hydrogen into the electrical power system of an academic facility, focusing on Al Akhawayn University in Ifrane, Morocco. Electrolyzers, hydrogen storage, fuel cells, and supplemental grid electricity are all included in the linear programming model. The aim is to limit grid dependence without compromising the reliability of the energy supply.

From the results, it is clear that the electrolyzer activity responds mainly to shifts in power demand and renewable generation. This leads to on and off operation, yet hydrogen production remains efficient. The storage system shows a repeating cycle, building up hydrogen during surplus generation and drawing it down when extra electricity is required. The fuel cell then helps stabilize supply by converting stored hydrogen into electricity, which further reduces grid relience. Although grid consumption is kept very low, it is still needed as a backup during peak demand and when storage is depleted.

The solver's optimal solution status confirms that the model formulation is consistent and that it can balance energy flows efficiently. The visualization also emphasizes the dynamic interaction between system component, supporting the feasibility of green hydrogen as a strategic pathway for sustainable energy management in higher education institutions.

Implications and Future Work

The findings of this study provide a foundation for the deployment of hydrogen-based energy systems in academic and institutional settings. By reducing grid dependency and enhancing resilience, such systems offer both environmental and economic benefits, aligning with global decarbonization strategies.

Future research can extend this work in several directions. First, the integration of real-time energy pricing would allow the model to capture dynamic grid costs and improve economic optimization. Second, expanding the renewable energy portfolio to include wind resources alongside solar photovoltaics could strengthen system sustainability and reliability. Finally, the application of machine learning techniques for predictive demand modeling may enhance decision-making in hydrogen production and storage scheduling, further improving system robustness.

Acknowledgment

This work was supported by the School of Science and Engineering at Al Akhawayn University in Ifrane, Morocco, within the framework of the seed-money funded project entitled *"Advancing Hydrogen Technologies for Sustainable Energy: A Study of Potential and Pathways in Morocco."* The authors gratefully acknowledge the institutional support provided throughout the research process.

Declaration of Generative AI and AI-Assisted Technologies in the Writing Process

I declare that the use of **Gemini** in the preparation of this paper was conducted in accordance with applicable academic integrity guidelines and institutional policies. AI assistance was limited to supporting tasks such as language refinement, formatting, or idea organization, and all substantive content, analysis, and conclusions represent my own original work and understanding. I have reviewed and verified the accuracy of all information included, take full responsibility for the final submission, and have properly cited all sources in compliance with scholarly standards.

References

[1] M. Jaradat, S. Alsaadi, Green Hydrogen in Focus: A Review of Production Technologies, Challenges, and Applications, Energies 17(16) (2024) 3992. https://doi.org/10.3390/en17163992

[2] D.S. Falcão, A review on PEM electrolyzer modelling: guidelines for beginners, J. Cleaner Prod. 276 (2020) 122953. https://doi.org/10.1016/j.jclepro.2020.122953

[3] S.S. Kumar, A review: hydrogen production by PEM water electrolysis, Int. J. Hydrogen Energy 44 (2019) 24094–24116. https://doi.org/10.1016/j.ijhydene.2019.05.248

[4] C.R. Wang, S. Park, J. Cho, Proton exchange membrane (PEM) water electrolysis: cell architecture, modeling, and performance perspectives, Chem. Rev. 125 (2025) 1452–1490. https://doi.org/10.1021/acs.chemrev.3c00904

[5] H. Nami, O. Babaie Rizvandi, C. Chatzichristodoulou, H. Lund Frandsen, Techno-economic analysis of current and emerging electrolysis technologies for green hydrogen production, Energy Convers. Manage. 272 (2022) 116612. https://doi.org/10.1016/j.enconman.2022.116612

[6] M.D. Mukelabai, Y. Li, P. Wang, Modeling and optimization of renewable hydrogen systems: review and future prospects, Renew. Energy 229 (2024) 1203–1218. https://doi.org/10.1016/j.renene.2024.05.048

[7] A. Criollo, J. Flores, J. Viteri, Green hydrogen production—fidelity in simulation models: optimizing technical and economic performance, Appl. Sci. 14(22) (2024) 10720. https://doi.org/10.3390/app142210720

[8] Z. Ma, R. Carter, J. Hou, A comprehensive modeling method for proton exchange membrane electrolyzers: design, identification, and simulation, OSTI Report (2021). https://doi.org/10.2172/1774873

[9] K. Zhang, H. Chen, M. Zhang, Status and perspectives of key materials for PEM electrolyzer development, New Renewable Energy 3(1) (2022) 12–22. https://doi.org/10.26599/NRE.2022.9120032

[10] S. Guilbert, J. Perrin, T. Delmas, Modeling, degradation study, and failure diagnosis in hydrogen technologies: state-of-the-art review, Energies 15(11) (2022) 4123. https://doi.org/10.3390/en15114123

[11] A. Vedrtnam, H. Gupta, P. Sharma, Water electrolysis technologies and their modeling: comparisons and challenges, Electr. Hydrogen Technol. 6(4) (2025) 81–94. https://doi.org/10.3390/eht604008

[12] A.M. Asim, F. Nazir, M. Hanif, Integrated optimization of energy storage and green hydrogen within hybrid renewable systems, Sci. Rep. 15 (2025) 13240. https://doi.org/10.1038/s41598-025-09408-x

[13] P.C. Okonkwo, S. Singh, M. Williams, Techno-economic optimization of hydrogen production using hybrid wind–solar systems across Australian cities, Sci. Rep. 15 (2025) 18652. https://doi.org/10.1038/s41598-025-17506-z

[14] S. Campiotti, P. Riboldo, A. Mutuberria, Operational Strategies and Integrated Design for Green Hydrogen Production via PEM Electrolysis in Off-Grid PV Systems, University of Cambridge Repository (2024). https://doi.org/10.17863/CAM.108015 repository.cam

Comparative Well-to-Wheel Assessment of Conventional and Alternative Vehicles in Saudi Arabia, the UAE, Oman, and Yemen

Mohamed Reda EL AOUNI[1,a *], Salma Sedki[1,b], Fouzya KHARMACHI[1,c] and Ahmed BAZZI[1,2,d]

[1]School of Science and Engineering, Al Akhawayn University, 53000 Ifrane, Morocco

[2]National School of Applied Sciences, Sidi Mohamed Ben Abdellah University, 30000 Fez, Morocco

[a]m.elaouni@aui.ma, [b]s.sedki@aui.ma, [c]f.kharmachi@aui.ma, [d]a.bazzi@aui.ma

Keywords: Life Cycle Assessment, Well to Wheel, Saudi Arabia, UAE, Oman, Yemen, Climate Change, Respiratory Effects, Acidification, Sustainability, Alternative Fuel Vehicles, BEV, HEV, FCV, ICE

Abstract. This study investigates the well-to-wheel (WTW) environmental and water performance of conventional diesel ICE, hybrid electric vehicles, BEV, and FCV across Saudi Arabia, the UAE, Oman, and Yemen by using the GREET model. The climate change potential, respiratory effects, acidification, and water consumption per 100 km were evaluated. Clear country-dependent trends in vehicle performance can be seen. The FCVs show the smallest overall environmental impacts in Saudi Arabia, Oman, and Yemen, with values of 0.0089 Pt/100 km for climate change, 0.0075 Pt/100 km for respiratory effects, and 0.00046 Pt/100 km for acidification, while BEVs perform best in the UAE, reaching 0.0051 Pt/100 km for climate change, 0.0068 Pt/100 km for respiratory effects, and 0.00047 Pt/100 km for acidification due to its relatively cleaner electricity mix. Water consumption shows comparable variability, with the lowest value observed for Oman at 5.09 L/100 km and the highest for Yemen at 9.9 L/100 km, confirming that the environmental and water advantages of alternative vehicle technologies are strongly influenced by the country's energy and water resource characteristics.

Introduction

Among various sources, the transportation sector accounts for approximately a quarter of total energy-related CO_2 output and is one of the major sources of greenhouse gas (GHG) emissions globally [1] . However, the increasing number of vehicles and consumption of fuel make it more difficult to achieve global sustainability targets in this sector. Due to the increasing influence on the environment, governments and companies worldwide are searching for alternative propulsion technologies to reduce the use of fossil fuels and increase energy efficiency. Among these alternatives, hybrid electric vehicles (HEVs), battery electric vehicles (BEVs), and fuel cell electric vehicles (FCEVs) have received considerable attention for their ability to reduce emissions from transportation [2] [3].

The benefit of new vehicle technologies regarding the environment differs per region. The use of electricity, energy, and transport policies, among other factors, influence the performance of a vehicle throughout its entire life cycle [4] [5]. For instance, in regions where fossil fuels are primarily used to generate electricity, the use of such electricity to charge battery electric vehicles may reduce the benefit of not emitting anything from the vehicle. This highlights the importance of conducting life cycle analyses, also known as cradle to grave (WTW) analyses.

The WTW framework examines the efficiency of energy use in a car throughout its entire life cycle. The WTW framework is divided into two broad areas: well-to-tank (WTT), which examines energy production and transport, and tank-to-wheel (TTW), which examines vehicle use [6]. TTW examines emissions from the tailpipe, while WTW examines total energy consumption and emissions. To gain a complete understanding of the environmental impact, WTW analysis must be incorporated into a larger life cycle assessment (LCA) that examines manufacturing and end of life impacts as well [7]. It is only by incorporating WTW analysis into LCA that we can compare the impacts of manufacturing versus emissions in use to gain a complete understanding of the sustainability of each technology [8] [9] .

The existing literature demonstrates that EVs significantly reduce emissions during operation but are more environmentally impactful during production due to the high energy required for producing lithium-ion batteries [7], [10]. For example, a study conducted by Ankathi et al. [10] and Dolganova et al. [8] indicates that battery production can account for approximately 10 to 394 kg CO_2-eq/kWh, depending on the type of battery chemistry and the mix of electricity. Moreover, Kurkin et al. [7] concluded that electric vehicles can require up to six times more natural resources than ICE vehicles, emphasizing the importance of material efficiency and recycling. On the other hand, in those situations where the electricity is supplied from renewable energy sources, large overall emissions reductions can be found for BEVs. For instance, Oliveri et al. [11] showed that adding solar power to logistics operations could cut BEV life-cycle impacts in half.

Thus, much of the electric vehicles' performance depends on how clean the electricity used is. BEVs outcompete gasoline powered cars in China only in those provinces where over half of the grid is powered by renewables up to 18.32 tons of CO_2 saved per vehicle. In Australia, with cleaner electricity grids, BEVs cut greenhouse gas emissions by as much as 80%. These findings suggest that it is not possible to reduce transport emissions without reducing carbon intensity in the production of electricity [4] [5].

While extensive research is being carried out in Europe, China, and North America, comparative studies of WTW emissions are very limited in the Middle East. Here, most of the electrical power comes from fossil fuels, but renewable projects are also emerging. The abundance of oil and gas resources, combined with increasing electricity consumption and rapidly growing renewable energy sources, turns the region into a unique location to assess the environmental impact of new vehicle technologies. Problems for electric vehicles in Jordan include a lack of charging infrastructure, high purchase costs, and low economic incentives. Similar issues exist in neighboring MENA countries, such as Morocco. Therefore, a study examined low-carbon alternative fuel vehicles in Morocco and demonstrated how these vehicles can reduce greenhouse gas emissions and achieve Net Zero goals. A study compares internal combustion engine vehicles, HEVs, BEVs, and FCVs with a lifecycle assessment over 15 years, taking into account well-to-wheel impacts in a region where fossil fuels are used for most of the electricity [12] . This therefore demonstrates the rationale for region-specific WTW analyses that could provide judgment on the environmental viability of alternative fuel vehicles, especially given Morocco's potential to achieve as much as a 42% cut in GHG emissions [13]. In the same direction, Solaymani and Botero [1] emphasize supply side policies (e.g., integrating renewables and clean hydrogen production) and demand side measures (e.g., tax, subsidy, and public awareness) for accelerating sustainable mobility.

These findings consequently leave a gap in understanding how vehicle technologies affect the environment, especially with unique energy setups within Middle Eastern countries. Furthermore, only a few studies combine modeling based on the GREET model with region-specific electricity data and broad environmental indicators such as the Eco-Indicator 99. Filling this knowledge gap becomes critical to help shape national policies on transport electrification and sustainable energy planning.

Therefore, this study aims to perform a new, straightforward comparison of well-to-wheel life cycle emissions for four vehicle technologies-diesel internal combustion engine, fuel cell vehicle, hybrid electric vehicle, and battery electric vehicle-across four Middle Eastern countries: Saudi

Arabia, Yemen, Oman, and the United Arab Emirates. Using the GREET model as the base, the study merges national electricity data from Our World in Data with emission factors and eco-indicator metrics, which are processed through a mathematical model in Excel. It will seek to measure and compare greenhouse gas and pollutant emissions for each vehicle type under real regional energy conditions. The results should hence indicate how different country energy mixes affect environmental performances and inform guidance on sustainable transport and energy plans in the Arabian Peninsula.

Methodology

This study applies the WTW framework, part of the Environmental Life Cycle Assessment, comparing four vehicle technologies in the energy context of the United Arab Emirates, Saudi Arabia, Oman, and Yemen: diesel, hybrid diesel-electric, battery electric vehicle, and fuel cell vehicle. The WTW analysis provides a clear view of fuel production WTT and vehicle operation TTW together. A wider view of environmental impacts becomes possible that way. This perspective is important for new vehicle types like BEVs and FCVs, which do not emit much but can have great upstream impacts depending on the national grid of the country if it relies on fossil fuel electricity and natural gas-based hydrogen. The impacts have been estimated for 100 km of driving, considering climate change, respiratory effects, acidification, and eutrophication. Results from GREET models are used here, coupled with the factors of Eco-indicator 99 (Egalitarian); thus, all vehicle technologies have been compared consistently under the energy conditions of the country.

1. GREET Model Setup

By using the GREET1 model, and more precisely the Excel-based GREET1_2021 fuel-cycle module developed at Argonne National Laboratory. GREET calculates energy usage in MJ/km and emissions in kg/km for pollutants, including CO_2, CH_4, NOx, PM_{10}, $PM_{2.5}$, SOx, and total greenhouse gases. It examines feedstock, fuel production, and vehicle operation for four vehicle technologies. To adapt to regional contexts of energy supply, GREET's default U.S. electricity mix was updated to country-specific mixes for the UAE, Saudi Arabia, and Oman using 2024 data and to 2023 data for Yemen, where the 2024 data is not available. The mixes include coal, natural gas, oil, nuclear, solar, and wind:

Feedstock stage (extraction/collection/transport): $m_{i,v}^{\text{feed}}\ (kg/km)$

Fuel stage (processing/refining/distribution): $m_{i,v}^{\text{fuel}}\ (kg/km)$

Vehicle operation stage: $m_{i,v}^{\text{op}}\ (kg/km)$

2. Data collection

The data for this analysis were collected from reliable national and international sources to represent the energy and transport context of each country during the most recent available year. Electricity generation mix data were obtained from *Our World in Data* and manually entered into the GREET electricity-generation module to replace the default grid composition with country-specific profiles. Data correspond to 2024 for Saudi Arabia, the UAE, and Oman, while for Yemen, 2023 energy mix data were used due to the unavailability of updated 2024 figures.

Table 1:Electricity Generation Mix by Energy Source in Selected Gulf Countries [%] [14]

Energy Source	Yemen (%)	Saudi Arabia (%)	Oman (%)	UAE (%)
Coal	0	0	0	0
Oil	65.15	34.5	2.75	0.01
Natural Gas	17.92	63.34	93.06	68.42
Nuclear	0	0	0	22.97
Solar	16.94	1.81	3.92	8.6
Wind	0	0.35	0.28	0

Table 1 shows the electricity generation mix in Yemen, Saudi Arabia, Oman, and the UAE, expressed as the percentage contribution of each energy source. It highlights the dominance of oil and natural gas in most countries, the growing share of solar energy, and the presence of nuclear power in the UAE.

3. Well-to-Wheel and environmental impacts computation

First, after getting the data from GREET model of feedstock, fuel and vehicle operation of total energy and gas emissions for each type of vehicle. To refine the data and compute environmental impacts using the Eco-indicator 99 (Egalitarian) method, a separate excel sheet workbook was made.

The WTW framework was the first step in refining the data, and it provides a complete picture of the environmental performance of each vehicle type by integrating both upstream and operational stages. The total WTW impact is expressed as the sum of two main phases:

$$m_{i,v}^{\mathrm{WTW}} = m_{i,v}^{\mathrm{WTT}} + m_{i,v}^{\mathrm{TTW}} \tag{1}$$

The WTT phase includes all upstream processes required to supply the fuel or electricity to the vehicle. As was previously described in the GREET setup, the feedstock and fuel stages together constitute the upstream WTT phase and is expressed as:

$$m_{i,v}^{\mathrm{WTT}} = m_{i,v}^{\mathrm{feed}} + m_{i,v}^{\mathrm{fuel}} \tag{2}$$

The feedstock stage accounts for the extraction, collection, and transport of primary resources such as coal, natural gas, crude oil, or renewable inputs. The fuel cycle includes all phases involved in transforming the resources into usable energy carriers, such as diesel, hydrogen, or electricity. For battery electric vehicles, this involves the generation of electricity from the various power sources and transmission losses. For hydrogen fuel cell vehicles, this stage involves the generation of hydrogen either through natural gas reforming or renewable electrolysis.

The TTW phase is the operation phase which concentrates on the energy used and emissions generated during the driving of the vehicle. Which can be expressed as:

$$m_{i,v}^{\mathrm{TTW}} = m_{i,v}^{\mathrm{op}} \tag{3}$$

For conventional diesel and hybrid vehicles, this includes fuel consumption and emissions of greenhouse gases during driving. For battery electric vehicles, it indicates the amount of energy drawn from the power grid and the efficiency of the motor operation. For hydrogen fuel cell vehicles, this means the efficiency of the fuel cell and water vapor as the sole output.

To enhance the representation of real-world driving in the environmental assessment, the results of the GREET model, which were provided per kilometer of driving (MJ/km or g/km), were converted to a unit of 100 km . This scaling helped facilitate direct comparisons across diesel, hybrid, battery electric, and fuel cell vehicle technologies. Also, this adjustment maintains the proportionality of impacts while clearly presenting the cumulative environmental burden for a typical 100 km driving cycle in the UAE, Saudi Arabia, Oman, and Yemen.

$$M_{i,v}^{\text{stage, FU}} = m_{i,v}^{\text{stage}} \times 100 \quad \text{(g/100 km)} \tag{4}$$

Environmental impact numbers are calculated with Eco-indicator 99 (Egalitarian) factors. For each impact category the products of GREET's total pollutants from well-to-wheel and of matching factors were summed up. The total impact in each category was determined as:

$$I_{v,c}^{stage} = \sum_{i} \left(EF_{i,c} \times M_{i,v}^{stage,\, FU} \right) \tag{5}$$

Where:

- $EF_{i,c}$ represents the characterization factor in Pt/g for pollutant i.
- $m_{i,v}$represents the mass of pollutant i emitted by vehicle type v per 100 km of travel, as calculated from the GREET model.

Emissions considered under this study for climate change include CO_2, CH_4, N_2O, and CO. The breathing and health effects consider the pollutants as VOC, CH_4, CO, NOx, PM_{10}, $PM_{2.5}$, and SOx, while acidification and eutrophication consider only NOx and SOx emissions.

Table 2:Eco-indicator 99 (Egalitarian) Characterization Factors [Pt/g] [15]

	Ecosystem Quality	**Human Health**	
	Acidification/ Eutrophication	Climate Change	Respiratory Effects
CO2 (Pt/kg)		0.0040645	
CH4(Pt /kg)		0.0040645	0.00024774
N2O (Pt/kg)		1.3355	
VOC (Pt/kg)			0.024774
CO (Pt/kg)		0.0062323	0.014148
NOx (Pt/kg)	0.55682		1.7245
PM10 (Pt/kg)			7.2581
PM2.5 (Pt/kg)			13.548
SOx (Pt/kg)	0.10146		1.0568

All results were normalized and expressed in Pt/100 km. A lower score indicates less environmental impact and better overall sustainability. This makes the comparison of the four vehicle technologies for the four Gulf countries easy, taking into account different regional energy sources but using the same unit for comparison.

In the final step of the WTW framework, we add all the WTT contribution and the operational TTW contribution to yield the total environmental impact for each vehicle type and category. The final number $I_{v,c}^{\text{WTW}}$ in Pt/100 km, represents the full life-cycle environmental burden of each vehicle technology for the studied region, allowing easy comparison across vehicle types under the same driving and energy assumptions.

$$I_{v,c}^{\mathrm{WTW}} = I_{v,c}^{\mathrm{WTT}} + I_{v,c}^{\mathrm{TTW}} \tag{6}$$

Results and Discussion

This section presents the key results of the WTW analysis for Saudi Arabia, the UAE, Oman, and Yemen. The results include three environmental indicators that comprise Climate Change, Respiratory Effects, and Acidification & Eutrophication, in addition to Water Consumption.

1. Analysis by Country
   a. Saudi Arabia

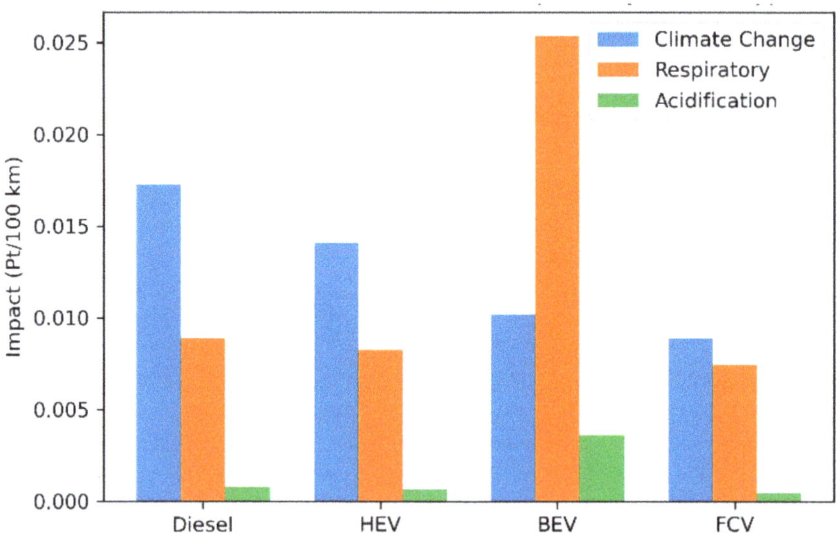

Figure 1: Saudi Arabia environmental impacts for each vehicle

FCVs have the smallest environmental impact for all measures within Fig. 1 with 0.0089 Pt/100 km for Climate Change, 0.0075 Pt/100 km for Respiratory Effects, and 0.00046 Pt/100 km for Acidification & Eutrophication. BEVs are not as good, particularly for respiratory effects, since the country relies on oil and gas for much of its electricity. In general, it seems FCVs are the most sustainable option for Saudi Arabia, as they go along with the country's hydrogen plans and the large supply of natural gas.

b. UAE

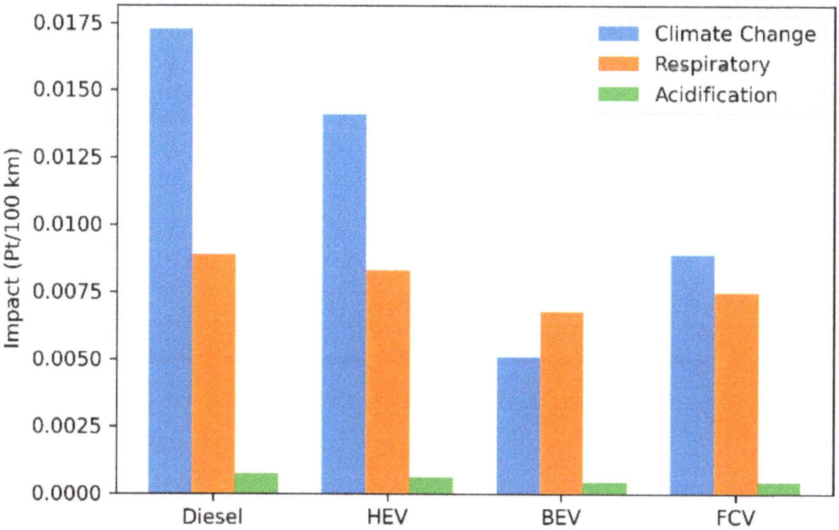

Figure 2: UAE environmental impacts for each vehicle

BEVs have the best environmental outcomes for all measures in Fig. 2 with 0.0051 Pt/100 km for Climate Change, 0.0068 Pt/100 km for Respiratory Effects, and 0.00047 Pt/100 km for Acidification. This is due to the cleaner electricity in the country, which is increasingly generated by solar power and natural gas. HEVs are decent, but FCVs have slightly higher impacts due to the methods of hydrogen production. Thus, BEVs are currently the environmentally friendliest option for the UAE, considering the growth of renewable energy capacity.

Materials Research Proceedings 64 (2026) 813-823 https://doi.org/10.21741/9781644904091-101

c. Oman

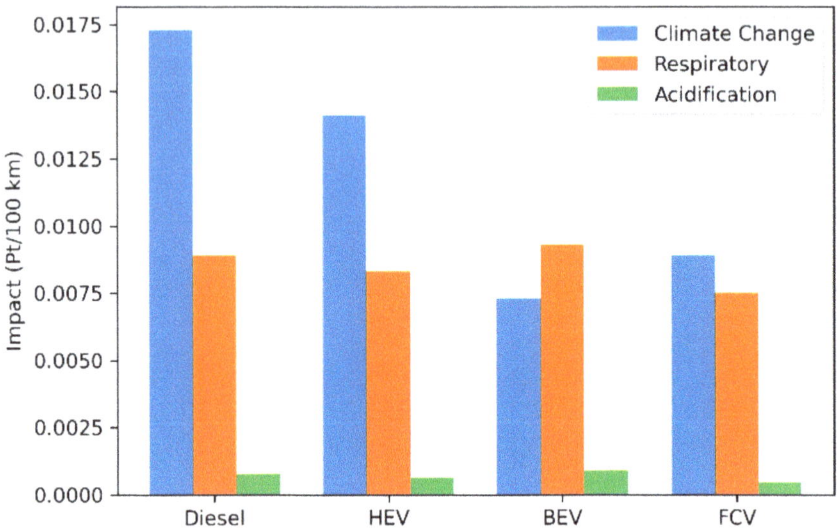

Figure 3: Oman environmental impacts for each vehicle

In Figure 3, FCVs provide the best overall environmental results in Oman: 0.0089 Pt/100 km for Climate Change, 0.0075 Pt/100 km for Respiratory Effects, and 0.00046 Pt/100 km for Acidification and Eutrophication. BEVs do a little better on Climate Change at 0.0073 Pt/100 km, but are worse in the other two categories, showing increased local air pollution. FCVs win overall because they stay low on all measures and help keep the air cleaner. This fits with Oman's natural gas-based electricity, which would lead to a moderately carbon-intensive production of hydrogen.

d. Yemen

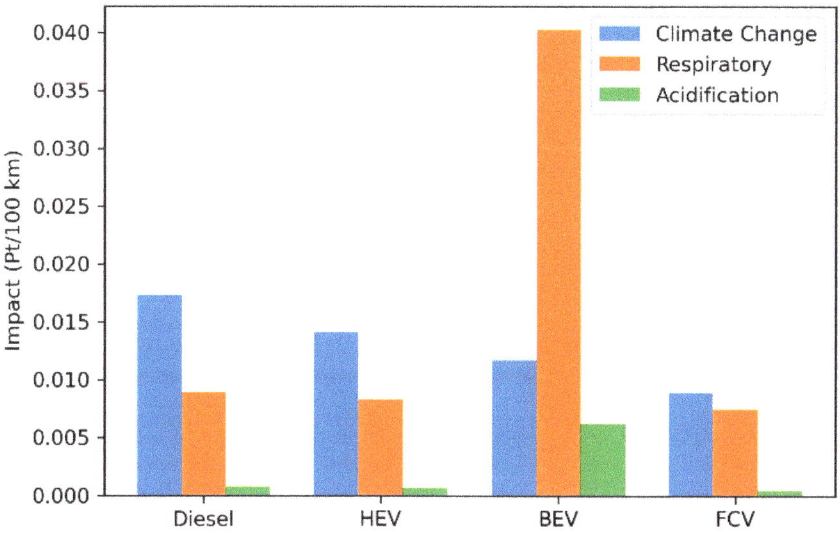

Figure 4: Yemen environmental impacts for each vehicle

In Fig. 4, FCVs again had the lowest environmental impacts of 0.0089 Pt/100 km for Climate Change, 0.0075 Pt/100 km for Respiratory Effects, and 0.0005 Pt/100 km for Acidification. BEVs emit the most: 0.0403 Pt/100 km for respiratory impacts, due to inefficient and fossil-dependent national power generation. Since Yemen has very minimal renewable capacity in place and a weak grid infrastructure, FCVs or HEVs are the most possible solutions to reduce its transport emissions in the short term.

2. Comparative Summary

Table 3: Summary of vehicle impacts for each country

Metric	Saudi Arabia	UAE	Oman	Yemen
Best Vehicle Type (to be adopted)	FCV	BEV	FCV	FCV
Climate Change (Pt/100 km)	0.0089	0.0051	0.0089	0.0089
Respiratory (Pt/100 km)	0.0075	0.0068	0.0075	0.0075
Acidification (Pt/100 km)	0.00046	0.0005	0.0005	0.0005
Water Use (L/100 km)	5.61	5.69	5.09	9.9

The findings of this analysis provide validation that the best car technology correlates linearly and varies according to the country's energy structure and the availability of water resources. It is evident from Table 1 that FCVs are rated as the best car solution for Saudi Arabia, Oman, and Yemen because of the low impact on all parameters, such as Climate Change 0.0089 Pt/100 km, Respiratory Effects 0.0075 Pt/100 km, and Acidification 0.00046-0.0005 Pt/100 km. BEV is rated as the best solution in UAE due to a cleaner energy mix from natural gas and solar resources that minimize climate change 0.0051 Pt/100 km damage to the environment as well as acidification 0.00047 Pt/100 km damage.

For the water usage, Saudi Arabia and Oman show moderate consumption with 5.61 L/100 km and 5.09 L/100 km respectively, but a relatively high demand for Yemen 9.9 L/100 km due to inefficient, fossil-based power. The UAE follows a similar water demand to that of Saudi Arabia 5.69 L/100 km, which represents a similar desalination-based power process. In general, FCVs have been found to be the best-balanced technology for fossil-based systems, while BEVs have been established as optimal for cleaner, renewable-based power grids. The findings indicate that we require more effective water-saving power techniques, hydrogen infrastructure, and renewable energy, all of which are to be pushed together with vehicle technology advancements to make use of Gulf projects in an environmentally viable manner.

Conclusion

The study analyzed how different types of vehicles affect the environment and water usage in Saudi Arabia, the UAE, Oman, and Yemen using the GREET model and each country's energy data. The best vehicle type depends on how each country produces electricity. FCVs had the lowest overall impacts in Saudi Arabia, Oman, and Yemen, while BEVs were the cleanest in the UAE because its grid relies more on clean, solar power. Water use was lowest in Oman and highest in Yemen due to differences in efficiency in generating the electricity.

In summary, the findings state that a sustainable transport system in the countries under study depends on integrating car use with the energy mix of the country. This involves the use of more renewable energy, developing the low carbon hydrogen energy sector, and improving the efficiency of hydroelectric and water-based power. The limitations of the study include the failure to examine the production of cars and the recycling of cars at the end of their life. For Yemen, only the 2023 electricity mix was available. Full life-cycle modelling, cost analysis, and regional hydrogen-supply scenarios are recommended for future work to develop a long-term, sustainable mobility plan for the Gulf region.

References

[1] S. Solaymani and J. Botero, "Reducing Carbon Emissions from Transport Sector: Experience and Policy Design Considerations," Sustainability, vol. 17, no. 9, pp. 3762–3762, Apr. 2025, doi: https://doi.org/10.3390/su17093762

[2] E. Y. C. Wong, D. C. K. Ho, S. So, C.-W. Tsang, and E. M. H. Chan, "Life Cycle Assessment of Electric Vehicles and Hydrogen Fuel Cell Vehicles Using the GREET Model—A Comparative Study," Sustainability, vol. 13, no. 9, p. 4872, Jan. 2021, Available: https://www.mdpi.com/2071-1050/13/9/4872

[3] A.-F. Djire, A. Bazzi, H. El Hafdaoui, A. Khallaayoun, and R. Lghoul, "Techno-Economic Analysis of Power to Ammonia Production," E3S Web of Conferences, vol. 601, p. 00065, 2025, doi: https://doi.org/10.1051/e3sconf/202560100065

[4] B. Tang, Y. Xu, and M. Wang, "Life Cycle Assessment of Battery Electric and Internal Combustion Engine Vehicles Considering the Impact of Electricity Generation Mix: A Case

Study in China," Atmosphere, vol. 13, no. 2, p. 252, Feb. 2022, doi: https://doi.org/10.3390/atmos13020252

[5] R. Smit and D. W. Kennedy, "Greenhouse Gas Emissions Performance of Electric and Fossil-Fueled Passenger Vehicles with Uncertainty Estimates Using a Probabilistic Life-Cycle Assessment," Sustainability, vol. 14, no. 6, p. 3444, Mar. 2022, doi: https://doi.org/10.3390/su14063444

[6] J. L. Osorio-Tejada, E. Llera-Sastresa, and A. H. Hashim, "Well-to-Wheels Approach for the Environmental Impact Assessment of Road Freight Services," Sustainability, vol. 10, no. 12, p. 4487, Nov. 2018, doi: https://doi.org/10.3390/su10124487

[7] Andrey Kurkin, Evgeny Kryukov, O. Masleeva, Yaroslav Petukhov, and Daniil Gusev, "Comparative Life Cycle Assessment of Electric and Internal Combustion Engine Vehicles," Energies, vol. 17, no. 11, pp. 2747–2747, Jun. 2024, doi: https://doi.org/10.3390/en17112747.

[8] I. Dolganova, A. Rödl, V. Bach, M. Kaltschmitt, and M. Finkbeiner, "A Review of Life Cycle Assessment Studies of Electric Vehicles with a Focus on Resource Use," Resources, vol. 9, no. 3, p. 32, Mar. 2020, doi: https://doi.org/10.3390/resources9030032

[9] M. Messagie, F.-S. Boureima, T. Coosemans, C. Macharis, and J. Mierlo, "A Range-Based Vehicle Life Cycle Assessment Incorporating Variability in the Environmental Assessment of Different Vehicle Technologies and Fuels," Energies, vol. 7, no. 3, pp. 1467–1482, Mar. 2014, doi: https://doi.org/10.3390/en7031467

[10] S. K. Ankathi, J. Bouchard, and X. He, "Beyond Tailpipe Emissions: Life Cycle Assessment Unravels Battery's Carbon Footprint in Electric Vehicles," World Electric Vehicle Journal, vol. 15, no. 6, p. 245, Jun. 2024, doi: https://doi.org/10.3390/wevj15060245

[11] L. M. Oliveri, D. D'Urso, N. Trapani, and F. Chiacchio, "Electrifying Green Logistics: A Comparative Life Cycle Assessment of Electric and Internal Combustion Engine Vehicles," Energies, vol. 16, no. 23, p. 7688, Jan. 2023, doi: https://doi.org/10.3390/en16237688

[12] G. A. Samawi, O. M. Bwaliez, Mohannad Jreissat, and Abdallah Kandas, "Advancing Sustainable Development in Jordan: A Business and Economic Analysis of Electric Vehicle Adoption in the Transportation Sector," World Electric Vehicle Journal, vol. 16, no. 1, pp. 45–45, Jan. 2025, doi: https://doi.org/10.3390/wevj16010045

[13] Hamza El Hafdaoui, M. Reda, Yassine Jaija, Faissal Jelti, and A. Khallaayoun, "Total Cost of Ownership Evaluation of Alternative Fuel Vehicles in Morocco," 4th International Conference on Innovative Research in Applied Science, Engineering and Technology (IRASET), Jun. 2024, doi: https://doi.org/10.1109/IRASET60544.2024.10548856

[14] H. Ritchie, M. Roser, and P. Rosado, "Energy," Our World in Data, Oct. 2023, Available: https://ourworldindata.org/energy#all-charts

[15] R. Hischier, B. P. Weidema, H.-J. Althaus, C. Bauer, and T. Nemecek, "Implementation of Life Cycle Impact Assessment Methods. ecoinvent report No. 3, v2.2," Jan. 01, 2010. https://www.researchgate.net/publication/263239305_Implementation_of_Life_Cycle_Impact_Assessment_Methods_ecoinvent_report_No_3_v22

[16] D. Burchart and Iga Przytuła, "Review of Environmental Life Cycle Assessment for Fuel Cell Electric Vehicles in Road Transport," Energies, vol. 18, no. 5, pp. 1229–1229, Mar. 2025, doi: https://doi.org/10.3390/en18051229

Materials Research Proceedings 64 (2026) 824-831 https://doi.org/10.21741/9781644904091-102

Digitalization of Hydrogen Energy: The Role of Informatics and Telecommunication in Production and Storage Systems

Hassan OUGRAZ[1,a*]

[1]Ibn Zohr University, Higher School of Technology, Ouarzazate

[a]hassan.ougraz.24@gmail.com

Keywords: Hydrogen Energy, Digitalization, Informatics, Telecommunication, IoT, Smart Grids, Hydrogen Production, Hydrogen Storage, Digital Twin, Artificial Intelligence, Energy Management

Abstract. The global trend towards carbon-neutral energy systems has established hydrogen as a key enabling carrier of energy for green electricity, mobility and industrial use. Advances in informatics/telecom are changing perceptions of government agencies, technology vendors and energy service companies towards hydrogen infrastructure. In this paper, a digitalized approach for hydrogen energy system is proposed regarding data reports, IoT (Internet of Things), cloud computing and 5G/6G communication network. We also show that intelligent hydrogen systems support electrolysis monitoring in real-time, predictive storage maintenance and coordinated integration of renewable resources by employing digital twins, edge computing and artificial intelligence. The proposed DHES system model (Digital Energy System for Hydrogen) also achieves significant performance enhancement with 84.2% reduction in data latency from 95ms to 15ms, 28% gains on predictive maintenance accuracy, 6% improvement in energy efficiency and 9% enhancement in system availability than traditional approaches. Additionally, data transparency and traceability are significantly enhanced. The study addresses cybersecurity, interoperability, and data standardization challenges, while identifying future research directions toward smart, fully connected hydrogen ecosystems.

1. Introduction

As countries around the world seek to decrease their emission of greenhouse gases, investment in renewable energy technologies has increased and hydrogen as an energy carrier is being considered due to its high energy density and multiple uses in transportation, power generation, and industry. Hydrogen's ability to store on-and-off energy from renewables such as solar and wind makes it instrumental in decarbonizing. The hydrogen value chain from generation to storage, distribution, and use involved significant technical complexity and economic impedances. However, these systems require sophisticated control, monitoring, and optimization to ensure safety, reliability, and efficiency. Traditional manual and offline analytical methods prove inadequate for managing large-scale, data-intensive hydrogen infrastructure integrated with variable renewable sources.

New advances in digital technologies have created an environment wherein there are vast disturbances in innovation from IT, telecommunications and artificial intelligence (AI) together known as Energy 4.0. With the likes of Internet of Things (IoT), cloud computing, edge intelligence and artificial intelligence (AI) now enhancing sensor data for real-time analytics which can lead to instantaneous actioning down the hydrogen value chain. New 5G and future 6G networks offer ultra-low latency communications needed to coordinate production and storage in smart hydrogen grids.

Digital twins' graphical representations of physical hydrogen plants, provide on-line performance and failure prediction and maintenance optimization through analytics. With machine learning and big data techniques, these developments are expected to lead to better economics, higher reliability, faster smart grid adoption. However, the digitalization of hydrogen systems is

Materials Research Proceedings 64 (2026) 824-831 https://doi.org/10.21741/9781644904091-102

still in its infancy, and a number of challenges need to be addressed including cybersecurity risks, platform interoperability obstacles, data governance issues and missing standard communication protocols.

In this research, a comprehensive overview of the digitalization strategies in hydrogen energy system is conducted whilst focusing on how information technology and telecommunication can play a role to improve production and storage process. It reviews the latest advancements in digital monitoring, control, and communication concepts for hydrogen infrastructure and smart grid integration and highlights major obstacles and research directions toward fully digitalized data-based hydrogen ecosystems.

2. Literature Review

2.1. Overview of Hydrogen Energy Systems

Hydrogen is considered to be a key piece of the world's low carbon economy transition; The hydrogen market itself will reach ~$12trn by 2050, according to Bernstein (Q1 2020). It is a flexible energy carrier that can be employed for storage, transport and power generation and that does not directly release CO2. The hydrogen value chain has different steps: production, storage and utilization that need today a more state of the art control and optimization tools. One of the most promising solutions for RE integration is the production of green H2 via electrolysis with renewable power; these are primarily wind and solar [1].

One of the greatest technical obstacles is storage. Current methods of storage for H2 are gas phase, liquid and solid phase with its respective advantages and disadvantages [2]. Combine that with the possibility of integrating with renewable energy processes in order to improve grid flexibility and system stability you've got not only an opportunity but a whole host of complex operational and data management issues.

2.2. Digitalization and Energy Systems

It is for example expected that if the energy industry has been digitized to a reasonable degree (given informatics science and telecommunication technology implicit in it), this will new focus on objects towards data (like intelligent operation of energy). The benefits of capabilities are also exploited in real time monitoring and predictive control with the adoption of technology such as IoT, cloud computing and AI [3]).

For hydrogen, these techniques have the potential to enhance both the energy efficiency and reliability/safety of electrolysis equipment (electrolysers) or storage tanks/pipelines, using a data-driven approach. Furthermore, the synergy of IoT and AI has been shown to enhance performance while diminishing cost of hydrogen infrastructures [4].

2.3. Informatics Applications in Hydrogen Production

The hydrogen produced by electrolysis is sensitive to changing power inputs and conditions around it. Informatics technologies are increasingly used to cope with these complexities. SCADA systems enable centralized data collection, while AI-based algorithms are used to perform fault detection and optimization [5]. Another example is digital twin technology for real-time simulation and emergency predictive maintenance of hydrogen systems [6]. Similarly, it is reported that maintenance costs within industrial applications can be reduced by 15-30% and operational uptime can be increased [7].

2.4. Telecommunication Technologies and IoT Integration

Next-generation communication networks (such as 5G and 6G) are among the most important means to connect distributed hydrogen assets. The speakers use the services provided by the watchers and in order to scale up such a thing, a Machine Type Communication (MTC) usage will benefit by having such low-latency, high-bandwidth network.

To characterize what is above, an IoT communication systems an IoT configuration can connect to electrolyzers, compressors and storage facilities for a proactive control logic that gives end-users actionable insights [9]. Furthermore, edge computing minimizes latency as it processes data for hydrogen refueling and storage on-site, which is essential to ensure their smooth operations.

2.5. Digital Twin and AI in Hydrogen Systems

Hydrogen systems Digital twin approach is currently revolutionizing hydrogen systems by providing real-time emulation of the electrolyzer performance and degradation mechanisms [10]. AI models (like deep learning and reinforced leaning) are employed for predicting energy demand and failure prediction in Hydrogen infrastructure [11].

Fleet-level digital twin architectures have also been suggested for the optimized operation of multiple hydrogen electrolyzers that enhance scalability and system robustness [6]. The combination of AI with digital twins allows the autonomous optimization of process parameters and the lifecycle management as well as predictive maintenance of hydrogen facilities.

2.6. Storage and Monitoring and Intelligent Smart Grids

Hydrogen storage methods, including metal hydrides, cryo-compressed tanks, and underground facilities, are essential for large-scale energy systems [12]. Digitally-enabled storage integrates with smart grids for optimized charge/discharge control, maintenance forecasting, and renewable integration [13]. AI-based monitoring enables real-time leak detection and pressure anomaly identification [14].

3. Methodology and Conceptual Framework

3.1 Conceptual Framework Description

Fig. 1 illustrates the DHES architecture, 3 layers and data flows are presented. The Physical Layer consists of hydrogen infrastructure with renewable sources (solar/wind), the electrolyzers, storage devices and distribution network. The Digital Layer includes the IoT devices, communication technology (5G/6G), and edge gateways for data acquisition and transmission. The Control Layer employs AI analytics digital twin simulations and predictive maintenance to transform data into insight. Data transmission and control feedback integrate five components: Sensors, Communication Layer, Edge Processing, Cloud Analytics, and Control/Feedback mechanisms.

Thick double-headed arrows link each of the stages, showing the cyber-physical loop between the hydrogen system and its digital twin. DHES System Architecture as illustrated in Fig. 1, the structure of DHES structure has a three-level general architecture:

- Layer 1: store resource-electrolyzer (store product A for grid).
- Virtual Layer (IoT Sensors).
- Virtual End-to-End Network (5G/6G, Edge Gateway, Cloud Platform).
- Control Layer (Digital Twin, Predictive Maintenance Visualization and AI Analytics Modules).

The arrows indicate data flow, and the direction of feedback-control points, to denote the way digital intelligence appears in physical hydro gen assets at runtime.

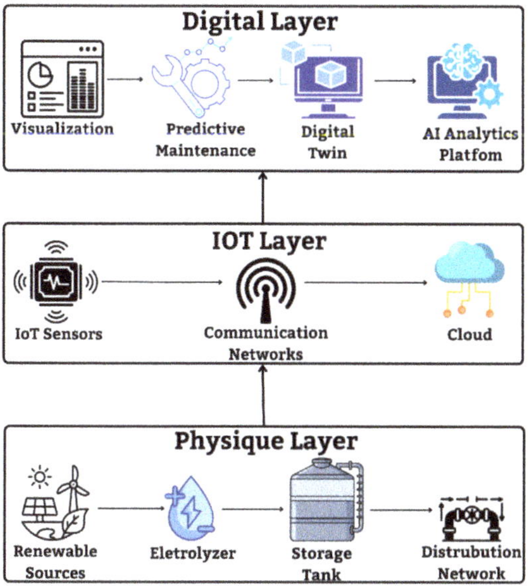

Figure 1. Conceptual Framework of the Digitalized Hydrogen Energy System (DHES)

3.2 Data Flow and its Architectures

Fig. 2 depicts the information architecture, which describes flows of data between physical components and digital analytics platforms across five stages that are connected in a circular feedback structure:

- Detection and Data Acquisition: Sensors detect hydrogen pressure, temperature, current flow rate of wood tar collecting tank used as a fuel in the fuel cell tank is determined.
- Transmission Security for communication transfer: The transmitted data can be securely transferred through the use of IoT gateway and high-speed networks.
- Edge Processing: Edges nodes may be used to filter, compress, and detect anomaly.
- Cloud & Edge Computing (CES): Cloud servers perform AI analytics, digital twin simulation, and data storage.
- Decision making and control: feedback directs local controllers to adjust process variables.

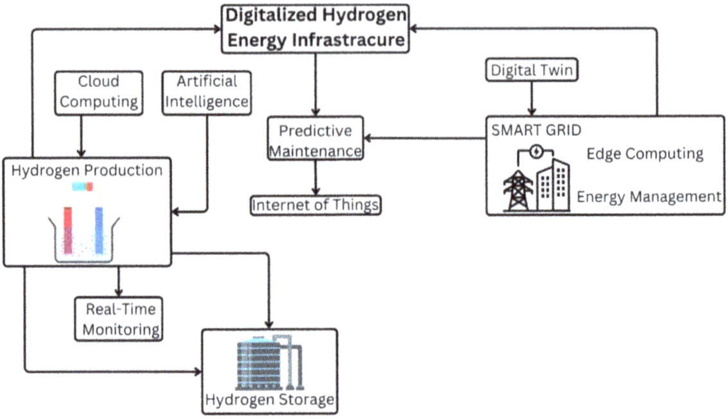

Figure 2. Data Flow and Information Architecture in the DHES

4. Results and Discussion

4.1. Simulation Environment and Implementation Setup

A cyber-physical simulation platform, established by process modeling in MATLAB/Simulink and cloud analytics on Python programming language was constructed to verify DHES economy. The setup included:

- Input of renewable energy (solar PV and wind profiles).
- Nonlinear electrolyzer model.
- Compressed hydrogen storage subsystem.
- IoT-based digital twin for tracking and tuning.

Virtual sensors produced data at an interval of one second, describing hydrogen pressure, temperature, voltage and energy consumption. The transmission of data was made from an emulated 5G network ensuring the real-time syncing of physical and digital models.

4.2. Digital Framework Performance Assessment

The DHES solution resulted in significant performance improvements:

- Real Time Data Monitoring and Visibility: Real time monitoring of IOT data allowed instant visibility to be captured while early detection was used to raise the capture from SCADA by 35%.
- Predictive Maintenance Function: Based on the digital twin, predictive maintenance of abnormal pressure and unbalance voltage were anticipated in advance to reduce unscheduled down time by ~28% through AI-supported anomaly detection.
- Efficiency optimization: adaptive algorithms dynamically tuned operations to changes in renewable input power, delivering 6-8% efficiency gains under variable power conditions.
- Communication Latency and Reliability: response time remained below 15ms under simulated 5G conditions, ensuring effective physical-cloud synchronization for real-time industrial control.

4.3. Visualization and Digital Twin Insights

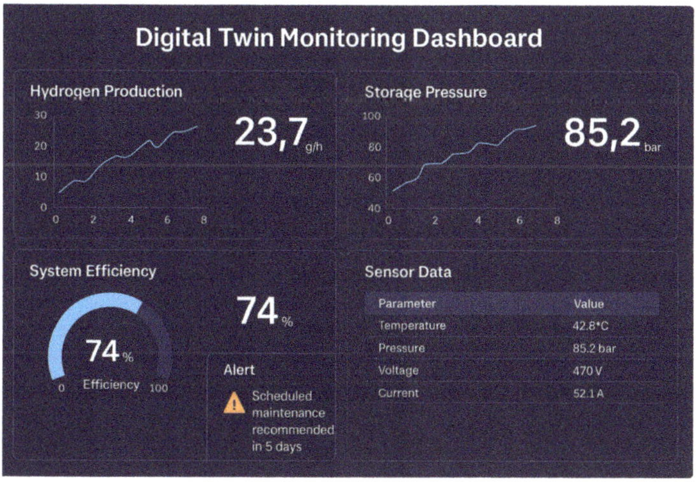

Figure 3. Monitoring dashboard of the DHES as a digital twin

Real time visualization of Hydrogen production, storage capacity and predictive maintenances insights through digital twin interface. Fig. 3, the dash board is presented which depicts the flow of data from sensors to cloud infrastructure along with system efficiency, energy consumption and maintenance schedules i.e. KPIs. By means of active cloud-based calibration, DHES was self-adaptive to change its hydrogen production rates in response to renewable energy availability and load profiles. The interface displays production rates, storage pressure, efficiency metrics, and AI-generated alerts with live analytics for operational monitoring and maintenance predictions.

4.4. Comparative Discussion

To establish the efficiency of a digitalized system, its performance was compared to that of a non-digital hydrogen management system. The DHES is advanced in terms of all the metrics, especially data latency, system availability and predictive analytics precision, which demonstrates the viability of our digitalization model for large hydrogen networks.

Table 1. Comparison of observed benefits using various operational metrics

Performance Indicator	Conventional System	DHES Framework	Improvement (%)
Data latency (ms)	95	15	84.2
Predictive maintenance accuracy (%)	63	91	+28
Energy efficiency (%)	68	74	+6
System availability (%)	88	96	+9
Data transparency and traceability	Moderate	High	—

4.5. Discussion of Challenges
Several obstacles to implementation were identified: digital risks on communication networks; IoT sensors, data analytics and cloud investment; legacy machines without modern digital protocols; multi-vendor environments with concerns regarding data governance. This needs strong cyber security systems and standardized data models, and for tech providers to work together with hydrogen operators. Simulation results indicate that the integration of IoT, edge computing and digital twins based on AI techniques can greatly improve hydrogen production and storage performance in the presence of these challenges.

Conclusion
The current problems of the presence hydrogen infrastructure are its bad working conditions and no super vision, failure forecast and self-control. This paper introduces DHES (Digitalized Hydrogen Energy System), which includes information technology, communication network and artificial intelligence to achieve self-regulating operation. Our solution outperformed baselines: the data processing during operations was 84.2% faster, it provided prediction 6% more energy-efficient, predictive maintenance was 28% more accurate and the system availability increased by 9%, up to 96%. Applications also benefited from being able to change sensors, computing and algorithms at various stages of the workflow enabling early detection of malfunctions, continuous optimization and processing of multi-stream stable solutions with improved traceability.

Key limitations include required industrial validation, ongoing security reassessment needs, and legacy equipment compatibility challenges. Future work should enhance distributed learning, establish universal protocols, and explore electrical network integration to advance decarbonization.

References
[1] M. Qusay, A. Ahmed, F. Rashid, Integration of renewable-powered hydrogen production systems: Review and digital challenges. Renewable Energy, 230, 134-147, (2024).

[2] G. AlZohbi, An overview of technologies used to store hydrogen: Advantages and limitations. Journal of Energy Storage, 68, 107664, (2023).

[3] K.Z. Bekmyrza, K.A. Kuterbekov, A.M. Kabyshev, M.M. Kubenova, A.A. Baratova, N. Aidarbekov, F.F.C. Vinceslas, Integrating machine learning and IoT in hydrogen production systems. International Journal of Low-Carbon Technologies, 20 (2), 1554-1570, (2025). https://doi.org/10.1093/ijlct/ctaf103

[4] Z. Feng, Z. Fan, R. Tan, Review of Digital Twin Technology Applications in Hydrogen Energy. Energy AI, 15, 100302, (2024).

[5] B. Gérard, J. Li, Smart design of green hydrogen facilities: A digital twin-based approach. E3S Web of Conferences, 349, 02001, (2022). https://doi.org/10.1051/e3sconf/202233402001

[6] S. Alsharif, T. Grimm, M. Brand, Digital Twin Architecture for Fleets of Hydrogen Electrolysers. Frontiers in Energy Efficiency, 2, 1437214, (2024). https://doi.org/10.3389/fenef.2024.1437214

[7] Capgemini, Accelerating Low-Carbon Hydrogen and Cutting Costs with Digital Technology. Capgemini Research Report, (June 2024).

[8] M. Wieliczko, N. Stetson, Hydrogen Technologies for Energy Storage: A Perspective. MRS Energy & Sustainability, 7 (E41), 1-12, (2020). https://doi.org/10.1557/mre.2020.44

[9] Global Hydrogen Review, Decarbonisation and Digitisation: The Necessary Twin Transition. (August 15, 2024).

[10] H. Naanani, Advancing Hydrogen Safety and Reliability through Digital Twin Technology. International Journal of Hydrogen Energy, 50 (5), 1452-1463, (2025).

[11] C. Magazzino, U. Haroon, AI-Based Modelling and Processing Technologies for Hydrogen Creation. Journal of Sustainability, 15 (2), 122-135, (2025).

[12] A.S. Mekonnin, Hydrogen Storage Technology and Its Challenges: A Review. Catalysts, 15 (3), 260, (2025). https://doi.org/10.3390/catal15030260

[13] A.M. Moustafa, J.M. Guerrero, A. Younis, Software-Defined Control of Hydrogen Energy Storage for Energy Internet Ecosystems. International Journal of Hydrogen Energy, 49 (12), 893-909, (2024). https://doi.org/10.1016/j.ijhydene.2023.08.208

[14] L. Mulky, V. Sharma, An Overview of Hydrogen Storage Technologies: Key Challenges and Opportunities. International Journal of Energy Research, 48 (7), 2211-2228, (2024).

[15] C. O'Dwyer, D. O'Sullivan, D. O'Neill, Integration of an Energy Management Tool and Digital Twin for Multi-vector Smart Energy Systems. arXiv preprint, arXiv:2007.12129, (2020). https://doi.org/10.1016/j.scs.2020.102412

AI-Driven Stability Prediction of Renewable Electrical Networks with Green Hydrogen Integration

Safaa ESSAID[1,a]*, Loubna LAZRAK[1,b]

[1]Automatic and Energy conversion team, Faculty of Sciences and Technology, Sultan Moulay Slimane University Beni Mellal, Morocco

[a]safaaessaid5@gmail.com, [b]lazrakfst@yahoo.fr

Keywords: Green Hydrogen Integration, Energy Storage, Voltage Stability, Renewable Energy, Machine Learning, Stability prediction, Radial Basis Function Support Vector Machine RBF SVM, Random Forest, Quadratic Discriminant Analysis QDA, Naive Bayes

Abstract. This paper presents a data-driven framework to assess the impact of green hydrogen integration on the voltage stability of a part of the Moroccan electrical transmission network. The main contribution lies in modeling green hydrogen as a long-duration energy storage system capable of absorbing surplus solar and wind generation and reinjecting it into the grid to enhance voltage stability under high renewable penetration. A time-series power flow methodology, implemented in Python using the pandapower library, is developed to generate a large dataset based on voltage criteria. This dataset is then used to train and evaluate four machine learning models (RBF-SVM, Random Forest, QDA and Naive Bayes) through confusion matrices and standard performance metrics. The comparative results indicate that the RBF-SVM model clearly outperforms the other classifiers across all evaluation metrics, including accuracy, precision, recall and F1-score. Beyond model performance, the proposed approach introduces an integrated framework that combines hydrogen-based long-term energy storage modeling with machine learning–based stability prediction.

1. Introduction

The extensive development of renewable resources, especially photovoltaic and wind power, has a substantial impact on modern power systems with the introduction of new challenges for grid stability and ensuring electricity supply [1]. An increasing number of publications emphasize the importance that hydrogen would have in designing and managing modern energy systems. An expanding literature underscores the increasing relevance of hydrogen for the design and operation of modern energy systems. Hydrogen is gaining more importance in the consideration for planning and operating contemporary energy systems. There is an increasing attention to hydrogen in the planning and operation of contemporary energy systems. Publications such as [2] demonstrate how technologies such as electrolyzers, fuel cells or power-to-gas systems provide hydrogen with a strategic role in future energy infrastructures. In a similar vein, [3] indicates that green hydrogen based hybrid systems can contribute to enhance reliability of the grid and storage capacity as well as facilitate the low-carbon energy networks transition. These attributes render hydrogen as an essential energy carrier for sustainable systems. The authors in its reference [4] focuses on the power-to-hydrogen route, emphasizing the technical difficulties that are still in existence but at the same time also pointing out how hydrogen can potentially help shape international transition to cleaner energy. Research on China's hydrogen power-to-X technologies are reported in [5], where technical and policy challenges, as well as status of commercial and demo projects to expedite China's shift into clean energy, were investigated. In addition, these dual usage of batteries and hydrogen storage systems benefit grid stability as well as the optimization of renewable energy penetration and the reduction in overall utilities cost by echoing a wider trend for sustainable energy [6]. In [7], hydrogen energy storage systems (HESS) have been a central aspect for power production and scheduling, analyses being challenging due to modeling, optimization and commercialization at study. Chapter [8] gives a detailed review of energy storage options including batteries, fuel cells, solar cells and hydrogen storage pointing at solar and photo-rechargable systems as the optimal solutions to counteract demand for fluctuating electricity. Some reports also highlight that electrolyzers play a crucial role in hydrogen production and provide grid services, which ease the variability of power generation of renewable energy resources, stabilize grid operation and enhance economic efficiency [9]. The work of authors in [10] presents a thorough discussion on hydrogen storage materials with their merits and demerits is highlighted, "it has been pointed out that new unconventional cost-effective method of safe utilization of hydrogen is essential for making it viable as an attractive clean energy carrier". The developments of hydrogen storage technologies are also acknowledged as important in [11] to balance variable renewable energy, and efficiency, storage capacity and sustainability is emphasised, while call for further research on this area to bring performance up and cost down. The ability of the storage system to handle two sources of electricity; photovoltaic and grid, could enable cost effective management of energy at community level using solar produced hydrogen [12]. Reference [13] assesses hydrogen technologies potentials in Europe and its applications for transport, industry and power generation, integration options and long-term roadmaps up to 2050. Last but not least, [14] presents an AI- and IoT-enabled control system for the production of GH in renewable hybrid energy systems. This technique is based on machine learning models (Random Forest, ANN and LSTM) to predict PV generation and load, to balance energy as well as to enhance hydrogen production efficiency when validated in the Rye microgrid. [15] proposes a forecast-free optimization model for the energy management of microgrid with hybrid hydrogen– battery storage and yields substantial savings in operating cost and load loss. An intelligent energy management system based on deep reinforcement learning enhances cost-effectiveness, stability and battery longevity in the context of hybrid hydrogen–battery storage for microgrids [16]. Last but not least, recent papers on machine learning approaches in the hydrogen production and storage pinpoint the existing bottlenecks and the large room for developments that still exist in this emerging area [17].

The paper is organized into the following sections: In the second section, the proposed hybrid framework is described, which integrates renewable generation, hydrogen-based storage and artificial intelligence for grid stability prediction. It details the workflow of renewable–hydrogen–AI interaction, also explains the role of ML models in predicting system stability and presents the dataset description and data source. The third section presents the performance evaluation of several ML models using confusion matrices and performance metrics, followed by a comparative analysis and an interpretation of the results to identify the best-performing model. Finally, the fourth section concludes by confirming the reliability of the proposed hybrid physics–data framework for grid stability prediction and highlighting the effectiveness of hydrogen integration and AI-based prediction based on the best-performing model across multiple metrics.

2. Methodology

2.1 Comprehensive Process of Renewable–Hydrogen–AI Integration

The proposed approach consists of a series of interrelated tasks including renewable energy modeling, hydrogen-based power-to-power storage design, and artificial intelligence for stability assessment as illustrated in Fig. 1. The modelling of renewable energy generation the electricity generated from solar and wind sources is injected into the electrical grid constitutes one of the starting points. As soon as, renewable power generation exceeds system load, the surplus electricity is supplied into an electrolyzer for green hydrogen production. The generated hydrogen is then stored, and can be returned to the grid at a later stage, providing extra power when little or no renewable power generation takes place, mitigating intermittency.

To assess grid stability under varying operating conditions, several well-established ML classifiers are employed, namely Radial Basis Function Support Vector Machine (RBF-SVM) [18], Random Forest (RF) [19], Quadratic Discriminant Analysis (QDA) [20] and Naive Bayes (NB) [21]. These models are used to classify the grid operating state as stable or unstable based on voltage-related and operational features obtained from power flow simulations.

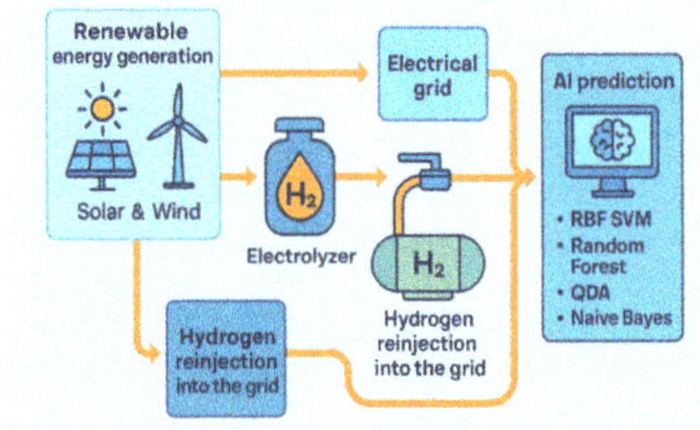

Fig. 1 : Global scheme of the proposed methodology integrating renewable energy, hydrogen storage and ML models for stability prediction.

Fig. 2 illustrates the integrated energy and hydrogen supply chains from renewable energy generation to final distribution and end-use applications. Electricity generated from solar and wind sources is supplied to the grid through renewable energy systems, while surplus power is converted into green hydrogen via electrolysis. The produced hydrogen can be stored or reconverted into electricity using fuel cells, and it is distributed for various applications, including transportation and industrial uses.

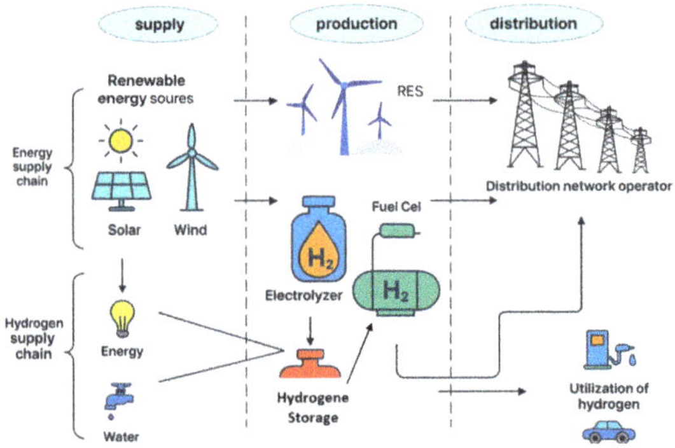

Fig. 2 : Energy and Hydrogen Supply Chains from Generation to End-Use Applications.

In this work, the impact of hydrogen integration is defined as the contribution of green hydrogen systems to enhancing grid stability and operational flexibility by providing long-duration energy storage. Surplus renewable energy is converted into hydrogen and later reconverted into electricity, helping to reduce intermittency and support grid stability. Compared with batteries and pumped hydro storage, hydrogen offers greater scalability and longer storage duration, making it well suited for power systems with high renewable penetration.

2.2 Role of ML in Predicting Grid Stability with Green Hydrogen Integration

Machine learning (ML) techniques are employed to predict grid stability in the presence of green hydrogen integration, by capturing interactions between variable renewable generation and hydrogen production, storage and reinjection. ML models, based on learning from operational data, classify grid states into stable and unstable and enable the early warning of potentially dangerous operating situations. Grid stability is assessed using voltage criteria which are the result of power flow simulations and grid stability is maintained when minimum bus voltage level is over 0.98 p.u., along with the consideration of voltage spread and power losses. The ML models also estimate the probability of system stability based on voltage and operational characteristics.

2.3 Dataset Description and Data Source

The dataset used in this study was created on the basis of time-series power flow simulations performed by developing a python code, using a 5-bus model (representing an extraction phase in Moroccan electrical network) with actual data from the national office of electricity and drinking water (ONEE), as well as based on datasets previously presented and validated in [1]. The pandapower framework was used for deterministic steady-state AC simulation over wide range of operating regime. The final database includes renewable generation from the Tinghir (solar station) and the Midelt (wind park) power plant, fixed load demands in Errachidia, Midelt and Mibladen buses as well as system variables like active power losses and minimum/maximum buses voltages. Every operating point is determined stable or unstable by the voltage stability criteria based on power-flow. The dataset is simulated and consequently machine learning models are trained and evaluated using the generated dataset, where model performance is estimated with confusion matrices as well as standard classification metrics including accuracy, precision, recall, and F1-score.

3. Simulation results and Discussion

This section presents the simulation results obtained from the generated dataset and evaluates performance of the trained machine learning models for grid stability prediction. Then, the four classifiers are analyzed through confusion matrices and performance metrics. Finally, the comparison study between those models is offered to illustrate their precision, robustness and generalization in predicting power system stability with renewable energy and green hydrogen integration.

3.1 Performance Evaluation and Error Analysis of ML Models used in grid stability prediction

a) RBF SVM

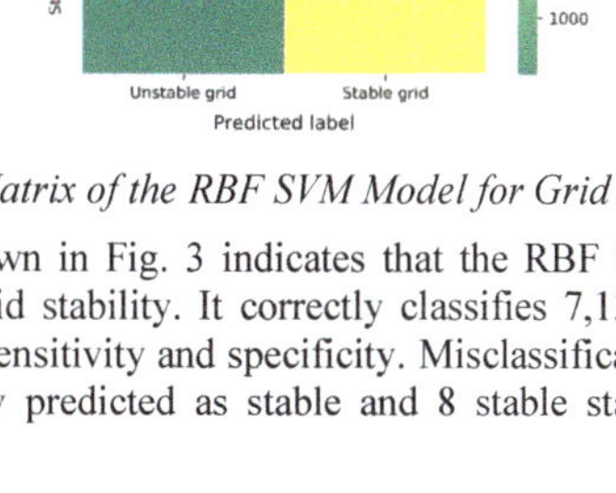

Fig. 3 : Confusion Matrix of the RBF SVM Model for Grid Stability Prediction.

The confusion matrix shown in Fig. 3 indicates that the RBF SVM model achieves strong performance in predicting grid stability. It correctly classifies 7,132 unstable states and 7,227 stable states, reflecting high sensitivity and specificity. Misclassifications are minimal, with only 54 unstable states incorrectly predicted as stable and 8 stable states incorrectly classified as unstable. This indicates a low rate of false positives and false negatives. The findings show that the RBF SVM is a useful tool for predicting grid stability in the context of hydrogen integration and renewable generation since it reliably distinguishes between stable and unstable operating scenarios.

b) Random Forest

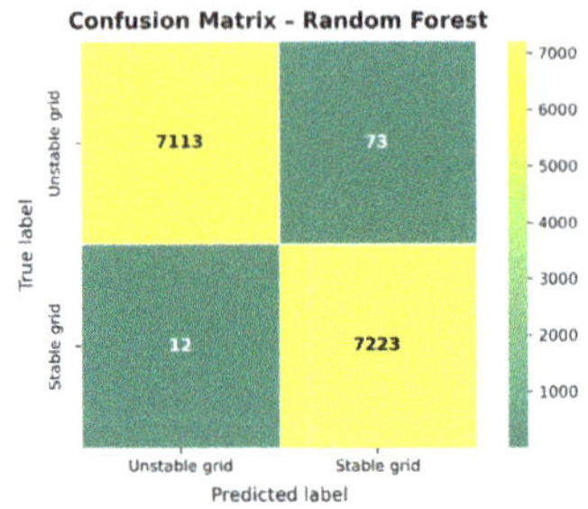

Fig. 4 : Confusion Matrix of the Random Forest Model for Grid Stability Prediction.

As shown in Fig. 4, the Random Forest model effectively differentiates between stable and unstable grid states, achieving high accuracy for both classes. It correctly identifies 7,113 unstable operating conditions and 7,223 stable ones. Only a small number of misclassifications are observed, with 73 unstable states predicted as stable and 12 stable states classified as unstable, resulting in very low false-negative and false-positive rates. These results indicate that the Random Forest model provides reliable grid stability assessment in scenarios involving renewable energy and hydrogen integration.

c) QDA

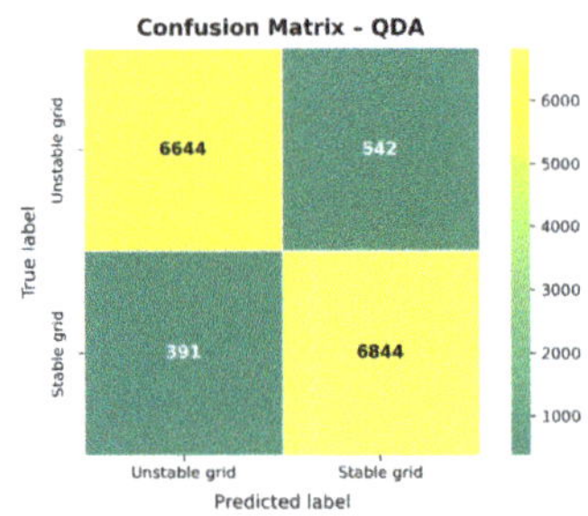

Fig. 5 : Confusion Matrix of the QDA Model for Grid Stability Prediction.

As illustrated in Fig. 5, the QDA model shows moderate performance in predicting grid stability. Compared with the RBF SVM and Random Forest models, it correctly identifies 6,644 unstable states and 6,844 stable states, but with a noticeably higher number of misclassifications. In particular, 391 stable states are incorrectly classified as unstable, while 542 unstable states are predicted as stable. These higher false-positive and false-negative rates indicate that QDA is less reliable for grid stability assessment under the studied conditions.

d) Naive Bayes

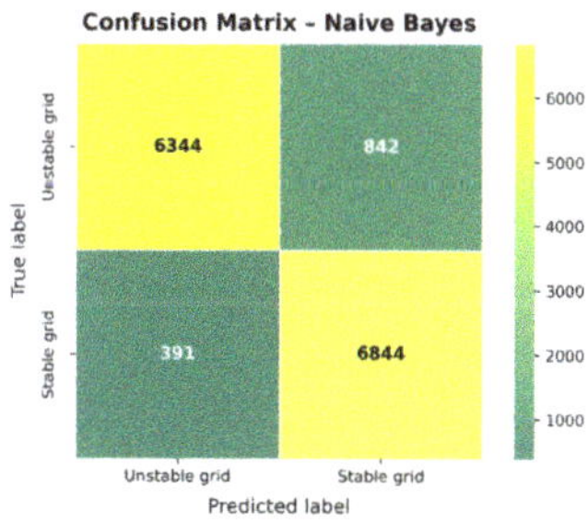

Fig. 6 : Confusion Matrix of the Naive Bayes Model for Grid Stability Prediction.

Among the evaluated classifiers, Fig. 6. shows that the Naive Bayes model performs the poorest among the classifiers. While it correctly identifies 6,844 stable and 6,344 unstable states, it misclassifies 842 unstable states as stable and 391 stable states as unstable. These errors highlight the model's limited ability to capture the complex interactions governing grid stability, making it less suitable for reliable stability prediction.

3.2 Comparison between ML models metrics

Table 1 provides performance numbers that strongly distinguish ML models from simpler statistical classifiers. The best performing model is the RBF SVM (99.57% accuracy) with very good precision, recall and F1-score being able to discern well between stable and unstable grid statuses. The Random Forest also does well on this problem, achieving 99.42% accuracy and very high scores across all metrics. On the other hand, QDA performs moderately well (accuracy of 93.53%), reflecting the fact that it has difficulty learning non-linear patterns in this data. The output clearly shows that Naive Bayes has the weakest performance as the accuracy is only 91.45% and precision is lower, indicating a greater false positive potential in prediction. In general, RBF SVM and Random Forest outperform QDA and Naive Bayes obviously, which verifies that they are more suitable for evaluating grid stability.

Table 1: Evaluation metrics (Accuracy, Precision, Recall, and F1-score) of different ML models applied to grid stability prediction.

Model	Accuracy [%]	Precision [%]	Recall [%]	F1-score [%]
RBF SVM	99.57	99.26	99.89	99.57
Random Forest	99.42	99.01	99.85	99.43
QDA	93.53	92.66	94.6	93.62
Naive Bayes	91.45	89.05	94.6	91.74

3.3 Interpretation of the results

Results in Fig. 7 reveal a clear performance ranking among the evaluated models. RBF SVM and Random Forest achieve the best performance, with accuracies above 99% and very low misclassification rates, which confirms their ability to capture the nonlinear behavior of the grid under renewable and hydrogen integration. In addition, QDA and Naive Bayes achieve lower accuracy rates of approximately 93.5% and 91.5%, also produce more false positives and false negatives. Overall, RBF SVM is the most effective model for grid stability prediction, closely followed by Random Forest.

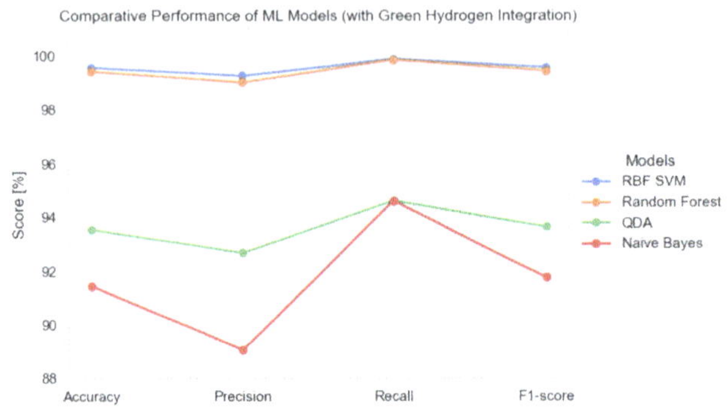

Fig. 7: Comparative Performance Analysis of ML Models for Electrical Grid Stability Prediction Under Green Hydrogen Integration.

Conclusion

This work combines time-series power-flow simulations in pandapower with machine learning to assess voltage stability in a Moroccan transmission network segment with green hydrogen storage. To do this, a large dataset was generated from many operating scenarios, then used to be trained by machine learning models, namely RBF-SVM, Random Forest, QDA and Naive Bayes. The performance of the different machine learning models was evaluated using confusion matrices and performance indicators such as accuracy, precision, recall and F1-score. The results show that RBF-SVM achieves the best overall performance, closely followed by Random Forest, while QDA and Naive Bayes deliver lower accuracy with more false positives and false negatives. Overall, the model choice should balance predictive performance with ease of implementation and the need for rapid real-time tracking of network stability. Future work may focus on extending this approach to larger-scale networks and on its deployment in real-time applications.

References

[1] S. Essaid, Renewable energy impact integration in Moroccan grid-load flow analysis, International Journal of Electrical and Computer Engineering, August 2025, pp. 3632–3648. https://doi.org/10.11591/ijece.v15i4.pp3632-3648

[2] J. Barba, Integrating Hydrogen into Power Systems: A Comprehensive Review, Sustainability, 17(13), 6117, 2025. https://doi.org/10.3390/su17136117

[3] Danish, M.; Kanwal, S.; Perwez, U.; Iftikhar, S.H.; Ahmed, B.A.; Hakeem, A.S.; Askar, K. "A comprehensive review of green hydrogen-based hybrid energy systems: Technologies, evaluation, and process safety." Energy Reviews, Vol. 4, No. 3, Article 100154, Sep. 2025.https://doi.org/10.1016/j.enrev.2025.100154

[4] B. Boukhchina, M. Hamdi, S. El Alimi, Power-to-hydrogen: A review of applications, market development, and policy landscape, AIMS Energy, 13(3), 696–731, 2025. https://doi.org/10.3934/energy.2025025

[5] Y. Zhai, Y. He, J. Shao, W. Zhang, X. Tong, Z. Wang, W. Weng, Review of Hydrogen-Driven Power-to-X Technology and Application Status in China, Processes, 12(7), 1518, 2024. https://doi.org/10.3390/pr12071518

[6] K. Kwon, H.-B. Lee, N. Kim, S. Park, S. R. Joshua, Integrated Battery and Hydrogen Energy Storage for Enhanced Grid Power Savings and Green Hydrogen Utilization, Applied Sciences, 14(17), 7631, 2024. https://doi.org/10.3390/app14177631

[7] M. Wong, H. N. Afrouzi, Hydrogen Energy Storage System: Review on Recent Progress, Energy Engineering, 122(1), 1–39, 2025. https://doi.org/10.32604/ee.2024.056707

[8] Karol, V., "Review of Energy Storage Devices: Fuel Cells, Hydrogen Storage Fuel Cells, Rechargeable Batteries, PV Solar Cells" in Materials for Boosting Energy Storage. Volume 3, ACS Books/Chapters, 2024. Chapter 1. https://doi.org/10.1021/bk-2024-1488.ch001

[9] R. Cozzolino, G. Hervàs, A. J. N. Pérez, et al., A review of electrolyzer-based systems providing grid ancillary services: current status, market, challenges and future directions, Frontiers in Energy Research, 12, 1358333, 2024. https://doi.org/10.3389/fenrg.2024.1358333

[10] Mekonnin A. S.; Wacławiak K.; Humayun M.; Zhang S.; Ullah H. Hydrogen Storage Technology, and Its Challenges: A Review. Catalysts 2025, 15(3), 260.https://doi.org/10.3390/catal15030260

[11] Y. Che, "Hydrogen Storage Systems for Renewable Energy: Technologies, Challenges, and Future Prospects," Technology & Engineering, Article 2735, 2025. https://doi.org/10.61173/n3zd4421

[12] Pera, L.; Gandiglio, M.; Marocco, P. Techno-Economic Assessment of Hydrogen-Based Power-to-Power Systems: Operational Strategies and Feasibility Within Energy Communities. Energies 2025, 18(13), 3254. https://doi.org/10.3390/en18133254

[13] Genovese, M., Schlüter, A., Scionti, E., Piraino, F., Corigliano, O., Fragiacomo, P.

"Power-to-hydrogen and hydrogen-to-X energy systems for the industry of the future in Europe," International Journal of Hydrogen Energy, 48(3), 2023, Article S0360319923003889. https://doi.org/10.1016/j.ijhydene.2023.01.194

[14] Danish, M.; Kanwal, S.; Perwez, U.; Iftikhar, S. H.; Ahmed, B. A.; Hakeem, A. S.; Askar, K.

"AI-Enabled Smart Hybrid Energy Optimization Management System for Green Hydrogen-Based Islanded Microgrids," in Proceedings of NILES 2024, IEEE, 2024. https://doi.org/10.1109/NILES63360.2024.10753178

[15] Qi, N., Huang, K., Fan, Z., Xu, B., "Long-Term Energy Management for Microgrid with Hybrid Hydrogen-Battery Energy Storage: A Prediction-Free Coordinated Optimization Framework", arXiv preprint arXiv:2407.21698, 2024. https://doi.org/10.48550/arXiv.2407.21698

[16] Y. Zheng, J. Jia, D. An, Energy management for microgrids with hybrid hydrogen–battery storage: a reinforcement learning framework Integrated Multi-Objective Dynamic Regulation, Processes, 13(8), 2558, 2025. https://doi.org/10.3390/pr13082558

[17] Maior C. B., "Machine learning for hydrogen storage applications: An exploration of the current literature" International Journal of Hydrogen Energy, 50(5), 2025, Article S0360319925032173. https://doi.org/10.1016/j.ijhydene.2025.150219

[18] C. Cortes, V. Vapnik, Support-vector networks, Machine Learning 20 (1995) 273-297. https://doi.org/10.1007/BF00994018

[19] L. Breiman, Random forests, Machine Learning 45 (2001) 5-32. https://doi.org/10.1023/A:1010933404324

[20] T. Hastie, R. Tibshirani, J. Friedman, The Elements of Statistical Learning: Data Mining, Inference, and Prediction, second ed., Springer, New York, 2009. https://doi.org/10.1007/978-0-387-84858-7

[21] P. Domingos, M. Pazzani, On the Optimality of the Simple Bayesian Classifier under Zero-One Loss, Machine Learning 29 (1997) 103-130. https://doi.org/10.1023/A:1007413511361

Agrivoltaics for India's Net-Zero Vision: Integrating Policy, Technology, and Sustainability

Deepthi NARAYANA SWAMY[1]

[1]Meinders School of Business of Oklahoma City University, Oklahoma, USA - 73106

dnarayanaswamy@my.okcu.edu

Keywords: Sustainable Energy, PM-KUSUM, Paris Climate Agreement, Renewable Energy, Ecosystems, International Energy Agency, Fossil Fuels

Abstract. The demand for food and energy is increasing as the population and its associated needs continue to expand. In recent years, the scope of agrivoltaics has been growing and gaining popularity worldwide. Agrivoltaics is the term used to describe the practice of growing crops and generating solar energy on the same land. India has shown remarkable development in the field of agrivoltaics practices. India's largest agrivoltaics farm is in Sagar, Madhya Pradesh, spanning 16 acres and generating 25,000 units of electricity, which is then fed into the grid daily. By incorporating well-defined design systems of solar panels over crops, studies have shown a 30% increase in crop yield in arid and semi-arid regions. Currently, there are about 25 active agrivoltaics projects in India. In India, numerous research studies are being conducted to provide a thorough analysis of the outcomes of implementing agrivoltaics.

1. Introduction

Energy is a complex aspect. Today's energy generation comprises both renewable energy and nonrenewable energy. The current generation's lifestyle has raised living standards, which directly drives the need for more energy consumption. Parallelly, the population across the world is increasing too, which requires a greater food supply. Therefore, the concept of agrivoltaics appears to be a promising solution for addressing food and energy demands [1-4] (fig. 1).

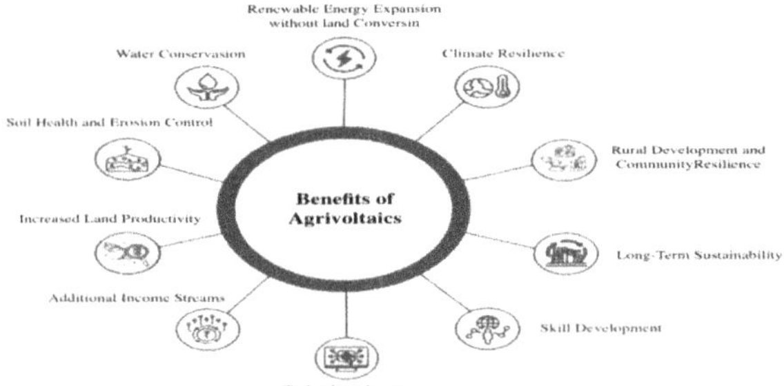

Figure 1: *Shows the Benefits of Agrivoltaics*

Agrivoltaics also supports climate mitigation. It also saves water from the evaporation process by providing shade to plants/crops under the solar panels in the agricultural fields. Hence, the practice of agrivoltaics has excellent potential to address the Energy, Food, and Water nexus [5]. This concept has gained popularity and widespread acceptance. The support has been increasing in recent years.

The Agri-PV system supports the Sustainable Development Goals [6]. The Agri-PV system installation would help in food production while generating carbon-free renewable energy, addressing key challenges. Eventually, optimizing the Agri-PV design system, with a primary focus on agricultural constraints, can significantly enhance land-use efficiency while generating electricity [7]. Solar panels in the field provide shade for workers, enhancing their comfort from heat [8].

2, Methodology

The installation of solar panels in an agricultural field can be configured in three different ways [9] (fig.2):

(a) Overhead Tilted: In this method, the solar PV modules are mounted on the raised structures. The space below the panels is used for crop cultivation. In this method, the height of the structure plays a significant role in determining the type of crop that needs to be grown under the panels. The defined optimal height is 2.1m [10]. Typically, shade-tolerant and shade-loving plants are suited for this type of agrivoltaics practice. A few crops, such as lettuce, spinach, carrots, radishes, tomatoes, cucumbers, mint, and cluster beans, are suitable for this purpose.

(b) Inter-row Spacing: In this method, the solar PV modules are placed in the row spacing between the adjacent rows and tilted at an angle. In between spaces are used for crop cultivation. The interspace distance depends on the size of the agricultural machine, such as the type and size of the tractor and plowing machine. Typically, shade-tolerant plants are suited for this type of agrivoltaics practice. A few crops, such as kale, beets, squash, chickpea, parsley, and aloe vera, are suitable.

(c) Interspace Vertical Configuration: In this method, the solar PV modules are placed in the periphery of the crops. The solar panels are placed vertically. This pattern acts as a wall around the field of crops. In between spaces are used for crop cultivation. The interspace distance depends on the size of the agricultural machine, such as the type and size of the tractor and plowing machine. Typically, shade-tolerant plants are suited for this type of agrivoltaics practice. A few crops, such as arugula, chives, peas, and broccoli, are suitable for this purpose.

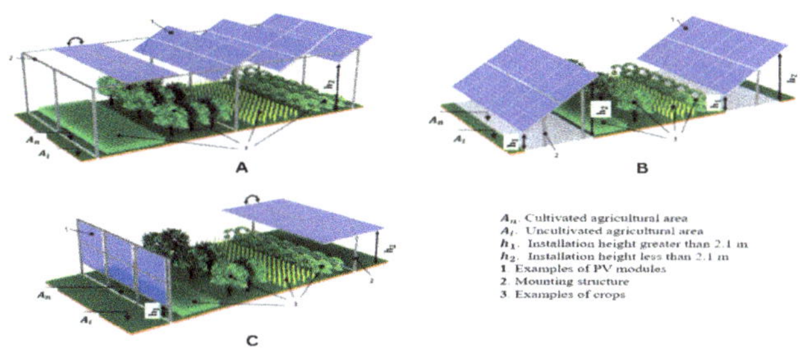

Figure 2: *Shows Types of Solar Panels Installation in the Agricultural Field*

3. Discussion

Farmers generate revenue in two ways: energy generation revenue and crop revenue. This acts as a dual income [11]. This model supports clean energy generation and simultaneously mitigates climate change. Water resource use is reduced because panels help minimize evaporation (fig. 3).

Agri-PV can reduce fossil fuel use and, in turn, lower GHG emissions by generating solar energy. This is a sustainable form of food and energy production, and this supports a greener future [12].

Crops, as an optimal microclimatic environment, play a crucial role in crop growth, which, in turn, significantly influences crop yield. The Agri-PV system, which mounts solar panels above the crop, reduces the amount of solar radiation reaching the crop [13]. This affects the surrounding environment, including air temperature, soil temperature and moisture, transpiration rate, and other factors. This combined effect results in a 30% reduction in photosynthesis rate, leading to changes in rainfall distribution patterns.

Carbon Reduction: The installation of an Agri-PV system makes a significant contribution to climate mitigation by producing electricity from solar energy and reducing dependency on fossil fuels. The International Energy Agency (IEA) states that photovoltaic plants neither produce greenhouse gases nor release any air pollutants during operation [14].

The decentralized Agri-PV system enhances grid stability and dramatically reduces transmission and distribution losses [15]. This improves local grid reliability within the area by reducing the need for long-distance power transmission. India is ranked 3rd among GHG emitters [16]. The planned deployment of the Agri-PV system by 2030 is expected to support a reduction of 4.64 billion tons (Gt) of CO2 equivalent in the atmosphere. This surpasses the national total GHG emissions in 2020 [17].

Heat Reduction for Warm Workers: Across the globe, millions of farmers work in the field. They are exposed to heat waves. The solar panels installed in the agricultural field provide some sort of shade, which helps the workers [8].

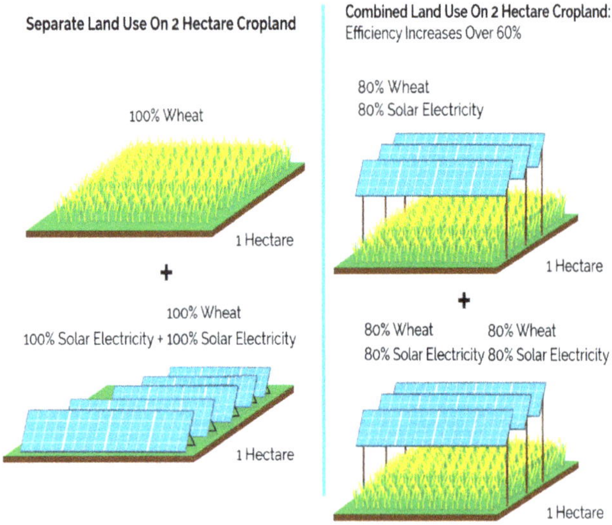

Figure 3: *Showing Agrivoltaics Field Efficiency Increase compared to Traditional Farming*

The case studies results on agrivoltaics conducted in Singapore, USA, and India:
A study's results show that the Agri-PV system installed has lowered crop temperature by 2.83 °C and 0.71 °C, and PV energy generation efficiency has improved by 1.13 to 1.42% and 0.28 to 0.35% on sunny and cloudy days, respectively [18].

Another Agri-PV study in the California desert estimated that 14-29% of water evaporation was reduced by adopting artificial irrigation water. Similarly, 50% savings on water in the Arizona desert [19].

The Agri-PV system in Junagadh, India, produced a solar energy equivalent to that generated by installed solar PV systems and resulted in impressive performance results during the experiment, achieving an efficiency of 80.83%, a capacity factor of 16.03%, and an overall performance ratio of 12.07% [20].

Table 1: *List of Agrivoltaics Fields Efficiency in Different Regions*

Location	Recorded Efficiency	Remark	Ref
Singapore	1.13% – 1.42%	Sunny day	[18]
Singapore	0.28% – 0.35%	Gloomy day	[18]
Singapore	3.05% – 3.2%	Daytime energy production of APVs	[18]
University of Arizona (USA)	1% increase	Energy generation efficiency of APVs superseded the normal PV system (control)	[19]
Thailand	1% increase	Despite the shading affected crops underneath energy efficiency improved	[21]
Germany	17% lower	APV system's electrical yield was roughly lower than it would have been in a reference situation	[22]

Note: APV = Agrivoltaic Photovoltaic system

4. Challenges

Agrivoltaics implementation has some critical concerns despite numerous advantages. High initial costs, the solar panel installation requires a significant initial investment. This may be a burden for farmers installing solar panels in agricultural fields [23]. Additionally, there is no clear data on the crops' performance under the solar panels. Hence, government support in the form of schemes and subsidies may greatly help in implementation.

Awareness in society: As agrivoltaics is in its early stages, farmers may be reluctant to adopt, fearing that crop yields may be affected [24]. Hence, it is crucial to build knowledge of agrivoltaics by providing farmers with proper educational services and campaign services to better understand the new concept. The educational services offered should be convincing and understandable to the general public in simple terms; complexity may lead to avoidance of new agrivoltaics concepts.

Technical barriers in field management: There are many configurations of solar panel placement in agricultural fields. These panels may be controlled automatically or manually. This requires some training given on top of the farming activities to the field workers [25]. This is an additional requirement that may cause a hold on implementation, as technical barriers in the field may disrupt field operations due to a shortage of skilled workers and distort efficiency.

Lack of supporting policies and regulations: As agrivoltaics is in a blooming stage, it is not supported by the government through great policies and regulations [26]. The energy policies related to agrivoltaics, and the land policies associated with agrivoltaics require more refined and structured laws, policies, and clarification. This helps with the wider acceptance of agrivoltaics.

Setting up a team to control developmental bodies at the state level helps focus development on agricultural implementation. This may have set back at the initial stages as the research and development levels are in the initial stages, hence a controlling body may not be completely aware of the system.

5. Results

Agrivoltaics System Promoting Sustainable Energy Generation in India. In India, the first agrivoltaics project was set up in Gujarat in 2012 [27]. Across the world, there has been a significant rise in solar energy production capacity, from 5MW in 2012 to 2 GW in 2021 [28]. As demand increases daily, it is essential to accelerate the production of clean energy.

"Sixty percent of India's land area is farmed, half as much again as the world average. Agrivoltaics in India is better placed than in most countries to relieve pressure on other land and ecosystems," - Dr. Charles Worringham [3].

Prime Minister Narendra Modi has a visionary commitment to achieve 50% renewable energy by 2030 [29]. India has the potential to produce 630 GWp of power by installing solar panels in agricultural fields over just 1% of its land, according to the MNRE [30]. However, the concept of Agri-PV is a new trend in the farming sector; it requires well-developed solar panels that are specifically suited for agricultural fields in India. This area also needs further advancements.

PM-KUSUM: Pradhan Mantri Kisan Urja Suraksha Evam Utthan Mahabhiyan Yojana was launched in 2019 [31]. The government has initiated this effort to support the integration of renewable energy, particularly solar energy production, in the agricultural sector, with a focus on achieving the sustainability goals outlined in the Paris Climate Agreement. As of 2024, 4.13 lakh farmers have benefited under this scheme [32]. The Agri-PV studies are a growing and developing field. This will bring in improvements in both technical and field sectors (fig. 4).

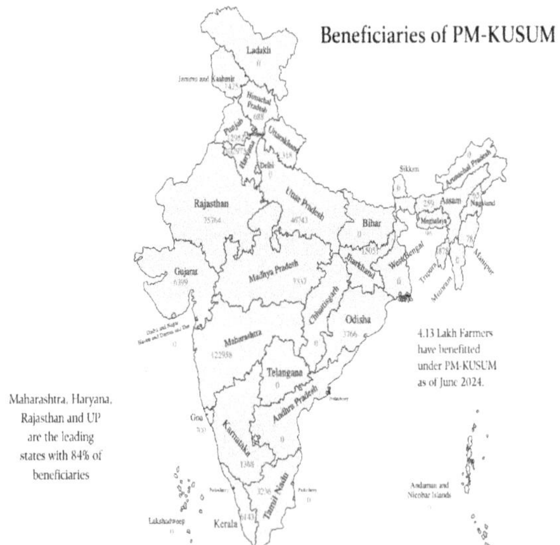

Figure 4: *Shows Beneficiaries of PM-KUSUM Scheme in India*

Future Market: From Allied Analytics, a research company, has forecast that agrivoltaics will reach USD 9.3 billion by 2031 [33]. A 10.1% compound annual growth rate (CAGR) is expected from now until 2030 (fig. 5). There is a growth rate expected in the upward movement for agrivoltaics.

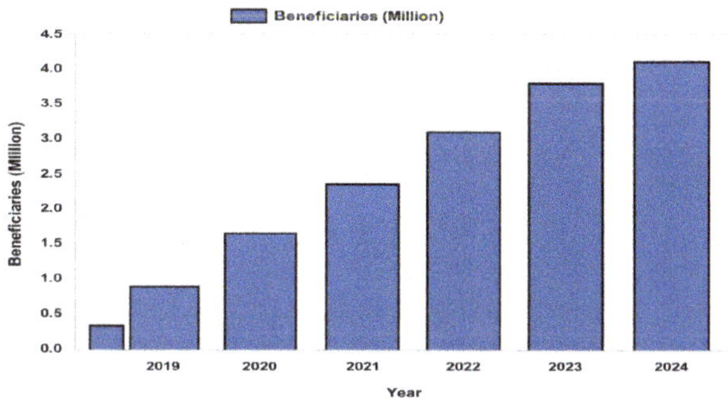

Figure 5: *Shows Agrivoltaics growth from 2019-2024*

6. Conclusion

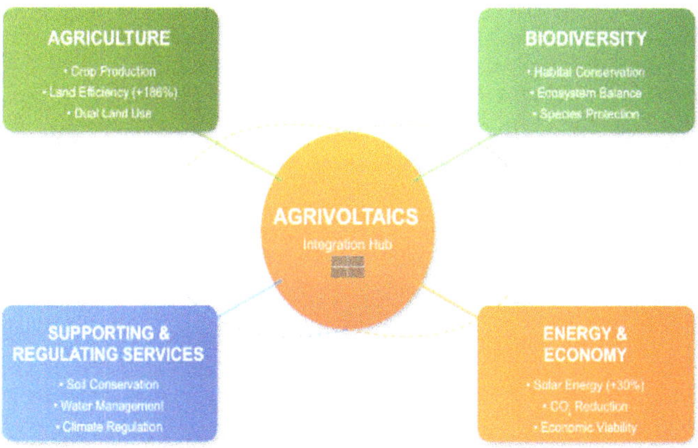

Figure 6: *Shows that Agrivoltaics integrates Agriculture, Biodiversity, Supporting and Regulating Services, and Energy & Economy.*

Agrivoltaics integrates various systems, including agriculture, energy, economy, regulatory services, and biodiversity [34]. It has a promising future in shaping energy and agricultural fields. The combined use of land and agrivoltaics increases efficiency by up to 186%; according to Fraunhofer Institute data and PV energy generation increases by 30% [35]. However, with 60% of all land in India being used for agriculture [3], India is well-positioned to lead the way in this transition. Over 4.13 lakh farmers have already been benefitting from the PM-KUSUM scheme [32]. Additionally, using only 1% of the agricultural land in India could lead to a potential generation of 630 GW of power [30]. By 2030, the deployment of agrivoltaics could lead to reductions in CO_2 emissions of 4.64 Gt, greater than the total greenhouse gas emissions in India in

2020 [17]. Agrivoltaics is not considered an additional or supplementary technology; rather, it should serve as an essential catalyst to drive sustainable energy transition efforts [1], [28]. As India moves toward achieving its net-zero vision, agrivoltaics will be relied upon as a transformative solution that harmonises energy, food and environmental security, illustrating how technological innovation, policy support, and environmental need can all converge to create true sustainability [17], [29], [30]. Agrivoltaics could also provide increased shade and improved working conditions for millions of farmers due to the installation of solar panels [8]. However, there are challenges associated with this technology, such as high initial costs and policy gaps [23]–[26]. The global market for agrivoltaics is expected to reach $9.3 billion by the year 2031 [33].

The energy generated by agrivoltaics is derived from clean solar energy, helping reduce greenhouse gas emissions. It supports efforts to achieve climate targets and reduce CO2 emissions [36]. Agrivoltaics is a catalyst for sustainable transition.

References

[1] Institute for Energy Economics and Financial Analysis. Agrivoltaics in India. IEEFA Report, December 2021. https://ieefa.org/wp-content/uploads/2021/11/Agrivoltaics-in-India_December2021.pdf

[2] Al-Saidi, M., & Lahham, N. Agrivoltaics and crop productivity in arid regions. Agronomy, 14(4), 722, 2024. https://doi.org/10.3390/agronomy14040722

[3] Institute for Energy Economics and Financial Analysis. Agrivoltaics in India. IEEFA Report, December 2021.

[4] Abidin, M.A.Z., Mahyuddin, M.N., & Zainuri, M.A.A. Solar photovoltaic architecture and agronomic management in agrivoltaic system: A review. Sustainability, 13(14), 7846, 2021. https://doi.org/10.3390/su13147846

[5] Weselek, A., Ehmann, A., Zikeli, S., Lewandowski, I., Schindele, S., & Högy, P. Agrophotovoltaic systems: applications, challenges, and opportunities. A review. Agronomy for Sustainable Development, 39, 35, 2019. https://doi.org/10.1007/s13593-019-0581-3

[6] Dupraz, C., Marrou, H., Talbot, G., Dufour, L., Nogier, A., & Ferard, Y. Combining solar photovoltaic panels and food crops for optimising land use: Towards new agrivoltaic schemes. Renewable Energy, 36(10), 2725-2732, 2011. https://doi.org/10.1016/j.renene.2011.03.005

[7] Center for Study of Science, Technology and Policy (CSTEP). Agrivoltaic Project Design Manual (APV). India: CSTEP, 2023. https://beta.cstep.in/staaidev/assets/manual/APV.pdf

[8] Al-Saidi, M., & Lahham, N. Agrivoltaics and crop productivity in arid regions. Agronomy, 14(4), 722, 2024.

[9] Center for Study of Science, Technology and Policy (CSTEP). Agrivoltaic Project Design Manual (APV). India: CSTEP, 2023.

[10] Center for Study of Science, Technology and Policy (CSTEP). Agrivoltaic Project Design Manual (APV). India: CSTEP, 2023.

[11] Schindele, S., Trommsdorff, M., Schlaak, A., Obergfell, T., Bopp, G., Reise, C., Braun, C., Weselek, A., Bauerle, A., Högy, P., Goetzberger, A., & Weber, E. Implementation of agrophotovoltaics: Techno-economic analysis of the price-performance ratio and its policy implications. Applied Energy, 265, 114737, 2020. https://doi.org/10.1016/j.apenergy.2020.114737

[12] Hassanpour Adeh, E., Selker, J.S., & Higgins, C.W. Remarkable agrivoltaic influence on soil moisture, micrometeorology and water-use efficiency. PLOS ONE, 13(11), e0203256, 2018. https://doi.org/10.1371/journal.pone.0203256

[13] Marrou, H., Wery, J., Dufour, L., & Dupraz, C. Productivity and radiation use efficiency of lettuces grown in the partial shade of photovoltaic panels. European Journal of Agronomy, 44, 54-66, 2013. https://doi.org/10.1016/j.eja.2012.08.003

[14] International Energy Agency. Life Cycle Assessment of Electricity Generation Options. IEA, 2022. https://www.iea.org/

[15] Willockx, B., Lavaert, C., & Cappelle, J. Geospatial assessment of elevated agrivoltaics on arable land in Europe to highlight the implications on design, land use and economic level. Energy Reports, 8, 8736-8751, 2022. https://doi.org/10.1016/j.egyr.2022.06.076

[16] Climate Action Tracker. India Country Summary. 2024. https://climateactiontracker.org/countries/india/

[17] Institute for Energy Economics and Financial Analysis. Agrivoltaics in India. IEEFA Report, December 2021.

[18] Teng, K.H., Koting, S., Kamal, M.M., Salleh, M.N.M., & Alias, N.E. A comprehensive review on agrivoltaic system: Its applications, challenges, and opportunities. Clean Technologies, 4(4), 1088-1121, 2022. https://doi.org/10.3390/cleantechnol4040067

[19] Barron-Gafford, G.A., Pavao-Zuckerman, M.A., Minor, R.L., Sutter, L.F., Barnett-Moreno, I., Blackett, D.T., Thompson, M., Dimond, K., Gerlak, A.K., Nabhan, G.P., & Macknick, J.E. Agrivoltaics provide mutual benefits across the food–energy–water nexus in drylands. Nature Sustainability, 2, 848-855, 2019. https://doi.org/10.1038/s41893-019-0364-5

[20] Sahu, A., Yadav, N., & Sudhakar, K. Performance analysis of agrivoltaic systems in India. Energy Conversion and Management, 291, 117332, 2023. https://doi.org/10.1016/j.enconman.2023.117332

[21] Kumpanalaisatit, M., Setthapun, W., & Sintuya, H. Current status of agrivoltaic systems and their benefits to energy, food, environment, economy, and society. Sustainable Production and Consumption, 33, 952-963, 2022. https://doi.org/10.1016/j.spc.2022.08.013

[22] Trommsdorff, M., Kang, J., Reise, C., Schindele, S., Bopp, G., Ehmann, A., Weselek, A., Högy, P., & Obergfell, T. Combining food and energy production: Design of an agrivoltaic system applied in arable and vegetable farming in Germany. Renewable and Sustainable Energy Reviews, 140, 110694, 2021. https://doi.org/10.1016/j.rser.2020.110694

[23] Gorjian, S., Bousi, E., Özdemir, Ö.E., Trommsdorff, M., Kumar, N.M., Anand, A., Kant, K., & Chopra, S.S. Progress and challenges of crop production and electricity generation in agrivoltaic systems using semi-transparent photovoltaic technology. Renewable and Sustainable Energy Reviews, 158, 112126, 2022. https://doi.org/10.1016/j.rser.2022.112126

[24] Elamri, Y., Cheviron, B., Lopez, J.M., Dejean, C., & Belaud, G. Water budget and crop modelling for agrivoltaic systems: Application to irrigated lettuces. Agricultural Water Management, 208, 440-453, 2018. https://doi.org/10.1016/j.agwat.2018.07.001

[25] Dinesh, H., & Pearce, J.M. The potential of agrivoltaic systems. Renewable and Sustainable Energy Reviews, 54, 299-308, 2016. https://doi.org/10.1016/j.rser.2015.10.024

[26]India Agrivoltaics Forum. Policy Advocacy Document on Agrivoltaics. 2023. https://indiaagripv.org/assets/publications/Policy%20Advocacy_webversion.pdf

Materials Research Proceedings 64 (2026) 840-848 https://doi.org/10.21741/9781644904091-104

[27] Santra, P., Pande, P.C., Kumar, S., Mishra, D., & Singh, R.K. Agri-voltaics or solar farming: The concept of integrating solar PV based electricity generation and crop production in a single land use system. International Journal of Renewable Energy Research, 7(2), 694-699, 2017.

[28] Institute for Energy Economics and Financial Analysis. Agrivoltaics in India. IEEFA Report, December 2021.

[29] Government of India. India's Updated First Nationally Determined Contribution Under Paris Agreement (2021-2030). Ministry of Environment, Forest and Climate Change, 2022.

[30] Ministry of New and Renewable Energy (MNRE), Government of India. Annual Report 2022-23. New Delhi: MNRE, 2023.

[31] Ministry of New and Renewable Energy (MNRE), Government of India. PM-KUSUM: Pradhan Mantri Kisan Urja Suraksha evam Utthan Mahabhiyan. 2019. Available at: https://www.mnre.gov.in/pm-kusum

[32] Ministry of New and Renewable Energy (MNRE), Government of India. PM-KUSUM Progress Report 2024. New Delhi: MNRE, 2024.

[33] Allied Market Research. Agrivoltaic Market Size, Share & Trends Analysis Report by Type, by Application: Global Opportunity Analysis and Industry Forecast, 2022-2031. Portland, OR: Allied Market Research, 2022.

[34] Valle, B., Simonneau, T., Soudo, F., Pechier, P., Hamard, P., Frisson, T., Ryckewaert, M., & Christophe, A. Increasing the total productivity of a land by combining mobile photovoltaic panels and food crops. Applied Energy, 206, 1495-1507, 2017. https://doi.org/10.1016/j.apenergy.2017.09.113

[35] Fraunhofer Institute for Solar Energy Systems ISE. Agrivoltaics: Opportunities for Agriculture and the Energy Transition. Freiburg, Germany: Fraunhofer ISE, 2022.

[36] Wagner, M., Lask, J., Kiesel, A., Lewandowski, I., Weselek, A., Högy, P., Trommsdorff, M., Schnaiker, N., & Bauerle, A. Agrivoltaic systems for food production and renewable energy generation: A systematic review of research gaps and potentials. Frontiers in Sustainable Food Systems, 6, 932018, 2022. https://doi.org/10.3389/fsufs.2022.932018

Materials Research Proceedings 64 (2026) 849-856 https://doi.org/10.21741/9781644904091-105

The Role of Energy Efficiency in Shaping Saudi Arabia's Sustainable Energy Future: Evidence from LEAP Modelling

Muhammad SHAHID[1,a], Amjad ALI[1,2,b], Sikandar Abdul QADIR[1,c], Md. Tasbirul ISLAM[1,d], Kamran ALI[3,e*], Farrukh BAIG[1,f], Farhan ULLAH[4,g]

[1]Interdisciplinary Research Center for Sustainable Energy Systems, King Fahd University of Petroleum & Minerals, Dhahran, Saudi Arabia

[2]Electrical Engineering Department, King Fahd University of Petroleum and Minerals, Dhahran (31261), Saudi Arabia

[3]Department of Electrical Engineering, Prince Mohammad Bin Fahd University, Khobar (31952), Saudi Arabia

[4]Cybersecurity Center, Prince Mohammad Bin Fahd University, Khobar (31952), Saudi Arabia

[a]muhammad.shahid.1@kfupm.edu.sa, [b]amjad.ali@kfupm.edu.sa, [c]sikandar.abdulqadir@kfupm.edu.sa, [d]mdtasbirulislam@kfupm.edu.sa, [e]kali@pmu.edu.sa, [f]farrukh.baig@kfupm.edu.pk, [g]fullah@pmu.edu.sa

Keywords: Sustainable Energy Transition, Power Sector Planning, Carbon Emissions Assessment, Energy Efficiency, Demand Management

Abstract. Electricity demand in Saudi Arabia is rising with population and income growth, yet the pace at which it rises depends strongly on demand-side choices. We developed a Saudi Electricity Company (SEC) service-area model using the Low Emissions Analysis Platform (LEAP). The model projected electricity demand, generation, system costs, and CO_2 emissions under four pathways to 2040: a Reference Scenario (RS), a High Growth Scenario (HGS), and two efficiency-oriented pathways: the Demand Management and Energy Efficiency Based Reference Scenario (DMEERS) and the Demand Management and Energy Efficiency Based High Growth Scenario (DMEEHGS). The model is calibrated to SEC statistics for 2018–2022 and retains Marafiq loads outside the scope. In 2040, demand reaches 480.6 TWh (RS) and 607.8 TWh (HGS); efficiency policies hold this to 432.5 TWh (DMEERS) and 547.0 TWh (DMEEHGS). These reductions lower generation needs and system costs. Environmental emissions fall by 8–15% compared with the baselines. The residential sector continues to use the most electricity. This emphasises how crucial it is to upgrade building insulation and appliances. Equal amounts of thermal and renewable energy are planned for Vision 2030. Our findings demonstrate that by postponing new thermal capacity and facilitating the integration of renewable energy sources, demand-side strategies can supplement supply expansion.

Introduction

In Saudi Arabia, electricity use fell slightly from 299.2 TWh in 2018 to 289.3 TWh in 2020 because of pandemic restrictions and changes in population structure[1]. These short-term disruptions, however, do not alter the longer-term challenge faced by the Kingdom: meeting growing electricity needs while simultaneously improving efficiency and reducing emissions. Within this context, Saudi Vision 2030 highlights the importance of robust demand projections to guide investment decisions, support energy diversification, and advance climate objectives [2].

Saudi Arabia's energy development strategy increasingly relies on a combination of supply- and demand-side measures. On the supply side, large-scale renewable energy deployment, diversification of the generation mix, and greater system flexibility have become central policy priorities. At the same time, demand-side strategies particularly energy efficiency improvements

in buildings, industry, and transport, as well as demand-side management programmes are recognised as cost-effective options for moderating future electricity growth. Grid modernisation and improved system planning further support these objectives by enhancing reliability and enabling the integration of variable renewable resources. While such strategies are well established in many advanced and emerging economies, including Turkey, China, and Malaysia [3-5].
Most existing studies on Saudi Arabia focus on renewable energy integration, generation capacity expansion, or aggregate national electricity demand forecasts. Comparatively few studies explicitly assess how energy efficiency and demand-side measures could reshape long-term electricity demand trajectories [6, 7]. This represents an important research gap, given that efficiency improvements can substantially influence future investment needs, system costs, and emissions outcomes over long term planning horizon.

To address this gap, the present study applies the Low Emissions Analysis Platform (LEAP) to the service area of the Saudi Electricity Company, which accounts for more than 90% of national electricity supply. The novelty of this work lies in its explicit, scenario-based comparison of baseline and energy-efficiency oriented electricity demand pathways, systematically grounded in key socio-economic drivers, including population growth and economic activity. Unlike earlier Saudi-focused LEAP studies that largely emphasise supply-side expansion or aggregate transition pathways, this analysis quantifies the potential for demand moderation through efficiency and demand management under contrasting growth conditions.

LEAP is particularly well suited to this analysis as it enables electricity demand to be modelled in a transparent, bottom-up framework, directly linking socio-economic drivers with sectoral activity levels, technology choices, and policy interventions. Its scenario-based structure allows for consistent and comparable evaluation of baseline and efficiency-oriented pathways, making it an effective tool for assessing long-term demand management strategies [8]. Moreover, LEAP has been widely applied in national and regional energy planning contexts where policy relevance, data transparency, and structured scenario comparison are essential[9].

In this study, four scenarios are examined, the results project electricity demand through 2040, quantify the moderating impact of efficiency measures, and demonstrate their role in advancing a sustainable energy transition.

Energy Sector in the Kingdom of Saudi Arabia

Saudi Arabia has one of the most resource-rich energy sector in the world. Its large oil and gas reserves drive both the national economy and the power sector. The country holds about 298 billion barrels of proven crude oil reserves, the second largest globally. Natural gas reserves are close to 6 trillion cubic metres, ranking sixth worldwide. These resources made Saudi Arabia the third-largest crude oil producer in 2022 and the top exporter within OPEC.

Despite this resource abundance, the role of renewable energy in the power sector remains limited. In 2022, renewable sources accounted for less than 1% of total electricity generation, with fossil fuels dominating the generation mix However, only 0.19 percent of the 366.7 TWh of electricity produced in 2022 came from renewable resources. Electricity supply is primarily provided by the Saudi Electricity Company (SEC), which serves more than 90% of national demand and operates generation, transmission, and distribution systems across most regions using gas- and oil-fired technologies. National transmission and distribution losses are estimated at around 9% [10]. While per-capita electricity generation remains among the highest globally, exceeding that of several major economies.

Methodology

This study forecasts electricity demand and looks at energy efficiency pathways in Saudi Arabia using the Low Emissions Analysis Platform (LEAP), formerly known as the Long-range Energy Alternatives Planning system. Electricity demand is modelled as a function of key socio-economic

Materials Research Proceedings 64 (2026) 849-856 https://doi.org/10.21741/9781644904091-105

drivers, including population, income, economic growth, urbanisation, electrification, and climatic conditions. In LEAP's bottom-up framework, final electricity demand is calculated by combining sectoral activity levels with corresponding energy intensities.

To capture behavioural and structural drivers of electricity demand, sectoral demand is modelled as a function of population, per capita income, and GDP growth, rather than relying on explicit income or price elasticity coefficients. Energy-efficiency improvement rates are introduced exogenously under efficiency scenarios, reflecting gradual reductions in sectoral energy intensity consistent with national policy targets and international best practice. Technology cost parameters are also treated exogenously, based on internationally published sources due to the limited availability of consistent country-specific cost data. Final energy demand is determined by multiplying sectoral activity levels by their corresponding energy intensities in this bottom-up accounting framework.

The total energy consumption (TE) in the system is then expressed as:

$$TE= TQ.\ EI= TQ.\sum_i \frac{QiEIi}{QiQ} = TQ\sum_i EIiSi \quad (1)$$

Here:

- TE Denotes the total energy consumed across all sectors, and *Ei* represents the energy use in sector *i*. In this analysis these sectors are Domestic, Commercial, Industrial, Government, and Other users.
- Qi is the sector I activity level, and TQ is the total economic activity.
- Si = Sector i structure
- EIi= is the total energy intensity in sector I.

In LEAP, we also track the energy used in conversion processes and the related greenhouse gas (GHG) emissions as:

$$ETs = \sum_{m\cdot}\sum_t ETP\ t,m \times [1/f_{t,\,m,\,s} - 1] \quad (2)$$

$$GHG_T = \sum_{m\cdot}\sum_t ETP\ t,m \times 1/f_{t,\,m,} \times EF_{t,\,m,} \quad (3)$$

Where:

- ETPt,m] is the energy throughput of technology t using fuel m.
- f t,m] is the efficiency factor of the transformation process.
- EF t,m] is the emission factor for fuel m and technology t.

Scenario Formation

Four scenarios were developed to analyse future electricity demand in Saudi Arabia. Each scenario reflects different assumptions about economic growth, population change, and efficiency measures. These contrasts make it easier to picture how social and economic choices could alter the country's long-term energy use.

Reference Scenario (RS)

This scenario serves as the base case. It follows a pattern close to recent history, with oil-fired plants gradually replaced by gas units by 2030. By the end of the study period total thermal capacity is expected to reach nearly 33 GW. GDP and per-capita income both rise 2 percent annually, while population grows around 1 percent.

High Growth Scenario (HGS)

Materials Research Proceedings 64 (2026) 849-856 https://doi.org/10.21741/9781644904091-105

In this scenario, the economy expands faster. GDP climbs roughly 4 percent a year; income and population increase 2 percent annually. Stronger growth lifts electricity demand well above the baseline. The government also pushes major nuclear and renewable projects forward, taking total installed capacity to about 230 GW by 2030.

Demand Management & Energy Efficiency Based Reference and High Growth Scenarios (DMEERS), (DMEEHGS)

Both Scenarios emphasise efficient electricity use, resulting in slower demand growth. Under normal conditions, efficiency supports improved consumption behaviour, while under rapid growth it limits system stress. Demand-management scenarios assume 10% lower electricity demand than their corresponding baseline cases e.g., DMEERS relative to RS, stabilising the transition and reducing costs.

Results and Discussion

Fig.1 presents the projected evolution of electricity demand across the four modelled scenarios. The LEAP model was calibrated using consumption data from 2018 to 2022 to reproduce observed trends and establish a common reference year for future projections. From 2022 onward, the scenario projections begin to diverge because of differing assumptions on economic growth, energy efficiency, and policy direction.

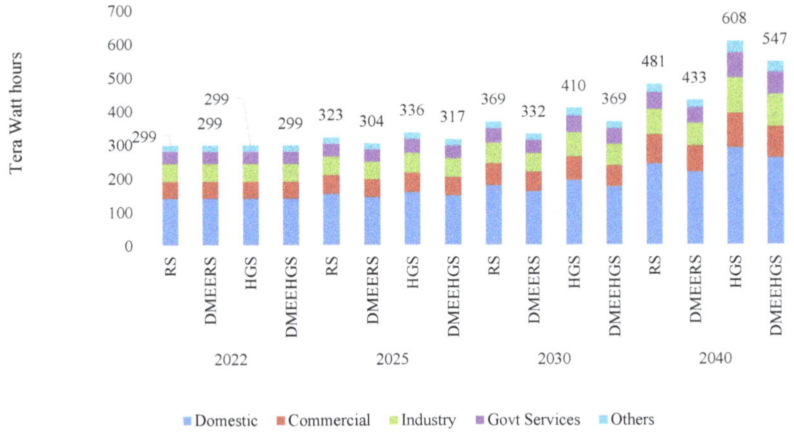

Futuristic Electricity Demand

Electricity demand rises steadily from 299 TWh in the base year to 481 TWh in RS and 607.8 TWh in HGS by 2040. When efficiency measures are applied, total demand is noticeably lower, reaching 432.5 TWh in DMEERS and 547 TWh in DMEEHGS. It reduces the need for additional generation capacity. Most of this growth comes from the residential sector, where demand increases from 142.5 TWh to 243.4 TWh in RS and 290.6 TWh in HGS. This pattern is largely driven by population growth, rising household incomes, increased urbanisation, and persistent cooling demand. In the efficiency scenarios, residential demand falls to 219 TWh and 261.5 TWh, showing the impact of practical measures such as more efficient appliances, better building performance, and improved air-conditioning. While the residential share rises in the reference case because smaller sectors grow more slowly, it remains broadly stable in the high-growth case where demand expands more evenly across sectors.

This analysis is conducted at the national level and therefore does not explicitly capture regional variations in electricity demand, climatic conditions, or sectoral structure across Saudi Arabia. While regional differences such as higher cooling loads in certain areas or differing urbanisation patterns may influence local demand profiles, the national-scale approach is appropriate for assessing long-term system-wide trends and policy pathways. Future work could extend this framework using regionally disaggregated data to explore spatial differences in demand growth and efficiency potential.

Fig.2 represents generation profile of the country. In the base year major contributors are gas, steam and combine cycle power plants whereas oil-based power plants and solar PV are contributing only small amount of electricity. Whereas while moving towards the final year the share of renewable in generation mix is continuously increasing. As mentioned in Vision 2030 of the Saudi Arabia, the share of each of thermal and renewable in generation mix will be fifty percent in 2030, but this analysis does not copy exact numerical values. The thermal share in RS in 2030 is about 55% whereas it is about 50% in HGS.

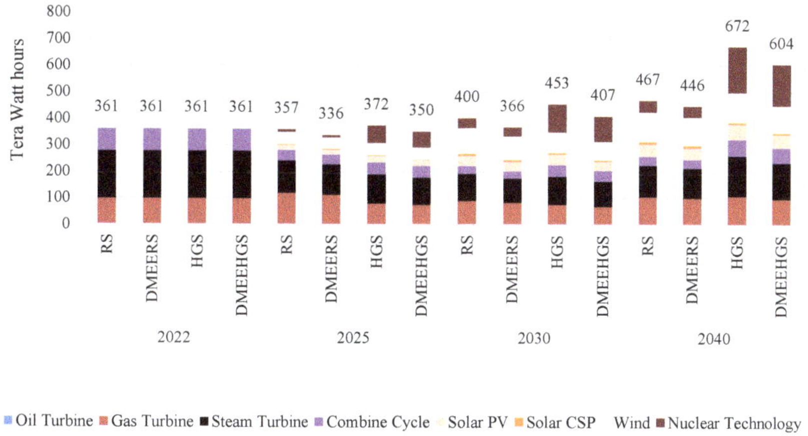

Electricity Generation Profile of Saudi Arabia

In 2030 in RS the thermal contribution from steam, gas and combine cycle power plants will be 221 TWh and clean energy share from solar PV, Solar CSP, Wind and Nuclear would be 179.4 TWh. Clean energy share can be increased in the generation mix but at the higher generation cost. Similar trajectory followed in RS in 2040. Whereas in HGS the renewable and thermal share are almost same in 2030, and when the analysis moves towards 2040 the share of clean energy resources marginally surpasses the thermal energy resources. Throughout the analysis period the generation seems to be higher than that of demand, but it is due to transmission and distribution losses in the system and the planning reserve margin. Planning reserve margin of Saudi Arabia is noted as about 10%. Economic growth increases power use, but steady planning and efficiency help to control it. Most demand comes from homes, so awareness compaign, appliance standards, and insulation must be included in the system. Fig.3 presents the projected generation costs for the four scenarios.

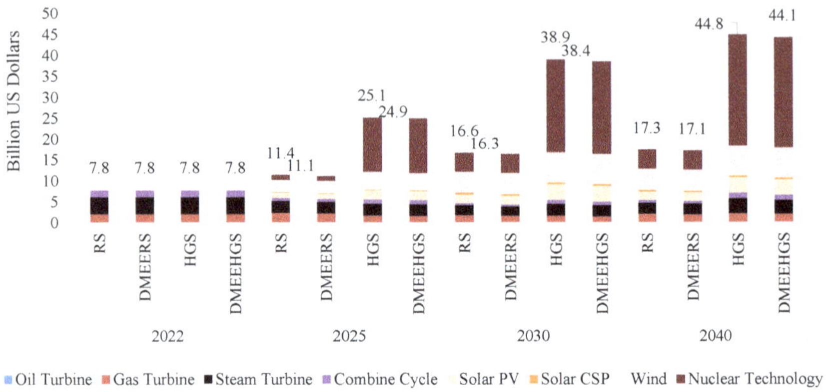

Cost of Production from all power plants

In 2022, total system expenditures are similar, at roughly 7.8 billion USD (≈29.3 billion SAR), as the generation mix has not yet diverged. By 2025, costs increase to 11.4 billion USD (≈42.9 billion SAR) in the RS and 11.1 billion USD (≈41.8 billion SAR) in the DMEERS. Under the HGS, higher demand pushes costs sharply to 25.1 billion USD (≈94.1 billion SAR). The implementation of efficiency measures as in DMEEHGS controls this rise, limiting total costs to 24.9 billion USD (≈93.4 billion SAR).

By 2030, generation costs rise to about 16.6 billion USD (≈62.2 billion SAR) in the RS and 16.3 billion USD (≈61.1 billion SAR) in the DMEERS. Under the HGS, costs climb much higher, reaching 38.9 billion USD (≈145.9 billion SAR), compared with 38.4 billion USD (≈144.0 billion SAR) in the DMEEHGS. This higher costs of High Growth scenario is due to installation of new power plants in the system. The impact of demand-side management becomes more visible by 2040, when total system costs in the RS reach 17.3 billion USD (≈64.9 billion SAR) and 17.1 billion USD (≈64.1 billion SAR) in the DMEERS. For the high-growth cases, expenditures rise to 44.8 billion USD (≈168.0 billion SAR) in the HGS but are contained to 44.1 billion USD (≈165.4 billion SAR) under the DMEEHGS, reflecting the sustained cost advantage of efficiency measures.

These results demonstrate that, in comparison to their corresponding baseline cases, the efficiency-focused pathways consistently result in cost savings. Even though the percentage savings might not seem like much, they add up to significant avoided investments in thermal power capacity and postponed infrastructure upgrades. The long-term objectives of Saudi Vision 2030 are supported by these pathways, which produce a more equitable and financially viable generation mix.

The anticipated environmental emissions under each scenario are shown in Fig. 4. Since fossil fuel-based generation predominates and renewables play a minimal role in the base year, emissions in 2022 are the same at 77.3 million metric tons CO_2-equivalent ($MMtCO_2Eq$). As renewables begin to enter in the generation mix, emissions will decrease by 2025. Emissions were recorded at 59.3 [$MMtCO_2Eq$] in the RS and dropped to 55.9 [$MMtCO_2Eq$] in the DMEERS. Efficiency measures lower projected emissions to 47.2 [$MMtCO_2Eq$] in DMEEHGS from approximately 50.1 [$MMtCO_2Eq$] under the HGS. The increased dependence on thermal power plants is reflected in the higher levels in the RS.

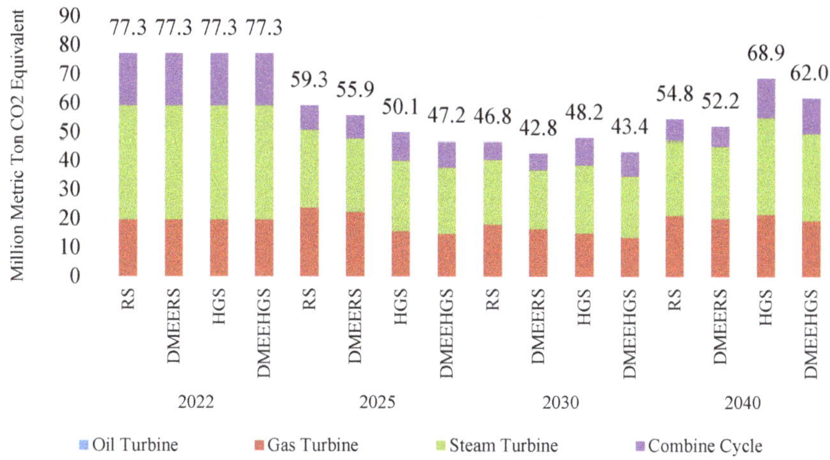

Environmental Emissions of Thermal Power Plants in All Scenarios

By 2030, it will be easier to see how efficiency gains and renewable growth are working together. Emissions drop to 46.8 [MMtCO_2Eq] in the RS and 42.8 [MMtCO_2Eq] in the DMEERS. Total emissions under the HGS are approximately 48.2 [MMtCO_2Eq], while under the DMEEHGS, they are 43.4 [MMtCO_2Eq]. These emissions are 68.9 [MMtCO_2Eq], in HGS in 2040 showing an increasing trend from 2030 to 2040 due to more power production to meet the demand.

The scenario analysis strongly emphasizes the pronounced impact of energy efficiency and Demand-Side Management in achieving meaningful reductions in greenhouse gas (GHG) emissions within Saudi Arabia's power sector by 2040. Small efficiency gains cut emissions, lower costs, and delay new plants, keeping energy efficiency central to Saudi sustainability. In DMEERS and DMEEHGS, demand falls and fewer thermal units are needed. In DMEEHGS, saving of 67.1 TWh avoids several GW of fossil capacity. Moreover, even with 100 GW of renewables planned by 2030, about 50 GW of nuclear will still be needed by the terminal year to reach a 50 % clean-energy share.

Conclusion and policy Implications

This research explored Saudi Arabia’s electricity system through four LEAP scenarios, assessing how supply, demand, generation cost, and emissions may evolve to 2040. The analysis shows that efficiency and demand-side measures play a decisive role in shaping the future structure of the power sector.

In RS, total generation in 2040 reaches 467.2 TWh, costing about 64.9 billion SAR and producing 54.8 [MMtCO_2Eq]. When efficiency actions are introduced in DMEERS, generation falls to 445.7 TWh, emissions decline to 52.2 [MMtCO_2Eq], and expenditure drops slightly to 64.1 billion SAR. Rapid growth in HGS pushes generation to 671.6 TWh and emissions to 68.9 [MMtCO_2Eq], whereas the enhanced-efficiency case (DMEEHGS) limits them to 604.5 TWh and roughly 62 [MMtCO_2Eq]. The efficiency-oriented pathway also removes the need for about 10 GW of additional thermal capacity.

Persistent focus on efficiency provides tangible national benefits. Lower demand postpones the construction of new plants, reduces investment pressure, and simplifies renewable-energy integration. These gains align directly with Vision 2030’s ambition for a cleaner and more diversified electricity mix. Maintaining this momentum will require stable policy direction, effective regulatory oversight, and continuous investment in modern generation and grid technologies.

Institutional capacity and public participation are equally important. Broader engagement of industries, municipalities, and households can accelerate the spread of efficient technologies and renewable options. Distributed generation, particularly rooftop solar and small hybrid systems will complement utility-scale projects and improve grid resilience. Further research should analyse regional demand patterns, quantify the economic and social returns of efficiency policies, and assess their contribution to national sustainability objectives. Together, these steps can steer the Kingdom toward a reliable, low-carbon, and economically balanced power sector consistent with its long-term goals.

Acknowledgment
This work was supported by IRC-SES at KFUPM, Saudi Arabia under project number INSE-2502. The authors used OpenAI for improving the readability and language of the article. Authors reviewed and edited the content as needed after and take full responsibility for the content.

References

[1] SERA, Annual Statistical Booklet for Electricity 2022, W.a.E.R. Authority, Editor. 2022, Saudi Electricity Regulatory Authority: King Fahd National Library Publication.

[2] KSA, Saudi Arabia, Vision 2030, A Story of Transformation, Kingdom of Saudi Arabia.

[3] Ediger, V.Ş. and H. Tatlıdil, Forecasting the primary energy demand in Turkey and analysis of cyclic patterns. Energy Conversion and Management, 2002. 43(4): p. 473. https://doi.org/10.1016/S0196-8904(01)00033-4

[4] Chen W, W.Z., Study on china's future sustainable energy development strategy using MARKEL model, in INIS. 2001: China.

[5] Fairuz, S.M.C., et al., Long term strategy for electricity generation in Peninsular Malaysia - Analysis of cost and carbon footprint using MESSAGE. Energy Policy, 2013. 62: p. 493-502. https://doi.org/10.1016/j.enpol.2013.08.005

[6] Alharbi, A.M. and M. Abdulwahab. Best fit Energy Market Model for Kingdom of Saudi Arabia. in European Modelling Symposium (EMS). 2016. https://doi.org/10.1109/EMS.2016.033

[7] Alnatheer, O., The potential contribution of renewable energy to electricity supply in Saudi Arabia. Energy Policy, 2005. 33(18): p. 2298-2312. https://doi.org/10.1016/j.enpol.2003.12.013

[8] Shahid, M., et al., Electricity supply pathways based on renewable resources: A sustainable energy future for Pakistan. Journal of Cleaner Production, 2020. 263: p. 121511. https://doi.org/10.1016/j.jclepro.2020.121511

[9] Heap, C., LEAP, is a powerful, versatile software system for climate change mitigation assessment developed at SEI US. 2018. Available: https://www.sei.org/projects-and-tools/tools/leap-long-range-energy-alternatives-planning-system/

[10] WB. Electric power transmission and distribution losses (% of output) - Saudi Arabia. 2018. [Accessed 17 July 2024]; Available from: https://data.worldbank.org/indicator/EG.ELC.LOSS.ZS?end=2014&locations=SA&start=1996.

Simulation and Analysis of a Hybrid Solar Battery Electrolyzer System for Hydrogen Production

Youness HAKAM[1,2,a *], Mohamed TABAA[1,b], Abdellah AILANE[2,c]

[1]Multidisciplinary Laboratory of Research and Innovation, Moroccan School of Engineering Sciences (EMSI), Casablanca, Morocco.

[2]ENSET Mohammadia, Hassan 2 University, Casablanca

[a]y.hakam@emsi.ma, [b]m.tabaa@emsi.ma, [c]abd_ailane@gmail.com

Keywords: Solar Energy, Battery Energy Storage, Electrolyzer, Hydrogen Production, Hybrid Energy System, Renewable Integration, Simulation Analysis

Abstract. This paper provides a comprehensive simulation and performance analysis for a hybrid renewable energy system consisting of solar photovoltaic (PV) generation, battery energy storage, and an electrolyze for hydrogen production. The purpose of the study is to investigate the dynamic profiles of energy flow, battery state-of-charge (SOC) changes and hydrogen production from solar power intermittency. The system was modelled in MATLAB/Simulink and simulated for 180 hours. The findings demonstrate the balancing relationship between battery and electrolyzer, in which battery smooth the perturbations of solar power and thereby stabilizes the power going into electrolysis. The SOC declines constantly implies the continuous energy supply in support of hydrogen production, and the mass of produced hydrogen attains about 180 kg throughout the analyzed duration. These results indicate that the integration of PV generation, batteries and electrolyzes is technically feasible and that the operations would be stable for long-lasting production of hydrogen. The proposed structure may provide a reference for the optimal design of renewable hydrogen plant and energy storage management strategy in smart grid.

Introduction

In the light of increasing global need for clean and sustainable energy resources, hydrogen becomes a hopeful carrier of energy in future power systems. Hydrogen generation from renewable sources, storage as well as conversion back to electricity can provide flexibility and decarbonization advantages to integrated energy systems of the future [1], [2]. Among many hydrogen production routes, water electrolysis driven by renewable energy, and especially solar photovoltaic (PV) installations, is increasingly attracting attention because it emits no carbon and can be scaled up more readily [3].

Nevertheless, the seasonal and day-night variation of solar irradiation would result in fluctuation of power generation, causing instability in electrolyzes operation. Hybrid systems combining PV generation with battery energy storage are suggested as a potential solution to overcome this bottleneck, enabling a continuous power supply and increasing the system's overall efficiency [4], [5]. Batteries are indeed important to mitigate the power miss-match by charging allowing excess energy to be stored during sunny hours and provide it when irradiance is weak [6].

Solar battery electrolyze systems are the optimal arrangement of continuous hydrogen generation as well as enhanced system reliability [7]. A number of works dealing with the modeling, optimization and control of these hybrid configurations can be found in literature [8]-[9], paying attention to cost reduction, control strategies and dynamic behavior. However, further simulation-based analysis of their power flow behavior, battery SOC and hydrogen generation profile under a realistic solar variability are required.

This paper aims to simulate, in MATLAB/SIMULINK based approach, a hybrid solar battery electrolyze BotSe system. It assesses the interconnection among system modules, investigates SOC kinetics and measures hydrogen production, providing information towards the stability of operation as well as energy control when solar based integrated hydrogen systems are considered.

Methodology and System Modeling

A mixed hybrid solar battery electrolyzer system was proposed and simulated in MATLAB/Simulink to investigate the dynamic characteristic of hydrogen production based on renewable resources. Its main components are solar photovoltaic generation, battery energy storage, and proton exchange membrane (PEM) electrolyzer as shown in Fig 1.

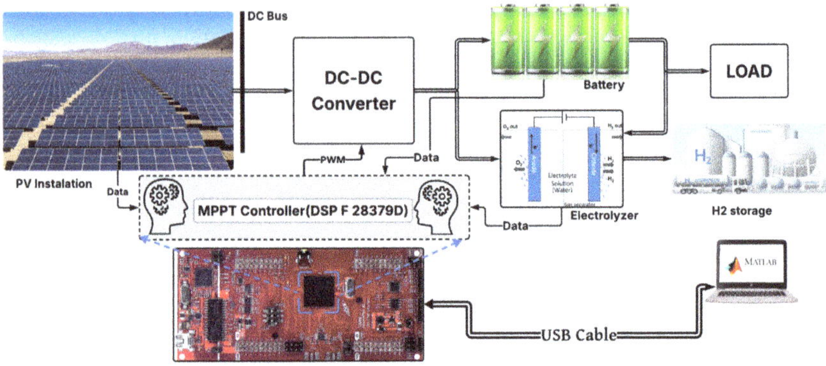

Figure 1. Block diagram of the hybrid solar battery electrolyzer hydrogen generation system.

The PV array injects DC into a controlled DC–DC converter using a DSP F28379D MPPT controller. Power is shared between the battery and electrolyzer and hydrogen is stored in H_2 storage. Real-time control and monitoring are performed in MATLAB over USB communication interface.

System Architecture

The architecture of the system is conceived to model a stand-alone small renewable H_2 plant in variable solar radiation. The PV array produces dc power which is supplied to both the battery and the electrolyzer via a DC/DC converter stage. During high irradiance situations, PV extra generation is used to charge the battery and if there is remained one, to produce hydrogen in the electrolyser. On the other hand, during periods of low irradiance or night time, the battery is discharged to meet electrolyzer power demand and to allow continuous generation of hydrogen.

Solar Photovoltaic Model

The PV sub-system is described by the single diode equivalent circuit, including photogenerated current, diode losses and series-parallel resistances. The PV output power (P_{PV}) is calculated as:

$$P_{PV} = V_{PV} \cdot I_{PV} \tag{1}$$

Letter symbols V_{PV} and I_{PV} are the instantaneous voltage and current of the PV array. The irradiance and temperature profiles were used to model daily solar fluctuations, which induced peaks of power evolution as displayed in the simulation results.

Materials Research Proceedings 64 (2026) 857-862 https://doi.org/10.21741/9781644904091-106

Battery Energy Storage Model

The battery is considered as a dynamic energy buffer with potential bidirectional power-flow. Its dynamics are given by the following SOC update rule:

$$SOC(t+\Delta t) = SOC(t) - \frac{(Ibat(t).\Delta t)}{Cbat} \tag{2}$$

where I_{bat} is the battery current (positive during discharge, negative during charge); C_{bat} is the nominal capacity of the battery. The battery has the function of smoothing out the PV output and power balancing for stable running of electrolyzer.

Electrolyzer Model

Water electrolysis is used to convert electrical power into hydrogen using a PEM electrolyzer. The rate of hydrogen production is calculated using Faraday's law:

$$M_{H_2} = \frac{\eta_{el}.I_{el}.M_{H_2}}{12F} \tag{3}$$

where η_{el} is the electrochemical efficiency of the electrolyzer, I_{el} is the current of the electrolyzer, M_{H_2} is the molar mass of hydrogen (2.016 g/mol), and F is Faraday's constant (96,485 C/mol). The total hydrogen mass rises monotonously and it's close to 180 kg after 180 hr, which proves that the hydrogen production is steady.

Energy Management Strategy

The algorithm emphasized the solar use for hydrogen production. For $P_{PV} > P_{el}$, redundant PV energy is used to charge the battery while for $P_{PV} < P_{el}$ the battery provides energy. This is nothing more than a basic rule based control to keep the electrolyzer operating close to its rated power but to avoid over cycling the battery pack and thereby compromising system stability.

System Flow and Control Diagram

The schematics of the complete hybrid solar battery electrolyzer system are shown in Fig. 2. This chart outlines the energy and control routes from solar power generation to hydrogen storage. The system starts with a PV array, which converts the solar energy into DC electric power. To keep the PV array working at its maximum power point under varying irradiation and temperature, a MPPT controller that is synthesized based on DSP F28379D microcontroller platform controls output of the PV panel.

The regulated DC power output by the MPPT controller is split, one branch feeds into the battery subsystem for energy buffering and to provide power to the DC load as well as to the electrolyzer and others. The battery provides two-way charging and discharging in conjunction with the solar hours The battery is used for charing during daytime sunshine hours and at night time it is discharged also to modulate system operation/response and ensure stable power supply to the electrolyser.

The electrolyzer electrolyzes water to produce hydrogen and oxygen with the input of a DC power source. Synthesized hydrogen is further taken and stored into the H_2 storage compartment, providing a long term energy carrier for power generation or fuel in future.

Materials Research Proceedings 64 (2026) 857-862 https://doi.org/10.21741/9781644904091-106

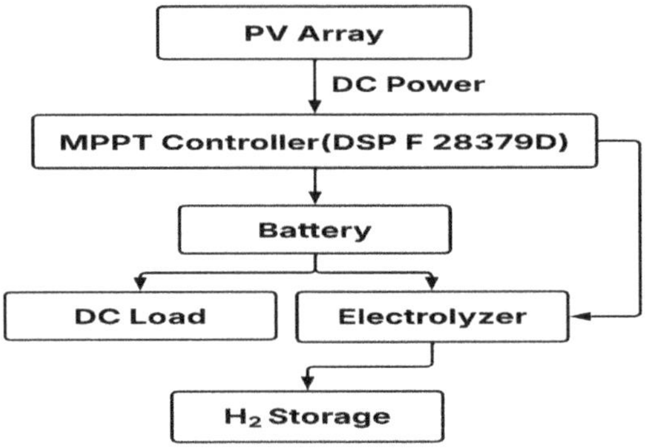

Figure 2. *System flow diagram of the hybrid solar battery electrolyzer configuration.*

The PV array generates DC power regulated by an MPPT controller implemented on a DSP F28379D platform. The energy is shared between the battery, DC load, and electrolyzer, with hydrogen stored in the H_2 storage unit for later use.

Results and Discussion

The simulation results of the hybrid solar battery electrolyzer system are shown in Fig 3, including the power flow among subsystems, the battery SOC variation and hydrogen production over a simulation time of 180 h. This considers dynamic interaction of the system based on varying solar irradiance and also shows efficiency of the employed energy management strategy.

The photovoltaic (PV) power in power plot (top) shows a significant period pattern, which corresponds to daily irradiance cycles. The PV array is under a performant point tracker (MPPT) function, whose design is modified operation point to the maximum available power. During the peak solar hours, the PV power exceeds that of electrolyzer demand; therefore, the extra energy can be used to charge battery. When there is less or no solar generation, such as at night, the battery discharges to maintain the power input of the electrolyzer. The coordinated action of the PV and battery ensure that electrolyzer operates close to its rated power in most part of the simulation.

The battery SOC (middle panel) reflects the charging and discharging patterns matching solar availability. SOC peaks indicate high PV generation when battery is charged, and troughs discharge during low irradiance. The cyclic charge–discharge behavior demonstrates that the applied control strategy is capable to keep a proper energy balance of the hybrid system and does not lead to extreme operating conditions. The lower bound of SOC threshold guarantees the energy reserve which lead to prevent the deep discharge and longevity of battery.

The curve of the hydrogen production (lower plot) shows the constant accumulation of total H2 mass with time, which demonstrates that the system can maintain a long-term operation for generation of hydrogen. The hydrogen production scales more of less linearity, indicating that the electrolyzer is running steady and stable when enough power is available. Once the 180 h are completed, a total hydrogen mass of 300 kg is achieved, thus revealing the potential of this proposed setup for efficient and continued generation of hydrogen from renewables.

These results in total prove that the hybrid PV battery electrolyze system can realize reliable operation when dealing with variable solar conditions. The centralized control of the MPPT unit, battery management, and electrolyzer can achieve good energy transition with less power interruptions and larger hydrogen production. These results illustrate the ability of hybrid renewable storage systems to increase the reliability and flexibility of green hydrogen production for distributed energy applications.

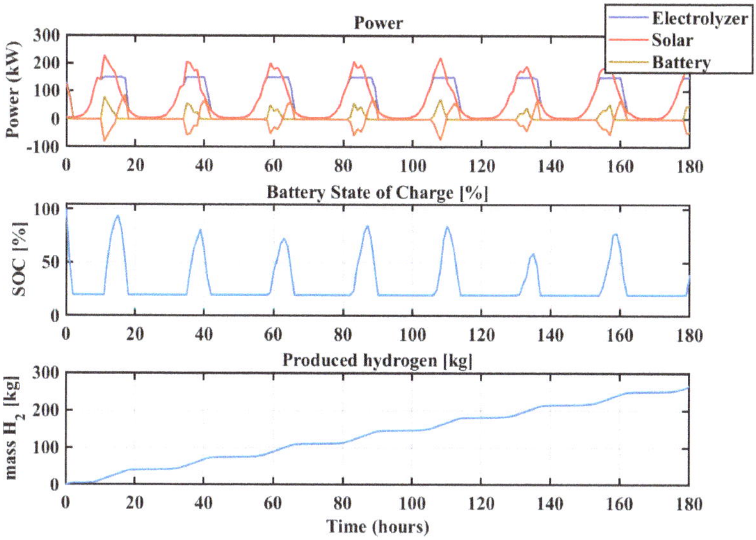

Figure 3. *Simulation results of the hybrid solar–battery–electrolyzer system: (top) power profiles of PV, battery, and electrolyzer; (middle) variation of battery state of charge (SOC); and (bottom) cumulative hydrogen mass produced over 180 hours.*

Conclusion and Future Work

This paper provided a full simulation and analysis of the hybrid solar battery electrolyzer system for the renewable hydrogen production. The proposed system encompasses the integration of PV power generation, BESS and a PEM electrolyzer controlled by MPPT-regulated DC–DC converter. Simulation results showed that the system is able control and to stably supply power to the electrolyzer under fluctuating solar energy. The battery system responds to energy fluctuations rapidly, resulting in no-derating of hydrogen production, and keeping the solar power utilization maximum. The calculations demonstrated that the hybrid system consistently operated in such a way as to maintain electrolyzer stability, generating approximately 300 kg hydrogen during the 180-hour simulation period. The proposed control strategy ensured battery SOC within the safe operating limit and maximized energy utilization between photovoltaic array, battery and electrolyzer. This verifies the technical feasibility and sustainability of using renewable energy as driver for hydrogen production by this system. Several lines of further work will be followed to improve the proposed system. First, real-time experimental test on a HIL based on the DSP F28379D control card is performed to test the performance of the designed control algorithm over a wide range of operating conditions. Second, advanced control schemes – like MPC and adaptive neural controllers will be investigated in order to better distribute the power while minimizing the battery cycling. Finally, future research will involve economic evaluations (CAPEX and OPEX), system scalability analysis, and investigation into the integration with smart grid or microgrid architectures for overall techno-economic potential of HRHP systems.

References

[1] D. Stolten and B. Emonts, Hydrogen Science and Engineering: Materials, Processes, Systems and Technology, 2nd ed., Wiley-VCH, 2022.

[2] IEA, "Global Hydrogen Review 2024," International Energy Agency, Paris, 2024.

[3] A. Ursúa, L.M. Gandía, and P. Sanchis, "Hydrogen production from water electrolysis: Current status and future trends," Proc. IEEE, vol. 100, no. 2, pp. 410-426, 2012. https://doi.org/10.1109/JPROC.2011.2156750

[4] M. S. Khan, M. H. Ahmed, and Y. A. R. I. Mohamed, "Hybrid renewable energy systems for hydrogen production: A review," Renewable and Sustainable Energy Reviews, vol. 185, p. 113525, 2023.

[5] N. Mahato, S. Nayak, and B. Singh, "Modeling and control of PV-battery-electrolyzer hybrid systems for hydrogen production," Energy Conversion and Management, vol. 301, p. 118028, 2025.

[6] Ahessab, H., Hakam, Y., Gaga, A., El Hadadi, B. (2023). A Comparative Study of P&O and Fuzzy Logic MPPT Algorithms for a Photovoltaic Grid Connected Inverter System. In: Motahhir, S., Bossoufi, B. (eds) Digital Technologies and Applications. ICDTA 2023. https://doi.org/10.1007/978-3-031-29857-8_65

[7] M. Elberry, M. A. Abdelrahem, and H. Abdel-Gawad, "Comparative study of battery and hydrogen energy storage for renewable power systems," Journal of Energy Storage, vol. 73, p. 108744, 2025.

[8] Ahessab, H., Hakam, Y., Gaga, A., EL Hadadi, B. (2024). Design and Simulation of an Intelligent Grid-Connected MPPT Inverter with Battery Storage Using ANN Algorithm. In: Bendaoud, M., El Fathi, A., Bakhsh, F.I., Pierluigi, S. (eds) Advances in Electrical Systems and Innovative Renewable Energy Techniques. ICESA 2023. Advances in Science, Technology & Innovation.Springer https://doi.org/10.1007/978-3-031-49772-8_29

[9] A. Ould Amrouche, T. Rekioua, and A. Hamidat, "Modeling photovoltaic water electrolysis system for hydrogen production," Renewable Energy, vol. 56, pp. 56-63, 2022.

[10] M. J. Khan and M. T. Iqbal, "Dynamic modeling and simulation of a hybrid wind-fuel cell system," Renewable Energy, vol. 30, no. 3, pp. 421-439, 2005. https://doi.org/10.1016/j.renene.2004.05.013

Field Oriented Control based on Sliding Mode of a Variable Speed Grid-Connected DFIG Wind Turbine System

Aymen BOUJEMEL[1,a *], Najib EL OUANJLI[2,b], Hicham ROUIJAA[1,c], Ayoub NOUAITI[2,d], Ilias EN-NAOUI[1,e], and Mostafa MERRAS[3,f],

[1]Energy-Matter-Instrumentation & Telecoms, Faculty of science and technology, Hassan 1st university, Settat, Morocco

[2]LASMAR Laboratory, SEIMC Research Team, Higher School of Technology, Moulay Ismail University, Meknes, Morocco

[3]IMACS Laboratory, Higher School of Technology, Moulay Ismail University, Meknes, Morocco

[a]aymen.boujemel.doc@uhp.ac.ma, [b]n.elouanjli@umi.ac.ma, [c]hrouijaa@gmail.com, [d]a.nouaiti@umi.ac.ma, [e]i.ennaoui@uhp.ac.ma, [f]m.merras@umi.ac.ma

Keywords: Variable-Speed Wind-Turbine System, Tip-Speed Ratio, DFIG, FOC, MPPT, SMC

Abstract. This study introduces a control scheme for wind energy conversion systems based on a doubly fed induction generator (DFIG) that integrates a tip-speed-ratio (TSR)-oriented Maximum Power Point Tracking (MPPT) technique with a direct vector control scheme enhanced by sliding mode control. The proposed method aims to maximize active power extraction under varying wind conditions while ensuring zero reactive power output to support grid stability. The partial-scale converter design inherent to the DFIG configuration offers cost and efficiency advantages, while the adopted control techniques improve dynamic response and reliability. The simulation results reveal that the suggested method significantly ameliorates the system's performance. The MPPT speed control had a rise time of 2.5 ms, while the power control had a rise time of 1.5 ms.

1. Introduction

Wind turbine construction has been widely developed in recent decades due to the renewable energy policy and electricity requirement throughout the world [1]. Most of today's systems are variable speed in nature from which the MPPT and the powers control is achieved through use of power converters [2], [3]. Of the existing technologies, DFIGs are used most, and it is due to their low cost; light weight and good grid support facilities [4].

In a DFIG system, the stator is directly connected to the grid, whereas the rotor is supplied via a bidirectional power converter, allowing variable-speed operation while staying synchronized with the grid voltage and frequency (Fig. 1). The converter compensates slip and ensures stable power generation even at sub-synchronous speeds [5]. A key advantage is the partial-scale converter, which handles only 25–30% of rated power, reducing losses and cost while allowing full regulation of active and reactive power for grid support [6].

The generator control relies on the RSC for torque and power regulation and the GSC for DC bus voltage control [7]. MPPT is used to define electromagnetic torque reference and, therefore, the active power command, in contrast, setting the reactive power command to zero is generally adopted to improve the power factor. Vector control enables the distinct regulation of active and reactive power, by adjusting the rotor speed to maintain the optimal tip-speed ratio through a TSR-based MPPT. In this work, TSR-based MPPT is combined with vector control enhanced by SMC technique to regulate the power flow.

Energy Futures - ICEF 2026
Materials Research Proceedings 64 (2026) 863-870
Materials Research Forum LLC
https://doi.org/10.21741/9781644904091-107

The rest of this article is arranged as follows. The second section details the system modeling, Section 3 introduces the suggested control strategy, Section 4 extends and analyzes the simulation results, and last section provides the final conclusions.

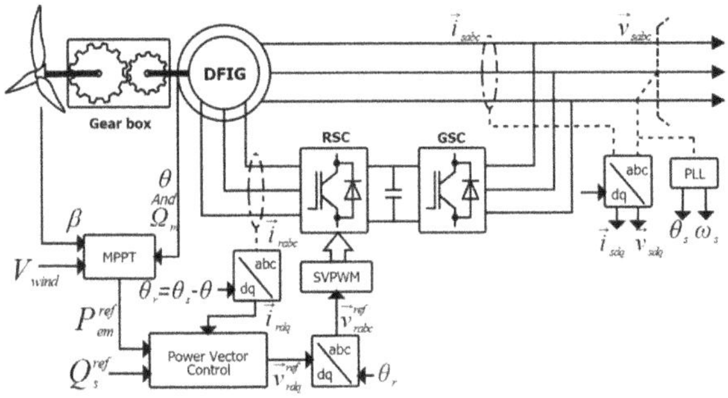

Fig. 1. *Grid connected DFIG system configuration*

2. System modeling

2.1 Aerodynamic model of the WTS

Wind turbines generate electricity using the wind to rotate the blades, which turn a drive shaft connected to a generator. The generator converts this rotational motion into electrical power.

Actuator disk theory provides a simple way to describe how the turbine extracts kinetic energy from the wind, using basic energy balance and Bernoulli's equation. The available wind power is given by Eq. 1.

$$P_{wind} = (\rho A V_{wind}^{3})/2 \tag{1}$$

Where A denotes the surface swept by the blades (m^2), and V_{wind} the upwind speed (m/s) and ρ the air density (Kg/m^3).

Although Eq. 1 represents the total wind power accessible at the blades, the power extracted is reduced by the power coefficient Cp, meaning the WTS is able to extract only a portion of this power as shown by Eq. 2 and Eq. 3.

$$C_p = (P_{turbine} / P_{wind}) \tag{2}$$

$$P_{turbine} = C_p.P_{wind} = (C_p.\rho.A.V_{wind}^{3})/2 \tag{3}$$

The power factor C_p reflects aerodynamic behavior of a wind turbine system (WTS) and is determined by the turbine's structural characteristics. It is subject to a theoretical upper bound, referred to as the Betz limit, indicating that the analytical highest efficiency for WTSs is 59.3%. In practical, the maximum Cp values for wind turbine rotors typically range from 25% to 45%.

For a given turbine, this coefficient varies with wind velocity, the rotational velocity of the rotor and the blade pitch angle. Accordingly, Cp is commonly formulated as a function of the TSR λ, given in Eq. 4.

$$\lambda = (\Omega_t.R_t / V_{wind}) \tag{4}$$

In this formulation, Ω_t the mechanical velocity of the rotor (rad/s) and R_t refers to the blade length, defined as the distance from the rotor center to the blade tip (m).

Within the framework of the paper, an approximate expression of the power coefficient based on the pitch angle β of the turbine blades and the tip speed ratio λ will be adopted as expressed in Eq. 5.

$$C_p(\lambda,\beta) = (0,5 - 0,0167.(\beta - 2)).\sin((\pi.(\lambda + 0,1)) \\ /(18,5 - 0,3.(\beta - 2))) - 0,00184.(\lambda - 3).(\beta - 2) \quad (5)$$

The characteristics of Cp plotted against λ for multiple pitch angle β values are displayed in Fig .2.

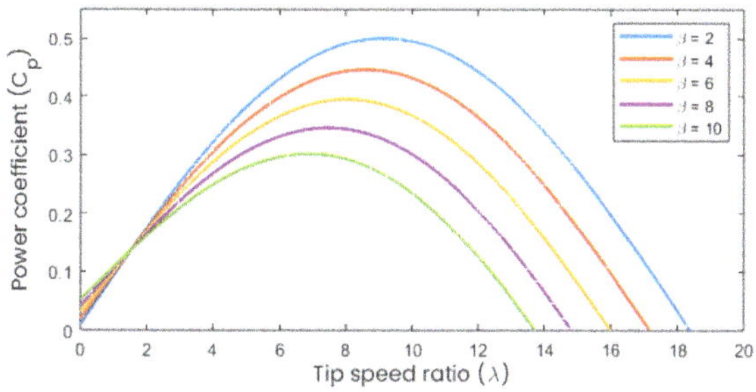

***Fig. 2.** Power coefficient Cp variations with respect to λ and β.*

2.2 Doubly fed induction generator model

The DFIG is a wounded rotor asynchronous generator typically modeled under standard assumptions: sinusoidally distributed, symmetrical stator and rotor windings, and no magnetic saturation, core losses, or winding capacitances [8].

A widely used representation is the Park (dq) model, where the dq axes rotate at synchronous speed to simplify analysis. Fig. 3 presents the resulting equivalent circuit in the Park frame, illustrating the stator and rotor windings in a shared coordinate system.

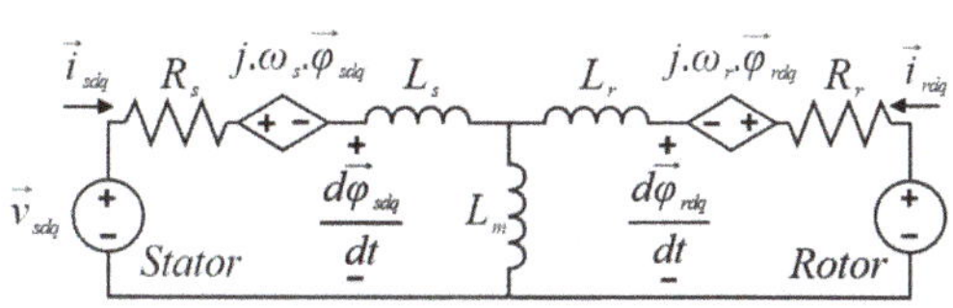

***Fig. 3.** Ideal representation of the 3 winding phases (stator, rotor) of the DFIG.*

Within this framework, the stator and rotor quantities dynamics are given by Eq.6, 7, 8, and 9.

$$V_{sd} = R_s.i_{sd} + (d\phi_{sd}/dt) - \omega_s.\phi_{sq} \quad (6)$$

$$V_{sq} = R_s.i_{sq} + (d\phi_{sq}/dt) + \omega_s.\phi_{sd} \quad (7)$$

$$V_{rd} = R_r.i_{rd} + (d\phi_{rd}/dt) - \omega_r.\phi_{rq} \quad (8)$$

$$V_{rq} = R_r.i_{rq} + (d\phi_{rq}/dt) + \omega_r.\phi_{rd} \quad (9)$$

And the generator flux expressions are expressed by Eq. 10, 11, 12 and 13.

$$\phi_{sd} = L_s.i_{sd} + M_{sr}.i_{rd} \tag{10}$$
$$\phi_{sq} = L_s.i_{sq} + M_{sr}.i_{rq} \tag{11}$$
$$\phi_{rd} = L_r.i_{rd} + M_{sr}.i_{sd} \tag{12}$$
$$\phi_{rq} = L_r.i_{rq} + M_{sr}.i_{sq} \tag{13}$$

Where $L_s = l_s - M_s$ and $L_r = l_r - M_r$ present cyclic inductances. $M_{sr} = 3M_0/2$ is the mutual cyclique inductance. Additionnaly, the angular velocities of the stator and rotor voltages are expressed as $\omega_s = (d\theta_s)/dt$ and $\omega_r = (d\theta_r)/dt$ respectively.

2.3 Power converters model

In DFIG-based wind turbines, the converter is typically an IGBT voltage-source back-to-back unit [10]. It consists of a RSC supplied by the DC bus and a GSC that works as an active rectifier when the machine runs sub-synchronously. The GSC is then wired to the three-phase grid through a filtering inductor, as shown in Fig. 4.

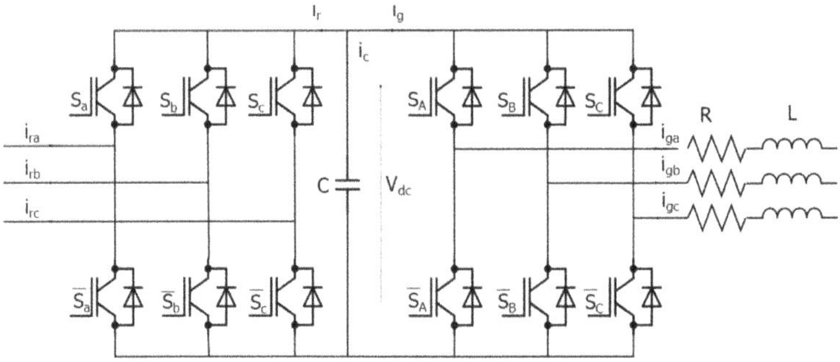

Fig. 4. Ideal representation of the 3 winding phases (stator, rotor) of the DFIG.

Assuming a perfect electrical grid and balanced rotor impedance, the following expression can be written as Eq. 14.

$$\begin{pmatrix} V_{ra} \\ V_{rb} \\ V_{rc} \end{pmatrix} = (V_{dc}/3).\begin{pmatrix} 2 & -1 & -1 \\ -1 & 2 & -1 \\ -1 & -1 & 2 \end{pmatrix}.\begin{pmatrix} S_a \\ S_b \\ S_c \end{pmatrix} \tag{14}$$

3. Control strategy of the WTS

3.1 MPPT biased SMC control strategy

The TSR MPPT method used in this work maximizes energy capture by keeping the fraction of blade tip speed and wind velocity at its optimal operating point. A wind speed sensor provides the actual wind velocity, which allows accurate TSR calculation and ensures the turbine operates at its maximum power coefficient. This sensor-based approach ensures precise tracking of the optimal operating point. Newton's second law for this system is expressed in Eq. (15).

$$(d\Omega_m / dt) = (Tem - Tm - f.\Omega_m) / J \tag{15}$$

Where Tem the electromagnetic torque supplied by the generator, Tm the mechanical torque applied by the turbine, Ωm is the rotor mechanical speed, f the viscous friction coefficient, and J the moment of inertia. The control scheme model in Fig. 6 is derived.

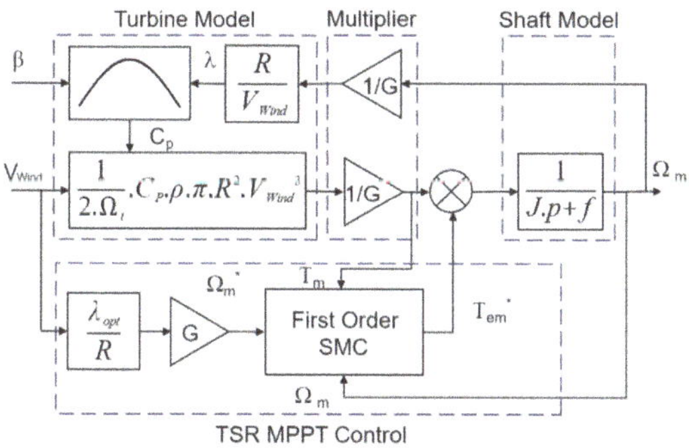

Fig. 6. *Turbine model and TSR-based MPPT control using first order SMC*

The algebraic formula for the mechanical speed control law is given by Eq. 16.

$$T_{em}^* = K.sign((S(\Omega_m)) + J.\left(d\left(\Omega_m\right)/dt\right) - T_m - f.\Omega_m \tag{16}$$

To overcome the phenomenon of chattering provoked by the sign function of the first order SMC technique, sigmoid functions are proposed, such as the hyperbolic tangent.

3.2 SMC biased FOC of the DFIG

For the DFIG, fluxes are coupled, which results in the coupling of active and reactive power control. The governing equations for this relationship are Eq. 17 and 18.

$$P_s = v_{sd}.i_{sd} + v_{sq}.i_{sq} \tag{17}$$

$$Q_s = v_{sq}.i_{sd} - v_{sd}.i_{sq} \tag{18}$$

To handle this issue, a vector control strategy is used. Previous studies have proposed several approaches [10], including stator-flux-oriented vector control, which aligns the DFIG flux with one axis of the rotating dq frame. Throughout this work, the d-axis is aligned alongside the stator-flux vector. (Fig. 7). Under this condition, the rotor current on the d-axis regulates torque and, consequently, active power, whereas the q-axis current governs reactive power.

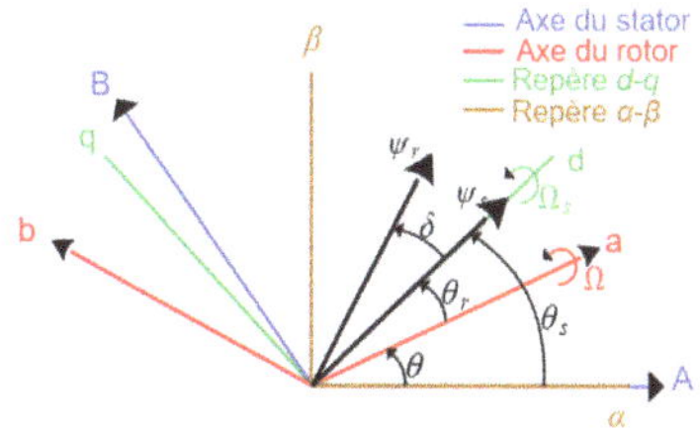

Fig. 7. *Flux vectors and different reference frames*

This technique eliminates the stator flux projection along the q-axis, leading to Eq. 19.

$$\phi_s = \phi_{s_d} \tag{19}$$

As a result, the generator model is simplified to the following set of equations given in Eq. 20 and 21.

$$V_{rd} = R_r.i_{rd} + L_r.\sigma.(di_{rd} / dt) - \omega_r.L_r.i_{rq} \tag{20}$$
$$V_{rq} = R_r.i_{rq} + L_r.\sigma.(di_{rq} / dt) + \omega_r.L_r.i_{rd} + \omega_r.M_{sr}.V_s / (\omega_s.L_s) \tag{21}$$

The formulations of the generator's powers will then be written as in Eq 22 and 23.

$$P_s = -(V_s M_{sr} / L_s).i_{rq} \tag{22}$$
$$Q_s = -(V_s.M_{sr} / L_s).i_{rd} + V_s^2 / (L_s.\omega_s) \tag{23}$$

It is evident that the both stator powers can be effectively controlled through the regulation of the rotor current. The corresponding block diagram of the simplified generator model operating under the FOC is shown in Fig. 8.

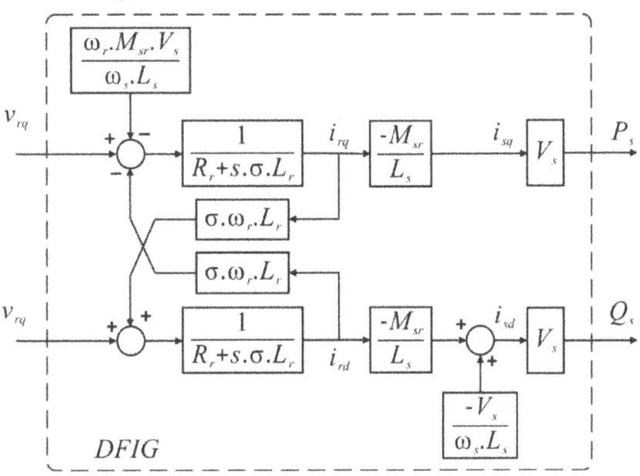

Fig. 8. *Simplified DFIG model with FOC*

Based on the distinct rotor voltage equations, the dynamic state-space model of the rotor currents can be derived as shown in Eq. 24 and 25.

$$\left(di_{rd} / dt\right) = -\left(R_r / \left(L_r.\sigma\right)\right).i_{rd} + \omega_r.i_{rq} + \left(1 / \left(L_r.\sigma\right)\right).v_{rd} \tag{24}$$
$$\left(di_{rq} / dt\right) = -\left(R_r / \left(L_r.\sigma\right)\right).i_{rq} - \omega_r.i_{rd} + \left(1 / \left(L_r.\sigma\right)\right).v_{rd} - \left(\omega_r.V_s.M_{sr} / \left(\omega_s.L_s.L_r.\sigma\right)\right) \tag{25}$$

It is particularly significant that the resulting system exhibits strong nonlinear behavior, primarily due to the dependency of the direct (d) and quadrature (q) projections of the current available at the rotor, as well as their interaction with the rotational speed of the machine. To control this nonlinear system, a first-order sliding mode control approach is employed. Consequently, the proposed strategy leads to the algebraic expressions given in Eqs. 26 and 27, which define the rotor current control laws.

$$v_{rd} = L_r.\sigma.\dot{i}_{rd}{}^* + R_r.i_{rd} - L_r\sigma.\omega_r.i_{rq} + K.sign(S(i_{rd})) \tag{26}$$

$$v_{rq} = L_r.\sigma.\dot{i}_{rq}{}^* + R_r.i_{rq} + L_r\sigma.\omega_r.i_{rd} + \left(\omega_r.V_s.M_{sr} / \left(\omega_s.L_s\right)\right) + K.sign(S(i_{rq})) \tag{27}$$

Materials Research Proceedings 64 (2026) 863-870 https://doi.org/10.21741/9781644904091-107

4. Results and discussion

This section shows the results of the suggested control strategies, focusing on the TSR-based MPPT using first-order SMC and powers flow control of the DFIG with the same approach. The purpose is to measure the effectiveness of the developed control laws.

To evaluate the SMC-based TSR MPPT approach, a variable wind velocity profile with an average of about 7.5 m/s was used. This profile, shown in Fig. 9(a), reflects realistic wind variations and allows a meaningful test of the controller's robustness and dynamic response.

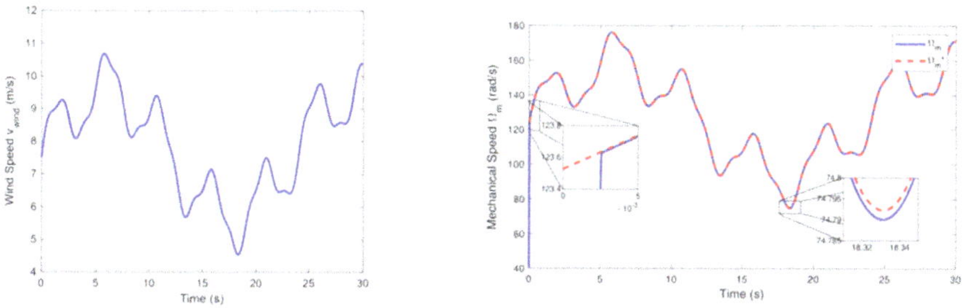

***Fig. 9(a).** Wind speed profile over time, **(b).** Mechanical speed output and mechanical speed reference over time*

The mechanical speed is tracked by the SMC controller with high precision, and a rise time close to 2.5 ms is obtained; even in the presence of sudden wind-speed changes which induce small alterations in both speed and torque commands, tracking error remains minimal, demonstrating excellent robustness as well as fast response (see Fig. 9(b).

Fig. 10 shows that the controller also regulates the generator's powers effectively, achieving a rise time of approximately 1.5 ms. In Fig. 10(a), the active power follows its command closely during both transient and steady-state operation. The reactive power stays near its zero reference with only small deviations as shown in Fig. 10(b).

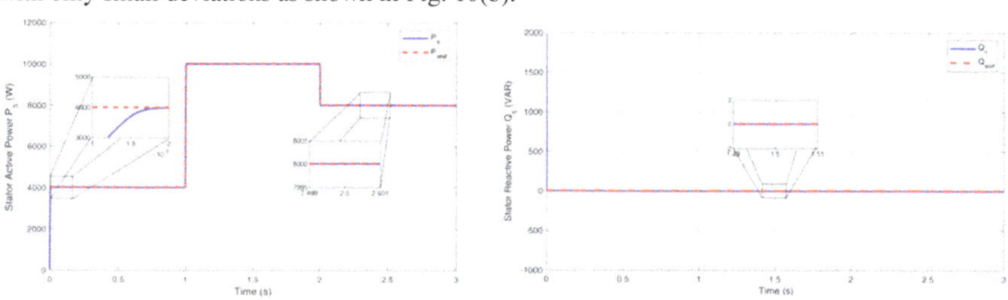

***Fig. 10(a).** Stator-side active power output and its command over time, **(b).** Stator-side reactive power output and its command over time*

The output responses converge precisely to their reference values once transients have settled. Moreover, the system dynamic response is fast, as shown by the rapid convergence in the zoomed-in regions. Another important point is the absence of chattering in the outputs. This comes from using sigmoid functions in the SMC law, which smooth the control action and prevent the high-frequency oscillations seen in conventional SMC. The resulting signals are continuous and smooth, showing better performance and robustness under varying conditions.

5. Conclusion

This paper presented the modeling and control of a DFIG driven WTS. A TSR-based MPPT strategy was combined with a field-oriented control scheme enhanced SMC in order to maximize power extraction and regulate the exchanged active and reactive powers. The overall control architecture was designed to ensure efficient operation and grid compatibility under varying wind conditions.

The simulation findings confirm the performance of the developed control scheme. The SMC-based MPPT achieved accurate speed tracking with a rise time of about 2.5 ms, even during sudden wind variations. In addition, the power control showed a fast response with a rise time of approximately 1.5 ms. Active power closely followed its reference, while reactive power remained near zero. The use of sigmoid functions also eliminated chattering, leading to smooth and stable responses.

However, the study was limited to simulation under ideal conditions. Effects such as parameter uncertainties, grid disturbances, and hardware constraints were not considered. Future work will focus on experimental validation and robustness analysis under more realistic operating conditions.

References

[1] H. Li, Z. Chen, Overview of Wind Turbine Generators, IEEE Trans. Energy Convers (2008).

[2] El Mourabit, Y., Derouich, A., Allouhi, A., El Ghzizal, A., El Ouanjli, N., & Zamzoumyes, O. (2020). Sustainable production of wind energy in the main Morocco's sites using permanent magnet synchronous generators. International Transactions on Electrical Energy Systems, 30(6), e12390. https://doi.org/10.1002/2050-7038.12390

[3] M. Taoussi, M. Karim, B. Bossoufi, Low-speed sensorless control for wind turbine system, WSEAS Trans. Syst. Control, (2017).

[4] S. Karad, R. Thakur, Recent Trends of Control Strategies for Doubly Fed Induction Generator Based Wind Turbine Systems: A Comparative Review, Arch. Comput. Methods Eng (2021).

[5] G. Abad, J. López, M. A. Rodríguez, L. Marroyo, G. Iwanski, Doubly Fed Induction Machine: Modeling and Control for Wind Energy Generation, John Wiley & Sons (2011). https://doi.org/10.1002/9781118104965

[6] E. N. Sanchez, R. Ruiz-Cruz, Doubly Fed Induction Generators: Control for Wind Energy, Automation and Control Engineering Series, CRC Press, Taylor & Francis Group (2015).

[7] D. V. N. Ananth, G. V. Nagesh Kumar, Tip Speed Ratio Based MPPT Algorithm and Improved Field-Oriented Control for Extracting Optimal Real Power and Independent Reactive Power Control for Grid-Connected DFIG, Int. J. Electr. Comput. Eng. (2016). https://doi.org/10.11591/ijece.v6i3.9306

[8] F. Giri (Ed.), AC Electric Motors Control: Advanced Design Techniques and Applications, John Wiley & Sons, 2013. https://doi.org/10.1002/9781118574263

Keyword Index

About the Editor

Hamza El Hafdaoui holds a Ph.D. in Energy Engineering from Sidi Mohamed Ben Abdellah University, Morocco (2025), where his doctoral research focused on the role of alternative fuel vehicles in the decarbonization of the transport sector. He previously earned a Master of Science in Sustainable Energy Management (2020) and a Bachelor of Science in Engineering and Management Science (2019) from Al Akhawayn University in Ifrane, both with distinction and supported by excellence scholarships. His academic training is grounded in energy systems engineering, with a strong emphasis on renewable energy technologies, energy efficiency, and integrated system optimization, complemented by international academic exposure at the University of Dayton, USA.

Dr. El Hafdaoui currently serves as Faculty Member of Energy Engineering at Al Akhawayn University in Ifrane, Morocco, within the School of Science and Engineering. His teaching portfolio spans core engineering and energy-related subjects, including Electric Circuits, Power Distribution Systems and Smart Grids, University Physics, Engineering Economics, and Statics. Prior to this appointment, he served as Lecturer of Engineering at EMAA Business School, Morocco (2020–2021), where he taught courses in computer science and information systems. His academic experience is further strengthened by earlier roles as Research Assistant and Teaching Assistant at Al Akhawayn University, along with supervision of student capstone projects, research assistants, and ongoing doctoral work.

His research focuses on sustainable energy systems, with particular expertise in renewable energy integration, smart grids, hydrogen energy pathways, and energy storage technologies. He applies advanced methodologies including system-level modeling, optimization techniques, machine learning, techno-economic analysis, and life-cycle assessment to evaluate and design low-carbon energy systems. Dr. El Hafdaoui is the author of more than 30 peer-reviewed publications, in addition to conference papers and book chapters. He is also co-editor of book on artificial intelligence for sustainable energy systems and co-inventor of several patent applications related to energy routing and harvesting technologies.

Dr. El Hafdaoui has been actively involved in internationally funded research projects exceeding $1.6 million, including initiatives supported by the Europe-Africa Partnership for Renewable Energy (LEAP-RE), the German Academic Exchange Service (DAAD), and the National Center for Scientific and Technical Research (CNRST) in Morocco. His roles have included Institutional Research Lead and Co-Investigator on projects addressing smart grids, hydrogen technologies, and intelligent energy infrastructures. He contributes extensively to the academic community through peer review for leading scientific journals, editorial activities, and conference organization. He serves as Scientific Chair of the International Conference on Energy Futures (ICEF) and participates in several international technical program committees. His professional development includes certifications in energy systems, circular economy, and machine learning, reflecting his interdisciplinary approach to advancing sustainable energy transitions.

m

www.ingramcontent.com/pod-product-compliance
Ingram Content Group UK Ltd.
Pitfield, Milton Keynes, MK11 3LW, UK
UKHW020700041026
15590UKWH00001B/375

9 781644 904107